LANDSLIDES AND ENGINEERED SLOPES

PROCEEDINGS OF THE TENTH INTERNATIONAL SYMPOSIUM ON LANDSLIDES AND ENGINEERED SLOPES, 30 JUNE–4 JULY, 2008, XI'AN, CHINA

Landslides and Engineered Slopes

From the Past to the Future

Editors

Zuyu Chen
China Institute of Water Resources and Hydropower Research, Beijing, China

Jianmin Zhang
Department of Hydraulic Engineering, Tsinghua University, Beijing, China

Zhongkui Li
Department of Hydraulic Engineering, Tsinghua University, Beijing, China

Faquan Wu
Institute of Geology and Geophysics, Chinese Academy of Sciences, Beijing, China

Ken Ho
Civil Engineering and Development Department, Hong Kong SAR, China

VOLUME 1

CRC Press is an imprint of the
Taylor & Francis Group, an **informa** business

A BALKEMA BOOK

CRC Press/Balkema is an imprint of the Taylor & Francis Group, an informa business

© 2008 Taylor & Francis Group, London, UK

Typeset by Vikatan Publishing Solutions (P) Ltd., Chennai, India
Printed and bound in Great Britain by Antony Rowe (A CPI-group Company), Chippenham, Wiltshire.

All rights reserved. No part of this publication or the information contained herein may be reproduced, stored in a retrieval system, or transmitted in any form or by any means, electronic, mechanical, by photocopying, recording or otherwise, without written prior permission from the publisher.

Although all care is taken to ensure integrity and the quality of this publication and the information herein, no responsibility is assumed by the publishers nor the author for any damage to the property or persons as a result of operation or use of this publication and/or the information contained herein.

Published by: CRC Press/Balkema
                   P.O. Box 447, 2300 AK Leiden, The Netherlands
                   e-mail: Pub.NL@taylorandfrancis.com
                   www.crcpress.com – www.taylorandfrancis.co.uk – www.balkema.nl

ISBN set: 978-0-415-41196-7 (2 volumes + CD)
ISBN Vol.1: 978-0-415-41194-3 (hbk)
ISBN Vol.2: 978-0-415-41195-0 (hbk)

Table of Contents

Preface	XXIII

VOLUME 1

Keynote lectures

Landslides: Seeing the ground N.R. Morgenstern & C.D. Martin	3
Limit equilibrium and finite element analysis – A perspective of recent advances Z. Chen & K. Ugai	25
Improving the interpretation of slope monitoring and early warning data through better understanding of complex deep-seated landslide failure mechanisms E. Eberhardt, A.D. Watson & S. Loew	39
Effects of earthquakes on slopes I. Towhata, T. Shimomura & M. Mizuhashi	53
Monitoring and modeling of slope response to climate changes H. Rahardjo, R.B. Rezaur, E.C. Leong, E.E. Alonso, A. Lloret & A. Gens	67
Soil nailing and subsurface drainage for slope stabilization W.K. Pun & G. Urciuoli	85

Special lectures

Loess in China and landslides in loess slopes Z.G. Lin, Z.J. Xu & M.S. Zhang	129
Advances in landslide continuum dynamic modelling S. McDougall, M. Pirulli, O. Hungr & C. Scavia	145
Deformation and failure mechanisms of loose and dense fill slopes with and without soil nails C.W.W. Ng	159
Capturing landslide dynamics and hydrologic triggers using near-real-time monitoring M.E. Reid, R.L. Baum, R.G. LaHusen & W.L. Ellis	179
The effects of earthquake on landslides – A case study of Chi-Chi earthquake, 1999 M.L. Lin, K.L. Wang & T.C. Kao	193
The role of suction and its changes on stability of steep slopes in unsaturated granular soils L. Olivares & P. Tommasi	203
Prediction of landslide movements caused by climate change: Modelling the behaviour of a mean elevation large slide in the Alps and assessing its uncertainties Ch. Bonnard, L. Tacher & M. Beniston	217

Geology, geotechnical properties and site characterization

Geotechnical appraisal of the Sonapur landslide area, Jainita hills, Meghalya, India R.C. Bhandari, P. Srinivasa Gopalan & V.V.R.S. Krishna Murty	231

The viscous component in slow moving landslides: A practical case — 237
D.A. González, A. Ledesma & J. Corominas

The systematic landslide investigation programme in Hong Kong — 243
K.K.S. Ho & T.M.F. Lau

General digital camera-based experiments for large-scale landslide physical model measurement — 249
X.W. Hu, H.M. Tang & J.S. Li

Shear strength of boundaries between soils and rocks in Korea — 257
S.G. Lee, B.S. Kim & S.H. Jung

Cracks in saturated sand — 263
X.B. Lu, S.Y. Wang & P. Cui

Some geomorphological techniques used in constraining the likelihood of landsliding – Selected Australian examples — 267
A.S. Miner, P. Flentje, C. Mazengarb, J.M. Selkirk-Bell & P.G. Dahlhaus

Rock failures in karst — 275
M. Parise

Geotechnical study at Sirwani landslide site, India — 281
V.K. Singh

Inferences from morphological differences in deposits of similar large rockslides — 285
A.L. Strom

Movements of a large urban slope in the town of Santa Cruz do Sul (RGS), Brazil — 293
L.A. Bressani, R.J.B. Pinheiro, A.V.D. Bica, C.N. Eisenberger & J.M.D. Soares

Geotechnical analysis of a complex slope movement in sedimentary successions of the southern Apennines (Molise, Italy) — 299
D. Calcaterra, D. Di Martire, M. Ramondini, F. Calò & M. Parise

Application of surface wave and micro-tremor survey in landslide investigation in the Three Gorges reservoir area — 307
A. Che, X. Luo, S. Feng & O. Yoshiya

A case study for the landslide-induced catastrophic hazards in Taiwan Tuchang Tribute — 313
C.Y. Chen & W.C. Lee

Pir3D, an easy to use three dimensional block fall simulator — 319
Y. Cottaz & R.M. Faure

Characterization of the fracture pattern on cliff sites combining geophysical imaging and laser scanning — 323
J. Deparis, D. Jongmans, B. Fricout, T. Villemin, O. Meric, A. Mathy & L. Effendiantz

In situ characterization of the geomechanical properties of an unstable fractured rock slope — 331
C. Dünner, P. Bigarré, F. Cappa, Y. Guglielmi & C. Clément

Properties of peat relating to instability of blanket bogs — 339
A.P. Dykes

Stability problems in slopes of Arenós reservoir (Castellón, Spain) — 347
J. Estaire, J.A. Díez & C. Olalla

The 22 August, 2006, anomalous rock fall along the Gran Sasso NE wall (Central Apennines, Italy) — 355
G.B. Fasani, C. Esposito, G.S. Mugnozza, L. Stedile & M. Pecci

New formulae to assess soil permeability through laboratory identification and flow coming out of vertical drains 361
J.C. Gress

Structure-controlled earth flows in the Campania Apennines (Southern Italy) 365
F.M. Guadagno, P. Revellino, G. Grelle, G. Lupo & M. Bencardino

Geotechnical and mineralogical characterization of fine grained soils affected by soil slips 373
G. Gullà, L. Aceto, S. Critelli & F. Perri

Vulnerability of structures impacted by debris flow 381
E.D. Haugen & A.M. Kaynia

Engineering geological study on a large-scale toppling deformation at Xiaowan Hydropower Station 389
R. Huang, G. Yang, M. Yan & M. Liu

Characterization of the Avignonet landslide (French Alps) with seismic techniques 395
D. Jongmans, F. Renalier, U. Kniess, S. Schwartz, E. Pathier, Y. Orengo, G. Bièvre, T. Villemin & C. Delacourt

Deformation characteristics and treatment measures of spillway slope at a reservoir in China 403
N. Ju, J. Zhao & R. Huang

Sliding in weathered banded gneiss due to gullying in southern Brazil 409
W.A. Lacerda, A.P. Fonseca & A.L. Coelho Netto

Experimental and three-dimensional numerical investigations of the impact of dry granular flow on a barrier 415
R.P.H. Law, G.D. Zhou, C.W.W. Ng & W.H. Tang

Temporal survey of fluids by 2D electrical tomography: The "Vence" landslide observatory site (Alpes-Maritimes, SE France) 421
T. Lebourg, S. El Bedoui, M. Hernandez & H. Jomard

Characteristics of landslides related to various rock types in Korea 427
S.G. Lee, K.S. Lee, D.C. Park & S. Hencher

Two approaches to identifying the slip zones of loess landslides and related issues 435
T. Li & X. Lin

Testing study on the strength and deformation characteristics of soil in loess landslides 443
H.J. Liao, L.J. Su, Z.D. Li, Y.B. Pan & H. Fukuoka

Failure mechanism of slipping zone soil of the Qiangjiangping landslide in the Three Gorges reservoir area: A study based on Dead Load test 449
X. Luo, A. Che, L. Cao & Y. Lang

Post-failure movements of a large slow rock slide in schist near Pos Selim, Malaysia 457
A.W. Malone, A. Hansen, S.R. Hencher & C.J.N. Fletcher

Characteristics of rock failure in metamorphic rock areas, Korea 463
W. Park, Y. Han, S. Jeon & B. Roh

Shape and size effects of gravel grains on the shear behavior of sandy soils 469
S.N. Salimi, V. Yazdanjou & A. Hamidi

Nonlinear failure envelope of a nonplastic compacted silty-sand 475
D.D.B. Seely & A.C. Trandafir

An investigation of a structurally-controlled rock cut instability at a metro station shaft in Esfahan, Iran 481
A. Taheri

Yield acceleration of soil slopes with nonlinear strength envelope 487
A.C. Trandafir & M.E. Popescu

Evaluation of rockfall hazards along part of Karaj-Chaloos road, Iran 491
A. Uromeihy, N. Ghazipoor & I. Entezam

Coupled effect of pluviometric regime and soil properties on hydraulic boundary
conditions and on slope stability 495
R. Vassallo, C. Di Maio & M. Calvello

Mechanical characters of relaxing zone of slopes due to excavation 501
H. Wang & X.P. Liao

Deformation characteristics and stability evaluation of Ganhaizi landslide in the Dadu River 507
Y. Wang, Y. Sun, O. Su, Y. Luo, J. Zhang, C. Zhou & S. Zhang

Landslide-prone towns in Daunia (Italy): PS interferometry-based investigation 513
J. Wasowski, D. Casarano, F. Bovenga, A. Refice, R. Nutricato & D.O. Nitti

Basic types and active characteristics of loess landslide in China 519
W. Wu, D. Wang, X. Su & N. Wang

Investigation of a landslide using borehole shear test and ring shear test 525
H. Yang, V.R. Schaefer & D.J. White

The importance of geological and geotechnical investigations of landslides occurred
at dam reservoirs: Case studies from the Havuzlu and Demirkent landslides
(Artvin Dam, Turkey) 531
A.B. Yener, S. Durmaz & B.M. Demir

An innovative approach combining geological mapping and drilling process
monitoring for quantitative assessment of natural terrain hazards 535
Z.Q. Yue, J. Chen & W. Gao

Types of cutslope failures along Shiyan-Manchuanguan expressway through
the Liangyun fracture, Hubei Province 543
H. Zhao, R. Wang, J. Fan & W. Lin

Advances in analytical methods, modeling and prediction of slope behavior

Probability limit equilibrium and distinct element modeling of jointed rock slope at northern
abutment of Gotvand dam, Iran 553
M. Aminpoor, A. Noorzad & A.R. Mahboubi

Rock block sliding analysis of a highway slope in Portugal 561
P.G.C. Santarém Andrade & A.L. Almeida Saraiva

Contribution to the safety evaluation of slopes using long term observation results 567
J. Barradas

Delimitation of safety zones by finite element analysis 573
J. Bojorque, G. De Roeck & J. Maertens

Laboratory and numerical modelling of the lateral spreading process involving
the Orvieto hill (Italy) 579
F. Bozzano, S. Martino, A. Prestininzi & A. Bretschneider

Albano Lake coastal rock slide (Roma, Italy): Geological constraints and numerical modelling 585
F. Bozzano, C. Esposito, S. Martino, P. Mazzanti & G. Diano

Superposition principle for stability analysis of reinforced slopes and its FE validation 593
F. Cai & K. Ugai

Soil suction modelling in weathered gneiss affected by landsliding 599
M. Calvello, L. Cascini, G. Sorbino & G. Gullà

Modelling the transient groundwater regime for the displacements analysis of slow-moving active landslides   607
L. Cascini, M. Calvello & G.M. Grimaldi

Numerical modelling of the thermo-mechanical behaviour of soils in catastrophic landslides   615
F. Cecinato, A. Zervos, E. Veveakis & I. Vardoulakis

Some notes on the upper-bound and Sarma's methods with inclined slices for stability analysis   623
Z.Y. Chen

Slope stability analysis using graphic acquisitions and spreadsheets   631
L.H. Chen, Z.Y. Chen & P. Sun

Efficient evaluation of slope stability reliability subject to soil parameter uncertainties using importance sampling   639
J. Ching, K.K. Phoon & Y.G. Hu

Prediction of the flow-like movements of Tessina landslide by SPH model   647
S. Cola, N. Calabrò & M. Pastor

Applications of the strength reduction finite element method to a gravity dam stability analysis   655
Q.W. Duan, Z.Y. Chen, Y.J. Wang, J. Yang & Y. Shao

Study on deformation parameter reduction technique for the strength reduction finite element method   663
Q.W. Duan, Y.J. Wang & P.W. Zhang

Stability and movement analyses of slopes using Generalized Limit Equilibrium Method   671
M. Enoki & B.X. Luong

Long-term deformation prediction of Tianhuangpin "3.29" landslide based on neural network with annealing simulation method   679
F. Zhang, C. Xian, J. Song, B. Guo & Z. Kuai

New models linking piezometric levels and displacements in a landslide   687
R.M. Faure, S. Burlon, J.C. Gress & F. Rojat

3D slope stability analysis of Rockfill dam in U-shape valley   693
X.Y. Feng, M.T. Luan & Z.P. Xu

3-D finite element analysis of landslide prevention piles   697
K. Fujisawa, M. Tohei, Y. Ishii, Y. Nakashima & S. Kuraoka

Integrated intelligent method for displacement predication of landslide   705
W. Gao

A new approach to *in situ* characterization of rock slope discontinuities: the "High-Pulse Poroelasticity Protocol" (HPPP)   711
Y. Guglielmi, F. Cappa, S. Gaffet, T. Monfret, J. Virieux, J. Rutqvist & C.F. Tsang

Fuzzy prediction and analysis of landslides   719
Y. He, B. Liu, W.J. Liu, F.Q. Liu & Y.J. Luan

LPC methodology as a tool to create real time cartography of the gravitational hazard: Application in the municipality of Menton (Maritimes Alps, France)   725
M. Hernandez, T. Lebourg, E. Tric, M. Hernandez & V. Risser

Back-analyses of a large-scale slope model failure caused by a sudden drawdown of water level   731
G.W. Jia, T.L.T. Zhan & Y.M. Chen

Effect of Guangxi Longtan reservoir on the stability of landslide at Badu station of Nankun railway   737
R. Jiang, R. Meng, A. Bai & Y. He

Application of SSRM in stability analysis of subgrade embankments over sloped weak ground with FLAC3D   741
X. Jiang, Y. Qiu, Y. Wei & J. Ling

Strength parameters from back analysis of slips in two-layer slopes 747
J.-C. Jiang & T. Yamagami

Development characteristics and mechanism of the Lianhua Temple landslide, Huaxian county, China 755
J.-Y. Wang, M.-S. Zhang, C.-Y. Sun & Z. Rui

Modeling landslide triggering in layered soils 761
R. Keersmaekers, J. Maertens, D. Van Gemert & K. Haelterman

Numerical modeling of debris flow kinematics using discrete element method combined with GIS 769
H. Lan, C.D. Martin & C.H. Zhou

Three dimensional simulation of landslide motion and the determination of geotechnical parameters 777
Y. Lang, X. Luo & H. Nakamura

Stability analysis and stabilized works of dip bedded rock slopes 783
J.Y. Leng, Z.D. Jing & X.P. Liao

A GIS-supported logistic regression model applied in regional slope stability evaluation 789
X. Li, H. Tang & S. Chen

The stability analysis for FaNai landslide in Lubuge hydropower station 795
K. Li, J. Zhang, S. Zhang & S. He

Numerical analysis of slope stability influenced by varying water conditions in the reservoir area of the Three Gorges, China 803
S. Li, X. Feng & J.A. Knappett

A numerical study of interaction between rock bolt and rock mass 809
X.P. Li & S.M. He

Macroscopic effects of rock slopes before and after grouting of joint planes 815
H. Lin, P. Cao, J.T. Li & X.L. Jiang

Two- and three-dimensional analysis of a fossil landslide with FLAC 821
X.L. Liu & J.H. Deng

Application of the coupled thin-layer element in forecasting the behaviors of landslide with weak intercalated layers 827
Y.L. Luo & H. Peng

Numerical modelling of a rock avalanche laboratory experiment in the framework of the "Rockslidetec" alpine project 835
I. Manzella, M. Pirulli, M. Naaim, J.F. Serratrice & V. Labiouse

Three-dimensional slope stability analysis by means of limit equilibrium method 843
S. Morimasa & K. Miura

Embankment basal stability analysis using shear strength reduction finite element method 851
A. Nakamura, F. Cai & K. Ugai

Back analysis based on SOM-RST system 857
H. Owladeghaffari & H. Aghababaei

Temporal prediction in landslides – Understanding the Saito effect 865
D.N. Petley, D.J. Petley & R.J. Allison

3D landslide run out modelling using the Particle Flow Code PFC^{3D} 873
R. Poisel & A. Preh

Double-row anti-sliding piles: Analysis based on a spatial framework structure 881
T. Qian & H. Tang

Centrifuge modeling of rainfall-induced failure process of soil slope *J.Y. Qian, A.X. Wang, G. Zhang & J.-M. Zhang*	887
A GIS-based method for predicting the location, magnitude and occurrence time of landslides using a three-dimensional deterministic model *C. Qiu, T. Esaki, Y. Mitani & M. Xie*	893
Application of a rockfall hazard rating system in rock slope cuts along a mountain road of South Western Saudi Arabia *B.H. Sadagah*	901
Model tests of collapse of unsaturated slopes in rainfall *N. Sakai & S. Sakajo*	907
Calibration of a rheological model for debris flow hazard mitigation in the Campania region *A. Scotto di Santolo & A. Evangelista*	913
Optical fiber sensing technology used in landslide monitoring *Y.X. Shi, Q. Zhang & X.W. Meng*	921
Finite element analysis of flow failure of Tailings dam and embankments *R. Singh, D. Mitra & D. Roy*	927
Landslide model test to investigate the spreading range of debris according to rainfall intensity *Y.S. Song, B.G. Chae, Y.C. Cho & Y.S. Seo*	933
Occurrence mechanism of rockslide at the time of the Chuetsu earthquake in 2004 – A dynamic response analysis by using a simple cyclic loading model *N. Tanaka, S. Abe, A. Wakai, H. Kawabata, M. Genda & H. Yoshimatsu*	939
Analysis for progressive failure of the Senise landslide based on Cosserat continuum model *H.X. Tang*	945
Large-scale deformation of the La Clapière landslide and its numerical modelling (S.-E. de Tinée, France). *E. Tric, T. Lebourg & H. Jomard*	951
A novel complex valued neuron model for landslide assessment *K. Tyagi, V. Jindal & V. Kumar*	957
Prediction of slope behavior for deforming railway embankments *V.V. Vinogradov, Y.K. Frolovsky, A. Al. Zaitsev & I.V. Ivanchenko*	963
Finite element simulation for the collapse of a dip slope during 2004 Mid Niigata Prefecture earthquake in Japan *A. Wakai, K. Ugai, A. Onoue, K. Higuchi & S. Kuroda*	971
Sensitivity of stability parameters for soil slopes: An analysis based on the shear strength reduction method *R. Wang, X.Z. Wang, Q.S. Meng & B. Hu*	979
Back analysis of unsaturated parameters and numerical seepage simulation of the Shuping landslide in Three Gorges reservoir area *S. Wang, H. Zhang, Y. Zhang & J. Zheng*	985
Slope failure criterion: A modification based on strength reduction technique *Y.G. Wang, R. Jing, W.Z. Ren & Z.C. Wang*	991
Unsaturated seepage analysis for a reservoir landslide during impounding *J.B. Wei, J.H. Deng, L.G. Tham & C.F. Lee*	999
A simple compaction control method for slope construction *L.D. Wesley*	1005

Numerical analysis of soil-arch effect of anti-slide piles *Y. Xia, X. Zheng & R. Rui*	1011
Determination of the critical slip surface based on stress distributions from FEM *D. Xiao, C. Wu & H. Yang*	1017
Effect of drainage facilities using 3D seepage flow analysis reflecting hydro-geological structure with aspect cracks in a landslide – Example of analysis in OODAIRA Landslide area *M. Yamada & K. Ugai*	1023
Back analysis of soil parameters: A case study on monitored displacement of foundation pits *B. Yan, X.T. Peng & X.S. Xu*	1031
3D finite element analysis on progressive failure of slope due to rainfall *G.L. Ye, F. Zhang & A. Yashima*	1035
Block-group method for rock slope stability analysis *Z. Zhang, Y. Xu & H. Wu*	1043
Quantitative study on the classification of unloading zones of high slope *D. Zheng & R.Q. Huang*	1051
Investigations on the accuracy of the simplified Bishop method *D.Y. Zhu*	1055
Author index	1059

VOLUME 2

Landslide mechanism, monitoring and warning

GIS-based landslide susceptibility mapping in the Three Gorges area – Comparisons of mapping results obtained by two methods: Analytical hierarchy process and logistic regression *S. Bai, J. Wang, G. Lu, P. Zhou, S. Hou & F. Zhang*	1067
Importance of study of creep sliding mechanism to prevention and treatment of reservoir landslide *J. Bai, S. Lu, J. Han*	1071
Stability prediction of landsides before and after impoundment for Lijiaxia hydropower station *J. Bai, S. Lu & J. Han*	1077
The technical concept within the Integrative Landslide Early Warning System (ILEWS) *R. Bell, B. Thiebes, T. Glade, R. Becker, H. Kuhlmann, W. Schauerte, S. Burghaus, H. Krummel, M. Janik & H. Paulsen*	1083
The Åknes rockslide: Monitoring, threshold values and early-warning *L.H. Blikra*	1089
DInSAR techniques for monitoring slow-moving landslides *D. Calcaterra, M. Ramondini, F. Calò, V. Longobardi, M. Parise, C.M. Galzerano & C. Terranova*	1095
Multitemporal DInSAR data and damages to facilities as indicators for the activity of slow-moving landslides *L. Cascini, S. Ferlisi, D. Peduto, G. Pisciotta, S. Di Nocera & G. Fornaro*	1103
The Serre La Voute Landslide (North-West Italy): Results from ten years of monitoring *M. Ceccucci, G. Maranto & G. Mastroviti*	1111
Onset of rockslide by the peak-residual strength drop *Q.G. Cheng & G.T. Hu*	1119
Analysis of mechanism of the K31 landslide of Changzhi-Jincheng express highway *Y. Cheng*	1127

A plane-torsion rockslide with a locked flank: A case study *Q. Cheng*	1133
Monitoring of natural thermal strains using hollow cylinder strain cells: The case of a large rock slope prone to rockfalls *C. Clément, Y. Gunzburger, V. Merrien-Soukatchoff & C. Dünner*	1143
Landslide hazards mapping and permafrost slope InSAR monitoring, Mackenzie valley, Northwest Territories, Canada *R. Couture & S. Riopel*	1151
Advanced monitoring criteria for precocious alerting of rainfall-induced flowslides *E. Damiano, L. Olivares, A. Minardo, R. Greco, L. Zeni & L. Picarelli*	1157
Investigation of slope failure mechanisms caused by discontinuous large scale geological structures at the Cadia Hill Open Pit *J. Franz & Y. Cai*	1165
Two approaches for public landslide awareness in the United States – U.S. geological survey warning systems and a landslide film documentary *L.M. Highland & P.L. Gori*	1173
Formation and mechanical analysis of Tiantai landslide of Xuanhan county, Sichuan province *R.Q. Huang, S. Zhao & X. Song*	1177
Development of wireless sensor node for landslide detection *H.W. Kim*	1183
Redox condition and landslide development *Y.H. Lang, S.Y. Liang & G.D. Zheng*	1189
Prepa displacement mechanism and its treatment measures for Hancheng landslide *T.F. Li & L.C. Dang*	1195
Investigation of the stability of colluvial landslide deposits *X. Li & L.M. Zhang*	1205
Choice of surveying methods for landslides monitoring *S.T. Liu & Z.W. Wang*	1211
No. 1 landslide on the eastern approach road to ErLang Mountain tunnel: Inference factors and controlling measures *H.M. Ma & Z.P. Zhang*	1217
Estimation of landslide load on multi-tier pile constructions with the help of a combined method *S.I. Matsiy & Ph.N. Derevenets*	1225
The use of PSInSAR™ data in landslide detection and monitoring: The example of the Piemonte region (Northern Italy) *C. Meisina, F. Zucca, D. Notti, A. Colombo, A. Cucchi, G. Savio, C. Giannico & M. Bianchi*	1233
Fill slopes: Stability assessment based on monitoring during both heavy rainfall and earthquake motion *T. Mori, M. Kazama, R. Uzuoka & N. Sento*	1241
The mechanism of movement of mud flows in loess soils, successful and unsuccessful cases of forecast *R.A. Niyazov, Sh.B. Bazarov & A.M. Akhundjanov*	1247
Influence of fine soil particles on excess of basal pore-fluid pressure generation in granular mass flows *Y. Okada & H. Ochiai*	1253
An early warning system to predict flowslides in pyroclastic deposits *L. Pagano, G. Rianna, M.C. Zingariello, G. Urciuoli & F. Vinale*	1259

Monitoring and modeling of slope movement on rock cliffs prior to failure 1265
N.J. Rosser & D.N. Petley

Active tectonic control of a large landslide: Results from Panagopoula landslide multi parametric analyses 1273
S. El Bedoui, T. Lebourg & Y. Guglielmi

A warning system using chemical sensors and telecommunication technologies to protect railroad operation from landslide disaster 1277
H. Sakai

Distributive monitoring of the slope engineering 1283
B. Shi, H. Sui, D. Zhang, B. Wang, G. Wei & C. Piao

Observational method in the design of high cutting slope around bridge 1289
S. Sun, B. Zhu, B. Zheng & J. Zhang

Ultrasonic monitoring of lab-scaled underwater landslides 1297
Q.H. Truong, C. Lee, H.K. Yoon, Y.H. Eom, J.H. Kim & J.S. Lee

Interaction between landslides and man-made works 1301
G. Urciuoli & L. Picarelli

Desiccation fissuring induced failure mechanisms for clay levees 1309
S. Utili, M. Dyer, M. Redaelli & M. Zielinski

Stability analysis by strength reduction finite element method and monitoring of unstable slope during reinforcement 1315
Z.Q. Wang, H.F. Li & L.M. Zhang

Displacement monitoring on Shuping landslide in the Three Gorges Dam reservoir area, China from August 2004 to July 2007 1321
F.W. Wang, G. Wang, Y.M. Zhang, Z.T. Huo, X.M. Peng, K. Araiba & A. Takeuchi

Deformation mechanism and prevention measure for strongly expansive soft-rock slope in the Yanji basin 1329
X. Wu, N. Xu, H. Tian, Y. Sun & M. He

Twenty years of safety monitoring for the landslide of Hancheng PowerStation 1335
M.J. Wu, Z.C. Li, P.J. Yuan & Y.H. Jiang

A time-spatial deterministic approach to assessment of rainfall-induced shallow landslide 1343
M.W. Xie, C. Qiu & Z.F. Wang

Introduction of web-based remote-monitoring system and its application to landslide disaster prevention 1349
M. Yamada & S. Tosa

Deformation mechanism for the front slope of the left bank deposits in Xiluodu hydro-electrical power station, China 1355
M. Yan, Z. Wu, R. Huang, Y. Zhang & S. Wang

Monitoring of soil nailed slopes and dams using innovative technologies 1361
J.-H. Yin, H.-H. Zhu & W. Jin

Application of multi-antenna GPS technique in the stability monitoring of roadside slopes 1367
Q. Zhang, L. Wang, X.Y. Zhang, G.W. Huang, X.L. Ding, W.J. Dai & W.T. Yang

Effects of earthquakes on slopes

Influences of earthquake motion on slopes in a hilly area during the Mid-Niigata Prefecture Earthquake, 2004 1375
S. Asano & H. Ochiai

The 1783 Scilla rock-avalanche (Calabria, southern Italy) — 1381
F. Bozzano, S. Martino, A. Prestininzi, M. Gaeta, P. Mazzanti & A. Montagna

Self-excitation process due to local seismic amplification and earthquake-induced reactivations of large landslides — 1389
F. Bozzano, S. Martino, G. Scarascia Mugnozza, A. Paciello & L. Lenti

Geological constraints to the urban shape evolution of Ariano Irpino (Avellino province, Italy) — 1397
D. Calcaterra, C. Dima & E. Grasso

Landslide zones and their relation with seismoactive fault systems in Azerbaijan, Iran — 1405
E. Ghanbari

Ground movements caused by lateral spread during an earthquake — 1409
S.C. Hsu, B.L. Chu & C.C. Lin

2-D analysis of slope stability of an infinite slope during earthquake — 1415
J. Liu, J. Liu & J. Wang

High-cutting slopes at Qingshuichuan electric power plant in the North of Shaanxi: Deformation and failure modes and treatment scheme — 1421
H. Liu, Z. Liu & Z. Yan

GIS-based real time prediction of Arias intensity and earthquake-induced landslide hazards in Alborz and Central Iran — 1427
M. Mahdavifar, M.K. Jafari & M.R. Zolfaghari

Geomorphology of old earthquake-induced landslides in southeastern Sicily — 1433
P.G. Nicoletti & E. Catalano

Coseismic movement of an active landslide resulting from the Mid-Niigata Prefecture Earthquake, Japan — 1439
T. Okamoto, S. Matsuura & S. Asano

Characteristics of large rock avalanches triggered by the November 3, 2002 Denali Fault earthquake, Alaska, USA — 1447
W.H. Schulz, E.L. Harp & R.W. Jibson

FE analysis of performance of the Lower and Upper San Fernando Dams under the 1971 San Fernando earthquake — 1455
C. Takahashi, F. Cai & K. Ugai

Reduction of the stability of pre-existing landslides during earthquake — 1463
B. Tiwari, I. Dhungana & C.F. Garcia

Probabilistic hazard mapping of earthquake-induced landslides — 1469
H.B. Wang, S.R. Wu, G.H. Wang & F.W. Wang

Investigation on stability of landfill slopes in seismically active regions in Central Asia — 1475
A.W. Wu, B.G. Tensay, S. Webb, B.T. Doanth, C.M. Ritzkowski, D.Z. Muhidinov & E.M. Anarbaev

Mechanism for loess seismic landslides in Northwest China — 1481
L. Yuan, X. Cui, Y. Hu & L. Jiang

Climate, hydrology and landslides

Evaluation of the landslide potential in Chahr Chay dam reservoir slopes — 1489
K. Badv & K. Emami

Effect of well pumping on groundwater level and slope stability in the Taiwan Woo-Wan-Chai landslide area — 1493
M. Chang, B.R. Li, Y.S. Zhang, H.S. Wang, Y.H. Chou & H.C. Liu

Case study: Embankment failure of Cable-Ski Lake development in Cairns *K. Chen*	1501
Analysis method for slope stability under rainfall action *X.D. Chen, H.X. Guo & E.X. Song*	1507
Hydrological modelling of the Vallcebre landslide *J. Corominas, R. Martín & E. Vázquez-Suñé*	1517
Landslides in stiff clay slopes along the Adriatic coast (Central Italy) *F. Cotecchia, O. Bottiglieri, L. Monterisi & F. Santaloia*	1525
Research on the effect of atomized rain on underground water distribution in Dayantang landslide *J. Ding*	1533
Landslide hydrogeological susceptibility in the Crati valley (Italy) *P. Gattinoni*	1539
Sustainable landslide stabilisation using deep wells installed with siphon drains and electro-pneumatic pumps *A. Gillarduzzi*	1547
Biological and engineering impacts of climate on slopes – learning from full-scale *S. Glendinning, P.N. Hughes, D.A.B. Hughes, D. Clarke, J. Smethurst, W. Powrie, N. Dixon, T.A. Dijkstra, D.G. Toll & J. Mendes*	1553
Some attributes of road-slope failure caused by typhoons *M.W. Gui, C.H. Chang & S.F. Chen*	1559
A small rock avalanche in toppled schist, Lake Wanaka, New Zealand *G.S. Halliday*	1565
NRCS-based groundwater level analysis of sloping ground *L.I. Ju, O.T. Suk, M.Y. Il & L.S. Gon*	1571
A numerical case study on load developments along soil nails installed in cut slope subjected to high groundwater table *A.K.L. Kwong & C.F. Lee*	1575
Landslides at active construction sites in Hong Kong *T.M.F. Lau, H.W. Sun, H.M. Tsui & K.K.S. Ho*	1581
Landslide "Granice" in Zagreb (Croatia) *Z. Mihalinec & Ž. Ortolan*	1587
Improvement of subsurface drainage provisions for recompacted soil fill slopes in Hong Kong *K.K. Pang, J.M. Shen, K.K.S. Ho & T.M.F. Lau*	1595
Biotechnical slope stabilization and using Spyder Hoe to control steep slope failure *P. Raymond*	1603
Rapid landslides threatening roads: Three case histories of risk mitigation in the Umbria region of Central Italy *D. Salciarini, P. Conversini, E. Martini, P. Tamburi & L. Tortoioli*	1609
Assessment of the slope stabilisation measures at the Cadas Pangeran road section, Sumedang, West Java *D. Sarah, A. Tohari & M.R. Daryono*	1615
Analysis of control factors on landslides in the Taiwan area *K. Shou, B. Wu & H. Hsu*	1621
Inclined free face riverbank collapse by river scouring *J.C. Sun & G.Q. Wang*	1627
Drainage control and slope stability at an open pit mine: A GIS-based hydrological modeling *C. Sunwoo, Y.S. Choi, H.D. Park & Y.B. Jung*	1633

Assessment of regional rainfall-induced landslides using 3S-based hydro-geological model   1639
C.H. Tan, C.Y. Ku, S.Y. Chi, Y.H. Chen, L.Y. Fei, J.F. Lee & T.W. Su

Investigation of a landslide along a natural gas pipeline (Karacabey-Turkey)   1647
T. Topal & M. Akin

Influence of extreme rainfall on the stability of spoil heaps   1653
I. Vanicek & S. Chamra

Behavior of expansive soil slope reinforced with geo-grids   1659
M.Y. Wang, X.N. Gong, M.Y. Wang, J.T. Cai & H. Xu

Geotechnical properties for a rainstorm-triggered landslide in Kisawa village, Tokushima Prefecture, Japan   1667
G. Wang & A. Suemine

Yigong rock avalanche-flow landslide event, Tibet, China   1675
Q. Xu, S.T. Wang, H.J. Chai, Z.Y. Zhang & S.M. Dong

Key issues of emergency measures and comprehensive remediation projects to control the Danba landslide, Sichuan province, China   1681
Q. Xu, X.-M. Fan, L.-W. Jiang & P. Liu

Enhanced slope seepage resulting from localized torrential precipitation during a flood discharge event at the Nuozhadu hydroelectric station   1689
M. Xu, Y. Ma, X.B. Kang & G.P. Lu

An issue in conventional approach for drainage design on slopes in mountainous regions   1697
Z.Q. Yue

Analysis of geo-hazards caused by climate changes   1703
L.M. Zhang

Slope stabilization and protection

Back experience of deep drainage for landslide stabilization through lines of siphon drains and electro-pneumatics drains: A French railway slope stabilization example   1713
S. Bomont

Experimental geo-synthetic-reinforced segmental wall as bridge abutment   1721
R.M. Faure, D. Rossi, A. Nancey & G. Auray

Rock slope stability analysis for a slope in the vicinity of Take-off Yard of Karun-3 Dam   1725
M. Gharouni-Nik

Stabilization of a large paleo-landslide reactivated because of the works to install a new ski lift in Formigal skiing resort   1731
J. González-Gallego, J. Moreno Robles, J.L. García de la Oliva & F. Pardo de Santayana

A case study on rainfall infiltration effect on the stability of two slopes   1737
M.W. Gui & K.K. Han

Consolidation mechanism of fully grouted anchor bolts   1745
S. He, Y. Wu & X. Li

Stability analysis for cut slopes reinforced by an earth retention system by considering the reinforcement stages   1751
W.P. Hong, Y.S. Song & T.H. Kim

Landslide stabilization for residential development   1757
I. Jworchan, A. O'Brien & E. Rizakalla

Influence of load transfer on anchored slope stability   1763
S.K. Kim, N.K. Kim, Y.S. Joo, J.S. Park, T.H. Kim & K.S. Cha

Review of slope surface drainage with reference to landslide studies and current practice in Hong Kong 1769
T.M.F. Lau, H.W. Sun, T.H.H. Hui & K.K.S. Ho

Analysis of dynamic stability about prestressed anchor retaining structure 1775
H. Li, X. Yang, H. Liu & L. Du

Safety analysis of high engineering slopes along the west approach road of ZheGu mountain tunnel 1781
T.B. Li, Y. Du & X.B. Wang

Landslide stabilizing piles: A design based on the results of slope failure back analysis 1787
M.E. Popescu & V.R. Schaefer

Landslides on the left abutment and engineering measures for Manwan hydropower project 1795
X. Tang & Q. Gao

Factors resulting in the instability of a 57.5 m high cut slope 1799
J.J. Wang, H.J. Chai, H.P. Li & J.G. Zhu

Orthogonal analysis and applications on anchorage parameters of rock slopes 1805
E.C. Yan, H.G. Li, M.J. Lv & D.L. Li

Waste rock dump slope stability for a gold mine in California 1811
H. Yang, G.C. Rollins & M. Kim

Properties of the high rock slope of Hongjiadu hydropower project and its engineering treatment measures 1817
Z. Yang, W. Xiao & D. Cai

Typical harbor bank slopes in the Three Gorges reservoir: Landslide and collapse and their stability control 1825
A. Yao, C. Heng, Z. Zhang & R. Xiang

Weighting predisposing factors for shallow slides susceptibility assessment on the regional scale 1831
J.L. Zêzere, S.C. Oliveira, R.A.C. Garcia & E. Reis

Analyses of mechanism of landslides in Tongchuan-Huangling highway 1839
L. Zhang & H. He

Treatments of Loess-Bedrock landslides at Chuankou in Tongchuan-Huangling expressway 1847
J.B. Zhao

Types, characteristics and application conditions of anti-slide retaining structures 1855
J. Zheng & G. Wang

The stabilization of the huge alluvial deposit on the left bank and the high rock slope on the right bank of the XiaoWan Hydropower Project 1863
L. Zou, X. Tang, H. Feng, G. Wang & H. Xu

Risk assessment and management

Malaysian National Slope Master Plan – Challenges to producing an effective plan 1873
C.H. Abdullah & A. Mohamed

Spatial landslide risk assessment in Guantánamo Province, Cuba 1879
E. Castellanos Abella & C.J. van Westen

Landslide risk management: Experiences in the metropolitan area of Recife – Pernambuco, Brazil 1887
A.P. Nunes Bandeira, & R. Quental Coutinho

Societal risk due to landslides in the Campania region (Southern Italy) L. Cascini, S. Ferlisi & E. Vitolo	1893
Landslide risk in the San Francisco Bay region J.A. Coe & R.A. Crovelli	1899
A first attempt to extend a subaerial landslide susceptibility analysis to submerged slopes: The case of the Albano Lake (Rome, Italy) G.B. Fasani, C. Esposito, F. Bozzano, P. Mazzanti & M. Floris	1905
Landslide susceptibility zonation of the Qazvin-Rasht-Anzali railway track, North Iran H. Hassani & M. Ghazanfari	1911
Assessment of landslide hazard of a cut-slope using linear regression analysis S. Jamaludin, B.B.K. Huat & H. Omar	1919
Global monitoring strategy applied to ground failure hazards E. Klein, C. Nadim, P. Bigarré & C. Dünner	1925
Regional slope stability zonation based on the factor overlapping method J.F. Liu, G.Q. Ou, Y. You & J.F. Lui	1933
Landslide hazard and risk assessment in the areas of dams and reservoirs of Serbia P. Lokin & B. Abolmasov	1939
The evaluation of failure probability for rock slope based on fuzzy set theory and Monte Carlo simulation H.J. Park, J.G. Um & I. Woo	1943
Macro-zoning of areas susceptible to flowslide in pyroclastic soils in the Campania region L. Picarelli, A. Santo, G. Di Crescenzo & L. Olivares	1951
Zoning methods for landslide hazard degree J. Qiao & L.L. Shi	1959
A proposal for a reliability rating system for fluvial flood defence embankments in the United Kingdom M. Redaelli, S. Utili & M. Dyer	1965
Simplified risk analysis chart to prevent slope failure of highway embankment on soft Bangkok clays A. Sawatparnich & J. Sunitsakul	1971
Determining landslide susceptibility along natural gas pipelines in Northwest Oregon, USA J.I. Theule, S.F. Burns & H.J. Meyer	1979
Landslide susceptibility assessment using fuzzy logic Z. Wang, D. Li & Q. Cheng	1985
Prediction of the spatiotemporal distribution of landslides: Integrated landslide susceptibility zoning techniques and real-time satellite rainfall H. Yang, R.F. Adler, G.J. Huffman & D. Bach	1991
Entropy based typical landslide hazard degree assessment in Three Gorges Z. Yang & J. Qiao	1995
The optimal hydraulic cross-section design of the "Trapezoid-V" shaped drainage canal of debris flow Y. You, H.L. Pan, J.F. Liu, G.Q. Ou & H.L. Pan	2001
Practice of establishing China's Geo-Hazard Survey Information System K. Zhang, Y. Yin & H. Chen	2005
A XML-supported database for landslides and engineered slopes related to China's water resources development Y. Zhao & Z. Chen	2011

Landslide and engineered slopes in China

Failure and treatment technique of a canal in expansive soil in South to North Water Diversion project Y.J. Cai, X.R. Xie, L. Luo, S.F. Chen & M. Zhao	2019
High slope engineering for Three Gorges ship locks G.J. Cao & H.B. Zhu	2027
Large-scale landslides in China: Case studies R.Q. Huang	2037
Early warning for Geo-Hazards based on the weather condition in China C.Z. Liu, Y.H. Liu, M.S. Wen, C. Tang, T.F. Li & J.F. Lian	2055
Slope engineering in railway and highway construction in China G. Wang, H. Ma, M. Feng & Y. Wang	2061
Mining slope engineering in China S. Wang, Q. Gao & S. Sun	2075
Structure and failure patterns of engineered slopes at the Three Gorges reservoir Y.P. Yin	2089
Slope engineering in hydropower projects in China J.P. Zhou & G.F. Chen	2101
A thunder at the beginning of the 21st century – The giant Yigong Landslide Z.H. Wang	2111
Author index	2119

Landslides and Engineered Slopes – Chen et al. (eds)
© 2008 Taylor & Francis Group, London, ISBN 978-0-415-41196-7

Preface

The city of Xi'an, China is privileged to have the honor of hosting the 10th International Symposium on Landslides and Engineered Slopes, following its predecessors: Rio de Janeiro, Brazil, 2004; Cardiff, U.K., 2000; Trondheim, Norway, 1996; Christchurch, New Zealand, 1992; Lausanne, Switzerland, 1988; Toronto, Canada, 1984; New Delhi, India, 1980; Tokyo, Japan, 1977; and Kyoto, Japan, 1972.

China is one of the countries in the world that suffer severely from landslide hazards. Statistics have shown that every year 700 to 900 people are killed by landslides. With the large scale infrastructure construction, failures of engineered slopes are increasing, and have become a serious concern of the government, various enterprises and technical societies. The Chinese geological and geotechnical communities look forward to this unique opportunity of exchanging and sharing technical know-how and experience of combating landslides disasters with our international peers.

Xi'an is a historical city of China. It has been the capital for China's twelve dynasties, spanning over 1200 years, and was a starting point of the famous Silk Road. It is a nice place for participants from all over the world to meet and review our past experience, and in the meanwhile, look forward to a productive future against landslides and slope failures.

From this book, readers will find that there are 7 Symposium Themes, as allocated by the Steering Committee, including 13 keynote and special papers. At the JTC1 meeting held at the 9th Symposium, it was decided that a special session entitled 'China Afternoon' would be organized, whose papers are arranged under a separate theme 'Landslides and Engineered Slopes in China' in this Proceedings.

This symposium was jointly organized by the Chinese Institution of Soil Mechanics and Geotechnical Engineering-CCES, Chinese National Commission on Engineering Geology, Chinese Society of Rock Mechanics and Engineering, and the Geotechnical Division of the Hong Kong Institution of Engineers. The Organizing Committee is grateful to the reviewers for reviewing more than 300 submitted papers. It is practically impossible to list the large number of these volunteers here. However, their contributions must be fully acknowledged, without which the quality of this book would not have been maintained.

Special thanks also go to the members of JTC1 for their constant attention and useful comments during the preparation of this symposium.

Zuyu Chen

Keynote lectures

Landslides: Seeing the ground

Norbert R. Morgenstern & C. Derek Martin
Dept. Civil & Environmental Engineering, University of Alberta, Edmonton, Canada

ABSTRACT: Landslide engineering requires the consideration of a number of complex processes ranging from geological and hydrogeological characterization to geomechanical characterization, analyses and risk management. This paper concentrates on recent advances that improve site characterization applied to landslide problems. It presents the view that one of the most exciting developments is the growing potential for application of Geographical Information Systems (GIS) and that making GIS goetechnically smart is a transformative development. Examples are given of integrating remote sensing data in GIS to improve visualization, mapping and movement characterization. Application of analysis of rockfall within GIS and complex slope stability evaluation with the aid of GIS are presented to illustrate recent developments and provide direction for future enhancements.

1 INTRODUCTION

A landslide, whether it occurs in a natural or an engineered slope, is a complex process. When Laurits Bjerrum, at the end of his Terzaghi Lecture (Bjerrrum, 1967), reminded us of the recognition in Japan of "a landslide devil who seems to laugh at human incompetence", he was reminding us of the complexity of the landslide process.

Managing complexity invariably requires simplification into a Process Model. A Process Model captures the essentials required to meet the objectives of using the model, without including details that are extraneous to these objectives. In geotechnical engineering these objectives can range from ensuring that an engineered structure will perform as intended to managing the risk associated with natural hazards over a larger scale. Establishing the appropriate process is both site and project dependent. It underpins the value associated with the practice of geotechnical engineering.

Understanding the landslide process and being able to simplify it effectively calls on interpreting a number of contributory processes and activities. The main ones are as follows:

- geomorphology—the multiplicity of physical and chemical processes that have affected the surface and near-surface of the site
- hydrology—the role of surface water in infiltration, erosion, etc.
- geology—the sequence and characteristics of the soils and rocks
- hydrogeology—the factors affecting the groundwater distribution
- geotechnical site characterization
- the geotechnical model, seepage, stability and deformation analyses
- risk assessment and risk mitigation

There has been very substantial progress in all of these areas in recent years. New tools are applied to site characterization. The range of geomechanical models that can be usefully applied in practice has grown substantially. The capacity to analyse often exceeds the capacity to characterize. Risk assessment and management of slopes is maturing quickly as a valuable tool for dealing with landslides both locally and regionally. Yet much uncertainty persists in geotechnical practice. The intrinsic presence of uncertainty in geotechnical practice was emphasized by Morgenstern (2000) who provided numerous examples of unanticipated behaviour of geotechnically engineered facilities, often with unfavourable results.

In developing the theme for this paper, we have drawn on our experience to conclude that the greatest uncertainties in the process modeling of landslides arise from inadequacies in site characterization, in the broadest sense, and therefore we concentrate on a discussion of recent advances that improve site characterization applied to landslide problems.

2 VIEWING THE GROUND SURFACE

2.1 Geographical Information Systems (GIS)

It is our view that one of the most exciting developments for landslide engineering is the growing potential for application of GIS. The power of GIS is that it enables us to ask questions of a database, perform spatial operations on databases and generate graphic output that would be laborious or impossible to do manually. Rhind (1992) observes that a GIS can answer five generic questions:

Question	Type of Task
1. What is at … ..?	Inventory
2. Where is. ….?	Monitoring
3. What has changed since…?	Inventory and monitoring
4. What spatial pattern exists..?	Spatial analysis
5. What if. ….?	Modelling

The first three questions are simple queries, while the last two are more analytical.

GIS on its own adds enormously to our capacity to see and interpret surface geospatial information which is essential for landslide engineering. A first fly-by experience in GIS soon convinces the landslide engineer of its potential. However, GIS has limitations in presenting three-dimensional (3D) geologic and geotechnical data since it was originally developed to deal with two dimensional plane problems. Some GIS systems, like ArcGIS, provide a functional developer kit which can be used to develop the 3D capability for geotechnical engineering problems. As pointed out by Lan & Martin (2007), the many current developments in 3D GIS are still not sufficient to meet the needs of the geotechnical engineer. Mining software such as Surpac Vision provides a comprehensive system for geological modeling, but not geotechnical modeling. For example, while three-dimensional solid modeling and two-dimensional sections can be easily created in Surpac, querying inclinometer or piezometer data is not easily accomplished. However an integrated system, which is illustrated in Figure 1, can be developed.

Even within Stage 1, limited ground behaviour can be modelled. The aim of Stage 2 is to establish a comprehensive ground model of the site. Stage 3 links geotechnical numerical analysis tools to conduct geotechnical analyses and assist in decision-making. Making GIS geotechnically smart is a transformative development for geotechnical engineering. Examples to illustrate this will follow in subsequent sections of the paper which will return to demonstrate the role of GIS in a number of slope related problems.

2.2 *Aerial and terrestrial photographs*

Aerial photographs are a well established resource for landslide studies and this is well-understood (Soeters & Van Westen 1996). Aerial photographs can be used for interpretation (API) for qualitative analysis and photogrammetry for extracting quantitative information. The former has been an essential tool for

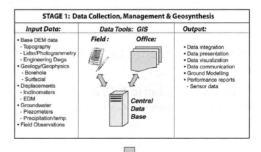

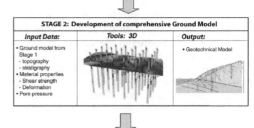

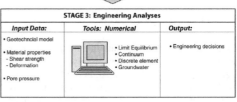

Figure 1. The architecture for an integrated system. It is composed of three different stages which required the implementation of specific tasks.

the landslide engineer for many years, although often neglected in many geotechnical curricula. The latter has usually been the preserve of specialists.

To recognize landslides, API relies on characteristic morphology, vegetation and drainage. Parise (2003) provides an example of how diagnostic surface features can be related to certain types of movement, the degree of activity and the depth of movement. The study of sequential photographs can provide information on the progressive evolution of landslides and can lead to a better understanding of their causes (Chandler & Brunsden, 1995; Van Westen & Getahun, 2003). GIS facilitates the application of API, the archiving of the photos and the production of geomorphological maps that arise from the interpretation.

While the application of API is common, more quantitative studies are rare, probably due to limited availability of good quality photographs, adequately fixed control points and cost. Modern photogrammetric software has been developed that should encourage greater use of photogrammetry for the construction of high quality digital elevation models (DEM). Differential DEMs will quantify landslide movements.

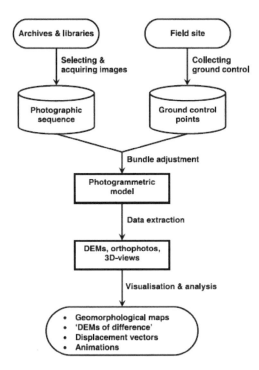

Figure 2. Flowchart of the working procedures used by Walstra et al (2007) for using historical aerial photographs in landslide assessment.

Walstra, Dixon & Chandler (2007) provide a flow chart for this process, reproduced here as Figure 2, and then illustrate its application to the Mam Tor landslide, looking back over photographs of adequate quality to 1953, and to the East Penwyn landslide over a comparable period. The need for accurate orthophotos are common to both the studies discussed in this paper as well as any spatial quantification and referencing within GIS. As the authors observe, the two case histories have demonstrated the value of the aerial photographic archive to extract spatial data necessary for assessing landslide dynamics.

Terrestrial-based digital photography is also gaining acceptance as an efficient method for creating DEMs as well as capturing geological structure of rock slopes (Pötsch et al. 2007). When designing remedial measures, the DEM's can be used to produce sections as well as estimate volumes. Findley (2007) describes the application of Sirovision technology using terrestrial-based digital photography to characterise a rock slope along Interstate 90 USA. He notes that the while the technology saves time and effort it is not a substitute for field mapping. It should also be noted that the technique works best when the slopes are bare of vegetation and, under good conditions, can produce DEM's that are comparable to those produced by terrestrial based LiDAR systems (Martin et al. 2007).

2.3 *Satellite sensing*

Van Westen (2007) has recently summarized the use of remote sensing imagery in creating landslide inventories and notes that medium resolution satellite imagery such as LANDSAT, SPOT, ASTER, IRIS-D etc. are used routinely to create landuse maps and inventories of landslides. At a broader scale, areas of global landslide susceptibility have been determined by correlating information on land surface features from the NASA Shuttle Radar Topography Mission (SRTM) which can resolve features up to 30 m in size. Correlations with other satellite-based information, such as that related to precipitation, results in a global landslide susceptibility map that is broadly supported empirically (Hong, Adler & Huffman 2007). While technically fascinating, the restricted resolution of the DEM and the lack of sub-surface information limits the value of this very remote sensing information for the landslide engineer.

While optical images with resolutions larger than 3 m (see medium resolution list above) have proven useful for interpretation of landslides in some individual cases (Singhroy 2005), Van Westen (op. cit) advises that very high resolution imagery (QuickBird, IKONOS, CARTOSAT-1, etc.) has become the best option for landslide mapping from satellite imagery and the number of earth observation satellites with stereo capabilities and resolution of 3 m or better is increasing rapidly.

Van Westen (op cit) also draws attention to the high resolution imagery that is available within Google Earth and notes that the 3-D capabilities and zooming functions that are available in Google Earth, together with the possibility of drawing polygons on the image, greatly facilitates the interpretation and mapping of slopes and landslides. These images can be transported into GIS for storage and mensuration.

2.4 *LiDAR technology*

LiDAR (Light Detection and Ranging) has become such an essential tool in the mapping and characterizing of landslides, that it is becoming difficult to imagine assessing all but the most local problems without it. Airborne LiDAR uses a powerful laser to map the ground surface in swathes. The literature on the application of LiDAR to landslide studies is growing rapidly and the procedures are becoming almost routine in professional practice in some countries. In our view, the definitive studies published so far are those carried out by the U.S. Geological Survey to assess landslide susceptibility in Seattle, Washington (Schulz 2004, Schulz 2007).

In this case, airborne laser pulses were uniformly spaced within a 600 m wide swath with an average pulse density of $1/m^2$. Up to four laser returns were collected for each pulse resulting in a vertical profile of ground features for each pulse location. Each pulse generates multiple returns due to reflections from features such as powerlines, buildings, trees, undergrowth and the ground surface. Simultaneous acquisitions of aircraft position and laser direction located laser returns with absolute vertical and horizontal accuracy of 15 cm and less than 1 m, respectively (Schulz 2004). Swathes are stitched together into a seemless DEM during processing.

All ground features that produce returns are represented in the laser survey, including buildings, trees and boulders. One of the most valuable developments is that the trees can be stripped away because some pulses penetrate the tree canopy and others are reflected off the forest floors. The latter can be separated from reflections from the trees to produce bald-earth DEM's. This processing for deforestation is a remarkable contribution but as pointed out by Haugerud & Harding (2001), there are some limitations in the algorithms that need to be recognized in interpreting the bare-earth DEM's. The technique has even been applied to faulting studies in high-relief Alpine landscapes, with spectacular results (Cunningham et al. 2006).

In the case of the Seattle bare-earth DEM, the vertical accuracy is typically about 30 cm, but is significantly less in areas of high vegetation. The data in the DEM have a grid cell size of 1.8 m. This DEM was entered into a GIS to produce a landslide map using derivatives of the DEM such as shaded relief maps (hill shades), a slope map, a topographic contour map and numerous ground surface profiles with a 2 m contour interval. This was supplemented by historical information and ground mapping.

The strength and weaknesses of LiDAR mapping are discussed in detail by Schulz (2007). It is of interest to note his conclusions that aerial photographs appeared to be more effective than LiDAR in the Seattle area for discerning boundaries of recently active landslides within landslide complexes. The resolution of the LiDAR data appeared inadequate to resolve landslide boundaries within landslide complexes. However, LiDAR was much more effective for identifying presumably older landslides and the boundaries of complexes in which recently active landslides occurred. Another recent example illustrating the value of high resolution DEM's provided by LiDAR for mapping landslides has been provided by Ardizzone et al. (2007).

Improvements in LiDAR technology, including processing, are rapidly leading to even more accurate bare earth DEM's. Examples of a bare earth DEM with a 1 m resolution applied to landslide studies and a bare-earth DEM with 25–50 cm resolution applied to faulting studies are citied by Carter et al. (2007).

Airborne LiDAR is increasingly being applied to map landslides and contribute to infrastructure locations such as pipelines and to develop landslide maps which contribute to risk analysis. The ability to penetrate forest cover, even with reduced DEM accuracy, is of enormous value. The increased accuracy of DEM's foresees the increasing use of differential LiDAR to measure ground deformations with time.

As an illustration of the use of LiDAR in current practice, Figure 3 shows the bare earth projection side by side with conventional aerial photography of a potential pipeline crossing of a river in central British Columbia. While most crossings are by horizontal directional drilling, design for conventional crossings by excavation methods are needed as a stand-by. Hence landslide identification is an important consideration in route selection. In this case the river has downcut into a deep deposit ($\sim$200 m) of glacio-lacustrine clays. The contrast between the information revealed by the bare earth imagery and conventional airphotos is striking.

While airborne LiDAR is commonly used for larger scale studies, it is also of value to enhance detailed geological studies at specific sites. Jaboyedoff et al. (2007) has suggested that the high resolution LiDAR DEM can be used to extract both regional and local scale geological structures. The advantages of such techniques are obvious when dealing with steep mountainous terrain. Figure 4 shows a portion of the famous Turtle Mountain-Frank Slide in Canada and the use of shading relief of a high resolution LiDAR DEM to portray the extent of tension cracks that still exist beyond the scarp of the slide (Sturznegger et al. 2007). The potential instability associated with these cracks is a matter of concern.

Figure 3. Example of a LiDAR bare-earth projection compared to the conventional aerial photograph.

Figure 4. Structural and tension crack mapping of the Frank Slide using LiDAR DEM.

In addition to airborne LiDAR, terrestrial-based LiDAR is also finding applications in slope stability studies. Examples of rock slope assessment, where LiDAR has been used to evaluate rock structure are given by Kemeny et al. (2006) & Martin et al. (2007). Using terrestrial based LiDAR portable scanners that can operate in the range of 50 m to 800 m greatly enhance our ability to map the slope discontinuities. Sturzenegger et al. (2007) combined both airborne and terrestrial-based LiDAR to map the structural features associated with Frank Slide (Fig. 4). Terrestrial-based LiDAR is also being used for direct monitoring of the process of hard rock coastal cliff erosion (Rosser et al. 2005).

3 GIS AND LANDSLIDE SUSCEPTIBILITY

As stated by Van Westen (op cit), GIS has determined, to a large degree, the current state of the art in landslide hazard and risk assessment, particularly for landslide studies that cover large areas. Chacón et al. (2006) have recently conducted a comprehensive general review of GIS landslide mapping techniques and basic concepts of landslide mapping. From this extensive investigation they identify three main groups of maps that have been propagated by means of GIS:

1. Spatial incidence of landslides
2. Spatial-temporal incidence and forecasting of landslides (hazard susceptibility)
3. Consequence of landslides.

Regional studies might characteristically have scales of 1:50,000 and smaller, while site specific studies will have larger scales ranging from 1:1000–1:25,000 depending on the project. At these larger scales one is characteristically merging from broader Engineering Geology or Geomorphology to Geotechnical Engineering.

The development of a landslide map, an essential for any hazard on risk assessment tool, relies primarily on the visualization techniques summarized previously. Aerial photo interpretation remains widely used and is increasingly enhanced by LiDAR imagery. GIS and image processing software facilitate the process. Soeters & Van Westen (1996) have summarized the geomorphic features that are diagnostic of landslides both recent and relict.

The assessment of landslide susceptibility goes beyond the cataloguing of past and current landslides by including areas that are susceptible to sliding. Ideally a susceptibility assessment is based on field reconnaissance to determine factors contributing to instability, utilizing the landslide inventory as a first step. Landslide susceptibility maps have been published for many decades (e.g., Radbruch & Crowther 1973). However the ability to manipulate geomorphic data within GIS has proliferated the number of landslide susceptibility studies and their associated methodology. Even prior to the use of GIS based techniques, relative landslide susceptibility in terms of simple bivariate analyses or more complex multivariate analyses had been developed. Early zonation methods based on these developments have been discussed by Varnes (1984). More recent GIS-based developments are listed in Chacón et al (op cit). Some highlights cited are:

- Franks et al. (1998) prepared detailed 1:1000 thematic maps in GIS for landslide hazards on Hong Kong Island, based on a very rich database.
- Wachal & Huduk (2000) used GIS to assess landsliding in a 1,500–2,000 km^2 area in the USA based on four factors—slope angle, geology, vegetation and distance to faults.
- Dai & Lee (2004) developed probabilistic measures of landslide susceptibility for Lantau Island using multivariate logistic regression of presence-absence of dependent variables relating landslides and contributing factors such as lithology, slope angle, slope aspect, elevation, soil cover, and distance to stream channels.

There is a tendency to incorporate increasingly complex statistical methods in these landslide susceptibility analyses. Spatial validation is essential for any practical application.

Temporal considerations most commonly enter into landslide susceptibility forecasting by coupling rainfall probability assessment as an important triggering factor (Lan et al. 2005). This can be undertaken empirically or on a more process-based consideration. The work of Mejia-Navarvo et al (1994) provides an example of the former while that of Dietrich et al. (1995) is an early example of the latter.

The inclusion of geomechanical and hydrological process considerations within GIS based modeling and landslide hazard analysis marks a convergence between the techniques for regional-based studies developed by Engineering Geomorphologists and Geologists and the inputs of the Geotechnical Engineer. Here the example offered by Delmonaco et al. (2003) is of interest. In this case the infinite slope analysis was applied at a river basin scale with basin scale characterization of all of the inputs to this classical equation. In order to calculate the likely pore pressure development, Green-Ampt infiltration analyses were also carried out over the basin scale, reflecting the variation of rainfall with different return periods. The relation between potential instability and return period was determined and the predicted scenarios of instability were found to correspond sensibly with observations made after an extreme rainfall event in 1966. Examples like this encourage the integration of process-based considerations into GIS-based hazard and risk analyses. The coupling of landslide susceptibility forecasts with earthquake effects have already been investigated in a GIS environment (Refice & Capolongo 2002).

The centrality of GIS-based processing has greatly advanced regional landslide hazard and risk analysis as summarized by Chacón et al. (2006). There has been some convergence between the tools used in regional studies and those used by geotechnical engineers in more site specific problems. As stated in Section 2, current GIS technology has significant limitations in truly three-dimensional problems. Chacón et al. (2006) concluded that "the use of three-dimensional GIS for large scale, detailed hazard or risk maps will be one of the significant developments in the near future". This, applied to landslide engineering, is the fundamental theme of this paper. Günther et al (2004) illustrate the kind of progress that is being made in their extension to GIS, designated RSS-GIS, that incorporates the deterministic evaluation of rock slope stability and is particularly useful for regional stability assessment. It incorporates grid-based data on rock structures, kinematic analyses for hard rock failure modes, some pore pressure effects and stability evaluations on a pixel basis. Other examples of integration of geotechnical considerations with GIS follow later in this paper, see Section 5, 6 and 7.

4 MAPPING GROUND MOVEMENT

4.1 InSar

Since the late 1990's the application of spaceborne Interferometric Synthetic Aperture Radar (InSAR) has slowly been incorporated into geo-engineering practice for mapping the rates and extents of ground deformations associated with landslides. As the potential applications and limitations of this tool are gradually being understood, the range of terrains and situations to which it may be applied are expanding. The strength of this technique is that either available archives of data can be utilized to better understand historical movements or new data can be acquired for go-forward monitoring for large areas (up to 2500 km^2) using a remote platform that can acquire data at night or through clouds.

Synthetic aperture radar (SAR) is an active sensor that can be used to measure the distance between the sensor and a point on the earth's surface. A SAR satellite typically orbits the earth at an altitude of approximately 800 km. The satellite constantly emits electromagnetic radiation to the earth's surface in the form of a sine wave, which reflects off the earth's surface and returns back to the satellite. The back-scattered microwave signal is used to create a SAR satellite image (a black and white representation of ground reflectivity) using SAR signal processing methods. SAR radar images are made up of pixels, with the specific size influenced by the SAR sensor resolution; the higher the resolution the smaller the pixel size. To measure differential ground movements over a specified time period, InSAR requires two SAR images of the same area taken from the same flight path, within typically 500 m laterally. During InSAR processing the phase of the corresponding pixels of both images are subtracted. The phase difference between the two SAR images can be used to determine the ground movement in the line-of-sight of the satellite.

Froese et al. (2004) discussed some of the limitations and applications of differential InSAR (D-InSAR) in mapping ground deformations associated with landslides. These included data availability, rate of motion, direction of movement, steep slope distortions and loss of coherence due to a variety of factors such as vegetation, ground moisture and atmospheric effects. Therefore the potential application of InSAR to landslide mapping and monitoring requires consideration on a case-by-case basis to determine the suitability of this method to a particular set of site conditions. Over the past few years a number of advances have lead to an increased reliability of InSAR for measuring ground motion in an ever increasing number of ground conditions.

PS-InSAR: In the last fifteen years, the available number of spaceborne SAR sensors (ERS 1/2, Radarsat 1, JERS, ALOS), has increased significantly. The capability of InSAR has been considerably improved by using large stacks of SAR images acquired over the same area, instead of the classical two images used in the standard configurations.

This multi-image InSAR technique was introduced as Permanent/Persistent Scatterer Interferometric Synthetic Aperture Radar (PS-InSAR) (Ferretti et al., 1999, 2000, 2001). With these advances the InSAR techniques are becoming more and more quantitative geodetic tools for deformation monitoring, rather than simple qualitative tools. Numerous recent projects in Europe (Farina et al. 2006, Colesanti & Waskowkski, 2006, Meisina et al. 2007) have shown good correlation between results obtained from PS-InSAR and traditional geotechnical instrumentation in urban areas impacted by landslide movements.

CR-InSAR: While the PS-InSAR technique is ideally suited to urban environments where buildings can be used as artificial reflectors or where suitable natural exposures exist, the application of this technique is more limited in northern boreal regions with sparse development and more dense vegetation cover. It is often in these remote northern areas where large slowly moving landslides whose size and rates of deformation are ideally suited for the InSAR technology are located. In order to overcome the issues associated with loss of coherence in vegetated and moist ground conditions the introduction of artificial, phase stable reflectors is emerging. This technique has been called either Corner Reflector InSAR (CR-InSAR) or Interferometric Point Target Analysis (IPTA). One of the first documented case histories of the use of artificial reflectors for monitoring of landslides was by Rizkalla & Randall (1999) where five corner reflectors were installed on the Simmonette River pipeline crossing as a trial to monitor slope movements. More recently Petrobras has utilized this technique along a pipeline crossing in Brazil (McCardle et al. 2007). Both of these applications have focused on the application of artificial reflectors along linear corridors in vegetated terrain. Perhaps the most complex landslide monitoring attempted utilizing CR-InSAR is for the Little Smoky River crossing of Highway 49 in northern Alberta, Canada. The application of D-InSAR to this site was first attempted in 2003 (Froese et al. 2004) but the heavy vegetation and ground moisture conditions limited the success of this application.

Both valley walls at the Highway 49 crossing of the Little Smoky River are subject to ongoing movements of deep seated, retrogressive slides in glacial materials and bedrock. The movements of each valley wall are very complex as there are a variety of zones of movement that differ in aspect and level of activity based on their proximity to the present day river. Since the completion of the bridge across the Little Smoky in 1957, there have been significant ongoing maintenance issues due to slope instability impacting on the highway. In order to provide a more stable long term solution to mitigate the impacts of slope movements on the highway, options were considered for stabilizing the existing road versus a re-route away from the area of most significant instability. As there is limited point source geotechnical instrumentation available in the areas that are easily accessible from the highway, larger portions of the valley slope do not have quantitative monitoring information.

In the fall of 2006, a series of 18 corner reflectors were installed on both the southwest and northeast valley walls in order to characterize the differential movements of the various portions of each valley walls (Figure 5). Between November 2006 and November 2007, scenes of Radarsat-1 ascending F2N scenes were obtained and processed by the Canadian Centre for Remote sensing using IPTA software (Froese et al. 2008). The preliminary results available at the time of the preparation of this paper indicate that for the reflectors that are situated on landslide blocks moving with the line-of-sight of the satellite, the movements observed from the CR-InSAR are greater than those found over the same time period as those observed on conventional slope inclinometers. As these slides are moving in colluvium, likely with a rotational component, the CR-InSAR results may be more representative of the actual deformations that are only represented in the horizontal plane by slope inclinometers. Evaluation of this data is currently on going.

Future Development: While the available resolution of the SAR sensors and the number of satellites has in the past been a limitation to the technique, the launching of new, higher resolution satellites provides the opportunities to overcome some of these limitations. With the launch of Radarsat-2 in December 2007, the ability of the satellite to look both right and left and obtain 3 m pixel resolution data will likely increase the directions of slope movements that can be measured and increase the amount of data that can be obtained.

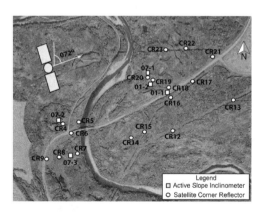

Figure 5. Layout of the corner reflector array in relation to recently installed instrumentation and profile locations (from Froese et al. 2008).

The introduction of 1 m pixel resolution available from the recently launched TerraSAR-X will also continue to increase the density of data that is available for target detection and monitoring.

As the quality and density of this data improves, three dimensional deformation information for landslides may become a reality. Recent studies by Farina et al. (2007) for the Ciro Marina village in Calabria, Italy have shown the potential for using data from different radar platforms to estimate the geometry of movement patterns, an essential step for defining the geometry of the three dimensional nature of the rupture surface.

4.2 Surface Radar (SSR)

The application of differential interferometry using synethic aperature radar has been recently applied to the monitoring of rock slopes. The technique is called Slope Stability Radar (SSR) and instead of using synthetic aperature radar from a moving radar platform, the SSR uses a real-aperture on a stationary platform positioned 50 to 1000 metres away from the foot of the slope (Harries & Roberts 2007). A major advantage of the technique is that it provides full coverage of the rock slope without the need to install reflectors. According to Harries & Roberts (2007) the technique offers sub-millimetre precision of slope wall movements without being affected by environmental conditions such as rain, dust, etc. The accuracy of this technique diminishes in areas of vegetative cover and hence the technique has been primarily used in open pit mines.

5 ROCK FALL PROCESS MODEL

Rock fall is the simplest of landslide processes and it is a surface phenomenon. If GIS can be made geotechnically smart, the development of rock fall simulation within GIS, a Stage 3 development in Figure 1, should form a starting point.

The Canadian railway industry has been exposed to various ground hazards since the first transcontinental line was constructed in the 1800s. One of the frequently occurring ground hazards is rock fall. These events in mountain regions occur as the result of ongoing natural geomorphologic processes (Figure 6). The slope and rock properties controlling the initiation and behaviour of these rock falls can vary widely and it is not practical to eliminate these rock fall hazards due to the extent and area of potential rock fall source zones. Nonetheless, reducing this hazard to an acceptable level of safety requires proactive risk management strategies.

Due to the linear corridor occupied by railways there is often a need to conduct a large number of

Figure 6. Example of rock fall hazard along a section of railway in British Columbia.

rock fall analyses at regular intervals and often the hazards come from inaccessible natural rock slopes well upslope of the track with previously undeterminable flow paths (Figure 6). GIS has been used as an effective tool in hazard delineation, but seldom is GIS used for rock fall process modeling (Dorren & Seijmonsbergen 2003). Stand alone computer software to assess rock falls have been developed to analyze trajectories, run-out distance, kinetic energies, and the effect of remedial measures (Pfeiffer & Bowen 1989, Guzzetti et al 2002, Jones et al. 2000). This software typically does not interact directly with existing GIS software. As a result to use these programs, one must first extract the digital elevation model (DEM) and then recompile it in a form that is suitable for the rock fall software.

RockFall Analyst, a three dimensional rock fall program that was developed as an extension to ArcGIS, is used to illustrate the added value GIS technology provides for hazard assessment for rock falls (Lan et al. 2007).

Rock fall hazard assessment for engineering purposes must capture as many variables as possible in relation to the rock fall process, kinetic characteristics and their spatial distribution (Dorren & Seijmonsbergen 2003). As a geomorphologic slope process, rock falls are characterized by high energy and mobility despite their limited volume. The dynamics of the rock fall process is dominated by spatially distributed attributes such as: detachment conditions, geometry features and mechanical properties of both rock blocks and slopes (Agliardi & Crosta 2003). Today accurate three dimensional morphology can be obtained from LiDAR data but the geotechnical parameters (the coefficient of restitution and friction) must be calibrated using historical rock fall events. The historical rock fall database records provides the information of past rock fall events including location of source and deposition, timing of events, size, influence on the railway operations and the effectiveness of existing barriers, should such barriers exist.

Lim (2008) showed that high resolution airphotos can also be used to aid in the assigning the geotechnical parameters to various regions of the slope and in the source zone characterisation process.

5.1 Rock fall modelling

Spatial modeling of rock fall hazards along a section of railway in the Rocky Mountains was carried out using RockFall Analyst, a GIS extension, which combines 3 dimensional dynamic modeling of the rock fall physical process with distributed raster-image modeling of rock fall spatial characteristics. The rock fall modeling process involves the following steps:

1. The potential instabilities areas (source areas) are evaluated using the DEM from LiDAR data and the spatial database of historical rock fall events. The available LiDAR DEM provided a spatial resolution of 0.15 m.
2. The rock fall hazard was assessed using Rock-Fall Analyst by taking into account the distributed geometry and mechanical parameters, and spatial pattern of rock fall characteristics. This step included two parts: (1) rock fall process modelling and (2) raster image modelling.
3. Raster image modelling of rock fall spatial frequency, flying height (potential energy) and kinetic energy was used to produce the rock fall hazard map.

Once the DEM was created from the LiDAR data and the spatial attributes assigned, potential rock falls were simulated from all of the possible source areas. Figure 7a shows the results from the simulation where the source is simulated as a horizontal line source. Using the ArcGIS tools, the number of rock fall trajectories occurring was calculated for each grid cell with an area of one square meter. Spatial geostatistical techniques within the GIS software were used to analyze the trajectories and determine the rock fall spatial frequency for the whole study area (Fig. 7b). The rock fall frequency occurring in the region of railway track was classified into 10 classes (0–9) and plotted in ArcGIS in three-dimensional view (Fig. 7c). Figure 7 demonstrates the advantage of using GIS technology to simulate the rock fall process model.

One of the concerns that frequently arises when conducting rock fall analysis is the level of resolution required for the DEM to provide reliable results. The resolution of the DEM controls the geometry of the rock fall impacts and trajectories and may control the physics of the impact. In British Columbia, where the case history is located, a DEM at a 10-m-grid spacing is freely available. For comparison purposes a 1-m-grid LiDAR survey was obtained to provide more accurate geometry of the slope. The rock fall simulations described above using the DEM from the

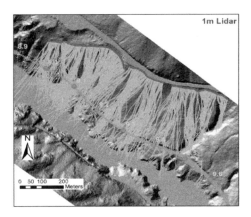

a) Rock fall trajectories from RockFall Analyst. Seeding of the rocks falls is carried out using a line source

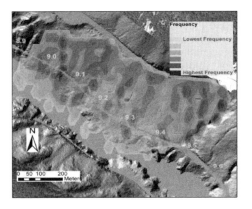

b) Raster image of rock fall frequency

c) Distribution of rock falls from the simulation

Figure 7. Spatial frequency analysis of rock falls using RockFall Analyst. All modeling and processing of the information was carried out in ArcGIS.

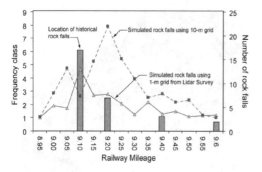

Figure 8. Comparison of the historical rock fall frequency impacting the railway tracks with the results from the Rock-Fall Analyst simulations using the 1-m and 10-m digital elevation models.

LiDAR survey were repeated using the 10-m DEM. Figure 8 compares the historical rock fall data base to the results from RockFall Analysts for both the 1-m and 10-m-grid. It is evident from Figure 8, that the rock fall simulations from the 1-m provides better agreement with the historical rock fall events (Lim 2008).

5.2 Hazard zoning

Once the spatial distribution of the rock fall has been computed the energy from such events is required to complete the hazard assessment. Two raster layers were created in ArcGIS to assess the spatial distribution of rock fall potential-energy (flying/bouncing height relative to ground) and the kinetic energy (velocity). The energy raster layers combined with the rock fall frequency is used to produce the rock fall hazard assessment shown in Figure 8. The rock fall hazard map clearly identifies the section of railway with the greatest risk, consistent with the historical evidence. Once the hazard has been identified, the energy raster layers can also be used to provide input to the design of protective barriers. Rock fall protection requires an assessment of both the height (bouncing/flying) and velocity of the rock falls. Without such information the design of protective measures is usually based on single 2 dimensional analysis or qualitative methods.

5.3 Summary

Rock falls are a significant hazard to Canadian railways. The assessment of such hazards over long sections of railway requires an efficient means for storing historical data and conducting rock fall simulations. The development of the three dimensional RockFall Analyst as an extension to ArcGIS, provides the framework for rapid assessment of rock fall hazards. The use of such tools requires a detailed DEM

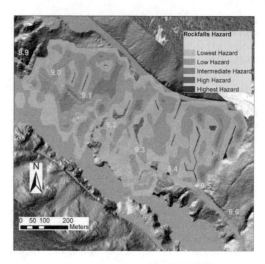

Figure 9. Rock fall hazard assessment based on rock fall frequency and kinetic and potential energy.

as well as historical data for calibration purposes. Once calibrated, the energy output from RockFall Analyst may be used in the design of protective measures.

6 INTEGRATING GIS INTO GEOTECHNICAL PRACTICE

Assessment of slope movement and associated hazards demands an understanding of the site characteristics and their spatial and temporal variability. Current geotechnical modelling tools are focused on numerical analyses and are not generally designed to facilitate the requirements of site investigation and characterization. Site characterisation must address key geospatial issues, e.g., complex geology, highly irregular porewater pressure, complex surface geometry and slip surface definition as appropriate. There is little doubt that capturing more complete geomorphological, geological and geotechnical information improves the quality of geotechnical site investigation, particularly when the site is geologically and geotechnically complex (Luna & Frost 1998, Tsai & Frost 1999, Parsons & Frost 2002, Jaboyedoff et al. 2004, Kunapo et al. 2005). Culshaw (2005) in the fifth Glossop lecture suggested that "the rapid development in technology over the last twenty years and the digitization of increasing amounts of geological data has brought engineering geology to a situation in which the production of meaningful three-dimensional spatial models of the shallow subsurface is feasible". Despite these advances there are very few spatial tools that help the geotechnical engineer achieve this goal.

While GIS is increasingly viewed as a key tool for managing spatial distribution of data (Nathanail & Rosenbaum 1998, Parsons & Frost 2000, Kunapo et al. 2005) it has significant limitations in presenting three-dimensional (3D) geologic and geotechnical data. Some GIS systems, like ArcGIS, developed by the Environmental Systems Research Institute, Inc. (ESRI), provide a functional developer kit which can be used to create 3D capability. However the current developments in 3D GIS are still not sufficient to meet the needs of the geotechnical engineer. Mining software such as Surpac Vision developed by Gemcom Software International Incorporated provide a comprehensive system for geological modelling but not geotechnical modelling. For example, while three-dimensional solid modelling and two-dimensional sections can be easily created in Surpac, querying inclinometer or piezometer data is not readily accomplished. In the following section we describe an integrated approach using ArcGIS, Surpac Vision and numerical modeling, to develop a three-dimensional spatial model of a shallow subsurface slide locally referred to as the Keillor Road Slide. This integrated approach illustrates the added value obtained when data and analyses are tightly integrated.

6.1 Development of an integrated approach

Nearly all slope site characterization efforts deal with surface mapping, geological information from borehole data and monitoring data. The work flow from data collection through to engineering analyses was outlined by Lan & Martin (2007) and can be summarized in three stages (see Figure 1). Stage 1 involves the data collection, management and geosynthesis of the data. Modern GIS software provides effective tools for the handling, integrating and visualizing diverse spatial data sets (Brimicombe 2003). Therefore, in Stage 1, the functionality of GIS provides an essential role in collecting, storing, analyzing, visualizing and disseminating geospatial information. This information could be basic site investigation data, such as geomorphology and geology conditions, and diverse, continually evolving geotechnical parameters, such as displacement and pore pressure readings from geotechnical instruments. Most GIS tools have limitations in representing time series data such as the displacement data from inclinometer or pore pressures from piezometers. Therefore additional functional tools were required for the standard ArcGIS software. These tools have been implemented using ArcObject, an ArcGIS developer kit, and Visual studio.net, a software developing package by Microsoft. These development tools provide capability for users to interact and communicate with various data sets.

The main aim of Stage 2 in Figure 1 is to establish a comprehensive ground model for the site. The construction of the ground model is enhanced using three dimensional geological modelling tools commonly available in the mining industry.

Finally, geotechnical analyses and engineering decisions are performed in Stage 3 (see Figure 1). In this step the ground behaviour is analysed using commercially available geotechnical numerical tools. It is essential that the tools used in Stage 2 communicate with the tools used in Stage 3 so that the data integration is maintained across all stages.

One of major issues in this integrated approach involves data input and output. In order to develop an appropriate easy-to-use input and output function, some industrial standard file formats are employed for data conversion and communication throughout the three Stages. Shape file (.SHP) from ESRI and Data Exchange File (.DXF) from AutoDesk are both industrial standard formats supported by almost all PC-based CAD and GIS products. The communication between different Stages in the system developed by Lan & Martin (2007) was implemented using these two file formats.

The integration of the tools described above offers effective digital tools to model heterogeneous geology, complex stratigraphy and slip surface geometry, and variable pore pressure conditions which are critical to complex slope stability problems or other analyses. The tools also provide for incorporating findings from monitoring data. In the following section the tools are demonstrated using a translational bedrock slide in Edmonton Alberta.

6.2 Case study: Keillor Road slide

A complete description of the Keillor Road bedrock slide was given by Soe Moe et al. (2005). The failure of the slope occurred over a number of years with the largest deformations occurring in 2002. The slide took place along the bank of the North Saskatchewan River valley in Edmonton, Alberta Canada (Fig. 10). The site investigations for the slide were conducted over a period of 15 years using traditional boreholes and monitoring systems.

Figure 11 shows a plan view of the site created in ArcGIS indicating the outline of the slide, the topography of the area, location of the tension cracks and location of the main scarp. Figure 11 also shows the location of the boreholes that had been used in the site investigations over the 15 year period. The major benefit of assembling the data in ArcGIS is that the borehole symbols are dynamically linked to the data base and instrumentation data installed in the boreholes.

To reconstruct the dynamics and kinematics of the processes acting on the slope and to determine their spatial and temporal distribution, Lan & Martin

Figure 10. Photograph of the Keillor Road Slide, from Soe Moe et al (2005).

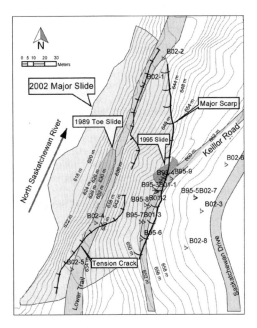

Figure 11. Plan view of Keillor Road slope showing location of the site investigation boreholes, tension cracks and outline of the slide. Contour elevations have been removed for clarity.

(2007) developed add-on tools for the processing of borehole information, plotting of time-series displacement data from slope inclinometers, pore pressure data from piezometers and relative geomorphological features, such as tension cracks. The add-on tools provide all of the standard types of plots for analysing slope inclinometer data. From these standard plots discrete movement zones can be defined by specifying the from-to-depths. The resultant time-displacement plots for these discrete zones show acceleration or deceleration of slope movement. In addition to the displacement versus time-plots, plots of displacement vector directions and displacement rate offer the ability to identify and evaluate the spatial and temporal character of the deformations, all in a user-friendly environment.

In addition to being able to process the data quickly in both a visual manner as well as conduct specific depth queries, the user can quickly assess the kinematics of the slide. Multiple movement zones at different depth are often detected in slope inclinometer plots. In this case, two movement zones were identified in the inclinometer readings for borehole B02-2 (Fig. 12). One zone was from depth 0.61 m to depth 2.44 m (zone 1) and the other zone was from depth 7.32 m to depth 8.53 m (zone 2). Their deformation histories can be rapidly shown on the plan map (Fig. 12). It can be seen clearly that zones 1 and 2 show different movement characteristics. The moving direction of the shallow zone 1 changed direction frequently while the direction of zone 2 was essentially unchanged.

Once the major rupture surface is identified, the spatial deformation pattern from all the inclinometer

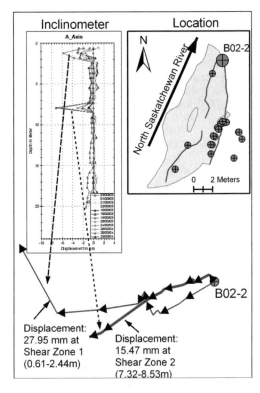

Figure 12. Displacement history at different shear zones of site B02-2. Shearing zone 1 and zone 2 are characterized by obviously different displacement vectors.

data can be shown (Fig. 13). This provides a consistency check within the data sets as well highlights the more active portions of the slide as both total displacements as well as displacement rate can be shown.

Geotechnical parameters are usually measured at points during site investigation by in-situ tests or by laboratory tests. Geostatistical kriging and simulation techniques in GIS offer powerful spatial modeling tools for visualising the spatial variability of these parameters (Nathanail & Rosenbaum 1998). Parsons & Frost (2002) argued that such statistical approaches improve the quality of site investigation data. Pore pressure is an essential parameter in slope stability studies. Lan & Martin (2007) used geostatistical techniques to interpret the point pore pressure data into a spatial pore pressure surface. When conducting such geostatistical analysis it is important to ensure that the pore pressures are being measured on the same geological unit which can be readily verified by comparing the borehole logs and piezometer installation locations. Lan & Martin (2007) also attached the three dimensional displacement curves obtained from the inclinometer data to the boreholes as lines to show a spatial relationship between displacement locations and pore pressure.

As mentioned earlier, Surpac Vision provides advanced tools for viewing and interpretation of geology data. Connecting to the same geological database as used by ArcGIS, the three-dimensional geology model for the Keillor Road Slide was created in Surpac Vision. Together with the other data imported from ArcGIS, such as the geomorphological surface, displacement and pore pressure readings, and tension crack planes, a comprehensive ground model for Keillor Road slope was created in Surpac Vision. From the model, the spatial extent of the slope which is at risk from instability can be immediately defined by the displacement data and the surface mapping information. Critical profile sections can be extracted in Surpac Vision along the section lines parallel to the displacement vectors. These sections now include all information managed and produced in GIS and Surpac. These section profiles can be exported to DXF files which can then be optimized for slope stability analysis. Nearly all modern slope stability software such as Slope/W or Slide can readily import DXF files. However, it is important that the user examine these DXF files to ensure that the relevant information is captured.

Slope/W is widely used in geotechnical engineering practice for analyzing the stability of slopes. It uses limit equilibrium theory to compute the critical factor of safety (Krahn 2003). The essential geometry elements in Slope/W include the ground surface, complex geological regions, and pore pressure line and tension crack lines. In many slope stability problems it is very important to establish an accurate representation of the slope surface geometry because small changes in the slope profile can have a significant impact on the calculated factor of safety especially when the rupture surface is relatively flat such as the translational slide at Keillor Road. Creating the section profile from LiDAR survey ensures that the most accurate surface geometry is captured. Figure 14 shows the final geometry and geology modeled using Slope/W. The integration of the

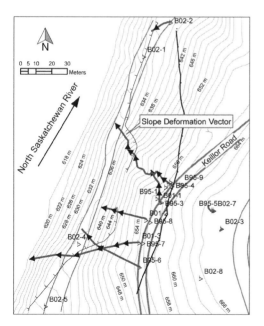

Figure 13. Spatial distribution of slope deformation from all the inclinometer data at the same discrete movement zone. Two major slope portions with different displacement evolution are divided by Keillor Road.

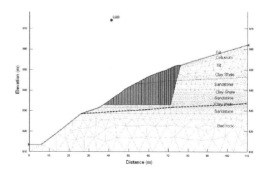

Figure 14. Slope stability analysis using Limit equilibrium and/or Finite-element analysis.

GeoStudio Software means that this model can also be used for conducting deformation or stress analyses.

7 ANALYSING COMPLEX LANDSLIDES

In the previous section we showed the benefit of integrating technologies when analyzing a single slide. In this section we demonstrate the added value when considering multiple complex landslides.

7.1 Background

Large translational landslides with rupture surfaces through glacial lake sediments in preglacial valleys are common hazards within river valleys of Western Canada (Evans et al. 2005). Eleven, retrogressive, multiple, translational earth slides have occurred along 10 kilometres of the Thompson River valley between the communities of Ashcroft and Spences Bridge in south-central British Columbia, Canada. The Canadian Pacific Railway (CPR) and Canadian National Railway (CN) main rail lines were constructed through the Thompson River valley in 1885 and 1905 respectively. Both have had recurring slope stability problems along this valley (Fig. 15). Given that the two national railroads traverse the same landslide prone area, the evaluation of risk at this location is a matter of considerable significance.

The Ashcroft area is part of the Thompson Plateau, a subdivision of the Interior Plateau of British Columbia. The Thompson River flows south and has cut through about 150 metres of glacial sediments (Porter et al. 2002). Quaternary sediments occur within the major valleys where deep valley fills have been dissected and terraced by postglacial downcutting of the trunk rivers. The landslides occurred on the steep walls of an inner valley that formed during the Holocene when Quaternary sediments filling the broader Thompson River valley were incised. The valley fill consists dominantly of permeable sediments, the exception being a unit of rhythmically-bedded silt and clay in the Pleistocene sequence (Clague & Evans 2003). The surficial materials in the area are tills, fluvial, fluvioglacial, lacustrine and colluvial deposits (Ryder 1976).

Individual investigations had been carried out for the six most active of these earth slides in the Thompson Valley since the early 1980s. A major effort was initiated in 2003 to re-analyse the data that had been collected over the past 20 years using the spatial capabilities inherent in GIS tools. Eshraghian et al (2007) completed a comprehensive study of the slides and concluded that the rupture surfaces that had been detected in the individual slides followed the highly plastic, overconsolidated clays within a Pleistocene stratigraphic unit consisting of up to 45 metres of rhythmically-bedded glaciolacustrine deposit of silt and clay couplets, ranging from less than 1 cm to several tens of centimetres thick (Fig. 16). These sediments may be several hundred thousand years old and

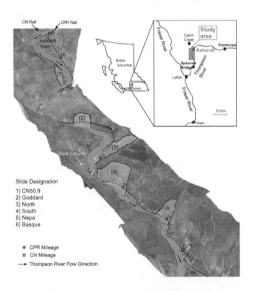

Figure 15. Major landslides south of Ashcroft, BC, (modified from Eshraghian et al 2007).

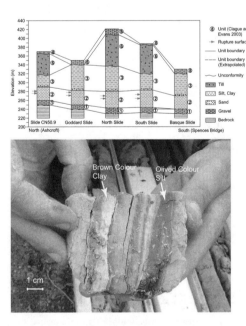

Figure 16. Geological units in the earth slides and highland terraces in Thompson Valley. The arrows indicate the rupture surfaces. Modified from Eshraghian et al 2007.

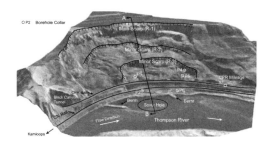

Figure 17. Example of an aerial photograph draped over a digital elevation model produced from an airborne Lidar survey (modified from Eshraghian et al 2007).

thus of Middle or Early Pleistocene age (Clague & Evans 2003). Samples of this unit from boreholes in South Slide show layers of brown, high-plastic clay 1 to 20 cm thick between thicker layers of olive silt (Figure 16).

The deposits of the three glacial sequences are separated by unconformities. Figure 16 shows the geological succession synthesized from borehole logs and outcrops in scarps and terraces in the Ashcroft area based on units proposed by Clague & Evans (2003). Eshraghian et al. (2007) used GIS technology to estimate the slide volumes which varied from 1.8 to 21.4 Mm^3 and spatially correlate the rupture surface. They concluded that the stratigraphic boundaries have tilts of 1.7 m/km similar to the glacial lake bottom and that sliding was occurring along essentially the same layers within the glaciolacustrine sediments (Figure 16). They then examined the surface exposure of the larger slides using LiDAR technology. The airborne LiDAR provided a vertical resolution of ±150 mm and when combined with the aerial photographs illustrated the multiple blocks associated with retrogressive slides of this nature (Figure 17).

7.2 Multiblock retrogressive model

Eshraghian et al 2007 examined Slide CN50.9 using traditional site investigation boreholes and the combined Lidar DEM and aerial photograph model to develop the geological and movement history since deglaciation (Fig. 18). During the first stage a braided Thompson River started cutting through the glacial sediments after deglaciation. Thompson River continued its down-cutting until it reached the first weak layer and potential rupture surface (stage 2, Fig. 18). Progressive failure within this weak layer caused sliding of blocks A and B on the shallower rupture surface. More down-cutting by Thompson River encountered a deeper weak layer (Stage 3, Fig. 18). This time, movement happened without retrogression on the deeper rupture surface. It caused more

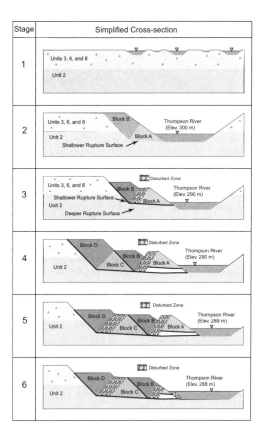

Figure 18. Simplified multiblock model illustrating the sliding process since deglaciation at the Slide CN50.9 (modified from Eshraghian et al 2007).

horizontal movement by block A and horizontal and vertical movement by block B. This sliding and also the Thompson River erosion caused progressive failure on the deeper rupture surface and the slide was ready for another retrogression (Stage 4, Figure 18). The most recent retrogression of the Slide CN50.9 happened in September 1897. During this stage, block D moved down on the main scarp and rest of slide material moved horizontally toward the river (Stage 5, Fig. 4). During this retrogression in the early morning of September 22, 1897, residents of Ashcroft were awakened by loud, thunder-like rumblings. The landslide constricted the Thompson River without completely blocking it (Clague & Evans 2003). From these explanations, the movement rate during this retrogression is estimated to be rapid. Following the slide, Thompson River removed part of the toe of the slide, mainly within block A (stage 6, Figure 18).

Eshraghian et al 2007 used the conceptual model developed in Figure 18 to develop the geological multiblock model for the retrogressive slide shown

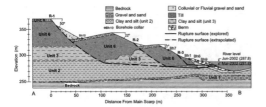

Figure 19. Slide CN50.9 cross-section showing the rupture surfaces, stratification, and borehole locations (see Figure 15 for location, modified from Eshraghian et al 2007).

in Figure 19. This slide is now moving on two rupture surfaces as a multiple translational earth slide. The positions of the rupture surfaces were determined by inclinometers and the rate of movement differs not only on the main rupture surfaces but between inclinometer measurements on the shallower rupture surface, suggesting possible more small blocks at the toe within R-3 block. Developing a single factor of safety for such a complex slide does not convey the geological complexity of the slide nor communicate the risk to the railways.

7.3 Trigger mechanisms

The location of the piezometers at the toe of Slide CN50.9 indicate upward gradients near the toe of the slope. All piezometers respond to changes in the river level but the shallower the piezometer and the closer it is located to the river, the greater the response. The piezometers also show a 7 to 10 days delay between the river level changes and the piezometer on the deeper rupture surface. The piezometers indicate that the slide portion near to the scarp, i.e. the slide head, is generally a recharge zone and the toe is a discharge zone when the Thompson River level is low. On the other hand, when the Thompson River starts rising, the water from the river seeps towards the slide mass and offsets the upward gradient condition at the toe. However, the river may not stay at these high levels for sufficient time to allow the flow system to reach equilibrium. Therefore, the top part of the rhythmically bedded silt and clay layer (unit 2) is more affected by river level changes than the lower part. In the years that the Thompson River stays at high levels for longer periods, the piezometers show the greatest increase in pore water pressures because the water has more time to seep through the soil mass.

The average rainfall in the area has been increasing since the 1920s from 150 mm/year to 240 mm/year (Porter et al. 2002). Despite this rainfall increase in the area, Eshraghian et al. 2005a did not find a correlation between slide movements and short term or long term rainfall. Eshraghian et al. (2005a, 2005b) examined the Thompson River levels and slide movements from 1970 to 2000. Data prior to 1970 are sparse. They presented a correlation between the cumulative river level difference from the average river level (CRLD) and active years. They concluded that the main trigger for reactivation of these slides was the discharge of Thompson River that produced above average river levels for prolonged periods (Figure 20).

Clague & Evans (2003) suggested "irrigation of the bench lands above the river, especially in the late 1800s, introduced large amounts of water into the valley fill. High pore pressures probably developed locally at the top of the rhythmically bedded silt-clay unit, triggering large landslides". They added "although high pore pressures generated by irrigation related groundwater discharge probably triggered most of the historical landslides in the Ashcroft area, the fundamental causes are geological". Morgenstern (1986) also concluded that the primary trigger for the movement in these complex slides was related to their retrogressive nature characterised by toe-initiated movements. Eshraghian (2007) examined the following possible trigger mechanisms: (1) rainfall, (2) irrigation, (3) Thompson River Level and (4) toe erosion. He concluded that the changes in the river level had the largest impact on the stability of the slides and that this occurred when the river remained high for a long period and then retreated to cause a drawdown effect on the slope toe blocks. Those slides which when combined with toe-erosion showed the greatest potential for rapid movements.

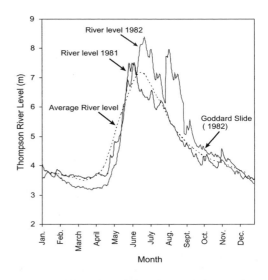

Figure 20. Illustration of the average Thompson River Level compared to 1981 when typical slow movements were recorded and 1982 when the Goddard Slide was reactivated, (modified from Eshraghian et al 2007).

7.4 Movement and risk

It is obvious from the discussion above that attempting to capture the risk from the movement associated with such landslides with a single number is not practical. Eshraghian et al 2007 used a quantitative hazard analysis in a framework that considered the different post-failure movement rates. They demonstrated the approach using probabilistic stability analyses that included material and trigger uncertainties as well as uncertainties from the groundwater modeling and toe erosion. The probabilistic rates of movement were calculated using the frequency of the trigger (the Thompson River flood, Figure 21), the historic movement rates, for each reactivation block.

The result of calculating the probability of movement for reactivation blocks are a movement probability distribution (Figure 22) which shows the probability distribution of different movement rates which may happen during the design life time of the project. They also reported the results for each reactivation block in the form of probability of different movement rates using the movement rate class suggested by Cruden & Varnes (1996) calculated for the designed life time of 100 years (Fig. 23).

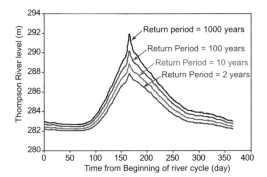

Figure 21. Thompson River level for different yearly discharge return periods.

7.5 Summary

The Introduction for this paper drew attention to the large number of contributory processes that have to be considered in developing an effective process model of a landslide and its consequences. All of these contributory processes have had to be considered in the example just presented; from geology and geomorphology through hydrological and geotechnical characterization and finally geotechnical and risk

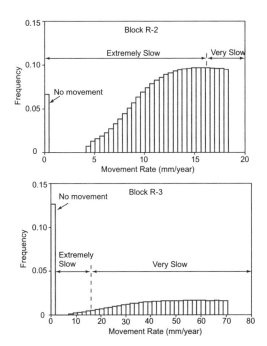

Figure 22. Histogram frequency distribution of movement rate for two translational blocks on shallower and deeper rupture surfaces.

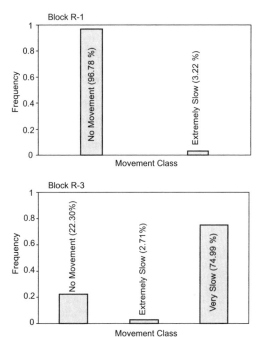

Figure 23. Frequency of different movement rate classes for reactivation blocks defined within Slide CN50.9 for a 100 year return period.

analyses, utilizing both deterministic and probabilistic considerations. The capacity for undertaking such complex landslide analyses was enhanced by the availability of recently developed tools such as LiDAR imagery. However, it is unlikely that the end product could have been achieved without utilizing GIS for spatial data management, correlations and analysis.

8 CONCLUDING REMARKS

A number of recent technical advances are leading to dramatic improvements in the study of landslides and the evaluation of appropriate risk mitigation measures. This paper draws attention to some, such as the application of LiDAR to delineate landslides more clearly than aerial photographs, and the role of InSAR to monitor ground movements over large areas with increasing accuracy. A number of other tools are entering practice that merit discussion but were beyond the scope of this paper.

It has been the central premise of this paper that the most important advances have been associated with improved visualization of landslides and related processes, both through surface and sub-surface features. To this end, our experience leads us to the view that GIS is capable of making transformative contributions.

Examples of use of GIS in geotechnical assessment, beyond its routine application of archiving surface information, have been provided. A general rockfall simulation model has been developed with GIS. Other examples illustrate the integration of GIS with subsurface modelling capability with interfaces to any kind of geotechnical analysis software.

Landslide data management and analysis of all kinds in GIS will be essential for future progress in landslide engineering as three-dimensional visualization and modelling capabilities improve.

ACKNOWLEDGMENTS

The authors wish to acknowledge the assistance of the following:

i. Dr. Hengxing Lan, Research Engineer, for his leadership in developing GIS based tools at the University of Alberta.
ii. Mr. Cory Froese, Team Leader—Geological Hazards, Alberta Geological Survey/Energy and Utilities Board, for assisting us with understanding the current status of InSAR applied to landslide studies.
iii. The graduate students who have collaborated with us in these and related studies over the past few years.
iv. The Gateway Pipeline Project for providing Figure 3.

The development of the ArcGIS tools and the case histories described in this paper was supported by the Canadian Railway Ground Hazard Research Program, a collaborative research program between Canadian National Railway, Canadian Pacific Railway, Transport Canada, Geological Survey of Canada, University of Alberta, Queen's University and the Natural Sciences and Engineering Council of Canada.

REFERENCES

Agliardi, F. & Crosta, G.B. 2003. High resolution three-dimensional numerical modelling of rockfalls. *International Journal of Rock Mechanics and Mining Science* 40 (4): 455–471.

Ardizzone, F., Cardinali, M., Gzzetti, F. & Reichenbach, P. 2007. Identification and mapping of recent rainfall-induced landslides using elevation data collected by airborne LiDAR. *Natural Hazards and Earth System Sciences* 7: 637–650.

Bjerrum, L. 1967. Progressive failure in slopes and overconsolidated plastic clay and clay shales. *Journal Soil Mechanics and Foundations Division, ASCE* 93 (SM5): 3–49.

Brimicombe, A. 2003. GIS, Environmental modelling and engineering, London; N.Y.: Taylor & Francis, 312 p.

Carter, W.E., Shresthe, R.L. & Slatton, K.C. 2007. Geodetic laser scanning. *Physics Today* 60: 41–47.

Chacón, J., Irigaray, C., Fernandez, T. & El Hamdouni, R. 2006. Engineering geology maps: landslides and geographical information systems. *Bulletin of Engineering Geology and the Environment* 65: 341–411.

Chandler, J.H. & Brunsden, D. 1995. Steady state behaviour of the Black Ven mudslide: the application of archival analytical photogrammetry to studies of landform change. *Earth Surface Process and Landforms* 20: 255–275.

Clague, J.J. & Evans, S.G. 2003. Geological framework of large historic landslides in Thompson River Valley, British Columbia. *Environmental and Engineering Geoscience* 9: 201–212.

Colesanti, C. & Wasowski, J. 2006. Investigating Landslides with Space-borne Synthetic Aperture Radar (SAR) Interferometry. *Engineering Geology* 88: 173–199.

Cruden, D.M. & Varnes, D.J. 1996. Landslide types and processes. *Transportation Research Board* Special report 247: 36–75.

Culshaw, M.G. 2005. The Seventh Glossop Lecture—From concept towards reality: developing the attributed 3D geological model of the shallow subsurface *Quarterly Journal of Engineering Geology and Hydrogeology* 38: 231–284.

Cunningham, D., Grebby, S., Tansey, K., Gosar, A. & Kastelic, V. 2006. Application of air borne LiDAR to mapping seismogenic faults in forested mountainous terrain, southeastern Alps, Slovenia. *Geophysical Research Letters* 33: L20308.

Dai, F.C. & Lee, C.F. 2004. A spatiotemporal probabilistic modelling of storm-induced shallow landsliding using aerial photographs and logistic regression. *Earth Surface Processes and Landforms* 28: 527–545.

Delmonaco, G., Leoni, G., Margottini, C., Puglisi, C. & Spizzichino, D. 2003. Large scale debris-flow hazard assessment: a geotechnical approach and GIS modelling. *Natural Hazards and Earth System Sciences* 3: 433–455.

Dietrich, W.E. & Montgomery, D.R. 1998. A digital terrain model for mapping shallow landslide potential. *Technical Report NCASI*. http://socrates.berkeley.edu/~geomorph/shalstab/

Dorren, L.K.A. & Seijmonsbergen, A.C. 2003. Comparison of three GIS-based models for prediction rock fall runout zones at a regional scale. *Geomorphology* 56: 49–64.

Eshraghian, A. 2007. Hazard analysis of reactivated earth slides along the Thompson River Valley, Ashcroft, British Columbia. *PhD Thesis Dept. Civil & Environmental Engineering, University of Alberta, Edmonton, Alberta, Canada*.

Eshraghian, A., Martin, C.D. & Cruden, D.M. 2005a. Landslides in the Thompson River valley between Ashcroft and Spences Bridge, British Columbia. In *Proceedings of the International Conference on Landslide Risk Management, Vancouver, Canada,* May 31 to June 4, 2005: 437–446.s

Eshraghian, A., Martin, C.D. & Cruden, D.M. 2005b. Earth slide movements in the Thompson River valley, Ashcroft, British Columbia. In *Proceedings of the 58th Canadian Geotechnical Conference, Saskatoon, Saskatchewan, Canada,* September 18–21, 2005.

Eshraghian, A., Martin, C.D. & Cruden, D.M. 2007. Complex Earth Slides in the Thompson River Valley, Ashcroft, British Columbia. *Environmental and Engineering Geoscience Journal* XIII: 161–181.

Evans, S.G., Cruden, D.M., Brobrowsky, P.T., Guthrie, R.H., Keegan, T.R., Liverman, D.G.E. & Perret, D. 2005. Landslide risk assessment in Canada; a review of recent developments. In *Proceedings of the International Conference on Landslide Risk Management, Vancouver, Canada,* 31 May–3 June 2005. A.A. Balkema. 351–434.

Farina, P., Casgli, N. & Ferretti, A. 2007. Radar-Interpretation of InSAR Measurements for Landslide Investigations in Civil Protection Practices. In Schaefer, V.R., Schuster, R.L., Turner, A.K. (eds), *Landslides and Society*. AEG Special Publication 23: 272–283.

Farina, P., Colombo, D., Fumagalli, A., Marks, F. & Moretti, S. 2006. Permanent scatterers for landslide investigations: Outcomes from the ESA-SLAM Project. *Engineering Geology* 88: 200–217.

Ferretti, A., Prati, C. & Rocca, F. 1999. Permanent scatterers in SAR interferometry. *International Geoscience and Remote Sensing Symposium, Hamburg, Germany,* 28 June-2 July, 1999. 1–3.

Ferretti, A., Prati, C. & Rocca, F. 2000. Nonlinear subsidence rate estimation using permanent scatterers in differential SAR interferometry. *IEEE Transactions on Geoscience and Remote Sensing* 38(5): 2202–2212.

Ferretti, A., Prati, C. & Rocca, F. 2001. Permanent scatterers in SAR interferometry. *IEEE Transactions on Geoscience and Remote Sensing* 39(1): 8–20.

Findley, D.P. 2007. Rock Stars. *Civil Engineering* July: 46–51.

Franks, C.A., Koor, N.P. & Campbell, S.D.G. 1998. An integrated approach to the assessment of slope stability in urban areas in Hong Kong using thematic maps. *Proc 8th IAEG Congress*. Vancouver: Balkema, 1103–1111.

Froese, C.R., Kosar, K. & van der Kooij, M. 2004. Advances in the application of InSAR to complex, slowly moving landslides in dry and vegetated terrain; in *Landslides: Evaluation and Stabilization,* W. Lacerda, M. Erlich, S.A.B. Fontoura & A.S.F. Sayao (ed.), *Proceedings of the 9th International Landslide Symposium, Rio de Janeiro, Brazil*: 1255–1264.

Froese, C.R., Poncos, V., Skirrow, R., Mansour, M. & Martin, C.D. 2008. Characterizing Complex Deep Seated Landslide Deformation using Corner Reflector InSAR (CR-INSAR): Little Smoky Landslide, Alberta. *Proceedings of the 4th Canadian Conference on Geohazards*. Quebec: In Press.

Günther, A., Carstensen, A. & Pohl, W. 2004. Automated sliding susceptibility mapping of rock slopes. *Natural Hazards and Earth Systems Sciences* 4: 95–102.

Guzzetti, F., Crosta, G., Detti, R. & Agliardi, F. 2002. STONE: a computer program for the three-dimensional simulation of rock-falls. *Computers & Sciences* 28: 1079–1093.

Harries, N.J. & Roberts, H. 2007. The use of slope stability radar (SSR) in managing slope instability hazards. In Eberhardt, E., Stead, D. & Morrison, T. (eds), *Proceedings 1st Canada-U.S. Rock Mechanics Symposium, Vancouver* 1: 53–59. London: Taylor & Francis Group.

Haugerud, R.A., & Harding, D.J. 2001. Some algorithms for virtual deforestation (VDF) of lidar topographic survey data. *International Archives of Photogrammetry and Remote Sensing* 34–3/W4: 211–217.

Hong, Y., Adler, R.F. & Huffman, G.J. 2007. Satellite remote sensing for global landslide monitoring, EDS. *Transactions of the American Geophysical Union* 88: 357.

Jaboyedoff, M., Ornstein, P. & Rouiller, J.-D. 2004. Design of a geodetic database and associated tools for monitoring rock-slope movements: the example of the top of Randa rockfall scar. *Natural Hazards and Earth System Sciences* 4: 187–196 (pdf, 5755 Ko).

Jaboyedoff, M., Metzger, R., Oppikofer, T., Coulture, R., Derron, M., Locat, J. & Turmel, D. 2007. New insight techniques to analyse rock slope relief using DEM and 3D-imaging cloud points: COLTOP-3D software, In Eberhardt, E., Stead, D. & Morrison, T. (eds), *Proceedings 1st Canada-U.S. Rock Mechanics Symposium, Vancouver* 1: 61–68. London: Taylor & Francis Group.

Jones, C.L., Higgins, J.D., & Andrew, R.D. 2000. Colorado Rock fall Simulation Program Version 4.0. Colorado Department of Transportation. *Colorado Geological Survey,* March 2000. 127 pp.

Kemeny, J., Norton, B. & Turner, K. 2006. Rock slope stability analysis utilizing ground-based LiDAR and digital image processing. *Felsbau* 24: 8–16.

Krahn, J. 2003. The 2001 R.M. Hardy Lecture: The limits of limit equilibrium analyses. *Canadian Geotechnical Journal* 40(3): 643–660.

Kunapo, J., Dasari, G.R., Phoon, K.K. & Tan, T.S. 2005. Development of a Web-GIS based geotechnical information system. *Journal of Computing in Civil Engineering* 19(3): 323–327.

Lan, H.X., Lee, C.F., Zhou, C.H. & Martin, C.D. 2005. Dynamic characteristics analysis of shallow landslides in response to rainfall event using GIS. *Environmental Geology* 47(2): 254–267.

Lan, H. & Martin, C.D. 2007. A digital approach for integrating geotechnical data and stability analyses in E.Eberhardt, D. Stead & Morrison, T. (eds), *Rock Mechnanics: Meeting Society's Challenges and Demands*, 45–2. London: Taylor and Francis Group.

Lan, H., Martin, C.D. & Lim, C.H. 2007. Rockfall Analyst: a GIS extension for three-dimensional and spatially distributed rockfall hazard modelling. *Computers & Geosciences* 33: 262–279.

Lim, C.H. 2008. A process model for rock fall, *PhD Thesis Dept. Civil & Environmental Engineering, University of Alberta, Edmonton, Canada.*

Luna, R. & Frost, J.D. 1998. Spatial liquefaction analysis system. *Journal of Computing in Civil Engineering* 12 (1): 48–56.

Maffei, A., Martino, S. & Prestininzi, A. 2005. From the geological to the numerical model in the analysis of gravity-induced slope deformations: An example from the Central Apennines (Italy). *Engineering Geology* 78(3–4): 215–236.

Martin, C.D., Tannant, D.D. & Lan, H. 2007. Comparison of terrestrial-based, high resolution, LiDAR and digital photogrammetry surveys of a rock slope. Eberhardt, E. Stead, D. & Morrison, T. (eds), *Proceedings 1st Canada-U.S. Rock Mechanics Symposium, Vancouver*. Taylor & Francis Group, London 1: 37–44.

McCardle, A., Rabus, B., Ghuman, P., Rabaco, L, M.L., Amaral, C.S. & Rocha, R. 2007. Using Artificial Point Targets for Monitoring Landslides with Interferometric Processing. *Anais XIII Simposio Brasileiro de Sensoriamento Remoto*. INPE: 4933–4934.

Meisina, C., Zucca, F., Conconi, F., Verri, F., Fossati, D., Ceriani, M. & Allievi, J. 2007. Use of Permanent Scatterers Technique for Large-scale Mass Movement Investigation. *Quaternary International* 171–172: 90–107.

Mejia-Navarro, M., Wohl, E.W. & Oaks, S.D. 1994. Geological hazards, vulnerability, and risk assessment using GIS: model for Glenwood Springs, Colorado. *Geomorphology* 43: 117–136.

Morgenstern, N.R. 1986. Goddard landslide of September, 1982, summary of the opinion of Norbert R. Morgenstern. *Report prepared for court between Canadian Pacific Limited and Highland Valley Cattle Company Limited.* Supreme Court of British Columbia: Vancouver registry No. C841694.

Morgenstern, N.R. 2000. Performance in geotechnical practice: Inaugural Lumb Lecture. *Transactions Hong Kong Institute of Engineers* 7: 1–15.

Nathanail, C.P. & Rosenbaum, M.S. 1998. Spatial management of geotechnical data for site selection. *Engineering Geology* 50(3–4): 347–356.

Parise, M. 2003. Observations of surface features on an active landslide and implications for understanding its history of movement. *Natural Hazards and Earth System Sciences* 3: 569–580.

Parsons, R.L. & Frost, J.D. 2000. Interactive analysis of spatial subsurface data using GIS-Based tool. *Journal of Computing in Civil Engineering* 14(4): 215–222.

Parsons, R.L. & Frost, J.D. 2002. Evaluating site investigation quality using GIS and geostatistics. *Journal of Geotechnical and Geoenvironmental Engineering* 128 (6): 451–461.

Pfeiffer, T.J. & Bowen, T. 1989. Computer simulation of rock falls. *Bulletin of the Association of Engineering Geologists* 26 (1): 135–146.

Porter, M.J., Savigny, K.W., Keegan, T.R., Bunce, C.M. & MacKay, C. 2002. Controls on stability of the Thompson River landslides. In *Proceedings 55th Canadian Geotechnical Conference: Ground and Water: Theory to Practice, Niagara Falls, Ontario*, Vol. 2, pp 1393–1400.

Pötsch, M., Schubert, W. & Gaich, A. 2007. The application of metric 3D images for the mechanical analysis of keyblocks. In Eberhardt, E., Stead, D. & Morrison, T. (eds), *Proceedings 1st Canada-U.S. Rock Mechanics Symposium, Vancouver* 1: 77–84. London: Taylor & Francis Group.

Radbruch, D.H. & Crowther, K.C. 1973. Map showing areas of estimated relative susceptibility to landsliding in California. *U.S. Geological Survey, Miscellaneous Geologic Investigation* Map I-747.

Refice, A. & Capolongo, O. 2002. Probabilistic modelling of uncertainties in earthquake-induced landslide hazard assessment. *Computing and Geoscience* 28: 735–749.

Rhina, D. 1992. Why GIS? *ARC News* 11: 1–4.

Rizkalla, M. & Randall, C. 1999. A Demonstration of Satellite-Based Remote Sensing Methods for Ground Movement Monitoring in Pipeline Integrity Management. *Proceedings 18th International Conference on Offshore Mechanics and Arctic Engineering*. St. John's, OMAE 99/PIPE-5004, ASME.

Rosser, N.J., Petley, D.N., Lim, M., Dunning, S.A. & Allison, R.J. 2005. Terrestrial laser scanning for monitoring the process of hard rock coastal cliff erosion. *Quarterly Journal of Engineering Geology and Hydrogeology* 38: 363–373.

Ryder, J.M. 1976. Terrain inventory and Quaternary geology, Ashcroft, British Columbia, *Geological Survey of Canada, Ottawa.*

Schulz, W.H. 2004. Landslides mapped using LiDAR imagery, Seattle, Washington. *U.S. Geological Survey, Open-File Report* 2004-1396.

Schulz, W.H. 2007. Landslide susceptibility revealed by LiDAR imagery and historical records, Seattle, Washington. *Engineering Geology* 89: 67–87.

Singhroy, V. 2005. Remote sensing of landslides. In Glade, T., Anderson, M. & Crozier, M.J. (eds), *Landslide Hazard and Risk* 469–492. Wiley.

Soe Moe, K.W., Cruden, D.M., Martin, C.D., Lewycky, D. & Lach, P.R. 2005. 15 years of movement at Keillor Road, Edmonton. *Proceedings 58th Canadian Geotechnical Conference & 6th Joint CGS/IAH-CNC Groundwater Specialty Conference, Saskatoon*, 1: 281–388.

Soeters, R. & Van Westen, C.J. 1996. Slope instability, recognition, analysis, and zonation in Turner, A.K. and Schuster, R.L. (eds). *Landslides: Investigation and Mitigation* 129–177. Washington, D.C.: National Academy Press.

Sturznegger, M., Stead, D., Froese, C., Moreno, F. & Jaboyedoff, M. 2007. Ground-based and airborne LiDAR for structural mapping of the Frank Slide. In Eberhardt, E., Stead, D. & Morrison, T. (eds), *Rock Mechanics: Meeting Society's Challenges and Demands* 925–932. London: Taylor and Francis.

Tsai, Y.C. & Frost, J.D. 1999. Using geographic information system and knowledge base system technology for real-time planning of site characterization activities. *Canadian Geotechnical Journal* 36 (2): 300–312.

Van Westen, C.J. & Getahun, F.L. 2003. Analyzing the evolution of the Tessina landslide using aerial photographs and digital elevation models. *Geomorphology* 54: 77–89.

Van Westen, C.J. 2007. Mapping landslides: Recent developments in the use of the digital spatial information. *Proceedings 1st North American Landslide Conference, Vail, CO* 221–238.

Varnes, D.J. 1984. International Association of Engineering Geology Commission on Landslides and Other Mass Movements on Slopes. Landslide Hazard Zonation. *International Association of Engineering Geology. UNESCO Natural Hazard Series* 3(63) pp.

Wachal, D.J. & Hudak. P.F. 2000. Mapping landslide susceptibility in Travis County, Texas, USA. *GeoJournal* 51: 245–253.

Walstra, J., Dixon, N. & Chandler, J.H. 2007. Historical aerial photographs for landslide assessment: two case histories. *Quarterly Journal of Engineering Geology and Hydrogeology* 40: 315–332.

Limit equilibrium and finite element analysis – A perspective of recent advances

Zuyu Chen
China Institute of Water Resources and Hydropower Research, China

Keizo Ugai
Department of Civil Engineering, Guma University, Japan

ABSTRACT: This paper gives a general review of the recent advances in the applications of limit equilibrium (LEM) and Finite Element Methods (FEM) for slope stability analysis. Accuracies of various LEM including Sarma's have been reviewed. Special attentions have been given to the strength reduction finite element method regarding its applicability, criteria for failure indications, and the treatment for modulus, Poisson ratio and dilation angles.

1 INTRODUCTION

Analyses play an important role in assessing the risks involved in a potential landslide and the design of an engineered slope.

The traditional approach based on the limit equilibrium methods (LEM) has found wide applications in slope stability analysis. The bench-mark reports given by Morgenstern (1992, 1991) and Duncan (1996) have covered almost all important aspects with regard to the use of LEM. Latest development in this area may not be substantial, but the improvement in enhancing its efficiency and widening its applicability deserve a special review. On the other hand, 3D LEM is an area that has received wide attentions whose review is also worthwhile.

The finite element method (FEM) offers an alternative that is more rigorous, being free of the assumptions regarding static indeterminacy, and descriptive, being able to provide both stress and deformation information. FEM normally does not give an explicit result for factor of safety, which limits its applicability in engineering judgments. The strength reduction finite element method (SRF), originally advocated by Zienkiewicz, et al (1975), has received warm response recently as it has similar theoretical background and the associated factor of safety to those of the conventional approach using LEM.

2 THE LIMIT EQUILIBRIUM ANALYSIS METHODS

2.1 General

When commenting the further efforts in updating the general limit equilibrium approaches, Morgenstern (1992) said that various LEM, at least for two-dimensional analyses, are well understood. The impact of new studies can be slight.

Indeed, new papers dealing with the improvement of the analytical aspects of LEM have not been many, compared to the early stages of the 1960s or 1970s. Some conclusive remarks on the accuracies of various LEM have been made (Duncan, 1996). More papers have focused on the search techniques for locating the critical slip surfaces, especially on the stochastic approaches of optimization methods.

2.2 The generalized formulations

A number of general formulations have been proposed in the literatures. The authors tried to fix various LEMs in a unified framework. Among these research outcomes (Zhu, et al 2003, Espinoza, et al 1994, Li, 1992), we wish to briefly review the work by Chen & Morgenstern (1983), whose analytical forms will greatly facilitate the calculations by spread sheets (Chen. et al., 2008).

The governing force and moment equilibrium equations provided by Chen and Morgenstern are

$$\int_a^b p(x)s(x)dx = 0 \tag{1}$$

$$\int_a^b p(x)s(x)t(x)dx - M_e = 0 \tag{2}$$

The conventional definition for factor of safety reduces the shear strength parameters c' and ϕ' by the following equations.

$$c_e = c/F \tag{3}$$

$$\tan\phi_e = \tan\phi/F \tag{4}$$

Eqs. (1) and (2) involve the following definitions:

$$p(x) = \frac{dW}{dx}\sin(\phi_e - \alpha) + q\sin(\phi_e - \alpha)$$

$$\quad - r_u \frac{dW}{dx}\cdot\sec\alpha\sin\phi_e + c_e\sec\alpha\cos\phi_e$$

$$\quad + \eta\frac{dW}{dx}\cos(\phi_e - \alpha) \tag{5}$$

$$s(x) = \sec(\phi_e - \alpha + \beta)$$

$$\quad \times \exp\left[-\int_a^x \tan(\phi_e - \alpha + \beta)\frac{d\beta}{d\zeta}d\zeta\right] \tag{6}$$

$$t(x) = \int_a^x (\sin\beta - \cos\beta\tan\alpha)$$

$$\quad \times \exp\left[\int_a^\xi \tan(\phi_e - \alpha + \beta)\frac{d\beta}{d\zeta}d\zeta\right]d\xi \tag{7}$$

$$M_e = \int_a^b \eta\frac{dW}{dx}h_e dx \tag{8}$$

where α = inclination of the slice base; β = the inclination of the inter-slice force. dW/dx = weight of the slice per unit width; q = vertical surface load; η = the coefficient of horizontal seismic force, h_e = the distance between the horizontal seismic force and base of the slice, r_u = pore pressure coefficient.

In order to avoid violating the Principle of complementary shear stresses, Chen and Morgenstern (1983) argued that β must be fixed at both ends of the sliding mass. They suggested [Figure 1(b) and (c)].

$$\tan\beta = f_o(x) + \lambda f(x) \tag{9}$$

in which $f(x)$ is a linear function that allows the values $f_o(a)$ and $f_o(b)$ to be equal to specified values of $\tan\beta$ at $x = a$ and $x = b$ respectively. $f(x)$ is another function

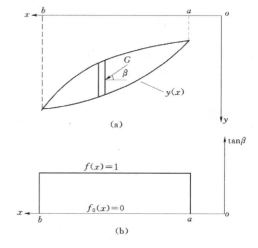

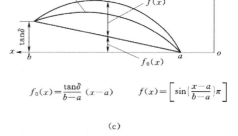

Figure 1. The generalized method of slices. (a) The failure mass, (b) Spencer method, (c) $\tan\beta$ to be fixed at both ends.

that is zero at $x = a$ and $x = b$. Figure 1(c) is an example that adopts a sine function for $f(x)$.

It is possible to find F and λ from Eqs. (1) and (2) by iterations.

2.3 The simplified methods

The various simplified methods in common use can be derived from Eqs. (1) and (2) as follows.

1. Spencer, 1966
   This method takes $f_o(x) = 0$ and $f(x) = 1$ in Eq. (9), which means [refer to Figure 1(b)],

$$d\beta/dx = 0 \tag{10}$$

Eqs. (1) and (2) thus respectively reduce to

$$s(x) = \sec(\phi_e - \alpha + \beta) \tag{11}$$

$$t(x) = \sin\beta(x - x_a) - \cos\beta(y - y_a) \tag{12}$$

The force and moment equilibrium equations (1) and (2) are simplified as:

$$\int_a^b p(x)\sec(\phi_e - \alpha + \beta)dx = 0 \quad (13)$$

$$\int_a^b p(x)\sec(\phi_e - \alpha + \beta)$$
$$\times (x\sin\beta - y\cos\beta)dx = M_e \quad (14)$$

2. U. S. Army, Corps of Engineers, 1967.
   This method assumes that the inclination of the inter-slice force of each slice is parallel to the average sloping of the slope surface, whose inclination is designated γ_a,

$$\beta = \gamma_a \quad (15)$$

3. Low, J. III and Katafiath, 1960.
   The authors assumed that the inclination of the inter-slice force of each slice is equal to the average of the slopings of the top and base of the slice.

$$\beta = \beta' = \frac{(\alpha + \gamma)}{2} \quad (16)$$

4. Janbu, 1954
   The simplified Janbu's method assumes

$$\beta = 0 \quad (17)$$

By specifying a particular value of β for each slice, it is possible to solve for F in Eq. (1) for the methods 2, 3 and 4.

5. Bishop 1952
   This method is concerned with circular slip surfaces whose center is taken to establish the moment equilibrium equation. The Bishop's simplified method assumes $\beta = 0$ that makes

$$s = \sec(\phi_e - \alpha) \quad (18)$$

and

$$t = -\int_a^x \tan\alpha \, d\xi$$
$$= -\int_a^x \frac{dy}{d\xi}d\xi = -y = -R\cos\alpha \quad (19)$$

where R is the radius of the circle. Substituting Eqs. (18) and (19) into Eq. (2), we have

$$\int_a^b \left[\frac{dW}{dx}\cos\alpha\tan(\phi_e - \alpha) - r_u\frac{dW}{dx}\sin\phi_e\right.$$
$$\left.\times \sec(\phi_e - \alpha)\right]dx + \int_a^b \left[c_e\sec(\phi_e - \alpha)\right.$$
$$\left.\times \cos\phi_e - \eta\frac{dW}{dx}R_d\right]dx = 0 \quad (20)$$

which can be demonstrated to be identical to the original formulation given by Bishop.

2.4 Applicability of various simplified methods

2.4.1 General remarks

The accuracies of various limit equilibrium methods have been conclusively summarized by Duncan as follows:

- The Ordinary (Fellunius) Method is highly inaccurate for effective stress analyses of flat slopes with high pore-pressure;
- Bishop's simplified method is accurate for all conditions;
- Factors of safety calculated by force equilibrium methods are sensitive with the assumed inclinations of the side forces between slices;
- Methods that satisfy all conditions of equilibrium are accurate for any conditions.

For years our profession has been puzzled by the fact that Bishop's simplified method always gives factors of safety in good agreement with those that satisfy complete equilibrium conditions. Zhu (2008) found that the omitted term on the summation of unbalanced shear forces on the interfaces in Bishop's simplified approach can be set to zero if a particular shear force distribution for inter-slices is assigned, which in the meanwhile allows the force equilibrium condition to be satisfied.

2.4.2 Illustrative examples

To illustrate the statements regarding the accuracies of the LEM methods in Section 2.3, we present the following two examples.

Example 1 Figure 2 shows a slope with simple geometry and material properties: $c = 5 \times 9.8$ kN/m^2, $\phi = 35°, \gamma = 1.7 \times 9.8$ kN/m^3. Table 1 summarizes the factors of safety associated with different arc angle α of the circle and pore pressure coefficient r_u. It can be found that the Bishop's simplified method in all cases gives basically the same results of Spencer's method. The discrepancies of Sweden's method, compared to Spencer's, increases rapidly as α and r_u enlarges. At $\alpha = 117.6$ and $r_u = 0.6$, the relative error is: $(F_S - F_F)/F_F = (1.381 - 0.769)/1.381 = 44.3\%$.

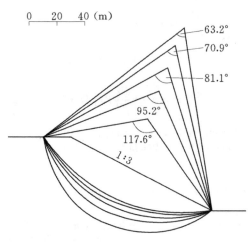

Figure 2. An illustrative example explaining the accuracies of Bishop, Fellunius and Spencer methods.

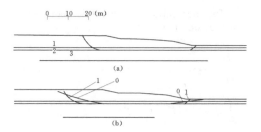

Figure 3. Back analysis of the Huaihexin Dike.

Table 2. Geotechnical parameters for example 2.

Soil layer number	γ (kN/m^3)	c (kPa)	ϕ(°)
1	19.11	11.0	26.0
2	13.03	2.67	0
3	17.12	27	2.84

Table 1. Factors of safety associated with various arc angle α of the circle and pore pressure coefficient r_u.

r_u	α(°)	F_b	F_F	F_s
0.0	117.6	3.020	3.009	2.544
	95.2	2.614	2.608	2.322
	81.1	2.451	2.446	2.245
	70.9	2.371	2.368	2.216
	63.2	2.332	2.329	2.209
0.2	117.6	2.444	2.444	1.953
	95.2	2.121	2.122	1.820
	81.1	1.994	1.995	1.782
	70.9	1.936	1.937	1.775
	63.2	1.910	1.910	1.783
0.4	117.6	1.876	1.895	1.361
	95.2	1.634	1.648	1.317
	81.1	1.542	1.552	1.318
	70.9	1.504	1.511	1.334
	63.2	1.490	1.496	1.354
0.6	117.6	1.325	1.381	0.769
	95.2	1.157	1.195	0.814
	81.1	1.098	1.126	0.855
	70.9	1.078	1.098	0.893
	63.2	1.076	1.091	0.929

NOTE: F_b, F_F, F_s are factors of safety obtained by methods of Bishop, Fellunius and Spencer respectively.

Example 2 Back analysis of the Huaihexin Dike

This example, shown in Figure 3, takes from a dike case in which the authors tried to use different methods to back analysis the failure (Chen, 1999). The strength parameters are shown in Table 2.

The factors of safety given by the approach of satisfying the force equilibrium method only, associated with different input of β, are shown in Table 3,

Table 3. Factors of safety associated with different values of β.

β(°)	0.0	1.55*	5.0	10.0	15.0	20.0
F	1.013	1.027	1.070	1.147	1.232	1.341

compared with that obtained by Spencer's method that gives $F = 1.027$ and $\beta = 1.55°$. It can be found that F varies with β considerably and the additional moment equilibrium method is indeed necessary to find a reasonable solution for F.

3 THE UPPER BOUND ANALYSIS

3.1 Sarma's (the upper-bound) method

Sarma presented a method that divides the failure mass into a number of slices with inclined interfaces. The limit equilibrium condition has been applied to both the base and inter-slice faces. This method is particularly applicable to rock slopes as advocated by Hoek (1983, 1987).

The original approach by Sarma (1979) is based on the force equilibrium conditions (Figure 4), which has complex recurrence formulations.

Donald and Chen (1997) presented an identical approach which is theoretically supported by the upper bound theorem and practically easy to handle.

This method starts with a kinematically admissible velocity field, in which the slice moves in a direction that inclined at a friction angle relative to its neighboring slice or the base (detailed discussion has been given in Chen (2008).

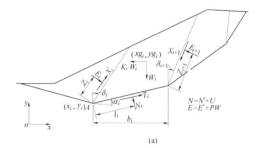

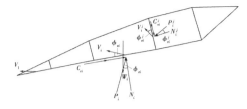

Figure 5. Sketch for the analyses by the energy approach of Sarma's method in finite difference forms.

Figure 4. Schematic illustrations for Sarma's method.

1. Formulations based on the finite differences

The velocity of a slice numbered i, designated V_i can be determined by (refer to Figure 5)

$$V_i = \kappa V_1 \quad (21)$$

where V_1 is the velocity of the first slice. κ is defined as

$$\kappa = \prod_{j=1}^{i} \frac{\sin(\alpha_i^l - \phi_{ei}^l - \theta_i^j)}{\sin(\alpha_i^r - \phi_{ei}^r - \theta_i^j)} \quad (22)$$

θ is the angle of the velocity with reference to the positive x axis. The superscript j refers to the variable on the interfaces, and l and r refer to the left and right sides of the interfaces.

The factor of safety, based on Eqs. (3) and (4) is obtained by the work-energy balance equation,

$$\sum_{i=1}^{n} \kappa[(c_e \cos\phi_e - u \sin\phi_e)\sec\alpha \, \Delta x$$

$$- \Delta W \sin(\alpha - \phi_e)]_i$$

$$- \sum_{i=1}^{n-1} \kappa(c_e^j \cos\phi_e^j - u^j \sin\phi_e^j)_i$$

$$\times \csc(\alpha^r - \phi_e^r - \theta_j)_i \sin(\Delta\alpha - \Delta\phi_e)_i L_i = 0 \quad (23)$$

The first term of the left-hand side of Eq. (23) refers to the work done by the external loads and the energy dissipation on the slip surface, while the second term is the energy dissipation on the interfaces between two contiguous slices.

2. Formulations based on integrals

Eq. (21) can be transformed to an integral if the width of the slice approaches to infinitesimally small (Figure 6),

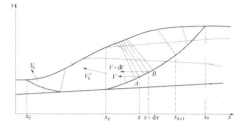

Figure 6. Sketch for the analyses by the energy approach of Sarma's method in integral forms.

$$V = \kappa \exp\left[-\int_{x_0}^{x} \cot(\alpha - \phi_e - \theta_j)\frac{d\alpha}{d\zeta}d\zeta\right]V_1 \quad (24)$$

Eq. (23) then becomes

$$\int_{x_0}^{x_n}\left[(c_e \cos\phi_e - u \sin\phi_e)\sec\alpha\right.$$

$$\left. - \frac{dW}{dx}\sin(\alpha - \phi_e)\right]E(x)dx$$

$$- \int_{x_0}^{x_n}(c_e^j \cos\phi_e^j - u^j \sin\phi_e^j)L\csc(\alpha - \phi_e - \theta_j)$$

$$\times \frac{d\alpha}{dx}E(x)dx + K_i = 0 \quad (25)$$

where K_i is a coefficient accounting for possible discontinuities in α, ϕ_e and c_e.

$$K_i = -\sum_{i=1}^{n}(c_e^j \cos\phi_e^j - u^j \sin\phi_e^j)_i$$

$$\times L_i \csc(\alpha^r - \phi_e^r - \theta_j)\sin(\Delta\alpha - \Delta\phi_e)_i E^l(x_i) \quad (26)$$

3.2 The optimization process

The conventional limit equilibrium methods and the upper-bound method (Donald and Chen, 1997) include an optimization process that finds the critical failure mode associated with the minimum factor of safety.

Early research work, such as Chen and Shao's (1988), has been continued recently by a number of researchers (Goh, 1999, Pham and Fredlund, 2003, Cheng, et al., 2003, Sarma and Tan, 2006).

The slip surface is discretized into a number of nodal points that are connected by either smooth curves or straight lines designated $A_1, A_2, \ldots, A_6$ (Figure 7). The variables defining the failure mode includes the co-ordinates of the nodal points and the inclinations of the interfaces if the upper-bound method is adopted. The optimization method will find these variables that give the minimum factor of safety designated $B_1, B_2, \ldots, B_6$.

3.3 Test examples

3.3.1 Theoretical verifications

A series of test problems based on the closed-form solutions provided by the slip-field method (Sokolovski, 1960) has been performed using the numerical approaches described in this Section (Donald and Chen, 1997; Chen, 1999). The results showed good agreements, demonstrating that the upper-bound method approach is more rigorous than the conventional method of vertical slices. As an extension, this method has been successfully applied to the calculation for bearing capacity analysis, in which the conventional method is generally not applicable (Wang et al, 2001). This means that various empirical coefficients involved in the conventional approaches accounting for the effect of soil weight, embedment of footing, complicated ground heterogeneities and water conditions are no longer necessary.

Example 3 Comparisons with the closed-form solution

Figure 8 shows a uniform slope subjected to a vertical surface load q. The weight of the soil mass is neglected. The closed-form solution for the ultimate vertical surface load has been provided by Sokolovski (1960). Chen (2008) gives a detailed description of the problem and demonstrated that Eq. (25) is reducible to the closed-form solution.

Figure 9 shows an example with the material property parameters $c = 98$ kPa, $\phi = 30°$. The inclination of the slope surface is $\gamma' = 45°$, the bearing capacity q calculated by the closed form solution (refer to Chen, 2008) is 10921.1 kPa. In Figure 9(a) the initially guessed slip surface is represented by 5 nodal points connected by straight lines. The inclinations of the interfaces are set arbitrarily. The factor of safety given by Eq. (23) is 1.047. Figure 9(b) shows the critical mode associated with $F = 1.013$. Figure 9(c) shows a more accurate result that employs 16 nodal points with $F = 1.006$. It can be seen that the upper-bound method gives an accurate result both in terms of the factor of safety and the critical failure mode, compared to the slip-line field method.

3.3.2 Comparisons with the conventional methods

Test examples have also shown that the upper-bound method is also able to give comparative results of factor of safety to those obtained by the conventional methods. The following two examples are taken from the ACADS slope stability programs review by Donald and Giam (1992) from which one may find the details including the material and geometry parameters.

Example 4 The ACADS test example EX1(a)

For a simple test example shown in Figure 10 the 'referee answer' based on the simplified Bishop's

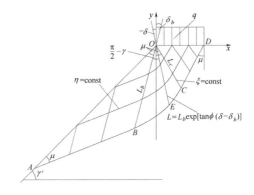

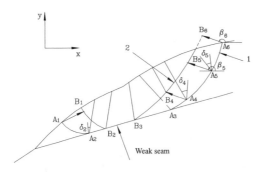

Figure 7. The optimization process for locating the critical failure mode.

Figure 8. Verifications of Eq. (24) compared to the theoretical solution provided by the slip-line field method, example 3.

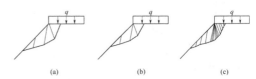

Figure 9. Example 3, an example describing the upper bound approach.

method for the critical slip surface is 1.00. The upper bound method defined 4 nodal points designated A, B, C, D with arbitrary interface inclinations as shown in the Figure 10(a). Factor of safety for this initial failure mode is 1.304. Figure 10(b) shows the critical failure mode associated with a minimum F of 0.997, which is very close to the 'referee answer'

Example 5 The ACADS test example EX1(c)

Using the similar algorithms the factor of safety for the initial failure mode was 1.630 as shown in Figure 11(a), and the minimum F for the critical failure mode shown in Figure 11(b) was 1.401. The upper bound results can be compared with those given by the conventional methods shown in Figure 12. The slip surfaces 1, 2, 3 are related to the methods of Spencer, Bishop, and Sarma respectively.

Chen (2008) illustrated that the limit equilibrium methods with the vertical and inclined slices can be approximately fixed in the theoretical framework of the lower and upper theorems of Plasticity.

Figure 10. The ACADS test example EX1(a). (a) The initial failure mode; (b) the critical one.

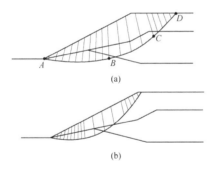

Figure 11. The upper bound solutions for ACADS test example EX1(c). (a) The initial failure mode; (b) the critical one.

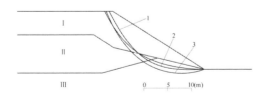

Figure 12. Comparisons of the critical slip surfaces obtained various methods. (1) Spencer, $F = 1.366$; (2) Bishop, $F = 1.378$; (3) Sarma, $F = 1.401$.

3.4 *Practical considerations with Sarma's method*

The experience of using Sarma's method shows that the following two issues are frequently encountered, which require proper treatments.

1. The alternative directions of shear force or relative velocity on the inter-slice surface

   It has been shown that there are two possible directions for a relative velocity between two contiguous slices. The conventional method only considers the condition that the left slice moves upward relative to the right one, as shown in Fig 7. However, it is sometimes likely that the left slice moves down ward relative to the right one. Failure to identify this alternative may occasionally yield wrong results.

2. Treatment when tension develops on the interfaces and/or the base of a slice

   Sarma's method assumes that shear failure develops along the slip surface and the interfaces. However the calculated results may show some tensile internal forces, which is contradictive to the original assumptions. In his program SARMA, Hoek gives a warning but no solution is offered. An approximate treatment is proposed by Chen (2008).

   Details working on the two issues deserve a special paper which is contained in this Proceedings (Chen, 2008).

4 THE 3D LIMIT EQUILIBRIUM AND UPPER BOUND ANALYSIS

4.1 *3D analysis based on the 'method of columns'*

In their keynote and state-of-the-art reports, Morgenstern (1992) and Ducan (1996) advocated the importance of the development of 3D limit equilibrium methods for slope stability analyses. A great number of papers on 3D slope stability analysis methods have emerged during the subsequent 10 years (e.g., Stark and Eid 1998, Chen et al. 2001, Huang and Tsai 2000, Jiang and Yamagami 2004). All these papers have dealt with the 'method of column' that can be considered to an extension of the "method of slices" in the two-dimensional area (Figure 13). On the other hand, the method employing columns or blocks with inclined interfaces have been developed (Michalowski, 1989, Chen et al. 2001a,b, Farzaneh and Askari 2003), which can be regarded as an extension of the 2D Sarma's method.

Review of the development of 3D limit equilibrium and upper bound analysis deserves a full paper, and indeed has been tried by Chen et al. (2006) in

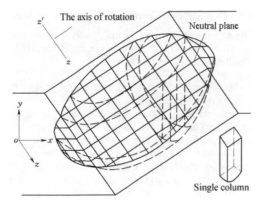

Figure 13. Sliding mass consisting of prisms with vertical interfaces.

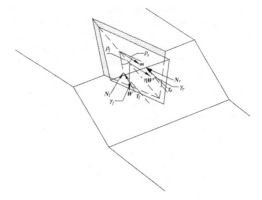

Figure 14. The rock wedge failure ρ_l, ρ_r = the dilatant angles. The subscripts 'l' and 'r' stand for the left and right planes respectively. For other parameters, refer to Chen (2004).

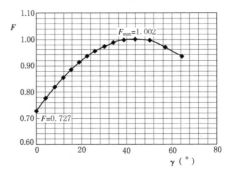

Figure 15. Factors of safety associated with various shear force directions on the failure planes.

their keynote paper of GeoShangai. Therefore no more elaboration will be made in this Paper.

4.2 The generalized solution to rock wedge analysis

Among a variety of three-dimensional stability problems of slopes, wedge is a special and also the simplest case that requires a special study.

It has been found that the limit equilibrium approach commonly used for tetrahedral rock wedge stability analysis actually involves an assumption that the shear forces applied on the failure planes are parallel to the line of intersection. It is because of this assumption that makes the solutions for normal forces applied on the left and right failure planes possible, as depicted in Figure 14.

To illustrate the impact of this finding, Chen took an example that has a symmetric geometry and material properties with respect to the line of intersection. The cohesion of the two failure surfaces is set to zero. The angle between the line of intersection and the shear force applied on the failure surface is denoted by γ. For this symmetric wedge with simple geometry, it is possible to establish a formulation to calculate F associated with different values of γ. The case $\gamma = 0°$ corresponds to the conventional method and gives a value $F = 0.727$. However F increases as γ becomes larger and eventually reaches a maximum of 1.002 at $\gamma = 42.5°$, as shown in Figure 15.

A new method that allows an input of various shear force directions has been presented by Chen (2004). The controlling equation is reducible to the conventional solution and permits a formal demonstration to confirm that when $\rho_l = \phi_{el}$ and $\rho_r = \phi_{er}$ factor of safety will obtain its maximum. Chen also discussed the theoretical implications of these findings regarding some fundamental understanding in Plasticity.

5 FACTORS OF SAFETY BY THE FINITE ELEMENT METHODS

5.1 Definition of the factor of safety

The finite element method normally gives information of stress and strain fields. Various approaches have been proposed to transfer them to the factor of safety that is a common concern in engineering practice.

1. Based on the stress levels

   Keep σ_3' of an element unchanged, draw a circle that is tangent to the Mohr-Coulomb's failure envelope with a corresponding diameter $(\sigma_1' - \sigma_3')_f$ and the stress level $(\sigma_1' - \sigma_3')/(\sigma_1' - \sigma_3')_f$. The factor of safety is defined as

$$F_{FE1} = \frac{\int dl}{\int \frac{\sigma_1' - \sigma_3'}{(\sigma_1' - \sigma_3')_f} dl} \tag{27}$$

where the integral represents the scalar summations along a potential slip surface

2. Based on the shear stress on an element

For a stress state $\sigma_x, \sigma_y, \tau_{xy}$, the normal and shear stresses σ'_n and τ on the slip surface can be determinated by:

$$\tau = \frac{1}{2}(\sigma_y - \sigma_x)\sin 2\alpha + \tau_{xy}\cos 2\alpha \quad (28)$$

$$\sigma'_n = \sigma_x \sin^2\alpha + \sigma_y \cos^2\alpha - \tau_{xy}\sin 2\alpha \quad (29)$$

where α the inclination of the slip surface to the x axis.

The shear strength that can be developed along the slip surface is

$$\tau_f = c' + \sigma'_n \tan\phi' \quad (30)$$

The factor of safety along the entire slip surface can be defined as

$$F_{FE2} = \frac{\int (c' + \sigma'_n \tan\phi')dl}{\int \tau dl} \quad (31)$$

3. Based on the weighted stress levels

This approach defines the factor of safety by the following equation:

$$F_{FE3} = \frac{\int (c' + \sigma'_n \tan\phi')dl}{\int \frac{\sigma'_1 - \sigma'_3}{(\sigma'_1 - \sigma'_3)_f}(c' + \sigma'_n \tan\phi')dl} \quad (32)$$

5.2 *Search for the critical slip surface*

Having given the definition of the factor of safety, a search technique, similar to that commonly used in the limit equilibrium analysis area, can be employed to find the critical slip surface associated with the minimum factor of safety (Zhou et al., 1995). Donald et al. (1985) used an algorithm called CRISS to calculate the factor of safety of a slope taken from ex1(a) of the test problems issued by ACADS.

Example 5 Reevaluations of Example 2 by FEM

The referee answer based on the conventional limit equilibrium method by the ACADS review program Ex1(a) is 1.00 associated with a critical slip surface passing through the toe of the slope as shown in Figure 16.

Table 4 summarizes the associated minimum factors of safety. It can be found that the results are close to one other.

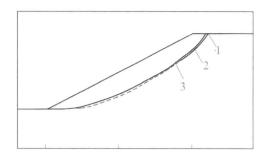

Figure 16. Factors of safety associated with various shear force directions on the failure planes.

Table 4. Comparisons of factors of safety obtained by different definitions.

No.	Types	Factor of safety		
		F_{FE1}	F_{FE2}	F_{FE3}
1	Embankment by layers	1.001	1.001	1.001
2	One-layer embankment	1.000	1.003	1.000
3	Excavation	1.082	1.044	1.044

6 THE STRENGTH REDUCTION FINITE ELEMENT METHOD

6.1 *The advances SRF has made*

SRF was used for slope stability analysis as early as 1975 by Zienkiewicz et al. In this method, the shear strength parameters are reduced by Eqs. (3) and (4), allowing the evolution of large area of plastic yielding.

Firstly, a gravity turn-on is implemented under elastic state to determine the initial stress distribution inside the slope. Then, stresses and strains are calculated by the elasto-plastic finite element method. The shear strength reduction factor, F, is then increased incrementally until the global failure of the slope reaches, which means that the finite element calculation diverges under a physically realistic convergence criterion (Refer to 6.3.1).

In a benchmark paper, Griiffiths and Lane (1999) provided a series of test examples that show good agreements both in terms of factor of safety and plastic zones with the conventional LE method.

SRF can be a powerful alternative to the traditional limit equilibrium methods. This technique has also been adopted in some well-known commercial software, such as FLAC, for practical applications.

The main advantages of the SRM can be summarized as follows:

- It requires no assumptions which have been commonly involved in LEM.
- The critical failure surface is found automatically.

- It offers much more detailed information such as the plastic zone, stress and deformation field, etc., compared to LEM.
- It is possible for SRF to include piles and anchors that produce the coupled stress fields for soil and structure simultaneously (Cai et al. 1998; Cai & Ugai 2000).

Perhaps, the most important contribution of SRF is that it makes geotechnical calculations by FEM self-checkable and reproducible. Since a large-scale non-linear finite element analysis involves complicated constitutive equations and iterations, it always happens that different computer programs cannot give same analytical results. Lack of unique and widely accepted solutions has discouraged the extensive use of FEM in geotechnical practice. Now SRF can be a tool to test the applicability of an EEM program which, as reliable software, should provide comparative results with LEM if SRF is performed using this program.

6.2 An illustrative example

A number of research work (Naylor, 1981; Donald and Giam, 1992; Matsui and San, 1992; Ugai and Leshchinsky, 1995; Dawson et al, 1999; Griffiths and Lane, 1999; Cheng et al, 2007) has confirmed good agreements between LEM and SRF in terms of factor of safety based on the definitions given by Eqs. (1) and (2). In this Proceedings, Duan et al. (2008) reported the calculated results by SRF for the gravity dam stability problems with weak seams, which are briefly summarized here.

Example 6 An example that compares the results of FEM and Sarma's method (Duan, et al., 2008).

Figure 17 shows the geometry of the example with material properties listed in Table 5. For the analysis along the weak seam ABC, Sarma's method gave a factor of safety $F = 2.12$.

Figure 18 shows the plastic zones from which one may find that the plastic zones along the weak seam extend as FOS increases. At $F = 2.20$, yielding dominates throughout the seam with a plastic zone near the dam toe shown in Figure 17, compared to the result $F = 2.12$ by Sarma's method.

Duan et al (2008) further investigated a case where section BC no longer exists as shown in Figure 17(b). The critical location is determined by an automatic search process in Sarma's method. The results obtained by Sarma and SRF were also in good agreement.

6.3 General issues with SRF

6.3.1 The failure criteria
There have been a number of criteria that define failure at which the calculation by SRF terminates: (1) nonconvergence of the numerical process; (2) rapid increase of displacement at some critical points; or (3) development of basically continuous plastic zones. The value of F at this moment is believed to be the solution for factor of safety of this problem.

Experience has shown that these criteria do not lead to substantially different values of F.

6.3.2 The constitutive laws
In SRF, both associated and non-associated elasto-plastic constitutive models can be adopted. The Mohr-Coulomb yield criterion is used to define the yield function if non-associated flow law is used.

$$f = -c' \cos\phi' - \frac{1}{3} I_1 \sin\phi'$$
$$+ \sqrt{J_2}\left(\cos\Theta - \frac{1}{3}\sin\Theta \sin\phi'\right) \quad (33)$$

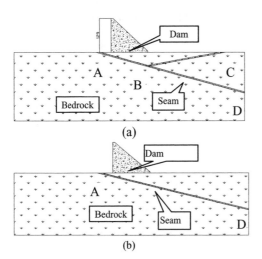

Figure 17. An example that compares the results of FEM and Sarma's method, the calculation by Sarma.

Table 5. Geotechnical properties for example 6.

Material	Density (kN/m^3)	Modulus (GPa)	Poisson ratio	Friction angle	Cohesion (MPa)	Tensile strength (MPa)
Dam	24.0	20	0.17	35	2.0	1.85
Bedrock	25.6	10	0.26	35	1.3	0
Seam	18.0	2.5	0.35	19.8	0.115	0

where

$$\Theta = \frac{1}{3}\sin^{-1}\left(-\frac{3\sqrt{3}}{2}\frac{J_3}{J_2^{3/2}}\right), \quad \left(-\frac{\pi}{6} \leq \Theta \leq \frac{\pi}{6}\right) \tag{34}$$

and the Drucker-Prager criterion is normally adopted to define the plastic potential function

$$g = -\alpha I_1 + \sqrt{J_2} - \kappa \tag{35}$$

where

$$\alpha = \frac{\tan \Psi}{\sqrt{9 + 12\tan^2 \Psi}}, \quad \kappa = \frac{3c'}{\sqrt{9 + 12\tan^2 \Psi}} \tag{36}$$

In the above equations $c, \phi,$ and Ψ are the effective cohesion, friction angle, and dilatant angle, respectively. $I_1, J_2,$ and J_3 are the first invariant of the effective stress, and the second and third invariants of the deviatoric stress, respectively.

6.3.3 Treatment of other parameters

The normal practice of SRF reduces the shear strength parameters during calculations while keeping other parameters constant. The necessity of treating these unchanged parameters has been discussed.

1. The Poisson ratio

    Zheng et al. (2002) found that if c and $\tan \phi$ are reduced considerably while Poisson ratio is still kept unchanged, it is likely that an element will inevitably yield, which is unrealistic. An approximate condition was suggested:

$$\sin \phi \geq 1 - 2\mu \tag{37}$$

where μ is the Poisson ratio. Therefore a better approach can be reducing the Poisson ratio simultaneously with the reduction of c and $\tan \phi$.

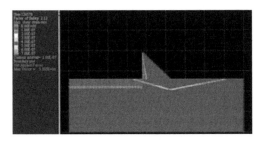

Figure 18. An example that compares the results of FEM and Sarma's method, the calculation by SRF.

Eq. (37) was derived on the assumption that c is negligible and the overburden h is very big. Duan et al. (2008) gave a more generalized criterion,

$$\sin(\phi + \alpha)/\cos \alpha \geq (1 - 2\mu) \tag{38}$$

where α is defined by

$$\sin \alpha = \frac{2c/\gamma h}{\sqrt{(1+K)^2 + \left(\frac{2c}{\gamma h}\right)^2}} \tag{39}$$

K is approximately taken to be coefficient of earth pressure at rest, which can be taken as:

$$K = \frac{\mu}{1 - \mu} \tag{40}$$

The detailed work is documented in a paper of this Proceedings (Duan et al., 2008).

2. Young's modulus

    From physical point of view, it is obviously advantageous to reduce both E and μ.

    Duan et al. (2008) also suggested a hyperbolic stress strain relationship, similar to that proposed by Duncan and Chang (1970) to derive the criterion for the reduced Young's modulus. Figure 19 illustrates how the modulus can be reduced based on the reduction of the strength envelop.

    As a matter of fact, the results shown in Example 4 and Figure 20 are based on the reduced values of μ. If μ keeps unreduced, a large area of

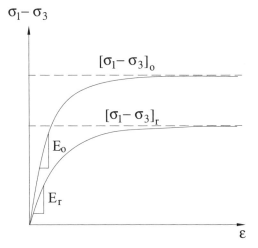

Figure 19. The concept of reducing Young's modulus E based on the hyperbolic stress strain relationship, the subscript 'o' and 'r' stand for the original and reduced variables respectively.

Figure 20. The large area of plasticity at the bottom by using unreduced values of μ by SRF.

plastic zone would develop at the bottom of the foundation, as shown in Figure 20. Although the factor of safety obtained was still $F = 2.12$, this unrealistic stress distribution would limit the credit and applicability of SRF.

3. The dilatant angles

A certain elasto-plastic constitutive law will be employed in the nonlinear finite element calculations, which can be associative or non-associative. The latter means a dilatant angle other than the friction angle can be assigned to an element. A number of papers (e.g. Cheng et al., 2007) investigated the influence of different values of dilatant angles to the final calculated results. The general conclusion is that adoptions of different values do not affect the final solution to factor of safety substantially in the 2D analysis.

7 CONCLUSIONS

This paper summarizes the recent advances in the traditional limit equilibrium and upper-bound methods, referred to as the method of slices with vertical and inclined interfaces respectively. Main findings in this area are:

- The analytical forms, such as Eqs. (1), (2) and Eqs. (13) and (14), can be derived for the generalized method of slices. They are particular useful for calculation by a spread sheet.
- Theoretical studies and test examples have shown that various LEM basically yield factors of safety close to one another, all lie on the lower bound side.
- Sarma's method with the slices of inclined interfaces can be formulated by an upper-bound approach represented by Eq. (23) or (25). This method enjoys a sound mechanical background. Some treatments for the alternative inter-slice shear force directions and internal tensions have been proposed.
- The three-dimensional limit equilibrium and upper-bound methods with vertical and inclined columns have been made possible. As a special case, the traditional rock wedge stability analysis method has been generalized allowing different input of shear force directions in the failure surfaces.

This paper also reviews the use of finite element method for slope stability analysis with particular attentions to the strength reduction approach (SRF). Main findings are:

- In the 2D areas, good agreements between LEM and SRF can be found. SRF can thus provide a useful approach for stability analysis;
- It would be advantageous to make appropriate reductions for other material parameters, such as Poisson ratio, Young's modulus, etc. Some associated criteria have been proposed.

REFERENCES

Bishop, A.W. 1955. The use of the slip circle in the stability analysis of slopes. Geotechnique 5(1):7–17.

Cai, F., Ugai, K., Wakai, A. & Li, Q. 1998. Effects of horizontal drains under rainfall by three-dimensional finite element analysis. Computers and Geotechnics 23:255–275.

Cai, F. & Ugai, K. 2000. Numerical analysis of the stability of a slope reinforced with piles. Soils and Foundations 40(1):73–84.

Chen, L.H., Chen, Z.Y. & Sun, P. 2008. Slope stability analysis using graphic acquisitions and spreadsheets. Proceedings of the 10th International Symposium on landslide and engineered slopes. Xi'an.

Chen, Z.Y. & Morgenstern, N.R. 1983. Extensions to the generalized method of slices for stability analysis. Canadian Geotechnical Journal 20(1):104–119.

Chen, Z.Y. & Shao, C.M. 1988. Evaluation of minimum factor of safety in slope stability analysis. Canadian Geotechnical Journal 25(4):735–748.

Chen, Z.Y. 1999. Discussions: Prior and back stability analysis of the Huaihongxin Dike. Chinese Journal of Geotechnical Engineering 21(4):518–519 (in Chinese).

Chen Z.Y., Wang xX.G., Haberfield C., Yin, J.H. & Wang, Y.J. 2001. A three-dimensional slope stability analysis method using the upper bound theorem-part I: theory and methods. International Journal of Rock Mechanics & Mining Sciences 38(3):369–378.

Chen Z.Y., Wang J., Wang Y.J., Yin, J.H. & Haberfield C. 2001. A three-dimensional slope stability analysis method using upper bound theorem-part II: numerical approaches, applications and extensions. International Journal of Rock Mechanics & Mining Sciences 38(3):379–397.

Chen, Z.Y. 2004. A generalized solution for tetrahedral rock wedge stability analysis. International Journal of Rock Mechanics & Mining Sciences 41:613–628.

Chen, Z.Y., Yin, J.H. & Wang, Y.J. 2006. Keynote lecture: The Three-Dimensional Slope Stability Analysis: Recent Advances and a Forward Look. Advances in Earth Structures, Research to Practice, Proceedings of Sessions of Geoshanghai, ASCE Special Publication No. 151:1–42.

Chen, Z.Y. 2007. The limit analysis in soil and rock: a mature discipline of geomechanics. Journal of Zhejiang University SCIENCE 8(11):1712–1724.

Chen, Z.Y. 2008. Some notes on the upper-bound and Sarma's methods with inclined slices for stability analysis. Proceedings of the 10th International Symposium on landslide and engineered slopes. Xi'an.

Cheng, Y.M. 2003. Locations of critical failure surface and some further studies on slope stability analysis. Computers and Geotechnics 30:255–267.

Cheng, Y.M., Lansivaara, T.B. & Wei, W.B. 2007. Two-dimensional slope stability analysis by limit equilibrium and strength reduction methods. Computers and Geotechnics 34:137–150.

Dawson, E.M., Roth, W.H. & Drescher A. 1999. Slope stability analysis by strength reduction. Geotechnique 49(6):835–840.

Donald, I.B., Tan, C.P. & Goh, T.C.A. 1985. Stability of geomechanical structures assessed by finite element method. Proc. 2nd Int. Conf. In Civil Engr. Hangzhou, 845–856. Beijing: Science Press.

Donald I.B. & Giam, S.K. 1988. Application of the nodal displacement method to slope stability analysis. In: Proceedings of the fifth Australia—New Zealand conference on geomechanics. 456–460. Sydney, Australia.

Donald, I.B. & Chen, Z.Y. 1997. Slope stability analysis by the upper bound approach: fundamentals and methods. Canadian Geotechnical Journal 34:853–862.

Donald, I.B. & Giam, P. 1992. The ACADS slope stability programs review. Proc. 6th International Symposium on Landslides. 3:1665–1670.

Duan, Q.W., Chen, Z.Y., Wang, Y., Yang, J. & Shao, Y. 2008. Applications of the strength reduction finite element method to a gravity dam stability analysis. Proceedings of the 10th International Symposium on landslide and engineered slopes. Xi'an.

Duan, Q. & Zhang, P.W. 2008. On the treatments for the deformation parameters in the strength reduction finite element method. Proceedings of the 10th International Symposium on landslide and engineered slopes. Xi'an.

Duncan, J.M. & Chang, C.Y. 1970. Nonlinear analysis of stress and strain in soils. Journal of Soil Mechanics and Foundation Engineering Division, ASCE 96(5):1629–1653.

Duncan, J.M. 1996. State of the art: Limit equilibrium and finite element analysis of slopes. Journal of Geotechnical Engineering 122(7):577–596.

Espinoza, R.D., Burdeau, P.L.P.C. & Mohunthan, B. 1994. Unified formulation for analysis of slopes with general slip surface. J. Geotech. Engng, ASCE 120(7):1185–1104.

Farzaneh, O. & Askari, F. 2003. Three-Dimensional Analysis of Nonhomogeneous Slopes. Journal of Geotechnical and Geoenvironmental Engineering 129(2).

Fellenius, W. 1927. Erdstatisch Berechnungen, Berlin W.Ernst und Sohn revised edition, 1939.

Goh, A.T.C. 1999. Genetic algorithm search for critical slip surface in multi-wedgestability analysis. Canadian Geotechnical Journal 36(2):383–391.

Griffiths, D.V. & Lane, P.A. 1999. Slope stability analysis by finite elements. Geotechnique 49(3):387–403.

Hoek, E. & Bray, J. 1977. Rock slope engineering. The Institute of Mining and Metallurgy.

Hoek, E. 1983. Strength of jointed rock masses. Geotechnique 33(3):187–223.

Hoek, E. 1987. General two-dimensional slope stability analysis-Analytical and Computational Methods in Engineering Rock Mechanics. Allen Unwin, London.

Huang, C.C. & Tsai, C.C. 2000. New method for 3D and asymmetric slope stability analysis. ASCE. Journal of Geotechnical and Environmental Engineering 126(9):917–927.

Janbu, N. 1954. Application of composite slip surfaces for stability analysis. Proceedings of European Conference on Stability of Earth Slopes. 3:43–49. Sweden.

Jiang, J.C. & Yamagami, T. 2004. Three-dimensional slope stability analysis using an extended Spencer method. Soils and Foundations 44(4):127–135.

Li, K.S. 1992. A unified solution scheme for slope stability analysis. Proceeding, 6th International symposium on landslides. 481–487. Christchurch.

Lowe, J. III. & Karaflath, L. 1960. Stability of earth dams upon drawdown. Proc. 1st Panamer. Conf. Soil Mech, 2:537–552. Mexico City.

Matsui, T. & San, K.C. 1992. Finite element slope stability analysis by shear strength reduction technique. Soils and Foundations 32(1):59–70.

Michalowski, R.L. 1989. Three-dimensional analysis of locally loaded slopes. Geotechnique 39:27–38.

Morgenstern, N.R. & Price, V. 1965. The analysis of the stability of general slip surface. Geotechnique 15(l):79–93.

Morgenstern, 1991. The evaluation of slope stability—a 25 year perspective. Proc. ASCE Conf. on Stability and Performance of Slopes and Embankments, 1:1–26. Berkeley.

Morgenstern, 1992. Keynote paper: The role of analysis in the evaluation of slope stability. Proceedings of 6th Internatioanl Symposium of Landslides: 1615–1629.

Naylor, D.J. 1981. Finite elements and slope stability. Numer. Meth. In:Geomech., Proceedings of the NATO Advanced Study Institute, Lisbon, Portugal:229–244.

Pham, H.T.V. & Fredlund D.G. 2003. The application of dynamic programming to slope stability analysis. Canadian Geotechnical Journal 40:830–847.

Sarma, S.K. 1979. Stability analysis of embankments and slopes. Journal of the Geotechnical Engineering Division, ASCE 105(GT12):1511–1524.

Sarma, S.K. & Tan, D. 2006. Determination of critical slip surface in slope analysis.Geotechnique 56(8):539–550.

Sokolovski, V.V. 1960. Statics of soil media. (Translated by Jones DH and Scholfield AN). London: Butterworth.

Spencer, E. 1967. A method of analysis of embankments assuming parallel inter-slice forces. Geotechnique 17:11–26.

Stark, T.D. & Eid, H.T. 1998. Performance of three-dimensional slope stability analysis method in practice. Journal of Geotechnical Engineering, ASCE 124: 1049–1060.

Tan, C.P. & Donald, I.B. 1980. Finite element calculation of dam stability. Proc. 11th Int. Conf. Soil Mech. and Fnd. Engr. San Francisco.

U.S. Army, Corps of Engineers. 1967. Stability of slopes and foundations, Engineering Manual, Visckburg, Miss.

Ugai, K. 1985. Three-dimensional stability analysis of vertical cohesive slopes. Soils and Foundations 25(3):41–48.

Ugai, K. & Leshchinsky, D. 1995. Three-dimensional limit equilibrium and finite element analysis: a comparison of results. Soils and Foundations 35(4):1–7.

Wang, Y.J., Yin, J.H. & Chen, Z.Y. 2001. Calculation of bearing capacity of a strip footing using an upper bound method. International Journal for Numerical and Analytical Methods in Geomechanics 25:841–851.

Whitman, R.V. & Bailey, W. 1967. Use of computers for slope stability analysis. Journal of Soil Mechanics and Foundation Engineering Division, ASCE 93(SM4).

Wright, S.G., Kulhawy, F.H. & Duncan, J.M. 1973. Accuracy of equilibrium slope stability analysis. Journal of Soil Mechanics and Foundation Division, ASCE 99(SM10):783–791.

Wright, S.G. 1978. Slope stability analysis. Proceedings on Analysis and Design in Geotechnical Engineering, Vol. 2. 153.

Yamagami, T. & Ueta, Y. 1988. Search for critical slip lines in finite element stress fields by dynamic programming. Proceedings of the 6th International Conference on Numerical Methods in Geomechanics. Innsbruck, Australia: 1335–1339.

Zhu, D.Y., Lee, C.F. & Jiang, H.D. 2003. Generalized framework of limit equilibrium methods for slope stability analysis. Geotechnique 53(4):377–395.

Zhu, D.Y., Dun, J.H. & Tai, J.J. 2007. Theoretical verification of rigorous nature of simplified Bishop's method. Chinese Journal of Rock Mechanics and Engineering 26(3):455–458 (in Chinese).

Zhu, D.Y. 2008. Investigation on the accuracy of the simplified Bishop method. Proceedings of the 10th International Symposium on landslide and engineered slopes. Xi'an.

Zienkiewicz, O.C. Humpheson, C. & Lewis R.W. 1975. Associated and nonassociated visco-plasticity and plasticity in soil mechanics. Geotechnique 25(4):671–89.

Zou, J.Z., Williams, D.J. & Xiong, W.L. 1995. Search for critical slip surfaces based on finite element method. Canadian Geotechnical Journal 32:233–246.

Zheng, H. & Li, C. 2002. Solution to the factor of safety by finite element method. Chinese Journal of Geotechnical Engineering 24(5):626–628 (in Chinese).

Improving the interpretation of slope monitoring and early warning data through better understanding of complex deep-seated landslide failure mechanisms

E. Eberhardt
Geological Engineering, University of British Columbia, Vancouver, Canada

A.D. Watson
BC Hydro, Burnaby, British Columbia, Canada

S. Loew
Engineering Geology, ETH Zurich, Switzerland

ABSTRACT: The past several years have seen significant advances in landslide monitoring technologies. Remote sensing techniques based on satellite and terrestrial radar can now provide high-resolution full area spatial coverage of a slope as opposed to relying on geodetic point measurements. Automation in the form of wireless data acquisition has enabled the collection of data with increased temporal resolution. These tools provide increased capacity to detect pre-failure indicators and changes in landslide behavior. Yet the interpretation of slope monitoring data, especially that for early warning, still remains largely subjective as geological complexity and uncertainty continue to pose major obstacles. This paper reviews several recent developments in landslide monitoring techniques but questions the phenomenological approach generally taken. Examples are then provided from several recent experimental studies involving "*in situ* laboratories" in which detailed instrumentation systems and numerical modeling have been used to better understand the mechanisms controlling pre-failure deformations over time and their evolution leading to catastrophic failure. Preliminary results from these studies demonstrate that by better integrating the different data sets collected, geological uncertainty can be minimized and better controlled with respect to the improved interpretation of slope monitoring and early warning data.

1 INTRODUCTION

Monitoring forms a key component of most landslide hazard assessments, providing data that may be used to quantify the nature of the hazard, its extent, kinematics and stability state, sensitivity to triggering mechanisms, response to mitigation works, etc., or to provide early warning of an impending failure especially those where lives or infrastructure may be at risk. In both cases, issues of uncertainty relating to the geological conditions, slope kinematics and failure mode provide major obstacles that contribute to a lack of definition of the problem.

Techniques used for forecasting impending failure (i.e. temporal prediction) are largely phenomenological, relying on surface-based point measurements of displacement monitored over time, which are then extrapolated or analyzed for accelerations that exceed set thresholds based on earlier patterns. Fukuzono's (1985) inverse velocity method is one such example.

Comprehensive reviews of these methods are provided by Bhandari (1988), Glastonbury & Fell (2002), Crosta & Agliardi (2003) and Rose & Hungr (2007). Inherently, these approaches are 'holistic', disregarding details pertaining to the underlying slope failure mechanism. Whether the displacement measurements are made using an extensometer positioned across a tension crack or a system of geodetic reflectors across a slope, the analysis is often carried out in the same way—surface displacements are recorded over time, which are then extrapolated or analyzed for accelerations in order to predict catastrophic/impending failure. Generally, the kinematics and causes of failure are not well defined, and instead, the surface manifestation of the instability (i.e. surface displacements) is relied upon for predictive analysis. Not surprisingly, only a few cases have been reported where these techniques have been successfully applied as part of a reliable forward prediction (e.g. Rose & Hungr 2007); most involve back analyses.

Numerical modeling offers a means to account for complex subsurface processes by breaking problems down into their constituent parts and analyzing the cause/effect relationships (and their evolution), which govern the behavior of the system as a function of changing environmental factors. However, these analyses require tight controls on the representation of geological heterogeneity and structure, soil and/or rock mass behavior, and special boundary conditions. This information may be derived in part from surface and borehole data, but more often it is limited to subsurface projections based on surface observations.

Slope monitoring data provides an important means to calibrate and constrain detailed numerical models. At the same time it must be recognized that most *in situ* measurements are affected by the same issues of rock mass complexity and variability as the numerical analyses they are meant to constrain. In many situations the interpretation of monitoring data is far from straight forward. In turn, it has been demonstrated that numerical modeling can be used to help constrain interpretations of complex field measurements (Eberhardt & Willenberg 2005, Watson et al. 2006). Thus it must be emphasized that a counterbalance and close association should exist between field measurements and analysis to develop a more complete understanding of the slope hazard problem (Sakurai 1991).

2 CONTINUING DEVELOPMENTS IN LANDSLIDE MONITORING

2.1 *Remote sensing and InSAR*

Although technological advancements continue with respect to traditional total station and prism monitoring systems (e.g. combined robotic total station and global navigation satellite systems; Brown et al. 2007), many of the key recent advances in slope monitoring are those related to remote sensing technologies, primarily satellite and ground-based radar but also non-radar variants like airborne and terrestrial laser scanning (e.g. Rosser et al. 2005).

At the regional scale, Differential Interferometric Synthetic Aperture Radar (DInSAR) is proving to be a useful means for identifying landslides within large coverage areas (e.g. 100 × 100 km using ERS data; Meisina et al. 2005) to help in the development of landslide inventories. InSAR uses satellite emitted electromagnetic signals to measure the phase difference resulting from the path length change between satellite passes of the same area taken from the same flight path. The difference in phase can be used to determine ground movement in the line of site of the SAR satellite, and with an emitted electromagnetic wavelength of a few centimeters, these ground movements can be measured to several millimeters accuracy (Froese et al. 2005). Temporal decorrelation due to vegetation coverage, however, dramatically affects interferometric coherence and limits the detection resolution, although multi-image based approaches like the Permanent Scatterers (Ferretti et al. 2001) and Small Baseline (Berardino et al. 2002) methods work to limit spatial decorrelation effects and topography errors.

Several studies have now been published where coherence for a study site could be maintained and DInSAR successfully applied. Most of these are from slides showing coherent movements over larger areas, often involving rock masses or colluvium with strong plastic deformations (e.g. Rott et al. 1999, Froese et al. 2005, Singhroy et al. 2005). Figure 1a shows deformations detected along a major fault and over old mine workings and colluvium towards the bottom

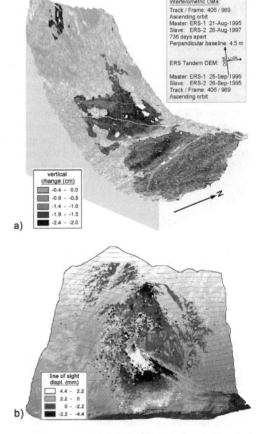

Figure 1. a) Vertical surface deformation map based on satellite DInSAR for Turtle Mountain for a 2-yr period (after Singhroy et al. 2005). b) Line of sight displacements derived from terrestrial radar for Randa for a 59-day period.

of Turtle Mountain in western Canada, location of the 1903 Frank Slide (Singhroy et al. 2005). These drew attention to a possible link between slope movements in the upper slope and de-stressing of the slope's toe due to the slow collapse of the old workings (Froese & Moreno 2007). In comparison, brittle rock masses showing smaller-scale complex block movements are more difficult to study with satellite-based DInSAR due to the strong spatial variability in block velocities. Instead, ground-based DInSAR (e.g. Tarchi et al. 2003) with a typical pixel size of a few meters may prove more suitable. Figure 1b shows preliminary results from Randa in southern Switzerland, location of the 1993 Randa rockslide, which are being used to provide important information about the area affected by slope movements and to support kinematic analyses.

2.2 Slope Stability Radar (SSR)

Other new developments in the use of radar have involved moving away from synthetic aperture radar and instead using real-aperture from a stationary platform, as in the case of Slope Stability Radar (SSR). With SSR, the system is typically set up 50 to 1000 m from the foot of the slope and the region of interest is continuously scanned, comparing the phase measurement in each image pixel with previous scans to determine the amount of movement. (Fig. 2; Harries & Roberts 2007). The combination of near real-time measurement, sub-millimeter precision and broad area coverage to quickly identify the size and extent of a developing failure is helping to establish SSR as a key tool for managing unstable rock slopes, especially in open pit mining (e.g. Harries et al. 2006, Little 2006, Day & Seery 2007). Further advantages of the system is that it is not adversely affected by rain, fog, dust or haze. Vegetation on the slope, however, may reduce the precision in pixels where there is low phase correlation between scans (Harries & Roberts 2007).

Again, the advantage of these remote sensing-type systems is that they provide fast and updatable data acquisition over broad areas as opposed to point measurements, coincident with prism placement, when using traditional geodetic monitoring systems.

2.3 Wireless data acquisition and data management

Conventional slope monitoring (e.g. inclinometers, total stations, tiltmeters, extensometers, crackmeters, etc.), continues to represent the favored means by which to measure and monitor slope deformation directions, magnitudes and rates, both on surface and at depth. Instrument reliability is of paramount importance, for which continuous improvement to the performance of vibrating-wire technology has led to it being widely recognized as the preferred choice for long-term *in situ* monitoring efforts. However, new technologies like fiber optics are also being tested to capitalize on the stability and insensitivity of fiber optic sensors to external perturbations. A review of new developments in fiber optic sensing technologies for geotechnical monitoring is provided in Inaudi & Glisic (2007).

Data reliability is equally a key issue. Recent studies involving cases where the displacements being measured are particularly small or where deep inclinometers are involved, have led to improved algorithms and procedures for identifying and correcting systematic errors (e.g. Mikkelsen 2003, Willenberg et al. 2003). Studies involving complex, deep-seated, rock slope instabilities have seen attempts to better integrate multiple geological and geotechnical data sets to improve data interpretation (e.g. Willenberg 2004, Watson et al. 2006, Bonzanigo et al. 2007, Hutchinson et al. 2007). These attempts at data "fusion" are moving towards the adoption of Virtual Reality (VR) technology, where the identification of hidden relationships, discovery and explanation of complex data interdependencies, and means to compare and resolve differing interpretations can be facilitated (Kaiser et al. 2002).

Concomitant with data integration is data management. Important new elements include Web GIS services integrated into the operational resources of decision makers. These services are linked to early warning systems through wireless data acquisition and transmission technologies, which enable real-time data from multiple remote monitoring sites to be accessed and viewed off-site by means of the internet. This is proving to be a highly valuable resource where an unstable slope threatens a community, critical facility or, in the case of large open pit mine slopes, worker safety. Furthermore, Hutchinson et al. (2007) propose that spatially and temporally distributed measurements should be combined with a knowledge engine and an evolving rule base to form the hub of a decision support system.

One of the more comprehensive systems in place is that installed by BC Hydro for its chain of hydroelectric dams on the Columbia River in British Columbia, Canada (Fig. 3). The system connects dataloggers at six large landslide sites along different dam reservoirs to a central monitoring computer using radio and microwave communication (Fig. 3c). The landslides sites are separated by up to 150 km and range in size from less than one million to over a billion cubic meters. These include: 731 Block, Checkerboard Creek, Downie Slide, Dutchmans Ridge, Little Chief Ridge and Little Chief Slide.

The extensive system in place is used for both investigative and predictive monitoring, and involves the continuous monitoring of a large number of piezometers, in-place inclinometers, extensometers, water weirs, load cells and tiltmeters. The basis of the data

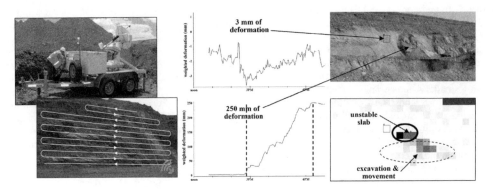

Figure 2. GroundProbe's Slope Stability Radar (SSR) system showing the continuous monitoring of millimeter-scale movements across the entire face of an unstable open pit mine slope (after Harries et al. 2006).

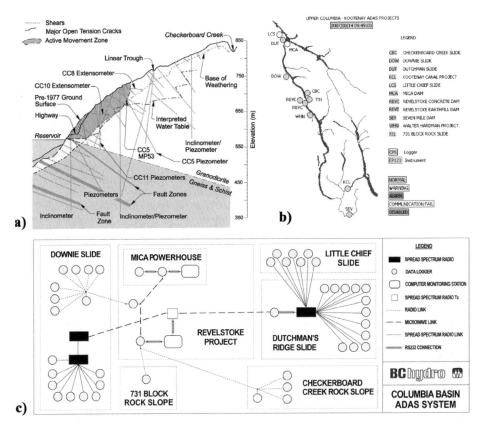

Figure 3. a) Example of the detailed instrumentation used by BC Hydro to monitor slope displacements, temperature and pore pressures at depth for a rock slope, Checkerboard Creek, above one of their dam reservoirs (after Watson et al. 2007). b) Monitoring and alarm interface used by BC Hydro to remotely monitor several reservoir slopes along their system of hydroelectric dams on the Upper Columbia River. c) Schematics of the wireless data acquisition/transmission system used for the remote monitoring system.

communication is a mix of UHF radio and spread-spectrum radio which allow the dataloggers to communicate with the central monitoring computer. The fundamentals of the communication system include a transmitter, receiver and surge arrestor which allows reliable communication over distances of a few kilometers, up to 100 kilometers with proper antennae. Advantages at these sites over satellite or cell phone communication include cost and ease of use. Disadvantages include the requirement to use repeaters if line of sight on long distances is not possible.

3 MONITORING OF LANDSLIDE BEHAVIOUR

3.1 Current state of practice

Landslide monitoring serves two important functions (Moore et al. 1991):

i. Investigative Monitoring: To provide an understanding of the slope and thus enable an appropriate action to be implemented.
ii. Predictive Monitoring: To provide a warning of a change in behavior and thus enable the possibility of limiting damage or intervening to prevent hazardous sliding.

Instrumentation typically includes: piezometers, in-place inclinometers, extensometers/crack monitoring, tiltmeters and surface geodetic monuments. Often for large landslides the most reliable instruments are those installed in deep subsurface exploratory boreholes (e.g. Fig. 3a).

Investigative monitoring can be used to obtain a greater understanding of the slope behavior, thus enabling the correct approach to be taken or to confirm that the approach taken was correct. Monitoring for investigative purposes can be as little as an annual visual inspection or as much as continuous measurement of a comprehensive instrumentation network (e.g. Fig. 3c). For the case of slow moving slides, many years of monitoring and annual cycles may be required to identify relationships between water levels, movement and other seasonal effects such as temperature. This was demonstrated by Bonzanigo et al. (2007) for the case of Campo Vallemaggia, an 800 million m^3 deep-seated landslide in strongly fractured and weathered crystalline rocks in southern Switzerland that threatened two villages founded on the landslide. Detailed inclinometer and piezometer measurements collected over a five year period were used to cross-correlate the stick-slip behavior of the landslide with pore pressures exceeding a threshold value tied to longer-term precipitation events (Fig. 4). These measurements were subsequently used in the decision making process to go forward with the construction of a deep drainage adit that successfully led to the stabilization of the landslide (Fig. 4, Eberhardt et al. 2007). A second example is that provided by Watson et al. (2006, 2007) for BC Hydro's Checkerboard Creek (Fig. 3a). Again, detailed monitoring of the slope revealed a similar persistent annual displacement cycle dominated by an active sliding phase in autumn to late winter and inactivity during the spring and summer. Although at first these periods of displacement activity appeared to correspond to periods of increased precipitation as in the case of Campo Vallemaggia, it wasn't until several years worth of measurements were collected that it was observed that the annual displacement cycle was repeated each year regardless of the amount of precipitation. Instead, the annual displacement cycle was more strongly correlated to seasonal temperature variations in the near surface bedrock and the deformation mechanism explained in terms of thermally induced slip along sub-vertical joints (Watson et al. 2006). In both cases, the assessments and subsequent approaches taken were completed based on the results of investigative monitoring used to develop reliable geological and hydrogeological models later aided by numerical modeling to more fully understand the deformation mechanisms involved.

Predictive monitoring systems usually evolve after investigative monitoring and assessment. Without this investigation phase it is unlikely that a predictive system could be designed, or early warning thresholds set, with confidence. Monitoring information must be assessed in the context of the physical setting and the conclusions of the investigation phase. Landslides can change their behavior within a few weeks or a few days. Thus, for a predictive monitoring system to be effective the frequency at which the instruments are monitored must be a fraction of this response time and the system must be in place to react to the instrumentation results. Furthermore, in order for decision makers to be in a position to react correctly to predictive monitoring data, the chance of faulty alarms or misleading instrument readings should be minimized by the use of the most reliable instrumentation possible.

3.2 In situ rockslide laboratories

Efforts to improve predictive monitoring have seen several recent multi-disciplinary studies focused on improving our understanding of complex landslide deformation mechanisms. These include the Randa *In Situ* Rockslide Laboratory in Switzerland, the Turtle Mountain Field Laboratory in western Canada and the Åknes/Tafjord Project in Norway.

The first of these, the Randa *In Situ* Rockslide Laboratory (Fig. 5), was a comprehensive experimental investigation into the spatial and temporal evolution of large rock slope failures in fractured crystalline rock.

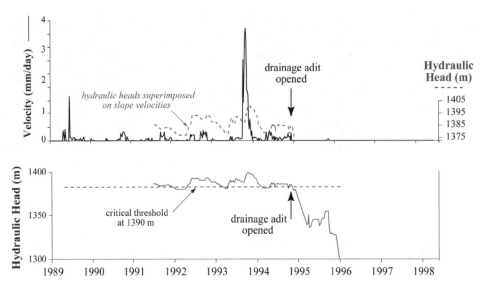

Figure 4. Correlation between the downslope velocities of the Campo Vallemaggia landslide and borehole pore pressures measured before and after the opening of a drainage adit to stabilize the slope. Slide velocities were measured using an automated geodetic station; pore pressures are expressed as the hydraulic head in the piezometer (after Eberhardt et al. 2007).

One of the prime motivating factors was to develop a better understanding of rock mass strength degradation (e.g. through the destruction of intact rock bridges between non-persistent discontinuities) and the progressive development of internal shear zones, and their accommodation of larger slope displacements, leading to increased extensional strains and eventually sudden collapse (e.g. Eberhardt et al. 2004b). Focus was also placed on improving early warning capabilities in the presence of persistent and non-persistent discontinuities, and multiple moving blocks and internal shear surfaces.

For this, a high-alpine facility was constructed above the scarp of the 1991 Randa rockslide (Fig. 5a), where ongoing movements of 1–2 cm/year are being recorded in gneissic rock for a volume of up to 10 million m^3. This facility included the installation of a variety of instrumentation systems designed to measure temporal and 3-D spatial relationships between fracture systems, displacements, pore pressures and microseismicity (Fig. 5). The monitoring was complimented by a detailed geophysical field campaign, which included 3-D surface seismic refraction and georadar surveys to resolve subsurface 3-D fracture distributions (Heincke 2005), and crosshole georadar and seismic tomography to identify key geological features (Spillmann 2007). These were then compared to those mapped on surface and in the boreholes to develop a 3-D geological model of the unstable rock mass (Willenberg 2004). This information helped to define the rock mass structure (e.g. Fig. 6), aided in the positioning of borehole instruments (e.g. in-place inclinometers) and was essential for reliably interpreting the monitoring data.

A similar multi-disciplinary field campaign was carried out for the Åknes/Tafjord Project, where up to 30–40 million m^3 of unstable rock moving with a mean rate of 2–4 cm/year has been identified as a potential threat and tsunami generating hazard for people and infrastructure living along the inner Storfjord (Blikra et al. 2005). Geological, geodetic and geophysical studies (including GPS, resistivity, georadar, reflection and refraction seismics, airborne laser scanning, and high-resolution air photography) were carried out to define the geometry and volume of unstable areas. Through these detailed studies it was found that a previously unmonitored section of the slide was moving at 15 cm/year, leading to revisions in the unstable volume and thus the magnitude of the potential hazard (Roth et al. 2006).

The fracture network, both existing and newly generated, was a central focus in the designs of the instrumentation networks at Randa, Åknes and Turtle Mountain. In different ways, displacement monitoring targeted resolving the complex displacement field generated by multiple moving blocks. At Turtle Mountain, a combination of crackmeters, wire-line extensometers and tiltmeters, were used to monitor surface tension cracks, enabling seasonal displacement patterns to be resolved (Moreno & Froese 2007). Similar

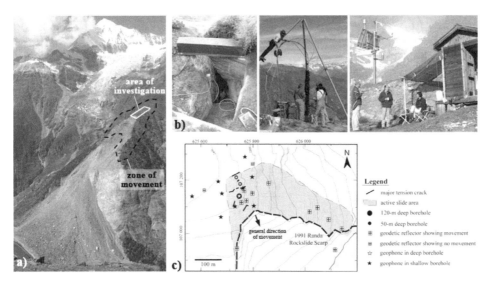

Figure 5. The Randa *In Situ* Rockslide Laboratory in southern Switzerland. a) Area of investigation (solid white line) and outline of present-day instability (black dashed line) above the scarp of the 1991 rockslide. Photo by H. Willenberg. b) Installation of surface and subsurface monitoring instruments and central data acquisition station housing batteries, power generation sources (solar and wind) and data acquisition and transmission hardware. c) Plan view map showing location of boreholes, geodetic reflectors and geophones relative to the active slide area and open tension cracks (after Willenberg 2004).

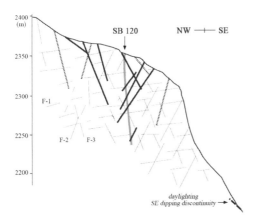

Figure 6. Cross-section through the Randa study area showing the network of discontinuities (F-1 to F-3) and faults (highlighted in black) mapped using geological and geophysical methods (after Loew et al. 2007).

monitoring was carried out at the Åknes and Randa project sites. This traditional focus on surface displacement monitoring addresses certain logistic and economic realities in terms of what may be feasible for on-site monitoring, yet it must also be asserted that only so much can be inferred on surface for a problem that develops at depth. For this, deep inclinometer measurements were added to the monitoring networks at Åknes and Randa. At Randa, the deepest inclinometer was also fitted with an Increx electromagnetic induction sliding extensometer system to enable the profile of 3-D displacement vectors at depth to be determined. Biaxial vibrating wire in-place inclinometers were installed and positioned at depth intervals coinciding with key fractures identified through the borehole televiewer surveys to provide continuous monitoring of subsurface deformations along these structures. Integration of these different displacement data sets with the 3-D geological model showed that the displacements recorded on surface and at depth were localized across active discontinuities (Fig. 7) and that the kinematic behavior of the slope was dominated by complex internal block movements rather than those of a coherently-sliding mass (Willenberg 2004). Willenberg et al. (2003) demonstrated that to resolve these complex displacement patterns, rigorous correction algorithms must be carried out to attain the requisite resolution.

The fracture network also plays a controlling role with respect to the distribution of pore pressures at depth and their coupled relationship with unstable rock slope movements. The design of the experimental monitoring networks at Randa, Åknes and Turtle Mountain each included borehole monitoring of pore pressures to correlate with measured displacements. At Randa, piezometers were positioned and packed off along zones indicating potentially higher fracture

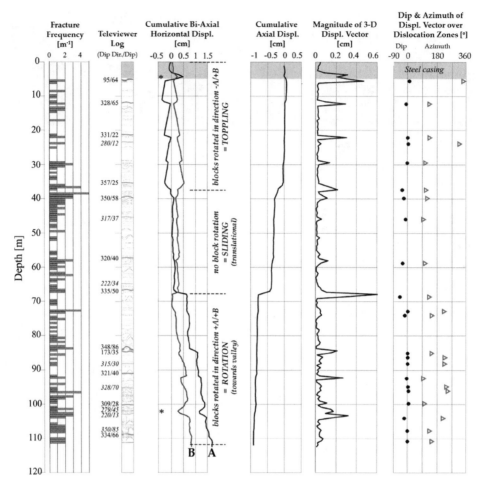

Figure 7. Integrated borehole data set for the 120-m deep borehole at Randa, showing from left to right: fracture frequency log, optical televiewer log (highlighting traces of major fractures), cumulative inclination changes for a two-year period (and corresponding preliminary kinematic interpretation), cumulative axial displacements for the same two-year period, and corresponding 3-D displacement vector magnitudes and orientations (after Willenberg 2004).

permeability as determined from borehole televiewer data. These data showed several water tables distributed within the rock mass and different types of pore pressure interactions with infiltrating surface water and atmospheric pressure variations (Willenberg 2004). This is a common feature for most deep-seated slides in crystalline rock, where preferential fracture permeability and hydraulic barriers (e.g. from fault gouge) result in isolated compartments of groundwater flow and reaction delays between surface precipitation and pressure responses at depth, making correlations between slope movements and precipitation events extremely difficult (e.g. Moore & Imrie 1992, Bonzanigo et al. 2007). A strong sensitivity of slope movements to changing groundwater recharge conditions at ground surface is observed in landslides showing artesian pressures at depth.

The final key monitoring strategy common to the Randa, Åknes and Turtle Mountain projects was the use of microseismic monitoring to detect and study subsurface brittle fracture processes. Spatially clustered microseismic events in numerous fields (e.g. mining, geothermal energy, nuclear waste disposal, etc.) have proven effective in providing critical information with respect to stress-induced tensile fracturing mechanisms and/or shear slip along internal fracture planes. The microseismic network at Randa was the most detailed of the three and included three geophones (28 Hz) mounted in deep boreholes, nine geophones (8 Hz) mounted in shallow boreholes, and

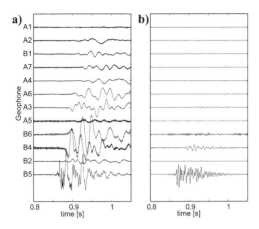

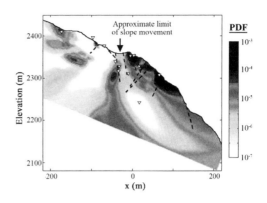

Figure 8. Vertical components of a locatable microseismic event: a) raw; b) 100–500 Hz bandpass filtered signals. Signals are sorted according to the source-receiver distance, with sen-sor A1 being the farthest and B5 the closest. Absolute time scale is arbitrary. After Eberhardt et al. (2004a).

Figure 9. Cross-section showing the relationship between microseismic activity (high PDF values indicate microseismogenic zones), faults (dashed lines) and the approximate limit of slope movements. The cumulative PDF represents the sum of the hypocenter probability density functions for all microseismic events. After Spillmann et al. (2007).

two 24-channel seismographs (Fig. 5c). The spatial distribution of the twelve triaxial geophones was chosen to ensure that the hypocenter parameters generated from the seismic sources could be reliably constrained within the area of interest (Spillmann et al. 2007). One of the initial findings from this system was that the higher frequency content of the recorded microseismic events was strongly attenuated (Fig. 8), pointing to the presence of large open fractures at depth. This was fully compatible with the geological model and borehole televiewer and pore pressure data (Willenberg et al. 2004). Larger low frequency events, such as those generated from natural seismic activity in the region, did not suffer as much from signal quality degradation. Interpretation of the recorded local microseismic activity involved comparing the cumulative hypocenter probability density functions, which incorporates uncertainties in the arrival times and travel times for each event (see Spillmann et al. 2007), with the results from the geological, geophysical and geotechnical investigations. From this, it was observed that the microseismic activity was concentrated in two main zones (Fig. 9): that near the scarp of the 1991 rock slide events and that coinciding with the highest density of faults (Spillmann 2007). These two zones are bound by the geodetically determined limits of the moving mass (e.g. Fig. 5c). A full interpretation of the microseismic activity recorded during the Randa experiment is reported in Spillmann et al. (2007).

The system at Turtle Mountain also combines surface- and borehole-mounted geophones, and builds on earlier experiences with seismic monitoring at Turtle Mountain carried out between 1983 and 1992. These identified several different sources including

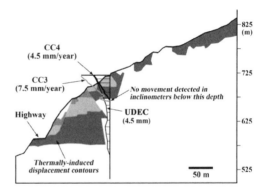

Figure 10. UDEC modeled thermal-induced horizontal displacements compared to inclinometer measured slope displacements (after Watson et al. 2006).

local earth tremors and rock falls, together with seismicity believed to be related to deformation and stress relief within Turtle Mountain and the ongoing collapse of mine workings at the base of the mountain (Read et al. 2005). Similar attempts will be made at Åknes to classify the microseismic events recorded by their system, based on rock falls, small-scale slides, and those directly related to the main slide body (Roth et al. 2006). As with Randa, signal quality has been noted as being a key limiting factor with respect to constraining source locations and mechanisms.

It is expected that the application of emerging monitoring technologies like those used in the Randa, Åknes and Turtle Mountain project studies, will provide an improved means to gain a better understanding of rock slope deformation kinematics and failure.

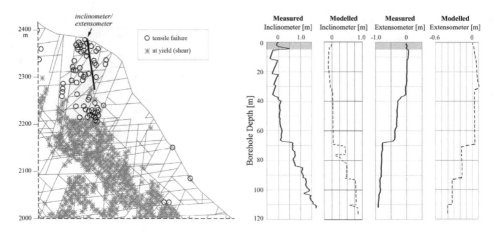

Figure 11. Distinct-element modeling of complex rock slope displacements at Randa and comparison between measured and modeled cumulative displacement profiles, assuming slip along discontinuities and elasto-plastic block deformation. Note that model boundaries extend beyond those shown (as indicated by the dashed boundary line).

At Turtle Mountain and Åknes, these are being integrated into early warning monitoring and response plans that comprise detailed monitoring procedures, threshold and alert level development, notification protocols and emergency response (e.g. Froese et al. 2005). Froese et al. (2005) conclude that the use of multiple systems that provide different spatial and temporal coverage, together with sufficient redundancy, provide a higher level of confidence in interpreting the kinematics of movement and impending failure than would be available from single sensor readings.

3.3 Use of numerical models to interpret slope monitoring responses

Numerical analysis can be a useful tool to provide confirmation of the geological model and/or conclusions drawn from investigative monitoring, as well as to explore possible future behavior. An example of this is the numerical analysis completed for the Checkerboard Creek rock slope to explore and confirm the indications from investigative monitoring that a thermal control existed for the deep-seated rock slope movements being measured (Watson et al. 2006). Modeling was initially carried out using the Itasca finite-difference code FLAC, but results suggested that a continuum approach could not capture the pattern of displacements and pore pressures measured in situ. Itasca's distinct element code UDEC was next used because of its capability to explicitly include joints and shear zones, with groundwater flow restricted to those joints, together with its ability to model thermal and dynamic loads. Models investigated the stability of the slope during a 1 in 10,000 year earthquake, as well as correlations between modeled displacement patterns induced during thermal cycling and those measured in situ by the instrumentation (Fig. 10). Results for the latter revealed that the cyclic nature of the displacements seen in the monitoring data was controlled by the thermal response of the rock mass to seasonal temperature changes in the upper 10 m of ground. This is similar to preliminary findings at Turtle Mountain, which suggest that thermal cycling (i.e. thermal induced stress changes) contribute more towards measured slope deformations than do heavy precipitation events (Moreno & Froese 2007).

Thus, not only does monitoring data provide an important means to constrain numerical models, but numerical modeling provides a means to better understand monitoring data (Eberhardt & Willenberg 2005). At both Randa and Åknes, simplified numerical modeling has been carried out together with the mapping and monitoring programs to help identify and constrain possible sliding surface/instability scenarios that would produce displacement patterns similar to those measured in situ. Figure 11 provides an example from a series of distinct element models generated for the Randa study, incorporating the key active geological structures identified through mapping and geotechnical monitoring. The blocks are modeled using a Mohr-Coulomb elasto-plastic constitutive model where the properties are scaled to those for an equivalent continuum to account for smaller scale discontinuities not explicitly included in the distinct-element model. Results show a correspondence between the measured and modeled block movements with toppling and translational movements in the upper part of the slide and outward rotation of the blocks at depth. The latter suggests that deep-seated yield together with shearing along persistent discontinuities may be an important contributing factor in the complex block deformation patterns measured. It should be noted that uncertainty does exist for the inclinometer readings over the lower portions of the borehole, and the modeling results incorporate the limitations inherent in a 2-D representation of a strongly 3-D problem. Still, any insights gained into the instability mechanism, either supporting or refuting a current interpretation, provides a means to better plan and design future *in situ* investigation, instrumentation and monitoring schemes for the site.

4 CONCLUSIONS

Issues related to geological complexity and uncertainty represent a significant obstacle to better predicting the spatial and temporal evolution of catastrophic rock slope failures. Advances in satellite and terrestrial radar technologies (e.g. Slope Stability Radar) and automation through wireless data acquisition are helping to address these issues. However, the findings summarized in this paper emphasize the additional need to not only improve landslide monitoring technologies but to also better integrate the various data sets collected during field-based investigations, monitoring and stability analyses in order to overcome these challenges. Through the integration of these data, a more reliable model of the controlling landslide mechanism can be incorporated into the engineering decision-making process.

Several examples are provided from recent studies that aim to improve our mechanistic understanding of deep-seated rock slope behavior. These include experimental efforts involving the construction of high alpine research facilities or "*in situ* rockslide laboratories", where the integration of state-of-the-art site characterization and instrumentation systems and numerical modeling are being used to better understand the mechanisms controlling pre-failure deformations over time. At depth, passive monitoring of microseismic activity offers a means to detect subsurface tensile fracturing and/or shear slip along internal fracture planes that may provide insights into the evolution of a progressively developing rock slope failure. Data from both surface geodetic and subsurface instrumentation systems can be integrated to obtain a description of the 3-D displacement field. When further integrated with subsurface geological information, derived through geophysical and/or borehole investigations, complex block movements can be resolved relative to major persistent fractures and shears.

The results presented also emphasize the importance of numerical modeling to provide support for and/or refute interpretations drawn from investigative monitoring as well as to explore possible future behavior. Focus is placed not only on using field mapping and instrumentation data to constrain numerical models but also in the use of numerical modeling to provide a means to better interpret and understand complex monitoring and early warning data. Thus, by better integrating the different data sets collected through all phases of an investigation, from mapping to monitoring to analysis, geological uncertainty can be minimized and controlled with respect to the comprehension of complex rock slope failure mechanisms, thereby improving our ability to effectively assess, monitor, mitigate and predict the potential for catastrophic rock slope failure and provide early warning to those endangered.

ACKNOWLEDGEMENTS

The authors wish to acknowledge and thank the different researchers, whose results were drawn on for the material presented in this paper, including those at BC Hydro, GroundProbe and those connected to the Randa, Åknes and Turtle Mountain rockslide research studies. Special thanks are extended to the Randa *In Situ* Rockslide Laboratory team, including Dr. Heike Willenberg, Dr. Keith Evans, Dr. Hansruedi Mauer, Dr. Tom Spillmann, Dr. Björn Heincke, Prof. Alan Green and Prof. Doug Stead.

REFERENCES

Berardino, P., Fornaro, G., Lanari, R. & Sansosti, E. 2002. A new algorithm for surface deformation monitoring based on Small Baseline Differential SAR Interferograms. *Transactions of Geoscience and Remote Sensing* 40 (11): 2375–2383.

Bhandari, R.K. 1988. Some practical lessons in the investigation and field monitoring of landslides. In Bonnard (ed.), *Proc., 5th Int. Symp. on Landslides, Lausanne*. Rotterdam: Balkema, v3, 1435–1457.

Blikra, L.H., Longva, O., Harbitz, C. & Løvholt, F. 2005. Quantification of rock-avalanche and tsunami hazard in Storfjorden, western Norway. In Senneset et al. (eds.), *Landslides and Avalanches, Proceedings of the 11th International Conference and Field Trip on Landslides (ICFL), Norway*. London: Taylor & Francis Group, pp. 57–64.

Bonzanigo, L., Eberhardt, E. & Loew, S. 2007. Long-term investigation of a deep-seated creeping landslide in crystalline rock—Part 1: Geological and hydromechanical factors controlling the Campo Vallemaggia landslide. *Canadian Geotechnical Journal* 44 (10): 1157–1180.

Brown, N., Kaloustian, S. & Roeckle, M. 2007. Monitoring of open pit mines using combined GNSS satellite receivesr and robotic total stations. In Potvin (ed.), *Proc. 2007 Int. Symp. on Rock Slope Stability in Open Pit Mining and Civil Engineering, Perth*. Perth: ACG, pp. 417–429.

Crosta, G.B. & Agliardi, F. 2003. Failure forecast for large rock slides by surface displacement measurements. *Can. Geotech. J.* 40 (1): 176–191.

Day, A.P. & Seery, J.M. 2007. Monitoring of a large wall failure at Tom Price Iron Ore Mine. In Potvin (ed.), *Proc. 2007 Int. Symp. on Rock Slope Stability in Open Pit Mining and Civil Engineering, Perth*. Perth: ACG, pp. 333–340.

Eberhardt, E., Bonzanigo, L. & Loew, S. 2007. Long-term investigation of a deep-seated creeping landslide in crystalline rock—Part 2: Mitigation measures and numerical modelling of deep drainage at Campo Vallemaggia. *Canadian Geotechnical Journal* 44 (10): 1181–1199.

Eberhardt, E., Spillmann, T., Maurer, H., Willenberg, H., Loew, S., and Stead, D. 2004a. The Randa Rockslide Laboratory: Establishing brittle and ductile instability mechanisms using numerical modelling and microseismicity. In Lacerda et al. (eds.), *Proc. 9th Int. Symp. on Landslides, Rio de Janeiro*. Leiden: A.A. Balkema, pp. 481–487.

Eberhardt, E., Stead, D. & Coggan, J.S. 2004b. Numerical analysis of initiation and progressive failure in natural rock slopes—the 1991 Randa rockslide. *Int. J. Rock Mech. Min. Sci.* 41 (1): 69–87.

Eberhardt, E. & Willenberg, H. 2005. Using rock slope deformation measurements to constrain numerical analyses, and numerical analyses to constrain rock slope deformation measurements. In Barla & Barla (eds.), *Proc. 11th Int. Conf. on Computer Methods and Advances in Geomechanics, Torino*. Bologna: Pàtron Editore, v4, pp. 683–692.

Ferretti, A., Prati, C. & Rocca, F. 2001. Permanent Scatterers in SAR interferometry. *Transactions of Geoscience and Remote Sensing* 39 (1): 8–20.

Froese, C.R., Keegn, T.R., Cavers, D.S. & van der Kooij, M. 2005. Detection and monitoring of complex landslides along the Ashcroft Rail corridor using spaceborne InSAR. In Hungr et al. (eds.), *Landslide Risk Management, Proc. of the Int. Conf., Vancouver*. Leiden: Balkema, pp. 565–570.

Froese, C.R. & Moreno, F. 2007. Turtle Mountain Field Laboratory (TMFL): Part 1—Overview and activities. In Schaefer et al. (eds.), *Proc., 1st North American Landslide Conference, Vail*. CD-ROM.

Fukuzono, T. 1985. A new method for predicting the failure time of a slope. In *Proc. IVth Int. Conf. and Field Workshop on Landslides, Tokyo*. Tokyo: NRCDP, pp. 145–150.

Glastonbury, J. & Fell, R. 2002. Report on the analysis of slow, very slow and extremely slow natural landslides. The University of New South Wales, UNICIV Report No. R-402.

Harries, N., Noon, D. & Rowley, K. 2006. Case studies of slope stability radar used in open cut mines. In *Stability of Rock Slopes in Open Pit Mining and Civil Engineering Situations*. Johannesburg: SAIMM, Symposium Series S44, pp. 335–342.

Harries, N.J. & Roberts, H. 2007. The use of Slope Stability Radar (SSR) in managing slope instability hazards. In Eberhardt et al. (eds.), *Proc. 1st Canada-U.S. Rock Mechanics Symposium, Vancouver*. London: Taylor & Francis, v1, pp. 53–59.

Heincke, B. 2005. Determination of 3-D fracture distribution on an unstable mountain slope using georadar and tomographic seismic refraction techniques. D.Sc. thesis, Applied and Environmental Geophysics, Swiss Federal Institute of Technology (ETH Zurich). 157 pp.

Hutchinson, D.J., Diederichs, M.S., Carranza-Torres, C., Harrap, R., Rozic, S. & Graniero, P. 2007. Four dimensional considerations in forensic and predictive simulation of hazardous slope movement. In Eberhardt et al. (eds.), *Proc. 1st Canada-U.S. Rock Mechanics Symposium, Vancouver*. London: Taylor & Francis, v1, pp. 11–19.

Inaudi, D. & Glisic, B. 2007. Overview of fiber optic sensing technologies for geotechnical instrumentation and monitoring. *Geotechnical News* 25 (3): 27–31.

Kaiser, P.K., Henning, J.G., Cotesta, L. & Dasys, A. 2002. Innovations in mine planning and design utilizing collaborative virtual reality (CIRV). *104th CIM Annual General Meeting, Vancouver*.

Little, M.J. 2006. Slope monitoring strategy at PPRust open pit operation. In *Stability of Rock Slopes in Open Pit Mining and Civil Engineering Situations*. Johannesburg: SAIMM, Symposium Series S44, pp. 211–230.

Loew, S., Willenberg, H., Spillmann, T., Heincke, B., Maurer, H., Eberhardt, E. & Evans, K. 2007 Structure and kinematics of a large complex rockslide as determined from integrated geological and geophysical investigations (Randa, Switzerland). In Schaefer et al. (eds.), *Proc., 1st North American Landslide Conference, Vail*. CD-ROM.

Meisina, C., Zucca, F., Fossati, D., Ceriani, M. & Allievi, J. 2005. PS InSAR integrated with geotechnical GIS: Some examples from southern Lombardia. In Sansò & Gil (eds.), *Geodetic Deformation Monitoring: From Geophysical to Engineering Roles*. Berlin: Springer, IAG Symposia, v131, pp. 65–72.

Mikkelsen, P.E. 2003. Advances in inclinometer data analysis. In Myrvoll (ed.), *Proc. 6th Int. Symp. on Field Measurements in Geomechanics, Oslo*. Lisse: A.A. Balkema, pp. 555–567.

Moore, D.P. Imrie, A.S. & Baker D.G. 1991. Rockslide risk reduction using monitoring. In *Proc., Canadian Dam Association Meeting, Whistler, BC*. Canadian Dam Safety Association.

Moore, D.P. & Imrie, A.S. 1992. Stabilization of Dutchman's Ridge. In Bel (ed.), *Proc. 6th Int. Symp. on Landslides, Christchurch*. Rotterdam: A.A. Balkema, pp. 1783–1788.

Moreno, F. & Froese, C.R. 2007. Turtle Mountain Field Laboratory (TMFL): Part 2—Review of trends 2005 to 2006. In Schaefer et al. (eds.), *Proc., 1st North American Landslide Conference, Vail*. CD-ROM.

Read, R.S., Langenberg, W., Cruden, D., Field, M., Stewart, R., Bland, H., Chen, Z., Froese, C.R., Cavers, D.S., Bidwell, A.K., Murray, C., Anderson, W.S., Jones, A., Chen, J., McIntyre, D., Kenway, D., Bingham, D.K., Weir-Jones, I., Seraphim, J., Freeman, J., Spratt, D., Lamb, M., Herd, E., Martin, D., McLellan, P. & Pana, D. 2005. Frank Slide a century later: The Turtle Mountain monitoring project. In Hungr et al. (eds.), *Landslide Risk Management: Proceedings of the International Conference on Landslide Risk Management, Vancouver*. Leiden: A.A. Balkema, pp. 713–723.

Rose, N.D. & Hungr, O. 2007. Forecasting potential rock slope failure in open pit mines using the inverse-velocity method. *Int. J. Rock Mech. Min. Sci.* 44 (2): 308–320.

Rosser, N.J., Petley, D.N., Lim, M., Dunning, S.A. & Allison, R.J. 2005. Terrestrial laser scanning for monitoring the process of hard rock coastal cliff erosion. *Quarterly*

Journal of Engineering Geology & Hydrogeology 38: 363–375.

Roth, M., Dietrich, M., Blikra, L.H. & Lecomte, I. 2006. Seismic monitoring of the unstable rock slope site at Åknes, Norway. In Gamey (ed.), *Proc. 19th Annual Symp. on the Application of Geophysics to Engineering and Environmental Problems (SAGEEP), Seattle*. CD-ROM.

Rott, H., Scheuchel, B. & Siegel, A. 1999. Monitoring very slow slope movements by means of SAR interferometry: A case study from mass waste above a reservoir in the Otztal Alps, Austria. *Geophys. Res. Lett.* 26 (11): 1629–1632.

Sakurai, S. 1991. Field measurements versus analysis in geotechnical engineering problems. In Sorum (ed.), *Field Measurements in Geomechanics, Proc. of the 3rd Int. Symp., Oslo*. Rotterdam: Balkema, v3, pp. 405–414.

Singhroy, V., Couture, R. & Molch, K. 2005. InSAR monitoring of the Frank Slide. In Hungr et al. (eds.), *Landslide Risk Management, Proc. of the Int. Conf., Vancouver*. Leiden: Balkema, pp. 611–614.

Spillmann, T. 2007. Borehole radar experiments and microseismic monitoring on the unstable Randa rockslide (Switzerland). D.Sc. thesis, Applied and Environmental Geophysics, Swiss Federal Institute of Technology (ETH Zurich). 205 pp.

Spillmann, T., Maurer, H., Green, A.G., Heincke, B., Willenberg, H. & Husen, S. 2007. Microseismic investigation of an unstable mountain slope in the Swiss Alps. *J. Geophys. Res.* 112(B07301): doi:10.1029/2006JB004723.

Tarchi, D., Casagli, N., Moretti, S., Leva, D. & Sieber, A.J. 2003. Monitoring landslide displacements by using ground-based synthetic aperture radar interferometry: Application to the Ruinon landslide in the Italian Alps. *J. Geophys. Res.* 108(B8): 2387, doi:10.1029/2002JB002204.

Watson, A.D., Martin, C.D., Moore, D.P., Stewart, T.W.G. & Lorig, L.L. 2006. Integration of geology, monitoring and modeling to assess rockslide risk. *Felsbau* 24 (3): 50–58.

Watson, A.D., Moore, D.P., Stewart, T.W. & Psutka, J.F. 2007. Investigations and monitoring of rock slopes at Checkerboard Creek and Little Chief Slide. In Eberhardt et al. (eds.), *Proc. 1st Canada-U.S. Rock Mechanics Symposium, Vancouver*. London: Taylor & Francis, v2, pp. 1015–1022.

Willenberg, H. 2004. Geologic and kinematic model of a complex landslide in crystalline rock (Randa, Switzerland). D.Sc. thesis, Engineering Geology, Swiss Federal Institute of Technology (ETH Zurich). 187 pp.

Willenberg, H., Evans, K.F., Eberhardt, E. & Loew, S. 2003. Monitoring of complex rock slope instabilities—correction and analysis of inclinometer/extensometer surveys and integration with surface displacement data. In Myrvoll (ed.), *Proc. 6th Int. Symp. on Field Measurements in Geomechanics, Oslo*. Lisse: A.A. Balkema, pp. 393–400.

Willenberg, H., Evans, K.F., Eberhardt, E., Loew, S., Spillmann, T. & Maurer, H.R. 2004. Geological, geophysical and geotechnical investigations into the internal structure and kinematics of an unstable, complex sliding mass in crystalline rock. In Lacerda et al. (eds.), *Proc. 9th Int. Symp. on Landslides, Rio de Janeiro*. Leiden: A.A. Balkema, pp. 489–494.

Effects of earthquakes on slopes

Ikuo Towhata
University of Tokyo, Tokyo, Japan

Tetsuo Shimomura
Ohbayashi Corporation, Tokyo, Japan

Masanori Mizuhashi
Public Works Research Institute, Tsukuba, Japan

ABSTRACT: This paper First addresses examples of large and small slope failures that were triggered by earthquakes. Their significance comes from the negative effects to the human community, which consist not only of the number of casualties but also of the difficulties in post-earthquake rescue and restoration. To mitigate the problems, identification of seismic instability and assessment of debris fun-out distance are important. Because the existing practical methods for these have several problems, the direction of their improvement is presented with examples and case studies.

1 INTRODUCTION

Slope failure is one of the most significant problems during strong earthquakes. It is important in that it might claim thousands of lives and destroy both public and private properties. When local transportation is blocked by slope failures, moreover, rescue and restoration become significantly difficult. It is true, on the other hand, that slope failure is a part of geomorphological processes and may be beyond the human control. Therefore, all what human beings can do is not to worsen the risk of slope failure, but to mitigate negative effects to the human community. The present paper addresses what happened in slopes during past and recent earthquakes, their effects to the public, and various kinds of efforts to mitigate the negative effects. Note that attention is focused mostly on natural slopes, but some events in manmade slopes will be introduced in relation to recent urban developments.

2 EXAMPLES OF ON-SHORE LARGE SLOPE FAILURES DURING EARTHQUAKES

This chapter addresses different kinds of seismically-induced slope failures that occurred both on shore and in the sea, in either big or small scale, and their consequences to the human communities. The discussion is initiated with the largest on-shore landslide in the world. Seimareh landslide is located in south-west Iran and measures 16 km in width (Fig. 1), 5 km in length of the failed slope, and 300 m in depth. The total volume of the failed mass is evaluated to be 24 billion m^3. This event occurred in a prehistoric time, probably 10,000 years ago, according to C_{14} dating (Watson & Wright 1969). Oberlander (1965) attributed this event to the foot erosion by river, but it is possible that the unstable slope was finally destroyed by seismic shaking. The important feature of this landslide lies in the enormous run-out length of debris, which traveled as long as 20 km, overtopping an anticline hill (Fig. 2).

Gigantic slope failures in more recent times often claimed many victims. Figure 3 demonstrates the ruin of Yungay City of Peru that was destroyed by seismic failure of Huascaran Mountain slope upon an earthquake of magnitude = 7.9 in 1970. The number of victims exceeded 17, 000. When a similar slope failure occurred in 1960, the debris flowed along a nearby Rio Santa channel, and the city was protected from it by a hill behind the city (Fig. 3). This experience gave people a wrong idea that debris would never hit the city in future. After the 1970 tragedy, the entire city was relocated to a safer place.

One of the consequences of a gigantic landslide is creation of a natural dam. The slope of Tsao-ling in Taiwan failed several times in a large scale in the past: in 1862 and 1941 due to earthquakes and in 1942 due to rainfall (Kawata 1943). Figure 4 shows a lake after the 1999 ChiChi earthquake.

Although a lake thus created may become a good tourist spot, problem is the possibility of the collapse of the dam and flooding in the downstream region.

Figure 1. View of Seimareh landslide from Pol-e-Doghtar.

Figure 2. Debris deposit in Seimareh.

Figure 3. Ruin of Yungay City in Peru.

Figure 4. Natural dam in Tsao-ling after 1999 ChiChi earthquake.

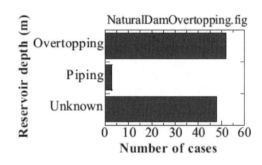

Figure 5. Statistics on cause of failure of natural dams (after Tabata et al., 2002).

Tabata et al. (2002) reported that the dam collapse due to overtopping is substantially more frequent than piping failure (Fig. 5). Another negative consequence is the long-term slope instability after the seismic failure. Fig. 6 shows a part of Ohya slide in the upstream area of Abe River, Japan. This slope collapsed upon the 1707 Hoei earthquake (M = 8.4). Since then, the destabilized slope has been producing debris flow at heavy rainfalls frequently (Imaizumi et al. 2005). Consequently, the river floor has raised significantly (Fig. 7) and the chance of flooding and overtopping river dikes has increased. To mitigate this situation, continuous efforts have been made to improve dikes and stabilize slopes.

3 PROBLEMS OF ON-SHORE MINOR SLOPE FAILURES

When the surface of a slope is subjected to weathering and disintegration, strong earthquake shaking triggers

Figure 6. Ohya slide in Shizuoka Prefecture.

Figure 7. Valley of Abe River filled with debris.

Figure 8. Surface failure of weathered slope (Kashmir of Pakistan in 2005).

falling of the surface soils. Figure 8 illustrates a slope of weathered limestone near Muzaffarabad, Pakistan, in 2005. The surface slide can easily destroy roads and prevent local transportation. Accordingly, emergency rescue and restoration become very difficult.

Figure 9. Seismic failure of road embankment during 2004 Niigata-Chuetsu earthquake, Japan.

Instability problems occur in artificial fills as well. Figure 9 illustrates an example in which a road embankment, which was constructed on a slope, failed upon an earthquake. Many local roads are subjected to a similar problem. The cause of the problem is that insufficient construction budget leads to the fill construction without removing the soft surface deposits. Those debris and organic materials at the bottom of the road embankment cannot resist against the seismic force, and the entire embankment collapses. What is important is the choice between earthquake resistance and money saving. In case the seismic failure would not be fatal, it may be reasonable to allow such a failure from the viewpoint of performance-based design and the policy of minimizing the life cycle cost. A similar but more serious problem lies in residential development. Around a sprawling mega city, inexpensive residential development is needed for low-income citizens, and such a development is conducted in hilly areas. Since cost has to be saved, soft valley deposits may not be removed, and, upon earthquakes, private lands and houses collapse. Such a loss of private properties is never allowed by limited budget of residents.

4 PROBLEMS OF SUBAQUEOUS SLOPE FAILURES

It is possible that earthquakes trigger failure of submarine slopes. In addition to the famous examples of Grand Banks in 1929 (Heezen & Ewing 1952) and off Orléansville in 1954 (Heezen & Ewing 1955), the 1964 Alaska earthquake of magnitude $= 9.2$ destabilized the sea bed deposit in front of Valdez (Fig. 10). A huge volume of soil mass (75 million m^3 according to Coulter & Migliaccio (1966)) collapsed upon shaking, and this mass movement caused tsunami in the

Figure 10. Aerial view of Valdez (photograph taken by Dr. K. Horikoshi).

Figure 11. Reinforcement of cliff behind house (Kamakura, Japan).

Valdez Bay. This tsunami destroyed the Valdez municipality and the town was later relocated in a safer place. The problem was that the submarine deposit of the Valdez Bay was most probably composed of fine nonplastic silt that was produced by the glacial erosion and was transported into the sea by the river in Figure 10. Being fine, silt grains deposit in water very slowly and the resultant density becomes very low. Since there is no cohesion, moreover, its undrained shear strength is very low. The possible excess pore water pressure during undrained shear does not dissipate during shaking, because the grain size is small. These situations made the bay deposit highly vulnerable to undrained failure and liquefaction. Moreover, it should be recalled that the highest tsunami so far recorded was triggered by rock slide impact induced by an earthquake (in Lituya Bay of Alaska: Miller 1960).

Stability of submarine slopes is reduced due to rapid rate of sedimentation (or human reclamation). If strong earthquake loading is superimposed on such a situation, slope failure is likely to occur. Note that a submarine landslide is one of the causative mechanisms of tsunami as was the case in North Sea (Long et al. 1989).

5 MITIGATION OF DAMAGE CAUSED BY EARTHQUAKE-INDUCED SLOPE

The previous sections introduced examples in which human communities were seriously damaged by seismic slope failures. Although it is advisable to stabilize all the potentially hazardous slopes (Fig. 11), financial restrictions do not allow it. Accordingly, there are three alternative options to be taken:

1. Reinforcement of slope by means of retaining walls, ground anchors, and other reinforcements
2. Relocation of human settlement from potentially hazardous areas
3. Emergency evacuation

The first choice is costly as stated above. The third choice may not be rapid enough during an earthquake. Both (1) and (2) require identification of potentially unstable slopes with practically reasonable cost. This assessment of failure risk may be made by determining shear strength of soil experimentally or by field investigation, and then running a stability analysis with reasonable seismic load taken into account. This procedure is unfortunately costly and can be conducted only in special situations. Hence, less expensive and time-efficient methodology is required, although the accuracy may be sacrificed to some extent.

The identification of potentially hazardous area for (2) further requires assessing the run-out distance of the failed mass. Fig. 12 compares two slope failures that were caused by the 2004 Niigata-Chuetsu earthquake. The one in front was of 110,000 m^3 in soil volume and traveled over some distance at the bottom of the valley. Consequently, this soil mass stopped the river flow and formed a natural dam. In contrast, the surface failure in the same figure had approximately 50 cm in thickness, suggesting that only the surface weathered soil fell down. This smaller soil mass did not travel laterally in the bottom of the valley and did not therein affect either the river flow or road traffic. Thus, the flow characteristics of soil mass depends at least partially on its volume as pointed out earlier by Hsü (1975). Furthermore, the effect of water content is illustrated in Fig. 13 in which a dry cliff in Muzaffarabad of Pakistan fell down upon the 2005 Kashmir earthquake. Noteworthy is that the soil did not translate laterally and houses in the bottom of the valley were

Figure 12. Failure of slope at Naranoki in Yamakoshi Village upon 2004 Niigata-Chuetsu earthquake (Mizuhashi et al., 2006).

Figure 13. Collapse of dry slope in Muzzafarabad, Pakistan.

not affected. This nature is different from the behavior of bigger soil mass in Fig. 12 that was of high water content.

6 ASSESSMENT OF SEISMIC RISK DUE TO SLOPE FAILURE

Any seismic risk depends primarily on the regional earthquake activities. Risk in seismic countries is higher than that in nonseismic countries. This feature is expressed in an empirical diagram in Figure 14 where the maximum distance of slope failure from an earthquake epicenter is plotted against the earthquake magnitude. In case a concerned slope lies within the critical epicentral distance, the failure risk has to be evaluated.

Financial reasons, however, eliminate in most cases the use of in-situ tests or laboratory shear tests on shear strength of undisturbed samples collected from a concerned slope. Accordingly, risk is judged on the basis of easily-available information. Table 1 shows an example in which various site factors are evaluated and weight points from them are added to make the final evaluation on the number of slope failures that

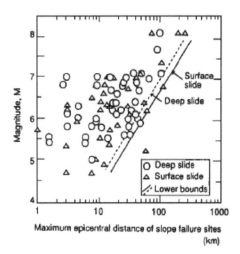

Figure 14. Maximum possible epicentral distance of failed slopes during past earthquakes (Yasuda, 1993).

would occur in a 500 m × 500 m grid during a future earthquake (Table 2).

The weight points in Table 1 were determined by regression analyses of past experiences. Therefore, some consistency is expected between evaluation and reality. There are problems, however, in Table 1. First, Table 1 is not intended to evaluate the safety of a particular slope. Second, the weight point for hardness of rock (W_4) increases as the rock becomes harder, implying more risk. This strange nature comes from the regression data in which failures in hard rock slopes were more numerous than those in soil, probably because slopes of weak soil did not exist or had failed during heavy rainfalls or in other circumstances prior to earthquakes. Consequently, Table 1 does not have a reasonable consideration of soil or rock material properties.

To solve those two problems mentioned above, an attempt was made to improve the method by taking into account the characteristics of soil. The weight in Table 1 was counted for individual slopes that failed during the 2004 Niigata-Chuetsu earthquake and their calculated risk was compared with reality (Mizuhashi et al. 2006). The results in Figure 16 are understandable to a certain extent. The good point is that the assessed rank of risk changes with the increasing volume of landslide mass. Since larger events are associated with greater risk assessment, the results are good. However, one of the biggest events at Naranoki is of lower assessed risk, and there is a need for further improvement.

The study puts emphasis on the effect of water. In particular, the deterioration of properties due to

Table 1. Weighting for factors related to slope instability (See Fig. 15 for W_7: Kanagawa Prefectural Government, 1986).

Factor	Category	Weight	Factor	Category	Weight
Maximum surface acceleration (Gal), W_1	0–200	0.0	Hardness of rock, W_4	Soil	0.0
	200–300	1.004		Soft rock	0.169
	300–400	2.306		Hard rock	0.191
	>400	2.754	Length of faults (m), W_5	No fault	0.0
Length of a contour line at mean elevation (m), W_2	0–1000	0.0		0–200	0.238
	1000–1500	0.071		>200	0.710
	1500–2000	0.320	Length of artificial slopes (m), W_6	0–100	0.0
	>2000	0.696		100–200	0.539
Difference between highest site and lowest site (m), W_3	0–50	0.0		>200	0.845
	50–100	0.550	Shape of slope, W_7	(1)	0.0
	100–200	0.591		(2)	0.151
	200–300	0.814		(3)	0.184
	>300	1.431		(4)	0.207

Table 2. Assessed number of slope instabilities per 500 m × 500 m grid (Kanagawa Prefectural Government 1986).

$W=W_1+W_2+W_3+W_4+W_5+W_6+W_7$		2.93	3.53	3.68	
Rank	A	B		C	D
Number of slope failures within 500m × 500m grid	0	1-3		4-8	>9

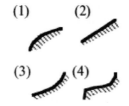

Figure 15. Vertical cross section of slopes.

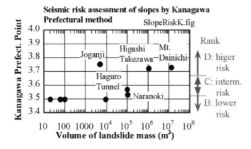

Figure 16. Seismic hazard assessment of slopes by using Kanagawa Prefectural method.

water submergence is focused on. Figure 17 illustrates drained shear behaviors of air-dry and water-submerged specimens of soil collected from the Urase site where rainfall-induced landslide occurred in July,

2004. The significant difference in strength was probably caused by deterioration or hydration of soil minerals upon submergence. Furthermore, Figure 18 compares both peak and residual strengths of different soils to confirm the submergence-induced deterioration. With these in mind, it was attempted to add a water-concerned index to risk assessment, which is obtained without running expensive shear tests.

The present study employed swelling characteristics and plasticity index. For swelling tests, in-situ soil grains were ground to the size less than 75 microns so that water effects would occur rapidly. Then, specimens for standard oedometer tests were reconstituted in the laboratory. Dry soil was compacted in two layers, with 20 impacts per layer, so that the dry mass density would be 1.061 g/cm^3 (60 g for a cylindrical sample). After compression under 20 kPa for 1,000 minutes in a dry state, specimens were submerged in water and the volume change was recorded (Fig. 19 for example). After some contraction occurred during the first 30 minutes, volume expansion (swelling) was observed in water-sensitive soils. Note that the extent of swelling was remarkably variable; some soils exhibited no volume expansion upon submergence.

The other index is plasticity index (liquid limit–plastic limit in %). Although this index is not directly related with mechanical properties of soils, it stands for the extent of mineral-water interaction and also is easy to measure in the laboratory. Figure 20 suggests

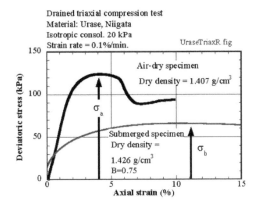

Figure 17. Drained triaxial compression tests on air-dry and submerged specimens.

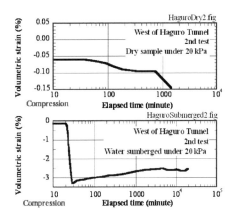

Figure 19. Water submergence of soil powder collected from Haguro Tunnel site in Niigata-Chuetsu.

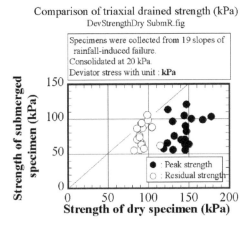

Figure 18. Reduction of shear strength of soils collected from rainfall-induced landslide sites.

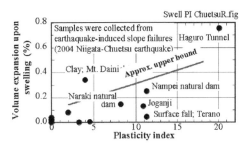

Figure 20. Correlation between swelling strain and plasticity index.

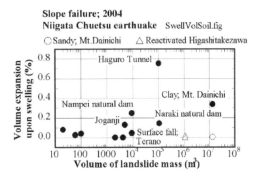

Figure 21. Correlation between landslide mass volume and swelling strain.

that there is a positive correlation between swelling strain and plasticity index. Figure 21 demonstrates that the swelling strain is greater for soils collected from sites of greater slope failure. However, the correlation is not clear in this figure because the volume of the soil mass is also affected by the topography and other geological conditions. Thus, swelling strain may be used as one of the indices to better assess the risk of earthquake-induced slope failure. Similarly, Fig. 22 illustrates that there is a positive correlation between the landslide volume and plasticity index, if such exceptional cases as sandy slopes and a reactivated landslide are eliminated from this figure.

On the basis of Figures 21, 22, the point, W, in Table 2 was adjusted by employing either $W+$ (swelling strain, %)/10 or $W+$ (plasticity index)/100. The results are shown in Figures 23, 24. The consistency with the damage extent and the determined rank are better than that in Figure 16.

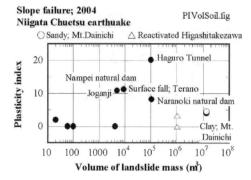

Figure 22. Correlation between landslide mass volume and plasticity index.

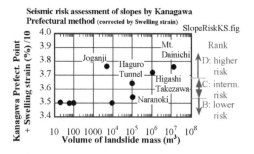

Figure 23. Correction of Kanagawa Prefectural method by swelling strain.

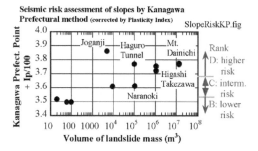

Figure 24. Correction of Kanagawa Prefectural method by plasticity index (I_p).

7 ASSESSMENT OF RUN-OUT DISTANCE OF DEBRIS

The extent of risk due to earthquake-induced landslide depends not only on the instability of a slope but also on the travel distance of a failed debris mass. In this regard, attention has to be paid to the dynamic characteristics of debris. Figure 25 illustrates the classical concept of the apparent friction angle by Hsü (1975).

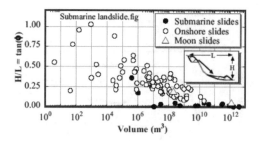

Figure 25. Apparent friction angle of landslides obtained from geometry.

Figure 26. Tsao-ling slide after 1999 ChiChi earthquake.

This diagram was drawn by using data collected by the present study. As is well known, the apparent friction angle, ϕ, or $H/L = \tan(\phi)$, decreases as the volume of landslide mass increases.

A second approach to assess the travel distance is the use of Newmark sliding block analogy (Newmark 1965). It is more advanced than the foregoing method because it employs the shear strength of soil and the time history of (design) earthquake acceleration. However, there is not yet a clear agreement on what kind of shear strength of soil should be used out of cyclic strength, undrained monotonic strength or else. In the present study, the Newmark method was used to reproduce the gigantic failure of Tsao-ling slope in Taiwan which was triggered by the 1999 ChiChi earthquake. The appearance of a part of the slide is shown in Figure 26, and the geological cross section is presented in Figure 27. It may be found that the slope is made of interbedding sandstone and mudstone (shale). To determine the shear strength of those materials, block samples were collected from the remaining part of the failed slope after the quake. Direct shear tests revealed that the mudstone was of the least shear strength and was likely to be the cause of the failure (Towhata et al. 2002).

Figure 28 indicates the stress-displacement relationship of mudstone specimens. It is therein found first that both peak and residual strengths increase

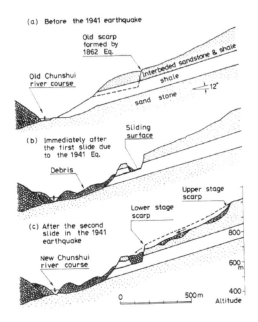

Figure 27. Geological cross section of Tsao-ling slope prior to 1999 event (Ishihara, 1985).

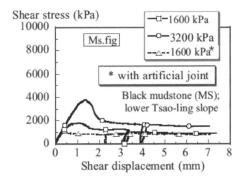

Figure 28. Direct shear tests on Tsao-ling mudstone specimen (Towhata et al. 2002).

with the confining pressure. A second finding is that a cycle of unloading and reloading after the post-peak softening does not accomplish the peak strength again. Since the Tsao-ling slope experienced instability several times in the past, the present study decided to use the residual strength for stability analyses. The employed strength parameters were $c = 417$ kPa and $\phi = 19.4$ degrees.

The first analysis was pseudo-static, in which the maximum acceleration (A_{max}) recorded at a nearby observatory (CHY028 at 8 km distance) of 748 cm/s^2 was converted to a seismic coefficient of $K_h = 0.305$

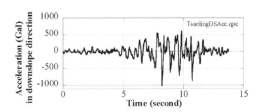

Figure 29. Acceleration time history for Tsao-ling analysis.

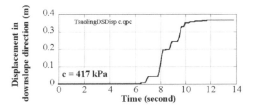

Figure 30. Calculated sliding displacement with $c = 417$ kPa.

by substituting A_{max} into an empirical formula by Noda (1975),

$$K_h = \sqrt[3]{A_{max}/g}/3 \qquad (1)$$

in which g stands for the gravity acceleration. By assuming the slope angle of 14 degrees and the thickness of the sliding rock slab to be 80 m, the pseudo-static factor of safety was obtained to be 0.73.

In order to conduct the Newmark-type analyses, the CHY028 acceleration records were converted to a component parallel to the slope (Fig. 29). Then the time history of sliding displacement was calculated. Figure 30, however, shows that the calculated displacement was far less than the reality. Furthermore, the displacement terminated at the end of strong shaking, which is contradictory to the complete failure of the slope in reality.

Sassa et al. (2004) used a ring shear device to study in more details the behavior of soil undergoing rapid flow.

The problem in underestimation of displacement is that the real failed rock mass was broken into pieces, as will be shown in the next chapter (Fig. 32), and the assumed strength parameters were not relevant. In particular, broken debris hardly had cohesion. Therefore, a second analysis was carried out by using $c = 0$ kPa, while using the same friction angle. The result in Figure 31 shows that displacement increases towards infinity even after the end of shaking.

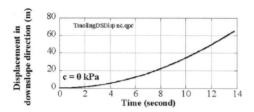

Figure 31. Calculated sliding displacement without cohesion.

Figure 32. Debris deposit at the bottom of Tsao-ling failed slope.

8 NUMERICAL ANALYSIS ON LAYERED MODEL OF SOIL MASS

The unsatisfactory performance of the Newmark-type analysis in the previous section implies that the calculation of debris over a long distance is beyond its capacity. This is because the Newmark method assumes a rigid block movement, while real debris flow consists of decomposed grains (Fig. 32). As an alternative, therefore, a more advanced analysis such as distinct element analysis is promising. The present paper, however, addresses a simpler choice.

To get an idea about the type of analysis, model tests were conducted on the nature of debris flow. Figure 33 shows the change of appearance of a model of flow. At 0.8 second, there are three targets near the front of the flow (in a white frame). Those targets are still visible at 1.0 second. However, they are overlain by following soil at 1.2 second, and have completely disappeared at 1.4 second. This suggests that there is a significant distortion and movement of grains inside the flow mass, which makes Newmark-type analysis totally irrelevant for the present type of problem.

Figure 34 illustrates the layered idealization of a debris mass in which the mass is composed of N-layers of finite thickness. The internal friction angle between layers is designated by ϕ, while the base friction by ϕ_b. When the bottom friction is greater than the internal friction, the bottom layer stops its movement after some translation, while overlying layers can continue their movement further. This idea seems to be consistent with the observation in Fig. 33.

The overall view of the analysis is illustrated in Figure 36 where the initial kinetic energy is supplied by free fall of H_i, and the velocity decreases gradually with the progress of down-slope movement. The apparent friction angle is assessed by the overall falling and lateral displacement (H/L).

An example analysis was made by assuming 100 layers ($N = 100$), the slope angle (θ) = 10 degrees, the base friction angle (ϕ_b) = 40 degrees, and varying international friction angles (ϕ). First, Figure 37 indicates the travel distance of the front of the debris that

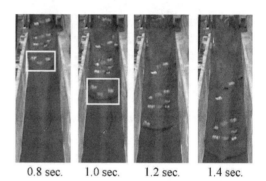

Figure 33. Distortion of debris flow with time.

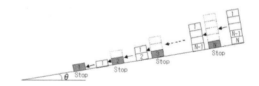

Figure 34. Layered model of debris flow.

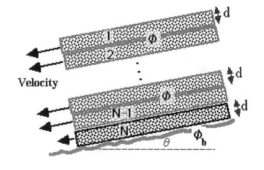

Figure 35. Internal and base friction angles.

increases with time. The lower the internal friction is, the longer is the run-out distance. Second, Figure 38 shows the number of layers (n) that arrive at different travel distances. As illustrated in Figure 34, lower

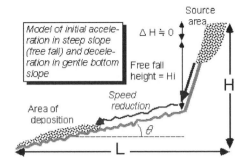

Figure 36. Initiation and termination of movement in layer analysis.

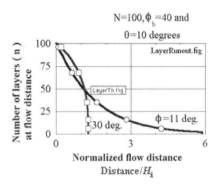

Figure 38. Number of layers that reach different travel distances.

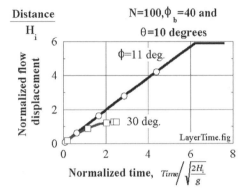

Figure 37. Travel distance of debris front increasing with time.

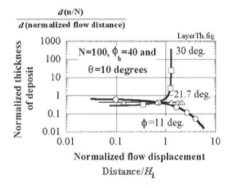

Figure 39. Height of debris deposit changing with horizontal distance.

layers stop their movement at shorter travel distance because of the greater bottom friction, while upper layers can travel longer. This feature makes a significant difference in the ultimate travel distance; the smaller internal friction allows longer travels.

Third, the thickness of debris deposit is assessed. When many layers stop within a short distance (case of $\phi = 30$ degrees in Figure 38), the accumulation of debris produces a high debris hill. This feature is somehow assessed by dn/d (*distance*). As shown in Fig. 39, the greater internal friction creates a high hill of accumulated debris at the front, and, in contrast, the lower internal friction allows more spread deposit over a greater area. The latter corresponds to a flow of liquefied (fluidized) debris with high water content in which the reduced effective stress makes the internal frictional resistance very small.

Finally an attempt was made to reproduce the empirical relationship in Figure 25. Because the thickness is represented by the number of layers, N, the volume is approximately represented by N^3. It is seen

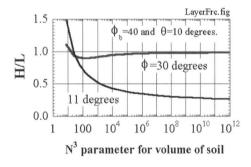

Figure 40. Reproduced apparent friction angle changing with debris volume.

in Figure 40 that H/L (tangent of the apparent friction angle) decreases with increasing volume when the internal friction angle is remarkably smaller than the bottom friction.

9 CONSIDERATION ON LAYER THICKNESS IN LAYERED FLOW SIMULATION

The numerical simulation in the previous section did not clearly indicate how to determine the layer thickness and, accordingly, the number of layers, N. At present, it is supposed that the thickness depends upon grain size of the debris. Figure 41 demonstrates a debris deposit in Niigata that was triggered by heavy rainfall. In this valley, the deposits developed a step in the surface and its height was approximately 3 m. Figure 42 shows a deposit of slope failure near Joganji in Nagaoka City that was caused by the 2004 Niigata-Chuetsu earthquake. The height of the step was approximately 50 cm. Finally, Figs. 43 and 44 reveal a fluidized slope failure in Tsukidate during the 2003 Miyagi-ken earthquake. The height of the step near the front was 58 cm.

Figure 43. Earthquake-induced slope failure at Tsukidate during the 2003 Miyagi-ken earthquake.

Figure 41. Rainfall-induced debris flow deposit Chu-ei Tunnel in Niigata-Chuetsu in 2005.

Figure 44. Step of deposit in earthquake-induced slope failure at Tsukidate.

Information was collected from case studies as well as literatures on the above-mentioned layer height and grain size distribution. The results are plotted in Figure 45. There is no good correlation between the layer thickness and the mean grain size, D_{50}. In contrast, D_{90} or probably the maximum grain size has some correlation. Tentatively, therefore, an empirical formula of

Layer thickness (m) = $30 \times D_{90}$ (m) (2)

is proposed. Consequently, the number of layers is determined by

Figure 42. Step of deposit in earthquake-induced slope failure near Joganji, 2005 Niigata-Chuetsu earthquake.

N = Total thickness of debris / layer thickness (3)

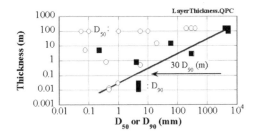

Figure 45. Empirical relationship between thickness of debris layer and grain size.

10 CONCLUSIONS

The present paper discussed knowledge collected from recent earthquakes. In mitigation of negative effects of slope failure to the human community, it is very important to identify potentially hazardous slopes and to assess the run-out distance. In particular, this has to be done with a reasonable cost. With regard to these, the following conclusions are drawn.

1. Problems of large slope failures are well known. Those of minor failures, however, have to be understood more, because they may stop local transportations and make rescue or restoration more difficult.
2. Seismically hazardous slopes should be identified with some consideration on characteristics of soils.
3. For the hazard assessment, a use of either swelling properties or plasticity index, both of which can be measured with minor efforts, is proposed.
4. Hazard assessment further needs to assess the travel distance of debris.
5. Newmark rigid block analogy is not relevant for this purpose. This study proposes a similarly simple layer idealization for the assessment of run-out distance.
6. To facilitate the determination of layer thickness, a correlation between the thickness and particle size was presented.

ACKNOWLEDGMENT

Direct shear tests on Tsao-ling rock block samples were conducted at the Central Research Institute of Electric Power Industries. Model tests on debris flow were carried out by Mr. Y. Nishimura for his undergraduate thesis. Field studies were made by T. Ito of Fudo-Tetra Corporation. Deep thanks are expressed to those supports and contributions.

REFERENCES

Coulter, H.W. & Migliaccio, R.R. 1966. *Effects of the earthquake of March 27, 1964 at Valdez, Alaska,* US Geological Survey Professional Report 542-C.

Heezen, B.C. & Ewing, M. 1952. Turbidity currents and submarine slumps, and the 1929 Grand Banks Earthquake, *Am. J. Science.* 250: 849–873.

Heezen, B.C. & Ewing, M. 1955. Orléansville earthquake and turbidity currents, *Bull. Am. Assoc. Petroleum Geologist* 39(12): 2505–2514.

Hsü, K.J. 1975. Catastrophic debris streams (Sturzstroms) generated by rockfalls. *Geol. Soc. Am. Bull.* 86: 129–140.

Imaizumi, F., Tsuchiya, S. & Ohsaka, O. 2005. Behaviour of debris flows located in a mountainous torrent on the Ohya landslide, Japan. *Can. Geotech. J.* 42: 919–931.

Ishihara, K. 1985. Stability of natural deposits during earthquakes, Theme Lecture. *11th ICSMFE, San Francisco,* Vol. 1: 321–376.

Kanagawa Prefectural Government. 1986. Report on Seismically induced Damage during future earthquake. pp. 13–63 (quoted by Manual for Zonation on Seismic Geotechnical Hazards, 1998, TC4 of ICSMGE).

Kawata, S. 1943. Study of new lake created by the earthquake in 1941 in Taiwan. *Bull. Earthq. Res. Inst., Univ. Tokyo* 21: 317–325 (in Japanese).

Long, D.E. & Dawson, A.G. 1989. A Holocene tsunami deposit in eastern Scotland. *J. Quaternary Science* 4(1): 61–66.

Miller, D.J. 1960. Giant Waves in Lituya Bay Alaska, Shorter Contributions to General Geology, US Geological Survey Professional Paper 354-C.

Mizuhashi, M., Towhata, I., Sato, J. & Tsujimura, T. 2006. Examination of slope hazard assessment by using case studies of earthquake- and rainfall-induced landslides. *Soils and Foundations* 46(6): 843–853.

Newmark, N.M. 1965. Effects of earthquakes on dams and embankments, *Geotech.* 5(2): 137–160.

Noda, S., Uwabe, T. & Chiba, T. 1975. Relation between seismic coefficient and ground acceleration for gravity quaywall. *Report of the Port and Harbor Research Institute* 67–111 (in Japanese).

Oberlander, T. 1965. *The Zagros streams, Syracuse Geographical Series, No. 1.* Syracuse University Press.

Sassa, K., Fukuoka, H., Wang, G.-H. & Ishikawa, N. 2004. Undrained dynamic-loading ring-shear apparatus and its application to landslide dynamics. *Landslides* 1(1): 7–19.

Tabata, S., Mizuyama, T. & Inoue, K. 2002. *Natural dam and disaster* Kokin Shoin Publ. pp. 50–53 (in Japanese).

Towhata, I., Yamazaki, H., Kanatani, M., Ling, C.-E., & Oyama, T. 2002. Laboratory shear tests of rock specimens collected from site of Tsao-ling earthquake-induced landslide. *Tamkang Journal of Science and Engineering* 4(3): 209–219.

Yasuda, S. 1993. Zoning for slope instability. *Manual for zonation of seismic geotechnical hazards, TC 4, Int. Soc. Soil Mech. Found. Eng.* 49–49.

Watson, R.A. & Wright Jr., H.E. 1969. The Saidmarreh landslide, Iran. *Geol. Soc. Am. Special Paper* 123: 115–139.

Monitoring and modeling of slope response to climate changes

H. Rahardjo
School of Civil and Environmental Engineering, Nanyang Technological University, Singapore

R.B. Rezaur
Department of Civil Engineering, Universiti Teknologi Petronas, Perak, Malaysia

E.C. Leong
School of Civil and Environmental Engineering, Nanyang Technological University, Singapore

E.E. Alonso, A. Lloret & A. Gens
Department of Geotechnical Engineering and Geosciences, UPC, Barcelona, Spain

ABSTRACT: Shallow slides are often triggered by climate effects. An understanding of the slope failure conditions and effective remedial measures can be achieved by comprehensive field monitoring of climatic and hydrologic changes and the consequent changes in slope responses. Two contributions from two different geographic regions are presented to gain understanding of the complex phenomena involved in slope failure studies. In the first part Alonso et al., contributes theoretical analysis of a stochastic model for the reliability of planar slides in a partially saturated soil, subjected to a rainfall history described as a time series and then presents a case history of shallow mudslides triggered by a Mediterranean climate, analyzed by means of a coupled hydro-mechanical modeling tool. The joint saturated-unsaturated consideration of the slide is necessary to understand field data. In the second part Rahardjo et al., contributes field monitored data from three residual soil slopes in Singapore and demonstrates how field monitored data on climatic, hydrologic, and slope variables were used to evaluate slope responses under subtropical Singapore climate.

1 INTRODUCTION

Global warming, rising sea level and climatic changes have become important issues of the world in recent decades. Climatic changes have affected rainfall patterns in many parts of the world, causing occurrences of numerous landslides. Many tropical areas are prone to frequent rainfall-induced slope failures (Poh et al. 1985, Pitts & Cy 1987, Tan et al. 1987, Chatterjea 1994, Toll et al. 1999). The problem is escalated by the increasing rate of hillside developments for engineered and fill slopes in many regions.

Rainfall of both event-based and antecedent, runoff, infiltration, and their contributions to pore-water pressure changes in a residual soil slope are the variables pertinent to slope responses. The dynamic flux boundary conditions are controlled by the physical properties of the soil, in particular the unsaturated soil in the vadose zone above the water table.

Attempts to relate climate conditions to the occurrence or reactivation of landslides are numerous (Wieczorek (1996) and Corominas (2000) provide a detailed account). Most methods combine antecedent rainfall, rainfall intensity and rainfall duration. However, it is impossible to dissociate climate effects from the type of slide. In general, large landslides require specific analysis. Shallow slides, however, react to rainfall conditions in a more predictable manner.

The objective of this paper is to present strategies for monitoring and modelling of slope response to climate changes. In this context two contributions from two different geographic regions are presented in two sections (PART-A and B). In the first part (PART-A) Alonso et al., presents a stochastic model for the analysis for the reliability of planar slides in a partially saturated soil, subjected to a rainfall history described as a time series and then presents a case history of shallow mudslides triggered by a Mediterranean climate analyzed by means of a coupled hydro-mechanical modeling tool. The theoretical analysis of the risk of slide associated with a given record of rainfall presented by Alonso et al., considers infiltration conditions through a partially saturated soil and integrates rainfall effects through a random function formulation of the concept of risk of failure. Closed form solutions were found for the probability of failure. The theoretical analysis highlights several aspects of rain-induced risk of failure. The case history for shallow instabilities in a clay formation analyzed by means of a coupled hydro-mechanical analysis presented by Alonso et al., is also valid for saturated and unsaturated slope conditions.

In the second part (PART-B) Rahardjo et al., presents field monitored data from three residual soil slopes in Singapore and demonstrates how field monitored data on climatic, hydrologic, and slope variables were used to evaluate slope responses in terms of; (i) pore-water pressure changes; (ii) runoff generation and infiltration amount. Part-I and II together provides understanding of the complex phenomena involved in slope failure studies.

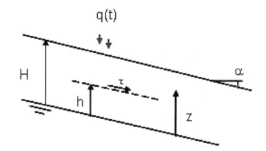

Figure 1. Geometry of the solved problem.

2 PART-A. MODELLING RAINFALL EFFECTS ON SHALLOW SLIDES (ALONSO ET AL.)

2.1 Rain infiltration and risk of sliding—a simple stochastic model

Soil moisture variation in time within slopes is recognized as a major source of uncertainty concerning the safety conditions of any slope. Any wetting process modifies several parameters which control the overall stability. In fact, the natural unit weight increases, the available shear strength decreases with a decrease in soil suction and, when the soil becomes saturated, the pore water pressure becomes positive thus modifying effective stresses. In addition, the internal geometry of different soil strata having varying permeabilities, plays a significant role in controlling the hydraulic regime within the slope. An example developed later (Villa Blasi slope) illustrates this comment. The variation of soil moisture conditions in the slope is controlled by the varying rates of infiltration which in turn depend on rainfall intensity, soil conditions and geometry. Except for the case of homogeneous masses of impervious clay soils or extremely pervious formations, experience shows that, in a large class of slopes, failures occur during periods of heavy rainfalls.

2.1.1 Safety margin of slopes in partially saturated soils

Fredlund et al. (1978) proposed the following expression for the strength of a partially saturated soil:

$$\tau = c' + (\sigma - p_a)\tan\phi' + (p_a - p_w)\tan\phi_b \quad (1)$$

where $(\sigma - p_a)$ and $(p_a - p_w)$ are the net normal stress and suction, (c', ϕ') are the effective cohesion and friction and $\tan\phi_b$ is a suction-related friction coefficient which takes the value of ϕ' for very low suctions and reduces progressively as suction increases. (However,

it will be assumed to be constant in the following). The safety margin of an infinite slope against failure along a plane at depth $(H - h)$ (Fig. 1) is given by

$$M = \left[\gamma_w n \int_h^H S_r dz + \gamma_s(1-n)(H-h)\right]$$
$$\times \cos\alpha(\cos\alpha\tan\phi' - \sin\alpha)$$
$$+ c' + (p_a - p_\omega)\tan\phi_b \quad (2)$$

Safety margin is defined as the difference between shear strength and shear stress and it provides a linear measure of safety. In Equation 2, S_r, n and γ_s are the degree of saturation, the porosity and the solid specific weight respectively.

2.1.2 Infiltration model

The classical theory assumes that the net infiltration, $q(t)$, results from a balance between rainfall, corrected by means of a runoff coefficient, and evapotranspiration. On the other hand the stochastic analysis of a time series of data regarding rainfall and evapotranspiration provides the necessary parameters to identify $q(t)$ from a stochastic point of view. The equations governing the flow of water in a rigid partially saturated soil are (Bear & Bachmat 1991)

$$\frac{\partial}{\partial t}(S_r n) + div(\bar{v}_\omega) = 0 \quad (3)$$

$$\bar{v}_\omega = -Kgrad\left(z + \frac{p_\omega}{\gamma_\omega}\right) \quad (4)$$

If a linear expression for the water retention curve is adopted,

$$S_r = S_{r0} + a_s(p_\omega - p_{\omega 0}) \quad (5)$$

where S_{r0} is the degree of saturation for a reference suction $p_{\omega 0}$ and a_s is a constant, Equations 3–5 become

$$\frac{\partial p_\omega}{\partial t} = C_s \frac{\partial^2 p_\omega}{\partial z^2}; \quad C_s = \frac{K}{na_s\gamma_\omega} \quad (6)$$

This equation may be normalized if,

$$Z = \frac{z}{H}, \quad T = \frac{C_s t}{H^2}, \quad u = [p_\omega/\gamma_\omega H(q_0/K - 1)] - z/H$$

q_0 being a constant under steady state conditions and it results in,

$$\frac{\partial u}{\partial T} = \frac{\partial^2 u}{\partial Z^2} \tag{7}$$

Boundary conditions for an infiltration $q(t)$ at the surface and a fixed ground water level at depth H are

$$u = 0; \quad Z = 0 \tag{8a}$$

$$\frac{\partial u}{\partial Z} = I(T) = \frac{q(t) - q_0}{q_0 - K} = \frac{q_1(T)}{q_0 - K} \tag{8b}$$

If q_0 is selected as the mean of the process $q(t)$, $I(T)$ becomes a stochastic process of zero mean, which, on account of the periodic nature of hydrologic events, will be expressed as:

$$\left.\frac{du}{dZ}\right|_{Z=1} = I(T) = \sum_{n=1}^{N} A_n \cos \lambda_n^2 T + B_n \sin \lambda_n^2 T;$$

$$\lambda_n^2 = \frac{2\pi n}{T_0} \tag{9}$$

where the coefficients A_n and B_n are non correlated Gaussian random variables of zero mean and common variance (σ_n^2) and T_0 is a dimensionless reference period.

Figure 2 shows a 6 year record of the process $I(t)$ for a meteorological station in the vicinity of Barcelona. Plotted values represent averages of 10 days. Evaporation data was obtained from actual monthly measurements, in a 1.25 m diameter tank corrected to reduce the free water surface to soil conditions. The solution of equation (7) with (8) and (9) is

$$u(Z,T) = \sum_{n=1}^{N} T_n(Z) \cos \frac{2\pi n T}{T_0} + W_n(Z) \sin \frac{2\pi n T}{T_0} \tag{10}$$

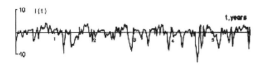

Figure 2. Infiltration record measured in a meteorological station in Barcelona.

Then, the safety margin (2) is found as:

$$M(Z,T) = \omega_0 + \sum_{n=1}^{N} (A_n \Omega_1 + B_n \Omega_2) \cos \frac{2\pi n T}{T_0}$$

$$+ (-A_n \Omega_2 + B_n \Omega_1) \sin \frac{2\pi n T}{T_0} \tag{11}$$

where the different coefficients have been defined in Appendix 1. The mean safety margin, ω_0, is a deterministic function of Z.

Figures 3a, b show the evolution of safety margin at different depths of a slope subjected to sinusoidal infiltrations of widely different frequencies. It can

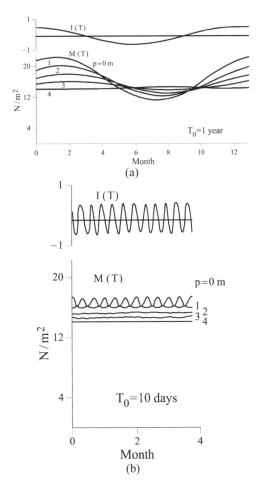

Figure 3. Evolution of safety margin at different depths (p = 0 to 4 m) of a slope ($\alpha = 32°$) subjected to sinusoidal infiltration: (a) Low frequency; (b) High frequency. $K = 0.7 \times 10^{-7}$ m/s, $\phi' = 30°$; $\phi_b = 25°$; $a_s = 10^{-5}$ m^2/N; $c' = 1000\, z$ N/m^2.

be observed the limited influence of high frequency components at points not very close to the surface. On the other hand, high period oscillations of infiltration have the capability of penetrating deep into the slope. Figure 3a shows also the delaying effect of infiltration into the soil.

2.1.3 *The safety margin as a random process in time—Correlation structure*

In order to analyze the slope reliability as a level-crossing problem associated with the evolution in time of the safety margin, it is first necessary to establish its correlation structure. It may be shown that the autocovariance function of the safety margin is given by

$$C_M(T) = \sum_{n=1}^{N} \sigma_n^2 (\Omega_1^2 + \Omega_2^2) \cos \lambda_n^2 \tau \quad (12)$$

The individual variances associated to each of these periodic terms reflect both the variability of the "external" infiltration terms (through σ_n^2) and the process of percolation, the modification of strength parameters of the soil and its weight and the (mechanical) definition of stability—(through the term $(\Omega_1^2 + \Omega_2^2)$, which is defined in the Appendix). Once the covariance is determined, standard procedures to analyze stochastic time records (see for instance Tretter (1976)) were used to determine the power spectrum density function of the safety margin. The analysis of a slope subjected to the Barcelona infiltration record shows how the influence of high frequencies of hydrologic variation is increasingly damped at depth.

2.1.4 *The level crossing problem*

For Gaussian processes, assuming that the rate of crossing a certain barrier follows a Poisson distribution and that the time intervals spent in the "safe" regions have a common exponential distribution whose mean is the inverse of the mean rate of crossings, Vanmarcke (1975) derived the following expression for the probability of no crossings in the interval 0 to t:

$$L_B(t) = [1 - \phi(r)] \exp\left[\frac{-v_a t}{1 - \phi(r)}\right] \quad (13)$$

where r is the normalized crossing level, $\phi(r)$ is related to the error function ($\phi(r) = 1/2[1 - erf(r\sqrt{2})]$) and v_a is the mean rate of crossings above the level r. In our case, if one works with the zero-mean safety margin process, $M^* = M - \omega_0$, the normalized barrier indicating slope failure is $r = -\omega_0/\sigma_M$, where σ_M is the standard deviation of safety margin. The mean rate of crossings above level r is given by

$$v_a = (\sigma_{\dot{M}}/2\pi \sigma_M) \exp(-r^2/2) \quad (14)$$

where $\sigma_{\dot{M}}$ is the standard deviation of the derivative in time of the safety margin. Knowing the expression (12) for the autocovariance function, both σ_M^2 and $\sigma_{\dot{M}}^2$ are found as

$$\sigma_M^2 = \sum_{n=1}^{N} \sigma_n^2 (\Omega_1^2 + \Omega_2^2);$$

$$\sigma_{\dot{M}}^2 = \sum_{n=1}^{N} \sigma_n^2 \lambda_n^2 (\Omega_1^2 + \Omega_2^2) \quad (15a, b)$$

All of these expressions are used subsequently to predict the reliability of an earth slope in a given hydrologic environmental condition (Barcelona area).

2.1.5 *Application of the developed formulation to the reliability of slopes in a given hydrologic environment*

In all the cases presented below the soil properties are maintained constant. They correspond approximately to a partially saturated silty soil and they are $\phi' = 30°$; $\phi_b = 25°$; $a_s = 10^{-5}$ m^2/N. The parameter c' of the strength envelope was assumed to increase slightly with depth ($1000p$ N/m^2, where p is the depth below the surface) in order to represent non homogeneous soil profiles, typical, for instance, in residual deposits. Failure probabilities correspond always to a yearly period.

Figure 4 shows the variation of probability of failure with the slope angle for different positions of the failure plane. It is interesting to note the transition from deep-seated surfaces, more likely to occur in steep slopes, to shallow failures in gentle slopes. In fact the influence of rainfall induced moisture changes is determinant to provoke shallow instabilities. On the other

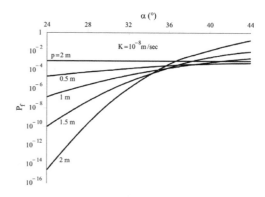

Figure 4. Probability of failure of an infinite slope in partially saturated soils for different slope angles and depths of failure plane.

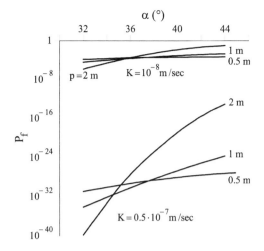

Figure 5. Influence of soil permeability. $H = 5$ m; $\phi' = 30°$; $\phi_b = 25°$; $a_s = 10^{-5}$ m^2/N; $c' = 1000$ p N/m^2.

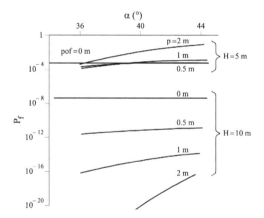

Figure 6. Influence of depth of ground water level in failure risk for different slope angles and position of failure plane.

hand steep slopes have small (deterministic) safety margins, decreasing with depth according to the stability model adopted, and their influence overrides the strong moisture induced effects close to the surface.

The marked influence of permeability is shown in Figure 5. Other conditions being equal the increase in permeability reduces fast the risk of failure. For a given infiltration history the developed pore pressure gradients are smaller for larger K values and therefore changes in soil strength are small.

The depth of groundwater level (Fig. 6) controls the mean "dryness" of the soil in the sense that both S_r and p_w are reduced. Larger strengths (implied by higher suctions) result in increased reliabilities. The important result is, however, that the slope is especially prone to shallow failure surfaces for a wide range of slope angles. In other words, the critical slope angle which marks the boundary between shallow or deep failure planes (approximately at 37° if $H = 5$ m in the example solved, see Figures 5 to 7) increases when H increases. It is finally noted that the above results were all obtained for rainfall records averaged over 10 days. No significant influence was found when the average period changed from 5 to 20 days.

2.2 A case history: ancona slides

2.2.1 Background
In December 1982, a large flow slide in overconsolidated clays destroyed the suburbs of the city of Ancona (Italy). Several years later, a research project was launched to investigate the behavior of shallow slides in clays in a Mediterranean climate. Villa Blasi slope was selected for the study and a field investigation was set out. Inclinometers, as well as piezometers, were installed and specimens were taken for specialized testing. Rainfall was recorded. A representative profile of the slope is shown in Figure 7. Three layers (α, β, γ) could be distinguished. The surface layer α (3–4 m thick) is made of remoulded brown Ancona clay with a significant proportion of organic matter. It overlies a 4 to 10 m thick layer (β) of stiff brown silty clay with sandy inclusions (brown Ancona clay). At depth, the Pliocene substratum consists of stiff blue silty clay with sandy inclusions (blue Ancona clay).

2.2.2 Soil properties
Dominant clay minerals are montmorillonite, illite and chlorite. Calcite amounts to 22% of the total mineral composition. Brown and blue Ancona clays are

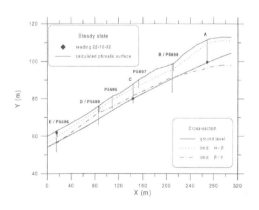

Figure 7. Longitudinal profile of slope and position of the phreatic level. The position of piezometers (A, B, C, D, E and P) is also shown.

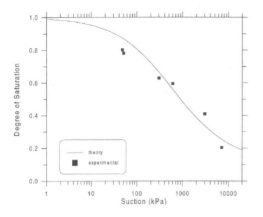

Figure 8. Water retention curve of brown Ancona clay.

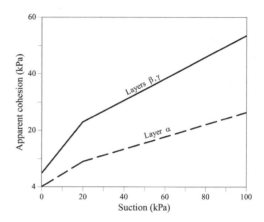

Figure 9. Variation of apparent cohesion with suction for Brown Ancona clay.

classified as CH or MH ($W_L = 52$–64%; PI: 25–34%; clay fraction: 40–55). Water content is low in these formations: 21%–32%. Oedometer tests indicated a significant overconsolidation stress: 2 to 3 MPa for layer β and 3 to 4 MPa for layer γ. In drained triaxial tests, the brown Ancona clay exhibited a brittle behaviour (brittleness index, $I_B = 0.40$).

Suction controlled oedometer tests were performed to determine the effect of suction on volumetric deformations and on permeability. Experimental data on the variation of permeability and degree of saturation with suction is shown in Figure 8.

A coupled finite element flow-deformation code (NOSAT) for saturated-unsaturated analysis, developed at the Department of geotechnical Engineering and Geosciences of the UPC, was used in calculations. NOSAT solves the balance equations for water and air, and the mechanical equilibrium.

2.2.3 Modelling Villa Blasi slope

The soil, in this case, was characterised by a non-linear elastic behaviour. Volumetric deformations induced by suction changes were introduced by means of state surfaces relating void ratio, suction and mean net stress. Figure 9 shows the variation of apparent cohesion with suction for the intermediate layer (brown clay).

A first step in modelling was to approximate the actual hydrogeologic conditions. Two flow regimes were identified. A deep flow regime was characterised by a phreatic level at nearly constant elevation in both wet and dry seasons (depth varies between 4 and 14 m). Figure 7 shows some piezometer readings and the calculated position of the phreatic level. In addition, a surface flow regime was directly controlled by rainfall. Positive and negative pore water pressures were measured during the year in the upper few meters. Calculations were performed in three phases. Equilibrium conditions were first defined. An idealized geological sequence was simulated. The soil was initially deposited under normally consolidated conditions. Then, erosion (unloading) took the slope to its actual geometry. For this phase on an elastoplastic hyperbolic model, having parameters $c' = 95$ kPa, $\varphi' = 27°$, was used. This sequence tried to reproduce the actual preconsolidation stresses measured in oedometer tests. The phreatic surface of the deep hydrogeological regime was also defined at this stage (see Fig. 7).

In the second phase of the analysis, rainfall infiltration is simulated. Average uniform monthly flow rates were imposed as flow boundary conditions at the slope surface. If a positive pore water pressure is calculated at the surface (which is an indication that the rainfall rate was larger than the infiltration capacity of the soil), the boundary condition is changed to zero suction, which is a flooding condition.

The third stage in the analysis concerns deformations and safety factors. Local safety factors were determined by comparing at some depths within the slope available shear strength and existing shear stress.

The rainfall record actually used in computations is shown in Figure 10. There was some uncertainty in the actual field permeabilities of layers α, β and γ. This is key information to understand stability conditions. Time records of pore pressure changes measured in piezometers could be used, however, to define a permeability profile. The idea was to perform a sensitivity analysis varying the permeabilities of layers α, β and γ.

A total of 10 different permeability profiles, shown in Table 1, were subjected to the same rainfall record in the manner outlined above. Then, the most probable permeability layering was identified by comparing piezometer measurements and calculated values.

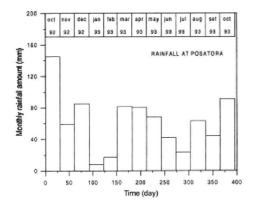

Figure 10. Rainfall records used in calculations.

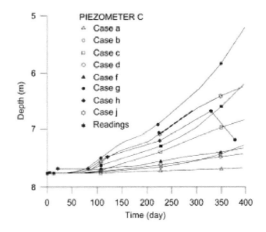

Figure 11. Comparison of measured and calculated piezometer readings (piezometer C). Depth of measuring chamber 9.80 m (open tube piezometer in substratum).

Table 1. Combination of soil layer permeability for sensitivity analyses.

Case	Saturated water permeability k_{ws} (m/s)		
	Layer α	Layer β	Layer γ
a	10^{-9}	10^{-9}	10^{-9}
b	10^{-9}	10^{-9}	10^{-9}
c	10^{-8}	10^{-9}	10^{-9}
d	10^{-7}	10^{-9}	10^{-9}
e	10^{-6}	10^{-9}	10^{-9}
f	10^{-8}	10^{-8}	10^{-9}
g	10^{-7}	10^{-8}	10^{-9}
h	10^{-6}	10^{-8}	10^{-9}
j	10^{-7}	10^{-7}	10^{-9}
k	10^{-6}	10^{-7}	10^{-9}

2.2.4 Computational results

Figures 11 and 12 show two examples of the comparison made between field data and calculated pore pressures. Figure 11 indicates the readings in an electrical piezometer located in α layer.

Some combinations of layer permeability lead to a consistent agreement with field data. (Case h, for instance). Suctions were often recorded. There is another interesting consideration about the relationship between failure risk and soil permeability. Examining the results of all computations, it became clear that some combinations of layer permeability and layering sequence led to the "strongest" response of the slope, in the sense of producing the highest pore pressures for a given sequence of rainfall. The general idea is that, given a climatic record, there are particular combinations of soil permeability which are critical in terms of pore water pressure development. This idea is illustrated in Figure 12 which is a plot of

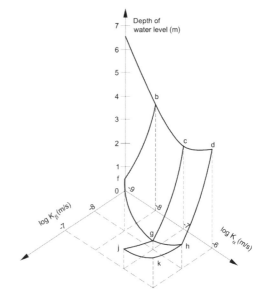

Figure 12. Computed depth of water level at piezometer P5699 for time $t = 350$ days after the beginning of the prediction exercise (October 1st, 1992). Depth of sensor: 2.80 m.

the computed response of piezometer P5699, 350 days after the initial date for the modelling.

The combination $K_\alpha = 10^{-7}$ m/s, $K_\beta = 10^{-8}$ m/s, leads to the maximum value of the (positive) pressure at the point considered. There are no easy rules to predict a critical permeability profile because the computed water pressure integrates many

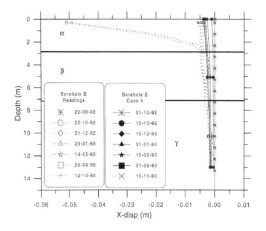

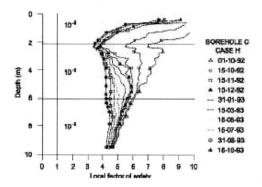

Figure 14. Profiles of local safety factors (based on FE computations) at a borehole C (Case h, homogeneous strength).

Figure 13. Computed and measured profiles of horizontal displacements at borehole B.

phenomena: surface flow, infiltration, storage capacity of the soil, previous wetting history of the soil etc.

The calculated and measured profile of horizontal displacements is shown in Figure 13. It is clear that a distinct shear surface has developed at the interface between α and β layers. Most probably, a previously existing shear surface was reactivated.

2.2.5 *Stability analysis—Local safety factors*

Field observations and the results of inclinometer readings indicate that a simple planar model could represent the shallow slides.

Since stresses were determined also in a theoretically more accurate analysis, an expression for the safety factor may be found. It uses stresses and water pressures calculated in the numerical model:

$$F_{(FEM)} = \frac{c' + \left[\dfrac{(\sigma_x + \sigma_y)}{2} - \dfrac{(\sigma_x - \sigma_y)}{2}\cos 2\alpha - \tau_{xy}\sin 2\alpha - p_w\right]\tan\varphi'}{\dfrac{(\sigma_x - \sigma_y)}{2}\sin 2\alpha - \tau_{xy}\cos 2\alpha} \quad (17)$$

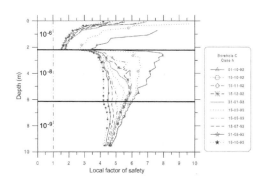

Figure 15. Profiles of FE-based local safety factors at borehole C (Case h, heterogeneous strength).

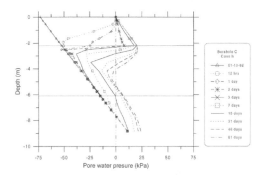

Figure 16. Profiles of water pressure for a permanent flooding of the slope. (Case h, heterogeneous soil strength).

Calculated variations of local safety factors with time and depth are shown in Figure 14. Absolute values of F are high, however. If the soil strength is assumed to be heterogeneous and remoulded parameters ($c' = 0$; $\phi' = 24°$) are assigned to the upper α layer, the calculated profiles of the local safety factor are shown in Figure 15. Minimum safety factors are now found at the α–β interface, where shear slips were detected by the inclinometer. However, the minimum calculated value ($F = 1.5$), is not able to explain the failure.

The analysis performed so far only includes the rainfall recorded in 13 months. Heavier rains certainly had occurred in the immediate past of the slope. An extreme rainfall event is to assume that the slope

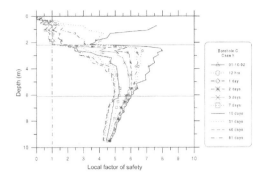

Figure 17. Profiles of FE-based local safety factor. Heterogeneous strength distribution. Case h.

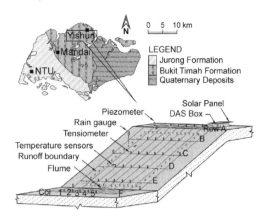

Figure 18. Location of the study slopes, generalized geological map of Singapore, schematic diagram of relative position and arrangement of field instruments.

surface is flooded during a certain time period. This condition is illustrated in Figure 16, which shows the calculated profiles of water pressure. Three days of permanent surface saturation leads to almost hydrostatic pore water pressure conditions in the upper remoulded α layer. This situation is compatible with suction values maintained in the lower β layer. This situation brings a different picture of the safety conditions of the slide (Fig. 17). Now surface instabilities are predicted after 12 hours of surface saturation of the slope. The entire upper α clay layer becomes soon unstable under extreme rainfall events.

Table 2. Description of the slopes.

Variable	Yishun	Mandai	NTU-CSE
Area (m^2)	165	180	140
Vegetation cover (%)	100	100	100
Slope angle (°)	23	31	27
Slope height (m)	7	11	7
Slope length (m)	16.5	18.0	14.0
Aspect	West	West	West
Slope form	Plain	Plain	Plain

3 PART-B: SLOPE HYDROLOGY AND RESPONSES TO CLIMATE CHANGES (RAHARDJO ET AL.)

3.1 Study area

Three slopes were selected for instrumentation to provide data for rainfall-induced slope failure studies. Out of these three slopes (see Figure 18) one was located in Yishun, one in Mandai and two in Nanyang Technological University (NTU) campus (hereafter called the NTU-CSE slope). The characteristics of the four slopes are shown in Table 2. These slopes were selected because they were located in two major geological formations in Singapore, subjected to frequent shallow landslides.

The climates at the study sites are hot and humid equatorial, with no marked dry season. The temperatures vary little throughout the year with an annual average temperature of 26.6°C and a mean relative humidity of 84% (Meteorological Service Singapore 1997). The average annual rainfall in Singapore varies between 2000 mm around the fringes of the island to about 2300 mm in the central region (Meteorological Service Singapore 1997). The three recorded highest rainfalls in Singapore are; 512 mm in 1978, 467 mm in 1969 and 366 mm in December 2006. The rainfall is usually greatest in the months of November to January (the north-easterly monsoon) but rain falls in all months of the year, with an average of 179 rainy days in a year. Rainstorms are short, intense and generally have a limited spatial extent, with intensities typically ranging between 20 and 50 mm/h, although short duration (5 min) rainfall intensities can exceed 100 mm/h (Sherlock et al. 2000). The potential evaporation rate in Singapore was calculated to vary between 5.16–7.53 mm/day (Gasmo 1997). All four slopes had grass as the vegetative cover.

The geology of the study sites consists of residual soils from two major geological formations (Public Works Department 1976); (a) the Bukit Timah Granite (Yishun and Mandai) which occupy the north and central-north region of Singapore; and (b) sedimentary rocks of the Jurong Formation (NTU-CSE), occupying the west and southwest region of Singapore (see Figure 18). These two residual soils comprise two-thirds of Singapore's land area.

3.2 Field instrumentation and data collection

The slope responses were characterised through measurement of the following variables: (i) rainfall inputs to the slopes (ii) runoff generation from the slopes (particularly NTU-CSE slope) and (iii) changes in pore-water pressures in response to rainfall.

Rainfall was recorded at each slope with a tipping-bucket rain gauge. Runoff was measured using a perspex flume (Rahardjo et al. 2004a). Corrugated zinc sheets, 300 mm high and driven about 100 mm into the ground bordered each instrumented slope (Figure 18). The boundaries guided the runoff into the perspex flume at the lower end of the plot where the surface runoff was measured using a capacitance water-depth probe installed in the flume. The water-depth probe was connected to a data logger that stored runoff data every 10 s during rainfall. Runoff measurements for both simulated and natural rainfalls were conducted particularly in NTU-CSE slope.

Pore-water pressure changes in response to climatic changes were recorded using jet-fill tensiometers. In addition to the tensiometers, piezometers and temperature sensors were installed at various locations in the slopes. Figure 18 shows the details of field installation of the instruments. All sensors within each instrumented slope were monitored continuously and automatically by a field data acquisition system (DAS). The DAS was programmed to scan the sensors at 4 h intervals during periods of no rain and at 10 min intervals during rainfall and continuing at the same rate until 30 min after cessation of rainfall. A rainfall event would trigger the rain gauge, and data for surface runoff were collected during rainfall events only. More details on field instrumentation can be found in Rahardjo et al. (2007).

3.3 Results and discussion

The results of the field tests, laboratory tests and field monitoring are presented in two forms. First, field test and laboratory test results used to characterise the engineering properties of the slopes are presented in the next section. Second, field monitoring and data analyses results are presented in the subsequent sections to show how slope responses were evaluated in terms of (i) pore-water pressure changes and (ii) runoff generation and infiltration amount.

3.3.1 Site observation

The geotechnical properties of the residual soils at the four slopes derived from laboratory tests and field tests are shown in Table 3. The information obtained from site investigation and laboratory tests was used to produce simplified soil profiles for the slopes as shown in Figure 19. The soil water characteristic curve (SWCC) of the slopes as obtained from pressure plate tests are shown in Figure 20. Elaborate discussion on the soil properties of these slopes are presented in Rahardjo et al. (2004b).

3.3.2 Evaluation of slope responses in terms of pore-water pressure changes due to rainfall

Slope responses to rainfall were assessed through the pore-water pressure changes in the slope. The tensiometer readings from the slope crest (row A) were plotted with the corresponding rainfall readings to indicate the seasonal pattern of pore-water pressures at Yishun slope (Figure 21). The tensiometers recorded increases in pore-water pressures at all levels on wet days. The deeper soil layers (3 m depth) at the slopes' crest and toe maintained positive pore-water pressures for all the 420 monitored days, while frequent changes in pore-water pressure from negative to positive were recorded for shallow depths (50 cm) in response to rainfall (Figure 21).

The soils at shallower depths are in close proximity to the atmosphere and slopes' vegetation. As a result, the soils at shallow depths are easily and frequently influenced by rainfall and evapotranspiration compared to deeper soil layers. During the monitoring period, negative pore-water pressure development as low as −57 kPa was observed at shallow depths (Yishun slope, Figure 21). Positive pore-water pressures were also observed at all soil depths after a significant rainfall and appear to be a common phenomenon in the monitored slope (Figure 21). Infiltration into the slope does not lead to a constantly wet soil condition, as is seen in the rapid matric suction recovery during dry periods (Figure 21). The average daily matric suction recovery rates were 5 kPa/day for shallow depths and 1 to 3 kPa/day for greater depths. The instant response of pore-water pressure to the infiltrating rainwater indicates considerable infiltration on the grass covered slopes.

To illustrate the sensitivity of pore-water pressures to rainfall and to reflect the hydrologic response of the slopes during a wet and a dry period, the record of pore-water pressures and rainfall for a period of 6 weeks was selected from the time series of the Yishun slope as shown in Figure 22, which illustrates the decreasing pore-water pressures during the dry period from February 12 to March 8, 1999. A small storm on March 8 only affected the pore-water pressures at a shallow depth (0.5 m) due to the small amount of infiltrated water. After the storm the pore-water pressures at 0.5 m depth started to decrease again due to the evapotranspiration processes. On the other hand, several rainfall events during the wet period from March 9 to March 16, 1999, caused the pore-water pressures at all depths (0.5, 1.7, and 2.9 m) to increase.

Table 3. Soil properties of the slopes.

Slope	Layer	Description	USCS	Index						Shear strength			Hydraulic
				w (%)	LL (%)	PL (%)	Fines (%)	ρ (Mg/m^3)	G_s	c' (kPa)	ϕ' (°)	C_c	k_s (m/s)
Yishun	1	Moderate to low plasticity silt	ML	35	45	31	58	1.88	2.688	12	33	0.30	7.2×10^{-7}
	2	Moderate plasticity silt	MH	52	57	40	78	1.76	2.714	13	29	0.64	–
	3	Silty sand	SM	19	42	26	34	2.05	2.667	35	31	–	–
Mandai	1	Silty sand	SM	25	55	31	41	2.02	2.684	12	30	0.19	9.5×10^{-6}
	2	Clayey sand	SC	26	88	34	40	1.94	2.686	14	31	–	–
	3	Moderate to low plasticity silt	MH–ML	36	52	34	60	1.89	2.680	10	29	0.24	–
	4	Silty sand	SM–MM	23	45	27	46	1.95	2.680	0	28	0.20	–
NTU-CSE	1	Moderate to high plasticity silt and clay	CH–MH	30	65	35	94	–	–	139	34	0.5	1.9×10^{-7}
	2	Low plasticity clay	CL to SC–CL	19	36	23	74	2.10	2.725	107	36	–	8.0×10^{-7}
	3	Silty sand	SM-ML	12	29	23	27	2.32	–	167	37	–	2.8×10^{-9}

USCS = Unified Soil Classification System (ASTM, 1997); w = Water content; LL = Liquid Limit, PL = Plastic Limit; ρ = Total density; c' = Effective cohesion; ϕ' = Angle of friction; G_s = Specific gravity; C_c = Compression index; and k_s = saturated permeability.
* Average of 2 to 9 samples, on some occasions there were only one sample.
– Results were not available due to sample damage or technical difficulty.

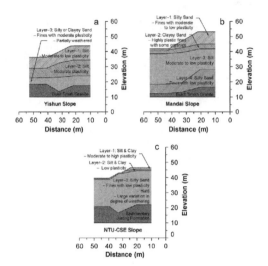

Figure 19. Generalized soil profile of the four slopes (a) Yishun slope, (b) Mandai slope, (c) NTU-CSE slope.

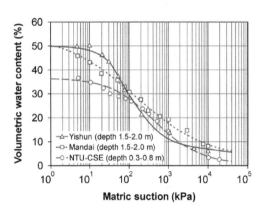

Figure 20. Soil-water characteristic curves of Yishun, Mandai and NTU-CSE slope.

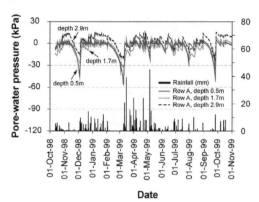

Figure 21. Time series of rainfall and pore-water pressure at Yishun slope.

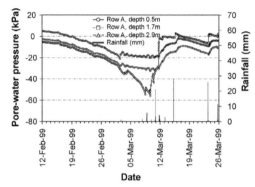

Figure 22. A trace of pore water pressure and rainfall record in Yishun slope showing tensiometer response to a dry period and wet periods.

Pore-water pressure records in row A at all depths of the Yishun slope are plotted for four different times during a dry period (Figure 23) and four different events during a wet period (Figure 24).

These figures show the pore-water pressure distribution across the soil depth and the progressive vertical movement of water (infiltration and evapotranspiration) during these two periods. Figure 23 indicates that the soil profile dries from the surface downwards, owing to evaporation and transpiration during dry periods. Figure 24, in conjunction with the rainfall events shown in Figure 22, shows the successive increase in pore-water pressure along the soil depths in response to cumulative rainfall amounts during the wet period.

Tensiometer readings at all depths and all locations (6 rows × 5 depths) in the Yishun slope for two selected periods (March 8 and March 16, 1999) were used to produce the pore-water pressure contours shown in Figures 25 and 26. These figures illustrate the magnitude and distribution of pore-water pressure within the entire slope profile and identify the movement of water in the subsurface regime.

These two specific events were chosen to observe the magnitude and distribution of pore-water pressures across the entire slope (that is slope response) at the end of a dry period (March 8) and a wet period (March 16) period (Figure 22). Figure 25 shows widespread negative pore-water pressure prevailing across the entire slope during the dry period. The groundwater table from the piezometric readings (near slope crest, midslope, and toe) for this period is also shown in Figure 25. However, the slope experiences more matric suction development at the crest than

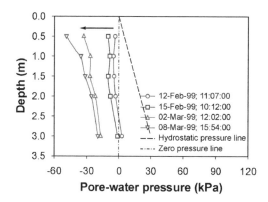

Figure 23. Pore water pressure profile in row A of Yishun slope showing the advancement of a drying event at different depths.

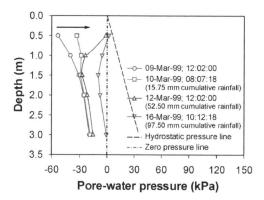

Figure 24. Pore-water pressure profile in row A of Yishun slope showing the advancement of the wetting at different depths during a wet period.

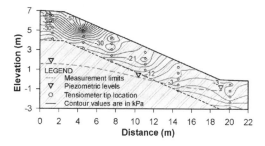

Figure 25. Magnitude and distribution of pore-water pressure along the soil profile in Yishun slope on 8 March 1999 at the end of a dry period.

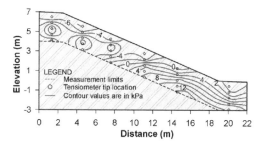

Figure 26. Magnitude and distribution of pore-water pressure along the soil profile in Yishun slope on 16 March 1999, after a wet period with 98 mm of cumulative rainfall.

at the toe. This reflects the variability in hydrologic response within the same slope and could be because, in addition to evaporation, there is preferential lateral movement of water downslope. Figure 26 shows the widespread development of positive pore-water pressure across the entire slope after the wet period, with about 98 mm of cumulative rainfall (from Figure 22). The groundwater table location from the piezometric readings for this period was not conclusive as they did not show any significant rise in the water table (instrument malfunction), and therefore the water table is not shown in Figure 26. The pore-water pressure distribution during this period is fairly uniform along the slope face, indicating that significant rainfall brings uniformity of pore-water pressures in the soil profile. The widespread development of positive pore-water pressures is a consequence of the increasing degree of saturation to near saturation conditions throughout the soil profile at higher rainfall rates.

The downslope movement of water becomes more obvious from the contours plots of the distribution of hydraulic heads shown in Figures 27 and 28.

As water flow is caused by the hydraulic head gradient, a comparison of total hydraulic heads (composed of the pressure head and elevation head, assuming velocity head to be negligible) at the slope crest and toe (both during dry condition, Figure 27, and wet condition, Figure 28) shows the existence of a hydraulic gradient during both dry and wet periods. The hydraulic gradient during the wet period is greater than during the dry period, which suggests there is more downslope flow during a wet condition than during a dry condition (assuming hydraulic isotropy).

Decreases in matric suction were observed during the analyses of pore-water pressure response to rainfall at almost all magnitudes of rainfall events. Therefore, the correlation between the increase in pore-water pressure Δu_w, (The difference in pore-water pressure before and after a rainstorm) and rainfall amount, R, was tested. This was done by splitting the rainfall data into daily totals and taking the algebraic difference in pore-water pressure before and after daily

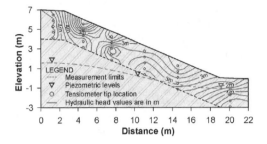

Figure 27. Magnitude and distribution of hydraulic head along the soil profile in Yishun slope on 8 March 1999 at the end of a dry period.

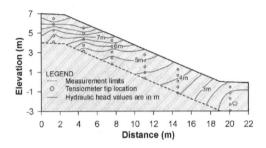

Figure 28. Magnitude and distribution of hydraulic head along the soil profile in Yishun slope on 16 March 1999, after a wet period with 98 mm of cumulative rainfall.

rainfall, in row A at the depth of 50 cm. A regression analyses was performed to fit the data sets of the increase in pore-water pressure, Δu_w, and daily rainfall amount, R. A semi-logarithmic equation of the form, $\Delta u_w = a_2 + b_2 \log R$, where a_2 and b_2 are coefficients, best described the data set. Figure 29 shows the relationships between the increase in pore-water pressure at 50 cm depth of row A, and the daily rainfall amount for the three slopes. The equations in Figure 29 indicate that both coefficients a_2 and b_2 increase with increasing particle sizes (see Table 2). Mandai, being rich in coarse particles, shows higher values of Δu_w than other slopes. This suggests that the increase in pore-water pressure is not only dependent on the rainfall amount, but it is also affected by soil properties and antecedent soil moisture. Although Δu_w values are different for Yishun, Mandai, and the NTU-CSE, they appear to increase with the rainfall amount, and the rate of increase tends to decline at a daily rainfall amount greater than about 10 mm.

3.3.3 Evaluation of slope responses in terms of runoff generation and infiltration

Rainfall and runoff data were collected for 27 natural rainstorms in the NTU-CSE slope. It is difficult

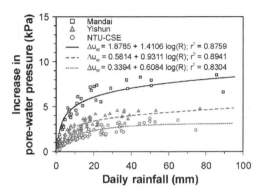

Figure 29. Relationship of increase in pore-water pressure at 50 cm depth to daily rainfall amount for all instrumented slopes.

to assess the infiltration effects on the slope due to natural rainfall events because natural rainfall intensity and duration vary within a rainfall event and from event to event. To understand infiltration effects on the residual soil slope under controlled rainfall conditions 10 simulated rainfall experiments (5 experiments during February of 1998 and another 5 experiments between 17 November 1998 and 5 January 1999) were conducted on NTU-CSE slope. The runoff data were analysed for the 27 natural and 10 simulated rainfalls to determine total runoff volume, total infiltration, peak intensity of each rainfall event. The total runoff amount was calculated from integration of the area under the runoff hydrograph. From the time series of rainfall and pore-water pressures shown in Figures 21 and 22 and similar plots for other slopes, rainfall amounts of less than 0.5 mm per day appear to have no significant effect on the pore-water pressures, even at shallow depths. Therefore, interception by the slope vegetation (grass) was estimated at 0.5 mm from the time series of pore-water pressures. Total infiltration was estimated for all runoff measurements by subtracting total runoff and an interception loss of 0.5 mm from total rainfall.

Figure 30 plots runoff amounts against rainfall amounts for the 27 monitored natural rainstorms. A linear regression fitted to the data points shows an intercept on the x-axis, indicating that a threshold rainfall amount of about 10 mm must be exceeded to produce a significant runoff. Hydrologic responses of slopes are locally and geographically variable, causing difficulty in comparing results. The limited references from Southeast Asia further add to this difficulty. The threshold rainfall observed in this study is comparable to that reported by Tani (1997) for hilly area in Okayama, Japan, but is about half of that reported by Premchitt et al. (1992) for Hong Kong in a sub-humid tropical climate.

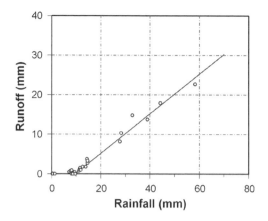

Figure 30. Relationship between storm rainfall (natural rainfall) and runoff amount at NTU-CSE slope.

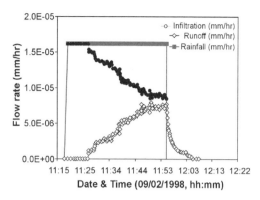

Figure 32. Rainfall and runoff hydrograph from a simulated rainfall event on 9 February 1998 in the NTU-CSE slope.

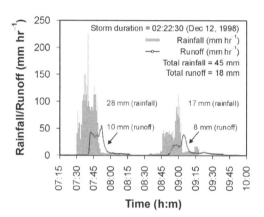

Figure 31. Rainfall and runoff hydrograph of a natural storm event on 12 December 1998 recorded at NTU-CSE slope.

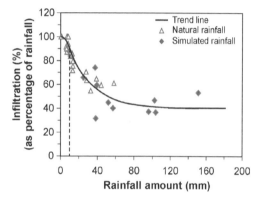

Figure 33. Percent infiltration as a function of rainfall amount.

Figure 31 shows the rainfall and runoff hydrographs of a natural composite storm recorded at the NTU-CSE slope on December 12, 1998, with a total rainfall of 45 mm, and two peak rainfall intensities of 240 and 120 mm/h, resulting 18 mm of total runoff. This composite storm is selected because it reflects the slope response to rainfall during dry and wet soil conditions. The first runoff hydrograph is characterized by a flash flood, which decreases quickly and represents overland flow under dry soil conditions. The first rainfall with 28 mm of rainfall produced 10 mm of runoff, which is about 35% of the rainfall. In contrast, the second rainfall event with similar rainfall characteristics and a total amount of 17 mm, but under wet soil conditions, produced about 8 mm of runoff, a consequence of increasing saturation condition of the slope. This runoff is about 47% of the rainfall.

Rainfall and runoff hydrograph from a simulated rainfall event during February 1998 on the NTU-CSE slope is shown in Figure 32. The infiltration rates shown in Figure 32 are derived by subtracting runoff rates from rainfall rates. The hydrograph (Figure 32) shows that there was no runoff during the early part of the simulated rainfall event. During this period all the rainfall water is lost as infiltration (a fraction may have also been retained by the slope vegetation as interception). It is noted that during the rainfall event of 9 February 1998 the runoff rate did not exceed the infiltration rate and the rise in runoff rate is slow. This is due to the relatively dry soil conditions that prevailed before the rainfall event.

In Figure 33, infiltration amounts (as a percentage of total rainfall) are plotted against rainfall amounts from 27 natural and 10 simulated rainfalls monitored in the NTU-CSE slope. It appears from Figure 16 that rainfall events producing small total amounts of rainfall may contribute fully to infiltration. This again suggests the existence of a threshold rainfall amount. Any rainfall below this amount will not produce any runoff, and the whole rainfall may end up as infiltration.

With reference to Figure 33 (broken line), this threshold appears to be about 10 mm of total rainfall. Beyond the threshold rainfall, the percentage of rainfall contributing to infiltration decreases with an increase in total rainfalls. The infiltration amount could decrease to about 40% of the rainfall (Figure 33) for rainfall events that produce a high total amount of rainfall. This, however, does not mean that the total infiltration amount is less during rainfall events with a higher total rainfall than the total infiltration during rainfall events with a smaller total rainfall. For example, with 40% of the rainfall contributing to infiltration, a 100 mm rainfall (higher total amount) would result in a total infiltration of 40 mm. With 100% of the rainfall contributing to infiltration, a 10 mm rainfall would produce only 10 mm of total infiltration. The data suggest that in residual soil slopes total infiltration could range between 40% and about 100% of the total rainfall depending on the rainfall amount. The relationship (Figure 33) derived from the rainfall records in the residual soil slope has practical significance. If the rainfall amount is known, Figure 33 could indicate the fraction of the rainfall that could become infiltration. This may be useful for seepage analysis that requires this information as flux boundary conditions. More rigorous interpretation on the hydrological responses of these slopes can be found in Rezaur et al. (2003); Rahardjo et al. (2005).

4 CONCLUSIONS

4.1 *PART-A: The stochastic model*

A method to analyze the risk of failure of slopes in partially saturated soils induced by climatic changes has been presented. Spectral and correlation functions for the safety margin of the slope were explicitly derived in terms of the spectral representation of the infiltration record. Some results of the theory of random processes were then used to find expressions for the failure risk in a given period (1 year). The method has been applied to the hydrologic conditions prevailing in the mediterranean coast close to Barcelona and to slopes in a partially saturated silty soil whose mean strength, permeability and suction characteristics were estimated from previous work and past experience. It was found that,

- Cyclic infiltration/evaporation records exhibiting a large time period of occurrence can penetrate deep into the soil. High frequency components only affect shallow depths and dampen fast at depth.

- In relatively gentle slopes shallow failures induced by moisture changes are more likely than deep seated failures.
- However, if the slope angle increases deep failures become more frequent
- For the cases analyzed here, increasing the permeability reduces the risk of failure
- In general, the deeper the water tables the lower the risk of sliding. Shallow failures tend to develop, with more probability, if the water table is deep.

4.2 *PART-A: The case history*

Two deformation mechanisms were identified in the slope: a creep-type displacement, which was detected along the full depth of the soil investigated (around 12 m) and a surface planar slide. The first mechanism is interpreted as a deformation associated with volume changes of the over-consolidated clays as the water pressures change in time as a reaction to rainfall events. Water pressures at the upper two weathered layers are controlled by the atmospheric weather.

A fairly good agreement between computed and measured water pressures is achieved when the three identified layers are characterized by three different permeabilities.

For a given climatic record, the critical situation of a given slope (in the sense of reacting with the maximum development of water pressures) is obtained for a particular combination of layer permeabilities. In other words, given a soil profile and geometry and its associated permeabilities and additional water flow parameters, there exist rainfall records which lead to a maximum "reaction" of the slope in terms of pore pressure development. Permeability and water retention are therefore fundamental properties in slope stability analysis.

As deformation accumulates, peak strengths are attained in surface layers and eventually remoulded and even residual strength conditions develop.

Weathering mechanisms result also in a change in permeability, which is stronger the closer to the surface. Water pressures recorded are consistent with a decrease of permeability with depth. (10^{-6} m/s, 10^{-8} m/s, 10^{-9} m/s).

Permeability transitions lead to peak pressures computed at the interfaces. Strength degradation (accumulated straining) and peak water pressures (positive or negative) result in the development of a sliding surface at the $\alpha-\beta$ interface, where minimum safety factors are consistently found.

Once a planar sliding surface has developed the conceptual model of slope motion is simple: an upper reworked layer slides on top of a critical surface at residual or near-residual conditions.

Periods of activity are dictated by suction changes in the upper few meters which depend critically on the slope geometry, flow boundary conditions, rainfall record, flow parameters (permeability, water retention properties) and its spatial variation.

4.3 PART-B: Slope hydrology and responses to climate changes

The pore-water pressure profile of the slopes presented in part-II shows distinct differences in slope hydrologic response during dry and wet conditions. The relationships between the increase in pore-water pressure and daily rainfall allows for an estimate of the rise in pore-water pressure due to rainfall. However, use of these relationships should be restricted to the field conditions under which measurements have taken place.

The results of natural and simulated rainfall—runoff experiments conducted on the test slope suggest that a large proportion of the rainfall contributes to infiltration in the residual soil slope. A rainfall may contribute from 40% to about 100% of its total rainfall as infiltration (assuming negligible interception losses) depending on the rainfall amount. This information is useful for seepage analyses that require the total infiltration amount as an input parameter. There appears to be a threshold rainfall of about 10 mm to generate runoff. The characteristics of infiltration processes, runoff generation, and pore-water pressure changes identified in this study have relevance for the assessment of rainfall-induced slope instability in residual soil slopes under similar climatic conditions in different geographic regimes.

ACKNOWLEDGEMENT

This work presented in Part-B was funded by a research grant from National Science and Technology Board, Singapore (Grant: NSTB 17/6/16). The authors gratefully acknowledge the field assistance of the Geotechnics Laboratory staff, School of Civil and Environmental Engineering, Nanyang Technological University, Singapore, during the field instrumentation, trouble shooting and data collection for this study.

REFERENCES

ASTM. 1997. Annual book of ASTM standards, Philadelphia, 04.08–04.09.

Bear, J. & Bachmat, Y. 1991. Introduction to Modeling of Transport Phenomena in Porous Media. Kluwer. Dordrecht.

Chatterjea, K. 1994. Dynamics of fluvial and slope processes in the changing geomorphic environment of Singapore. *Earth Surface Processes and Landforms* 19: 585–607.

Corominas, J. 2000. Landslides and climate. *Proc. 8th Int. Conf. Landslides. Cardiff.* pp. 1–31.

Fredlund, D.G., Morgenstern, N.R., & Widger, R.A. 1978. The shear strength of unsaturated soils. *Canadian Geotechnical Journal* 15(3): 313–321.

Gasmo, J.M. 1997. Stability of Unsaturated Residual Soil Slopes as Affected by Rainfall. Master of Engineering Thesis. School of Civil and Structural Engineering, Nanyang Technological University, Singapore.

Meteorological Service Singapore. 1997. Summary of Observations (annual publication), Singapore.

Pitts, J., & Cy, S. 1987. In situ soil suction measurements in relation to slope-stability investigations in Singapore. E.T. Hanrahan, T.L.L. Orr, and T.F. Widdis, eds., *Proc., 9th European Conf. on Soil Mechanics and Foundation Engineering.* Vol. 1, Balkema, Rotterdam, The Netherlands, 79–82.

Poh, K.B., Chuah, H.L. & Tan, S.B. 1985. Residual granite soils of Singapore. *Proceedings of the 8th Southeast Asian Geotechnical Conference, Kuala Lumpur, Malaysia. 11–15 March, 1985.* 1(3):1–9.

Premchitt, J., Lam, T.S.K., Shen, J.M. & Lam, H.F. 1992. Rainstorm runoff on slopes. GEO Rep. 12, Geotechnical Engineering Office, Hong Kong.

Public Works Department. 1976. The geology of the Republic of Singapore, Singapore.

Rahardjo, H., Aung, K.K., Leong, E.C. & Rezaur, R.B. 2004b. Characteristics of residual soils in Singapore as formed by weathering. *Engineering Geology* 73: 157–169.

Rahardjo, H., Lee, T.T., Leong, E.C. & Rezaur, R.B. 2005. Response of a residual soil slope to rainfall. *Canadian Geotechnical Journal* 42: 340–351.

Rahardjo, H., Lee, T.T., Leong, E.C. & Rezaur, R.B. 2004a. A flume for assessing flux boundary characteristics in rainfall-induced slope failure studies. *Geotechnical Testing Journal* 27(2): 145–153.

Rahardjo, H., Leong, E.C. & Rezaur, R.B. 2007. Effect of antecedent rainfall on pore-water pressure distribution characteristics in residual soil slopes under tropical rainfall, Hydrological Process. 21: (in press).

Rezaur, R.B., Rahardjo, H., Leong, E.C. & Lee, T.T. 2003. Hydrologic behavior of residual soil slopes in Singapore. *Journal of Hydrologic Engineering, ASCE* 8(3): 133–144.

Sherlock, M.D., Chappell, N.A. & McDonnell, J.J. 2000. Effects of experimental uncertainty on the calculation of hillslope flow paths, Hydrological Processes. 14(14): 2457–2471.

Tan, S.B., Tan, S.L., Lim, T.L. & Yang, K.S. 1987. Landslide problems and their control in Singapore. *Proc., 9th Southeast Asian Geotechnical Conf., Southeast Asian Geotechnical Society, Bangkok, Thailand.* Vol. 1, 25–36.

Tani, M. 1997. Runoff generation processes estimated from hydrological observations on a steep forested hillslope with a thin soil layer. *Journal of Hydrology* 200(1–4): 84–109.

Toll, D.G., Rahardjo, H. & Leong, E.C. 1999. Landslides in Singapore. 2nd Int. Conf. on Landslides, Slope Stability and the Safety of Infra-structures, Singapore. 269–276.

Tretter, S.A. 1976. *Introduction to Discrete Time Signal Processing.* Wiley. NY.

Vanmarcke, E. 1975. On the Distribution of the First Passage Time for Normal Stationary Random Processes. *Journal of Applied Mechanics Division. A.S.M.E.* March: 215–220.

Wieczorek, G.F. 1996. Landslide triggering mechanisms, In A.K. Turner & R.L. Schuster (eds.) Landslides: investigation and mitigation. TRB Special Report, 247. National Academy Press, Washington, 1996, 76–90.

5 APPENDIX: COEFFICIENTS FOR EQUATIONS (10) AND (11)

$T_n = \omega_{21} C_n + \omega_{22} D_n$

$W_n = -\omega_{22} C_n + \omega_{21} D_n$

$C_n = \dfrac{f_1 (B_n - A_n) + f_2 (B_n + A_n)}{2\alpha_n (f_1^2 + f_2^2)}$

$D_n = \dfrac{f_2 (B_n - A_n) - f_1 (B_n + A_n)}{2\alpha_n (f_1^2 + f_2^2)}$

$\alpha_n = \sqrt{1/2 \lambda_n}$

$f_1 = \cos \alpha_n \left(e^{-\alpha_n} + e^{\alpha_n} \right)$

$f_2 = \sin \alpha_n \left(-e^{-\alpha_n} + e^{\alpha_n} \right)$

$\omega_{21} = \cos \alpha_n Z \left(e^{-\alpha_n Z} - e^{\alpha_n Z} \right)$

$\omega_{22} = \sin \alpha_n Z \left(-e^{-\alpha_n Z} - e^{\alpha_n Z} \right)$

$\omega_0 = \left[S_{r0} - a_s \left[p_{\omega 0} - \gamma_\omega (q_0 / K^{-1}) \left(\dfrac{H+h}{2} \right) \right] \right] \gamma_\omega n + \gamma_s (1-n)$

$\times (\tan \phi' \cos \alpha - \sin \alpha) \cos \alpha (H - h)$

$+ c' + \tan(\phi_b) \left[p_a - h \gamma_\omega \left(\dfrac{q_0}{K} - 1 \right) \right]$

$\Omega_1 = \Omega_{11} \beta_{11} - \Omega_{12} \beta_{12}$

$\Omega_2 = \Omega_{12} \beta_{11} + \Omega_{11} \beta_{12}$

$\Omega_{11} = \dfrac{\omega_1 H(-\omega_{11} + \omega_{12})}{2\alpha_n} - \omega_2 \omega_{21}$

$\Omega_{12} = \dfrac{\omega_1 H(\omega_{11} + \omega_{12})}{2\alpha_n} - \omega_2 \omega_{22}$

$\omega_{11} = \cos \alpha_n (e^{-\alpha_n} + e^{\alpha_n}) - \cos \alpha_n Z \left(e^{-\alpha_n Z} + e^{\alpha_n Z} \right)$

$\omega_{12} = \sin \alpha_n (e^{-\alpha_n} - e^{\alpha_n}) - \sin \alpha_n Z (e^{-\alpha_n Z} - e^{\alpha_n Z})$

$\beta_{11} = \dfrac{f_2 - f_1}{2\alpha_n (f_1^2 + f_2^2)}$

$\beta_{12} = \dfrac{f_2 + f_1}{2\alpha_n (f_1^2 + f_2^2)}$

$\omega_1 = \gamma_\omega^2 n a_s H \left(\dfrac{q_0}{K} - 1 \right) \cos \alpha (\cos \alpha \tan \phi' - \sin \alpha)$

$\omega_2 = \tan \phi_b \gamma_\omega H \left(\dfrac{q_0}{K} - 1 \right)$

Soil nailing and subsurface drainage for slope stabilization

W.K. Pun
Geotechnical Engineering Office, Civil Engineering Department, Government of the Hong Kong Special Administrative Region, China

G. Urciuoli
Department of Geotechnical Engineering, Via Claudio, Napoli, Italy

ABSTRACT: A wide range of slope stabilization and protective measures are available. Soil nailing and subsurface drainage are amongst the very commonly used techniques for slope stabilization. The concept of soil nailing involves creating a stable block of composite material by strengthening the insitu ground with soil nails. The interaction between the ground and the soil nail is complex, and the mobilization of forces in the soil nail is dependent on many factors. The mechanism of subsurface drains in slopes involves a decrease in pore water pressures in the subsoil, and consequently an increase in effective stresses and soil shear strength in the whole drained domain. This paper gives an overview of the mechanism of soil-nailed system and subsurface drainage measures and presents some geotechnological developments related to their applications.

1 INTRODUCTION

Landslides have resulted in the loss of human lives and properties in many parts of the world. To combat landslide risk, a wide range of risk mitigation measures are available. These range from hard engineering measures of slope stabilization and landslide protective works to soft community means of public education. Stabilization works aim at reducing the likelihood of failure of a slope whereas the other measures reduces the risk by minimising the consequences of slope failures.

The range of slope stabilization works may be categorized as follows (Ho 2004):

a. surface protection and drainage,
b. subsurface drainage,
c. slope regrading,
d. retaining structures,
e. structural reinforcement,
f. strengthening of slope-forming material,
g. vegetation and bioengineering,
h. removal of hazards, and
i. special materials and techniques.

Further detail breakdown of the various categories of slope stabilization works and landslide protective measures are depicted in Figure 1.

The development, characteristics, application and performance of various types of slope stabilization works have been reported by many researchers and practitioners (e.g. Hutchinson 1977, Veder 1981, Zaruba & Mencl 1982, Leventhal & Mostyn 1987, Schuster 1992, Hausmann 1992, Fell 1994, Holtz & Schuster 1996, Perry et al, 2003, Ho 2004). This paper focuses on the developments in geotechnology in the two of the common stabilization measures, viz. soil nailing and subsurface drainage.

2 SOIL NAILING TECHNOLOGY

2.1 Introduction

The soil nailing technique was developed in the early 1960s, partly from the techniques for rock bolting and multi-anchorage systems, and partly from reinforced fill technique (Clouterre 1991, FHWA 1998). The New Austrian Tunnelling Method introduced in the early 1960s was the premier prototype to use steel bars and shotcrete to reinforce the ground. With the increasing use of the technique, semi-empirical designs for soil nailing began to evolve in the early 1970s. The first systematic research on soil nailing, involving both model tests and full-scale field tests, was carried out in Germany in the mid-1970s. Subsequent development work was initiated in France and the United States in the early 1990s. The result of this research and development work formed the basis for the formulation of the design and construction approach for the soil nailing technique in the subsequent decades.

The concept of soil nailing involves creating a stable block of composite material by strengthening the

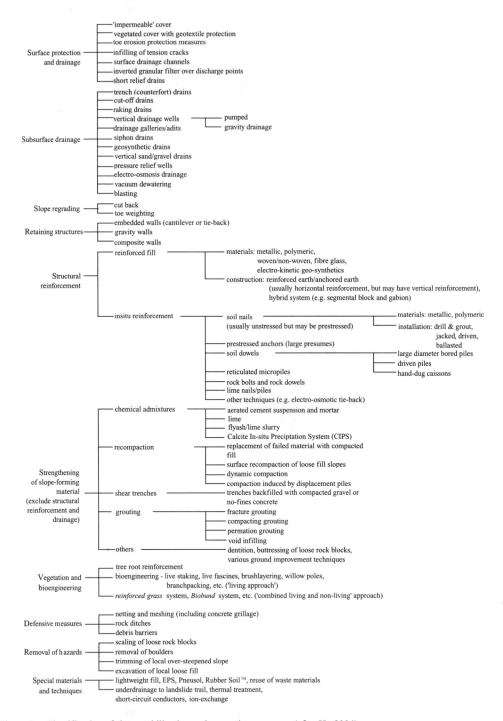

Figure 1. Classification of slope stabilization and protective measures (after Ho 2004).

insitu ground with soil nails. This requires that the soil nails are installed at close spacing, both horizontally and vertically. The following are typical merits of adopting the soil nailing technique in respect of ease of construction, cost and performance (GEO 2008):

- It is suitable for sites with difficult access because the construction plant required for soil nail installation is small and mobile.
- It can more easily cope with site constraints and variations in ground conditions encountered during construction, e.g., by adjusting the location and length of the soil nails to suit the site conditions.
- During construction, it causes less environmental impact than cutting back and retaining wall construction as no major earthworks and tree felling are needed.
- There could be time and cost savings compared to conventional techniques of cutting back and retaining wall construction which usually involve substantial earthworks and temporary works.
- It is less sensitive to undetected adverse geological features, and thus more robust and reliable than unsupported cuts. In addition, it renders higher system redundancy than unsupported cuts or anchored slopes due to the presence of a large number of soil nails.
- Its failure mode is likely to be ductile, thus providing warning signs before failure.

Like every other stabilization technique, soil nailing has its limitations:

- The presence of utilities, underground structures, or other buried obstructions poses restrictions to the length and layout of soil nails.
- The zone occupied by soil nails is sterilised and this site poses constraint to future development.
- Permission has to be obtained from the owners of the adjacent land for the installation of soil nails beyond the lot boundary. This places restrictions on the layout of soil nails.
- The presence of high groundwater levels may lead to construction difficulties in hole drilling and grouting, and instability problem of slope surface in the case of soil-nailed excavations.
- The effectiveness of soil nails may be compromised at sites with past large landslides involving deep-seated failure due to disturbance of the ground.
- The presence of permeable ground, such as ground with many cobbles, boulders, highly fractured rocks, open joints, or voids, presents construction difficulties due to potential grout leakage problems.
- The presence of ground with a high content of fines may lead to problems of creeping to soil nails.

- Long soil nails are difficult to install, and thus it renders the use of the soil nailing technique more difficult in dealing with deep-seated landslides and sizeable slopes.
- As soil nails are not prestressed, mobilisation of soil-nail forces will be accompanied by ground deformation. The effects on nearby structures, or services may have to be considered, particularly in the case of soil-nailed excavations.
- Soil nails are not effective in stabilising localised steep slope profiles, back scarps, overhangs or in areas of high erosion potential. Suitable measures, e.g., local trimming, should be considered prior to soil nail installation.

2.2 Mechanisms of soil-nailed system

2.2.1 Load transfer mechanism

The soil nailing technique improves the stability of slopes, retaining walls and excavations principally through the mobilisation of tension in the soil nails. The tensile forces are developed in the soil nails primarily through the frictional interaction between the soil nails and the ground as well as the reactions provided by soil-nail heads and the facing (Fig. 2). The tensile forces in the soil nails reinforce the ground by directly supporting some of the applied shear loadings and by increasing the normal stresses in the soil on the potential failure surface, thereby allowing higher shearing resistance to be mobilised. Soil-nail heads and the facing also provide a confinement effect by limiting the ground deformation close to normal to the slope surface. As a result, the mean effective stress and the shearing resistance of the soil behind the soil-nail heads will increase (Fig. 3). Soil-nail heads and the facing also help preventing local failures near the surface of a slope and promote an integral action of

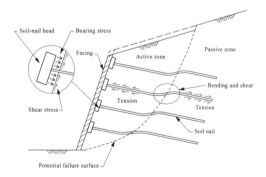

Figure 2. Load transfer mechanism of soil nailed structure.

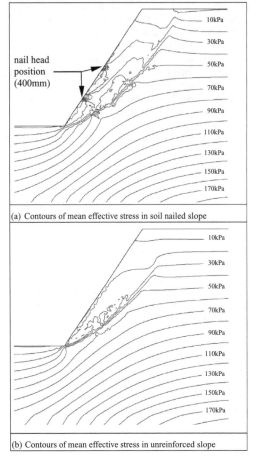

Figure 3. Contour of mean effective stress in (a) soil-nailed slope, and (b) unreinforced slope.

the reinforced ground mass through redistribution of forces among soil nails. The resistance against pullout failure of the soil nails is provided by the part of soil nails that is embedded into the passive zone.

When there is a tendency of ground movement in the active zone, the soil nail will experience both axial and lateral strains through two fundamental mechanisms of nail-ground interaction. They are: (i) the nail-ground friction that leads to the development of axial strains in the soil nails, and (ii) the soil bearing on the soil nails and the nail-ground friction on the sides of soil nails that lead to the development of lateral stains in the soil nails. In these two mechanisms, the interactions between the ground and the soil nails are complex and the forces developed in the soil nails are influenced by many factors such as the bearing capacity of the ground to resist reaction force from the soil nail, the relative stiffness of the soil nail and ground, the nail inclination, and the tensile strength, shear strength and bending capacity of the soil nail. Generally speaking, the axial strain will mobilize tensile or compressive forces, and the lateral strain will mobilize shear force and bending moment in the soil nail. Due to relatively slender dimensions of soil nails, the reinforcing actions from shear and bending are limited by the small flexural strength, and they are usually negligible (FHWA 1998). The effect of inclination and bending stiffness of soil nail are discussed further in the subsequent sections of this paper.

Compressive and shear strains are developed in the soil behind a soil-nail head in response to the ground deformation in the active zone (Fig. 2). If the resultant strain is close to the direction perpendicular to the base of soil-nail head, the head-ground interaction will be predominantly in the form of a bearing mechanism. However, if the resultant strain is in a direction that deviates significantly from the normal to the base of the soil-nail head, the head-ground interaction will be a combination of bearing and sliding mechanisms. These interaction, particularly the bearing mechanism, gives rise to tensile loads at the heads of soil nails. The tensile loads at the soil-nail heads are taken up by the soil-nail reinforcement. The interaction increases as the size of the soil-nail heads or the coverage of facing increases, resulting in larger tensile loads. Further discussion on the effect of soil-nail heads is given in Section 3.5 below.

The mobilization of pullout resistance along a soil nail in the passive zone depends on many factors. Theoretically, the bond strength between the soil nail and the ground depends on the contact stress and the interface coefficient of friction. Where a soil nail is installed by the drill-and-grout method, the process of drilling reduces significantly the radial stress at the circumference of the drillhole. The drillhole remains stable by soil arching. Subsequent grouting will restore a certain level of the radial stress in the soil around the drillhole. The contact pressure at the drillhole face is generally small compared to the overburden pressure except where pressure grouting is adopted. This seems to imply small bond strength at the ground/grout interface. In reality, as the drillhole face may be irregular and rough, the mechanical interlocking between the cement grout and the ground also contributes a significant portion of the bond strength. Upon pulling of the soil nail, shearing may occur within the ground mass in a finite zone surrounding the soil nail. If the soil is dilative, the effect of restrained soil dilatancy will come into play. The effect of this can be significant and can lead to high soil-nail friction (Pun & Shiu 2007).

Soil nails are considered to tie the active zone to the passive zone. It should be noted that the two-zone concept is only an idealisation for design purpose. In reality there is a complex shearing zone subject to shear distortion, unless the failure is dictated by joint settings where the failure surface is distinct. The effect of the shear zone on the mobilization of forces in soil nails is discussed in Section 3.4 below.

2.2.2 *Effect of nail inclination*

Unlike the reinforcements in reinforced fill structures, which are placed in horizontal direction, soil nails can be installed in the ground at various inclinations. In cramped sites, soil nails are sometimes installed at large inclinations. Different nail inclinations may produce different effects on the behaviour of soil-nailed structures.

In this paper, nail inclination, α, is the angle of a soil nail made with the horizontal; and nail orientation, θ, is the angle between a soil nail and the normal to the shearing surface. The typical relationship between α and θ is presented in Figure 4.

Jewell (1980) investigated the fundamental behaviour of reinforced soil by carrying out a series of direct shear box tests on sand samples reinforced with bars and grid reinforcements. One of the significant findings of his work was that the shear strength of the reinforced soil is dependent on the orientations of the reinforcements. Jewell's investigation shows that reinforcement significantly modifies the state of stress and strain in soil, and that by varying the orientation of the reinforcement, the reinforcement can either increase or decrease the shear strength of the soil. Figure 5 compares the pattern of strain in soil between the unreinforced and reinforced tests. The presence of the reinforcement causes a significant reorientation of the principal axes of strain increment of the soil. The soil strains close to the reinforcement are small because the reinforcement inhibits the formation of the failure plane. When the reinforcement is orientated in the same direction of the tensile strain increment of the soil, tensile forces are induced in the reinforcement through the friction between the soil and the reinforcement. Likewise, compressive forces are induced in the reinforcement if the reinforcement is placed close to the compressive strain increment of the soil.

Figure 6 shows the orientations of the reinforcement in which compressive or tensile strain increments

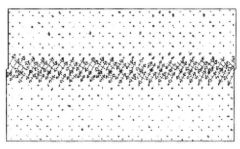

(a) Principal Strains in Unreinforced Sand after Peak

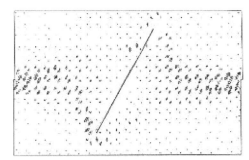

(b) Principal Strains in Sand Reinforced by a Grid at an Orientation $\theta = +30°$ (legend: double arrow represents principal tensile strain)

Figure 5. Incremental strains at peak shearing resistance in unreinforced and reinforced sand (after Jewell 1980).

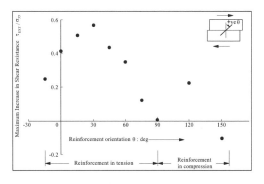

Figure 6. Increase in shear resistance for reinforcement placed at different orientation (after Jewell & Wroth 1987).

are experienced. The shear strength of the soil starts to increase when the reinforcement is placed in the direction of tensile strain increment, and it reaches a maximum when the orientation of the reinforcement is close to the direction of the principal tensile strain increment. When the reinforcing elements are oriented in a direction of a compressive strain increment, there is a decrease in shear strength of the reinforced soil.

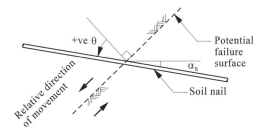

Figure 4. Relationship between nail inclination and orientation.

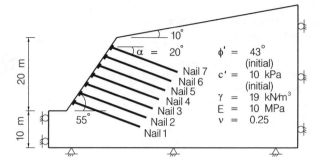

Figure 7. Geometry and material parameters of model slope.

This shows that in order to optimise strength improvement of soil, the reinforcement should be placed in the directions of principal tensile strain in the soil. When the reinforcement deviates from its optimum orientation, strength improvements decrease.

Results of laboratory investigation of the behaviour of soil reinforced with bars were also reported by Marchal (1986), Hayashi et al (1988), and Palmeira & Milligan (1989). They showed that the orientation of reinforcements played a significant role in the improvement of the shear strength of the reinforced soil mass. A negative orientation can lead to the development of compressive force in the reinforcement and consequently loss of shear strength of the soil. There exists an optimum reinforcement orientation in terms of strengthening of the soil. These findings are consistent with that of Jewell (1980).

The effects of nail inclination on the safety margin of a slope was studied by Shiu & Chang (2005) in Hong Kong by means of numerical analysis using the two-dimensional finite difference code, Fast Lagrangian Analysis of Continua (FLAC) (Itasca 1996). A simulated slope of 20 m in height, standing at an angle of 55°, and with an up-slope of 10° in gradient was adopted for the analysis. Figure 7 shows the geometry of the slope and the material parameters used in the numerical analysis. Each soil nail was 20 m long with a 40 mm diameter steel bar in a 100 mm grouted hole. A Mohr Coulomb model was assumed for the soil. A cable element was used to represent the soil nail as the bending stiffness of the soil nail was not considered. Developments of the tensile forces in the soil nails were governed either by the tensile strength of the nail or the peak shear strength at the soil-grout interface.

Slope stability analysis was first carried out on the unreinforced slope. From the results of the analysis, the unreinforced slope has a minimum factor of safety (FoS) close to 1.0 for the initial soil strength parameters of $c' = 10$ kPa, and $\phi' = 43°$. In slope engineering, the FoS is conventionally defined as the ratio of the actual soil shear strength to the minimum shear strength required for equilibrium. As pointed out by Duncan (1996), FoS can also be defined as "the factor by which the shear strength of the soil would have to be divided to bring the slope into a state of barely equilibrium". FoS can therefore be determined simply by reducing the soil shear strength until failure occurs. This strength reduction approach is often used to compute FoS using finite element or finite difference programs (Dawson et al 1999, Krahn 2003).

Figure 8 shows that for soil nails with a small inclination of 20°, tensile forces are developed in all the soil nails. On the other hand, when the soil nails are inclined steeply at an inclination of 55°, compressive forces are developed in the top four rows of soil nails whereas tensile forces are mobilized only in the bottom three rows of soil nails. Tensile forces in the soil nails can improve the slope stability whereas compressive forces can have opposite effect. Increases in FoS (ΔFoS) due to the soil nails were calculated for different nail inclinations. Figure 9 shows the relationship between the calculated ΔFoS and nail inclinations (α) for the model slope. The ΔFoS is close to 1 with little variations for the range of α between 0° and 20°. The ΔFoS decreases quickly as α increases beyond 20°, reflecting that the reinforcing effects of the soil nails reduce rapidly with increasing nail inclinations. At $\alpha = 55°$, the value of ΔFoS is almost zero.

These studies show that the nail inclination can significantly affect the reinforcing action of the soil nails. Increase in nail inclination would decrease the efficiency of the reinforcing action of the soil nails. For steeply inclined soil nails, axial compressive forces may be mobilized in the soil nails. The compressive forces would reduce the stability of the soil-nailed structure.

2.2.3 *Effect of bending stiffness of soil nail*

Steel reinforcements can sustain shear forces and bending moments, and thus this ability of steel soil

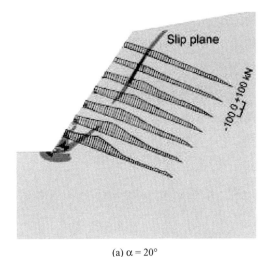

(a) α = 20°

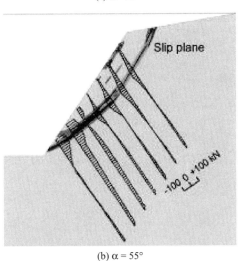

(b) α = 55°

Figure 8. Axial force distribution in soil nails in model slope.

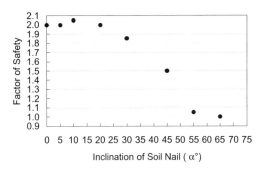

Figure 9. Variation of increase in factor of safety against inclination of soil nails.

nails may enhance the shear strength of soil. The development of shear force in soil nails involves a mechanism which is dependent on the relative stiffness of the soil nail and the ground, the soil bearing strength, the orientation and shear deformation of reinforcement, and the thickness of shear zone.

The effect of the bending stiffness of a soil nail on nail forces and displacements has been investigated by many researchers, e.g. Schlosser (1982), Marchal (1986), Gigan & Delmas (1987), Pluemelle et al (1990), Pedley (1990), Jewell & Pedley (1990, 1992), Bridle & Davies (1997), Davies & Le Masurier (1997), Smith & Su (1997) and Tan et al (2000). The most notable and comprehensive investigation was the laboratory and theoretical study reported by Pedley (1990) and Jewell & Pedley (1990, 1992).

In the study of Pedley (1990), a series of direct shear tests were carried out in a large-scale direct shear apparatus (1 m × 1 m × 1 m). Three different types of circular elements were tested. They were solid steel bars (16 to 25.4 mm in diameter), metal tubes (15.88 to 25.4 mm in external diameter and 13.19 to 22.36 mm in internal diameter) and grouted bar (50.8 mm in diameter with steel bar diameters 6.71 to 16 mm). Figure 10 shows the distributions of: (a) the reinforcement bending moment (M) normalised by the plastic moment capacity (M_p); (b) the reinforcement shear force (P_s) normalised by the full plastic axial capacity (T_p); and (c) the lateral stress on the reinforcement (σ_l) normalised by the limiting soil bearing stress (σ'_b). These distributions were the stress conditions of the reinforcement at a shear displacement of soil of 60 mm. The test results show that even when the measured bending moments were close to the fully plastic moment ($M/M_p = 1$) in all the tests, the maximum shear force (P_s) in the reinforcement was less than 6% of the plastic axial capacity (P_p).

Pedley (1990) confirmed the laboratory test results by theoretical analysis. He derived elastic and plastic models for determining the maximum shear force mobilised in the reinforcement bar. These models were also reported in Jewell & Pedley (1990, 1992). The elastic analysis simply defines the stress condition before the reinforcement reaches plasticity; it does not represent the failure condition. As such, only the plastic analysis is discussed here. The plastic analysis will always give a larger reinforcement shear force than the elastic analysis.

The limiting plastic envelope for a bar of rectangular cross-section is given by (Calladine 2000):

$$\frac{M}{M_p} + \left(\frac{T}{T_P}\right)^2 = 1 \qquad (1)$$

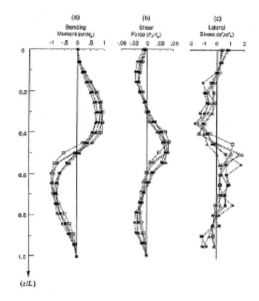
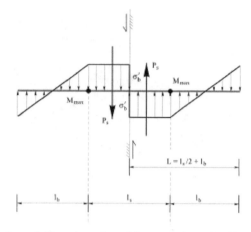

Legend: Ps = shear force; Mmax = maximum bending moment in soil nail; I_b = minimum required length beyond point of maximum bending moment; Is = distance between points of maximum moment on either side of shear plane; σ_b' = limiting bearing stress between soil and reinforcement

Figure 11. Plastic analysis of soil nail interaction (after Jewell & Pedley 1992).

Legend:

	Bar diameter (mm)	Nail orientation°
○	15.88	0°
□	15.88	15°
△	15.88	25°
●	25.4	0°
■	25.4	15°
▲	25.4	25°

Figure 10. Profiles of bending moment, shear force and lateral stress in reinforcement (after Pedley 1990).

Equation 1 is slightly conservative for circular bar. Since no simple relationship can be derived for circular bar, this equation has been adopted by Jewell & Pedley (1990).

The relation between maximum shear force P_s and the maximum moment M_{max} depends on the magnitude and distribution of the lateral loading on the reinforcement bar. For plastic analysis for a soil nail under the lateral loading shown in Figure 11, the equations for this distribution of lateral loading are:

$$(P_s) = \frac{4M_{max}}{l_s} \qquad (2)$$

$$\left(\frac{l_s}{D}\right) = \sqrt{\frac{4\sigma_y}{3\sigma_b'}} \qquad (3)$$

where l_s is the distance between the points of maximum moment on either side of the potential shear surface, (Fig. 11), D is the diameter of reinforcement bar, σ_y is the yield stress of the bar and σ_b' is the limiting bearing stress between soil and reinforcement.

The limiting bearing pressure (σ_b') between the soil and the reinforcement required to achieve the plastic equilibrium is:

$$(\sigma_b')_{max} = \frac{8M_{max}}{l_s^2 D} \qquad (4)$$

The theoretical plastic limiting maximum shear force P_s that can be generated in an ungrouted round bar that also supports axial force, P, is:

$$\frac{P_s}{T_p} = \frac{8}{3\pi(l_s/D)}\left\{1 - \left[\frac{T}{T_p}\right]^2\right\} \qquad (5)$$

Details of explanations and derivation of Equations (1) to (5) can be found in Pedley (1990).

Jewell & Pedley (1992) computed and presented envelopes of limiting combinations of shear force P_s and axial force T for a grouted reinforcement bar of 25 mm diameter with typical soil parameters and showed that the magnitude of the limiting shear force in the reinforcement was only a small proportion of the axial force capacity. This was the case even when the reinforcement was oriented so as to mobilize the maximum shear force.

Pedley (1990) also back-analysed an instrumented 6 m high soil-nailed wall, which was loaded to failure. He found that the highest contribution of reinforcement shear force to soil strength improvement was less than 3% to that due to reinforcement axial force. This is in agreement with the theoretical study result that only very small amount of shear force can be mobilised in a soil nail.

The effect of bending stiffness of soil nails on nail forces and displacements and the safety margin of soil-nailed slopes has also been studied by Shiu & Chang (2005) in Hong Kong by means of numerical simulations using the finite element code PLAXIS. The slope model is the same as that shown in Figure 7. The FoS of the soil-nailed slope, the tensile forces, shear stresses and bending moments developed in the soil nails at different inclinations were computed.

The maximum axial force developed in a soil nail is T_{max}. Figure 12 shows the sum of the maximum tensile forces mobilised in all the soil nails (ΣT_{max}) at limit equilibrium condition of the slope model. The maximum shear force in a soil nail at the location where the shear plane intersects the soil nail is Ps_{max}. The sum of the maximum shear forces (ΣPs_{max}) mobilized in the soil nails at limit equilibrium condition of the model are also plotted in Figure 12. The value of ΣPs_{max} rises steadily with increasing nail inclination. The rise is small, from 31 kN/m at $\alpha = 10°$, to 76 kN/m at $\alpha = 55°$. In contrast, the value of ΣT_{max} decreases rapidly with increasing nail inclination. For small nail inclinations, ΣT_{max} is much larger than ΣPs_{max}. Comparing Figures 9 and 12, it can be seen that both ΔFoS and ΣT_{max} decrease with increasing nail inclinations. This similarity illustrates that ΔFoS is strongly influenced by the nail axial force. The ΔFoS is not sensitive to the mobilized shear resistances in the soil nails. The modeling results show that small shear forces are mobilized in soil nails and they have little effect on the factor of safety of the slope, except perhaps at very steep nail inclination where dowel action starts to play a role. However, soil nails are not effective in providing dowel action. For that purpose, other types of structural element should be considered, e.g. large diameter piles.

The above studies show that the contribution of shear force and bending stress of soil nails on enhancing the shear strength of the soil mass is very small. Large soil displacements are required to mobilise shear and bending forces in the soil nails. When failure conditions are approached, the contribution of shear and bending action may be more significant but is still small. For these reasons, the soil nail design practice in most countries, such as the USA (FHWA 1998), the UK (Department of Transport 1994), Japan (JHPC 1998) and Germany (Gässler 1997), ignores any beneficial effects from the mobilisation of shear force or bending stress in the soil nails. An exception to this is the French design approach (Clouterre 1991) in which the contributions of shear and bending of the nails are considered. Clouterre (1991) emphases that shear forces are mobilised in soil nails only when the structures are near failure.

Although the bending stiffness of soil nail has little contribution to the shear resistance of a soil-nailed system, the beneficial effect of shear ductility of soil nail reinforcement should not be ignored. For example, steel has large shear ductility. As a result of the mobilization of shear and bending ductility at large deformations, a soil-nailed system comprising steel reinforcement tends to exhibit ductile rather than brittle failure.

2.2.4 Effect of thickness of shear zone

The reported laboratory investigation and numerical studies generally considered sliding along a shear plane. To investigate the influence of the thickness of shear zone on the mobilisaion of shear force and bending moment in soil nails, a study involving numerical simulations of large shear box tests was conducted by the Geotechnical Engineering Office (GEO) in Hong Kong. The two-dimensional finite element code PLAXIS was used. Three cases were considered, ranging from a well-defined shear plane to a wide shear zone (Fig. 13).

The model shear box was 3 m deep and 6 m long. A reinforcement was placed in the middle of the box across the vertical slip surface. The box was assumed filled with homogeneous sand. The model and the assumed parameters are shown in Figure 14. A Mohr-Coulomb model was assumed for the soil.

In the numerical simulation, the shear box was initially restrained to move. An overburden pressure of 80 kPa was applied on the top of the box to model a 5 m high earth pressure. In the case of sliding along a shear plane, an imposed downward uniform displacement δ of 5 mm, 10 mm, 25 mm and 50 mm was applied in sequence at the right-half part of the shear box.

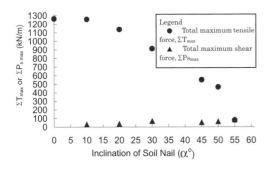

Figure 12. Variation of total maximum tensile force (T_{max}) and total maximum shear force ($P_{s\ max}$) with nail inclination (α).

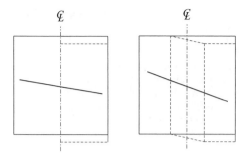

(a) Direct Shear Box Test   (b) Zone Shear Box Test

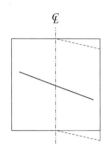

(c) Simple Shear Box Test With Half Part Fixed

Figure 13. Cases considered in numeral study of effect of shear zone.

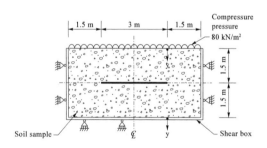

Steel bar size: 25 mm diameter steel bar x 3 m (length)

Steel bar parameters
σ_y = 460 N/mm²
E_{steel} = 205 kN/mm²

Soil parameters:
c' = 0 kN/m²
ϕ' = 40°
γ = 16 kN/m³
E_{soil} = 60 MN/m²
ν = 0.4

Figure 14. Model for numeral study of effect of shear zone.

In the case of a narrow shearing zone, a linearly varying downward displacement was applied across the shear zone of width z during the loading phases. An imposed uniform displacement of 5 mm, 10 mm,

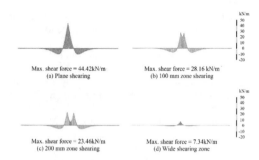

Figure 15. Shear force along steel bar in shear box model.

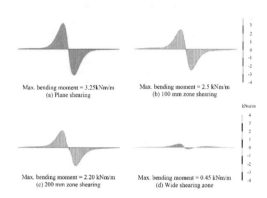

Figure 16. Bending moment along steel bar in shear box model.

25 mm and 50 mm was again applied in sequence in the downward direction beyond the shear zone along the top and bottom of the right-half part of the box. Two values of z, 100 mm and 200 mm respectively, were considered. In the case of wide shearing zone, the shear zone was extended to cover the entire right-hand part of the shear box.

The results of the numerical simulations are shown in Figures 15 and 16. The mobilisation of shear force and bending moment in soil nails is affected by the thickness of the failure shear zone. The narrower the shear zone, the higher is the shear force and bending moment in the soil nails.

2.2.5 Effect of soil nail head and facing

Our understanding of the magnitude and distribution of loadings developed at soil-nail head is not as good as our knowledge of the development of tensile forces in soil nails (FHWA 1998). This is because of the lack of good quality field monitoring data. The available data from instrumented soil nails are generally difficult to interpret in the vicinity of soil-nail heads, where bending effects of soil nail tend to be more significant arising from the weight of the soil-nail head (Thompson & Miller 1990). There has been little field monitoring data obtained using load cells at soil-nail heads probably because of the difficulties in placing load cells between soil-nail heads and soil (Stocker & Reidinger 1990).

Despite the lack of good quality field monitoring data, a number of studies including model tests (e.g. Muramatsu et al 1992, Tei et al 1998), full-scale field tests (e.g. Gässler & Gudehus 1981, Plumelle & Schlosser 1990, Gutierrez & Tatsuoka 1988, Muramatsu et al 1992) and numerical simulations (e.g. Ehrlich et al 1996, Babu et al 2002) have been carried out. The results of these studies provide useful insight into the role and behaviour of soil-nail heads. Many of the studies are related to soil-nailed retaining walls where soil-nail heads are integrated into a concrete facing.

Gutierrez & Tatsuoka (1988) reported loading tests performed on three model sand slopes: (i) unreinforced slope, (ii) slope reinforced with metal strips but without a facing, and (iii) slope reinforced with metal strips and with a facing (Fig. 17). The slopes were loaded at the crest by a footing with a smooth base. Result of the tests is shown in Figure 18. It indicates that the reinforced slope with facing can sustain a higher load than the reinforced slope with no facing, and a much higher load than the unreinforced

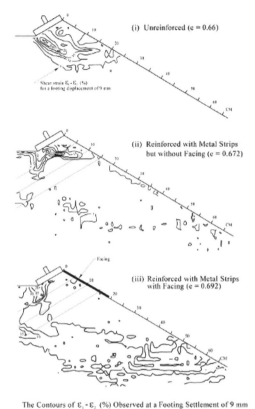

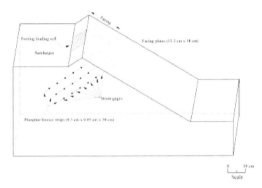

Figure 17. Model slope reinforced with metal strips and with a facing tested by Gutierrez & Tatsuoka (1988).

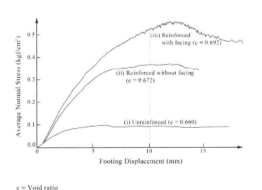

e = Void ratio

Figure 18. Results of tests of Gutierrez & Tatsuoka (1988).

The Contours of $\varepsilon_1 - \varepsilon_3$ (%) Observed at a Footing Settlement of 9 mm
e = Void ratio

Figure 19. Shear stain contour in model tests by Gutierrez & Tatsuoka (1988).

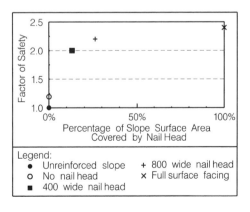

Figure 20. Relationship between factor of safety and soil-nail head size.

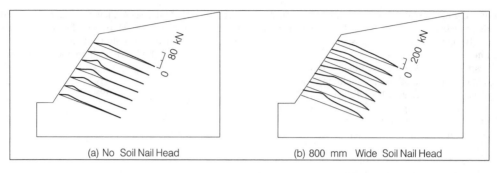

Figure 21. Variation of axial nail forces for (a) 800 mm soil head and (b) no nail head.

slope. Figure 19 shows the failure planes of the slopes when loaded at the crest. Deep and shallow failure planes were observed in the unreinforced slope. For the reinforced slope with no facing, failure took place close to the slope face. This was because the reinforcement alone was not effective at retaining the active zone. For the reinforced slope with facing, the failure was observed at a greater depth. The maximum tensile force generated in the reinforcement was larger than that in the reinforced slope with no facing. Furthermore, substantial tensile force was induced in the reinforcement at the connection to the facing. The full-scale field tests by Plumelle & Schlosser (1990) and Muramatsu et al (1992) also showed that the presence of a facing could enhance the stability of a reinforced slope and helped prevent shallow failures.

Soil-nail heads used in slope stabilization works in Hong Kong are usually in the form of isolated reinforced concrete pads. To investigate the effect of soil-nail heads on stability of nailed slopes, a series of numerical simulations using the two dimensional finite element code FLAC was conducted (Shiu & Chang 2004). Figure 7 shows the slope model. Strength reduction technique (Dawson et al 1999) was employed to compute the factor of safety (FoS) of the model slope. In the simulations, soil-nail heads of different sizes were considered. The slope without any soil nails (i.e. unreinforced) had a minimum FoS close to 1. Based on the FLAC analysis, Figure 20 shows the relationship between the calculated FoS of the model slope and soil-nail head sizes. The FoS increases from 1 for the unreinforced slope to 1.2 for the soil-nailed slope with no soil-nail heads. Substantial increases in the FoS are observed when soil-nail heads of sizes ranging from 400 mm wide to 800 mm wide are provided. The trend of increase in FoS levels off for soil-nail head sizes larger than 800 mm wide. It shows that soil-nail heads can significant enhance the stability of a soil-nailed slope.

Figure 21 compares the axial tensile forces developed in soil nails without soil-nail heads with those in nails with heads of 800 mm wide. For the soil nails without soil-nail heads, no tensile force is developed at the front end of the soil-nail whereas for the soil nails with soil-nail heads, large tensile forces are mobilised in the soil nails at the connections to the soil-nail heads. The maximum tensile forces mobilized along the soil nails are much larger in the latter case.

A series of centrifuge tests has also been conducted in the Geotechnical Centrifuge Facility of the Hong Kong University of Science and Technology to investigate the reinforcing effect of soil nails and soil-nail heads (Ng et al 2007). Figure 22 shows an instrumented model used in one of the nailed slope centrifuge tests. The test results support the results of the numerical simulations that soil-nail heads can substantially improve the stability of a soil-nailed slope.

Results of the above model tests, field measurements and numerical simulations highlight the importance of soil-nail heads in the soil nailing applications. They show that soil-nail heads, whether in the form of individual concrete pads or as part of concrete facing, greatly enhance the stability of a soil-nailed slope.

2.3 *Modelling and design*

2.3.1 *Design approach and standard*
A soil-nailed structure is required to fulfil fundamental requirements of stability, serviceability and durability during construction and throughout its design life. Other issues such as cost and environmental impact are also important design considerations.

The design for stability generally entails the setting up of ground and design model, consideration of potential failure mechanisms, stability analyses, determination of soil-nail design capacity, soil-nail head and facing design, and detailing.

The failure mechanisms of nailed structures can broadly be classified as external failure and internal failure. External failure refers to the development of

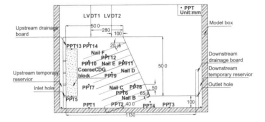

(a) Set-up of a nailed-slope model in centrifuge test.

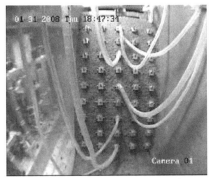

(b) Front view of the slope model in centrifuge test

Figure 22. Centrifuge test to study soil-nailed slope.

potential failure surfaces essentially outside the soil-nailed ground mass. The failure can be in the form of sliding, rotation, bearing, or other forms of loss of overall stability (see Fig. 23a). Internal failure refers to failures within the soil-nailed ground mass. Internal failures can occur in the active zone, passive zone, or in both of the two zones of a soil-nailed system.

In the active zone, internal failure modes could be:

- failure of the ground mass, i.e., the ground disintegrates and 'flows' around the soil nails and soil-nail heads
- bearing failure underneath soil-nail heads
- structural failure of the soil nail under combined actions of tension, shear and bending
- structural failure of the soil-nail head or facing, i.e., bending or punching shear failure, or failure at head-reinforcement or facing-reinforcement connection
- surface failure between soil-nail heads, i.e., washout, erosion, or local sliding failure

In the passive zone, the failure mode is mainly pull-out failure of soil nail along soil-grout interface or reinforcement-grout interface.

The various internal failure modes are illustrated in Figure 23b.

The approach of limit state design incorporating partial factors is adopted in many soil nail design codes. Driven by the Eurocode, all the European countries under the European Union use the partial factors approach in soil nail design. In the USA, it appears that it is at a transition stage of changing over from global safety factor approach to partial safety factors approach. The American design code permits the use of either load and resistant factor design approach (which is similar to partial factors approach) or service load design approach. In Hong Kong, the soil nail design approach is essentially a combination of global safety factor approach (permissible stress design) and partial safety factor approach. Table 1 summarises the design approaches recommended by the different design codes.

Calculation methods involving trial wedges (single-wedge or double-wedge) and limit equilibrium methods (LEM) of slices on circular, spiral, or other non-circular slip surfaces are commonly used. While these methods are good enough for design purpose, none of them can account for the actual behaviour of a soil-nailed structure, which is a strain compatibility problem. It is possible to define a wide variety of nail length patterns that satisfy stability requirements but that may not satisfy serviceability requirements.

It is essential to have a good understanding of the principles behind the calculation methods so that the appropriate method is used and the results are interpreted correctly. For instance, the factor of safety of a soil-nailed slope computed using the simplified Janbu method is insensitive to the location of the applied soil nail force. This is an inherent limitation of the method, and it may give rise to an over- or under-estimation of the true safety margin. In light of this, GEO (2008) recommends that only stability analysis methods that satisfies both moment and force equilibrium should be used in soil nail design.

In Hong Kong, more than 3,000 slopes and retaining walls have been stabilized using soil nails. The vast amount of soil nail designs had allowed the development of a prescriptive design approach for the design of soil-nailed soil cut slopes and retaining walls. Prescriptive measures are pre-determined, experience-based and suitably conservative modules of works prescribed to a slope or retaining wall to improve its stability or reduce the risk of failure, without detailed ground investigation and design analyses. Using prescriptive measures has the technical benefits of enhancing safety and reducing the risk of failure, by incorporating simple, standardised and suitably conservative items of works to deal with uncertainties in design that are difficult to quantify. There would also be savings on time and human resources, by eliminating detailed ground investigation and design analyses.

The prescriptive soil nail design guidelines in Hong Kong comprise standard soil nail layouts and a set of qualifying criteria for the application of the prescriptive measures to ensure that the prescriptive design

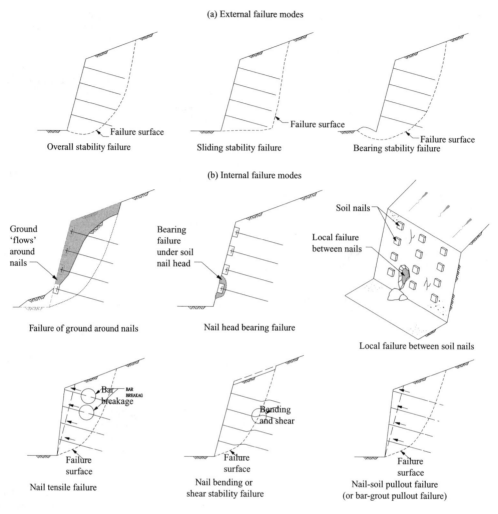

Figure 23. Principal modes of failure of soil-nailed system.

approach was applied safely within the bound of past experience. The guidelines were developed based on the findings of review of several hundreds of soil nail designs (Pang & Wong 1997, Pun et al 2000, Lui & Shiu 2004). Soil nail designs using the prescriptive approach have been successfully applied to many soil cut slopes and retaining walls since the promulgation of the design guidelines in Wong et al (1999) and Lui & Shiu (2005).

Soil nailing is also a feasible option for the stabilization of loose fill slopes. However, as loose granular fill material exhibits contractive behaviour upon shearing, there is concern that the loose fill may lose strength at such a rate that the forces mobilised in the soil nails will not be able to compensate for the loss of shear strength in the fill. Following comprehensive literature reviews, laboratory investigations (e.g. Law et al 1998, Ng & Chiu 2003) and numerical modelling (e.g. Cheuk 2001), specific guidelines on the use of soil nails in fill were developed by HKIE (2003). The following are salient points of the design recommendations:

- Large-strain steady-state undrained shear strength should be assumed for loose fill in the design.
- Global stability should be provided for by bonding soil nails into a competent stratum.
- Local surface stability should be enhanced by the provision of a concrete grid structure covering not less than 50% of the slope surface and connecting soil nail heads.
- Soil nails should be closely spaced horizontally and vertically.

Table 1. Summary of approach and method of soil-nailed slopes design.

Place	Document	Design approach	Common method of analysis
China	Geotechnical Engineering Handbook 1994	Partial factors approach	Limit equilibrium method using single wedge mechanism
Europe	Eurocode 7 : Geotechnical Design, 1995	Limit state approach incorporating partial safety factors	No specific method defined for soil nail design
France	Soil Nailing Recommendations Clouterre 1991	Limit state approach incorporating partial safety factors	Limit equilibrium method based on Bishop and two-part wedge mechanism
Hong Kong	Guide to Soil Nail Design and Construction (Geoguide 7)	Global safety factor approach combined with separate factors of safety on tensile strength of steel and pull-out resistance	Limit equilibrium method of slices on non-circular slip
Japan	Japanese Design Guide: Design and Construction Guidelines for Reinforced Cut Slopes	Global safety factor approach combined with safety factors for tensile strength of steel and pull-out resistance	Calculation models based on conventional circular or linear slip surface analyses
South Africa	Lateral Support in Surface Excavations: Code of Practice 1989	Global factor of safety combined with separate factors of safety on strength of steel and pull-out resistance	Limit equilibrium method using trial sliding wedges
UK	Design Methods for the Reinforcement of Highway Slopes by Reinforced Soil and Soil (HA 68/94)	Limit state approach incorporating partial safety factors	Limit equilibrium method using two-part wedge mechanism
UK	BS 8006: 1995 Code of Practice for Strengthened/Reinforced Soils and Other Fills	Limit state approach incorporating partial safety factors	Limit equilibrium method with reference two-part wedge, circular slip and log-spiral methods
USA	Manual for Design & Construction Monitoring of Soil Nail Walls	Limit state approach incorporating partial safety factors or Service Load Design	Design models based on Load and Resistance Factor Design and Service Load Design Limit equilibrium based on two-part wedge and slip circle method is adopted

- The grid structure should be designed to withstand bending moments and shear forces generated by the loose fill it is retaining; it should be adequately founded on a competent stratum.
- The potential of leakage from water-carrying services should be duly considered.

2.3.2 Soil-nail head design

Guidelines on the design of soil-nail head are available in the design codes of the UK (Department of Transport 1994), France (Clouterre 1991), USA (FHWA 1998), Japan (JHPC 1998) and Germany (Stocker & Riedinger 1990). All these documents recognize the soil-nail head or facing as a significant component of the overall soil nail system, and they provide specific recommendations for design pressures. They also recognize that the magnitudes of pressures induced in the soil-nail heads are controlled by many factors such as the density and length of the nails and the stiffness of the soil-nail head. Both the UK and the Japanese practice require that the pull-out failure at the active zone is to be checked. The UK practice also requires the checking of the bearing capacity failure in soil.

The French, Japanese and German methods use empirical earth pressures which are related either to the maximum tension developed in the soil nail (T_{max}) or Coulomb earth pressure. The U.S. and Japanese practice consider directly the strength of soil-nail head when determining the magnitude and distribution of nail forces along the length of the soil nails. If the beneficial effect of the soil-nail head is not considered, the pull-out resistance of the soil nail at the active zone would be significantly reduced. This may lead to more number of soil nails being required.

In all the design methods, the size, thickness and reinforcement details of soil-nail heads are determined on the basis of the earth pressure acting on the soil-nail heads. Two main design aspects are considered: the bearing capacity of the soil beneath the soil-nail head and the structural strength of the soil-nail head itself. Many of the design methods (such as those used in France and U.S.) were developed mainly for soil-nailed walls, where the soil-nail heads form part of the concrete facing. In these cases, bearing failure of the soil beneath soil-nail heads or facing is unlikely to occur and as such little guidance has been provided in respect of soil bearing failure. A method on the design against bearing failure of the soil behind isolated nail heads is given in the UK guidance document HA68/94.

For the development of soil-nail head design guidelines in Hong Kong, a series of numerical analysis was carried out. In the study, FLAC analysis was performed to examine the bearing failure beneath square soil-nail heads. A small slope model of 5 m in height was used and various slope angles were considered (Fig. 24). In the analysis, the soil-nail head was pushed into the ground by a nail force to simulate the situation of soil moving out from a slope and pressing against the soil-nail head. The nail forces used are determined from the allowable tensile strength of steel bars. Figures 25 and 26 show the shear strains and the displacement vectors respectively at the point of bearing failure for a 600 mm x 600 mm soil-nail head on a 45° slope. Typical results of the analyses in terms of c'-ϕ' envelope for limit equilibrium (i.e. when bearing failure occurs) are plotted in Figure 27. In this plot, the soil-nail head loads are expressed as diameters of steel bars. A number of the plots have been developed for different slope angles and soil-nail head sizes.

Knowing the shear strength parameters of the soil, the steel bar diameter and the slope angle, a designer can determine the size of soil-nail head from these plots. A design table has been derived from the plots for different combinations of slope angle and angle of shearing resistance of soil (Table 2).

2.3.3 Pullout resistance

Pull-out capacity is a key parameter for the design of soil nails. At present, methods for estimating pullout capacity are not unified as reflected by the many approaches used in different technical standards and codes of practice, such as effective stress method (GEO 2008), empirical correlation with SPT N values (JHPC 1998), correlation with pressuremeter tests (Clouterre 1991), and correlation with soil types (FHWA 2003). The merits and limitations of the various methods are summarized in Table 3.

The effective stress method is adopted in Hong Kong. The allowable pullout resistance provided by the soil-grout bond strength in the passive zone, T_P, is given by (Schlosser & Guilloux 1981):

$$T_P = \frac{c' P_c L_c + 2D\sigma'_v \mu^* L_c}{F_P} \quad (6)$$

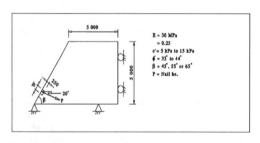

Figure 24. Slope model for bearing capacity analysis.

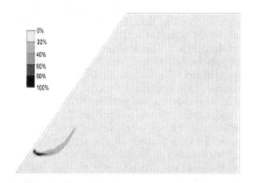

Figure 25. Typical shear strain plot.

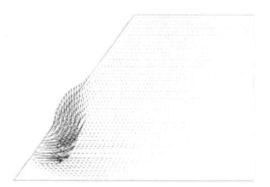

Figure 26. Typical displacement vector plot.

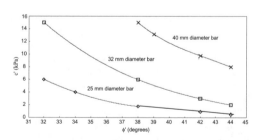

Figure 27. Shear strength required for 600 mm × 600 mm nail heads on a 45° slope to mobilise allowable tensile strength of nails of specified diameters.

Table 2. Recommended sizes of isolated soil-nail heads in Geoguide 7 (GEO, 2008).

Soil shear strength parameter near the slope surface		45° ≤ Slope angle < 55°			55° ≤ Slope angle < 65°			Slope angle ≥ 65°		
		Diameter of soil-nail reinforcement (mm)			Diameter of soil-nail reinforcement (mm)			Diameter of soil-nail reinforcement (mm)		
φ'	c' (kPa)	25	32	40	25	32	40	25	32	40
34°	2	800	800	800	600	600	800	600	600	800
	4	600	800	800	600	600	800	600	600	800
	6	600	800	800	400	600	800	400	600	600
	8	600	600	800	400	600	800	400	600	600
	10	400	600	800	400	600	600	400	600	600
34°	2	600	800	800	600	600	800	600	600	800
	4	600	800	800	400	600	800	400	600	800
	6	600	600	800	400	600	800	400	600	600
	8	400	600	800	400	600	600	400	600	600
	10	400	600	800	400	600	600	400	400	600
38°	2	600	800	800	400	600	800	600	600	600
	4	600	600	800	400	600	800	400	600	600
	6	400	600	800	400	600	600	400	600	600
	8	400	600	800	400	600	600	400	400	600
	10	400	600	800	400	400	600	400	400	600
40°	2	600	600	800	400	600	800	600	600	600
	4	400	600	800	400	600	600	400	400	600
	6	400	600	800	400	600	600	400	400	600
	8	400	600	600	400	400	600	400	400	600
	10	400	600	600	400	400	600	400	400	600

Notes: (1) Dimensions are in millimetres unless stated otherwise. (2) Only the width of the square soil-nail head is shown in the Table.

Table 3. Merits and limitations of the methods for determining ultimate pull-out resistance.

Method	Merits	Limitations
Empirical correlation	Related to field performance data; can better account for influencing factors.	Need a large number of field data and take a long time to establish a reasonable correlation; a general correlation may not be applicable to all sites.
Pull-out test	Related to site-specific performance data.	Need to carry out a considerable number of field pull-out tests during the design stage; not feasible for small-scale project; time consuming.
Undrained shear strength	Based on soil mechanics principles; easy to apply.	Generally not suitable for Hong Kong; many factors that affect the pull-out resistance are not accounted for.
Effective stress	Based on soil mechanics principles; easy to apply.	Many factors that affect the pull-out resistance are not accounted for.
Pressuremeter	Related to field performance data; can better account for influencing factors.	Need a large number of field data to establish a reasonable correlation; a general correlation may not be applicable to all sites; pressuremeter test is not common in Hong Kong.

where c' is effective cohesion of the soil, P_c is outer perimeter of the cement grout sleeve, L_c is bond length of the cement grout sleeve in the passive zone, D is outer diameter of the cement grout sleeve, σ'_v is vertical effective stress in the soil, μ^* is coefficient of apparent friction of soil (μ^* may be taken to be equal to tan ϕ', where ϕ' is the effective angle of shearing resistance of the soil), F'_p is factor of safety against pullout failure at soil-grout interface.

For design purpose, the vertical effective stress in the soil is calculated from the overburden pressure, which implies that the contact pressure at the soil-grout interface is governed by the overburden pressure. This assumption is not necessarily true because the normal stress at the face of the drillhole is reduced to zero after drilling due to arching effect and the grouting pressure is generally so low that only a small contact pressure can be restored. The contact pressure is likely much less than the overburden pressure.

The effects of hole drilling process, overburden pressure and grouting pressure on pullout resistance was investigated by means of laboratory pullout tests by Yin & Su (2006). The test set up is shown in Figure 28. Compacted fill of completely decomposed granite was used in the tests. The study showed that (a) the drilling process during soil nail installation led to stress reduction in the soil around the drillhole and the pullout resistances of the nails were not dependent on the amount of vertical surcharge applied if gravity grouting was adopted (Fig. 29); and (b) pullout resistances of the soil nails increased with an increase of grouting pressure (Fig. 30).

Pullout tests are routinely carried out in sacrificial test nails in Hong Kong for the verification of design assumptions. In order to examine the significance of the potential stress reduction due to the arching effect, the results of about 900 pullout tests were reviewed. The pullout resistance measured in the field was compared with the theoretical values estimated by the effective stress method. About 84% of the tests were conducted in granite or volcanic saprolite. The rest were conducted in other types of material such as fill, colluvium and moderately decomposed rock.

Many of the test nails were not loaded to bond failure because the ultimate pullout resistance (T_{ult}) of the bonded section was higher than the yield strength of steel. The pull-out tests were stopped when the test load reached 90% of the yield strength of steel to avoid tensile failure of the steel reinforcement. Figure 31 shows the plot of the ratio of the field pull-out resistance to that estimated using the effective stress method against the overburden pressure. The field pull-out resistances were generally several times higher than those estimated, but the safety margin (i.e. $T_{ult(field)}/T_{ult(estimate)}$) gradually decreases when overburden pressure increases.

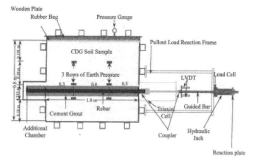

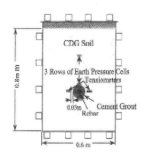

Figure 28. Schematic set-up of the laboratory pullout test.

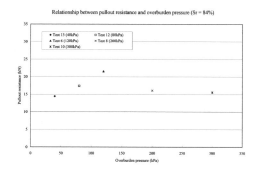

Figure 29. Plot of pullout resistance against overburden pressure.

The difference between the measured and the estimated pullout resistance is due to many factors including soil arching, restrained soil dilatancy, soil suction, roughness of drillhole surface, and over-break, which are hard to quantify in design. All these factors except soil arching tend to result in higher pullout resistance than the design value. The finding of the review gives assurance on the adequacy of the effective stress method.

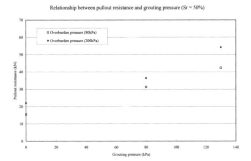

Figure 30. Plot of pullout resistance against grouting pressure.

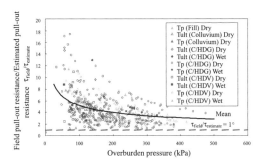

Figure 31. Plot of field ($T_p + T_{ult}$) to estimated pull-out resistance against overburden pressure.

2.3.4 Potential effect of blockage of subsurface drainage by soil nailing works

Soil nails installed in the ground may impede groundwater flow and as a result dam up the water level. To study the significance of this effect, a number of numerical models were set up in both 2-D and 3-D for various geological settings, subjected to infiltration (Halcrow China Limited 2007). Typical nail spacings of 1 m to 2 m were adopted in the models. Figure 32 illustrates an example of computed flow nets and water table distributions for a slope under three conditions: (a) without soil nails; (b) soil nails with excessive grout loss, and (c) soil nails with no grout loss.

Results of the numerical modeling show that under typical conditions where there is little grout loss during the grouting operation, there should be no significant blockage of the drainage paths. It is also found that the influence of soil nails on groundwater flow can be significant if excessive grout escapes laterally to affect large volumes of the country rock. Therefore, measures should be taken to avoid excessive grout loss. Where excessive grout loss occurs during installation of soil nails, the cause should be investigated and, if necessary, measures taken to monitor rises in hydraulic head and to take action to drain the ground upstream of the nails.

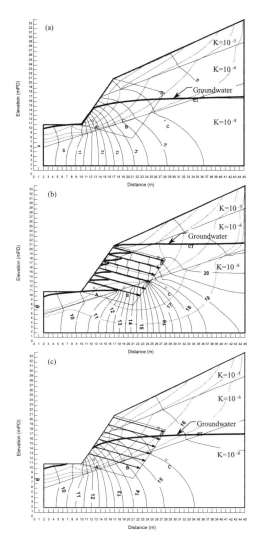

Figure 32. Flow Patterns in a slope (a) without soil nails, (b) soil nails with excessive grout loss, (c) soil nails with no grout loss (after Halcrow, 2005).

2.3.5 Long-term durability of soil nails

Durability is an important aspect of soil nailing system. The long-term performance of soil nails depends on their ability to withstand corrosion attack from the surrounding ground. Soil nails of different ages exhumed from the ground in Hong Kong revealed that localized corrosion could occur even if hot dip galvanization was provided, particularly in areas where voids existed

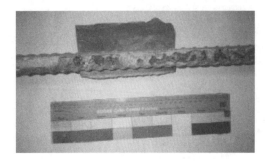

Figure 33. Localized corrosion in exhumed soil nail.

in the cement grout (Shiu & Chang 2003). Figure 33 shows the corrosion on an exhumed soil nail.

The design for durability of a steel soil nail entails the assessment of the corrosivity of soil at the site and the provision of corrosion protection measures. Different soil corrosivity assessment schemes are available in the world, and different countries may adopt different corrosion protection measures.

Eyre & Lewis (1987) developed two soil corrosivity assessment schemes, one for general assessment purposes and another for reinforced earth and culvert application. Two types of assessment scheme were subsequently developed by others in the UK. The first was that of Murray (1993), who incorporated the scheme for general assessment purposes into the specifications for soil corrosivity assessment in respect of soil nailing in the UK. Tests are conducted on soil samples only. The soil aggresivity classification scheme in Hong Kong (GEO 2008) was developed based on that of Murray (1993).

The UK Department of Transport made reference to the scheme for reinforced earth and culvert application and the recommendations of Brady & McMahon (1993) and developed a corrosivity assessment scheme for corrugated steel buried structures under roadways (Department of Transport 2001). In this scheme, both soil and water samples are collected and tested. CIRIA (2005) has recommended a corrosivity assessment scheme based on that developed by the Department of Transport (2001).

Depending on the soil aggresivity of a site, the required design life and the intended degree of protection, different measures may be adopted to protect steel bars against corrosion. The common corrosion protection measures are cement grout, sacrificial thickness to the steel, sacrificial metallic coating to the steel (e.g. hot-dip galvanizing with zinc coating), sacrificial non-metallic coating to the steel (e.g. epoxy coating), and corrugated plastic sheathing.

Cement grout can prevent corrosion by forming a physical and a chemical barrier. The cement grout physically separates the steel from the surrounding soil. Due to the alkalinity of cement grout, a tight oxide film is also formed on the surface of the steel bar. This further protects the steel from corrosion. However, micro-cracks will occur in the cement grout when the soil nail is subject to tensile stress. Shrinkage cracks may also be formed during the setting of the cement grout. Once cracked, the function of the cement grout in corrosion protection is not reliable.

The provision of sacrificial steel thickness is a simple and widely used method. It allows for corrosion of the steel by over-sizing the cross-section of the steel bar. Products of corrosion also form a protective coating between the steel and its surrounding. Whilst this coating offers no physical protection to the steel, it may slow down the rate of corrosion by changing the kinetics of the chemical reactions.

Zinc is the most common type of metal used to provide corrosion protection to steel bars. The galvanizing zinc coating is strongly resistant to most corrosive environments. It provides a barrier protection and a cathodic protection function to the steel.

Non-metallic coatings in the form of fusion-bonded epoxy have been used in the USA to protect steel bars from corrosion. The epoxy coatings do not conduct electricity and they isolate the steel bars from the surrounding environment. To be effective, the coating has to be impermeable to gases and moisture and free from cracks. The interface between the steel and the coating has to be tight.

When a high level of corrosion protection is needed, corrugated plastic sheaths can be used in conjunction with cement grout. The sheath prevents ingress of water and corrosive substances even if the cement grout is cracked.

To overcome the problem of corrosion of metallic reinforcement, non-metallic soil nails may be used. An alternative to steel reinforcement is composite material made of fibres embedded in a polymeric resin. It is generally known as fibre-reinforced polymers (FRP). FRP is highly corrosion resistant. The common types of fibre used in composites for civil engineering works are carbon FRP (CFRP), glass FRP (GFRP) and aramid FRP (AFRP). Shanmuganathan (2003) gave a state-of-the-art review of the development and application of FRP composites in civil and building structures. There are reported cases of using CFRP reinforcements to slope stabilization works in the USA, the UK, Spain, Greece, Japan and Korea (e.g. Unwin 2001, Ground Engineering 2004).

Carbon fibres are the primary load-carrying component in CFRP reinforcement, which are characterized by low weight, high strength and high stiffness. The primary function of the resin is to provide a continuous protection medium to the fibres and to transfer stresses among fibres. CFRP reinforcement is anisotropic in nature and is characterized by high tensile strength in the direction of the fibres. It is non-corrosive and has a much better strength-to-weight ratio than steel reinforcement. However, CFRP reinforcement does not exhibit yield behaviour. The lack of ductility necessitates special consideration in its application as a soil nail.

Figure 34 illustrates schematically the stress-strain behaviour of CFRP reinforcement in comparison to steel. According to ACI (2001), the tensile strength of CFRP reinforcement ranges from 600 MPa to 3,690 MPa (c.f. 460 MPa for high yield steel bar). Specific test results on tensile testing of CFRP reinforcement strips (Fig. 35) which were used in a field trial by the GEO in Hong Kong are summarized in Table 4. The tests gave tensile strength ranging from 1,990 MPa to 2,550 MPa, with an average value of 2,280 MPa. This average tensile strength is about five times that of high yield steel.

The shear strength of CFRP reinforcement is generally much lower than its tensile strength. Benmokrane et al (1997) reported that the shear strength of some CFRP reinforcements is only about 11% of its tensile strength.

While there are a number of national design and construction guides on the use of CFRP reinforcement in concrete structures (e.g. Japanese Ministry of Construction 1997, JSCE 1997, IStructE 1999, ACI 2001), international standard on the use of CFRP reinforcement as soil nails is lacking. An interim design and construction guideline for CFRP soil nails has been developed by the GEO for use in its slope upgrading programme. The following are salient points of the GEO guideline:

- A suitably conservative estimate of the design tensile strength is made using a partial material safety factor of 3.3 to cater for the uncertainty in material properties and to compensate for the lack of ductility of CFRP.
- A partial safety factor of 1.4 on bond strength between CFRP reinforcement and grout is adopted following the recommendation of IStructE (1999).
- The inclination of the CFRP soil nails to limited to within 15° from the horizontal so as to optimize the reinforcing efficiency of the soil nails and to limit the slope movement for the mobilization o the tensile force in the soil nails.
- Only CFRP reinforcements with a shear strength equal to or greater than that of steel are used.
- CFRP reinforcements with polyester resins should not be used because of their relatively ease of degradation in highly alkaline environment.

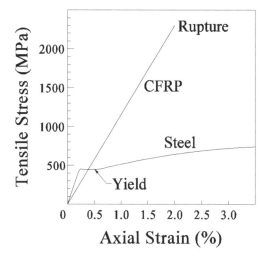

Figure 34. Stress/strain curves of typical high yield steel bar and CFRP bar.

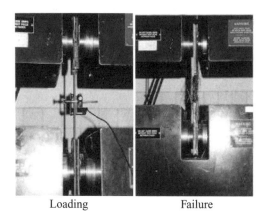

Figure 35. Tensile testing of CFRP strips.

2.3.6 *Aesthetic considerations*

A good soil nail design should give due attention to the aesthetic aspects in addition to safety and serviceability considerations. GEO Publication No. 1/2000 (GEO 2000) provides comprehensive guidance on the landscape treatment and bio-engineering for slopes and retaining walls. The publication contains general principles and good practice for enhancing the appearance of engineered slopes and illustrates these by a large number of case examples. While the principles of aesthetics and landscape treatment are given for unreinforced slopes, they are generally applicable to soil-nailed systems. The following design principles are worth considering for enhancing the appearance of soil-nailed slopes and retaining walls:

- Make the appearance of soil-nailed systems compatible with and minimize visual impact to the existing environment.
- Identify and preserve, wherever practical, mature trees on slopes and near their crests and toes.

Table 4. Results of tensile test on CFRP reinforcement bars.

Specimen no.	Width (mm)	Thickness (mm)	Rupture load (kN)	Tensile stress at rupture (MPa)	Modulus of elasticity (GPa)	Rupture strain (%)
1	29.0	4.3	318	2,548	120.2	2.12
2	29.0	4.5	330	2,522	119.4	2.11
3	28.9	4.5	262	1,996	111.6	1.79
4	29.0	4.4	253	1,985	114.9	1.73
5	28.7	4.6	314	2,363	103.5	2.28
			Average	2,283	113.9	2.01

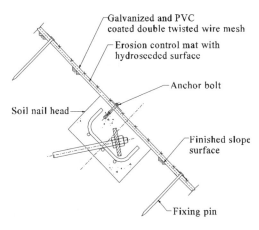

Figure 36. Fixing details of steel wire mesh and erosion control mat on slope face with soil nail heads.

- Locate soil nails and other engineering features away from tree trunks and roots.
- Pay attention to the design and location of man-made features such as surface drainage channels, stairways and catchpits in order to minimise their visual impact, e.g. concrete aprons on either side of drainage channels can be designed using geotextiles or other bioengineering techniques.
- Route maintenance stairways with care to minimise visual impact and paint railings in sympathetic unobtrusive colours.
- Place soil nails in a regular, rather than a random pattern.
- Recess isolated soil-nail heads and treated them with a matt paint of a suitable colour to give a less intrusive visual appearance.

A large variety of greening techniques are available in the market. The Hong Kong experience shows that a simple method which involves the use of an erosion control mat in conjunction with a steel wire mesh allows the provision of vegetation covers to steep slopes (Fig. 36). Many steep soil-nailed cut slopes in granitic or volcanic saprolite of a gradient up to 60° have been successfully vegetated using this technique.

Where the provision of vegetated surface cover on a slope is practically not feasible, hard landscape treatment can be provided to improve its appearance. Possible methods are masonry block facing, ribbed or other patterned concrete finishes, toe planters, colouring and planter holes. More fancy techniques such as decorative artwork and artificial rock may also be used.

2.4 *Construction technology*

2.4.1 *Construction method*
The choice of installation method depends on a number of factors such as cost, site access, working space, and ground and groundwater conditions. A brief description of the commonly available soil-nail installation methods is summarised below.

a. *Drill-and-Grout*. This is the most common installation method. In this method, a soil-nail reinforcement is inserted into a pre-drilled hole, which is then cement-grouted under gravity or low pressure. Various drilling techniques, e.g., rotary, rotary percussive and down-the-hole hammer, are available to suit different ground conditions. The advantage of this method is that it can overcome underground obstructions, e.g., corestones, and the drilling spoil can provide information about the ground. The size and alignment of the drillholes can be checked before the insertion of reinforcement, if needed. Potential construction difficulties are hole collapse and excess grout loss. The drilling and grouting process may also cause disturbance and settlement to the adjacent ground.

b. *Self-Drilling*. This is a relatively new method when compared with the drill-and-grout method. The soil-nail reinforcement is directly drilled into the ground using a sacrificial drill bit. The reinforcement, which is hollow, serves as both the drill rod and grout pipe. The installation process is rapid as the drilling and grouting are carried out simultaneously. Instead of using air or water, cement grout is used as the flushing medium, which has the benefit of maintaining hole stability. No centralisers nor grout pipes are needed and casing is usually not required. However, self-drilling soil nails may not be suitable for the ground containing corestones as they cannot penetrate through rock efficiently. It may be hard to ensure the alignment of long soil nails due to the flexibility of reinforcement. Durability may also be a concern if it relies on the provision of grout cover and corrosion protective coatings to steel reinforcement as corrosion protection measures.

c. Driven. In this method, soil-nail reinforcement is directly driven into the ground by the ballistic method using a compressed air launcher, by the percussive method using a hammering equipment, or by the vibratory method using a vibrator. During the driving process, the ground around the reinforcement will be displaced and compressed. The installation process is rapid and it causes minimal ground disruption. However, due to the limited power of the equipment, this method can only be used to install soil nails of relatively short length. Moreover, the soil-nail reinforcement may be damaged by the excessive buckling stress induced during the installation process, and hence it is not suitable for sites that contain stiff soil or corestones. Since the soil-nail reinforcement is in direct contact with the ground, it is susceptible to corrosion unless non-corrodible reinforcement is used.

The experience in Hong Kong shows that soil nails can generally be constructed by means of the drill-and-grout method without many difficulties. However, under some unfavourable ground conditions, construction problems may be encountered. The following geological conditions are susceptible to excessive grout leak during soil nail installation: f

- fill, containing a significant proportion of coarse materials, i.e., boulders, cobbles, gravel, and sand;
- colluvium and fluvial deposits with a high proportion of coarser material;
- erosion pipes which may be partly infilled by porous and permeable material;
- material boundaries within colluvium, and between colluvium and in-situ material, and within corestone-bearing saprolite, especially at the margins of corestones, open joints, faults and shear zones, and other discontinuities (e.g., zones of hydrothermal alternation, etc.) that are weathered and eroded, and so are open;
- landslide scars, tension cracks, and other features related to slope deformation, as these may include voids within transported and in-situ materials; and
- drainage lines intersecting slopes, within which colluvium may be present, erosion pipes may be developed, and preferred groundwater throughflow indicated by seepage locations/horizons may also occur.

2.4.2 Use of non-destructive testing method for quality control

Like other buried works, it is difficult to verify the quality of an installed soil nail. In the context of this paper, the quality of an installed soil nail refers to the as-built length and the integrity of cement grout. In order to enhance the quality control of soil nailing works, non-destructive testing (NDT) methods could be carried out on installed soil nails. With the help of NDT, the overall picture of the quality of installed soil nails can be built up, which facilitates the identification of the areas for follow-up actions. A number of NDT methods including sonic echo method, Mise-a-la-Masse method, magnetometry, electromagnetic induction method and time domain reflectometry (TDR) have been examined in Hong Kong. Amongst these, the TDR method was found to be reliable, simple and not expensive (Cheung 2003, Lee & OAP 2007).

The principle of TDR technique was derived in 1950s from that of radar. Instead of transmitting a 3-D wave front in radar, the electromagnetic wave in the TDR technique is confined in a waveguide (O'Connor & Dowding 1999). TDR is commonly used in the telecommunications industry for identification of discontinuities in transmission lines. In the 1980s, the application of the technique was extended to many other areas such as geotechnology, hydrology, material testing, etc (Dowding & Huang 1994, Siddiqui et al 2000, Liu et al 2002, Lin & Tang 2005). TDR is based on transmitting electromagnetic pulses through a transmission line, which is in the form of coaxial or twin-conductor configuration, and receiving reflections at the locations of discontinuities. By measuring the time for the pulses to travel from the pulse generator to the point of discontinuity, one can determine its location using Equation (7).

$$L = v_p t \quad (7)$$

where L is the distance between the pulse generator and the point of discontinuity, and t is the respective pulse travel time. The pulse propagation velocity, v_p, is related to the electrical properties of the material in the close proximity to the pair of conductors by the following expression (Topp et al 1980):

$$v_p = \frac{v_c}{\sqrt{\varepsilon}} \quad (8)$$

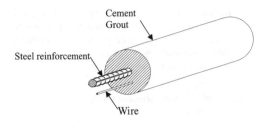

Figure 37. Analogy of a soil nail with pre-installed wire as a twin-conductor transmission line.

where v_c is the speed of light in vacuum (3×10^8 m/s) and ε is the dielectric constant which measures how a material reacts under a steady-state electric field (for air $\varepsilon \approx 1$, for cement grout $\varepsilon \approx 10$, for water $\varepsilon \approx 80$).

If a wire is pre-installed alongside a soil-nail reinforcement, which is generally a steel bar, as shown in Figure 37, the configuration becomes analogous to a twin-conductor transmission line and the end of the reinforcement-wire pair becomes a discontinuity. This suggests that TDR can be used to determine the length of installed steel soil nails.

As indicated in Equation (1), the two key parameters that have to be known for the estimation of soil-nail length are (i) the time for a pulse to travel from the reinforcement head to its end, t, and (ii) the pulse propagation velocity, v_p. Equation (8) further suggests that the pulse propagation velocity, v_p, along a reinforcement-wire pair in air will be much greater (2 to 3 times) than that in cement grout. Hence, the pulse travel time along a soil nail with voids in grout will be less than that in a fully grouted soil nail of the same length.

Apart from the effect on pulse propagation velocities, a reflection will be induced whenever an electrical pulse reaches the location of discontinuity in the grouted reinforcement-wire pair (e.g. the end of a soil nail or a void). The magnitude and polarity of the reflection depend on the amount of changes in electrical impedance at the location of discontinuity, which can be expressed in terms of the reflection coefficient, Γ (Hewlett Packard 1998):

$$\Gamma = \frac{V_r}{V_i} = \frac{Z - Z_o}{Z + Z_o} \qquad (9)$$

where V_r is the peak voltage of the reflected pulse, V_i is the peak voltage of the incident pulse, Z is the electrical impedance at the point of reflection and Z_o is the characteristic electrical impedance of the grouted reinforcement-wire pair.

Figure 38 shows a theoretical TDR waveform of a cement grouted reinforcement-wire pair with void section in the middle. There will be reflections at the location of the void as well as the end of the pair.

According to Equation (9), a positive reflection will be returned at the discontinuity when there is an increase in electrical impedance (e.g. reflection 1 at the interface of grout/void and reflection 3 at the end of the pair), whereas a negative reflection will be returned otherwise (e.g. reflection 2 at the void/grout interface). Moreover, the pulse travel time is less than that in the fully grouted pair. In other words, one can in-principle determine the quality of an installed soil based on a TDR waveform.

In order to investigate the feasibility of applying TDR technique in the estimation of soil-nail length, TDR tests were conducted on prefabricated soil nails of various known lengths. Figure 39 shows the TDR test results where reflections are returned from the respective soil-nail ends and the time of pulse propagation is found to be proportional to the length of the soil nail.

Based on the contrast in pulse propagation velocity in air and grout and the occurrence of reflections where these is a change in impedance along the reinforcement-wire pair, TDR results in-principle can be interpreted to infer the grout integrity of a soil nail. To examine this, TDR tests were conducted on prefabricated soil nails with built-in grout defects of varying void sizes at different locations along the soil nails as shown in Figures 40 and 41. These model test results indicate that soil nails with significant grout defects will result in shorter TDR-deduced length with some characteristic patterns in the TDR waveform. These patterns depend on the location as well as size of the defects.

Up to the end of 2007, over 10,000 soil nails at about 850 sites have been tested using TDR in Hong Kong (Fig. 42). In general, the percentage difference between the TDR-deduced length and design length of the soil nails is small and lies within the uncertainty limit of the test. There were a small number of soil nails (less than 1%) with such difference exceeding an alert limit and displaying anomalous TDR wave forms. Further investigation was conducted where needed. The anomalies encountered so far were found to be due to grout defect in the soil nails.

The experience shows that the TDR technique can be an effective tool to supplement site supervision in the quality control of soil nailing works, which cannot be checked easily after construction. While TDR, like any other NDTs, does not give definitive answer to the cause of anomalous test results, it flags up soil nails that warrant further examination and coupled with appropriate NDTs, the conditions of the soil nails can be ascertained (Pun et al 2007).

2.5 *Performance of soil-nailed system*

Since the introduction of the soil nailing technology to Hong Kong in late-1980's (Watkins & Powells 1992),

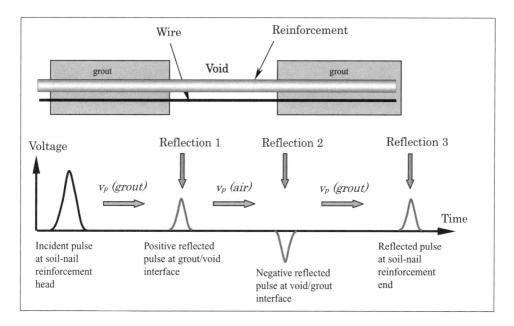

Figure 38. Theoretical TDR waveform of a soil nail with defect in grout sleeve.

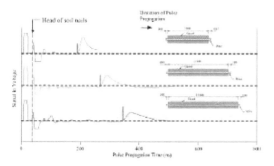

Figure 39. TDR results on soil nails of various known length.

Legend:
↓ Inferred grout/void interface
| Inferred soil nail end

Figure 40. TDR waveforms for soil nails with a void section between grouted sections.

more than 3,700 soil cut slopes have been upgraded by means of soil nailing in Hong Kong. The performance of these soil-nailed slopes gives an indication of the reliability of the soil nailing technique.

In the period between 1993 and 2006, a total of 25 landslide incidents on permanent soil-nailed slopes were reported to the GEO. The landslides were generally of small scale, involving local shallow sliding failure or washout, with failure volumes ranging from less than 1 m^3 to a maximum of about 35 m^3. The average annual failure rate of such relatively minor landslides on soil-nailed slopes between 1997 and 2006 is 0.078%. This average annual failure rate is of a similar order of magnitude as that of engineered, unsupported soil cut slopes in Hong Kong.

Amongst the 25 reported landslide incidents, 15 cases were reviewed in detail. All the 15 landslide incidents involved active zone failures. There was no report of external failure or passive zone failure. The slopes were steep and had a vegetation cover before failure, four of which had a gradient equal to or exceeding 45° and the other 11 exceeding 50°. Twelve of the 15 cases were associated with surface

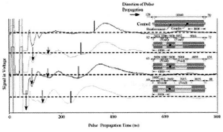

Note: All dimensions are in mm.
Legend
↓ Inferred grout/void interface
| Inferred soil nail end

Figure 41. TDR waveforms for soil nails with two void sections between grouted sections.

Figure 42. Conducting TDR test at a soil nail.

Figure 43. Shallow failure at a soil-nailed slope.

(a) General view

(b) View of soil-nailed head

Figure 44. Failure of a soil-nailed slope undermining soil-nail heads.

erosion and detachments from the near-surface materials between the nail heads (see example in Fig. 43). In two cases, the soil-nail heads were partially exposed but the nail reinforcements and grout sleeves remained intact. In the remaining case, the soil-nail heads were undermined and the soil nail reinforcement bars were bent (see Fig. 44). The common factors contributing to the landslides were inadequate slope protection, inadequate drainage provisions or presence of adverse geological or hydrogeological conditions.

The review shows that steep soil-nailed slopes with vegetated covers are fairly vulnerable to minor failures, as the potential for shallow small detachments between soil-nail heads within the active zone of the soil nail system cannot be prevented effectively by means of the soil nails. The efficiency of the soil nails, which are typically shallowly inclined for the upgrading of substandard soil cuts in Hong Kong, is not high in so far as prevention of shallow detachment is concerned. This is because the nails are not orientated at an optimal inclination in relation to the steep slip surfaces with regard to the mobilization of tension forces in the soil nails. In case of shallow failures on a steep vegetated soil-nailed slope, there is little horizontal displacement to mobilize the tensile forces in the nail bars and hence their effectiveness in stabilizing potential vertical/subvertical failure surfaces may be limited.

No failure with a volume larger than 50 m^3 on soil-nailed slopes has been reported in Hong Kong so far.

As a comparison, the average failure rate of landslides with a volume larger than 50 m³ for engineered, unsupported soil cuts in Hong Kong is 0.018%. Soil nails appear to be effective in preventing large-scale failures.

It is also worth noting that no landslides have been reported on soil-nailed slopes with a hard surface cover in Hong Kong. A hard surface cover is effective in minimizing surface infiltration and provides a better protection against surface erosion than a vegetated cover. However, a hard surface cover may not be acceptable from the environmental and aesthetic points of view. A variety of flexible structural facing such as tensioned wire mesh may be used to enhance the stability of vegetated slopes. This would however increase the construction cost. A proper design has to balance between the risk of possible minor failure, cost and environmental considerations. In this regard, risk mitigation measures, such as debris traps, toe barriers or buffer zones, may be considered as an integral part of the slope design to cater for possible minor detachments from vegetated slopes.

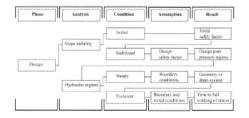

Figure 45. Approach to design drains to stabilize a slope.

3 SUBSURFACE DRAINAGE

3.1 General aspects

In saturated soils, subsurface drains are widely used as control measures against slope instability, as they are less costly than other types of stabilization works and suitable for a large number of cases, even when the landslide is very deep and structural measures are not effective.

The mechanism of drains inside slopes involves a decrease in pore pressures in the subsoil and consequently an increase in effective stresses and soil shear strength in the whole drained domain. In particular, the increase in soil shear strength along the potential sliding surface of the landslide body, due to the function of drains, is responsible for the slope stability improvement. Therefore the first step in the design of a drainage system is the determination of the pore pressure change that is required to increase the factor of safety of the slope to the design value (Fig. 45). The next step is to design the geometric configuration of drains that will result in the required pore pressure change.

The effect of the drainage system is usually analyzed for the steady-state condition, which is attained some time after drainage construction (i.e. in the long term). After drain installation, a transient phenomenon of equalization of pore pressures occurs, provoking subsidence of the ground surface. The magnitude of subsidence depends on (i) the compressibility of the soils concerned, (ii) the thickness of the drained domain, and (iii) the amount of lowering of the water table. Problems related to excessive ground settlements are expected when the drained soil is very thick, as in the case of deep drains.

As regards the transient phase, two aspects have to be evaluated in the design:

a. whether the delay until the drains are completely effective is affordable,
b. whether settlements associated with de-watering will damage buildings and infrastructures at the ground surface.

The steady-state condition is usually analyzed by assuming continuous infiltration of water at the ground surface to recharge the water table. In the literature, results of steady-state analysis are often presented in non-dimensional design charts, that practitioners generally use to design drainage systems.

The water flow captured and discharged by drains depends largely on the permeability of the drained soils. In steady-state condition the permeability of the ground does not affect the amount of lowering of the pore pressures in the subsoil, which depends on the hydraulic conditions at the boundaries of the examined domain and the geometry of the drainage system. Thus the quantity of discharge is not an indicator of the performance of drains, which has to be investigated by means of piezometers to measure the change in the level of water table by drains. Indeed, pore pressure changes are the most direct and useful indicators of drains being in good working condition. Measurements of surface and deep displacements are good indicators of overall slope stability. These measurements complete the instrumentation framework (Fig. 46).

3.2 Drain types

Among the measures for slope stabilization, drains are probably the most commonly used. Because of their widespread application in very different ground conditions and geomorphology, the technology in this field

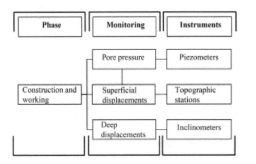

Figure 46. Measurements to evaluate that drains are in good working condition.

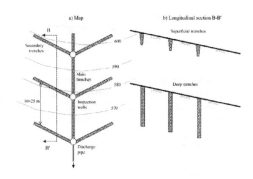

Figure 48. Superficial and deep trenches, with main and secondary branches: a) map, b) longitudinal section.

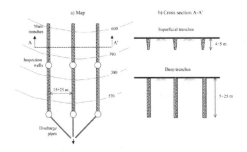

Figure 47. Superficial and deep trenches, with only main branches: a) map, b) cross section.

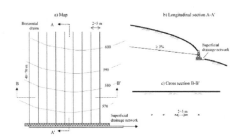

Figure 49. Horizontal drains inserted from the ground surface: a) map, b) longitudinal section, c) cross section.

is advancing continuously. Nowadays the most common types of drains used in geotechnical engineering applications are as follows:

- Superficial and deep trenches, with main (Fig. 47) and possibly secondary (Fig. 48) branches,
- Horizontal drains (Fig. 49), executed from the ground surface,
- Wells, with or without horizontal drains (Fig. 50),
- Tunnels, with or without horizontal drains.

Drain trenches should be excavated deep enough to intercept the regions of positive pore pressures. Superficial trenches can be excavated by means of an excavator up to a depth of approximately 5 m from the ground surface. The width of the trench is dependent on the type of excavator being used and may vary from 0.5 to 1.0 m.

In open areas, trenches can have sloping sides, the gradient of which is based on stability consideration. Where there is not enough space, trench sides have to be formed to vertical and should be properly supported. Guidelines on the design of lateral support to excavation are given in many publications, e.g. BS 6031:1981 (BSI 1981). Problems of trench instability can be reduced by opening up trenches in short

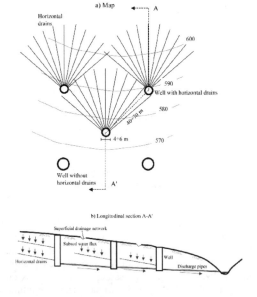

Figure 50. Drain wells with and without horizontal drains: a) map, b) longitudinal section.

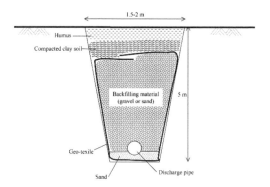

Figure 51. Scheme of a superficial trench.

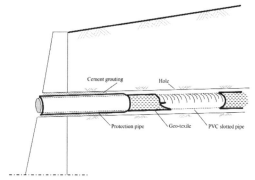

Figure 52. Scheme of a horizontal drain.

lengths and backfilling the trench within a short time after excavation.

Deep trenches can be excavated by means of grab shells. The sides of the trenches, being vertical, should be supported by slurry, e.g. polymeric mud.

Trenches need to have a high discharge capacity to avoid the saturation of the backfilling material or of the lower portion of it. This can be achieved providing a drainage layer of gravel materials or installing at the bottom of the trench a perforated pipe (with slots on the upper part). The perforated pipe should be wrapped with a geotextile to prevent the clogging of the slots by fine soil particles (Fig. 51). A compacted clay cover should be placed on the top of the trench to prevent ingress of surface water, which should be drained by means of a system of surface drainage network. The impermeable cover should have a minimum thickness of 0.5 m and should be compacted in layers.

Trenches should be constructed starting from the lowest point in the area to be drained, so that they can drain water during construction. Inspection wells that intercept the trenches should be installed to allow:

- monitoring of the working condition of the drainage system, possibly by measuring the flow;
- maintenance, possibly flushing of the perforated pipe.

Horizontal drains involve the drilling of holes in the ground. The diameter of the hole is usually 100-120 mm, and it is drilled with a tricone or drag bit. A PVC slotted pipe, protected by a geo-textile, is inserted in the hole (Fig. 52). The maximum length of horizontal pipes is around 100 m, but in some cases it has been possible to reach 300 m.

Deposits of calcium salts and iron oxide can block horizontal drains; regular maintenance by flushing the pipes with a high pressure water jet, should therefore be programmed. In the absence of maintenance, drain pipes cannot remain functional for a long time.

To reduce precipitation of calcite it is good practice to drill the hole at an inclination slightly above horizontal, such that the pipe is not continuously submerged. Conversely, there are other chemical phenomena, favored by bacterial activity, that are due to aeration (Walker & Mohen 1987).

At the portion of a horizontal drain near to the slope surface, it is recommended to use a 3–6 m long un-perforated pipe, grouted all around with cement, to prevent the penetration of tree roots into the pipe, which could block the water flow.

Wells and tunnels are costly and complex to construct; for this reason they tend to be used in special circumstances, e.g. in deep landslides where other types of drain are unable to reach the sliding surface.

Drain trenches and horizontal drains are the most commonly used drain systems in slope stabilization. Therefore these are treated in detail in the next part of this paper.

3.3 *Influence of groundwater on landslides in saturated soils*

Investigation carried out in many parts of the world (Kenney & Lau 1984, Urciuoli 1998) showed that pore pressures at shallow depths are strongly influenced by seasonal atmospheric conditions. Urciuoli (1998) showed, on the basis of piezometer measurements, that a critical line can be drawn inside clay formations, separating the zone in which groundwater regime is transient, due to variation in atmospheric conditions, from the deep zone, in which pore pressures remain essentially constant throughout the year (Fig. 53). The division of the saturated soil domain into two zones is useful to interpret the mechanism of landslides and its relation with pore pressures, as is described below.

Active shallow landslides are usually characterized by the presence of a sliding surface in the zone affected

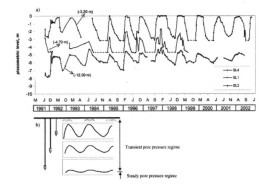

Figure 53. Pore pressure in clayey formation: a) measurements at Casagrande piezometers installed at different depths (Basento Valley, Italy), b) scheme of pore pressure regime in the subsoil.

by a transient groundwater regime. In this case, the safety factor of the landslide body varies with time along with pore pressure fluctuations. Pore pressures attain their maximum values during wet seasons: shallow landslides can reactivate, as a direct consequence of reduction in shear strength along the sliding surface. The rate of slope movement is characterized by a seasonal trend (Pellegrino et al. 2004b).

In active deep landslides, slope movement may well be due to intense plastic strains, occurring in some regions of the subsoil where shear stresses are close to the failure envelope of the soil. In other words, displacement points of the landslide body may not be localized on a sliding surface. The rate of movement is generally very small and constant (i.e. not characterized by a seasonal trend), according to the steady groundwater regime.

In both cases drains can play an important role in contributing to slope stability. In potential or active shallow landslides drains might prevent the rise of the water table, which is a consequence of atmospheric changes, up to the critical level that endangers slope stability. Nonetheless, methods of analyzing the stabilization effect of drains commonly available in the literature (e.g., Hutchinson 1977, Desideri et al. 1997) model the groundwater regime as a steady-state phenomenon (seepage) and assume the presence of a film of water on the ground surface. In areas where the weather is not very rainy, such as in southern Europe, this assumption underestimates the effects of drains on slope stability.

In potential or active deep landslides drains reduce the steady-state pore pressures, which in turn increase the difference between mobilized shear stresses and the soil shear strength. In this way the plastic strain rate in the overstressed zone is drastically reduced. The effectiveness of drains is correctly analyzed by considering the groundwater regime as a steady-state phenomenon.

3.4 Groundwater regime

3.4.1 Equation

If the subsoil is assumed to be a saturated porous medium, characterized by isotropic permeability K, the equation governing transient flow can be obtained by imposing the conservation of the liquid mass. Darcy's Law is inserted therein. In this way the subsequent equation is obtained:

$$-K \cdot (h_{xx} + h_{yy} + h_{zz}) = \dot{\varepsilon}_v \qquad (10)$$

where h is the piezometric head:

$$h = \xi + \frac{u}{\gamma_w} \qquad (11)$$

ξ is the geometric head, u is the pore pressure and γ_w is the water unit weight; $\dot{\varepsilon}_v$ is the volumetric strain rate of the soil.

Assuming the solid skeleton is an isotropic linearly elastic medium, with Young's modulus E and Poisson index ν, volumetric strain is expressed as:

$$\varepsilon_v = dp' \cdot \frac{3(1-2\nu)}{E} \qquad (12)$$

where dp' is the variation in the mean effective stress during the transient phenomenon.

Under the assumption that mean total stresses $p (p = p' + u)$ remain constant in all the volume subjected to the analysis we obtain:

$$dp' = -du = -\gamma_w dh \qquad (13)$$

Substituting eq. (13) in eq. (12) and eq. (12) in eq. (10), we obtain:

$$h_t - c_v \left(h_{xx} + h_{yy} + h_{zz} \right) = 0$$
$$c_v^{3D} = \frac{KE}{3(1-2\nu)\gamma_w} \quad (x,y,z) \in \Omega \ 0 < t \leq T \qquad (14)$$

where c_v^{3D} is the coefficient of consolidation in 3D condition; Ω and T are respectively the spatial and time integration domains.

The first of equations (14) can be simply specialized to 2D and 1D cases; contextually the expression of the coefficient of consolidation has to be modified, according to table d5.

Once the hydraulic boundary conditions are fixed, the transient solution tends to a steady distribution

Table 5. Consolidation equation and coefficient of consolidation for 2D and 1D conditions.

Conditions	2D	1D
Equation	$h_t - c_v^{2D}(h_{xx} + h_{zz}) = 0$	$h_t - c_v^{1D} h_{zz} = 0$
p' (mean effective stress)	$\dfrac{(\sigma'_x + \sigma'_z)(1+v)}{3}$	$\dfrac{\sigma'_z(1+v)}{3(1-v)}$
ε_v (volume strain)	$\dfrac{(1+v)(1-2v)}{E} d(\sigma'_x + \sigma'_z)$	$\dfrac{1-v-2v^2}{(1-v)E} d\sigma'_z$
assumption	$d(\sigma_x + \sigma_z) = 0$	$d\sigma_z = 0$
c_v (coefficient of consolidation)	$c_v^{2D} = \dfrac{KE}{2 \cdot \gamma_w (1+v)(1-2v)}$	$c_v^{1D} = \dfrac{KE \cdot (1-v)}{\gamma_w (1-v-2v^2)}$

of pore pressures $u(\infty, x,y,z)$, that can be obtained directly by integrating equation (15):

$$(h_{xx} + h_{yy} + h_{zz}) = 0 \qquad (15)$$
$$(x,y,z) \in \Omega \; 0 < t \leq T$$

From eq. (15), it is clear that the steady solution does not depend on the properties of the soil.

At equations (10) or (15), hydraulic conditions at boundaries must be added. Among them, water flow through the ground surface affects pore pressures in subsoil more than others. To take this aspect into account, two different hydraulic conditions at ground surface are considered in this paper:

- a film of water continuously present,
- a flux of water varying with a seasonal trend.

When, in practice, drainage works is analysed by means of numerical codes (DEM or FEM), the problem may be solved by taking soil stratigraphy and heterogeneity into account. Pore pressures can be calculated all along the critical sliding surface; then they can be used in slope stability analysis.

Practitioners very often prefer to estimate pore pressure, lowered by drains, by means of non-dimensional charts obtained for homogeneous soil and very simple geometric schemes. Design charts are a general tool: they cannot consider hydraulic conditions at ground surface with a seasonal trend, which necessarily depends on typical climatic features of the region being considered. Hence the design charts presented in this paper are obtained under the most generic assumption of a water film at the ground surface.

3.4.2 First solution (film of water at ground surface)
In this section steady-state solutions are presented for drains operating in 3D conditions, assuming a film of

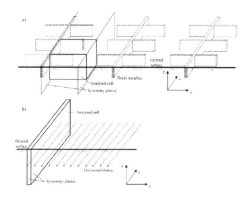

Figure 54. Usual geometry of drain systems: a) drain trenches, b) horizontal drains. The cells analysed in this paper are delimited with bold lines.

water fixed at ground surface. From this configuration 2D solutions are derived as particular cases.

Cells are considered, each containing a single drain and confined on lateral boundaries with impermeable planes (Fig. 54). The scheme represents a case in which an infinite distribution of drains is working, so that it is possible to isolate the soils drained by a single drain, between two consecutive planes of symmetry, that behave as impermeable surfaces. Hydraulic conditions along cell boundaries are indicated in Figure 55. At ground surface the following condition, which represents the film of water at the ground surface and is able to recharge the water table, is imposed:

$$h = z \qquad (16)$$

Under this assumption all the examined domain is submerged.

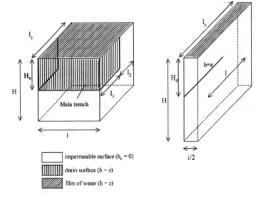

Figure 55. Analysed cells, with indication of hydraulic boundary conditions.

Condition (16) is imposed along the drain boundaries[1]; it physically means that inside the drains the pore air pressure is atmospheric.

The presence of drains inside soils modifies the water level in a complex way. In the analysed cell represented in Figure 56 this situation is shown for the case of drain trenches, by means of pressure head envelopes on three planes parallel to the ground surface at three different depths. The lowering of the water table caused by drains is not homogeneous in Ω: it depends upon the distances of the examined point from the drain boundaries and from the ground surface. The drainage effect is weaker in the deepest zone of the slope.

Drains cause a non-hydrostatic pore pressure distribution, involving a flux downwards. The vertical component of head gradients are larger near the drain boundaries where pore pressures can be much lower than those of hydrostatic ones.

The effect of drain surface on pore pressure distribution is also evident in Figure 57, where the case of three trenches all around the cell is compared to the case with two trenches positioned in front of one another.

The action of horizontal drains is represented in Figure 58, through the pore pressure distribution and its evolution in time, along three vertical axes and three horizontal ones.

To use design charts it is useful to refer the problem to a simple framework that describes pore pressure regime in slopes, as indicated in the following.

---

[1] For horizontal drains, Marino (2007) has shown that the hole can be schematized as a segment coincident with its axis. With this simplification she obtained results very close to the more rigorous ones in which the hole is correctly schematized as a cylinder.

Groundwater regime in natural conditions, before the construction of drains, can be very often schematized through the model of infinite slope (1D), which is a very simple case to analyse, the pore pressure regime being described by the relationship below:

$$u = \frac{\gamma_w \cdot (D - H_w)}{1 + \tan\alpha \cdot \tan\beta} \quad (17)$$

where D and H_w are the depths of a generic point and the water table surface respectively, both measured along the vertical from the ground surface (Fig. 59); α and β are the slopes of the ground surface and of the water flux respectively.

Supposing the water flux is parallel to the ground level and using the quantity z_w, which expresses the depth below the water level measured along the normal to the slope, eq. (17) is changed into eq. (18):

$$u = \gamma_w z_w \cos\alpha \quad (18)$$

As has been shown above, distribution of pore pressure caused by drains is complex. Therefore simplification is required to handle the problem more manageably. Accordingly, 3D pore pressure distribution resulting from the action of drains can be schematized as a 1D distribution, equivalent to 3D distribution as regards its influence on slope stability. 1D distribution can be obtained by replacing the pressure head envelope on each plane parallel to the ground surface with a uniform one. In Figure 60, on the longitudinal section of the slope, is represented the equivalent pore pressure regime that reconnects the problem to that of an infinite slope.

Assuming 1D equivalent condition in groundwater regime, pore pressure is easily calculable. On a planar sliding surface parallel to the ground level, pore pressure is uniform and is calculable by means of eq. (17). Along a circular sliding surface Γ, pore pressure $u_\Gamma(t,x)$ is a function of the depth of the point of Γ. To simplify the analysis mean pore pressure on Γ, $\bar{u}(t,\Gamma)$, can be defined as follows:

$$\bar{u}(t,\Gamma) = \frac{\Gamma \int u_\Gamma(t,x)dx}{l} \quad (19)$$

where l is the length of the unstable mass on the ground level, between the two ends of the sliding surface. Because a film of water is assumed at ground surface, sliding surface is completely submerged, and the average pore pressure $\bar{u}(t,\Gamma)$ along it can be evaluated at the depth $2/3z_\Gamma$, where z_Γ is the maximum depth below the water level in direction z, normal to the slope.

The effect of drainage is often evaluated by means of the efficiency $E(t,x,y,z)$, which in each point of Ω expresses the difference between the initial and current

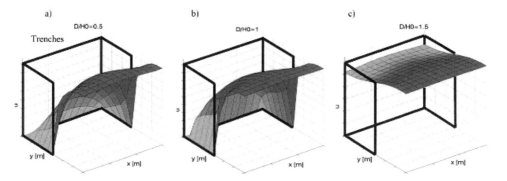

Figure 56. Pressure head envelope between trenches on three planes at different depths: a) $D/H_0 = 0.5$, b) $D/H_0 = 1$, c) $D/H_0 = 1.5$.

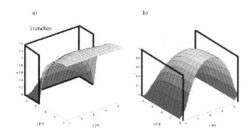

Figure 57. Pressure head envelope on the plane at the base of trenches for the case of 3 drain surfaces (a) and 2 drain surfaces (b), adjacent to the analysed cell.

value of pore pressure (at a generic time t), normalized to the initial value:

$$E(t,x,y,z) = \frac{u(0,x,y,z) - u(t,x,y,z)}{u(0,x,y,z)} \quad (20)$$

When design charts are used, the function E, defined in each point of Ω, is not particularly useful for evaluating the overall stability of slopes. We need a function that expresses the global evolution of pore pressure along the critical sliding surface Γ.

By means of $\bar{u}(t,\Gamma)$ we can define the average efficiency $\bar{E}$ along the sliding surface:

$$\bar{E}(t,\Gamma) = \frac{\bar{u}(0,\Gamma) - \bar{u}(t,\Gamma)}{\bar{u}(0,\Gamma)} \quad (21)$$

Finally, with a view to representing the steady-state solution, the function $\bar{E}_\infty$ can be used:

$$\bar{E}_\infty(\Gamma) = \frac{\bar{u}(0,\Gamma) - \bar{u}(\infty,\Gamma)}{\bar{u}(0,\Gamma)} \quad (22)$$

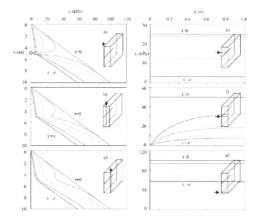

Figure 58. Pore pressure distribution determined by horizontal drains, along 3 vertical axes (a, b, c) and 3 horizontal axes (e, f, g).

The function $\bar{E}_\infty$ plays a key role in designing slope stabilization by drains, because it represents the final distribution of pore pressure $[\bar{u}(\infty,\Gamma)]$, used in the calculation to obtain the desired improvement in slope stability.

In practice, $\bar{E}_\infty(\Gamma)$ is calculated, after determining the function $\bar{u}(\infty,\Gamma)$ from slope stability analysis, as the pore pressure distribution that guarantees the safety factor chosen by the designer. From $\bar{E}_\infty(\Gamma)$, by means of non-dimensional charts, the designer can determine the geometric characteristics of the drain system.

On the basis of the analysis presented in this paper, the value of $\bar{E}_\infty(D)$ has been calculated and represented in design charts (see Appendix A) as a function of the following parameters:

H = depth of analysed volume Ω,
H_0 = depth of drain,

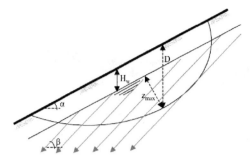

Figure 59. Infinite slope with a generic water flux.

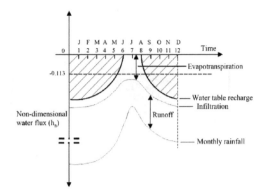

Figure 61. Water flux recharging water table during the year, obtained from monthly rainfall, from which runoff and evapotranspiration are subtracted (from D'Acunto and Urciuoli 2006, modified).

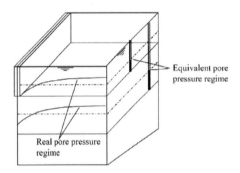

Figure 60. Transformation of the real pressure regime (represented in the cross section) in a equivalent one (in the longitudinal section).

D = depth of the plane on which efficiency is evaluated (correspondent to sliding surface),
L_y = longitudinal length of the analysed volume Ω (in the case of trenches it is the spacing between principal branches of drain trenches),
S = spacing between secondary branches of drain trenches,
i = spacing between horizontal drains,
l_2 = length of secondary branches of drain trenches,
$l_1 = L_y - l_2$.

3.4.3 Second solution (variable water flux at ground surface)

The solution presented in this section regards the effect of an unsteady hydraulic condition at the ground surface, where a function of time, representing water flux which recharges the water table, is applied (Fig. 61). Water recharging the water table is calculated by subtracting runoff and evapotranspiration from rainfall, using a study on groundwater regimes in relation to climatic features in southern Italy (D'Acunto & Urciuoli 2006). The adopted function expresses the concept that when rainfall is abundant a film of water forms at the ground surface and an amount of rain infiltrates. If the climate is cool, evapotranspiration is low and most of the infiltrated water recharges the water table. If the climate is warm, rain evaporates and the quantity that infiltrates is removed by evapotranspiration. This case is solved only for a system of drain trenches analysed in 2D conditions so as to quantify the influence of hydraulic conditions at the ground surface on drainage efficiency.

Having fixed a flux at the ground surface, instead of a head, the position of the water table is unknown and is obtainable from analysis. In the cases studied, the ground water level is always located a few metres from the ground level. In this case, the sliding surface is partially submerged. In order to calculate efficiency (E), in all points above the water table it is assumed u = 0 (neglecting suction); thus efficiency varies from 0 to 1 during the phenomenon of pore pressure equalization. As regards the depth at which efficiency has to be calculated if the sliding surface is circular, mean pore pressure is first calculated along the submerged part of the sliding surface. It corresponds to the pore pressure at an equivalent depth $z_w = 2/3 z_\Gamma$. Hence this value has to be adjusted to take into account the part of the sliding surface above the water level, along which pore pressure is zero. To this aim, in Figure 62 a coefficient [η] is proposed to make this adjustment, based on simple geometric considerations on the examined problem that are not illustrated here to save space. By means of η the equivalent depth must be calculated as $2/3 \cdot \eta \cdot z_\Gamma$.

While at work, drain trenches are partially submerged. This situation influences hydraulic conditions on the contacts between drains and soils. Given that backfilling material has a degree of saturation more or less equal to zero, water can flow only from the subsoil to the trenches and not vice versa. In this case, a double boundary condition is necessary on the trench:

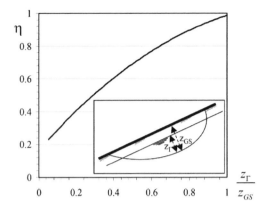

Figure 62. Corrective coefficient to evaluate the equivalent depth at which efficiency must be calculated in the case of partially submerged circular sliding surface.

$h_n = 0$, if $u < 0$ in Ω (the trench boundary must be considered impermeable),
$h = z$, if $u > 0$ in Ω,

where n is normal to the trench boundary.

In Figure 63 the solution of a case with a flux applied at the ground surface, taken from D'Acunto et al. (2007), is reported. The piezometric surface is compared to the pressure head envelope on the plane through the bases of trenches at different times: at the initial condition, one year after, eight years after the beginning of trench work (data on soil and geometry are reported in the figure). In each diagram two axes are reported: on the left there is the height above the $z = 0$ plane (positioned at 10 m from the ground surface), which is useful to locate the water table; on the right there is the pore pressure scale, to represent the envelope of pore pressure on the plane through the base of the trenches. The position of the water table allows us to determine the height of the trench which is working at the time considered (it is the height below the water table). The envelope of pore pressure on a deep plane allows the efficiency of the trench to be assessed on that plane. It can be seen that the water table (whose position is essential to fix the boundary condition on the trench wall) is well above the pore pressure envelope. This means that much of the height of the trench works long after the beginning of drainage, and trenches are always ready to discharge the water flow of heavy rains. This explains the ability of drain trenches to avoid pore pressure peaks (as is shown in Figure 53 for natural regimes) during wet seasons. Measurements carried out on instrumented sites where drains were constructed show that the water table is not subject to seasonal fluctuations where drains are in good working condition. This is clearly shown in the next section.

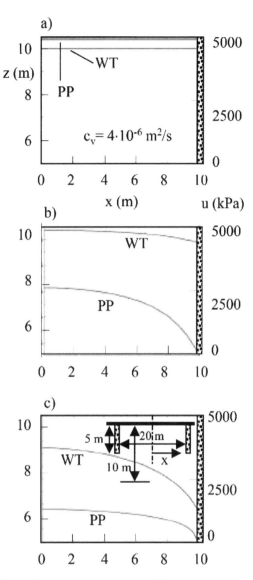

Figure 63. Evolution of water table (WT) and pore pressure (PP) on the plane at the base of trenches in the case of a water flux applied at the ground surface: a) initial condition, b) 1 year after the beginning of the drainage process, c) 8 years after (from D'Acunto et al, 2007). In figure a) the water table and pore pressure are coincident; the difference is due to different scales used for the representation of WT and PP.

3.5 *Case history*

3.5.1 *The site*

In 1990, in southern Italy, during highway works along the Sele river some ancient landslides were reactivated as a consequence of trench excavations. After these events further works was needed to assure the safety of the highway. In particular, in the urban district of Contursi, the highway crosses an extensive unstable zone affected by several landslides. The unstable area was stabilized by drains, consisting of trenches in some zones and wells with horizontal drainage pipes in others (Fig. 64).

The groundwater regime was investigated before and after drain construction, using Casagrande piezometers installed inside boreholes in the whole area. The investigation program was fairly extensive, aiming to measure pore pressure fluctuations inside and around the landslide body and in the substratum, in order to determine the hydraulic conditions at the boundary of the unstable area. Piezometer measurements were carried out continuously from 1991 onwards (Fig. 65). Data are grouped according to zones. As regards zones far away from drains (Fig. 65a), where the groundwater regime is affected by natural conditions, pore pressure undergoes seasonal fluctuations of the order of some metres, more significant in the upper cells but noticeable also in the lower cells. As regards the drained zone, measurements at piezometers installed between trenches show that pore pressures are well below those measured far from the drained area. Moreover, seasonal fluctuations more or less vanish. This is more evident in zone B, which is the closest to the drain system, while in zone A the behaviour is intermediate.

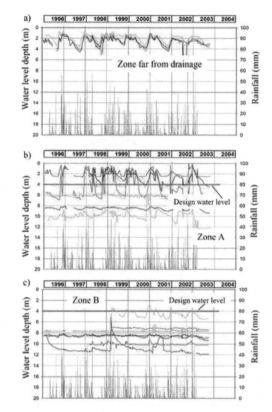

Figure 65. Pore pressure measurements at Contursi in zones: a) far from drainage, b) intermediate, c) very close to drainage (from Pellegrino et al. 2004a).

These observations show the greater stabilizing effect of drains, since the maximum pore pressure, attained during winter and early spring, is the cause of reactivation of seasonal landslides. This aspect of drain action consisting in the "lamination" of the seasonal trend has not been stressed in the literature despite its great importance in slope stability.

At the Contursi site, soils were investigated extensively both in the laboratory and in the field: the most reliable value of permeability k was obtained from fallen head tests carried out inside piezometers. Results were very scattered, with an average close to 10^{-8} m/s. As regards the coefficient of consolidation c_v, a value of $3.8 \cdot 10^{-5}$ m^2/s was calculated from the permeability obtained by in situ tests and oedometer modulus E_{oed}, by laboratory tests ($c_v = \frac{k \cdot E_{ed}}{\gamma_w}$, $\gamma_w =$ unit weight of water).

3.5.2 Analysis of groundwater regime in the site of Contursi

Referring to the described site the slope was modelled as a trapezium (Fig. 66), constituted by homogeneous

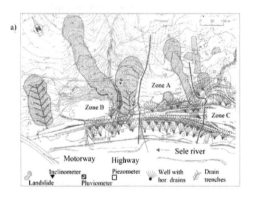

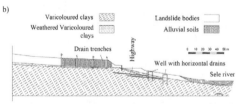

Figure 64. The slope uphill of the highway along the Sele river in Contursi; stabilized with drain trenches and wells: a) map, b) stratigraphic section (from Pellegrino et al. 2004a).

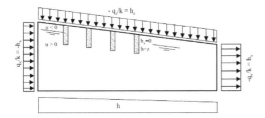

Figure 66. Analysed domain with hydraulic conditions adopted at the boundaries (D'Acunto and Urciuoli 2006).

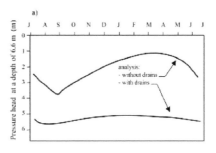

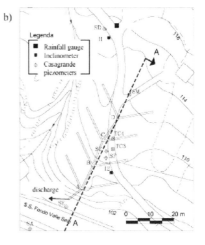

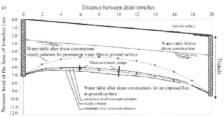

soil, whose bases are 105 m (H) and 70 m, and length 280 m(L). The slope α is 7°1'. Numerous parametric analyses were performed in order to study the response of the water table to different conditions at the boundaries. It emerged that water flow through the ground surface was the hydraulic condition that most influenced pore pressures in subsoil around the slip surface. This result coincides with the fact that the slip surface is relatively shallow (its depth is always less than 9 m). Hence major efforts were made to examine the hydraulic condition at the ground surface, which allowed the function in Figure 61 to be obtained.

The vertical boundaries on the uphill side were considered a section of an almost infinite slope, since there was no perturbation either of the geometry or soil stratigraphy nearby. It was then supposed that around the boundary the pore pressure was constant along the direction, parallel to the ground surface. All along the base of the trapezium, the piezometric head was assigned according to a linear function, starting at the downhill side where the water level was at 65 m (in the Sele River). As regards the hydraulic condition on the downhill vertical side of the examined volume it should be noted that this boundary is very close to the alluvial soils of the Sele river. Hence it is considered to have a greater capability of draining water from the analysed subsoil.

The analysis was developed in transient conditions, regulated by the function in Figure 61. Thus results were obtained that represent the pore pressure fluctuations without and with drains as a consequence of the variable hydraulic conditions at ground surface. When drains are working, fluctuations are drastically reduced: Figure 67a shows the seasonal trend of pressure head in a point at a depth of 6.6 m from the ground surface. Numerical results are compared (Fig. 67c) with measurements at piezometers inside two secondary branches of trenches (piezometers TC3, Sc and SF in Fig. 67b). It can be seen that the solution obtained for a permanent film of water at the ground surface overestimates pore pressure in the subsoil. Measurements are well fitted by the analytical solution with flux imposed at the ground surface.

Figure 67. Groundwater regime at Contursi. Comparison between measurements and numerical results (from D'Acunto and Urciuoli, 2006): a) calculated pore pressure without and with drains, b) position of piezometers with respect to drain trenches, c) comparison between measurements and numerical results for different boundary conditions.

3.5.3 Influence on efficiency of the hydraulic condition at the ground surface

By considering the imposed flux at the ground surface, instead of the water film, the following results were obtained:

- the water table is strongly depressed compared with the conditions without drains,
- seasonal fluctuations are lowered by drains.

To evaluate efficiency, reference should be made to maximum values assumed by functions u_o and u_∞ during the year. It is thus possible to calculate that efficiency is up to 40% greater than that corresponding to a film of water fixed at the ground surface (D'Acunto & Urciuoli 2006).

4 CONCLUSION

Soil nailing is a robust slope stabilization measures and is generally applied to stabilize man-made slopes, retaining walls or excavations. The interaction between the ground and the soil nails is complex. Nevertheless, the findings of numerous researches have provided some understanding of the mechanism and behaviour of soil-nailed systems, particularly on the effect of nail inclination, soil-nail heads and bending stiffness of reinforcement bars. Design practice is well established. In contrast, there is room for advancement of the construction technology and the application of innovative materials as soil nail reinforcement. The use of non-destructive technique can greatly enhance the confidence of quality control of soil nail construction.

Subsurface drainage stabilizes the ground by preventing significant rise in porewater pressure. Compared with other types of engineering works, it is a very cost-effective stabilization measures, especially at large site. Different types of drains have been developed and they serve different purposes. Trench drains and horizontal drains are more commonly used than other types of drains, and simple design charts have been developed to facilitate designers to assess the effects of the drains on pore pressures.

ACKNOWLEDGEMENT

This paper is published with the permission of the Head of the Geotechnical Engineering Office and the Director of Civil Engineering and Development, Government of the Hong Kong Special Administrative Region.

The part of this paper on soil nailing technology was based on the review and research work carried out in the Geotechnical Engineering Office in Hong Kong. The soil-nail related studies was mainly led by the first author with contributions made by many colleagues, particularly Mr Y.K. Shiu, Dr W.M. Cheung, Dr G.W.K. Chang and Dr D.O.K. Lo. All contributions are gratefully acknowledged.

The part of this paper on drainage was based on the research work carried out in the Department of Geotechnical Engineering in Naples. The related studies was mainly led by the second author with contributions made by Prof. Berardino D'Acunto, Ms. Nunzia D'Esposito (Eng.) and Ms. Roberta Marino (Eng.), that are gratefully acknowledged.

REFERENCES

ACI 2001. *Guide for the Design and Construction of Concrete Reinforced with FRP Bars (ACI 440.1R-01)*. American Concrete Institute. USA.

Babu, G.L.S., Murthy, B.R.S. & Srinivas, A. 2002. Analysis of construction factors influencing the behaviour of soil-nailed earth retaining walls. *Ground Improvement* 6(3): 137–143.

Benmokrane, B., Xu, H. & Nishizaki, I. 1997. Aramid and carbon fibre-reinforced plastic prestressed ground anchors and their field applications. *Canadian Journal of Civil Engineering* 24: 968–985.

Brady, K.C. & McMahon, W. 1993. *The Durability of Corrugated Steel Buried Structures*. Transport and Road Research Laboratory, Department of Transport.

Bridle, R.J. & Davies, M.C.R. 1997. Analysis of soil nailing using tension and shear: experimental observations and assessment. *Geotechnical Engineering, Proceedings of the Institution of Civil Engineers*, July, 155–167.

BS: 6031 1981. *Code of Practice for Earthworks*. British Standard Institution.

Calladine, C.R. 2000. *Plasticity for Engineers: Theory and Applications*. Horwood Publishing Limited.

Cheuk, C.T., Ng, C.W.W. & Sun, H.W. 2001. Numerical analysis of soil nails in loose fill slopes. *Proc. 14th Southeast Asian Geot. Conf.*, Hong Kong, 1: 725–730.

Cheung, W.M. 2003. *Non-Destructive Tests for Determining the Lengths of Installed Steel Soil Nail (GEO Report No. 133)*. Geotechnical Engineering Office, Hong Kong.

CIRIA 2005. *Soil Nailing—Best Practice Guidance (CIRIA C637)*. CIRIA, UK.

Clouterre 1991. *French National Research Project Clouterre—Recommendations Clouterre*. (English Translation 1993). Federal Highway Administration, FHWA-SA-93-026, Washington, USA.

D'Acunto, B. & Urciuoli, G. 2006. Groundwater regime in a slope stabilised by drain trenches. *Mathematical and Computer Modelling. Pergamon-Elsevier Science LTD*, 43(7–8): 754–765.

D'Acunto, B., Parente, F. & Urciuoli, G. 2007. Numerical models for 2D free boundary analysis of groundwater in slopes stabilized by drain trenches. *Computers & Mathematics with applications*, 53(4): 1615–1626.

Davies, M.C.R. & Le Masurier, J.W. 1997. Soil/nail interaction mechanism from large direct shear tests. *Proc. Third Int. Conf. on Ground Improvement GeoSystems*, London, 493–499.

Dawson, E.M., Roth, W.H. & Drescher, A. 1999. Slope stability analysis by strength reduction. *Géotechnique* 49(6): 835–840.

Department of Transport 1994. *Design Manual for Roads and Bridges: Design Methods for the Reinforcement of*

Highway Slopes by Reinforced Soil and Soil Nailing Techniques, HA68/94, Department of Transport, UK.

Department of Transport 2001. *BD 12/01—Design of Corrugated Steel Buried Structures with Spans Greater than 0.9 Metres and up to 8 Metres*. (Design Manual for Road and Bridges). Department of Transport, UK.

Desideri, A., Miliziano, S. & Rampello, S. 1997. *Drenaggi a Gravità per la Stabilizzazione dei Pendii*. Hevelius Edizioni, Benevento.

D'Esposito, N. 2007. *Analisi 3D Dell'efficienza diTtrincee Drenanti Utilizzate per la Stabilizzazione dei Pendii*. Graduate thesis, University of Naples Federico II.

Dowding, C.H. & Huang, F.C. 1994. Early detection of rock movement with time domain reflectometry. *Journal of Geotechnical Engineering* 120: 1413–1427.

Duncan, J.M. 1996. State of the art: limit equilibrium and finite element analysis of slopes. *Journal of Geotechnical Engineering ASCE* 122(1): 577–596.

Ehrlich, M., Almeida, M.S.S. & Lima, A.M. 1996. Parametric Numerical Analyses of Soil Nailing Systems. In H. Ochiai, N. Yasufuku & K. Omine (eds.), *Earth Reinforcement* 747–752. Balkema, Rotterdam.

Eisbacher, G.H. & Clague, J.J. 1984. Destructive mass movements in high mountains: hazard and management. *Geol. Surv. Can., Pap.* 84–16. 230 p.

Eyre, D. & Lewis, D.A. 1987. *Soil Corrosivity Assessment*. Contractor Report 54. Transport and Road Research Laboratory, Department of Transport, UK.

Fell, R. 1994. Stabilization of soil and rock slopes. *Proc. of East Asia Sym. and Field Workshop on Landslides and Debris Flows*, Seoul, 1: 7–74.

FHWA 1998. *Manual for Design & Construction monitoring of Soil Nail Walls. Federal Highway Publication No. SA-96-069R*. U.S. Department of Transportation, Federal Highway Administration, Washington, D.C.

FHWA 2003. *Geotechnical Engineering Circular No. 7: Soil Nail Walls*, Report No. FHWA0-IF-03-017, Federal Highway Administration, Washington, USA.

Gässler, G. 1997. Design of reinforced excavations and natural slopes using new European Codes. In H. Ochiai, N. Yasufuku & K. Omine (eds.), *Earth reinforcement* 943–961. Balkema, Rotterdam.

Gässler, G. & Gudehus, G. 1981. Soil nailing—some aspects of a new technique. *Proc. Tenth Int. Conf. Soil Mechanics and Foundation Engineering*, Stockholm, 3: 665–670.

GEO 2000. *Technical Guidelines on Landscape Treatment and Bio-engineering for Man-made Slopes and Retaining Walls (GEO Publication No. 1/2000)*. Geotechnical Engineering Office, Hong Kong.

GEO 2008. *Guide to Soil Nail Design and Construction (Geoguide 7)*. Geotechnical Engineering Office, Hong Kong.

Gigan, J.P. & Delams, P. 1987. Mobilisation of stresses in nailed structures. English translation. *Transport and Road Research Laboratory, Contractor Report* 25.

Ground Engineering 2004. Soil-nailing—called to the bar. *Ground Engineering*, May, 14.

Gutierrez, V. & Tatsuoka, F. 1988. Roles of facing in reinforcing cohesionless soil slopes by means of metal strips. *Proc. Int. Geot. Sym. on Theory and Practice of Earth Reinforcement*, (IS Kyushu '88), Fukuoka, 289–294. Balkema, Rotterdam.

Halcrow China Limited 2007. *Study on the Potential Effect of Blockage of Subsurface Drainage by Soil Nailing Works—Study Report (GEO Report No. 218)*. Geotechnical Engineering Office, Hong Kong.

Hausmann, M.R. 1992. Slope remediation. (Invited lecture). *Proc. Specialty Conf. on Stability and Performance of Slopes and Embankments—II*, Berkeley, Geotechnical Special Publication No. 31, American Society of Civil Engineers, 2: 1274–1317.

Hayashi, S., Ochiai, H., Yoshimoto, A., Sato, K. & Kitamura, T. 1988. Functions and effects of reinforcing materials in earth reinforcement. *Proc. Int. Geot. Sym. on Theory and Practice of Earth Reinforcement*, (IS Kyushu '88), Fukuoka, 99–104. Balkema, Rotterdam.

Hewlett Packard 1998. *Time Domain Reflectometry Theory (Application Note 1304-2)*. Hewlett Packard Company, USA.

HKIE 2003. *Soil Nails in Loose Fill Slopes—A Preliminary Study*. Hong Kong Institution of Engineers Geotechnical Division, Hong Kong.

Ho, K.K.S. 2004. Keynote paper: Recent advances in geotechnology for slope stabilization and landslide mitigation—perspective from Hong Kong. *Proc. Ninth Int. Sym. on Landslides*, Rio de Janeiro, 2: 1507–1560.

Holtz, R.D. & Schuster, R.L. 1996. Stabilization of soil slopes. In A.K. Turner & R.L. Schuster (eds.), *Landslides Investigation and Mitigation* 439–473. Transportation Research Board, Special Report 247, National Research Council, National Academy Press, Washington.

Hungr, O., Morgan, G.C. & Kellerhals, R. 1984. Quantitative analysis of debris torrent hazards for design of remedial measures. *Can. Geotech. J.* 21: 663–677.

Hungr, O., Morgan, G.C., VanDine, D.F. & Lister, D.R. 1987. Debris flow defences in British Columbia. In J.E. Costa & G.F. Wieczorek (eds), *Debris flows/avalanches: process, recognition and mitigation. Reviews in Engineering Geology*. Geol. Soc. Am., 7: 201–222.

Hutchinson, J.N. 1977. Assessment of the effectiveness of corrective measures in relation to geological conditions and types of slope movement (General Report). *Bulletin of the Int. Association of Engineering Geology* 16: 131–155.

Itasca 1996. *Fast Lagrangian Analysis of Continua (FLAC) Manual*, Version 4.0. Itasca Consulting Group, Inc., Minnesota.

IStructE 1999. *Interim Guidance on the Design of Reinforced Concrete Structures Using Fibre Composite Reinforcement*. The Institution of Structural Engineers, UK.

Japan Highway Public Corporation (JHPC) 1998. *Design & Works Outlines on the Soil-Cutting Reinforcement Soilworks*. (English Translation). Japan Highway Public Corporation.

Jewell, R.A. 1980. *Some Effects of Reinforcement on the Mechanical Behaviour of Soils. PhD thesis*, University of Cambridge.

Jewell, R.A. & Pedley, M.J. 1990. Soil nailing design: the role of bending stiffness. *Ground Engineering*, March, 30–36.

Jewell, R.A. & Pedley, M.J. 1992. Analysis for soil reinforcement with bending stiffness. *Journal of Geotechnical Engineering, ASCE*, 118(4): 1505–1528.

Japanese Ministry of Construction 1997. Design guidelines of FRP reinforced concrete building structures. *Journal of Composites for Construction*, August, 90–115.

JSCE 1997. *Recommendations for Design and Construction of Concrete Structures Using Continuous Fiber Reinforcing Materials*. Concrete Engineering Series, Japan society of Civil Engineers, No. 23.

Kellerhas, R. 1970. Runoff routing through steep natural channels. *ASCE Journal of Hydraulics Division*, 96: 2201–2217.

Kenney, T.C. & Lau, K.C. 1984. Temporal changes of groundwater pressure in a natural slope of non fissured clay. *Can. Geotech. J.* 21(1): 138–146.

Kenney, T.C., Pazin, M. & Choi, W.S. 1977. Design of horizontal drains for soil slopes. *ASCE Journal of Geotechnical Engineering Division*, 103(GT11): 1311–1323.

Krahn, J. 2003. The 2001 R.M. Hardy Lecture: The limits of limit equilibrium analyses. *Canadian Geotechnical Journal* 40: 643–660.

Law, K.T., Shen, J.M. & Lee, C.F. 1998. Strength of a loose remoulded granitic soil. In K.S. Li, J.N. Kay & K.K.S. Ho (eds.), *Slope Engineering in Hong Kong, Proc. HKIE Annual Seminar*, Hong Kong, 169–176. Balkema.

Lee, C.F. & OAP 2007. *Review of Use of Non-destructive Testing in Quality Control in Soil Nailing Works (GEO Report No. 219)*. Geotechnical Engineering Office, Hong Kong.

Lin, C.P. & Tang, S.H. 2005. Development and calibration of a TDR extensometer for geotechnical monitoring. *Geotechnical Testing Journal* 28(5): 464–471.

Liu, W., Hunsperger, R., Chajes, M., Folliard, K. & Kunz, E. 2002. Corrosion detection of steel cables using time domain reflectometry. *Journal of Materials in Civil Engineering*, ASCE, 14(3): 217–223.

Leventhal, A.R. & Mostyn, G.R. 1987. Slope stabilization techniques and their application. In *Slope Stability and Stabilisation*, 183–230. Balkema, Rotterdam.

Lui, B.L.S. & Shiu, Y.K. 2004. Prescriptive soil nail design for concrete and masonry retaining walls. *Proc. the HKIE Geotechnical Division Annual Seminar 2004—Recent Advances in Geotechnical Engineering*, Hong Kong, 185–197.

Lui, B.L.S. & Shiu, Y.K. 2005. *Prescriptive Soil Nail Design for Concrete and Masonry Retaining Walls (GEO Report No. 165)*. Geotechnical Engineering Office, Hong Kong.

Marchal, J. 1986. *Soil Nail—Experimental Laboratory Study of Soil Nail Interaction*. (English Translation). Transport and Road Research Laboratory, Department of Transport, Contractor Report No. 239.

Marino, R. 2007. *Analisi 3D Dell'efficienza di Aste Drenanti per la Stabilizzazione dei Pendii*. Graduate thesis, University of Naples Federico II.

Mears, A.I. 1981. *Design Criteria for Avalanche Control Structures in the Runout Zone*. United States Department of Agricolture, Forest Service General Technical Report RM-84.

Muramatsu, M., Nagura, K., Sueoka, T., Suami, K. & Kitamura, T. 1992. Stability analysis for reinforced cut slopes with facing. *Proc. Int. Sym. on Earth Reinforcement Practice*, Fukuoka, Kyushu, Japan, 503–508.

Murray, R.T. 1993. *The Development of Specifications for Soil Nailing. Research Report 280*, Transport Research Laboratory, Department of Transport, UK.

Ng, C.W.W. & Chiu, A.C.F. 2003. Laboratory study of loose saturated and unsaturated decomposed granitic soil. *Journal of Geotechnical and Geoenvironmental Engineering*, ASCE 129(6): 550–559.

Ng, C.W.W., Pun, W.K., Kwok, S.S.K., Cheuk, C.Y. & Lee, D.D.M. 2007. Centrifuge modelling in engineering practice in Hong Kong. *Geotechnical Advancements in Hong Kong since 1970s, Proc. HKIE Geotechnical Division Annual Seminar*, Hong Kong, 55–68.

O'Connor, K.M. & Dowding, C.H. 1999. *GeoMeasurements by Pulsing TDR Cables and Probes*. CRC Press.

Palmeira, M. & Milligan, G.W.E. 1989. Large scale direct shear tests on reinforced soil. *Soils and Foundations*, 29(1): 18–30.

Pang, L.S. & Wong, H.N. 1997. Prescriptive design of soil nails to upgrade soil cut slopes. In K.S. Li, J.N. Kay & K.K.S. Ho (eds.), *Slope Engineering in Hong Kong, Proc. HKIE Annual Seminar*, Hong Kong, 259–266.

Pedley, M.J. 1990. *The Performance of Soil Reinforcement in Bending and Shear. PhD thesis*, University of Oxford.

Pellegrino, A., Ramondini, M. & Urciuoli, G. 2004a. Regime delle pressioni neutre in un pendio in Argille Varicolori stabilizzato con trincee drenanti. *International Workshop: "Living with landslides: effects on structures and urban settlements. Strategies for risk reduction"*, 141–149. Anacapri.

Pellegrino, A., Ramondini, M. & Urciuoli, G. 2004b. Interplay between the morphology and mechanics of mudslides: field experiences from Southern Italy. *IX International Symposium on Landslides*, 2: 1403–1409. Rio de Janeiro.

Perry, J., Pedley, M.J. & Reid, M. 2003. *Infrastructure embankments- condition appraisal and remedial treatment. (Second edition)*. CIRIA Report C592. CIRIA, UK.

Plumelle, C., Schlosser, F., Delage, P. & Knochenmus, G. 1990. French National Research Project on Soil Nailing. *Proc. Conf. of Design and Performance of Earth Retaining Structures*, ASCE Geotechnical Special Publication No. 25, 18–21 June, 660–675.

Plumelle, C. & Schlosser, F. 1990. A French National Research Project on Soil Nailing: Clouterre. In A. McGown, K.C. Yeo & K.Z. Andrawes (eds.), *Performance of Reinforced Soil Structures* 219–223. British Geotechnical Society.

Pun, W.K., Cheung, W.M., Lo, D.O.K. & Cheng, P.F.K. 2007. Application of time domain reflectometry for quality control of soil nailing works. *Proc. Int. Forum on Landslide Disaster Management*, Hong Kong (in press).

Pun, W.K., Pang, P.L.R. & Li, K.S. 2000. Recent development in prescriptive measures for slope improvement works. *Proc. Sym. on Slope Hazards and Their Prevention*, Hong Kong, 303–308.

Pun, W.K. & Shiu, Y.K. 2007. Design practice and technical developments of soil nailing in Hong Kong. *Proc. HKIE Geotechnical Division Annual Seminar*, May, Hong Kong, ??.

Shanmuganathan, S. 2003. Fibre reinforced polymer composite materials for civil and building structures—review of the state-of-the-art. *The Structural Engineer*, July, 26–33.

Schlosser, F. 1982. Behaviour and design of soil nailing. *Proc. of Sym. on Recent Developments in Ground Improvements*, Bangkok, 399–413.

Schlosser, F. & Guilloux, A. 1981. Le forttement dans les sols. *Revue Francaise de Gèotechnique*, 16: 65–77. (in French)

Schuster, R.L. 1992. Recent advances in slope stabilization. (Keynote paper). *Proc. of Sixth Int. Sym. on Landslides*, Christchurch, 3: 1715–1745.

Shiu, Y.K. & Cheung, W.M. 2003. *Long-term Durability of Steel Soil Nails (GEO Report No. 135)*. Geotechnical Engineering Office, Hong Kong.

Shiu, Y.K. & Chang, G..W.K. 2004. *Soil Nail Head Review (GEO Report No. 175)*. Geotechnical Engineering Office, Hong Kong.

Shiu, Y.K. & Chang, G.W.K. 2005. *Effects of Inclination, Length Pattern and Bending Stiffness of Soil Nails on Behaviour of Nailed Structures (GEO Report No. 197)*. Geotechnical Engineering Office, Hong Kong.

Siddiqui, S.I., Drnevich, V.P. & Deschamps, R.J. 2000. Time domain reflectometry for use in geotechnical engineering. *Geotechnical Testing Journal*, ASTM, 23(1): 9–20.

Smith, I.M. & Su, N. 1997. Three-dimensional FE analysis of a nailed wall curved in plan. *International Journal for Numerical and Analytical Methods in Geomechanics* 21: 583–597.

Stocker, M.F. & Riedinger, G. 1990. The bearing behaviour of nailed retaining wall. *Proceedings of Design and Performance of Earth Structure*, Geotechnical Special Publication No. 25, ASCE, 613–628.

Tan, S.A., Luo, S.Q. & Yong, K.Y. 2000. Simplified models for soil-nail lateral interaction. *Ground Improvement*, Thomas Telford, 4(4): 141–152.

Tei, K., Tayor, R.N. & Milligan, G.W.E. 1998. Centrifuge model tests of nailed soil slopes. *Soils and Foundations* 38(2): 165–177.

Thompson, S.R. & Miller, R. 1990. Design, construction and performance of a soil nailed wall in Seattle, Washington. *Proc. Conf. on Design and Performance of Earth Retaining Structures*, Geotechnical Special Publication No. 25, June 18–21, 1990, New York, 629–643.

Topp, G.C., Davis, J.L. & Annan, A.P. 1980. Electromagnetic determination of soil water content: measurement in coaxial transmission lines. *Water Resources Research* 16(3): 574–582.

Thurber Consultants Ltd. 1983. *Debris Torrent and Flooding Hazards, Highway 99, Howe Sound*. Report to B.C. Min. Transportation and Highways, Victoria, B.C.

Unwin, H. 2001. Carbon fibre soil nailing for railway embankments. *Proc. Int. Exhibition and Sym. of Underground Construction*, London, 697–706.

Urciuoli, G .1998. Pore pressures in unstable slopes constituted by fissured clay shales. *2nd Int. Symp. on The Geotechnics of Hard Soils—Soft Rocks*, 2: 1177–1185. Napoli.

Van Dine D.F. 1996. *Debris Flow Control Structures and Forest Engineering*. Working paper, Ministry of Forests, British Columbia.

Veder, C. 1981. *Landslides and their stabilization*. Springer-Verlag, New York.

Walker, B.F. & Mohen, F.J. 1987. Ground water prediction and control and negative pore water pressure. In Walker & Fell (eds.), *Soil Slope Instability and Stabilisation*, 121–181.

Watkins, A.T. & Powell, G.E. 1992. Soil nailing to existing slopes as landslip preventive works. *Hong Kong Engineer*, March, 20–27.

Wong, H.N., Pang, L.S., A.C.W., Pun, W.K. & Yu, Y.F. 1999. *Application of Prescriptive Measures to Slopes and Retaining Walls (GEO Report No. 56). (Second Edition)*. Geotechnical Engineering Office, Hong Kong.

Yin, J.H. & Su, L.J. 2006. An innovative laboratory box for testing nail pull-out resistance in soil. *Geotechnical Testing Journal* 29(6): 451–461.

Zaruba, Q. & Mencl, V. 1982. *Landslides and their control*. New York, Elsevier.

Appendix A. Example of design charts.

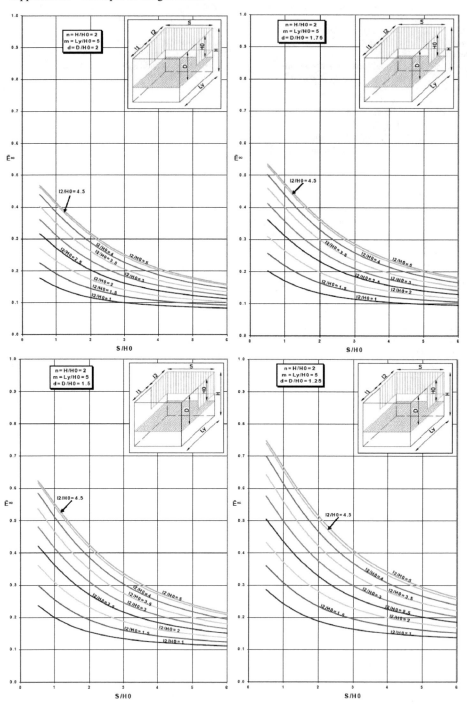

Special lectures

Loess in China and landslides in loess slopes

Z.G. Lin & Z.J. Xu
Northwest Research Institute of Engineering Investigations and Design, Xian, China

M.S. Zhang
Xian Geological and Mineral Resources Research Institute, Xian, China

ABSTRACT: Loess in china is outstanding for its stratigraphical intactness, huge thickness, vast expanse of distribution and metastable geotechnical properties. During the Quaternary Period the arid, semi-arid climatic environment provided the stage on which material sources came into being, grain particles were wind-transported and deposited and loess strata were thus formed. The paper puts emphasis on the granulometry and engineering properties peculiar to loessial deposits as well as the role they might play in the initiation and evolution of loess landslides. From this, the paper proceeds to such problems as geomorphological zonation, landslide distribution, morphological and structural features of loess landslides as well as landslide classifications, giving due attention to seismic landslides. This is followed by a summing-up of the methods of landslide prevention and remediation. Finally, the issues that need more attention and further in-depth research are raised.

1 INTRODUCTION

Loess covers large and wide parts of Northwest, North and Northeast China, its distribution being concentrated on the so-called Central Loess Plateau which comprises Shaanxi, Gansu and Shanxi provinces as well as most of Ningxia Autonomous Region. Here the loess deposit formed in the unique Quaternary climatic environment is outstanding for its stratigraphical intactness, sustained distribution and huge total thickness. The granulometric composition of loess is remarkable for stableness, maintaining in the meanwhile a well-oriented gradual change from Northwest to Southeast. In engineering properties the soil is characterized by its state of consolidation (underconsolidated in the upper part and slightly overconsolidated in its lower part). These properties are likewise remarkable for their regular and oriented change in the general direction from Northwest to Southeast all across the Loess Plateau.

In such climatic, sedimentation and stratigraphical environment conditions, geomorphological evolution has followed correspondingly unique paths resulting in the formation of loess landslides peculiar to such conditions. In the push for the development of the West, it is important to have a correct understanding of loess landslides and use proper methods of investigation, prevention and remediation. This paper tries to give a brief summing-up of the research results achieved and experiences accumulated up to now so as to contribute to the development of new research programs and new prevention and remediation methods.

2 DISTRIBUTION OF LOESS IN CHINA AND ITS ENGINEERING PROPERTIES

2.1 Distribution of loess in China

Loess in China is mainly distributed in Gansu, Shaanxi, Shanxi, Henan and Qinghai provinces, the Ningxia Autonomous Region and the Inner Mongolian Autonomous Region, blanketing the Central Loess Plateau and its neighboring areas. Next in distribution are Hebei, Shandong, Liaoning and Heilongjiang provinces, and the Xinjiang Autonomous Region. The land surface covered by loess in China is estimated to be 6.3×10^5 km^2 which is equal to about 6.6% of the total land area of the country. See Figure 1.

2.2 Stratigraphical features

Stratigraphical intactness and huge total thickness are prominent features of loess in China, owing to which almost complete records of the geographical, climatic, depositional and biological changes and evolution over the entire Quaternary Period have been preserved. What is shown in Figure 2 is a typical stratigraphical profile of the loess in China.

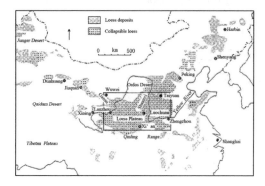

Figure 1. Distribution of loess in China.

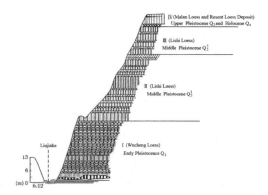

Figure 2. A typical sketch profile of loess in China.

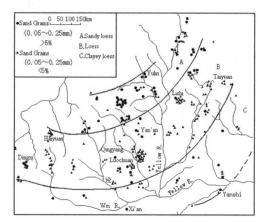

Figure 3. Granulometric zoning of the loess plateau.

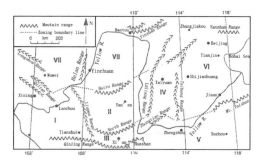

Figure 4. Engineering geological zonation of the loess of China.

2.3 *Granulometry*

Throughout the Quaternary Period the granulometric composition kept stable. However against this background a gradual change from coarse to fine in the general direction from NW to SE is discernible (Fig. 3). This stableness in composition is reflective of the geographical, climatic and depositional environment conditions of that period and evidences the eolian origin of loessial deposits. Of course the existence of the numerous reddish brown paleo-soil layers bears evidence of the climatic fluctuations in the generally stable environment.

2.4 *Engineering properties*

Of the engineering properties of loess four deserve special attention. They either play important roles in the initiation and evolution of landslides or are contributive to the behavior of man-made slopes.

2.4.1 *Collapsibility*
Loessial terrains (or construction sites) fall into two types, i.e. the type which collapses under the pressure from its overburden weight when saturated and that which collapses under the combined pressure from its own overburden and additional loads (fills, foundations, embankments etc) when saturated. In Figure 4 is shown the engineering geological zonation of the loess in China and in Table 1 are given the soils laboratory test results (mean values of physico-mechanical indices) for Zone I, Zone II and Zones I, II, III, IV, V and VII as a whole. Figure 5 is a generalization of the afore-mentioned regularity of oriented gradual change (amelioration) in engineering properties of loess in the direction from NW to SE all across the Central Loess Plateau and its neighboring regions. And Figure 6 is a photo which tells of the grave consequences that could be brought about by flooding in overburden-collapsible loessial terrains (Qian et al. 1988, Lin 1994, Wang et al. 1990).

In the meanwhile, it should be noted that in step with increase of water content the soil sample invariably tends to increase in compressibility with corresponding decrease in collapsibility, as shown in Figure 7.

Table 1. Soils laboratory test results (mean values of physico-mechanical indices) for different zones.

Zone	Age	Nat.wat. cont. w(%)	Unit wt. γ(kN·m⁻³)	Void rat. e	Liquid lim. w_L(%)	Plast. index I_p(%)	Coef.of overburd. collaps. δ_{zs}	Init. pressure P_{sh}(MPa)	Precons. press.(nat.) P_c(MPa)	Precons. press.(sat.) P_c(MPa)	Sand grains (%)	Slit cont. (%)	Clay cont. (%)
Gansu Prov. (Zone I)	Q_3	8.98	14.58	1.032	27.60	9.37	0.052	0.135	0.508	0.137	11.20	78.75	10.05
	Q_2^2	6.79	15.37	0.868	26.87	7.34	0.054	0.270	2.047	0.287	10.00	76.00	14.00
	Q_2^1	4.90	16.20	0.742	26.40	8.90	0.051	1.000	3.800	1.040	9.00	77.00	14.00
North. Shaanxi and	Q_3	13.38	14.70	1.105	29.9	11.2	0.023	0.092	0.835	0.097	13.20	65.80	21.00
	Q_2^2	15.30	16.40	0.924	30.9	12.2	0.014	0.416	1.224	0.473	15.00	68.10	16.90
Longdong (Zone II)	Q_2^1	18.10	18.30	0.760	30.6	12.3	0.008	0.684	1.680		12.00	64.00	24.00
	Q_1	20.10	18.50	0.773	31.1	11.7	0.003		1.856	1.615	13.00	70.00	17.00
Six Zones (Zone I-V)	Q_3	13.21	15.26	1.016	28.62	10.86	0.015	0.137	0.580	0.158	10.96	70.70	18.34
	Q_2^2	12.85	16.26	0.884	29.11	10.79	0.018	0.419	1.380	0.453	15.19	68.62	16.19
and Zone VII	Q_2^1	13.09	17.23	0.783	29.01	11.25	0.026	0.981	2.148	1.036	12.25	69.87	17.88
	Q_1	20.10	18.55	0.773	31.05	11.65	0.003		1.855	1.615	13.00	70.00	17.00

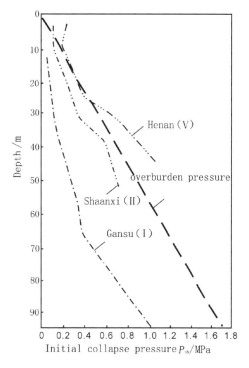

Figure 5. Characteristic generalization of the initial collapse pressure-overburden pressure relationships.

Figure 6. General view and close-up of ground subsidence in and around test pit caused by flooding (Q_4 and Q_3 loess).

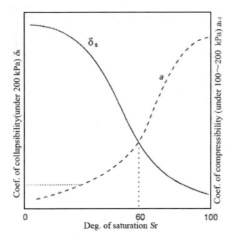

Figure 7. Interrelationship between the coefficient of collapsibility, δ_s, and the coefficient of compressibility, a, as a function of the degree of saturation, S_r.

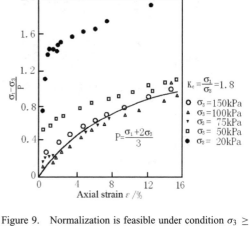

Figure 9. Normalization is feasible under condition $\sigma_3 \geq 75$ kPa.

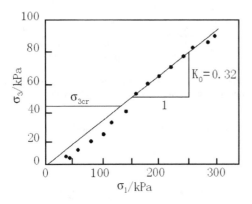

Figure 8. K_0 becomes constant when consolidation pressure $\sigma_3 \geq \sigma_{3cr}$.

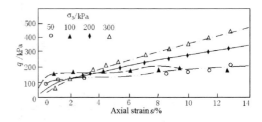

Figure 10. Stress-strain curves of soil samples from Xian (depth ≤ 10 m).

2.4.2 *Structurality and structural strength*

Loess is outstanding for its structurality which is due to its environment of deposition and the diagenetic process it underwent, its micro-structural features, the existence of soluble salts and the soil matrix suction. As shown in Figure 8 the coefficient of lateral pressure K_0 turns constant only after the consolidation pressure reaches and exceeds σ_{3cr}. Figure 9 shows that normalization by $P(=(\sigma_1 + 2\sigma_3)/2)$ of the $(\sigma_1 - \sigma_3)/P$ vs axial strain ε_1 curves of the soil sample is feasible only after the consolidation pressure exceeds 75 kPa. Figure 10 and Figure 11 respectively depict the $q(=(\sigma_1 - \sigma_3)/2)$ vs ε_1 relationships of loess samples from two different depths and consolidated under different pressures σ_3 or, in other words, the soil

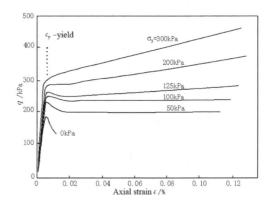

Figure 11. Stress-strain curves of soil samples from Xian (depth: 13.2–17.4 m).

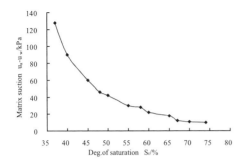

Figure 12. Soil-water characteristic curve of a loess (Q_3) sample from Yangling, Xi'an, Shaanxi Province.

Table 2. Classification of sliding bed soil types in loess slides.

Sliding bed soil type and age	Physical properties of bed soil
Loess — Malan Loess (Q_3)	Silty bed soil
Loess — Lishi Loess (Q_2)	
Loess — Wucheng Loess (Q_1)	Clayey bed soil
Mixtures of loess and underlying bedrock material (Q_4)	
Weak intercalations in bedrock	Weak rock-turned bed soil

structurality vs soil age relationships (Qian et al. 1988, Lin 1994, Wang et al. 1990).

Structural strength of loess invariably weakens as water content increases until the state of saturation is reached. In this state the structural strength of loess can be termed the "residual structural strength" or, usually in subsoil treatment and foundation design in China, the "initial (collapse) pressure". It is to be pointed out that for loess the terms "residual structural strength", initial (collapse) pressure" and "preconsolidation pressure" are technical synonyms, having the same numerical value.

2.4.3 Suction or matrix suction

As a kind of unsaturated clayey soils and like swelling clays, loess distinguishes itself by particularly high suction when it is in a state of low moisture content. This suction in loess manifests itself as a kind of negative pore pressure effect and enables the soil to manifest high strength which however decreases in step with increase of water content. Usually, this suction-dependent strength becomes very low or even vanishes when the soil water content is over 25% or degree of saturation exceeds 65%. Figure 12 (Xing 2001) gives an example depicting this soil-water characteristic.

2.4.4 Shear strength of sliding bed soil types

In addition to its suction-dependency and like many other soils, the shear strength of loess is also remarkable for the large drop from peak to residual parameter values. Table 2 gives the classification of sliding bed soil types in loess slides and in Table 3 are given the geotechnical properties of those soil types (Wang 2005).

3 DISTRIBUTION AND TYPES OF LOESS LANDSLIDES

3.1 Loess landslides as a peculiar geomorphological element of the Loess Plateau

Figure 13 (Derbyshire et al 1994) is a schematic representation of the geomorphological divisions (zoning) of the Central Loess Plateau. Each division has its characteristic and representative landform element or elements such as Yuan (platforms or residual platforms) as shown in Figure 14, Liang (ridges) as shown in Figure 15, Mao (mounds, hills) also shown in Figure 15, fluvial plains and fluvial terraces as shown in Figure 16. Loess landslides as peculiar landform elements are scattered among them.

3.2 Distribution of loess landslides

Figure 17 (Jin et al. 1996) is a schematic map of landslide distribution in the Loess Plateau and its contiguous areas, on which the clusterly nature of landslide distribution is discernible. Each large cluster has its combination of natural conditions and slide-causative factors. For examples, the NS trending lengthy cluster near the right side margin of the map is just where the Yellow River gorge and the west slope of the Lüliang mountain range are; the linear assemblage at the lower middle of the map is representative of the 180 or more old landslides which are closely spaced one after the other and extend eastward from Baoji along the north shore the Wei River; and the clusters of landslides west of the Lüliang Range on the map came into being during the 1920 M = 8.5 Haiyuan Earthquake.

Landslides and their distribution can be located and delineated with precision by means of satellite images and air photos. The successful use of infrared color imagery for Xi-Yu expressway is an example, as shown in Figure 18.

Table 3. Geotechnical properties of different sliding bed soil.

	Geotechnical properties					
			Strength parameters			
			Cohesion c(kPa)		Angle of int. friction $\varphi(°)$	
Sliding bed soil type	Permeability k(cm·s^{-1})	Coef. of collaps. δ_s	Peak	Residual	Peak	Residual
Silty bed soil	$(3–7) \times 10^{-4}$	0.01–0.08	15–30	3–15	12–22	8–12
Clayey bed soil	$(5–6) \times 10^{-6}$	0	20–130	5–25	10–20	2.5–8
Weak rock-turned bed soil	$<6 \times 10^{-6}$	0	50–200	5–25	10–20	2.5–8

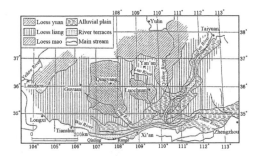

Figure 13. Geomorphological divisions of the Loess Plateau of China.

Figure 15. Liang (ridges) and Mao (hilly mounds), Yanan, Shaanxi Province.

Figure 14. Part of Xifengyuan, Ningxian, Gansu Province.

Figure 16. Loess terraces, Malianhe Valley, Gansu Province.

3.3 Structural types of loess landslides

Based on the strata involved in sliding and the structural features of the slide, loess slides can be viewed as falling into three types, i.e. intra-loess slides in which the sliding plane confines itself within the loessial soil mass and sliding takes place following an internal weak zone or weak layer; loess-bedrock surface contact slides in which the sliding bed and the soil-bedrock interface coincide; and bedrock slides which again

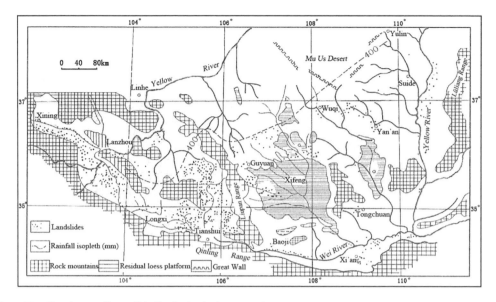

Figure 17. Sketch map of loess slide distribution in the Loess Plateau.

Figure 18. Infrared color converted to color image of Xi-Yu expressway.

can be divided into two subtypes, i.e. slides whose sliding bed is bedrock-conformative and slides whose sliding bed is bedrock-nonconformative.

3.3.1 Intra-loess landslides

What Figure 19 (Wang et al. 2005) shows is an intra-loess slide very common in the Loess Plateau. As shown, the sliding plane (or bed) is at the top of a paleo-soil layer where perched water existed. The non-conformative contact between the "new loess" (Q_3) and "old loess" (Q_2) shown in Figure 20 is also a causative factor for intra-loess slides. It is to be pointed out that the volume of slide débris involved is usually less than 100,000 m^3.

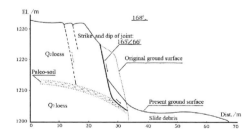

Figure 19. Profile of Changqingqiao landslide, Xifeng, Gansu Province.

Figure 20. Loess profile, Zhangzigou, Qinghuabian, Yan'an, Shaanxi Province.

Figure 21. A loess landslide in Tongchuan, Shaanxi Province.

3.3.2 *Loess-bedrock surface contact landsides*

Shown in Figure 21 and Figure 22 are loess-bedrock surface contact slides, also common in the Loess Plateau, The sliding plane (or bed) is at the surface of contact between loess and the underlying bedrocks of the Tertiary, Jurassic or Triassic periods. In such slides the volume of slide débris involved is mostly larger than 100,000 m^3, exceeding 1,000,000 m^3 sometimes.

3.3.3 *Loess-bedrock landslides*

What is shown in Figure 23 is a loess-bedrock slide. It can be seen that the sliding plane passed all the way through the loess strata, incised at a certain dip angle into the underlying Tertiary strata and finally levelled off in conformative, near-horizontal sliding.

4 EVOLUTION MECHANISM OF LOESS LANDSLIDES

4.1 *Causative factors of loess landslides*

From the above-mentioned pattern of landslide distribution and geologic structural features of loess slides, four kinds of causative factors can be summed-up as follows:

4.1.1 *Steeping of loess banks by downward erosion in gullies and intensified lateral erosion of loess platforms and terraces by river streams*

This kind of causative factors is very common. They are related to regional periodical deposition-erosion changes. See Figure 24.

4.1.2 *Sustained rainfalls, rainstorms and frequent irrigation involving large areas of loessial terrain*

Here, the former two are natural processes, while the latter is human. These processes inevitably lead to increased water content in the loess strata and lowered stability of loess slopes and are therefore slide-causative. Figure 25 shows the Jiangliu landslide on the south bank of the Jing River as an example. That slide was an old one. It was reactivated on a large scale at the end of the rainy season of 1984. During that year rainfalls were much more abundant than usual and the rainy season was also much longer. In consequence, more landslides took place in Shaanxi Province during that year and the following one.

The phenomena mentioned above can also be expounded to a certain degree in the context of the relationship between the strength and matrix suction of unsaturated soils. In Figure 26 is shown the progressive failure of a "typical loess cut slope" (Wang et al. 2005). In pace with water content increase, the matrix suction decreases. The corresponding effect is lowered strength and weakened stability. If this process goes on unremedied, the slope will fail and eventually turn into a slide, as shown in Figure 27.

4.1.3 *Human activities*

In step with economic development in the Loess Plateau and the contiguous regions, landslides caused or triggered by irrigation, highway and railway construction, mining and underground space development activities are becoming more frequent. This has raised serious concern.

4.1.4 *Seismic landslides*

This problem will be discussed in Section 5.

4.2 *Evolution track of loess landslides*

Except for slides caused by earthquakes, it usually takes time, sometimes long time for slides to evolve from initiation to general sliding. In other words, the process is progressive. An incisive summing-up is given in Figure 28.

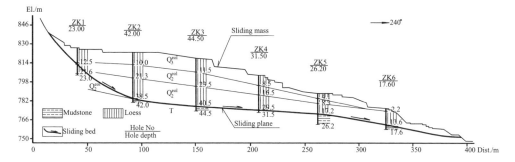

Figure 22. Profile of a landslide at K10 Wubu, Northern Shaanxi.

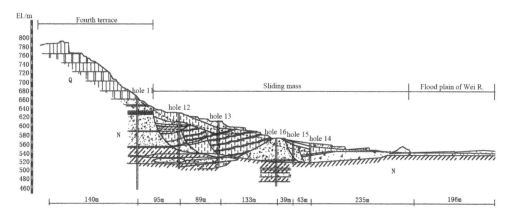

Figure 23. Longitudinal section of Pojishan landslide, Baoji, Shaanxi Province.

Figure 24. Wangjiaao landslide, Tongchuan, Shaanxi Province.

Figure 25. Jiangliu landslide on south bank of the Jing River, Guanzhong Basin.

5 LOESS LANDSLIDES IN HIGHLY SEISMIC REGIONS

The east, south and west parts of the Loess Plateau and some neighboring areas covered by loess are highly seismic regions where at least six earthquakes of magnitude 8 and larger took place in history.

5.1 *Distribution of earthquake landslides*

Figure 29 (Lin et al 2000) shows the distribution of landslides caused by historical earthquakes (1500~1949). Please note the lower middle left of

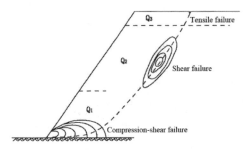

Figure 26. Progressive failure mode for a "typical loess slope".

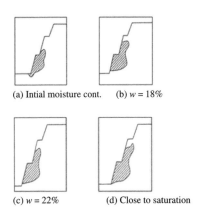

Figure 27. In the slope the plastic state zone enlarges in pace with water content increase (at Jiangzhang).

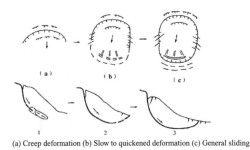

Figure 28. Landslide evolution (after Wang Gongxian).

Figure 29 where single large slides and slide clusters that came into being during the 1920 M = 8.5 Haiyuan earthquake are shown.

5.2 Landslides in the course of whose initiation and evolution soil liquefaction played an inalienable part

During the 1920 Haiyuan earthquake (M = 8.5) numerous slides and slide clusters came into being. They are characterized by very small sliding bed angles and unusually long slide distances like Huihuichuan slide shown in Figure 30. A lot of such slides are very large and characterized by panoramic wavy terrain as can be seen in Figure 31. And what Figure 32 shows is part of the results obtained in studying the stratigraphic structures before and after sliding uncovered by trenching. These results bear witness to the presence of soil (sand) liquefaction.

It is to be noted that soil (sand) liquefaction during strong earthquakes is not exclusive. Often, sudden heavy loading on alluvial deposits by a falling or down-sliding soil mass can cause instantaneous excessive pore water pressure build-up in the alluvial deposits and result in liquefaction of the underlying alluvium or the soil (sand) layers in the vicinity of the sliding bed as demonstrated in Figure 33 and Figure 34. Because of greatly reduced resistance the slide-mass can not but drift a much longer distance before full energy dissipation. The 1983 Sale mountain landslide in Gansu Province (Fig. 35) and the old Pojishan landslide in Shaanxi Province (Fig. 23) are examples of such events.

6 LANDSLIDE PREVENTION AND REMEDIATION

6.1 Measures and methods commonly used for landslide prevention and remediation in loessial regions

6.1.1 Modification of the slide's geometrical configuration

Examples of this approach are: improvement of slope configuration like partial removal of slide débris with a view to reducing the pushing force; back-loading where the slope of the sliding bed turns reversed so as to increase anti-sliding resistance.

6.1.2 Drainage

Examples of this approach are: use of a system of open or covered ditches to drain rainwater out of the slide area and the contiguous ground surfaces; use of underground drainage ditches and underground slide-resistant drainage ditches; drainage of the débris mass by means of large diameter boreholes; drainage of the débris mass by means of inclined boreholes; use of underground galleries inside as well as outside of the landslide.

6.1.3 Retaining structures

Examples of this approach are: gravity retaining walls including gabion baskets; anchored retaining structures; anti-sliding piles, pretensioned cable-tied anti-sliding piles; soil-nailed walls.

The measures and methods enumerated above are often used in a joint manner.

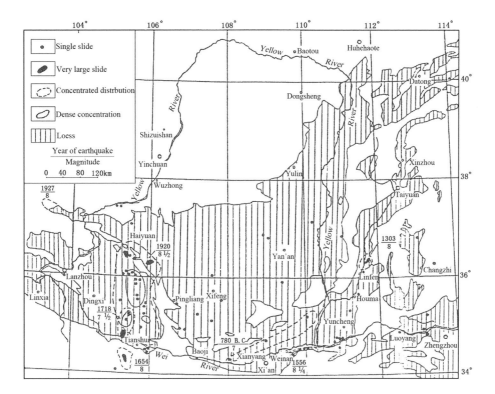

Figure 29. Location and distribution of historical earthquake-induced landslides and slump failures (1500–1949).

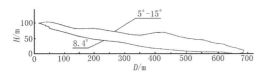

Figure 30. Longitudinal section of Huihuichuan slide.

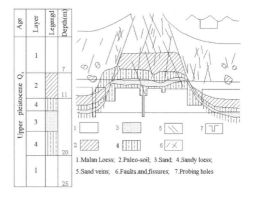

Figure 32. Original stratigraphic structure of the slide area and evidence of thrust faulting and shearing as well as traces of liquefaction.

6.2 Examples of remediation projects

6.2.1 Phoenix mountain slide, Yan'an

This slide is an old yet reactivated one. The remedial efforts include drainage of rainwater, backloading at the slide toe and anti-sliding piles with evident, satisfactory effects. See Figure 36 and Figure 37.

Figure 31. Wavy terrain characterizing low-angle long distance slide.

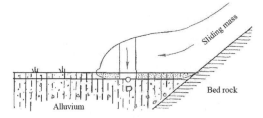

Figure 33. Schematic diagram of undrained loading upon alluvial deposits by a moving landslide mass.

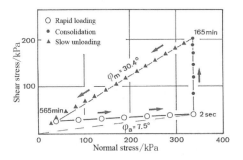

Figure 34. Rapid loading test on a saturated alluvial deposit.

Figure 35. Sale mountain slide, 1983 (location: Dongxiang, Gansu Province; slide volume: 3100×10^4 m^3; slide distance: in excess of 1000 m; casualties: 220 deaths) (Wu et al. 2006).

Figure 36. Mount Phoenix slide, Yan'an, Shaanxi Province.

6.2.2 *Fengjiahe loess landslide, Tongchuan-Huangling 1st class highway*

Being a large, very thick old landslide, it is 300–440 m long, 1300 m wide and 25–45 m thick, reaching 60 m where the slide mass is most thick. Road cutting led to the appearance of two 150–220 m long tensile cracks in the upper reach of the slope which kept on widening. The main design approach adopted for remediation was drainage of the slide-mass, including 29 inclined drainage boreholes each 32.5 m long and inclined at an angle of elevation of 5°, an underground drain 320 m long and positioned along the up-slope side of the road cut, unloading by soil removal at the top of the slope with a rainwater interception ditch on the cut platform, and a 5 m high anti-sliding wall at the foot of the cut slope. See Figure 38 and Figure 39. According to the results of surface movement monitoring and down-hole inclinometer measurements over the past 6 years, the slide has proved to be stable (Wang et al. 2002).

6.2.3 *Mt. Lishan landslide, Lintong, Xi'an*

Figure 40 gives a plane view of Mt. Lishan landslide. On the right of the Figure is the landslide No. 2 which is a set of retrogressive intra-loess or loess-bedrock slides. A deep gully named Lao Ya gully separates landslide No. 2 from landslide No.1 The latter falls into 3 blocks of which Block I has been the focus of research, monitoring and remediation.

One of the distinguishing features of the research study of Mt. Lishan landslide is the completeness of the set of instruments used and the extended time span of monitoring. It should be pointed out that the "full-length" extensometers and extensometers of various lengths played important roles in delimiting and delineating different deforming parts of the slide. Down-hole inclinometering was of great help in determining the sliding bed. All such results formed a reliable data base on which subsequent specific remedial measures were worked out.

The target of remediation of the first phase was Block I of slide NO. 1. The structural features of Block I and measures taken as shown in Figure 41 (Lin 1997) are representative of that block. As can be seen, Block I was taken as composed of three diffentiated slides, i.e. slides A, B and C. For them the measures taken were anti-sliding piles, cable-tied slide-resistant walls and cable-tied anti-sliding piles respectively.

6.2.4 *Pojishan landslide, Baoji, Shaanxi*

It is a very large old landslide situated close to the downtown district of the city of Baoji. To the rear of the slide is the high fourth terrace of the Wei River and to the front is the flood plain of the latter. See Figure 23. In the 80's of the last century, several east-west extending long cracks appeared above the

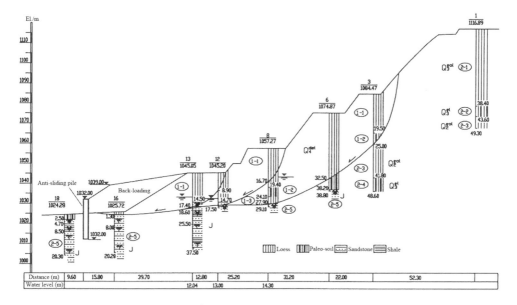

Figure 37. Engineering geological section 2–2′ of Mt. Phoenix landslide, Yan'an Shaanxi Province (Zhang et al 2004).

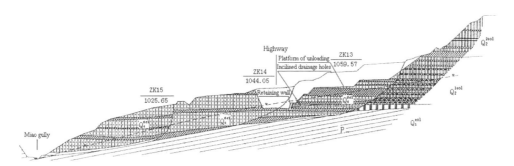

Figure 38. Longitudinal section of Fengjiahe landslide and remedial works.

Figure 39. Low angle inclined draining boreholes at Fengjiahe landslide.

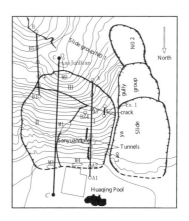

Figure 40. Lishan landslide and layout of extensometers and inclinometering hole BZ4.

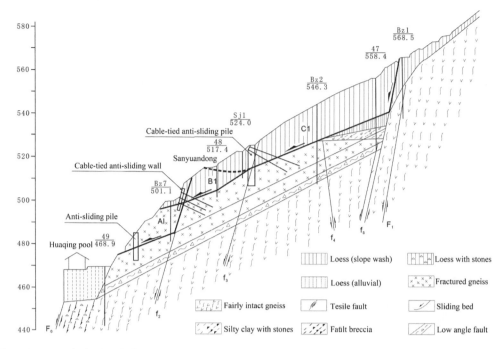

Figure 41. Longitudinal geological section of Block I of Lishan landslide and remedial works.

waist of the slide, widening year by year. As there is a large irrigation canal which runs from west to east along the down-slope side of the slide waist and also as the landslide overhangs the densely populated downtown district of the city, a decision was made to remedy the situation and the main approach was interception of water feeding the slide from outside and drainage of the slide-mass. To that end a system of underground galleries was adopted and constructed with satisfactory results.

7 PROBLEMS IN NEED OF FURTHER IN-DEPTH STUDY

This paper is intended as a concise summing-up of the research results and engineering experiences with respect to loess and loess landslides in China. It is obvious that much more needs to be done. Here, the authors would like to mention four issues. Advancement in these respects will be of much help to the development of the "West", especially in disaster prediction and disaster prevention and remediation.

7.1 Strength variability of loess

In loess slope stability study, one should not only know the peak and residual strengths of the soil and carefully select testing methods, especially the loading-drainage path to be followed in the course of test, but also pay attention to the important factor of matrix suction. In other words, loess strength is very sensitive to change in soil water content (or degree of saturation), so it can be said that in the geomorphological evolution of the Loess Plateau and initiation of loess landslides soil suction has been playing an important part.

7.2 Stability assessment and computation of the "slide push"

In domestic engineering practice it is usual to use either of the following two methods:

For a segmented sliding bed, the stability factor can be computed by means of the following Equation. See also Figure 42 (GB50330–2002).

$$K_s = \frac{\sum_{i=1}^{n-1}\left(R_i \prod_{j=i}^{n-1} \Psi_j\right) + R_n}{\sum_{i=1}^{n-1}\left(T_i \prod_{j=i}^{n-1} \Psi_j\right) + T_n} \quad (1)$$

Where ψ_j is the coefficient of force transfer expressed as follows:

$$\psi_j = \cos(\theta_i - \theta_{i+1}) - \sin(\theta_i - \theta_{i+1})\tan\psi_{i+1}$$

Although it is not without deficiencies, most relevant codes currently in effect in China are for use of

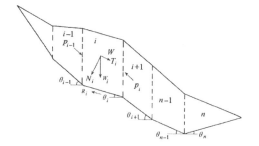

Figure 42. Stability factor computation for a non-circular slide plane.

this method in engineering practice, with some points of notice and advised care, of course.

For slides in a relatively homogeneous soil medium, the sliding bed can be regarded as circular in shape, and either the total stress or effective stress method can be used to compute the stability factor as follows (TB 10027–2001, J 125–2001):

$$K_s = \frac{\sum N \tan \varphi + cL}{\sum T} \quad (2)$$

or

$$K_s = \frac{\sum (N - u) \tan \varphi' + \sum c'L}{\sum T} \quad (3)$$

Besides these equations, Bishop's Simplified Method is often used in China.

It is to be reiterated that the foregoing methods are far from being perfect and much more research work remains to be done, including sustained systematic and instrumented monitoring on landslides of various types, structures and scales, as well as development of new methods of data analysis.

7.3 Forecast of disaster extent

Over the past years, few research efforts have been made to forecast the extent of destruction of a landslide disaster. At present, the advised and feasible approach is data collection and reduction, back-computation and selection of relevant parameters for regression analysis. for example, Wang & Zhang (2005) started from the Factual data of "typical" loess landslides, selected thickness of loess (h), slope angle (θ), length of slide previous to sliding (l) and previous thickness (d) of soil mass involved in sliding as parameters for regression analysis and proposed the following models to be used in forecasting the extent of a "typical" loess landslide disaster.

For intra-loess landslides

$$L = -124.96 - 1.86\,h + 3.13\,\theta + 3.24\,d + 2.871 \quad (4)$$

For loess-bedrock surface contact landslides

$$L = 14.72 - 5.55\,h + 2.49\,\theta + 20.05\,d + 2.571 \quad (5)$$

Where L is the forecast slide distance.

These models are results of tentative efforts only and thus have limitations in practical application. Especially as it is difficult to estimate the slide length (l) previous to sliding and the previous thickness (d) of soil mass involved in sliding with fair accuracy.

7.4 High cut slopes in railway, highway and deep excavation design

Efforts to develop the "West" call for large-scale construction of railways and highways and utilization of underground space in urban areas. In loess regions this inevitably entrains the problem of high cut slopes. In recent years slopes as high as 60 m and more are no longer rare in highway building. In face of a deficient theoreretical basis and insufficient experiences, the approach people presently take is that of a coordinated combination which comprises a safe general slope rate, an appropriate number of steps and platforms, a system of drainage ditches on slope top and along platforms including torrent troughs, and proper slope (step) face framing, planting and slope foot protection (Li et al. 2005).

8 CONCLUSIONS

This paper is a concise description and analysis of a natural phenomenon that has been coming into being, evolving and often threatening to cause havoc in regions of unique environmental conditions, i.e. loess landslides in provinces of loessial terrain. Northwest and North China is typical in this respect.

Description and analysis of the behavior peculiar to loess landslides is made not only in view of the climatic, depositional, stratigraphical, geomorphological and in many cases seismic features characteristic of lands of loessial terrain, but also in the light of the inalienable granulometry and engineering properties of the soil, collapsibility and matrix suction in particular.

Methods of landslide prevention and remediation currently in use in China are enumerated and actual representative examples given.

The four issues mentioned in Part 7 of the paper are only a few of the problems that need in-depth study. Typical and urgent examples are disaster extent forecast and stability of high railway and highway cuts which are no longer rare issues as China's West quickens its pace of development.

REFERENCES

Derbyshire, E., Meng, X.M. & Wang, J.T. 1994. Collapsible loess on the loess plateau of China. In Derbyshire, E., Dijkstra, T.A. & Smally, I.J. (eds), *Genesis and Properties of Collapsible loess; Proc. of the NATO Advanced Research Workshop, Loughborough, UK 1994*: 267–293. Nether lands: Kluwer.

Jin, Z.X. 1996. Factors governing the distribution of landslide hazards (in Chinese). In *Gansu Science Bulletin, 1996 (Supp.)*: 123–128.

Lin, Z.G. 1994. Variation in collapsibility and strength of loess with age. In *Genesis and properties of collapsible soils; Proc. NATO Advanced Research Workshop, Loughborough, UK 1994*: 247–263. Nether lands, Kluwer.

Lin, Z.G. 1997 Huaqing Palace, Xian, China and landslide hazard. In *Natural disaster prediction and mitigation; Proc. intern. Symp., Kyoto, 1997*: 299–307. Kyoto: Kyoto Univ. Press.

Lin, Z.G., Sassa, K. & Bai M.X. 2000. Undrained loading and soil liquefaction in the evolution of low angle long travel distance landslides in regions of loessial terrain. In *Proc. of the Third Multi-lateral Workshop on Development of Earthquake and Tsunami Disaster Mitigation Technologies and their Integration for the Asia-Pacific Region, Manila, Philippines, 2000*: 39–43.

Li, Z.Q. & Zhao Z.S. 2005. *Research on highway high slope protection technology in loess regions* (in Chinese). Xi'an: Highway Investigation and Design Institute, Shaanxi Province.

Qian, H.J. & Lin, Z.G. 1988. Loess and its engineering problems in China. In *Engineering Problems of Regional Soils; Proc. Intern. Conf., Beijing, 1988*: 136–153. Beijing: International Academic Publishers.

Professional Standards Compilation Group, PRC. 2001. *Code for unfavorable geological condition investigation of railway engineering (TB 10027–2001, 125–2001)* (in Chinese). Beijing: China Railway Press.

National standards compilation group, PRC. 2002. *Technical code for building-and construction-related slope engineering (GB50330–2002)* (in Chinese). Beijing: China Architecture and Building Press.

Wang, Y.Y. & Lin, Z.G. 1990. *Structural features and physico-mechanical properties of loess in China* (in Chinese). Beijing: Science Press.

Wang, X.Y., Zhao, Z.S. & Zhao, J.B. 2002. *Analysis of landslides along Tongchuan-Huangling Highway and research on their multi-approach remediation*. Xi'an: Highway Investigation and Design Institute, Shaanxi Province.

Wang, N.Q & Zhang, Z.Y. 2005. *Study on loess landslide disasters* (in Chinese). Lanzhou: Lanzhou University Press.

Wu, W.J. & Wang, N.Q. 2006. *Landslide hazards in Gansu Province* (in Chinese). Lanzhou: Lanzhou University Press.

Xing, Y.C. 2001. *Research on effective stress and deformation-strength characteristics unsaturated soils* (in Chinese). Xi'an: Xi'an University of Science and Technology.

Zhang, Y.Z. & Xu, Z.J. 2004. Investigation and remediation of Mt. Phoenix landslide (in Chinese). In *Proc. of the 6th National Symp. on Geotechnical Engineering case histories, Beijing, 2004*: 284–290. Beijing: Weaponry Industry Press.

Advances in landslide continuum dynamic modelling

S. McDougall
BGC Engineering Inc., Vancouver, Canada

M. Pirulli
Politecnico di Torino, Torino, Italy

O. Hungr
University of British Columbia, Vancouver, Canada

C. Scavia
Politecnico di Torino, Torino, Italy

ABSTRACT: Landslide continuum dynamic models have improved considerably in the past three decades, especially within the past few years, but a consensus on the best method of determining the input resistance parameter values for predictive runs has not yet emerged. A calibration-based approach is one possible method, although different calibrated values are often obtained using different models to back-analyze the same event. These differences may be amplified by the different input geometrical assumptions made in each case. With the authors' own models, consistent ranges of calibrated parameter values have been found for specific classes of events, and these ranges can be used for parametric forward-analyses. A long-term goal of this work is to build a calibration database large enough to permit probabilistic input parameter selection.

1 INTRODUCTION

Continuum dynamic modelling has emerged as a useful tool for landslide runout analysis and risk assessment. With increasing attention and coinciding advances in computational capabilities, a large number of models have been developed or are currently in development. Several of these models have included innovations that have significantly advanced both our ability to simulate real events and our fundamental understanding of rapid landslide processes. Contributions have been made by a number of researchers with a wide variety of perspectives and goals, making this topic truly multidisciplinary.

Still, some significant challenges remain as we move towards more accurate and objective runout prediction using continuum dynamic models. One of the most important challenges is how to select the input parameters for forward-analysis, and in the case of calibration-based parameter selection, transferability of parameters between different models is a key issue.

This paper begins with a detailed review of previous developments in landslide continuum dynamic modelling. The latter part of the paper addresses the issues of input parameter selection and transferability and is based mainly on the authors' experiences with our own models.

2 REVIEW OF EXISTING MODELS

Landslide continuum dynamic models evolved from established lumped mass and hydrodynamic methods. Subsequent innovations have included various methods to account for the behaviour of real landslides, including the effects of 3D terrain, internal strength, mass and momentum changes due to entrainment and spatial and temporal rheology variations.

The following review of existing models is not exhaustive. Instead, an emphasis has been put on the most important innovations with time. Models that are closely related in this context have been grouped together and, within these groupings, a chronological order has been attempted. Although the focus of this review is on landslide-specific models, some innovative snow avalanche models have also been included.

In the following review, a distinction has been made between 2D and 3D models. Although there are some inconsistencies in terminology related to depth-averaging (which essentially eliminates one dimension from the governing equations), for the purposes of this paper, 2D and 3D simply denote models that simulate motion across 2D and 3D paths, respectively.

2.1 Extension of hydrodynamic methods

The earliest continuum dynamic models for landslides and snow avalanches were essentially dam-break or flood routing models with modified basal rheologies. Lang et al. (1979) modified an existing 2D hydrodynamic model to include frictional resistance in addition to classical kinematic viscosity. The frictional component in the model (AVALNCH) was increased at low speeds to simulate the "fast-stops" that had been observed during snow avalanche deposition. Appropriate values for the resistance parameters were investigated by back-analyses of actual events (Lang & Martinelli 1979).

Dent & Lang (1980) modified another 2D hydrodynamic model, based on the Simplified Marker-and-Cell numerical method, to include various combinations of frictional, viscous and turbulent resistances. Dent & Lang (1983) modified the same model to include a biviscous resistance, similar to the Bingham rheology (e.g., Johnson 1970). The model (BVSMAC) has also been used to back-analyze full-scale rock and debris avalanches (Trunk et al. 1986, Sousa and Voight 1991, Voight & Sousa 1994). Sousa & Voight (1991) and Voight and Sousa (1994) used progressively-decreasing resistance parameters to simulate changes along the path.

The Bingham rheology has been implemented by other workers. Jeyapalan (1981) and Jeyapalan et al. (1983a) presented a 2D model (TFLOW) based on the Bingham rheology, which was used to back-analyze mine waste flow slides (Jeyapalan 1981, Jeyapalan et al. 1983b). Schamber & MacArthur (1985) and MacArthur & Schamber (1986) developed 2D and 3D models for routing mud flows which were also based on the Bingham rheology. Pastor et al. (2002) presented a 3D model for the simulation of flow slides. The Bingham rheology was used and the consolidation model proposed by Hutchinson (1986) was implemented to simulate pore pressure dissipation during the course of motion. Several cases of tailings dam and mine waste dump failures were back-analyzed.

Fread (1988) developed a dam break model based on a power law rheology. The model (DAMBRK) was used by Costa (1997) to back-analyze historic and prehistoric lahars in the Cascade Volcanic Belt.

O'Brien et al. (1993) presented a 3D model for routing floods, mud flows and debris flows using a generalized quadratic rheology, which is capable of accounting for various combinations of plastic/frictional, viscous and turbulent/dispersive resistances. The model (FLO-2D) was demonstrated using a full-scale mudflow case study. FLO-2D is notable because it has been commercially available for several years and has been used extensively in practice.

Laigle & Coussot (1997) presented a 2D model for mud flows based on the Herschel-Bulkley rheology (e.g., Coussot 1994). The input parameters were measured independently using standard rheometric techniques. A 3D version of the model was presented by Laigle (1997).

2.2 Incorporation of path-dependent rheology

Landslide mobility is related to the volume and character of the source material, but often more importantly to the extent, depth and character of the surficial material encountered along the path. This material can strongly influence the shear resistance at the basal interface (Figure 1).

With noted exceptions (i.e., Lang et al. 1979, Sousa & Voight 1991, Voight & Sousa 1994), the aforementioned models used constant bulk rheological parameters and neglected the influence of internal strength. Sassa (1988) presented a 3D model that included a method to account explicitly for changes in resistance along the path using a frictional resistance model. Instead of using a constant friction angle for the duration of motion, Sassa (1988) used a spatially-variable effective basal friction angle, which accounts for the influence of pore water pressure implicitly.

Figure 1. Oblique aerial view of the May/June 2002 McAuley Creek rock avalanche near Vernon, British Columbia. Most of the rock debris (about 7 Mm^3 total) deposited immediately below the source area (A), but about 1 Mm^3 continued another 1.6 km down the narrow valley (B). The mobility of this part of the landslide may have been enhanced by the presence of fine-grained, relatively weak and possibly saturated glaciolacustrine deposits in the path. (Photograph taken on August 26, 2002, courtesy of Dr. Réjean Couture, Geological Survey of Canada, Ottawa.)

He argued that the basal resistance is governed by the properties of the bed material, which can fail under rapid loading (Hutchinson & Bhandari 1971, Sassa 1985), and proposed that the effective basal friction angle at different locations could be measured using high speed ring shear tests on samples taken from the path.

2.3 Incorporation of earth pressure theory

The concept of strain-dependency of stresses within a deforming granular mass is well established in geotechnical earth pressure theory (e.g., Rankine 1857). Although the internal stresses in an extremely rapid landslide have never actually been measured, strain-dependency can be inferred from the observed behaviour of experimental granular flows as well as real landslides. In particular, granular flows and landslides do not appear to spread out or contract as readily as fluids (Gray et al. 1999, Hungr 1995).

Adopting an established soil mechanics technique, Sassa (1988) normalized the internal horizontal normal stresses by the total vertical normal stress using pressure coefficients (depth-averaging was performed in the vertical direction). He suggested that the stress coefficients could range between 0 and 1 (representing solid and fluid extremes, respectively) and used a value based on Jaky's (1944) equation for the static or "at rest" (Terzaghi 1920) earth pressure coefficient, which he assumed to be valid during motion and horizontally-isotropic. The model was demonstrated using a back-analysis of a real rock avalanche.

Hutter & Savage (1988) presented a 2D model for the simulation of dry granular flows that incorporated a more advanced method to account for internal strength. A detailed description of the model was presented subsequently by Savage and Hutter (1989). Significantly, they recognized that the internal stresses in a deforming granular mass are strain-dependent and coupled with the basal shear strength. They assumed that both the internal and basal rheologies were frictional but could be governed by distinct friction angles. Using classical Rankine earth pressure theory, they suggested that the internal stresses in a deforming granular mass tend to either active or passive states as the material diverges or converges, respectively. In contrast to Sassa's (1988) assumptions, this implied that the internal stresses could be greater than hydrostatic in directions of converging motion. They used only the theoretical limiting values for the stress coefficients in their model, independent of spreading rate or magnitude, an approach that assumes instantaneous stress response and therefore implies infinite stiffness of the granular mass.

The so-called Savage-Hutter Theory has been the basis for a large body of work (c.f., Pudasaini & Hutter 2006). Savage & Hutter (1991) modified their original model to account for centripetal acceleration due to curvature of the path, which influences the magnitudes of the internal and basal stresses. They tested the model using their own series of laboratory flume experiments involving dry plastic beads. The best simulations were achieved by incorporating velocity-dependent resistance into the model, which they justified on the basis of previous tests on glass beads. Hutter & Koch (1991) and Greve & Hutter (1993) compared model results to additional confined chute experiments involving the motion of various materials through both concave and convex path segments. Velocity-dependent resistance was not implemented in these tests and differences between measured and calibrated basal friction angles were observed.

Hutter et al. (1993) extended the Savage-Hutter theory for the simulation of 3D motion down an inclined plane. To maintain compatibility of the multi-dimensional internal stresses, they assumed that the stresses in the downslope direction dominate, consistent with the original Savage and Hutter (1989) model. Instantaneous stress response was still assumed and, in contrast to Sassa's (1988) 3D model, anisotropic stress states were possible. Greve et al. (1994) and Koch et al. (1994) developed an improved 3D extension of the Savage-Hutter theory, which they solved using a Lagrangian finite difference scheme that employed a triangular mesh.

Gray et al. (1999) extended the Savage-Hutter theory further for the simulation of motion across irregular 3D terrain. The assumption of stress dominance in the general downslope direction, as measured in a global reference coordinate system, was maintained and a similar Lagrangian method was employed. The model was tested using partially confined chute experiments involving deposition on a flat surface. Additional tests with various materials were presented by Wieland et al. (1999). Tai and Gray (1998) proposed a modification to the model to account for gradual transitions between limiting stress states, which reduced artificial shocks and instabilities that were caused by the original numerical method. Various alternative shock-capturing methods, including Eulerian schemes, have been investigated by Gray et al. (2003), Wang et al. (2004) and Chiou et al. (2005). Pudasaini and Hutter (2003) and Pudasaini et al. (2005) proposed additional modifications for the simulation of motion down highly irregular channels.

2.4 Incorporation of entrainment capabilities

Surficial deposits (e.g., colluvium, till, residual soil, alluvium, organics) in the path of a landslide may fail under rapid loading. Entrainment of this material increases the volume and alters the composition of a landslide. Momentum transfer accompanies volume change, as the initially stationary path material is accelerated to the landslide velocity by a combination of solid collisions and fluid thrust. This process results in a velocity-dependent inertial resistance, which is additional to the basal shear resistance (e.g., Perla et al. 1980). Entrainment can be an important characteristic of rapid landslides at any scale, including debris avalanches, debris flows, flow slides and rock avalanches (Figure 2).

Entrainment capabilities within a continuum framework were introduced by Takahashi (1991), who developed 2D and 3D models for simulating debris flows in which the solid-fluid mixture was modelled as a dilatant fluid (Bagnold 1954). The internal stresses were assumed to be hydrostatic and isotropic. Mass and momentum transfer due to erosion and deposition of material were accounted for explicitly using semi-empirical relationships for erosion and deposition rates. Simulation results were compared with laboratory experiments and observations of real debris flow deposits.

An alternative to Takahashi's (1991) erosion/ deposition rate formulas was proposed by Egashira and Ashida (1997) and implemented in the 2D models of Brufau et al. (2000) and Egashira et al. (2001). Brufau et al. (2000) separated the mass balances of the solid and fluid constituents in order to simulate spatial and temporal variations in flow density. This two-phase model was extended to 3D by Ghilardi et al. (2001) and was used to back-analyze two real debris flows. Egashira et al. (2001) pointed out that entrainment is affected by the grain size of the bed material and that large boulders often cannot be entrained, even in a reach where erosion dominates. Papa et al. (2004) described entrainment characteristics for cases in which bed material is different from the solid material in a debris flow. Other modifications to Takahashi's (1991) methods were proposed by Lo & Chau (2003) and Chau & Lo (2004).

Entrainment capabilities have also been incorporated into snow avalanche models. Entrainment of a finite depth of snow cover by plowing at the flow front was included in the 3D model (SAMOS) presented by Sailer et al. (2002). Sovilla and Bartelt (2002) used a 2D model to investigate plowing at the front as well as erosion at the base of dense snow avalanches, and their simulations were compared with field mass balance measurements. Entrainment of snow cover was also enabled in simulations presented by Turnbull & Bartelt (2003) using a 2D model. Naim et al. (2003) presented a 3D model that accounted for erosion and deposition explicitly using kinematic criteria similar to the semi-empirical erosion/deposition rate formulas proposed by Takahashi (1991) and Egashira & Ashida (1997).

2.5 Towards comprehensive models

Hungr (1995) presented a 2D Lagrangian model (DAN) that synthesized much of the earlier work by including features to account for internal strength, entrainment and rheology variations. He formalized the concept of "equivalent fluid", used tacitly by previous workers, by modelling the landslide as a hypothetical material governed by simple internal and basal rheological laws. As in the Savage-Hutter theory, the bulk of the material was assumed to be frictional and therefore capable of sustaining strain-dependent internal stresses that could range between active and passive states. The influence of internal stiffness was accounted for by incrementing the stresses in proportion to the magnitude of the prevailing strain, the same concept proposed later by Tai & Gray (1998). Compatibility of the internal and basal stresses was neglected in the original model, but the Savage-Hutter coupled stress approach has since been implemented. Instead of imposing a single basal rheological law, Hungr (1995) included a selection of rheologies, which could

Figure 2. The Tsing Shan debris flow, which started as a relatively small slide of 400 m^3 but grew to a total volume of 20,000 m^3 through entrainment. (Photograph taken on September 14, 1990, courtesy of J. King, Geotechnical Engineering Office, Hong Kong.)

be assigned to different zones within the landslide and/or different segments along the path. The user could also specify the depth of erodible path material in each segment, and mass and momentum transfer during entrainment were accounted for explicitly. The model was used to back-analyze several cases of mine waste flow slides. DAN has since been used by many workers to back-analyze a variety of landslides (e.g., Hungr & Evans 1996, Hungr et al. 1998, Ayotte & Hungr 2000, Jakob et al. 2000, Evans et al. 2001, Hungr et al. 2002, Pirulli et al. 2003, Hürlimann et al. 2003, Revellino et al. 2004, Pirulli 2005). It has also been used to demonstrate conceptual theories on the dynamics of rapid landslides, such as the influence of longitudinal sorting on the surging behaviour of debris flows (Hungr 2000) and the influence of entrainment on landslide mobility (Hungr & Evans 2004).

A 3D extension of DAN, based on a Lagrangian finite element method, was proposed by Chen and Lee (2000). Internal stiffness and compatibility of the multi-dimensional internal stresses were neglected. Instead, instantaneous stress response was implemented in orthogonal, but otherwise arbitrarily-oriented, horizontal directions (depth-averaging was performed in the vertical direction). The frictional basal rheology was implemented to back-analyze two real landslides. Bingham and Voellmy rheologies were implemented in subsequent back-analyses (Chen & Lee 2002, Chen & Lee 2003, Crosta et al. 2004). Entrainment capabilities were incorporated by Chen et al. (2006) and applied to the back-analysis of a debris avalanche.

An alternative 3D extension of DAN was proposed by McDougall and Hungr (2004). The model (DAN3D) was based on Smoothed Particle Hydrodynamics (SPH), a Lagrangian numerical method capable of handling large deformations. As in DAN, the internal stresses were coupled to the basal shear stress and incremented in proportion to the magnitude of the prevailing strain. Stress dominance in the direction of motion was assumed in order to maintain stress compatibility. A material entrainment feature, based on the concept of natural exponential volume growth, was presented by McDougall and Hungr (2005). The user-specified growth rate and maximum erosion depths along the path could be constrained by field observations and a change in rheology could be implemented at the onset of entrainment. The method was demonstrated through back-analysis of a rock slide-debris avalanche involving substantial entrainment. DAN3D has subsequently been used to back-analyze a large number of case studies and good correspondence with DAN results using the same set of input parameters has been repeatedly demonstrated (McDougall 2006). Using the growing database of calibrated parameters, the model is currently being used in practice for runout prediction.

Iverson (1997) challenged the use of bulk rheological relationships, especially in the dynamic modelling of debris flows, whose dynamics are strongly influenced by the interaction of relatively distinct solid and fluid components. He proposed a generalization of the Savage-Hutter theory, based on grain-fluid mixture theory, to account for viscous pore fluid effects explicitly, and incorporated the theory into a 2D Lagrangian model. The original Savage-Hutter assumptions of coupled internal and basal stresses and instantaneous stress response to deformation were retained. Iverson (1997) imposed a longitudinal pore pressure distribution, based on experimental evidence, and used the model to simulate a large-scale flume experiment. Viscous effects were neglected in the simulations and the required frictional input parameters were based on independently measured static values. Significant differences between the measured and simulated travel distance and travel time were observed, which were attributed to multiple surges that developed in the experiment but could not be modelled.

Iverson and Denlinger (2001) and Denlinger and Iverson (2001) presented a 3D extension of the theory, which they implemented in a conventional Eulerian framework. The theory was demonstrated using analytical solutions of simplified forms of the governing equations and the numerical model was tested using both small-scale and large-scale flume experiments. A kinematic criterion was used to identify relatively high resistance zones near frictional flow margins, allowing the spatial distribution of pore pressure to evolve automatically throughout each simulation. The internal stress distribution computed in the model was dependent on the orientation of the local reference frame, which was itself arbitrarily-oriented.

Denlinger and Iverson (2004) presented a revised 3D model for dry flows based on a unique Eulerian-Lagrangian hybrid numerical method. The governing equations were solved in an Eulerian framework, similar to the previous model, while a Lagrangian finite element method was simultaneously used to track deformations and redistribute the internal stresses accordingly. The model was demonstrated using hypothetical experiments and tested using analytical solutions of the dam-break problem. Iverson et al. (2004) presented additional experimental tests involving the motion of granular materials across an irregular surface.

Mangeney-Castelnau et al. (2003) presented a 3D model for simulating dry granular flows using a kinetic scheme. The model (SHWCIN) was based largely on the Savage-Hutter theory but used the multi-parameter velocity and depth-dependent basal resistance relationship proposed by Pouliquen (1999). A hydrostatic, isotropic internal stress state was assumed. Pirulli (2005) proposed modifications to SHWCIN to combat observed mesh-dependency problems, permit simulation of motion across irregular 3D terrain, incorporate the influence of internal strength and allow the selection of more than one possible basal resistance relationship. The modified model (RASH3D) employed an unstructured finite volume mesh and Iverson and Denlinger's (2001) frame-dependent method to account for non-hydrostatic internal stresses (Pirulli et al. 2007). RASH3D has been used to back-analyze a large number of case studies.

Pitman et al. (2003a) presented a 3D model based on the Savage-Hutter theory. The model (TITAN2D) was designed to provide high resolution using a parallel, adaptive mesh numerical method, which runs on distributed memory supercomputers. Iverson and Denlinger's (2001) frame-dependent method was employed to account for non-hydrostatic internal stress states. Entrainment capabilities, based on an empirical erosion rate, were incorporated by Pitman et al. (2003b) and demonstrated with a back-analysis of a historic rock avalanche. TITAN2D is currently available as an open source program.

Naef et al. (2006) presented a 2D model (DFEM) which allowed the selection of a number of resistance relationships. A 3D version of the model was presented by Rickenmann et al. (2006), in which DFEM, FLO-2D and the 2D model developed by Laigle and Coussot (1997) were applied to the back-analysis of two different field cases.

A true 2D (i.e., non-depth-averaged) finite element model based on the program TOCHNOG (Roddeman 2001) was presented by Crosta et al. (2006). Entrainment of weak path material was incorporated and a number of resistance relationships could be implemented. The model was demonstrated using a series of hypothetical cases.

Alternative solution methods have also been explored. Sampl (1993) used the Particle-in-Cell method to solve the 3D Savage-Hutter equations and model the dense flow component of snow avalanches. Frenette et al. (1997) presented a 3D model based on a pseudoconcentration function. Quecedo and Pastor (2001) presented a two-phase 3D model based on the level set method. True 3D SPH has been applied to the simulation of pyroclastic flows (Nagasawa and Kuwahara 1993) and dam-breaks (Cleary and Prakash 2004). Bursik et al. (2003) coupled the numerical methods of Cellular Automata and SPH in a depth-averaged form to model erosion and consequent surface evolution by debris floods in 3D.

3 INPUT PARAMETER SELECTION

Landslide dynamic models have advanced incrementally in the past three decades to the point where, when used in combination with careful engineering and geoscience judgement, first-order runout prediction is possible. However, there is still considerable room for improvement. In particular, the problem of how to choose the input parameters for forward-analysis remains a major challenge.

3.1 *Input parameter measurement vs. calibration*

A distinction is sometimes made between models that require the input of measured rheological parameters and models that must be calibrated by back-analysis. The difference between the two classes is not always clear because parameter measurement and parameter calibration are not necessarily characteristics of the models themselves, but rather the modelling approach adopted by the model developers and/or users.

Parameter measurement is advocated by workers who maintain that rapid landslide dynamics can be described by constitutive relationships that are functions of intrinsic material properties, that these relationships can be incorporated into dynamic models and that the required material properties can be measured using independent methods. Measurement-based models are typically tested using established analytical solutions and back-analyses of controlled experiments before being applied to the analysis of real cases. This could be considered the traditional scientific approach to dynamic analysis.

In contrast, in the calibration-based approach, rheological parameters are constrained by systematic adjustment during trial-and-error back-analysis of full-scale prototype events. Simulation is typically achieved by matching the simulated travel distance, velocities and extent and depth of the deposit to those of the prototype. Calibration-based models may utilize the same physically-justifiable constitutive relationships as measurement-based models and should similarly be tested using analytical solutions and controlled experiments before being calibrated to specific classes of field problems. The calibration-based approach is rooted in the empirical methods of classical hydraulic engineering.

Critics of the calibration-based approach argue that model calibration, especially without controlled testing, is equivalent to tuning or curve-fitting (i.e., successful simulation can be achieved by arbitrary adjustments of the right type and number of variables), and that model adaptability can therefore be mistaken for model accuracy (Iverson 2003). Parameter selection for the purposes of prediction can also be difficult because, while simulation of a single event can be performed quite efficiently, back-analysis of a significant number of similar prototype events is required for the calibrated parameters to be physically and/or statistically justifiable. Accurate data for several suitable events can be hard to acquire and the calibration process can be time-consuming.

On the other hand, given the extreme complexity of landslide dynamics, the measurement-based approach could be considered idealistic. Although it is scientifically appealing to be able to measure the input parameters independently, no standard tests are available to measure, for example, the properties of coarse rock avalanche debris travelling at extremely rapid velocities. Such properties, even if measurable, may change significantly during the course of motion, along with the rheology itself, and may be scale-dependent. True measurement-based models must be able to account for any evolution in behaviour without the need for user-imposed changes and without using constitutive laws that include ad hoc assumptions. These requirements are extremely challenging.

3.2 *The "equivalent fluid" approach*

In our own modelling, the authors have adopted the semi-empirical "equivalent fluid" approach (Figure 3), defined by Hungr (1995) but used for a long time by many other workers (e.g., Sousa and Voight 1991, O'Brien et al. 1993, Rickenmann & Koch 1997). In this framework, the heterogeneous and complex landslide material is modelled as a hypothetical material governed by simple internal and basal rheological relationships, which may be different from each other. The internal rheology is typically assumed to be frictional, based on the method developed by Savage & Hutter (1989). In contrast, a unique basal rheology is not imposed. Instead, the basal rheological model and its associated parameters, generally only one or two, are selected based on an empirical calibration procedure, in which actual landslides of a given type are subjected to trial-and-error back-analysis. The results are judged in terms of their ability to reproduce the bulk external behaviour of a prototype event, including the travel distance and duration and the spatial distribution of velocities and flow/deposit depths (wherever comparable estimates are available from field observations).

The calibrated parameters are considered apparent, rather than actual, material properties and cannot, in general, be measured in the laboratory. As mentioned previously, extensive back-analysis is required to build a database of input parameters that can be used for prediction. However, this approach reduces reliance on laboratory-derived material properties and constitutive relationships that may not be valid at full-scale. Because the equivalent fluid rheologies are simple and need only a limited number of controlling parameters, the models are generally easy to constrain. Both velocity and deposit distributions are strongly affected by the chosen rheology.

The premise of the equivalent fluid approach is perhaps summarized best by Voight & Pariseau (1978), who wrote, "Any model that allows the slide mass to move from its place of origin to its resting place in the time limits that bound the slide motion is likely to be consistent with the principal observable fact—that of the slide occurrence itself." In the same sense, continuum dynamic models that can accurately simulate/predict the extent and duration of a landslide and the distribution of intensity (e.g., flow depth and velocity) within the impact area, regardless of the underlying micro-mechanics, should be considered useful. For the practical purposes of landslide risk assessment, this is the only information that is relevant.

As noted above, a common criticism of this approach is that, with a sufficiently flexible model, nearly any event can be simulated by choosing a certain set of parameters. What is necessary is to seek patterns of rheological type and parameter ranges that reproduce the behaviour of groups of events of similar description. Predictive capability will result only once such consistent patterns have been identified. This work is now in progress.

3.3 *Calibrated parameters*

The three models developed by the authors (DAN, DAN3D and RASH3D) have been used to back-analyze a number of landslides of various types and scales, including several of the same cases. It should be noted again that, although there has been extensive cooperation between the authors, RASH3D was developed largely independently as an extension of the model SHWCIN, originally introduced by Mangeney-Castelnau et al. (2003). In keeping with the equivalent fluid approach, several alternative rheological relationships, including frictional, Voellmy and Bingham models, have been used for each case study.

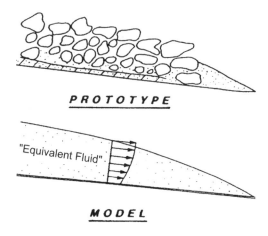

Figure 3. Schematic illustration of the equivalent fluid approach (after Hungr 1995). The complex landslide material is modelled as a hypothetical material governed by simple internal and basal rheologies.

Rock avalanches are an interesting subset of the combined case study database. Hungr & Evans (1996) and Pirulli (2005) used DAN to back-analyze 34 different rock avalanches using the frictional rheology with a constant bulk basal friction angle, ϕ_b (i.e., assuming a constant pore pressure ratio). Total runout distance was the single calibration criterion in each case. The calibration results are summarized in the histogram in Figure 4. Note that nine duplicate events and two outlier events analyzed by Pirulli (2005) were excluded from the data set. The results approximately obey a normal distribution with a mean ϕ_b of 16° and a standard deviation of 4.3°.

The rock avalanche data set can be further subdivided based on event volume, path morphology, source and path composition and many other factors, providing further constraint to the calibration results. For example, Pirulli (2005) suggested that the data set could be split into three groups, those with elongated, tongue and T-shaped runout areas (after Nicoletti & Sorriso-Valvo 1991), and noted three distinct ranges of calibrated friction angles.

Overestimation of velocities and often unrealistic forward-tapering deposits are well-known characteristics of the frictional model. The Voellmy model, which includes frictional and velocity-dependent terms (cf., Hungr 1995), typically produces better simulations of velocity and deposit distribution, but requires the input of two parameters: a friction coefficient, f, and a turbulence parameter, ξ. Since these two parameters are not necessarily independent, statistical analysis of calibration results is not straightforward.

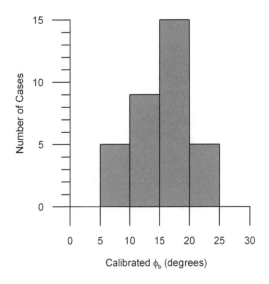

Figure 4. Histogram of calibrated bulk basal friction angles based on DAN back-analyses of 34 different rock avalanches by Hungr and Evans (1996) and Pirulli (2005).

Twenty-three rock avalanches in the data set presented above were also back-analyzed by Hungr and Evans (1996) using DAN with the Voellmy model. Total runout distance and observed/estimated velocity and duration were used as calibration criteria to constrain the two-parameter model. Calibrated friction coefficients ranged between 0.03 and 0.24, while calibrated turbulence parameters ranged between 100 and 1000 m/s^2. Using $f = 0.1$ and $\xi = 500$ m/s^2, Hungr & Evans (1996) found that the total runout distance in 70% of the cases could be simulated within an error of approximately 10%.

Similar ranges of calibrated frictional and Voellmy parameters have been found in more recent back-analyses of rock avalanches using DAN3D and RASH3D (McDougall & Hungr 2004, McDougall 2007, Pirulli 2005, Pirulli & Mangeney 2007). For example, the 1987 Val Pola rock avalanche in the Italian Alps (cf., Govi et al. 2002) has been back-analyzed using both of the 3D models, with similar results. Using the frictional model, for both DAN3D and RASH3D the best simulation of the extent of the impact area was achieved using $\phi_b = 16°$ (McDougall 2006, Pirulli & Mangeney 2007). The same calibrated friction angle was obtained by Hungr and Evans (1996) in their DAN back-analysis of the event (included in the data set shown in Figure 4). The DAN3D and RASH3D simulation results are compared in Figure 5. While some differences are evident, the results are qualitatively very similar.

Pirulli and Mangeney (2007) also presented a systematic calibration of RASH3D for the Val Pola case study using the Voellmy model. The best simulation of the event, in terms of the extent of the impact area and the distribution of deposits, was achieved using $f = 0.24$ and $\xi = 1000$ m/s^2. Figure 6 shows Pirulli and Mangeney's (2007) results, along with the results of a corresponding DAN3D simulation using the same pair of input parameters. Relatively higher runup against the opposite slope, particularly by the northern lobe of the slide mass, was simulated by DAN3D.

While some of the differences between the DAN3D and RASH3D results shown in Figures 5 and 6 are no doubt due to differences in the governing equations and numerical methods used in each model, it is important to note that different assumptions about the geometry of the source failure and the sliding surface were also made in each case. Significantly, the DAN3D simulations were based on a bulked source volume of approximately 50 Mm^3 (McDougall 2006), while the RASH3D simulations were based on a source volume of only 30 Mm^3 (Pirulli & Mangeney 2007). This difference has an important influence on the results when the Voellmy model is used, since the flow resistance relative to the driving forces is inversely related to the flow depth; the relatively

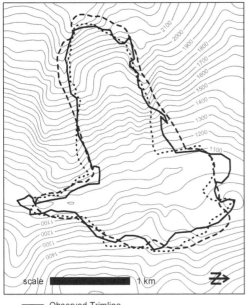 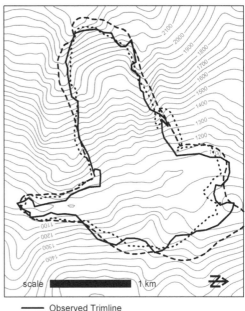

Figure 5. Comparison of DAN3D and RASH3D simulations of the Val Pola rock avalanche using the frictional model with $\phi_b = 16°$. The sliding surface shown is based on the DAN3D analyses. Elevation contours are shown at 50 m intervals.

Figure 6. Comparison of DAN3D and RASH3D simulations of the Val Pola rock avalanche using the Voellmy model with $f = 0.2$ and $\xi = 1000$ m/s^2 (note that the DAN3D case is not calibrated and uses a bulked volume of the source mass). The sliding surface shown is based on the DAN3D analyses. Elevation contours are shown at 50 m intervals.

higher volume and flow depths in the DAN3D case contributed significantly to the higher runup against the opposite slope. With DAN3D, a better simulation of the runup was achieved using $f = 0.2$ and $\xi = 500$ m/s^2.

Similar to the frictional case, in both the DAN3D and RASH3D analyses, the calibrated Voellmy parameters are within the range of values reported by Hungr and Evans (1996). However, in this case they do not correspond exactly with Hungr and Evans' (1996) calibrated values of $f = 0.1$ and $\xi = 500$ m/s^2 for the Val Pola event. This discrepancy may simply be due to the inherent limitations of the 2D model, which requires the input of a path profile. In the 2D case, the deflection of the southern part of the slide mass down the Adda River valley was not modelled. Since the mean channel slope in this area is greater than 0.1, use of $f = 0.1$ in the 3D case should allow the southern lobe to continue travelling downstream, beyond the observed deposition zone. Nevertheless, using the 3D model developed by Chen & Lee (2000), Crosta et al. (2004) obtained calibrated values of $f = 0.1$ and $\xi = 200$ m/s^2 in their back-analysis of the Val Pola rock avalanche. Comparative simulations using DAN3D and RASH3D with these input parameter values did not produce satisfactory results.

4 DISCUSSION

As shown above, unique values of calibrated resistance parameters do not necessarily apply to each event. The calibrated values depend on the governing equations and numerical solution methods that are used, which vary between different models and even between different versions of the same model. As a result, calibrated values are not necessarily transferable, and it may be necessary to revisit previous back-analyses after any significant modifications are made to a particular model. Perhaps more importantly, the calibrated values depend on the input geometrical assumptions made in each case. For this reason, the collection of more accurate field and remote observations of case studies and the development of more objective methods of setting up and running the numerical models are key factors in the future success of the calibration-based approach. In particular, improved methods of estimating flow velocities are needed to help constrain multi-parameter models. These goals may be somewhat conflicting, however; as better observations become available there is a tendency towards more detailed, and sometimes correspondingly more subjective, numerical analyses.

In spite of these challenges, using DAN, DAN3D and RASH3D, consistent ranges of input parameter values have emerged for certain classes of events. Rock avalanches have been used as an example in this paper, but similarly well-defined (although typically region-specific) calibration results have also been presented for debris avalanches and debris flows in Hong Kong (Hungr et al. 1998, Ayotte and Hungr 2000), Italy (Revellino et al. 2004) and California (Bertolo and Wieczorek 2005) and flow slides of coal mine waste in British Columbia (Hungr et al. 2002). In each of these cases, constant or "bulk" rheologies with only one or two adjustable input parameters were implemented and objective calibration criteria were defined, so model tuning has been kept to a minimum.

These ranges of calibrated input values can be used for first-order, parametric runout prediction. Reasonable confidence limits can often be applied, but the calibration database is still too small to be used for true probabilistic forward-analysis. With continued expansion of the database, a probabilistic approach, which would fit well within the framework of quantitative landslide risk assessment, may eventually be possible.

In any case, landslide continuum dynamic models should never be used as "black boxes", and experience and judgement will always play important roles in runout prediction. Back-analyses of real landslide case studies using dynamic models can serve to enhance this experience.

5 CONCLUSIONS

As demonstrated in this paper, different calibrated values may be obtained using different models to back-analyze the same event. These differences may be amplified by the different input geometrical assumptions made in each case. With the authors' own models, consistent ranges of calibrated values have been found for specific classes of events. These ranges are reasonably well-defined and can be used for parametric forward-analyses. Eventually, the calibration database may be large enough to permit probabilistic input parameter selection.

ACKNOWLEDGEMENTS

The authors gratefully acknowledge Prof. Giovanni Crosta (University of Milan), who provided the DEM for the DAN3D analyses of the Val Pola rock avalanche, and Dr. Franco Godone (CNR-IRPI) and Dr. Luca Mallen (ARPA Piemonte) for their contributions to the development of the DEM for the RASH3D analyses of the Val Pola rock avalanche.

REFERENCES

Ayotte, D. & Hungr, O. 2000. Calibration of a runout prediction model for debris-flows and avalanches. In G.F. Wieczorek & N.D. Naeser (eds.), *Proc. of the 2nd Int. Conf. on Debris-Flow Hazards Mitigation, Taipei*: 505–514. Rotterdam: Balkema.

Bagnold, R.A. 1954. Experiments on a gravity-free dispersion of large solid spheres in a Newtonian fluid under shear. *Proc. of the Royal Society of London A* 225: 49–63.

Bertolo, P. & Wieczorek, G.F. 2005. Calibration of numerical models for small debris flows in Yosemite Valley, California. *Natural Hazards and Earth System Sciences* 5: 993–1001.

Brufau, P., Garcia-Navarro, P., Ghilardi, P., Natale, L. & Savi, F. 2000. 1D mathematical modelling of debris flow. *Journal of Hydraulic Research* 38(6): 435–446.

Bursik, M., Martínez-Hackert, B., Delgado, H. & Gonzalez-Huesca, A. 2003. A smoothed-particle hydrodynamic automaton of landform degradation by overland flow. *Geomorphology* 53: 25–44.

Chau, K.T. & Lo, K.H. 2004. Hazard assessment of debris flows for Leung King Estate of Hong Kong by incorporating GIS with numerical simulations. *Natural Hazards and Earth System Sciences* 4: 103–116.

Chen, H. & Lee, C.F. 2000. Numerical simulation of debris flows. *Canadian Geotechnical Journal* 37: 146–160.

Chen, H. & Lee, C.F. 2002. Runout analysis of slurry flows with Bingham model. *Journal of Geotechnical and Geoenvironmental Engineering* December: 1032–1042.

Chen, H. & Lee, C.F. 2003. A dynamic model for rainfall-induced landslides on natural slopes. *Geomorphology* 51: 269–288.

Chen, H., Crosta G.B. & Lee, C.F. 2006. Erosional effects on runout of fast landslides, debris flows and avalanches: a numerical investigation. *Geotechnique* 56: 305–322.

Chiou, M.C., Wang, Y. & Hutter, K. 2005. Influence of obstacles on rapid granular flows. *Acta Mechanica* 175: 105–122.

Cleary, P.W. & Prakash, M. 2004. Discrete-element modelling and smoothed particle hydrodynamics: potential in the environmental sciences. *Phil. Trans. of the Royal Society of London A* 362: 2003–2030.

Costa, J.E. 1997. Hydraulic modeling for lahar hazards at Cascades Volcanoes. *Environmental and Engineering Geosciences* 3(1): 21–30.

Coussot, P. 1994. Steady, laminar flow of concentrated mud suspensions in open channel. *Journal of Hydraulic Research* 32(4): 535–559.

Crosta, G.B., Chen, H. & Lee, C.F. 2004. Replay of the 1987 Val Pola Landslide, Italian Alps. *Geomorphology* 60(11): 127–146.

Crosta, G.B., Imposimato, S. & Roddeman, D.G. 2006. Continuum numerical modelling of flow-like landslides. In S.G. Evans, G. Scarascia Mugnozza, A. Strom & R.L. Hermanns (eds.) *Landslides from Massive Rock Slope Failure*: 211–232. Springer.

Denlinger, R.P. & Iverson, R.M. 2001. Flow of variably fluidized granular masses across three-dimensional terrain, 2. Numerical predictions and experimental tests. *Journal of Geophysical Research* 106(B1): 553–566.

Denlinger, R.P. & Iverson, R.M. 2004. Granular avalanches across irregular three-dimensional terrain: 1. Theory and computation. *Journal of Geophysical Research* 109: F01014.

Dent, J.D. & Lang, T.E. 1980. Modeling of snow flow. *Journal of Glaciology* 26(94): 131–140.

Dent, J.D. & Lang, T.E. 1983. A biviscous modified Bingham model of snow avalanche motion. *Annals of Glaciology* 4: 42–46.

Egashira, S. & Ashida, K. 1997. Sediment transport in steep slope flumes. *Proc. of the Roc Japan Joint Seminar on Water Resources*.

Egashira, S., Honda, N. & Itoh, T. 2001. Experimental study on the entrainment of bed material into debris flow. *Physics and Chemistry of the Earth (C)* 26(9): 645–650.

Evans, S.G., Hungr, O. & Clague, J.J. 2001. Dynamics of the 1984 rock avalanche and associated distal debris flow on Mount Cayley, British Columbia, Canada; implications for landslide hazard assessment on dissected volcanoes. *Engineering Geology* 61: 29–51.

Fread, D.L. 1988. *The National Weather Service DAMBRK Model: Theoretical Background and User Documentation.* Hydrologic Research Laboratory, National Weather Service, Silver Spring, MD, 320 pp.

Frenette, R., Eyheramendy, D. & Zimmermann, T. 1997. Numerical modelling of dam-break type problems for Navier-Stokes and granular flows. In C.L. Chen (ed.) *Debris-flow hazards mitigation: mechanics, prediction and assessment*: 586–595.

Ghilardi, P., Natale, L. & Savi, F. 2001. Modeling of debris flow propagation and deposition. *Physics and Chemistry of the Earth* 26(9): 651–656.

Govi, M., Gullà, G. & Nicoletti, P.G. 2002. Val Pola rock avalanche of July 28, 1987, Valtellina (Central Italian Alps). In S.G. Evans & J.V. DeGraff (eds.) *Catastrophic Landslides: Effects, Occurrence and Mechanisms* 15: 71–89. Geological Society of America.

Gray, J.M.N.T., Wieland, M. & Hutter, K. 1999. Gravity-driven free surface flow of granular avalanches over complex basal topography. *Proc. of the Royal Society of London A* 455: 1841–1874.

Gray, J.M.N.T., Tai, Y.C. & Noelle, S. 2003. Shock waves, dead zones and particle-free regions in rapid granular free-surface flows. *Journal of Fluid Mechanics* 491: 161–181.

Greve, R. & Hutter, K. 1993. Motion of a granular avalanche in a convex and concave curved chute: experiments and theoretical predictions. *Phil. Trans. of the Royal Society of London, Physical Sciences and Engineering* 342(1666): 573–600.

Greve, R., Koch, T. & Hutter, K. 1994. Unconfined flow of granular avalanches along a partly curved surface, Part 1: Theory. *Proc. of the Royal Society of London A* 445: 399–413.

Hungr, O. 1995. A model for the runout analysis of rapid flow slides, debris flows, and avalanches. *Canadian Geotechnical Journal* 32: 610–623.

Hungr, O. & Evans, S.G. 1996. Rock avalanche runout prediction using a dynamic model. In K. Senneset (ed.) *Proc. of the 7th Int. Symp. on Landslides, Trondheim*: 233–238. Rotterdam: Balkema.

Hungr, O., Sun, H.W. & Ho, K.K.S. 1998. *Mobility of selected landslides in Hong Kong—Pilot back-analysis using a numerical model. Report of the Geotechnical Engineering Office.* Hong Kong SAR Government.

Hungr, O. 2000. Analysis of debris flow surges using the theory of uniformly progressive flow. *Earth Surface Processes and Landforms* 25: 1–13.

Hungr, O., Dawson, R., Kent, A., Campbell, D. & Morgenstern, N.R. 2002. Rapid flow slides of coal-mine waste in British Columbia, Canada. In S.G. Evans & J.V. DeGraff (eds.) *Catastrophic Landslides: Effects, Occurrence and Mechanisms* 15: 191–208. Geological Society of America.

Hungr, O. & Evans, S.G. 2004. Entrainment of debris in rock avalanches; an analysis of a long run-out mechanism. *Geological Society of America Bulletin* 116(9/10): 1240–1252.

Hürlimann, M., Corominas, J., Moya, J. & Copons, R. 2003. Debris-flow events in the eastern Pyrenees: preliminary study on initiation and propagation. In D. Rickenmann & C.L. Chen (eds.) *Proc. of the 3rd Int. Conf. on Debris-flow Hazards Mitigation: Mechanics, Prediction and Assessment, Davos*: 115–126. Rotterdam: Millpress.

Hutchinson, J.N. 1986. A sliding-consolidation model for flow-slides. *Canadian Geotechnical Journal* 23: 115–126.

Hutchinson, J.N. & Bhandari, R.K. 1971. Undrained loading, a fundamental mechanism of mudflow and other mass movements. *Géotechnique* 21: 353–358.

Hutter, K. & Savage, S.B. 1988. Avalanche dynamics: The motion of a finite mass of gravel down a mountain side. In C. Bonnard (ed.) *Proc. of the 5th Int. Symp. on Landslides, Lausanne*: 691–697. Rotterdam: Balkema.

Hutter, K. & Koch, T. 1991. Motion of a granular avalanche in an exponentially curved chute: experiments and theoretical predictions. *Phil. Trans. of the Royal Society of London A* 334: 93–138.

Hutter, K., Siegel, M., Savage, S.B. & Nohguchi, Y. 1993. Two-dimensional spreading of a granular avalanche down an inclined plane, Part 1: Theory. *Acta Mechanica* 100: 37–68.

Iverson, R.M. 1997. The physics of debris flows. *Reviews of Geophysics* 35(3): 245–296.

Iverson, R.M. & Denlinger, R.P. 2001. Flow of variably fluidized granular masses across three-dimensional terrain, 1. Coulomb mixture theory. *Journal of Geophysical Research* 106(B1): 537–552.

Iverson, R.M., Logan, M. & Denlinger, R.P. 2004. Granular avalanches across irregular three-dimensional terrain: 2. Experimental tests. *Journal of Geophysical Research* 109: F01015.

Jakob, M., Anderson, D., Fuller, T., Hungr, O. & Ayotte, D. 2000. An unusually large debris flow at Hummingbird Creek, Mara Lake, British Columbia. *Canadian Geotechnical Journal* 37: 1109–1125.

Jaky, J. 1944. The coefficient of earth pressure at rest. *Journal of the Society of Hungarian Architects and Engineers* October: 355–358.

Jeyapalan, J.K. 1981. *Analyses of flow failures of mine tailings impoundments.* Ph.D. thesis, University of California, Berkeley.

Jeyapalan, J.K., Duncan, J.M. & Seed, H.B. 1983a. Analyses of flow failures of mine tailings dams. *Journal of Geotechnical Engineering* 109(2): 150–171.

Jeyapalan, J.K., Duncan, J.M. & Seed, H.B. 1983b. Investigation of flow failures of tailings dams. *Journal of Geotechnical Engineering* 109(2): 172–189.

Johnson, A.M. 1970. *Physical Processes in Geology*. Freeman, Cooper and Co., San Francisco. 577 pp.

Koch, T., Greve, R. & Hutter, K. 1994. Unconfined flow of granular avalanches along a partly curved surface, Part 2: Experiments and numerical computations. *Proc. of the Royal Society of London A* 445: 415–435.

Laigle, D. 1997. A two-dimensional model for the study of debris-flow spreading on a torrent debris fan. In C.L. Chen (ed.) *Debris-flow hazards mitigation: mechanics, prediction and assessment*: 123–132.

Laigle, D. & Coussot, P. 1997. Numerical modeling of mudflows. *Journal of Hydraulic Engineering* 123(7): 617–623.

Lang, T.E., Dawson, K.L. & Martinelli Jr., M. 1979. Application of numerical transient fluid dynamics to snow avalanche flow, Part 1: Development of computer program AVALNCH. *Journal of Glaciology* 22(86): 107–115.

Lang, T.E. & Martinelli Jr., M. 1979. Application of numerical transient fluid dynamics to snow avalanche flow, Part 2: Avalanche modeling and parameter error evaluation. *Journal of Glaciology* 22(86): 117–126.

Lo, K.H. & Chau, K.T. 2003. Debris-flow simulations for Tsing Shan in Hong Kong. In D. Rickenmann & C.L. Chen (eds.) *Proc. of the 3rd Int. Conf. on Debris-flow Hazards Mitigation: Mechanics, Prediction and Assessment, Davos*: 577–588. Rotterdam: Millpress.

MacArthur, R.C. & Schamber, D.R. 1986. Numerical methods for simulating mudflows. In *Proc. of the 3rd Int. Symp. on River Sedimentation, Mississippi*: 1615–1623.

Mangeney-Castelnau, A., Vilotte, J.P., Bristeau, O., Perthame, B., Bouchut, F., Simeoni, C. & Yerneni, S. 2003. Numerical modeling of avalanches based on Saint Venant equations using a kinetic scheme. *Journal of Geophysical Research* 108(B11): 2527.

McDougall, S. & Hungr, O. 2004. A model for the analysis of rapid landslide motion across three-dimensional terrain. *Canadian Geotechnical Journal* 41: 1084–1097.

McDougall, S. & Hungr, O. 2005. Dynamic modelling of entrainment in rapid landslides. *Canadian Geotechnical Journal* 42: 1437–1448.

McDougall, S. 2006. *A new continuum dynamic model for the analysis of extremely rapid landslide motion across complex 3D terrain*. Ph.D. thesis, University of British Columbia, Canada. 253 pp.

Naaim, M., Faug, T. & Naaim-Bouvet, F. 2003. Dry granular flow modelling including erosion and deposition. *Surveys in Geophysics* 24: 569–585.

Naef, D., Rickenmann, D., Rutschmann, P. & McArdell B.W. 2006. Comparison of flow resistance relations for debris flows using a one-dimensional finite element simulation model. *Natural Hazards and Earth System Sciences* 6: 155–165.

Nagasawa, M. & Kuwahara, K. 1993. Smoothed particle simulations of the pyroclastic flow. *Int. Journal of Modern Physics B* 7(9/10): 1979–1995.

Nicoletti, G. & Sorriso-Valvo, M. Geomorphic controls on the shape and mobility of rock avalanches. *Geological Society of America Bulletin* 103: 1365–1373.

O'Brien, J.S., Julien, P.Y. & Fullerton, W.T. 1993. Two-dimensional water flood and mudflow simulation. *Journal of Hydraulic Engineering* 119(2): 244–261.

Papa, M., Egashira, S. & Itoh, T. 2004. Critical conditions of bed sediment entrainment due to debris flow. *Natural Hazards and Earth System Sciences* 4: 469–474.

Pastor, M., Quecedo, M., Fernandez Merodo, J.A., Herrores, M.I., Gonzalez, E., & Mira P. 2002. Modelling tailings dams and mine waste dumps failures. *Geotechnique* 52(8): 579–591.

Perla, R., Cheng, T.T. & McClung, D.M. 1980. A two-parameter model of snow-avalanche motion. *Journal of Glaciology* 26(94): 197–207.

Pirulli, M., Preh, A., Roth, W., Scavia, C. & Poisel, R. 2003. Rock avalance run out prediction: combined application of two numerical methods. In *Proceedings of the Int. Symp. on Rock Mechanics, Johannesburg*: 903–908.

Pirulli, M. 2005. *Numerical modelling of landslide runout: A continuum mechanics approach*. Ph.D. thesis, Politecnico di Torino, Italy. 204 pp.

Pirulli, M. & Mangeney, A. 2007. Results of back-analysis of the propagation of rock avalanches as a function of the assumed rheology. *Rock Mechanics and Rock Engineering* (published online June 21, 2007).

Pirulli, M., Bristeau, M.O., Mangeney, A. & Scavia C. 2007. The effect of the earth pressure coefficients on the runout of granular material. *Environmental Modelling & Software* 22: 1437–1454.

Pitman, E.B, Nichita, C.C., Patra, A., Bauer, A., Sheridan, M. & Bursik, M. 2003a. Computing granular avalanches and landslides. *Physics of Fluids* 15(12): 3638–3646.

Pitman, E.B., Nichita, C.C., Patra, A.K., Bauer, A.C., Bursik, M. & Webb, A. 2003b. A model of granular flows over an erodible surface. *Discrete and Continuous Dynamical Systems Series B* 3(4): 589–599.

Pouliquen, O. 1999. Scaling laws in granular flows down rough inclined planes. *Physics of Fluids* 11(3): 542–548.

Pudasaini, S.P. & Hutter, K. 2003. Rapid shear flows of dry granular masses down curved and twisted channels. *Journal of Fluid Mechanics* 495: 193–208.

Pudasaini, S.P., Wang, Y. & Hutter, K. 2005. Rapid motions of free-surface avalanches down curved and twisted channels and their numerical simulation. *Phil. Trans. of the Royal Society of London A* 363: 1551–1571.

Pudasaini, S.P. & Hutter, K. 2006. *Avalanche Dynamics: Dynamics of rapid flows of dense granular avalanches*. Springer. 626 pp.

Quecedo, M. & Pastor M. 2001. Application of the level set method to the finite element solution of two-phase flows. *Int. Journal of Numerical Methods in Engineering* 50: 645–663.

Rankine, W.J.M. 1857. On the stability of loose earth. *Phil. Trans. of the Royal Society of London* 147.

Revellino, P., Hungr, O., Guadagno, F.M. & Evans, S.G. 2004. Velocity and runout simulation of destructive debris flows and debris avalanches in pyroclastic deposits, Campania region, Italy. *Environmental Geology* 45: 295–311.

Rickenmann, D., Laigle, D., McArdell, B.W. & Hübl, J. 2006. Comparison of 2D debris-flow simulation models with field events. *Computational Geosciences* 10: 241–264.

Roddeman, D.G. 2001. *TOCHNOG User's Manual*. FEAT, 177 pp.

Sailer, R., Rammer, L. & Sampl, P. 2002. Recalculation of an artificially released avalanche with SAMOS and validation with measurements from a pulsed Doppler radar. *Natural Hazards and Earth System Sciences* 2: 211–216.

Sampl, P. 1993. Current status of the AVL avalanche simulation model—Numerical simulation of dry snow avalanches. In L. Buisson & G. Brugnot (eds.) *Proc. of the Pierre Beghin Int. Workshop on Rapid Gravitational Mass Movements, Grenoble*: 269–276.

Sassa, K. 1985. The mechanism of debris flows. In *Proceedings of the 11th Int. Conf. on Soil Mechanics and Foundation Engineering, San Francisco* 1: 1173–1176.

Sassa, K. 1988. Geotechnical model for the motion of landslides. In C. Bonnard (ed.) *Proc. of the 5th Int. Symp. on Landslides, Lausanne*: 37–56. Rotterdam: Balkema.

Savage S.B. & Hutter, K. 1989. The motion of a finite mass of granular material down a rough incline. *Journal of Fluid Mechanics* 199: 177–215.

Savage S.B. & Hutter, K. 1991. The dynamics of avalanches of granular materials from initiation to runout, Part 1: Analysis. *Acta Mechanica* 86: 201–223.

Schamber, D.R. & MacArthur, R.C. 1985. One-dimensional model for mud flows. In *Proceedings of the ASCE Specialty Conf. on Hydraulics and Hydrology in the Small Computer Age* 2: 1334–1339. American Society of Civil Engineers, New York.

Sousa, J. & Voight, B. 1991. Continuum simulation of flow failures. *Géotechnique* 41(4): 515–538.

Sovilla, B. & Bartelt, P. 2002. Observations and modelling of snow avalanche entrainment. *Natural Hazards and Earth System Sciences* 2: 169–179.

Tai, Y.C. & Gray, J.M.N.T. 1998. Limiting stress states in granular avalanches. *Annals of Glaciology* 26: 272–276.

Takahashi, T. 1991. *Debris flow*. International Association for Hydraulic Research monograph. Rotterdam: Balkema.

Terzaghi, K. 1920. Old earth pressure theories and new test results. *Engineering News Record* 85: 632.

Trunk, F.J., Dent, J.D. & Lang, T.E. 1986. Computer modeling of large rock slides. *Journal of Geotechnical Engineering* 112: 348–360.

Turnbull, B. & Bartelt, P. 2003. Mass and momentum balance model of a mixed flowing/powder snow avalanche. *Surveys in Geophysics* 24: 465–477.

Voight, B. & Sousa, J. 1994. Lessons from Ontake-san: a comparative analysis of debris avalanche dynamics. *Engineering Geology* 38: 261–297.

Wang, Y., Hutter, K. & Pudasaini, S.P. 2004. The Savage-Hutter theory: A system of partial differential equations for avalanche flows of snow, debris, and mud. *Zeitschrift fur Angewandte Mathematik und Mechanik* 84(8): 507–527.

Wieland, M., Gray, J.M.N.T. & Hutter, K. 1999. Channelized free-surface flow of cohesionless granular avalanches in a chute with shallow lateral curvature. *Journal of Fluid Mechanics* 392: 73–100.

Deformation and failure mechanisms of loose and dense fill slopes with and without soil nails

C.W.W. Ng
Department of Civil Engineering, The Hong Kong University of Science and Technology, HKSAR

ABSTRACT: Although fill slopes have been constructed worldwide to meet the needs and development of various human activities for years, the effects of fill density, material type, including gap-graded or well-graded soils, various stabilisation methods, including soil nails, and destabilising agents, including rainfall infiltration, rising ground water and earthquakes, are still not well understood. In this paper, various studies of deformation and failure mechanisms of unreinforced and nailed fill slopes using a geotechnical centrifuge are described and key findings are reported and explained.

1 INTRODUCTION

Fill slopes have been constructed across the world for highway and railway embankments, road widening projects, earth dams, landfills and levees for years. The construction of fill slopes generally involves borrowing fill materials from local sources or from foreign sites and compaction. Catastrophic failures of uncompacted and improperly compacted fill slopes resulting in loss of lives and severe damage to property have been documented (Hong Kong Government 1977, Sun 1999). In addition, excessive settlement and surface erosion failure to densely compacted new slopes due to rainfall infiltration have been reported. It is clear that the deformation and failure mechanisms of fill slopes subjected to rainfall infiltration and earthquakes are not fully understood. In this paper, the performance of both loose and dense fill slopes is investigated wih the aid of centrifuge modelling technology. Both gap-graded Leighton Buzzard (LB) Fraction E fine sand and well-graded completely decomposed granite (CDG), i.e. silty sand, are used as model fill materials. The model fill slopes are destabilised by rainfall infiltration, rising ground water and simulated earthquakes using a hydraulic shaker installed on the geotechnical centrifuge at the Hong Kong University of Science and Technology (HKUST; Ng et al. 2001, 2004b). The 400 g-ton centrifuge is 8.4 m in diameter and is equipped with a hydraulic bi-axial shaker that can simulate earthquakes in two horizontal directions (Ng et al. 2004b) with a maximum input acceleration of 50 g (model scale). In addition, a four-axis robotic manipulator is installed in the centrifuge for simulating some construction activities in-flight such as soil nailing (Ng et al. 2002a).

The effects of using soil nails for stabilising both loose and dense fill slopes are also investigated in this paper. Failure and deformation mechanisms of slopes, which cannot be obtained easily and economically in the field or reliably from numerical analysis are provided and highlighted in this paper. All test results are reported in prototype scale unless stated otherwise.

2 ROLES OF FIELD WORK AND NUMERICAL AND CENTRIFUGE MODELLING

Geotechnical centrifuge modelling has become an alternative modelling tool to investigate failure mechanisms and deformations of slopes and soil structures (Kimura 1998, Ng et al. 2003, 2006a). The basic principle of geotechnical centrifuge modelling is to recreate the stress conditions that would exist in a full-scale construction (prototype), using a model on a greatly reduced scale. This is done by subjecting the model's components to an enhanced body force, which is provided by a centripetal acceleration of magnitude, ng, where g is the acceleration due to the earth's gravity (i.e. 9.81 m/s^2). Stress replication in an nth scale model is achieved when the imposed "gravitational" acceleration is equal to ng. Thus, a centrifuge is suitable for modelling stress-dependent problems. Moreover, a significant reduction of testing duration such as consolidation and seepage can be achieved by using a reduced size model (Ng 1999).

Figure 1 shows the inter-relationships between field investigation, numerical modelling and physical (including centrifuge) modelling. These three approaches are definitely not mutually exclusive. On the contrary, they are complementary as no

Figure 1. The inter-relationships between field investigation, numerical modelling and centrifuge modelling.

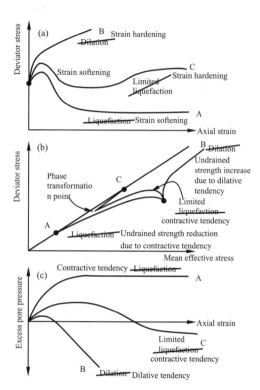

Figure 2. Liquefaction, limited liquefaction, and dilation in monotonic loading tests (modified from Castro 1969, Kramer 1996).

approach is perfect in solving every geotechnical or geo-environmental problem in terms of the quality of results, time and cost. Ideally, two or three approaches should be utilized to calibrate their results against each other and to verify any conclusion drawn. Small-scale laboratory model tests without correctly considering stress effects or using an incorrect constant stress distribution with depth can be very misleading as soil behaviour is stress dependent. These tests should therefore be treated with caution.

3 FAILURE MECHANISMS OF LOOSE SAND FILL SLOPES WITH AND WITHOUT SOIL NAILS

Liquefaction failure due to cyclic and earthquake loading has received much attention and significant progress has been made to improve our understanding of the failure mechanisms and viable stabilisation measures. On the contrary, the mechanisms of static liquefaction failure of loose fill slopes are not yet fully understood. It is clear that an under-design of stabilisation measures for existing loose fill slopes may lead to catastrophic results. In contrast, an over-design of upgrading methodologies could result in excessive costs and adverse environmental impacts, such as unnecessarily removing trees and shrubs that have been growing on existing fill slopes for years. It is clear that more research work is needed.

3.1 Clarification of the terminology relating to static liquefaction

Figure 2 shows some typical results from monotonic triaxial tests on saturated, anisotropically consolidated sand specimens. The figure shows that the very loose sand specimen, A, exhibits a peak undrained shear strength at a small shear strain and then "collapses" to large strains at a low effective confining pressure and a low shear strength at large strain. This behaviour is often loosely called "liquefaction" or "flow liquefaction". No matter whether it is called "flow liquefaction" or "liquefaction", the terminology to describe the material behaviour observed in the laboratory is rather confusing and, strictly speaking, incorrect. Would it be clearer and more precise to describe the material behaviour of loose specimen, A, and dense specimen, B, as "strain-softening" and "strain-hardening", respectively, in the deviator stress-axial strain space (see Fig. 2a)? In the mean effective stress-deviator stress space (see Fig. 2b), would it be more precise to use the terms "undrained strength reduction" (or so-called collapse (Sladen et al. 1985)) and "undrained strength increase" to describe the strength changes of specimen A and specimen B, respectively? Of course, it is well-recognised that a reduction and an increase in the undrained shear strength are caused by the respective tendency of sample contraction and dilation, leading to a respective increase and a reduction in the pore water pressure (Δu) for specimens A and B during undrained shearing (see relationship between

Δu and axial strain in Fig. 2). It must be pointed out that these are simply material element behaviour that do not capture the global behaviour of an entire fill slope or an earth structure.

3.2 Investigation of the failure mechanism of liquefied flow in the centrifuge

3.2.1 Model material

Centrifuge model tests were previously carried out to investigate the failure mechanisms of static liquefaction flow of loose fill slopes subjected to rainfall and a rising ground water table at HKUST (Zhang 2006, Zhang et al. 2006, Ng et al. 2007). Leighton Buzzard (LB) Fraction E fine sand was selected as the fill material for the model tests. Figure 3 shows the gap-graded particle size distribution of LB sand. D_{10} and D_{50} of the sand were 125 µm and 150 µm, respectively. Following BS1377 (1990), the maximum and minimum void ratios of the LB sand were found to be 1.008 and 0.667, respectively (Cai 2001). The estimated saturated coefficient of permeability was 1.6×10^{-4} m/s. LB sand was chosen because of its pronounced strain-softening characteristics with its high liquefaction potential, LP, i.e. it experiences substantial strength reduction (see Fig. 4a). The results from four loose specimens with different initial void ratios (e_o) shown in the figure are obtained from isotropically consolidated undrained compression triaxial tests. The loose sand clearly shows pronounced strain-softening behaviour and substantial strength reduction in the deviator stress and shear strain ($q - \varepsilon_q$) space and contractive responses in the mean effective stress (p') and deviator stress (q) space, i.e. p' decreases continuously as q increases until a peak state is attained (see Fig. 4b), where p' and q are equal to $(\sigma_1' + 2\sigma_3')/3$ and $(\sigma_1' - \sigma_3')$, respectively. After the peak state, q drops (the soil collapses) with a large deformation develops until the quasi-steady state (a shear strain of about 15%) or the critical state (shear strain = 30%) is reached. The critical state friction angle (ϕ_c') of the

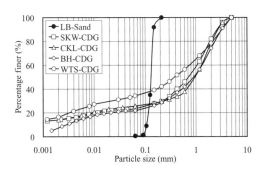

Figure 3. Particle size distributions of LB sand and CDG samples.

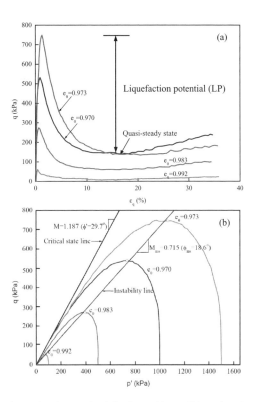

Figure 4. Contractive behaviour of loose LB sand under consolidated undrained tests (a) in the $\varepsilon_q - q$ and (b) in $p' - q$ planes (modified from Zhang 2006; data from Cai 2001).

sand is 30° (Cai 2001). Following the approach proposed by Lade (1992), the angle of instability (ϕ_{ins}') determined for the sand is 18.6°. It is well-known that ϕ_{ins}' is dependent on void ratio and stress level (Chu & Leong 2002). For engineering assessment and design of remedial work for loose fill slopes, it may be reasonable to assume this angle is a constant as the first approximation.

3.2.2 Model package and test procedures

Figure 5 shows an instrumented loose sand fill slope model together with the locations of the pore water pressure transducers (PPTs) (Zhang & Ng 2003, Ng et al. 2008). The model slope was prepared by moist tamping. The initial slope angle and relative compaction were 29.4° and 68%, respectively.

The body of the sand slope was instrumented with seven PPTs and arrays of surface markers were installed for image analysis of soil movements. Linear variable differential transformers (LVDTs) and a laser sensor were mounted at the crest of the slope to monitor its settlement.

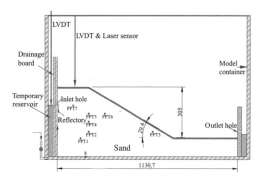

Figure 5. Centrifuge model of a loose sand fill slope subjected to rising ground water table at 60 g (Zhang & Ng 2003). The dimensions are given in model scale.

The ratio of the model height to the average particle size was on the order of 1000. According to Goodings & Gillete (1996), particle size effects would be small if the slope height were higher than 158 times the grain diameter of the model material.

3.2.3 *Observed static liquefaction mechanism*

Although the initial angle of the loose slope at 1 g was 29.4°, the slope was densified to 80% of the maximum relative compaction due to self-weight compaction at 60 g. The slope angle was therefore flattened to 24° (see Fig. 6a), which is steeper than the angle of instability of 18.6°. This implies that the slope was vulnerable to instability, which could lead to liquefaction (see Fig. 4). At 60 g, the 18 m-height (prototype) slope was de-stabilised by rising ground water from the bottom of the model (Zhang 2006, Ng et al. 2008). The loose sand slope liquefied statically and flowed rapidly (see Fig. 6b), i.e. it followed a process in which the loose slope was sheared under the undrained condition, lost its undrained shear strength as a result of the induced high pore water pressure and then flowed like a liquid, which we call "liquefied flow". The liquefaction of the loose sand slope was believed to be initially triggered by seepage force (Ng et al. 2008).

Figure 7 shows the measured rapid increases in the excessive pore water pressure ratio ($\Delta u/\sigma'_v$) within about 25 seconds (prototype) at a number of locations in the slope during the test. The maximum measured $\Delta u/\sigma'_v$ was about 0.6, which would be much higher if a properly scaled (i.e. more viscous) pore fluid were used to reduce the rate of dissipation of the excess pore pressure correctly in the centrifuge test. This means that the slope would liquefy much more easily. As shown in Figure 6b, the completely liquefied slope inclines at about 4° to 7° to the horizontal after the test. The observed fluidization from in-flight video cameras and the significant rise in excessive pore

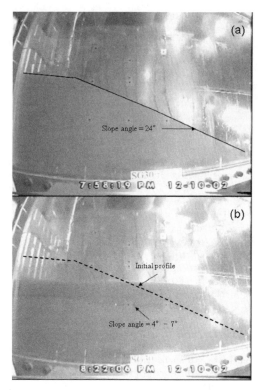

Figure 6. Slope profile in a loose sand fill test (a) before rising ground water table; (b) after static liquefaction (Zhang & Ng 2003).

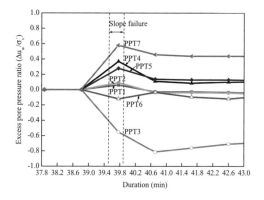

Figure 7. Measured sudden and substantial increases in pore water pressure at seven locations inside the slope (Zhang & Ng 2003).

water pressures during the test clearly demonstrated the static liquefaction of the loose sand fill slope. It should be noted that measurements of the sudden and significant rise of excessive pore water pressures are

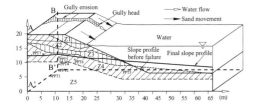

Figure 8. Postulated failure zones during the liquefaction of slope SG30 (Ng et al. 2008).

essential to "prove" or verify the occurrence of static liquefaction of loose fill slopes if no video recording is available.

Figure 8 shows five postulated zones, Z_1-Z_5, corresponding to the sequence of the failure and liquefaction process of the slope (Ng et al. 2008). Z_1 is the failure region de-stabilised by the loss of its toe due to the seepage force in the gully. The soil mass at the toe of Z_1 slid with the soil at the gully head to trigger the failure of Z_2. The soil mass in Z_2 collapsed rapidly which was then followed by the collapse of Z_3 without inducing obvious deformation in the lower part. The collapses of Z_2 and Z_3 were due to the strain-softening associated with the significant strength reduction (i.e. the high liquefaction potential) of the loose LB sand as illustrated in Figure 4. The rapid undrained collapses of Z_2 and Z_3 were evident from the measured large excess positive pore pressures at PPT7 (see Fig. 7). The collapse of the soil mass induced by the collapse of the other parts of the slope inclined at an angle of 10° was also reported by Moriwaki et al. (2004). They observed a large-scale landslide in loose sandy soil induced by artificial rainfall at 1 g.

Subsequently, Z_4 collapsed as a result of the strain-softening associated with the significant strength reduction (high liquefaction potential) of the loose LB sand (see Fig. 4). In this zone, relatively smaller excess positive pore pressure ratios were recorded at PPT4 and PPT5 (see Fig. 7) as compared to that at PPT7. This likely indicates the smaller extent of shearing in Z_4. The fluidised soil was carried by the water wave and finally deposited at an angle of 4° to the horizontal. The dotted line in this figure represents the upper boundary of the stable region (Z_5), monitored by markers and the small excess pore pressures at PPT1 and PPT2 (see Fig. 7) during the liquefaction process.

Based on the centrifuge results described above and other centrifuge model tests on unreinforced and reinforced loose sand fill slopes subjected to rainfall infiltration and shaking (Ng et al. 2006b, Zhang et al. 2006), it is fairly clear that the strain-softening characteristic of a material is a necessary but not a sufficient condition to induce liquefied flows. A trigger such as the seepage force in the gully is required.

3.3 The role of soil nails in liquefiable flow (Zhang et al. 2006)

The soil nailing technique has been used worldwide to stabilise slopes because of its ease of construction and robustness. In Hong Kong, soil nailing has been used for improving soil slope stability for many years and is currently the predominant method used for upgrading the stability of existing cut slopes.

3.3.1 Model preparation, instrumentation and test procedures

Three dynamic centrifuge model tests using LB sand were carried out to investigate the effect of soil nails on the failure mechanism of a liquefiable loose sand fill slope (Zhang et al. 2006). The tests were subjected to severe groundwater conditions and then large dynamic loading in the centrifuge. To create reliable liquefaction, the in-flight hydraulic shaker was used in the centrifuge tests (Ng et al. 2001). Figure 9 shows the initial geometry and instrumentation of a nailed model slope (Zhang et al. 2006). Due to page constraints here, only two tests are summarised in Table 1. LB sand was used to construct the model slopes. In the two tests, i.e. DS25 and DSN25 (D stands for dynamic, S for sand, N for nail and 25 is the initial slope angle), the slopes were initially subjected to severe seepage controlled by the boundary conditions before shaking.

All the slope models had an initial slope height of 250 mm and an initial angle of 25°, as shown in Figure 9. Sand with water content of about 5% was compacted into 50 mm thick layers at a void ratio of about 1.5. To minimize side friction, the container wall was covered with a thin layer of smooth plastic membrane. The upstream drainage board (Fig. 9) was covered by a rubber plate to decrease the wave reflection effects. A thin layer of sand-glue mixture was used on the inner bottom surface of the container to increase the roughness.

As shown in Figure 9, soil nails were installed in slope DSN25. Each modelled nail was equivalent to a steel bar 35 mm in diameter with a grouting layer of 210 mm on the outer diameter of the prototype. The nails were arranged in four rows and five columns with a horizontal distance of 67 mm (2.1 m in the prototype) and a vertical distance of 42.3 mm (1.3 m in the prototype). The inclined angle of each nail was 40° to the horizontal. Depending on the locations of the nails, the lengths of the nails (i.e. A, B, C and D) varied. The soil nails located in the centre column were instrumented with five sets of semiconductor strain gauges.

The instrumentation layout for all tests included five miniature accelerometers (ACCs) installed in the slope to measure soil acceleration in the X-direction, eight PPTs installed at the base of the slope or near the accelerometers to record excessive pore pressures

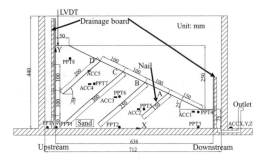

Figure 9. Initial geometry and instrumentation of model slope for tests DS25 (without nails) and DSN25 (nailed slope) (Zhang et al. 2006).

Table 1. Slope geometries, soil properties and input motion (data from Zhang et al. 2006).

Model identity	DS25	DSN25
Total height at 30 g before shaking (m)	8.13	8.16
Initial density (kg/m³)	1373	1314
Initial water content (%)	4.8	5.6
Void ratio: initial	1.022	1.125
at 30 g before shaking	0.858	0.976
Duration of shaking (s)	15.2	66.2
Peak amplitude (g)	0.02	0.16

Note: The initial slope angle and the initial total height was 25° and 9.75 m, respectively.

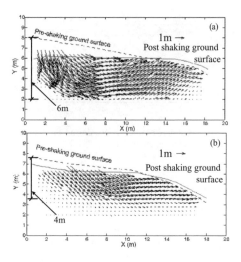

Figure 10. Measured displacement vectors of slopes during shaking by image analysis (a) for DS25, (b) for DSN25 with the slope profiles before the earthquake (Zhang et al. 2006).

during the shaking, and two LVDTs installed to measure the crest settlement.

The deformation of the model slopes was captured using a texture-driven image analysis system, GeoPIV, which combines the techniques of digital photography, multi-threshold centroiding, particle image velocimetry (PIV), and close-range photogrammetry (White et al. 2003, Zhang et al. 2006). The accuracy of the measurements in the centrifuge tests was estimated to be about 0.1 mm.

When the acceleration in the centrifuge reached a nominal value of 30 g, the slopes were subjected to transient seepage controlled by an upstream water supply. A high hydraulic gradient was created between the upstream and the downstream. The outlet located at the downstream was kept open throughout the tests. During an increase in the hydraulic head at the upstream, the slope deformed and some seepage failure occurred (Zhang et al. 2006). Subsequently, a sinusoidal earthquake was triggered by applying acceleration along the base of the slope. Some details of each shaking are given in Table 1.

3.3.2 The role of soil nails during slope liquefaction (comparisons of tests DS25 and DSN25)

Due to the page length constraints, only some results for DS25 and DSN25 are reported in this paper. Other test details and results are described by Zhang et al. (2006).

Figure 10 shows localised seepage failure induced by the high hydraulic gradient created between the upstream and the downstream prior to shaking. The pre-shaking ground surface profiles are also shown. No static liquefaction was observed at this stage. After shaking, displacement vectors obtained from image analysis of PIV data of DS25 (see Fig. 10a) indicated a deep-seated rotational failure of the liquefied soil, reducing the once sloping ground to an almost horizontal surface. Figure 11a shows the slope profile captured by a digital camera after the shaking. The sand mass was shaken down in the left of the slope (or the part close to the slope crest), whereas the sand mass at the right (or the sand close to the slope toe) moved horizontally. The estimated maximum depth of the liquefied zone was up to about 6.0 m.

In slope DSN25, the soil nails restricted the development of this deep-seated failure mechanism, with soil deformations being smaller in magnitude and nearer to the surface (Fig. 10b). The sand slid downslope along the failure surface (see Fig. 11b). The estimated maximum thickness of the failure zone was up to about 4.0 m. Based on the comparison of slope profiles between the two tests, the final lateral displacement of the sand mass at the slope toe was estimated to be 0.71 m, which was 41% less than that of DS25. The estimated maximum settlement at the

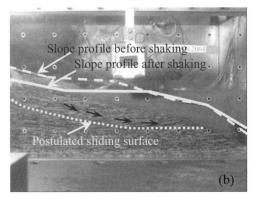

Figure 11. Slope profiles after the shaking for (a) DS25 and (b) DSN25 (Zhang et al. 2006).

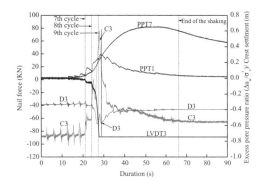

Figure 12. Typical mobilized nail forces with excess pore pressure ratios and crest settlement in DSN25 (Zhang et al. 2006).

Table 2. Maximum excess pore pressure ratios at PPTs 1 to 8 for the two tests (data from Zhang et al. 2006).

Test ID	PPT1	PPT2	PPT3	PPT4*
DS25	0.06	0.51	0.39	0.22
DSN25	0.30	0.12	0.36	0.22
	PPT5	PPT6	PPT7	PPT8*
DS25	0.41	0.20	0.80	0.96
DSN25	0.28	0.02	0.66	1.35

* Estimated locations of PPT4 and PPT8 may be questionable.

slope crest was 0.76 m, which was 62% less than that of DS25, even though the magnitude and duration of the shaking were substantially larger and longer, respectively, for DSN25 than that for DS25, (see Table 1). Hence, the presence of soil nails greatly reduced the deformation of the sand mass during the shaking.

Figure 12 shows the typical responses of nail forces recorded by strain gauges C3 and D3, which were located at the middle of nail C and nail D, respectively (see Fig. 9). Tension (+ve) and compression forces (−ve) in the nails increased when large deformation was initiated and excess pore pressure started to build up at the eighth cycle of the shaking. All forces reached their peaks at the end of the eighth cycle (or at the beginning of the ninth cycle) when the excess pore pressure ratio, $\Delta u_w/\sigma_v'$, was about 0.27 and 0.23 at PPT1 and PPT7, respectively, where σ_v' is the estimated effective overburden stress just before failure and Δu_w is measured excess pore water pressure. At this time, LVDT3, located at the crest of the slope, also reached the end of its travel. As the shaking continued, the two nail forces decreased and eventually reached their respective ultimate values but the excess pore pressure ratio at PPT7 (close to the crest, see Fig. 9) continued to rise and reached its peak of 0.66 after about 50 cycles of shaking. The sudden reduction of nail force might be attributed to the destruction of bond strength between the soil and the nail, resulting from the significant reduction of shear strength after the peak as shown in Figure 4.

Table 2 summarises the measured excess pore pressure ratios at PPT1-8 in the two tests. For both slopes, peak ratios were measured at PPT7 and the values were 0.80 and 0.66 for slopes DS25 and DSN25, respectively. The measured ratios would be larger if a scaled viscous liquid were used in the tests. With the exception of PPT1, the estimated peak ratios for DS25 were generally larger than those at the corresponding locations (PPT2~PPT7) of DSN25 by up to 100%, despite the facts that a denser soil, a much smaller peak amplitude and a shorter duration of shaking were adopted for DS25 (see Table 1). These differences demonstrate the role of soil nails in reducing the volume contraction of the sand mass during earthquake-induced liquefaction by resisting the relative displacement between the soil and the nails.

165

4 EXCESSIVE SETTLEMENTS OF THICK LOOSE CDG FILL SLOPES

4.1 Monotonic and cyclic behaviour of CDG from Beacon Hill (Ng et al. 2004b)

Prior to the centrifuge model tests, a series of undrained monotonic and cyclic triaxial tests on normally consolidated CDG specimens 70 mm in diameter and 140 mm in height were performed to assist in the interpretation of the centrifuge test results. Figure 3 shows the particle size distribution of the well-graded CDG samples obtained from Cha Kwo Ling (CKL), Kowloon. In the figure, the well-graded CDG taken from Beacon Hill (BH) is also included for comparison. The mean particle size, D_{50}, of the CDG from CKL is 1.18 mm and the sample contains about 15% fines content. According to the British Standard, BS1377 (1990), CDG can be classified as well-graded silty sand. The maximum dry density and the optimum water content are 1.82 g/cm³ and 14.4%, respectively.

The triaxial specimens were prepared by moist tamping at the optimum moisture content. The initial relative compaction of the specimens was 70% before saturation. Enlarged lubricated end platens were used in the tests to reduce the end constraints on the soil specimens. In the undrained monotonic triaxial compression tests, the soil specimens were consolidated isotropically to different initial mean effective stresses before shearing. Figure 13a shows the effective stress paths of five isotropically consolidated undrained compression tests with the initial p' ranging from 50 kPa to 400 kPa (corresponding to void ratios varying from 1.05 to 0.78). The effective stress path of each loosely compacted specimen is characterized by its initial increasing q with decreasing p', due to an increase in pore water pressure during undrained shearing resulting from the contractive tendency of the soil. After a peak is reached, q decreases with a further reduction in p' until the critical state (M = 1.54, $\phi' = 37.8°$) is reached, illustrating the unstable nature of the specimen. By joining the stress origin and the peak of each stress path, an instability line (Lade 1992) can be identified with its slope equal to 1.12, corresponding to ϕ'_{ins}=28.2°. Strain-softening behaviour with very small liquefaction potential but without any phase transformation phenomenon was noted in these tests (see Fig. 13b).

During the cyclic tests, a cyclic deviator stress of equal magnitude in compression and extension was applied to the specimens. Figure 13c shows a typical result of CDG (e = 0.821) from a cyclic triaxial test with a cyclic stress ratio (CSR) of 0.1, where CSR is defined as the single amplitude cyclic shear stress (σ_d) divided by twice the initial effective confining pressure (σ'_3), i.e. CSR = $\sigma_d/(2\sigma'_3)$. In the test, p' decreased monotonically but the rate of the pore water

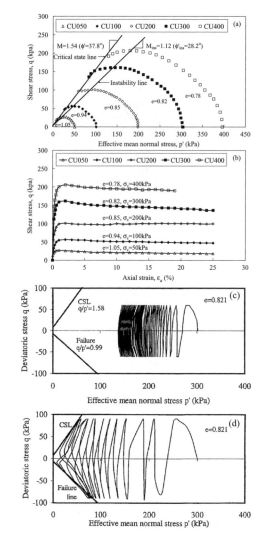

Figure 13. Triaxial tests on loose CDG: (a) Stress paths of static triaxial tests; (b) Stress-strain relationships of static triaxial tests; (c) Cyclic triaxial test with CSR = 0.1; (d) Cyclic triaxial test with CSR = 0.15.

pressure build-up decreased as the number of cycles increased, due to the relatively low CSR. Eventually, the pore water pressure ceased to develop further as the contractive and dilative tendency of the soil specimen balanced out. The total deviator strain developed was less than 0.2% at the end of the test. On the other hand, for a cyclic test on CDG with CSR = 0.15 (e = 0.821) as shown in Figure 13d, the pore water pressure accumulated continuously and resulted in a continuous decrease in p', illustrating a typical cyclic mobility phenomenon (Castro 1969).

4.2 Response to rainfall infiltration (Take et al. 2004)

With the support of the Geotechnical Engineering Office (GEO) of the Civil Engineering and Development Department of HKSAR, collaborative and complementary centrifuge model tests were carried out at the University of Cambridge and HKUST. Bulk samples of CDG taken from BH were delivered to the two universities for centrifuge model tests.

Figure 14 shows an initially 45° loose fill model slope. The model was constructed by moist-tamping with only a minimal compaction effort. To reduce particle size effects, the fill material was first sieved to remove all particles in excess of 5 mm in diameter. To simulate effects of rainfall infiltration by controlling water contents (or moisture), the fill slope was installed in an atmospheric chamber, which was sealed from the external environment (Take et al. 2004). It is well known that suction is related to the water contents in soil pores (Ng & Menzies 2007). In the test, positive and negative pore water pressures were measured using a network of miniature PPTs and pore pressure tension transducers (PPTT), respectively, at each of locations indicated by open circles in the figure. The deformation of the model fill slopes was captured by PIV (White et al. 2003).

During the centrifuge test at Cambridge University, the model was slowly brought to the testing acceleration of 60 g in 10 g increments. The response of the loose fill slope during the self-weight consolidation and rainfall (see Fig. 15a) at $PPTT_1$ and displacements of the crest region (a 32 × 32 pixel patch PIV_1) are shown in Figures 15b and c, respectively.

As the loose fill material became incrementally heavier, a cumulatively larger percentage of the loose fill no longer supported the increase in total stress and

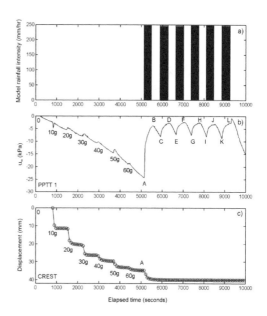

Figure 15. Observed behaviour of a loose CDG model slope (Take et al. 2004).

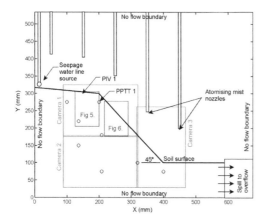

Figure 14. Model geometry of a CDG fill slope (Take et al. 2004).

rapidly decreased its void ratio. The biggest pores in the soil shed their pore water vertically downwards into the fill (consolidation), creating an initial suction distribution that increased with elevation from a value of approximately zero at the toe to −25 kPa at the crest ($PPTT_1$ in Fig. 15b). Figure 16 illustrates the changes of the initially moist-tamped structure of the model fill at the crest during the test. At 1 g, the very loose soil had an initially very open structure (see Fig. 16a), which consisted of large voids supported by capillary suction. One such void is circled in the figure. At 60 g, many of these macro-voids were observed to collapse (Fig. 16b). However, not all the voids collapsed. In particular, the voids at low stress levels (i.e. shallow depths) such as the highlighted void in Figure 16a simply settled along with the fill. The observations of the collapse and the mechanisms shown in these two figures cannot be easily obtained from the field or numerical analyses even with large-strain formulations.

After the initial self-weight consolidation, the fill slope was subjected to the equivalent of six weekly periods of rainfall infiltration (Fig. 15a). As shown in Figure 15b, the arrival of rainfall on the slope surface at time A destroyed a significant portion of the soil suction very rapidly at the shallow location of $PPTT_1$. The loose model fill responded immediately to this loss of surface tension by collapsing the macro-voids that had survived self-weight consolidation (Fig. 16c).

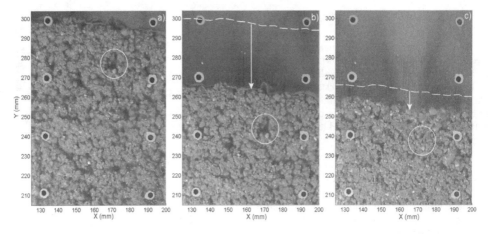

Figure 16. The soil structure of the crest region (Take et al. 2004).

The resulting settlement of the crest was purely vertically downwards.

As the rainfall infiltration continued, the rate of suction loss decreased (see Fig. 15b), as did the rate of settlement (Take et al. 2004). Further rainfall infiltration resulted in the development of a water table at the toe of the slope. By the time the model had been subjected to rainfall for an equivalent duration of 1 week (Time B in Fig. 15b), the rate of the pore water pressure increase diminished to almost zero, becoming asymptotic to a small negative value. At this point, the mist nozzles were turned off and the fill slope was allowed to consolidate for an equivalent period of one week.

Despite being subjected to an additional five infiltration events of identical severity, the model slope was observed neither to achieve positive pore water pressures nor to experience any significant additional deformation. Figure 17 shows the spatial distribution of the settlement vectors, revealing the region of instability recorded during the first rainfall event. Although the fill experienced instability, a liquefied flow was not triggered. The finding is consistent with that reported by Ng et al. (2002b). This was probably because of the small liquefaction potential of the CDG (see Fig. 13b).

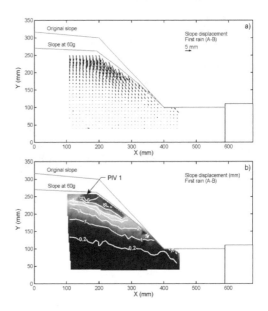

Figure 17. Observed wetting collapse of the loose model fill slope (Take et al. 2004).

4.3 Response to rising ground water (Ng et al. 2002b)

To complement the rainfall infiltration tests carried out at Cambridge, a series of centrifuge model tests on loose CDG fill slopes with and without soil nails was subjected to rising ground water at HKUST (Ng et al. 2002b, Zhang 2006). The CDG fill material used for the tests in Hong Kong was also from BH. A model slope was initially prepared to incline at 45° to the horizontal and the initial relative compaction of the fill was less than 80%. At 60 g, a 300 mm high model slope was equivalent to an 18 m high prototype slope. Figure 18 shows the measured displacement vectors of an unreinforced loose CDG fill slope destabilised by the rise of the ground water. Excessive settlement was measured but no sign of liquefied flow or slide of the slope was observed during and after the test. This was probably because of the small liquefaction potential of the CDG (Ng et al. 2004a). Similar findings are also

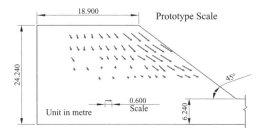

Figure 18. Displacement vectors in unreinforced slope (CG45) (Ng 2007).

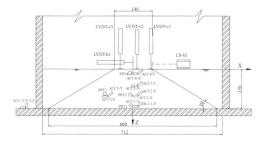

Figure 19. Configuration of the model slope and instrumentation (Ng et al. 2004b).

reported by Take et al. (2004) as described in Section 4.2 of this paper.

Tang & Lee (2003) report on a large-scale field trial on a loose CDG fill slope compacted with an initial angle of 33° to the horizontal. Again, the bulk fill material was taken from BH. The height and width were 4.75 m and 9 m, respectively. It was constructed by the end-tipping method and resulted in a loose state with an initial dry density ranging from 70% to 75% of the maximum dry density. It is considered that the stress state of this slope would represent reasonably well that of many existing fill slopes in Hong Kong. The fill slope was subjected to a rise in the ground water and rainfall infiltration. Only excessive settlement of the slope was recorded and no sign of liquefied flow or non-liquefied slide was observed (Tang & Lee 2003). Consistent results are obtained between the independent centrifuge model tests at Cambridge and at HKUST and the field trial.

4.4 *Response to earthquake loading*

4.4.1 *Centrifuge model and test procedures (Ng et al. 2004b, Ng 2007)*

To further investigate the possibility of flow liquefaction of loose CDG fill slopes, uni-axial and bi-axial dynamic centrifuge tests were carried out using soil samples taken from BH (Ng et al. 2004b). Model embankments were subjected to shaking ranging from 0.08 g to 0.28 g (prototype) in the centrifuge at HKUST. All the models were essentially the same in geometrical layout and made of loose CDG with the same initial dry density. Figure 19 shows a typical model slope (6 m in prototype) initially inclined at 30° to the horizontal with its instrumentation. A rigid rectangular model box was used to contain the CDG samples compacted to an initial dry density of about 1.4 g/cm^3 (or 77% of relative compaction). The insides of the container were lined with a lubricated plastic membrane to minimize friction of the soil with the container walls. For two-directional (X-Y) horizontal shaking tests, a layer of 5 mm soft modelling clay (putty) was sandwiched between a silicon rubber sheet and the sidewall of the model container. This set-up was used to reduce reflected stress waves at the boundaries in the Y-direction. Five pairs of miniature accelerometers were installed in the embankment. Each pair was arranged to measure soil accelerations in two horizontal directions (i.e. the X- and Y-directions). Four miniature pore pressure transducers were installed in the soil near the accelerometers to record pore water pressures during shaking. On top of the embankment, three LVDTs were mounted to measure the crest settlement, and one LVDT and one laser sensor (LS) were used to measure horizontal movement of the crest.

To simulate the correct dissipation rate of excessive pore pressures in the centrifuge tests, sodium carboxy methylcellulose (CMC) powder was mixed with distilled deionized water to form the properly scaled viscous pore fluid and to saturate the loose CDG model slopes.

After model preparation, the speed of the centrifuge was increased to 38 g. Once steady state pore pressure condition was reached at all transducers, a windowed 50 Hz (1.3 Hz prototype), 0.5 s (19 s prototype) duration sinusoidal waveform was then applied (Ng et al. 2004b). After triggering each earthquake, the centrifuge acceleration was maintained long enough to allow the dissipation of any excess pore pressure. Due to page limits, only some results from one biaxial shaking test are discussed here. Other details of all the tests are presented in Ng et al. (2004b).

4.4.2 *Measured responses of the loose CDG fill slope subjected to bi-axial shaking (M2D-0.3) (Ng et al. 2004b)*

Figure 20 shows some measured horizontal acceleration time histories in the X- and Y-directions together with their normalized amplitudes in the Fourier domain. In the biaxial shaking test, the base input accelerations (recorded by ACC-T-X & ACC-T-Y as shown in the figure) were 11.26 g (0.28 g prototype) and 7.77 g (0.19 g prototype) in the

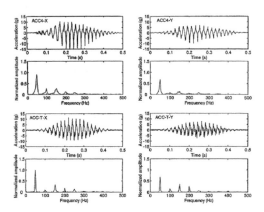

Figure 20. Seismic acceleration history and Fourier amplitude spectrum (M2D-0.3) (extracted from Ng et al. 2004b).

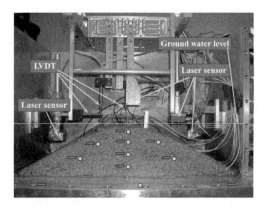

Figure 22. A typical profile of a loose fill slope after shaking (Ng et al. 2004b, Ng 2007).

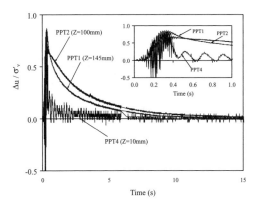

Figure 21. Measured excess pore-water pressure ratios in bi-axial shaking test M2D-0.3 (Ng et al. 2004b).

X-direction and Y-direction, respectively. The windowed sinusoid waveform applied in the Y-direction lagged the X-direction input signal by 90°. Recorded by the accelerometer near the crest, the peak acceleration in the X-direction increased 45% at ACC4-X, higher than that measured in a corresponding uni-axial shaking test (Ng et al. 2004b). A similar trend of variations in the acceleration was also found in the Y-direction. The normalized spectral amplitudes of acceleration at the predominant frequency of 50Hz decreased about 9% in the X-direction but increased about 4% in the Y-direction in the upper portion of the embankment.

Figure 21 shows the time history of the excess pore pressure ratios along the height of the model embankment during shaking. Peak acceleration occurred at about 0.25 s after the start of shaking. The maximum pore pressure ratio occurred at about 0.33 s at each of the three transducers (PPT1, PPT2 & PPT4).

PPT1 and PPT2 recorded about the same maximum pore pressure ratio of 0.87, whereas PPT4 registered the smallest of 0.75. These measured values were less than the theoretical value of 1.0 for liquefaction, even though the pore fluid was correctly scaled in the test. The excess pore pressures dissipated to zero at about 12 s (6.8 minutes in prototype) after the start of shaking.

Figure 22 is a photograph of the model taken after the completion of a shaking test. The deformation profile for the slope was similar in both the uni-axial and bi-axial shaking tests. The observed profile of the deformed slope clearly illustrates that no liquefied flow and non-liquefied slide took place during the shaking. The significant difference between the observed physical test results from the loose LB sand and CDG fill slopes may be attributed to the significant difference in liquefaction potential of the two materials (see Figs 4a & 13b).

5 NON-LIQUEFIED SLIDE MECHANISMS OF SHALLOW CDG FILL SLOPES

5.1 Destabilisation of loose shallow CDG fill slopes near the crest (Ng et al. 2007)

The Housing Department of HKSAR has been actively looking for innovative methods to preserve the environment by minimizing the need for felling trees when improving the stability of existing shallow loose CDG fill slopes. Centrifuge model tests were commissioned to investigate possible failure mechanisms of loose fill slopes. Figure 23 shows an instrumented centrifuge model created to study the potential static liquefaction of a loose shallow CDG fill slope subjected to a rising ground water table. The particle size distribution

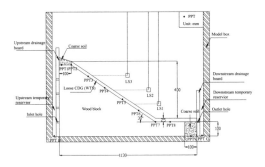

Figure 23. Model package of an instrumented shallow fill slope (Ng et al. 2007).

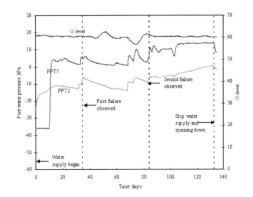

Figure 25. Variations in the measured pore water pressure at the crest (PPT2) and at the toe (PPT7) of the slope over time (Ng et al. 2007).

Figure 24. Top view of the model showing a non-liquefied slide (Ng et al. 2007).

of the CDG used is denoted as WTS in Figure 3. The initial fill density was 66%. This model was used to simulate a 1.5 m thick, 24 m high layered fill slope when tested at 60 g. In addition to laser sensors (LSs) installed for monitoring soil surface movements, PPTs were installed to measure excess pore water pressures during the tests. Effects of layering were considered by titling the model container during model preparation. The slope was destabilised by downward seepage created by a hydraulic gradient, which was controlled by the water level inside the upstream temporary reservoir and the conditions of the outlet hole located downstream (see Fig. 23).

Figures 24 and 25 show the occurrence of a non-liquefied slide and the measured excessive pore water pressure during two failures, respectively. The slide was initiated near the crest. Based on the observed failure mechanisms and the small excessive pore water pressures measured, it was concluded that non-liquefied slide of loose shallow CDG fills slopes could occur but static liquefaction was very unlikely to happen in the slopes.

5.2 Destabilisation of loose shallow CDG fill slope at the toe (Take et al. 2004)

Take et al. (2004) also carried out centrifuge model tests to investigate the possible slide-flow failure mechanism of a loose thin CDG fill layer. The CDG used was taken from Beacon Hill. Figure 26 shows the geometry adopted. The slope angle was 33°. At 30 g, the model corresponded to a fill slope of 9 m in height, with a vertical depth of fill of 3 m. The chosen soil profile for the model fill also represents an idealized case of layering in which the CDG fill material has been sieved and separated into its coarse and fine fractions and placed one on top of the other to form a layered backfill. The layer ends blindly at the toe of the slope to generate elevated transient pore pressures (Take et al. 2004). This ensures that the rate of arrival of the seepage water at the toe greatly exceeds that of the leakage, thereby ensuring a more rapid local transient build up of pore water pressures in this region than would have existed in the absence of layering. In this experiment, the impermeable bedrock layer was modelled by a solid wooden block, the top surface of which was coated with varnished coarse decomposed granite to ensure a high interface friction angle.

The density of the fill material in the first layered slope model was very loose, with an approximate relative compaction of 77%. After preparation, the model fill slope was installed on the centrifuge and slowly brought to the testing acceleration of 30 g.

Figure 27 shows the arrival of the transient pore water at the toe of the slope. Once the line source of seepage water was activated, the high transmissivity of the coarse layer quickly delivered water to the toe

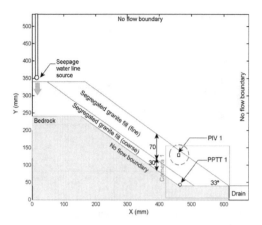

Figure 26. A slide-to-flow landslide triggering mechanism model (Take et al. 2004).

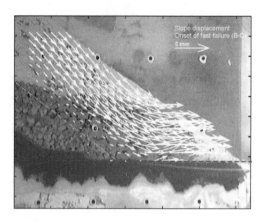

Figure 28. Displacement field prior to final acceleration of loose fill model (Take et al. 2004).

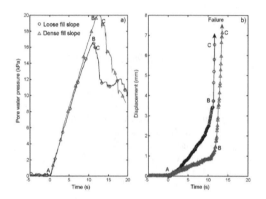

Figure 27. Observed behaviour of slide-to-flow models (Take et al. 2004).

of the fill slope. As intended, the rate of water transfer into the toe region exceeded the seepage velocity through the model fill, causing a transient increase in the pore water pressure at the toe. The local pore water pressure was observed to increase at a nearly constant rate reaching a maximum value of 16 kPa at point B in Figure 27a. As this seepage front progressed towards the toe, the slope was slowly creeping (Fig. 27b).

After time B, the slope mass is observed to accelerate (points B–C on Fig. 27b). Analysis of this pair of images indicates that, at the onset of more rapid failure, the toe accelerated horizontally with an average velocity of approximately 6 mm/s (Fig. 28). The observed displacement field over this time interval indicates that the surface of the model fill moved down-slope at a slower velocity. When the fill material finally came to rest, it formed a low-angle run-out. This failure mechanism differs from that of the slope destabilised by downward seepage in the test for the Housing Department in which the slope was not blinded hydraulically at the toe (see Fig. 23). The initiation of the non-liquefied slides differed in these two slopes.

5.3 Destabilisation of a dense shallow CDG fill slope at the toe (Take et al. 2004)

Unlike the static liquefaction mechanism of loose sand fill slopes, the non-liquefied slide triggering mechanism is argued to be independent of soil density (Take et al. 2004, Ng 2007). In order to verify this hypothesis, the experiment was therefore repeated with a fill compacted to 95% maximum Proctor density while all other factors remained constant (Take et al. 2004).

As before, seepage water was introduced to the crest of the model fill slope and it was quickly transmitted to the toe of the slope, building up localized transient pore water pressures at an identical rate as in the loose fill model (Fig. 29a). Since the slope material was dry, the position of the wetting front could be observed. This dense slope exhibited a much stiffer response to the build up of pore water pressures, with less than one half of the pre-failure displacements signaling the onset of failure (see Fig. 27b). Just before reaching the failure pore water pressure, the brittle fill material cracked and water rapidly entered the fill. As high-pressure water entered the crack, the acceleration of the slide increased. The extent to which this crack injected water into the fill material at time B is shown in Figure 27a. After time B, the slope mass accelerated, although at a slower slide velocity than observed in the loose fill slope (points B–C in Fig. 27b). The subsequent behaviour of the model fill slope is laid out pictorially in the remainder of Figure 29. As the toe continued to accelerate horizontally, the surface of the

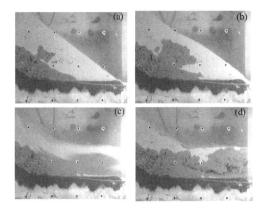

Figure 29. Failure mechanism in the dense fill model (modified from Take et al. 2004).

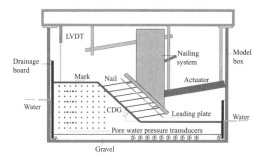

Figure 30. A typical centrifuge model slope package equipped with an in-flight soil nailing system (Ng et al. 2002b).

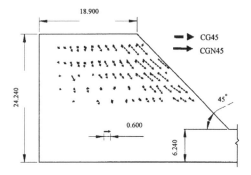

Figure 31. Comparisons of measured soil displacements without (CG45) and with soil nails (CGN45) in two centrifuge tests using CDG loose fill at 60 g (dimensions in metres at prototype scale) (Ng et al. 2002b).

model fill accelerated towards the toe (Fig. 29b), with the velocity increasing to such a point that it exceeded the shutter speed of the camera (Fig. 29c). Eventually, the slope came to rest (Fig. 29d). Similarly to in the shallow loose fill slope, the landslide event triggered from localized transient pore water pressures formed a low-angle run-out. The densification of the fill slope slightly increased the pore water pressure required to initiate failure (see Fig. 27a), but it made the failure more brittle (Take et al. 2004).

6 THE ROLE OF SOIL NAILS IN CDG FILL SLOPES

6.1 Performance of soil nailed loose fill slopes

Centrifuge model tests on CDG fill slopes subjected to rising ground water and rainfall were conducted by Ng et al. (2002b) and reported by Ng et al. (2003). The main objective of these tests was to investigate static liquefaction in loose CDG fill slopes and the ability of soil nails to stabilise these loose slopes. Figure 30 shows a model package of a loose CDG fill slope equipped with an in-flight soil nailing system. The relative compaction of the loose fill model slope was less than 80%. At 60 g, the 300 mm high, 45° model slope was equivalent to an 18 m high prototype slope. Figure 31 compares the displacement vectors of the loose CDG fill slopes obtained from two model tests, one without and one with soil nails. The soil nails were installed in-flight at 60 g and it can be seen that the soil nails substantially reduced soil movements by at least a factor of 5. No sign of static liquefaction of the slopes was observed during and after the tests. Similar findings are also reported by Take et al. (2004) from independent centrifuge model tests using the same loose CDG fill at Cambridge University and by Tang & Lee (2003) from large-scale field tests conducted at Hong Kong University.

6.2 Effects of soil nails on steep dense CDG fill slopes (Zhou et al. 2006)

6.2.1 Centrifuge model and test procedures

Figure 32 illustrates an instrumented model of a nailed slope centrifuge test (CGN65_30) on model scale. A similar model was used for an unreinforced slope (CG65_30) except that no soil nails were installed. The soil used in the experiment was sieved CDG from BH with particles larger than 2 mm removed. D_{10} and D_{50} were found to be 8 μm and 600 μm, respectively. The specific gravity of the CDG was 2.62. The maximum dry density was 1845 kg/m^3 and the optimum water content was 14.2% as established by the standard Proctor compaction test.

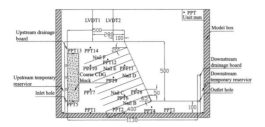

Figure 32. A nailed slope model on the model scale (CGN65_30) (Zhou et al. 2006).

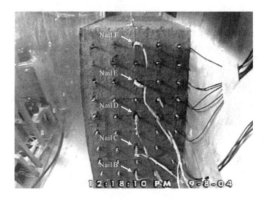

Figure 33. Positions of instrumented soil nails (front view) (Zhou et al. 2006).

A horizontal bed of CDG was first formed by layer-by-layer moist tamping. The mass of soil required in each layer was calculated based on a dry density of 1753 kg/m^3 (relative compaction 95%) at a water content of about 16% (wet of the optimum). A rectangular block of coarse CDG was used to facilitate the seepage of water into the slope mass. Silicon grease was used on both sides of the container to reduce side friction and preferential water flow at the interfaces. After the model was formed, the horizontal bed of soil was cut to form a slope that was 500 mm high with a slope angle of 65°. The dimensions of a 1/30th-scale steep slope model were chosen to represent the prototype behaviour of a 15 m high steep slope when tested at 30 g.

Holes were drilled for inserting nails into the slope mass at 1 g. The inclination of each nail was 20° below the horizontal. The nails were arranged in nine rows and in five columns with horizontal spacing of 67 mm (2.0 m in prototype) and vertical spacing of 50 mm (1.5 m in prototype) (Fig. 33). Five nails (Nail B to Nail F) of the middle column were instrumented with strain gauges (Fig. 33). The length of the nails was 400 mm (12 m in the prototype). The modelled nail was equivalent to a steel bar 32 mm in diameter with a grouting layer of 205 mm in the prototype. Seven sets of full-bridge semiconductor strain gauges were mounted on each nail to monitor the axial nail forces.

Two LVDTs were used to monitor the settlement of the modelled slope (Fig. 32). In addition, the test process and slope deformations were monitored in the control room by four digital cameras and four video cameras.

During a test, the g-level was slowly increased to the target value of 30 g. Then, the rising water table was simulated by allowing water to flow into the standing pipe outside the model container on the upstream side. On the downstream side, the water level was kept within 3 m (in the prototype) of the model base. The supply of groundwater was stopped when no further failure was observed after over four months of seepage in prototype time.

6.2.2 *Measured settlement responses and crack formations due to rising of the ground water*

Due to space limitations, only selected results are presented here. Figure 34 shows the relationships between measured settlement at LVDT2 and pore pressure at PPT4 for both the unreinforced and nailed slopes.

In the unreinforced slope (CG65_30), initially, settlement at the crest was not sensitive to the increase in pore water pressure below the slope toe (O-A). At point A, the positive pore water pressure indicates a complete loss of suction. Settlement increased suddenly at point A, corresponding to the occurrence of the first crack near the slope toe (see Fig. 35). Then, the settlement gradually increased with the pore water pressure. When the pore water pressure increased to about 14 kPa (point B), the settlement significantly increased again. Point B exactly corresponds to the observed second crack at the crest. From point B to point C, a slight increase in pore water pressure induced a significant vertical deformation at the crest. This observed deformation-suction relationship is consistent with results from field stress path triaxial tests reported by Ng & Chiu (2003). At point C, the third crack occurred below the crest. Then, instability from the slope toe quickly developed upwards with increased pore water pressure (PPT4). At point D, the third crack spread to the slope surface as indicated by the formation of a continuous slip plane. After point D, the settlement at LVDT2 could not represent the deformation of the slipped mass because of the detachment of the soil mass below the crest (see Fig. 35).

In the reinforced slope (CGN65_30), the initial decrease at PPT4 indicated that consolidation was possibly still in process before the water flowed near the toe. Once the water from the upstream reached the slope toe, the pore water pressure started to increase at point A'. Localized failure between nails near the toe occurred (see Fig. 36). Then, the crest settlement

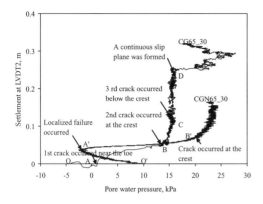

Figure 34. Settlements at LVDT2 versus pore water pressure (PPT4) in two models (CG65_30 & CGN65_30) (Zhou et al. 2006).

Figure 36. Side view of the nailed slope (CG65_30) (Zhou et al. 2006).

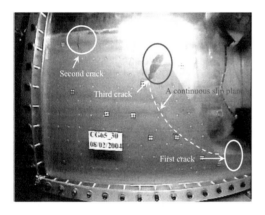

Figure 35. Side view of the unreinforced model (CG65_30) (Zhou et al. 2006).

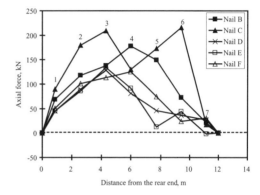

Figure 37. Distribution of axial forces at each instrumented nail when the groundwater stopped rising (CGN65_30) (Zhou et al. 2006).

increased with the pore water pressure at PPT4 slowly to about 20 kPa (point B′). After point B′, there was a significant increase of crest settlement, which corresponded to the appearance of a crack at the crest. In comparison to the measured pore water pressure of 14 kPa in the unreinforced slope, the critical pore water pressure at PPT4, corresponding to the initiation of the slope instability, increased by about 6 kPa in the nailed slope. This corresponded to a longer duration of seepage prior to the formation of the crack at the crest. After that, the crest settlement increased significantly with a minor change at PPT4. However, unlike in the unreinforced model, a continuous slip plane in this test was not observed. When the seepage flow was stopped in the nailed slope, the crest settlement in the reinforced model was 0.16 m in the prototype, which was only about half of that in the unreinforced model. This implies that the scale of failure is reduced with the use of soil nails.

6.2.3 *Distribution of nail forces*

Figure 37 shows the measured axial nail forces when the groundwater stopped rising. The maximum axial tensile force (216 kN) was mobilised at Nail C, which was 31% larger than that when the g-level stopped rising. The axial force mobilised at Nail B was about 20% smaller than that at Nail C. The largest mobilized maximum tensile force did not occur at Nail B, which was located in the lowest row. This is in good agreement with numerical predictions reported by Shiu & Chang (2005). Nails D, E and F mobilised similar but the smallest axial forces. The measured maximum tensile force at Nail F increased by about 220% as compared with that recorded when the g-level stopped rising.

7 CONCLUSIONS

By making use of geotechnical centrifuge modelling technology, the deformation and failure mechanisms of sand and CDG fill slopes with and without soil nails were investigated. Both static and dynamic centrifuge tests were carried out and in-flight rainfall infiltration and rising ground water were simulated. Based on the static centrifuge tests, it can be concluded that static liquefaction/fluidization of a loose sand fill slope due to a rising ground water table was successfully created in the centrifuge. The occurrence of liquefaction in sand was observed by in-flight video cameras and verified by the significant and sudden buildup of excessive positive pore water pressures measured at various locations in the slope. It is found that strain softening of the material is a necessary but not a sufficient condition to cause flow liquefaction. A trigger such as seepage force or additional loading is needed.

Dynamic centrifuge model tests on loose sand fill slopes with and without soil nails reveal that soil nails increase the stability of the slope against dynamic liquefaction failure (a larger dynamic loading is needed to trigger failure). This is because soil nails can resist soil mass movements and reduce the amount of contraction of the soil. Measured settlements and lateral movements of the sand mass were reduced at least by 62% and 41%, respectively, due to the presence of soil nails. The maximum excess pore pressure ratio at the corresponding locations could be reduced by up to 100% at certain locations.

No liquefied flow or slide was observed in thick loose CDG fill slopes when they were subjected to rising ground water tables and/or heavy rainfall infiltration. Only excessive settlements were measured. The centrifuge test results are consistent with the field observations from a large-scale field trial of loose CDG fill slope. With the use of the two-dimensional shaking table at HKUST, both one-dimensional and two-dimensional earthquakes with up to 0.3 g peak ground acceleration (prototype) were imposed on loose CDG fill slopes saturated with correctly scaled pore fluid. The earthquakes did not liquefy the loose slopes and they also did not induce any slide but resulted in excessive positive pore pressures and settlements. The significant difference between the observed physical test results on the sand and CDG models may be attributed to the difference in the fine contents, gradation and liquefaction potential of the two materials.

Although static and dynamic liquefaction is very unlikely to occur in loose CDG fill slopes, because of CDG's small liquefaction potential, non-liquefied shallow slides were observed in both loose and dense shallow fill slopes. Depending on the boundary conditions, different initiations of non-liquefied shallow slides were captured in the centrifuge. The landslide event triggered by high localized transient pore water pressures at the toe results in a low-angle run-out in both shallow loose and dense CDG fill slopes. A denser CDG fill slope shows stiffer and more brittle responses than does a looser fill slope.

The use of soil nails in steep dense CDG fill slopes can prevent the slope from forming a continuous slip plane and reduce settlements at the crest by about 50%. In addition, the presence of soil nails delays and minimizes crack formation at the crest and localized failures between nails near the toe.

ACKNOWLEDGEMENTS

The work presented here was supported by research grants HKUST6053/97E, HKUST6046/98E and CA99/00.EG01 provided by the Research Grants Council of the Hong Kong Special Administrative Region (HKSAR) and DAG00/001.EG36 from HKUST. The author is grateful for research contracts provided by the Geotechnical Engineering Office of the Civil Engineering and Development Department and the Housing Department of the HKSAR. Moreover, the author thanks Mr Chen Rui for assisting in checking and formatting the paper.

REFERENCES

British Standards Institution (BSI). 1990. *BS1377: Methods of tests for soils for civil engineering purposes.* BSI, London.

Cai, Z.Y. 2001. *A comprehensive study of state-dependent dilatancy and its application in shear band formation analysis.* PhD thesis, HKUST.

Castro, G. 1969. Liquefaction of sands. *Harvard Soil Mechanics Series 87*, Harvard University, Cambridge, Massachusetts.

Chu, J. & Leong, W.K. 2002. *Géotechnique* 52(10): 751–755.

Goodings, D.J. & Gillette, D.R. 1996. Model size effects in centrifuge models of granular slope instability. *Geotech. Testing J.* 19(3): 277–285.

Hong Kong Government. 1977. *Report on the Slope Failures at Sau Mau Ping 25th August 1976.* Vol. 1–3, Hong Kong Government Printer, Hong Kong.

Kimura, T. 1998. Development of geotechnical centrifuge in Japan. *Proc. Centrifuge 98*, Tokyo, Pre-print volume: 23–32.

Kramer, S.L. 1996. *Geotechnical earthquake engineering*, Prentice-Hall, Inc. New Jersey, USA.

Lade, P.V. 1992. Static instability and liquefaction of loose fine sandy slopes. *J. Geotech. Eng.* 118(1): 51–71.

Moriwaki, H., Inokuchi, T., Hattanji, T., Sassa, K., Ochiai, H. & Wang, G. 2004. Failure processes in a full-scale landslide experiment using a rainfall simulator. *Landslide* 1: 277–288.

Ng, C.W.W. 1999. Applications of Geotechnical Centrifuge Modelling Techniques for Engineering Designs. *Proc.*

Construction Challenges into the Next Century. Dec., 1999, Hong Kong Institution of Engineers: 241–252.

Ng, C.W.W. 2007. Keynote: Liquefied flow and non-liquefied slide of loose fill slopes.*Proc. 13thAsian Regional Conference on Soil Mechanics and Geotechnical Engineering, 10–14, Kolkata.* Vol. 2. In press.

Ng, C.W.W & Chiu, C.F. 2003. Laboratory study of loose saturated and unsaturated decomposed granitic soil. *Journal of Geotechnical and Geoenvironmental Engineering, ASCE* 129(6): 550–559.

Ng, C.W.W., Fung, W.T., Cheuk, C.Y. & Zhang, L.M. 2004a. Influence of stress ratio and stress path on behaviour of loose decomposed granite. *J. Geotech. and Geoenviron. Eng., ASCE* 130(1): 36–44.

Ng, C.W.W., Kusakabe, O. & Leung, C.F. 2003. Theme lecture: Applications of centrifuge modelling technology in geotechnical engineering practice. *Proc. of 12th Asian Regional Conference on Soil Mechanics and Geotechnical Engineering,* August, Singapore, Vol. 2: 1277–1285.

Ng, C.W.W., Li, X.S., Van Laak, P.A. & Hou, Y.J. 2004b. Centrifuge modelling of loose fill embankment subjected to uni-axial and bi-axial earthquake. *Soil dynamics and earthquake Engineering* 24(4): 305–318.

Ng, C.W.W. & Menzies, B. 2007. *Advanced Unsaturated Soil Mechanics and Engineering.* Publisher: Taylor & Francis. ISBN: 978-0-415-43679-3 (Hardcopy). 687p.

Ng, C.W.W., Pun, W.K., Kwok, S.S.K., Cheuk, C.Y. & Lee, D.M. 2007. Centrifuge modelling in engineering practice in Hong Kong. *Proc. of the Geotechnical Division's Annual Seminar, Hong Kong Institution of Engineers (HKIE),* 55–68.

Ng, C.W.W., Van Laak, P., Tang, W.H., Li, X.S. & Zhang, L.M. 2001. The Hong Kong Geotechnical Centrifuge. *Proc. 3rd Int. Conf. Soft Soil Engineering*: 225–230.

Ng, C.W.W., Van Laak, P.A., Zhang, L.M., Tang, W.H., Zong, G.H., Wang, Z.L., Xu, G.M. & Liu, S.H. 2002a. Development of a four-axis robotic manipulator for centrifuge modeling at HKUST. *Proc. Int. Conf. on Physical Modelling in Geotechnics,* St. John's, Canada: 71–76.

Ng, C.W.W., Zhang, M., Pun, W.K., Shiu, Y.K. & Chang, G. W.K. 2008. Investigation of static liquefaction mechanisms in loose sand fill slopes. Re-submitted to *Géotechnique.*

Ng, C.W.W., Zhang, M. & Shi, X.G. 2002b. Keynote (In Chinese): An investigation into the use of soil nails in loose fill slopes. *Proc. of the 1st Chinese Symposium on Geoenvironment and Geosynthetics,* 17–19 Nov., Hangzhou, Zhejiang, China: 61–80.

Ng, C.W.W, Zhang, L.M. & Wang, Y.H. 2006a. *Proceedings of 6th Int. Conf. on Physical Modelling in Geotechnics.* Volumes 1 and 2. Publisher: Taylor & Francis. ISBN: 978-0-415-41587-3 and 978-0-415-41588-0.

Ng, C.W.W., Zhang, E.M. & Zhou, R.Z.B. 2006b. Centrifuge modelling of use of soil nails in loose and dense fill slopes. *GEO Report,* CEDD of HKSAR.

Shiu, Y.K. & Chang, G.W.K. 2005. *Nail head review.* Geotechnical Engineering Office, HKSAR.

Sladen, J.A., D'Hollander, R.D. & Krahn, J. 1985. The liquefaction of sands, a collapse surface approach. *Canadian Geotechnical Journal* 22: 564–578.

Sun, H.W. 1999. *Review of fill slope failures in Hong Kong. GEO Report No. 96.* Geotechnical Engineering Office, Hong Kong.

Take, W.A., Bolton, M.D., Wong, P.C.P. & Yeung, F.J. 2004. Evaluation of landslide triggering mechanisms in model fill slopes. *Landslides, Japan,* Vol. 1: 173–184.

Tang, W.H. & Lee, C.F. 2003. Potential use of soil nails in loose fill slope: an overview. *Proceedings of the International Conference on Slope Engineering,* Hong Kong, China: 974–997.

White, D.J., Take, W.A. & Bolton, M. D. 2003. Soil deformation measurement using particle image velocimetry (PIC) and photogrammetry. *Géotechnique* 53(7): 619–631.

Zhang, M. 2006. *Centrifuge modelling of potentially liquefiable loose fill slopes with and without soil nails.* PhD Thesis, the Hong Kong University of Science and Technology.

Zhang, M. & Ng, C.W.W. 2003. *Interim Factual Testing Report I—SG30 & SR30.* Hong Kong University of Science & Technology.

Zhang, M., Ng, C.W.W., Take, W.A., Pun, W.K., Shiu, Y.K. & Chang, G.W.K. 2006. The role and mechanism of soil nails in liquefied loose sand fill slopes. *Proc. of 6th Int. Conf. Physical Modelling in Geotechnics (TC2),* Hong Kong. Vol. 1: 391–396.

Zhou, R.Z.B., Ng, C.W.W., Zhang, E.M., Pun, W.K., Shiu, Y.K. & Chang, G.W.K. 2006. The effects of soil nails in a dense steep slope subjected to rising groundwater. *Proc. of 6th Int. Conf. Physical Modelling in Geotechnics (TC2), Hong Kong.* Vol. 1: 397–402.

Capturing landslide dynamics and hydrologic triggers using near-real-time monitoring

M.E. Reid
U.S. Geological Survey, Menlo Park, California, USA

R.L. Baum
U.S. Geological Survey, Denver, Colorado, USA

R.G. LaHusen
U.S. Geological Survey, Vancouver, Washington, USA

W.L. Ellis
U.S. Geological Survey, Denver, Colorado, USA

ABSTRACT: Near-real-time monitoring of active landslides or landslide-prone hillslopes can provide immediate notification of landslide activity, as well as high-quality data sets for understanding the initiation and movement of landslides. Typical components of ground-based, near-real-time landslide monitoring systems include field sensors, data acquisition systems, remote telemetry, and software for base-station data processing and dissemination. For the last several decades, we have used these monitoring tools to investigate different landslide processes. Some of our field applications have determined the groundwater conditions controlling slow-moving landslides, detected 3-D displacements of large rock masses, and characterized the transient near-surface hydrology triggering shallow landsliding.

1 INTRODUCTION

Most landslide investigations are autopsies of inactive slides or inventories of past slope failures. Such studies, however, reveal little about the dynamics of active landslides. Reliable landslide warning systems require accurate short-term forecasts of landslide activity, which in turn demand a detailed understanding of current field conditions and a quantitative framework for interpreting those conditions. This knowledge is difficult to extract solely from landslide postmortem studies.

Real-time or near-real-time landslide monitoring can provide insight into the dynamics of landslide initiation and movement. In addition, this monitoring can provide immediate notification of landslide activity that may be critical to protecting lives and property. Displaying current landslide conditions on the Internet can be extremely valuable to a wide variety of end users, including emergency responders, land managers, geotechnical engineers, researchers, teachers, and the general public. These groups may have very different uses for near-real-time monitoring data. Near-real-time systems also tend to ensure high-quality data sets about landslide behavior by helping to maintain continuity of monitoring during critical periods. If the systems or field sensors malfunction, they can be quickly repaired to minimize data interruption. In addition, such systems tend to promote the evolution of better landslide monitoring by identifying the need for additional or different sensors to better detect changing field conditions. The resulting data sets are valuable for improved geotechnical designs or emergency actions aimed at mitigating landslide hazards. They are also crucial for advancing scientific understanding of active landslide behavior.

The term real-time monitoring has become common in many settings, from finance to computer performance to environmental conditions. However, remote monitoring systems are not truly real time; there is always some delay between sampling conditions and displaying those conditions to users. Here we use the term near-real-time monitoring to designate observations that are delayed slightly (typically minutes to hours) but still close enough in time to represent the current status of field conditions. The degree to which a remote system approaches real time depends on the frequency of 1) data sampling in the field, 2) data transmission, and 3) data updates available to users.

Near-real-time monitoring systems have been used throughout the world to detect or forecast landslide activity. In Hong Kong, the USA, and Brazil, regional warning systems have been operated to forecast conditions for rainfall-induced shallow landslides, using near-real-time rainfall observations (Finlay et al. 1997, Ortigao & Justi 2004, Wilson 2005, Chleborad et al. 2006). Frameworks for similar systems have been developed for mountainous regions of Italy, New Zealand, and Taiwan (Aleotti 2004, Chien-Yuan et al. 2005, Schmidt et al. 2007). Site-specific, near-real-time systems have been applied in many countries to monitor critical structures, such as dams, or hazardous landslides (e.g. Angeli et al. 1994, Berti et al. 2000, Husaini & Ratnasamy 2001, Froese & Moreno 2007). Since 1985, researchers with the U.S. Geological Survey (USGS) have used near-real-time monitoring systems for regional warning systems (Keefer et al. 1987, NOAA-USGS Debris Flow Task Force 2005) and for recording the dynamics of hazardous active landslides or landslide-prone hillslopes (e.g. Reid & LaHusen 1998, e.g. Baum et al. 2005).

In this paper, we discuss some of the design considerations and components typical of ground-based, site-specific, near-real-time landslide monitoring systems. We then discuss some USGS applications of such monitoring systems. Finally, we present three brief case studies that illustrate monitoring system configurations and that document landslide dynamics or hydrologic triggering in very different geologic settings. These studies include: 1) identifying the groundwater pressures controlling a slow-moving, coastal landslide, 2) detecting 3-D displacement of a large rock block using inexpensive GPS receivers, and 3) capturing the transient, rainstorm-induced, soil-moisture conditions triggering a shallow landslide.

2 COMPONENTS OF NEAR-REAL-TIME LANDSLIDE MONITORING SYSTEMS

2.1 System design considerations

The technologies used in ground-based, near-real-time landslide monitoring systems have evolved rapidly in recent years; the development of new sensors, low-cost methods of telemetry, and new software for data dissemination have made this type of monitoring readily available and affordable. However, there is no standard setup that will work for all landslide monitoring. The design and implementation of near-real-time systems depends on considerations that vary from site to site, including: 1) The end purpose of the monitoring. Systems for public safety often differ from those intended to record data for scientific research studies.

Automated warning systems may need redundant field sensors, power supplies, and data serving computers to help ensure continuous operation during landslide triggering events. The desired frequency of data updating, i.e. how close to real time the system is, can be greatly influenced by the end purpose of the system. 2) The type of landslide to be monitored. Instrumentation techniques and sampling frequencies to detect rapid debris flows differ greatly from those used to monitor slow-moving landslides. Moreover, designs to monitor displacement of a currently active slide can differ from those monitoring the hydrologic conditions the might trigger future sliding. 3) The physical setting of the field site. Landslides in urban settings may have access to AC power and readily available telecommunications, whereas very remote sites may need multiple radio repeaters or satellite links to relay data to a secure base-station computer.

Although designs and configurations can vary considerably, most ground-based, near-real-time landslide monitoring systems have certain components in common. These include: 1) sensors on or within the landslide mass or landslide-prone area, 2) data acquisition systems to sample and control the sensors, 3) a communication system to relay data from the field to base-station computers or the Internet directly, and 4) software for data analysis and visualization. An example system configuration is shown in Figure 1. Below, we provide brief overviews of each component with an emphasis on techniques we have found successful in USGS landslide monitoring.

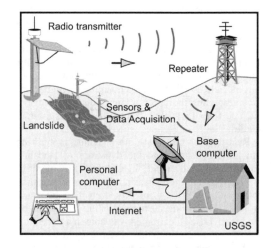

Figure 1. Example of a ground-based, near-real-time landslide monitoring system. Typical components include field sensors, field data acquisition system, remote communications (here via radio telemetry), data processing (on a base computer) and data dissemination (via the Internet).

2.2 Field sensors

For ground-based monitoring, detecting and measuring landslide activity typically requires an array of sensors placed on the ground surface and/or in the subsurface at the slide. For USGS monitoring, we have used both commercially available and USGS custom-built sensors. A wide variety of commercial-off-the-shelf electronic and/or mechanical geotechnical and hydrologic sensors exist (Dunnicliff 1993, Mikkelsen 1996). For remote sites, sensors need to be rugged, weather resistant, portable, and have low power consumption; power is often the most critical issue. Sensors also need to easily interface with data acquisition systems and have adequate sensitivity and resolution. Very-high precision instruments, common in scientific laboratories, are seldom needed for field monitoring. Instead, relatively inexpensive, yet precise, sensors are preferred because they may be destroyed by landslide activity. In addition, low-cost sensors can allow more units to be deployed. Often, sensor arrays with more complete spatial coverage, both on the ground and in the subsurface, are more useful than a few high-precision sensors.

Monitoring sensors typically provide two types of information: 1) actual displacement of a landslide or debris flow or 2) environmental conditions that affect slide activity. Slow-moving landslide displacement can be measured using surface or subsurface extensometers, tiltmeters, ultrasonic or laser distance meters, radar (Tarchi et al. 2005), digital cameras, videos cameras, or downhole inclinometers. Although surface cable extensometers are relatively inexpensive, they are particularly subject to disturbance by weather, animals, and local ground failure. Subsurface deformation can also be detected using grouted-in TDR (Time Domain Reflectometry) cables (Kane & Beck 1996). Measurements from Global Positioning System (GPS) receivers located on a slide can be processed relative to a known reference station to provide sub-cm positions and 3-D displacements (Kramer & Rutledg 2000, LaHusen & Reid 2000). Rapidly moving debris flows can be detected with tripwires, geophones, or flow-height sensors (LaHusen 2005).

Because most landslide and debris flow activity is triggered by hydrologic conditions, it is often important to detect changes in subsurface pore-water pressures. Furthermore, it may be desirable to understand the hydrologic processes modifying those pore pressures, including the infiltration of rainfall or snowmelt. Commonly used hydrologic sensors include precipitation gauges, tensiometers and dielectic soil-moisture probes for measuring unsaturated soil conditions, and piezometers for recording positive pore pressures. For automated recording, tensiometers and piezometers can be fitted with transducers for measuring fluid pressure. Piezometers can be directly buried in a sand pack, grouted in place (Mikkelsen & Green 2003), or installed in a cased borehole, whereas tensiometers and moisture sensors typically need direct soil contact.

2.3 Data acquisition systems

Remote sites usually require on-site data acquisition systems to power and control sensor sampling, log sensor data, and interact with the remote communication system. Site conditions and system configurations often dictate the number of data acquisition systems required; most systems can handle multiple sensors. Seismic and GPS instruments can require relatively high frequency sampling (0.1 to 100 Hz), whereas most geotechnical and hydrologic sensors are sampled at relatively low frequency (minutes to hours). USGS landslide monitoring has used both commercially available data loggers and USGS systems custom-developed for monitoring active volcanoes (Hadley & LaHusen 1995). A reliable power supply is an absolute necessity for each acquisition system, as well as for the remote communication system. We commonly use batteries and solar panels to supply power at remote sites; AC power is sometimes available at urban sites. For sites without AC power or adequate solar exposure, we often use air-alkaline batteries.

2.4 Remote communications

To provide near-real-time data updates from remote monitoring systems, some form of dependable communication system is needed. There are many options using either dedicated telemetry or commercially available services. Typically, the crucial link is between the remote monitoring stations and a secure base station with Internet access or a dedicated telephone line. This remote link can be provided by radio transceivers, satellite uplinks, or telephone services. The choice of a communication component depends, in part, on site remoteness, power availability, the frequency of data transmission desired, reliability, data throughput, and recurring expense. Low-power radio transceivers can use either a fixed frequency or license-free spread-spectrum technology, but line-of-site transmission may require repeater stations. Meteor burst radio communications can be used over long distances for low data-rate applications. Satellite uplinks may use a dedicated service such as GOES (Geostationary Operational Environmental Satellite). Commercial vendors can provide satellite phone modems or a fixed Internet address accessible through a low-cost VSAT (Very Small Aperture Terminal) satellite ground station. Although satellite uplinks may reach many remote areas, their transceivers can require more power than line-of-sight radios and their use can incur service charges. Telephone services, either land line, cellular or satellite, can be reliable options for low data-rate transmission.

However, they entail recurring service charges and may not be available at remote sites. For monitoring where close to real-time response and/or high data-rate transmissions are required, we often use dedicated radio telemetry. Importantly, battery-powered radio links are usually very reliable during stormy weather when landslides may be active. AC power or telephone communications may fail during these stormy periods.

2.5 *Data processing and dissemination*

After remote monitoring data are collected and relayed to a protected base-station computer, additional actions are needed to provide information to end users. Base-station computer actions typically include receiving the data, processing the data if needed, creating graphs and/or tables, and archiving the data. Remote data transmission may be controlled by the field acquisition system or by the base-station computer. There are a variety of software options for performing these tasks, including commercially available software packages, some using Open Process Control (OPC) protocols (http://opcfoundation.org), that can handle real-time data flow and processing. For many USGS systems, we use custom-written base-station software controlled by automated batch processing. Graphs can be generated, using commercial or license-free software, at specified intervals or in response to user requests. Once graphs and tables are created, they can be disseminated via a user's local computer network, or commonly, on web pages with public or password protected access. We typically use USGS web servers to disseminate monitoring information.

3 USGS APPLICATIONS OF NEAR-REAL-TIME LANDSLIDE MONITORING

Over the last several decades, researchers with the USGS have used monitoring systems to understand both the dynamic behavior of individual slides and the hydrologic conditions triggering widespread landsliding. Many of these efforts involved remote data acquisition and some invoked periodic transfer of data via cellular telephone service. (Ellis et al. 2002). USGS automated, near-real-time landslide monitoring sites are listed in Table 1 with a brief summary of their field sensors, data acquisition systems, and remote communication set-ups. Publicly accessible monitoring data from USGS systems currently in operation can be viewed at http://landslides.usgs.gov/monitoring.

Early USGS monitoring efforts at La Honda, California contributed to a San Francisco Bay regional landslide warning system that operated between 1985 and 1995 (Keefer et al. 1987, Wilson 2005). Here, near-real-time observations of rainfall, shallow pore pressures, and soil suction were transmitted using ALERT system radio telemetry. Starting in 1997, the Cleveland Corral landslide, threatening U.S. Highway 50 in California, was our first monitoring site with automated data dissemination via publicly accessible web pages on the Internet (Reid & LaHusen 1998, Reid et al. 2003). Since then, the USGS has operated many other near-real-time monitoring sites. With the exception of La Honda, all of the systems listed in Table 1 use or used USGS web servers to disseminate data over the Internet, typically with updates at a frequency similar to that of the listed data transmission. Our sampling, transmission, and update frequencies were selected to capture changes in the physical processes occurring in the field (e.g. movement, rain infiltration) while minimizing field station power usage. Sites with geophones, such as the Cleveland Corral landslide, scan data every second and transmit immediately if ground vibrations exceed a chosen threshold; thus these sites are closer to true real-time monitoring. USGS researchers have also played key roles in designing and installing near-real-time landslide monitoring systems to monitor alpine debris-flow activity in Italy (Berti et al. 2000) and volcanic debris flows at Ruapehu Volcano in New Zealand.

Below, we briefly present three USGS case studies using near-real-time monitoring that illustrate some of the advantages and complexities involved. Each study examines a different type of slide, uses different monitoring instrumentation and communication telemetry, and addresses different scientific questions. In particular, we focus on how near-real-time monitoring provides crucial insight into different landslide triggering and behavior. Each of the three cases is located near the Pacific Coast of the USA (Fig. 2) where rainfall-induced landslide activity occurs primarily during the winter/spring-wet season. Brand names for sensors and data acquisition systems are provided for descriptive purposes only and do not imply endorsement by the USGS; other vendors can provide similar equipment.

3.1 *Case study 1: Identifying groundwater controls on the motion of a slow-moving landslide, Newport, Oregon*

Most landslide movement is activated or reactivated by increased pore-water pressures acting on a slide's slip surface (Terzaghi 1950, Sidle & Ochiai 2006). These pressure increases can result from many processes. (e.g. Reid & Iverson 1992, e.g. Iverson 2000); understanding the timing and pathways of subsurface water flow leading to landslide movement is crucial to forecasting future slide behavior, developing warning strategies, and designing effective mitigation measures. Our first brief case study illustrates the use

Table 1. USGS near-real-time landslide monitoring sites.

Location and period of operation	Type of slide	Field sensors	Data acquisition system***	Communication system and transmission frequency
La Honda, California (1985–1995)	Shallow earth slide**	Rain gauges, piezometers, tensiometers, extensometers	Sierra Misco ALERT system	Radio network with repeater (15 minutes)
Cleveland Corral landslide, U.S. Highway 50, California (1997-present)	Translational earth slide	Rain gauges, geophones, piezometers, extensometers	USGS custom system	Radio network with repeater (15 minutes)
Woodway, Washington (1997–2006)	Rotational debris slide	Rain gauge, piezometers, extensometers	Campbell CR10X data logger	Telephone (15 minutes)
Rio Nido, California (1998–2001)	Earth slide	Rain gauge, geophones, piezometers, extensometers	USGS custom system	Radio network with repeater (10 minutes)
Headscarp of Mission Peak landslide, Fremont, California (1998-present)*	Rock block slide	L1-GPS receivers, extensometers, air temperature sensor	Environmental Cellular initially, then USGS custom system	Cellular telephone initially, then spread-spectrum radio network (30 minutes or hourly)
Edmonds, Washington (2001–2006)*	Shallow translational earth slide**	Rain gauges, soil-temperature probe, soil-moisture profilers, tensiometers, piezometers	Campbell CR10X data logger	Radio network (hourly and 15 minutes)
Everett, Washington (2001-2006)	Shallow earth slide**	Rain gauge, water-content reflectometers, piezometers	Campbell CR10X data logger	Radio network (hourly and 15 minutes)
State Route 20, Newhalem, Washington (2004–2005)	Rock block slide	Geophones, tiltmeters, extensometers	USGS custom system	Radio network repeater with (15 minutes)
Johnson Creek landslide, Newport, Oregon (2004-present)*	Translational slide	Rain gauge, downhole extensometers, piezometers, soil-moisture sensors, air and ground temperature sensors	Campbell CR10X data logger	Cellular telephone (daily)
Florida River landslide, Durango, Colorado (2005-present)	Ancient translational slide and recent debris slides in wildfire burn area	Rain gauge, extensometers, tiltmeters, piezometers, air temperature sensor	Campbell CR1000 and CR200 data loggers with radio network	Cellular telephone (hourly)
Ferguson rockslide, near Yosemite Natl. Park, California (2006-present)	Rock block slide	L1-GPS receivers, geophones	USGS custom system	Spread-spectrum radio network with repeater (hourly)
Portland, Oregon (2006-present)	Shallow earth slide**	Rain gauges, tensiometers, piezometers, soil-moisture sensors	Campbell CR1000 data logger	Cellular telephone (15 minutes)

*Monitoring at this site is discussed further in a case study.
**Instruments monitor(ed) hydrologic conditions in landslide-prone hillslope. Slide occurred at end of monitoring at Edmonds site.
***Brand names are provided for descriptive purposes only and do not imply endorsement by the USGS.

Figure 2. Map of western USA showing the locations of our three case studies (Newport, Fremont, Edmonds) illustrating near-real-time landslide monitoring.

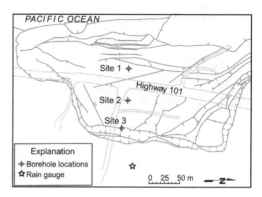

Figure 3. Map of the Johnson Creek landslide showing major structural features, location of rain gauge, and the three sites of grouped instrumentation boreholes. Slide motion is west towards the Pacific Ocean. Modified from Landslide Technology (2004).

of near-real-time monitoring at a slow-moving landslide to identify the relations between rainfall, pore pressure, and slide movement.

3.1.1 Background and setting

Many large, episodically active landslides disrupt U.S. Highway 101, the major north-south transportation corridor that links towns along the Pacific Ocean coast of Oregon, USA. The Johnson Creek landslide, near Newport, Oregon, has a history of repeated movements during winter rainy seasons, and frequently impacts the highway. This translational slide, about 200 m long, 360 m wide, and 26 m thick, occurs in seaward dipping (15–20°) siltstone, sandstone, mudstone and tuffaceous claystone of the Astoria Formation and is located on a nearly flat Pleistocene marine terrace. Total landslide displacement is about 28 m horizontal and 6 m vertical. (Priest et al. 2006). The largest recent movement episode occurred between January 2002 and February 2003, when the central part of the slide moved about 25 cm horizontally and dropped several cm vertically. (Landslide Technology 2004).

3.1.2 Near-real-time monitoring

Beginning in November 2004, the USGS installed near-real-time monitoring at this site, in cooperation with the Oregon Department of Geology and Mineral Industries and the Oregon Department of Transportation. Our efforts were focused on understanding the subsurface pore pressures controlling movement as well as the timing of and processes creating elevated pore pressures on the basal slip surface. Earlier investigators installed a rain gauge and pairs of borings at three sites along a longitudinal section of the slide (Fig. 3). At each site, a vibrating-wire piezometer (Slope Indicator) had been installed just above the slide plane in one boring and inclinometer casing in the companion boring. Wire-rope extensometers, anchored beneath the landslide's basal slip surface had also been installed in the inclinometer borings. (Landslide Technology 2004). We added electronic cable extensometers (Celesco) to the down-hole, wire-rope extensometer cables for automated recording. In November 2006, we installed two vertical arrays of six vibrating-wire piezometers each (Slope Indicator) between depths of 3 m and 26 m at sites 1 and 2 within the central and upper parts of the landslide (Fig. 3) so that both the lateral and vertical distribution of pore pressures could be monitored. These were installed in grout and are capable of measuring unsaturated soil suctions as well as positive pore pressures. We also installed two sets of dielectric soil-moisture content sensors (Decagon Devices) at shallow depths of 1.5 m and 3 m at sites 1 and 3 to assess the contribution of vertical rainfall infiltration to pore pressure changes at depth within the slide (Schulz & Ellis 2007). Two data loggers (Campbell Scientific), powered by batteries and solar panels, record data in 15-minute intervals. These data are transmitted automatically every 24 hours using cellular-telephone telemetry, graphed on a USGS base-station computer, and placed on a USGS website for viewing.

3.1.3 Results

Our monitoring between 2004 and 2007 (Ellis et al. 2007), spanning both dry and wet years, showed that basal-shear pore pressures begin to increase within just a few hours following rainfall events. Monitoring also indicated that landslide movement initiates when pore pressures exceed a threshold, with some minor variability (Fig. 4). Our monitoring also shows that rainfall-induced pore-pressure increases travel from near the headscarp (site 3) westward toward the toe of the slide (site 1), and that the travel time of the pore-pressure pulses decreases significantly with increased antecedent pore-pressure conditions (Fig. 4). Following rainstorms, there are almost simultaneous increases in pore pressures at all depths within each vertical array of piezometers located beneath the water table while shallower unsaturated zone responses lag (Fig. 5). This suggests that rapid pore-pressure increases at depth within the slide do not result directly from vertical infiltration of rainfall, but are likely due to lateral pore-pressure response from the headscarp graben area. Our observations demonstrate that enhanced forecasting of slow-moving landslide activity requires detailed knowledge of the links between pore-pressure response and movement, and that inferences based on other landslide studies may be inadequate. Near-real-time monitoring can provide the required information.

3.2 Case study 2: Detecting 3-D movement using inexpensive GPS receivers, Fremont, California

Predicting the timing of rapid, catastrophic failure of landslides and rockslides is a long sought after goal of landslide science (e.g. Saito 1965, Varnes 1983, Voight 1989). Most forecasting approaches rely on detecting the acceleration associated with a transition

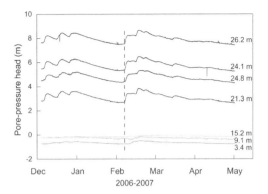

Figure 5. Pore-pressure head at various depths in the Johnson Creek landslide at site 1 between December 2006 and May 2007. Grey lines represent negative pore pressures from three different depths in the unsaturated zone; black lines represent pore pressures at four different depths in the saturated zone. Vertical dashed line shows timing of similar response in the saturated zone; response is delayed in the unsaturated zone.

from slow creep to rapid movement. For some failures, the timing of rapid movement can be predicted by projecting the trend of 1/velocity (Fukuzono 1990, Petley et al. 2002, Petley et al. 2005). Thus, detecting slide displacement in near real time is crucial to most forecasting efforts. The following brief case study examines the use of inexpensive, single-frequency (L1) GPS receivers to detect the 3-D displacement of a large rock block having the potential to fail rapidly.

3.2.1 Background and setting

More than 75,000 ancient and dormant landslides are scattered throughout the hills of the San Francisco Bay region, California, USA (Pike 1997). During the wet, El Niño influenced winter and spring of 1998, the large (35 hectare) Mission Peak landslide reactivated, and moved more than 5 m (Geolith Consultants 2000). The slip surface of this 1.2 km long, 30–55 m thick, historically dormant earthflow is primarily in the clay-rich Orinda Formation (Geolith Consultants 2000). The reactivated slide is a small part of a much larger ancient landslide complex located above the City of Fremont in a tectonically active region, with on-going vertical uplift. Near the head of the slide, the seismically active Mission Fault crosses the slope. Upslope of this fault, the steep (45°) headscarp is composed of relatively competent Briones Sandstone (Graymer et al. 1995) that dips backward into the slope and contains several persistent joint sets. The headscarp area shows geomorphic evidence of prior large massive rock block failures (Fig. 6), as well as small sackung features, suggesting that both slow and rapid movements are possible in this setting.

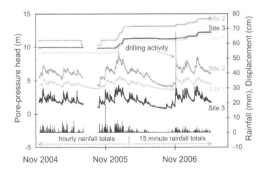

Figure 4. Basal shear zone pore-pressure head, rainfall, and landslide movement (from downhole extensometers) at the Johnson Creek landslide between November 2004 and May 2007. Site locations are shown in Figure 3.

Figure 6. Photograph of active rock mass in headscarp of the Mission Peak landslide, showing large tension crack and adjacent older failure scar. Locations of two extensometers and two GPS stations also shown. Photo: Phil Stoffer, USGS.

In March 1998, the main earthflow was active and threatened homes at the toe of slide. During this time, we observed disturbance of a pre-existing, prominent tension crack (previously open about 1.5 m) in the headscarp area (Fig. 6). Our quadrilateral measurements of 3-D displacement across the crack showed continued movement and prompted concerns about potential catastrophic failure of a large rock block, partially bordered by this tension crack. The estimated volume of this block ranges between 50,000 and 170,000 m^3, depending on inferred thickness (Geolith Consultants 2000). Rapid failure of the remaining entire rock mass might result in a rockfall avalanche. Subsequent analyses of such an avalanche show potential maximum runouts of about 500 m along slope (Jurasius 2002).

3.2.2 *Near-real-time monitoring*

USGS monitoring at this site focuses on measuring surface displacement of rock in the headscarp region where acceleration might be a possible precursor to rapid failure. Over time, our monitoring tools have evolved to better identify the 3-D strain across the entire rock block as well as the time history of movement. Initially in March 1998, we used manually surveyed quadrilateral monuments located across the large tension crack. Over the next several months, we installed two surface cable extensometers (UniMeasure) to record downslope displacement across tension cracks (Fig. 6), first using a data logger (Campbell Scientific) and later using cell-phone communications (Environmental Cellular).

We then developed a low-cost, single-frequency (L1) GPS receiver system designed for automated data acquisition, rapid deployment, and prolonged operation in remote hazardous areas (LaHusen & Reid 2000). In February 2000, we installed a working prototype of this system in the headscarp area, using NovAtel GPS receivers, Micropulse GPS antennas, and a USSG-designed data acquisition and controller system. To obtain sub-cm measurements, we utilize very short baseline, static differential processing of GPS observations from two antenna/receiver stations, one located on the moving rock block and another located off the block, about 67 m away. Power for these two remote stations is supplied by solar panels and batteries. The GPS antenna on the moving rock mass is located near the outer edge of the block (Fig. 6), just upslope of the headscarp, to measure strain across the entire block (relative to the stable GPS receiver) and to increase the likelihood of detecting a rapid failure. Instead of continuously operating the GPS receivers, which use significant power, we employ a novel scheme of powering the receivers on and off with a variable duty cycle controlled by base-station computer software. Typically, we collect 30 minutes of GPS observations at 10-second intervals, transmit these data using 900 MHz spread-spectrum radio transceivers, and then power down the system for the next 30 minutes. Independent, high-precision, static GPS solutions, with fixed ambiguity resolutions, for each 30-minute observation period are automatically computed on the base-station computer using GPS processing software (Waypoint). Results are then automatically graphed and placed on a USGS website for viewing.

3.2.3 *Results*

Although our monitoring between 1998 and 2007 did not record rapid, catastrophic failure, it did demonstrate the ability of our L1-only GPS system to detect 3-D, sub-cm movement and accelerations of the rock mass during wet seasons. More than 40 cm of downslope motion of the block, measured by our lower extensometer crossing the large tension crack prior to installation of the GPS system, occurred during the wet 1998 season. Subsequently, between 2000 and 2007 our differential GPS system showed long-term northward, westward, and downward creep of the block, resolved into 3-D components in Figure 7. Creep might be expected because the GPS antenna is located near the headscarp free face. During the relatively wet springs of 2000 and 2006, the block accelerated slightly but then slowed during the following summers (Fig. 7). It did not exhibit creep to rapid failure. Nevertheless, our observations indicate that significant movement of the block is related to wet years or sequential wet years.

When the block was active during the spring of 2000, we measured movement with both extensometers and the differential GPS system (Fig. 8). The upper extensometer recorded about 5 cm of downslope

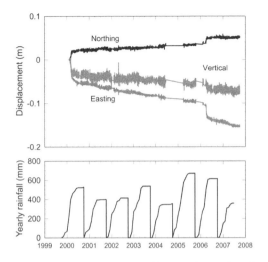

Figure 7. 3-D components (northing, easting, and vertical) of the high-precision, differential GPS solutions for the Mission Peak active rock block between February 2000 and September 2007. Points shown are 5-point medians of the independent static solutions for each 30-minute satellite observation period. Cumulative displacement is since installation in 2000; overall the block is moving northward, westward, and downward. Cumulative yearly rain begins October 1 of each water year.

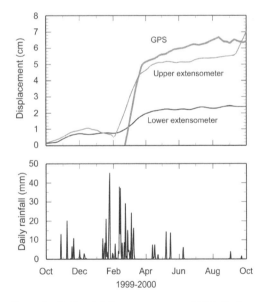

Figure 8. Daily rainfall, extensometer and GPS displacement measurements at Mission Peak during the 1999-2000 wet season. Extensometers have cables oriented downslope; therefore smoothed GPS displacement is computed along slope using the 3-D components. GPS data start in February 2000.

movement, the lower extensometer about 2 cm, and the GPS about 6 cm of cumulative slope displacement. Detrended GPS solutions have a standard deviation of about 2 mm in the horizontal and 4 mm in the vertical. Our monitoring illustrates the advantage of using high-precision GPS position solutions to detect 3-D strain across larger areas than can be readily measured using extensometers.

3.3 Case study 3: Capturing shallow landsliding triggered by rainstorms, Edmonds, Washington

Shallow landslides and accompanying debris flows pose a frequent and often devastating hazard worldwide (Iverson et al. 1997, Jakob & Hungr 2005, Sidle & Ochiai 2006). These slides typically occur in thin colluvium and are often induced by intense rainstorms or rapidly melting snow. Although such occurrences are exceedingly common, well-documented field examples of the subsurface hydrologic conditions controlling slide initiation are rare. Some studies have demonstrated that transient positive pore pressures in shallow saturated zones trigger failure (e.g. Sidle & Swanson 1982, Reid et al. 1988), whereas others have inferred that suction changes in unsaturated materials might instigate failure (e.g. Wolle & Hachich 1989, Collins & Znidarcic 2004). Understanding the near-surface transient hydrologic conditions controlling shallow failure is crucial to developing reliable forecasting or warning systems. In our final brief case study, we illustrate how near-real-time monitoring can identify the transient, shallow subsurface hydrologic conditions triggering a shallow landslide.

3.3.1 Background and setting

Shallow, rapidly moving landslides occur almost every winter along steeper sections (45–60°) of the coastal bluffs of Puget Sound between Seattle and Everett, Washington, USA (Baum et al. 2000). Although most of these slides are less than 1000 m^3 (Baum et al. 2000), they pose a continuing threat to public safety in this area, including disruption of a railway at the base of the bluffs and destruction of homes, other structures, and utilities on the bluffs. Rainstorm events producing one or more landslides have an average recurrence of six times per year (Chleborad et al. 2006). In cooperation with the Burlington Northern Santa Fe (BNSF) Railway, we selected a coastal bluff near Edmonds, Washington (20 km north of Seattle) for detailed monitoring of subsurface hydrologic conditions (Fig. 9). The purpose of our monitoring in this area was not to provide warning of individual landslides, but rather to determine when subsurface conditions are wet enough to make the slopes highly susceptible to landslides. During heavy rains in 1996–1997, several shallow landslides occurred near this site in weathered glacial deposits and colluvium (Baum et al. 2000). The 50-m-high bluff that we selected for monitoring is underlain by subhorizontally bedded glacial and interglacial sediments. A 3-m-thick layer of glacial till caps the bluff; beneath the till is a layer of glacial advance outwash that overlies dense glaciolacustrine silt (Minard 1983). Mechanical weathering of the dense, uniform, medium outwash sand produces a loose sandy colluvium mantle that covers much of the lower bluff.

3.3.2 Near-real-time monitoring

USGS monitoring at this site, which operated on AC power, focused on identifying the transient subsurface hydrologic conditions triggering shallow failure. We experimented with various kinds of sensors here in an effort to find a combination that provided hydrologic monitoring data of sufficient quality, reliability, and relevance to be suitable for forecasting landslide activity (Baum et al. 2005). Between September 2003 and January 2006, our remote station at the Edmonds site was equipped with sensors to monitor both unsaturated and saturated volumetric soil-moisture contents and pore-water pressure or suction (Fig. 9). We installed two adjacent tipping bucket rain gauges, two water-content profilers (Sentek EnviroSMART) equipped with eight (soil capacitance) sensors each at depths ranging from 20 cm to 200 cm, and two nests of six tensiometers (Soil Moisture Equipment Corp.), ranging in depth from 20 cm to 150 cm, to measure soil suction (Baum et al. 2005). The soil-water instruments were installed in dense glacial outwash sand and colluvium about 25–35 m above sea level (Fig. 9). Data were relayed every hour using line-of-sight radio telemetry to a server at Meteor Communications, then received, reduced, and graphed on a USGS base-station computer, and finally placed on a USGS website for viewing.

3.3.3 Results

On January 14, 2006, a shallow landslide occurred at the Edmonds site, destroying much of the instrumentation. However, we measured the near-surface hydrologic conditions through the previous three wet seasons and just prior to failure. Our near-real-time monitoring revealed several relations between rainfall, soil moisture, pore pressure, and the occurrence of shallow landslides in the Seattle area, including: 1) The timing and magnitude of soil moisture/pore pressure response from rainfall is highly dependent on antecedent soil moisture. For example, during October 2003, the soil was dry and wetting fronts moved slowly in response to rainfall (Fig. 10). As soil wetness increased throughout the winter season, pore pressure and soil wetness at depth responded much more rapidly to heavy rainfall. For example, heavy rainfall in mid-October produced an increase in soil moisture at 2 m depth after 6 days, whereas in mid-November, heavy rainfall resulted in a similar increase after only 1 day (Fig. 10). 2) The pattern of soil-moisture response was consistent with vertical downward infiltration,

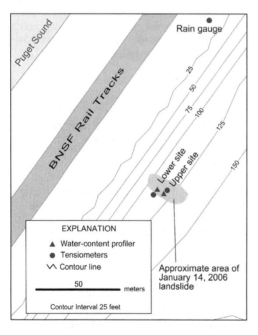

Figure 9. Map of Edmonds monitoring site near Seattle, Washington showing instrument locations and extent of January 2006 shallow landslide.

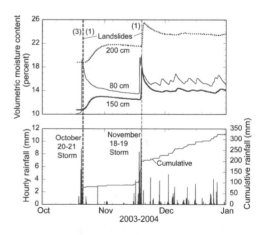

Figure 10. Soil-moisture response to rainfall at the Edmonds site between October 2003 and January 2004. Sensor depths (cm) are indicated next to response curves. Numbers of landslides that occurred on each date are indicated in parenthesis beside dashed vertical lines.

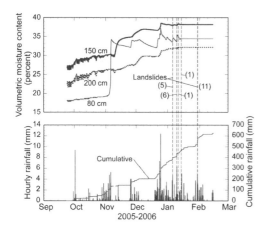

Figure 11. Soil-moisture response to rainfall at the Edmonds site between October 2005 and January 2006. A shallow landslide destroyed the instrumentation on January 14, 2006. Numbers of landslides that occurred on each date are indicated in parenthesis beside dashed vertical lines.

rather than lateral flow. 3) Landslide occurrence was strongly correlated to wet antecedent soil conditions. Intense rainfall that occurred in October 2003, November 2003, and November 2005 (Fig. 11) caused very few landslides because soil was relatively dry prior to these storms. However, storms of moderate intensity during January 2006, when soil was relatively moist after many successive days of rainfall, caused moderate numbers of landslides, even though the rainfall did not exceed our empirical rainfall threshold. (Baum et al. 2005, Godt et al. 2006). After several weeks of rain, the landslide at our monitoring site occurred during light rainfall on January 14, 2006 (Fig. 11), when overall soil moisture was at the highest since the beginning of the 2005–06 rainy season. Overall, our monitoring illustrates that soil-moisture conditions exert a strong control over the timing of shallow landsliding. The ability to measure these conditions in near real time makes it possible to determine when heavy rainfall is likely to cause landslides.

4 CONCLUSIONS

Near-real-time systems for monitoring active landslides or landslide-prone hillslopes have advanced rapidly in recent years and offer many advantages for understanding landslide processes. They can provide current information about remote landslide conditions, help ensure high-quality data sets, and capture transient and dynamic processes. Their configurations vary widely depending on the style of landsliding being monitored and the end purpose of the monitoring system, however most ground-based systems are composed of field sensors, field data acquisition systems, remote communications, and base-station data processing and dissemination, often over the Internet. We have used near-real-time monitoring to understand the dynamic behavior and hydrologic conditions triggering different types of landslides. Our investigations include identifying the groundwater conditions controlling slow-moving landslides, detecting 3-D displacements of large rock masses, and documenting the transient near-surface hydrology triggering shallow landsliding. Knowledge of both current field conditions and likely future behavior are crucial to developing better landslide forecasting and warning systems. Near-real-time monitoring systems can provide the current field conditions, but more work is needed on techniques to rapidly forecast future landslide behavior based on these near-real-time observations.

ACKNOWLEDGEMENTS

Many people and agencies contributed to the landslide monitoring described in this paper. We particularly thank Dianne L. Brien, Jonathan W. Godt, Jonathan P. McKenna, and William H. Schulz of the USGS for their multi-faceted assistance.

REFERENCES

Aleotti, P. 2004. A warning system for rainfall-induced shallow failures. *Engineering Geology* 73 (3–4): 247–265.

Angeli, M.G., Gasparetto, P., Menotti, R.M., Pasuto, A. & Silvano, S. 1994. A system of monitoring and warning in a complex landslide in northeastern Italy. *Landslide News* 8: 12–15.

Baum, R.L., Godt, J.P., Harp, E.L., McKenna, J.W. & McMullen, S.R. 2005. Early warning of landslides for rail traffic between Seattle and Everett, Washington, USA. In O. Hungr, R. Fell, R. Couture & E. Bernhard (eds), *Landslide Risk Management, Proc. of the 2005 International Conference on Landslide Risk Management*: 731–740. New York: A.A. Balkema.

Baum, R.L., Harp, E.L. & Hultman, W.A. 2000. Map showing recent and historic landslide activity on coastal bluffs of Puget Sound between Shilshole Bay and Everett, Washington. *U.S. Geological Survey Miscellaneous Field Studies Map MF 2346*.

Berti, M., Genevois, R., LaHusen, R.L., Simoni, A. & Tecca, P.R. 2000. Debris flow monitoring in the Acquabona watershed on the Dolomites (Italian Alps). *Physics and Chemistry of the Earth, Part B: Hydrology, Oceans and Atmosphere* 26 (9): 707–715.

Chien-Yuan, C., Tien-Chien, C., Fan-Chieh, Y., Wen-Hui, Y. & Chun-Chieh, T. 2005. Rainfall duration and debris-flow initiated studies for real-time monitoring. *Environmental Geology* 47: 715–724.

Chleborad, A.F., Baum, R.L. & Godt, J.P. 2006. Rainfall thresholds for forecasting landslides in the Seattle, Washington, Area—exceedance and probability. *U.S. Geological Survey Open-File Report 2006–1064*.

Collins, B.D. & Znidarcic, D. 2004. Stability analyses of rainfall induced landslides. *Journal of Geotechnical and Geoenvironmental Engineering* 130 (4): 362–372.

Dunnicliff, J. 1993. *Geotechnical Instrumentation for Monitoring Field Performance*. New York: John Wiley & Sons.

Ellis, W.L., Kibler, J.D., McKenna, J.W. & Stokes, R.L. 2002. Near real-time landslide monitoring using cellular telephone telemetry—some recent experiences. *Geological Society of America Abstracts with Programs* 34 (6): 49.

Ellis, W.L., Priest, G.R. & Schulz, W.H. 2007. Precipitation, pore pressure, and landslide movement—detailed observations at the Johnson Creek landslide, coastal Oregon. In V.R. Schaefer, R.L. Schuster & A.K. Turner (eds), *Conference Presentations: 1st North American Landslide Conference, AEG Special Publication 23*: 921–934. Vail, Colorado: Assoc. of Engineering Geologists.

Finlay, P.J., Fell, R. & Maguire, P.K. 1997. The relationship between the probability of landslide occurrence and rainfall. *Canadian Geotechnical Journal* 34 (6): 811–824.

Froese, C.R. & Moreno, F. 2007. Turtle Mountain Field Laboratory (TMFL): Part 1—overview and activities. In V.R. Schaefer, R.L. Schuster & A.K. Turner (eds), *Conference Presentations: 1st North American Landslide Conference, AEG Special Publication 23*: 971–980. Vail, Colorado: Assoc. of Engineering Geologists.

Fukuzono, T. 1990. Recent studies on time prediction of slope failure. *Landslide News* 4: 9–12.

Geolith Consultants. 2000. *Executive Summary, Mission Peak Landslide, Fremont, California*.

Godt, J.G., Baum, R.L. & Chleborad, A.F. 2006. Rainfall characteristics for shallow landsliding in Seattle, Washington, USA. *Earth Surface Processes and Landforms* 31: 97–110.

Graymer, R.W., Jones, D.L. & Brabb, E.E. 1995. Geologic map of the Hayward fault zone, Contra Costa, Alameda, and Santa Clara Counties, California: A digital database. *U.S. Geological Survey Open-File Report 95–587*.

Hadley, K.C. & LaHusen, R.L. 1995. Technical manual for an experimental acoustic flow monitor. *U.S. Geological Survey Open-File Report 95–114*.

Husaini, O. & Ratnasamy, M. 2001. An early warning system for active landslides. *Quarterly Journal of Engineering Geology and Hydrogeology* 34: 299–305.

Iverson, R.M. 2000. Landslide triggering by rain infiltration. *Water Resources Res.* 207: 59–82.

Iverson, R.M., Reid, M.E. & LaHusen, R.G. 1997. Debris-flow mobilization from landslides. *Annual Review of Earth and Planetary Sciences* 25: 85–138.

Jakob, M. & Hungr, O. 2005. Introduction. In M. Jakob & O. Hungr (eds), *Debris-flow Hazards and Related Phenomena*: 1–7. Berlin: Praxis, Springer.

Jurasius, M. 2002. Rock slope kinematics of the Mission Peak Landslide, Fremont, CA (M.S. thesis), San Jose State University.

Kane, W.F. & Beck, T.J. 1996. Rapid slope monitoring. *Civil Engineering-ASCE* 66 (6): 56–58.

Keefer, D.K., Wilson, R.C., Mark, R.K., Brabb, E.E., Brown, W.M., Ellen, S.D., Harp, E.L., Wieczorek, G. F., Alger, C.S. & Zatkin, R.S. 1987. Real-time landslide warning during heavy rainfall. *Science* 238 (4829): 921–925.

Kramer, J. & Rutledg, D. 2000. Advances in real-time GPS deformation monitoring to landslides, volcanoes, and structures. *Annual Meeting—Assoc. of Engineering Geologists* 43 (4): 97.

LaHusen, R.L. 2005. Debris-flow instrumentation. In M. Jakob & O. Hungr (eds), *Debris-flow Hazards and Related Phenomena*: 291–304. Berlin: Praxis, Springer.

LaHusen, R.L. & Reid, M.E. 2000. A versatile GPS system for monitoring deformation of active landslides and volcanoes. *Eos, Transactions of the American Geophysical Union* 81 (48): F320.

Landslide Technology. 2004. Geotechnical investigation Johnson Creek Landslide, Lincoln County, Oregon. *Oregon Department of Geology and Mineral Industries Open-File Report O-04-05*.

Mikkelsen, P.E. 1996. Field instrumentation. In A.K. Turner & R.L. Schuster (eds), *Landslides: Investigation and Mitigation, Special Report 247*: 278–316. Washington, DC: National Academy Press.

Mikkelsen, P.E. & Green, G.E. 2003. Piezometers in fully grouted boreholes. In F. Myrvoll (ed.), *Field Measurements in Geomechanics, Proc. of the 6th International Symposium, Oslo*: 545–554. Lisse: Swets & Zeitlinger.

Minard, J.P. 1983. Geologic map of the Edmonds East and part of the Edmonds West Quadrangles, Washington. *U.S. Geological Survey Miscellaneous Field Studies Map MF-1541*.

NOAA-USGS Debris Flow Task Force. 2005. NOAA-USGS Debris-Flow Warning System—Final Report. *U.S. Geological Survey Circular 1283*.

Ortigao, B. & Justi, M.G. 2004. Rio-Watch: the Rio de Janeiro landslide alarm system. *Geotechnical News* 22 (3): 28–31.

Petley, D.N., Bulmer, M.H. & Murphy, W. 2002. Patterns of movement in rotational and translational landslides. *Geology* 30 (8): 719–722.

Petley, D.N., Higuchi, T., Petley, D.J., Bulmer, M.H. & Carey, J. 2005. Development of progressive landslide failure in cohesive materials. *Geology* 33 (3): 201–204.

Pike, R.J. 1997. Index to detailed maps of landslides in the San Francisco Bay region, California. *U.S. Geological Survey Open-File Report 97-745D*.

Priest, G.R., Allen, J., Niem, A., Christie, S.R. & Dickenson, S.E. 2006. Interim Report: Johnson Creek Landslide Project, Lincoln County, Oregon. *Oregon Department of Geology and Mineral Industries Open-File Report O-06-02*.

Reid, M.E., Brien, D.L., LaHusen, R.L., Roering, J.J., de la Fuente, J. & Ellen, S.D. 2003. Debris-flow initiation from large slow-moving landslides. In D. Rickenmann & C. Chen (eds), *Mechanics, Prediction and Assessment: Proc. of the Third International DFHM Conference, Davos, Switzerland*: 155–166. Rotterdam: Millpress Science Publishers.

Reid, M.E. & Iverson, R.M. 1992. Gravity-driven groundwater flow and slope failure potential, 2, Effects of slope morphology, material properties, and hydraulic heterogeneity. *Water Resources Res.* 28 (3): 939–950.

Reid, M.E. & LaHusen, R.L. 1998. Real-time monitoring of active landsides along Highway 50, El Dorado County. *California Geology* 51 (3): 17–20.

Reid, M.E., Nielsen, H.P. & Dreiss, S.J. 1988. Hydrologic factors triggering a shallow hillslope failure. *Assoc. Engineering Geologists Bulletin* 25 (3): 349–361.

Saito, M. 1965. Forecasting the time of occurrence of slope failure. *Proc. of the Sixth ICSMFE, Montreal* 2: 537–541.

Schmidt, J., Turek, G., Clark, M. & Uddstrom, M. 2007. Real-time forecasting of shallow, rainfall-triggered landslides in New Zealand. *Geophysical Research Abstracts* 9: 5778.

Schulz, W.H. & Ellis, W.L. 2007. Preliminary results of subsurface exploration and monitoring at the Johnson Creek landslide, Lincoln County, Oregon. *U.S. Geological Survey Open-File Report 2007-1127*.

Sidle, R.C. & Ochiai, H. 2006. *Landslides: Processes, Prediction, and Land Use*. Water Resources Monograph: American Geophysical Union.

Sidle, R.C. & Swanson, D. A. 1982. Analysis of a small debris slide in coastal Alaska. *Canadian Geotechnical Journal* 19: 167–174.

Tarchi, D., Antonello, G., Giuseppe, C., Farina, P., Fortuny-Guasch, J., Guerri, L. & Leva, D. 2005. On the use of ground-based SAR interferometry for slope failure early warning—the Cortenova rockslide (Italy). In K. Sassa, H. Fukuoka, F. Wang & G. Wang (eds), *Landslides—Risk Analysis and Sustainable Disaster Management*: 337–342. Berlin: Springer.

Terzaghi, K. 1950. Mechanism of landslides. In S. Paige (ed.), *Application of Geology to Engineering Practice (Berkey Volume)*: 83–123. New York: Geological Society of America.

Varnes, D.J. 1983. Time-deformation relations in creep to failure of earth materials. *Proc. of the 7th Southeast Asian Geotechnical Conference* 2: 107–130.

Voight, B. 1989. A relation to describe rate-dependent material failure. *Science* 243: 200–203.

Wilson, R.C. 2005. The rise and fall of debris-flow warning system for the San Francisco Bay region, California. In T. Glade, M. Anderson & M.J. Crozier (eds), *Landslide Hazard and Risk*: 493–516. West Sussex, United Kingdom: John Wiley & Sons.

Wolle, C.M. & Hachich, W. 1989. Rain-induced landslides in southeastern Brazil. In *Proc. of the 12th International Conference on Soil Mechanics and Foundation Engineering*: 1639–1642. Rio de Janeiro: A.A. Balkema.

The effects of earthquake on landslides – A case study of Chi-Chi earthquake, 1999

M.L. Lin, K.L. Wang, & T.C. Kao
Department of Civil Engineering, National Taiwan University, Taiwan, China

ABSTRACT: The Chi-Chi earthquake struck central region of Taiwan on September 21, 1999, with a local magnitude of 7.3, which induced extensive landslides covering a total area of more than 8000 ha. In this paper, the effects of the ground motion of the Chi-Chi earthquake on the landslides are examined and discussed. The relationship between the critical acceleration and threshold displacement to the landslides and ground motion is analyzed based on the sliding block analysis for evaluation of the seismic stability of slope. Finally, the effects of the earthquake to triggering of subsequent landslides are discussed.

1 INTRODUCTION

Taiwan is situated at the juncture zone of the Euroasia Continental Plate and the Philippine Sea Plate with a northwest tectonic movement of the Philippine Sea Plate. As the consequences of the tectonic actions, the Central Mountain Range as well as the Coastal Mountain Ranges is produced, and the prevailing geological formations and structural patterns align approximately parallel to the longitudinal axis of the island. The topography and geological conditions are highly related to the tectonic activity, where the mountain area composes about 77% of the whole area with highly fractured geological conditions. As the results of the tectonic activity, Taiwan is among the most active seismic districts in the world. The Chi-Chi earthquake struck central region of Taiwan on September 21, 1999, with a local magnitude of 7.3, had caused severe ground failures and loss of lives and properties. The earthquake was triggered by the faulting action of the Chelungpu fault, with a fault rupture length of 105 km. The Chelungpu fault is a shallow thrust east-dipping fault which moved westward. The focal depth of the earthquake was only 8 km, which meant a tremendous amount of energy was released near the ground surface. The maximum peak horizontal ground acceleration recorded was about 1 g, and the maximum peak vertical ground acceleration recorded was 0.7 g. The PGA contour of motion caused by the main shock is shown in Figure 1. During the Chi-Chi earthquake, it was found that the ground motion is strong for area next to the fault, and decreases rapidly with distance as illustrated in Figure 1. The distribution of the measured ground motion along with the seismic intensity was as shown in Figure 2 (Shin, 1999). Noted that three triggered events with magnitude larger than 6 were also recorded which caused higher seismic intensity for some areas far away from the Chelungpu fault. The final ground displacements caused by the thrusting of the fault were surveyed using global positioning system (GPS) (Huang et al., 1999), and the distribution of the displacements was as shown in Figure 3. The maximum horizontal displacement was about 9 m, and the maximum vertical displacement was about 6 m near the north-end of the fault where it turned eastward. With such strong ground motion and large displacements, various types of ground failure occurred causing tremendous amount of damages and loss. Among all types of failure, landslides covered

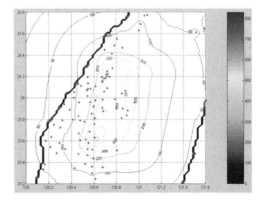

Figure 1. The distribution of horizontal peak ground acceleration of the Chi-Chi earthquake.

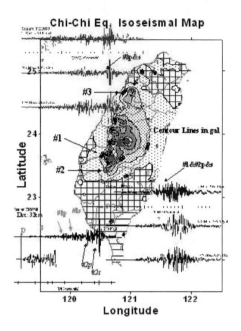

Figure 2. The isoseismal map with the three triggered events of the Chi-Chi earthquake (Shin, 1999).

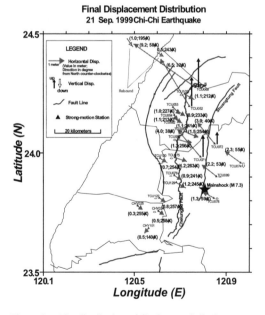

Figure 3. The distribution of final ground displacements caused by Chi-Chi earthquake. The arrows indicate displacement quantity and its direction whereas values appeared in parenthesis. (Huang et al., 1999).

the most significant extents and caused tremendous loss and profound effects (NCREE, 2000). The effects of the earthquake on the landslides and the subsequent effects of such incident will be discussed in this paper.

2 LANDSLIDES CAUSED BY THE CHI-CHI EARTHQUAKE

The ground-based field reconnaissance called by the NCREE for slope failure has reported a total number of 436 items of slope failure, and the distribution of slope failure along with the surface rupture of Chelungpu fault and strong motion stations are as shown in Figure 4. A follow-up investigation was conducted based on the aero-photo and SPOT satellite images taken before and after the earthquake. More than 21,900 items of pixel variation were identified with a total area of more than 8,600 hectare as shown in Figure 5. The slope failures distributed from Miao-Li to Jia-Yi Counties as shown in the figures, and almost all items located to the right or hanging wall of the thrust fault, where the mountain terrain located. The slope failures in Tai-Chung County, and Nan-Tou County were most widely spread. In Tai-Chung County, the most severe slope failure is along the trans-island highway. In Nan-Tou County, a large scale slope failure of Juo-Juo peaks covered an area as large as 950 hectare, which was major shallow sliding and spalling of gravelly material. In Juo-Feng-Err mountain area, a dip slope sliding occurred with an area of 200 hectare. In Yun-Lin County, a massive dip slope failure occurred in the Tsao-Ling area which covered an area of 400 hectare and with an amount of 120 million cubic meter of deposit material. The large amount of material deposited in the valley of Ching-Sui river and formed a dammed-up lake.

Based on the data from the reconnaissance and field observations, most of the landslides induced by the earthquake were with small to medium scales, and

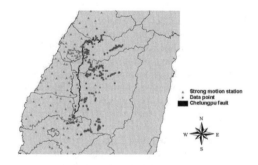

Figure 4. The distribution of the slope failure cases and the strong motion stations (Lin & Tung, 2004).

about 70% of the landslides had area smaller than 4000 m². The distribution of different types of failure of all items reported is as shown in Figure 6. In Figure 6 the debris slide is the most encountered failure type which accounts for 63% of the failure, while toppling and rock fall is the second with 22%. Distribution of the slope angle of the failure slopes is as shown in Figure 7. The landslides with slope angle in the range of larger than 45° accounts for 90% of all cases, reflecting the fact that the slope with high slope angle would have high potential of failure. The results of the slope angle compared relatively well with the distribution of the types of failure of the slope, because typically the debris slide and toppling/rock fall would occur on steep slope. The field observation also confirm that the most common type of slides were shallow slides on the steep slope occurring near the crest. The weather condition before the Chi-Chi earthquake was fairly dry without much precipitation as shown in Figure 8. Thus it is unlikely to have significant cases of deep-seated slides for lacking of ground water pressure effects.

The horizontal peak ground acceleration (PGA-H) and vertical peak ground acceleration (PGA-V) of each slope failure were determined by interpolation of the strong motion station records. The results were as shown in Figures 9 and 10, respectively, and it was found that number of the slope failures increased significantly when the vertical component of the peak ground acceleration reached 200 gal. There were 74% of the slope failures with the vertical component of PGA-V larger than 200 gal. As for the horizontal peak ground acceleration, about more than 81% of slope failure were with mean PGA-H within the range from 150 to 450 gal. A cross-examination of the distribution of the rock formations and topography suggested that the effects of the geological formation and topography did not appear to be significant. It was suggested that

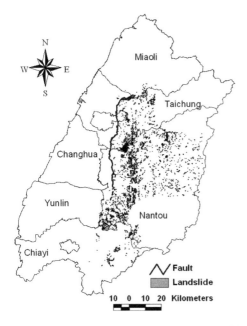

Figure 5. Distribution of identified landslide caused by Chi-Chi earthquake using SPOT images (Lin et al., 2002).

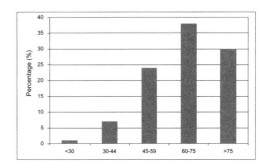

Figure 7. The distribution of slope angle of landslides.

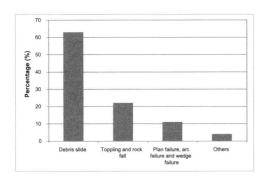

Figure 6. The distribution of failure types of landslides.

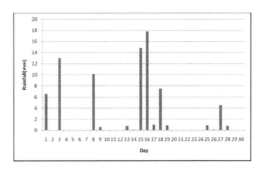

Figure 8. Rainfall record of the Sun Moon Lake station in September, 1999.

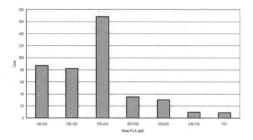

Figure 9. The distribution of mean horizontal peak ground acceleration.

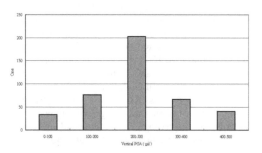

Figure 10. The distribution of vertical peak ground acceleration.

the ground motion might be the most important factor for earthquake induced slope failures (Lin et al. 2000).

3 THE EFFECTS OF GROUND MOTION ON THE LANDSLIDES

In order to understand the effects of the ground motion on the landslides, the pixel variation data by COA (2000) combined with SPOT images and aerial photos covering more areas were used. The data were carefully checked to ensure accuracy and representative of landslides. Each identified ground surface variation was screened in accordance to the variations of elevation and slope angle (Lin & Tung, 2004). After the screening process, the number of data points in this study is 31702, and the study area with distribution of landslide events are as shown in Figure 11. The epicenter of Chi-Chi earthquake is also shown in Figure 11 with the marked time 01:47:16 (local time, GMT +8) as the main shock and the other three are triggered events with magnitude larger than 6.

The distribution of the mean horizontal peak ground acceleration and vertical peak ground acceleration of each event were determined and plotted with respect to the slope angle in Figures 12 and 13, respectively. It was found that the effects of the ground motion were significant and relatively independent of the

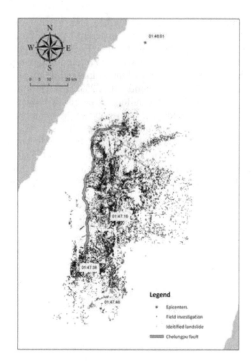

Figure 11. The distribution of selected landslide cases.

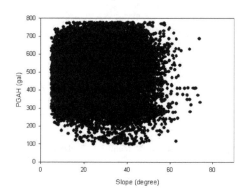

Figure 12. The distribution of horizontal peak ground acceleration versus slope angle.

slope angle, and there appeared to be threshold accelerations for both the vertical and horizontal ground motions. Observing Figures 12 and 13, the threshold vertical peak acceleration is approximately 70 gal for slope angle larger than 20°, and gradually increases as the slope angle decreases. The threshold horizontal peak acceleration is approximately 100 gal for slope angle larger than 20°, and also gradually increases as the slope angle decrease. In both cases, the threshold peak ground accelerations are constants and being

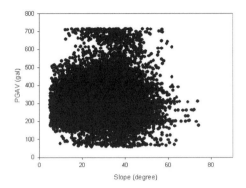

Figure 13. The distribution of vertical peak ground acceleration versus slope angle.

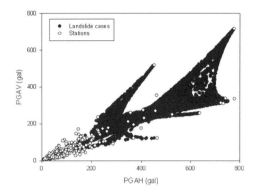

Figure 15. The distribution of vertical versus horizontal peak ground acceleration of the free-field strong motion stations and landslide events.

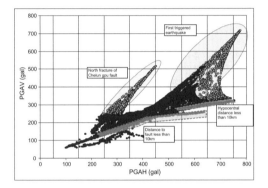

Figure 14. The distribution of horizontal peak ground acceleration versus vertical peak ground acceleration.

independent of the slope angle for the range of slope angle larger than 20°. To examine the effects of the two components of ground motion, the vertical peak acceleration versus horizontal peak acceleration of each landslide was plotted in Figure 14. The distribution of the ground motions of the landslide cases can be separated into few groups. The group of events approximately parallel and to the left of the main group was landslides located close to the north end of the fault where the maximum ground displacement occurred. This may be caused by the energy concentration where the fault rupture turned eastward, which also induced large vertical and horizontal displacements. While the second group of events with the higher vertical peak ground acceleration distributed in the right upper portion of the main group was actually affected by the first triggered earthquake. For the first triggered earthquake, large vertical and horizontal ground accelerations were also recorded. Both groups appeared to have higher vertical acceleration comparing to the main group of events, and it appeared that the

large vertical motion could have significant effects on the landslides. The main group of events distributed in a bi-linear form with events located within about 10 km to the epicenter and to the surface fracture of the fault all appeared to have about constant vertical and higher horizontal peak ground accelerations with events close to the epicenter displaying larger ground motion. As moving further away from the epicenter and fault rupture, the vertical and horizontal ground motion of the events decreased more rapidly. Comparing the ground motion of landslide events to the ground motion recorded by the strong motion station in Figure 15, the distribution of the main group of events fell in the similar trend as ground motion recorded by the strong motion stations. However, it was found that the landslide events appeared to locate in the ranges with higher ground motion, and typically with high vertical acceleration.

The attenuation of the peak ground acceleration with respect to the hypocentral distance for vertical motion and horizontal motion were plotted in Figures 16 and 17, respectively. Generally, the attenuation of ground motion in both directions of the landslide events followed the same trend of attenuation recorded by the free field strong motion stations, and again located in the range with larger ground motion and closer to the epicenter. The attenuation curve proposed by Wen et al. (2004) was also plotted in Figure 16. The curve provided satisfactory results but tended to underestimate the ground motion of the landslide case within 10 km of the hypocentral distance. The two small peaks observed in the figures were the groups affected by the large ground displacement in the north part and the first triggered event as described previously. Again the effects appeared to be more significant on the vertical ground motion than the horizontal ground motion. Based on the previous discussions, the ground motion

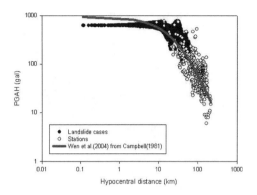

Figure 16. The distribution of horizontal peak ground acceleration versus hypocentral distance.

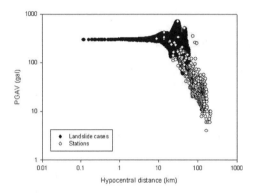

Figure 17. The distribution of vertical peak ground acceleration versus hypocentral distance.

appeared to be the most important factor causing landslides during the Chi-Chi earthquake, and the effects of the vertical peak ground acceleration were quite significant.

4 THE THRESHOLD DISPLACEMENTS VERSUS CRITICAL ACCELERATION

Among the various seismic slope stability analysis methods, the sliding block method proposed by Newmark (1965) was often used for description of seismic slope behavior. In Newmark's method, the displacement of the slope is integrated from the acceleration record when ground acceleration exceeds the critical acceleration determined from stability condition of the slope based on material strength and geometry of slope, and the stability of the slope is evaluated accordingly. Therefore, the determination of the critical acceleration and threshold displacement is vital for assessment of slope stability. In order to understand the effects of ground motion with respect to the critical acceleration and threshold displacement, cases of sliding failure caused by Chi-Chi earthquake are analyzed, and the displacements of the slopes are calculated based on the sliding block method. The conditions of simple slope and homogeneous soil properties are assumed and the log-spiral failure surface method developed by Huang & Lin (2003) is used. The study cases are selected and screened based on the assumptions and requirements of data for the analysis.

It was suggested that the slope angle is the most important factor affecting the critical acceleration of the slope, and the method was appropriate for slope with slope angle of 20 to 45 degrees of the range most sliding occurred, while for slope with angle larger than 45 degree rock fall and toppling were more likely to occur (Huang & Lin, 2003). A cross comparison of the sliding-block method and the numerical modeling method was conducted, and it was suggested that the displacement calculated in either method was consistent for landslides with ratio of the critical acceleration to peak ground acceleration smaller than 0.5. It implied that the peak ground acceleration was quite large and more than two times of the critical acceleration when landslides occurred. Analyses were performed on landslide cases satisfying this requirement, and the resulting toe displacements of 63 such cases were plotted versus the ratios of critical acceleration to peak ground acceleration in Figure 18 (Lin & Kao, 2005). The calculated toe displacements appeared to increase with decreasing ratio of the critical acceleration to the peak ground acceleration and followed a consistent trend. Noted that in Figure 18 most of the ratios of the critical acceleration to the peak ground acceleration are much smaller than 0.5, and typically the toe displacements thus calculated are much larger than 10 centimeters. Such conditions indicated that strong ground motion could induce large displacements and failure of slopes. However, in Figure 18 there are a few cases with displacements smaller than 10 centimeters. For those cases the ratios of the critical acceleration to the peak ground acceleration are close to 0.5, implying that the ground motion is not much larger than the critical acceleration. Still for these cases, the calculated toe displacements are in the range of 5 to 10 cm, which are consistent with the threshold displacement proposed by Jibson (1993), Keefer & Wilson (1989), and Wieczorek et al. (1985). Therefore, it is suggested that the peak ground acceleration being two times larger than the critical acceleration and a threshold displacement of 5 to 10 centimeters could be critical for initiation of landslide in the Chi-Chi earthquake, and the ratio of the

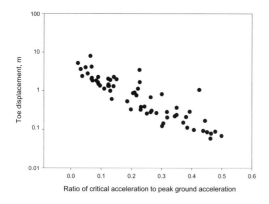

Figure 18. The toe displacement of landslide cases calculated using sliding block method versus the ratio of critical acceleration to the peak ground acceleration.

critical acceleration to the peak ground acceleration could provide information on the stability of the slope subjected to earthquake load.

5 EFFECTS ON SUBSEQUENT LANDSLIDES

Due to the severe slope failures caused by the Chi-Chi earthquake, the slope material was loosened and with fissures and cleavages. New landslides, rock falls, and debris flow may be easily triggered by other earthquakes or rainfall. In 2001, typhoon Toraji caused severe landslide and debris flow hazard, and it was found that most of the landslides and debris flows occurred at locations identified with previous landslides caused by the Chi-Chi earthquake as shown in Figure 19. In 2004, typhoon Mindule again caused severe landslide and debris flow hazard in the Central Taiwan area. The reconnaissance (Lin, et al., 2004) again indicated a close relationship between the hazard induced by the typhoon Mindule and landslides induced by the Chi-Chi earthquake as shown in Figure 20. For both events, it was found that the magnitude and extend of failures of many reactivated landslides and debris flows increased significantly. Case histories of 14 debris flow torrents with known significant recurring hazards were documented for major typhoon events from 1985 to 2004 as shown in Figure 21. Observing Figure 21, it was noted that the number of debris flow occurrence increased significantly after the Chi-Chi earthquake, and the recurring period of severe events decreased.

A close examination of the two most severely impacted watersheds- the Ta-Chia river watershed, and the Chen-You-Lan river watershed was performed. The landslide ratios caused by events of the Chi-Chi earthquake, typhoon Toraji, and typhoon Mindule

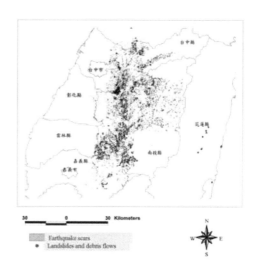

Figure 19. Landslides and debris flows induced by typhoon Toraji in 2001 versus landslides induced by the Chi-Chi earthquake.

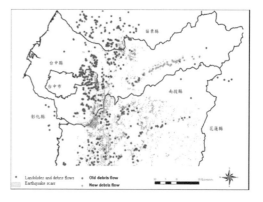

Figure 20. Landslides and debris flows induced by typhoon Mindule in 2004 versus landslides induced by the Chi-Chi earthquake (Lin, et al. 2004).

together with the reactivated landslide ratio by subsequent typhoon were listed in Table 1. It was found that the landslides reactivated by subsequent typhoon events were significantly. In Ta-Chia river watershed the reactivated landslides took up about 30% of landslides induced by both typhoons Toraji and Mindule. In Chen-You-Lan river watershed the reactivated landslides took up about 55% and 33% of landslides induced by typhoons Toraji and Mindule, respectively. Comparison of the subsequent landslides and landslides induced by the Chi-Chi earthquake revealed that the landslides caused by the earthquake occurred near

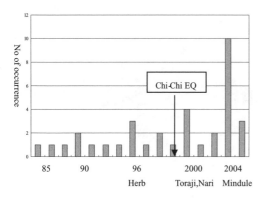

Figure 21. Number of debris flow occurrences of 14 torrents during major typhoon events (data from SWCB 2002).

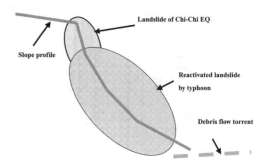

Figure 22. Illustration of reactivation of landslides and triggering of debris flows following the earthquake induced landslides.

Table 1. The landslide ratio of the Ta-Chia river watershed and Chen-You-Lan river watershed triggered by three major events.

Triggering event	Ta-Chia river watershed	Chen-You-Lan river watershed
Chi-Chi EQ	0.48%	2.46%
Toraji	1.63%	3.95%
Toraji & Chi-Chi	0.48%	2.16%
Accumulated rainfall of typhoon Toraji, mm	480	634
Mindule	3.19%	4.83%
Mindule & Chi-Chi	0.94%	1.57%
Accumulated rainfall of typhoon Mindule, mm	1658.5	1418

the crest of the slope due to the characteristic of seismic load and ground amplification. Subsequently, the landslide scars and open cracks would easily lead to landslide reactivation in the lower part of the slope profile and triggering of debris flow caused by heavy rainfall of typhoons as illustrated in Figure 22.

Thus, a large magnitude earthquake such as Chi-Chi earthquake could lead to extensive ground failures with landslides, cracks and fissures, which could lead to activation of the subsequent landslides and triggering of debris flows. It is expected that the effects would be prolonged but in a diminishing rate, as suggested in the Table 1 that the accumulated rainfall of typhoon Mindule was much higher than that of typhoon Toraji for Chen-You-Lan watershed without causing significant increase of reactivated landslides. However, efforts are still required in order to reduce such effects of secondary hazard in the near future.

6 CONCLUSIONS

In 1999 the Chi-Chi earthquake struck central Taiwan and caused extensive landslide hazard. Based on the analysis of ground motion data of the identified landslide events, it is suggested that the ground motion is the most important factor for causing landslides, and the vertical peak ground acceleration has a significant effect. The threshold peak ground acceleration observed is approximately 70 gal in vertical direction, and 100 gal in horizontal direction. Furthermore, the stability of slope can be evaluated using the ratio of the critical acceleration to the peak ground acceleration based on the sliding block method, and the threshold displacement of 5–10 cm. The follow up investigation of the landslides caused by the typhoon events suggested that the landslides caused by the earthquake could be easily reactivated in the subsequent triggering events. Such effects could be significant and prolonged, but it would be in a diminishing rate. However, efforts are still required in order to reduce such effects of secondary hazard in the near future.

REFERENCES

Huang, C.Y., Lin, C.W., Chen, W.S., Chen, Y.G., Yu, S. B., Chia, I.P., Lu, M.D., Hou, C.S. & Wang, Y.S. 1999. Seismic Geology of the Chi-Chi Earthquake, *Proceedings, International Workshop on the September* 21, 1999 Chi-Chi Earthquake, 25–42.

Huang, G.J. & Lin, M.L. 2003. Dynamic Slope Analysis Using Sliding Block Method, *Journal of Chinese Institute of Civil and Hydraulic Engineering*, 15(4), 655–665.

Jibson, R.W., 1993. Predicting Earthquake-Induced Landslide Displacements Using Newmark's Sliding Block Analysis, *Transportation Research Record*, 1411, 9–17.

Keefer, K.K. & Wilson, R.C. 1989. Predicting earthquake-induced landslides with emphasis on arid and semi-arid environments, *in Sadler, P. M., and Morton, DD. M., eds.,*

Landslides in a semi-arid environment. California, Iland Geologicial Society, 2, 118–149.

Lin, M.L., Chen, T.C. & Chen, L.C., 2002. The Geotechnical Hazards and Related Mitigation after Chi-Chi Earthquake, *Proceedings, Disaster Resistant California Conference, City of Industry, California, USA.*

Lin, M.L., Chen, C.H., Lin, M.L., Chen, H.Y., Lin, C.C. & Hsue, M.H. 2004. Reconnaissance of the Hazard Induced by Typhoon Mindule in Ta-Chia River Watershed, *the magazine of the Chinese Institute of Civil and Hydraulic Engineering*, 31(4), 19–25.

Lin, M.L. & Kao, J.J. 2005. The Threshold Displacement of Landslides Caused by Chi-Chi Earthquake, *International Symposium on the Potential, Risk, and Prediction of Earthquake-induced Landslides, Taipai, Taiwan.*

Lin, M.L. & Tung, C.C. 2004. A GIS-Based Potential Analysis of the Landslides Induced by the Chi_Chi Earthquake, *Engineering Geology*, 71, 63–77.

Lin, M.L., Wang, K.L., & Chen, T.C., 2000. Characteristics of the Slope Failure Caused by Chi-Chi Earthquake, *Proceedings of International Workshop on Annual Commemoration of Chi-Chi Earthquake, III-Geotechnical Aspect,* 199–209.

NCREE, NAPHM, and Taiwan Geotechnical Society, 1999. *Reconnaissance report of the geotechnical hazard caused by Chi-Chi earthquake,* National Research Center on Earthquake Engineering, Taiwan, 111p.

Newmark, N.M. 1965. Effects of Earthquake on Dams and Embankments, *Géotechnique*, 15(2), 139–160.

Shin, T.C., 1999. Chi-Chi Earthquake-Seismology, *Proceedings, International Workshop on the September* 21, 1999 Chi-Chi Earthquake, 1–14.

Wen, K.L., Jan, W.Y., Chang, Y.W., Chen, K.T & Jian, J.S. 2004. Site effects on strong motion of Taiwan area, *Report*, Central Weather Bureau, Report No. MOTC-CWB-93-E-09.

Wieczorek, G.F., Wilson, R.C. & Harp, E.L. 1985. Map showing slope stability during earthquakes of San Mateo County, California. *U.S. Geological Survey Miscellaneous Geologic Investigations* Map 1 1257E, scale 1:62500.

The role of suction and its changes on stability of steep slopes in unsaturated granular soils

L. Olivares
Dept. of Civil Engineering, Second Univ. of Naples, Italy

P. Tommasi
Institute for Environmental Geology and Geo-Engineering, National Research Council, Rome, Italy

ABSTRACT: A significant part of Italian mountain areas are covered by pyroclastic deposits resting at slope angles higher than 40–50°. The stability of these steep slopes in loose or poorly cemented pyroclastic materials is essentially guaranteed by the positive effects of matrix suction on shear strength until an increase in saturation (and hence a decrease in suction) is induced by seepage initiated by different processes. The Cervinara flowslide (Campania, Italy) is a typical case where rainfall infiltration increased saturation and hence led to failure of shallow pyroclastic deposits. At La Fossa Crater (Vulcano Island, north of Sicily, Italy) seepage induced by condensation of vapor produced by the degassing activity of the active volcano seems instead to be correlated to slope movements. In the paper the two case studies are examined by means of monitoring data, laboratory investigations and numerical analyses.

1 INTRODUCTION

Unsaturated cohesionless or slightly-bonded pyroclastic materials can form relatively steep slopes whose stability suffers from wetting, which reduces suction and hence shear strength.

In this paper two cases of slope instability initiated by wetting occurring through different seepage processes are described and discussed.

The first is the Cervinara slide, which is representative of flows triggered by rainfall infiltrating in the shallow pyroclastic deposits that overlie the limestone massifs of the Campanian Apennines (southern Italy).

The second case is a landslide on the flank of the pyroclastic cone of *La Fossa* volcanic edifice in Vulcano Island (North of Sicily). Its initiation is believed to have been influenced by a sharp increase in condensed vapour produced by the degassing of the active volcano.

1.1 Landslides induced by rainwater infiltration

In recent decades, a large number of landslides of flow type (flowslides) have been triggered in pyroclastic deposits in the southern Italian region of Campania by rainwater infiltration, causing loss of life and property (Olivares & Picarelli, 2003; Calcaterra & Santo, 2004). In Figure 1 a typical example of a landslide is shown as it evolves into a flowslide.

In most rainfall-induced flowslides it is possible to recognize some recursively similar elements:

– flowslides take place essentially in shallow cohesionless air-fall deposits in primary deposition lying on fractured limestone;

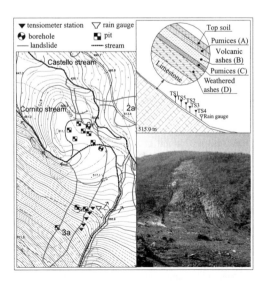

Figure 1. Cervinara flowslide: plan-view and schematic cross-section.

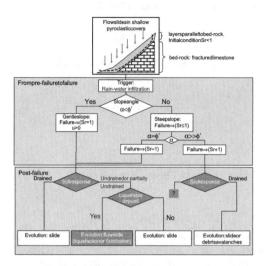

Figure 2. Conditions for rainfall-induced flowslide in unsaturated soils (infinite slope) (Olivares and Damiano, 2007).

- deposits are constituted by alternating ashy and pumiceous layers with an average thickness in the order of a few decimeters;
- the slope angle is almost constant and high (35–45°) and the shallow failure surface is parallel to the limestone formation;
- the water table is absent and the deposits are initially unsaturated;
- the soil is susceptible to liquefaction.

Starting from these similarities, Olivares & Damiano (2007) proposed a framework (Fig. 2) to predict the development of landslides induced by rainfall infiltration in homogeneous deposits under the assumption of an infinite slope.

On steep slopes (i.e. slope dip α greater than the saturated friction angle ϕ' of the soil) failure can occur in almost complete saturation or in states still far from saturation, conditioning the failure mechanisms (drained or undrained) and post-failure evolution (flowslide, slide or debris avalanche).

Hence, in this schematization the elements controlling different failure and post-failure mechanisms are represented by the state properties and in particular by the degree of saturation S_r (or water content w) at the onset of instability.

The weakness of this framework is its ability to predict the hydrologic slope response (i.e. of linking rainfall to the space-time distribution of S_r or w and that of the state of stress within the soil deposits). In this complex linkage a fundamental role is played by the boundary conditions, by the intrinsic hydraulic characteristics of each layer of the deposit and, above all, by the state variables at the beginning of rain.

To remove this uncertainty at the Geotechnical Laboratory of the Second University of Naples a complex experimental program was conducted. It included:

- mechanical and hydraulic characterization of the ashy soils in both saturated and unsaturated conditions (see section 2.1);
- infiltration tests performed on both homogeneous and layered small-scale slopes in a well instrumented flume (FL tests) to investigate the mechanics of rainwater infiltration leading to slope failure (see section 2.2);
- calibration of a numerical model through back-analysis of infiltration tests (see section 2.3.1);
- in situ monitoring of the slope involved in the 1999 flowslide, including suction measurements at different depths in the ashy layers and pluviometric measurements (see section 2.3.2);
- back-analysis of in situ suction measurements, by means of numerical simulations carried out with boundary and initial conditions derived from in situ monitoring (see section 2.3.2).

1.2 Landslides induced by volcanic activity

In active volcanoes hydraulic conditions are affected not only by infiltrating rainwater but also by volcanic activity, which produces complex changes in the state variables of pore fluids (i.e. pore fluid pressure). In particular, volcanic activity can modify pore fluid pressure as far as to induce slope instability.

In the geotechnical and geological literature this effect is well documented only for lava domes with reference to gas overpressures induced by rainfall infiltration (see e.g. Elsworth et al. 2007).

In relatively steep pyroclastic cones, slope instability may also be induced by the increase in water content produced by vapour condensation, whose main effect is the decrease in suction up to possible development of positive pore pressures.

This mechanism seems to apply to slope stability conditions of the cone of *La Fossa* at Vulcano Island where, during the most intense well documented volcanic unrest, large slope movements developed on the NE flank of the cone.

In order to verify the validity of the proposed mechanism, geotechnical investigations including laboratory testing and in situ suction measurements were conducted and used to set up simple seepage models and run stability analyses (see section 3). Continuous monitoring of suction and soil/air temperature at two selected sites was recently started with the aim of ascertaining whether suction changes can be correlated to temperature variations induced by an evolution of the volcanic activity.

The research currently involves the following institute: CNR-IGAG, CIRIAM and INGV-Palermo.

2 CERVINARA FLOWSLIDE

The Cervinara flowslide (3a in Fig. 1) was a typical rainwater-induced flowslide that occurred along a fairly regular slope formed by a primary deposit of unsaturated layered air-fall pyroclastites overlying fractured limestone. Figure 1 presents a plan-view and a schematic cross-section obtained from boreholes and pits carried out along the landslide. At the main scarp, the slope is around 40° and the average thickness of the pyroclastic cover is about 2–2.5 m. From top to bottom under a top soil formed by remoulded volcanic ashes (60 cm thick) the following layers were identified: A) an upper layer (20 cm) of coarse pumices; B) a layer (75 cm) of volcanic ashes; C) a horizon of finer pumices mixed with ash (20 cm); D) a bottom layer (40 cm) of weathered ash in contact with the fractured limestone bedrock. The total thickness of the cover increases at the slope toe due to the presence of colluvium.

Table I reports the mean values of physical properties of the different materials (Olivares 2001; Olivares & Picarelli 2003; Picarelli et al. 2006). Porosity is very high and the soils are unsaturated with S_r depending on environmental conditions. Within each layer the grain size distribution is quite uniform (Fig. 3) indicating a highly selective deposition mechanism (air-fall) (Picarelli et al., 2006). In particular, volcanic ashes (B) display a high sandy component but also a significant amount of non-plastic silt; the altered ash in layer D is finer than the ash in layer B and pumices fall in the domain of sandy gravel (A) and gravelly sand (C). SEM observations show that pumices are vesicular and present small pores (intragranular) separated by glassy diaphragms caused by quick quenching in air after explosion (Fig. 4a). Soil B is characterized by intergranular and intragranular pores (Fig. 4b, c). The intergranular pores and have a size comparable to that of the coarsest particles.

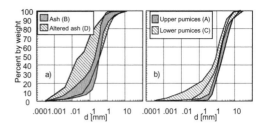

Figure 3. Grain size distribution of the tested material.

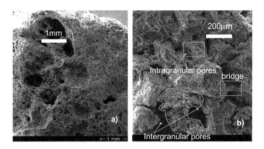

Figure 4. SEM photographs: a) pumice; b) fabric of layer B showing large (from Picarelli et al, 2006) intergranular pores, often filled with silt.

2.1 Mechanical and hydraulic characterization

Mechanical and hydraulic characterization of soils was carried out on undisturbed and remoulded specimens recovered from pits. As the layers are cohesionless, undisturbed sampling was successfully carried out only in the ash (layer B), and altered ash (layer D). Data on shear strength of pumices are not available. The shear strength of saturated volcanic ashes was measured through CID and CIU triaxial tests on saturated natural samples (Olivares and Picarelli, 2001). The ash (B) is characterized by a friction angle ϕ' of 38° and a nil value of cohesion c', while weathered ash has a peak and critical state-strength characterized by ϕ' and c' equal to 31° and 11 kPa and 35° and 0 respectively.

Table 1. Average values of properties of pyroclastic soils.

Layer	Material	d_{max} mm	U	G_s	γ kN/m³	w [%]	n [%]	S_r [%]
A	Coarse pumices	20	5	2.5	13	25	52	36
B	Volcanic ashes	10	5	2.6	14	67	67	84
C	Pumices	20	42	2.6	14	40	50	40
D	Weathered ashes	10	25	2.6	16	54	54	75

* w and S_r were obtained from samples taken at the end of wet season.

Since suction largely controls strength and hydraulic properties of these materials, a significant part of the testing program was carried out using apparatus especially suited to unsaturated soils. The shear strength in unsaturated conditions was measured only for soil B by means of suction-controlled triaxial tests (SCTX) using a mean net stress $(p - u_a)$ between 20 and 200 kPa and a suction $(u_a - u_w)$ between 10 and 80 kPa (Olivares and Picarelli 2003). The shear strength envelope of the saturated material ($c' = 0$; $\phi' = 38°$) and strength values of the unsaturated material measured at different values of suction are reported in Figure 5b. All data show the significant role of suction, even at small values: in particular, for a suction of 4–8 kPa cohesion is 2–6 kPa (Fig. 5a). These values are sufficient to justify the stability of steep slope during dry period.

The coefficient of permeability at saturation measured on natural samples taken from layers B and D through constant head tests for different effective stresses (40–70 kPa) ranges between 1.04×10^{-7} and 5.46×10^{-7} m/s for layer B and between 8.5×10^{-8} and 6.1×10^{-7} m/s for layer D. The saturated permeability of reconstituted pumices by constant head permeability tests assume the values of 1.1×10^{-5} m/s (layer A) and 5.1×10^{-6} m/s (layer C).

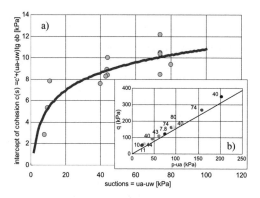

Figure 5. Unsaturated shear strength of Cervinara ashes (Olivares, 2001).

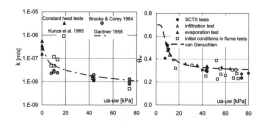

Figure 6. Unsaturated permeability and retention curve on Cervinara ashes.

The unsaturated coefficient of permeability of ashes B obtained from the interpretation of the transient phase of suction equalization (Kunze et al., 1965) in SCTX tests is reported in figure 6a. In the same figure the permeability values versus suction estimated using the expression proposed by Brooks and Corey (1964) and the permeability function obtained by Gardner's equation (1958) are plotted. In both cases in the range of suction between 0 and 80 kPa, permeability decreases by about two orders of magnitude as suction increases. Figure 6b reports the corresponding data in terms of volumetric water content θ_w versus (u_a-u_w) from SCTX tests at the end of suction equalization together with the corresponding retention curve obtained by means of conventional long-term evaporation and infiltration tests on natural and reconstituted samples. These results are in good agreement with the expression proposed by van Genuchten (1980) ($\theta_r = 0.3; \theta_s = 0.7; m = 0.2; n = 7; \alpha = 1.7$).

2.2 Results of infiltration tests and comparison with numerical analyses

Two types of infiltration tests performed on a small scale of slope on pyroclastic soils taken from the Cervinara site are presented (Tables 2 and 3).

The tests were performed on both homogeneous and layered slopes: in the first case the infiltration process in a deposit with a slope of 40° lying on an impervious base is presented; by contrast, in the second case, the infiltration process on layered deposit lying on a pervious base is analyzed. In all tests, to simulate a hydraulic boundary condition of free flow at the toe of the slope a geotextile drain is positioned (side 3 in figure 7), with conductivity more than one order of magnitude higher than that of pumices. In figure 7 a sketch of the model slope and monitoring system is given. Dimensions of a small scale slope are selected as regards the thickness/length ratio to consider the assumption of an infinite slope valid. During the infiltration (constant rainfall with intensities varying from 18 to 105 mm/h) and evaporation stages, the following were monitored: water content (by the TDR technique), displacements (by laser transducers and the PIV technique), suction (by mini-tensiometers) and positive pore pressures (by miniaturized pore pressure transducers). More details about tests results, the characteristics of the apparatus and the monitoring system are reported in Olivares & Damiano (2007) and Olivares et al. (2008).

2.2.1 Results of infiltration tests on homogeneous artificial deposit

In figure 8 (a and b) results of FL16 and FL10 tests are select which show, for state of stresses corresponding to 10 cm of thickness, the typical response

Table 2. Tests on homogeneous slopes in volcanic ashes (B): initial and boundary conditions.

test	slope angle α	initial conditions n_i	w_i	rainfall characteristics i [mm/h]	Δt	boundary conditions* side 1	side 2	side 3
FL7	40	0.71	0.31	80	27'	b	b	a
				0	4'			
				80	19'			
FL8	40	0.70	0.30	60	31'	b	b	a
FL9	40	0.73	0.37	60	30'	b	b	a
FL10	40	0.65	0.43	60	37'	b	b	a
FL15	40	0.75	0.50	40	36'	b	b	a
FL16	40	0.75	0.40	80	30'	b	b	a

* a = pervious; b = impervious.

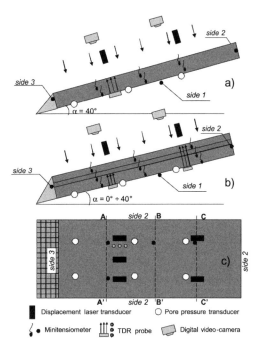

Figure 7. Sketch of the instrumented flume: a) test on homogeneous slopes; b) test on layered slopes; c) plan-view.

Table 3. Tests on layered slopes: geometry, initial and boundary conditions.

slope	layer	soil	L [m]	h_i [m]	n_i [%]	w_i [%]
layered	upper	ashes (B)	1.7	0.10	76.0	35.0
	medium	pumices (A)	1.7	0.04	72.0	12.6
	lower	ashes (B)	1.7	0.06	75.0	35.0

test FL20 Stage	slope angle α	rainfall characteristics i [mm/h]	Δt	boundary side 1	side 2	side 3
I_a	0	50	61'	a	b	a
I_b	0	0	6days 20h 10'	a	b	a
II_a	0	27	2h 32'	a	b	a
II_b	0	0	2days	a	b	a
III_a	20	27	21'	a	b	a
III_b	20	0	12days 21h	a	b	a
IV_a	40	18	4h 50'	a	b	a
IV_b	40	0	18days 17h 58'	a	b	a
V_a	40	85	1h 45'	a	b	a
V_b	40	85	5h 7'	a	b	b
V_c	40	0	3days 17h 24'	a	b	b
VI_a	40	105	7h 25'	a	b	b

* a = pervious; b = impervious.

respectively of a loose and a dense soil. The geometry, the initial and boundary conditions of the slope and the intensity and duration of artificial rainfall are summarised in Table 2. In both the tests during infiltration a marked suction decrease occurs starting from the ground surface towards the base of the model as confirmed by the delay (about 6–8 minutes) of the deep tensiometers installed at the bottom of the model slope (figure 8a, b). As revealed by settlement measurements recorded by laser sensors, the volumetric collapse is significant (about 8%) in the case of loose soil (Figure 8a; FL16 test) and negligible in the case of dense soil (Figure 8b; FL10 test). In both cases, about 10 minutes before instability, the deepest tensiometers record a practically nil value of suction and the pore pressure transducers record a positive value at the base, while those superficial maintain constant values of about 2–3 kPa.

Soil water content was measured by an innovative TDR technique (inverse profiling methods; Greco, 2006) to extract a non-homogeneous moisture profile along the probe axis. Measurements are based on the correlation between the bulk dielectric permittivity of wet soil (ε_r) and volumetric water content θ_w appropriately determined on Cervinara volcanic ashes (Damiano et al., 2008) since it differs significantly from the relationship proposed by Topp et al. (1980).

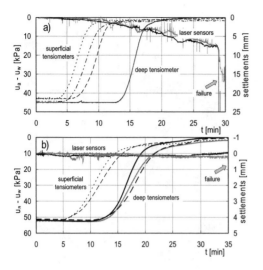

Figure 8. Suction and settlement measurements during: a) FL16 test; b) FL10 test.

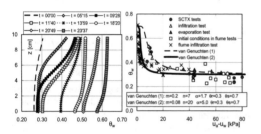

Figure 9. Volumetric water content profiles retrieved by TDR during experiment FL16 (a) and Water Retention Curve derived from TDR compared with WRC obtained from laboratory tests.

As an example of this application, in Figure 9 are plotted the results of the FL16 test are plotted in terms of θ_w profiles at different times extracted along a probe installed normally to the ground surface. At 11'40 from the beginning of test, the θ_w profile shows a strong curvature due to the wetting of the uppermost soil layer, in line with a decrease in suction recorded by superficial tensiometers (fig. 8a). Subsequently, as the deep tensiometers start to record significant decreases in suction, the form of the profile changes (t = 18'20 min) due to the progressive wetting of the lower part of the slope. As the deep tensiometers record a suction of 1 kPa the θ_w profiles (23'37 min) show that an almost uniform moisture distribution is reached, highlighted by a value of θ_w approaching the porosity of soil.

Coupling the measurements of the suction with the θ_w at the same depth, it is possible to extract the water retention curves experienced by the soil during the infiltration process. In Figure 9b these curves are compared with those obtained through conventional long-term laboratory techniques (see section 2.1). The two experimental results were fitted through the van Genuchten (1980) expression: the characteristic curve extrapolated by flume tests (curve 2 in fig. 9b) clearly shows that for $\theta_w > 0.4$, suction is lower than the values obtained by conventional laboratory tests (curve 1 in fig. 9b). This effect is probably related to the different rate of suction decrease imposed in the two procedures accounting for the pyroclastic nature of the particles. The size of the pores between the particles (intergranular pores) is significantly higher than that of internal pores (intragranular pores) (see fig. 4). Consequently the amount of retained water required to cause a given suction decrease is a function of the exposure time. Indeed only long exposure times (as in the conventional tests) enable the intragranular pores to absorb water.

2.2.2 Results of infiltration tests on layered artificial deposits

In this section we discuss the results of phase IVa of the layered FL20 test with a pumiceous layer of soil A interbedded between two ashy layers of soil B. In this test the slope angle was of 40° and a pervious boundary was reproduced at the base (side 1); rainfall intensity was kept constant at 18 mm/h for 4 hours and 50 minutes.

The results in terms of suction against time are shown in figure 10. As in the case of homogeneous deposits a marked suction decrease appears throughout the soil mass, but in this case an instability condition was not reached. The comparison between the measurements in the two ash layers across contact with

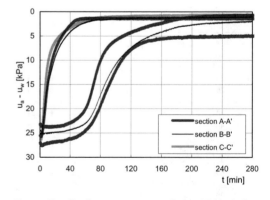

Figure 10. Suction measurements during FL20 at three different depths.

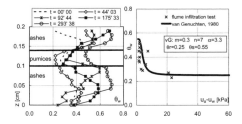

Figure 11. Volumetric water content profiles retrieved by TDR during experiment FL20 (a) and Water Retention Curve of pumices derived from TDR.

the pumiceous layer shows an immediate response in the top layer with a different trend and a marked delay in the bottom layer. Furthermore, the comparison between the two suction measurement performed in the same bottom ash layer highlights a negligible difference despite the different depth.

Figure 11a gives the θ_w profiles at different times obtained with the TDR inverse profiling methods using a single probe installed through the three layers. Since the suction assumes the same value at the ash-pumice interface, coupling the measurements of the suction in the ash with the θ_w evaluated in the pumices, it is possible to extract the water retention curves of pumices experienced by the soil during the infiltration process (Figure 11b). The experimental points were fitted by the expression proposed by van Genuchten (1980) ($\theta_r = 0.25; \theta_s = 0.55; m = 0.3; n = 7; \alpha = 3.3$).

2.3 Analysis of infiltration

The complexity of infiltration process on unsaturated soils requires the use of a numerical model. The reliability of the model can be improved through a preliminary calibration (given the variability in soil parameters) and a validation through a back-analysis of in situ monitoring. In this paper the former one was obtained using data from laboratory tests on natural samples and simulating infiltration tests on the slope model; the latter through back-analysis of in situ suction measurements using initial and boundary conditions derived from monitoring.

Numerical analysis was carried out using an I-MOD3D (volume finite method) program developed in VBA application for ARCOBJECT™/ARCGIS 9.2™ to automate the mesh-generation starting from a Digital Terrain Model.

The analysis was performed under isothermal conditions for a undeformable unsaturated porous medium neglecting the flux of the gas phase. The saturated permeability of volcanic ashes (B and D) and pumices (A and D) was assumed equal to the maximum values obtained in laboratory tests and reported in section 2.1. The unsaturated permeability functions are described by the Brooks and Corey (1964) expression. The adopted retention curves are described by the van Genuchten (1980) expressions: for both topsoil and ash (B) the retention curves of figure 9b were adopted; for pumices (A and C) the retention curve of figure 11b; for altered ash (D) data from the literature (Fredlund and Rahardjo, 1993).

2.3.1 Calibration of numerical model trough a back-analysis of infiltration tests

Calibration of numerical model was carried out on the basis of results of infiltration tests on homogeneous (FL10) and layered slopes (FL20). A 3D analysis was performed schematizing the slope model with a mesh characterized by $dx = dy = dz = 1$ cm, considering the boundary conditions indicated in tables 2 and 3 and assuming for the initial condition a constant value of suction equal to the mean value recorded at the beginning of tests.

As an example of this calibration stage, in figure 12 the experimental data of FL10 test are compared with the results of two numerical simulations, considering for the ashes B the van Genuchten parameters obtained either in conventional laboratory tests (simulation 1 with curve 1 in Fig. 9b) or in the flume infiltration tests (simulation 2 with curve 2 in Fig. 9b). As expected, the best fitting was obtained by using retention curve 2 that allows both the trend and the final values of suction at different depths to be captured. Conversely, simulation 1 cannot reproduce the trend of the deepest tensiometers and the value of suction at the end of the test. Retention curve 2 probably leads to higher suction than measured since it was obtained for equalization times longer than the test duration, which was insufficient to involve the water inside the intragranular pores.

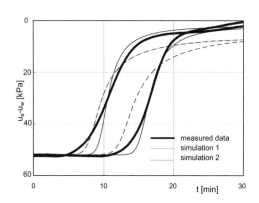

Figure 12. Test FL10: comparison between suction measurements and numerical simulation at section A-A' in fig. 7.

209

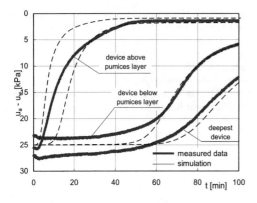

Figure 13. Test FL20: comparison between suction measurements and numerical simulation at three depths at section A-A′ in fig. 7.

In the figure 13 the results of the layered FL20 test are compared with the numerical simulation considering, for both the soils, pumices and ashes, the retention curves obtained in flume tests. Numerical analysis correctly simulated the response of both points located across the pumice layer and points located within the ash layer below the pumices, reproducing the different trend recorded above and below the pumice. The results show the importance of model parameter calibration based on flume infiltration test results, since infiltration rates and boundary conditions applied to the model slope resemble real slope conditions.

2.3.2 Validation of model: comparison between field data and numerical analysis

At 2002 a complex monitoring system was installed on the natural grassed area surrounding the catastrophic flowslide of Cervinara (on the right side of the landslide 3a in fig. 1). The equipment consists of a pluviometer and five tensiometer stations. The tensiometers (jet-fill type manufactured by Soil-Moisture) were installed at different dephts within the top soil and the volcanic ashes (B and D). Figure 14 shows the suction measured at different depth from April 2002 to April 2003. During the winter season, suction reached very low values, ranging between 2 kPa (in the top soil) and 15 kPa (in layer B). The dry season started in June as indicated by the lack of rainfall during the period June-July. In this dry period suction increased up to values of some tens of kPa with a peak in middle July, when the shallow device of nest 1, installed at depth of 60 cm in the evapotranspiration zone, indicated a maximum value near to 50 kPa.

Subsequently the summer had an abnormal rainfall regime for the Mediterranean area. On August 7th the daily rainfall was of 165 mm only slightly lower than

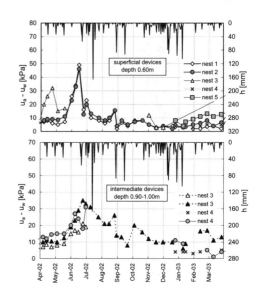

Figure 14. Suction measured at the five tensiometer stations as a function of daily rainfall.

the previous 30 years maximum (180 mm in December 19th 1968). Furthermore the total rainfall during the summer (550 mm) was almost half the annual precipitation. Due to the summer rains, suction fell from the end of July with a faster decrease at the end of September. In the months of October and November, it fluctuated around values of a few kPa.

Monitoring also shows that during summer a significant reduction in suction occurred as a consequence of prolonged rainfall, as after July, 19th and not during the most intense and short events, as on August 7th. This is a consequence of the low hydraulic conductivity of the shallow unsaturated soils, which prevents water infiltration during short precipitations. In contrast, infiltration is facilitated by the increase of degree of saturation occurring for long lasting events.

The period between 13/7/2002 to 8/8/2002, characterized by abnormal rainfall after a prolonged dry period, was selected to validate the numerical model. The slope has been schematized with a 3D mesh derived by DEM (with dx=dy=0.5 m) and by stratigraphy of fig. 1 using a dz=0.12 m. At the ground surface two conditions have been used: average daily rainfall intensity from monitoring or evaporation flux during dry period considering the minimum value suggested by Wilson (1990) for unsaturated soil (ranging between 0.33 mm/h (first day) and 0.012 mm/h (after three days)). For the lateral and base surfaces have been considered a condition of free flow. The initial condition in terms of suction has been established from field measurements.

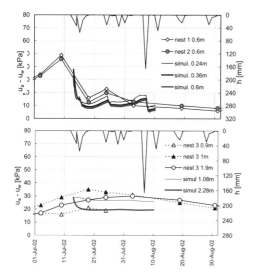

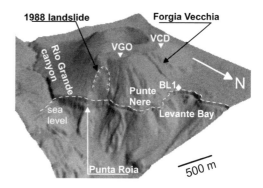

Figure 16. Shadow relief of La Fossa edifice.

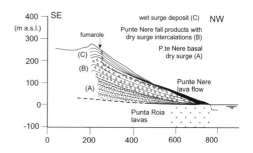

Figure 15. Comparison between in situ measurements and numerical analysis.

A comparison between the monitoring results and numeric simulations at three depths is shown in figure 15. During this wet period a adequate agreement has been obtained in both top soil and volcanic ash layer B, although the measured values are discontinuos and a higher decrease of suction is predicted during rainfall. As observed by field measurements and confirmed by numerical analysis prolonged rainfall (13/07–19/07) have an effect on suction higher than a shorter and heavier rainstorm (7/08) in the shallowest layers. In layer D the suction variation does not seem to be strictly influenced by rainfall events and the inadequate prediction of numerical analysis is probably related to the assumption on the initial condition. During dry period the predict trend is qualitatively in agreement with measurements even if characterized by lower value. This is probably due to the effect of transpiration at the upper boundary due to the presence of a grassed surface that has not been considered in the analysis.

3 LANDSLIDES AT "LA FOSSA" CONE

Instabilities occur along the flanks of *La Fossa* pyroclastic cone that has a base diameter of 1200 m at sea level and is 400 m high (Fig. 16).

Major instabilities affected the NE sector (where investigations concentrated) that is formed by a sequence (Fig. 17) consisting from bottom to top of (Dellino et al. 1990):

– laminated sands from dry surge deposition, up to 150 m thick (unit A);

Figure 17. Schematic SW–NE profile of La Fossa cone based on data from Dellino et al. (1990). The dashed line indicates the approximated location in the sequence of the 1988 slide.

– layers of coarse grained (up to 200 mm in diameter) fall products with thin intercalations of sands from dry surge deposition (unit B);
– a lava flow closing the eruptive cycle.

This sequence is overlaid by a wet-surge deposit (varicoloured tuffs) up to 20 m thick (unit C).

At present volcanic activity is essentially hydrothermal and concentrates at *La Fossa* edifice. Degassing is both concentrated at fumaroles (i.e. particular points where structural, stratigraphic and stress conditions created preferential paths) or is diffused throughout the whole cone, favoured by the high permeability of the pyroclastic formations.

Around fumaroles pyroclastic materials are altered, often up to the complete argillification; volumes and geometry of the argillified material (material D) are highly variable, depending on flow paths and physico-chemical characteristics of the hydrothermal fluids.

Small sheet slides occur in the Varicoloured tuffs but the largest documented instability was the translational slide occurred along a partly altered dry surge horizon within the Punte Nere fall deposit (unit B).

Table 4. Physical properties of pyroclastic materials.

Mat.	γ kN/m^3	w_n%	ρ_s Mg/m^3	S_r %	e_0	CF %	w_L, I_P%
C	14.70	13.52	2.52	36–45	0.9	10	–
A	13.84	13.00	2.67	28.2	1.13	–	–
B	14.62	20.8	2.66	47.5	1.16	–	–
D1	15.70	74.5	2.74	100	2.04	57	75, 23
D2	14.91	68.7	2.61	95.0	1.89	63	85, 41

D1: 1988 slip surface; D2: outer cone rim.

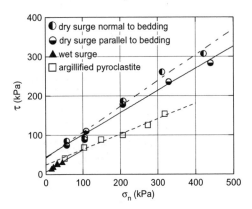

Figure 18. Direct shear tests on pyroclastites.

Figure 19. View of the 1988 slide. At the top on the right the main scarp white altered materials is apparent.

3.1 Geotechnical characterization

Undisturbed samples of dry surge materials (from units A and B), varicoloured tuffs (unit C) and argillified pyroclastites (material D) were taken from cuts. The average values of the physical properties are reported in Table 4.

The material forming the *varicoloured tuffs* is a silt with sand and traces of clay. It has low saturation, apart from local increases around fumaroles. The deposit is homogeneous and has the mechanical behaviour of a non-cohesive soils. Direct shear tests on saturated specimens performed at normal stresses lower than the fragile-ductile transition provide, similarly to triaxial CIU tests, a friction angle φ' of 36° with negligible cohesion (Fig. 18).

Dry surge deposits at the base of the Punte Nere sequence and those intercalated in the overlying fall products are both uniform medium sands formed by extremely irregular, clean particles at times bonded together.

Strength data from direct shear tests, reported in Figure 18, show slight non-linearity and anisotropy (φ' from linear regressions forced through the origin varies between 35° and 38° depending on shearing direction). Otherwise if a cohesion intercept c' is considered regression yields values of $c' = 30$ kPa and $\varphi' = 30°$.

Hydrothermal alteration around fumaroles has completely changed mineralogy, texture and hence mechanical behaviour of the parent pyroclastites. Argillified materials are fully saturated and have medium-to-high plasticity deriving from a high smectite content (D1 and D2 in Tab. 4). Materials sampled at different sites show unexpected similarity, from both mineralogical and geotechnical point of view, which indicates that alteration has the same effect on pyroclastites with different texture. Argillification drastically reduces shear strength: friction angle decreases to 26 or 21° depending on whether null intercept is imposed or not (in the latter case a cohesion of 21 kPa is obtained). Alteration also affects hydraulic conductivity which reduces to 3.5×10^{-10} m/s.

3.2 Instability phenomena

Small slips occur around the cone within the Varicoloured tuffs but the largest instability recorded at Vulcano is the 1988 slide (Fig. 19). On April 20th about 2×10^5 m^3 of pyroclastic material detached on the eastern side of the cone along a dry surge horizon of the fall deposit (B) (Tinti et al., 1999). The main scarp developed at an active fumarole and the uppermost part of the sliding surface was altered up to argillification. Seepage of water condensed from fumarole vapours

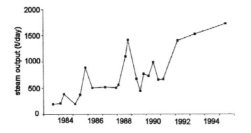

Figure 20. Steam output at fumaroles located at La Fossa crater (after Bukumirovic et al. 1997, re-drawn).

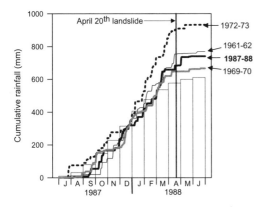

Figure 21. Average cumulative rainfall between 1965 and 1995 (bar chart) and actual cumulative rainfall in particularly wet years (line plots).

was observed along fractures forming the scarp even some years after the slide.

The slide occurred during a major period of unrest, characterized by intensification of seismic activity and degassing (Fig. 20). However failure does not seem to have been triggered by dynamic actions because the strongest seismic event (M = 4, Neri et al. 1991) occurred 20 days before the slide with epicentre at 20 km from the island. Rasà & Villari (1991) invoke also reiterate inflation-deflation cycles and localized action of high-temperature hydrothermal fluids as general causes of strength reduction in the volcano flank.

The analysis of hydraulic conditions (groundwater circulation and suction), degassing and rainfall regime at La Fossa cone suggest that failure could be initiated by a sharp change in pore fluid pressures due to the volcano unrest which was possibly favoured by a preceding rainy wet season. The sole rainwater infiltration seems instead to be not sufficient to induce instability. Analysis of hydrologic data between 1958 and 1988 indicate that other three histories of cumulative rainfall were equal or higher than that recorded during the months preceding the slide (Fig. 21).

3.3 Monitoring

In order to collect experimental evidence of the link between soil suction and degassing activity, in 2007 continuous monitoring of soil temperature and suction was started by INGV-Palermo and 2nd University of Naples at two stations (figure 16). Increase in soil temperature revealed, in fact, to be associated to the rise of steam output. Probes were installed in altered pyroclastites, close to a degassing point (VGO station), and far from fumaroles, in fresh pyroclastites (VCD station). During the preceding years, suction was also measured manually by means of portable tensiometres at other locations.

Soil temperature and suction are plotted versus time in Figure 22. In order to separate the effect of degassing from those of evaporation and rainwater infiltration, air temperature and rainfall plots were also included.

Data allow to draw some preliminary considerations. Soil temperature markedly increases in middle July 2007 and maintains at the same level over all summer. During this period of more intense volcanic activity suction at VGO station progressively decreases in spite of the absence of rainfall and of high air temperature (i.e. higher evaporation). In this respect short suction increments only occur at air temperature peaks. The effect of the increased degassing lasts until the end of November as the insensibility of suction to the increased rainfall demonstrates. Successively suction slightly increases as an effect of scarce rainfall and normalization of degassing.

Finally, far from degassing points (station VDC), where soil and air temperature are similar, suction is virtually influenced only by rainfall.

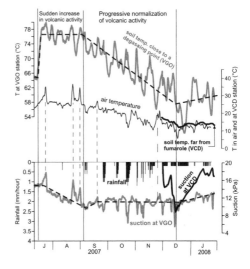

Figure 22. Soil/air temperature, suction and daily rainfall versus time.

3.4 Hydraulic conditions and slope stability

Monitoring data seem to support the hypothesis that condensed vapour increases water content of pyroclastites. Furthermore, observations around fumaroles and in wells drilled at the cone foot (Madonia, pers. com.) suggest that condensed vapour could change seepage in the cone and hence shallow groundwater circulation. In fact continuous seepage of warm water from fractures forming the 1988 slide scarp was noticed even during the driest season of the last fifty years (Sept. 2003). Furthermore in fboreholes at the cone foot (i.e. BL1 borehole in Figure 16) high pore water temperature was measured in the dry season within the first 15 m, accompanied by intense alteration.

These two elements seem to indicate that warm water from the cone permanently circulates in the shallower layers. The complete displacement of the colder sea water (the borehole top is at sea level) also suggests that flow is continuous and seepage velocity is significant.

Evaluating amount and distribution of saturation changes and seepage processes in the pyroclastic deposits is extremely difficult. In order to make preliminary considerations on flow paths and distribution of pore pressures, steady-state and transient 2D seepage analyses were carried out under simplified hypotheses using water retention curves determined in the laboratory. Analyses wee conducted through the SEEP/W code (Geo-Slope International, 2004). Stratigraphy and boundary conditions of the model are reported in Figure 23: the seepage domain was delimited by the ground surface at the top, the slip surface at the bottom, and the fumarole fracture at the upper side where inflow of condensed vapour is imposed (Fig. 23). Along the slip surface the argillified portion is fully saturated whilst downslope suction is equal to average values measured in situ. Inflow is evaluated from average flow measurements in the shallow groundwater (Madonia pers. com).

Seepage parallel to layering establishes throughout the partially saturated varicoloured tuffs and in most of the underlying B unit, being more defined at the interface with the argillified horizon. The influence of the fumarole inflow is greatly increased accounting for material anisotropy ($k_h/k_n = 10$); under this assumption a maximum matric suction of 70 kPa at the slope top is calculated. If during the unrest the inflow is increased (four times according to the steam output reported by Bukumirovic, 1997), suction decreases by some 20 kPa in the upper third of the slope and a continuous zone of positive pore pressures forms within the varicoloured tuffs, which extends downslope for few tens of metres.

Limit equilibrium analyses run using strength parameters reported in section 3.2. indicate that the

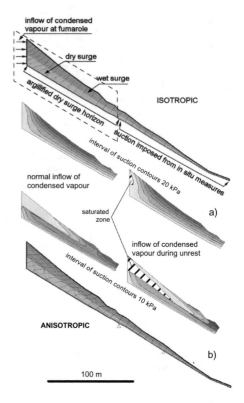

Figure 23. Increase in concentrated degassing on seepage and suction distribution in the 1988 slide area (a) considering or (b) otherwise anisotropy of the pyroclastic materials; suction contours start at zero value on left boundary.

1988 slide mobilized a shear strength which is intermediate between that of the fresh and of the argillified dry surge material. In the case of dry slope the percentage of argillified slip surface that causes failure does not match in situ observations. Therefore positive pore pressures and/or partial filling of the tension crack at the main scarp are to be considered.

Finally the drop in suction evaluated from seepage analyses in the partially saturated Varicoloured tuff blanket could be sufficient to trigger the small slips and sheet slides observed around fumaroles.

4 CONCLUSIVE REMARKS

Comparison between the results of model tests, numerical simulations and in situ measurements allowed us to set up a reliable tool for predicting slope response to rainfall infiltration for shallow layered pyroclastic deposits. The analysis of tests on Cervinara ash showed that analysis is correct if the highest saturated permeability obtained from laboratory tests is adopted and the retention curve is extracted from the results of infiltration test characterized by boundary conditions and infiltration rates similar to real slope conditions.

Comparison between homogeneous and layered slope models indicates the strong influence of the pumice layer interbedded between ash layers that substantially modify the flow regime in the lowermost ash layer, preventing soil wetting. This result is confirmed by field monitoring.

On the flanks of pyroclastic cones of volcanoes characterized by active hydrothermalism, intensification of degassing can result in a decrease in suction due to localized or diffuse release of condensed vapour. At La Fossa crater the phenomenon was evidenced by in situ monitoring of soil suction and soil temperature. In situ observations and measurements indicate that seepage of condensed vapour is appreciable. Simple models based on the geotechnical characterization of pyroclastic materials suggest the hypothesis that variations in suction can be significant to stability of volcano slopes when these are very close to limit conditions and if material hydraulic anisotropy is considered. The validation of this hypothesis requires further monitoring data during periods of intense unrest and more comprehensive models that account for non-isothermal multiphase pore fluid pressure and groundwater circulation, influencing the state of stress and hence stability.

ACKNOWLEDGEMENTS

The research was supported by the INGV-DPCI contract V5/12 (2005–2007) and PRIN 2006 Project.

Special thanks are due to P. Madonia (INGV-Palermo) for monitoring data and suggestions on volcanic activity, to E. Damiano (2nd Univ. of Naples) for model slope tests and to V. Savastano (AMRA) and V. Grana (CNR-IGAG) for numerical simulations.

REFERENCES

Brooks, R.H. & Corey, A.T. 1964. Hydraulic properties of porous media. *Hydrology Paper No. 3*, Colorado State Univ., Fort Collins, Colorado

Bukumirovic, T., Italiano, F. & Nuccio, P.M. 1997. The evolution of a dynamic geological system: the support of a GIS for geochemical measurements at the fumarole field of Vulcano, Italy. *J. Volcanol. Geotherm. Res.*, 79: 253–262.

Calcaterra, D. & Santo, A. 2004. The January, 10, 1997 Pozzano landslide, Sorrento Peninsula, Italy. *Engineering Geology*, 75: 181–200.

Dellino, P., Frazzetta, G. & La Volpe, L. 1990. Wet surge deposits at La Fossa di Vulcano: depositional and eruptive mechanisms. *J. Volcanol. Geotherm. Res.*, 43: 215–233.

Elsworth, D., Voight, B. & Taron, J. 2007. Contemporary views of slope instability on active Volcanoes. *Volcanic Rocks, Proc. Workshop W2—11thISRM Congress*, Malheiro A.M. & Nunes J.C. (Eds.), Azores, 3–9. Taylor & Francis.

Fredlund, D.G. & Rahardjo, H. 1993. Soil Mechanics for Unsaturated Soils. In *Wiley-Interscience Pubblication*, John Wiley & sons, inc.

Gardner, W.R. 1958. Some steady state solutions of the unsaturated moisture flow equation with application to evaporation from water table. *Soil Sci.*, 85(4): 228–232.

Geo-slope International 2004. SEEP/W Finite-element code for groundwater seepage analyses.

Greco, R. 2006. Soil water content inverse profiling from single TDR waveforms. *Journal Hydrol.*, 317: 325–339.

Kunze, R.J., Uehara, G. & Graham, K. 1968. Factors important in the calculation of hydraulic conductivity. *Proc. Soil Sci. Soc. Amer.*, 32: 760–765.

Neri, G., Montalto, A., Patanè, D. & Privitera, E. 1991. Earthquake space-time-magnitude patterns at Aeolian Islands (Southern Italy) and implications for the volcanic surveillance of Vulcano. *Acta Vulcanologica*, 1: 163–169.

Olivares, L. & Picarelli, L. 2001. Susceptibility of loose pyroclastic soils to static liquefaction—Some preliminary data. *Proc. int. conf. Landslides—Causes, countermeasures and impacts. Davos*

Olivares, L. & Picarelli, L. 2003. Shallow flowslides triggered by intense rainfalls on natural slopes covered by loose unsaturated pyroclastic soils. *Géotechnique*, 53(2): 283–288.

Olivares, L. & Damiano, E. 2007. Post-failure mechanics of landslides: laboratory investigation of flowslides in pyroclastic soils. *Journal of Geotechnical and Geoenvironmental Engineering ASCE*, 133(1): 51–62.

Olivares, L., Damiano, E., Greco, R., Zeni, L., Picarelli, L., Minardo, A., Guida, A. & Bernini, R. 2008. An instrumented flume for investigation of the mechanics of rainfall-induced landslides in unsaturated granular soils. *Submitted to ASTM Geotechnical Testing Journal*.

Picarelli, L., Evangelista, A., Rolandi, G., Paone, A., Nicotera, M.V., Olivares, L., Scotto di Santolo, A., Lampitiello, S. & Rolandi, M. 2006. Mechanical properties of pyroclastic soils in Campania Region. *Proc. 2nd int. work. on Characterisation and Engineering Properties of Natural Soils*, Singapore, 3: 2331–2383.

Rasà, R. & Villari, R. 1991. Geomorphological and morphostructural investigations on the Fossa cone (Vulcano, Aeolian Islands):a first outline. *Acta Vulcanologica*, 1: 27–133.

Tinti, S., Bortolucci, E. & Armigliato, A. 1999. Numerical Simulation of the landslide-induced tsunami of 1988 on Vulcano Island, Italy. *Bulletin of Volcanology*, 61: 127–137.

Topp, G.C., Davis, J.L. & Annan, A.P. 1980. Electromagnetic determination of soil water content: measurement in coaxial transmission lines. *Water Resour. Res.*, 16: 574–582.

van Genuchten, M. Th. 1980. A closed-form equation for predicting the hydraulic conductivity of unsaturated soil. *Soil Sci. Soc. Am. J.*, 44: 615–628.

Prediction of landslide movements caused by climate change: Modelling the behaviour of a mean elevation large slide in the Alps and assessing its uncertainties

Ch. Bonnard
Formerly Soil Mechanics Laboratory, Swiss Federal Institute of Technology, Lausanne, Switzerland

L. Tacher
Engineering and Environmental Geology Laboratory, Swiss Federal Institute of Technology, Lausanne, Switzerland

M. Beniston
University of Geneva, Caroug, Switzerland

ABSTRACT: The consideration of predicted climate change conditions in the hydrogeological and geomechanical modelling of a large landslide allows the assessment of its future behaviour in case of crisis. This application shows that the predictions are not necessarily pessimistic, despite of the uncertainties of the needed assumptions.

1 INTRODUCTION AND OBJECTIVES

For many decades there has been a clear consensus within the scientific community to express various quantitative or semi-quantitative relations between climatic conditions and general landslide movements of different kinds, as well as to use these experimental or empirical relations to try to predict future movements (Terzaghi 1950, Wieczorek 1996). However such relations have often proved to be deceiving as many short-term or long-term complex factors influence the crises of landslides.

One of the possible predictive approaches with neural networks has tried to combine observed past movement data and climatic information to predict future movements (Vulliet et al. 2000) Such a prediction is nevertheless reliable only in a short-term perspective and when no critical situation is likely to occur (Bonnard 2006), which is indeed the case when a reliable forecast is wished!

On the other hand, because of the numerous impacts that a changing climate can have on many elements of the planetary environment, it is of essence to predict the future course of climate forced by enhanced concentrations of greenhouse gases in the atmosphere. Predicting the speed and amplitude of climatic change can thus provide a measure of guidance to decision makers and climate-impact specialists.

In terms of land-surface processes, and particularly slope instability events, climatic factors are often assessed as a key factor in the triggering and/or the amplification of various forms of landslides, rock-falls, debris flows, etc. Precipitation is certainly today the dominant driving mechanism for many forms of slope instabilities, through water loading in soils beyond a critical threshold, or through excessive runoff that will lead to rapid surface erosion and debris flows. Both heavy but short-lived precipitation or moderate but continuous rainfall can thus provoke various forms of slope response, either in a natural or a man-made surrounding. Extremes of temperature can also contribute to slope instability, notably through repetitive freeze-thaw mechanisms at high elevations that tend to weaken rocks by progressively enlarging fractures. In addition, permafrost degradation in high mountains resulting from milder atmospheric temperatures can also contribute to slope instabilities by reducing the cohesion of slope material currently embedded within subsurface ice.

In a changing global climate, it is thus of interest to know how temperature and precipitation patterns may change, both in space and time, and also in terms of mean and extreme conditions. Such changing conditions may result in increased or even new forms of slope hazards compared to those encountered under current climate.

It is also essential to model the complex infiltration process of rainfall and snowmelt in large slides, in order to be able to establish a transient distribution of groundwater pressure at any point of the landslide mass and within the slip surface, so as to model the movements induced by these pressure changes in a FEM.

The objective of this paper is thus to present the global trend of climate change and then to illustrate its possible long-term effect in the case of a large slide. The obtained results will show that in some cases, the so-called evidences are not granted for sure. It is also important to trace and quantify all kinds of possible uncertainties in this multiple process in order to assess the reliability of long-term predictions.

2 TREND OF GLOBAL CLIMATE CHANGE

Perhaps the most exhaustive source of information concerning future climatic change is provided by the Intergovernmental Panel on Climate Change (IPCC). In the succession of reports published in 1996, 2001 and 2007 (IPCC 2007), a number of global climate simulation models have been applied to assess the response of the climate system to anthropogenic forcing in the 21st century, based on a number of greenhouse-gas emission scenarios developed by Nakicenovic et al. (2000). According to the scenario, itself a function of assumptions on population growth, economic growth, technological choice, and policy decisions, the global mean temperature change over present ranges from 1.5–5.8°C, as illustrated in Figure 1. This represents an amplitude of change that is probably one order of magnitude larger than changes reconstructed for the past 20, 000 years (i.e., since the last glacial maximum), and a speed of change that is up to two orders of magnitude greater than typical natural fluctuations of climate.

Climate model solutions suggest that the change in temperature will be stronger in the high latitudes compared to the equatorial region. This is because of the strong positive feedback that can be expected as a result of smaller areas covered by snow in the northern continents, a shorter winter period, and reduced sea-ice cover in the Arctic Ocean. Reduced snow and ice cover will substantially modify the surface energy balance, particularly through increased absorption of solar energy. Temperature change will also be greater over the continents than over the oceans, because of the larger heat capacity of the oceans.

While a warmer climate will enhance the hydrologic cycle, precipitation will not necessarily increase everywhere. The latest climate models published by the IPCC (2007) suggest that the northern latitudes may experience greater precipitation than currently, but that rainfall in Mediterranean and arid climates

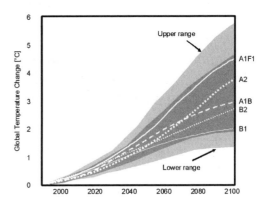

Figure 1. Range of climate futures according to greenhouse-gas emission scenarios; the dark gray zone represents the more likely range of global temperature change (IPCC 2007). For the emission scenarios, see Nakicenovic et al. (2000).

may decrease (i.e., many semi-arid and arid regions could become even drier in the future). Precipitation totals will probably increase by 2100 in the Monsoon climates of India and China, and in the inter-tropical convergence zone around the equatorial belt.

Temperature in Europe will increase on average by 4–6°C, with strong regional and seasonal differences; for example, summertime warming in southern Europe is expected to be greater than during the winter, because of the positive feedback effect of dry soils during this season (Seneviratne et al. 2006). In the Alps, wintertime temperatures will rise by 3–4°C by 2100 compared to current climate, according to the level of greenhouse gases. Summer temperatures may rise by more than 6°C during the same period, as a result of positive feedback effects from dry soils and reduced snow and ice cover in the Alps (Beniston et al. 2007). Figure 2 shows the difference between summer temperatures for current (1961–1990) and future climates in Basel, Switzerland, not only for mean conditions, but also in terms of the upper quantiles that essentially represent heat-wave conditions.

Simulated results for the low emissions B2 scenario and the high emissions A2 scenario are shown; interestingly, the difference in temperature between the high and low emissions scenarios is less than between the B2 scenario and current climate. This implies that even with rather stringent policies to abate greenhouse gas emissions, the increase in temperatures as seen for the B2 scenario will result in summer heat waves that are as intense, or even stronger, than the 2003 heat wave, with an even greater potential for strong heat waves in the A2 scenario. Indeed, statistically-speaking, the 2003 heat wave could occur one summer out of two in a future climate (Schär et al., 2004).

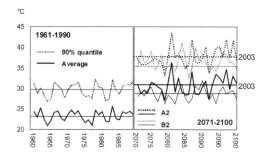

Figure 2. Observed summer temperatures for 1961–1990 (means and 90% quantile values), and simulations for 2071–2100 for the low emissions B2 and the high A2 emissions scenarios established by the IPCC (2007). See text for further details.

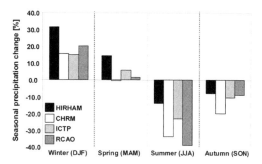

Figure 3. Percentage change in average seasonal precipitation for the control and scenario climates as simulated by 4 different regional climate models; see text for further details.

In a future climate, whatever the emissions scenario considered, the freezing level will thus rise by about 500–600 m in winter and close to 1, 000 m in summer, thereby accelerating glacier retreat and exacerbating the natural hazards associated with deglacierized landscapes. Positive temperatures at increasingly high elevations will penetrate into permafrost layers, leading to its progressive melting and thus reducing the cohesion of soils.

Christensen & Christensen (2003) have shown that northern Europe will experience more precipitation on average, while in a large band stretching from France to the Black Sea and beyond, summer precipitation is projected to diminish by as much as 40%. Simultaneously, many regional climate models point to a strong increase of short-lived but very intense precipitation events in certain regions that are already prone to such hazards, such as parts of central Europe, the Alps, southern France, and northern Spain.

In Switzerland, simulations of climate forced by high greenhouse-gas emissions for the period 2071–2100 compared to the reference 1961–1990 period shows a marked shift of the seasonality of mean precipitation (e.g., Beniston 2006), with strong increases in winter and spring, and substantial reductions in summer and fall, as seen in Figure 3 for four different regional climate model simulations (the Danish HIRHAM; the Swiss CHRM; the Italian ICTP; and the Swedish RCAO models); while the absolute value of change differs from one model to another, all models agree on the seasonal sign of change. The principal cause of these changing patterns is related to the strong summer warming and drying in the Mediterranean zone that would also affect the Alps and regions to the north, and the enhanced winter precipitation that a milder climate may bring to the region, rather under the form of rain than of snow. As a result of the change in mean precipitation, the frequency of extreme rainfall events also changes in seasonality compared to current climate. Model simulations suggest that springtime extremes will increase the most, while summer events may decline by as much as 50%. However, when heavy precipitation occurs in a warmer climate, it may be even more intense because of the additional energy provided by a warmer atmosphere than today.

In terms of the potential for floods, natural hazards and damage that the changes in means and extremes of precipitation may trigger in the Alps, it should be emphasized that heavy precipitation is a necessary but not sufficient condition for strong impacts. For example, Stoffel and Beniston (2007) have shown that while debris-flows of the past have occurred mostly during wet summers, it is conceivable that in a greenhouse climate the frequency of such events could decrease because of the shifts in the occurrence of extreme precipitation from summer to spring or fall by 2100. The response of slopes and watersheds to high precipitation levels varies from one event to another for a number of reasons, in particular the prior history of precipitation, evaporation, permeability of soils and the buffering effect of snow during an event. That may lead to decrease at mean altitude. The more elevated the freezing level, the greater the potential for strong runoff and high intensity of erosion since there is a larger surface area upon which water can flow off the slopes. Under current climate, the most intense events are observed to occur during the summer months, where the freezing level is higher than 3,500 m above sea level. In a future climate, on the other hand, the freezing level associated with heavy rainfall will be on average 500 m lower because many events will take place either in spring or in autumn, at a time of the year where conditions will be cooler than for current summers. However, even if their frequency is likely to decrease, the magnitude and impacts of future summertime debris flows, mudslides or rock-falls could be greater than currently because of warmer temperatures and higher precipitation intensities.

3 TYPES OF LANDSLIDES AND SLOPES LIKELY TO BE AFFECTED BY CLIMATE CHANGE

It is evident that all kinds of shallow slides and improperly drained engineered slopes, as well as potential debris flow creeks are likely to present a higher hazard level in the future, as one of the major characteristics of climatic conditions in the 21st century is to display more intense storms occurring probably with an increased frequency. Another reason for this increased hazard is the always extending area of impervious zones in the concerned watersheds, due to the development of urbanization and roadways. This situation causes higher peak floods in streams or excessive discharges in inappropriate sewage or drainage ducts that are likely to divert sudden flows at the surface of slopes through manholes or pipe failure and thus generate destructive mud flows, if specific retention works are not foreseen.

In the case of high mountain slopes, the increasing melting of the permafrost zone due to higher summer temperatures can also be a cause of unexpected debris flows, even outside of a rainfall event. The loosened fan material at the toe of mountain cliffs provides more sediments for the debris flows that can generate more extensive impacts in the valley floor; this critical situation is especially due to the intense tourist development of chalets in zones providing space with a view and easy access.

In all these cases there is a nearly simultaneous occurrence between the storm triggering the landslide and the development of the landslide process leading to severe visible impacts on buildings or agricultural land, so that there is no doubt about the relation between the climatic conditions and the consecutive damage. However in the cases of larger slides such a correlation is not evident to demonstrate and a nearly similar rainfall pattern can cause a slight increase of the movements of a slide during one winter and a severe crisis during the next winter, as it was observed and monitored at La Chenaula landslide in Switzerland in 1982 and 1983 (Noverraz & Bonnard 1992).

In the case of large to very large slides, extending on an area of one to several km^2, with differences of elevations between scarp and toe that can reach several hundreds of meters, the situation is even more complex, as several factors may influence the reaction of the landslide mass, namely:

- variation of rainfall amount with elevation
- offset of snowmelt episodes along the slope
- capacity of snow cover to absorb a large amount of rainfall before infiltration occurs
- variation in vegetation implying different interception and evapotranspiration patterns

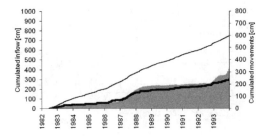

Figure 4. Relationships between horizontal displacements and computed flux entering the La Frasse landslide in a cumulated representation. Thin line: infiltration issued from COUP model (considering temperature, snowmelt, vegetation, soil, sun exposure...). Thick line: weighted and truncated infiltration over several years in the past. The best fit of the thick curve with the displacements (grey area) is concluded from the shape of the curves rather than from the value of the corresponding points.

- distribution of permeability fields at the surface of the slide and in the landslide mass
- possibilities of differentiated infiltration along streams flowing on the slide as well as in cracks
- occurrence of floods in the stream flowing at the toe of the slide and likely to cause erosion
- development of urbanization and collecting water pipes and ducts
- differences of reaction of the landslide mass along the slope (swelling and depletion zones).

In such cases it is also necessary to consider the effect of climatic impact not only for one specific event, but for a crisis period that can be caused by antecedent conditions extending to several years before the observed crisis. In the case of La Frasse Landslide, in Switzerland, it has thus been shown (Tacher et al. 2005) that several factors like interception of modest rainfall and evapotranspiration had to be considered in order to express a good correlation between infiltration flow and movements (Fig. 4).

It is then necessary to analyse the changing climatic situation through the formulation of scenarios implying different episodes along the yearly cycle. It is of course essential to gather enough long-term data on the movement of the slide as well as continuous displacement information on so as to understand its global behaviour and to confront the observed and modelled displacement vectors.

4 LONG-TERM MONITORING OF THE IMPACT OF CLIMATE CHANGE ON LANDSLIDE BEHAVIOUR

In order to determine the possible effect of climate change conditions on the behaviour of large slides, it is first necessary to gather available data providing information on their long-term movements. Such an investigation has been carried out in Switzerland within a national research project (PNR31), considering a dozen of very large slides, extending over areas from one to some 40 km^2, and for which ancient geodetic survey data had been collected (Noverraz et al. 1998).

In the specific case of Lumnez Landslide, in the Canton of Graubünden in Eastern Switzerland, the position of the spires of 7 village churches has been regularly monitored for more than a century (up to 17 monitoring campaigns beginning in 1887). The results showed in general a very constant average velocity, varying from 3 to 20 cm/year (Fig. 5).

Only one point located near the toe of the slide (village of Peiden) displayed a clear reduction of velocity after the years 1940, which can be partially explained by a series of dry years and then by the construction of a dam on the river Glenner flowing at its toe, upstream of the slide. As far as the annual rainfall is concerned, the long-term trend is not so marked in this region (average value of 950 mm/year—Figure 5) as in western Switzerland, in which a clear increase of annual rainfall by some 10% has been observed since the years 1980 (this fact has induced the Swiss hydrological service to change the long-term reference rainfall value from the period 1901–1960 to that extending from 1961–1990 for all rain gauge stations).

It can thus be observed that most of the very large slides monitored display a fairly constant velocity even if the are affected by long periods of higher precipitation. This fact can of course be due to their size and depth, inducing a certain mechanical inertia, as well as to the large storativity of their hydrogeological conditions. But this observation is not necessarily valid for all slides.

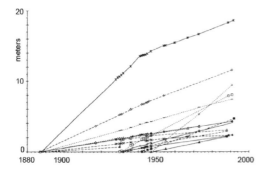

Figure 5. Cumulated displacements of several topographic points at Lumnez landslide over a period of more than one hundred years.

In a few cases, indeed, some monitored points of these large slides have displayed an acceleration phase that may last from several months to a year, like in the case of La Frasse landslide, for which crisis periods of a few months duration have been recorded in 1966, 1982–83 and 1993–94 (Tacher et al. 2005). In most of the duly monitored slides, this major acceleration phase does not imply the whole landslide mass, but a part of it, generally located at its toe or eventually in an area in which the depth of the slide is reduced. Such a situation was clearly put forward in the case of Chlöwena Landslide in Switzerland, in 1994: the crisis lasted for 4 to 5 months, with a peak velocity of 6 m/day at the end of July that was reduced to a few cm/year at the end of September (Vulliet & Bonnard 1996).

A comprehensive approach of such complex phenomena therefore requires first a long-term monitoring and then a detailed modelling in order to understand the hydrogeological and geomechanical conditions that explain the crisis episodes. In a second step it is possible to determine the probable effects of climate change in a quantitative way, considering several crisis scenarios. Such an approach has been applied to various slides and in particular to the Triesenberg landslide.

5 MODELLING THE CLIMATE CHANGE CONDITIONS : THE TRIESENBERG LANDSLIDE

The Triesenberg landslide extends over a significant part (i.e. 5 km^2) of the Principality of Liechtenstein (160 km^2), located to the East of Switzerland (Figure 6). It also includes two villages, Triesen at its toe and Triesenberg at mid-slope, the infrastructures of which incur occasional damage, in particular during crisis episodes.

The movements of this landslide are quite ancient and date back to the end of the Wurmian period; presently they are generally slow (i.e. some mm/year to cm/year) in normal conditions and locally may reach velocities of a few dozens of cm/year during severe crisis periods. As the slide displays a relatively slow movement, many buildings have spread on the slope in particular during these last decades, due to the real estate development.

The objectives of the research were the following:

– to determine the critical hydrogeological conditions that cause an acceleration of the slide and that may justify the triggering of an alarm system;
– to foresee the behaviour of the slide under the possible effect of climate change, so as to establish bases for the sustainable development of the slope.

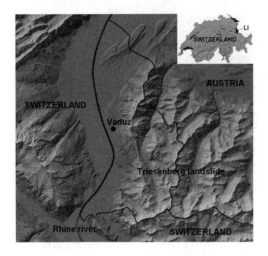

Figure 6. Location of the Principality of Liechtenstein and of the Triesenberg landslide.

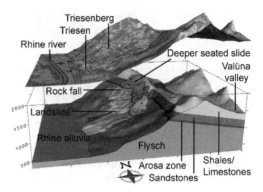

Figure 7. Geological model of the Triesenberg landslide. The draped topographic map is lifted to display the geological units.

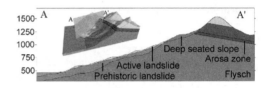

Figure 8. Geological vertical cross-section.

The first aim of the models developed does not consist in determining the possibilities of stabilizing the overall slope, as it can be expected that such works would by far pass the planned investments by the authorities of the Principality. What is aimed at is to live with the slide, and not to slow it down.

The specific difficulties presented by the Triesenberg landslide mainly refer to its large area, to its essentially unsaturated hydrogeological conditions and to the slow movement velocities.

After calibrating the model parameters with respect to the crisis of 2000, the impact of climate change has been analysed by modifying the boundary conditions of the hydrogeological model, on the basis of the relevant climatic scenarios, as set up by the Swiss Commission for the assessment of climate change (OcCC, 2004). Then the respective computed groundwater pressures have been introduced in the geomechanical model, as it was done for the year 2000.

5.1 Main features of the landslide

5.1.1 Morphology and geology

The slope is oriented from North-East (up) to South-West (down). It presents some small undulations but is generally fairly regular. Based on a digital terrain model, the mean slope is 24° (Figures 7 and 8). Three parts are distinguished:

– In the upper part, deep-seated slope movements occurred, probably at the end of the Wurmian glacial retreat (14,000 years); they are now underlined by a terrace in the topography at the top of the slope (Figure 8) and were triggered by a deep landslide, the so-called Prehistoric landslide. This upper zone, largely inactive, is not considered to cause a driving force on the slope. Approximately, it covers 1.7 km^2 with a volume of 74 million m^3.

– The prehistoric landslide is known by some boreholes. It is more than 80 m deep and is made of flysch (clayey shales). This zone is today stabilised; moreover, no movement has been observed at the toe of the landslide, where it lies under the Rhine river alluvia (gravels).

– The active landslide (Table 1) covers the prehistoric one. It is also composed of flysch and takes place on a slip surface located at an average depth of 10–20 m. According to the inclinometer data, available from 1995 to 2002, the slip surface is approximately one meter thick.

The analysis of the observed intensities and directions of the movements (Figure 9) showed that the area is indeed composed of three instability zones that can be considered as independent.

This is confirmed by the reduced depth of the slip surface close to the assessed boundaries of the three areas and by the spatial distribution of damage to infrastructures (Frommelt AG 1997).

The practical consequence of the decomposition of the landslide into three distinct systems is to allow defining three different modelling areas for the 3D

Table 1. Main features of the Triesenberg active landslide.

Aspect	Characteristics
Area	3.1 km²
Altitudes	min. 460 m, max. 1500 m a.s.l.
Length	2300 m
Width	1500–3200 m
Mean depth	10–20 m
Volume	37 millions m³
Mean slope	24°
Mean velocity	0 to 3 cm/year
Soil	Flysch (clayey shales) including elements of limestone and sandstone
Vegetation	Pasture land and some wooded zones
Investigations	Hydrogeology, boreholes with inclinometers, GPS, RMT geophysical methods, laboratory tests, modelling
Possible damage	Infrastructures of two villages

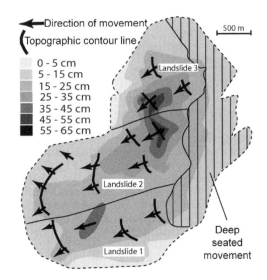

Figure 9. Total displacements between 1976/1981 and 1996/1997 (i.e. during some 20 years) in the whole instability area, and boundaries of the three so called independent landslides.

mechanical modelling, supposing negligible kinematic interactions between them (the hydrogeological model involves a single regional area).

5.1.2 Hydrogeology

Besides supplying the geomechanical models with hydraulic pressures, the hydrogeological analysis and modelling aims at a better understanding of the particular hydrogeological behaviour of the slope.

The yearly observation of displacements shows a close dependence of the movements on the seasons. A reactivation is generally perceived in the spring, which corresponds to the snowmelt period. This indicates that the main driving force of the movements is the variation of pore water pressure in the slope. However, a reactivation may also occur following a storm event.

The tectonic Arosa zone (Fig. 8) is a very important feature of the hydrogeological system due to its low permeability: a part of the Valüna Valley groundwater (Fig. 7) flows on the Arosa zone and feeds the basal surface of the landslide, causing the Triesenberg groundwater basin to be much larger than its topographic watershed. This mechanism is proven by several observations (Tacher & Bonnard 2007).

Such a double feeding is also effective outside intensive infiltration periods. Both a hydraulic balance of the Triesenberg slope (Bernasconi 2002) and a numerical model calibration suggest that about one half of the inflow in the landslide is supplied by a base flow from the Valüna Valley through the sandstones covering the Arosa zone (ca. 9 mio m³/year). Groundwater discharge occurs through some one hundred springs distributed over the landslide, as well as at its toe, in the River Rhine alluvia. The water table is located about 20 to 30 m below the soil surface at the top of the landslide, whereas at the bottom, it almost reaches the ground surface.

5.2 Hydrogeological modelling of the year 2000

The year 2000 was chosen to perform the modelling; during this year, a critical phase with a reactivation of the movements was observed, showing a good correlation with the snowmelt phase in April. A violent thunderstorm also occurred on August 6th. This year has a return period of the annual rainfall of 42 years, according to the Gumbel law. Another reason of this choice is the availability of calibration data for both hydrogeological and geomechanical models. As the slide is very thin, the unsaturated zone is of relatively high importance, which justifies computing groundwater flows in unsaturated regime, i.e. the flows are governed by Richards' equation (Hillel 1980).

From the model results, in terms of volumes, the direct infiltration reached 7.52 mio m³ in 2000, while the inflow through the Arosa zone was about 9.86 mio m³ (Figure 10). The cumulated rate of the springs reached 1.06 mio m³, which represents only a few percent of the total outflow; the balance flow seeps in the Rhine river alluvia.

The outflow curve is smoother than the inflow events, due to the capacitive function of the landslide. Typically, the August 6th storm response was absorbed and delayed. In May, the snowmelt episode did not lead to spectacular changes in the hydraulic balance because, due to the slope topography, the melting occurred progressively from the bottom to the top. For example, when the snowmelt occurred in the Valüna Valley, it had already finished on the landslide several days to weeks earlier.

More relevant from the geomechanical point of view is the piezometric behaviour. It is illustrated by piezometer B8 (Fig.11). The respective calibration was carried out by comparing the water table data with the hydraulic head computed at the bottom node at this site, i.e. at the slip surface. Both main events of the year 2000 led to a peak more than 2 m high. Just after these peaks, the head decrease was slower in the model than in the reality. This can be explained by the relative smoothing of the parameter field, mainly over depth.

Heterogeneities are also responsible for another observation: during the snowmelt event, the model reacted with a delay of some days with respect to the monitoring data. Local pervious heterogeneities that were not considered in the model accelerated the piezometer response to inflows in the Valüna Valley. Such a delay did not occur at the beginning of August since inflows concern both the Triesenberg and Valüna basins.

The numerical results suggest that the model globally fits with reality, despite a simplification of the parameter fields, a rough estimation of the unsaturated parameters and a minimal knowledge of the real hydraulic balance. Computed hydraulic pressures are thus suitable as an input in the hydro-mechanical models in order to describe the direct causes of the movements during crises.

5.3 Geomechanical modelling of the year 2000

The effect of the hydraulic head variation with time, as determined by the hydrogeological modelling, on the mechanical behaviour of the whole slide, has been modelled by a FE code, Z-SOIL (2-D and 3-D), using a Biot-type formulation, implying the conservation of mass and momentum of both fluid and solid phases (François et al. 2007).

In the 3-D model, the maximum displacement values are in general slightly lower, but they appear within zones where damage has been reported (Fig. 12).

Parametric studies have also been carried out to evaluate the effect of the selected friction angles (between 30 and 21°) and of the range of water pressure variation (the computed data through the hydrogeological model were multiplied by 1.25 and 1.5 respectively).

Both simulations display nearly linear variations and prove that, even in extreme conditions, it is not expected that the movements will lead to a catastrophic behaviour of the whole slope.

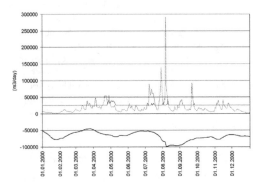

Figure 10. Hydraulic balance of the model for the year 2000. Thin line: Feeding by the Arosa zone. Dot line: Direct infiltration on the slope. Thick line: Outflow rates through springs and the River Rhine valley.

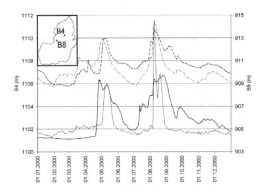

Figure 11. Measured and computed hydraulic heads in piezometers B4 and B8. Thin line: B4 measured. Thick line: B4 computed. Thin dotted line: B8 measured. Thick dotted line: B8 computed.

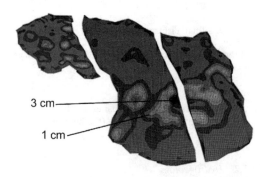

Figure 12. Distribution of the obtained displacements after 291 days of simulation, from January 1 to October 18, 2000 (water pressure data multiplied by 1.25) and location of the more active zone in the central slide (3 cm).

5.4 Modelling of Climate change impact and related uncertainties

According to (IPCC 2007), the air temperature should increase in the medium term, especially in summer, and the rainfall should increase in winter, but decrease in summer. The climatic scenario for 2050 used in this study is issued from the Swiss "Organe consultatif sur les Changements Climatiques" (OcCC 2004), more specific to the North of the Swiss Alp context (Fig. 13).

According to this scenario, it can thus be expected that:

- In winter, the total infiltration would increase and rain would partly replace snow accumulation. On the other hand, snowmelt at the beginning of the spring would be less important.
- In summer, the storm events would remain similar, if not slightly worse, but the total infiltration would be smaller than today because of higher evapotranspiration.

In this study, the rainfall in 2050 is considered to increase by 2 mm/day in winter and to decrease by 2 mm/day in summer. Those values are added as a one year sinusoidal transformation to the records for the year 2000. Similarly, the temperature curve for 2050 is obtained by adding a one year sinusoidal function to the records of the year 2000, considering a warming of 1.5°C in winter and 3.5°C in summer (Figure 14).

Considering the impact on landslides, such a scenario is not obviously more severe, mainly for the landslide zones in altitude. Indeed, besides the total infiltration, the groundwater pressure fluctuations have a major effect on the movements. By diminishing the rather massive infiltration period of snowmelt, the 2050 scenario smoothes out the groundwater head curve at spring time. In particular in the Valüna valley, the fast snowmelt at the beginning of May 2000 might be replaced by a succession of less important episodes of rain, falling on a thin accumulation of snow.

The target of the models is here to consider the most unfavourable scenario as far as the landslide movements are concerned. Thus these worst case infiltration conditions for 2050 are as follows, even if they are not the most plausible:

- No consideration of the decreasing of gross rainfall in summer. The infiltration curve is left intact from May 1st,
- Keeping the snowmelt event of the end of April,
- In winter, adding infiltration periods without decreasing the accumulated snow height.

In practice, the 2050 infiltration scenario implies to add infiltration days between January 1 and April 20 to the year 2000 conditions. For all altitude classes of infiltration, a 5 mm/day event is introduced each ten days (Fig. 15). This represents an additional infiltration of 55 mm/year.

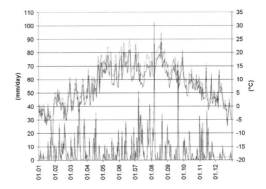

Figure 14. Climatic scenario for 2050 in the Valüna valley. Upper curves: Temperatures in 2000 (solid line) and in 2050 (dotted line). Lower curves: Gross rainfall in 2000 (solid line) and in 2050 (dotted line).

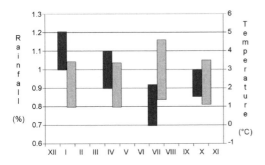

Figure 13. Climatic scenario for 2050 for the North of the Alps, after OcCC. Horizontal axis: months of the year. Black bars: Rainfall change in % with respect to present average seasonal values. Grey bars: Temperature change in °C.

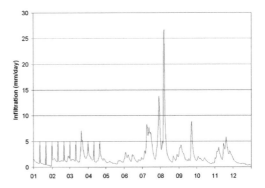

Figure 15. Scenario 2050. Infiltration conditions for altitudes below 625 m.

225

The results of both hydrogeological and geomechanical models with such modified boundary conditions are very similar to those obtained for the year 2000. Typically, the hydraulic heads in piezometer B8 (Fig. 11) are changed by some centimetres only.

Considering uncertainties on climatic changes, the modelled scenario appears to be probably the worst case. In such conditions, the computed velocity field for 2050, if the parameter calibration on year 2000 is considered as reliable, is a rather pessimistic global assessment of the slope. However, the transition to a stormier climatic regime may have local consequences (hectometric slides, mudflows) not considered in this regional modelling.

6 CONCLUSIONS

A detailed hydrogeological and geomechanical modelling as it was recently applied at the Triesenberg and La Frasse landslides allows a significant modelling of large landslide movements during crises, provided sufficient information is available. The application of predicted climatological conditions in the future then supplies quantitative values of possible movements, considering appropriate scenarios. However, extremely rare conditions with a very remote probability cannot be modelled reliably, as the boundary conditions may significantly differ from the ones considered in the original model.

The analysis of several large landslides in other contexts (Bonnard et al. 2004) also shows that the effect of climate change on landslides within the next 50 years or so must not be overemphasized. Indeed, as shown here, the progressive snowmelt that will begin earlier than before tends to reduce the occurrence of critical situations in the spring or summer. On the contrary, it is clear that the expected increase of storm intensity, as foreseen by some climatologists, may produce more violent and frequent small slides and debris flows; but this specific prediction is not relevant for large landslides and cannot justify a development of more severe disasters related to this type of phenomena. Indeed, due to the heterogeneity of the material at a large scale, to the increased range of altitude where infiltration occurs, to the capacitive function of the landslide mass and to the more complex hydraulic relationships with the bedrock, the response to climatic events may be significantly smoothed and delayed, which explains this relatively optimistic vision.

ACKNOWLEDGMENTS

The authors wish to thank the authorities of the Principality of Liechtenstein for supporting this research, Dr Riccardo Bernasconi geological office for supplying data and advice, as well as all the colleagues who participated to the modelling.

REFERENCES

Beniston, M. 2006. The August 2005 intense rainfall event in Switzerland: not necessarily an analog for strong convective events in a greenhouse climate. Geophysical Research Letters 33, L5701.

Beniston, M., Stephenson, D.B, Christensen, O.B., Ferro, C.A. T., Frei, C., Goyette, S., Halsnaes, K., Holt, T., Jylhä, K., Koffi, B., Palutikof, J., Schöll, R., Semmler, T., and Woth, K., 2006. Future extreme events in European climate; an exploration of regional climate model projections. Climatic Change, 81, 71–95.

Bernasconi, R. 2002. Tiefbauamt des Fürstentums Liechtenstein - Hangsanierung Triesenberg—Hydrogeologische Überwachung—Ergebnisse der Markierversuche Valünatal 1999/2000, Hydrogeologischer Bericht Nr. 1124–04, August 2002.

Bonnard, C., Forlati, F. & Scavia, C. 2004. Identification and mitigation of large landslide risks in Europe : advances in risk assessment. IMIRILAND Project. 317 p. Leiden: Bal-kema, ISBN 90 5809 598 3.

Christensen, J.H. & Christensen, O.B. 2003. Severe summertime flooding in Europe. Nature, 421, 805–806.

COUP model software, Per-Erik Jansson, Department of Land and Water Resources Engineering, Royal Institute of Technology, Stockholm.

François, B., Tacher, L., Bonnard, Ch., Laloui, L. & Triguero, V. 2007. Numerical modelling of the hydrogeological and geomechanical behaviour of a large slope movement: The Triesenberg landslide (Liechtenstein). *Canadian Journal of Geotechnics* 44:840–857.

Frommelt, A.G., Ingenieurbüro, Vaduz. & Hangbewegungen, Triesenberg. Verschiebungsmessungen 1978/81–1996/97, 1997.

Hillel, D. 1980. *Fundamentals of soil physics*. Academic Press, New York.

IPCC, 2007. Climate Change. The IPCC Fourth Assessment Report. Cambridge University Press, Cambridge, UK.

Nakiæenoviæ, N., et al., 2000: IPCC Special Report on Emissions Scenarios, Cambridge University Press, Cambridge, UK.

Noverraz, F. & Bonnard, Ch. 1992. Le glissement rapide de la Chenaula. Proc. Symp. INTERPRAEVENT, Berne. 2:65–76.

Noverraz, F., Bonnard, Ch., Huguenin, L., Dupraz, H. 1998. Grands glissements de versants et climat. Projet VERSINCLIM, Comportement passé, présent et futur des grands versants instables subactifs en fonction de l'évolution climatique, et évolution en continu des mouvements en profondeur. Rapport final PNR 31, FNRS, Berne, 1998. Zürich : V/d/f, 314 p.

OcCC, 2004. Die Klimazukunft der Schweiz—Eine probabilistische Projektion, Christoph Frei, Institut für Atmosphäre und Klima, ETH Zürich.

Schaer, C., et al. 2004. The role of increasing temperature variability in European summer heat waves. *Nature* 427:332–336.

Seneviratne, S.I., et al. 2006. Land-atmosphere coupling and climate change in Europe. *Nature* 443:205–209.

Stoffel, M. & Beniston, M. 2006. On the incidence of debris flows in the Swiss Alps since the early Little Ice Age and in a future climate. Geophysical Research Letters, 33, L16404.

Tacher, L., Bonnard, C., Laloui, L. & Parriaux, A. 2005. Modelling the behaviour of a large landslide with respect to hydrogeological and geomechanical parameter heterogeneity. *Landslides Journal* 2 (1):3–14.

Tacher, L. & Bonnard, Ch. 2007. Hydromechanical modelling of a large landslide considering climate change conditions. Lecture at International conference on 'Landslides and Climate Change—Challenges and Solutions'. Ventnor, Isle of Wight, UK. 21–24 May 2007.

Vulliet, L. & Bonnard, Ch. 1996. The Chlöwena landslide: Prediction with a viscous model. *Proc. VIIth Int. Symp. on Landslides, Trondheim* Vol. 1:397–402.

*Geology, geotechnical properties
and site characterization*

Geotechnical appraisal of the Sonapur landslide area, Jainita hills, Meghalya, India

R.C. Bhandari, P. Srinivasa Gopalan & V.V.R.S. Krishna Murty
Intercontinental Consultants and Technocrats Private Limited, New Delhi, India

ABSTRACT: The occurrence of landslides particularly on cut slopes along the roads with in Jainita hills in Northern Eastern part of Himalayas of Meghalaya state in India are common features. These slope failures causes considerable loss of life and property along with many inconveniences such as disruption of traffic along highways. The paper present deals with geotechnical synthesis of slides as per recommendations of RHRS system and stabilization measures suggested. The landslide is located on North South trending ridge on eastern bank of river Sonapur. The rocks involved in sliding are highly jointed sandstone shale of Oligocene series. The total inclined length of the affected slopes is around 800 meter. The jointing in rock is attributed to the nearness of the area to a major thrust. These studies indicate the failure along the slopes is "Rock fall-cum debris flow".

1 INTRODUCTION

1.1 Occurrence of land slide

A land slide occurs when due to gravity forces the rock/soil mass moves down wards due to heavy precipitation, run off or ground saturation. The flow occurs generally during period of intense rainfall, on steep hill slopes where the rocks are tectonically disturbed. The flow/fall from many different sources can combine in channels, and their destructive power is greatly increased. They continue flowing down hills and through channels, growing in volume in addition of water sand, mud, boulders, up rooted trees and other material. When the flows reach flatter ground, the debris spread over a broad area thus affecting the considerable length of road.

1.2 Location of area

The occurrence of land slides, particularly along the road in the Jaintia hills of Meghalaya state is common feature during rainy season. This cause considerable loss of life and property along with many inconveniences such as disruption of traffic between Shillong and Silcher every year. Sonapur land slide occurring near vicinity of township and is located between KM 141.100 to KM 141.350 on Shillong—Silcher highway.

1.3 Climate

The climate of the area is moderate. The rainfall occur during monsoon period which extends from May to October. The month of June and July experiences maximum precipitation. The annual precipitation varies from 4000 mm to 8000 mm during the year maximum precipitation of 1200 mm is recorded in month of June and July. Maximum temperature reaches to 24°C in month of September while minimum temperature recorded is 10°C in month of February.

2 REGIONAL GEOLOGY OF THE AREA

2.1 Rock types

The rock types in the region available, varies in age from Archeans to Tertiary the general sequence of the rocks available in the area is as follows (Figure 1).

3 LOCAL GEOLOGY AND GEOMORPHOLOGY

3.1 Geology

The Sonapur landslide is located on North–South trending ridge, on east bank of river Sonapur/Lubah. The river Lubah flows at elevation 33.0 m from mean sea level, and the present road elevation 44.0 meter. The highest elevation at top of the crown of slide is 481.25 meters. Thus the total height affected by slide along the slopes are also observed slopes is around 440 meters. The rock exposed in the area are sand stones and shales of Oligocene age. The rocks are highly jointed and local folding along the slopes are also observed.

Table 1. Regional geology.

Rock types	Group	Age
Feldspathic sandstones, pebbles and conglomerates	Brail	Oligocene
Shales Sandstone and marls (Kopiliformation)	Garo	Miocene
Limestone and Sandstone (Sylhet formation)	Jaintia	Eocene
Gneissses and granites	Gniessic complex	Archeans

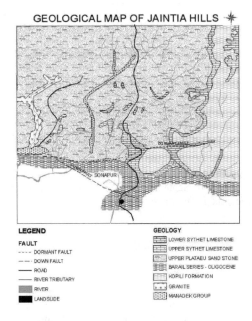

Figure 1. Regional geological map of the area.

3.2 Geomorphology

The area is occupied by undulating topography characterized by presence on hills and valleys. The area is dissected by number of streams and network of their tributaries. The drainage pattern in the entire district represent a most spectacular feature revealing extra ordinary straight course of rivers and streams, evidently along master joints and faults which are impressions of major geological activity in the area. The magnificent gorges scooped out by the river in the southern part on Jainitia district are result of massive headword erosion by antecedent streams along joints of sedimentary rocks, exposed in the area.

4 HISTORY OF SLIDE

For information see table on next page.

5 STRUCTURAL ANALYSIS

5.1 Geometry of slide

The geometry of the slide is as under. (For plan and section of land slide area please refer to Figures 2 and 3).

The area is being drained by two prominent drainages. The drainage, which is located on southern side is being continuously recharged through minor

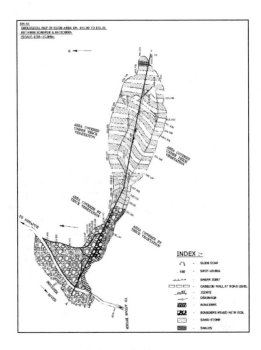

Figure 2. Geological plan of slide area.

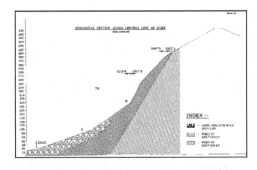

Figure 3. Section along the central line of slide area.

Table 2. History of slide.

Date of occurrence	Possible cause	Remedial steps taken	Remarks
1998 July	i) Loss of cohesion in soil along slopes due to high precipitation ii) Earth quake of 1987, possibly resulted loosening of shear strength between jointed blocks	Clearance of debris Construction of gabion walls and road level diversion the drain	Slide of high intensity Road closed for more than one month
1989 September	Flow of debris material from higher slopes	Clearance of debris and maintenance of remedial measures taken were same as in 1998	Intensity was less; road was blocked daily for 1–2 hours for 15 days
1999 June–July	Rolling down of boulders from top 200 m above road level along slopes	Clearance of debris Diversion of water at road level ii) Repairs of gabion structures at road level	Road blocked 2–3 days Land slide of less magnitude but size of boulders was 5.5 m × 1.8 m × 4.5 m
2000 Aug–Sep	Rolling down of boulders from top 200 m above road level along slopes	Clearance of debris Diversion of water at road level ii) Repairs of gabion structures at road level	Land slide frequently occurred. The road was blocked for total period of 22 days
2001 01 to 15 June	Loss of cohesion in soil along slopes due to heavy rains	Construction of Retaining wall, Check wall water chute Concrete pavement and geotextiles	High-magnitude slide, Boulders of 3.75 m × 2.5 m × 8.8 m rolled down along slopes Road blocked for 10 days one oil tanker, truck with loaded goods, one passenger bus buried down. Number of casualties not known
2004 July	Rolling down of boulders from top 200 m above road level along slopes	Construction of culvert Retaining wall, Wire crated wall with jute mesh	Total number of incidence occurred. Four road was closed for 5–6 days. Size of boulders 2 × 2 × 3 and 7 × 10 × 3

Table 3. Geometrical parameters of slide.

Base—Width at road level	300–350 m
Width at center	180 m
Crown—Width	70 m
Inclined Length (from River Bed to Crown of slide)	800 m (approx)
Slope from river bed level to point A	35°
Slope from Point A to B	25°
Slope from point B to C (Crown)	>60°

Table 4. Wedge analysis.

Wedge No.	Intersection of Joint set	Amount and direction of plunge	Relationship with slope stability (on southern face)
1	1 and 2	50° S 50° W	Forms slope most unstable as the axis of wedge plunge towards open face
2	2 and 3	44°: S 33°W	Forms unstable slopes as axis of wedge plunge towards face by angle of <45°
3	1 and 3	50°: S	Forms stable slopes as wedge is dipping inside the face

springs emerging from the escarpments on southern face of slide. The spring mostly emerges from the contact plane between shale and sand stone.

5.2 Discontinuity analysis

The prominent sets of geological discontinuities observed at road level are:

Joint set (1) Strike—N 50° W–S 50° E—Dip 50°: S 40° W

Joint set (2) Strike—N 20° E–S 20° W—Dip 65°: N 70° W

Joint set (3) Strike—N 70° E–S 70° W—Dip 50°: S 20° E

The simple stereographic projection method used for "Wedge—analysis" along slopes for evaluating slopes stability due to intersection of above mentioned joint planes indicates, that the following wedges are formed.

6 SLOPE STABILITY ANALYSIS AND ROCK FALL MECHANISM

The geological structures of the study area are characterized by strongly jointed rocks of variable strength mainly sand stone, and shales of Oligocene age. The synthesis of geological date indicates that the slope stability problems in the area are associated with surcharging of slopes during heavy precipitation and with inflow of ground water from fissures. A great loss of strength of the rock mass results, particularly in zone of weathering and causes its subsequent displacement along slopes.

1. The dislodged blocks from higher reaches rolls down along the slopes are mixed with debris and moves down along slopes, thus forming the slides as "Complex Slide". The schematic geological plan and sections are shown as Figures 2 and 3 respectively.
2. The slope analysis is divided in two parts

   i. Rockfall along the exposed rock faces of slide above 200 meter high along slopes with respect to road level.
   ii. Movement of pre-existing slide debris along the slopes due to high degree of saturation during high rainfall from June to Oct every year in the area. The total rainfall in the year 2005 was 1662.05 mm with maximum rainfall alone in the month of Aug 2005 718.

3. The Southern and Eastern part of slide at higher levels are having vertical to sub vertical slopes. At places escarpments of 20–30 meter high are seen. Along the vertical scarp faces, the sand stone with minor bands of shale are exposed. The shale bands are having a thickness of 25 cm to 50 cm and are very soft and fragile, where as sand stone are hard and massive.

   The rock falls in the area are guided by following factors.

   i. Orientation of geological discontinuities, and their relation with respect to exposed faces.
   ii. Increase in pore—water pressure, the heavy rainfall in area, and infiltration of water through open joints—causes changes in forces acting on a rock mass.
   iii. Erosion of surrounding materials during heavy rainfall, chemical degradation, or weathering of rock.

4. The field observations shows that the wedges formed by intersection of Joint set 1and 2, 2 and 3 have plunge towards SW, i.e, towards exposed face,

these joints are clay filled and are prominent with shales.

The blocks rolling from top of slopes, particularly from southern face causes rock fall hazards. The rock fall occur when a block is suddenly released from apparently sound face, of sand stone with thick layers of shale this is due to small deformities on surrounding rock mass. This phenomenon is anticipated when the forces acting across discontinuity planes, changes due to pore water pressure developed on account of presence of water with in joint planes which isolates a block from its original position due to reduction of shear strength between the contact planes of shale and sand stone. This phenomenon is attributed to long term detoriation due weathering along joint planes, dipping towards slopes and results the release of "Key-Blocks" causing rock falls of significant size, at some time larger blocks having diameters 7 × 3 × 3 m are also released from higher slopes.

Once the movement of a rock, on top of crown of slide scar, has been initiated, its trajectory is controlled by geometry of the slope. As the slopes are almost vertical, the southern face acts as "SKY JUMP", and imparts a high horizontal velocity to falling rock, causing it to bounce a very long way.

The downward slopes from 250 m up to river bed are covered by boulders of gravels, mixed with clay and this bed absorbs a considerable amount of energy of falling rock, and movement of boulders is reduced in many cases are even stops completely.

The slopes from river bed level to 200 m high are covered with loose pre-existing slided materials with gravels and boulders mixed with clay. Such type of materials is having very low cohesion and angle of internal friction. The material when charged with water forms a slurry, and has a tendency to flow along the slope with significant velocity, thus causing damage to protection measures provided at road level.

7 STUDY OF AREA AS PER ROCK FALL HAZARD SYSTEM (RHRS)

The Highway and railway construction in mountainous regions present a special challenge to engineers and geologists. The rock fall hazard rating system "RHRS" was developed by Pierson et al 1990. The following analysis of Sonapur slide as has been made as per Rock Fall Hazard Rating System.

The RHRS system does not include recommendation on actions to be taken for different rating this is because decisions on remedial action for specific slope depends upon many factors such as budget allocation for highway work. However slopes with rating

Table 5. Analysis as per—RHRS—parameters.

Parameter	Observation	Points
Slope height	100 ft	81
Ditch effectiveness	Good catchment	3
Average vehicle risk	25% of time	3
Percent decision of sight distance	Adequate sight distance 100% of low design value	3
Road way width widening paved shoulders	36 feet	9
Geological character	Discontinuous joints with random orientation	9
Rock friction	Clay-filling or slickensided joints	81
Structural condition	Occasional erosion features	9
Difference in erosion rates	Moderate difference	9
Block size	>4'	81
Quantity of rock fall/event	12 cubicyard	27
Climate and presence of water in slope	High precipitation period or continual water on slope	27
Rock Fall history	Many falls	27
Total score		369

less than 300 are assigned, a very low priority where as the slopes with rating excess 500 are identified as areas requiring urgent remedial measures. Since in the present case the total rating is more than 300 but less then 500, the area falls under category requiring long term planning for slope stabilization after detailed geological investigation.

8 REMEDIAL MEASURES ADOPTED

The following remedial measures have been adopted so sorto control the damage at road level time to time.

1. Construction of retaining wall at road level
2. Construction of culvert at road level
3. Diversion of drain at road level
4. Construction of check dams across perennial drains along slope.
5. Concrete pavement
6. Construction of chute along the slope

To keep the road free for vehicular movement through-out the year, as a part long term planning the road maintenance Engineers are proposing construction of twin landslide gallery, Alternatively construction of approx. 500 m long tunnel by passing slide—zone has also been recommended as a measure to the traffic smoothly by passing the slide zone.

9 CONCLUSION AND RECOMMENDATIONS

The rock falls, are due to falling of blocks of sand stone/shale from top i.e. near crown area of slide. Which are generated due to intersection of joints, with in sand stone and shale.

The pre-existing old landslide debris material lying from river bed level to a height of around 200 meters along slopes is mixed with clay and boulders. The loose blocks resting on saturated slopes increases the load on soil. Under these circumstances when the soil gets saturated with water, looses its cohesion and angle of internal friction added with increase of overlying weight, starts flowing with velocity along slopes. When such material with great momentum hits out at remedial structures such as retaining wall, breast wall, water chute concrete pavement and Geo-textile causes complete damage of these structures.

The studies carried out earlier has suggested construction of "Twin Slide Shed" at the road level, allow to pass the debris material over it, alternatively construction of approximately 500 meter long tunnel has been also recommended.

By studying the survey data and detailed section of land slide area. The protection measures which is appears to be more appropriate are

i. Construction catch pit at RL ± 200 m of (bigger dimension 25 × 10 × 5) to arrest the movement of water, boulders coming from up slopes
ii. Construction of cascading chute from catch pit to road level along perennial drainage.
iii. Construction of gabion wall at road level.
iv. Construction of culvert at road level.
v. Construction of chute from Invert level of culvert at road level to river bed for an inclined length of 50 m.

ACKNOWLEDGEMENT

Authors are thankful to Chairman and Managing director ICT Shri. K.K. Kapila for encouraging the authors to write the manuscript of this paper.

REFERENCES

Hoek & Bray J.W. 1981. *Rock slope engineering.* Revised third edition. The institution of Mining and Metallurgy, London.

Pierson L.A., Davis S.A. & Van Vickle R. 1990. *Rock fall hazard Rating system—implementation manual.* Federal highway administration (FHWA) Report FHWA OREG 90-91. FHWA Department of transportation.

The viscous component in slow moving landslides: A practical case

D.A. González, A. Ledesma & J. Corominas
Department of Geotechnical Engineering & Geosciences, Tech. University of Catalonia (UPC), Barcelona, Spain

ABSTRACT: The availability of continuous records of both velocities and groundwater table in monitored landslides has increased the interest of the scientific community for the dynamics of the slow moving landslides. In this paper we analyse the role of the viscous component in large landslides by using ten years record of monitoring data from the Vallcebre translational landslide, located north of Barcelona (Spain). Previous research showed that a viscosity term should be considered in order to reproduce the measured displacements in this landslide. We discuss here the hysteretical behaviour of the landslide velocity records observed during some acceleration events produced by the rise and withdraw of the groundwater table. The conceptual model we have used shows that a constant viscous component of the movement is not able to explain the hysteretic behaviour and that consequently, other mechanisms should be searched to explain such a behaviour.

1 INTRODUCTION

In many slow moving landslides a close relationship between landslide velocity and position of the groundwater level has been observed. Acceleration changes in these landslides are usually controlled by water table. Despite this evidence, research works describing in detail the dynamics of such kind of relations are still scarce. This is partly due to the difficulty in the past to measure, in a continuous manner, both the velocities and the position of the groundwater table.

Nakamura (1984) showed for a particular landslide in Japan that the amount of landslide movement in the rising limb of groundwater level is larger than that observed in the lowering limb for the same groundwater level. Figure 1 shows that points indicating the groundwater level at the time of measurement line up in a circle counterclockwise with time. A number at the side of each point indicates the month, and "e", "m" and "l" indicate early part, middle part and the last part of the month, respectively. The author compares the velocity of the landslide (amount of displacement in 10 days) in different events for one year. The landslide response is influenced by the different soil conditions and particular characteristics of each event.

Bertini et al (1984, 1986) and Picarelli (2004) performed a similar analysis, based on measurements in the Fosso San Martino landslide (Italy) suggesting different velocities for the rising and lowering limb of the piezometric levels (Figure 2).

Van Asch et al (2007) coincides with the latter authors saying that for a given groundwater level,

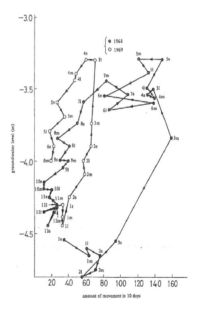

Figure 1. Groundwater level in relation to amount of landslide movement for a landslide in Japan. After Nakamura (1984).

velocities are higher when water table is increasing than when it is decreasing.

In this paper we analyse the relationship between groundwater level changes and the landslide displacements and velocities using data from the Vallcebre

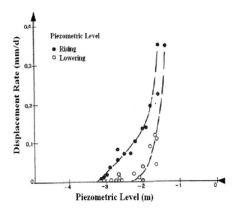

Figure 2. Displacement rate versus piezometric level for Fosso San Martino landslide. After Bertini et al (1986).

landslide. The analysis is restricted to acceleration events occurring in the period between November 1996 and August 1998. To check whether these results are consistent, finally a basic conceptual model is presented as well.

2 THE VALLCEBRE LANDSLIDE

2.1 General setting

The Vallcebre landslide is a large, active slope failure located in the upper Llobregat river basin, in the Eastern Pyrenees, 140 km north of Barcelona, Spain. The landslide is situated on the western slope of the Serra de la Llacuna.

The mobilised material consists of a set of shale, gypsum and claystone layers of continental origin gliding over a thick limestone bed, all of which are of Upper Cretaceous—Lower Palaeocene age. The dimensions of the slide mass are 1200 m long and 600 m wide. The entire landslide involves an area of 0.8 km^2 that shows superficial cracking and distinct ground displacements.

The toe of the landslide extends to the Vallcebre torrent bed, and is pushing it towards the opposite bank. As a result of this, the Vallcebre torrent has been shifted to the west more than ten meters and the foot of the landslide has overridden the opposite slope to form a back tilted surface. The torrent undermines the landslide toe during floods, causing erosion and local rotational failures which decrease the overall stability. A comprehensive description of the landslide is found in Corominas et al (2005).

2.2 Monitoring of the landslide

Since 1996, systematic recording of rainfall, groundwater level changes, and landslide displacements has been carried out every 20 minutes. Piezometric readings have indicated that changes in groundwater levels occur quickly. In-hole wire extensometers have recorded sudden changes in displacement rates that can be directly related to the fluctuations of the water table which is governed by rainfall (Corominas et al 1999).

The wire extensometer measurements show that the landslide has never stopped completely. It has been moving since we started the continuous monitoring in November 1996, although velocities slow down significantly during dry periods (Corominas et al 2000). On the other hand, the history of displacement of the extensometers reflects that different parts of the landslide mass move synchronically but with a different rate of displacement.

2.3 Hydrological changes and landslide response

The data show that groundwater reacts almost immediately to rainfall inputs, suggesting that water infiltration is controlled by fissures and pipes rather than by soil porosity. The role of the karstic network in the gypsum lenses is unclear but all the observed field features are very shallow (up to 3 m depth), which is well above of the normal groundwater level fluctuation. Because of this, we have assumed that karstic network (piping) play only a secondary role.

A close relationship between the groundwater level changes and landslide activity was observed at borehole S2 (Figure 3). There exists a strong level of synchronism between the two records.

Figure 4 shows an interesting relationship between observed velocities and the depth of water table at borehole S2 for the period considered. A cubic curve may be fitted to the data. These data are going to be analyzed thoroughly by considering the landslide acceleration events separately. Each event is defined by the rising and lowering groundwater level limb.

Note in Figure 4, that when depth of groundwater table is close to 6 m, velocities tend to be nil. That is, there is a level of water table below which landslide stops.

In a previous analysis (Corominas et al 2005), we have considered that beside frictional resisting forces, additional resisting forces (i.e. viscous forces) were necessary to explain the rate of displacement of the landslide.

3 ANALYSIS OF THE DATA

3.1 Field data observations

The data shown in Figure 4 correspond to different rainfall events occurred in the period considered without discrimination whether the groundwater level is in a rising or lowering limb.

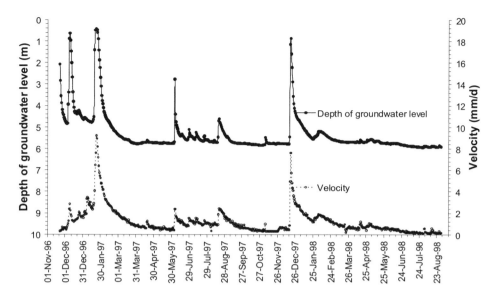

Figure 3. Velocities and water table depths for January, February and March 1997.

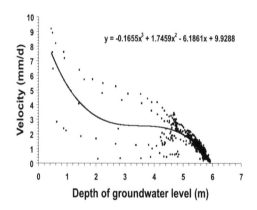

Figure 4. Velocities versus water table depths for November 1996 to August 1998 period. Data correspond to mean daily values.

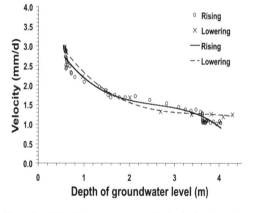

Figure 5. Velocities versus water table depths for December 1996.

Figure 5 shows the relationship between groundwater level changes and landslide velocities for one of the biggest rainfall events of the study period, that of December 1996. This figure shows that velocities in both limbs (rising and lowering) are very close. The difference between two points with the same groundwater level is negligible.

However, other events of the same period of data show a completely different behaviour. Figures 6 and 7 show the changes in velocities and groundwater table for the events of January–March 1997 and December 1997–January 1998. Figure 6 shows that the landslide velocities of the rising limb and those of the lowering limb do not coincide. Velocities of the rising limb are slower than the velocities in the lowering limb which is opposite to what has been found by authors mentioned previously (in particular, Nakamura, 1984, Bertini et al., 1986).

This particular behaviour was also observed in other events (Figure 8) mainly for small rainfall events where the difference between the values of velocity for the same groundwater level is very small.

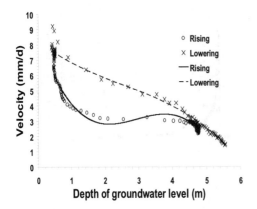

Figure 6. Velocities versus water table depths for January, February and March 1997.

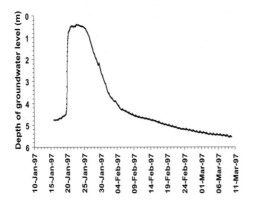

Figure 7. Groundwater table depths versus time. Event for January, February and March 1997.

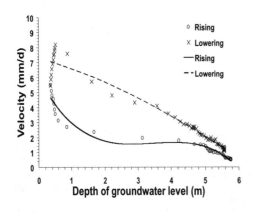

Figure 8. Velocities versus water table depths for December 1997 and January 1998.

At the beginning of the event, velocities increase gradually. Close to the maximum level of water table, velocities increase very quickly. Velocities are still high when water table starts to withdraw but eventually they decrease as well.

Once the rainfall has stopped the pore water pressures start to decrease and also the velocity with a decreasing rate which is different to the one during the rising limb.

Therefore, a preliminary conclusion for this landslide is that the landslide-velocity response to the rise and withdraws of water table depends on the initial hydrological conditions of the ground and on the magnitude of the event.

3.2 Theoretical Analysis

An attempt to simulate the landslide hydromechanical behaviour was considered, and for that purpose, a basic conceptual model based on the classical equilibrium differential equation was developed:

$$\sum F = m \cdot \frac{d^2u}{dt^2} + C \cdot \frac{du}{dt} + k \cdot u \quad (1)$$

where F = Equilibrium forces; m = mass; u = displacement; t = time; C = damping coefficient related to viscosity; and k = stiffness.

The analysis is based on the equilibrium of a infinite slope, using the basic equations described in Corominas et al (2005). A representative soil block with unit length and width and the initial parameters presented in Table 1 were used for the analyses. The stiffness was assumed nil and the soil parameters were obtained from laboratory experiments simulating residual conditions on ring shear equipments allowing for large displacements.

Two different situations were considered: i) Model including a viscous component; and ii) Model without a viscous component (i.e., C = 0).

In Figure 9 the relationship between velocities and depth of groundwater level taking into account a viscous component is shown. Note that with this model,

Table 1. Initial model parameters from the Vallcebre landslide.

Setting	Value
Damping Coefficient (C)	2.46×10^{11} Ns/m
Stiffness (k)	0.0
Slope angle (α)	10°
Friction angle (ϕ)	14°
Cohesion (c')	0.0

the response of velocities to the rising and lowering limb of groundwater level is the same.

The results corresponding to the other case analyzed is shown in Figure 10. Now the viscosity term has been neglected. Note that in the rising limb the velocity is increasing and in the lowering limb continues increasing without deceleration of the mass movement when the groundwater level is low. That is, the viscosity provides with a mechanism for dissipating energy in the system, and eventually helps reducing velocity and stabilising the landslide.

This is the result of a simple analysis and a simple model, but it may be useful to understand the dynamics of the landslide and the effect of the viscosity term in the movement.

According to this conceptual model, a constant viscosity will give a "reversible" response in terms of velocity versus groundwater table. Therefore, a hysteretic behaviour seems to be related either with a non-constant viscosity or with another effect not considered in the conceptual model. For instance, in Vallcebre the toe of the landslide is eroded by the Vallcebre torrent in the case of heavy rains and this may affect the equilibrium conditions. In such particular situations, one may expect a different behaviour of the landslide in the rising limb or in the lowering limb of the velocity curve.

4 CONCLUSIONS

The recorded data available for Vallcebre landslide has been used in this work to analyse the dynamics of the movement during rising and lowering of the ground water table. The velocity of the movement has been considered as main variable to be analysed as dependent of the ground water table level.

In some rainfall events the velocities in the rising limb are very close or the same that in the lowering one and comparing this case with the theoretical analysis implies that the mechanism of slope movement has to include a viscous component. However, other events have shown a different behaviour with velocities in the rising limb lesser than velocities in the lowering limb.

A simple conceptual model based on the infinite slope analysis was considered to understand the relationship between velocities and groundwater table. It was found that for a single event one should expect a "reversible" behaviour of velocities, and that has been measured in many situations in Vallcebre.

However, sometimes a hysteretic effect has been observed in single events, a situation that has been reported by other authors as well. That may be due to a non constant value of the viscosity, or to changes in the conditions acting on the landslide, i.e. the toe of the moving mass has been eroded by the torrent and therefore the conditions when rising or lowering the water table can not be compared directly.

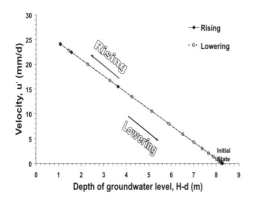

Figure 9. Results from the model: velocities, u' versus water table depths, H-d, including a viscous component.

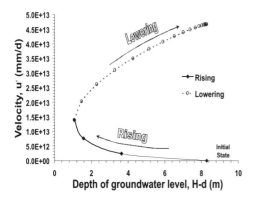

Figure 10. Results from the model: velocities, u' versus water table depths, H-d, without a viscous component.

ACKNOWLEDGEMENTS

First author is on leave from University Centre-occidental "Lisandro Alvarado" from Barquisimeto, Venezuela and express his thanks for the financial support. Moreover, grant provided by Gran Mariscal de Ayacucho Foundation (FUNDAYACUCHO) is also recognized.

This research work has been supported by the Spanish Science and Education Ministry (MEC), project number CGL2005-05282 (MODEVALL) and by the Institute of Geomodels (UPC-UB-CSIC).

REFERENCES

Bertini, T., Cugusi, F., D'Elia, B. & Rossi-Doria, M. 1984. Climatic conditions and slow movements of colluvial covers in central Italy. Proceeding of the 4th International Symposium on Landslides. Toronto, Canada. 1: 367–376.

Bertini, T., Cugusi, F., D'Elia, B. & Rossi-Doria, M. 1986. Lenti movimenti di versante nell'Abruzzo Adriatico: Caratteri e criteri di stabilizzazione. Proc. XVI Convegno Nazionale di Geotecnica, Bologna maggio 1986: 91–100.

Corominas, J., Moya, J., Ledesma, A., Rius, J., Gili, J.A. & Lloret, A.1999. Monitoring of the Vallcebre landslide, Eastern Pyrenees, Spain. Proceedings Intern. Symp. on Slope Stability Engineering: IS-Shikoku'99. Matsuyama. Japan, 2: 1239–1244.

Corominas, J., Moya, J., Lloret, A., Gili, J.A., Angeli, M.G. & Pasuto, A. 2000. Measurement of landslide displacements using a wire extensometer. Engineering Geology, 55: 149–166.

Corominas, J., Moya, J., Ledesma, A., Lloret, A. & Gili, J.A. 2005. Prediction of ground displacements and velocities from groundwater level changes at the Vallcebre landslide (Eastern Pyrenees, Spain). Landslides 2: p. 83–96.

Nakamura, H. 1984. Landslides in silts and sands mainly in Japan. Proc. IV Int. Symp. On Landslides, Toronto 1984, 1: 155–185.

Picarelli, L., Urciuoli, G., & Russo, C. 2004. Effect of groundwater regime on the behaviour of clayey slopes. Canadian Geotechnical Journal, 41: 467–484.

Van Asch, Th.J.W., Van Beek, L.P.H. & Bogaard, T.A. 2007. Problems in predicting the mobility of slow-moving landslides. Engineering Geology 91: 46–55.

The systematic landslide investigation programme in Hong Kong

K.K.S. Ho & T.M.F. Lau
Geotechnical Engineering Office, Civil Engineering and Development Department, Hong Kong SAR, China

ABSTRACT: The Geotechnical Engineering Office (GEO) has been collecting data and conducting annual reviews of rainfall and landslides since the 1980s. During this time, significant landslide incidents have been selected for detailed study for the purposes of advancing the understanding of landslides. Under the GEO's systematic landslide investigation programme which was implemented in 1997, all reported landslides are examined and significant landslide cases selected for study to document the failure, establish the probable causes and identify the lessons to be learnt and the necessary follow-up actions. This paper provides an overview of the systematic landslide investigation programme in Hong Kong.

1 THE GEOTECHNICAL ENGINEERING OFFICE

Hong Kong is vulnerable to landslides due to its hilly terrain with dense urban development, the presence of a large number of substandard man-made slopes mostly formed before the 1970s without adequate geotechnical input and control, deep weathering profiles and high seasonal rainfall. In the aftermath of several serious landslides with multiple fatalities, the Geotechnical Control Office (renamed Geotechnical Engineering Office (GEO) in 1991) was established by the Hong Kong Government in 1977 to regulate the planning, investigation, design, construction, monitoring and maintenance of slopes in Hong Kong.

Much of the enhanced slope engineering practice in recent years has originated from an improved understanding of landslides in Hong Kong. In particular, the systematic landslide investigation programme of the GEO, which was implemented in 1997, has played a key role in advancing the state of knowledge on slope performance and facilitated a better understanding of the causes and mechanisms of slope failures (Wong & Ho, 2000a).

2 THE SYSTEMATIC LANDSLIDE INVESTIGATION INITIATIVE

Between the 1980s and the early 1990s, significant landslides were selected for detailed studies by GEO's in-house professionals as research and development projects to enhance the understanding of causes and mechanisms of landslides. The technical findings from these selected landslide studies provided insights for improvement to slope engineering practice in Hong Kong (e.g. Hencher et al. 1984; Wong & Ho, 1995; Wong et al. 1998a, b).

As a result of the Kwun Lung Lau landslide on 23 July 1994, a new systematic landslide investigation (LI) programme was implemented by the GEO. This fatal landslide occurred on a 100-year old masonry wall located within a public housing estate, resulting in 5 fatalities and 3 serious injuries, and temporary evacuation of more than 3,900 residents (Figure 1).

The GEO carried out a comprehensive investigation into the causes and mechanism of the landslide (GEO, 1994) and an international geotechnical expert, Professor N R Morgenstern, was engaged by the Government to conduct an independent review of the technical investigation. The investigation established that thin masonry walls are liable to fail in a brittle manner without appreciable prior warning. The landslide also highlighted the adverse effects of leakage from buried water-carrying services on slope stability.

Figure 1. 1994 Kwun Lung Lau landslide.

These findings led to the issue of guidelines to rationalize the assessment of the stability of masonry walls and a Code of Practice on Inspection and Maintenance of Water-carrying Services Affecting Slopes (ETWB, 2006).

In his independent review (Morgenstern, 1994), Professor Morgenstern concluded that "Practice in Hong Kong with respect to evaluation of slope stability is excessively influenced in a restricted manner by the slope catalogue and is not sufficiently responsive to indications of potential problems on a project or development scale". One of his recommendations was for Government to introduce a more integrated approach into the slope stability assessment process through review of landslides. In response to this, a systematic LI programme was launched by the GEO in 1997.

Following a 3-year trial implementation to develop a new LI methodology for long-term use, the systematic LI work has been integrated with the Landslip Preventive Measures (LPM) Programme since 2000. The average annual cost of the systematic LI work is about HK$25 million (about US$3 million).

3 OBJECTIVES OF SYSTEMATIC LANDSLIDE INVESTIGATIONS

The main goals of the LI programme are illustrated in Figure 2 and described as follows:

a. identification of slopes in need of early attention before the situation deteriorates to result in a serious problem;
b. improvement in knowledge on the causes and mechanisms of landslides so as to formulate new ideas for reducing landslide risk and enhancing the reliability of landslide preventive or slope remedial works;
c. provision of data for reviewing the performance of the Government's slope safety system and identifying areas for improvement;
d. provision of evidence in forensic studies of serious landslides that may involve coroner's inquest, legal action or financial dispute.

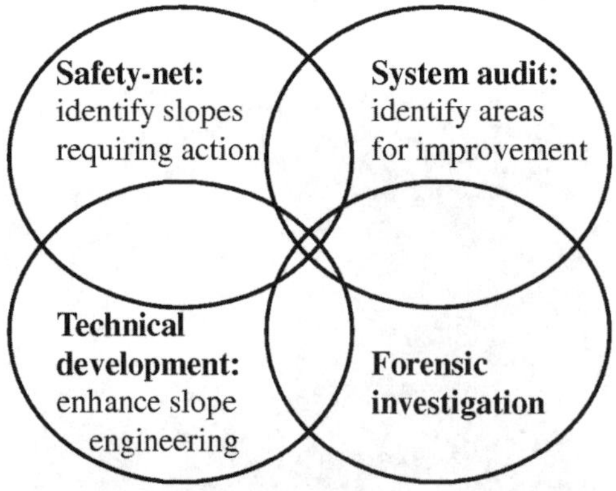

4 IMPLEMENTATION OF THE SYSTEMATIC LANDSLIDE INVESTIGATION PROGRAMME

In undertaking the investigation of significant landslides, it is important to attend to the sites as soon as practicable in order to collect crucial field evidence that could otherwise be destroyed or removed as part of the debris clearance operation or emergency repair works. Since a large number of landslides may occur within a short period of time during severe rainstorms (e.g. over 250 landslides were reported to the GEO during and immediately following the 19–22 August 2005 rainstorm), an adequate supply of standby resources is essential for the prompt mobilization of a sufficient number of investigation teams. The GEO has been outsourcing the LI work to consultants under a standby arrangement, which has worked well in meeting the operational needs. The advantage of engaging consultants to review the performance of Government's slope safety system through study of landslides is the impartiality of an independent party. This is especially important for forensic investigations of fatal landslides from a public accountability point of view. Also, overseas landslide experts can be mobilized as members of the landslide investigation teams on a need basis for serious landslides.

5 METHODOLOGY OF SYSTEMATIC LANDSLIDE INVESTIGATIONS

Under the LI programme, all reported landslide incidents are examined to collate data for analysis. The landslides are screened by a panel of experienced geotechnical professionals to identify cases that warrant follow-up inspections and detailed investigations. On average, about 300 landslides are reported to the GEO every year. About 20% of the cases would be selected for inspection by the LI consultants (the vast majority of the reported cases would be inspected by the GEO under the emergency system in providing advice to government departments). Typically, about 10% of the landslides are found to deserve follow-up studies.

Figure 2. Main goals of the landslide investigation programme.

The following are some of the relevant considerations in screening the landslide incidents for inspections and follow-up studies:

- large-scale failures;
- failures with serious consequences, e.g. casualties, major evacuation and significant social disruption;
- failures with technical interest, e.g. sites with special geological or hydrogeological features;
- failures involving slopes which were previously designed and checked to the required safety standards;
- failures of special engineered slopes, e.g. soil-nailed slopes;
- slopes with major signs of distress;
- slopes with landslide clustering or a history of repeated failures.

The following types of landslide studies are carried out under the LI programme:

a. Landslide Examination—all the available information on landslide incidents are examined shortly after they are reported to collate data for analysis and the identify cases which deserve further studies.
b. Landslide Review—these cover salient aspects of selected landslide and focus on the most important elements of the incident. This type of study is particularly relevant where the incident in itself does not warrant a detailed landslide investigation, and will enhance cost-effectiveness and ensure more effective use of resources.
c. Landslide Study—these comprise in-depth studies of selected landslides examination of the history of the failed slope and identification of the causes and mechanisms of failure.
d. Forensic investigation—these comprise detailed investigations of fatal or serious landslides to the highest possible rigour of proof in order to prepare a report that can be presented as evidence in legal proceedings.

In order to retain in-house expertise in landslide investigations, a small number of landslides continue to be studied by the GEO. In addition to the studies of individual landslides, a diagnostic review of all the landslide data and findings from landslide studies is carried out by the GEO every year to consolidate experience and make recommendations to enhance slope engineering practice and landslide risk management. Integrated thematic studies (e.g. review of slope surface drainage with reference to landslide studies, review of landslides at active construction sites, review of landslides involving slopes affected by water-carrying services, review of soil-nailed slope failures, etc) are also conducted.

6 KEY FINDINGS FROM SYSTEMATIC LANDSLIDE INVESTIGATIONS

6.1 Performance of engineered slopes

There was a perception in the past that the slope safety problem in Hong Kong was dominated by failure of old substandard slopes formed before 1977 and that the stability of engineered slopes built after 1977 to a high safety standard should be of little concern. An important development in slope engineering practice in Hong Kong over the more recent years has been the realization of the fact that even engineered slopes have quite a high failure rate and that there is a need to further improve the practice in order to reduce the rate of failure.

From the systematic review of the landslide data between 1997 and 2006, the annual average failure rate for engineered slopes is about 0.015% for 'major' landslides (defined as landslide with a failure volume $\geq 50 \text{ m}^3$), and about 0.068% for minor landslides (viz. $<50 \text{ m}^3$). The overall failure rate of engineered slopes in terms of major failures is lower than that for the pre-1977 slopes (i.e. slopes formed before the establishment of the GEO) by a factor of about 2 to 3 (Figure 3).

6.2 Causes of failures of engineered slopes

Wong & Ho (2000a) have summarised the common mechanisms of failures of man-made slopes with fast-moving debris (which include fill slopes, soil cuts and rock cuts) and some technical observations on notable generic factors contributing to landslides in Hong Kong, based on the findings of landslide studies. The key contributing factors to failure on engineered soil cuts have been discussed by Wong (2001). These factors comprise adverse groundwater, weak geological materials and inadequate slope maintenance.

A total of 56 landslides that occurred on engineered soil cut slopes between 1997 and 2006 have been investigated in detail to diagnose the key contributory

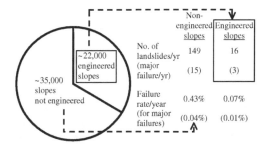

Figure 3. Landslide data in Hong Kong (1997–2006).

causes of the failures. The findings are summarized in Table 1 below.

About 80% of the landslides on engineered slopes from 1997 to 2006 were minor failures. The main problems with respect to minor failures of engineered slopes are as follows:

– uncontrolled surface runoff,
– inadequate slope maintenance,
– poor detailing of slope surface cover and surface drainage provisions (Figure 4), and
– local weaknesses in the ground mass, or a combination of the above factors.

With regard to major failures of engineered slopes, the main contributing factors are related to:

a. inadequate consideration of 'stability-critical' features in the geological models, such as relict discontinuities, sheeting joints, kaolin-rich seams in the weathered profiles, and kaolin and manganese oxide infill with discontinuities (Figure 5),

Table 1. Key contributory factors of engineered soil cut failures.

Contributory factor	All failures (56 nos.)	Minor failures (<50 m³) (40 nos.)	Major failures (≥50 m³) (16 nos.)
Adverse groundwater condition	30% (17 nos.)	38% (15 nos.)	75% (12 nos.)
Weak geological materials	46% (26 nos.)	40% (16 nos.)	67% (10 nos.)
Inadequate slope maintenance	45% (25 nos.)	55% (22 nos.)	19% (3 nos.)
Inadequate surface drainage provisions	5% (3 nos.)	7.5% (3 nos.)	–
Uncontrolled, concentrated surface water flow	2% (1 nos.)	2.5% (1 nos.)	–

Figure 4. Surface runoff overflowing off road.

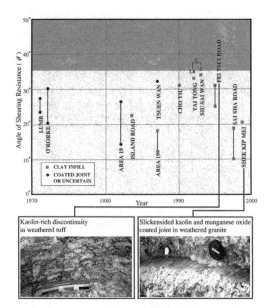

Figure 5. Selected landslides involving relict-jointed saprolite.

b. adoption of inadequate hydrogeological models in slope design leading to under-prediction of groundwater pressures,
c. progressive slope deterioration and displacement as evident by signs of distress.

The above diagnosis emphasizes the need to further improve the design practice and detailing so as to further enhance the reliability of engineered slopes. Key areas that warrant attention are highlighted, which include inadequate consideration of local weaknesses in the groundmass, inadequate engineering geological input during investigation, design and construction leading to the adoption of inadequate geological and hydrogeological models, insufficient attention to history of instability, inadequate consideration of overall site setting in an integrated perspective, uncontrolled surface runoff, poor detailing in slope drainage provisions and inadequate slope maintenance.

7 BENEFITS OF THE SYSTEMATIC LI WORK

7.1 Landslide data

The comprehensive and good quality landslide data collected from the systematic LI work have provided a valuable source of information to facilitate the GEO to make major technical advancements in the following areas:

a. assessment of landslide debris mobility of man-made slope and natural terrain failures (Wong et al. 1997; Wong, 2001)

b. the application of quantitative risk assessment (QRA) to quantify the landslide risk (Wong & Ko, 2006), and
c. continual refinement of the rainfall-landslide frequency correlation model for landslide risk management actions (e.g. criteria for issue of landslip warnings) (Yu, 2004).

All of the above have contributed to the enhancement of the slope safety system in Hong Kong.

7.2 Enhancement of technical knowledge on landslides

The major advances made as a result of landslide studies include:

a. improved understanding of the progressive nature of some of the slope failures,
b. consideration of failure mechanisms and debris movement mechanisms in the assessment of mobility of landslide debris (Figure 6),
c. importance of subsurface water (e.g. perched water) as well as surface runoff in triggering landslides,
d. the significance of the range of adverse geological features and the need to account for the potential for local as well as large-scale failures in slope design,
e. importance of robustness in slope design to combat uncertainties in geological and groundwater conditions.

The findings from landslide investigations highlighted the following points as indicators of potentially difficult sites that may have complex geological/hydrogeological conditions (Wong & Ho, 2000b):

- sites with relict massive failures,
- evidence of progressive slope movement and deterioration,
- slopes with a history of failure despite having been assessed or designed to the required geotechnical standards,
- planar geological features (such as joints, faults, weak seams, bedding, foliation, planar soil-rock interface), especially where they are dipping out of the slope, laterally persistent, show evidence of previous movement, associated with weak materials such as kaolin, and affect groundwater flow,
- evidence of high groundwater, or seepage at high levels, associated with drainage valleys, subsurface drainage concentrations (e.g. depression in weathering front), dykes or persistent sub-vertical discontinuities,
- complex groundwater conditions with a significant response or delayed response to rainstorms,
- large cuttings in a deep weathering profile.

7.3 Improvements to the slope safety system and engineering practice

Based on lessons learnt from landslide studies, the following improvement measures have been developed:

a. improved detailing for subsurface drainage provisions for recompacted soil fill slopes,
b. improved detailing for surface drainage provisions in slope designs,
c. improved detailing for soil-nailed slopes,
d. technical guidelines for the enhancement of rock slope engineering practice, and
e. technical guidance to enhance the reliability and robustness of engineered soil cut slopes.

8 OUTPUT OF THE SYSTEMATIC LANDSLIDE INVESTIGATION PROGRAMME

Examples of work completed under the LI consultancies to date include about 700 landslide inspections (out of more than 3,000 records examined), about 200 landslide studies, including six forensic investigations.

The findings of all landslides studies are published in a series of reports, which are distributed widely to the local profession. The key findings and lessons learnt from landslide studies are also made available on the Government's slope safety website http://hkss.cedd.gov.hk.

9 CONCLUSIONS

Based on the lessons learnt from landslide investigations, improvements have been made to the slope engineering practice to mitigate small and large-scale failures of engineered slopes (Ho et al. 2003). The

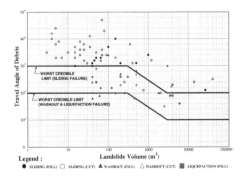

Figure 6. Data on debris mobility for different mechanisms and scale of landslides in Hong Kong.

systematic LI programme has resulted in improved reliability of engineered slopes, improved slope detailing and enhanced geotechnical and engineering geological input for slopes. It has served as an important asset management tool and will continue to form an integral part of the Government's long-term slope stabilisation programme.

ACKNOWLEDGEMENTS

This paper is published with the permission of the Head of the Geotechnical Engineering Office and the Director of Civil Engineering and Development, Government of the Hong Kong Special Administrative Region.

REFERENCES

Environmental, Transport and Works Bureau. 2006. *Code of Practice on Monitoring and Maintenance of Water-carrying Services Affecting Slopes*. Environmental, Transport and Works Bureau. The Government of the Hong Kong Special Administrative Region. Second Edition. November 2006, 93 p.

Geotechnical Engineering Office. 1994. *Report on the Kwun Lung Lau landslide of 23 July 1994. Vol. 2—Findings of the Landslide Investigation*. Geotechnical Engineering Office, Hong Kong, 379 p.

Hencher, S.R., Massey, J.B. & Brand, E.W. 1984. Application of back analysis to some Hong Kong landslides. *Proceedings of the Fourth International Symposium on Landslides*, Toronto, vol. 1, pp 631–638.

Ho, K.K.S., Sun, H.W. & Hui, T.H.H. 2003. *Enhancing the Reliability and Robustness of Engineered Slopes*. Geotechnical Engineering Office, Civil Engineering and Development Department, HKSAR Government, 63 p. (GEO Report No. 139).

Morgenstern, N.R. 1994. *Report on the Kwun Lung Lau landslide of 23 July 1994. Vol. 1—Causes of the Landslide and Adequacy of Slope Safety Practice in Hong Kong*. Report prepared for the Hong Kong Government, 21 p.

Wong, H.N. 2001. Invited Paper—*Recent Advances in slope engineering in Hong Kong*. Geotechnical Engineering, Ho & Li (Eds), 2001, pp 641–659.

Wong, H.N. & Ho, K.K.S. 1995. *General Report on Landslips on 5 November 1993 at Man-made Features in Lantau*. Geotechnical Engineering Office, Hong Kong, 78 p. & 1 drg, (GEO Report No. 44).

Wong, H.N. & Ho, K.K.S. 1996. Travel distance of landslide debris. *Proceedings of the Seventh International Symposium on Landslides*, Trondheim, Norway, vol. 1, pp 417–422.

Wong, H.N. & Ho, K.K.S. 2000a. Learning from slope failures in Hong Kong. *Proceedings of the 8th International Symposium on Landslides*, Cardiff, Bromhead et al (Eds.), Thomas Telford.

Wong, H.N. & Ho, K.K.S. 2000b. *Review of 1997 and 1998 Landslides*. Geotechnical Engineering Office, Hong Kong, 53 p (GEO Report No. 107).

Wong, H.N. & Ko, F.W.Y. 2006. *Landslide Risk Assessment—Application and Practice*. Geotechnical Engineering Office, Hong Kong, 277 p (GEO Report No. 195).

Wong, H.N., Ho, K.K.S. & Chan, Y.C. 1997. Assessment of consequence of landslides. *Proceedings of the International Workshop on Landslides Risk Assessment*, edited by D.M. Cruden & R. Fell, Honolulu, February 1997, pp 111–149.

Wong, H.N., Ho, K.K.S., Pun, W.K. & Pang, P.L.R. 1998a. Observations from some landslide studies in Hong Kong. *Proceedings of the HKIE Geotechnical Division Seminar on Slope Engineering in Hong Kong, 1998*, edited by Li, K.S., Kay, J.N. & Ho, K.K.S., Hong Kong Institution of Engineers, pp 277–286.

Wong, H.N., Ho, K.K.S. & Ho, K.K.S. 1998b. *Diagnostic Report on the November 1993 Natural Terrain Landslides on Lantau Island*. Geotechnical Engineering Office, Hong Kong, 98 p. plus 1 drg. (GEO Report No. 69).

Yu, Y.F. 2004. *Correlations Between Rainfall, Landslide Frequency and Slope Information for Registered Man-made Slopes*. Geotechnical Engineering Office, Hong Kong, 109 p. (GEO Report No. 144).

General digital camera-based experiments for large-scale landslide physical model measurement

X.W. Hu
Engineering Faculty, China University of Geosciences, Wuhan, China
School of Civil Engineering and Architecture, Wuhan University of Technology, Wuhan, China

H.M. Tang
Engineering Faculty, China University of Geosciences, Wuhan, China

J.S. Li
School of Remote Sensing and Information Engineering, Wuhan University, Wuhan, China

ABSTRACT: Physical model experiment is one of the most effective methods for studying deformation and failure mechanisms of landslides. The traditional displacement sensors, which must be in contact with or be buried in the landslide model, can only monitor limited points and limited displacement range, while digital camera close-range photogrammetry can obtain 3D surface information by means of no-contact. In this paper, on the basis of the Direct Linear Transformation theory, a large number of general digital camera close range photogrammetric experiments on a 2D large-scale landslide model were conducted. The experiment results show that (1) for the same camera with about 2 to 4 Mega-Pixels resolution, the measure accuracy hardly increase with resolution of images; (2) the different resolutions of images also hardly affect measure accuracy when the pixel size of images from different type cameras is between about 2 to 4 mega; (3) the precision is high when convergent angle of two cameras is beyond 10 degree, which dose not change clearly with increasing of the angle, but it is low when the angle is below 10 degree; (4) the increasing of station numbers and image number per station does not improve the precision, but much more time and money to be spent; and (5) the best two-dimensional measure accuracy ranges between 0.56 mm and 0.77 mm when resolutions of cameras are between about 2 and 4 Mega-Pixels, but the accuracy is low when the resolutions are below 1.38 Mega-Pixels.

1 INTRODUCTION

In the last decades, landslides in China are increasing with the rapid development of engineering practices. Physical model test is one of the most effective methods for studying deformations and failure mechanisms of landslides. Presently, more and more researchers in China are using it to study landslides so as to well understand and control landslides. Deformation and failure process of a landslide model must be monitored, which are commonly measured by conventional sensors such as strain gauge and dial indicator (Hu, 2004). These sensors or some fittings must be positioned in the model and can only measure the deformation or displacement of one direction of the measuring point, and likely influence model test results.

General digital camera close-range photogrammetry can be used to get the surface information of a 2D landslide model (Hu, 2006). This photogrammetric technique measures the physical model by using two general digital cameras or images taken from at least two different positions of one general digital camera at close range, which has several advantages over the conventional surveying methods. The first advantage is the non-contact measuring, and the tested landslide model is not affected by the measuring device and the model do not have to be accessible. Secondly, all surface information of the 2D physical model can be achieved almost immediately and conserved enduringly along with digital camera images. Data processing can be implemented at any time thereafter. Thirdly, the measuring range can be adjusted to the measuring task. Finally, photogrammetry can provide several kinds of products such as images, 2D displacement and velocity, and displacement vector.

Li (2003) investigated close-range photogrammetric network with variant interior elements of general digital camera. But his studies did not combine with a large-scale physical model. On the condition of large-scale landslide physical model experiment, it is very

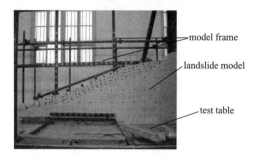

Figure 1. 2-D landslide physical model.

Figure 2. Layout of photogrammetry control and check points with white number mark.

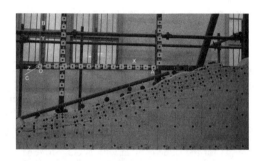

Figure 3. Spatial reference frame and target points for general digital camera close-range photogrammetry.

important to use general camera to achieve deformation characters of the landslide model. In this paper, the goal of the studies is to investigate the optimal parameters of general digital camera close-range photogrammetry monitoring a large-scale landslide physical model and verify the accuracy potential of this technique.

2 LANDSLIDE PHYSICAL MODEL

A 2D landslide physical model (Fig.1) based on the prototype of Zhaoshuling landslide in the Three Gorges Reservoir in China, is about 118.75 cm in height, 300 cm in length and 20 cm thick. The model was laid on a test table which could be raised. The model frame was fixed on the test table.

The landslide model was made of mixtures. The sliding mass was comprised of barite, fine sand, manganese carbonate, gypsum, water and engine oil. The sliding bedding was comprised of barite, fine sand, manganese carbonate and engine oil. Polyethylene membranes and polytetrafluoroethylene membranes were introduced as the sliding surface.

3 CLOSE-RANGE PHOTOGRAMMETRIC PHILOSOPHY

General digital camera close-range photogrammetric method is different from conventional displacement measure methods. Procedures of photogrammetry are as follows:

1. Measuring points, pasted on the model frame and the landslide model side (Fig. 2), were signalized with white papers consisting of a black annular target, white background and a small reticle target for total station instrument reference measurements. Enough control points and check points shown in Fig. 2, of which 3D coordinates (x, y, z) were surveyed with a SOKKI∧ SET-2C and a SOKKI∧ SET-3B, were located on the model frame and the sliding bedding side, respectively. Some of the high precise control and check points were made into the practical control points locating on the model frame and the model side to set up a 3D reference frame, and the others were made into both check points and target points so as to verify the precision and the reliability of the results from the general digital camera close-range photogrammetry. Figure 3 shows that the coordinate plane XY and the coordinate Z are in parallel with and perpendicular to the landslide model side, respectively.
2. Convergent photography was taken to gain a series of images with general digital cameras.
3. JPEG format of images is transformed into BMP format, then target points in the images are identified and their 2-D image coordinates (x, y) are obtained.
4. Based on the Direct Linear Transformation (DLT) analytical method, object spatial coordinates (X, Y, Z) of the measuring points are calculated.

The object spatial coordinates (X, Y, Z) of a target point can be calculated by the fundamental expression (1) and (2) (Feng, 2002) below:

Table 1. Photogrammetry with two stations of different convergent angles by a Kodak DC 4800 Zoom.

Serial of images	φ (°)	ω (°)	k (°)	x_0 pixel	y_0 pixel	f mm	XS mm	YS mm	ZS mm	Mean errors of measuring points (mm)		
										Mx	My	Mx,y
1	−18.81 / 28.55	−12.45 / 9.43	−4.98 / 1.33	905.04 / 928.16	−570.02 / −560.53	2837.38 / 1712.65	2539.12 / −238.77	704.84 / −575.44	4244.26 / 2335.74	0.36	0.47	0.57
2	−19.27 / 3.92	0.15 / 3.96	−2.32 / −7.49	908.40 / 912.93	−563.87 / −562.06	2229.76 / 1996.54	2305.42 / 947.20	−108.24 / −483.26	3307.48 / 3467.82	0.40	0.41	0.57
3	−19.24 / −7.48	6.85 / 5.22	−9.48 / −10.32	902.62 / 910.64	−564.98 / −563.64	2116.25 / 2117.87	2422.98 / 1711.27	−628.38 / −521.84	3336.21 / 3530.37	0.41	0.39	0.57
4	−18.80 / −0.57	−12.45 / −13.25	−4.98 / −2.34	905.04 / 912.89	−570.02 / −568.03	2837.38 / 2650.95	2539.12 / 1245.91	704.84 / 722.95	4244.26 / 4103.93	0.34	0.51	0.62
5	−0.57 / 3.92	−13.25 / 3.96	−2.34 / −7.49	912.89 / 912.93	−568.03 / −562.06	2650.95 / 1996.54	1245.91 / 947.20	722.95 / −483.26	4103.93 / 3467.82	0.40	0.49	0.64
6	28.55 / −29.80	9.43 / 9.98	1.33 / 2.32	928.16 / 905.41	−560.53 / −563.31	1712.65 / 1966.78	−238.77 / 2677.64	−575.44 / −634.09	2335.74 / 2749.34	0.42	0.49	0.65
7	−19.27 / −0.57	0.15 / −13.25	−2.32 / −2.34	908.40 / 912.89	−563.87 / −568.03	2229.76 / 2650.95	2305.42 / 1245.91	−108.24 / 722.95	3307.48 / 4103.93	0.41	0.51	0.65
8	−18.80 / 12.16	−12.45 / 10.08	−4.98 / −0.03	905.04 / 921.16	−570.02 / −561.75	2837.38 / 1509.74	2539.12 / 532.41	704.84 / −685.84	4244.26 / 2568.83	0.43	0.52	0.68
9	0.81 / −0.57	10.76 / −13.25	0.25 / −2.34	914.63 / 912.89	−559.73 / −568.03	1470.66 / 2650.95	1092.53 / 1245.91	−699.20 / 722.95	2624.98 / 4103.93	0.44	0.53	0.68
10	28.55 / 0.81	9.43 / 10.76	1.33 / 0.25	928.16 / 914.63	−560.53 / −559.73	1712.65 / 1470.66	−238.77 / 1092.53	−575.44 / −699.20	2335.74 / 2624.98	0.52	0.48	0.71
11	12.16 / −29.80	10.08 / 9.98	−0.03 / 2.32	921.16 / 905.41	−561.75 / −563.31	1509.74 / 1966.78	532.41 / 2677.64	−685.84 / −634.09	2568.83 / 2749.34	0.47	0.55	0.73
12	−19.27 / 0.81	0.15 / 10.76	−2.32 / 0.25	908.40 / 914.63	−563.87 / −559.73	2229.76 / 1470.66	2305.42 / 1092.53	−108.24 / −699.20	3307.48 / 2624.98	0.59	0.50	0.77
13	−19.27 / −29.80	0.15 / 9.98	−2.32 / 2.32	908.40 / 905.41	−563.87 / −563.31	2229.76 / 1966.78	2305.42 / 2677.64	−108.24 / −634.09	3307.48 / 2749.34	1.27	0.63	1.41
14	3.92 / 0.81	3.96 / 10.76	−7.49 / 0.25	912.93 / 914.63	−562.06 / −559.73	1996.54 / 1470.66	947.20 / 1092.53	−483.26 / −699.20	3467.82 / 2624.98	1.59	0.51	1.67

251

$$x + \frac{l_1 X + l_2 Y + l_3 Z + l_4}{l_9 X + l_{10} Y + l_{11} Z + 1} = 0 \quad (1)$$

$$y + \frac{l_5 X + l_6 Y + l_7 Z + l_8}{l_9 X + l_{10} Y + l_{11} Z + 1} = 0 \quad (2)$$

where $l_i (i = 1, 2, \cdots, 11)$ are factors. During calculating, optical aberrance should be taken into account, and the image coordinates (x, y) must be corrected (Li, 2003).

4 CLOSE-RANGE PHOTOGRAMMETRIC EXPERIMENTS AND RESULTS

4.1 Experiment preparation

The model table and the landslide physical model were static. The general digital cameras used to test involved a Canon Powershot45 Zoom camera with the images of 3.87, 1.92 and 0.78 M-Pixels, a Kodak DC290 Zoom camera with the resolutions of 3.36, 2.15 and 1.38 Mega-Pixels and a Kodak DC4800 Zoom camera with 2.16 Mega-Pixels resolution. At test, every camera was fixed on a tripod with the distance about 2 m∼4 m to the model.

4.2 Influence of the convergent angle of cameras on the measuring accuracy

A Kodak DC4800 Zoom camera with the image resolution of 1800 × 1200 (2.16 Mpixels) and a Kodak DC290 Zoom camera with the image resolution of 2240 × 1500 (3.36 Mpixels) were adopted. Two photographic stations with one image every station were applied to obtain 19 pairs of images with different convergent angles.

The results shown in Table 1 and Figure 4 from Kodak DC4800 Zoom camera indicate that when one of the angle element difference $\Delta\varphi$ and $\Delta\omega$ of every pair of images is excess of 10 degrees, the accuracy values of coordinates X and Y both reach about 0.34 mm∼0.59 mm and that of plane XY is 0.57 mm∼0.77 mm. But the measuring accuracy of coordinates X and Y is hardly improved with the increase of the angle element difference $\Delta\varphi$ and $\Delta\omega$. While the angle element differences $\Delta\varphi$ and $\Delta\omega$ are below or equal to 10 degrees, the accuracy value of coordinate X is over 1.00 mm and that of plane XY is over 1.40 mm.

The results shown in Table 2 and Figure 5 from the combination of a Kodak DC290 Zoom camera and a Kodak DC4800 Zoom camera indicate that when the angle element difference $\Delta\varphi$ of a pair of images is between 13 and 28 degrees and the angle element difference $\Delta\omega$ is between 3 and 5 degrees, the accuracy values of coordinates X and Y are about

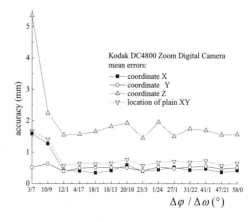

Figure 4. Relationship between measuring accuracy and the angle element difference $\Delta\varphi(\Delta\omega)$ of pairs of images from a Kodak DC4800 Zoom camera.

0.35 mm∼0.46 mm and 0.44 mm∼0.47 mm, respectively, and that of plane XY is 0.56 mm∼0.66 mm. While the angle element differences $\Delta\varphi$ and $\Delta\omega$ both are below or equal to 4 degrees, the accuracy value of coordinate X is over 0.90 mm and that of plane XY is over 1.00 mm.

4.3 Influence of the image resolution from same digital cameras on the measuring accuracy

A Canon Powershot45 Zoom camera and a Kodak DC290 Zoom camera were respectively used to make three team tests with two photographic stations, and one image every station was taken with convergent photogrammetry.

Figure 6 shows that for two images with same pixel of 1.92∼4.00 Mega, the accuracy value of coordinate X is about 0.3 mm∼0.4 mm, that of coordinate Y is about 0.5 mm∼0.6 mm and that of coordinate plane XY is about 0.65 mm. But, when two image resolutions are 0.78∼1.38 Mega-Pixels, the accuracy values of coordinates X and Y is about 0.54 mm∼0.65 mm and about 0.82 mm∼2.04 mm respectively, and that of coordinate plane XY is about 0.98 mm∼2.14 mm. The results illuminate that when the image pixel of general digital camera is about 1.92∼4.00 mega, the photogrammetric measuring accuracy value of coordinate plane XY is about 0.65 mm and hardly influenced by the change of image resolution, but when the image pixel is under about 1.5 mega, the measuring accuracy is low.

4.4 Influence of the image resolutions from different digital cameras on the measuring accuracy

The combination of a Canon Powershot45 Zoom camera and a Kodak DC290 Zoom camera was used to

Table 2. Photogrammetry with two stations of different convergent angles by the combination of a Kodak DC 290 Zoom camera and a Kodak DC 4800 Zoom camera.

Serial of images	φ (°)	ω (°)	k (°)	x_0 pixel	y_0 pixel	f mm	XS mm	YS mm	ZS mm	Mean errors of measuring points (mm)		
										Mx	My	Mx,y
1	8.26 / −19.96	2.06 / 7.11	−7.32 / −9.37	1139.405 / 900.915	−758.04 / −558.13	2455.87 / 2237.45	684.17 / 2431.52	−424.64 / −630.05	3224.58 / 3311.47	0.35	0.44	0.56
2	4.23 / −20.32	2.55 / 7.17	−7.35 / −9.377	1139.90 / 903.35	−755.68 / −560.08	2463.44 / 2232.13	932.45 / 2421.32	−462.24 / −629.76	3446.11 / 3287.34	0.42	0.44	0.60
3	4.24 / −8.98	2.46 / 5.63	−7.346 / −10.12	1140.75 / 906.04	−759.71 / −559.40	2455.14 / 2234.04	932.8 / 1874.13	−462.08 / −547.74	3435.54 / 3682.97	0.46	0.47	0.66
4	0.32 / −3.97	3.27 / 4.70	−7.64 / −10.38	1134.00 / 909.51	−746.60 / −561.94	2454.64 / 2232.14	1215.11 / 1536.39	−503.40 / −497.55	3526.49 / 3745.83	0.93	0.50	1.08
5	0.26 / −2.51	2.19 / 2.63	−1.86 / −1.036	1133.04 / 905.61	−755.79 / −563.66	2461.77 / 1906.43	1115.88 / 1368.97	−256.38 / −247.45	3514.34 / 3435.12	1.66	0.59	1.77

conduct five team tests with two photographic stations, and one image every station was taken with convergent photogrammetry.

Figure 7 shows that when the place and the pose of two cameras are constant and the two image pixels are 1.38~3.87 mega, the accuracy values of coordinate X, Y and plane XY are about 0.54 mm~0.64 mm,

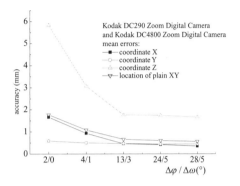

Figure 5. Relationship between measuring accuracy and the angle element difference $\Delta\varphi(\Delta\omega)$ of pairs of images from the combination of a Kodak DC290 Zoom camera and a Kodak DC4800 Zoom camera.

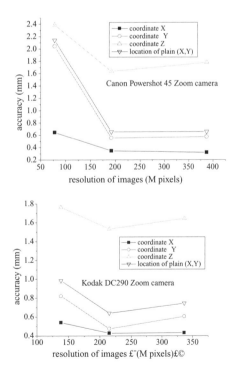

Figure 6. Relationship between measuring accuracy and image resolution from a camera.

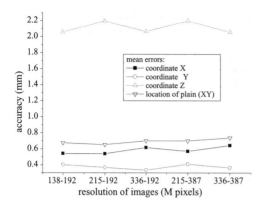

Figure 7. Relationship between measuring accuracy and image resolution from two-station different digital cameras.

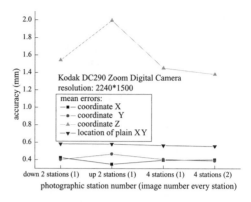

Figure 9. Relationship between measuring accuracy and the number of station and image.

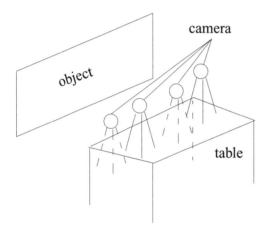

Figure 8. Sketch of 4 photographic stations.

0.33 mm~0.41 mm and 0.65 mm~0.74 mm, respectively, and hardly influenced by the change of two image resolution.

4.5 Influence of station and image number on the measuring accuracy

A Kodak DC290 Zoom camera was used to obtain images with the resolution of 2240 × 1500 (336 M pixels) and 1792 × 1200 (225 Mpixels). Four photographic stations (Fig. 8), two of which were set up on ground and a table respectively, were taken with convergent photogrammetry.

Figure 9 shows that when the place and the pose of camera are nearly constant and the image pixels are 3.36 mega, the accuracy values of coordinate X, Y and XY are about 0.35 mm~0.40 mm, 0.39 mm, 0.44 mm and 0.57 mm, respectively, and hardly influenced by photographic station number and image number every station.

5 DISCUSSION AND CONCLUSION

From the above studies on general digital camera close-range photogrammetry based on DLT and convergent photography for a large-scale landslide physical model, some conclusions drawn are as follows.

1. When the resolutions of images from general digital camera range between 1.92 M-pixels and 4.00 M-pixels, the measure accuracy is high and hardly influenced by the improving of image resolution. However, image resolutions below 1.38 M-pixels evidently reduce the measuring accuracy. And the accuracy is independent of the type of general digital camera.
2. Photogrammetric station numbers and image numbers per station hardly influence the measuring accuracy, which is different from the studied results by Dr. Li (2003).
3. When the angle element differences $\Delta\varphi$ and $\Delta\omega$ of images are below or equal to 10 degrees, the 2-D measure accuracy is low. While one of angle element differences is in excess of 10 degrees, the 2-D measuring accuracy is high and its value

reaches about 0.56 mm~0.77 mm, but it does not get higher because of the increase of angle element differences.

REFERENCES

Feng, W.H. 2002. *Close-range photogrammetry: measuration of shape and movement state of a object.* Wuhan: Wuhan University Publish Company.

Hu, X.W. 2004. Studies on similar material model experiment for landslide stability of Zhoushuling in the Three Gorges Reservoir. PhD thesis, China University of Geosciences.

Hu, X.W., Tang, H.M. & Li, J.S. 2006. Reliability of landslide model monitoring based on digital camera. *In proceedings of the Sixth International Conference on Physical Modelling in Geotechnics (6th ICPMG '06), Hong Kong, 4–6 August 2006,* Vol.1: 193–197.

Li, J.S. 2003. Research of digital close range photogrammetric network with variant interior elements. PhD thesis, Wuhan University.

Shear strength of boundaries between soils and rocks in Korea

S.G. Lee, B.S. Kim & S.H. Jung
Department of Civil Engineering, The University of Seoul, Seoul, Korea

ABSTRACT: In Korea, heavy rainfall each summer triggers many landslides with consequent loss of life and considerable damage to properties. Most of the landslides in mountain regions occur along the boundary between an overlying soil layer of weathered rock and colluviums and the underlying less weathered rock, regardless of rock type. This paper presents the results of a series of direct shear tests was performed to investigate the shear strength of the boundary between soils and rocks as well as the intact soil above the boundary under natural and saturated moisture conditions. Direct shear tests were conducted for a variety of igneous, sedimentary, and metamorphic rocks and the results are compared to identify the difference related to various rock types. In general it was found that the shear strength of boundaries between soil and rock are lower than the shear strength of the intact soil for all rock types. Furthermore shear strength was generally lower for the saturated condition compared to the natural moisture condition.

1 INTRODUCTION

Seventy percent of Korea is mountainous. Numerous landslides occur during the rainy season from June to September every summer resulting in on average 60 persons killed and hundreds of millions US dollars of property damage and traffic disruption (Lee et al, 2007). In the mountainous areas soil is typically thin around 1 m thickness, overlying rock. The majority of landslides occur along this boundary between the thin soils and rocks (Lee, 1987, 1988 & 1995; Lee et al., 2008).

Numerous landslides occurred during the passage of Typhoons Rusa in 2002 and Maemi in 2003 and, as is usually the case, most of the landslides that were triggered by these storms occurred on the boundary between soil and rock (Lee et al. 2008).

The purpose of the study reported in this paper was to investigate the shear strength of the boundaries between soil and rocks which is an important element for slope stability analysis and the geotechnical engineering of landslides.

In general, when analyzing slope stability, the shear strength of intact soil is adopted in calculations on the assumption that that shear strength is the minimum. This is despite other workers suggesting that the shear strength may be lower at the soil-rock contact surface compared to the intact soil (Patton, 1968). Similarly Kanji (1974) experimentally demonstrated that the shear strength on the boundary between weathered granitic soil and granitic rocks is lower than that of the soil itself.

As the landslides occur frequently at the boundary between soil and rock in Korea, irrespective of the kind of rock, shear strength of contacts between soil and rocks were studied using a variety of rock types.

2 SELECTION OF SURVEY AREAS AND METHOD OF STUDY

One hundred and seventy one landslides were surveyed in various areas listed in Table 1, where numerous landslides during rainfall associated with Typhoons Rusa in 2002 and Maemi in 2003 (Table 1). The general geology of the studied areas are shown in Figure 1 (Um & Reedman, 1975). It can be seen that the studied landslides occurred in terrain underlain by a variety of igneous, sedimentary and metamorphic rocks.

Table 1. Areas and numbers of landslides surveyed.

Surveyed areas		Numbers	Time of production
Igneous rock areas	Gangreung	80	1 Sep 2002
	Donghae	40	1 Sep 2002
	Taebaeg	6	30 Aug 2003
Metamorphic rock areas	Hamyang	26	1 Sep 2002
	Samcheog	6	30 Aug 2003
Sedimentary rock areas	Habcheon	6	30 Aug 2003
Total		171	

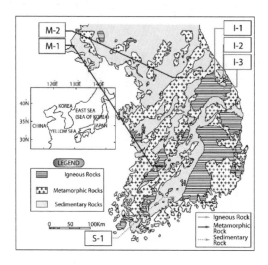

Figure 1. Locations and geological characteristics of studied areas.

Figure 3. Shear surface almost parallel to sheeting joint in igneous rock.

Figure 4. Shear plane mismatched with the direction of sheeting joint developed in igneous rocks.

Figure 5. Landslides produced parallel with the bedding plane of sedimentary rock area.

Figure 6. Landslide discordant with bedding discontinuities in sedimentary rock area.

Figure 2. Survey of landslides.

3 PROPERTIES OF SHEAR STRENGTH OF THE AREAS IN LANDSLIDE AREAS

3.1 Landslide Studies

In the study areas, the original slope angles, the length, depth, slope, and width of landslides, location of landslide, and strength characteristics of failure surfaces were recorded for more than 170 landslides (Figure 1) (Lee et al, 2008).

In granitic areas many landslides occurred parallel to unloading (sheeting) joints (Figure 3) but in some cases the shear surface was mismatched with sheeting joints (Figure 4). In sedimentary rocks, mainly comprising sandstone and shale, in some cases the shear plane is parallel to bedding (Figure 5) but in other cases there is a mismatch (Figure 6). Metamorphic rocks in the study areas mostly consist of gneiss; in these areas the interface between the detached soil horizon and underlying rock is generally not associated with any specific geological structure (Figure 7).

3.2 Preparation of rock and soil samples

About 400 samples were collected for each rock type using sampling rings to minimise disturbance. Shear tests were then conducted on the intact soil from above the slip surface and on samples representing the boundary between soil and rocks at different normal

Figure 7. In metamorphic rock areas slip surfaces generally are not associated with pre-existing geological structures.

Figure 8. Collection of rock and soil samples.

stress levels, for natural moisture content (as collected) and in a soaked condition. To perform the shear tests on the boundary between soil and rocks, rock samples were collected as representative as possible of small scale roughness of the areas where the landslides occurred (Figure 8).

3.3 Soil property test

Soil samples collected in each area were wrapped to prevent changes in moisture content and to minimise disturbance. The soil collected from each site was tested by graded and various index tests conducted including specific gravity, liquid limit, plastic limit, and water content (Head, 1980). Mean values were determined for each rock type.

3.4 Shear strength test

Shear tests were carried out at moisture content as recovered and in a soaked state. For soaked tests samples were packed with filter papers and then immersed in water for 5 days prior to testing. Shearing rate was set as 0.2 mm/min. Normal stress was given was varied over five steps from 30 kPa to 150 kPa and maximum shear displacement was set as 9 mm, i.e. 15% of the whole length of a sample. The tests on the boundary between soils and rocks were essentially the same as those on intact soils but in these cases a rock specimen was placed inside the lower ring.

3.5 Result of shear property tests and shear strength tests

Based on the result of soil property tests (Table 2, Figures 10, 11 & 12), unit weight of rock was $17.06 \sim 18.04 \, kN/m^3$, mean moisture content was 19%

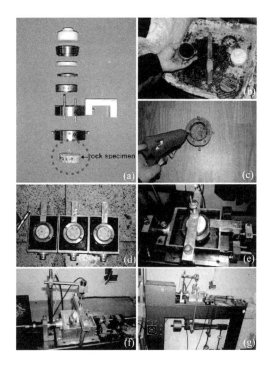

Figure 9. Details of direct shear strength test on boundary between soil and rock (a) Components of shear box (b) Rock sample placed in the lower part of shear box (c) Intact soil placed in the upper part (d) Porous plate placed above soil (e) Application of normal load (f) LVDTs for measuring vertical and horizontal displacement (g) Shear test underway.

in case of igneous rock, 18% in case of sedimentary rock, and 17% in case of metamorphic rock; the contents were high. The range of specific gravity was $2.56 \sim 2.71$ and soil classification through sieve analysis showed well-graded result. As shown in Table 2, as the result of direct shear strength tests, the c and Φ' values of boundaries between soils and rocks were smaller in all rock types compared to the values obtained from soil and the c and Φ' values were lower when the tests were performed natural moisture content condition compared to saturated condition. The fact that the shear strength of boundaries is lower than that of soil indicates the proof of the landslide collapse on the boundaries.

Also it could be identified that, when the internal friction angle of soil is assumed as 100% and when the internal friction angle lowering of boundaries is expressed in percent, the internal friction angles were lowered up to $71 \sim 94\%$ of natural moisture content and $82 \sim 98\%$ of saturated moisture content in case of igneous rocks, $74 \sim 96\%$ of natural moisture content and $89 \sim 99\%$ of saturated moisture content in case of

Table 2. Results of physical properties and shear strength tests on residual soils derived from various rock types.

Rock type				γ (kN/m³)	w/t (%)	G_s	LL (%)	PI (%)	USCS	C (kPa)	Φ (%)
Igneous rock	Breaking down by sheeting joints	Soil	dry	17.75	19.18	2.625	42.55	30.57	SP	18.5	24.3
			wet	18.04	23.51	2.624	42.55	30.57	SP	17.3	20.5
		Soil/Rock boundary	dry	17.95	21.56	2.606	42.55	30.57	SP	17.6	19.6
			wet	18.04	21.77	2.561	42.55	30.57	SP	19.6	18.8
	Breaking with no influence from sheeting joints	Soil	dry	17.26	17.82	2.607	31.88	11.56	SM	28.1	20.9
			wet	17.55	20.75	2.607	31.88	11.56	SM	27.1	20.5
		Soil/Rock boundary	dry	17.46	16.18	2.646	31.88	11.56	SM	26.7	19.1
			wet	17.85	19.12	2.705	31.88	11.56	SM	28.1	18.8
Sedimentary rock	Landslides produced parallel with bedding plane	Soil	dry	17.65	18.82	2.654	35.81	11.25	SM	28.6	22.4
			wet	17.65	19.04	2.614	35.81	11.25	SM	34.2	21.1
		Soil/Rock boundary	dry	17.65	19.87	2.647	35.81	11.25	SM	27.5	19.6
			wet	17.65	18.72	2.597	35.81	11.25	SM	29.7	18.4
	Landslides produced with quite mismatch with bedding plane	Soil	dry	17.75	18.94	2.603	45.54	13.24	SW	30.2	23.6
			wet	17.75	20.14	2.641	45.54	13.24	SW	27.4	22.1
		Soil/Rock boundary	dry	17.75	18.97	2.647	45.54	13.24	SW	29.7	20.5
			wet	17.65	19.37	2.640	45.54	13.24	SW	26.4	19.2
Metamorphic rock	Landslides produced regardless joins or foliation are the geological structure of metamorphic rocks	Soil	dry	17.65	18.68	2.598	40.30	19.76	SM	32.2	17.7
			wet	17.65	19.08	2.597	40.30	19.76	SM	28.8	15.8
		Soil/Rock boundary	dry	17.55	14.45	2.592	40.30	19.76	SM	32.7	17.4
			wet	17.65	14.72	2.640	40.30	19.76	SM	34.9	16.2
		Soil	dry	17.46	18.67	2.664	39.64	21.54	SM	31.6	21.5
			wet	17.65	18.98	2.564	39.64	21.54	SW	27.8	20.7
		Soil/Rock boundary	dry	17.06	18.28	2.578	39.64	21.54	SW	28.7	19.5
			wet	17.65	19.57	2.642	39.64	21.54	SW	27.4	18.3

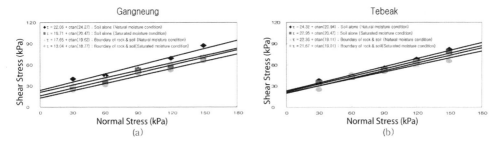

Figure 10. Shear strength of Igneous rocks: (a) Landscapes having the collapse matching with sheeting joint; (b) Landscapes having the collapse quite mismatching with sheeting joint orientation.

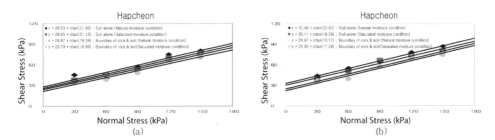

Figure 11. Shear strength of sedimentary rocks: (a) Landscapes having the collapse matching with bedding plane; (b) Landscapes having the collapse quite mismatching with bedding plane.

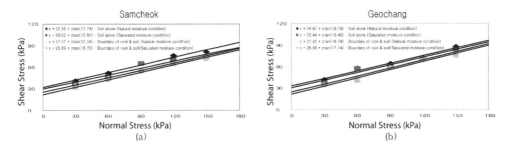

Figure 12. Shear strength of metamorphic rocks: (a) Landslides in metamorphic rocks in Samcheog area of Gangwon-do; (b) Landslides in metamorphic rocks in Geochang area of Gyeongnam.

metamorphic rocks, and 77~94% of natural moisture content and 82~98% of saturated moisture content in case of sedimentary rocks.

4 CONCLUSION

1. A survey of 117 landslides in Korea triggered by Typhoons Rusa in 2002 and Maemi in 2003 showed that most of the landslides involved sliding on the boundaries of soil and rocks.

2. As the result of direct shear tests, it was found that, when the internal friction angles of soil was assumed as 100% and the internal friction angles of boundary to the internal friction angles of soil were expressed in percent, the internal friction angles were lowered to 72%~96% in natural moisture content condition and to 82~94% in saturated moisture content condition.

3. The shear strength of boundaries was lower in all rock types compared to the shear strength of soil and, in both cases, lower shear strength was found in saturated moisture condition compared to natural

moisture condition. The fact that the shear strength of boundaries is lower than that of soil indicates the proof of the landslide collapse on the boundaries.

ACKNOWLEDGEMENTS

This research was partially supported by a grant (NEMA-06-NH-05) from the Natural Hazard Mitigation Research Group, National Emergency Management Agency.

REFERENCES

Head, K.H. 1980. Manual of Soil Laboratory Testing. VI, Soil Classification and Compaction Tests. V2, Permeability, Shear Strength and Compressibility Tests. Publ, London. Pentach Press.

Kanji, M.A. 1974. Unconventional Laboratory Tests for the Determination of the Shear Strength of Soil-Rock Contacts. *The 3rd Congress of ISRM*. Denver 2, pp. 241–247.

Lee, S.G. 1987. Weathering and geotechnical characterization of Korean granites, *PhD thesis*, Imperial College, University of London.

Lee, S.G., 1988. A study on landslide in Korea, Researches on geological hazards, *Research report of Korea* Institue of Geoscience and Mineral Resources (KIGAM), KR-88-(B)-7, 145–148.

Lee, S.G., 1995. Natural hazard in Korea. *Proc. Of the int. Forum on Natural hazard mapping, Geological Survey of Japan Report*, 281, 145–148.

Lee, S.G., Kim, M.S. & Park, D.C., 2008. A study on the characteristics of landslides related to various rock types in Korea. *Proc. 10th Int. Symp. on Landslides and Engineered Slopes*, Xi'an, China (in press).

Patton, F.D. 1968. The determination of shear strength of rock masses, Paper presented to the terrametric course on measurement systems of control of construction and mining, Denver: 37.

Um, S.H. & Reedman, A.J., 1975. *Geology of Korea*. Korea Institue of Geoscience and Mineral Resources (KIGAM), Seoul.

Cracks in saturated sand

X.B. Lu & S.Y. Wang
Institute of Mechanics, Chinese Academy of Sciences, Beijing, China

Peng Cui
Institute of Mountain Hazard and Environment, Chinese Academy of Sciences, Chengdu, China

ABSTRACT: The formation mechanism of water film (or crack) in saturated sand is analyzed numerically. It is shown that there will be no stable "water film" in the saturated sand even if the strength of the skeleton is zero and no positions are choked. The stable water films initiate and grow if the choking state keeps unchangeable once the fluid velocities of one position decreases to zero in a liquefied sand column. A simplified method for evaluating the thickness of water film is presented according to a solidification wave theory. The theoretical results obtained by the simplified method are compared with the numerical results and the experimental results of Kokusho.

1 INTRODUCTION

It is often occurred on the ground slope that sand deposit translates to lateral spreading or even landslide or debris flow not only during, but also after earthquakes. If the sand deposit on a slope are composed of many sublayers, there will be a water film forms once it liquefied (Kokush et al. 1998) which may serves as a sliding surface for postliquefaction failure. As a result, landslide or debris flow may happen on a slope with very gentle slope-angle. Seed (1987) was the first to suggest that the existence of water film (crack) in sand bed is the reason of slope failures in earthquakes. Some researchers (Fiegel, G.L. & Kutter, B.L. 1994, Kokusho, T. 1999, Zhang Junfeng 1998) performed some experiments to investigate the formation of crack in layered sand or in a sand containing a seam of non-plastic silt. Nevertheless, the mechanism of the formation of cracks or "water film" in a sand layer with the porosity distributed continuously is not very clear.

In the viewpoints above, Firstly, we present a pseudo-three-phase model describing the moving of liquefied sand and numerically simulates. Secondly, we present a simplified method to analyze the evolution of the water film.

2 FORMULATION OF THE PROBLEM

It is considered here a horizontal sand layer, which is water saturated and the porosity changes only vertically. The fine grains may be eroded from the skeleton and the eroding relation is assumed to be proportional to the velocity difference of grains and pore water (Cheng, C.M. et al. 2000). The x axis is upward.

$$\frac{1}{\rho_s}\left(\frac{\partial Q}{\partial t} + u_s\frac{\partial Q}{\partial x}\right) = \frac{\lambda}{T}\left(\frac{u - u_s}{u^*} - q\right)$$

if $-\varepsilon(x,0) \leq \dfrac{Q}{\rho_s} \leq \dfrac{Q_c(x)}{\rho_s}$ (1)

$$\frac{1}{\rho_s}\left(\frac{\partial Q}{\partial t} + u_s\frac{\partial Q}{\partial x}\right) \leq 0 \quad \text{otherwise} \quad (2)$$

in which Q is the sand mass eroded per unit volume of the sand/water mixture, ρ_s is the grain density, u and u_s are the velocities of pore fluid and sand grains, q is the volume fraction of sand carried in percolating fluid, T and u^* are physical parameters, λ is a small dimensionless parameter, $\varepsilon(x,t)$ is the porosity, $Q_c(x)$ is the maximum Q that can be eroded at x.

3 MODEL OF THE PROBLEM

The mass conservation equations are as follows:

$$\begin{cases} \dfrac{\partial(\varepsilon - q)\rho}{\partial t} + \dfrac{\partial(\varepsilon - q)\rho u}{\partial x} = 0 \\ \dfrac{\partial q\rho_s}{\partial t} + \dfrac{\partial q\rho_s u}{\partial x} = \dfrac{\partial Q}{\partial t} + u_s\dfrac{\partial Q}{\partial x} \\ \dfrac{\partial(1-\varepsilon)\rho_s}{\partial t} + \dfrac{\partial(1-\varepsilon)\rho_s u_s}{\partial x} = -\dfrac{\partial Q}{\partial t} - u_s \end{cases} \quad (3)$$

in which ρ is the density of water. A general equation may be obtained by eq. (3)

$$\varepsilon u + (1-\varepsilon)u_s = U(t) \quad (4)$$

in which $U(t)$ is the total mass of fluid and grains at a transect. The momentum equations are:

$$\begin{cases} [(\varepsilon-q)\rho + q\rho_s]\left(\dfrac{\partial u}{\partial t} + u\dfrac{\partial u}{\partial x}\right) \\ \quad = -\varepsilon\dfrac{\partial p}{\partial x} - \dfrac{\varepsilon^2(u-u_s)}{k(\varepsilon,q)} - [(\varepsilon-q)\rho + q\rho_s]g \\ \quad \times [(\varepsilon-q)\rho + q\rho_s]\left(\dfrac{\partial u}{\partial t} + u\dfrac{\partial u}{\partial x}\right) \\ \quad + (1-\varepsilon)\rho_s\left(\dfrac{\partial u_s}{\partial t} + u_s\dfrac{\partial u_s}{\partial x}\right) \\ \quad = -\dfrac{\partial p}{\partial x} - \dfrac{\partial \sigma_e}{\partial x} - [(\varepsilon-q)\rho + q\rho_s]g \\ \quad -(1-\varepsilon)\rho_s g - \left(\dfrac{\partial Q}{\partial t} + u_s\dfrac{\partial Q}{\partial x}\right)(u - u_s) \end{cases} \quad (5)$$

in which p is the pore pressure.

Here k is assumed as following

$$k(\varepsilon,q) = k_0 f(q,\varepsilon) = k_0(-\alpha q + \beta\varepsilon) \quad (6)$$

in which α, β are parameters and $1 < \beta << \alpha$, which indicates that changes in q overweighs that of ε.

The mass conservation equation (4) yield assuming both u and u_s are zero at $x = 0$.

$$\varepsilon u + (1-\varepsilon)u_s = U(t) = 0 \quad (7)$$

Taking T as characteristic time. u_t the characteristic velocity and L the characteristic length of the problem. We make eq. (1) non-dimensional. Letting

$$\bar{u} = \dfrac{u}{u_t}, \quad \tau = \dfrac{t}{T}, \quad \xi = \dfrac{x}{Tu_t} \quad (8)$$

Instituting equ. (1), (2), (4), (7) (8) into eqs. (3), we may obtain

$$\dfrac{\partial \varepsilon}{\partial \tau} + \dfrac{\partial \varepsilon \bar{u}}{\partial \xi} = \bar{u}\dfrac{u_t}{u^*(1-\varepsilon)} - q$$

$$\dfrac{\partial q}{\partial \tau} + \dfrac{\partial q\bar{u}}{\partial \xi} = \bar{u}\dfrac{u_t}{u^*(1-\varepsilon)} - q \quad (9)$$

For $Tg/u_t >> 1$, the inertia terms are negligible, the last equation of eq. (5) becomes

$$\bar{u} = \left(\dfrac{1-\varepsilon}{\varepsilon}\right)^2 (\varepsilon-q)f(q,\varepsilon)\dfrac{k_0\rho_s g(1-\rho/\rho_s)}{u_t}$$

$$= \left(\dfrac{1-\varepsilon}{\varepsilon}\right)^2 (\varepsilon-q)f(q,\varepsilon) \quad (10)$$

$$u_t = k_0\rho_s g(1-\rho/\rho_s) \quad (11)$$

The initial conditions are:

$$\varepsilon(\xi,0) = \varepsilon_0(\xi), \quad q(\xi,0) = 0. \quad (12)$$

4 NUMERICAL RESULTS

Numerical simulation is carried out based on eq. (8). The parameters adopted in simulation are as follows: $\Delta t = 9 \times 10^{-4}$, $\Delta \zeta = 5 \times 10^{-3}$, $\beta = 46 \sim 56$, $\kappa = 50.0$, $a = 0.08$, $\rho_s = 2400$ kg/m^3, $\rho_w = 1000$ kg/m^3, $u^* = 0.04$, $k_0 = 4 \times 10^{-6}$ m/s, $\alpha = 1$. The boundary conditions:

1. The initial porosity distribution is $\varepsilon_0(x) = \bar{\varepsilon}_0(1 - a\tanh((x-0.5L)/2)\cdot\kappa)$, in which $\bar{\varepsilon}_0 = 0.4$, $L = 1$, $0 \leq x \leq 1$, L is the length of sand column. There is an assumption that u keeps zero once it drops to zero.

2. The distribution of initial porosity is the same as that in condition 1, there is no assumption.

Figure 1 shows that if we assume that once the sand column at some point is jammed, they keep this state forever, then the sand above the jammed position will be prevented to drop cross the jammed point and so the porosity becomes smaller and smaller, while the sand below the point will settle gradually and makes

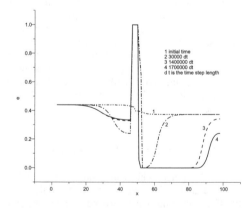

Figure 1. The evolution of cracks in the condition 1.

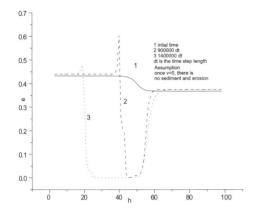

Figure 2. The evolution of cracks in condition 2.

the crack extends gradually. But if we do not adopt the assumption as in Figure 1, the crack will form first and then disappear gradually (Figure 2). The results show that the forming conditions of stable water film are: (1) the porosity of the upper part of the sand column must be smaller than that of the lower. (2) The keeping of the jamming state to prevent the free dropping of the grain or the skin friction.

5 A SIMPLIFIED EVALUATION METHOD

A simplified method is presented here for analyze the evolution of cracks fast and practically. Florin and Ivanov (1961) pointed out that when the settling particles reach solid material, such as the unliquefied underlying sand, or the container base in a experiment, they accumulate to form a solidified zone which increases in thickness with time. A solidification front therefore moves upward until it reaches the surface, or overlying unliquefied material. Scott et al. (1986) had analyzed the development of the solidification.

Assuming that the whole mass reaches its terminal velocity, k, which is the permeability, instantaneously at the end of liquefaction, Florin have given an expression for the constant velocity $\dot{z}$, of the solidification front:

$$\dot{z} = \frac{\rho}{\rho_w} \frac{1-n_1}{n_0-n_1} k \qquad (13)$$

in which $\rho' = \rho_s - \rho_w$ is the buoyant unit weight of the liquefied sand n_0 is the porosity of the liquefied sand, n_1 is the porosity of the solidified sand.

From eq. (13), we can obtained the duration of liquefaction and subsequent excess pore pressure decline for any point in the sand column.

$$t = \frac{\rho_w}{\rho} \frac{n_0-n_1}{1-n_1} \frac{h}{k} \qquad (14)$$

in which h is the height of any point in the sand column.

The final settlement of the top surface of the sand layer is

$$\Delta L = \frac{n_0-n_1}{1-n_0} H \qquad (15)$$

in which H is the maximum height of sand layer.

The rate of settlement is

$$\dot{s}_a = \frac{\gamma' k}{\gamma_w} \qquad (16)$$

The settlement at any time is

$$s_a = \frac{\gamma' k}{\gamma_w} t \qquad (17)$$

The settlement velocity v_e of the elements above the water film is determined by the combined permeability k_{es} of the middle layer and the upper layer as (Kokusho, 2002)

$$k_{es} = \sum_1^m L_i \bigg/ \sum_1^m L_i/k_i \qquad (18)$$

The upward seepage flow and the settlement of grains have the following Velocities:

$$v = k_{es} i_e; \quad u = -\frac{nv}{1-n} \qquad (19)$$

i_e is the average hydraulic gradient, n is the porosity.

The deform of the skeleton by the geostatic stress in the solidification zone after the solidification may be expressed as[5]

$$s_2 = \frac{1}{2} \frac{\rho' g}{m_s} z^2(t) \qquad (20)$$

in which m_s is the compressible modulus. The total deformation is:

$$\Delta s = \Delta L + \Delta s_2 \qquad (21)$$

Instituting eqs. (15) and (20) into eq. (21), considering that the initial height is equal to maximum height of the liquefied zone and the solidified zone and the water layer above the sand surface:

$$\Delta s = \frac{n_0-n_2}{1-n_2}(\Delta z + \Delta s) + \frac{\rho' g}{m_s} z \Delta z \qquad (22)$$

because $h = h_0 + L + s$, which yields $\Delta h = (\Delta L + \Delta s)$.

Eq. (16) may be written as

$$\frac{\Delta s}{\Delta t} = k\frac{\rho'}{\rho_w} \quad (23)$$

According to eqs. (22) and (23), the increase velocity of the solidification thickness:

$$\frac{\Delta z}{\Delta t} = \frac{k\rho'}{\rho_w} \Big/ \left(\frac{n_0 - n_1}{1 - n_1} + \frac{1 - n_1}{1 - n_0}\frac{\rho' g}{m_s}z\right) \quad (24)$$

Then we can obtain the duration of any location that the solidification front arrives at as follows:

$$t = \frac{\rho_w}{k\rho'}\left(\frac{n_0 - n_1}{1 - n_1}z + \frac{1}{2}\frac{1 - n_1}{1 - n_0}\frac{\rho' g}{m_s}z^2\right) \quad (25)$$

Side friction may be expressed if it should be considered

$$\sigma_s = \mu K_0 \sigma_z \quad (26)$$

The effect of the changes of porosity on the permeability k is considered as a linear relation:

$$k = k_0[1 - \alpha(n_0 - n)] \quad (27)$$

in which k_0 is the initial porosity, α is a parameter, n_0 is the initial porosity.

By considering eqs. (26) and (27) in the pore pressure gradient and considering the consolidation of the solidification zone, we can compute the development of cracks at these conditions.

6 COMPARISON WITH EXPERIMENTAL RESULTS

It is shown that the results computed by numerical method and the simplified method are close to the experimental results (Figure 3, parameters used in computing is the same given in literature 7). The simplified method presented in this paper may be used to compute the evolution of the water film.

7 CONCLUSIONS

Numerical simulations show that there are stable water films only in the conditions that: (1) the porosity of the upper part of the sand column must be smaller than that of the lower. (2) The keeping of the jamming state. A simplified method for evaluating the thickness of water film is presented. It is shown that the simplified method are agree with the experimental results.

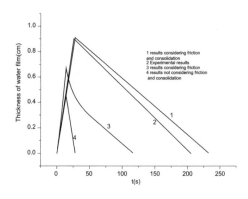

Figure 3. The comparison of the results computed by simplified method with the experiment of Kokusho.

ACKNOWLEDGEMENTS

This paper is supported by Chinese Academy of Sciences and The Key Program of the Academy of Sciences (KZCX2-YW-302-02), Key Laboratory of Mountain Hazards and Surface Process.

REFERENCES

Cheng, C.M., Tan, Q.M. & Peng, F.J. 2000. On the mechanism of the formation of horizontal cracks in a vertical column of saturated sand. *ACTA Mechanica Sinica* (English Serials) 17(1): 1–9.

Fiegel, G.L. & Kutter, B.L. 1994. Liquefaction mechanism for layered sands. *ASCE J Geotech Engrg.* 120(4): 737–755.

Florin, V.A. & Ivanov, P.L. 1961. Liquefaction of saturated sandy soils. In British National Society (ed.), *Proc. 5th int. Conf. On Soil Mech. Found. Engrg.*, 17–22 July, 1961. Paris, DUNOD, 1: 107–111.

Kokusho, T. 1999. Water film in liquefied sand and its effect on lateral spread. *J Geotech and Geoenviron Engrg.* 10: 817–826.

Kokusho, T., Watanabe, K. & Sawano, T. 1998. Effect of water film on lateral flow failure of liquefied sand. *Proc. 11th European Conf. Earthquake Engrg.*, Paris, CD publication, ECEE/T2/kokeow.pdf.

Scott, R.F. 1986. Solidification and consolidation of a liquefied sand column. *Soils and Foundations* 26(4): 23–31.

Seed, H.B. 1987. Design problems in sand liquefaction. *ASCE J Geotech Engrg.* 113(8): 827–845.

Zhang Junfeng 1998. *Experimental study on the strengthening of percolation and the damage of structure under impact loading.* dissertation for Ph.D, Institute of mechanics, Chinese Academy of Sciences.

Some geomorphological techniques used in constraining the likelihood of landsliding – Selected Australian examples

A.S. Miner & P. Flentje
Faculty of Engineering, University of Wollongong, Wollongong, NSW, Australia

C. Mazengarb & J.M. Selkirk-Bell
Mineral Resources Tasmania, Department of Energy, Infrastructure and Resources, Hobart, Australia

P.G. Dahlhaus
Department of Geology, University of Ballarat, Ballarat, Victoria, Australia

ABSTRACT: Techniques for landslide risk management in Australia have evolved considerably since the publication of the first formal process in 1985. The Australian Geomechanics Society recently published the next generation of updated landslide risk documents in 2007. The estimation of landslide likelihood is fundamental to the outcome of the landslide risk management process. However, experienced practitioners still regard this component as one of the most difficult and challenging aspects of the assessment as it requires information about the age of landslides, an understanding of landscape processes and the rate of slope evolution. Such information is difficult to obtain and is often not a core competency among practitioners undertaking landslide risk assessment. In order to provide insight into the methods of estimating and constraining landslide likelihood, a number of different geomorphological approaches are herewith reviewed through a series of selected Australian cases studies. Whilst the case studies highlight inherent limitations and uncertainties they also demonstrate how geomorphological studies can provide validation and constraints to a quantification of likelihood and ultimately risk.

1 INTRODUCTION

Techniques for landslide risk management in Australia have evolved considerably since the publication of the first formal process in 1985 (Walker et al 1985). In 2007 the Australian Geomechanics Society released a significant set of updated guidelines that will help shape the nature of slope stability investigations in Australia including the need for quantitative risk assessment (AGS 2007a & c). The adoption of these guidelines as a requirement for development approval by regulators, pose considerable challenges for stakeholders. There are several reasons for this, including a paucity of suitable published geological information for the urban environment and a skills shortage for suitably qualified practitioners.

However, one of the most significant issues impeding the estimation of risk is an underinvestment of hill country geomorphological research; hence landscape age, process rates and likelihood are often unknown or poorly constrained. Without an adequate understanding of likelihood, a risk assessment may require a highly conservative approach or worse be unreliable.

2 LANDSLIDE FREQUENCY

The AGS 2007 lists various techniques for assessing landslide frequency. These include:

- Landslide inventory compilation by gathering local historical records.
- Correlation between landslide events and rainfall.
- Relationships to geomorphology and geology.
- Simulation modeling and probabilistic analyses.
- Knowledge based expert judgment.

The first two techniques involve periods of record rarely spanning longer than 140 years. This typically allows estimation of annual frequencies of the order of 0.01 (10^{-2}) for a single event in a particular location. As such, these techniques have greatest application in the first two categories of the AGS likelihood classification system as shown in Table 1 (AGS 2007c, Appendix C).

Whilst the latter two methods are commonly used to provide validation to inventory and observational techniques, they are more importantly used to provide estimates of longer frequency occurrences which are evident in the landscape. To further highlight this

Table 1. Qualitative and quantitative measures of likelihood (AGS 2007c).

Approximate annual probability						
Indicative value	National boundary	Implied indicative landslide recurrence interval		Description	Descriptor	Level
10^{-1}	5×10^{-2}	10 years	20 years	The event is expected to occur over the design life.	ALMOST CERTAIN	A
10^{-2}	5×10^{-3}	100 years	200 years	The event will probably occur under adverse conditions over the design life.	LIKELY	B
10^{-3}	5×10^{-4}	1000 years	2,000 years	The event could occur under adverse conditions over the design life.	POSSIBLE	C
10^{-4}	5×10^{-5}	10,000 years	20,000 years	The event might occur under very adverse circumstances over the design life.	UNLIKELY	D
10^{-5}	5×10^{-6}	100,000 years	200,000 years	The event is conceivable but only under exceptional circumstances over the design life.	RARE	E
10^{-6}		1,000,000 years		The event is inconceivable of fanciful over the design life.	BARELY CREDIBLE	F

• This table should be used from left to right; use approximate annual probability or description to assign descriptor, not vice versa.

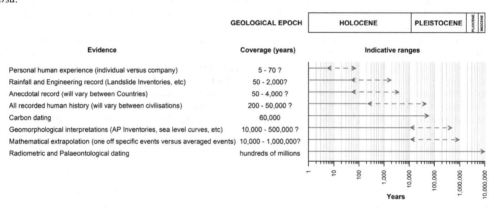

Figure 1. Areas of human knowledge and historical interpretation displayed with indicative time ranges.

fact the application of various evidence based methods, including indicative time ranges over which these evidence sources may extend, is shown in Figure 1.

The case studies provided in this paper specifically focus on the use of geomorphological techniques to assist in the classification, validation and constraint of landslide frequency.

3 GEOMORPHOLOGICAL PROCESSES

Geomorphology is the study of landforms, including their origin and evolution, and the processes that shape them. Geomorphology focuses on the analysis of interconnected physical processes which shape or have shaped the landscape. Integral to the study of geomorphology is Hutton's 18th century principle of uniformitarianism that assumes the natural processes operating in the past are the same as those that can be observed operating in the present. Unfortunately, uniformitarianism only tells part of the tale in that we can now see that the rates of geomorphological processes have changed over geological time due to such factors as plate tectonics and changing sea levels and climate. With good research and site based observations, we are increasingly able to place some constraints on these issues and factors. This is well demonstrated in the following three case studies. The location of the three case study sites are shown in Figure 2.

4 SOUTH WESTERN VICTORIA

On March 28 1990, two high school students were tragically killed by a rockfall at the Lal Lal Falls Reserve. The accident was the subject of a coroner's inquest

Figure 2. South Eastern Australia showing the location of three case study sites discussed in this paper.

which has resulted in detailed geological and geomorphological investigations at the site (Cooney 1990; Dahlhaus and Miner 2000).

The site is a small recreation and scenic reserve surrounding a waterfall, about 18 km southeast of the City of Ballarat. The waterfall, approximately 35 m high, is located at the head of a gorge bordered by steep cliffs along the Lal Lal Creek (Figure 2).

Dense, olivine-rich basalt of Pliocene - Pleistocene age is exposed in the gorge. The basalt comprises two separate flows, both exhibiting columnar jointing, with the columns in the upper flow being narrower (1/2 – 1 m) than in the bottom flow (1 1/2 – 2 m) (Dahlhaus and Miner, 2000).

4.1 Geomorphic evolution

The falls have formed by headward erosion of the Lal Lal Creek over the past 3 million years. Below the falls the creek follows a deeply incised valley that reflects the original meandering course of the stream. Above the falls the valley of the meandering Lal Lal Creek is very shallow.

4.1.1 Calculation of erosion rates
The volume of material eroded from the landscape since the emplacement of the basalt was calculated using a digital terrain model. It has been estimated (Dahlhaus and Miner, 2005) that the Lal Lal Falls have retreated a distance of approximately 1650 m in the past 3 million years which averages 0.55 mm/year. By comparison, the adjacent Moorabool Falls have retreated about 1400 m over the same time period— an average of 0.46 mm/year. Assuming a constant erosion rate, the valley slopes have receded away from the streams at 0.015 mm/m^2/year. Based on these calculations, a volume of 0.04 m^3/year has eroded from the upper columnar basalt cliff of the entire southern slope over the past 3 million years. Based on the observed average column diameter of 0.8 m, this equates to an average column height of 80 mm per year or more realistically a 0.8 m diameter column of 0.8 m length eroding every decade (an ARI of 1 in 10 years and a classification of *Almost Certain* as per AGS 2007). Despite the unavoidable assumption of a uniform rate of landscape evolution the calculated rate serves as a useful estimate.

4.2 Rock fall frequency assessment

4.2.1 Hazard types and historical inventory data
The entire cliff face surrounding the falls is prone to both block toppling and rockfall. Open joints, large columns leaning out from the cliff face and undercut columns are evident at the site.

Information on previous accidents at the Lal Lal Falls Reserve as sourced from local newspaper and historical records is poor. However historic photographs, taken on previous geology student excursions, proved to be a useful record of rockfall events. Comparison of photographs from 1975 to 1990 shows few changes in the cliff faces, and major rockfalls are not evident. However the actual site of the 1990 fatality was photographed in 1972, with both the column involved in the accident and a 0.8 m length that had fallen sometime prior, still in place. Both columns represent approximately 0.2 m^3 volume.

Another rockfall event at a different site on the upper cliff was also observed in May 1992. This fall involved a toppling failure of a column 1.5 m high by 0.5 m diameter (approx 0.3 m^3).

4.2.2 Calculation of rockfall frequency
Based on the calculated average erosion rates, the observed volumes of the rockfalls at the fatality site each represent five year's average erosion volume, suggesting an average frequency of 1 in 5 years. The 0.3 m^3 rockfall observed in 1992 is 7 1/2 year's average erosion volume, or a 1 in 7 1/2 year event. All three events are of a similar size.

A total of 0.7 m^3 is known to have fallen from the southern cliff face in three separate events between 1972 and 1992 indicating an observed annual rate of (0.035 m^3/y). The photographic record and historical records suggest that no other significant falls occurred during this time. The observed annual rate compares well to the calculated erosion rate (0.04 m³/y) with the additional volume probably made up from smaller falls and erosion.

4.3 Likelihood estimates

Calculation of rockfall likelihood for the Lal Lal site was achieved through the use of the calculation of the erosion rate in combination with site observations and the establishment of a site inventory through anecdotal and historical evidence.

Rockfalls within the study area with an estimated frequency as described fall into the *Almost Certain* category as per the AGS classification system. It is however very important to note that the techniques only allow for such a likelihood classification if it is applied to the entire study area as average rates of erosion have been used. Hence a better statement about rockfall likelihood from this study is that rockfalls of approximately 0.2 to 0.3 m³ could be expected to occur somewhere within the 1.65 km length of the study area every 5 to 7.5 years.

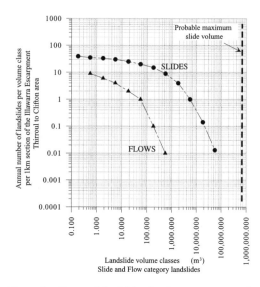

Figure 3. Proposed landslide size frequency curves for slides and flows normalised for 1 km of escarpment slopes between the suburbs of Thirroul and Clifton. It is not appropriate to use these curves for slopes outside this area.

5 ILLAWARRA REGION, NEW SOUTH WALES

Landsliding is a common occurrence along the Illawarra Escarpment. The University of Wollongong Landslide Inventory (LI) identifies almost 600 landslide sites on the escarpment. Most of these landslides have occurred since 1950. Susceptibility zoning (Flentje et al, 2007) identifies 12% of the land as either high or medium susceptibility whilst containing 92% of the known landslides.

5.1 Landslide process rates for the thirroul to Clifton escarpment

The Illawarra escarpment is a variable geomorphological feature. In the southern half of the study area it is quite a mature feature with a wide coastal plain and well developed and incised drainage lines.

North of the suburb of Thirroul, to Clifton the escarpment coastline comprises a series of small bays and headlands which merge to the north into a cliff line up to 100 m in height. This 10 km section of escarpment is currently being undercut at or near sea level by marine erosion and is relatively over steepened when compared to the escarpment further to the south. It is not surprising then to note that it contains a significantly higher density of landslides.

This area includes 155 landslide sites, of which 123 are slide category landslides, 16 are debris flows and a further 16 are rock falls (from various different source areas). The annual size frequency distributions derived from the actual mapped slides and flows have been used as a guide in the development of the interpretive size-frequency curves shown in Figure 3 below (as discussed in Moon et al 2005).

The area under each curve (the landslide process rate) represents the annual average volume 'mobilised' by slide and flow landsliding respectively per one km of escarpment in this area. This approximate total volume is 29,000 m³. Of course, the material mobilised does not mean this amount of material is directly removed from the escarpment slopes. Not withstanding the inherent uncertainties involved, if an 'average' rate of movement of 1 cm per year is adopted for all the landslides combined, the volume removed annually from the escarpment reduces to approximately 300 m³.

The annual number of landslides per volume class per 1 km of escarpment slope can then be used to assign a broad range of landslide likelihoods as per the AGS classification system. However, each volume class must be considered independently as the rate of displacement and associated consequences for each can greatly affect the risk.

5.2 Comparison of process rates with estimates derived from rates of sea-floor spreading

The Illawarra Escarpment is an erosional escarpment that has evolved on a passive continental margin. This margin developed following the rifting and continental breakup between Australia and the Lord

Howe Rise/Dampier Ridge approximately 80 million years ago (Brown et al, 2003).

The outer edge of the continental shelf is approximately 30 km offshore north of Wollongong. If it is assumed that the outer edge of the continental shelf represents the point from which the escarpment has retreated since the onset of sea floor spreading, then an overall rate of escarpment retreat can be estimated. These limitations suggest 30 km of escarpment retreat in 80 Million years (Ma), or approximately 0.375 m per 1000 years, on average during the geological life of the escarpment. However it must be acknowledged that this does not consider changes in sea level during this period, other tectonic influences and climatic variations which will also impact upon rates of escarpment retreat. Such influences will have had marked impacts on this 'average' rate of retreat, at times causing it to be higher and at others forcing it lower.

This rate indicates an average annual volume of approximately 600 m^3 of material is removed from a 1 km width of the escarpment (given an escarpment slope length of 1700 m, such as north of Thirroul). A percentage of this can occur as landslides in one form or another (such as slides, flows or falls) whilst some will also be lost by alluvial processes.

Hence this volume as determined from a broad geomorphological approach is of the same order of magnitude (albeit double the value) as the figure determined above on the basis of data from the LI.

5.3 Discussion

Clearly, the LI does not identify all the landslides and the reported volumes within the LI are not precise. In addition the rate of average movement is uncertain. However, despite these 'averages' it is encouraging to see that the estimates from the two approaches to process rate determination are of comparable magnitude. This compatibility helps validate the modeling assumptions and adds substantial credibility to the outcomes, despite the uncertainty which is implicit. Such assessments, founded in sound site based observations and a rational geomorphological understanding of the processes active in slope formation, are fundamental in achieving realistic assessment of landslide frequency required under the AGS classification system.

6 NORTH WEST TASMANIA

Landslides are a widespread feature in the Tasmanian landscape and there is much pressure to develop landslide affected terrain. Mineral Resources Tasmania (MRT) is progressively undertaking regional landslide mapping of urban areas and their surrounds to assist stakeholders such as local councils in addressing landslide risk, particularly in terms of the new AGS guidelines and documents.

The overall approach being used by MRT includes geological and geomorphological mapping and the compilation of a landslide inventory. When combined within a GIS framework these information layers are then used to produce landslide susceptibility maps. While this form of a landslide assessment is useful, the future production of hazard maps would be a significant advance allowing, for example, tolerable risk levels to be determined.

Whilst the MRT landslide mapping project is yet to be completed the following example provides an insight into landslide likelihood as gained from geomorphic evaluation of the landscape.

6.1 Devonport-Burnie, NW Tasmania

The Devonport-Burnie area is located on the northwest coast of Tasmania, overlooking Bass Strait which separates the island from mainland Australia to the north.

The dominant features of the coastal strip are a hinterland of deeply dissected sub horizontal Tertiary basalt, an escarpment (fossil sea cliff) and associated coastal plain as well as contemporary coastal cliffs on headlands. Fluvial terraces occur in major waterways. Based on dating by Murray-Wallace & Goede (1995) the escarpment is believed to have formed during the last interglacial sea level high at about 125,000 years (125 Ka) BP, and tectonically uplifted by about 22 m. The uplift is not only reflected in the elevated coastal plain (marine terrace), but also in incised catchments and the presence of fluvial aggradational terraces at similar elevations above the channels adjacent to the elevated coastal plain.

The fossil sea cliff and incised valley walls have collapsed in numerous places and in two major styles (deep seated landslides and earthflows) after sea levels retreated. Landslides that runout over the marine terrace and matching aggradational fluvial terraces are therefore stratigraphically less than 125 Ka old (Figures 4 and 5). Such features are relatively common.

The age constraints provided brings the understanding of landslide age from millions of years (basalts ages are over 10 Ma old) down almost three magnitudes (<125,000 years). This would initially tend to suggest a conservative estimate of *Rare* likelihood under the AGS system of classification if age is simply equated to annual probability in the absence of known triggering events.

However whilst rainfall and elevated groundwater are likely to be key triggering mechanisms, if climate was considered to be cool and dry during the glacials (these last in the order of 100,000 years), it may imply

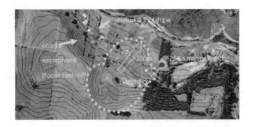

Figure 4. Orthophotograph of relic (prehistoric) landslide that has runout nearly 200 m over the 125 Ka bench. Other terrain features are also depicted to provide indication of mapping approach. It is believed that a significant component of the colluvial footslope is composed of earth flow deposits derived from the basaltic soils in the escarpment. Orthophoto from Department of Primary Industries and Water, Hobart.

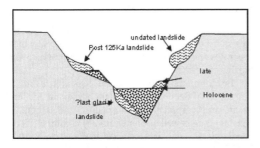

Figure 5. Conceptual model of how landscape chronology can be used to constrain the age of landslides in an incised valley situation.

that landslide activity was greater during the interglacials. Given that we are currently in an interglacial period, likelihood may not be simply calculated as an average over 125 Ka but could be higher by as much as 2 orders of magnitude. Hence the current likelihood might more reasonably be assumed to be *Unlikely* to *Possible* under the AGS system when considering individual landslides.

6.2 Discussion of likelihood estimate

While there are many landslides in the Tasmanian landscape, almost all of these features are undated which makes it difficult for estimating likelihood of future landslides and understanding risk at a regional or local level with any confidence. The popular opinion by the geotechnical community, as expressed in geotechnical reports supporting development applications, is that the deep seated landslides such as the one described at Burnie are "fossil" features (more correctly referred to as relict landslides) and formed under differing climatic conditions. This implies AGS classifications of *Rare* and *Barely Credible* and suggests that they would not occur in the present day. However as the discussion indicates, likelihoods may well be at least an order of magnitude higher. Without considered opinion of geomorphic processes initial unsupported assumptions may be dangerous as they could significantly underestimate landslide hazard.

7 DISCUSSION AND CONCLUSIONS

The case studies provide insight into how an understanding of long term geomorphological processes can validate inventory and observational techniques and allow estimates of low frequency landslide events. Techniques such as scarp retreat rate and regional surface dating can provide opportunities in constraining landslide likelihood as defined in the AGS system of classification. The case studies also clearly highlight the limitations and uncertainties with such estimations and emphasize the importance of using all available techniques when assessing landslide likelihood as required under the AGS system of landslide risk management.

REFERENCES

AGS, 2007a. Australian Geomechanics Society Working Group. Guideline for landslide susceptibility, hazard and risk zoning for land use planning. Australian Geomechanics Journal, Volume 42, No. 1 March, pp 13–37.

AGS, 2007c. Australian Geomechanics Society Working Group. Practice note guidelines for landslide risk management. Australian Geomechanics Journal, Volume 42, No. 1 March, pp 63–114.

Brown B.J. Muller, R.D. Gaina, C. Struckmeyer, H.M. Stagg, H.M.J. & Symonds, P.A., 2003. Formation and evolution of Australian passive margins: implications for locating the boundary between continental and oceanic crust. In Hillis and Muller (eds.) Evolution and Dynamics of the Australian Plate. Geol Soc of Aust Special Pub. 22.

Cooney, A.M. 1990. Geologocaical report on the rock fall at Lal Lal. Report to the Deputy Coroner. Unpublished report 1990/13 geological survey of Victoria.

Dahlhaus, P.G. & Miner A.S. 2000. Estimating the occurrence of Rockfalls in columnar Basalt. GeoEng 2000. (Reprinted in Australian Geomechnaics Vol 36, No. 2 June 2001).

Flentje, P., Stirling, D. & Chowdhury, R.N., 2007. Landslide Susceptibility and Hazard derived from a Landslide Inventory using Data Mining—An Australian Case Study. Proceedings of the First North American Landslide Conference, Landslides and Society: Integrated Science, Engineering, Management, and Mitigation. Vail, Colorado June 3–8, 2007. CD, Paper number 17823–024, 10 pages.

Moon, A.T., Wilson, R.A. & Flentje, P.N., 2005. Developing and using landslide size frequency models. Joint Technical Committee on Landslides and Engineered Slopes, JTC-1, in association with Vancouver Geotechnical Society. Proceedings of the International Conference on Landslide Risk Management/18th Annual Vancouver Geotechnical Society Symposium, Vancouver. May 31 to June 4, pp 589–598.

Murray-Wallace, C.V. & Goede, A. 1995. Aminostratigraphy and electron spin resonance dating of Quaternary coastal neotectonism in Tasmania and the Bass Strait islands. Australian Journal of Earth Sciences, 42: 51–67.

Walker, B.F., Dale, M., Fell. R., Jeffrey, R., Leventhal, A., McMahon, M., Mostyn, G. & Phillips, A. 1985. Geotechnical Risk Association with Hillside Development. Australian Geomechanics News, No10, pp 29–35.

ced by Chen et al. (eds)

Rock failures in karst

Mario Parise
National Research Council, IRPI, Bari, Italy

ABSTRACT: Rock failures in karst environments present peculiar features related to the typical setting of karst, characterized by presence of caves created by chemical solution of soluble rocks, direct connection between the surface and the subsurface, and an overall high fragility of the environment. In addition to landslides *s.s.* and sinkholes, breakdown processes within caves are extremely common in karst, and may represent a geohazard even to the built-up environment, due to possibility of void migration toward the surface. They deserve, therefore, great attention by scientists, and should be carefully examined through an integrated approach which has necessarily to include direct surveying in natural karst caves.

1 INTRODUCTION

Karst processes affect soluble rocks, creating peculiar landforms at the surface, and caves and drainage systems underground, through chemical solution of the bedrock. The main rock type interested by karst development is carbonate. Nevertheless, evaporites are greatly involved too, and other rock types such as sandstones and quartzites, even though very resistant, may be also affected by development of karst and/or pseudokarst caves.

An important, often underestimated, process in karst is represented by rock failures, with the production of huge amount of breakdown deposits in caves (White & White 1969). Slope instability in karst plays actually a twofold role: it occurs upon it, at the surface, as in many other settings where steep to vertical rock walls crop out. In addition, rock failures in karst significantly shape the underground systems. Some of the largest rooms in caves all over the world are situated at the sites of intersections of faults and/or fractures, that is at the main weakness zones in the rock mass. Caves, as we see them today, have been produced by the combined action of dissolution (karst *sensu strictu*) and breakdown processes. The latter become particularly significant once the cave passages are abandoned by the flowing water, which level is lowered below the passage horizon (White 1988). Lack of the support exerted by water on the cave walls and roof, together with the degree of fracturing in the rock mass, determines the continuous detachment of blocks and slabs of rock, and their later fragmentation into chips, thus enlarging the cave size. Upward, the process may proceed until reaching the ground surface, to produce collapses of variable size and depth.

Rock failures in caves represent a danger to cavers and karst scientists, and, at those sites where show cave exploitation is carried out, even to tourists. Further, the instabilities that occur underground may have indirect effects at the surface, producing likely damage to the built-up environment, and creating a very subtle geohazard to engineers.

2 TYPES OF FAILURES

Karst environments present peculiar features that create the possibility of different types of rock failures:

1. landslides *sensu strictu*, in those karst areas characterized by slope relief and geological settings prone to slope movements;
2. sinkholes related to the presence of subterranean cavities;
3. rock failures in the underground environment.

In the following, I will briefly deal with categories 1 and 2 above, mostly focusing the attention on the features within natural karst caves.

2.1 *Landslides*

The phenomena belonging to category 1 are studied and analyzed following the usual approaches defined and well established for slope movements, in particular for those affecting rock slopes (De Freitas & Watters 1973; Hoek & Bray 1981; Cruden & Varnes 1996).

Even in flat karst regions, where slope relief is generally low, occurrence of rock failures may be common along the walls of the main valleys, as shown by the topples in figure 1. In addition to the usual factors that generally influence or predispose to slope movements,

Figure 1. Topple failures long the walls of a deep karst valley (Laterza, southern Italy).

Figure 2. Rockslide that occurred in January, 2002 in the Amalfi coast (Campania, southern Italy).

Figure 3. Collapse sinkhole in carbonate rocks of the Gargano Promontory (Apulia, southern Italy).

the presence of karst voids and conduits, and the deriving modality of water circulation may further favour in karst the occurrence of failures, as illustrated by Santo and co-workers (2007) in their study of carbonate rock failures in the Campania region of southern Italy (Figure 2).

2.2 Sinkholes

Sinkholes are rock failures typical of karst environments, being related to the presence of subterranean cavities (Figure 3), both natural and man-made (Waltham et al. 1986; Culshaw & Waltham 1987). In the last decades they have been object of many studies in different parts of the world, and their research has become an important and specific field of engineering geology in many countries (Delle Rose et al. 2004, Gutierrez et al. 2004, Dogann 2005).

Sinkholes are a subtle type of failure, since they may occur even on very gentle slopes, and potentially affect inhabited areas and man-made infrastructures.

Sinkholes may be classified into six main types, distinguished by their genetic process and morphology (Waltham et al. 2005). These include: solution sinkhole, created by the slow process of dissolutional lowering of the surface; collapse sinkhole, caused by rock roof failure into an underlying cave; caprock sinkhole, involving failure of insoluble rock into cave in soluble rock below; dropout sinkhole, when the soil collapses into soil void formed over bedrock fissure; suffosion sinkhole, consisting of down-washing of soil into fissures in bedrock; and, eventually, buried sinkhole, when an old sinkhole in rock becomes soil-filled after environmental changes have occurred.

3 ROCK FAILURES IN KARST

Caves are a notable karst geohazard, due to the potential for gravitational collapse of rock and/or soil into them, either naturally or under induced load (Waltham & Lu 2007).

Karst caves formed in phreatic environment ideally present a cylindrical shape, produced through slow development in geological times, and may have reached a stable configuration with equilibrium roof arches. However, reduction in the thickness of the roof might determine further instability processes. In fractured rock masses, the setting is more complicated depending upon frequency and orientation of the main discontinuity systems. In any case, instability phenomena may become the main evolution process in caves initiated by classical karst evolution. Through natural stoping and upward cavity migration, the progressive rock failures can go on until reaching the ground surface.

Rock failures in underground environments are represented by falls, topples, and slides that affect the walls and roofs of karst caves. Since inspection of the underground world is generally restricted to experienced cavers, it becomes clear that these types of failures have to be studied by scientists-cavers, or alternatively by a team at least partly composed by cavers. Besides the inherent difficulties of the hosting environment, rock failures within caves often differs in some ways from the analogue types observed at the earth's surface, due to greater importance of weathering processes (Fookes & Hawkins 1988, Anon 1995, Zupan Hajna 2003, Forti & Parise 2008), and groundwater chemistry and circulation (Harmon & Wicks 2006, Ford & Williams 2007).

Cave roof failure is the type of great worrying, since its evolution, and upward stoping, may eventually lead to ultimate surface collapse. Mechanisms and rates of cave roof failure is essentially dependant upon geological features of the rock mass, namely attitude and bedding (Waltham & Lu 2007).

Even though the matter is extremely complicated, and a number of other factors might play crucial roles, a commonly accepted rule of thumb is that an individual bed of strong limestone is going to fail when the cave's unsupported span exceeds 10–20 times the bed thickness. Besides the aforementioned attitude and bedding of the rock mass, however, other factors must be considered, including presence and typology of discontinuities, weathering, and water flow.

Discontinuities weaken the rock mass, breaking its continuity, and constitute preferential ways to water flow. In karst, the latter means essentially a combination of mechanical erosion and chemical solution that may result in creating voids in the rock.

Over a void, a zone of ground compression in the form of an arch naturally develops (Figure 4). In a cave, the arch is produced through progressive detachment of beds from the tension zone beneath the compression arch. In this way, caves may reach a stable configuration, with arched roof profiles approaching the stability of a voussoir arch in uncemented masonry (an arch formed of shaped blocks designed to be stable in compression). When jointing becomes extensive, and the rock mass is heavily to extremely

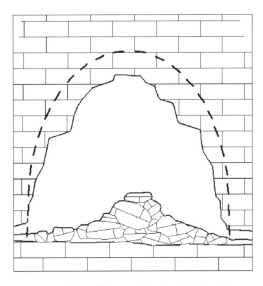

Figure 4. The dotted line marks the "tension dome" or zone of maximum shear stress induced by the presence of a cavity (modified after White 1988).

fractured, stability of the arched roof profiles is much less guaranteed.

Among the different approaches to analyze such an evolution, numerical modelling has been advanced by defining fractured rock masses in terms of their "rock mass ratings" (RMR). The geomechanic system derives RMR by summing rating values ascribed on the basis of RQD (rock quality designation, based on fracture intersections in borehole core), mean fracture spacing, fracture conditions, fracture orientation, unconfined compressive strength of the intact rock and groundwater state (Bieniawski 1973; Hoek & Brown 1980).

RMR values range from more than 80 for very good rock of rock mass class I to less than 20 for very poor rock of rock mass class V; they may be correlated with Q values derived from the alternative Norwegian classification scheme (Barton et al. 1974).

RMR for typical cavernous karst in strong limestones is taken conservatively as between 30 and 40, whilst in chalk and other weak or thinly bedded limestones it may be estimated as nearer 20 (Waltham & Lu 2007).

However, direct observations and surveys in natural caves show a great variety in the range of values of the stability parameters which are strongly a function of the rock mass quality (Figure 5).

The instability process within stratified hard rock masses has been widely discussed by Diederichs & Kaiser (1999), according to the traditional Voussoir beam theory (Evans 1941), suitably revisited by them.

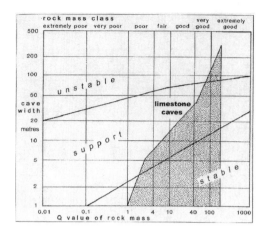

Figure 5. Cave stability related to cave width and rock mass quality (after Waltham, 2002). Q value is after Barton et al. 1974.

In particular, they suggest that timing for roof collapse is often controlled by residual tensile strength available along the existing rock bridges (Diederichs & Kaise, 1999). According to the Authors, the tensile strength degradation due to humidity and chemically assisted stress corrosion can be responsible for failure of spans that remained safe for very long term. Therefore, the transition from systems of stratified continuous beams to systems of stratified voussoir beams due to time-dependant tensile strength degradation as an effect of weathering processes, and the following collapse of the stratified blocky roof, is thought to be among the main processes acting in underground karst systems. Such a process is locally favoured by water infiltration from the ground surface, which may produce significant weathering effects.

3.1 The importance of sub-surface investigations

High variability is probably the main feature of karst landforms and features. At a same site, part of a cave may be stable and present sound roofs, whilst another may be characterized by weak zones with high percentage of joints and/or broken rocks. This translates into the need to pursue as much as possible direct observations (including those from caving activities) in order to collect useful databases to implement numerical modelling aimed at assessing the stability conditions in underground karst environments.

Investigating extensive subsurface cave systems, and mapping the main morphological features related to breakdown processes, including those not yet manifested at the surface, may provide a significant contribution to the understanding of the different stages of breakdown development. As shown by Klimchouk &

Andrejchuk (2002) in their study on the gypsum caves in Western Ukraine (a region where the five largest known gypsum caves in the world are present), the determination of the degree of propagation of the breakdown structures toward the surface through the cover, allows to assess the related geohazard with a precision and certainty unavailable by the approaches of conventional engineering geology (Klimchouk & Andrejchuk 2002).

In order to best evaluate the gravity-related processes in the cave, there is the necessity to map any outlet features that may indicate considerable breakout in the vault: from domes and cupolas, to domepits in cave passages. In addition, mapping of rockfall deposits, together with determination of their size, and possibly the origin of detachment (vault, wall, etc.), and of taluses deposits is also important (Parise & Trisciuzzi 2007). Stage and activity of the breakdown may be ascertained through recording several features in the cave such as maximum height of the cupolas, presence of open cracks and fissures at its margin, water seepage or flow in the breakdown taluses, dampness of breakdown sediments, freshness of fallen blocks.

Klimchouk & Andrejchuk (2002) observe, in their study on the breakdown structures in the Zoloushka cave, that the largest density of breakdown structures was observed in regions of not very large passages; this finding strongly contrasts with one of the most established views, which suggests that breakdown formation is controlled primarily by passage size. The Authors above identify two different mechanisms of breakdown formation, with mechanism A leading only to gradual subsidence and not collapse, which, on the other hand, is produced through mechanism B (Fig. 6).

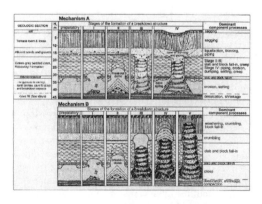

Figure 6. The two mechanisms of breakdown formation hypothesized by Klimchouk & Andrejchuk (2002) in their study of gypsum caves of Western Ukraine.

Even the presence of smooth dolines at the surface may be related to past events of collapse, likely within still unexplored caves. The example of Dan'kivsky shaft (a 22 m deep shaft that opened on January 11, 1998 in Ukraine) illustrates how the shape of a surface form is not necessarily indication of its sudden or gradual origin (in other words, of a collapse or subsidence formation). The original deep shaft, in fact, within few years transformed into a gently sloping doline, due to the presence of soft sediments within the overburden (Klimchouk & Andrejchuk 2002).

In stratified soluble rocks, especially in evaporites showing a cover beds (Figure 7), most of the breakdown features are generally originated through more or less prolonged multi-stage development rather than through a single massive collapse of the roof (Iovine et al. 2007). This is also indicated by connection through the different storeys by large pits, which represented hydraulic communication at different levels during the cave formation.

4 DISCUSSION AND CONCLUSIONS

Due to their intrinsic characteristics, karst environments are extremely susceptible to a number of geohazards (White, 1988; Parise & Pascali 2003; Parise & Gunn 2007), including rock failures. These may involve the landscape both at the surface and in the subterranean world, which in karst are strongly and intimately linked. Changes induced by man in the surface karst environment may produce in fact rapid events underground, contributing to accelerate the generally slow process of karst solution. For example, external loading due to road construction and traffic (Kambesis & Brucker 2005), or high water flow deriving from diversion of the natural surface flow, clogging of original swallets, and/or impermeabilization of natural surface drainage by man (White & White 1984;

Figure 7. Rockfall deposits at the entrance of a gypsum cave in Calabria, southern Italy.

Parise 2003) may work to strongly change the natural setting, with likely severe consequences and damage for the anthropogenic environment.

Studying rock failures in karst is therefore a challenging but highly stimulating issue. In many cases, however, traditional approaches are not sufficient to fully appreciate the phenomena. Lacking detailed data on discontinuity systems in the rock mass, and/or direct observations in caves, an estimate of the overall conditions of a site can be derived from the engineering classification of the karst, assessed from broad visual inspection (Waltham & Fookes 2003). However, such estimates should be taken into account only as a first approximation, whilst collection of further data (possibly through direct inspection and observations) is strongly encouraged.

An interesting point is that availability of surface and subsurface investigation may allow a double geomechanical survey, which has to be carried out according to the internationally accepted and codified standards (ISRM 1978). Through this approach, it will be possible to identify the main discontinuity sets in the rock mass, their continuity within the cave systems, and the control they have on the occurrence of slope failures. The data so collected can be used to identify the main processes that trigger breakdown in caves, to assess the related hazard to man and the anthropogenic structures, and to implement numerical models dedicated to the comprehension of the modality of evolution of the rock mass (Lollino et al. 2004; Waltham & Swift 2004).

REFERENCES

Anon, 1995. The description and classification of weathered rocks for engineering purposes: Geological Society Engineering Group Working Party Report. *Quart. J. Eng. Geol.* 28: 207–242.

Barton, N., Lien, R. & Lunde, J. 1974. Engineering classification of rock masses for tunnel design. *Rock Mechanics* 6: 189–236.

Bieniawski, Z.T. 1973. Engineering classification of jointed rock masses. *Trans. South Afr. Inst. Civil Eng.* 15: 335–343.

Cruden, D.M. & Varnes, D.J. 1996. Landslide types and processes. In A.K. Turner & R.L. Schuster (eds) *Landslides. Investigation and Mitigation.* Transp. Res. Board, sp. rep. 247: 36–75.

Culshaw, M.G. & Waltham, A.C. 1987. Natural and artificial cavities as ground engineering hazards. *Quart. J. Eng. Geol.* 20: 139–150.

De Freitas, M.H. & Watters, R.J. 1973. Some field examples of toppling failure. *Geotechnique* 23 (4): 495–514.

Delle Rose, M., Federico, A. & Parise, M. 2004. Sinkhole genesis and evolution in Apulia, and their interrelations with the anthropogenic environment. *Natural Hazards and Earth System Sciences* 4: 747–755.

Diederichs, M.S. & Kaiser, P.K. 1999. Tensile strength and abutment relaxation as failure control mechanisms in underground excavations. *Int. J. Rock Mech. Min. Sc.* 36: 69–96.

Dogann, U. 2005. Land subsidence and caprock dolines caused by subsurface gypsum dissolution and the effect of subsidence on the fluvial system in the Upper Tigris basin (between Bismil-Batman, Turkey). *Geomorphology* 71: 389–401.

Evans, W.H. 1941. The strength of undermined strata. *Trans. Inst. Min. Metall.* 50: 475–500.

Fookes, P.G. & Hawkins, A.B. 1988. Limestone weathering: its engineering significance and a proposed classification scheme. *Quart. J. Eng. Geol.* 21: 7–31.

Ford, D. & Williams, P. 2007. *Karst hydrogeology and geomorphology.* John Wiley & Sons, 562 pp.

Forti, P. & Parise, M. 2008. The role of weathering in favouring instability processes in natural karst caves. In D. Calcaterra, D. Campbell & M. Parise (eds) *Weathering as predisposing factor to slope movements.* Geol. Soc. London, Engineering Series sp. publ.

Gutierrez, F., Guerrero, J. & Lucha, P. 2004. Paleosubsidence and active subsidence due to evaporite dissolution in the Zaragoza area (Huerva river valley, NE Spain): processes, spatial distribution and protection measures for transport routes. *Engineering Geology* 72: 309–329.

Harmon, R.S. & Wicks, C.M. (eds) 2006. *Perspectives on karst geomorphology, hydrology, and geochemistry.* Geol. Soc. Am., sp. paper 404: 366 pp.

Hoek, E. & Bray, J.W. 1981. *Rock slope engineering.* Institution of Mining and Metallurgy, London, 358 pp.

Hoek, E. & Brown, T. 1980. *Underground excavations in rock.* Institution of Mining and Metallurgy.

Iovine, G., Parise, M. & Trocino, A. 2007. Looking at instability phenomena from different perspectives: an experience from the Verzino area (Calabria, southern Italy). Epitome 2: 209.

ISRM 1978. Suggested methods for the quantitative description of discontinuities. *Int. J. Rock Mech. Min. Sc.* 15: 319–368.

Kambesis, P. & Brucker, R. 2005. Collapse sinkhole at Dishman Lane, Kentucky. In T. Waltham, F. Bell & M. Culshaw *Sinkholes and subsidence: karst and cavernous rocks in engineering and construction.* Springer: 277–282.

Klimchouk, A. & Andrejchuk, V. 2002 Karst breakdown mechanisms from observations in the gypsum caves of the Western Ukraine: implications for subsidence hazard assessment. *Int. J. Speleol.* 31 (1/4): 55–88.

Lollino, P., Parise, M. & Reina, A. 2004. Numerical analysis of the behavior of a karst cave at Castellana-Grotte, Italy. In H. Konietzky (ed), *Proc. 1st Int. UDEC Symp. "Numerical modeling of discrete materials"*, Bochum (Germany), 29 september—1 october 2004: 49–55.

Parise, M. 2003. Flood history in the karst environment of Castellana-Grotte (Apulia, southern Italy). *Natural Hazard and Earth System Sciences* 3: 593–604.

Parise, M. & Pascali, V. 2003. Surface and subsurface environmental degradation in the karst of Apulia (southern Italy). *Environ. Geol.* 44: 247–256.

Parise, M. & Gunn, J. (eds) *Natural and Anthropogenic Hazards in Karst Areas: Recognition, Analysis and Mitigation.* Geol. Soc. London, sp. publ. 279, 202 pp.

Parise, M. & Trisciuzzi, M.A. 2007. Geomechanical characterization of carbonate rock masses in underground karst systems. *Kras i speleologia.*

Santo, A., Del Prete, S., Di Crescenzo, G. & Rotella, M. 2007. Karst processes and slope instability: some investigations in the carbonate Apennine of Campania (southern Italy). In M. Parise & J. Gunn (eds) *Natural and Anthropogenic Hazards in Karst Areas: Recognition, Analysis and Mitigation.* Geol. Soc. London, sp. publ. 279: 59–72.

Waltham, A.C. 2002. The engineering classification of karst with respect to the role and influence of caves. *Int. J. Speleol.* 31 (1/4): 19–35.

Waltham, A.C. & Fookes, P.G. 2003. Engineering classification of karst ground conditions. *Quart. J. Eng. Geol. and Hydrog.* 36: 101–118.

Waltham, T. & Lu, Z. 2007. Natural and anthropogenic rock collapse over open caves. In M. Parise & J. Gunn (eds) *Natural and anthropogenic hazards in karst areas: recognition, analysis and mitigation.* Geol. Soc. London, sp. publ. 279: 13–21.

Waltham, A.C., Vandenver, G. & Ek, C.M. 1986. Site investigations on cavernous limestone for the Remouchamps Viaduct, Belgium. *Ground Engineering* 19 (8): 16–18.

Waltham, T., Bell, F. & Culshaw, M. 2005. *Sinkholes and subsidence: karst and cavernous rocks in engineering and construction.* Springer, Berlin.

White, W.B. 1988. *Geomorphology and hydrology of karst terrains.* Oxford Univ. Press, 464 pp.

White, E. & White, W. 1969. Processes of cavern breakdown. *Bull. Natl. Speleol. Soc.* 31 (4): 83–96.

White, E.L. & White, W. 1984. Flood hazards in karst terrains: lessons from the Hurricane Agnes storm. In A. Burger & L. Dubertret (eds) *Hydrogeology of karst terrains* 1: 261–264.

Zupan Hajna, N. 2003. *Incomplete solution: weathering of cave walls and the production, transport and deposition of carbonate fines.* Carsologica, Postojna-Ljubljana, 167 pp.

Geotechnical study at Sirwani landslide site, India

V.K. Singh
Scientist, Slope Stability Division, Central Mining Research Institute, Dhanbad, Jharkhand, India

ABSTRACT: The paper deals with the landslide study at Sirwani, Sikkim. It was also aimed to know the influence of water on the safety factor by sensitivity analysis. The rock discontinuities were mapped along the landslide and road cutting exposures as per the norms of International Society of Rock Mechanics. After identifying kinematically possible failure modes, detailed stability analysis is carried out by limit equilibrium. Landslides occur frequently in the Himalayas. This is due to the high intensity of rainfall that contributes to rapid erosion and weathering of the rock mass, which leads to reduction in the stability of natural slope. Remedial measures were suggested for better stability in the landslide prone zone.

1 INTRODUCTION

The research work was aimed for the landslide study of Sirwani landslide, Sikkim. It was also aimed to know the influence of water on the safety factor by sensitivity analysis, which tells the importance of the parameter in the slope. A more justified and suitable remedial measure can be planned for any critical slope after sensitivity analysis.

The Sirwani landslide is located at a distance of about 28 km from Gangtok and at about 1.7 km from Singtam on the road of Singtam to Dikchu. The yearly precipitation in the capital city of Gangtok is relatively high with an annual average of 3500 mm.

The stability of slopes depends on the geological structures, geomechanical properties of the slope materials, groundwater/rainwater condition. The rock discontinuities were mapped along the exposures of landslide and road cuttings as per the norms of International Society of Rock Mechanics (ISRM 1978). Geotechnical mapping was undertaken to determine the critical orientation of structural discontinuities. After identifying kinematically possible failure modes, detailed slope stability analysis was carried out by limit equilibrium method.

2 GEOLOGY OF THE AREA

The area near landslide consists of metamorphic arenaceous and argillaceous rocks intruded by basic sills that have been metamorphosed to epidiorite and talcose phyllites. A layer of weathered rock material consisting of sand and silt covers most of these phyllite formations. The area is characterized by steep slope.

The Top most lithology in the area is soil, which is underlain by shales and phyllites.

3 PHYSICAL CHARACTERISATION OF DISCONTINUITIES

The geotechnical mapping was done along the exposures of road cuttings. The mapping was done as per the norms of ISRM (1978). The general structural pattern in the rock mass of the phyllite is presented in Table 1. It will be appropriate to mention here that the characteristics of discontinuities are varying in different parts of the landslide zone because of the local folding and the area is geologically complex.

4 GEOMECHANICAL PROPERTIES

The samples of the slope mass in which the failure had taken place, were tested for various geo-mechanical properties at CMRI geo-mechanical testing laboratory. The mineralogical analysis was done with help

Table 1. Orientation of major joint sets in phyllite.

Joint sets	Mean Orientation of Joint Sets			
	Dip direction (°)	Dip amount (°)	Spacing (cm)	Persistence (m)
J1	N251	44	20 to 40	0.5 to 10
J2	N186	48	30 to 50	0.2 to 0.4
J3	N320	72	200 to 400	0.4 to 0.7

Table 2. Geo-mechanical properties of the slope material mass at Sirwani landslide site, Sikkim.

Swell Pressure (kg/cm²)	angle (degree)	Bulk Density (kN/m³)	Cohesion (kPa)	Friction
0.02		2.39 (DC)	0.92 (DC)	35 (DC)
		2.64 (UC)	0.32 (UC)	26 (UC)

Abbreviations: DC—Drained condition, UC—Undrained condition.
Major minerals in slope material: Quartz, Illite, Muscovite, Clinochlore, Phlogopite.
Slake durability test on shale: First cycle: 90.00 %, second cycle: 58.00 % (Medium durability).

of DB Advance X-ray Diffraction system (from M/s Bruker AXS, Germany). The relevant geotechnical parameters of slope material have been presented in Table 2.

5 SLOPE STABILITY ANALYSIS

Initially the kinematic analysis is done to identify the types of failure. The detailed stability analysis was done with the help of GALENA software, which is based on Limit Equilibrium Method.

5.1 Kinematic Analysis

The average orientations of the discontinuity sets (Table 1) determined from the geologic structural mapping were analysed to assess kinematically possible failure modes involving structural discontinuities (Figure 1). The critical discontinuity must lie within 20 degree of the slope face for plane failure to occur.

The kinematic analysis was done to determine the types of failure in the rock mass (Figure 1). The joint sets (J1, J2 and J3) present in the area are dipping obliquely to the slope face. So a large-scale plane failure is unlikely in the rock mass. The failure along the wedges (W1 and W3) is unlikely because these wedges are dipping obliquely to the slope face. Small-scale failures are possible along W2 wedge. But it is a general phenomenon in the hilly terrain.

5.2 Stability analysis by limit equilibrium method

The kinematic analysis shows that large scale plane or wedge failures are unlikely in the rock mass. So, the failure was limited to the topsoil mass along with

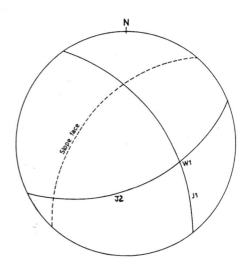

Figure 1. Kinematic analysis for types of failure.

the upper weathered-highly fractured rock mass and the same was observed in the field also. The failure in this type of slope is always characterised by circular type of failure. The earthquake effect is simulated in the GALENA software by assigning a value for basic horizontal seismic coefficient for the concerned area.

The presence of rain/surface water may decrease the shear strength of the clays, so sensitivity analysis was done separately to determine the effect of water on factor of safety. The drained slope mass condition means that the slope mass has been provided proper drainage in and around the landslide site. The undrained condition means that either the drains are not provided or the provided drains are not effectively maintained.

The contour plan (Fig. 2) has been prepared based on the data obtained with the help of sophisticated & precise Electronic Distance Meter (WILD DI4L, Swiss make) during the fieldwork of the landslide area. The typical section of the landslide area shows the existing steepest slope face direction. The stability analysis was done along this section in drained (Fig. 3) and undrained (Fig. 4) conditions.

The basic horizontal seismic coefficient for Gangtok is 0.05 (IS: 1893–1984. 2000). This value was used during stability analyses along with the other geotechnical parameters of slope mass (Table 2). The results of stability analyses are presented in Table 3.

The stability analysis shows a factor of safety of 1.5 in drained condition, so the slope is likely to be stable in drained condition. But the factor of safety under undrained condition reduces to less than the cut-off value of factor of safety (1.5). So, every attempt should be made to keep the slope in drained condition for the slope stability.

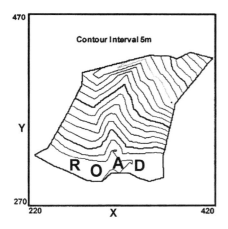

Figure 2. Contour plan.

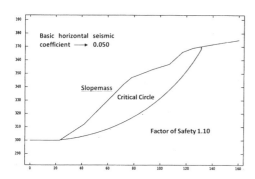

Figure 3. Stability analysis in drained condition.

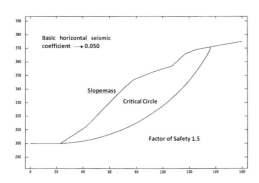

Figure 4. Stability analysis in undrained condition.

Table 3. Slope stability analysis of Sirwani landslide site.

Sr. No.	Slope mass condition	Factor of safety
1.	Drained slope mass	1.48
2.	Undrained slope	0.90

6 DISCUSSION

Sensitivity analysis shows that the influence of water is alarming. The slope may become unstable if the drainage in and around the failed zone is not effectively maintained. Every attempt should be made to minimize the entry of surface/rainwater to the landslide area.

There are main two reasons of slides in this area. First reason is the formation of deep gullies along the flow channel of the failed slope. In rainy season, a huge quantity of surface water flows down through these nala with a great speed. The current of water flow used to be very high due to steep gradient of the topography. The flowing of water with great current undercuts the existing/overlying slope mass. When the undercutting is deep enough then the undercut overlying slope mass losses equilibrium and fails down the slope. The failed mass is again washed away down the slope after subsequent rains. The failed mass used to be top debris and under cut highly weathered phyllitic rock mass. The slide area is being provided by a garland drain to check the entry of rainwater in to the slide zone. The slide zone is being terraced by providing stepped concrete steps to check the momentum of the running water during rainstorm.

The second reason of sliding is the under cutting at the toe position of the hill slope below road by the high current of Tista river. Due to this undercutting, the overlying slope mass becomes unstable. The heavy gabions are being provided near the undercut to check the undercutting by the speedily flowing river water during the rainy season.

Further, it is likely that topsoil and highly weathered-fractured phyllitic formation became saturated and lost shear strength due to the intense infiltration of water during the rainstorm. The water could not percolate to the underlying unweathered phyllite. It resulted in high water pressure in the top fragile formation. Pore pressure reduces the available shear strength and the weight of the water adds to the forces that induce sliding.

The slope movements in the landslides involve a combination of earth slide and debris flow. These landslides are characterised as sliding of earth followed by flow of the displaced material. The convergence of surface run-off of all the catchments area in a small area provides water for an initial failure. If a landslide occurs during a run-off producing storm, then surface run-off water will flow down the catchments area and underground water flowing from slope face will spill into the initial failure. Deformation accompanying an initial failure may allow further incorporation of water emanating from bedrock springs and surface runoff into the failed material, thus increasing debris mobility.

The amount of water available for mixing with landslide debris and the gradient of the down slope channel contribute to the transition of an initial landslide into a mobile debris flow. Incorporation of excessive volumes of water may dilute landslide debris and increase its mobility.

7 CONCLUSIONS

An assessment of the engineering and structural geology, strength properties and related geotechnical controls show that the existing slope is likely to be stable in drained condition but the same slope becomes unstable in undrained condition.

Sensitivity analysis shows that the influence of water is alarming. The slope may become unstable if the drainage in and around the failed zone is not effectively maintained. The landslide is occurring mainly due to the formation of deep gullies and subsequent undercuts formed by the currents of uncontrolled flowing rainwater down the hill slope. The second reason of sliding is the under cutting of the hill slope at the toe position by the flowing water of Tista river. Due to this undercutting, the overlying slope mass becomes unstable. Proper terracing of the slide area and heavy gabions are being provided to check the slides triggered by these two factors.

The topsoil and highly weathered-fractured phyllitic formation becomes saturated and loose strength due to the intense infiltration of water during the rainstorm. The water could not percolate to the underlying unweathered phyllite resulting in high water pressure in the top fragile formation. So every attempt is now being made to keep the slope in drained condition and to minimize the entry of surface/rainwater to the landslide area by providing suitable drainage in and around the slide area.

REFERENCES

IS:1893–1984. 2000. *Indian Standard Criteria for Earthquake Resistant Design of Structures*. Bureau of Indian Standards, pp. 58–59.

ISRM. 1981. Suggested methods for the quantitative description of discontinuities in rock masses. *Rock Characterization, Testing and Monitoring, ISRM Suggested Methods*: 3–45. E.T. Brown (ed.), Published for the Commission on Testing Methods, International Society of Rock Mechanics, Pergamon Press.

Inferences from morphological differences in deposits of similar large rockslides

A.L. Strom
Institute of Geospheres Dynamics, Russian Academy of Sciences, Moscow, Russia

ABSTRACT: Case-by-case comparisons of otherwise similar rockslides and rock avalanches that differ in morphology can provide reliable and important data for better understanding rockslide motion. Several groups of rockslide examples with more or less similar failure conditions but different deposit shapes or runouts are described. A common theme of differences in mode of momentum transfer is offered in explanation of the observations.

1 INTRODUCTION

A number of rockslides having similar failure conditions, but differing deposit morphologies are described briefly. Events are compared that have similar volumes, occur on slopes of similar height and steepness, are composed of similar lithologies, and moved over more or less similar terrain, but which formed deposits with differing morphologies and/or have different runout lengths. Such comparisons help to reveal the factors influencing rockslide morphology and, thus, provide better understanding of their motion.

2 COMPARATIVE CASE STUDIES

2.1 *Differing debris distributions along rock avalanche paths*

Significant difference in debris distribution along rock avalanche paths can be illustrated with examples from the Kokomeren River valley in the Central Tien Shan. Several rockslides in granite and gneiss, 10 to 20 Mm^3 in volume had fell from slopes of similar height and steepness. Nevertheless, they formed rock avalanches with widely different debris distributions in their transition-deposition zones, forming different shaped debris accumulations.

One of these examples—the Seit rock avalanche—entailed the entire rock mass in the avalanche-like motion (Figure 1-A).

In contrast, the Southern Karakungey and the Chongsu rock avalanches both formed a compact body at the foot of their source slopes and a secondary rock avalanche extending beyond it. Both compact and long runout parts include significant portions of the collapsed rock mass. In each example, the boundary between the components is defined by a concave 'secondary scar' on the downstream slope of the compact body (Figure 1-B).

The Northern Karakungey and the Kashkasu rockslides jumped from benches and struck their valleys' bottoms nearly at a right angle. These also formed compact bodies accompanied by avalanche-like tongues. But here, the compact portions have convex slopes, while their mobile debris portions gradually thin towards the distal margins (Figure 1-C).

It was proposed (Strom 1996, 2006) that these variations in rockslide debris distribution and morphology depends on the shape of the collapsing slope foot where the initial acceleration ceases and motion continues due to the momentum of the rockslide mass.

2.2 *Assymmetry of up- and down-valley flow*

Rockslide-deposit shape differs also in the up- and down-valley direction when rockslides spread both up- and down-valley—transverse to the initial debris motion vector. This variability in spreading can be observed at rockslides falling from one valley side and striking the opposite slope at a nearly right angle to form a high natural dam. Some examples are shown in Figure 2.

It is hypothesized that the extent of the spreading depends on the presence or absence of a tributary valley at the collision zone to act as a "trap" for the rapidly moving rockslide mass. Its presence at a rockslide axis prevents transverse debris spreading and leads to a symmetric dam with steep up- and downstream slopes (Figure 2-A). If such "trap" is positioned up- or downstream from the rockslide axis, spreading along the main valley is much more pronounced just on the side of the blockage farther from the tributary

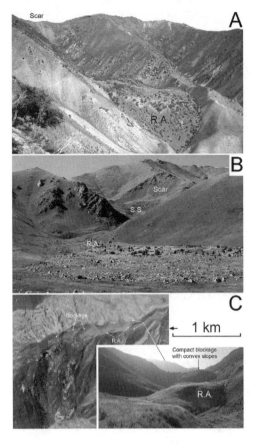

Figure 1. Rock avalanches featuring different debris distribution along their path. A—the Seit rock avalanche ~15×10^6 m^3 in volume, which debris accumulated at the distal part of the 3-km long rock avalanche path (R.A.). B—The Chongsu rock avalanche ~6×10^6 m^3 in volume that formed compact body with well-expressed secondary scar (S.S.) and 1.5-km long rock avalanche (R.A.) that involves about 65% of debris. C—Aerial and hand-made photographs of the Northern Karakungey rock avalanche ~10×10^6 m^3 in volume that jumped from a bench and formed a compact blockage with convex slopes; its portion turned right and traveled ~0.5 km downstream (R.A.).

mouth, irrespective of whether is it located on the up- (Figure 2-B) or downstream side (Figure 2-C).

2.3 Runout anomalies

Another interesting phenomenon is variation in runout distance for rock avalanches originating from the slopes of nearly similar height and steepness. This contradicts the "standard" H/L vs. rockslide

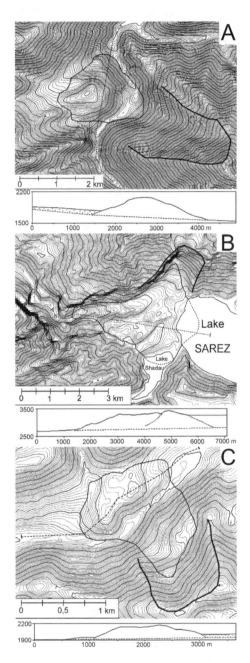

Figure 2. Large rockslide dams (maps and along-valley cross-sections). A—the symmetric Aksu rockslide dam, Tien Shan; B—the Usoi rockslide dam, Pamirs, that demonstrates distinct downstream spreading while its upstream slope is very steep; C—the Djashilkul rockslide dam, Tien Shan, with more pronounced upstream spreading rather than downstream one.

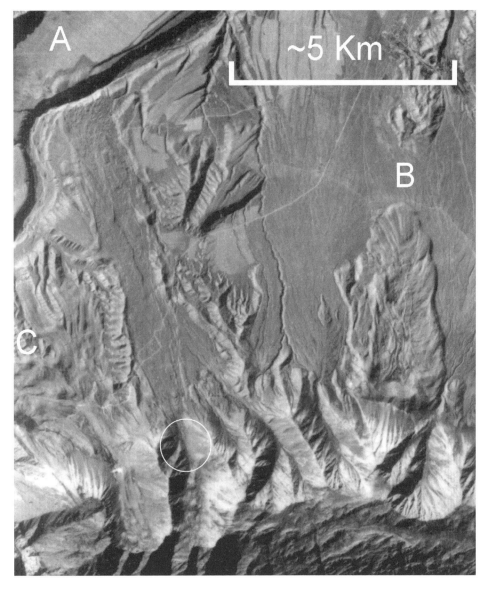

Figure 3. The Chaartash rock avalanches, Tien Shan. **A**—the Chaartash-2; **B**—the Chaartash-3, **C**—the Chaartash-1. White circle marks narrowing of the valley through which the Chaartash-2 rock avalanche passed.

volume relationship. The latter implies that increasing rockslide volume results in decreasing H/L ratio (Scheidegger 1973, Hsu 1975, Shaller 1991).

Such anomaly can be exemplified by two rock avalanches that descended from the slopes of the Western Akshiyriak range in the Central Tien Shan. Both are composed of granite and moved over the slightly inclined intermontane depression bottom (Strom1998). One—the Chaartash-2 rock avalanche of about 120×10^6 m^3 in volume—initiated from the 600-m high slopes of a deep gully, so that collapsing masses collided each other at the gully bottom. This collision formed a secondary rock avalanche, which passed beyond the narrow gully mouth to travel 7 km

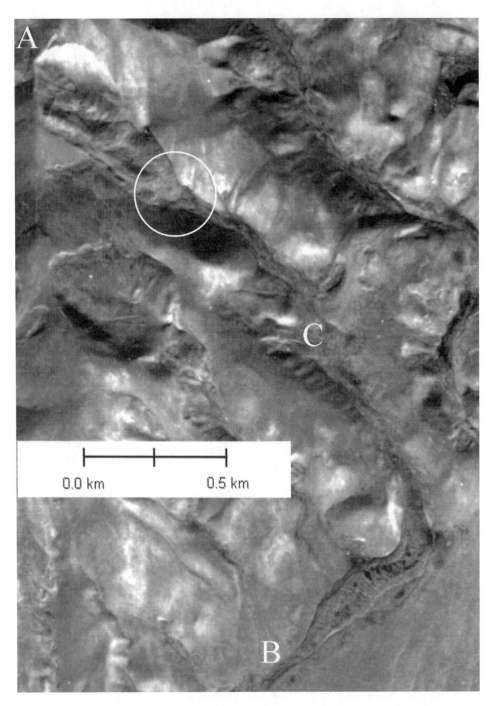

Figure 4. Long runout Snake-head rock avalanche. The initial massive rock slope failure at **A** was accompanied by rock avalanche that traveled along the narrow gorge about 2.2 kilometers up to **B**. **C**—part of the transitional zone where debris of additional landslides "fell into" the main rock avalanche. Circle marks the funnel-shape narrowing shown on Figure 5.

Figure 5. Funnel-shape narrowing after which the initial rock slope failure of the Snake-head rockslide converted into long runout rock avalanche (see Figure 4).

(!) across the depression on a mean slope of 6°–7° (Figure 3-A). At the end of its path it crossed the Kugart River valley and stopped at the opposite bank. Its total runout is 9 km and H/L ratio is 0.18. The second example—the Chaartsh-3 rock avalanche of about 200×10^6 m^3 in volume—descended a 800 m high slope rising just above the unconfined plain, and traveled 4 km across the depression on a mean slope of 8°–9° (Figure 3-B). Here, the total runout is 4.96 km and the H/L ratio is 0.27.

Seemingly, the Chaartash-3 conditions should be more favorable for long runout—there is no confinement, whereas the Chaartash-2 had to pass through a "bottleneck" of a confining valley. Moreover, the Chaartash-3 event had the greater volume and initial fall height, however the Chaartash-2 rockslide travelled farther.

3 DISCUSSION

The qualitative model was proposed to explain the phenomena described above (Strom, 1996, 2006). It was hypothesize that debris distribution and, correspondingly, its motion mechanism is governed by the shape of the collapsing slope foot, where the initial acceleration ceases and motion continues due to rockslide-mass inertia. A smooth transition from the steep slope onto a valley bottom leads to the formation of a "primary" rock avalanche where most of debris mass retains its mobility (see Figure 1-A).

In contrast, a sharp slope-to-foot transition or significant narrowing of the debris path results in an abrupt deceleration of part of the debris so that it forms a compact body and, as I assume, the momentum is transferred to the frontal or downvalley portion of debris, which retains possibility of further motion. This part moves as a "secondary" rock avalanche (see Figure 1-B). The same phenomenon may explain the asymmetric along-valley profiles of the large rockslide dams (see Figure 2-B, C). It seems that friction consumes the momentum of rapidly moving debris entering the tributary valley, while after direct impact on an opposite slope, some momentum transfer occurs, producing transverse spreading and, in the extreme case, formation of a secondary rock avalanche.

Momentum transfer from abruptly decelerating debris to a still moving portion may also explain the abnormally high mobility of the secondary rock avalanches that passed through the funnel-shape narrowing in comparison with other types of rock avalanches. Such a "funnel effect" likely took place at the Chaartash-2 secondary rock avalanche (see Figure 3-A) and at the much smaller Snake-head rock avalanche (Figures 4 & 5) in the Kokomeren

River basin (Central Tien Shan) (Strom et al., 2006). Considering the rather small volume of latter rockslide ($5 \times 10^6 - 7 \times 10^6$ m^3 only) its 2815 m runout is very long, and its H/L ratio is an abnormally low 0.12. Since only ~10% of the collapsed rock mass was involved in the avalanche-like motion the mobility of this particular part is extraordinary.

Transformation to rock avalanches in rockslides that fall from benches and strike the valley bottom nearly at a right angle could be governed by another mechanism supporting debris mobility. We should consider the linear dimensions of the catastrophically collapsing slopes in question; these are large—hundreds of meters, up to kilometer or so—and their complete descend should last from dozens of seconds to several minutes. Therefore the first part of the moving debris to reach the valley bottom could be over-ridden by subsequent debris. This may result in some type of fluidization of the over-ridden portion of debris which than spreads from under the main body as a viscous liquid. Perhaps the process is analogous to concrete released from a dumper.

4 CONCLUSIONS

Case-by-case comparisons of otherwise similar rock avalanches that differ in morphology can provide reliable and important data for better understanding rockslide motion. Such qualitative comparisons provide a basis for predicting future styles of debris motion and runout from potential large-scale rockslides, if we can compare their situations with studied events of similar geological and, above all, geomorphic conditions.

Besides clarification of some aspects of rockslide-motion mechanisms, case-by-case comparative studies allow more strict selection of case studies that can be used to derive relationships between various parameters characterizing rockslide size and mobility. Only those events should be compared quantitatively that have occurred under similar 'basic' conditions. For example, the Chaartash-3 unconfined rock avalanche can be compared with the Blackhawk (Shreve 1968, Johnson 1978) and the Chaos Jumbles (Eppler, et al. 1987) rock avalanches in California, the Bayan-Nur rock avalanche in Western Mongolia (Strom 2005) and 2 other old rock avalanches in the same region at 47°42′36.83″N, 92°35′40.90″E and 47°45′22.64″N, 92°33′33.41″E, with the 1st rock avalanche of the Sierra Laguna Blanka complex (Argentina)—(Hermanns et al. 2001), with rock avalanche in Chinese Tien Shan at 42°16′48.95″N, 87°18′41.59″E, and with some other Terrestrial and, probably, Martial (Shaller 1991) rock avalanches that had moved along the similar unconfined terrain. The Chaartash-2 event should be excluded from this sampling, since it occurred in a significantly different geomorphic situation.

Many more rockslide case studies of this type will provide more representative sampling and allow more strict selection criteria, thus providing more reliable relationships between basic parameters.

ACKNOWLEDGEMENTS

I want to express my gratitude to anonymous reviewers for critical analysis of the manuscript and to Mauri McSaveney for his kind assistance in improving the text and useful comments.

REFERENCES

Eppler, D.B., Fink, J. & Fletcher, R. 1987. Rheologic properties and kinematics of emplacement of the Chaos Jumbles rockfall avalanche, Lassen Volcaic National Park, California, *Journal of Geophysycal Research,* B95: 3623–3633.

Hermanns, R.L., Niedermann, S., Garsia, A.V., Gomes J.S. & Strecker M.R. 2001. Neotectonics and castasrophic failure of mountain fronts in the southern intra-Andean Plateau, Argentina, *Geology* 29: 619–623.

Hsü, K.J. 1975. Catastrophic debris streams (sturzstroms) generated by rockfalls. *Geological Society of America Bulletin* 86:129–140.

Johnson, B. 1978. Blackhawk landslide, California, U.S.A, In B. Voight (ed), *Rockslides and Avalanches*, 1, *Natural Phenomena*, 481–504. Amsterdam: Elsevier.

Shaller, P.J. 1991. *Analysis and implications of large Martian and Terrestrial landslides.* Ph.D. Thesis. California Institute of Technology.

Sheidegger, A.E. 1973. On the prediction of the reach and velocity of catastrophic landslides. *Rock Mechanics* 5: 231–236.

Shreve, R.L. 1968. *The Blackhawk landslide*, Geological Society of America Special Paper 108.

Strom, A.L. 1996. Some morphological types of long-runout rockslides: effect of the relief on their mechanism and on the rockslide deposits distribution, In: K. Senneset (ed) *Landslides; Proc. of the Seventh International Symposium on Landslides, 1996, Trondheim, Norway*: 1977–1982. Rotterdam, Balkema.

Strom, A.L. 1998. Giant ancient rockslides and rock avalanches in the Tien Shan Mountains, Kyrgyzstan, *Landslide News.* 11: 20–23.

Strom, A.L. 2005. Gigantic rockslides and rock avalanches in the Central Asian region, In: Senneset, K., Flaate, K. & Larsen, J.O. (eds.), *Landslides and Avalanches, ICFL 2005 Norway* : 343–348. Taylor & Francis Group, London.

Strom, A.L. 2006. Morphology and internal structure of rockslides and rock avalanches: grounds and constraints for their modelling, In S.G. Evans, G. Scarascia Mugnozza, A. Strom, and R.L. Hermanns (eds.), *Landslides from Massive Rock Slope Failure, NATO Science Series: IV: Earth and Environmental Sciences,* 49: 321–346. Springer, Dordrecht.

Strom, A.L., Djumabaeva, A.B., Dyikanalieva, J.K. & Ormukov, Ch.A. 2006. The Snake-head rock avalanche: rock slope failure caused by horizontal seismic acceleration, In: *Quantitative Geology from Multiple Sources, Liège, Belgium, September 3–8, 2006.* CD-ROM. S08–12.

Movements of a large urban slope in the town of Santa Cruz do Sul (RGS), Brazil

L.A. Bressani
PPGEC/Federal University of RGS, Porto Alegre, RS, Brazil

R.J.B. Pinheiro
PPGEC/Federal University of Santa Maria, Santa Maria, RS, Brazil

A.V. D. Bica
PPGEC/Federal University of RGS, Porto Alegre, RS, Brazil

C.N. Eisenberger & J.M.D. Soares
PPGEC/Federal University of Santa Maria, Santa Maria, RS, Brazil

ABSTRACT: This paper presents a study of an unstable urban slope situated in the city of Santa Cruz do Sul, RGS, Brazil. The slope is located in the northern part of the city, where several houses have been affected by slope movements. Its overall size is around 8 hectares. The slope consists of a layer of colluvium and man made fill. Instruments for field monitoring have been installed and they showed a relationship between slope displacements and peaks of piezometric level. Slope stability analyses were carried out using shear strength parameters from direct shear tests. Factors of safety obtained in the analysis were typically close to one, presenting some variability. An analysis of building pathologies caused by slope movements was also carried out, using data obtained with site inspection and interviews, which indicated the lateral extension of the movement.

1 INTRODUCTION

The town of Santa Cruz do Sul is located in the central part of Rio Grande do Sul State, with 107.000 inhabitants. The town is surrounded by colluvium slopes, especially to the North and East. Its average altitude is about 122 m. The region has a sub-tropical climate with humid and hot summers. The rains have low intensity and long durations during winter and high intensity but short durations during summer. In the period of 1988 to 2003 the accumulated precipitation measured in the town during the year varied between 1419 mm and 2100 mm.

The town is known to have problems of slope instability since the 70's (Grehs 1976). This paper describes the study carried out in a region with some houses damaged at various degrees. By chance, the work was carried out in a period with large rainfall and the soil mass movements were clearly seen (and measured).

2 GEOLOGY AND SITE INVESTIGATION

The geology of the North and East parts of the town is formed by siltstones of the Santa Maria Formation at the base, followed by Botucatu sandstone and basaltic layers of the Serra Geral Formation. In some places the Botucatu is not present, as in the case described here. The Santa Maria siltstones are quite red in colour which makes its field identification very easy. The basaltic layers present on the top of the sequence are heterogeneous and with varying thickness. These rocks present an intense vertical and horizontal fracturing (Grehs 1976, Wenzel 1996).

The Santa Maria Formation is formed basically by clayey-siltstones with a massive aspect, reddish in colour and with smectite type clays. This Formation is widespread in the town on altitudes of 30–100 m. Most of the central portion of the town is located on this material. Its permeability is much lower than the permeability of the basalts and this has important

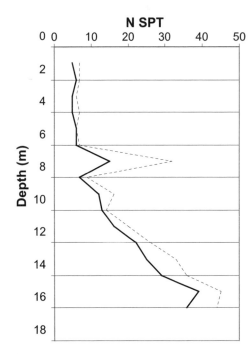

Figure 1. Typical geotechnical profile (some presence of cobbles at the materials contact, 6–8 m).

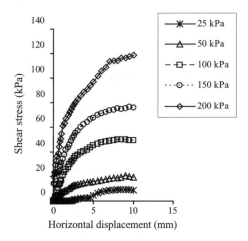

Figure 2. Results of direct shear tests on colluvium.

consequences. Grehs (1976) has indicated that regions where saprolitic soil of this Formation are present have slope problems.

2.1 Site investigations

A number of SPT borings and manual auger borings have been made by Pinheiro et al. (2002) to obtain the geotechnical profile of the region. Some samples for laboratory testing were also collected. From these results the geotechnical profile was described as a deposit of colluvium soil and fill with varying thickness (3–10 m). The soil can be described as a sandy silt-clay mixture having boulders of basalt and sandstone and brown to reddish colours. Underneath this layer, the red siltstones of Santa Maria Formation were present. Figure 1 shows a typical sounding profile of the region. The change in materials was clearly seen in the field by the difference in colour and blow counts. Presence of cobbles was sometimes noticed at the contact.

2.2 Laboratory tests

Pinheiro et al. (2002) and Eisenberger (2003) carried out a series of characterization tests with the materials taken from the borings. The scatter in the results was significant on grading, plasticity limits and physical indexes of samples taken from 12 borings.

Direct shear test on colluvium and red siltstone specimens soaked in water were carried out (5×5 cm e 10×10 cm sizes). The colluvium samples were taken from 2 positions on the slope and the void ratio varied between 1.2 and 1.5. Drained triaxial (CID) tests in the colluvium soil were also carried out on specimens of 5 cm in diameter. The results of direct shear tests have not shown peak strength on any of the specimens. The colluvium specimens generally maintain a constant level of shear strength after some deformation. The volumetric variation only presented compressive behaviour.

The resulting envelopes obtained have shown some variation but they seem to pass through the origin. The internal friction angle varied from 23° to 29°, with $c' = $ zero. Figure 2 presents the curves of a set of direct shear tests carried out on colluvium.

3 FIELD INSTRUMENTATION

The instrumentation for monitoring of this slope was first installed in 2001 and during 2002 other instruments were installed. The monitoring included Casagrande piezometers, inclinometer tubes, an automatic pluviometer and physical markers to measure opening of fissures in buildings and walls. The installation procedures generally follow recommendations of Dunniclif (1988). Most of the data presented here is base on a time scale of 707 days, which cover the period March 30, 2001 to March 08, 2003.

3.1 Automatic pluviometer and piezometers

Four Casagrande type piezometers were installed on auger borings of 100 mm diameter previously prepared. The piezometers were made of plastic tubes with sand filter of 60 cm and a bentonite seal on top of it. One of the piezometers (P4) had an electric transducer connected to a data acquisition system installed inside of the tube. This piezometer recorded the pore water pressure every hour. The other piezometers (P1, P2 and P3) were read manually by conventional means (a probe inserted inside of the plastic tube).

An automatic pluviometer was installed in the same place as one of the inclinometers and close to the piezometers. The main purpose was to obtain the rain intensity as real as possible. The system registered the rains with 1h intervals during the period of June 24, 2002 to February 15, 2003. The records collected during this period showed monthly average values larger than the average of the last 16 years and in some cases being even larger than the maximum previously recorded. During October 2002 the precipitation reached 369 mm and that was the maximum of the historical series (Figure 3). The rains during 2002 have shown larger variation than normal and the town had a number of street floodings as a result. The coincidence of this natural phenomenon with the instrumentation timing was an unexpected opportunity for the work.

The measurements obtained from the manual piezometers from May 2001 indicated a similar trend of pore water pressure in all of them. The piezometers have shown variation of 2 to 3 m in the water level. No artesian pressures were observed.

3.2 Inclinometers

The inclinometers tubes installed were positioned close to the position were most of the damages in buildings were first observed. The results obtained are presented in Figure 4 for the two inclinometers.

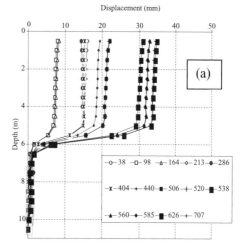

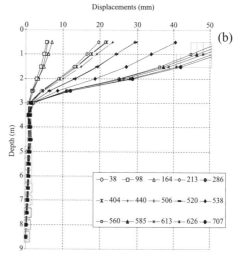

Figure 4. Results obtained from inclinometers; (a) inclinometer I1; (b) inclinometer I2.

The results presented in Figure 4(a) shows a very clear slip surface at 6 m depth, at the contact colluvium-siltstone. The accumulated displacement of the soil mass was of 35 mm during 22 months.

The second tube (I2) was situated above the first inclinometer and crosses a smaller colluvium deposit. The results presented in Figure 4(b) have shown a slip surface at 3 m depth and so well defined as the previous one. The results show a gradient of displacements reducing with depth, but the maximum accumulated displacement measured during the 22 months was 63 mm.

The displacements measured with the inclinometers have shown clear periods of acceleration which

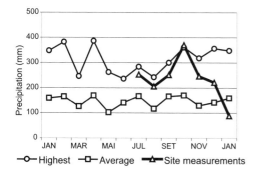

Figure 3. Precipitation records during the period compared to historical averages.

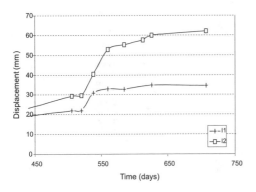

Figure 5. Displacement at the top of the inclinometer tubes versus time.

coincide with periods of intense rains. The larger displacements were after periods of intense rains, especially on Oct 30, 2001 and Oct 12, 2002. Figure 5 shows the horizontal displacements measured at the top of the tubes versus time of monitoring (days measured from the beginning of monitoring). It is evident that the soil mass have periods of no-displacements and periods of acceleration. This correlates well with the observed rains.

3.3 Displacement pins

In order to monitor the opening of existing fissures caused by the slope movements in buildings, three sets of pins have being installed on damaged structures in the area. The arrangement of pins formed a vertical triangle which allowed the measurement of horizontal and vertical components of movement across the fissure. The resolution of readings was of the order of 1 mm. The sections called T1 and T2 were established in one house which showed considerable damages. The total horizontal displacement was of 2.4 cm and the vertical displacement of 0,6 cm. The other set, 12 m away, showed 2.7 cm and 3.9 cm, respectively, in the same period of around 6 months (Sept 2002 until March 2003).

3.4 Buildings damage

In order to have a better mapping of the movements, a series of inspections were made on buildings in the region and interviews with the owners. This allowed the construction of a house damage map. The damages were classified by type, origin and severity, following Alexander (1986) and Chiocchio et al. (1997).

The residences inspected generally follow the same standard: the structures were made of reinforced concrete supported by concrete or metal piles. There was generally some earth moving in the plots. The center of the slope studied presented the most severe damages. Using evidences taken from fissures in a garden paved floor it was possible to infer movements of 40 cm in the horizontal direction from 1997 to 2002. The damages reduce in severity as the residences are situated further from the center line of the main movement, but as the damages are a bit scattered in these regions the limits could not be defined accurately. The total area with movements has around 7 to 9 hectares.

4 DATA ANALYSIS

The analysis of data obtained gave rise to some interesting findings. The measurements of automatic piezometer P4 sometimes gave responses to precipitation which were quite fast in relation to the precipitation. This peaks were generally correlated with movements of the slope measured by the inclinometers and displacement pins, especially during September 2002.

Figure 6 shows the water level (depth below surface) during the critical period of rain. Comparing it with Figure 5 it can be seen that the peaks of water pressure correlate well with the inclinometers movement acceleration. Those peaks have not been shown by the manually read piezometers, as the velocity of increase and decreased of water level is quite fast (some hours in most cases).

Figure 7(a) shows the results of precipitation during the month of October 2002, one of the wettest in the 16 year period and Figure 7(b) shows the measurements of water level obtained from piezometer P4. As can be seen, the first precipitation of 45 mm (day 3) caused a pronounced peak of 44 cm on it, but the larger amount of subsequent rain did not appear to show any influence, in particular the precipitation of 60 mm on day 7. So, sometimes it responds quite well to heavy rains, other times showed almost no response. This may be a consequence of the movement of the slope as this may change the internal mass permeability.

In general, the horizontal displacements measured at the surface through the pins positioned at the

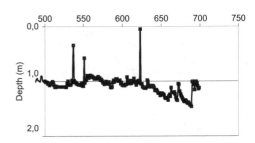

Figure 6. The water pressure measured by piezometer P4 during September and October 2002.

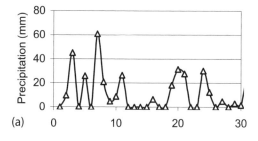

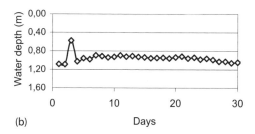

Figure 7. (a) Precipitation during October, 2002; (b) water depth measured by Piezometer 4 on the same days.

buildings showed the same trend but much larger displacements than the inclinometers. The reason is that they were installed in the fissures where most of the movements are concentrated. The inclinometers and the pins showed the larger movements around the day 550. It is interesting to note that the peak of water pressure on day 625 had effects on inclinometer I2 and pin T3 but no significant influence on inclinometer I1 and pins T1 and T2. This data confirms other evidences that the movement of the slope is a sequence of small retrogressive movements of individual blocks.

5 STABILITY ANALYSES

Two representative cross-sections of the slope were analyzed through the SLOPE/W program. The sections were coincident with the areas where scars are visible at the surface. Various simulations were carried out using the observed pore-water pressures and measured internal friction angles. In general the factors of safety obtained were close to 1.0 but the results were quite sensitive to small values of cohesion. The most critical surfaces were found at the contact between the siltstone and the colluvium soil. The critical area pointed out by the software agreed quite well with the main scar found in the region but it was interesting to observe that the analysis showed not one critical surface covering the whole length, but a series of smaller surfaces. These surfaces reproduce quite well the movements observed in place and also shown by the monitored displacements.

6 CONCLUSIONS

The area studied was known to have some damages in the houses but little more was known. The area is formed by a gentle slope composed by a colluvium and man-made fill of siltstone with 3 to 10 m depth. A comprehensive damage survey in the slope buildings revealed that the damages varied between negligible to severe (following Alexander, 1986) depending on their position on the slope (there is a central region were the worse cases are found).

Borings were opened through the area and a number of soil samples were taken. The thickness of the soil mass varied between 3 and 10 m. Tests showed that the soil has some variability and this affect the shearing parameters as well.

The monitoring of the slope was carried out with piezometers, inclinometers and fissure measurements. The use of an automatic piezometer and pluviometers (1 hour interval readings) showed that the pore-water pressure has a fast response to the rains after a threshold value. The automatic piezometer showed peaks of up to 1 m (10 kPa) for intense daily rains and some response for accumulated rains over 3 days periods. The manual readings of the other piezometers showed no such peaks (as they were not read at the proper times). The rains which caused peaks of pore-water pressures were those above a 50 mm/day during wet periods.

The monitoring covered a period in which the monthly rains were more intense than the average of the last 16 years. This caused large displacements of the mass and the possibility to observe clearly these movements. The critical period, with larger inclinometer readings and surface displacements, was September and October of 2002 coincident with larger precipitations. Two inclinometer readings showed (a) a failure well defined in the contact of materials and (b) a distortion zone up to 3 m depth. Both instruments showed an acceleration of movements in that period.

The peaks of pore-water pressures have a good correlation with the movements of the slope as measured by inclinometers and fissure measurements. The movements seem to happen in retrogressive manner following a temporal sequence.

Stability analysis carried out on typical sections showed factors of safety around 1.0 in a number of critical surfaces which were limited in extension. It is interesting to observe that the composition of these individual surfaces appears to represent well the observed movement of the overall slope. In general the movements are slow (40 cm in 5 years is a good estimate) but with acceleration during the periods of pore-water pressure peaks, as was the case during monitoring.

Further work still in progress has shown that other areas of the town are also affected by movements of similar magnitudes.

ACKNOWLEDGMENTS

L.A. Bressani and A.V.D. Bica have been supported by CNPq. The MSc dissertation of C. Eisenberger was also supported by CNPq. The municipal local government gave support for the field work. CAPES provided funds for some of the instrumentation.

REFERENCES

Alexander, D. 1986. Landslide damage to buildings. *Environ. Geol. Water Sci.*, vol. 8, no. 3, pp. 147–151.

Chiocchio, C., Iovine, G. & Parise, M. 1997. A proposal for surveying and classifying landslide damage to buildings in urban areas. *Engineering Geology and the Environment*, Rotterdam, pp. 553–558.

Dunniclif, J. 1988. *Geotechnical Instrumentation for Monitoring Field Performance*. John Wiley & Sons Inc., New York, 577p.

Eisenberger, C.N. 2003. *Study of an urban colluvium slope in Santa Cruz do Sul-RS*. Porto Alegre: UFRGS, 2003. 111p. Master dissertation. Escola de Engenharia, UFRGS (in Portuguese).

Grehs, S.A. 1976. Geological Mapping of Santa Cruz do Sul—Basic information for Integrated Planning, *ACTA GEOLÓGICA LEOPOLDENSIA*, v.1, Universidade Federal do Vale do Rio dos Sinos, p. 121–152 (in Portuguese).

Pinheiro, R.J.B., Soares, J.M.D., Bica, A.V.D. Bressani, L.A. & Eisenberg, C.N. 2002. Geotechnical investigation of an urban slope in Santa Cruz do Sul—RS. *Proc. XII Brazilian Cong. Soil Mech. Geotech. Engng*, ABMS, Sã Paulo, Vol.,2, 1247–1257 (in Portuguese).

Wenzel, J.A. 1996. *Structura, geological mapping of the town of Sant Cruz d Su fo urba planning u U NISC Sant Cruz d Sul*. (in Portuguese).

Geotechnical analysis of a complex slope movement in sedimentary successions of the southern Apennines (Molise, Italy)

D. Calcaterra, D. Di Martire, M. Ramondini & F. Calò
Department of Geotechnical Engineering, Federico II University of Naples, Naples, Italy

M. Parise
National Research Council, IRPI, Bari, Italy

ABSTRACT: The results of a geological and geotechnical study on a complex landslide occurred in Molise, Italy, are described. The slope movement, a roto-translational slide evolving in an earth flow, was firstly re-activated in January 2003 by intense rainstorms, causing serious damage to man-made structures and the evacuation of 15 families from their houses. After a thorough archival research, a geological field survey was performed, integrated by interpretation of multi-temporal air-photos and by two site investigation and monitoring campaigns. Inclinometer measurements revealed some slip surfaces at depths between 5 and 30 metres. The geotechnical laboratory tests, aimed at characterizing the upper, weathered portion of the sequence and at comparing it with the unsheared materials, were used for a preliminary back-analysis, to verify the shear strength mobilized under the stability limit conditions, in view of developing a more suitable kinematic model for the landslide.

1 INTRODUCTION

The southern Italian Apennines are well known as areas highly susceptible to landslides (Cotecchia & Melidoro 1974; Pellegrino et al. 2003). The complex geological setting, where a great variety of different lithologies are present within the framework of a recently-built mountain chain, makes these territories among the most affected by slope movements in Italy. Predominance of clays and marls in the Miocene and Pliocene formations cause flow-type landslides to be the most diffuse typology. Usually induced by rainfall, earth flows can even be triggered by human activities, whilst the role played as triggering factor by earthquakes has also to be noted, in particular as regards the reactivation of large phenomena.

Notwithstanding the limited extension, Molise, the smallest region of central-southern Italy, is severely affected by landslides and erosional processes. In the last decade, several events brought to the attention of the public and the scientific community some cases of slope movements, triggered either by seismic shocks (Bozzano et al. 2004) or by heavy and/or prolonged rainfall events (Corbi et al. 1999; Picarelli & Napoli 2003).

In January 2003, a complex landslide was reactivated as a consequence of an intense rainstorm in the territory of Agnone, a town situated in the Isernia province. Known as the Colle Lapponi—Piano Ovetta landslide (CL-PO landslide in the following), after the localities where it develop, the phenomenon is part of an historic landslide, which already affected in the past the right slope of the Verrino Torrent. The present study analyzes the CL-PO landslide, illustrating the results of a geological, geomorphologic and geotechnical characterization, aimed at better comprehending the landslide kinematics.

2 SLOPE MOVEMENTS AT AGNONE

2.1 Historical landslides

Analysis of historical data on landslides, as well as on any other type of natural hazards, has repeatedly proved to be a necessary phase in the process of the hazard assessment. In particular as regards slope movements, availability of information on past events, including location, date of occurrence, and triggering factor, is extremely important for a proper comprehension of the recent geomorphologic evolution of the slopes. Regions where a long history of documentation is available, as in southern Italy, are particularly suitable for this type of research, as demonstrated in many cases (e.g. Calcaterra & Parise 2001; Calcaterra et al. 2003).

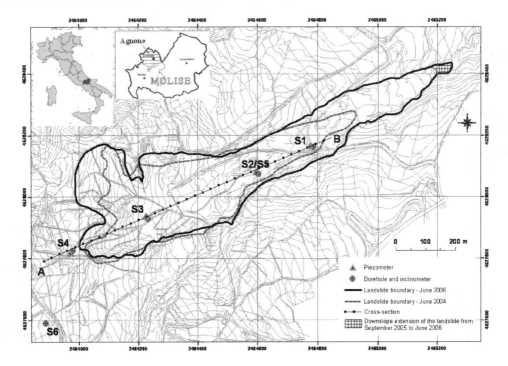

Figure 1. Plan view of the CL-PO landslide and its evolution in the time-span June 2004–June 2006.

As for many other districts in Molise, landslides in the municipality of Agnone have been known for long times. The oldest report of gravity-related phenomena, in fact, dates back to March 1905, when a slope movement was registered in the Vallone S. Nicola, due to snow-melting and a very intense rainfall period. The landslide damaged the bridge of one of the main roads entering the historic centre of Agnone (Almagià 1910). It is interesting to note that the locality involved in the 1905 event is the same dealt with here.

More recently, the Agnone area has been affected by several landslides, including some large phenomena covering many hectares. The historical and archival research on landslides carried out within the framework of a national project (AVI, an acronym which stands for Areas Vulnerated in Italy; Guzzetti et al. 1994) has shown the occurrence of at least 60 landslides at Agnone and the surrounding territory in the time-span 1970–1988. Among these, it is worth to remind of the landslide which occurred at Colle Lapponi in February 1994, and which caused a country road to be interrupted. A few hundreds of metres east of the site where the 2003 event took place, another landslide was triggered on February 1984 (Guadagno et al. 1987). Known as the Fonte Griciatta landslide, this rotational slide evolving to earth flow affected two abutments of a viaduct along the State Road no. 86 "Istonia".

2.2 *The Colle Lapponi—Piano Ovetta landslide*

The CL-PO landslide (Fig. 1) is a complex slope movement, consisting of a roto-translational slide which evolves in an earth flow. The landslide took place in the catchment of Vallone S. Nicola, a tributary to the Verrino Torrent, which runs west of Agnone. It was the consequence of the intense pluviometric event that occurred on January 23–27, 2003, in southern Italy. The event was characterized by a maximum rainfall height of about 200 mm over 72 hours, with less than 50 mm registered at Agnone. The landslide caused serious damage to rural buildings and the local road network. The main social consequence was the cautional evacuation of 15 families from their houses, located in the areas directly affected by the slope movement or in its immediate surroundings.

The geological setting of the landslide area is characterized by the presence of the Agnone Flysch, a structurally complex formation dated to Upper Miocene (Vezzani et al. 2004). In the study area, the lower member of the Agnone Formation is present, made up of marly clays, clayey marls, silty-sandy clays and subordinate sandy levels, with intercalated carbonate beds (thickness between 10 cm and 2 m). The clayey-marly terms show features typical of a weak rock and a scaly structure; along the scales, weathering evidences can be found, represented by reddish coatings. The formation is overlain by a weathered mantle, constituted by brownish-yellowish clays, silty clays, often iron-oxidated, rich in organic matter and including heterometric carbonate fragments.

A recent site investigation allowed an in-depth analysis of the lithostratigraphical features of the terrains, explored down to 40 m from the ground surface. In view of the eventual geotechnical slope stability analysis, four homogenous horizons have been recognized, of which their characters are as follows (Fig. 2):

– Complex A: chaotic, plastic, remoulded greyish to brown clay deposits, directly involved in the CL-PO reactivations. Thickness: 6.70–11.20 m;
– Complex B: light- to dark grey clays, silty clays, sandy clays, silty sands. Thickness: 4–12 m;
– Complex C: calcareous levels. Thickness variable from dm to m;
– Complex D: dark grey, scaly marly clays, clayey marls and marls, which minimum depth is at between 15.30 and 20 m from ground surface.

As regards the relative position of the various complexes, A, B and D are usually found in such sequence from the ground surface downward, while C can be found interbedded with A, B and D. Complex C is represented by lithoid, highly permeable beds, and assumes a fundamental role with respect to groundwater circulation. Following the above litostratigraphical scheme, a cross section has been reconstructed (Fig. 3) where layers belonging to complex C are not visible due to the scale adopted.

The rotational character is shown in the source area by the steeply inclined main scarp and by the overall spoon-like shape. Due to strata attitude (which, even though with several folds and irregularities, mostly

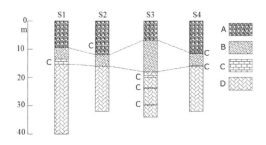

Figure 2. Lithological complexes recognized in the boreholes.

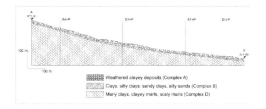

Figure 3. Geological cross-section through the CL-PO landslide.

dips toward the east and the north-east) the slope movement evolves from rotational to translational slide and successively into an earth flow.

As evident from Figure 1, the landslide has shown both a retrogressive and an advancing tendency, which resulted in an overall increase in the unstable area, evaluated in about 80,000 m^2 from 2004 to 2006, reaching a global value of about 240,000 m^2. From September 2005 to June 2006 the foot has progressed of about 70 m; consequently, the landslide has reached a total length of about 1500 m.

The CL-PO has completely altered the local hydrographic network. In fact, during the main phases of reactivations, the San Nicola torrent has been repeatedly obstructed; in addition, a number of ponds were noted, partly related to groundwater circulation hosted by the uppermost weathered horizons and often coinciding, during the rainy periods, with the ground surface. To this regard, it is important to note that the presence of carbonate rocks in contact with the structurally complex formation involved in the landslide provide a huge amount of water from the above. Further evidence of the active slope morphodynamics are given by several minor scarps, located upslope of the main crown of the CL-PO body, and by a number of small flows triggered by the side-slopes undercutting of the San Nicola torrent.

A source of major concern is represented by the toe zone. In fact, a possible further remobilization of the landslide could rapidly reach the intersection with the Verrino torrent, which at present is about 350 m apart. In such case, a landslide dam could build up resulting in adverse consequences to downriver man-made structures, among which the above cited viaduct, already damaged by the 1984 Fonte Griciatta landslide.

3 THE MONITORING CAMPAIGNS

With the aim of identifying the viable remedial measures, the CL-PO landslide has been under observation since 2003, through three site and laboratory investigation campaigns (2003, 2004, 2006). The 2004 and

2006 campaigns consisted of 26 boreholes, 11 inclinometers, 14 Casagrande and 1 open-pipe piezometers, optical levelling (58 benchmarks) and 36 samples for geotechnical laboratory analyses.

Measurements on a network of 58 benchmarks have shown that horizontal movements at the surface of the landslide ranged from 20 to 76 cm between April 2006 and April 2007, hence showing a displacement rate between 1.5 and 5.8 cm/month. Such rate is lower than that empirically evaluated during the March 14–15, 2004 reactivation, when the bridge over the San Nicola torrent was overtopped; on that occasion velocities between 0.5 and 1 m/hr were calculated. Comparing two series of air-photos, the landslide toe showed a total displacement of about 280 m over the June 2004–September 2005 period, corresponding to an average rate of 18.6 m/month. Field surveys in the toe zone allowed to recognize a further downslope extension of the landslide of about 70 m between September 2005 and June 2006 (Fig. 1), corresponding to 7.7 m/month.

The borehole inclinometers of the 2006 campaign (Figure 1) have recorded, from January 2006 to January 2007, similar rates of movement, with maximum displacements of about 30 cm. The depths of the major active shear surfaces range from 7 m to about 27 m. Accordingly, a longitudinal profile of the landslide has been reconstructed (Fig. 4). From these data, it is estimated that the active landslide has a volume of about 3.5 million m^3.

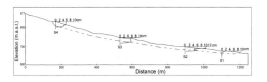

Figure 4. Slip surface by inclinometer measurements.

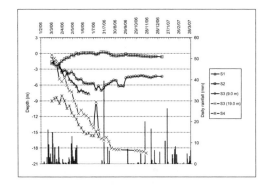

Figure 5. Groundwater levels compared with rainfall measurements.

From Figure 5, it is clear that the groundwater levels have shown variations in a depth range between −19 m and the ground surface. Different reasons can be invoked to explain this situation, such as a marked morphological disorder of the slope, caused by the CL-PO landslide and hydrogeological heterogeneities among the lithological complexes. In particular the gauges located in the C complex or near to it show a behavior independent by the rainfalls measured at the local raingauge, probably depending on the upslope connection of these levels to the carbonate aquifer.

4 GEOTECHNICAL DATA

The geotechnical characterization of the CL-PO terrains derives from the re-interpretation of the 2004 and 2006 laboratory tests, which, in turn, can be usefully compared with the literature data available for the Agnone Fmn. (Cotecchia et al. 1977; Guadagno et al. 1987). In particular, Guadagno et al. (1987), studying the Fonte Griciatta landslide, geotechnically characterized the clayey terms of the formation, distinguishing intact clays from the mudslide weathered clays.

Following the already recalled subdivision, quite different parameters can be referred to the A, B, and D complexes (Table 1 and Fig. 6). In the following, complex D parameters are derived from Cotecchia et al. (1977) and Guadagno et al. (1987).

Complex A is composed by lapideous fragments and shales plunged into a softened, fine grained matrix, consisting of a mix of clay and thin hard lumps. The clay fraction varies from 48% to 63%; dry density ranges between 14.0 and 16.5 kN/m^3, with a saturation ratio always high, in agreement with the water table

Table 1. Mean values of the geotechnical parameters of the complexes A, B, D.

Parameter	Complex A	Complex B	Complex D
γ (kN/m^3)	19.4	19.7	22.5
γ_d (kN/m^3)	15.2	16.0	21.2
γ_{sat} (kN/m^3)	19.6	20.1	22.9
γ_s (kN/m^3)	27.2	27.4	27.0
w (-)	0.27	0.23	0.10
e (-)	0.79	0.71	0.28
n (-)	0.44	0.41	0.22
Sr (-)	0.93	0.89	0.99
wl (-)	0.59	0.62	0.44
wp (-)	0.26	0.26	0.20
Ip (-)	0.33	0.35	0.24
ϕ' (°)	19	23	22
c' (kPa)	20	28	59
ϕ'_{res} (°)	17	22	9
c'_{res} (kPa)	0	0	0

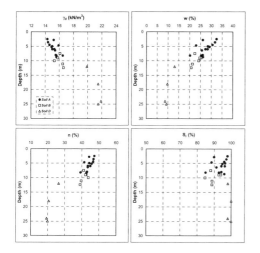

Figure 6. Physical properties of the investigated complexes.

position, near the ground surface. The porosity values vary between 40% near the bottom of the complex and 47% in its uppermost portion. As regards plasticity, complex A falls in the field of high plasticity clays, characterized by low activity. These values are based on the measurements of the "overall" water content; however, these results depend on the low water content of the lithorelicts included in the matrix. In fact, when the lithic inclusions are separated from the remaining part of the matrix, a higher value of the water content and of the liquid index can be obtained (Picarelli 1993). The shear strength, detected by means of direct shear tests, is characterized by a value of the peak friction angle near to 19° and a cohesion around 20 kPa.

Complex B has a grain distribution as clays, silty clays, sandy clays, silty sands. with a clay fraction varying between 50% and 55%, and a silt fraction slightly higher than complex A. Physical properties show a limited variability: dry density is between 15.0 and 16.5 kN/m³ while S_r varies from 85% to 95%. Such values, apparently contrasting with the boundary hydraulic conditions, depend on the structure of the complex and its sandy fraction. Porosity is lower than in complex A, with n values from 40% to 44%. In terms of plasticity, complex B can be referred to medium plasticity clays, again showing a low activity. The shear strength offers a value of the peak friction angle near 23° and a cohesion value around 28 kPa.

Complex D has a grain distribution as marly clays and clayey marls, with a clay fraction varying between 25% and 35%, and a silty fraction varying between 50% and 60% (Cotecchia et al. 1977; Guadagno et al. 1987). Physical properties are significantly different from the upper complexes: dry density is in fact definitely higher (20.0–22.0 kN/m³), porosity n is on average equal to 22%, showing a slightly higher value towards the top of the horizon. Low porosity also explains the high S_r values, even though the water content is quite low (Table 1). While plasticity and activity values are comparable to those belonging to complex B, the shear strength results higher: peak friction angle is in fact near 22° with a cohesion around 60 kPa.

The significant difference between the peak strength values in the three complexes disappears in the residual strength; in fact the friction angles ϕ'_{res} are near 17°, 22° and 9° referring to complex A, B and D respectively. These results are probably due to better mechanic characteristics of the complex B if compared to complex A. The lower value measured for complex D is to correlate with its scaly structure, characterized by low friction strength between each scale after the rupture of the cohesive bond.

5 SLOPE STABILITY BACK-ANALYSIS

Aimed at defining a suitable kinematic model, the best parameters governing the CL-PO landslide have been sought. Hence, a preliminary stability back-analysis has been performed, adopting the classical infinite slope model, that can be usefully applied to the case-study, considering the section of the longitudinal profile (depth/length ≪ 1–Fig. 4).

In the back-analysis the original pre-landslide profile has been considered, subdividing the slope into segments with constant slope angle; accordingly, the safety factor (SF) has been calculated for each segment. By doing so, slope sectors characterized by a SF lower than 1 have been defined, for different positions of the water table, considered with a flow parallel to the soil profile. Three hypotheses have been made as concerns the surface of rupture, placed at 5, 10 and 20 m below the ground surface respectively. Both peak and residual shear strength have been considered.

Figure 7 shows that the upper portion of the slope is under limit equilibrium conditions, while downslope SF is higher than 1. It can be also noted that, to explain the observed displacements, residual strength should have been available along some slope segments, since for $\varphi = \varphi'_p$ FS is always higher that 1. As regards the groundwater table, the critical conditions for the slope are reached when H_w is close to the ground surface for $\varphi = \varphi'_p$. On the other hand, when φ'_{res} is taken into account, SF is close to 1 when H_w is between 2 and 8 m below the ground surface.

These results, referred to the measured Hw piezometric data (Fig. 5), show some interesting behaviors (Fig. 8). In the upper zone of the slope, in fact, the SF values are lower than or close to 1 even to peak friction, at variance with the actual displacement regime; such

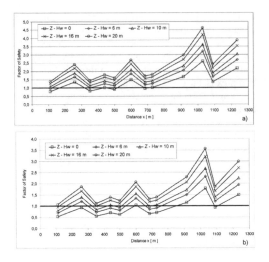

Figure 7. SF variations along the slope as a function of the groundwater table position (surface of rupture = 20 m). a = peak strength; b = residual strength.

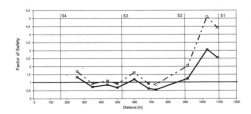

Figure 8. SF variations along the slope (surface of rupture = 20 m) in relation to the piezometer data. a = peak strength; b = residual strength.

evidence has probably to be correlated to some reconsolidation phenomena developed in this zone after the landslide occurrence. On the other hand, in the lowermost segments of the slope, the SF values are quite high with residual strength too: it can therefore be hypothesized that in this zone, during its occurrence, the landslide has changed its rheological features due to its velocity, evolving from a slide to a flow.

6 DISCUSSION AND CONCLUSIONS

Evidence from archival sources, field survey, airphoto interpretation and monitoring campaigns clearly demonstrate that the 2003–2007 CL-PO landslide is the reactivation of a pre-existing mass movement, known since 1905 at least, showing a behavior which can be ascribed to an intermittent kinematics.

In fact, following the Cruden & Varnes (1996) velocity scale it can be stated that the CL-PO landslide has reached values of displacements typical of a rapid movement (10^{-4} m/s) during the March 2004 crisis, slowing down to moderate velocities (10^{-6} m/s) in the June 2004–June 2006 period. In the last months the optical levelling indicates a further reduction of the displacements with a mean rate between 10^{-8} and 10^{-9} m/s.

As for the geotechnical characterization, the CL-PO data are in good agreement with the literature, which indicates the weathering processes as responsible for a number of mudslides (Chandler 1972; Taylor & Cripps 1987; Cafaro & Cotecchia 2001; Picarelli et al. 2005). In the CL-PO case-study, the presence of three overlying soil complexes provided with different properties supports the importance of the weathering processes as a key factor in predisposing the local instability. To this respect, Guadagno et al. (1987), studying the Fonte Griciatta landslide, recognized two main complexes, distinguishing intact clays from weathered clays, based on significant differences between the two complexes.

The preliminary slope stability back-analysis highlighted the complex behavior of the CL-PO slope, which can be defined as a "long-extension" slope (Pellegrino et al. 2003; Picarelli & Russo 2004), characterized by a length in the order of several hundred metres and, consequently, by a high length-thickness ratio. In these kinds of slopes it is very difficult to hypothesize the occurrence of the same geotechnical characteristics and the same behavior along the whole slip surface. It will be therefore necessary to use more complex geotechnical models (i.e. constitutive models including creep behavior) to perform a best-fit of the slope evolution for variations of the rain-water level regime.

REFERENCES

Almagià, R. 1910. Studi geografici sulle frane in Italia. L'Appennino centrale e meridionale. Conclusioni generali. *Mem. Soc. Geogr. It.* 14 (2): 1–431.

Bozzano, F., Martino, S., Naso, G., Prestininzi, A., Romeo, R.W. & Scarascia Mugnozza, G. 2004. The large Salcito landslide triggered by the 2002 Molise, Italy earthquake. *Earthquake Spectra* 20 (2): 1–11.

Cafaro, F. & Cotecchia, F. 2001. Structure degradation and changes in the mechanical behaviour of a stiff clay due to weathering. *Géotechnique* 51 (5): 441–453.

Calcaterra, D. & Parise, M. 2001. The contribution of historical information in the assessment of landslide hazard. In T. Glade, A. Albini & F. Frances (eds), *The use of historical data in natural hazard assessment*: 201–217. Dordrecht: Kluwer Acad. Publ.

Calcaterra, D., Parise, M. & Palma, B. 2003. Combining historical and geological data for the assessment of the landslide hazard: a case study from Campania, Italy. *Nat. Hazards and Earth System Sc.* 3 (1–2): 3–16.

Chandler, R.J. 1972. Lias clay: weathering processes and their effects on shear strength. *Géotechnique* 22: 403–431.

Corbi, I., De Vita, P., Guida, D., Guida, M., Lanzara, R. & Vallario, A. 1999. Evoluzione geomorfologica a medio termine del vallone in località Covatta (Bacino del Fiume Biferno, Molise). *Geografia Fisica e Dinamica Quaternaria* 22: 115–128.

Cotecchia, V. & Melidoro, G. 1974. Some principal geological aspects of the landslides of southern Italy. *Bull. Int. Ass. Eng. Geology* 9: 23–32.

Cotecchia, V., Monterisi, L., Salvemini, A., Spilotro, G. & Trisorio Liuzzi, G. 1977. Geolithological, structural and geotechnical aspects of some arenaceous-marly formations cropping out in Central-Southern Apennines. *Proc. Int. Symp. on Structurally Complex Formations, Capri*: 71–78.

Cruden, D.M. & Varnes, D.J. 1996. Landslide types and processes. In: A.K. Turner & R.L. Schuster (eds), *Landslides. Investigation and mitigation*. Transp. Res. Board, spec. rep. 247: 129–177. Nat. Acad. Press, Washington, D.C.

Guadagno, F.M., Palmieri, M., Siviero, V. & Vallario, A. 1987. La frana del febbraio 1984 in località Fonte Griciatta nel comune di Agnone (Isernia). *Mem. Soc. Geol. It.* 37: 127–134.

Guzzetti, F., Cardinali, M. & Reichenbach, P. 1994. The AVI Project: a bibliographical and archive inventory of landslides and floods in Italy. *Environ. Management* 18 (4): 623–633.

Pellegrino, A., Picarelli, L. & Urciuoli, G. 2003. Experiences of mudslides in Italy. *Proc. Int. Works. on Occurrence and Mechanisms of Flow-like Landslides in Natural Slopes and Earthfills, Sorrento:* 191–206. Bologna: Pàtron Editore.

Picarelli, L. 1993. Structure and properties of clay shales involved in earthflows. *Proc. Intern. Symp. on The Geotechnical Engineering of Hard Soils-Soft Rocks, Athens:* 2009–2019. Rotterdam: Balkema.

Picarelli, L. & Napoli, V. 2003. Some features of two large earthflows in intensely fissured tectonized clay shales and criteria for risk mitigation. *Proc. Int. Conf. on Fast Slope Movements. Prediction and Prevention for Risk Mitigation, Naples:* 431–438. Bologna: Pàtron Editore.

Picarelli, L. & Russo, C. 2004. Remarks on the mechanics of slow active landslides and the interaction with man-made works. *Proc. 9th Intern. Symp. on Landslides*, Rio de Janeiro, vol. 2: 1141–1176. Rotterdam: Balkema.

Picarelli, L., Urciuoli, G., Ramondini, M. & Comegna, L. 2005. Main features of mudslides in tectonised highly fissured clay shales. *Landslides* 2: 15–30.

Taylor, R.K. & Cripps, J.C. 1987. Weathering effects: slopes in mudrocks and overconsolidated clays. In M.G. Anderson & K.S. Richards (eds), *Slope Stability: Geotechnical Engineering and Geomorphology*: 405–445. New York: John Wiley & Sons.

Vezzani, L., Ghisetti, F. & Festa, A. 2004. *Carta geologica del Molise*. Firenze: S.EL.CA.

Application of surface wave and micro-tremor survey in landslide investigation in the Three Gorges reservoir area

Ailan Che & Xianqi Luo
China Three Gorges University Key Laboratory of Geological Hazards on Three Gorges Reservoir Area, Ministry of Education, China

Shaokong Feng
Chuo Kaihatsu Corporation, Tokyo, Japan

Oda Yoshiya
Xi'an Jiaotong University, Xi'an, China

ABSTRACT: The surface wave and micro-tremor survey are new methods of the engineering geophysical investigation. They have been widely used in recent years because they are of low cost and easy operation and they give the dynamic property of the ground, i.e. the shear wave velocity. This paper discusses the application of micro-tremor and surface wave in the Qianjiangping and Shuping landslides of the Three Gorges reservoir area. For surface wave survey, we analyzed the dispersion of the recorded data and then estimated the shear wave velocity structure by phase velocity inversion. For micro-tremor survey, we observed the ground micro-tremor signals, and analyzed them by fast Fourier transform method (FFT). The soil response to frequency of ground micro-tremor is revealed and the function with frequency-dependence and frequency-selection of micro-tremor for different foundation strata are investigated. The vertical to horizontal spectra ratio (H/V, Nakamura technique) of micro-tremor observed at the surface was used to evaluate the site's predominant period. All of these have provided a more dependable basis for the landslide material classification and for analysis of these hazardous phenomena. The experimental foundation and the deduction process of the method were described in detail. Comparison of the geophysical investigation results with the geological data shows their good agreement, thus proving the applicability of the methods in landslide study.

1 INTRODUCTION

The Three Gorges dam on the Yangtze River in China is the largest hydroelectric project in the world. After the first impoundment in June 2003, many landslides in her upstream area occurred or were reactivated. For example, in the early morning at 00:20 July 13, 2003, the Qianjiangping landslide occurred on the left bank of the Qinggan River, a tributary of the Yangtze River, three kilometers away from its confluence with the Yangtze mainstream (Figure 1). 14 people were killed and another 10 listed as missing (Wang et al. 2004). Direct economic losses caused by the landslide were about 7 million USD, and the asset value of Shazhenxi town was reduced by 40%. The Shuping landslide that is 3.5 km from the previous one is one of the most active landslides at the south bank of Yangtse River near the Shazhenxi town (see Figure 1) (Gan et al. 2004). It is anticipated that the frequency of landslides will increase due to the impoundment of the Three Gorges reservoir, and this has prompted geotechnical researchers to pay special attention to the problem.

It is well known that the ground structure is great important for geotechnical engineering. However, the study by boreholes is too expensive, especially in complex geological conditions, which is the typical case of the Three Gorges area landslides. The Surface wave survey and micro tremor observation are one of the most convenient methods to investigate the dynamic characteristics of the surface ground (Yamanaka, et al., 1996, Bard, 1998). They were applied for investigation of Qianjiangping— rockslide and Shuping landslide.

2 GEOLOGICAL CONDITIONS IN RESEARCH AREAS

2.1 *The Qianjiangping landslide*

The Qianjiangping rockslide is located about 50 km upstream from the Three Gorges Dam, China

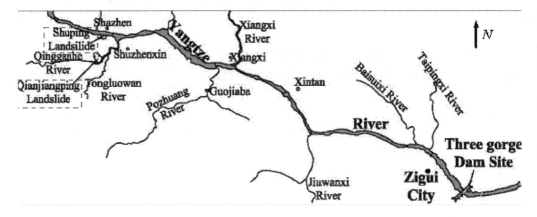

Figure 1. Location of the Qianjiangping landslide and Shuping landslide in Three Gorges area.

(Figure 2) (Wang et al. 2004). The landslide is a tongue-shaped in plan, 1200 m long, and 1000 m wide. It moved about 250 m in the main sliding direction S45°E. The average thickness of the sliding mass was about 20 m, thinner in the upper part and thicker at the lower part. It was a typical rockslide in weathered shale and sandstone, a block of which slipped down along the bedding plane. The exposed sliding surface at the upper part was very straight, sub-parallel to the bedrock strata.

The site is composed of quartzo-feldspathic sandstone, fine sandstone with carbonaceous siltstone, siltstone with mudstone, and silty mudstone of the Shazhenxi group of Late Triassic age. Three units can be roughly found in the cross-section — the slide mass, sliding zone and the bedrock, respectively, as shown in Table 1 (Luo, et al., 2001).

Figure 2. The Qianjiangping landslide.

Table 1. Physico-mechanical parameters of main units.

Serial number	Name of soil -layers	Specific gravity (Gs)	Natural water content (%)	Natural density (g/cm³)	Void ratio (e°)
1	Slip mass	2.63	19.89	1.63	0.93
2	Slip band	2.70	15.76	2.02	0.55

2.2 The Shuping landslide

The Shuping reactivated landslide occurred in shale and sandstone of Tertiary period. As soon as the first impoundment of the Three Gorges reservoir was conducted, serious deformation occurred in the Shuping landslide. It formed a large dangerous factor for the local residents and the shipment in the main stream of the Yangtze River (Gan et al. 2004).

The Shuping landslide is an old landslide composed of two blocks. The sliding mass consists of red muddy debris of old landslide, and of sandy mudstone, muddy siltstone of Badong group of Triassic period (Wang, et al., 2005). Three units can be roughly selected in the cross-section — loam with gravel, clay and sandy clay with gravel, respectively, as shown in Table 2.

3 OBSERVATIONS IN RESEARCH AREAS

3.1 Rayleigh wave prospecting

At Qianjiangping landslide, the surface wave survey is conducted from SK4 boring hole, which are about 360 m away from the sliding cliff (Figure 4).

Table 2. Physico-mechanical parameters of main units.

Serial number	Name of soil-layers	Specific gravity (Gs)	Natural water content (%)	Natural density (g/cm³)	Void ratio (e°)
1	Loam with gravel	2.63	19.89	1.63	0.93
2	Clay	2.70	15.76	2.02	0.55
3	Sandy clay with gravel	2.70	15.76	2.02	0.55

Figure 3. Shuping landslide consisting of two blocks at the main stream of the Three Gorges area.

At Shuping landslide, the surface wave survey is conducted from ZG86 GPS point on II-II″ line of Block-A, its elevation is about 355 m downward Shahuang high way (Figure 5). The data are acquired on a 58-m survey line with a receiver interval of 2 m for each site. Sledgehammer is used as a vertical source for data acquisition, and five shot gathers with source offsets for 4 m, 8 m, 12 m, 16 m and 20 m were recorded. Because of limited space, source is only placed on one end of the survey line. Data were acquired with 1 ms sample rate, and total of 2048 measurements are recorded.

3.2 Micro tremor observations

The micro tremor observations were conducted at both sites — from SK4 borehole at the Qianjiangping landslide (Figure 4) and from ZG86 GPS point on II-II″ line of Block-A of the Shuping landslide (Figure 5). The total length of survey line for the latter is about 200 m with an interval of about 10 m, thus summing up to 40 observation points. The data were recorded for 5 minutes with 1 ms sample interval. The records of the horizontal and vertical components of short-period micro tremors are obtained using a three components highsensitive seismometer, which has a one second natural period. The potential noise sources such as machinery, vehicles traffic or pedestrians, near the seismometer are avoided during the measurement time.

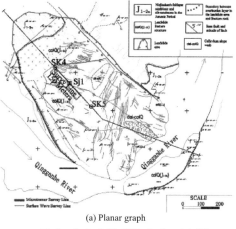

(a) Planar graph

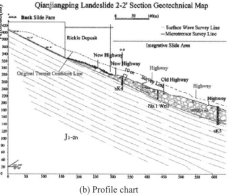

(b) Profile chart

Figure 4. Measurement points at Qianjiangping landslide.

4 ANALYSES AND RESULTS

4.1 Rayleigh wave prospecting

For dispersion analysis, first, the cross-spectrum of each trace with a reference trace is calculated to form a cross-spectrum gather for each shot location, and these are then stacked over the five shots to improve the signal-to-noise level.

Then, in order to construct a 2D profile over the line, spatial windows of 12 traces are analyzed and moved along at 2 m intervals, with the nominal position taken as the center of the trace window. Figure 6 is the example of the dispersion analysis. These are interpreted as the fundamental and higher modes. As the figure shows, we can clearly see the higher-mode dispersion curves, as well as the fundamental-mode curve.

Finally, we estimate the Vs structure by inverting the phase velocity data (Feng et al., 2005). Figure 7 shows the borehole log results and the estimated 2D shearwave velocity model from the phase velocity inversion.

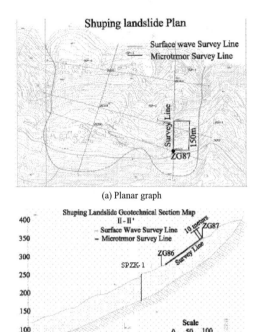

Figure 5. Measurement points at Shuping landslide.

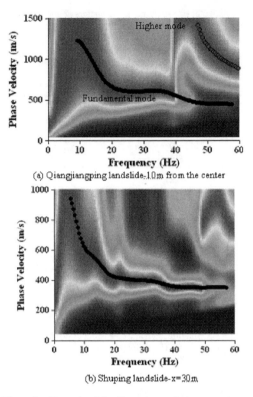

Figure 6. Example of the dispersion analysis at test site.

The model given by the multi-mode inversion is in better agreement with soil types. At the Qiangjiangping landslide Vs section (Figure 7a), there are 3 main shear wave velocity (Vs) boundaries, corresponding to the main geological boundaries according to the drilling core. Especially, the boundary between Vs = 600 m/s and Vs = 800 m/s units on the Vs section, agrees very well with the slide surface. At the Shuping landslide (Figure 7b), the Vs section shows 2 clear velocity boundaries inferring that the slide mass consists of 3 layers, which shear wave velocities (Vs) are 400 m/s, 600 m/s, and 800 m/s, respectively.

4.2 Micro tremor observations

The micro tremor is a kind of geophysics information with abundant intension, which is related to site structure, corresponding to engineering geological condition. The frequency characteristics of micro tremor can be obtained by spectral analyses of its signals, which can be used to probe the dynamic characteristics of a studying area ground. From the recorded data of micro tremor measurements, five sets of 2048 digital data at lower noise periods are selected to use for FFT analysis. The velocity Fourier amplitude spectra, spectra ratio and relative variation amplitude are computed. The spectra ratio of vertical component to the horizontal one (H/V, Nakamura technique) of micro tremor observed at the surface ground are widely used to evaluate the site dynamic response (Nakamura, 1989), which is considered the transfer function $S_r = SH_s/SH_B = SH_s/SV_s$.

In this study, the spectral ratio (H/V— average type suggested by Japanese Education Ministry was used) from the five sets of 2048 digital data, proposed by Nakamura was used to identify the predominant frequency (f_p: Hz) of the research ground (Iwatate, et al., 1996). A sample H/V of micro tremor measurements is shown as Figure 8.

In order to verify the results obtained from the micro tremor data, ground structures based on the actual borehole logs (SK-4 and SPZK-1) were revealed. One-dimensional seismic response analyses of horizontally layered soils using Multiple Refraction Theory were performed (Enomoto, et al. 2002). The values of shear-wave velocity were determined from the Rayleigh wave prospecting. The transfer functions of the model ground were calculated. Compared with the predominant frequency values from H/V of micro-tremor measurements and analytical results, it shows a good

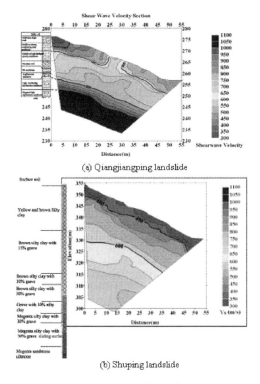

Figure 7. 2D shear-wave velocity model from the phase velocity inversion compared with boring logs.

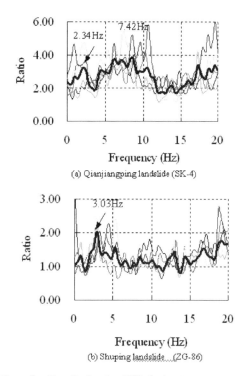

Figure 8. Transfer function H/V of micro tremor measurements.

agreement (Figure 9). Therefore, it can be considered that the dynamic characteristics of the ground in the research area can be evaluated by the spectral ratios (H/V) from micro-tremor measurements. The predominant frequencies obtained from the Qianjiangping ground model clearly show the effect of slip band that are relatively identical to that as obtained from the micro tremor results (Figure 9a).

Figure 10 shows the H/V of all the measurement points. The result shows that the predominant frequencies obtained from the micro tremor data show clear three peaks at 2 Hz, 4 Hz and 7 Hz. Compared with the predominant frequency values from H/V of micro tremor measurements and analytical results (Figure 9), it shows a response of the sliding zone in the Qianjiangping landslide. The change of predominant period of ground micro tremor is closely related to the formation of site ground structure, and mutually corresponds to the change of engineering geological conditions of the site. Therefore, it can be considered that the dynamic characteristics of the ground can be evaluated by the spectral ratios (H/V) from micro tremor measurements.

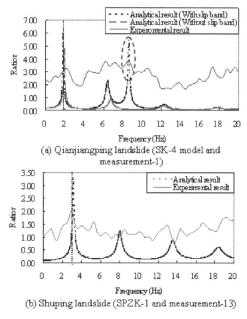

Figure 9. Compared with the transfer function from H/V of micro-tremor measurements and analytical results.

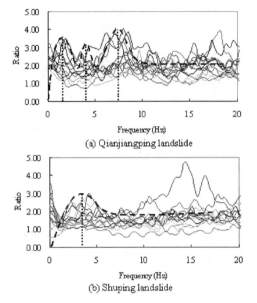

Figure 10. Predominant frequencies from H/V of micro tremor measurements.

5 CONCLUSIONS

The surface wave survey and micro tremor observation are conducted on the ground of the typical sites in the Qianjiangping landslide and Shuping landslide. The phase velocity of surface wave is calculated from the observed surface wave survey, and the subsurface shear wave velocity structure is estimated by phase velocity inversion. The microtremors are observed simultaneously on the ground surface. To study the application of this method to landslide geotechnical investigation, a series of field experiments and analyses are carried out. The main results are as follows.

1. The Vs structure from the phase velocity inversion is in good agreement with soil structure.
2. In Qiangjiangping landslide, the result shows that there are 3 main shear wave velocity (Vs) boundaries on the Vs section, corresponds to the main geological boundaries of the drilling core. Especially, the boundary between Vs = 600 m/s and Vs = 800 m/s on the Vs section, agrees very well with the slide surface.
3. In Shuping landslide, the Vs section shows 2 clear velocity boundaries infers that the slide mass consists of 3 layers.
4. The predominant frequencies obtained from the Qianjiangping clearly show the effect of slip band in both analytical results and the micro tremor results. The change of predominant period of ground micro tremor is closely related to the formation of site ground structure, and mutually corresponds to the change of engineering geological conditions of the site. So it can be considered that the ground structure in the landslide can be identified by the micro tremor.

REFERENCES

Wang, F.W., Zhang, Y.M., Huo, Z.Y., Matsumoto T. & Huang, B.L. 2004. The July 14, 2003 Qianjiangping Landslide, Three Gorge Reservoir, China, *Landslide*, 1(2):157–162.

Gan, Y.B., Sun, R.X., Zhang, Y.Q. & Liao, S.Y. 2004. *Urgent investigation report on the Shuping landslide in Shazhenxi town, Zugui Country, Hubei Province (in Chinese)*.

Bard, P. Y. 1998. Micro tremors measurements, a tool for site effect estimation. *ESG 98 symposium*.

Yamanaka, H. & Ishida, H. 1996. Application of genetic algorithm to an inversion of surface-wave dispersion data: *Bulletin of the Seismological Society of America*, 86:436–444.

Luo Xianqi, Jiang Qinghui, Ge Xiurun & Liu Sifeng, 2001. Study of the stability of Qianjiangping slope in Huanglashi slope group, *Chinese Journal of Rock Mechanics and Engineering*, 21(1):29–33.

Wang Fawu et al. 2007. Landslide caused by water level changes in Three Gorges Reservoir Area, China, *Chinese Journal of Rock Mechanics and Engineering*, Vol. 26, No. 3, pp. 509–517.

Fawu Wang, Gonghui Wang, Kyoji Sassa, Kiminori Araiba, Atsuo Takeuchi, Yeming Zhang, Zhitao Huo, Xuanming peng & Weiqun Jin. 2005. Deformation monitoring and the exploration on Shuping landslide induced by imm-poundment of the Three Gorge Reservoir, China, Annuals of disas. Prev. Res. Inst. Kyoto Univ., No. 48B.

Shaokong Feng, Takeshi Sugiyama & Hiroaki Yamanaka, 2005. Effectiveness of multi-mode surface wave inversion in shallow engineering site investigations, Butsuri-Tansa (Vol. 58, No. 1), Mulli-Tamsa (Vol. 8, No. 1), *Exploration Geophysics* 36, 26–33.

Nakamura, Y. 1989. A method for dynamic characteristics estimation of subsurface using micro tremor on the ground surface. *OR or RTR1*, 30, 25–33.

Iwatate, T., Akira, O. & Koji, A., 1996. Surface Ground Motion Characteristics of Zushi-Site. *Eleventh World Conference on Earthquake Engineering (11 WCEE)*, Acapulco, MEXICO, June 23–28.

Enomoto, T. & Iwatate, T. 2002. Site-effects evaluation H/V spectra comparing micro-tremor with strong motion records observed at ground surface and basement using borehole. *12th European Conference on Earthquake Engineering*.

A case study for the landslide-induced catastrophic hazards in Taiwan Tuchang Tribute

C.Y. Chen
National Chiayi U., Dept. of Civil & Water Resources Eng., Chiayi City, Taiwan, China

W.C. Lee
National Science & Technology Center for Disaster Reduction, Sindian City, Taipei County, Taiwan, China

ABSTRACT: A landslide-induced flow failure caused 15 residents dead, and 24 houses buried during a Typhoon Aere in 2004. The landslide mechanism by a GIS spatial analysis was studied for clarifying the reasons attributed to the catastrophic hazards for geological, hydrological, and topographical characteristics. The evacuation and rescue response for the hazards are being discussed for enhancing the emergency response in mountainous areas during harsh climatic conditions. After analysis, it was later found that the landslide was associated with post-seismic behaviour, torrential rainfalls, steep slopes and unfavorable geological conditions. More attention to the earthquake-induced slopeland ground surface cracking is needed during post-seismic rainfall events.

1 INTRODUCTION

Landslides have frequently occurred during recent years with the visibly abnormal climate change. One of these catastrophic landslide hazards occurred in the Leyte Island, Philippines, on 17 Feb., 2006. The rainfall induced a giant landslide which caused more than 1000 people and 300 houses to be buried in the mass of debris. In Taiwan, the climate-change induced landslide increased after the M7.6 Chi-Chi earthquake in 1999. The landslide hazard in Tuchang Tribute was speculated as having occurred on 25 Aug., 9:00 AM, 2004, followed by a landslide dam formed and breached at 10:00 AM according to the communication record from the buried police station. There were 15 residents dead and 24 houses buried by the landslide masses, including 3 policemen and the police station. More attention is needed for the reaction of historical landslides induced by earthquakes after torrential rainfalls in the mountainous areas.

2 SITE LOCATION

The Tuchang Tribute is located at Wufeng Village, Hsinchu County, northern Taiwan (Figure 1). The tribute has a population of approximately 100 people and more than 20 buildings. The landslide site is an immediate neighbor to the center of the tribute. The tribute is

Figure 1. Site location of the Tuchang Tribute in Wufeng Village, Hsinchu County.

situated in a mountainous area with an elevation about 780 m. The tribute is a gate to the high mountainous area and a police station for personnel management.

3 LANDSLIDE MAGNITUDE

The landslide covered an area of 0.58 km², with a length of 490 m, and a base width of 200 m. The landslide area included in the affected area was about 1.5 km² by the measure of the post-landslide aerial photo, shown in Figure 2. The estimated landslide volume was about 1.0 million m³. Figure 3 depicts the buried area of the tribute by the debris masses, which

Figure 2. Aerial Photos for the Tuchang Tribute pre- and post-landslides (scale: 1/5000, photo by the Forestry Bureau, http://www.forest.gov.tw/).

4.1 Historical hazards

There were 77 historical landslides (source: Soil & Water Conservation Bureau, called SWCB hereinafter, taken in 2003 by Spot 5 Satellite), 22 potential debris flow torrents, as show in Figure 5 (source: SWCB, 2003). There were 21 debris flow hazard spots in the village from the Typhoon Area (source: NCDR, http://www.ncdr.nat.gov.tw/). The landslide is a reactive site for a visible landslide scarp after the M7.6 Chi-Chi earthquake, as shown in Figure 6.

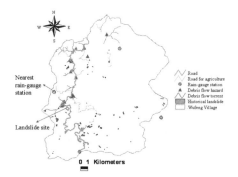

Figure 5. Historical Hazards in the Tuchang Tribute.

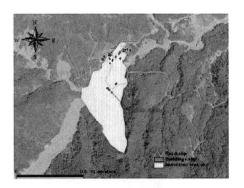

Figure 3. Buried area by the debris masses.

Figure 4. Landslide deposition area in the Tuchang Tribute (taken on 30 September, 2004).

includes the position of the police station. Field investigation was made after the traffic was available on 30 Sep., 2004, as shown in Figure 4.

4 REASONS ATTRIBUTE TO THE HAZARDS

The reasons attributed to the landslide hazards were studied for their landslide history, geology, hydrology, and topographic characteristics, as shown in the following.

Figure 6. Historical scarp in the landslide site (scale: 1/5000, photo by the Forestry Bureau, http://www.forest.gov.tw/).

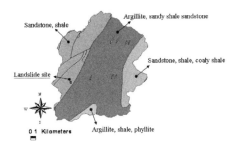

Figure 7. Geology map for the Tuchang Tribute.

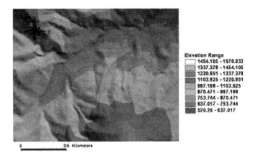

Figure 10. Elevation map for the landslide area (unit: m).

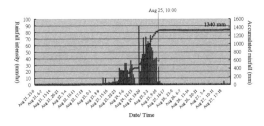

Figure 8. Rainfall time history at the nearby rain-gauge station (source Central Weather Bureau, http://www.cwb.gov.tw/).

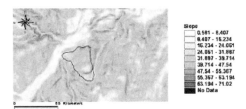

Figure 11. Slope map for the landslide area (unit: %).

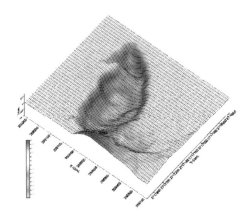

Figure 9. Surface map for the landslide area (unit: m).

4.2 Geologic condition

The geological map shows that the landslide is located at the margin of bedrocks for argillite, sandy shale sandstone, and sandstone and shale, as shown in Figure 7.

4.3 Hydrological condition

The accumulated rainfall was 1340 mm and the rainfall intensity was 27 mm/hr at the nearest rain-gauge station at the time in which the landslide dam breached, shown in Figure 8. The rains were over the debris-flow warning threshold of 250 mm in this area, coded by the Soil & Water Conservation Bureau in 2007.

4.4 Topographic characteristics

The averaged slope for the landslide area was about 35.5° and an elevation changed 285 m from the top at 1100 m to the base of 815 m as shown in Figures 9, 10 and 11. Its slope length was 490 m (source: the 9 m DTM from the Forestry Bureau, http://www.forest.gov.tw). The aspect of the landslide was north-northwest.

5 EMERGENCY RESPONSE FOR THE LANDSLIDE HAZARDS

Four stages of the responses for the landslide hazards relief are separated herein as (1) actions before typhoon landfall, (2) actions during typhoon landfall, (3) emergency relief before disaster, and (4) emergency rescue after disaster. The following documented statements were sourced from the National Disaster Prevention and Protection Commission (http://www.ndppc.nat.gov.tw/), and the National Fire Agency (http://www.nfa.gov.tw/).

5.1 Actions before typhoon landfall

The Central Emergency Operation Center was established on 23 Aug. for preparing the Typhoon Aere landfall in Taiwan. The Council of Agriculture on 17:40 23 Aug., gave information to those residents who live near to the potential debris flow torrents to pay attention to the increased rainfall and to prepare to evacuate for the mountainous areas in Hsinchu County.

5.2 Actions during typhoon landfall

At 08:00 AM on 24 Aug., the commander of the Central Emergency Operation Center for the Typhoon Aere asked the Council of Agriculture to announce to the public the warning areas for debris flows before 15:00 PM, especially for the tributes in the mountainous areas. The information was provided to the Council of Indigenous Peoples and the Department of Social Welfare for the relief and preparedness of residents' evacuation in the debris-flow warning areas for the commander at the local governmental emergency operation center. Then, at 14:40, the commander asked the Council of Indigenous Peoples, the Council of Agriculture, and the Department of Social Welfare to enforce the local government to evacuate the residents at the debris-flow warning areas before 17:30. On 24 Aug. 16:23, the Wufeng Village was listed as a warning area for debris-flow, and the commander asked those residents who live nearby the potential debris flow torrents to evacuate. There was a strong suggestion given to the village head of Wufeng Village on 24 Aug. 17:33 to evacuate the residents, and three times to inform the local government to pay more attention to this area. On 24 Aug. 22:10, the village head rendered that he had asked the policemen to help evacuate the residents, and three shelters in the village were opened for resident accommodation.

Then, at 08:00 AM on 25 Aug., the commander asked the Council of Indigenous Peoples, the Soil & Water Conservation Bureau, and the Department of Social Welfare help the local governments to force residents to evacuate in debris-flow warning areas and to make a record for responsibility.

5.3 Emergency relief before disaster

The communication records for the sacrificed policemen before the disaster are list below:

- 24 Aug. 16:30. The policemen got the command to force the residents in the Tuchang Tribute and in other lower land areas to evacuate to a place of safety at 17:00.
- 24 Aug. 17:40. The police station was inside the warning areas. Three policemen left their station and sought safety at a nearby house.
- 25 Aug. 06:10. The bridge connected to the police station was inundated by a flash flood (it was speculated as to the time of flood).
- 25 Aug. 09:10. The police station was covered by the landslide (it was speculated as to the time the landslide was initiated).
- 25 Aug. 09:50. The policemen replied to the local emergency operation center in Wufeng Village by wireless phone and reported that they had evacuated to nearby houses.
- 25 Aug. 09:55. Connection failed with the three policemen by the wireless phone (it was speculated as to the time the landslide dam breached and debris flow initiated).

5.4 Emergency rescue after disaster

On 25 Aug. the commander of the Central Emergency Operation Center asked the vice-commander, delegates in the Council of Indigenous Peoples, and the Soil & Water Conservation Bureau to set up the Forward Command Post near the village to aid the Hsinchu County Government in the emergency rescue.

Starting on 26 Aug., the Central Emergency Operation Center asked the National Search and Rescue Command Center, the rescuers at the Fire Departments in local areas, the non-governmental emergency relief organizations, and the operators for engineering machines, to help the local emergency operation center at the Hisnchu County Government for rescue actions. The excavation by heavy machines began on 1 Sep. 2004 for the dead in the disaster. The rescue actions were performed as follows:

- Seven special search and rescue teams from the National Fire Agency were assigned to seven tributes in the mountainous areas in Hsinchu County, for the aids of the rescue actions.
- Seven international maritime satellite telephones for emergency communication to the seven tributes in the mountainous areas were provided.
- The Ministry of National Defense sent soldiers to Hsinchu County to help in the rescue actions.
- The National Search and Rescue Command Center sent helicopters to deliver foodstuffs and transportation for residents.
- The Council of Agriculture, the Council of Indigenous Peoples, and the Ministry of Transportation & Communications were responsible for debris flow field investigations and emergency road repair.

Three bodies were found at the landslide site on 5 Sep. 08:00. The search actions were continued and supported by 174 men, 19 hawks and 19 trucks, on 11 Sep. 2004.

6 DISCUSSION AND CONCLUSION

A landslide dam breach, which induced debris flow hazards, occurred on 25 Aug. 2004. The landslide mechanism was examined for its historical hazards, geology, hydrology, and topographical characteristics. The aids of field investigation, aerial photos, and final communication records with the dead from the disaster help to clarify the reasons attributed to the hazards. The landslide mechanism was speculated as a flood-induced riverbank erosion resulting in a landslide, followed by a landslide dam, which formed and breached a flow of debris.

There seems to have been a weak link during the emergency operation for the central departments and local government before the landslide, causing an inability by authorities to evacuate the residents to a place of safety early on. There also were numerous difficulties during the emergency rescue actions because of cut-off roads from landslides, causing an inability to communicate with the landslide-initiated mountainous area, which resulted in lack of information to this region in time. It seems need more efforts on enhancing the hazard management and a more efficient field mechanism for improving road blockage in the mountainous areas is needed (Chen et al., 2006; Chen, 2007). More attention to the seismic-induced ground cracking is also needed to prevent a similar case hereafter.

ACKNOWLEDGEMENTS

The author appreciates the National Science & Technology Center for Disaster Reduction (NCDR), the Central Geology Survey, the National Fire Agency, the Soil & Water Conservation Bureau, and the Council of Agriculture for providing valuable materials for the study.

REFERENCES

Central Geology Survey, Ministry of Economic Affairs, retrieved date August 1, 2007, from http://www.moeacgs.gov.tw/.

Central Weather Bureau, retrieved date August 1, 2007, from http://www.cwb.gov.tw/.

Chen, C.Y., Lee, W.C. & Yu, F.C. 2006. Debris flow hazards and emergency response in Taiwan. First International Conference on Monitoring, Simulation, Prevention and Remediation of Dense and Debris Flows, 7–9 June 2006, Rhodes, Greece, 311–320.

Chen, C.Y. 2007. Landslide Characteristics and rainfall distributions in Taiwan. Landslide and Climate Change-Challenge and Solution, 20–24 May 2007, Ventnor, Isle of Wight, UK, 35–40.

Council of Agriculture, Executive Yuan, retrieved date August 1, 2007, from http://www.coa.gov.tw/.

Council of Indigenous Peoples, Executive Yuan, retrieved date August 1, 2007, from http://www.apc.gov.tw/.

Department of Social Affairs, Ministry of Interior, retrieved date August 1, 2007, from http://www.moi.gov.tw/.

Forestry Bureau, retrieved date August 1, 2007, from http://www.forest.gov.tw/.

Ministry of Transportation and Communications, retrieved date August 1, 2007, from http://www.motc.gov.tw/.

National Disaster Prevention and Protection Commission, retrieved date August 1, 2007, from http://www.ndppc.nat.gov.tw/.

National Fire Agency, Ministry of Interior, retrieved date August 1, 2007, from http://www.nfa.gov.tw/.

National Science and Technology Center for Disaster Reduction (NCDR), Sindian City, Taiwan, retrieved date August 1, 2007, from http://www.ncdr.nat.gov.tw/.

Soil & Water Conservation Bureau, Council of Agriculture, Executive Yuan, retrieved date August 1, 2007, from http://www.swcb.gov.tw/.

The Ministry of National Defense, R.O.C., retrieved date August 1, 2007, from http://www.mnd.gov.tw/.

Pir3D, an easy to use three dimensional block fall simulator

Y. Cottaz
Geociel, Décines, France

R.M. Faure
Centre d'Etude des Tunnels, Bron, France

ABSTRACT: We present here the last release of a falling block simulator, called Pir3D. The main advance is the user friendly front-end that allows a fast definition of the digital elevation model, from usual map to an accurate 3D representation of the ground. New bouncing algorithms are now available and give the possibility of comparisons of models. With these enhancements fence position is easy to define using mouse drag when trajectories are stopped in real time.

1 INTRODUCTION

Populated areas are increasing and in mountainous regions falling blocks from cliffs or steeped slopes are a real threat for buildings, roads, tunnel portals and lives. For risk assessment the new release of the 3D simulator Pir3D, is really a jump in the use of it, with a new user-friendly front end and new algorithms for bouncing calculation. (Faure et al, 1995).

The paper will present, in a first part, the simulator following a real study, showing practical steps and giving information about the consumed time.

In a second part, focus is put on bouncing aspect, developing the two algorithms used for the geotechnical approach. (Faure et al, 1995).

Comparisons between these two approaches and others codes are given as conclusion.

2 GEOMETRY OF PIR3D

Three steps can be identified in the process: giving topography, starting lines of trajectories and fence position.

2.1 Topography and ground zones

The topography of the studied area is obtained from a usual map or from an "Autocad"® set of points, that topographer usually use.

When one uses a map, the interesting area is scanned as an image. The scale of the picture is given through two identified points on the map. On the computer screen, the operator gives the altitude of a level line, and with the mouse clicks along this line.

The computer stores all this points with 3D coordinates. The ground model is so very quickly determined: it takes less than 60 minutes for 1000 validated points. The point density is chosen by the operator that can add points where it is necessary, and any point can be added if necessary. When clicking on the "mesh" icons the convex contour of the set of points is calculated and a Delaunay algorithm builds a triangular mesh and immediately the 3D model is displayed. The Delaunay algorithm ensures that each circle built on the three points of a triangle does not include any other point. With this algorithm the operator can tighten the points where necessary and have more large triangle in other part of the ground model. (see figure 1).

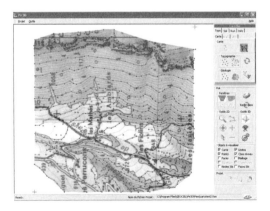

Figure 1. Following the level lines the operator introduce a set of points (1000 points in 60 minutes).

Using a kind of 'brush', the operator paints the triangles as to determine the kind of soil they represent. Figures 2, 3 and 4 describe this sequence.

When the set of points is an "Autocad"® file the operator only gives the name of the file and the 3D model appears.

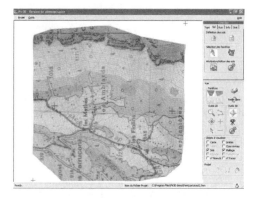

Figure 2. The triangular mesh appears when Delaunay algorithm is called, and the operator can determine the type of grounds that are shown in different colors.

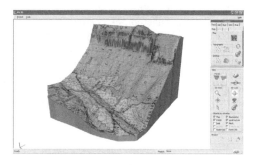

Figure 3. A block diagram of the studied area. The map is mapped on the ground.

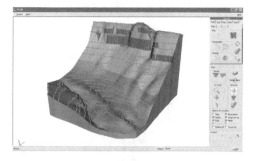

Figure 4. The same block diagram with soils differentiation and level lines.

2.2 *Starting lines*

For block fall simulation, the starting point of trajectories is usually not well known as it may occur somewhere in the cliff.

A segmental line defined (mouse clicking) by the operator determines the starting points of the blocks. Along this line a random algorithm gives the initial point of each trajectory.

The easy to use front-end of Pir3D gives the possibility to define several lines and move these lines with a drag of the mouse.

When ready, Pir3D asks the operator for the number of blocks, usually, between 1000 to 10000, and the initial speed. Mainly the initial speed is a vertical falling of 1 meter high above the starting point, or an accurate value giving the three speed components, v_x, v_y, v_z, in case of seismic shock or blasting (Figure 5).

2.3 *Fences*

One of the most powerful possibilities in Pir3D, is the simulation of fences, that will stop the block. The position of the fence is given by mouse dragging and its efficiency is computed in real time and on the block diagram, trajectories intercepted by the fence don't appear beneath the fence. The operator can change also the height, and very quickly, he can select on the map the best position of the fence and determine its height.

With this possibility the operator quickly determines the best position and the height of the fence and for a better knowledge of the fence efficiency two other graphics are drawn by Pir3D. The first one is the fence elevation with the position of all the block impacts, the colour of them indicate the energy of the block following a colour scale. It is so possible to determine the kind of fence we can use. The second graphic is an histogram of the energy impact, or the height of the impact on the fence. Figures 6, 7, and 8 illustrate the use of fence, and on figure 8, one can see the zoom effect, available at any time when using Pir3D.

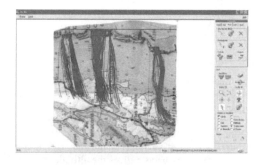

Figure 5. Simulation with three starting lines.

Figure 6. The role of the fence (blue line with transparency and height).

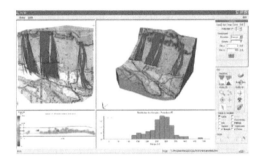

Figure 7. The two diagrams giving more information on the fence system.

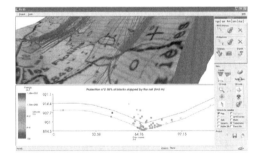

Figure 8. A zoom for a better view on the simulation.

3 BOUNCING SIMULATION

The bouncing model is based on point movement, but with uncertainties on soils, on block shape, on vegetation, we introduce uncertain parameters giving to the operator the choice between uniform distribution and normal distribution. Soil parameters are presented on

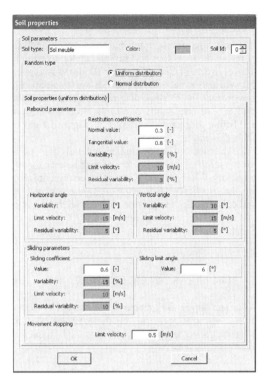

Figure 9. Soil parameters showing uncertainty use on bouncing and sliding, for uniform distribution (The code can be used in French and English).

Figure 9. Lot of soils can be stored inside Pir3D, so we have for identifying a soils its name and its color.

The uncertain on bounce parameters is defined as a function of the speed. Usually variations are greater at low speed and less important at high speed. It is why when speed is close to zero we have "maximum variability" and over a "speed limit", the variability called "residual variability" is used. Between these two speeds a linear function gives the value to be used. (User manual, 2007).

The same table is defined for normal distribution.

The bouncing parameters are: energy restitution with a distinction between vertical and horizontal restitution, vertical and horizontal angle, and if the angle after the bounce is less than a value, the motion is a slide on the ground, with constant deceleration.

The computer code is written with object representation and optimized, so 10000 trajectories are computed within a minute. Pir3D appears as a very efficient tool that can be easily compared to other similar tools as results, but can't be compared as use, its user friendly interface give to it some advance.

4 USE FOR SPECIFIC STUDIES

In some case, for research purposes (back analysis of registered trajectory and soil parameter determination) or in case of comparisons Pir3D allows studies on 2D profiles. In fact, these profiles are automatically transformed in 3D block diagram by extension of the profile. The uncertainty due to the ground vanishes, and fitting parameters is simplified. Figure 10 and 11 show this way.

Each trajectory can be studied in details; the operator can reach all the parameters of each bounce for any trajectory. When some hundred of trajectories are plotted, the selection is made using the mouse, and with up and down touch the operator analyses all the bounces of the trajectories. (Labiouse et al, 2001) (Aliardi, Crosta, 2003).

5 RISK AREAS DETERMINATION

After simulating the trajectories of more than 10000 blocks, the risk areas, at the foot of the slope can be determined in several manners and maps are automatically drawn.

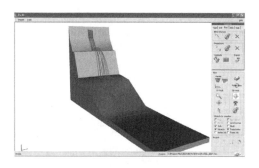

Figure 10. A extended profile showing the influence of lateral uncertainty.

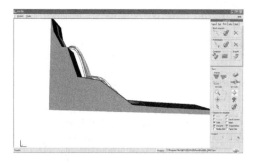

Figure 11. Trajectories.

Each time a block flies over a pixel of a given area at the foot of the slope, a counter notices it. For each pixel of the given zone, the operator obtains the probability of being reached by a block, and maps of probability, can be drawn. This representation is full of interest when discussions with populations occur. For engineer the display of energy received by each facet (total or maximum) is an appreciated information for determining the best protection.

6 CONCLUSIONS

As a conclusion we summarize the hypothesis used in Pir3D following:

Air resistance is not taken into account.

Rocks have not a form. The influence of the form is represented by random parameters.

Rock rotation on itself is not taken into account in the calculations.

Rolling is not directly taken into account in calculations. Our software ignores this aspect as it depends on the shape of the rock or stone in question. We have replaced this by a notion of rock sliding on the ground, below a minimum bounce angle.

There is no interaction between rocks. Only isolated rock fall situations and independent mass landslides are considered.

Rock fragmentation is not considered. This has as a consequence the dissipation of a large quantity of energy; the worst scenario is thus represented by a rock remaining intact.

Conditions relating to rock face breaking are not taken into account. Initial rock fall conditions are determined by the user.

These simplifications, used with uncertainty as defined in Pir3D, give results very closed of reality, and the use of Pir3D increases.

REFERENCES

Faure R.M. Fayolle G. Tartivel F., 1995, PIR3D, a code which simulate falling blocks in three dimensions. *European Conference of soils mechanics, Copenhagen.* Vol 6 pp 27–32.

Labiouse V., Hendereich B., Desvarreux., Viktorovitch M., Guillemin P., 2001. Confrontation de logiciels trajectographiques dans le cadre d'un programme Interreg IIC consacré aux instabilités de falaises. *Convegno Internazionale su opera di diffesa da caduta massi, Siusi, Italia,* Ottobre 2001. pp 13–23.

Agliardi F., Crosta G.B., 2003. High resolution three dimensional numerical modeling of rockfalls. *Int. J. of Rock Mechanics and Mining Sciences* 40, pp 455–471.

User manual, 2007, www.geociel.fr/pir3d.php.

Characterization of the fracture pattern on cliff sites combining geophysical imaging and laser scanning

J. Deparis & D. Jongmans
LGIT, Université Joseph Fourier Grenoble, Grenoble Cedex 9, France

B. Fricout & T. Villemin
LGCA, Université de Savoie, Le Bourget du Lac Cedex, France

O. Meric & A. Mathy
SAGE, Gières, France

L. Effendiantz
CETE Lyon, Bron Cedex, France

ABSTRACT: Fracture characterization of potentially unstable cliffs is a crucial problem which can only partly be solved by geological measurements at the surface. In this study we combine Laser Scanning (Lidar) and geophysical techniques for obtaining the best possible image of the fracture pattern at the surface and inside the rock mass. Two limestone cliff sites around Grenoble (French Alps), exhibiting different geometrical and geotechnical features, were investigated in order to show the potential and the limits of the methods. Processing of the Dense Digital Surface Models (DDSM) derived from Lidar data allowed fracture analysis which compares well with field observations. For the site where abseiling was possible, performing Ground Penetrating Radar (GPR) on the cliff face turned out to be the most effective method for accurately imaging the fractures inside the rock mass. On the other site, seismic and electrical imaging techniques were used on the plateau for mapping the fractures seen on the cliff.

1 INTRODUCTION

Rock fall hazard assessment is a complex problem in high cliff areas, due to the difficulty of measuring and mapping geological structures on the cliff face, and to the little information about the internal discontinuities provided by these conventional mapping techniques. Within the framework of a French project funded by the "Ministère de l'Ecologie et du Développement Durable", we have developed a new approach which aims at giving a more detailed description of the 3D structure of the potentially unstable rock mass, for volumes from a few thousands to a few hundred of thousands of cubic metres. This methodology combines two types of investigations: a structural analysis based on remote measurements and geophysical imaging.

Airborne and terrestrial laser scanning techniques are increasingly used for mapping in landslide detection and rock engineering (Lemy & Hadjigeorgiou 2004; Feng & Röshoff 2004; Schulz et al. 2005). Bornaz & Dequal (2003) developed the concept of Solid Image combining laser scanning data and co-registered images. In this approach, the point clouds obtained from laser scanning are projected on the image which keeps its original geometry and resolution. Each pixel can be localized and the 3D orientation of fractures observed on the image can be

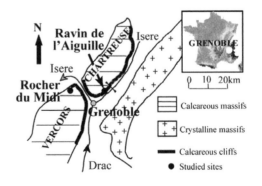

Figure 1. Location map of the two cliff sites: Le Rocher du Midi and Le Ravin de l'Aiguille.

automatically computed. Solid images are a new tool offered to structural geologist for quickly, easily and safely having access to measurements.

The geophysical methods have been increasingly used for cliff investigation, both on the plateau above the cliff (Busby & Jackson 2006) and on the cliff face itself (Dussauge et al. 2003; Roch et al. 2006; Jeannin et al. 2006; Deparis et al. 2007). The use of GPR (Ground Penetrating Radar) on the cliff face was found to be the most valuable tool in terms of resolution for investigating a rock mass (Jongmans & Garambois 2007). Two limitations of GPR for cliff investigation are safety requirements for abseiling and the penetration depth which was found to be lower than 30 m with 100 MHz antenna in the limestone rocks around Grenoble (France).

This study aims at combining the laser scanning and geophysical imaging techniques for characterizing the fracture pattern on two cliff sites located in the French sub-alpine limestone Massifs of Vercors and Chartreuse (Figure 1). On each site, the geometrical and mechanical characteristics of the cliffs allow adaptation of the investigation survey.

2 STUDIED SITES CHARACTERISTICS

The first site (Le Rocher du Midi, Figures. 2a and b) is a 100 m high column made of massive limestone, dated back to Lower Cretaceous (Urgonian limestone,
Arnaud et al. 1978, Philippe et al. 1998). It is separated from the mass on one side by a 1 m wide NW-SE fracture F1. The potential unstable volume was estimated at 50,000 m^3. A structural study of all the nearby outcrops and of the vertical cliff has evinced the presence of three sets of discontinuities: the near-horizontal bedding and a system of two conjugate near vertical fracture families, F_a and F_b, striking N20°E and N125°E, respectively. Two electrical tomography profiles ($E1$ and $E2$, Figure 2a) were conducted on the plateau. As the rock offers the required quality for abseiling, four vertical GPR profiles (P1 to P4) were performed on the cliff face.

The second site (Le Ravin de l'Aiguille, Figs. 2c and d), composed of Tithonian limestone (Philippe et al. 1998), is a 100 m wide and 170 m high tetrahedron, down pointed and limited by two large fractures, F_c and F_d striking N65°E and N130°, respectively. The tetrahedron exhibits current signs of instability, with frequent rock falls. The potential unstable volume was originally estimated at 2×10^5 m^3. Due to the danger of abseiling, GPR acquisition is impossible on this cliff face and geophysical prospecting methods (seismic and electrical profiles) were applied on the plateau in order to delineate the internal limits and fractures of the unstable rock mass.

3 STRUCTURAL ANALYSIS OF THE CLIFF USING SOLID IMAGES

For both sites, terrestrial laser scanning was made from two different locations in order to have the most complete coverage of the cliff (Figure 2).

For the "Ravin de l'Aiguille" site, the two acquisitions points are located north of the cliff face. The first one looked at the scarp with a very low incidence angle while the second one had an incidence of 30° on average. Figure 3a shows a view of the "Ravin de l'Aiguille" scarp from the first scanning point. The dots have a color computed from digital images of the site.

For the "Rocher du Midi" site, the first scanning was performed from the North of the site at a distance of about 150 m on average (Figure 2b). From this point the scanner looked at the outcrop in a N140°E ± 10° direction with a dip ranging from +5° to −50° relative to the horizontal plane. The second scanning point, south of the cliff, looked at the outcrop in a N30°W ± 20° direction with a dip of 0° to −60°.

The process chain of the laser raw data was as follows:

- Orientation of each point cloud: we placed on the cliff high reflectivity targets that were easily distinguished in the point cloud. The location of these targets was measured in the field and used

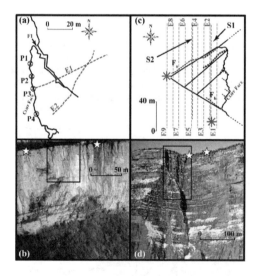

Figure 2. Helicopter views of the two sites and schematic surface maps showing the location of the main observed fractures and of the geophysical profiles. (a) and (b) Rocher du Midi, (c) and (d) Ravin de l'Aiguille. The scanning points are shown with white stars.

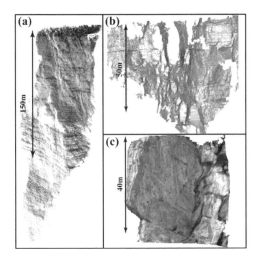
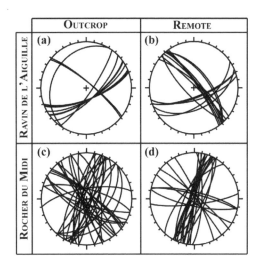

Figure 3. Illustrations of the process chain for laser data a) View of the Ravin de l'Aiguille site after orientating of point clouds and filtering b) Triangular Irregular Network (Rocher du Midi) and c) Synthetic 3D view with photographic draping (Upper south corner of the Ravin de l'Aiguille). The large rock face corresponds to the N130°E limit of the tetrahedron.

Figure 4. Stereographic projection (lower hemisphere) of the fracture planes measured at the "Ravin de l'Aiguille" (top row) and the "Rocher du Midi" (bottom row). The left column represent the data collected directly in the field while data remotely measured on the solid images are on the right.

to transform the laser coordinates to geographical coordinates. After this step the point clouds were northerly and vertically oriented.
- Filtering of outliers for eliminating isolated points that are significantly far from the scarp.
- Filtering of points corresponding to the vegetation in order to keep only the points on the rock face itself.
- Co-registration of points clouds with the images of the sites. This allows development of solid images. All points of the clouds are re-projected in the images, allowing the field location of each pixel to be fixed.
- Triangulation of the point clouds. Each point cloud was triangulated independently in a spherical geometry (Figure 3b), using only laser (azimuth, dip) coordinates (Alberts 2004). This processing chain results in a TIN (Triangular Irregular Network), a model of the site close to the real surface at a resolution of a few cm. The TIN was then used for viewing in combination with the images of the site (Figure 3c).

A structural analysis was first performed on the accessible outcrops in the vicinity of each site. The fractures were measured with a clinometers-compass and are shown in Figure 4 (a and c). A remote analysis was done using the solid images on which fracture planes were manually delineated. If the selected areas are planar, the best fit plane is computed. The selected fracture planes have a surface ranging between 10^{-1} m^2 to a few m^2. Most of them correspond to fractures or to small faults.

On the "Rocher du Midi" site, our measurements show two main families of near-vertical fractures (Figures. 4c and d). One strikes N10°E to N40°E while the second one has a direction ranging between N110°E to N140°E. The two techniques (direct and remote measurements) yield similar results although the remote technique gives a less scattered diagram (Figure 4d). The second family has an orientation similar to the one of the large fissure F1 behind the site. On the "Ravin de l'Aiguille" site, results from both methods are similar (Figs. 4a and b) although with slightly different orientations. Outcrop data are rare. The measurements also show two families of vertical fractures, N65°E ± 15° and N140°E ± 20°, the orientation of which correspond to the ones of the two large fractures F_c and F_d defining the tetrahedron, north and south respectively.

4 GEOPHYSICAL MEASUREMENTS

Geophysical experiments were conducted on the plateau of the two sites, while GPR vertical investigation on the cliff face was only performed for the "Rocher du Midi" site. Indeed, the cliff face of the Ravin de l'Aiguille did not meet the safety requirements for abseiling, due to the presence of numerous unstable blocks.

4.1 Rocher du Midi

Geophysical experiments included two electrical tomography profiles on the plateau and four GPR vertical profiles on the cliff face, the location of which is given in Figure 2a.

4.1.1 Electrical Resistivity Tomography (ERT)

Electrical profiles ($E1$ and $E2$, Figure 5) were carried out perpendicularly to the cliff face (Figure 2a) in order to check the continuity of the open fracture F1. The Wenner alpha array configuration was chosen for its robustness (Dahlin & Zhou 2004), with an electrode spacing of 1 m and 2 m for profiles $E1$ and $E2$, respectively. Inversion of apparent resistivity values was made using the software RES2DINV with the L1 norm (Loke & Barker 1996). The influence of the 150 m high cliff on apparent resistivity measurements was not corrected, as this effect would regularly increase the resistivity values by a factor between two near the cliff edge and one at the farthest distance (Sahbi et al. 1997). Thus, strong lateral resistivity contrasts, which are the targets of this study, are little affected by the presence of the cliff. The two profiles (Figure 5) shows electrical resistivity values ranging from 50 Ω.m in the highly weathered clayey zones, to more than 5000 Ω.m in the open fractures, with a mean resistivity of 500 to 1500 Ω.m within the rock mass. On profile $E1$, the $F1$ fracture (0.5 – 1 m wide) is clearly displayed by a continuous 3 m wide vertical resistive band, the thickness of which increases with depth. On profile $E2$, the fracture whose aperture is less than 10 cm at the surface, appears as a resistive spot at the surface. These results highlight the limited resolution of the electrical tomography method for low-aperture fractures and the decrease of resolution with depth.

4.1.2 GPR profiles

Four vertical GPR profiles ($P1$ to $P4$), 30 m to 60 m long, were recorded on the cliff wall (Figure 2). These Transverse Electric mode profiles were acquired with a trace spacing of 20 cm, using unshielded antennas of 100 MHz which offer a good compromise between resolution and penetration in this limestone (Jeannin et al. 2006). Reflecting targets were placed on the cliff during the acquisition in order to locate precisely the GPR profiles. Due to higher reflectivity of these points compared to the limestone rock, their location is easily derived during the Lidar data processing. GPR data processing consisted of bandpass filtering followed by a zero-phase band-pass filter and a time to depth conversion using velocity of 10 cm/ns, which is derived from CMP (Common Midpoint Profile) analysis. To amplify the late reflected events which were highly attenuated, an AGC (Automatic Gain Control) process was used. Finally, a static correction was performed to take into account the topography of the cliff face. Non migrated filtered profiles are displayed in Figure 6 for the 100 MHz antenna. As expected, the penetration depth of 100 MHz antenna is about 25 m, except along profile P3 (Figure 6c), located in a weathered conductive zone where waves are strongly attenuated (Reynolds 1997). The theoretical resolution at 100 MHz of an open fracture location is about 25 cm but the detection power of a thin bed could reach about 1 cm for air filling (Jeannin et al. 2006). The GPR profiles display numerous reflected waves, corresponding to near vertical or inclined fractures affecting the rock mass.

Two horizontal GPR profiles acquired at the bottom of two vertical ones ($P1$ and $P2$) give the fracture orientation. The near vertical fractures mainly exhibit two orientations (N20° and N130°) corresponding to the two families F_a and F_b observed at the cliff surface. A strong continuous reflector (labelled F_2) dipping inside the massif and parallel to the cliff face was shown on the four GPR profiles. This major discontinuity affecting the rock mass was not detected during the initial surface investigation.

To validate the GPR results, a detailed investigation of the karstic network affecting the rock mass was performed. The vertical sketch made from the ground observations (Figure 7) shows the existence of a major open fracture dipping to 45° at a depth of 20 m and to 70° between 25 m and 40 m depth. These observations are consistent with the geometry of the reflector F_2 shown on the close $P2$ profile (compare Figures. 6b and 7). This fracture is parallel to the well-known and large Montaud fault located 250 m from the site. These results highlight the interest and the power of GPR methods for characterizing the discontinuity pattern inside a rock mass.

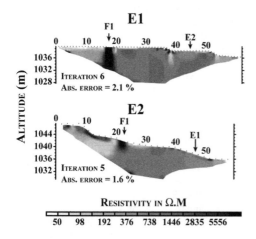

Figure 5. Rocher du Midi. Electrical profiles $E1$ and $E2$.

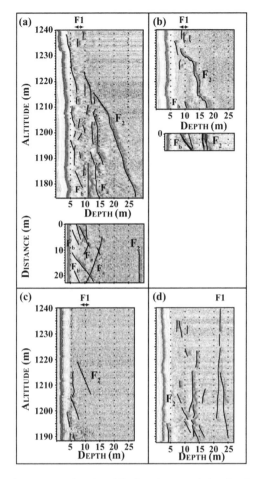

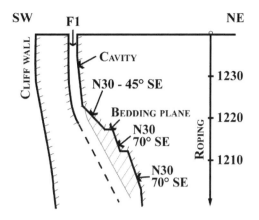

Figure 7. Schematic SW-NE cross-section drawn from observations in the karstic network (Rocher du Midi site).

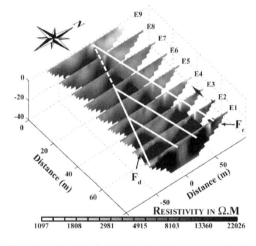

Figure 6. Rocher du Midi. GPR interpreted data for the four profiles acquired with 100 MHz antenna a) P1, b) P2, c) P3 and d) P4. Two horizontal profiles were conducted at the bottom of P1 and P2. Thin black lines show the interpreted fractures.

Figure 8. Ravin de l'Aiguille. Block diagram showing the 9 ERT profiles (E1 to E9). White lines delineate the dihedrons (see text).

4.2 Ravin de l'Aiguille

The size of the Ravin de l'Aiguille site is far greater than the Rocher du Midi. As the cliff face conditions do not allow abseiling, nine ERT profiles (labelled $E1$ to $E9$ in Figure 2c) and two seismic profiles ($S1$ and $S2$) were conducted on the plateau, parallel to the cliff face. Electrical images (Figure 8) exhibit resistivity values ranging from 1000 to more than 14000 Ω.m. The dihedron affected by open fractures is characterized by high resistivity values of a few thousands Ω.m bounded by conductive zones (less than 400 Ω.m). To the north, the narrow conductive zone coincides with the major fracture F_c limiting the dihedron and filled with clay. This fracture is clearly mapped by the electrical profiles down to 20 m deep, particularly on the $E1$ to $E5$ profiles. To the south and in the middle, the resistivity variations show the presence of smaller dihedrons imbricate into the main one (Figure 8). These results are validated by field observations. Two seismic profiles were carried out along the $E2$ and $E5$ electrical profiles. We used 48 vertical geophones at 3 m apart with a natural frequency of 4.5 Hz. Seismograms and first arrival picking are shown in Figure 9 for an end shot (a) and a fan shot (b) for the $S1$ profile.

Figure 9a clearly shows a change of the wave field characteristics, as well as a strong decrease of the signal to noise ratio, at a distance of about 40 m, which

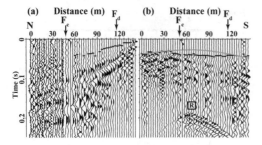

Figure 9. Seismic profile $S1$ recorded for (a) an end shot at 141 m and (b) a fan shot (b) (Figure 2). Black bars are the picked first arrivals and grey bars (b) are the theoretical times computed for a weak anisotropic medium. R: reflection on the cliff face.

corresponds to the trace of fracture F_c. A more subtle modification of the signal frequency is observed at 120 m, close to the F_d trace. A fan shot was made at the dihedron edge (Figure 2c). P-waves (Figure 9b) exhibit a nearly constant arrival time (around 0.04 s), in spite of the distance variations. The observed time values are compared in the same figure with theoretical times computed in a weak anisotropic medium (Thomsen 1986) characterized by a N65° orientated fracture. P-waves are clearly delayed between the main fractures F_c and F_d, with a maximum value of about 0.01 s. These results highlight the fracturing effect within the dihedron, the extension of which is approximately delineated by the time anomaly. Similar results were obtained along the $S2$ profile.

5 DISCUSSION AND CONCLUSION

Fracture characterization on cliffs is a complex problem due to the difficulty of performing geological observations and structural mapping on the cliff face, and to the lack of information about the rock mass provided by these techniques. This study aimed at evaluating the potential of applying laser scanning and geophysical imaging techniques. Two sites showing potential instability problems with different characteristics of volume (50,000 and 200,000 m^3) and of shape (column and wedge) were chosen in the cliffs surrounding the town of Grenoble (French Alps) as test sites. In such high cliffs, terrestrial laser scanning is only appropriate when the cliff topography allows the site to be correctly seen from one point or more, located along the crest at a short distance (typically <300 m) and covering the cliff with a significant incidence angle. The obtained point clouds can be combined with optical images to obtain a solid image on which a structural analysis can be performed. The comparison of the fracture pattern deduced from the remote analysis and the one determined from direct field measurement is good. These results highlight the potential of using laser scanning techniques on high cliffs, which yield a gain in time and in safety. For the site where abseiling was possible, the Ground Penetrating Radar (GPR) on the cliff face turned out to be the most effective method for imaging the fractures inside the mass, with a detection power of a few cm with an antenna of 100 MHz and a penetration of maximum 30 m. The traces of the profiles were located on the solid images and the 3D coordinate of the trace computed and used for topographical corrections. Combining vertical and horizontal GPR profiles on the cliff face allowed the 3D geometry of the fractures to be mapped in homogeneous resistive limestone rock mass. In other geological formations (e.g in shale, micaschist and marly limestone), penetration can be limited by the high rock electrical conductivity which attenuates radar waves. When abseiling was impossible due to the instability of rock blocks, seismic and electrical imaging techniques were used on the plateau for mapping the fractures seen on the cliff. Only the ones with an opening larger than 1 m were detected using these methods, with a penetration limited to a few tens of meters. On the "Ravin de l'Aiguille" site, these techniques helped to delineate the geometry of the potential unstable rock masses and to determine its volume.

ACKNOWLEDGMENTS

This work was funded by the French national project RDT (Risques, Décisions, Territoire, http://www.rdtrisques.org/projets/camus/) from the Ministry of the Ecology and Sustainable Development. We thank the federal organization VOR (Vulnérabilité des Ouvrages aux Risques) for its financial support.

REFERENCES

Alberts, C.P. 2004. Surface reconstruction from scan paths. *Future Generation Computer Systems*, 20:1285–1298.

Arnaud, H., Bravard, Y., Fournier, D., Gidon, M. & Monjuvent, G. 1978 Carte géologique à 1/50000, feuille de Grenoble. Tech. rept. BRGM Ed.

Bornaz, L. & Dequal, S. 2003. The solid image: a new concept and its applications. In *The International Archives of the Photogrammetry, Remote Sensing and Spatial Information Sciences*, volume XXXIV.

Busby, J. & Jackson, P. 2006. The application of time-lapse azimuthal apparent resistivity measurements for the prediction of coastal cliff failure. *Journal of Applied Geophyiscs*, 59:261–272.

Dahlin, T. & Zhou, B. 2004. A numerical comparison of 2d resistivity imaging with ten electrode arrays. *Geophysical Prospecting*, 52:379–398.

Deparis, J., Garambois, S. & Hantz, D. In press. On the potential of Ground Penetrating Radar to help rock fall hazard assessment: a case study of a limestone slab, Gorges de la Bourne (French Alps). *Engineering Geology.*

Dussauge-Peisser, C., Wathelet, M., Jongmans, D., Hantz, D., Couturier, B. & Sintes, M. 2003. Seismic tomography and ground penetrating radar applied on fracture characterization in a limestone cliff, Chartreuse massif, France. *Near surface geophysics*, 1:161–172.

Feng, Q.H. & Röshoff, K. 2004. In-situ mapping and documentation of rock faces using full coverage 3d laser scanning technique, *International Journal of Rock Mechanics and Mining Sciences*, 41, 1, 139–144.

Jeannin, M., Garambois, S., Jongmans, D. & Grégoire, C. 2006. Multiconfiguration gpr measurements for geometric fracture characterization in limestone cliffs (alps). *Geophysics*, 71:B85-B92.

Jongmans, D. & Garambois, S. 2007. Surface geophysical characterization and monitoring: a review. *Bull. Soc. géol. France*, 178:101–112.

Lemy, F. & Hadjigeorgiou, J. 2004. A Field application of laser scanning technology to quantify rock fracture orientation. In *EUROCK 2004 & 53rd Geomechanics*, 435–438.

Loke, M.H. & Barker, R.D. 1996. Rapid least-squares inversion of apparent resistivity pseudosections by a quasi-newton method. *Geophysical Prospecting*, 44:131–152.

Philippe, Y., Deville, E. & Mascle, A. 1998. *Thin-skin inversion tectonics at oblique basin margin: example of the western Vercors and Chartreuse Subalpine massifs*. Mascle A., Puigdefµabregas C., Luterbacher H., Fernµandez M. (eds) Geological Society, London, Special Publications. Pages 239–262.

Reynolds, J.M. 1997. *An introduction to applied and environmental geophysics*. John Wiley & Sons, Chichester, England.

Roch, K.H., Chwatal, E. & Brückl, E. 2006. Potential of monitoring rock fall hazards by GPR: considering as example of the results of salzburg. *Landslide*, 3:87–94.

Sahbi, H., Jongmans, D. & Charlier, R. 1997. Theoretical study of slope effects in resistivity surveys and applications. *Geophysical prospecting*, 45(5):795–808.

Schulz, T., Lemy, F. & Yong, S. 2005. Laser scanning technology for rock engineering applications. In *Optical 3-D Measurement Techniques VII (Eds: Grün, Kahmen), Vienna.*

Thomsen, L. 1986. Weak elastic anisotropy, *Geophysics*, 51, 1954–1966.

In situ characterization of the geomechanical properties of an unstable fractured rock slope

C. Dünner & P. Bigarré
INERIS, Ecole des Mines de Nancy, Nancy, France

F. Cappa & Y. Guglielmi
Géosciences Azur, CNRS-UNSA-IRD-UPMC, Sophia-Antipolis, France

C. Clément
LAEGO ~ INERIS, Ecole des Mines de Nancy, Nancy, France

ABSTRACT: Rockfalls and landslides are recognized as a major natural hazard across the Alps with strong economical and social impacts on regional land settlement and transportation policies. While numerous Alpine zones and valleys are prone to gravitational risks, climate change is becoming an important issue on whether or not the situation facing landslide hazards could become worse for decision makers. Along the *Tinée* Valley, a steep fractured rock slope known as the *Rochers de Valabres* is currently being investigated as a Pilot Site Laboratory (PSL), aiming to develop scientific knowledge on the physical processes of rock slope instabilities. An original field characterization protocol coupling stress measurements and acoustic logging in a borehole was conducted in order to improve site geomechanical properties and potential instability estimations. The authors present an overview of this research site and objectives, as well as some preliminary results collected from subsurface investigations and an assessment of the knowledge gained at this stage.

1 INTRODUCTION

Rockfall is a slope failure mechanism for which both the preparatory and triggering processes are still poorly understood. Identification and quantification of numerous predisposing factors—based on reliable field data—is a challenging task while early-warning signs of failure, or short term precursory mechanisms, still require research and development work.

At present, it is difficult to correctly characterize, on a deterministic and exhaustive basis, the rock mass fabric and the whole set of physical properties needed, mainly due to the variety of heterogeneities usually encountered and to the differing scales of such heterogeneities. Correct assessment of rock mass initial volumetric conditions as internal boundaries, *in situ* stresses, stiffness and moreover strength degradation and/or stress build up mechanisms with time requires considerable effort and patience.

In order to improve knowledge about all of these issues relating to rock slope failure mechanisms, a large-scale fractured rock slope—the *Rochers de Valabres* located in the Southern French Alps—is being investigated as a Pilot Site Laboratory (PSL), by INERIS and GEOSCIENCES AZUR Laboratory, among other partners.

The recent history of the PSL destabilization marked by two important rockfalls that occurred in May 2000 (around 2,000 m^3 of blocks and fragments) and October 2004 (about 30 m^3 of material) and caused much damage. The first incident seriously damaged one pile of a bridge over the *Tinée* River and caused a road traffic interruption, lasting several few weeks, due to the considerable renovation works required. Subsequent geological reconnaissance was urgently undertaken and provided clear evidence of other potential rockfalls. Since then, a research program has been initiated and progressively developed as a long-term, extensive and multi-disciplinary program, the description of which is beyond the scope of this paper.

The VAL-STRESS3D specific field experiment is part of this program. It is based on subsurface investigations coupling *in situ* stress measurements and acoustic logging in a shallow borehole set in the lower part of the slope. The geological context of the site, the preliminary results of stress measurements and acoustic measurements are firstly presented.

Then, we discuss a better way to estimate slope properties and stability from a comparison between the three approaches.

2 SETTING OF THE ROCHERS DE VALABRES PSL

2.1 Geological features

The *Rochers de Valabres* PSL is located on the right bank of the *Tinée* Valley, which is situated at the northwestern edge of the *Argentera Mercantour* crystalline massif (French Alps). The basal portion of the PSL is at an elevation of 900 m and the top of the slope culminates at an elevation of 2,254 m. The topographic surface is characterized by considerable roughness and has a slope angle ranging from 40° to 60° (Figure 1).

This basement unit consists of metamorphic rocks that have undergone a polyphased tectonic and metamorphic evolution during the Variscan and Alpine orogeneses. The slope is mainly composed of two hard mica gneisses with a sub-vertical foliation striking from N110° to N140°. The metamorphic foliation appears undulated and micro-folded. Four sub-vertical fault zones of decametric thickness cut the slope in three zones. In these zones, the rock mass is intensively fractured near the fault zones, with fracture density decreasing rapidly while moving away from the faults (Figure 2). These fractures can be seen to be partially closed or opened with millimetric to centimetric apertures.

Considering both the topographic and structural complexities of the rock slope, subsurface investigations were undertaken to provide more quantitative information regarding stability analysis and the hypotheses to be constrained.

2.2 Geological borehole logging

At the end of 2005, a 19 metres deep sub-horizontal over-cored borehole was sunk in a compartment block of interest, underlying a large planar failure flagstone (Figure 1) to measure stresses, obtain samples and run surveys for refined characterization of the rock fabric. Except for the very first metres, RQD resulted in excellent, numerous collected cores exceeding 0.5 metre in length. A geological analysis of the drill cores as well as a borehole endoscopy were performed (Clément et al., 2006; Marignac et al., 2006).

The lithological analysis of the drill cores (Figure 3) shows rather homogeneous intact rock corresponding to highly migmatized gneisses with rough foliation planes. For depths ranging from 9 m to 11 m, foliation planes appear more regular and characterized by a significant increase in biotite content.

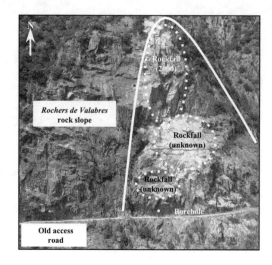

Figure 1. Frontal view of the *Rochers de Valabres* PSL.

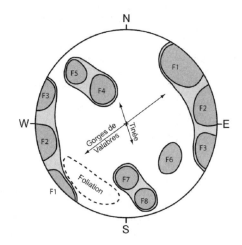

Figure 2. Stereographic projection of the different sets of fractures (lower hemisphere). Sub-vertical fractures are clearly dominant. Dip directions of the *Tinée* and the *Gorges de Valabres* are indicated (after Gunzburger et al., 2004, 2005).

Figure 3. Drill core from over-coring test N° 5, at a depth of 15.75 m.

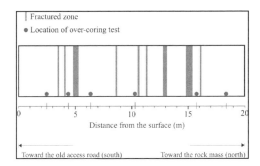

Figure 4. Log of location of over-coring tests and fractured zones in the investigation borehole.

Table 1. Synthesis of the elastic parameters used for inversion of the gneisses of the *Rochers de Valabres* PSL.

Over-coring test	1 to 4 and 6	5
Transverse Isotropy	Marked	–
E_1 (GPa)	25.2	28.6
E_2 (GPa)	35.9	28.6
ν_{12}	0.14	0.06
ν_{23}	0.15	0.06
G_{12} (GPa)	12.4	13.5
G_{23} (GPa)	15.7	13.5
V_P (m/s)	4500	3900
R_C (MPa)	76.4	88.8

Endoscopy reveals nine fault zones (Figure 4), six of these as a few millimetres-thick fault planes and three 0.5 to 0.8 metre-thick breccia zones, located respectively at depths of 5 m, 13 m and 15 m.

Faults are oriented sub-vertical and N40° to N70° dip oriented, in full accordance with previous geological analyses (Figure 2). A significant number of fracture planes display a rough and oxidized surface (calcite infillings). Signs of water circulation were apparent although no actual water was found when visual inspections were carried out.

3 IN SITU STRESS MEASUREMENTS

3.1 Experimental protocol

If rockfalls are disruptions originating from unbalanced forces exceeding rock strengths, the local stress field in steep fractured rock slopes remains difficult to assess. One objective of the VAL-STRESS3D experiment was to perform a 3D *in situ* stress profile based on the over-coring method.

The over-coring method is based on CSIRO cell measurements that make it possible to measure a full 3D strain tensor estimation and thus to back-calculate a 3D stress tensor $(\sigma_1, \sigma_2, \sigma_3)$. Successive incremental over-coring tests while drilling is progressing enable sharp stress gradients and locally heterogeneous stress fields with good accuracy to be assessed.

Six tests were conducted ranging from 2.5 m down to 18.5 m depth along the borehole (Figure 4). More details of the experimental protocol can be found in Dünner et al., 2007.

3.2 Laboratory testing of the intact rock

Two cores from over-coring tests, located at depths of 2.5 m and 15.75 m, were completed by laboratory biaxial tests in order to characterize the rock's local elastic parameters that had to be considered in stress determinations.

Although biaxial testing does not allow the transverse isotropic rock parameters to be determined, it can qualitatively highlight such parameters and, in any case, provides average values. Laboratory measurements on intact rock samples selected from drill cores were thus conducted to complete and refine the determination of elastic properties related to the pre-identified transverse isotropy of the gneiss. The two zones sampled for this were those nearest to the tests zones where a biaxial test could be carried out, in order to allow an easy quantitative correction of expected scale effects.

Geomechanical results, along with scrutiny of the drill cores and lithological analysis, confirmed a rather homogeneous transverse isotropy of the deformability and the shear modulus due to the foliation (Table 1) except in the fifth test zone, where foliation tapers off.

3.3 Results of stress measurements

The over-coring data was processed with SYGTGEOstress® software, developed by INERIS, featuring data inversion including anisotropic materials. Figure 5 presents the results of stress measurements with $(\sigma_1, \sigma_2, \sigma_3)$ magnitudes plotted as a function of borehole depth. The results show essentially:

- a first zone between 0 and 1 m, which appears as rather fractured. Note that the first target zone was intended to be placed at 1 m depth from the free surface, but disking to very shallow and dense fissuring meant the first measure had to be moved back to 2.5 m. This pre-existing fissuration is assumed to be most likely blast-induced;
- a second relatively homogeneous zone, between 2 and 6.5 m (over-coring tests N° 1, 2 and 3). This zone is characterized by a major stress σ_1 of approximately 6 MPa and an intermediate stress σ_2 of 5.4 MPa. Principal stresses (σ_1, σ_2) are contained in a sub-vertical plan. This orientation—which does

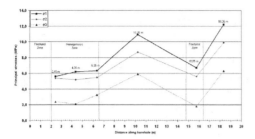

Figure 5. Profile of principal stresses (σ_1, σ_2) versus depth.

not correspond in the outcrop to that of the geometry of the average slope of the rock slope—is most likely influenced by the vertical discontinuities identified afterwards. Minor stress σ_3 presents a value of about 2.6 MPa; it is horizontal, close to the drill hole axis, in other words oriented sub-perpendicularly to the free surface at the borehole collar level;
- a third zone beyond 6.5 m (over-coring tests N° 4, 5 and 6). This zone has a stronger level in term of average stresses, with a shift in the orientation of the principal stresses. It presents a sharp stress gradient at 10 m depth, for all three components ($\sigma_1, \sigma_2, \sigma_3$);
- a local singularity at 15 m depth, with stresses in full continuity to those measured in the first 6 m, followed by a maximum stress level at 18 m depth with σ_1 exceeding 12 MPa;
- except for measurement points 4 and 6 located at 10 m and 18 m depth, (σ_1, σ_2) is located in a sub-vertical plane, whereas locally the slope is 60°. This is driven probably by the different sets of vertical fractures cutting through the rock mass.

4 ULTRASONIC BOREHOLE LOGGING

4.1 Ultrasonic acquisition set up

In a second phase, a geophysical survey was carried out, based on high resolution ultrasonic logging, in order to investigate the geophysical properties of the rock fabric.

The probe used for this logging was equipped with three ultrasonic transducers (1 transmitter and 2 receivers) operating in a 50–700 kHz frequency range centred at 150 kHz. Working with a dominant wavelength of a few centimetres offers a high resolution in detecting slight changes in physical properties and mineralogy, but most of all in detecting joint alteration states and opened fractures. Three transducers, two receivers respectively 9 cm and 18 cm distant from the emitter, were placed in contact with the smooth borehole walls using a pneumatic clamping device and some coupling agent, while a push-rod system enabled the probe's depth position to be controlled. Measurements provided two values of apparent velocity of P-waves associated respectively with each receiver, V_{P1} and V_{P2} (Contrucci et al., 2007).

Data acquisition was based on automatic stacking while analyses consisted in manually picking first break arrival times in order to evaluate the P-wave velocity profile versus depth. Eventually, a total of 156 measurements were collected.

4.2 Results of velocity logging

It is worth noting that the ultrasonic logging was carried out more than one year after the borehole drilling. Visual inspection showed a significant alteration of the borehole wall-sides with visible slight breakouts and disseminated rock fragments, requiring a cleaning operation. This was quite unexpected when considering the very smooth cores obtained initially and high values of the strengths of the intact rock compared to the very moderate level of stresses measured as described here-above. As previously noted, no trace of water was found in the borehole.

Two loggings were carried out along the borehole 90° from each other, first in the vertical plane and secondly in the horizontal plane.

Vertical logging was not conclusive since recurrent problems of coupling related to significant localized and superficial alteration of the borehole wall-side impeded correct emission of the mechanical stress wavefront and introduced much dispersion in the data. Data was found to be emergent, with low signal to noise ratio, and not repeatable, while intensive use of coupling agent could not improve measurement quality and reliability.

A second successful attempt was run 90° clockwise towards the right walls of the borehole. The depth of investigation focused on the first 8 metres, showing good signal-to-noise ratio and repeatability of the measurements.

This profile is presented in Figure 6, featuring:
- in situ velocities $V_{P1} \sim 3{,}650$ m/s and $V_{P2} \sim 4{,}150$ m·s^{-1}, with V_{P2}/V_{P1} ratio equal to 1.13. These values are consistent with the velocities obtained from laboratory testing (Table 1);
- a first zone between 0 and 2.5 m, where V_{P1} is slightly increasing while V_{P2} is almost constant. This means that the first couple of metres of borehole walls may even be slightly damaged also along this generatrix while free-surface and stress-release effects, measured by V_{P2}, are not perceptible. This low V_{P1} zone, measuring a damage zone induced around the borehole, most likely originates from the construction of the overhanging road;

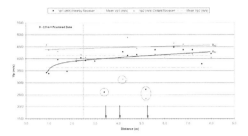

Figure 6. Horizontal plane velocity profile versus depth. The arrows indicate velocity drops respectively at 3.5 m, 4.1 m and 5.3 m corresponding to fault zones located in Figure 3.

- beyond 2.5 m, V_{P1} and V_{P2} are almost constant and $V_{P2} \sim V_{P1}$. This trend indicates fairly homogeneous rock properties except for some specific zones;
- three thin zones affected by velocity drops respectively at 3.5 m, 4.1 m and 5.3 m where the borehole is intersecting fault zones.

5 DISCUSSION

Observations from the three approaches allow the borehole to be divided into the following geomechanical zones:

- a first zone between 0 and 2 m, which is significantly altered and fractured. Disking, lithological analysis of drill cores video analysis and lower velocity confirmed this important set of fracturing, which made it necessary to conduct the first over-coring test at a depth of about 2.5 m. This altered zone was most likely induced by artificial blasting when the road was built and by natural long term surface effects (stress release and some weathering);
- a second zone between 2.5 and 7 m depth that appears as rather homogeneous, for intact rock, in terms of geology, stress tensor and wave velocity, despite the presence of three fractures at 3.5 m, 4.1 m and 5.3 m shown by endoscopy and distinctly identified through velocity drops;
- a clear singularity at about 15 m depth highlighted by stress measurements. The geological survey carried out on borehole cores and the endoscopy showed a breccia zone between 15.5 m and 16.5 m. This zone presents a strongly altered quartzic and hydrochloric matrix. The presence of calcite in the fractures and fluid inclusions proves the occurrence of temporary meteoric fluid circulation;
- the excellent precision of the deformation data input made it possible to highlight stress variations—in

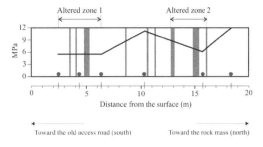

Figure 7. Profile of stresses and location of fractured zones and over-coring tests.

term of modulus and orientation—along the profile. Even when considering the topography and fractured setting of the rock slope, the *in situ* stresses found in the first 0–18 m range of depth may be considered unexpectedly high and discontinuous. These results confirm that the local stress field is rather complex, characterized by both an overall pattern and variability revealing the largely predominant role of pre-existing fractures and local heterogeneities.

As shown in Figure 7, both the measured principal stress and acoustic velocities variations appear closely related to fault zones. Both stress magnitudes and V_P values of intact rock are a factor of 1.6 higher than in fault zones. These results show that fault zones clearly display very different mechanical properties compared to intact rock zones and that such contrasts in the properties induce large stress variations in the subsurface of the rock slope.

The low V_P values in macroscopic fault zones intersected in boreholes are in general related to low Young's modulus values of the zone materials compared to intact rock Young's modulus values (Barton et al. 1992; Kuwarahara et al. 1995; Malmgren et al. 2006). Young's modulus can be related to values from the following Equation (1):

$$E = V_P^2 \rho (1 + \upsilon)(1 - 2\upsilon)/(1 - \upsilon) \qquad (1)$$

Where E is the Young's modulus, V_P is the velocity, ρ is the rock density and υ is the Poisson's ratio.

If the Poisson's ratio and the density are assumed constant, Equation 1 shows that a factor of 1.6 increase of V_P between intact rock and fault zones corresponds to a factor of 2.5 increase in the Young's modulus. Then, for an average value of $E_{\text{Intact rock}} = 30$ GPa, it provides a rough estimation of $E_{\text{Fault zone}} = 12$ GPa for a thickness of the zones of a few 0.5 centimetres, much less for a thickness of a few centimetres.

Moreover, at low stress, the elastic parameters of fault zones depend on the Uniaxial Compressive Strength (UCS) of the infillings and the fault walls, the stress applied normal to the fault walls and the shape of fault pores (Barton 1976 and 1990). Moisture effects in fault pores can be neglected because the pores were dry when the experiments were performed. In our case, fault zones that are filled with breccia and altered by temporary water flow clearly display a much higher porosity and lower UCS compared to intact rock that could well fit with the average elastic modulus value estimated from Equation 1.

Finally, such a method based on coupled accurate acoustic logging and stress measurements through over-coring appears promising since it allows the *in situ* rough estimation at shallow depths of macroscopic fault properties (stiffness and strength) from two independent approaches respectively based on velocity variations and stress-strain measurements.

6 CONCLUSIONS

Complementary to surface investigations, subsurface surveys including *in situ* stress measurements, removal and laboratory tests of rock samples and geophysical logging have been undertaken at the *Rochers de Valabres* PSL, France. As part of a long term research program related to rock slope failure mechanisms, these surveys aim to quantify the predominant role of the geological structure, through the presence of discontinuities, faults and fractures in the failure mechanisms.

In the zone of investigation, the *in situ* stress field was quantified accurately and found much higher than expected when considering analytical or numerical modelling. The *in situ* stress profile exhibits significant variability and complexity, most likely governed both in terms of orientation and amplitudes by the pre-existing sub-vertical discontinuities. Normal stress to the discontinuities is clearly reduced. It must be kept in mind that recent rockfalls just overlying the investigation zone may have a key role in the readjustment of stresses and could explain the complex stress field. Measured stresses are still found one order of magnitude lower than UCS estimated from laboratory tests on intact rock.

Video and ultrasonic logging show that discontinuities contain infilling materials between intact rock walls that generate very sharp contrasts in stiffness and shear strength that can be coarsely estimated.

Some temporary water flows were apparent in the fracture zones, which could also correspond to preferential groundwater paths. Further *in situ* experiments are planned to refine these results and estimate these fracture zone permeabilities and the relationships between permeabilities and fracture mechanical properties.

ACKNOWLEDGEMENTS

Special thanks are extended to the French Ministry of Ecology, Sustainable Development and Planning for financial support, as well as to the French National Research Agency, to the Provence-Alpes-côte d'Azur Region and the Alpes Maritimes Département.

All authorizations and help from the *Mercantour National Park* and *Electricité de France* are fully acknowledged.

REFERENCES

Barton, N. 1976. Rock Mechanics Review—The Shear Strength of Rock and Rock Joints. *International Journal of Rock Mechanics and Mining Sciences, Vol. 13, pp. 255–279.*

Barton, N., Bandis, S. 1990. Review of predictive capabilities of JRC-JCS model in engineering practice. *Rock Joints. Barton & Stephansson (eds). Balkema, Rotterdam. ISBN 90 6191 109 5, pp. 603–610.*

Barton, N., Zoback, M.D. 1992. Self-Similar distribution and properties of macroscopic fractures at depth in crystalline rock in the Cajon Pass scientific drill hole. *J. Geophys. Res. 97, 5181–5200.*

Clément, C., Laumonier, B., Brouand, M. 2006. Mineralogical study of gneisses of the *Rochers de Valabres* PSL. *Internal memo referenced INERIS DRS-06-75754LN07, 10 p.*

Contrucci, I., Cabrera, J., Klein, E., Ben Slimane, K. 2007. Excavation Damaged Zone characterized by ultrasonic borehole logging in drifts of different ages excavated in the argillaceous formation of the Tornemire experimental station (Aveyron, France), *3rd International Meeting on Clays in natural & Engineered barriers for radioactive waste confinement, 17–20th September 2007, Lille, France.*

Dünner, C., Bigarré, P., Clément, C., Merrien-Soukatchoff, V., Gunzburger, Y. 2007. Natural and thermomechanical stress field measurements at the "Rochers de Valabres" Pilot Site Laboratory in France. *In: Ribeiro E Sousa et al. Editors. Proc. of the 11th Congress of the International Society for the Rock Mechanics (ISRM), 9–13th July 2007, Lisbon, Portugal. Leiden, The Netherlands: Taylor & Francis Group; 2007, vol.1, pp. 69–72.*

Gunzburger, Y., Merrien-Soukatchoff, V., Senfaute, G., Piguet J.-P., Guglielmi, Y. 2004. Field investigations, monitoring and modeling in the identification of rock fall causes. *In: Lacerda et al. Editors. Landslides: evaluation and stabilization. Proc. of the 9th International Symposium on Landslides (ISL), 28th June—2nd July 2004,*

Rio de Janeiro, Brazil. London: Taylor & Francis Group; 2004, pp. 557–563.

Gunzburger, Y., Merrien-Soukatchoff, V., Guglielmi, Y. 2005. Influence of daily surface temperature fluctuations on rock slope stability: case study of the Rochers de Valabres slope (France). *International Journal of Rock Mechanics and Mining Sciences, vol. 42, no.3, April 2005, pp. 331–349.*

Kuwarahara, Y., Ito, H., Ohminato, T., Nakao, S., Kiguchi, T. 1995. Size characterization of in situ fractures by high frequency s-wave vertical seismic profiling (VSP) and borehole logging. *Geotherm. Sci. Technol. 5, 71–78.*

Malmgren, L., Saiang, D., Töyrä, J., Bodare, A. 2006. The excavation disturbed zone (EDZ) at Kiirunavaara mine, Sweden—by seismic measurements. *Journal of Applied Geophysics 61, 1–15.*

Marignac, C., Clément, C., Laumonier, B. 2006. Petrographic analysis of the drill cores of VAL-STRESS3D experiment. *Internal memo referenced INERIS DRS-06-75754LN06, 7 p.*

Properties of peat relating to instability of blanket bogs

A.P. Dykes
School of Earth Sciences and Geography, Centre for Earth and Environmental Science Research, Kingston University, UK

ABSTRACT: Instability and failure of hillslopes mantled with blanket peat occurs naturally in cool temperate regions that favour the development of blanket bogs. Such peat failures are most common in the UK and Ireland, where they pose potential hazards to properties and infrastructure. These blanket bogs are also particularly susceptible to engineering-induced instability. This paper identifies an urgent need for new approaches to the assessment of blanket peat stability, particularly through appropriate geotechnical characterisation, and examines some of the principal issues identified during recent research that now need to be addressed including validation or modification of existing methods of quantitative stability analyses for use with peat.

1 INTRODUCTION

Peat is a material formed from the accumulation of waterlogged remains of dead and decaying plant matter. It has traditionally been investigated by engineers primarily concerned with settlement following loading associated with construction of roads or other infrastructure across peat deposits such as the Irish raised bogs (e.g. Hanrahan 1964) or the Canadian Muskeg (MacFarlane 1969). However, natural instability of peat deposits had not been systematically investigated until the last ten years, when initial detailed research on specific peat landslides (Kirk 2001, Dykes & Kirk 2001) and on the general topic of peat failures (Mills 2002, Warburton et al. 2004) was followed by an extensive research programme arising from the peat landslide events of 2003 in the UK and Ireland and founded on a new classification scheme for peat failures (Dykes & Warburton 2007a).

On 19 September 2003, extreme rainfall triggered 40 landslides on a small mountain in western Ireland that included 12 large peat failures and had severe impacts on the rural communities below the mountain (Dykes & Warburton 2007b). On the same day, a separate extreme storm triggered around 35 landslides in Shetland, northern Scotland, including 20 large peat slides, which combined with floodwaters to cause significant damage to local infrastructure (Dykes & Warburton, in press). Then, a month later, engineers preparing the site for a wind turbine in Ireland triggered a 450,000 m^3 peat flow that also had environmental and socio-economic consequences. All of these events involved failures of upland blanket bog, a type of peat deposit not previously of concern to engineers.

Properties of peat were examined and reviewed specifically for engineers by Hobbs (1986). However, data presented in his review were primarily obtained from fen or transitional (fen-bog) peat, and the shear strength of peat was not addressed. More recent reviews have provided little additional material. In other words, properties and characteristics of some peatlands that are particularly susceptible to natural or engineering-induced failure have not previously been reported in the mainstream engineering or geomorphological literature.

The upland blanket bogs of the UK and Ireland in the northern hemisphere and subantarctic islands in the south fall into this category of susceptible peatlands, although the latter locations are largely uninhabited and peat failures generally do not constitute a hazard. In fact, over 80% of all known natural peat mass movements occurred in the British Isles (60% in Ireland and 20% in England and Scotland: Dykes & Kirk 2006), most commonly involving upland blanket bog on hillslopes with gradients typically in the range 2–12°. Windfarm construction and commercial forestry operations on British and Irish uplands create additional potential hazards from failures of blanket peat.

The aim of this paper is to highlight the problems of assessing and analysing the stability of hillslopes mantled with blanket peat. Specific issues are identified, arising from the nature of peat as an engineering material, that need to be addressed if the hazard from natural blanket bog failures is to be adequately assessed and engineering-triggered failures avoided.

2 TYPES AND MECHANISMS OF PEAT FAILURES

Several distinct types of peat failures have been identified and defined by Dykes & Warburton (2007a) primarily according to field morphology as controlled by apparent failure position and mechanism, but with additional divisions for types of peat deposit (e.g. 'bog bursts' exclusively refer to raised bogs) and engineering-induced failures ('peat flows' being used to identify head-loaded failures). The latter designation makes no assumption about failure mechanism: for example, the 450,000 m^3 Irish windfarm peat flow began as a translational shear failure within the basal peat caused, at least in part, by placement of excavated peat up to 1 m thick on the in-situ blanket bog (AGEC 2004). It is not known if any of the other head-loaded peat flows failed by translational shearing or some other mechanism, but for any stability assessment it would be reasonable to initially assume a shearing component.

Large and potentially highly damaging natural 'bog slides' involve translational shearing within the lower peat, whereas 'bogflows' are thought, on the basis of eyewitness accounts in the early 20th century (e.g. Delap et al. 1932), to begin due to in-situ loss of strength (analagous to sensitive 'quick' clay) caused by high water pressures. However, there is as yet no evidence for any specific failure mechanism for bogflows so an initial shearing component cannot be ruled out. The potential for severe impacts from these events is perhaps demonstrated by their scale: whilst bog slides tend to involve several tens of thousands of cubic metres of peat moving from 5–8° slopes, bogflows of up to 375,000 m^3 have been recorded from 2–5° slopes in Ireland. Both types can produce extensive debris runout with a high capacity to cause physical damage and pollution.

3 GEOTECHNICAL ISSUES IN BLANKET BOG STABILITY ASSESSMENT

Investigations of twelve separate peat landslides at seven locations in northern and western Ireland have generated a set of data that describe the index properties and some geotechnical properties of the failed blanket bogs. These failed peats possessed much higher water contents and liquid limits than fen or transitional peats at the same degree of humification as reviewed by Hobbs (1986) (Fig. 1). Furthermore, although the limited data do not appear to show any differences between the different failure types, they do include preliminary geotechnical results that highlight the need for new or modified techniques in order to be able to reliably assess the stability condition of undisturbed or engineered blanket bog covered hillslopes.

There are four principal difficulties associated with stability assessments of blanket peat. These will be addressed in turn.

1. All peats are highly heterogeneous, making the classification of peat for engineering purposes difficult. Small-scale variations in the botanical composition and stratigraphy undoubtably give rise to widely ranging values for some geotechnical and index properties (although these groups of peat properties have never yet been correlated), and in some cases the matrix properties may be less useful than the 'peat mass' (c.f. rock mass) properties. For example, the saturated hydraulic conductivity (k_{sat}) of the lower 'catotelm' peat, i.e. below the level of maximum seasonal water table draw-down 0.2–0.5 m below the surface, could be between 10^{-3} and 10^{-10} m s^{-1}, higher in the horizontal plane due to compression of plant fibres (Hobbs 1986). On the other hand, laboratory measurements of k_{sat} of undisturbed 'small' and 'large' core samples (50 or 100 mm long × 50 or 100 mm diameter) of basal catotelm peat from several Irish bogflows and bog slides, using a standard constant-head technique, varied between 10^{-4} and 10^{-11} m s^{-1} with either higher vertical values or no significant anisotropy (e.g. Yang and Dykes 2006, Dykes et al., in review, Fig. 2). Dissipation of excess pore water pressures due to loading, for example by placement of fill or spoil on intact peat over a relatively short time (as occurred at the Irish windfarm), will be limited not only because of the low natural permeability but because the permeability falls rapidly as the void ratio is reduced by loading (Hobbs 1986): the resulting potential for instability is clear. In natural blanket bogs, the permeability of the peat matrix is much less important than that of the peat mass, with inputs and subsequent drainage of rainwater occurring by means of natural pipes, cracks and other macropores (Holden and Burt 2003). Natural instability has commonly been associated with heavy rainfall, the assumption being that rainwater supply may be sufficiently rapid to exceed the natural drainage capacity so that excess water pressures were generated within the basal peat (Warburton et al. 2004). Indeed, at some sites, the low permeability peat matrix is thought to have caused artesian water pressures in soil pipes beneath thin blanket bog, contributing to widespread peaty-debris slides (e.g. Dykes and Warburton 2007b).
2. There are almost no data to show how peat strength relates to any other index physical properties or indeed botanical properties. Long (2005) and Dykes & Kirk (2006) reviewed published peat

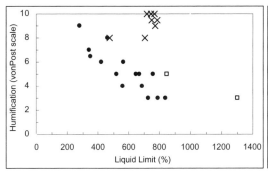 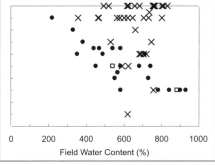

Figure 1. Variations in liquid limit and water content with humification (after Hobbs, 1986). Black circles represent data from fen-transitional peat in Shropshire, UK (50–70% organic content), open squares refer to a drained raised bog in County Durham, UK (99% organic), and crosses indicate properties of Irish upland blanket bogs (>95% organic). Both plots use the same y-axis.

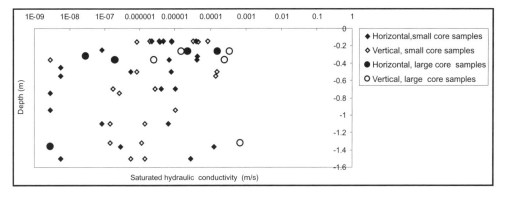

Figure 2. Saturated hydraulic conductivity of blanket peat samples from four landslides on Dooncarton Mountain, Ireland, in 2003. No upper acrotelm layer could be distinguished in any of the peat profiles, so all samples are assumed to represent catotelm peat.

strength data: Table 1 presents data from peat deposits of similar type and/or similar strength characteristics to blanket bog. All of the results were obtained from samples consolidated under specified normal stress (within the range 5 to 50 kPa) prior to application of the compressive (triaxial) or shearing (direct shear) stress. The final three entries in Table 1 high-light two important points: firstly, that application of identical (direct shear) test procedures on the same equipment demonstrates real differences between different Irish blanket bogs, and secondly, that these shear tests overestimated the cohesion. In fact, back-analyses of peat failures using, for example, the infinite slope model, consistently require much lower shear strengths than laboratory tests produce (e.g. Dykes et al., in review). It is now thought that peat does not fully (if at all) conform to the Mohr-Coulomb failure model, but it has not yet been established that conventional slope stability analyses using M-C parameters cannot adequately represent shear failures in peat. However, the underlying difficulty is that shear strength of peat cannot yet be reliably estimated from other properties. The shear strength of peat obtained from vane tests has been demonstrated to be unreliable because of the influence of fibres in the peat (Helenelund 1967, Landva 1980), although vanes may be useful for estimating relative strength variations. Helenelund (1967) considered the tensile strength to be a potentially more reliable indicator of peat strength, and found tensile strength values of some Canadian and Finnish peats to be approximately half of the corresponding vane strengths.

Table 1. Published values for the shear strength of peat (after Dykes & Kirk, 2006).

Original source	Peat type/characteristics	Cohesion, c (kPa)	Internal friction angle, ϕ
Adams (1965)	Peat with low moisture contents	0	48°
Hanrahan et al. (1967)	Remoulded H_4 *Sphagnum* peat	5.5–6.1	36.6–43.5°
Hollinshead & Raymond (1972)	No information	4.0	34°
Landva & La Rochelle (1983)	*Sphagnum* peat (H_3, mainly fibrous)	2.4–4.7	27.1–35.4°
Farrell & Hebib (1998)	Raised bog, 98% organic; undr. triax. compression: direct *and* ring shear:		55° 38°
Kirk (2001)	Ombrotrophic blanket peat, 0.35–1.25 m deep	2.7–8.2	26.1–30.4°
Long (2005)	Review of triaxial test data from studies of Indonesian, Brazilian and Japanese peat		32–58°
Long & Jennings (2006)	Unspecified (blanket?) peat, 1–2.5 m deep, H_{4-5} to H_{7-8}	3–5	35°
Dykes & Warburton (2008)	Ombrotrophic blanket peat, 1.4–1.7 m deep, H_{10}	8–11	21–25°
Dykes & Warburton (2008)	As above, calculated from back-analyses of failures	2.0–6.2	21°
Dykes (unpublished data)	Ombrotrophic blanket peat, 2 m deep, H_{5-6}	5.2	33.4°

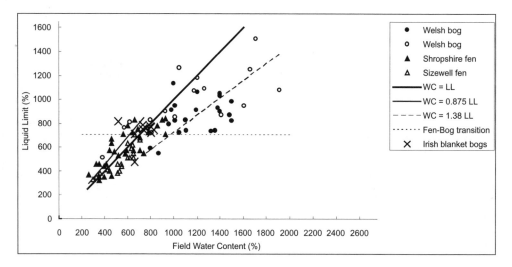

Figure 3. Variations in liquid limit with water content (after Hobbs, 1986). Hobbs identified fen peat below the horizontal transition line, 'amorphous bog peat' immediately above this line where WC > 1.38 LL, and 'fibrous peat' higher up the 'WC = 1.38 LL' line.

This measure of strength may be worth revisiting in the context of blanket peat instability.

3. The liquid limit (LL) of peat is known to sometimes exceed the natural water content (*wc*) (Fig. 3) but it is difficult to determine. The 'loss on ignition' is often used as an alternative indicator of the susceptibility of the peat, but this is unreliable as LL and *wc* vary with humification (and usually depth) (Fig. 1) *and* with type of peat (Fig. 4). The LL may be a more fundamental indicator of peat properties and behaviour than is currently thought (Hobbs, 1986), and issues relating to its determination have been examined using samples of Irish blanket peat (Yang and Dykes 2006). Hobbs demonstrated clear relationships between LL and organic content in fen and transitional peats, but that within any narrow range of organic contents, the LL falls with increasing humification (Figs 1, 4). In Figure 4, the latter control is best exemplified by the results from bog peat (organic content > 90%), the highly humified Irish blanket peat having the lowest LLs among this group (c.f. Fig. 1). Hobbs also suggested that bog peats can be readily distinguished from fen and transitional peats according to whether the natural water content is greater or less than the liquid limit, with LL = 700% being

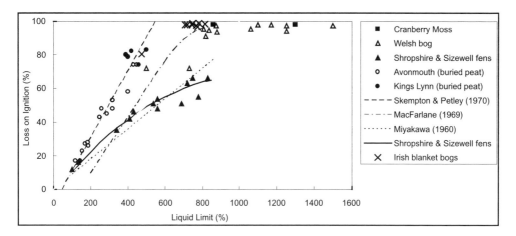

Figure 4. Variations in liquid limit with organic content as indicated by ignition loss for different UK peatlands, including summary relationships obtained from Japan (Miyakawa) and Canada (MacFarlane) (after Hobbs, 1986, Figs 16 & 17b).

identified as the fen-bog boundary (Fig. 3, also applicable to Fig. 4). Hobbs explicitly excluded blanket peat from this analysis, although the data from Irish blanket bogs (Yang & Dykes 2006) do plot within the 'amorphous bog peat' zone. It can therefore be seen that engineering works involving blanket peat deposits cannot rely on published general relationships between index peat properties because of their characteristically high degree of humification, negligible mineral content and botanical composition associated with acidic plant species (c.f. generally alkaline fen peats: Hobbs, 1986). Analysis and interpretation of natural failures in blanket bogs will also therefore need to be primarily based on data obtained from site-specific peat samples.

4. Blanket bog is typically fully saturated all year, thus in its natural state it exerts a negligible (or even zero) normal stress on its basal layer. If the bog is slightly unsaturated, for example if the water table falls up to 0.5 m below the surface, this representing an exceptional degree of drying in western Ireland, the basal peat may experience an effective normal stress of up to around 5 kPa. The low to zero effective normal stresses that apply within natural blanket bogs raise serious doubts about the validity of any shear strength data obtained using standard procedures, in which the peat consolidates significantly under the applied normal loads. Preliminary experiments were therefore undertaken using a direct shear apparatus to provide results for comparison with data obtained previously from the same blanket peats using standard procedures. Small block samples of undisturbed basal peat extracted from near the margin of an Irish bog slide were fully saturated then sheared in a 100 mm square shearbox. These samples were not consolidated prior to shearing, and the normal loads of between 1 and 10 kPa were applied to each respective sample as shearing commenced, possibly generating some matrix pore water pressures broadly similar to the in-situ field condition. The samples were sheared rapidly, at 0.2 or 0.5 mm min^{-1}, to reflect the field evidence of extremely rapid failure with the associated likelihood of some undrained shearing effects, again better representing field conditions. The problems of this approach are recognised, including the absence of pore water pressure monitoring or control and the use of normal loads far below the generally accepted minimum for this type of equipment. However, the results from two replicated sets of experimental tests were sufficiently similar to allow preliminary interpretation and to guide subsequent research. What the results appear to show is: (i) the natural basal peat has very little in-situ strength in its undisturbed state with minimal effective stresses, consistent with back-analyses of stability and with field experience of extracting samples for testing; and (ii) further low-stress direct shear and triaxial strength tests are warranted to investigate the validity and potential utility of this approach for use with existing or improved stability assessments and analyses.

4 CONCLUSIONS

The possibility of more frequent natural peat failures as a consequence of climatic changes, and the

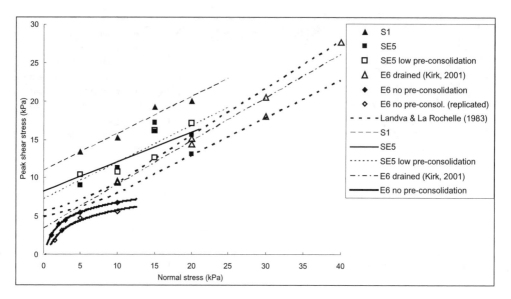

Figure 5. Shear strength results obtained from direct shear tests of Irish upland blanket bog peat (S1, SE5—Dykes & Warburton, 2008; E6—Dykes & Kirk, 2006), also showing the range of ring shear test results (upper and lower envelopes) obtained from Canadian H_{3-4} *Sphagnum* peat by Landva & La Rochelle (1983). All samples were normally consolidated prior to rapid first-time shearing except those identified as 'no pre-consolidation' (in which the normal loads were applied as shearing commenced, with consolidation occurring during the initial shearing), 'low pre-consolidation' (consolidated under 5 kPa normal loads prior to shearing, with the full normal loads applied as shearing commenced) and 'drained' (consolidated-drained tests using very low shear rates).

increasing numbers of windfarms and other activities proposed for upland environments that require engineering such as construction of access roads, suggest that the need for improved methods for analysing and modelling the stability of upland blanket bogs is urgent. Conventional stability analyses of blanket bog covered slopes may be appropriate, as has thus far been assumed, but there are as yet insufficient data to verify this or to permit the development of more reliable failure models for peat deposits. However, conventional geotechnical analyses of this material are clearly inappropriate. The research priorities are therefore to investigate the botanical controls on the geotechnical properties of peat, to establish a reliable method for determining the shear strength of peat, and to identify or develop a reliable method for analysing the stability of blanket bog covered slopes.

ACKNOWLEDGEMENTS

The Irish blanket peat data were obtained during projects supported financially or otherwise by the UK Natural Environment Research Council (Grant Refs. NER/A/S/2003/00888–9), the Limestone Research Group, Fermanagh District Council & Huddersfield University. The contributions of many colleagues to this work over several years are gratefully acknowledged.

REFERENCES

Adams, J.I. 1965. The engineering behaviour of a Canadian muskeg. *Proceedings, 6th International Conference on Soil Mechanics and Foundation Engineering* 1: 3–7.

AGEC. 2004. *Reports on the Derrybrien Windfarm—Final Report on Landslide of October 2003*. Unpublished report, Applied Ground Engineering Consultants Ltd., Ireland.

Delap, A.D., Farrington, A., Praeger, R.L. & Smyth, L.B. 1932. Report on the recent bog-flow at Glencullin, Co. Mayo. *Scientific Proceedings of the Royal Dublin Society* 20: 181–192.

Dykes, A.P. & Kirk, K.J. 2001. Initiation of a multiple peat slide on Cuilcagh Mountain, Northern Ireland. *Earth Surface Processes and Landforms* 26: 395–408.

Dykes, A.P. & Kirk, K.J. 2006. Slope instability and mass movements in peat deposits. In I.P. Martini, A. Martínez Cortizas & W. Chesworth (eds), *Peatlands: Evolution and Records of Environmental and Climate Changes*: 377–406. Amsterdam: Elsevier.

Dykes, A.P. & Warburton, J. 2007a. Mass movements in peat: A formal classification scheme. *Geomorphology* 86: 73–93.

Dykes, A.P. & Warburton, J. 2007b. Geomorphological controls on failures of peat-covered hillslopes triggered by extreme rainfall. *Earth Surface Processes and Landforms* 32: 1841–1862.

Dykes, A.P. & Warburton, J. 2008. Failure of peat-covered hillslopes at Pollatomish, Co. Mayo, Ireland: analysis of topographic and geotechnical influences. *Catena* 72: 129–145.

Dykes, A.P. & Warburton, J. In press. Characteristics of the Shetland Islands (UK) peat slides of 19 September 2003. *Landslides*.

Dykes, A.P., Gunn, J. & Convery (née Kirk), K.J. In review. Landslides in blanket peat on Cuilcagh Mountain, Northern Ireland. *Geomorphology*.

Farrell, E.R. & Hebib, S. 1998. The determination of the geotechnical parameters of organic soils. *Proceedings of the International Symposium on Problematic Soils, IS-TOHOKU 98, Sendai, Japan*: 33–36.

Hanrahan, E.T. 1964. A road failure on peat. *Géotechnique* 14: 185–202.

Hanrahan, E.T., Dunne, J.M. & Sodha, V.G. 1967. Shear strength of peat. *Proceedings of the Geotechnical Conference, Oslo*, Vol. 1: 193–198.

Helenelund, K.V. 1967. Vane tests and tension tests on fibrous peat. *Proceedings of the Geotechnical Conference, Oslo*, Vol. 1: 199–203.

Hobbs, N.B. 1986. Mire morphology and the properties and behaviour of some British and foreign peats. *Quarterly Journal of Engineering Geology* 19: 7–80.

Holden, J. & Burt, T.P. 2003. Runoff production in blanket peat covered catchments. *Water Resources Research* 39: 1191, doi:10.1029/2002 WR001956.

Hollingshead, G.W. & Raymond, G. 1972. Field loading tests on Muskeg. *Canadian Geotechnical Journal* 9: 278–289.

Kirk, K.J. 2001. *Instability of blanket bog slopes on Cuilcagh Mountain, N.W. Ireland*. Unpublished PhD thesis, University of Huddersfield, UK.

Landva, A.O. 1980. Vane testing in peat. *Canadian Geotechnical Journal* 17: 1–19.

Landva, A.O. & La Rochelle, P. 1983. Compressibility and shear characteristics of Radforth peats. In P.M. Jarrett (ed.), *Testing of Peats and Organic Soils*: 157–191. Philadelphia: ASTM Special Technical Publication 820.

Long, M. 2005. Review of peat strength, peat characterisation and constitutive modelling of peat with reference to landslides. *Studia Geotechnica et Mechanica* XXVII: 67–90.

Long, M. & Jennings, P. 2006. Analysis of the peat slide at Pollatomish, County Mayo, Ireland. *Landslides* 3: 51–61.

MacFarlane, I.C. (ed.) 1969. *Muskeg Engineering Handbook*. Toronto: University of Toronto Press.

Mills, A.J. 2002. *Peat slides: morphology, mechanisms and recovery*. Unpublished PhD thesis, Durham University, UK.

Miyakawa, I. 1960. *Some aspects of road construction in peaty or marshy areas in Hokkaido*. Sapporo, Japan: Civil Engineering Research Institute, Hokkaido Development Bureau.

Skempton, A.W. & Petley, D.J. 1970. Ignition loss and other properties of peats and clays from Avonmouth, King's Lynn and Cranberry Moss. *Géotechnique* 20: 343–356.

Warburton, J., Holden, J. & Mills, A.J. 2004. Hydrological controls of surficial mass movements in peat. *Earth Science Reviews* 67: 139–156.

Yang, J. & Dykes, A.P. 2006. The liquid limit of peat and its application to the understanding of Irish blanket bog failures. *Landslides* 3: 205–216.

Stability problems in slopes of Arenós reservoir (Castellón, Spain)

J. Estaire & J.A. Díez
Laboratorio de Geotecnia, CEDEX, M° de Fomento, Madrid, Spain

C. Olalla
ETSI Caminos, Canales y Puertos, Universidad Politécnica de Madrid, Madrid, Spain

ABSTRACT: This paper presents the analysis of the stability conditions of two slopes situated in Arenós reservoir, taking too into account that a town, named Puebla de Arenoso, is located on the top of one of the two slopes. With the information provided by the geotechnical investigation carried out, some soil profiles were elaborated that made it possible to identify the types of soils existing in the slopes: sandy coluvial sediments on a 10 m thick layer of altered marls. The interpretation of all the data available made it possible to deduce that the slope was in a quite strict equilibrium situation that produced movements in the coluvial sediment layer amplified by the successive increases and decreases of water level in the reservoir. The slope situated in front of Puebla de Arenoso was also studied. This slope suffered a great dimension slide at the beginning of Quaternary, whose stability should be verified. In this case, the main problem that can be originated by a reactivation of that huge slide was the formation of a wave that could affect some buildings situated in low parts of the town or even the dam. To face these problems, some soil treatments were carried out.

1 INTRODUCTION

Arenós dam and reservoir are situated in Castellón, at the east of Spain, on Mijares river. The slopes of the reservoir have presented, since dam construction, instability problems and slides. Among those slides, there are some of great dimensions that limit the maximum volume of water that can be stored in the reservoir. The slopes studied in this paper, as shown in Figure 1, are slopes 1 and 2, that suffered a sliding at the beginning of Quaternary and a small reactivation in 1989, and slope 4 in whose top Puebla de Arenoso is situated. In the paper those slopes are referred as "paleoslide" (slopes 1 and 2 and "Puebla de Arenoso slope" (slope 4). Slope 3 was also studied but it does not have any important problem of stability.

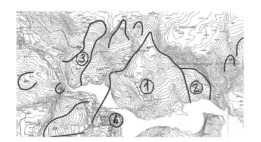

Figure 1. Diagram showing potential slides in the studied area.

2 DESCRIPTION OF PROBLEMS

The potential instability of those slopes make necessary to analyze the following aspects:

a. Determination of the possibility of a reactivation of the great slide occurred in Quaternary that could affect almost all the surface of slopes 1 and 2.
b. Determination of the possible maximum volume of the slides with highest probability of occurrence, taking into account the present geomorphological and hydrological conditions of the slopes.

c. Determination of the velocity of the material flow in its way to the reservoir, in case of a slide occurrence. That velocity is one of the factors with more influence in the height of the waves produced in the water stored in the reservoir.
d. Determination of the possibility that the material fallen from the slopes could form an artificial dam, whose failure could produce very big waves with a great destructive potential, affecting mainly the low parts of the town or the dam.
e. Determination of the stability condition of slope 4 and the definition of the remedial treatments, taking into account its importance as there are some houses inhabitant in its top.

3 GEOTECHNICAL INVESTIGATION

The works performed to analyze the stability of the different slopes are described below.

3.1 Paleoslide (slopes 1 and 2)

a. Geological study: it consisted in the interpretation of stereoscopic pairs of photos and "in situ" ground investigation, whose data were used to elaborate geomorphological maps at 1:10.000 and 1:5.000 scales.
b. Campaign of geotechnical investigation: five boreholes were drilled, of more than 100 m length each. Inclinometric tubes were installed in all the boreholes. Before installation, narrow and lengthwise incisions were made in the tubes to be used also as piezometric tubes to analyze the variations in water level. The surrounding hole was filled with fine gravel to allow water go into the tubes.
c. Control of water levels in the boreholes.
d. Search of the eleven boreholes drilled in 1989, after the reactivation of a slide in the toe of slope 1 and analysis of their present state.
e. Campaign of laboratory tests made mainly with samples obtained in the altered rock substratum.

3.2 Puebla de Arenoso slope

a. Control of the present state of the 21 boreholes drilled in 1972, in the town, before the construction of the dam. The position of those boreholes can be seen in Figure 2.
b. Drill of other six boreholes equipped with inclinometric tube to be used also as piezometric tube. Its position can also be seen in Figure 2.
c. Inclinometric measurements were made, one per month, during six months.
d. A map of cracks existing in the buildings and streets of the town was made to determine the main directions of the ground movements.

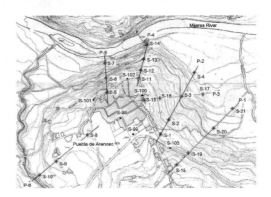

Figure 2. Position of boreholes drilled in Puebla de Arenoso.

4 SOIL PROFILES

Based on the data collected in the geological and geotechnical investigation, some geological soil profiles were made.

4.1 Paleoslide (slopes 1 and 2)

In slopes 1 and 2, situated in front of Puebla de Arenoso and in the left margin of Mijares river, there is a great slide formed by a principal, old and of great dimension slide and other minor slides produced in its toe. The principal slide possibly occurred at the beginning of Quaternary epoch, during Pleistocene, that means it is between 1, 8 and 0, 07 million of years old. Afterwards, and during recent Holocene, other minor slides were generated in the slope toe due to the erosive action of Mijares river. Nowadays that zone is covered by water stored in Arenós reservoir. In 1989 a rotational slide occurred in the slope toe after a period of heavy rains. Those slides can be seen in Figure 3.

The upper part of the great paleoslide is situated between 750 and 850 m, above sea level, and its toe is in the bottom of Mijares river valley, at about 540 m, above sea level, as can be seen in Figure 4.

It is interesting to remark that, according to different studies, the flow of this slide closed the valley and went up in the other margin. The rests of this flow form the area in which Puebla de Arenoso is situated, about 50 m above the bottom of the valley.

The slope is formed by different slide masses that suffered several readjustments. The slide material is formed by an up to 90 m thick disorganized mass with metric and decimetric loose limestone blocks, embedded in sandy-clay matrix.

The base of the paleoslide is formed by Aptiense marls and, in the toe area, by Weald clays, that they are in contact with the marls by a fault. In some boreholes, metric and decimetric thick breccia zones were

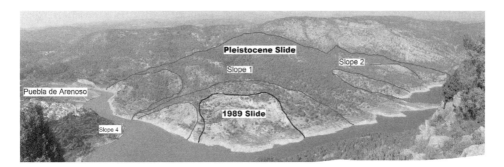

Figure 3. General view of the paleoslide.

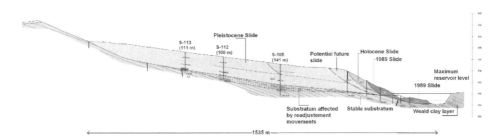

Figure 4. Soil profile of the paleoslide.

Figure 5. Aspect of the breccia material found in boreholes.

Figure 6. General view of the town near the top of the slope.

detected, near the contact between the slide material and the substratum. The slide movement very probably occurred through those breccia zones which are formed by angled fragments of marly limestone, of milimetric and centimetric size, embedded in a marly-clay, blackish grey matrix, as it can be seen in Figure 5.

The piezometric level in the upper part of the slope was in the contact zone between slide material and substratum. In the boreholes situated in the low part of the slope, the piezometric level coincided with the reservoir level.

4.2 Puebla de Arenoso slope

As it was said previously, Puebla de Arenoso is located on slide materials belonging to the toe of the paleoslide occurred in the opposite margin of Mijares river, as it can be seen in Figures 3 and 6.

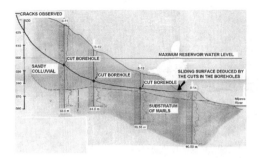

Figure 7. Soil profile of Puebla de Arenoso slope.

Table 1. Material geotechnical parameters deduced by stability back analysis.

Material	Cohesion (kPa)	Friction angle (°)
Slide materials	15	28.0
Substratum altered layer	0	12.5
Substratum Weald clays	0	25

Table 2. Different hypothesis made in the stability analysis.

Reservoir level (m)	Phreatic level	Safety factor
575	Flat	1.06
575	High	1.08
600	Flat	1.05
600	High	1.08
600	Fast drawdown	1.01

Figure 7 shows the geological soil profile of Puebla de Arenoso slope. The average thickness of slide materials is about 50 m, with a detected maximum value of 56 m. Great blocks and rocky fragments in a chaotic disposition, embedded in a marly-clay matrix, form the slide materials. The substratum, formed by marls that are altered in its upper part, appears below.

The points where the boreholes made in 1972 are cut are also drawn. Those points form a possible cinematically sliding surface, whose top coincides with some of the cracks observed in that slope area. This sliding surface goes through the mass of slope materials and it has its toe some meters above the river level.

5 DATA ANALYSIS

5.1 Paleoslide (slopes 1 and 2)

The first step was to perform 2D stability back analysis of the different slides occurred in the slope: the great paleoslide occurred in Quaternary, the middle sized slide produced in Holocene and the last and smallest one, occurred in 1989.

In these calculations, using Morgenstern-Price method, the sliding surface of the great paleoslide was supposed to develop through the altered rock substratum layer. In the other slides, the sliding surface can be seen in Figure 4.

The results of those calculations were the geotechnical parameters of the materials existing in the slope whose values are given in Table 1.

The main conclusion of these first analyses is that the reactivation of the paleoslide is not probable as the strength values of the altered substratum layer deduced from stability calculations are clearly lower than the ones determined in laboratory ($c = 10\,kPa$; $\phi' = 18°$).

The second step was to analyse the present slope stability situation, in the calculation hypothesis shown in Table 2.

The first results of this second set of calculations is that the most probable future sliding surface is similar to one corresponding to a partial reactivation of the slide occurred in Holocene.

The slight difference between the results obtained is due to the fact that the volume of materials affected by an increase in the reservoir water level is relatively small.

The main conclusion of this second set of calculations is that present stability situation of the toe of the slope is quite precarious, as the maximum safety factor is below 1.1.

5.2 Puebla de Arenoso slope

The potential sliding surface drawn in Figure 7 seems to indicate that the slope is in a very precarious equilibrium situation that produces readjustment movements in the slide mass, increased by the successive oscillations of the reservoir water level. However, it is important to remark that, since the drilling of the boreholes in 1972, no important slide has occurred in this slope.

The stability back analysis performed with that surface show that the slide material has the following strength parameters: cohesion between 0 and 10 kPa, combined with friction angles ranging between 20 and 22°. Those values are in correspondence with the characteristics of the matrix of the slide mass formed by sandy-clay coluvial with a slight cohesive component. Those values could not be contrasted with laboratory tests due to the great size of the limestone blocks embedded in the sandy-clay matrix.

On the other hand, in the inclinometric measurements some movements were detected in three of the boreholes, at depths corresponding with zones near the contact between the slide materials of the paleoslide and the grey marls of the substratum. These measurements were performed during a period of draught and with a slow and progressive decrease in the reservoir water level.

6 CONCLUSIONS OF THE STUDY

6.1 Paleoslide (slopes 1 and 2)

a. It is not probable that the great paleoslide has a sharp and sudden reactivation, without some slow movements appear previously in the slope that make it possible to take some remedial treatments.
b. However small movements at the toe of the great paleoslide can be produced, as it occurred in 1989. Taking into account the precarious stability equilibrium of the toe of the slope, a period of heavy rains can be an important activation factor for those small movements to be produced.
c. The estimated volume of that future potential slide at the toe is about 2,1 millions of m^3. Based on geological considerations, it is estimated that the volume of material that could reach the reservoir would be 1,55 millions of m^3.
d. Taking into account the characteristics of the materials existing in the slope, it is not foreseeable that its flow into the reservoir was going to be quick enough not to be detected by a intensive system of auscultation, before a sharp and sudden movement was produced.

6.2 Puebla de Arenoso slope

a. The analysis performed with the boreholes drilled in 1972 makes it possible to draw a possible cinematically sliding surface, whose top coincides with some of the cracks detected in that zone of the slope. This sliding surface goes through the mass of slide material existing in the slope and it has its toe some meters above the reservoir water level.
b. The interpretation of the inclinometric measurements performed in the boreholes of the town, during a period of draught and with a slow and progressive decrease in the reservoir water level, made it possible to determine that the slope was in strict equilibrium. Those measurements indicated that movements are mainly produced in the contact between the slide material and the marly substratum.

7 DESCRIPTION OF SOLUTIONS

Taking into account the previous conclusions, some remedial measurements were taken to try to solve the stability problems of the slopes.

7.1 Installation of auscultation

14 GPS stations were installed to monitor the slope movements, as it can be seen in Figure 8, (Solanes, 2007). Besides, the inclinometers, installed during the geotechnical investigation campaign, are read periodically. By now, the movements detected do not exceed 5 mm and are quite slow.

7.2 Analysis of wave height

Some calculations were performed to determine the wave height and the transient elevation of the reservoir water level due to a more or less sudden entrance of material, coming from some of the potential unstable slopes of the reservoir, (Segura, 2007).

As the velocity of the potential slopes depends on some physical parameters that are difficult to quantify, calculations were made with a wide range of times of material flow (3, 6, 10 and 30 minutes).

Before that, a data recompilation from reference literature was made. The values of velocity of material flow in landslides similar to the one studied here are represented in Figure 9 (Segura, 2007).

It can be seen that the velocities used in this study are clearly greater to the ones measured in most of the real cases referenced.

Figure 8. View of one of the GPS station installed in the slope.

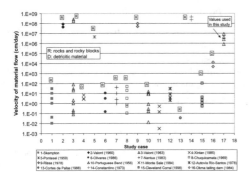

Figure 9. Reference values of velocity of material flow in landslides (Segura, 2007).

In spite of it, the obtained results indicate that the maximum reservoir water level is not greater, in any case, to 3 m, in Puebla de Arenoso slope, and to 1 m, in the dam. These values must be considered as acceptable as the existing clearance are greater in both zones.

7.3 Remedial works

The solution for the stabilization of Puebla de Arenoso slope consisted in the construction of fills made with granular soils, protected by rockfills, in both margins of the river and of a concrete structure to channel the river (Solanes, 2007). A sketch of the solution can be seen in Figures 10 and 11.

These fills, with stabilization effects in the toes of both margins, are between 30 and 40 m high. Their inclination is 2H:1V. They have draining layers to dissipate water pressures in case of a fast drawdown.

In the riverbed, a layer of 5.5 m of granular material was placed as a foundation layer of the concrete structure.

This 190 m long structure is formed by a slab of a variable thickness between 1.6 and 2.0 m, and vertical walls of height between 8 and 10 m and thickness between 1.2 and 2.0 m. Their main function is to avoid water erosion of the stabilization fills placed in both margins. To design this structure some numerical hydraulic simulations were performed and a scaled physical model was tested.

Figures 12 and 13 show a general view of the works performed in the Puebla de Arenoso slope.

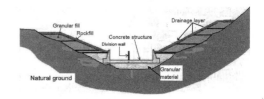

Figure 11. Section of the works performed in Puebla de Arenoso slope.

Figure 12. General view of the works performed in Puebla de Arenoso slope.

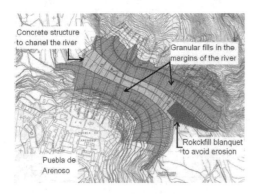

Figure 10. Plan of the works performed in Puebla de Arenoso slope.

Figure 13. General view of the slope after the end of the works.

ACKNOWLEDGEMENTS

The authors of this paper want to acknowledge Mr. Julián Cuesta, geologist of Eptisa, due to his effort during the geological study, and Mr. Fernando Solanes and Miss María Irles of Spanish Ministerio de Medio Ambiente (Ministry of Environment) due to their support and help during the execution of the works.

REFERENCES

Solanes F., Calderón P., Escuder I. & Martínez J. 2007. *Estabilización de la ladera de Puebla de Arenoso en el embalse de Arenós (Castellón)*. Jornadas Técnicas sobre Estabilidad de Laderas en Embalses. pp 295–332. Zaragoza. Conf. Hidrográfica del Ebro.

Segura N. & Fernández L. 2007. *Estudio de la variación del nivel de embalse producida por la entrada de material de laderas en el embalse de Arenós*. Jornadas Técnicas sobre Estabilidad de Laderas en Embalses. pp 455–476. Zaragoza. Conf. Hidrográfica del Ebro.

The 22 August, 2006, anomalous rock fall along the Gran Sasso NE wall (Central Apennines, Italy)

G. Bianchi Fasani, C. Esposito, G. Scarascia Mugnozza & L. Stedile
"Sapienza" University of Rome, Department of Earth Sciences and "CERI" Research Center on Geological Risks, Rome, Italy

M. Pecci
IMONT, National Mountain Institute of Italy, Rome, Italy

ABSTRACT: It is described the rock fall event occurred along the Gran Sasso massif (Central Apennines, Italy) on 22 August, 2006, when a limestone block, with an estimated volume of about 30,000 m^3, fell from the sub-vertical NE wall nearby the Corno Grande peak, the highest peak of the Italian Apennines. Despite the small rock volume involved in the landslide, the rock fall deposits covered an area of about 35,000 m^2, a giant and abrasive dust cloud was generated by the atmospheric pressure waves (air blasts) induced by the rockfall impact and determined destructive effects over an area of about 110,000 m^2 at the base of the slope. Moreover the dust cloud covered a distance of about 3 km, thus reaching the village of Casale San Nicola and the A24 motorway that was temporarily closed for security reasons. The seismic noise generated by the rock fall was recorded by the National Institute of Nuclear Physics seismometric devices located in the Gran Sasso underground laboratories (LNGS).

1 INTRODUCTION

On August 22nd, 2006 a limestone block of about 30,000 m^3 detached from the NE slope of the Gran Sasso d'Italia (the Apennines' highest peak, 2912 m a.s.l.) at an elevation of 2,800 m asl and was involved in a rock fall along the 1,500 m high cliff, also known as the "Paretone" (big wall). Because of the huge dust cloud generated by the rock fall, the motorway A24, located few kilometres downhill, was closed for security reasons. The seismic noise generated by the rock fall until the final impact over the cliff base was recorded by the seismometric devices located in the underground laboratories (LNGS) run by the National Institute of Nuclear Physics. The rock block involved in the fall consists of Triassic dolomitic-limestone which thrust marly-calcareous and marly-arenaceous formations of Cretaceous-Tertiary age within a very complex structural setting. Prompt site investigations allowed to estimate the volume of the felt rock block by means of laser telemetry measurements from the nearest peak; the effects induced by the impact of the rock fall were also observed and the run out area at the steep cliff toe measured as well. In this paper we analyse the possible causes which led to the slope failure, the dynamics of the fall and the mechanisms which generated the giant dust cloud. We also compare the present case history with previous unusual rock falls (Morrissey et al. 1999; Wieczorek et al. 2000) in order to predict hazard scenarios potentially induced by future rockfall events along the Gran Sasso NE wall.

2 GEOLOGY AND CLIMATE

The unique landscape of this zone within the Apennines is dominated by the Gran Sasso Massif that lies just like a "huge boulder" (from which the name itself) (Fig. 1) made of Triassic and Jurassic massive limestone and dolomite over relatively smooth valley slopes incised in bedded marly-limestone and sandy-marlstone.

The geomorphological setting strictly reflect the structural setting of the area: it is featured by numerous thrust sheets, superimposed one each other during the Pliocene compressive tectonic phases (Ghisetti & Vezzani 1990, Calamita et al. 2002).

The more evident sheet crossing the sub-triangular wall of the Corno Grande (Fig. 1) is also the easier mountaineering route to reach the top. The so called "Jannetta route" overlaps the main overthrust, bringing into contact the stratigrafical-structural Corno Grande sub unit (sector above the Jannetta route, containing the rock failure) with the same formation, but

Figure 1. Virtual 3D view of the Gran Sasso NE wall (from the website http://maps.live.com).

Figure 2. Frontal view of the landslide paths along the rock slope.

belonging to the stratigrafical-structural Corno Piccolo sub unit in the sector below (Adamoli 1992). This structural element cuts all the eastern face and corresponds to an original normal faults system. Such system elevated the Corno Grande structural high during the extensional phase at the beginning of medium Lias and have been re-activated during the Messinian by the Apennine compressive tectonic phase.

Glacial and periglacial processes typify the summit area of the massif which hosts the southernmost glacier (Calderone) in Europe, in a fast reduction phase (Pecci 2007).

Extensive historic and prehistoric rockfall and debrisflow deposits have accumulated at the base of the wall (D'Alessandro et al. 2003), where some villages and important lifelines are located. In fact, earthquakes, snowmelt, freezing-melting cycles and frost-wedging effects have caused rock falls in the area, like the historically recorded 1897 event which made a scar known as "Farfalla" (Butterfly) due to its shape (see the circle in Figure 2). The NE wall is a very popular and very challenging climbing cliff, thus imposing further risk conditions (Amanti et al. 1994)

3 DESCRIPTION OF THE EVENT

The event has been observed just after the release until the impact (Fig. 3), while time histories (Fig. 4) recorded by the seismic antenna at LNGS (Laboratori Nazionali del Gran Sasso—National Laboratories of the Gran Sasso) and field investigations helped in better frame the dynamics of the rock fall. It occurred at 7.30 am GMT, when a 30,000 m^3 rock block detached from the "Guglia Bambù" (Bamboo Pinnacle) along a slickensided fault surface.

After the failure, the already jointed rock mass underwent a crumbling process while experiencing a fall as high as 1,500 m along the steep cliff which features this Gran Sasso slope. Along this section the debris motion took place mainly by free fall and bouncing. In addition, shortly after the detachment the falling mass split into two parts: a first one fell down along an almost straight, ballistic trajectory, while a second one was channelled within a chute (Vallone Jannetta; see the white arrows in Figure 2) following a more winding path and entrapping rock fragments and debris along the path.

In both cases, the fallen rock fragments reached the base of the steep slope where they spread over an area of about 35,000 m^2; a part of the debris was then channelled within the upper section of a deeply cut valley. In addition, immediately after the rock impact at the wall base, a huge and dense dust cloud developed and rapidly moved downhill covering a distance of about 3 km until the A24 motorway and the village of Casale San Nicola (Fig. 3); because of this dense and rapidly moving dust cloud the motorway was temporarily closed for security reasons.

The rock fall occurred on a sunny day without any apparent trigger event such as an earthquake or

Figure 3. Photographics sequence of the rock fall event, showing the spreading of the dust cloud.

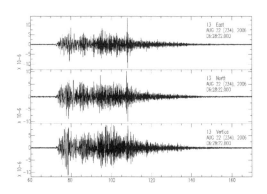

Figure 4. Seismogram recorded by the LNGS in concomitance with the rock fall event.

Figure 5. Rock free face of the detached block from different points of view.

heavy rainstorm. During the preceding days some rainfalls were recorded, due to some afternoon thunderstorms, a typical phenomenon during this hot period of the year. Dark streaks observable on the detachment free face indicated paths where water had seeped through the jointed rock mass (Fig. 5). As regards the geologic controls on the failure mechanism, the shape and volume of the detached block is clearly guided by the presence of tectonic slickensided surfaces (Fig. 5); the jointing conditions within the rock mass are on their turn connected with the local structural setting, featured by tectonic lines of regional importance.

Laser-GPS telemetry measurements from the nearest peak also allowed to determine the actual detached rock volume (30,000 m³), once compared with images showing the shape of the Bamboo Pinnacle before the fall. The remainder of the pinnacle still in place is an overhanging rock dihedral estimated to be 20,000 m³ in volume. Site investigations along the steep cliff pointed out the numerous joints within the rock mass and the presence of unstable rock blocks.

4 RECORDED EFFECTS

The most significant feature of the 22 August rock fall is definitely represented by the high energy dissipation derived from the impact and rock fragmentation at the slope toe and the so generated dust cloud (Fig. 3). Actually, according to the above reconstructed kinematics of the event and as inferred from the time histories recorded at LNGS (Fig. 4), we hypothesize two distinct impacts due to a gap in the arrival times by the rock fragments which followed the two different paths. As regards the mechanisms involved in such highly dissipative process, it can be assumed that the energy of the fallen rock volume was transferred to both ground and atmosphere, thus creating air pressure waves (airblast effect) and a consequent dense dust cloud.

The latter has been driven by the airblast with high velocity and has been able to preserve an abrasive effect over a quite large area (110,000 m²) (Fig. 6), and reached the village of Casale San Nicola located 3 km downhill the impact area. Such an abrasive effect is shown by the uprooting and snapping of hundreds of trees and bushes as well as by the debarking (Fig. 7). Airblasting is usually generated by landslides involving rock volumes in the order of 10⁶ m³. Only few cases have been previously reported (e.g. Yosemite, 10 July 1996 event by Morrissey et al. 1999 and Wieczorek et al. 2000) regarding relatively small magnitude landslides able to generate air pressure waves.

In both cases the airblast generation is due to the high impact velocity and the two separated impacts, as inferred from the seismograph time histories. During the time interval between the impacts, the initial atmospheric conditions were changed after the formation of a first dust cloud. Such high density fluid could have favoured the sudden variation in air pressure (Morrissey et al. 1999).

The rock debris within the landslide deposit shows grain size within fine sands and boulder (1 m³), and

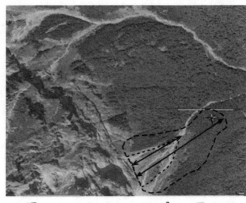
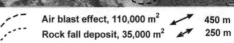

Figure 6. Aerial view showing the extension of the rock fall deposit and the zone affected by air blast.

Figure 7. Abrasive air blast effect on the vegetation.

a reduction in rock element dimensions versus distance from impact point are recorded. The larger landlside debris channeled within a stream and came to halt 250 m far from the impact point, while the finer material blanketed areas within 3 km distance.

4.1 Considerations about further risk scenarios

In addition, during the survey phases that were carried out to better reconstruct the discussed rock fall, it was possible to preliminarily define the overall geomorphic, geomechanical and, finally, slope stability conditions along the whole NE wall of the Gran Sasso. For this purpose, data derived from direct surveys in the accessible areas were coupled with data inferred by remote techniques (such as laser telemetry), performable through devices easy to transport on steep and quite unstable tracks. As a result, other blocks within the rock mass prone to detachment have been recognized. Even if most of the so identified blocks have dimensions quite comparable with those of the hereby presented rock fall, it was possible to observe also some huge rock "pillars" with volumes ranging between $10^5 m^3$ and $10^6 m^3$ (Fig. 8). Furthermore, field surveys pointed out the presence of even large-sized (tens of cubic meters) limestone blocks also in some areas downslope of Casale San Nicola village and the motorway, thus testifying the occurrence of past massive and catastrophic rock slope failure.

Based on this evidence, it is possible to hypothesize risk scenarios for both the motorway and the village. Even if the gravity-induced landscape evolution under the present boundary conditions seems to be characterized by frequent, small-sized rock fall events, the potential occurrence of massive rock slope failures involving the "pillars" under specific conditions, could evolve in a dry granular flow. As a matter of fact, the presence of deeply incised channels at the base of the wall can represent the geomorphic "constraint" able to convey the highly fragmented debris, allowing for a long run-out (Fig. 9).

5 CONCLUDING REMARKS

As before the event no earthquake was recorded, no seismic trigger can be invoked for the failure. The only possible causes can be ascribed to the intense jointing within the rock mass and to the frost (thermal) wedging and to the possible permafrost degradation phenomena. The latter is justified by the progressive retreat of the Calderone glacier nearby the source area at the same elevation (Pecci 2007). According to recent studies in the Alps (Davies et al. 2001, Fischer et al. 2006), rockfall events can be referred to climate changes in high mountain environment, thus determining a significant permafrost or seasonal ice pattern modification. Annual mean temperature can influence the stability of rock masses with ice bearing joints. This can be related to the water seepage evidences observed on the free face of the 22 August event. As a consequence, the presented case history could be considered as a further example of rock slope failure induced by climatic change in a very sensitive environment such as the high peaks of the Gran Sasso Massif, the southernmost glacial and periglacial environment in Europe.

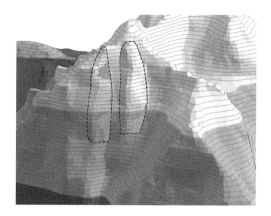

Figure 8. DEM adopted for the volume calculations of the unstable blocks.

Figure 9. Possible morphologic constrain able to produce the channelling of granular flow deriving from larger sized landslide event.

REFERENCES

Adamoli, L. 1992. Evidenze di tettonica d'inversione nell'area del Corno Grande-Corno piccolo (Gran Sasso d'Italia). *Bollettino della Società Geologica Italiana* III: 53–66.

Amanti, M., Pecci, M., Scarascia Mugnozza, G. & Vallesi, R. 1994. Comparison and critical review of quick field data collection methods on rock slopes: a contribution from climbing techniques and experiences. *Atti del Convegno "Man and mountain '94"*, 20–24 giugno 1994, Ponte di Legno (bs), pp. 189–198.

Calamita, F., Scisciani, L., Adiamoli, M., Ben M'Barek, M. & Pelorosso, M. 2002. Il sistema a thrust del Gran Sasso d'Italia (Appennino Centrale). *Studi Geologici Camerti* 1/2002: 19–32.

D'Alessandro, L., De Sisti, G., D'orefice, M., Pecci, M. & Ventura, R. 2003. Geomorphology of the Summit Area of The Gran Sasso d'Italia (Abruzzo Region, Italy). *Geografia Fisica e Dinamica Quaternaria* 26: 125–141.

Davies, M.C.R., Hamzal, O. & Harris, C. 2001. The effect of rise in mean annual temperature on the stability of rock slopes containing ice-filled discontinuities. *Periglac. Process* 12: 137–144.

Fischer, L., Kaab, A., Huggel, C. & Noetzli, J. 2006. Geology, glacier retreat and permafrost degradation as controlling factors of slope instabilities in a high-mountain rock wall: the Monte Rosa east face. *Nat. Hazards Earth Syst. Sci* 6: 761–772.

Ghisetti, F. & Vezzani, L. 1990. Stili strutturali nei sistemi di sovrascorrimento della Catena del Gran Sasso (Appennino Centrale). *Studi Geologici Camerti* vol. spec. 1990, 37–50.

Morrissey, M.M., Savane, W.Z. & Wieczorek, G.F. 1999. Air blast generated by rockfall impacts: Analysis of the 1996 Happy Isles event in Yosemite National Park. *Journal of Geophysical Research* 104, n° B10: 23189–23198.

Pecci, M. 2007. The shrinkage of the central Mediterranean cryosphere in a changing mountain environment. *Mountain Forum Bullettin* VII Issue 2, ISSN 1029–3760 (http://www.mtnforum.org/rs/bulletins/mf-bulletin-2007-07.pdf).

Wieczorek, G.F., Snyder, J.B., Waitt, R.B., Morrissey, M.M., Uhrhammer, R.A., Harp, E.L., Norris, R.D., Bursik, M.I. & Finewood, L.G. 2000. Unusual July 10, 1996, rock fall at Happy Isles, Yosemite National Park, California. *GSA Bullettin* 112: 75–85.

New formulae to assess soil permeability through laboratory identification and flow coming out of vertical drains

J.C. Gress
Ecole Nationale des Travaux Publics de l'Etat, Lyon, France

ABSTRACT: Mastering a lot of landslides through deep dewatering drainage by one or more lines of drains, being pumped either by siphoning pipes or by electropneumatic pumps®, it has appeared to us that water tests were not reliable for different reasons. We suggest here a procedure through laboratory identification tests and new formulae to assess the permeability, we can wait for, and a new formula to assess the flow which will come out of a line of vertical drains in a slope.

1 INTRODUCTION

For past twenty years, the mastering of landslides, in France, has taken great benefit of the experience of a great number of works of deep drainage through siphon drains® or electropneumatic drains®.

But it has clearly demonstrate that it was difficult to have a good estimation of the flow, we could wait for. After having analysed why, we propose hereunder new formulae, in order to have a better approach of deep drainage efficiency through lines of vertical drains, formulae to assess soil permeability through identification tests and a formula to assess flow coming out of vertical drains in a slope.

2 BASIC PRINCIPLES OF VERTICAL DRAINS

Lines of vertical drains are placed, on the site, in order to dewater the landslide and stop the instability.

These drains, with an average spacement of 5 meters must:

– reach the aquifers to be drained,
– be bored with the good tools, in order to avoid the decrease of permeability,
– must be equipped like wells (proper slotted pipes, filter).

The efficiency between adjacent drains must be designed; this will lead to a better assessment of the spacing between drains.

If the wanted drawdown, under the soil surface, is less than eleven meters, then the drain will be pumped through siphoning pipes, a hydraulic accumulator,

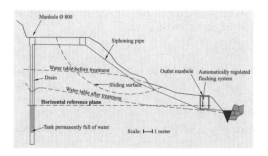

Figure 1. Cross section through a siphon drain®.

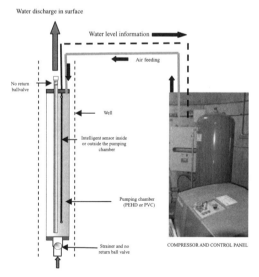

Figure 2. Electropneumatic pump® and view of the compressor chamber.

equipped with a flushing system, regulating the flow above the critical one.

If it is greater, then we use electropneumatic pumps®, air under pressure being fed by a compressor placed in a chamber, the pump being equipped with a sensor, analysing if the pump is empty or full and then regulating the feeding with air.

Then, we can pump up to depth of 100 meters.

3 CLASSICAL MEASUREMENTS OF THE PERMEABILITY

In order to study the feasibility of such dewatering scheme, there is always a preliminary geotechnical and hydrogeological study with borings, samplings, laboratory tests and water tests on site and in the laboratory.

It appears very often that water tests didn't give a good order of the real permeability, because of:

- bit to do the boring being not the good one,
- too small diameter,
- water test run in an injection way and not in a pumping way,
- had we to run a Lefranc or Nasberg test?

These difficulties can be mastered. But it stays the major difficulty, that is to say, that the permeability is not homogeneous and can vary vertically through the same geological layer, due to the variation of the clayey fraction or to the superposition of layers of sand and clay, the thickness of these layers being maybe less than decimetric.

To put this in evidence, the best thing is to do intact continuous samplings and to analyse the variation of density, granulometry and qualify the activity of clay through Atterberg limits or methylene blue tests. We have then tried to obtain formulae giving a rough assessment of permeability through the different parameters measured.

4 ASSESSMENT OF THE PERMEABILITY THROUGH SOIL IDENTIFICATION PARAMETERS

There are not much formulae allowing us to have a good assessment of the permeability, these formulae working for gravely soils to clay.

The Hazen formula is well known:

$$k = K(D_{10})^2 \quad (1)$$

k in cm/s and D_{10} being the diameter in cm of the screen allowing ten per cent in weight of the soil to go through. But it works only for sands and sandy gravels.

The oedometer formula:

$$\Delta \log k = Ck \Delta e \quad (2)$$

with $Ck \approx 0.5 e_o$ (Tavenas, F. et al. 1983) is giving only a variation of k with the void index e for clayey soils.

We have worked on correlations, we had through water tests and identification tests of different soils and on the works of Nagaraj et al. 1986 and Sivapullaiah P.V., et al. 2000.

It appears that the formulae proposed hereunder fits relatively well, correlating $log_{10}k$ with W_L = liquidity limit, when the particle of the soil have a maximum size of 400 µm; otherwise, VBS = methylene blue value of the total soil; e = void index, %2µ = percent finer than 2 µm.

The methylene blue value is very frequently measured in France. The methylene blue value of the 0/400 µm fraction is well correlated to the plasticity index Ip and to the liquidity limit W_L:

$$Ip \approx 0.045 \, VB0400\mu \quad (3)$$

$$WL \approx 0.14 + 0.063 \, VB0400\mu \quad (4)$$

the methylene blue value of a granulometric portion o/d being linked to the percent finer than d through the formulae:

$VBod \, x \, \% \, od$
$$= VB0400\mu \, x \, \% \, 400\mu = VB02\mu \, x \, \% \, 2\mu \quad (5)$$

When the maximum size of the soil particle is 400 µm, we propose the formulae hereunder, in order to have a rough estimation of soil permeability:

if $WL < 0.25$

$$\log k = -(1.41 + 25.55 \, W_L) + (4.46 - 3.5 \, W_L)e \quad (6)$$

if $0.25 \leq WL < 0.80$

$$\log k = -(5.23 + 9.2 \, W_L) + (4.6 - 4.11 \, W_L)e \quad (7)$$

When the maximum size D of the particles is greater than 400 µm, then we propose:

if $VBS < 1.5$

$$\log k = -(4.99 + 1.61 \, VBS) + (3.97 - 0.22 \, VBS)e \quad (8)$$

if $1.5 \leq VBS < 10$

$$\log k = -(6.52 + 0.58 \, VBS) + (4.03 - 0.259 \, VBS)e \quad (9)$$

where VBS is the methylene blue value of the total soil.

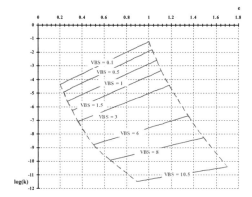

Figure 3. Values of $\log_{10} k$ with e and VBS.

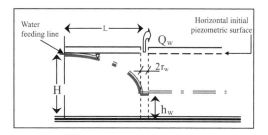

Figure 4. Cross section after dewatering (horizontal initial piezometric surface).

We can then with these formulae, through simple laboratory tests, check the in-situ water tests and have a relatively better knowledge of the different levels of permeability.

5 NEW FORMULA TO ASSESS THE FLOW COMING OUT OF THE DRAINS

For a line of drains, dewatering an horizontal piezometric surface, for an unconfined aquifer, the flow of each drain is given by:

$$H^2 - h_w^2 = \frac{2 Q_w L}{ka} + \frac{Q_w}{\pi k} Ln\left(\frac{a}{2\pi r_w}\right) \quad (10)$$

where Q_w = flow coming out of each well; r_w = radius of each well; k = permeability of the aquifer; a = distance between each drain.

Each drain penetrating totally the aquifer, and a feeding line being located at a distance L:

In case of an inclined piezometric surface having an initial p_o slope, we propose this formula:

$$q_w = \frac{H_o^2 - h_w^2}{\frac{H_o + h_w}{a p_o k} + \frac{1}{\pi k} Ln\left(\frac{a}{2\pi r_w}\right)} \quad (11)$$

where H_o = total initial height of the sheet of water; h_w = thickness of the dewatered sheet; p_o = initial slope of piezometric surface; a = distance between each drain; k = permeability; r_w = radius of the drain.

The initial upstream flow q_o is equal to $p_o H_o k$. The flow coming out of the vertical drains per meter of line is equal to $q_1 = \frac{Q_w}{a}$. The downstream flow q_2 is equal to $q_0 - q_1$.

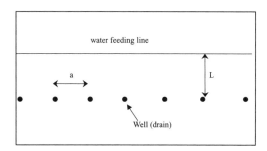

Figure 5. Overview of a set of drains.

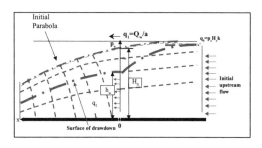

Figure 6. Cross section after dewatering (surface of drawdown).

The substratum is supposed to be horizontal.
Example if:

$H_o = 20$ m
$h_w = 10$ m
$p_o = 8\%$
$a = 5$ m
$k = 5 \times 10^{-6}$ m/s
$r_w = 0.08$ m

then $Q_w = 71$ liters per hour.

6 CONCLUSIONS

Vertical drains, either drained by siphoning pipes® or electropneumatic pumps®, have allowed the stabilization of more than too hundred lanslides, these last twenty years.

The hereabove proposed formulae of the permeability and of the flow coming out of the drains should lead to an improvement in the design of the scheme of drains, these formulea having to be adjusted to the further experience of new works.

REFERENCES

Bomont, S. 2002. Drainage with electropneumatic drains®. Conference JNGG 2002. Nancy, France.

Bomont, S. 2004. Back experience from four landslides stabilized through lines of siphon drains® in Normandy, France. 9th International Symposium on Landslides 2004. ISL Rio.

Bomont, S. et al. 2005. Two applications for deep drainage using siphon and electropneumatic drains®. Slope works for Castlehaven Coast Protection Scheme, Isle of Wight (UK) and slope stabilisation for the Railways Agency, France. In, Proceedings of the International Conference on Landslide Risk Management. 18th Annual Vancouver Geotechnical Society Symposium.

Clark, A.R. et al. 2002. The planning and development of a coast protection scheme in an environmentally sensitive area at Castlehaven, Isle of Wight. Proc. Int Conf on Instability, Planning & Management, Thomas Telford, 2002.

Clark, A.R. et al. 2007. Allowing for climate change; an innovative solution to landslide stabilisation in an environmentally sensitive area on the Isle of Wight International Conference on 'Landslides and Climate Change-Challenges and Solutions' Ventnor, Isle of Wight, UK.

Gress, J.C. 1996. Dewatering a landslip through siphoning drain®. Ten years experiences. Proc 7th International Symposium on Landslide. Trondheim.

Gress, J.C. 2002. Two sliding zones stabilized through siphon drains®. International conference on Landslide, slope stability of infrastructures. Singapor.

Nagaraj, T.S. et al. 1993. Stress state—permeability relation for fine grained soils. *Geotechnical* (43): 333–336.

Mitchell, J.K. 1993. *Fundamentals of soil behavior*. 2nd ed. John Wiley & Sons, Inc., New York.

Pandian, N.S. et al. 1995. Permeability and compressibiliy behavior of bentonite—sand/soil mixes. Geotechnical Testing Journal (18): 86–93.

Sivapullaiah, P.V. et al. 2000. Hydraulic conductivity of benton. Canadian Geotechnical Journal (37): 406–413.

Tavenas, F. et al. 1987. State of the Art on Laboratory and in situ stress strain time behavior of soft clays. LAVAL University.

Structure-controlled earth flows in the Campania Apennines (Southern Italy)

F.M. Guadagno, P. Revellino, G. Grelle, G. Lupo & M. Bencardino
Department of Geological and Environmental Studies, University of Sannio, Italy

ABSTRACT: Slow-velocity landslides predominate in the area of the Province of Benevento, due to the prevalent clay nature of its outcropping deposits. An analysis of these instabilities, via detailed inventory mapping, has shown a pervasive diffusion of earth flows, characterized by a reactivation tendency. Their evolution, in terms of activity and kinematic mechanisms, is structurally-controlled and can be generally connected to three principal controls: i) bedding; ii) stratigraphic or tectonic contact between lithologically differentiated sequences; and iii) zones of intense fracturing linked to folds and faults. Differences in pattern and controlling factors led to their grouping into recurrent types characterised by a different style of evolution.

1 INTRODUCTION

Like other sectors of the Italian Apennines, the area of the province of Benevento (Campania region, Southern Italy) is characterised by recurring events of slope instabilities, in time and space, determining conditions of high risk. The geological and structural setting of the Campania Apennines is the basis of landsliding processes. Moreover, recent occurrences demonstrate and confirm that human activities can have a determinant role upon the morphologic evolution of the slopes. The changes linked to increasing urbanization of morphologically complex settings and agricultural and forestry practices seem to have an important effect.

Further problems connected to landslide and erosion phenomena are linked to the responses of natural slopes to rainfall regime modifications induced by climatic changes, which could produce changes in the spatial-temporal recurrence.

The principal objective of this paper is to highlight the influence of the geological and structural settings on the occurrence of earth flows in the study area, by using data from a "Landslide Inventory Map" (Guadagno et al., 2006) of the province. These controls affect the evaluation of the susceptibility required in hazard assessment.

2 THE GEOLOGICAL ENVIRONMENT

The territory of the province of Benevento is located in one of the most geologically complex areas in Italy. The Southern Apennines consists of thrust-belt structures of carbonate terrain and clayey and stony flysches, resulting from the deformation of paleogeographic domains (Patacca & Scandone 1989).

As a consequence, lithologically differentiated sequences belonging to marine sedimentary rocks, ranging from Cretaceous to Pliocene, and continental deposits of Pleistocene, outcrop in the area. They can be grouped as follows (Pescatore et al. 2000): 1) *Platform carbonate Units*: limestone and dolomitic limestone forming the higher slopes; 2) *Pre-orogen basin Units*: basinal facies and clayey, quarzarenite and arenite deposits; 3) *Synorogen and lateorogen Units*: arenaceous and arenaceous-clayey flysch-like successions; 4) *Pyroclastic deposits*: a) generally lithoid and coherent, grey-yellowish flow deposits (the *Campanian Ignimbrite*); and b) incoherent or weakly cemented air-fall deposits; 5) *Continental, fluvial and detrital deposits*: debris and colluvial fans at the toe

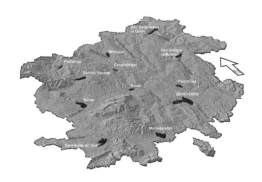

Figure 1. DEM of the province of Benevento.

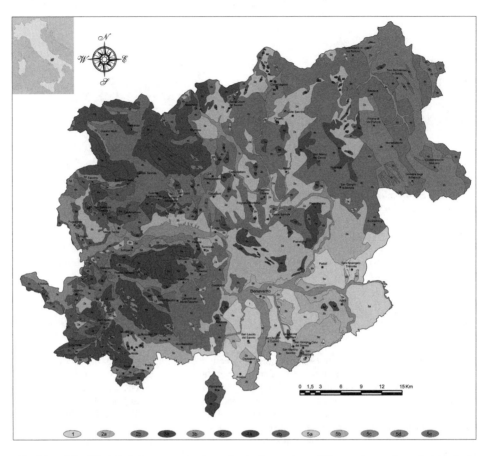

Figure 2. Map of the lithotechnical sequences outcropping in the province of Benevento, as shown in table 1. Legend: 1) Clayey-silty sequences; 2a) Clayey-marly sequences; 2b) Clayey sequences; 3a) Calcareous sequences; 3b) Conglomeratic sequences; 3c) Tuffaceous sequences; 4a) Calcareous sequences (calcareous s.s. and calcareous-clayey); 4b) Arenaceous-clayey and arenaceous-conglomeratic sequences; 5a) Sandy- arenaceous sequences; 5b) Alluvial sequences; 5c) Fluvial sequences; 5d) Cemented detrital sequences; 5e) Incoherent detrital sequences.

of the carbonate slopes; generally terraced, Quaternary alluvial fan deposits along the river valleys.

The structural setting, resulting from the tectonic phases, strongly influences the morphological configurations on the area. The western portion of the province is characterized by the presence of calcareous mountains, while the eastern sector is characterized by a hilly morphology (Figure 1). In contrast, the central area corresponds to the depression where marine and continental clastic Pleistocene deposits outcrop and along which the main rivers flow.

Bearing in mind the aims of this paper, it was felt opportune to group the deposits outcropping in the province into successions, characterized by a likely lithotechnical homogeneity. In other words, these successions are constituted by similar lithotypes according to geological-technical and geomechanical features (Figure 2). This procedure permitted the obtainment of data regarding the relationships between landslides and involved deposits and consequently on the principle causes of landsliding.

The tectogenetic phases have induced complex tectonic settings in the area, testified by typical structures of a ductile-type and fragile-type tectonics, which completely deform and displace the sequences. Sets of joints are connected to these, which pervasively affect the masses and show high frequency near the most important tectonic lineations.

Therefore, most of the above-indicated sequences can be defined as *structurally complex formations* (AA.VV. 1985, Picarelli 1986) as consequence of lithostratigraphical and tectonical features.

Table 1. Geomechanical and lithological characteristics of the sequences outcropping in the province of Benevento.

	Group of sequences	Lithotechnical sequences		Competence	Setting
1	Prevalently pelitic—from low-degree to medium-degree of tectonization	Clayey-silty (Ag-L)		Incoherent	Stratified, generally monoclinalic
2	Prevalently pelitic—high-degree of tectonization	Clayey-Marly (Ag-M) Clayey (Ag)		Complex. Prevalently incoherent	From mildly to intensely folded. Scaly clay.
3	Stony—from low-degree to medium-degree of tectonization	Calcareous (Ca)		Stony	Well stratified strata and banks (from 30–40 cm to 5–10 m)
		Conglomeratic (Cg)		Stony	Sometimes stratified
		Tuffaceous (Tf)		Weakly lithified	Up to 30 m banks. Columnar jointing
4	Stony and complex—high-degree of tectonization	Calcareous (Ca)	Calcareous s.s.	Stony	Well stratified and highly tectonized
			Calcareous-clayey	Complex, prevalently stony	Well stratified strata and banks Intensely jointed and folded
		Arenaceous-calyey Arenaceous-conglomeratic (Ar-AgCg)		Complex	Arenaceous and conglomeratic bank, thin layers of marl, clay and conglomer ate Intensely jointed and folded.
5	Coarse clastic and/or non-homogeneously lithificated	Sandy- arenaceous (S-Ar)		Non-homogeneous lithification	Generally stratified and well stratified
		Alluvial (Al) Fluvial (Fl)		Incoherent or weakly cemented	Unclear bedding
		Cemented detrital (Dc)		Non-homogeneous lithification	Irregular bedding
		Incoherent detrital (Ds)		Incoherent or locally weakly cemented	Irregular bedding

On the basis of qualitative and quantitative observations, together with data collected from site and laboratory tests (Guadagno et al. 2006), the masses were classified from a geo-mechanical point of view. Table 1 shows the grouping of the various successions as a function of their common mechanical and lithological characteristics. The main groups identified can be listed as below: i) Group 1, Prevalently pelitic sequences (from low-degree to medium-degree of tectonization); ii) Group 2, Prevalently pelitic sequences (high-degree of tectonization); iii) Group 3, Stony sequences (from low-degree to medium-degree of tectonization); iv) Group 4, Stony and complex successions (high-degree of tectonization); v) Group 5, Coarse clastic and/or non-homogeneously lithificated sequences.

The map in Figure 2 shows the areal distribution of the above described sequences. It is possible to observe that wide areas are characterised by the presence of both clayey and stony tectonization formations. These areas correspond to the sectors which are the most involved in erosion and landslide processes.

3 LANDSLIDES OF THE BENEVENTO PROVINCE

In order to investigate the distribution and type of failures and their correlation with the lithostructural and morphological setting, a landslide inventory map of the province was compiled on a 1:75,000 scale (a copy of the map can be requested by e-mail—guadagno@unisannio.it—from the authors). The surveys were performed on the basis of a 1:25,000 scale topographic map, whereas, specific areas, characterized by the presence of typical landslides and towns or infrastructures, were surveyed on more detailed maps (greater than 1:10,000 scale).

Analyses were carried out by means of an interpretation of historical aerial photos from different time periods, dating from 1954, together with geological surveys carried out between the years 2001 and 2005. It should be noted that, 2003 and 2005, two heavy rainfall events triggered the reactivation of hundreds of landslides.

The classification criteria adopted refer to well-know classification systems (Varnes 1978, Hutchinson 1988, Cruden & Varnes 1996, Hungr 2001). Furthermore, they are based on some specific characteristics of landsliding processes which are worth pointing out.

3.1 Classification criteria

Six different typological groups (numbered from 1 to 6 below) were distinguished, taking into account the survey results: i) (1) falls and topples; ii) (2) rock and debris avalanches, and debris flows; iii) (3) translational, rotational and composite slides, and; iv) earth flows distinguished in (4) single (Figure 3), (5) multi-source (Figure 4) and (6) coalescent (Fig. 5).

This latter typological choice derives from the necessity to provide morpho-evolutive elements for the earth flows, addressed, on the one hand, to understanding the landslide mechanisms and, on the other, to defining hazard management. The use of diversified typological contexts permits to stress the

Figure 5. Example of basin affected by coalescent earth flows.

wide differences in the landsliding processes involving the slopes.

In relation to *coalescent earth flows*, the surveys highlighted that different sectors, mainly characterized by the outcropping of pelitic and complex sequences from a medium to high degree of tectonization, have been and are still affected by several landslide generations that mobilise interacting among themselves. These groups of landslides, which can be considered as a *unicum* from a morphological point of view, are localized inside unitary sub-catchments (Figure 5).

Multi-source earth flows reflect the complex lithostructural settings. In particular, in specific areas, lithololegically differentiated sequences alternate or they are in contact, inducing differentiated erosion zones. The result is the formation of articulated source areas, composed of several branches.

The structural setting may also influence the morphology of the channels, as well as that of the accumulation zones.

These aspects are important in defining the distribution of activity, which contributes to the assessment of the evolution of the landslide body and, therefore, to the obtainment of data on landslide susceptibility evaluation.

Figure 3. Oblique aerial view of the Sant'Agata de' Goti landslide of 11 January '97.

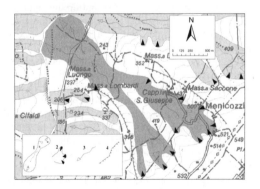

Figure 4. Multi-source earth flow from a segment of the Landslide Inventory Map (Guadagno et al. 2006). Legend: 1) Distribution of the activity in the landslide area: a. Retrogressing; b. Enlarging; c. Advancing; 2) Source area; 3) Movement direction; 4) Lithostructural element controlling the evolution of the landslide mass.

3.2 Areal distribution of the earth flows

In the study area, 3,160 landslides were inventoried, covering an approximate area of 358 km^2, equal to about 18% of the whole provincial surface. This value refers to parts of the territory already affected by landslides and mappable on the scale of representation. It does not include those areas where predisposing conditions have been verified and which indicate possible future landslide evolutions.

Moreover, the Landslide index, calculated as the percentage of area affected by landslide events per grid of 1 km^2, reaches values even greater than 75% where clayey sequences are outcropping.

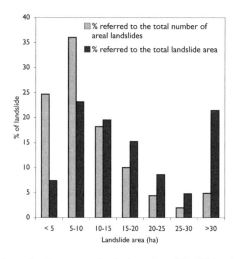

Figure 6. Percentage distribution of areal landslide phenomena, computed for number and area, in dimensional classes.

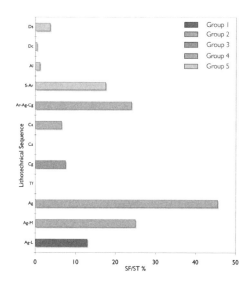

Figure 7. Percentage ratio between area involved in landsliding (SF) and total outcropping area (ST) for each lithotechnical sequence. Refer to Table 1 for the symbology used.

Figure 6 shows the percentage distribution of the landslides in dimensional classes. The data refers to the single events, even if they make up part of the coalescent groups. It is interesting to note that landslides <10 ha constitute over 50% of the inventoried phenomena, whereas the few larger events (ha >25) represent over 20% of the areas affected by landslides.

3.3 Controlled evolution of earth flows

The litho-structural setting determines favourable conditions for the development of landslides. To verify from a quantitative point of view, analyses of data were carried out both for the 5 groups of sequences and for the single lithotechnical sequences (cf. Table 1). Considering the landslide type, analyses were carried out evaluating the formations specifically involved in the source areas.

As shown in Figure 7, the percentage of outcropping surface affected by landslides is up to 46% of the total. The prevalently pelitic and complex sequences at a high-degree of tectonization are those mostly affected by landslides, totalling more than 80% of the inventoried phenomena. Additionally, over 40% of the areas where clayey formations crop out, are also affected by landslides. In particular, lithotechnical sequences named *clayey*, *clayey-marly* and *arenaceous-clayey—arenaceous-conglomeratic* are those in which most of the landslides occur.

Once the initial failure is identified, landslide development can be guided by the presence and orientation of the structural control elements, which influence the style and the distribution (cf. WP/WLI 1993, Cruden & Varnes 1996) of the landslides and therefore their evolution in time (Figures. 8, 9). In particular, the evolution of earth flows, in terms of activity and kinematic mechanisms, can be generally explained by: i) bedding of the homogeneous and complex sequences; ii) stratigraphic or tectonic contact between deposits with a different competence; iii) zone of intense fracturing linked mainly to the presence of axes of folds and faults.

These settings constitute fundamental geological elements in understanding the control mechanisms in the earth flow source areas and channels. The hazard analysis and, in particular, aspects connected to the spatial prediction, imply a careful evaluation of the geostructural conditions that become a fundamental element in the areas where first-order events are developing.

Where geological bodies with different competence outcrop, the lithostructural control is recurrent. A typical example is the instability in Figure 9. The presence of a bedding plane (N125°/40°) of a stony sequence directs the shape, the orientation and the evolution of both the source area and the channel of the landslide, inducing a lateral retrogression of the source in comparison with the flow channel.

Areas of severe structural control are the source areas of multi-source earth flows (more than 150), where separation into source branches is a direct consequence of the local lithological and structural setting. Figures 4 and 10 show the morphological characters of multi-source instabilities and the distribution of the

Figure 8. Earth flow structurally controlled by a marly-clayey hill. During motion, the landslide body affected some houses: A) Image of November 2003; B) Image of March 2005.

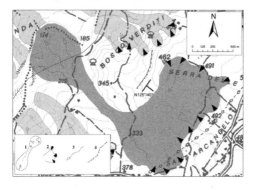

Figure 9. Earth flow from a segment of the Landslide Inventory Map (Guadagno et al. 2006). For the legend see Figure 4.

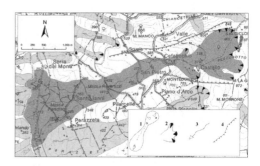

Figure 10. Example of multi-source earth flow. For the legend see Figure 4.

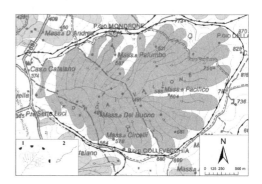

Figure 11. Examples of basin affected by coalescent flows; Legend: 1) Coalescent earth flows and basin boundary; 2) Flow direction.

activity by means of the empty triangles. The peculiar shaping of the source areas are fixed by the presence of stony sequences overlapping highly-tectonized, pelitic terrains.

In contrast, conditions of relative lithological homogeneity can be attributed to the basins affected by coalescent earth flows. In particular, about 50 basins (Figure 11) have been identified that include over 390 events (size <10 ha). For each single basin, about 60% of the surface is involved in mass movements on average, with extreme values of 72%, whereas the number of events with interconnected evolution varies from 5 to 36.

4 CONCLUSION

The structural control of the earth flows of the Province of Benevento was investigated by means of a landslide inventory map. A morphological evolution conditioned by active landslide processes affects a large part of the territory. Considering the prevalent clayey nature of the outcropping deposits, slow-velocity landslides predominate, having in some areas a pervasive diffusion.

Landslides are characterized by a reactivation tendency. They are linked to the repetition of triggering conditions which are correlated to the occurrence of seismic events and rainfall with specific characteristics, in terms of intensity and duration.

The diffusion of the earth flows led to their grouping into three types. On this basis, the definition of the distribution of the activity seems to be relevant in order to obtain basic data with the aim of ascertaining landslide susceptibility.

This distinction seems to be a basic aspect for the comprehension of landslide evolutive mechanisms and, consequently, for evaluations related to prediction of future events. It also opens a classification strategy for earth flows.

ACKNOWLEDGEMENTS

The research was financially supported by the PRIN 2005 project (prot. 2005047032_003, resp. F.M. Guadagno) and by the Provincial Council of Benevento and Benevento Chamber of Commerce. We would like to thank Italo Abate and Donato Tornesiello for the photographs in Figures 3 and 8a.

REFERENCES

AA.VV. 1985. Geotechnical Engineering in Italy. An overview. Published on the occasion of the ISSMFE Golden Jubilee. Roma, *A.G.I.-Ass.Geotecnica Italiana*: 414 pp.

Cruden, D.M. & Varnes, D.J. 1996. Landslide Types and Processes. In "Landslides: Investigation and Mitigation", Ed. Turner A.R. e Shuster *R.L.Sp.Rep. 247, Transportation Research Board*, National Research Council, National Academy Press, Washington D.C.: 36–72.

Guadagno, F.M., Focareta, M., Revellino, P., Bencardino, M., Grelle, G., Lupo, G. & Rivellini, G. 2006). *La carta delle frane della provincia di Benevento*. Sannio University Press.

Hungr, O., Evans, S.G., Bovis, M. & Hutchinson, J.N. 2001. Review of the classification of landslides of the flow type. *Environmental and Engineering Geoscience*, 7(3): 1–18.

Hutchison, J.N. 1988. General Report: Morphological and Geotechnical Parameters of Landslide in Relation to Geology and Hydrogeology.- *Proc., Fifth International Symposium on Landslide* (C. Bonnard, ed), A.A. Balkema, Rotterdam, Netherlands, 1, 3–35.

Patacca, E. & Scandone, P. 1989. Post-Tortonian mountain building in the Apennines. The role of the passive sinking of a relic lithospheric slab. In A. Boriani, M. Bonafede, G.B. Piccardo, G.B Vai (Eds.): *The lithosphere in Italy. Advances in Earth Science Research. It. Nat. Comm. Int. Lith. Progr., Mid-term Conf.* (Rome, 5–6 May 1987), Atti Conv. Lincei, 80: 157–176.

Pescatore, T.S., Di Nocera, S., Matano, F. & Pinto, F. 2000. L'Unità del Fortore nel quadro della geologia del settore orientale dei Monti del Sannio (Appennino Meridionale). *Boll., Soc., Geol., It.* 119, 587–601.

Picarelli, L. 1986. Caratterizzazione geotecnica dei terreni strutturalmente complessi nei problemi di stabilità dei pendii. *Proc. 16th Conv. Ital. di Geotecnica*, Bologna, 3, 155–170.

Varnes, D.J. 1978. Slope movements, type and processes. In: Schuster R.L. & Krizek R.J. (Eds.), *Landslides analysis and control. Washington Transportation Research Board, Special Report 176.* National Academy of Sciences, WA, 11–33.

WP/WLI 1993. A Suggested Method for Describing the Activity of a Landslide. *Bulletin of the I.A.E.G*, 47, 53–57.

Geotechnical and mineralogical characterization of fine grained soils affected by soil slips

G. Gullà & L. Aceto
CNR-IRPI_Sede di Cosenza, Rende (CS), Italy

S. Critelli
Università degli Studi della Calabria, Dip. Scienze della Terra, Rende (CS), Italy

F. Perri
Università degli Studi della Basilicata, Dip. Scienze Geologiche, Potenza (PZ), Italy

ABSTRACT: Soil slips affect essentially the degraded or weathered covers of soil, and could be particularly dangerous because of their kinematic mechanism and their spreading over wide areas during rainstorms. The high incidence of soil slips in an area of Central Calabria (Southern Italy) prompted a research aimed at geotechnical characterisation of fine-grained soils involved in this kind of instability. This paper illustrates the geotechnical and the mineralogical characterization carried out on samples coming from three sites representative of the wider study area, where sedimentary terrain (Plio-Pleistocene) crops out. Classification and direct shear tests have been conducted on undisturbed specimens and artificially degraded specimens. The results supply a reference frame of the soils physical-mechanical characteristics. Correlating the geotechnical analysis with the mineralogical-geochemical investigations allows a wider characterization of the sediment properties.

1 INTRODUCTION

Shallow and fast sliding-flow instabilities, soil slips in this paper (Campbell 1975), are dangerous phenomena produced by frequently shallow landsliding events related to rainstorms (Govi & Sorzana 1980, Brand 1984, Antronico & Gullà 2000, Wieczorek et al. 2001, Antronico et al. 2002, Antronico et al. 2004, Sorriso-Valvo et al. 2004).

The presence of many structures and infrastructure creates very high risk conditions for wide areas affected by shallow landsliding events.

Furthermore all lithotypes involving thick weathered or degraded soil profiles can be affected by soil slips. Shallow landslide density, and then the magnitude of the shallow landsliding events, depends on soil cover saturations conditions before the triggering rainstorms (Antronico et al. 2004), and to the geomaterial characteristics (Antronico et al. 2002).

In particular, the density of soil slips in fine-grained soils is probably related to the saturation-desaturation cycles that affect their degradation and reduce their shear strength (Gullà et al. 2004, Gullà et al. 2006).

To better understand this predisposing-triggering effect, we discuss in this paper the geotechnical and mineralogical characteristics of some silt and clay samples collected in a study area affected by shallow landslides in Calabria (Southern Italy). Above all we compare in this paper the geotechnical and mineralogical characters of natural (intact) silt and clay soils with the same soils artificially degraded in the laboratory (Gullà et al. 2004, Gullà et al. 2006).

2 GEOLOGICAL SETTING

The Calabria-Peloritani Terrane (CPT) is a fault-bounded, allochthonous terrane located between the NW-SE-trending Southern Apennines and E-W-trending Sicilian Maghrebides (Bonardi et al. 2001). The CPT is characterized by a pre-Mesozoic crystalline basement and it shows evidence of pre-Neogene tectonism.

The CPT comprises several Oligocene-to-Recent sedimentary units covering about half of its emerged bedrock area. These sedimentary deposits are mostly equally distributed along the Ionian and Tyrrhenian margins of Calabria and in mostly subsided areas in the central portions of the Calabrian terranes, one of these latter subsided areas is the Catanzaro Graben oriented orthogonal to the overall structural grain of the CPT. The Catanzaro graben is filled with deposits of Tortonian to Quaternary age, composed of marginal fanglomerates, evaporites and shallow marine facies,

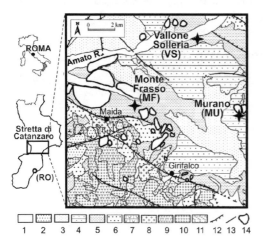

Figure 1. Geological outline of the study area and location of the sample sites. 1) alluvial deposits; 2) continental conglomerates and sand; 3) marine conglomerates and sands; 4) mostly sandstone deposits; 5) mostly clay deposits; 6) mostly conglomerate deposits; 7) evaporitic level; 8) Castagna Unit (two mica gneiss); 9) Polia-Copanello Unit (garnet and sillimanite-bearing gneiss); 10) sedimentary deposits which overlay Polia-Copanello Unit; 11) Stilo Unit (phyllites); 12) normal fault; 13) transcurrent fault; 14) landslide; RO = Roccella site. From Gullà et al. 2004, modified.

characterized by fine grained sediments (Fig. 1). Fine-grained sediments are widespread in the Catanzaro Graben and are mostly Pliocene to Pleistocene in age.

3 EXPERIMENTAL PROGRAM AND MEASURING TECHNIQUES

Laboratory tests were carried out to detect the characteristics of the fine-grained soils present in the study area, and to investigate the variations in mechanical characteristics due to the wetting-drying-freezing-thawing cycles (Antronico & Gullà 2000, Gullà & Antronico 2001, Gullà et al. 2004, Gullà et al. 2006).

The experimental program, that has been carried out on samples collected in three different sites of the study area (Monte Frasso, Vallone Solleria and Murano; Fig. 1), is based on classification tests (particle size distribution analysis and index test), direct and ring shear tests on intact, degraded and reconstituted specimens. Twenty-nine fine grained soil samples were analyzed by X-ray diffraction (XRD) and X-ray fluorescence spectrometry (XRF).

For the chemical and mineralogical analyses, the whole-rock samples were first dried and then crushed by hand in an agate mortar. Randomly-oriented whole-rock powders were run using a Scintag X_1 apparatus

equipped with a solid-state Si (Li) detector. The mineralogical composition of the <2 μm grain-size was determined from a thin section of highly oriented aggregate. The presence of expandable clays was determined after treatment with ethylene glycol at 25°C for 15h. Elemental analyses for major and some trace elements concentrations were obtained by X-ray fluorescence spectrometry (Philips PW 1480). Total loss on ignition (L.O.I.) was determined, after heating the samples for three hours, at 900°C.

The morphological change and secondary mineral formation were examined by field emission scanning electron microscopy (ESEM Philips Electronics QUANTA 200F with EDX GENESIS 4000). Secondary mineral compositions were determined qualitatively by energy dispersive X-ray spectrometer (EDS).

4 GRAIN SIZE DISTRIBUTIONS AND INDEX PROPERTIES

Based on the grain size distribution envelope of 54 clay samples collected in the studied area, it is possible to classify the soils of this area as clay with silt to silt with clay and sand (Fig. 2a). The grain size distribution curves of the clay samples investigated by mineralogical analyses (Monte Frasso site) are shown in Figure 2a.

The grain size distribution envelope of 99 silt soil samples (Vallone Solleria and Murano sites) show a composition that varied from sandy silt with clay to slightly gravelly clayey sand with silt (Fig. 2b). Figure 2b shows the grain size distribution curves of the silt samples used for mineralogical analyses.

Based on some index properties, it is possible to observe that there are differences between clay and silt soils, even if the ranges of overlap values are comparable. The average values of all clay samples are: unit weight of solid particles 26.7 kN/m^3, unit weight of dry soil 15.1 kN/m^3, void ratio 0.778; for the clay samples used for mineralogical analyses the average values are: unit weight of solid particles 26.6 kN/m^3, unit weight of dry soil 14.9 kN/m^3, void ratio 0.786. The average values of all silt samples are: unit weight of solid particles 26.6 kN/m^3, unit weight of dry soil 16.0 kN/m^3, void ratio 0.674; the average values of the silt samples used for mineralogical analyses are: unit weight of solid particles 26.7 kN/m^3, unit weight of dry soil 16.3 kN/m^3, void ratio 0.639.

Plasticity and activity characteristics of the examined soils are illustrated in Figure 3. It is possible to observe that the clay soils are classifiable as inorganic inactive, with medium-high plasticity, whereas the silt soils are classifiable as inorganic inactive or normally active, with medium-low plasticity. Clay and silt samples used for mineralogical analyses are representative of the total analysed samples. Plasticity and activity characteristics of the clay and silt soils are quite similar.

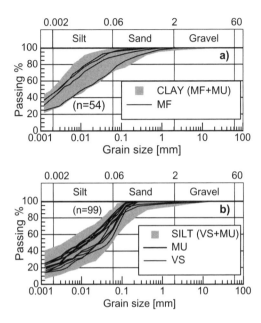

Figure 2. Grain size distribution envelopes for clays (a) and silts (b) in the study area, and grain size distribution curves for the samples used for present study (MF = Monte Frasso, VS = Vallone Solleria; MU = Murano).

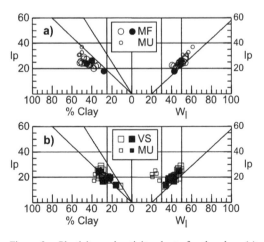

Figure 3. Plasticity and activity charts for the clays (a) and silts (b) in the study area (solid symbols), and for the samples used for mineralogical analyses (empty symbols) (MF = Monte Frasso, VS = Vallone Solleria; MU = Murano).

5 MINERALOGY AND CHEMISTRY

The XRD patterns of whole-rocks show that the analyzed samples are rich in clay minerals associated with significant amounts of calcite, quartz and feldspars, whereas minor concentrations of dolomite have been identified in some XRD patterns (Figs. 4–5). The <2μm grain-size fraction of all the samples is composed predominantly of illite, followed by illite-smectite mixed layers, chlorite and kaolinite.

The differences among silts and clays are related to the clay mineral contents (Fig. 4). The clay soils show a higher percentage of clay minerals and lower content of feldspars, dolomite and quartz. These differences are minor, and do not significantly affect the plasticity and activity features (Fig. 3).

The XRD patterns of natural (intact) silt samples and silt samples degraded in the laboratory do not show substantial mineralogical differences (Fig. 5); only the <2 μm fraction of these samples shows a little variation related to the illite and illite-smectite mixed layers (I-S) content (Fig. 6). These slight differences are probably related to the laboratory degradation treatments (cycles of saturation, drying, wetting and freezing).

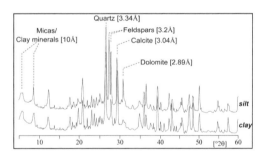

Figure 4. The XRD patterns of silt and clay whole-rocks fine grained soil samples.

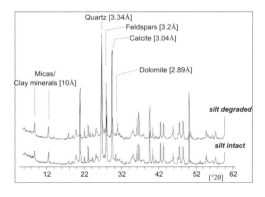

Figure 5. The XRD patterns of silt samples degraded in laboratory and natural (intact) silt samples whole-rocks.

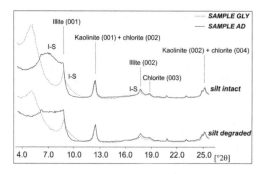

Figure 6. The XRD patterns of the air dried (AD) and ethylene-glycol solvated (GLY) <2 μm fraction specimens of silt samples degraded in laboratory and natural (intact) silt samples.

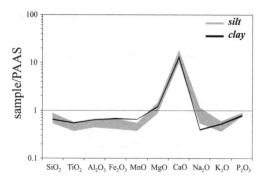

Figure 7. Major element distributions normalized to PAAS (Post-Archean Australian Shales; Taylor and McLennan, 1985) of silt and clay samples.

The <2 μm XRD pattern of natural (intact) silt samples and silt samples degraded in the laboratory shows variations in the expandable clays, in which a different peak shape is observed in the 5–10 °2θ range, related to different content of the I/S mixed layers. This is probably associated with chemical and mineralogical variations in response to different cycles of degradation. The silt samples show a content of smectite layers in the I-S mixed layers in a range of 40–50% (R ordering $= 0$; Reickeweite number), suggesting a shallow diagenetic stage. The same characteristic seems detectable too in the clay samples.

Major and trace element variations and chemical weathering indices are related to a different behaviour and distributions of chemical elements, principally controlled by the degree of artificial degradation and weathering. The elemental distributions of the studied samples have been normalized to the PAAS standard (Post-Archean Average Shale; Taylor and McLennan 1985).

Major element composition of the samples is quite homogeneous, whereas the trace element concentration is relatively variable.

Chemical analyses show similar variations between silts and clays (Fig. 7) and between natural (intact) silt samples and silt samples degraded in the laboratory (Fig. 8).

The soil samples are characterized by narrow compositional changes for all elements which have concentrations rather weakly depleted relative to the PAAS. Mg alone has concentrations close to those of the PAAS, whereas Ca is strongly enriched. Sr, similar to Ca, is enriched relative to the PAAS. In standard-normalized variation diagrams, for the natural (intact) silt samples and silt samples degraded in the laboratory, only the Na ratio weakly decrease during the course of artificial degradation.

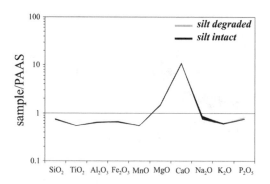

Figure 8. Major element distributions normalized to PAAS (Taylor and McLennan, 1985) of silt samples degraded in laboratory and natural (intact) silt samples.

Al_2O_3, Fe_2O_3, TiO_2, MnO, MgO and K_2O have a positive correlation whereas CaO has a negative correlation. In fine-grained sediments Al_2O_3 commonly monitors (indicates) clays and this behaviour may thus account for the competition between mica-like clay minerals and carbonate phases. Most of the trace elements show positive linear trends against Al_2O_3, Fe_2O_3, K_2O and TiO_2, indicating clay mineral control on their contents.

6 MINERALOGICAL SIGNATURES OF DEGRADATION AND WEATHERING

The scanning electron microscope (SEM) and application of X-ray diffraction (XRD) have been used to better understand the behaviour of grain minerals and their morphological variation during the course of sample degradation. Among common rock-forming minerals, quartz is the most resistant mineral, while plagioclase and mica (usually biotite) are easily replaced by secondary phases.

In Figure 9 SEM photomicrographs show the grain feature, coating products and interlayer sites in the mica minerals.

The difference between Figure 9a and 9b are related to the grain-size particles; Figure 9a shows the silt features characterized by more feldspar and quartz grains, whereas the clay sample (Fig. 9b) is clearly marked by abundant clay minerals.

The slightly weathered mica (Fig. 9c) of the natural (intact) silt sample shows a smooth basal surface with etch pits of submicrometer size, whereas secondary minerals are attached to the edge. During the course of weathering and the degradation processes, mica minerals have a tendency to open the reticular sheets and gradually lose K^+ ions toward the top part. This result indicates that illitization has occurred and such K-depleted inter-layers in the mica minerals will be referred to hereafter as illite/smectite-like inter-layers. The mica mineral of the silt sample degraded in the laboratory (Fig. 9d) is slightly deformed, with secondary minerals present in the lower part and at the edge of the grain. One of the secondary minerals is identified qualitatively as an illite and/or smectite-like silicate, as indicated by its morphology and composition (revealed by SEM-EDS).

These deformation mechanisms is the mica minerals are probably partially related to the artificial degradation carried out in the laboratory.

7 SHEAR STRENGTH

The direct shear tests (peak and residual) carried out on specimens used for the present study gave results that are congruent with the shear strength envelopes proposed by Gullà et al. (2004).

In order to investigate the mechanisms that produced a reduction of the shear strength, we have been reproduced in the laboratory a lot of wetting-drying-freezing-thawing cycles with a period that range from one to 90 days (Gullà et al. 2004, Gullà et al. 2006); using this technique, we have been studied the degradation of the soils induced by saturation-desaturation cycles.

The results of the direct shear tests carried out on natural (intact) and artificially degraded specimens, at the same vertical stress, are shown in Figure 10. In both cases, for clay and silt, we can see the same failure mechanism, and find significant reductions of shear strength for the degraded specimens relatively to the shear strength of the natural (intact) specimens: for clay 74% and 35% respectively after one day and two days of degradation; for silt 42% after one day, 32% after 14 days and 28% after 30 days (after 60 and

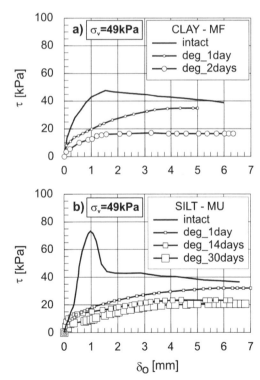

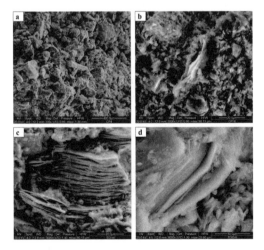

Figure 9. SEM images from grains of four different fine grained soil samples: a) silt sample, b) clay sample, c) natural (intact) silt sample, d) silt sample degraded in laboratory.

Figure 10. Stress-displacement curves of intact and artificially degraded specimens for clay (a) and silt (b) soils.

90 days of degradation there is no further reduction in shear strength).

For the silts, the previously illustrated results allow us to assume that the decrease of shear strength for degradation is not due to mineralogical and/or chemical changes, according to the results obtained from Gullà et al. (2006) for the soils outcropping in an other area of the Calabria region (Roccella site, RO in Fig. 1). The reduction of shear strength for the silts tested in this study, according to the studies by Gullà et al. (2006), is probably related to the soil structure changes induced from the degradation cycles. However, other enquiries are necessary in the study case in order to confirm this hypothesis.

Figure 11 gives a first reference frame for the shear strength data related to artificially degraded specimens. Figure 11 shows the ratio values among shear strength of the degraded specimens and maximum peak shear strength of the undisturbed specimens, related to the time of degradation and for all examined vertical stresses.

The examined vertical stresses are: from 13 kPa to 98 kPa for clays, with cycles of degradation until seven days; from 24.5 kPa to 147 kPa for silts, with cycles of degradation until 30 days. After seven days of degradation it is possible to observe shear strength decreases between 55% and 20% for clays, and between 50% and 30% for silts (Fig. 11).

Figure 11 shows an asymptotic trend of the curves for silts, that indicate a reduction to 27% of the maximum peak shear strength after thirty days of degradation. The same feature is found for the reduction curve of the clay studied by Gullà et al. (2006) at the Roccella site (Fig. 11). It is possible to assume similar features for the other envelope curves for the clay and silt studied in this paper (reduction of the maximum peak shear strength after thirty days of degradation: between 20% and 45% for clay and between 27% and 35% for silt).

8 CONCLUSIONS

This paper has illustrated an integrated study, based on mineralogical and geotechnical investigations, aimed to improve the characterisation of fine grained soils involved in soil slip phenomena. This approach is crucial to better understand the magnitude and distribution of shallow landsliding events in such soils.

The degradation cycles simulated in the laboratory do not produce mineralogical and chemical changes in tested clay. However cycles of wetting-drying-freezing-thawing have disturbed the natural structure of the soil, producing changes in fabric and bonding.

The stress-displacement curves obtained from direct shear tests on intact silts and clays show a well-defined peak strength in contrast to similar tests on the degraded specimens. Peak shear strength shows a reduction during one month of degradation; after this period no further meaningful change is observed. The reduction of the peak shear strength due to the degradation induced in the laboratory is comparable to the residual conditions (Gullà et al. 2004).

In conclusion, the approach used and the results proposed in this paper represents valuable support for the study of shallow landslide susceptibility in fine grained soils.

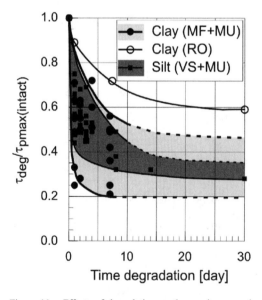

Figure 11. Effects of degradation on the maximum peak shear strength. MF = Monte Frasso; MU = Murano; VS = Vallone Solleria; RO = Roccella.

REFERENCES

Antronico, L. & Gullà, G. 2000. Slopes affected by soil slip: validation of an evolutive model. *Proc. of the 8th Intern. Symp. on Landslides, Cardiff* 1: 77–84.

Antronico, L., Gullà, G. & Terranova, O. 2002. L'evento pluviometrico dell'8–10 settembre 2000 nella Calabria Ionica Meridionale: Dissesti sui versanti e processi in alveo. *Proc. of the Symp. "Il dissesto idrogeologico: inventario e prospettive", XIX Giornata dell'Ambiente, Accademia Nazionale dei Lincei, Roma, 5 Giugno 2001*: 67–79.

Antronico, L., Gullà G. & Borrelli, L. 2004. Shallow instabilities for sliding flow: regional influence and area effects. *Proc. of the 9th Intern. Symp. on Landslides, Rio de Janeiro* 2: 1381–1387.

Bonardi, G. Cavazza, W. Perrone & V. Rossi, S. 2001. Calabria-Peloritani terrane and Northern Ionian Sea. *In: Vai G.B., Martini I.P., Eds. Anatomy of an orogen: the Apennines and Adjacent Mediterranean Basins*, 287–306. Dordrecht/Boston/London, Kluwer Academic Publishers.

Brand, E.W. 1984. Landslides in Southeast Asia: A State-of-the-Art Report. *Proc. of the 4th Intern. Symp. on Landslides, Toronto*: 17–59.

Campbell, R.H. 1975. Soil slip, debris flow and rainstorms in the Santa Monica Mountains and vicinity, Southern California. U.S. *Geological Survey Professional Paper* 851: 1–51.

Govi, M. & Sorzana, P.L. 1980. Landslide susceptibility as a function of critical rainfall amount in Piemont Basin (North-Western Italy). *Studia Geomorph. Carpatho-balcanica* 14: 43–61.

Gullà, G. & Antronico, L. (eds.) 2001. Le instabilità superficiali per scorrimento-colata nella Stretta di Catanzaro. In *"Linee Guida per Interventi di Stabilizzazione di Pendii in Aree Urbane da Riqualificare"*, CNR-IRPI, Regione Calabria-UE, POP 1994/99.

Gullà, G. Aceto, L. & Niceforo, D. 2004. Geotechnical characterization of fine-grained soils affected by soil slips. *Proc. of the 9th Intern. Symp. on Landslides, Rio de Janeiro* 1: 663–668.

Gullà, G. Mandaglio, M.C. & Moraci, N. 2006. Effect of weathering on the compressibility and shear strength of natural clay. *Can. Geotech. J.* 43: 618–625.

Sorriso-Valvo, M., Antronico, L., Gaudio, R., Gullà, G., Iovine, G., Merenda, L., Minervino, I., Nicoletti, P.G., Petrucci, O. & Terranova, O. 2004. Carta dei dissesti causati in Calabria meridionale dall'evento meteorologico dell'8–10 settembre 2000. *CNR-GNDCI, Pubblication 2859, Geodata* 45, Rubbettino Publisher, Soveria Mannelli, Italy.

Taylor, S.R. & McLennan, S.M. 1985. *The Continental Crust: Its Composition and Evolution*. Oxford, Blackwell Scientific.

Wieczorek, G.F., Larsen, M.C., Eaton, L.S., Morgan, B.A. & Blair, J.L. 2001. Debris-flow and flooding hazards associated with the December 1999 storm in coastal Venezuela and strategies for mitigation. *U.S. Geological Survey, Open File Report* 01–144, pp 40.

Vulnerability of structures impacted by debris flow

E.D. Haugen & A.M. Kaynia
Norwegian Geotechnical Institute, Oslo, Norway

ABSTRACT: As debris flow impacts a structure it sets the structure in vibratory motion. This article presents a simple method for prediction of damage in a structure impacted by a debris flow of known magnitude. The method uses the principles of dynamic response of simple structures to earthquake excitation and fragility curves proposed in HAZUS for estimation of the structural vulnerability, by the damage state probability. The method was tested by applying it to a debris flow site in Italy where the reported structural damages ranged from light to complete devastation. The vulnerability of six of the impacted structures was assessed by the proposed model and compared with the real damage states.

1 INTRODUCTION

In areas susceptible to landslides, it is becoming more and more common to assess the risk in two parts: 1) Assessing the hazard, which is defined by: "The probability that a particular danger (threat) occurs within a given period of time." 2) Assessing the vulnerability, which is defined by: "The degree of loss to a given element or set of elements within the area affected by a hazard" (ISSMGE, 2004).

The area of risk management in geohazards is fairly complex. Much work has been done on making adequate models for the probability of occurrence of an adverse event and developing landslide hazard maps (e.g. Remondo et al. 2005). But within the area of vulnerability many conflicts exist; vulnerability analysis has commonly been integrated directly in the risk analysis and no unique or simple method is found for landslide vulnerability assessment (Glade, 2003).

Many suggestions have been made to vulnerability assessments of structures struck by landslides. Cardinali et al. (2002) assessed the vulnerability as a function of qualitative landslide intensity and type. Leone et al. (1996) made use of the damage intensity as a vulnerability measure. Heinimann (1999) estimated rock fall vulnerability by using the building type and rock fall magnitude as input.

As one may see, none of these methods cover the whole scope of the vulnerability problem. Each of the methods has used important parameters, such as landslide intensity, volume and type, but not one of them incorporates completely all the aspects of vulnerability. Glade (2003) states that most approaches do not distinguish between types of processes or magnitudes, that absolute values of vulnerability vary significantly and these values are spread over a wide range; this makes the comparison of approaches difficult. In addition, many vulnerability or risk assessment models depend on empirical data from past damages (e.g. Wong et al. 1997, Remondo et al. 2005) and on local historical databases which contain the socio-economic damage values.

This paper presents a suggestion to a quantitative model for vulnerability assessment of structures hit by debris flow, where the vulnerability is measured by the probability of reaching a certain damage state. The method is based on the premises that the debris flow is not so large as to completely wipe out the structure, but impacted by a force, which leads to structural vibration and results in a specific state of damage.

2 THEORY

2.1 HAZUS damage state probabilities

The vibrations from an impact will damage a structure in approximately the same way as a ground vibration will damage a structure standing on the ground. From this it may be assumed that a debris flow impact force, modelled as an impulse, may be compared to ground acceleration from an earthquake. This allows one to try to incorporate existing literature on structural damage from earthquake vibrations in landslide vulnerability assessment.

The Federal Emergency Management Agency (FEMA) of the United States together with the National Institute of Building Sciences (NIBS), have developed the HAZUS programme (short for Hazard US). HAZUS is used in USA to map the expected losses due to earthquake events. HAZUS quantitatively estimates the losses in terms of direct cost for repair, direct cost associated with loss of function, casualties, displacement from residences, debris quantity and regional economic impact (HAZUS, 2006).

In contrast to the conventional vulnerability grading from 0–1 for a given element at risk, HAZUS defines five damage states: None, slight, moderate, extensive and complete.

These damage states are displayed by fragility curves, which portray the damage-motion relationship. The fragility curve for a given building type describes the probability of a specific damage state arising as a function of the spectral displacement. The spectral displacement is the maximum displacement under earthquake of a single-degree-of-freedom system, with the same natural frequency as the building relative to the ground motion. HAZUS has classified 36 building classes, and the vulnerability for each of these may be assessed with the fragility curve of each class. An example of fragility curves for unreinforced old masonry buildings (class 34) is seen in Figure 1, which shows how the damage state probability may be found from the spectral displacement of the building.

Each fragility curve is a log-normal curve, characterised by a median spectral displacement ($\hat{S}_{d,ds}$), which corresponds to the threshold of the damage state, and a log-normal standard deviation (β_{ds}), the variability associated with the damage state.

The probability of a certain damage state, ds, given a spectral displacement, u_{max}, is defined by:

$$P(ds|u_{max}) = \Phi\left[\frac{1}{\beta_{ds}}\ln\left(\frac{u_{max}}{\hat{S}_{d,ds}}\right)\right] \quad (1)$$

where Φ is the standard normal cumulative distribution function.

The objective of this study is to provide a framework for assessment of structural vulnerability to debris flow using HAZUS principles. This means that a structure's maximum lateral displacement caused by debris flow impact may be taken as the spectral displacement and used in the fragility curves. The result would be a quantitative estimation of vulnerability—that is; probability of certain damage states occurring for a given debris flow impact.

2.2 Spectral displacement

A debris flow hits a structure with an impulse force, the dynamic impact force, P_{dy}. This will set the structure in vibrations and give it a lateral displacement, which is increased by the hydrostatic force, P_{st}. Assuming a single-degree of freedom, the maximum displacement of a structure, which may be set as the spectral displacement, is:

$$u_{max} = u_{st} + u_{dy,max} \quad (2)$$

The displacement from the hydrostatic force, u_{st}, may be found directly from the static force, P_{st}, and the structure stiffness, k:

$$u_{st} = \frac{P_{st}}{k} = \frac{p_{st}A}{k} = \frac{\rho_{df}gbh^2}{2k} \quad (3)$$

where p_{st} is the average hydrostatic pressure, $A = bh$ is the area of the structure hit by the debris flow, ρ_{df} is the debris flow density, g the gravitational constant (9.81 m/s^2), b is the width of the flow and h is the height of the flow.

The maximum dynamic displacement, $u_{dy,max}$, must be found from the impulse force, P_{dy}, and structure parameters such as the natural period of the structure, T_n, and the stiffness, k. Using a shock spectrum, see Figure 2, the displacement may be found via R_{max}. R_{max} is the response ratio which is the ratio of the dynamic displacement to the static

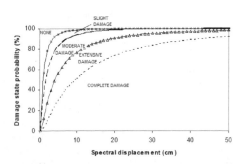

Figure 1. HAZUS fragility curves for building class 34, unreinforced masonry buildings.

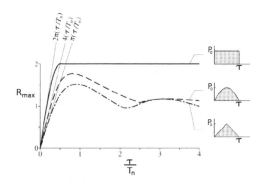

Figure 2. Shock spectrum for three force pulses of equal amplitude (Chopra, 2001).

displacement, caused by the impulse force amplitude $P_{dy,max}$ (Clough, 1993):

$$R_{max} = \frac{u_{dy,max}(P_{dy,max})}{u_{st}(P_{dy,max})} = \frac{u_{dy,max}}{P_{dy,max}/k} \quad (4)$$

In the shock spectrum the shape of the impulse force, $P_{dy}(t)$, is assumed with a amplitude $P_{dy,max}$. Three impulse shapes, rectangular, half-cycle sine and triangular, are plotted for τ/T_n versus R_{max}, where τ is the impulse duration. For debris flow impact a conservative assumption is a rectangular impulse force shape, so that $P_{dy,max} = P_{dy}$.

As seen from Equation (4) the maximum lateral displacement imposed by a dynamic impulse impact may be found from the response ratio for a known amplitude of the dynamic force and stiffness of the building.

The dynamic force must be estimated from parameters which may be approximated. A function of the debris flow velocity at impact, v_{df}, the flow density, ρ_{df}, and the area hit by the flow, A, is amongst others proposed by Hungr et al. (1984), Armanini and Scotton (1992) and Lin et al. (1992). The dynamic force suggested by Hungr et al. is used:

$$P_{dy} = p_{dy} \cdot A = \rho_{df} \cdot v_{df}^2 \cdot A \quad (5)$$

Most debris flow pulses will last longer than $T_n/2$, which gives $\tau/T_n > 1/2$. For these pulses the maximum deformation occurs during the impulse, and the pulse shape plays a bigger role than for pulses shorter than $T_n/2$ (Chopra, 2001). In these cases the shock spectrum may be used directly, inserting τ/T_n, and finding the response ratio from the corresponding pulse shape. For structures with less than 20 stories and impulse duration equal to that of a flood impact, which FEMA (1995) proposes to be $\tau = 1$ s, rectangular impulse forces give a response ratio with a constant value of $R_{max} = 2$.

The stiffness of the building is also needed to find the lateral displacement. It may be approximated by:

$$k = \frac{4\pi^2 m_{str}}{T_n^2} \quad (6)$$

where m_{str} is the mass of the structure.

Using Equations (5) and (6) in Equation (4) and rearranging it, gives the maximum lateral displacement caused by the dynamic impact:

$$u_{dy,max} = R_{max}\frac{P_{dy,max}}{k} = \frac{\rho_{df}v^2 A}{2\pi^2 m_{str}}T_n \quad (7)$$

The natural period of the structure, T_n, can be estimated by the structure's height, H_{str} (NS 3491-12, 2004) as follows:

For concrete frames:

$$T_n = 0.075 \cdot H_{str}^{3/4} \quad (8a)$$

For wooden frames:

$$T_n = 0.050 \cdot H_{str}^{3/4} \quad (8b)$$

3 APPLICATION AND IMPLEMENTATION OF METHOD

On 5–6 May 1998 several devastating debris flows occurred in the Sarno area of the mountain Pizzo d'Alvono, in southern Campania, Italy. The area was hit by tens of debris flows. 160 people were killed, over 150 buildings were totally destroyed and over 500 damaged seriously or partially. The value of the damage to structures was estimated to approximately 34 million Euros (European Commision, 2003). Zanchetta et al. (2004) mapped the extent of the debris flows in the Sarno area.

The debris flows originated from shallow soil slips on slopes with an angle of 30–35 degrees. Nine days of continuous rainfall had saturated the volcaniclastic soil, which failed on the boundary between soil and bedrock. Progressive failure and liquefaction enlarged the mass as it flowed downhill. The debris flows followed the stream channels until they reached the alluvial fans, with an inclination less than 12 degrees, and spread out into the inhabited areas and caused damage and casualties.

Zanchetta et al. have estimated the physical and dynamical properties of the debris flow which are needed to find the impact pressure and mapped the structural damage. This data is very convenient for using in the proposed vulnerability model, as all the input factors are available, and the damage is mapped.

Figure 8 in Zanchetta et al. gives the variation of the dynamic and hydrostatic pressure and the damage done to the buildings by three debris flows named Ep-6, Lav-2 and Ep-3b. The damage varies from light to complete.

Using the data found from the Sarno debris flows in the vulnerability model, the vulnerability of two buildings in each debris flow are assessed; these buildings are named Ep-6-1, Ep-6-2, Lav-2-1, Lav-2-2, Ep-3b-1 and Ep-3b-2.

3.1 Impact force

For the three debris flows the impact pressures of the structures 1 and 2 in every debris flow are found directly from Zanchetta et al. (2004). Table 1 shows

Table 1. Debris flow impact and damage at different structures.

	Dynamic force MN	Static force MN	Structural damage
Ep-6-1	3.32	1.03	Complete
Ep-6-2	1.01	0.45	Heavy
Lav-2-1	6.12	1.99	Complete
Lav-2-2	1.08	0.81	Moderate-light
Ep-3b-1	3.93	1.14	Complete
Ep-3b-2	0.43	0.34	Complete

the slide properties, the dynamic force and the static force and the damage of each structure.

3.2 Natural period and stiffness of structure

The natural period of the structure depends on the type and mass of the structure. The buildings in the area hit by debris flows were both old country houses and new, reinforced concrete houses, mostly one- and two-story masonry residences with basement. Since it has not been possible to find out the structure type of the six structures selected, the vulnerability assessment is performed for both types of structures. Assuming the structures in question all have two stories, which gives them a height $H_{str} = 5$ m, the natural period is found from Eq. (8) to be $T_n = 0.25$ s.

Making assumptions on the structural dimensions and using a concrete density, 2300 kg/m^3, gives the approximated mass, m_{str}, in Table 2. The stiffness, found by Eq. (6) with the structure mass and the natural period, is also found in Table 2.

3.3 Spectral displacement

Using the impulse duration $\tau = 1$ s, one sees that $\tau > T_n/2 = 0.13$ s, and therefore the maximum displacement ratio may be found directly from the shock spectrum as a function of impact force shape and τ/T_n. By assuming a rectangular impulse force shape and using $\tau/T_n = 4$ in Figure 2, gives the maximum displacement ratio $R_{max} = 2$. For the less conservative triangular and sinusoidal impulse shapes, the maximum displacement ratio is 1.0 and 1.15 respectively.

Continuing with $R_{max} = 2$, the displacement of each structure due to the dynamic impact may be found by Eq. (7), and the displacement from the static force from Eq. (3). The displacements for the various structures are calculated and listed in Table 3, where the total displacement is taken as the spectral displacement in the following steps.

3.4 HAZUS fragility curves

The displacements found in Table 3 are used as input in the HAZUS fragility curves. But first the appropriate fragility curves must be selected by finding the FEMA building class of the buildings in the area, and the belonging curve parameters. The two types of buildings in the area may be classified in the FEMA system as URML (class 34) and RM2L, which respectively refer to unreinforced old masonry buildings and new reinforced masonry buildings, both with 1–2 stories and basement (HAZUS, 2006b).

As the buildings have not been seismically designed, the fragility curve parameters may be found from Table 5.9d of the HAZUS Earthquake Technical Manual (HAZUS, 2006b). These are listed in Table 4.

Having found the parameters for the standard cumulative normal distribution of the fragility curves, and the maximum displacement of the structures, the probability of the different damage states may be found by Eq. (1).

The tables in the following section give the damage state probabilities for the spectral displacements found in Table 3 for both un-reinforced masonry houses (URML) and reinforced masonry houses (RM2L).

Table 2. Approximated mass and stiffness of structures in flow paths.

	Flat area m²	Mass kg	Stiffness MN/m
Ep-6-1	105	$97 \cdot 10^3$	61
Ep-6-2	170	$156 \cdot 10^3$	98
Lav-2-1	160	$147 \cdot 10^3$	92
Lav-2-2	320	$294 \cdot 10^3$	185
Ep-3b-1	140	$129 \cdot 10^3$	81
Ep-3b-2	140	$129 \cdot 10^3$	81

Table 3. Displacements of the structures 1 and 2 in debris flows Ep-6, Lav-2 and Ep-3b.

	Dynamic displacement m	Static displacement m	Total displacement m
Ep-6-1	0.110	0.017	0.127
Ep-6-2	0.021	0.005	0.026
Lav-2-1	0.133	0.022	0.155
Lav-2-2	0.012	0.004	0.016
Ep-3b-1	0.097	0.014	0.111
Ep-3b-2	0.011	0.004	0.015

Table 4. Fragility curve parameters (the mean $\hat{S}_d$, ds in meters).

	URML		RM2L	
	$\hat{S}_{d,ds}$	β_{ds}	$\hat{S}_{d,ds}$	β_{ds}
Slight damage	0.0081	1.15	0.0147	1.14
Moderate damage	0.0165	1.19	0.0234	1.10
Extensive damage	0.0411	1.20	0.0587	1.15
Complete damage	0.0960	1.18	0.1600	0.92

4 RESULTS: DAMAGE STATE PROBABILITIES

Tables 5–10 give the computed vulnerability of the six different buildings in terms of the probability of reaching certain damage states. For unreinforced masonry buildings the damage state probabilities exceed those of the reinforced buildings, which is closer to the actual damage.

Table 5. Damage state probabilities for building Ep-6-1, where complete structural damage was observed.

	Slight	Moderate	Extensive	Complete
URML	0.992	0.957	0.826	0.593
RM2L	0.970	0.938	0.748	0.400

Table 6. Damage state probabilities for building Ep-6-2, where there was heavy damage and partial devastation.

	Slight	Moderate	Extensive	Complete
URML	0.837	0.639	0.341	0.128
RM2L	0.681	0.527	0.231	0.002

Table 7. Damage state probabilities for building Lav-2-1, where complete structural damage was observed.

	Slight	Moderate	Extensive	Complete
URML	0.995	0.970	0.864	0.656
RM2L	0.980	0.957	0.799	0.484

Table 8. Damage state probabilities for building Lav-2-2, where moderate to light structural damage was observed.

	Slight	Moderate	Extensive	Complete
URML	0.722	0.490	0.216	0.065
RM2L	0.530	0.365	0.130	0.062

Table 9. Damage state probabilities for building Ep-3b-1, where complete structural damage was observed.

	Slight	Moderate	Extensive	Complete
URML	0.989	0.946	0.797	0.550
RM2L	0.962	0.922	0.711	0.347

Table 10. Damage state probabilities for building Ep-3b-2, where complete structural damage was observed.

	Slight	Moderate	Extensive	Complete
URML	0.700	0.465	0.198	0.057
RM2L	0.503	0.340	0.116	0.005

Building Ep-6-1 was completely damaged. The probability of complete damage from the HAZUS fragility curve for unreinforced masonry buildings (URML) is 59%. Building Ep-6-2 was heavily damaged and partially devastated, which is predicted by the URML curve by a probability of 34%.

Building Lav-2-1's was completely damaged. The computations predict a 66% probability of complete damage based on the URML fragility curve. Building Lav-2-2 was, on the other hand, moderately to lightly damaged. The computations indicate a 53% probability of moderate damage.

Buildings Ep-3b-1 and Ep-3b-2 were both completely damaged, the computations from the URML fragility curves give respectively 55% and 6% probability of this damage state occurring.

5 DISCUSSION AND LIMITATIONS

Assuming un-reinforced structures (URML), the HAZUS fragility curves give the probability of reaching the correct damage level by 34–66%, with the exception of building Ep-3b-2 with 6%. Building Ep-3b-2 was completely destroyed even though it was hit by less than a third of the force which hit building Ep-3b-1, therefore it is appropriate to believe that this building is not assigned the correct class and must be left out of the further discussion.

The damage state probabilities for the remaining five buildings must be evaluated, so that one may judge whether they are applicable or not. The Sarno debris flow used to implement the model is a real case; therefore the actual damage states are known with certainty. When using the vulnerability model to estimate the damage from the exact landslides which hit Sarno in

1998, one expects the probability for the actual damage state to be sufficiently high, so that the model may be used.

The question is: What is a sufficiently high probability? For instance, for the case of the completely damaged building Ep-6-1, the probability of complete damage using the vulnerability model was 59%. Is a 59% probability of complete damage high enough to lead to the necessary interventions to prevent damage which may happen (as seen in the Sarno case)?

One may argue that the model gives imprudently low values for the actual damage state arising. If so the probabilities should be adjusted, so that, for instance, a 40% probability of a certain damage state is equivalent to a higher probability in reality. This may also be seen from the values in the fragility curves. Table 11. shows how large the spectral displacement of an un-reinforced building must be to obtain certain probabilities of reaching a given damage state. The spectral displacements which are required for a two-story building to get 80% or higher probability of complete damage are unrealistically large.

The reasons for this discrepancy may either lie in the model itself or in the input, which leaves the following options:

1. The model gives non-conservative values. The output must be correlated to realistic values, if it is possible.
2. The building classes chosen to find the fragility curve parameters do not reflect the actual behaviour of buildings subject to debris impact, more exact data must be found on the structures in the area.
3. The displacement calculation models are erroneous and must be refined.

Using the model in practice a sensitivity analysis of the input should be implied to see how the result is affected by variations. The impact force, for instance, may be up to twice the magnitude suggested in Eq. (5) (Armanini & Scotton, 1992, Lin et al. 1997). Using double dynamic force for the heavily damaged building Ep-6-2 gives a 54% probability of being extensively damaged compared to 34% with the former force. The impulse force shape (see Figure 2) also varies, but assuming $R_{max} = 2$ is conservative for any case.

6 CONCLUSION

The proposed model for assessing the vulnerability of structures hit by debris flows has been tested for the debris flows in Sarno area, Italy, 5–6 May, 1998. Six different structures with varying damage levels were tested. The model gave the probabilities between 34% and 66% for reaching the damage levels which actually occurred for five of the six structures. For example for a building which was completely damaged in reality, the model gave 66% probability of being completely damaged for the same debris flow.

The uncertainties of the model are fairly large, which makes it difficult to see where the possible discrepancy lies, whether it is in the input parameters or the model itself. Although the case had much information, it did not give the full information needed to limit the input uncertainty. To decide whether the method of assessing vulnerability to debris flows is worth using, input from more case studies must be tested, preferably using input more accurate than the Sarno case, to locate the error sources. A sensitivity analysis should also be done. For a more detailed damage state prediction the structural model may be extended to more degrees of freedom, which would show the effect of the loading point and structure height.

The method could be included in a statistical model, to give the degree of reliability of the output. To increase the applicability, the model may be extended to consider punching and rupture of structural elements, which will be the case for inhomogeneous debris flow (i.e. with embedded rocks) and may be more critical. By using GIS software the model could be used to assess the vulnerability of several structures in an area exposed to debris flow.

Table 11. Spectral displacements (in meters) for different damage state probabilities for URML structures.

Damage state probability for URML structures	50%	80%	95%
S.D. giving slight damage	0.008	0.021	0.053
S.D. giving moderate damage	0.016	0.045	0.115
S.D. giving extensive damage	0.041	0.111	0.292
S.D. giving complete damage	0.095	0.255	0.659

REFERENCES

Armanini, A. & Scotton, P. 1992. Experimental analysis on the dynamic impact of a debris flow on structures. In *Proceedings, International Symposium, INTRAPREVENT 1992, Bern*, volume 6, pages 107–116.

Cardinali, M., Reichenbach, P., Guzzetti, F., Ardizzone, F., Antonini, G., Galli, M., Cacciano, M., Castellani, M. & Salvati 2002. A geomorphological approach to the estimation of landslide hazards and risks in Umbria, Central. *Natural Hazard and Earth System Sciences* 2(1):57–72.

Chopra, A. 2001. *Dynamics of Structures, Theory and Applications to Earthquake Engineering*, 2nd edition. New Jersey: Prentice Hall.

Clough, R. & Penzien, J. 1993. *Dynamics of Structures*, 2nd edition. Singapore: McGraw-Hill, Inc.

European Commission 2003. Lessons Learnt from Landslide Disasters in Europe. In Hervás, J. (ed.), *Technical report, Joint Research, NEDIES PROJECT*.

FEMA 1995. *Engineering Principles and Practices for Retrofitting Flood Prone Residential Buildings. FEMA 259*.

Glade, T. 2003. Vulnerability Assessment in Landslide Risk Analysis. *Die Erde (Beitrag zur Erdsystemforschung)* 134:121–138.

HAZUS 2006. *Multi-hazard Loss Estimation Methodology, Earthquake Model, HAZUS MH-MR2, User Manual. FEMA and NIBS*.

HAZUS 2006b. *Multi-hazard Loss Estimation Methodology, Earthquake Model, HAZUS-MH-MR2, Technical Manual. FEMA and NIBS*.

Heinimann, H.R. 1999. Risikoanalyse bei gravitativen Naturgefahren—Fallbeispiele und Daten. *Umwelt-Materialien* 107/II. Bern.

Hungr, O., Morgan, G. & Kellerhalls, R. 1984. Quantitative Analysis of debris Torrent Hazard for Design of Remedial Measures. *Canadian Geotechnical Journal* 21: 663–667.

ISSMGE TC32 2004. Glossary of Risk Assessment Terms, Technical Committee on Risk Assessment and Management. www.engmath.dal.ca/tc32/2004Glossary_Draft1.pdf.

Lin, P., Chang, W. & Liu, K. 1997. Retaining Function of Open-Type Sabo Dams. *Debris-Flow Hazards Mitigation: Mechanics, Predication and Assessment*, 636–645.

Remondo, J., Bonachea, J. & Cendrero, A. 2005. A statistical approach to landslide risk modelling at basin scale: from landslide susceptibility to quantitative risk assessment. *Landslides*, 2:321–328.

Wong, H.N., Ho, K.K.S. & Chan, Y.C. 1997. Assessment of Consequences of Landslides. In: Cruden, D.M. & Fell, R. (eds.), *Landslide Risk Assessment. Proceedings of the Workshop on Landslide Risk Assessment, Honolulu, Hawaii, USA*, 19–21 February 1997. Rotterdam: Balkema.

Zanchetta, G., Sulpizio, R., Pareschi, M., Leoni, F. & Santacroce, R. 2004. Characteristics of May 5–6, 1998 volcaniclastic debris flows in the Sarno area (Campania, southern Italy): Relationships to structural damage and hazard zonation. *Journal of volcanology and geothermal research* 133:377–393.

Engineering geological study on a large-scale toppling deformation at Xiaowan Hydropower Station

Runqiu Huang, Genlan Yang, Ming Yan & Ming Liu
State Keyl Laboratory of Geohazards Prevention, Chengdu University of Technology, Chengdu, China

ABSTRACT: Toppling failure is one of deformation and failure types of bedded rock mass with high dip angle. It usually develops near the ground surface. Based on the field investigation, we discovered a large-scale toppling deformation occurred at the dam site, which developed to a great depth of about 200 m. In this paper, the deformation phenomena revealed by surface excavation and underground galleries are described in detail including its dip angle changes and rock mass structure variation. In terms of them, a mechanism model of the slope deformation is established and its formation process is analyzed.

1 INTRODUCTION

Toppling is a common kind of deformation and failure for rock slopes. Toppling deformation mainly develops in the slopes of platy rock mass formed by steep-dip-angle cracks and fissures, and the angle between structure plane strike and slope surface one is less than 30 degrees. Its formation mechanism is that due to bending action caused by the deadweight the steep layered rock mass faces are free, much like cantilever bending, and gradually develops in the inner slope. Toppling deformation finally happens (Huang et al, 1996, 2000; Wang et al, 1996; Rui et al, 2001; Wang et al, 2001; Sun, 2002). Toppling deformation usually occurs at the after-edge of the slope, especially margin of the steep cliff. Limited development depth is a characteristic, commonly less than 50 m, and vertical depth is more than 60 m. The type of toppling deformation occurs in the local area of shallow earth surface, and after-edge part is easily destroyed, and local collapse and landslip are caused.

However, in recent years we researched the accumulation on the left bank of Xiaowan Hydropower Station, and found large-scale deep-layer bending toppling deformation failure phenomena. Its maximum horizontal growth depth is near 200 m, and present most vertical developing depth is 160 m (EL.1460 m~EL.1600 m). It is meaningful to understand deformation failure mechanism of high rock slopes under complex conditions that the characteristics of a large-scale toppling failure and its formation conditions are disclosed.

2 GEOLOGICAL CONDITIONS

Xiaowan Hydropower Station is located in the southwestern part of Yunnan Province, the position between Fengqing County and Nanjian County. It lies at the south of famous Sanjiang Structure Belt in southwest region with complex geological structure. Lantsang have been cut into deep vale, and magnificent deep cut gorge landform was formed (Chen, et al., 2004; Xu, et al., 2004; Cheng, et al., 2005). The dam site of Xiaowan Hydropower Station lies at the position that is 3.85 m far from the joining point between Lantsang and its branch, Heihui River. Both sides' valley in the area of the dam is basically symmetrical, and V shape gorge is presented. There are grand hills and steep slope in both sides, and average slope gradient ranges from 35 degrees to 50 ones. The river elevation is 987.92 m in the region of the dam, and the peak elevation is 2168.8 m. The difference between them is more than 1000 m.

The paper studied the top part of Yinshui Gully accumulation before the dam on the left bank of the Xiaowan Hydropower Station, Lantsang (See Figure 1). The elevation difference between deformation's lower boundary and the riverbed is about 500 m. Its north is the gentle-slope flat of Longtai Road, No 2 hill girder's top lie in its south, and there is Malutang Gully in its northwestern direction. The growth depth of the large-scale deformable body varies from 150 m to 200 m, maximum level growth depth across the river direction is up to 180 m, and vertical growth depth is about 160 m.

The bedrock below the deformation body is mainly composed of biotite granite gneiss including lamellar lentiform schist. The trend of the granite gneiss is nearly EW direction with vertical dip angle, and its depth is about 400 m. It distribute along Longtangan Gully and Gouzi Gully.

The most fracture developed in the deformable body is Fault F7, which belongs to second-class structure plane with the altitude of N80~90° E/SE∠80~82°. The width of the crash zone ranges from 3.8 m to 5.0 m, and the effect width varies from 20 m to 30 m. The fault spread out in the center of Yinshui Gully, which is located in the northern part of the toppling deformation failure zone. Due to the debris covered the fault crops out in the excavated slope surface. The small fault in the original rock underlies the deformable body, and extruding face well developed. According to growth scale, the structure planes most belong to IV-class structure planes. They usually developed along the schist, thickness and space length of which greatly varied. The thickness commonly ranges from 20 cm to 50 cm, and the maximum thickness arrives at 2 to 3 m. Their space length is usually from 9 m to 14 m, and the maximum length is up to 18 m.

During excavating the rock slopes the different altitudes between the bedrock and the top rock mass of Yinshui Gully is found, as indicates that there is perhaps a special deformation failure mechanism for the slope.

In order to investigate deformation failure mechanism of the deformable body, we added a series of exploration tunnels in terms of construction of the drainage tunnels. These tunnels, which is seen in Figure 2, discloses the multiform deformation and failure phenomena of the rock mass.

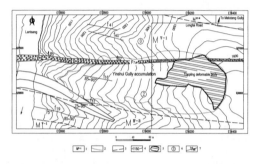

Figure 1. Plan view of Yinshui Gully deformable body with large-scale toppling deformation: 1. rock layer symbol; 2. accumulation boundary; 3. stratum boundary; 4. contour line; 5. toppling deformable body; 6. ridge number; 7. small fault and its number.

3 GEOLOGICAL PHENOMENA OF A LARGE-SCALE TOPPLING DEFORMATION FAILURE

3.1 Deformation failure phenomena revealed by surface excavation

During excavating the slopes of No. 2 ridge and Yinshui Gully, altitude of the slope surface between EL. 1600 and 1460 m is different that of the original rocks, i.e., there is an evident trend to topple northward for the rock layers (See Figure 3).

The rock stratum's dip angle in the deformable body and changing characteristics of the rock mass structure on the surface are analyzed on the basis of

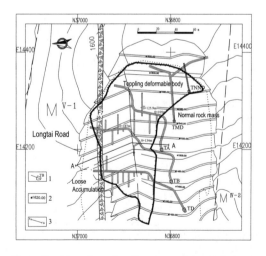

Figure 2. Plan view of surface excavation and drainage tunnel distribution: 1. drainage tunnel and its number; 2. excavated berm line and its elevation; 3. profile line.

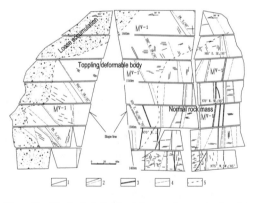

Figure 3. Geological sketch of slope surface excavation for Yinshui Gully and No. 2 Ridge from the altitude 1460 m to the altitude 1600 m above sea level: 1. cracks of original rock; 2. schist interlayer; 3. small fault; 4. extruding plane; 5. boundary of the deformable body.

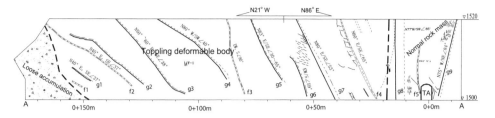

Figure 4. Geological sketch of profile A-A from the altitude 1500 m to the altitude 1520 m above sea level g1~g9. Schist interlayer; f1–f4. small fault, extruding belt.

geological investigation and sketch figures from the most typical excavated profile between elevation 1500 m and 1520 m in the deformable body.

3.1.1 *Change of the altitude*

It can be seen from Figure 4 that the dip angle of the rock stratum gradually decreases along the berm from south to north. Figure 5 shows that the dip angle varied with the distance from south to north from the initial position of TA drainage tunnel (EL. 1500 m). In Figure 5 the rock stratum's dip angles between 0 m to 15 m keep unchanged, and are all more than 80 degrees, which belong to normal rock mass altitude. From 15 m to 55 m the dip angles show great changes, and the dip angle decreases to 50 degrees from 85 degrees. In Figure 5 the curve between 55 m to 150 m is smooth, and the dip angles are about 40 degrees. There is no obvious layer structure to the north of 150-m point, and the layer structure is basically even.

3.1.2 *Rock mass characteristics*

Figure 4 indicates that there are normal rock masses to the south of TA drainage tunnel exit. In Figure 5 g9 to the north of the tunnel exit, which is 4.5 m to 5 m wide, is the southern boundary of toppling failure. There occur normal rock masses between 0 m to 15 m from the initial point of the tunnel exit. Toppling-falling rock is distributed from 15 m to 150 m. There is the accumulation of soil including block stone (toppling failure after-edge) to north of the 150 m point.

The geological information disclosed by the surface excavation indicates that large-scale toppling deformation phenomena mainly occur above the elevation of 1460 m at Yinshui Gully. No marked toppling phenomena are found in the slope surface between El. 1380 m and El. 1460 m.

3.2 *Phenomena disclosed by drainage tunnel and sub-tunnel construction*

Five drainage tunnels were constructed for excavated slope surface of Yinshui Gully and No. 2 Ridge above

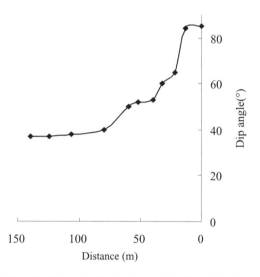

Figure 5. Curve showing the change of the dip angles along the berm at the altitude 1500 m.

the elevation of 1380 m. The serial number is separately TD, TB, TA, TMD and TNND from bottom to top, the elevation of which is separately 1420 m, 1460 m, 1500 m, 1540 m and 1580 m. Toppling deformation mainly developed in the drainage tunnel TB, TA and TMD above the elevation of 1460 m. In terms of field survey, the rock stratum's dip angle, rock mass structure characteristics and crack development of the above drainage tunnels are as follows.

3.2.1 *Dip angle change of the rock strata*

Figure 6 describes the relationship between rock stratum's dip angle of different altitude and tunnel depth. For TD drainage tunnel from elevation 1420 m the dip angles show no obvious change, and don't vary with depth. This indicates that a large-scale toppling failure hasn't occurred below elevation 1420 m or TD drainage tunnel. According to data statistic the dip angles of the rock strata are all less than 60 degrees in TB drainage tunnel (elevation 1460 m). At the position of 0+187 m the dip angles begin to well varies with

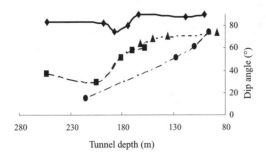

Figure 6. Graph of dip angle varying with the drainage tunnel depths at different levels: 1. TD drainage tunnel (1420 m); 2. TB drainage tunnel (1460 m); 3. TA drainage tunnel (1500 m); 4. TMD drainage tunnel (1540 m).

Table 1. The relationship between the crack aperture and the horizontal depth.

Distance apart from the slope surface (m)	Opening degree (cm)
100	2~3
107	0.5~1
109	1~2
110	5
113	1~1.5
118	2~5
120	80~100

the tunnel depth, and decrease to less than 40 degrees. The boundary between toppling loose rock mass and normal one is determined by the cracks and fissures of the drainage tunnels, which is the position of main drainage 0+134 m. The rock stratum's dip angles in TA drainage tunnel (elevation 1500 m) investigated exhibit approximate linear fall, as shows typical characteristics of toppling failure. In TMD drainage tunnel (elevation 1540 m) there is a trend to decrease for the dip angles, however, the dip angles are more than those of another two tunnels, and are more than 60 degrees.

3.2.2 Crack development characteristics

In the most typical TB drainage tunnel (elevation 1460 m) the cracks' width at the position of main tunnel 0+170 m almost are more than 2 cm (See Table 1). From Table 1 opening degree of the cracks is related with the length apart from the slope surface. As a whole, from 100 m to 125 m far from the slope surface, opening degrees of the cracks increase with depth. At the position 120 m far from the slope surface, the cavity formed by SN-direction unloading cracks is 80 to 100 cm long, and show an inverse wedge shape. There occur rock mass unloading phenomena to the south of the point, and a few SN-direction unloading cracks grow to the north with small opening cracks, width of which commonly ranges from 1 mm to 5 mm. Distributing characteristics of these cracks shows that main tunnel 0+170 m of TB drainage tunnel is the most bending plane of rupture of toppling failure at the elevation of 1460 m.

3.3 Total characteristic analysis of toppling deformation

According to the geological information disclosed by surface investigation and tunnels, large-scale toppling and deformation rock mass researched presents the characteristics as follows.

The altitude of the rock strata: the altitude is dispersed as compared with the original rock. The strike of the original rock strata is commonly NWW direction. However, in the local part of collapse accumulation the strike is a NEE direction, even NE direction. The altitude of the rock strata is SW direction, and the dip angles range from 40 degrees to 60 ones. And there is a trend to gradually decrease from south to north. These show that the local part of the rock mass exhibited deflexion to some degrees during toppling-falling-covering.

In the toppling deformable body the joints mainly include SN-direction steep joints and EW-direction bedding joints. Secondly there occur gentle SN-direction joints and EW-direction joints. The SN-direction joints' dip angles vary from 20 degrees to 35 degrees, and the mean value is 30 degrees. It is found in terms of joint growth contrast between the original rock and the accumulation that there is difference between them. One is that joints' trace length in the accumulation is greater than that of the original rock. The other is that the difference from the joint growth frequency is not greater than that from the original rock. These indicate that there is an even trend for toppling accumulation as compared with the original rock.

In the toppling deformable body the physical properties are different for different parts. As a whole there are greater deflection degrees, smaller blocks, and stronger weathering from south to north. At the after-edge of the falling-covering accumulation the rock blocks usually show full weathering, and the lamination of the accumulation is difficult to identify.

4 GROWTH MODEL AND FORMATION CONDITION ANALYSIS OF THE LARGE-SCALE TOPPLING DEFORMATION

4.1 Growth model analysis of a large-scale toppling deformation

According to geological information of surface and drainage tunnels deformation failure at the top part of

Yinshui Gully, Xiaowan Hydropower Station, can be generalized into a toppling failure model described in Figure 7. In the region of large-scale toppling deformation failure the granite gneiss is cut by the schist interlayer, and thus stratoid rock mass is presented. The rock layer is approximately vertical with near EW direction of the rock layer surface. During releasing the stress with the river valley incision, ductile bending toppling deformation is generated due to the long-term self-weight stress from the approximate layered rock mass. The schist interlayer g9 is the boundary of toppling deformation, which is 4.5 m to 5 m wide. It is difficult to observe the whole failure deformable body due to Occurrence of the gentle slope platform.

In terms of the ground and underground deformation failure we can obtain the deformation failure models described in Figure 7, i.e., toppling failure models. The models have the characteristics as follows

1. Great deformation area. In the horizontal range the deformation's depth ranges from 150 m to 200 m in the slope, and vertical growth depth is about 200 m. Therefore, this is a kind of toppling deformation that is different from common large-scale deformable bodies.
2. Marked ductile deformation characteristics. The rock layer shows big deformation range, and toppling-bending is very strong. The dip angle of the rock strata becomes approximate level from being vertical. This is obviously related with weak schist growth from local parts of the slope, and great decrease of the whole rock strength.
3. Obvious zoning. In terms of strong degrees of deformation, the area can be divided into A, B, C and D four regions from inner to outside, where, A is the vertical rock mass that no deformation happens; B is the toppling loose rock mass; C is the toppling-falling accumulation; D is the colluvial deposit.

4.2 Forming condition of large-scale toppling deformation

Formation of the large-size toppling deformation failure at last has two basic conditions. One is the slope structure, i.e., stratoid structure and the structure growth with gentle dip angle, the other is the favorable landform condition in the failure region, i.e., gently platform developed at the after-edge of the toppling deformable rock mass.

4.2.1 The conditions about the slope structure

Underlying rockbed of the Yinshuigou accumulation has a strike of EW-direction with vertical dip angle. The schist in the rockbed well develops. Its width much varies, and the rock layer is only several cm thin, and even 2 m to 3 m deep. The space between the schist gradually increases from north to south. In the rockbed slope the average space between the schist ranges from 9 m to 14 m, and the most one is up to 18 m. However, in Yinshuigou accumulation, the space is smaller, and the least is 2 m. Growth of the schist interlayer makes the rock mass in the deformation failure zone being layered structure, as provides structural conditions for the large-size toppling deformation failure.

4.2.2 Landform conditions

In the region of the large-size toppling deformable body, underlying rock bed plane is the U-shape trough that developed by F7 fault (See Figure 8). In the trough above the elevation of 1460 m the slope is gentle in the longitudinal direction, and that below the elevation of

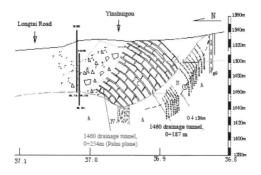

Figure 7. Yinshuigou toppling deformable body structure of Xiaowan Hydropower Station in Lantsang: A—Erect rock mass; B—Toppling loose rock mass; C—Falling accumulation; D—Drift bed; ①—Plane of maximum bending rupture.

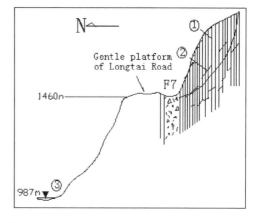

Figure 8. The sketch of toppling failure initial form before toppling failure: 1. EW-direction bedding structure plane; 2. EW-direction low-angle structure plane; 3. Lantsang.

1460 m is steep. The degree of slope from the original rock plane ranges from 25 to 30. Underlying rock plane below gently platform of Longtai Road, the top part of No. 0 Ridge, is about the elevation of 1460 m occurrence of the gentle platform from the rockbed from Yinshuigou to Longtai Road Village provides a deformation space for the large-size toppling failure.

5 CONCLUSION

According to geological information about surface excavation and drainage tunnel construction, knowledge on the deformation failure rock mass at the top Yinshuigou can be obtained as follows: the dip angle of the rock layer plane gradually changes from being steep to being gentle. The rock structure shows a trend to change massive texture into scattered structure from inner to outside. In the deformable rock mass the unloading cracks well develop neat the inner boundary that opening degree of the rock mass is great. The deformable body can be considered as the product of the large-size toppling deformation in terms of dip angle change of the deformable rock mass, rock mass structure and unloading crack growth characteristics. This large-size toppling deformation usually occurs in the high slopes that flexible rock layers well develop or there are many flexible rock layers. And the landform has accumulating conditions of deformable toppling bodies with favorable free faces.

REFERENCES

Huang Runqiu, Wang Shitian, Hu Xiewen et al. 1996. Study on the main Engineering Geological Problems of Xiaowan high arch dam. Chengdu: Southwest Jiao-Tong University Press (in Chinese).

Huang Runqiu. 2000. Time-dependent deformation of a high rock slope and its engineering-geological significance. *Journal of Engineering Geology*, 8 (2): 148–153 (in Chinese).

Wang Jianfeng, Wilson H Tang & Cui Zhengquan. 2001. Stability analysis of Toppling Failure of block rock slopes. *The Chinese Journal of geological hazard and control*, 12 (4): 1–8 (in Chinese).

Rui Yongqin, He Chunning, Wang Huiyong et al. 2001. Analysis of Deformation and Failure Development of the Large-scale Toppling-sliding Slope under Mining. *Journal of Changsha Jiao-tong University*, 17 (4): 8–12 (in Chinese).

Wang Xiaogang, Jia Zhixin, Chen Zuyu et al. 1996. The Research of Stability Analyses of Toppling Failure of jointed rock slops. *Journal of Hydraulic Engineering*, (3): 7–15 (in Chinese).

Sun Dongya, Peng Yijiang & Wang Xingzhen. 2002. Application of DDA method in stability analysis of topple rock slope, *Rock Mechanics and Engineering*, 21 (1): 39–42 (in Chinese).

Chen Hongqi & Huang Runqiu. 2004. Stress and flexibility criteria of bending and breaking in a countertendency layered slope. *Journal of Engineering Geology*, 12 (3): 243–246 (in Chinese).

Xu Peihua, Chen Jianping, Huang Runqiu & Yan Ming. 2004. Deformation mechanism of Jiefanggou high steep dip slope in Jinping Hydropower Station. *Journal of Engineering Geology*, 12 (3): 247–252 (in Chinese).

Cheng Dongxing, Liu Da' an, et al. 2005. Three-dimension numerical simulation of deformation characteristics of toppling rock slope. *Journal of Engineering Geology*, 13 (2): 222–226 (in Chinese).

Characterization of the Avignonet landslide (French Alps) with seismic techniques

D. Jongmans, F. Renalier, U. Kniess, S. Schwartz, E. Pathier & Y. Orengo
LGIT, Université Joseph Fourier, Grenoble Cedex 9, France

G. Bièvre
LGIT, Université Joseph Fourier, Grenoble Cedex 9, France
CETE de Lyon, Laboratoire Régional d'Autun, Autun cedex, France

T. Villemin
LGCA, Université de Savoie, France

C. Delacourt
Domaines Océaniques, UMR6538, IUEM, Université de Bretagne Occidentale, France

ABSTRACT: The large Avignonet landslide (40×10^6 m^3) is located in the Trieves area (French Alps) which is covered by a thick layer of glacio-lacustrine clay. The slide is moving slowly at a rate varying from 1 cm/year near the upper scarp to over 13 cm/year at the toe. A preliminary geophysical campaign was performed in order to test the sensitivity of geophysical parameters to the gravitational deformation. In the saturated clays where the landslide occurs, the electrical resistivity and P-wave velocity are little affected by the slide. On the contrary, S-wave velocity (Vs) values in the first ten meters were found to be inversely correlated with the measured displacement rates along the slope. These results highlight the interest of measuring Vs values in the field for characterising slides in saturated clays and of developing techniques allowing the 2D and 3D imaging of slides.

1 INTRODUCTION

Slope movements in clay formations are world widespread and usually result from complex deformation processes, including internal strains in the landslide body and slipping along rupture surfaces (Picarelli et al. 2004). Such mass movements are likely to generate changes in the geophysical parameters characterizing the ground, which can be used to map the landslide body. Since the pioneering work of Bogoslovsky & Ogilvy (1977), geophysical techniques have been increasingly but still relatively little used (or referenced) for landslide investigation purposes (McCann & Forster 1990, Jongmans & Garambois 2007).

The recent emergence of 2D and 3D geophysical imaging techniques, easy to deploy on slopes and investigating a large volume in a non-invasive way, has made more attractive the geophysical methods for landslide applications. One of the key factors controlling the success of geophysical techniques is the existence of a contrast differentiating the landslide body to be mapped. In the past (Caris & van Asch 1991, Schmutz et al. 2000, Lapenna et al. 2005, Grandjean et al. 2006, Méric et al. 2007) seismic and electrical methods were successfully used in clay deposits for distinguishing the mass in motion from the unaffected ground.

The aim of this study is to test the sensitivity of the main geophysical parameters (with a focus on the shear wave velocity) to the clay deformation generated by the landslide of Avignonet where geotechnical and geodetic data are available.

2 THE AVIGNONET LANDSLIDE

The large Avignonet slide (40.10^6 m^3) located in the Trieves region (French Alps, Figure 1) was studied. This 300 km^2 area is covered by a thick Quaternary clay layer (up to 200 m) deposited in a glacially dammed lake during the Würm period (Giraud et al. 1991). These clayey deposits overlay compact old alluvial layers and marly limestone of Mesozoic age, and are covered by thin till deposits. After the glacier melting, rivers have cut deeply into the geological

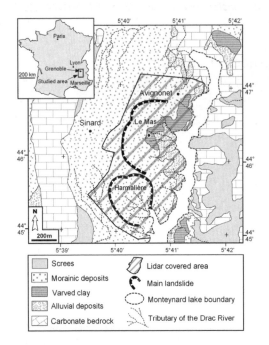

Figure 1. Geological map with the location of the Avignonet and Harmalière landslides, and of the area investigated by Lidar.

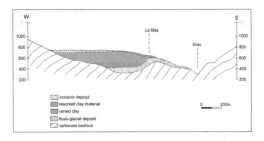

Figure 2. EW geological cross-section over the hamlet of Le Mas (Figure 1.).

formations, triggering numerous landslides (Giraud et al. 1991). Figure 1 shows the simplified geological map of the studied area, in the northern part of the Trièves region, as well as the two main landslides (Avignonet, Harmalière) occurring in the clay deposits. An EW synthetic geological section over the Avignonet landslide is presented in Figure 2, showing the thickness variation of the clay layer, from 0 m to more than 200 m. The translational Avignonet slide, whose first signs of instability were noticed between 1976 and 1981 (Lorier & Desvarreux 2004), affects a surface of about 1×10^6 m^2, with a global eastward motion to the Drac valley which is dammed downstream. South of this slow moving slide, a quick mudslide (Harmalière, Figure 1) occurred in March 1981, creating a head scarp of 30 m and affecting a surface of about 450,000 m^2 (Moulin & Robert 2004) in the same material. Between 1981 and 2004, the head scarp has continuously regressed with an average of 10 m/year in a north-eastward direction. It now intersects the limit of the Avignonet landslide. The source zone displays complex deformation patterns, including rotational slips, cracks, slumps, and translational failures. In the track and at the toe, the slide evolves into a flow during heavy rainfalls, contributing to the depletion of the landslide mass and the southward erosion process.

3 GEOTECHNICAL AND GEODETIC INVESTIGATION

The Avignonet landslide which affects the hamlet of Le Mas (Figures 1 and 2) was investigated by four boreholes (T0 to T3), equipped with inclinometers (see Figure 4 for location and Table 1 for the main results). The contact with the alluvial deposits was found at 14.5 m, 44.5 m and 56 m in T2, T3 and T1 respectively, whereas T0 was still in the clay deposits at 89 m, in agreement with the westwards thickening of this formation (Figure 2). Inclinometer data revealed several rupture surfaces, at a few m depth (T0 and T2), between 10 m and 17 m (T0, T1, T2, T3) and up to 42 to 47 m (T1 and T0) (Table 1, Lorier & Desvarreux 2004). Of particular importance is the presence of a major active slip surface found at 13 m depth in borehole T2 which is located in the more active area.

Piezometric measurements showed the presence of a very shallow water table (1 to 3 m below the ground level). No geotechnical investigation was performed

Table 1. Borehole and inclinometer results.

	Geological formations	Depth of rupture surface
T0	0–5 m: morainic deposits 5–89 m: varved clays	5 m 10 m 47 m
T1	0–5 m: morainic deposits 5–56 m: varved clays 56–59 m: alluvial deposits	15 m 34 m 42.5 m
T2	0–4 m: morainic deposits 4–14.5 m: varved clays 14.5–17 m : alluvial deposits	1.5 m 4 m 12 m
T3	0–4 m: morainic deposits 4–44.5 m: varved clays 44.5–59 m: alluvial deposits	16 to 17 m

on the Harmalière landslide which does not threaten any property in the short term.

3.1 Lidar acquisition

A Lidar (Light Detection and Ranging) acquisition, covering the zone displayed in Figure 1, was performed in November 2006 with the handheld airborne mapping system Helimap system® (Vallet & Skaloud 2004). As regards the Lidar, the measurement unit is composed of 3 sensors: a GPS receiver, providing the position of the unit, an inertial measurement unit which provides the orientation of the system, and a laser scanner unit measuring a point cloud of the surface. The height of flight was of 500 m above the ground and allowed to acquire a density of one point by square meter in average. The system displays a high accuracy of $\sim$10 cm both in horizontal and vertical. The interpolated Digital Elevation Model (DEM) is shown in Figure 3 at a resolution of 2 m. The DEM enlightens several landslide indicators. It clearly displays the crescent-shaped front scarp of the Avignonet landslide which intersects the Harmalière one to the South and another minor one to the North.

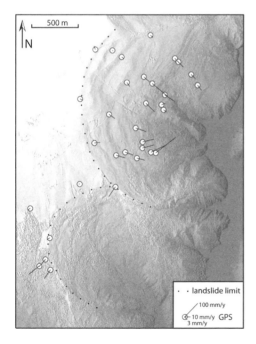

Figure 3. Shaded Lidar-derived DEM of the Avignonet and Harmalière areas acquired from Helicopter. White circles are campaign GPS stations. Black lines indicate mean velocity values measured by GPS from 1995 to present. Dashed lines show the limits of the two landslide.

Inside the Avignonet slide, the DEM shows the presence of multiple parallel scarps down the slope with a spacing of about 100 m. Scarp heights are higher within the slide, suggesting that this latter regresses toward the plateau at the west and that the motion could be greater at the toe than at the crown. In the lower part of the hill, the slope increases due to the presence of consolidated alluvial layers over which the clay material flows (Fig. 2). The geometry of the Harmalière landside is more elongated than the Avignonet one, with a the presence of multiple curved scarps in the source area and a funnel shaped track zone through which the material flows to the lake with a regular slope. To the south-west, the Harmalière landslide also intersects another landslide. This difference in the morphology and in the mechanical behaviour between the two landslides, developing in the same material, probably results from the disappearance of the crest made of bedrock and alluvial layers to the south, removing the buttress that prevents the deep sliding of the Avignonet landslide.

3.2 GPS measurements

The Avignonet slide has been monitored by biannual GPS measurements at 26 geodetic points since 1995, while only 6 points were installed around the Harmalière landslide, due to the strong deformation inside the mass in motion. The locations of the GPS points are shown on a DEM (Figure 3).

The velocity values at the surface of the Avignonet landslide, averaged from the GPS measurements available (Figure 3) increase downhill, varying from 0 to 2 cm/y at the top to more than 13 cm/y in the most active part of the toe. Most of the area is sliding southeastward, parallel to the general slope. In detail, the deformation pattern is complex and velocity and direction of the ground movement are influenced by local geological and morphological features. The concave shape of the river bank below the landslide clearly controls the slope orientation and consequently the slide direction in the lower part of the hill where the displacement vectors rotate. This morphology seems to be linked to the presence of old and consolidated alluvial deposits overlying the bedrock, around which the clay slides (compare Figures 1 and 3). At lower altitudes, the higher slide velocities measured with GPS are accommodated by scarps and bulges spaced by less than 20 m. Strong velocity contrasts are observed along the landslide toe, which seem to be linked to slope angles which are higher in the southern part. The GPS measurement on the ridge between the Harmalière and Avignonet landslides shows that the Harmalière lateral head scarp still actively moves backwards, involving material belonging to the Avignonet landslide. On the contrary, GPS measurements northwest of the Harmalière landslide exhibit little displacement. Finally, the three GPS points located along the south-western limit shows the presence of another active slide, south of the Harmalière one.

4 GEOPHYSICAL PROSPECTING

A preliminary geophysical campaign was performed in 2006–2007 in order to test the sensitivity of three geophysical parameters (the electrical resistivity ρ, the P-wave seismic velocity Vp and the S-wave seismic velocity Vs) to the deformation resulting from the slide. It turned out that, in such saturated clays, ρ and Vp are strongly influenced by the water level and are little affected by the landslide activity. On the contrary, Vs showed significant variations both vertically and laterally. For this reason, we have focused our study on the Vs measurements.

Shear wave velocity Vs can be measured by a relatively large number of methods including active source techniques (borehole tests, SH-wave refraction tests, surface wave inversion) and ambient vibration techniques (Jongmans 1992, Socco & Jongmans 2004). In the present study we apply the SH seismic refraction tomography method and the surface wave (SW) inversion for deriving Vs values. Seismic refraction tomography consists in inverting the first arrival times picked on all the signals recorded at all geophones for different shots spread along the profile. In the SH case, transverse horizontal ground motions are generated using a sledge hammer hitting laterally a loaded plank as a source. The picked first arrival times are inverted using the Simultaneous Iterative Reconstruction Technique (SIRT, Dines & Lyttle 1979) and provide 2D Vs images.

In the SW method, Rayleigh waves are generated by a vertical shock and are recorded by vertical geophones, together with P waves. SW dispersion curves are computed using the f-k method, which assumes plane wave propagation and no lateral seismic velocity variations (Socco & Strobbia 2004). Dispersion curves are then inverted to get 1D Vs profiles, using the Geopsy software (http://www.geopsy.org).

For this study, we performed five 115 m long seismic profiles (P1 to P5) and two 470 m long profiles (P6 and P7). The profile location is given in Figure 4. For the short profiles, we used 24 vertical geophones (4.5 Hz) and 24 horizontal geophones (14 Hz) spaced by 5 m, for recording Rayleigh waves and SH waves, respectively. Shots were made every 15 m, with two offsets for SW recordings. Figure 5 shows the seismograms generated along profile P1 by a vertical source and a horizontal SH source. In Figure 4a, one can distinguish the P waves from the Rayleigh waves which are inverted for retrieving

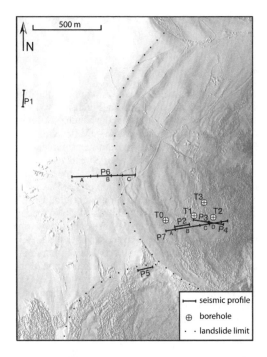

Figure 4. Shaded Lidar-derived DEM with location of the seismic profiles (P1 to P7) and of the boreholes (T0 to T3).

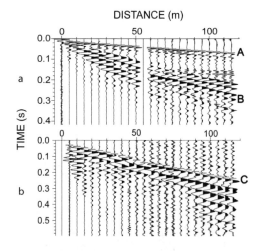

Figure 5. Seismograms along profile P1 for a shot at 0 m. a) Vertical motions generated by a sledge hammer. A: P-wave first arrivals, B: Rayleigh waves. b) Horizontal motions for a SH source. C: SH waves.

the Vs structure. The two long profiles (P6 and P7) were conducted with 48 vertical 4.5 Hz geophones, using explosive sources 80 m and 50 m apart, respectively.

4.1 SH tomography

As both SW and refraction travel-time data inversions are non unique problems, we performed a joint analysis of refraction data with surface waves, looking for a common solution in the 1D part of the profiles, with local validation by borehole data (Renalier et al. 2007). Figure 6 shows the Vs images obtained by the SH refraction tomography method for profiles P1, P3 and P5, which are respectively located outside the landslides, on the Avignonet landslide and on the Harmalière landside (Fig. 4). Below profile P1, located on the Sinard plateau, Vs quickly increases to 550 m/s at 5 m depth. Profile P3 exhibits a low-velocity layer (Vs around 250 m/s) with a thickness of 14 m to 19 m from E to W, overlying a more compact layer (Vs > 600 m/s). The depth of this velocity contrast coincides with the slip surface at 13 m found in borehole T2 (Fig. 4). The low Vs values above the slip surface are probably linked to internal strains in the mass resulting from the slide. On the Harmalière landslide (P5), the Vs parameter delineates three distinct layers: a very slow (Vs between 80 and 200 m/s) 5 m thick layer, overlying a 15 m thick layer around 250 m/s, over a more compact layer (Vs > 500 m/s). These results suggest the presence of a slip surface at 20 m deep. SH refraction tomography thus enlightens the evolution of the Vs shallow structure depending on the state of the ground—from undisturbed at P1 to highly disturbed at P5.

4.2 Surface wave interpretation

Dispersion curves (Rayleigh fundamental mode) for profiles P1, P5, P6 and P7 are plotted in Figure 7. For

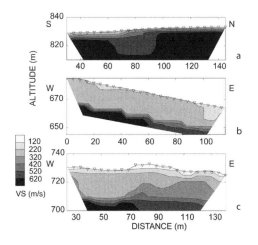

Figure 6. Vs seismic images (SH refraction tests) along three seismic profiles. a) P1, b) P3 c) P5. RMS values are below 3%.

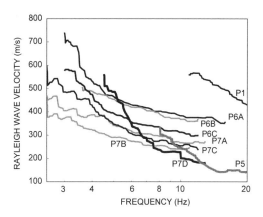

Figure 7. Comparison of the phase velocity dispersion curves calculated along profiles P1, P5, P6 and P7.

the 115 m long profiles P1 and P5, all geophones were considered together. Dispersion curves were computed for both direct and reverse offset shots in order to check the 1D hypothesis. For the 470 m long profiles P6 and P7 seismogram examination and P-wave tomography images were used to define groups of geophones which roughly fit the requirement of 1D media: 3 groups along P6 (from West to East: A: 1–15, B: 19–30, C: 38–48) and 4 groups along P7 (from West to East: A: 1–9, B: 9–28, C: 29–35, D: 35–44). For these groups, dispersion curves were also computed for the two offset shots. All the dispersion curves are gathered on Figure 7.

The two different types of profiles can be easily recognised by their frequency range: dispersion curves for P6 and P7 (explosive sources) exhibit lower frequencies than the curves for P1 and P5 (hammer source). Despite this difference, the Rayleigh wave velocity at high frequency (over 8 Hz) exhibits a significant decrease according to the position of the profile on the two landslides, with the exception of P7C. Rayleigh wave velocities at 14 Hz are divided by 3, from 500 m/s out of the landslides (P1) to 150 m/s on the Harmalière landslide (P5). These results agree with the Vs images of Figure 6. At lower frequencies, the curves do not display such decrease because they are also influenced by the underlying higher velocity alluvial layers and bedrock, which come up from more than 200 m deep on the western part of the studied area, to the surface at the eastern end of the landslides (see Fig. 2).

In a second step, these dispersion curves were inverted using the Neighbourhood Algorithm (Wathelet et al. 2004), giving a two layer Vs model estimate for each group of geophones.

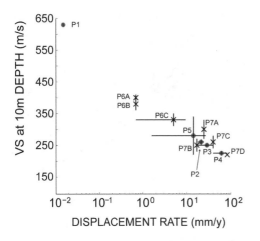

Figure 8. Evolution of shear wave velocity values at 10 m depth with slide velocities. Crosses: surface wave data. Dots: SH refraction data. Vertical error bars on the Vs values are indicated. Horizontal bars indicate the uncertainty range on the displacement rate when no data are available close to the seismic profiles. An arbitrary low displacement rate value of $1\cdot 10^{-2}$ mm/y has been assumed for profile P1 (Vs = 630 m/s).

4.3 Vs value interpretation

All Vs values obtained by the two methods for each profile or group of geophones are correlated with the displacement rates measured at the GPS points (Fig. 3). For each SH refraction image, two Vs values were extracted at the third and the two thirds of the profile. When no GPS point is close to the seismic profile, the displacement rate is averaged using the two closer data. Figure 7 shows the Vs value with error bars at 10 m depth versus the ground displacement rate in a semi-logarithmic scale. The striking feature is the regular decay of Vs values with the displacement rate, from 630 m/s far from the slide, to 225 m/s at the slide toe, where the slope surface is strongly deformed. These results show that the gravitational deformation strongly affects the shear wave velocity within the clay, which could be used as a parameter for mapping the slide activity.

5 CONCLUSIONS

On the Avignonet landslide, Vs values at shallow depth (10 m) were found to be inversely correlated with displacement rates measured by GPS, with a division by a factor of almost 3 between the zones unaffected and the ones strongly deformed by the landslide. This strong decrease of Vs values is probably linked to internal strains in the mass above the main rupture surface.

Such variations were not observed on P-wave velocity and resistivity values. These results highlight the interest of in-situ measuring Vs values for characterising slides in such saturated clays and of developing techniques allowing the 2D and 3D imaging of landslides. The relationship between Vs values, deformation and pore pressure should be investigated through laboratory tests in order to allow a quantitative interpretation of the field results. Combining Vs imaging with multitemporal remote sensing (satellite and aerial) giving a continuous image of the displacement rates at different times would also allow a deeper insight into the 3D deformation processes and pattern of the landslide. Aerial image archive and new Lidar acquisition planned in autumn 2008 on the Avignonet and Harmalière landslides will supply a global view on the past and present day slide velocity.

ACKNOWLEDGMENTS

This work was financially supported by regional, national and European funds coming from the Conseil Général de l'Isère, the region Rhône-Alpes, the Cluster VOR (Vulnérabilité des Ouvrages aux Risques) and the European project NERIES. The authors thank all the people who participated to the field investigation, as well as RTM (Restauration des Terrains en Montagne) and SAGE (Société Alpine de Géotechnique) for providing the geotechnical data.

REFERENCES

Bogoslovsky, V. & Ogilvy, A. 1977. Geophysycal methods for the inverstigation of landslides. *Geophysics* 42: 562–571.

Caris, J.P.T. & van Asch, T.W.J. 1991. Geophysical, geotechnical and hydrological investigations of a small landslide in the French Alps. *Engineering Geology* 31 (3–4): 249–276.

Dines, K. & Lyttle, J. 1979. Computerized geophysical tomography. *Proceedings of the Institute of Electrical and Electronics Engineers* 67: 1065–1073.

Giraud, A., Antoine, P., van Asch, T.W.J. & Nieyuwenhuis, J.D., 1991, Geotechnical problems caused by glaciolacustrine clays in the French Alps: Engineering Geology 31, 185–195.

Grandjean, G., Pennetier, C., Bitri, A., Méric, O. & Malet, J.P. 2006. Caractérisation de la structure interne et de l'état hydrique de glissements argilo-marneux par tomographie géophysique: l'exemple du glissement-coulée de Super-Sauze. *Comptes Rendus Geosciences* 338 (9): 587–595.

Jongmans, D. 1992. The application of seismic methods for dynamic characterization of soils in earthquake engineering. *Bulletin of the International Association of Engineering Geology* 46 (1): 63–69.

Jongmans, D. & Garambois, S. 2007. Geophysical investigation of landslides: A review. *Bulletin Société Géologique de France* 178 (2): 101–112.

Lapenna, V., Lorenzo, P., Perrone, A., Piscitelli, S., Rizzo, E. & Sdao F. 2005. 2D electrical resistivity imaging of some complex landslides in Lucanian Apennine chain, southern Italy. *Geophysics* 70: B11–B18.

Lorier, L. & Desvarreux, P. 2004. Glissement du Mas d'Avignonet, commune d'Avignonet. *Proceedings of the workshop Ryskhydrogeo, Program Interreg III, La Mure (France)*.

McCann, D.M. & Forster, A. 1990. Reconnaissance geophysical methods in landslide investigations. E*ngineering Geology* 29 (1): 59–78.

Méric, O., Garambois, S., Malet, J-P, Cadet, H., Guéguen P. & Jongmans, D. 2007. Seismic noise-based methods for soft-rock landslide characterization. *Bulletin Société Géologique de France* 178 (2): 137–148.

Moulin, C. & Robert, Y. 2004. Le glissement de l'Harmalière sur la commune de Sinard. *Proceedings of the workshop Ryskhydrogeo, Program Interreg III, La Mure (France)*.

Park, C.B., Miller, R.D. & Xia, J. 1999. Multi-channel analysis of surface waves. *Geophysics* 64 (3): 800–808.

Picarelli, L., Urciuoli, G. & Russo, C. 2004. The role of groundwater regime on behaviour of clayey slopes. *Canadian Geotechnical Journal* 41: 467–484.

Renalier, F., Jongmans, D., Bièvre, G., 'Schwartz, S. & Orengo, Y. 2007. Characterisation of a landslide in clay deposits using Vs measurements. 13th *European Meeting of Environmental and Enginering Geophysics, Istanbul, 3–4 September 2007.*

Schmutz, M., Albouy, Y., Guerin, R., Maquaire, O., Vassal, J., Schott, J.J. & Descloitres, M. 2000. Joint inversion applied to the Super Sauze earthflow (France). *Surveys in Geophysics* 21: 371–390.

Socco, L.V. & Jongmans, D. 2004. Special issue on Seismic Surface Waves. *Near Surface Geophysics* 2: 163–258.

Socco, L.V. & Strobbia, C. 2004. Surface-wave method for near-surface characterization: a tutorial. *Near Surface Geophysics* 2: 165–185.

Vallet, J. & Skaloud J., 2004. Development and Experiences with A Fully-Digital Handheld Mapping System Operated From A Helicopter, The International Archives of the Photogrammetry, Remote Sensing and Spatial Information Sciences, Istanbul, Vol. XXXV, Part B, Commission 5.

Wathelet, M., Jongmans, D. & Ohrnberger, M. 2004. Surface wave inversion using a direct search algorithm and its application to ambient vibration measurements. *Near surface geophysics* 2: 211–221.

Deformation characteristics and treatment measures of spillway slope at a reservoir in China

Nengpan Ju, Jianjun Zhao & Runqiu Huang
State Laboratory of Geo-hazard Prevention and Geo-Environment Protection, Chengdu University of Technology, Chengdu, China

ABSTRACT: This paper presented the failure model of some oblique slope consisting of silt mudstone, argillaceous sandstone and carbon shale interlayers, by virtue of analysis of the slope deformation character. According to engineering geological conditions and analysis of cracks, rock mass relaxation and bedding displacement in the slope, a conclusion was drawn that the deformation of the slope was mainly controlled by weak interlayers and steep unloading cracks dipping out of slope. Owing to tensile stress and shear stress concentration at the middle and upper of the slope along the bedding displacement belt, and the compressive deformation of soft layer at the bottom of slope, rock blocks deform out of slope or along dip direction of bedding planes. And the deformation develops upwards along rock beds and unloading cracks, which results in step creeping and cracking deformation. Because it is cataclastic rock mass intensely weathered and strongly off-loaded with many weak interlayers downstream slope, the potential landslide may slide along the weak layers. So the control of the slope deformation should be concentrated on reduce-loading, together with pre-stressed cable framework to control the deformation of weak layers.

1 INTRODUCTION

Oblique slope means that strike of strata obliquely intersects with strike of slope surface. Such slope is often high, steep and steady, and its deformation and failure is mainly local falling (Yin Yueping 2005). Under rapid basal sapping of river or excavation, stress will change inside high and steep slope. Owing to release stress, rock mass has to occur unloading rebound deformation toward the free face of slope, and stress field of valley will be adjusted. With this process, the shallow slope within a certain range can form steep and inclined unloading cracks in the same direction as dip of slope. When slope consists of soft and hard rocks interlayers, along the contact planes between soft rock and hard rock strata, shear dislocation will happen due to released stress, with successional tension cracks (Huang Runqiu et al. 2001, Ding Xiuli et al. 2005). Such weak structure planes and unloading cracks often become controlling structure planes which can control stability of high steep slope. When slope accrues deformation, such cracks having the same strike as controlling structure planes commonly are formed on the surface of slope. So study of relation between slope deformation and structure planes will be helpful to know deformation and failure mode of slope, which is very important to ensure reasonable and effective slope treatment (Yang Yonghong & Lv Dawei 2006, Sun Hongyue & Shang Yuequan 1999, Fan Wen et al. 2000, Huang Zhengjia & Wu Aiqing 2001). Taking slope of spillway of a hydropower station as an example, under detailed analysis of geological conditions and phenomena of deformation, this paper studied function of weak structure planes and unloading cracks, which plays a controlling role in stability of slope. On the basis of analysis of deformation and failure mode of slope, targeted treatment measures is proposed.

2 ENGINEERING GEOLOGICAL CONDITIONS OF SLOPE

The study area lies in the southeastern part of Sichuan basin, steep terrain with asymmetrical "V" valley. Slope is on the right bank of Taoxi River and the natural angle of slope is 35° to 55°. There are two gullies in excavated slope, which are located at 0+000 and 0-045 respectively. Excavated slope is discount linear and dip direction is 39° to 80°; Along dip of slope, the slope is divided into A, B, C, of which dip is 74°, 39°, 40° to 80° respectively. Height of slope is 88 m, and 6 steps are designed in the process of excavation. The first level slope is sidewall slope of lock chamber which is a vertical slope. Gradient of the second and

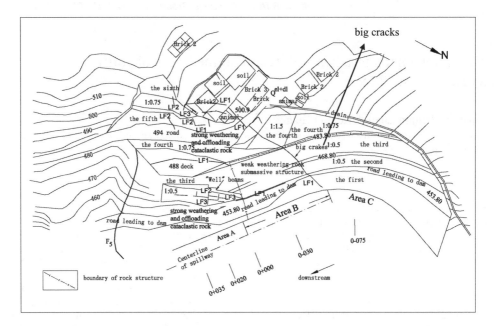

Figure 1. Plane sketch map of distribution of cracks.

the third level slopes is 1:0.5, and gradient of the rest slopes above the fourth level is 1:0.75, see Figure 1.

The inclined slope consists of soft and hard interlayers. Exposed bedrock is the middle Jurassic Qianfuyan Formation silt mudstone, pelitic siltstone, carbonaceous shale, muddy siltstone, and carbonaceous shale. They has a characteristic of inter-layered distribution, occurrence of strata N 60° to 80° E/NW ∠ 20° to 40°.

The study area lies in northwest wall of Wenquanjing anticline and southeast wall of Pingloushan syncline, with simple geological structure. F5 reversed fault is found in a gully, on downstream side of Area A, and its occurrence is N30°~70° E/NW 37°~45°, and it tends to taper. Shale interlayers mostly develop into bedding displacement zone, generally about 10 cm thick. The total amount is more than 10; these weak interlayers often control stability of slope.

Weak interlayers exist in the inclined slope which consists of soft and hard interlayers. Aclinic depth of unloading rock mass is about 22 m; aclinic depth of strong weathering is about 10~25 m, and aclinic depth of weak weathered is 33 m. The original terrain is small ridges at 0+00~0+35 in Area A. Aclinic depth of excavation is less than 15 m, so rock mass of excavation plane is in the strong weathering and offloading zone. Coupled with the impact of F5 faults, rock mass was broken into cataclastic structure, and most of structure planes are open and filled with secondary mud. The fourth grade slope of Area B, C is also in strong weathering and unloading zone. Slope from the first to the third grade in Area B, C is in weak

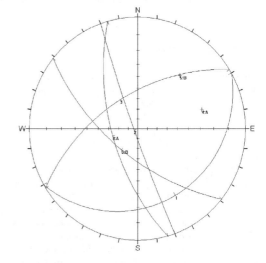

Figure 2. Stereographic polar projection of dominant structural planes and slope surface.

weathering, and rock mass structure is the sub-massive structure, where joint planes are partial opening without filling. Groundwater of slope develops, which is locally found in this slope, and it seeps along bedrock cracks.

Main three groups of dominant structure planes developed in this slope: (1) contact planes along the carbonaceous shale, mudstone and sandstone can develop into weak interlayers, 0.1 m to 1 m thick, occurrence N60° E/NW ∠ 21°, which can form potential sliding surface of block; (2) Unloading cracks along river, occurrence N20° W/NE ∠ 86°, can form cutting plane of rear edge of block; (3) occurrence of joints is N58° E/SE ∠ 56°, which is nearly vertical to strike of slope and dips downstream. They can form lateral edge of block. From Figure 2, the above three groups of structure planes made up block in Area B, and it has conditions of deformation and failure. It may generate deformation along weak interlayers which is dip out of slope.

3 DEFORMATION CHARACTERISTICS AND MECHANISM ANALYSIS OF SLOPE

3.1 Deformation and failure characteristics of slope

Figure 1 shows that deformation of the slope occurs mainly in Area A and B, and there are no obvious signs of deformation and failure in Area C. Successional and big cracks form the upstream border of the block in Area B. Gully forms the downstream border of it in Area. Cracks of house at the back of the fifth grade slope in Area I form the rear boundary edge, and the exposed carbonaceous shale layer at the sidewall of lock chamber forms the bottom boundary. Confined block generates deformation. Deformation of the slope mainly includes the following three aspects:

3.1.1 A large number of cracks in the slope

According to strike of cracks, they can be divided into three groups. LF1 nearly is parallel to excavation plane, occurrence N30° to 50° W. It is a big cracks which develops along the river, and its opening width 2 cm to 10 cm. Unloading cracks control formation of the group of cracks which are N20° W/NE ∠ 86°; LF2 cuts excavation plane with a large angle, occurrence N50° to 70°E, while length of extension is not long, and it has close relation with joints which is N58° E/SE ∠ to 56°; LF3 also cuts excavation plane with a large angle. As a whole it inclines upstream. The group of cracks has a large quantity and short extension. When block creeps along the layer, these cracks are formed.

Cracks intensively distribute in two zones: below 488 platform in Area A and above Altitude 494. Cracks of Area A are mainly big LF1 at 488 platform and step cracks which are formed by LF2 and LF3 below this platform. Its opening width is 5 to 10 cm. There are 3 big cracks LF1 above Altitude 494. Their extending length is 15~30 m. They cross houses at the top of the slope, and their opening width is 5 to 10 cm. Extending length of LF2 and LF3 is smaller, about 0.5~1 m. The opening width is less than 2 cm. These cracks

Figure 3. Relaxation of rock mass on the slope surface.

mainly distribute in the excavation plane and gullies, the amount more than 10. Moreover, two main cracks exist in Area B. Extending length of them is more than 30 m, and their opening width is 10 to 30 cm.

3.1.2 Relaxation of rock mass in the slope

Relaxation of rock mass shows that rock mass becomes loose, so these original filled structure planes can further open, and local rock mass generates deformation (Figure 3). Relaxation of rock mass can generally attribute to two reasons: one type of relaxation is caused by rebounding and unloading function (Huang Runqiu 2000). The relaxation shows that it is strong in the shallow part of slope, and gets gradually weak inwards slope so as to disappear. The other relaxation is caused by deformation of slope. Potential slip surface becomes relaxation boundary. Inside it, relaxation phenomenon is not existence and rock mass is integrity; Outside it, repeated unloading function makes relaxation phenomenon of mantle rock mass become more strong, and the most strong relaxation phenomenon of rock mass is close to potential shear edge, and then local position of slope causes bulging deformation, perhaps resulting in collapse. The slope below 488 platform in Area A has a smaller excavation depth. It is in a strong weathering and unloading zone. Open structure planes are filled with plenty of mud and debris. Hole can be found in local position. Instability of local block can be found near the road which leads to dam. This explains that relaxation of rock mass below 488 platform in Area A in the slope is mainly caused by repeated unloading function.

3.1.3 Interbedded shearing deformation along the carbonaceous shale layer

From Figures 4 and 5, when the first grade slope was excavated in Area B, weak weathered sandstone containing thin silt mudstone was exposed, thickness of which ranges from 30 to 50 cm; Furthermore, there is black carbonaceous shale zone, thickness of which is about 1 m. Deformation is always along plane and carbonaceous shale seam, which moves towards outside of the slope and upstream. It shows that displacement deformation of rock mass developed towards outside of the slope. Displacement distance upstream is 3~6 cm (dip direction), and displacement distance towards outside of the slope is 5 cm. Meanwhile, fresh tension cracks are found at the back of the second grade berm. Its extension length is about 30 m, and its width is 10~20 cm. LF2 and LF3 constitute step plane, which separates first grade slope from the upper rock mass. It shows that it is possible for the block to get instable along potential sliding surface, which is formed by carbonaceous shale layer and cracks.

Clearly, deformation of the slope can be divided into three levels: (a) width of cracks above Altitude 494 is smaller, and extending length is also short. Detritus soil is exposed near houses. Cracks running along slope have long length, which damage foundation of houses, so that wall cracks; (2) below 488 platform in Area A, the slope has strong unloading and weathering rock mass. Deformation shows that a larger longitudinal crake is formed at the 488 platform, and relaxation of rock mass is in front of the slope. Local instability can be found; (3) Excavation depth of the first grade slope in Area B is about 60 m rock mass is weak weathering, and it forms creep sliding deformation along carbonaceous shale intercalation. Under certain conditions the three levels of deformation above can cause overall instability of slope.

Figure 4. Slipping along carbonaceous shale.

3.2 Analysis of deformation and failure mode of slope

A number of weak interlayers and steep dip outwards off-load cracks develop in the slope. The combination of the two structure planes can form stepladder-like sliding plane, as shown in Figure 6. It is extensive that the stepladder-like creep and crack deformation distributes in hard and thick strata, and the development of this deformation is often from up to down. It creeps along structure planes dipping out of slope with middle-low angles to the dip direction of the slope (Huang Runqiu 2004, Yang Xuetang et al. 2004). The deformation is gradually transferred by steep cracks. The stepladder-like creep and crack develops from down to up, namely: exposed mudstone and carbonaceous shale in the first grade of the slope form unloading rebound and compressive deformation, and block consists of 3 dominant structure planes can form creep deformation along weak interlayers dipping out of slope and strata dip direction. In addition fracture happens along off-load cracks upwards.

From Figure 6, it is clear that deformation of the slope controlled by weak interlayers is the deformation of block comprised by dominant structure planes. If supporting intensity in not enough, it can develop upwards along stepladder-like sliding plane comprised by weak layers and steep dip offload cracks. Under certain conditions, Area A and B may form stepladder-like creep and crack, resulting in massive landslide.

In Area C slope was curved. Combination of structure planes can not form potential instable block controlling stability of slope, but higher step and steeper slope may form local falling. So overall stability of slope in Area C is better, but protection of slope surface has to be done.

4 STUDY ON TREATMENT

Deformation of the slope is mainly caused by surficial excavation in Area A. In certain range of depth, the excavated surface is filled with cataclastic rock in which joints are extremely developed, and inner weak interlayers of the slope are also developed. Combination of structure planes is harmful for the stability of the slope. It can easily make deformation of block, resulting in a large-scale instability of slope. Therefore the emphasis of slope treatment should be put on following aspects:

1. A majority of rock mass in Area A is located in the strong weathering and off-load zone, and excavated depth is not enough to form relative protuberant terrain between 0+00 and 0+35. 488 platform is 20 m wide, and the slope has a very good condition to be cut below it. The slope above road

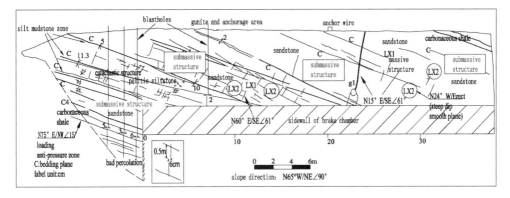

Figure 5. Geological sketch map of the first slope in Zone B.

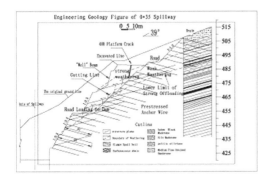

Figure 6. Distribution sketch map of deformation and controlling measurements of 0+35 profile.

leading to dam in Area A, is cut by ratio of 1:1, and it's beneficial to improving the stress condition of the slope. But after cutting, there still exist some parts of strong weathering and off-load rock mass. Height of the step is 15 m and a lot of weak interlayers of slope are well developed, so cutting and offloading can't satisfy the request of keeping stability of the slope in a long run.

2. The slope has many weak interlayers, and all of them can form bottom sliding plane of creep and crack deformation. Through the deformation and failure model of slope, it can be concluded that holistic deformation of the slope may be caused by creep deformation of carbonaceous shale interlayer. And along combination of steep dip off-loading cracks and weak layers, it may gradually develop upwards. Because above the tear line, slope is very tall and the weak layers are extremely developed, after cutting, there is the possibility of large-scale creep deformation in Area A. So the emphasis of controlling slope deformation is to prevent the deformation of weak layers.

3. For slope of the Area A and B, the support project of combination of the pre-stressed anchor cable and framework beams is chosen. Length of anchor cable is 20 m to 25 m. After cutting and offloading slope above the third step, there is still a certain depth of rock mass which is strong weathering and offload. Direction of anchor should be perpendicular to slope. The first and the second step are weak weathered rock mass. So preventing rock mass from creeping along the weak layers is key task. It is why direction of anchor wire should be perpendicular to the slope. Anchor framework is used to slope protection in Area C and length of anchor ranges from 8 m~10 m.

4. Rock mass which is incised by structural planes is fragmentation. The groundwater of slope is also developed. So a row of acclivitous drainage holes of 15 m deep are set at the bottom.

5 CONCLUSIONS

1. The oblique slope is an inclined slope which consists of interbedded mudstone, siltstone, carbonaceous shale. And interbedded displacement zone is also well-developed. The deformation of slope is showed with offloading rebound and compressive deformation of soft rock. Blocks transfer deformation towards top, which are formed by weak layers and off-load cracks. The shallow surficial rock mass of the slope is quite crashed, and joints are well developed. The slope has a large number of cracks, of which strikes are in the same direction with these dominant structural planes.

2. Combination of structural planes forms a large amount of potential instable blocks. Deformation and failure model of slope is stepladder-like creep and crack, which develops from bottom to top. Controlling of the slope deformation should be set

on cutting and offloading, cleaning out cataclastic rock in Area A, and in order to control the deformation of weak layers, anchor measures should be taken.
3. Because of spatial difference of rock structure, deformation of slope in different parts is usually different, therefore, study of deformation and failure model of slope should analyze the mutual relationship of rock structure and deformation in details.

REFERENCES

Ding Xiuli, Fu Jing & Liu Jian. 2005. Study on creep behavior of alternatively distributed soft and hard rock layers and slope stability analysis. *Chinese journal of rock mechanics and engineering*. 24(19):3410–3418.
Fan Wen, Yu Maohong & Li Tonglu. 2000. Failure pattern and numerical simulation of landslide stability of startified rock. *Chinese journal of rock mechanics and eng.* 19(supplement):983–986.
Huang Runqiu. 2000. Time-dependent deformation of a high rock slope and its engineering-geological significance. *Journal of engineering geology.* 8(2):148–153.
Huang Runqiu, Lin, Feng & Chen Deji. 2001. Formation mechanism of unloading fracture zone of high slopes and its engineering behaviors. *Journal of engineering geology* 9(3):227–232.
Huang Zhengjia & Wu Aiqing, 2001. Usage of Block Theory in Three Gorges Projects. *Chinese journal of rock mechanics and engineering* 20(5):648–652.
Huang Runqiu. 2004. Mechanism of large scale landslides in western China. *Advances in earth science* 19(3):443–450.
Yang Xuetang, Ha Qiuling, Gao Xizhang. 2004. Research on unloading deformation and support of high slope rock interlaced with hard rock. *Chinese journal of rock mechanics and engineering*, 23(16):2681–2686.
Sun Hongyue & Shang Yuequan. 1999. Study on deformation and failure characteristics of inclined slope. *Journal of engineering geology* 9, 7(2):141–146.
Yang Yonghong & Lv Dawei. 2006. Study on treatment of high-cut carbonaceous shale slope in expressway. *Chinese journal of rock mechanics and engineering*, 25(2):392–398.
Yin Yueping. 2005. Human-cutting slope structure and failure pattern at the three gorges reservoir. *Journal of engineering geology* 13(2):145–154.

Sliding in weathered banded gneiss due to gullying in southern Brazil

W.A. Lacerda
COPPE/UFRJ, Rio de Janeiro, Brazil

A.P. Fonseca
CEFET-RJ, Rio de Janeiro, Brazil

A.L. Coelho Netto
IGEO/UFRJ, Rio de Janeiro, Brazil

ABSTRACT: This geotechnical study of a case of sliding associated with gullying is of great importance for understanding the various mechanisms of this type that are frequently found in the Middle Paraiba do Sul Valley in the southeast of Brazil. The evolution of erosive canals has been widely studied in this region (Coelho Netto, 2003). Once an erosive canal reaches a drainage basin with a steeper topography, it can trigger landslides. These landslides frequently intensify erosion (the soil is left more exposed to surface drainage) and significantly increases the volume of soil carried to rivers. In this paper landslides caused by gullying in Três Barras were studied. In this concavity two independent mechanisms for the movement of masses exist: the erosion of the more friable material (recent residual material) and the sliding of overlying soil masses.

1 INTRODUCTION

In the area of study it was noted that the propagation of gullying was associated with the expansion of the canal network, which can rupture and force back topographical divides. The advance of the gully undermines the stability of the two masses with different slide directions, one along the erosion axis and the other approximately perpendicular, which is dislocated as if it were a plate. These masses are called A and B respectively, and can be seen in Figure 1.

The two unstable masses studied (A and B) were measured with inclinometers and piezometers.

2 AREA LOCATION

The area of study is the Três Barras concavity, located in the southeastern plateau of Brazil, on the northern watershed of Serra da Bocaina, in the middle Paraiba do Sul valley, close to the frontier between the states of Sã Paul an Ri de Janeiro.

In the area of gullying there was a vein of pegmatite at the latitude N 55°E/35°NW, in other words parallel to the strike of the Pre-Cambrian rock foliation.

The actions of a temporary aquifer in contact with the altered rock on top of the pegmatite vein were also observed. This aquifer is solely supplied by local rainfall.

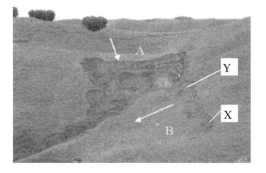

Figure 1. Photo of gullying in study. Mass B moves along a planar surface, with zero movement at point X and maximum movement at point Y. Mass A moves along a conchoidal surface.

3 GEOTECHNICAL PROFILE OF THE AREA

14 holes were drill using percussion drilling and 3 using mixed drilling techniques in the area of study. Piezometers and inclinometers were placed in these holes. Figures 2, 3 and 4 show the geotechnical sections. The topographic plan with the instrumentation points and the section sketches are shown in Figure 6. The subsoil essentially consists of layers of

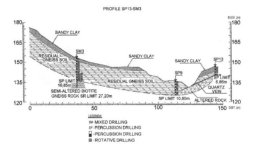

Figure 2. Geotechnical section SP6-SP12 of the area of study.

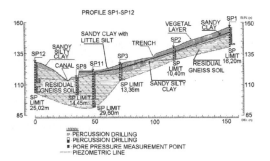

Figure 3. Geotechnical section SP6-SP3 of the area of study.

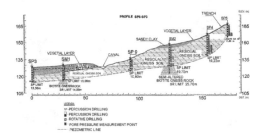

Figure 4. Geotechnical section SM3-SP13 of the area of study.

sandy clay in a process of laterization over the residual sandy-silty gneissic soil.

4 INSTRUMENT READINGS

14 Casagrande type piezometers and 3 inclinometers were set up in the area. The location of these instruments is shown in Figure 5. It should be noted that piezometers (Pz) were placed in the holes dug by percussion drilling (SP) and the inclinometers (I) in the holes dug by mixed drilling (SM).

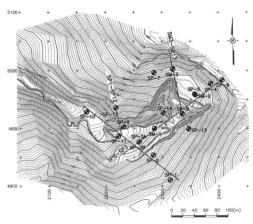

Figure 5. Topographical plan with the location of the drilling, the instrumentation and the indication of the subsoil sections (elevations in meters).

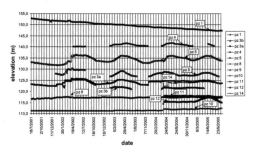

Figure 6. Average piezometric elevations.

4.1 Piezometers

Figure 6 shows the Piezometric elevations for all the piezometers.

Piezometers 2a, 2b, 6, 7 and 13 remained continually dry. Some instruments remained dry for sustained periods of time, which is represented by the discontinuities in the chart.

The piezometric level of Pz1 declined continually, as can be seen in Figure 5. In relation to this, it should be noted that the nearest piezometers, Pz2 a and b, remained continually dry and that Pz1 is at the top of the slope, on a elevation 15 meters above Pz2. Therefore, it can be assumed that the instrument in question had some sort of defect. As a result Pz1 was not considered in the analysis.

It can also be noted that the piezometers that contained water during the monitoring period are located in the direction of the erosion axis, favored by the probable geological fractures in the area of study. Avelar and Coelho Netto (1992) obtained similar data when monitoring the Bom Jardim Concavity, also in Bananal.

Figure 7 shows the accumulated rain over a 25 day period measured in the Bananal pluviometric station (SIGRH/SP).

The variation in the piezometric level of instruments Pz3a and b, Pz4, Pz5 and Pz9 bore some relation to the rainfall measured. Looking at Figures 7 and 8, it can be noted that elevations in the piezometric levels occurred approximately two months after the respective accumulated rainfall over a 25 day period measured in the Bananal station.

This variation in the piezometric levels following a long delay in relation to the measured rainfall is common in the region. Rocha Leão (2005) found an interval of time similar to those found in this paper when he compared the piezometric levels and the local rainfall in an area of gullying in Bela Vista, approximately 7 km from the area of study.

It can also be seen that the piezometric levels in instruments placed in the lowest elevations (pz8, pz10 and pz14) did not present any relation with the rainfall that occurred, though pz10 and 14, which are located at the bottom of the canal, along the erosion axis, presented artesianism.

4.2 Inclinometers

Table 1 shows the depths at which the inclinometers were installed and the date of the initial reading taken two days after installation.

Figures 8, 9 and 10 show the inclinometer readings. Figure 11 shows the topographic plan with the vector direction of the displacements measured at the depth of the rupture surface. Figure 12 shows the geotechnical profiles in the direction of movement of each inclinometer with the maximum displacement measured (resultant displacement).

Table 2 shows the maximum horizontal displacements measured and the depth of the rupture surface.

It is evident that part of the rupture surface is practically situated on the contact between the soil and the rock.

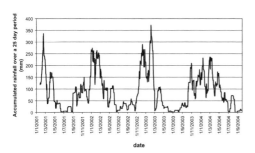

Figure 7. Accumulated rainfall over a 25 day period.

Table 1. Depth of installation and date of initial reading of inclinometers.

Inclinometer	Date of initial reading	Installation depth (m)
I 01	05/05/2004	15.0
I 02	01/05/2004	24.5
I 03	29/04/2004	21.0

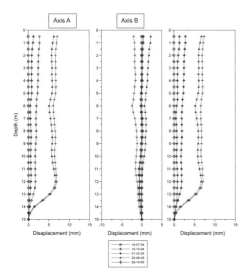

Figure 8. Readings of inclinometer I1.

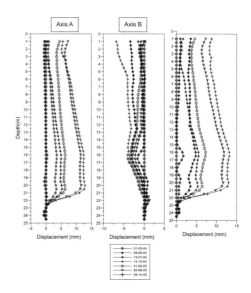

Figure 9. Readings of inclinometer I2.

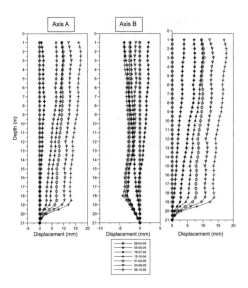

Figure 10. Readings of inclinometer I3.

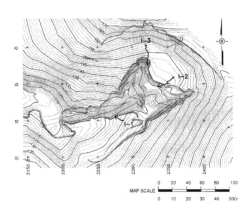

Figure 11. Displacement vectors measured at rupture surface.

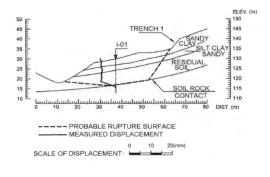

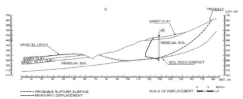

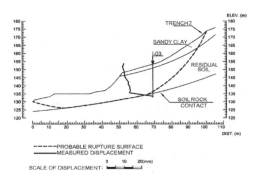

Figure 12. Geotechnical profiles in the direction of movement of each inclinometer with the maximum displacement measured (resulting) and the indication of the probable surface slide.

Table 2. Maximum horizontal displacement and the depth of the rupture surface between June 2004 and October 2005.

Inclinometer	Maximum displacement (mm)	Depth (m)
I 01	7.1	14.0
I 02	13.2	21.5
I 03	17.7	20.0

5 INSTABILITY MECHANISMS

A hypothesis that can explain the phenomenon is that soil mass A becomes periodically unstable due to the gradual erosion that occurs around its base. When this takes place mass A moves, which in turn also permits the movement of mass B.

Stability analyses have shown that the shear strength along the slide surface can be classified as residual. Skempton (1964), Lupini et al. (1981) and Leroueil (2001) have shown the importance of studying the residual condition in the case of slides that may have already occurred.

6 CONCLUSIONS

It was noted that the underground water in the area is fed by the local regional water table and not by local rain. Some points of artesianism were measured in the area, all of which were at the bottom of the erosive channel. The flow of water has been eroding the saprolitic soil, loosening the underlying levels which then rupture and slide. In this way a continuous process of erosion and sliding was started throughout the entire area.

In the large majority of cases the development of concave units is associated with rock fracturing, which, according to Avelar and Coelho Netto (1992), allows the exfiltration of ascendant underground water flows (artesianism). Due to the excess pore-pressure on the exfiltration rock faces located on the gully walls, these flows can unleash erosive mechanisms.

It was also noted through the inclinometer readings and the field observations that the shearing surface of mass A is conchoidal while that of mass B is planar, fan-shaped in plan.

ACKNOWLEDGEMENTS

The authors would like to register their gratitude to CNPq and FAPERJ, through the PRONEX Program, for the financial support they provided for this research project and to MSc students Raquel Maciel, Vitor Aguiar and Marina Duarte for support during the field work.

REFERENCES

Almeida, J.C.H., Eirado Silva, L.G. & Avelar, A.S. (1991) "Coluna tectono-estratigráfica da parte do Complexo Paraíba do Sul na região de Bananal, SP" Anais do Simp. Geol. Do Sudeste/SBG, São Paulo.

Avelar, A.S. & Coelho Netto, A.L. (1992a) "Fraturas e desenvolvimento de unidades côncavas no médio vale do rio Paraíba do Sul". Rev. Bras. De Geociências, vol. 22 no. 2.

Avelar, A.S. & Coelho Netto, A.L. (1992b) Fluxos d'água subsuperficiais associados a origem das formas côncavas do relevo. Anais da 1a conferência Brasileira de Estabilidade de Encostas/COBRAE, ABMS and SBGE, Rio de Janeiro; vol. 2: 709–719.

Coelho Netto, A.L. (2003) "Evolução de cabeceira de drenagem no Médio Vale do Rio Paraíba do Sul (SP/RJ): A formação e o crescimento da rede de canais sob controle estrutural". Revista Brasileira de Geomorfologia, vol. 4, no. 2, pp. 118–167.

Fonseca, A.P. (2006) Analise de mecanismos de voçoramento associados a escorregamentos na Bacia do rio Bananal (SP/RJ). Doctoral Dissertation, COPPE/UFRJ.

Heilbron, M. (1995) O Segmento Central da Faixa Ribeira: síntese geológica e ensaio de evolução geotectônico. Livre Docência. DGeoUERJ, 110p.

Leroueil, S. (2001) "Natural slopes and cuts: movement and failure mechanisms". Géotechnique, vol. 51, no. 3, pp. 197–243.

Lupini, J.F., Skiner, A.E. & Vaughan, P.R. (1981) "The drained residual strength of cohesive soils". Géotechnique, 31(2), pp. 181–213.

Rocha Leão, O.M. (2005) Evolução regressiva da rede de canais por fluxos de água subterrânea em cabeceiras de drenagem: bases geo-hidroecológicas para recuperação de áreas degradadas com controle de erosão. Doctoral Thesis, IGEO-UFRJ, Rio de Janeiro.

SIGRH/SP—Homepage: http: www.sigrh.sp.gov.br/cgi-bin/bdhm.exe/plu

Skempton, A.W. (1964) "Long-Term stability of clay slopes." Fourth Rankine Lecture, Géotechnique, vol. 14, no. 2 pp. 77–101.

Experimental and three-dimensional numerical investigations of the impact of dry granular flow on a barrier

R.P.H. Law, G.D. Zhou, C.W.W. Ng & W.H. Tang
Department of Civil Engineering, Hong Kong University of Science and Technology, HKSAR

ABSTRACT: In this study, the impact behaviour of dry granular flows on a barrier was investigated both experimentally and numerically. Physical model tests were carried out using a 3.8 m long flume whereas numerical simulations were conducted using a three-dimensional particulate code (PFC^{3D}). Uniform clean sands (Leighton Buzzard sands fraction C and E) were used for the flume tests. During each test, impact profiles were recorded using a high speed camera and impact force was measured by recording the deformation of a barrier with known stiffness. Results obtained from experimental and numerical tests are compared and analyzed.

1 INTRODUCTION

Many landslides with long travel distance are flow-like in character. Some, such as debris flows, have distributed velocity profiles resembling the flow of fluids (Hungr 1995). Debris flow poses a costly risk for lives and properties in mountainous and rainy areas. Understanding of the flow and impact characteristics of debris flow would facilitate decision and implementation of debris flow mitigation strategy.

In previous work by Law et al (2007), a series of flume model tests and numerical simulations were conducted to investigate the flow characteristics of dry granular material. Pore fluid was not considered in the tests to simplify the experiments and numerical analysis. Physical model test results revealed that small plugs were observed in all dry flow tests. Plugs did not grow in size significantly during the flow process. Reverse segregation was clearly observed in all flow tests using non-uniform silty sand (completely decomposed granite). Coarse particles were observed at the front of the flow, whereas fine particles were located at the rear and at the bottom of the flow. Reverse segregation was also simulated in 3D numerical codes (PFC^{3D}). In the previous publication, impact characteristics of granular flow on a barrier are not studied.

In the field, a flexible barrier may be composed of rings, ropes, nets, springs and dashpots. Anchors may be installed to fix the barrier in position. Anchorage may break under heavy impact load, leading to the failure of the barrier. Impact force analysis is therefore important for barrier design.

According to Wendeler (2006), flexible barriers were originally designed to protect villages, highways and railway lines from rockfalls. Their main loadbearing principle is to restrain the falling rocks using a long braking distance and therefore producing a soft stop, reducing the peak loads in the barrier components and the anchors. The same principle also works for a variety of other problems such as snow slides, tree falls, floating woody debris during flooding and debris flows. According to Rorem (2004), several natural debris flow events in the USA and around the world have demonstrated that flexible barriers can be effective at stopping debris flows. However, as far as the authors are aware, impact behaviour of debris on a barrier is not fully understood yet.

This study therefore aims at investigating the impact characteristics of dry granular flow through flume model tests (see Figure 1) and three-dimensional numerical simulations by PFC^{3D}. Dry Leighton Buzzard sands (LB sands) fraction C and E (see Figure 2) were used for the experimental tests. Granular flow simulations started with the release of dry granular materials from the straight flume. Impact process was observed through high speed image and the force exerted on the barrier was recorded. Emphasis was made to study the effect of grain size on the measured force. Numerical simulation of dry granular flow was conducted by using Particle Flow Code in Three Dimensions (PFC^{3D}) (Itasca 2005). Elastic balls were used to model granular materials and the barrier. Emphasis was made on the influence of static load of deposited material on impact force. Computed results are compared and analyzed with experimental results in this paper.

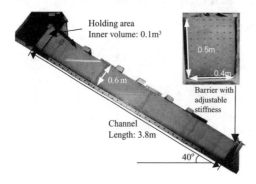

Figure 1. The experimental setup for study of granular flow.

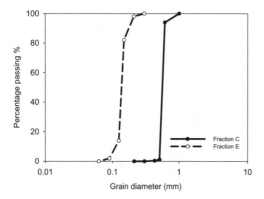

Figure 2. Particle size distributions of dry LB sands (fraction C and E).

2 DETAILS OF PHYSICAL MODEL TESTS

Figure 1 shows the setup of flume model. Two tests were carried out using the flume of 3.8 m long, 2.8 m high, 0.4 m wide and 0.6 m in depth. The slope angle was adjusted to be 40°. The base of the slope was smoothened with polythene to ensure uniform roughness along the slope. The flume was mainly made of plywood, except that one side of channel was made of transparent Perspex. The channel was held in position by a heavy duty crane system. A MotionScope® PCI Series high speed imaging system was installed to view the impact behaviour. The recording speed was 125 frames per second and the resolution was 480 × 420 pixels.

For modelling a barrier, an aluminium plate was installed to stop the flow at the lower end of the channel. The plate was 0.5 m in depth, 0.4 m in width and 0.01 m in thickness. The aluminium plate was supported by a set of shafts and springs of known stiffness. Load was transferred to the springs during an impact. The deformations of the springs were recorded by laser sensor. The measured force is calculated by the measured deformations of the springs times the stiffness of the springs. The installation of springs serves to provide flexibility of the aluminium plate upon impact. The shafts and the springs are supported by an aluminium structure. The structure restricts the movement of the shaft such that the aluminium plate could move parallel to the channel only. The barrier was calibrated by static load test, in which loads were added successively onto the barrier and the displacement was recorded. The calibrated stiffness of the springs was 29.3 N/mm. Materials to be tested were placed in a holding area at the highest point of the slope. The inner volume of the holding area is 0.1 m^3. The prepared dry granular materials were placed in the holding area just before tests started to minimize possible consolidation and segregation.

Two impact tests were conducted. In each test, 30 kg of dry LB sands (either fraction C or E) was released on the channel. LB sand (fraction C) composed of fairly uniform grains with grain diameter ranging from 300 μm to 600 μm whereas LB sand (fraction E) composed of fairly uniform grains with grain diameter ranging from 150 μm to 90 μm. Figure 2 shows the particle size distribution of the LB sands (fraction C and E) used. The two tests aim to examine the effect of grain size on measured impact forces.

3 INTERPRETATIONS OF MODEL TEST RESULTS

3.1 Effects of grain size on measured impact force

Figure 3 shows the observed time history of measured impact forces by LB sands (fraction C and E). The measured forces were deduced by using the average calibration constants obtained from the best fit

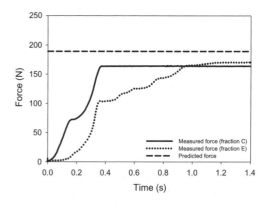

Figure 3. The time history of measured impact forces using LB sands (fraction C and E).

line. The moment at which the fraction C sand first impacted the barrier is taken as time zero. As shown in the figure, the measured force rises continuously in two stages with a similar rate of increase until the time reaches approximately 0.35 s for the test using fraction C sand. After 0.35 s, the measured force does not change noticeably as time elapses. The measured maximum force is 165 N. Regarding the test using fraction E sand, the measured force rises continuously in multi stages with different rates of increase until the time reaches approximately 0.95 s. After 0.95 s, the measured force does not change significantly as time elapses. The measured maximum force by the static load of deposit is 171 N for fraction E sand. It is observed that no peak force was measured for both tests using fraction C and E sands. The duration of impact was much longer for fraction E sand than that for fraction C sand.

According to the research work done by Shinohara (2000), the angle of internal friction of dry powders increases with a smaller porosity. As the mean grain size of uniform fraction C sand is larger than that of fraction E sand, it is expected that fraction E sand should have a smaller porosity and hence a higher dynamic angle of internal friction to resist to flow. Thus, it may be reasonable to deduce that fraction E should flow slower because of its larger internal frictional resistance and so it take a longer duration to complete the impact process. The time history of the measured forces suggests that the duration of impact decreases with an increase in grain size of uniform granular materials.

In Figure 3, the measured forces are compared with a predicted static force. The predicted force represents the total static load exerted on the barrier by deposited granular material. The predicted force is calculated by the following simple equation:

$$F = mg \sin \theta_c \sin \theta_i \qquad (1)$$

where F is the predicted impact force, m is mass of the granular material, g is the gravitational acceleration, θ_c is an angle of the slope, θ_i is an angle between the barrier and the channel. Based on the set up of the flume, m, θ_c and θ_i are taken to be 30 kg, 40° and 90° respectively. It should be noted that the rate of change of momentum of the granular material and the basal frictional resistance of the channel are not considered in the equation. The predicted force therefore does not vary with time. The dashed line represents the predicted static force by Equation 1 and the predicted value is 190 N, which is greater than the measured static load acting on the barrier in both tests. It could be explained by the reduction of the measured static load by the basal frictional resistance, which is not considered in the equation.

3.2 The influence of deposition process on measured force

Figure 4 shows the images recorded during the deposition processes in the two tests. Each dotted line represents the surface of deposited granular material at each given time interval. The time increment (Δt) between each dotted line is 0.08 s. Since the shutter speed of the camera is 0.008 s, distances between two dotted lines therefore represent the observed surfaces obtained between every ten camera shots. In each of the Figures 4a and 4b, the rightmost dotted line is the deposited surface 0.08 s after the granular material first impacted on the barrier in each test. It can be seen in both figures that that the horizontal distribution between every two dotted lines is fairly constant and uniform, thus suggesting a progressive accumulation of the deposit of the granular material. Progressive accumulation of the deposit involves a relatively longer

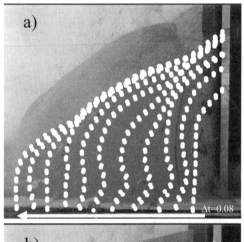

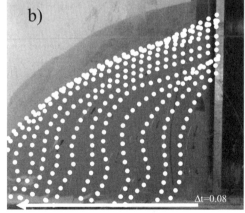

Figure 4. Recorded images of deposited surfaces in tests using a) LB sand (fraction C) and b) LB sand (fraction E).

impact process than impact which is caused by rigid object, such as a boulder impact. A longer impact process reduces the rate of change of momentum of the granular material, leading to smaller dynamic impact. The smaller dynamic impact force helps to explain why no peaks of the measured force were observed.

According to Wendeler (2006), the debris deposit on a ring and net barrier (one form of flexible barrier) serves as a dam. The dam is strong enough to retain the remaining debris. In the process of the progressive accumulation of the deposit, the deposit is expected to act like a dam to reduce the dynamic impact of the upcoming granular material. As a result, the existence of accumulated deposit during the impact process is expected to reduce the impact load.

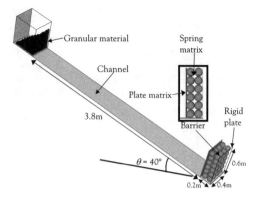

Figure 5. A typical setup of numerical model.

4 NUMERICAL MODELING OF THE IMPACTS OF GRANULAR MATERIAL ON A BARRIER

Particle Flow Code in Three Dimensions (PFC^{3D}) (Itasca 2005) is a numerical simulation tool that models the motion and interaction of spherical particles by the distinct element method (DEM) (Cundall & Strack 1979). In this study, it is used to study the impact behaviour of dry granular materials on a barrier. It should be pointed out that the main purpose of the numerical simulations is to investigate the impact behaviour qualitatively, not quantitatively due to the complexity of dry granular flows, uncertainty in model parameters and limitations of the numerical code such as handling the presence of pore fluid.

Figure 5 shows the setup of the numerical model. The two side walls and the bottom floor of the flume (as shown in Figure 1) are idealized as elastic "walls" with normal and tangential (shear) stiffness of 10^8 N/m. For simplicity and reduction of computational time, LB sand is modelled by a cluster of 21 mm-diameter identical spheres. Obviously, the diameter of each ball is much larger than that of soil grains used in the flume tests. Hence, computed results should be treated with caution. In order to reduce the computation time, stiffness of the spheres is limited. Therefore, the stiffness of the spheres is set to be 10^6 N/m. The density of spheres is chosen as 1480 kg/m^3, which is the measured bulk density of fraction C sand used in laboratory tests. According to Brown (2000), the internal friction angle of LB sand fraction C is 35.4°. The internal friction angle of the granular material is therefore set to be 35.4°. The interface friction angle between the granular material and the channel are set to be 22.5°. The value was obtained by testing a hollow paper cylinder sliding down the channel. The hollow paper cylinder was filled with LB sand fraction C.

Two matrices of spherical particles were used to simulate the barrier, namely the plate matrix and the spring matrix. The aluminium plate of the barrier is modelled by the plate matrix. The plate matrix composes of 2400 spheres. Each of the spheres is 0.005 m in radius and 10^6 N/m in stiffness. Based on the density of the aluminium plate used in experimental tests, the density of the spheres is set to be 2700 kg/m^3. The springs of the barrier is modelled by the spring matrix. The plate matrix composes of 48 spheres. Each of the spheres is 0.05 m in radius. Based on the barrier stiffness used in experimental tests, the stiffness of the simulated barrier is set to be 29.3 N/mm. Correspondingly, the normal stiffness of the spheres that compose the spring matrix is set to be 3960 N/m. Since a ball with zero density is not allowed in the code, the density of spheres that compose the spring matrix is set to be 100 kg/m^3. Based on the barrier design in experimental tests, each of the spheres in the plate matrix and the spring matrix could move parallel to the channel only. The local damping ratio of the barrier is roughly taken as 0.7. All spheres that constitute the barrier are frictionless. The normal and shearing bonding in the barrier is set to be 10^7 N/m. The arrangement of the spheres in the plate matrix and the spring matrix is shown in Figure 5.

During impact, the spring matrix is compressed by approaching granular material. As the stiffness of the plate matrix is 250 times higher than the spring matrix, the compression of the barrier is mostly contributed by the spring matrix. The aluminium structure in the barrier is modelled by a rigid wall. A rigid wall is installed at the back of the barrier to support the barrier and record the impact force. The computed impact force is calculated by the force acting on the rigid wall minus the static load by the weight of the barrier. Table 1 summarizes the parameters used for the numerical analysis. Details of the contact behaviour of the model could be found in Law et al. (2007).

Table 1. Parameters adopted for numerical simulations.

Parameter	Magnitude	
Slope angle	40°	
Length of the channel	3.8 m	
Stiffness of walls	Normal:	10^8 N/m
	Tangential:	10^8 N/m
Number of balls	Granular material:	4000
	Plate matrix:	2400
	Spring matrix:	48
Ball radius	Granular material:	0.01 N/m
	Plate matrix:	0.05 m
	Spring matrix:	0.03 m
Density of each ball	Granular material:	1450 kg/m^3
	Plate matrix:	2700 kg/m^3
	Spring matrix:	100 kg/m^3
Mass of granular material	30 kg	
Normal and tangential ball stiffness	Granular material:	10^6 N/m
	Plate matrix:	10^6 N/m
	Spring matrix:	3960 N/m
Internal friction angle	35.4°	
Interface friction angle	22.5°	
Contact bond	Normal:	10^7 N/m
	Shear:	10^7 N/m
Local damping	Granular material:	0.05
	Barrier:	0.7
Viscous normal damping	Granular material:	0.2
	Barrier:	0
Viscous shear damping	Granular material:	0.2
	Barrier:	0

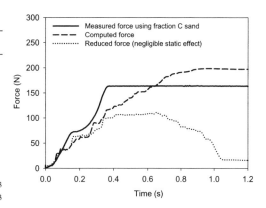

Figure 6. Comparisons of measured and computed time history of impact forces.

5 THE ROLE OF STATIC LOAD ON IMPACT FORCE

Figure 6 compares the measured and computed time histories of impact forces acting on the barrier. The solid line represents the measured force in the test using fraction C sands. The time zero represents the moment at which the granular material first hit the barrier, whereas the dotted line represents the computed force time history. It can be seen that the computed time history is generally consistent with the measured one, although the duration of computed impact (about 0.9 s) is almost twice as the measured value (about 0.35 s). In addition, it takes numerous stages in the numerical prediction, rather than two stages in the experiment, to reach the maximum computed impact force of 198 N, which is about 17% larger than the measured value of 171 N. Considering the approximations involved and simplifications made in the numerical simulations and the uncertainties encountered during the experiment, the fairly consistent results obtained are quite fortuitous.

In order to understand the components of an impact force, a reduced force is calculated by reducing the mass of balls from 1480 kg/m^3 to 100 kg/m^3 whenever they rest on the barrier. This is represented by the dashed line in Figure 6. Thus, the difference in magnitude between the dotted and the dashed lines is the contribution of static load of the granular material. It can be observed that the computed and the reduced force, in general, rise in magnitude to 101 N at about 0.35 s. While the computed force continues to rise to a maximum of 198 N at the elapsed time is 0.9 s, the reduced force, however, falls to 15 N at time equal to 1.05 s. This "residual" 15 N force is due to the static load of the deposited granular material specified with the reduced density of 100 kg/m^3 in the PFC analysis. The ratio between the maximum computed reduced force of 101 N and the maximum residual reduced force (i.e., 15 N) is around 15%. Thus, the static effect of the granular material is considered to be partially eliminated from the reduced force.

By comparing the difference in magnitude between the dotted and the dashed lines, the contribution of static load is observed to increase during the impact process. The computed force at the end of impact is primarily composed of static load. It is explained by the accumulation of deposited granular material during the impact process. On the other hand, it is observed that the maximum measured and computed forces were recorded at the end of impact. The results suggest that the maximum force for dry granular material is substantially contributed by static load.

6 CONCLUSIONS

In order to investigate the impact characteristics of dry granular materials, flume model tests and numerical simulations were conducted. Physical model test results reveal that the duration of impact was longer for fraction E sand (smaller particles) than that of fraction C sand. It may be explained by the difference in grain size between the fraction C and E sands. No peak force was measured for both fraction C and E sands. Progressive deposition was observed in tests using fraction C and E sands.

Numerical simulations indicate that the contribution of static load increases during the impact process. Static load is illustrated and verified to contribute the maximum impact force substantially.

ACKNOWLEDGEMENTS

The authors would like to acknowledge the financial support from the Research Grants Council of HKSAR (grant no. HKUST6294/04E) and research grants DAG05/06.EG39 and CA-MG07/08.EG01 provided by Hong Kong University of Science and Technology (HKUST) and GEL05/06.EG01 by Geotech Engineering Ltd. The assistance and contribution by Messrs T.T.L. Yeung, Y. Kwok & T.H.N. Ho and technicians at the Geotechnical Centrifuge Facility, HKUST are also gratefully acknowledged.

REFERENCES

Brown, C.J., Lahlouh, E.H. & Rotter, J.M. 2000. Experiments on a square planform silo. *Chem. Eng. Sci.* 55(20): 4399–4413.

Chau, K.T., Wong, R.H.C. & Wu, J.J. 2002. Coefficient of restitution and rotational motions of rockfall impacts. *International Journal of Rock Mechanics & Mining Science* 39: 69–77.

Cundall, P.A. & Strack, O.D.L. 1979. A Discrete Numerical Model for Granular Assemblies. *Geotechnique,* 29, 1: 47–65.

Hungr, O. 2000. A model for the runout analysis of rapid flow slides, debris flows, and avalanches. *Canadian Geotechnical Journal* 32: 610–623.

Itasca, 2005. Itasca Consulting Group Inc. PFC3D (Particle Flow Code in 3 Dimensions), Version 3.1. Minneapolis: ICG.

Law, R.P.H., Zhou, G.D., Chan, Y.M. & Ng, C.W.W. 2007. Investigations of fundamental mechanisms of dry granular debris flow. *Proc. 16th Southeast Asia Geot. Conf. 8–11 May, Malaysia*: 781–786.

Rorem, J. 2004. Debris Flow Remediation: Case study San Bernardino Mountains near Crestline and Lake Arrowhead, CA/USA. Geobrugg North America, LLC.

Shinohara, K., Golman, B. & Oida, M. 2000. Effect of particle shape on angle of internal friction by triaxial compression test. *Powder Technology* 107: 131–136.

Wendeler, C., McArdell, B.W., Rickenmann, D., Volkwein, A., Roth, A. & Denk, M. 2006. Field testing and numerical modeling of flexible debris flow barriers. *Proc. 6th International Conference on Physical Modelling in Geotechnics, Hong Kong.* Vol. 2: 1573–1578. ISBN: 0-0415-41586-1.

Wu, S.S. 1985. Rockfall evaluation by computer simulation. *Transportation Research Record.,* volume: issue: 1031: 1–5.

Temporal survey of fluids by 2D electrical tomography: The "Vence" landslide observatory site (Alpes-Maritimes, SE France)

T. Lebourg, S. El Bedoui, M. Hernandez & H. Jomard
UMR Géosciences Azur, CNRS-UNSA-IRD-UPMC, Sophia-Antipolis, France

ABSTRACT: A best knowledge about landslide processes requires the characterisation of the triggered factors and their impact on the process cinematic. These factors are often time dependent and that a reason why it is very complex to have a quantitative approach. We propose a temporal imagery of water circulation in a moving mass by Electrical Resistivity Tomography (ERT). The purpose is to be able to quantify the coupling between the rainfalls and the water inflows in a sliding mass, and the evolution of resistivity values acquired by ERT. Our work was based on a multi-scale and multi-survey device approach on the "Vence" landslide (South-eastern France). It is considered like a sandy mass sliding on a sedimentary rock substratum (limestone and calcareous) and largely controlled by rainfall.

1 INTRODUCTION

In most of failure landslide processes fluids are considered as the most important triggered factor. In order to constrain the risk associated to the landslide, it appears necessary to localise and understand the fluids draining network inside the soil mass. Since ten years, the geophysical methods were developed and applied to the study of instabilities of ground. These methods are characterized by their non-destructive aspect and the important investigation depth. They allow obtaining some quantitative data about the slip surfaces, the faults networks and the water flows in depth.

Indeed, combination of fractures, breccia, gouge and the fluids can produce large contrasts in observable geophysical properties (Lebourg et al., 2003). The electrical resistivity is one of them and it is particularly sensitive to the water content of investigated areas.

Precedent authors (Eberhart-Philipps and al., 1995; Saarenketo, 1998; Fukue and al., 1999, and Friedel and al, 2006) shown that the clay resistivity values are strongly dependent with the water content. They conclude that there is a critical water content (depending with the pore distribution) producing a strong resistivity value modification (by a factor 1000 in their case). Consequently a strong resistivity gradient can be associated with the presence of water in the case of a clay ground.

During the last few years, a large number of electrical surveys have been conducted to acquire information about fault geometry or sliding surface

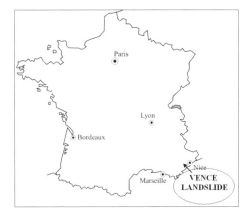

Figure 1. Localisation of the field study.

in various areas (Ritz et al., 1999; Jongmans et al., 2000; Demanet et al., 2001; Lebourg et al., 1999; 2001; 2005; Wise et al., 2003; Lapenna et al., 2003). However, very few studies have been realised in order to obtain data on the water spatiotemporal distribution and its role in the sliding dynamic. Recently a first tentative has been successfully realised on the La Clapière landslide (Alpes-Maritimes, SE France) (Lebourg et al., 2001, Jomard et al. 2005). Their preliminary results allowed the identification of the sliding surface and the network water drainage strongly dependent with the inherited tectonic structures.

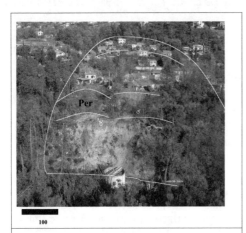

Figure 2. Photography of the landslide (A) and the mains scarps (C) of the studied area.

This study is based on an approach using resistivity method, with a temporal follow-up, on a landslide located in Vence (Alpes Maritimes, SE France), The aim is to characterized the drainage channel system and its evolution in time and space by the resistivity variations.

The choice of this landslide was motivated by: 1) its size is well adapted to such dense geophysics researches, 2) its geological framework is well known and present the advantage to illustrated the possible role of folded and fractured structure in the draining channel network, and, 3) raining conditions can be concentrated on a short time with an abundant volume of water and clearly control the landslide activity.

In a first time, our work has been focused on geotechnical, geological and hydrogeological investigations. After that, we include the fluids propagation survey by ERT in order to demonstrate: 1) the precise geometry of the potential sliding surface and, 2) the possibility to prove a high and fast correlation between an external water injection and resistivity changes. Finally, we performed an original survey device based on resistivities, pluviometric and piezometric data acquisitions (Figure 3).

2 HISTORICAL, GEOLOGICAL AND HYDROGEOLOGICAL SETTINGS

The "Pra of Julian" landslide, more commonly called the "Vence" landslide, is regarded like a rotational one including 1.2×10^6 m^3 of material. It affects an area about 250 m by 350 m, with a slope around 12° to 20°. The current landslide activity is underlined by some surface expressions appearing in the landscape morphology: tension cracks, scarps, disorders affecting constructions and particularly the deviation of the "Lubiane" river at the foot of the slope. The historical activity of the landslide is related since 1950/1960, with the first constructions, but was restricted to minor disorders (little cracks in the soil). During the 70's years, urbanization increased (9 houses in 1970 to 17 in 1980) and more disorders appeared a few days after strong rainfalls (250 mm in 2 days). The acceleration of the landslide kinematic induced the obstruction of the "Lubiane" river bed. In October 1981, an intense rainfall episode (250 mm of 24 hours) generated mud flows. In 1989, the number of houses was about 26. From 1990 to present, the area was more and more urbanized and the sliding phases occur always with

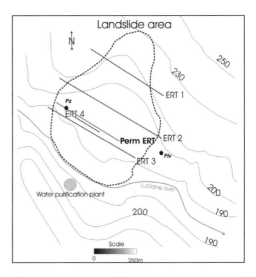

Figure 3. Localisation of the electrical tomography profiles (ERT1, ERT2 and ERT3) and the temporal tomography profile (ERT4 & PERM_ERT).

intense rainfall episodes. At the end of 2000, the landslide activity became more intense with the opening of important tension cracks in the upper part and the destabilisation of the front with a scarp about several meters.

Geology and geomorphology: The studied area is localized within the NW-SE trending the Oligocene "Vence" fold related to the alpine tectonic deformations (Laurent, 1998).

The hydrogeology of the sector shows the presence of a free water table in the sand and clay layer and a captive water table in marly limestones. The presence of faults in the limestones creates a connection between the Cretaceous calcareous permeable reservoir and the overlying sandy-clay layer.

3 A CONCEPTUAL MODEL OF THE "VENCE" LANDSLIDE

The conceptual model consisted in the definition of the geometrical limits observed on the field and also limits in depth of the sliding surface. In order to describe it, we realised several ERT in two dimensions (not developped in this short paper): three profiles perpendicular to the slope. They enable us to approach the morphology of the active slide surface, and to highlight drained faults in the landslide.

To understand these complex relations between water and structure of a landslide we carry out and compare the three types of measurements:

– Large scale profiles to complete the geometrical/conceptual model of the Vence landslide, and to calibrate the resistivity values with the boreholes and field observations (not developed in this paper, ERT1, 2, 3),
– temporal measurements (ERT4), based on a controlled/artificial water injection (controlled quantity) coupled with an ERT survey. ERT4 was performed on the top of the basal scarp, during 2 days with the geophysical survey of the evolution of the resistivities at a regular temporal follow-up (short period of about 1 hour at the beginning and 3 times 24 h after the first day), but not developed in this paper.
– permanent measurement device (Perm_ERT) to analyse the correlation between daily piezometric and rainfall measurements.

4 RESISTIVITY TEMPORAL FOLLOW UP

4.1 *ERT 4: to analyse the resistivity soil evolution versus water injection*

The precedent approach shows that there are several origins of water supply in the solid mass: the rain fall (external factor) and the interconnection between

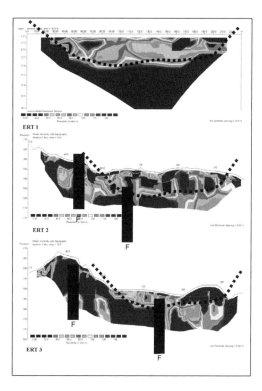

Figure 4. ERT 1, 2, 3 (resistivity in Ohm.m), vertical faults (**F**) and landslide surface in dotted points).

the sliding mass and the rock substratum by draining structures. In this study, we want analyze the water circulation during the time in the sliding mass (ERT4 by injection) and the rate between the higher watertable and the drained faults system. This correlation can be obtained by the coupled analysis between temporal resistivity variations (Perm_ERT), the elevation of the piezometric level and the analysis of the rainfall.

We use two profiles at different scales. The first one is parallel to the middle active scarp (ERT4), during a 2 days period and coupled with a controlled water injection. The second one is longer and parallel to the active scarp of the slip and crossing the landslide (Perm_ERT). It is performed during three months and coupled with the rainfall measurement.

The first phase of measurement was realized during summer 2005. We imposed a controlled quantity of water close to the electrical profile (ERT4, Figure 3, (Jomard et al, 2006)). In order to explore the resistivity variations assign to the injected water, we carried out a measurement before (reference), during and after the injection of water (Figure 9).

The water was injected between the electrodes 4 and 5. The interpretation of the first profile (T0) correlated with the field information (sliding surface at

423

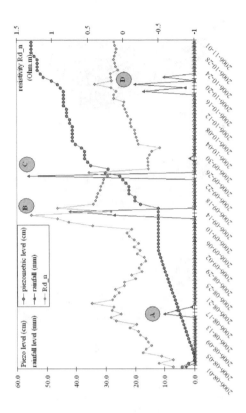

Figure 5. Temporal evolution of resistivity (Ohm.m)/ rainfall (mm)/piezometric level (cm), (Rho_n average of the 574 points of measurement Rd_n) and A, B C and D are four rainfalls events.

12 m) confirm us the presence of the sliding surface towards 12 m of depth in the first right half part of the tomography and an increase of the sliding surface on the level of the electrode 24, which follows the calcareous substratum until its appearance on the field to the level of electrode 32.

4.2 Perm_ERT : 3 month of follow up to

We propose to focus on 3 months of monitoring with a daily measurement:

– ERT with 574 points of measurement,
– Measurement of the piezometric level (with a centimetre resolution).
– Measurement of rainfall events with a pluviometre (with a millimetre resolution).

On the basis of the ERT measurement, we try to extract a signal representing the resistivity variations from external conditions. As we have just shown it (4.1), a temporary water contribution on the surface involves a modification of the resistivity, but what can be the response resistivity measurement at a natural scale?

The acquisition period presents here (resistivity, rainfall and piezometric level) is from July 2006 to November 2006.

The first treatments (calibration of the repeatability of measurement), which are not the aim of this paper, enabled us to show that there was a daily repetitivity of the measurement: a maximum average variation of $+/-0.2$ Ohm.m/day without climatic changes.

For each day we consider:

– The average of the 574 points of measurement Rho_day (Rd_n),
– The standard deviation of the 574 points (SdDay_n).

After different studies of the data distribution, we choose to represent one acquisition day by the data average: Rd_n (n from day 1 st July the first to the last day November the first).

To highlight the differences from a day to another, we calculate the difference between the resistivity value of a day (Rd_n) and the average of the resistivity day considering a three month period (Mean_Rho_day: Rd3 months).

4.3 Discussion and interpretation of data

The observations carried out over the 3 months period show that there is a very fast response of the field to the rainy episodes 4.3, as described with the ERT4. We observe a several centimetres increasing of the piezometric level and a period of 24/48 h of stabilization.

According to the resistivity survey, we measured:

– in a drilled hole of 20 m of depth near ERT_perm the evolution of the piezometric level. Measure carried out manually with a centimetric resolution,
– near the site (200 m from ERT_perm) we daily quantify the rainfall with a pluviometre characterized by a millimetric resolution.

When we observe the evolution of the curves resistivities/rainfalls/piezometry we can analyse:

– that important precipitations (events A, B, C and D) make decreasing resistivity,
– at the same periods we could observe a strong increase in the piezometric levels (an increase of 5 cm after the event C (2006/09/25) measured the 2006/09/27 due to strong precipitations). This event involved a very important reduction in the daily resistivity average. Thus, the daily average falls of

39.19 Ohm.m to 38.35 Ohm.m, which represents a signal/noise ratio of 3. At the same time we observed an increase in a factor 5 of the piezometric level.

These observations on the event C (September 25 2006) are the same for event A, B and D, but to a lesser extent.

Several remarks which will be the subject of a new work are to be brought:

Notice 1: The traditional analysis by inversion of the data (RES2Dinv, Loke, 1997) and the subtraction of the matrices of inversion (Jomard and al., 2007) do not allow highlighting an effective signal.

Notice 2: The ERT survey seems to be a powerful/ adapted tool in the study of the rainfall impact on the moving mass. . ... But, it appears necessary to develop an adapted data treatment related to the temporal aspect of the survey.

Notice 3: The distribution of the 574 points of measurement is carried out according to a bimodal distribution. Indeed we can observe that the data are distributed around a first peak located between 20 and 25 Ohm.m, then that the other data are distributed according to another mode from 50 Ohm.m. The average of the resistivity is thus not inevitably the tool most representative of the variation of the electrical measurements, nevertheless this mathematical variable is for the moment the most representative of the daily group of 574 measurements.

5 CONCLUSION

The electrical resistivity tomography is a powerful tool to constraint geometrical and geological boundaries in landslide areas. A complex problem in landslide survey is related to the temporal survey of triggered factors as the water.

On the basis of the electrical sensitivity toward the soil water content, we show the possibility to have a quantitative approach of soils response from external water solicitations, using electrical resistivity measurement.

It seems to be accessible to clearly associate a rainfall measurement with a piezometric variation and with a electrical signal. In this case, another mathematical data treatment must be considered in order to have higher signal accuracy and to locate the ground areas where the variations are concentrated.

In the case of the "Vence" landslide, two aspects must be more improved: 1) the impact of the drained fault in the water flow inside the slide mass, 2) the mechanical parameters evolution toward the soil water content. The results of hydrogeological investigations and the influence of weather conditions will allow establishing an evolutive slope stability calculus. Finally, it could possible to related an electrical signal with a slope stability value.

REFERENCES

Demanet Donat, Eric Pirard, François Renardy & Denis Jongmans, 2001. Application and processing of geophysical images for mapping faults. *Computers & Geosciences*, Volume 27, Issue 9, 1 November 2001, Pages 1031–1037.

Eberhart-Phillips D., William D. Stanley, Brian D. Rodriguez & William J. Lutter, 1995. Surface seismic and electrical methods to detects fluids related to faulting. *Journal of geophysical research*, vol. 100, NO. B7, pages 12, 919–12, 936.

Friedel S., Thielen A. & Springman S.M. 2006. Investigation of a slope endangered by rainfall-induced landslides using 3D resistivity tomography and geotechnical testing. *Journal of Applied Geophysics*, Volume 60, Issue 2, Pages 100–114.

Fukue M., T Minato., Horibe H. & Taya N. 1999. The micro-structures of clay given by resistivity measurements. *Engineering Geology*, Volume 54, Issues 1–2, September, pp 43–53.

Griffiths, D. H. & Barker, R.D. 1993. Two-dimensional resistivity imaging and modelling in areas of complex geology: *Journal of applied geophysics*, v. 29, p. 211–226.

Jomard, H., Lebourg T., Binet S., Tric E. & Hernandez M. 2006. Characterization of an internal slope movement structure by hydrogeophysical surveying: Terra Nova.

Jongmans, D., Hemroulle P., Demanet D., Renardy F. & Vanbrabant Y., 2000. Application of 2D electrical and seismic tomography techniques for investigating landslides: Eur. *J. Environ. Eng. Geophys.*, v. 5, p. 75–89.

Lapenna, V., Lorenzo, P., Perrone, A. & Piscitelli, S. 2003. High resolution geoelectrical tomographies in the study of Giarrossa Landslide (Southern Italy): *Bull. Eng. Geol. Env.*, v. 62, p. 259–268.

Laurent O., Stephan J-F. & Popoff. M. 2000. Modalités de la structuration miocène de la branche sud de l'arc de Castellane (Chaînes subalpines méridionales). *Géol. France*, 3, p.33–65.

Lebourg, T., Frappa M. & Sirieix C. 1999. Reconnaissance des surfaces de rupture dans les formations superficielles instables par mesures éléctriques: *PANGEA*, v. 31, p. 69–72.

Lebourg, T., Binet, S., Tric, E., Jomard, H. & El Bedoui, S. 2005. Geophysical survey to estimate the 3D sliding surface and the 4D evolution of the water pressure on part of a deep seated landslide: *Terra Nova*, v. 17, p. 399–407.

Lebourg, T. & Frappa, M. 2001. Mesures géophysiques pour l'analyse des glissements de terrain: *Revue Française de Géotechnique*, v. 96, p. 33–40.

Loke, M.H. 1997. Res2Dinv software user's manual.

Loke, M.H.& Barker, R.D., 1996, Rapid least square inversion of apparent resistivity pseudosection by a quasi Newton method: *Geophysical research letter*, v. 44, p. 131–152.

Ritz Michel, Parisot Jean-Claude, Diouf S., Beauvais A., Dione F. & Niang M. 1999. Electrical imaging of lateritic weathering mantles over granitic and metamorphic basement of eastern Senegal, West Africa. *Journal of Applied Geophysics,* Volume 41, Issue 4, June 1999, Pages 335–344

Rizzo E., Colella A., Lapenna V. & Piscitelli, S. 2004. High-resolution images of the fault.

Saarenketo Timo. 1998. Electrical properties of water in clay and silty soils. *Journal of Applied Geophysics*, Volume 40, Issues 1–3, pages 73–88.

Characteristics of landslides related to various rock types in Korea

S.G. Lee, K.S. Lee & D.C. Park
Department of Civil Engineering, The University of Seoul, Seoul, Korea

S. Hencher
Halcrow China Ltd, Department of Earth Sciences University of Leeds, UK

ABSTRACT: Korea experiences numerous landslides every summer due to intense rainfall and typhoons which result in considerable damage to property, loss of life and disruption to traffic. This paper concerns the nature of landsliding related to rock type in Korea and is based on empirical studies of landslides together with sensitivity analysis of slopes with respect to groundwater levels. The vast majority of natural terrain landslides are very shallow and associated with the sliding of thin weathered soil horizons on underlying stronger rock, irrespective of rock type.

1 INSTRUCTION

Many landslides occur each summer in Korea due to heavy rainfall. These result in 60 deaths on average, hundred of millions US dollars damage to property and considerable disruption to traffic flow (Lee & Hencher, 2007).

More than 70% of Korea is mountainous. Typically in these hilly areas, only a thin layer of soil overlies rock and the majority of landslides occur at the boundary between rock and the overlying soil. In terms of scale, most natural terrain landslides are less than 20 to 30 m in length, less than 10 m wide and less than 1 m deep (Lee, 1987, 1988, 1995; Lee et al., 2008).

Rainfall is an important trigger of landslides but this is not the only factor. Other important susceptibility conditions are underlying geology and topographic situation. This paper reports on a study to investigate the relationships between landslide occurrence and landslide characteristics and geological factors.

Furthermore, back analysis has been conducted to look at the sensitivity of different geological and topographic settings to changes in groundwater conditions.

2 SELECTION OF SURVEY AREAS

Areas were selected for study that had numerous landslides triggered by heavy rainfall in the summers of 2001–2003 and where the landslide damage is still evident and amenable to measurement and analysis. Areas were also selected for study with the aim of discerning the influence of geological factors. The distribution of those areas finally selected areas finally selected for study are shown in Table 1 and Figure 1 (Um & Reedman, 1975).

Table 1. Landslide survey areas and numbers of landslides.

Rock types	Location		2001	2002	2003	Total
Igneous rocks	Hongcheon	(I-1)	36	–	–	36
	Gangneung	(I-2)	–	46	44	90
	Donghae	(I-3)	–	46	–	46
	Gimcheon	(I-4)	–	25	–	25
	Samcheok	(I-5)	–	–	8	8
Metamorphic rocks	Hamyang	(M-1)	–	26	–	26
	Geochang	(M-2)	–	–	4	4
	Samcheok	(M-3)	–	–	7	7
Sedimentary rocks	Pohang	(S-1)	–	2	–	2
	Hapcheon	(S-2)	–	19	4	23
Total			36	164	67	267

3 METHOD OF FIELD STUDY

At each landslide location various aspects were recorded. Overall slope angle was measured using a clinometer; depths, lengths, and widths of each failure were measured physically using tapes. The nature of the slip surface was recorded using Schmidt Hammers (SHV value) to provide quantitative data on the nature of underlying rock, where appropriate (ISRM, 1978).

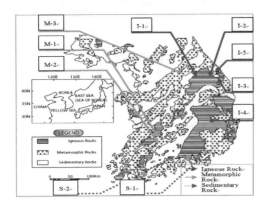

Figure 1. Locations and geological feature of survey areas.

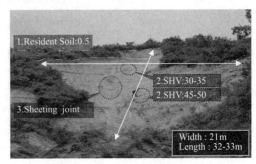

Figure 3. View and geological feature of igneous rock area landslides (Site I-2).

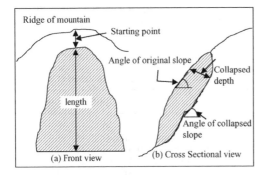

Figure 2. Schematic representation of surveyed features.

Figure 4. View and geological features of typical landslide in metamorphic rock area (Site M-2).

4 GROUND CHARACTERISTICS OF LANDSLIDES

4.1 Igneous rock areas

Most of the surveyed areas underlain by igneous rock have a thin surficial layer of about 0.5 m of weathered granite soil (residual and completely weathered). The underlying rock is typically highly to moderately weathered ("soft rock" in Korean terminology) with Schmidt Hammer rebound values of the exposed surface in the range of 30 to 50 (Hencher & Martin, 1982) (Figure 3).

4.2 Metamorphic rock areas

Most of the metamorphic rock areas are underlain by gneiss and, as in the granitic areas, there is a thin layer of soil with a thickness of 1–2 m underlain by highly, moderately and slightly weathered rock with Schmidt Hammer rebound values on exposed failure surfaces typically of the order of 35–55.

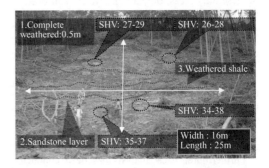

Figure 5. View and geological feature of sedimentary rock area landslides (Site S-2).

4.3 Sedimentary rock areas

In areas underlain by sedimentary rock there is, as in the igneous rock and metamorphic rock areas, a thin soil horizon of perhaps 1 m developed, underlain by relatively strong sandstone and weak shale. The Schmidt Hammer rebound values of the relatively strong sandstone were in the range of 25–35 which is indicative of a *moderately weathered* state for a rock that is moderately strong (unconfined compressive strength >12.5 MPa) or stronger in its fresh state (Anon, 1995). The shale was much weaker with no rebound value from a Schmidt Hammer.

5 GEOMETRICAL CHARACTERISTICS OF LANDSLIDES

5.1 Original slope angles

The slope angles of the surveyed areas were measured using a clinometer. The average angle of those slopes in which landslides occurred was 39°. By rock type, landslides were generally developed in areas underlain by igneous rock in relatively steeper terrain with an average angle of 43°; for metamorphic rock areas the average was 39°; for sedimentary rock the average was 35° with 15% occurring in terrain with an original slope angle of less than 30 degrees (Figure 6).

5.2 Location of landslides relative to mountain crest

The locations of collapse starting points were classified into 10 kinds between the peak points and lowest points starting from mountain tops or mountain ridges. In terms of the location where the collapse started, most collapses frequently started at the 7th–8th ridges.

In case of igneous rock areas and metamorphic rock areas, the frequency of occurrences of landslides at 8th or lower ridges was low; however, in case of sedimentary rock areas, the landslides were occurred at 5th or lower ridges at fairly high frequencies.

5.3 Longitudinal section types of slopes

Collapse type was linear in terms of longitudinal section type regardless of rock types (Figure 8). Especially, 99% of the landslides on sedimentary rock areas were occurred on parallel slopes and the cause was that the longitudinal direction slopes in sedimentary rock areas were almost linear.

5.4 Cross section types of slopes

As shown in Figure 9, the cross section types at the landslide points were classified into three types; the majority of landslides were developed in concave, depressed topography in all rock types although some landslides occurred in flat terrain. The occurrence of landslides in depressed terrain is probably related to surface flow concentration allowing saturation of the

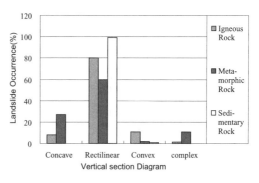

Figure 8. Longitudinal section types by rock types.

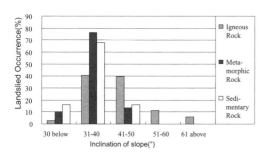

Figure 6. Original slope angles by rock types.

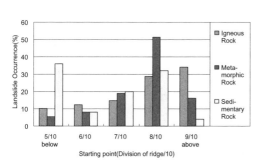

Figure 7. Location of collapse starting points by rock types.

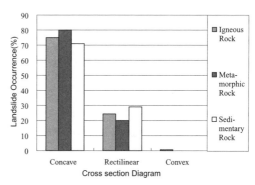

Figure 9. Cross section types by rock types.

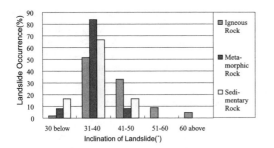

Figure 10. Slope angles by rock types after collapse.

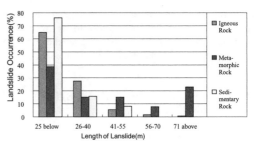

Figure 11. Collapse lengths by rock types.

surficial soil layer, the development of flows along the boundaries between rocks and soil and elevated ground water pressures above those boundaries resulting in reduced shear strength. Surficial and internal erosion (piping) may also be important factors (Hencher et al. 2006).

5.5 Slope angles after collapse

The average angle of all slopes after occurrence of landslides was 37° and in terms of the average values by rock type, igneous rock areas showed 41°, metamorphic rock areas showed 36°, and sedimentary rock areas showed 32° (Figure 10).

The average slope angle after collapse was 37° and was smaller than the original average slope angle, 39°, by 2°. In terms of the change in slope angles by rock type, the angle was changed from 43° to 41° in case of igneous rocks, from 39° to 35° in case of metamorphic rocks, and from 35° to 33° in case of sedimentary rocks.

5.6 Length of collapse

The lengths of collapse were physically measured in the filed with tape measures. The average length was 30 m: in terms of rock types, igneous rock landslides were on average 29 m, metamorphic rock landslides 41 m, and sedimentary rock landslides 13 m. Most of the igneous rock and sedimentary rock landslides were less than 25 m of but landslides in metamorphic terrain were 40 m or longer in 46% of cases (Figure 11).

5.7 Width of collapse

Generally the width of landslides were constant from the upper part to the lower part as might be expected for such shallow detachments of soil above rock, typical of most of the studied landslides. In some cases however, where the width was different in upper, central and lower parts, average values were recorded as the collapse widths. The average of widths of landslides

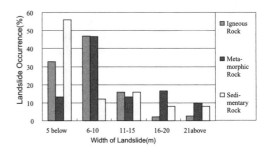

Figure 12. Collapse widths by rock types.

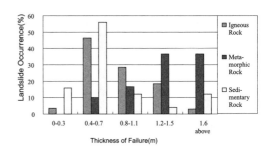

Figure 13. Collapse depths by rock types.

was 10 m; igneous rocks showed 11 m, metamorphic rocks showed 12 m, and sedimentary rocks showed 8 m (Figure 12).

Although the collapse widths differed slightly according to rock type, most of the landslides were 10 m or smaller regardless of rock type.

5.8 Collapse depth

The average depth of landslides was 1.0 m: igneous rocks showed 0.8 m, metamorphic rocks showed 1.1 m, and sedimentary rocks showed 0.9 m (Figure 13).

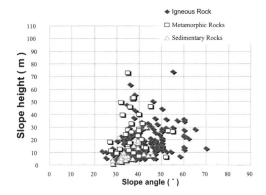

Figure 14. Original slope height versus original slope angle.

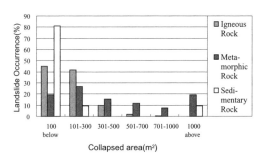

Figure 15. Collapse areas by rock types.

Landslides in areas underlain by sedimentary and igneous rocks were typically shallower than 1 m but in metamorphic rock areas they were deeper than 1.5 m in 37% of cases which is indicative of a greater developed thickness of weathered soil in areas underlain by metamorphic rock.

5.9 Angles and heights of original Slopes

It is shown in Figure 14 that most landslides occurred in slopes with original angles of 30–50° of angle and heights of 10–30 m regardless of rock type although the distribution was different depending upon rock type. Especially, sedimentary rocks showed the landslides of which the slope angles and heights were same with the original angles and heights, compared to other rock types.

5.10 Collapse areas

Surface areas of landslides were 500 m² or smaller in 97% of cases in igneous rock areas, 61% in metamorphic rock areas, and 91% in sedimentary rock areas.

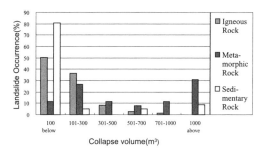

Figure 16. Collapse volume by rock types (Back analysis).

5.11 Collapse volumes

Collapse volumes were compared by rock types as shown in Figure 16. Landslides in sedimentary rock and igneous rock areas were typically 100 m^3 or less in volume in many cases. However, in metamorphic rock areas landslide volumes were sometimes rather larger.

6 CRITICAL GROUNDWATER PRESSURES FOR FAILURE BY BACK ANALYSIS

In Korea, it is recommended cut-slope design standards that ground water is assumed up to ground surface, i.e. 100% (KEC, 2001).

The Slope/W program was used to carry out back analysis of trial landslides using the Janbu (1954) method for interpretation. The shear strength of boundaries between soil and rocks for different rock types were measured under saturated conditions using laboratory direct shear tests (Lee et al. 2008).

When back analysis was performed in consideration of the properties of each rock type and with changing the ground water table, the time of collapse occurrences was regarded as Fs = 1; and then, this assumption was applied to 4 areas to which the back analysis was to be applied and the result was arranged as shown in Table 3 and Figure 17. The underground level on the upper part of rocks at creation of landslides was assumed by back analysis; the distribution at such level was 25–39% in case of igneous rock areas, 20–44% in case of metamorphic rock areas, and 35–49% in case of sedimentary rock areas.

The reason for such difference in underground water levels by rock types in back analysis was deemed that the original slope angle influenced in formation of underground water levels in natural slopes.

Table 2. Percentages of underground water levels by rock types when the factor of safety by back analysis is 1.

Rock types	Site no. 1	2	3	4
Igneous rocks	39%	39%	25%	36%
Metamorphic rocks	44%	20%	36%	20%
Sedimentary rocks	35%	40%	44%	49%

Table 3. Result of back analysis of ground wuawater level, case of igneous rock area no.1 in Table 2.

Ground water level (%)	Factor of safety (FS)	Ground water level (%)	Factor of safety (FS)
100	0.738	50	1.056
90	0.819	40	1.109
80	0.878	33	1.163
70	0.937	30	1.163
60	0.994	20	1.216
		10	1.269

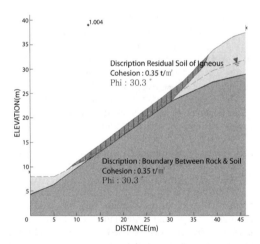

Figure 17. Estimated result of ground water level by back analysis (Slope/W Interpretation cross section).

7 CONCLUSION

Landslides triggered by rainfall in 2001 to 2003 in Korea have been studied to examine their typical characteristics and to consider any variation with fundamental geology.

The vast majority of natural terrain landslides studied, irrespective of rock type, are associated with the sliding on thin, 1–2 m thick, weathered soil horizons overlying stronger, less weathered rock. The boundary is often well defined and acts as an aquiclude allowing the development of perched water tables and through flow. The thickness of the weathered soil layer is sometimes greater in the case of metamorphic rocks but this may be partly related to the location of landslides in metamorphic terrain in valley bottoms rather than side slopes (this may be a study-specific conclusion rather than generally applicable). Failed slopes typically occurred in terrain with original slope angle of 30–50°. Landslides do however occur in gentler terrain in areas underlain by metamorphic and sedimentary rock.

Landslides were typically 20–30 m long, and less than 10 m wide; 85% of collapses occurred at 7th–8th ridges. In terms of longitudinal section type, 80–90% of landslides were created at parallel slopes; in terms of cross section type, 70–80% of landslides occurred in depressed areas, probably associated with surface water concentration.

Also, the underground water level when the factor of safety by back analysis for each rock types is 1 was obtained; igneous rocks took 25–39%, metamorphic rocks took 20–44%, and sedimentary rocks took 35–49% in whole soil layers; such small difference was deemed to be caused by the difference in outflow of underground water depended upon the difference in slope angles of original slopes.

ACKNOWLEDGEMENT

This research was partially supported by a grant (NEMA-06-NH-05) from the Natural Hazard Mitigation Research Group, National Emergency Management Agency.

REFERENCES

Anon 1995. The description and classification of weathered rocks for engineering purposes. Geological Society Engineering Group Working Party Report. *Quarterly Journal of Engineering Geology*, 28, pp 207–242.

Korea Expressway Corporation (KEC). 2001. Road design handbook (II). 406–410.

Hencher, S.R. & Martin, R.P. 1982. Description and classification of weathered rocks in Hong Kong for engineering purposes. *Proceedings of the 7th Southeast Asian Geotechnical Conference*, Hong Kong, 1, pp 125–142.

Hencher, S.R., Anderson, M.G. & Martin, R.P. 2006. Hydrogeology of landslides. *Proceedings of International Conference on Slopes*, Malaysia, pp 463–474.

International Society for Rock Mechanics (ISRM). 1978. Suggested methods for the quantitative description of discontinuities in rock masses. *Int. J. Rock Mech. Mining Sci. Geomech. Abstr.*, 15: 319–368.

Janbu, N. 1954. Application of composite slip circles for stability analysis. *Proc. European Conference on Stability of Earth Slopes*, Stockholm, 3: 43–49.

Lee, S.G. 1987. *Weathering and Geotechnical characterization of Korean Granites*. PhD thesis. Imperial Collage. University of London.

Lee, S.G. 1988. A study on landslide in Korea. Researches on geological hazards, *Research report of Korea Institute of Geoscience and Mineral Resources (KIGAM)*, KR-88-(B)-7: 145–148.

Lee, S.G. 1995. Natural hazard in Korea. *Proc. of the int. Forum on Natural hazard mapping, Geological Survey of Japan Report* (281): 145–148.

Lee, S.G., Hencher & Kim, B.S. 2008. A study on the shear strength of boundaries between soils and rocks in Korea. *Proc. 10th Int. Symp. on Landslides and Engineered Slopes,* Xi'an, China (in press).

Lee, S.G. & Hencher, S.R. 2007. Slope safety and landslide risk management practice in Korea. *Proc. of 2007 Inter. Forum on Landslide Disaster Management,* Hong Kong (in press).

Um, S.H. & Reedman, A.J. 1975. Geology of Korea. *Korea Institute of Geoscience and Mineral Resources (KIGAM)*, Seoul.

Two approaches to identifying the slip zones of loess landslides and related issues

Tonglu Li & Xiaoyan Lin
Department of Geological Engineering, Chang'an University, Xi'an, China

ABSTRACT: With the rapid development of highways on the loess area in recent decades, a large number of loess landslides are encountered and some of them have to be controlled. Unfortunately, the invalid technical methods for landslide exploration generally employed often led to misjudgment of their geological conditions and stability that make stabilizing construction either failure or extra investment. For landslide exploration, it is a key problem to locate the slip zone, so we are going to discuss some practical skills in loess landslide exploration after the research project we have finished which funded by Communication Bureau of Shaanxi Province, China. The skills include drilling technique in loess landslides, identification of loess slip zone and use of electric resistivity for inspecting the slip zone. From in situ and laboratory tests and observation, we have suggested a series of easy operating and effective ways for loess landslide exploration.

1 INTRODUCTION

Loess covers about 631 thousands square kilometers of Chinese land surface, occupies about 4.4 percent of the total land area of China (Sun, 2005). It mainly distributes in the central area of China, the middle reach of the Yellow River in where the source of Chinese civilization is located. The loess region has been rising up by wind blown deposit while cutting down by the tree-like river system. The two reverse processes consequently produce a dense distributed loess slopes in the sides of the river valleys. As they are going on, the slopes are becoming higher and steeper, finally to form a large number of landslides. Statistics of the investigation data suggests that there are 16616 landslides and soil falls developed in the north of Shaan'xi, the density exceeds 5 ones in each square kilometer. In the eastern and western Gansu Province, 14109 loess landslides had developed since the end of 1950s to 1992, the density exceeds 6 ones each square kilometer (Lei, 2001). Loess landslides have been the main geological hazards in the traffic and civil engineering construction. Therefore, some of the landslides which the engineering encountered have to be controlled. In-situ Investigation and exploration of the landslides should be taken before design of the controlling construction. But the engineers now still use some unreasonable methods to give a unreliable geological information, which led to the design being either high risk or high invest, such as the general used continuous-flight auger or direct rotary drilling bit disturb the soil seriously, so the soil from borehole can seldom be used to identify the slip zone. Therefore, we have researched on the aspects of drilling skills of loess landslides, identification of the loess slip zone, and application of electric resistivity, which consists of composite methods for loess landslide exploration.

2 THE DRILLING SKILLS OF LOESS LANDSLIDE

It is common concerned to determine the slip zone of landslides in their exploration, because the features of loess slip zone is changeable. It may be thin, few millimeters to even a contact plane and may be thick, from several centimeters to meters, may has typical shearing traces and may has not, may be single layer or several layers. Therefore, we need continuous undisturbed soil core from boreholes to identify the slip zone, but as mentioned above, the present used drilling skills is not satisfactory for the purpose. The general used method of continuous flight auger or percussion hammer in foundation exploration disturbs the soil completely in the process of driving down, so the sampler is used to collect undisturbed soils in certain depth in this case. While for landslide drilling, the depth of slip zones is difficult to predict and it is not practical to sample in a long range continuously, so the methods is not available. Generally, the direct-rotary method is used to drill rock mass and sometimes to drill soil under water level. This method usually is carried out with circulation-liquid for lubricating and cooling the bit. In the most codes and specifications of China, it is prohibit to use circulation-liquid above the water level in landslide drilling because the liquid will soak the soil core and make it dispersed easily. As drilling without circulation-liquid by this method, the sample is distorted seriously as showing in Figure 1, so it is not available too.

In ASTM standard (2003), an hollow stem auger is introduced to drilling in soils as an conventional methods in American, but it is seldom applied in China. Referring this technique, we design a similar auger which has dual tubes with the inner tube static, the outer one rotated. The inner tube can be spited for easily taking the soil core out and the outer one has a continuous flight auger round its out surface. The tip of the outer tube attaches three hard steel blades as cutter heads whose width is in the range from the out side of the auger to the inner wall of the inner tube. The out side of the auger attaches vertical steel rids on each side which plays a role in smoothing the borehole wall and keeping vertical of the hole. Fig. 2A shows the feature of the designed auger.

(A)

(B)

Figure 1. Loess samples taken by direct rotary method without circulation fluid. (A) Brown-yellow Loess and red paleosol mixed together, completely disturbed; (B) The white concretions in brown-red paleosol was crushed, also completely disturbed.

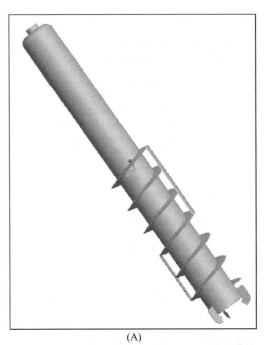

(A)

(B)

Figure 2. A dual tube auger (A) A sketch of the auger. (B) Drilling with the auger, the soil cohered on the flight of the outer tube was taken off, then the soil core in the inner tube was taken out while.

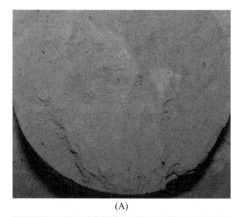

(A)

(B)

(C)

Figure 3. Soil cores taken by dual tube auger (A) From slip bed, the loess the perfect original structure of root holes being clearly seen; (B) From slip zone, loess become schistose and dense; (B) From slip zone, paleosol become hard and compacted.

The drilling test with the auger was carried out in a large loess landslide, the Taiping landslide to the north of Xi'an City. With rotation of the drilling rod, the cutter head would cut down the outer ring part of the soil first and keep the inner rounded part without disturbing; meanwhile the cuttings of the soil outside would go along the flight auger and the soil core inside goes into the inner wall up. The soil cohered on the flight of the outer tube was taken off (Figure 2B) while the soil core in the inner wall was taken out. The soil core keeps dry and intact. Figure 3A is the sample taken by the auger, which has really perfect original structure of the loess. The samples could be used to not only identify the slip zone, but to do the laboratory tests for physical and mechanical properties. Figures 3B and 3C show the core soils taken from the slip zone of which the original material is loess and paleosol respectively; they are schistose and compacted apparently.

3 IDENTIFICATION OF THE LOESS SLIP ZONE

The above technique provides us a continuous intact soil core, but how to recognize the slip zone from the core is also an essential work in landslide exploration.

Megascopically, as observing the structures of loess from slip mass through slip zone to slip bed in the excavated investigation pits or the soil core from boreholes, the differences among them are apparent. The slip mass has loose structure and the red paleosol generally is enclosed by yellow-brown loess or inter-enclosed each other. The root holes have been damaged by compression and extension in most portions, but locally they can be seen in the centre of some soil blocks. These appearances are not difficult to be observed. The soil of slip zone shows compact and homogonous structure, the root holes are crushed completely by extensive compressing and shearing. The foliations and striates within the zone could be seen times but not always. The soil of slip bed has the typical structure of loess that is homogonous with vertical root holes and inter-bedded loess-paleosol sequences. Both in excavated pit and soil cores, looking for the slip zone should be from lower to upper. Because the slip bed is easy to be recognized, as the soil structure going to complex, it implies that the soil was disturbed by slipping. Even we could not found the typical features of slipping movement as foliations and striates, variation of the soil structures is helpful to judge the position of slip zone.

Microscopically, the soil of slip zone shows apparent compacting and distorting effect relative to the natural soil. Comparing the SEM photos under same amplified times between the loess undisturbed and that disturbed in slip zone, it is found that the undisturbed loess has typical hollow-skeletons structure,

(a)

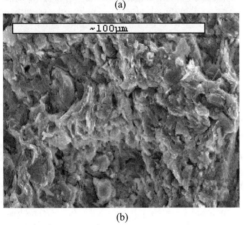

(b)

Figure 4. Comparing of the microstructures between the undisturbed loess and the loess in slip zone. (A) Microstructure of intact loess (500 times); (B) Microstructure of the loess in slip zones (500 times).

while the loess in slip zone is extensively compacted and the scales of clay minerals are distorted. However, the directed features megascopically do not reveal in microscopic structures (Figure 4).

Quantitatively, to find differences of the physical indexes among the soils from slip mass to slip bed, we sampled with short intervals of 5 cm on the four investigation pits in which the slip zone outcropped out. The pits was excavated near the back of the four large loess landslides of Dongfeng, Taiping, Shutangwang and Xiuchidu which located to the north side of Jinghe Rive and in the south side of a loess platform, to the north of Xi'an City. The sampling section is nearly perpendicular to the slip zone and at least one sample in slip zone was collected. Then the samples were tested in lab for the parameters of water content, plastic limit, liquid limit, particle components.

Plot these parameters and their relevant results verse to the position relative to slip zone. It reveals that the water content of slip zone is higher than that of slip mass and slip bed, but there is one exception that the water content above the slip zone is higher in Taiping landslides (Figure 5A); plastic limit is lower in slip zone without exception (Figure 5B); liquid limit is higher in slip zone with one exception that the Dongfeng landslide has a lower liquid limit in slip zone (Figure 5C), plastic index is higher in slip zone, but in which Dongfeng landslide is not so apparent (Figure 5D), and liquid index is higher in slip zone, in which Taiping landslide is not so apparent (Figure 5E) and average particle diameter is higher in slip zone in all the four landslides (Figure 5F). It implies that abrasion between slip mass and slip bed has made the soil finer and compact, which correspondingly changed its water content and plasticity. Even so the indexes have exceptions in some of the landslides which may be caused by the processes of sampling and testing, they are helpful to determine the slip zones of the loess landslides.

4 APPLICATION OF ELECTRIC RESISTIVITY IN LOESS LANDSLIDES

Geophysical methods are often used to prospect the slip zone of landslide, such as seismic method, electric method, geological radar and twinkling alteration electric-magnetic waves etc. However, the methods based on wave transition may be suitable to inspect the large scale and deep sited geological body. In landslide prospecting, the slip zone is generally no more than 100 meters depth, so the error of these methods is not suitable to locate the exact position of slip zone. Electric method has the similar difficulties, so it should not be available alone. While combining with in-situ exploration, an effective result may be produced.

Therefore we apply the electric method in investigation pits and boreholes for further confirming the slip zone. First, the electric resistivity was measured on the section crossing slip zone in the pits. They were measured in three states of natural, freshwater saturated and salt liquid saturated to the sections. The results shown in Fig. 6 clearly reflect that resistivity of both natural and freshwater saturated decline sharply as they cross the slip zone; the resistivity of slip mass is much higher than that of slip bed. That of the freshwater saturated soil is a little lower than natural soil. As the soil is saturated by salt liquid, the resistivity of slip mass and slip bed has no much difference, the electric property mainly controls by dielectric other than soil properties. The high resistivity of slip mass may be because of its loose structure. The results suggest that we can directly measure the resistivity to confirm

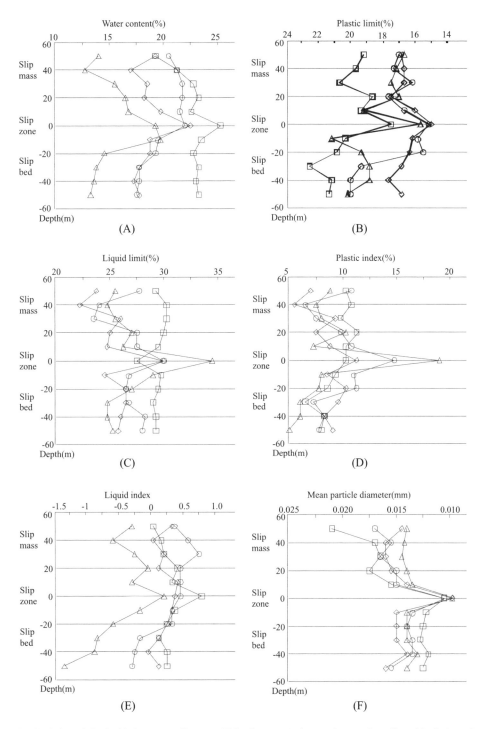

Figure 5. Variation of physical indexes near slip zone. Of the figures, -△-denotes the samples collected in the investigation pit on the back of Shutangwang landslide; -◇-denotes that of Xiuchidu landslide; -□-denotes that of Dongfeng Landslide; -O-Denotes that of Taiping landslide.

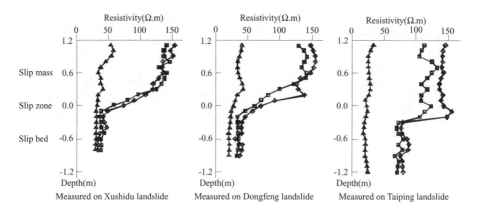

Figure 6. Variation of eclectic resistivity near slip zone in the exploration pits. Of the figures, -◇-denotes measuring on natural state; -□-denotes measuring after water saturated; -△-Denotes measuring after salt liquid saturated.

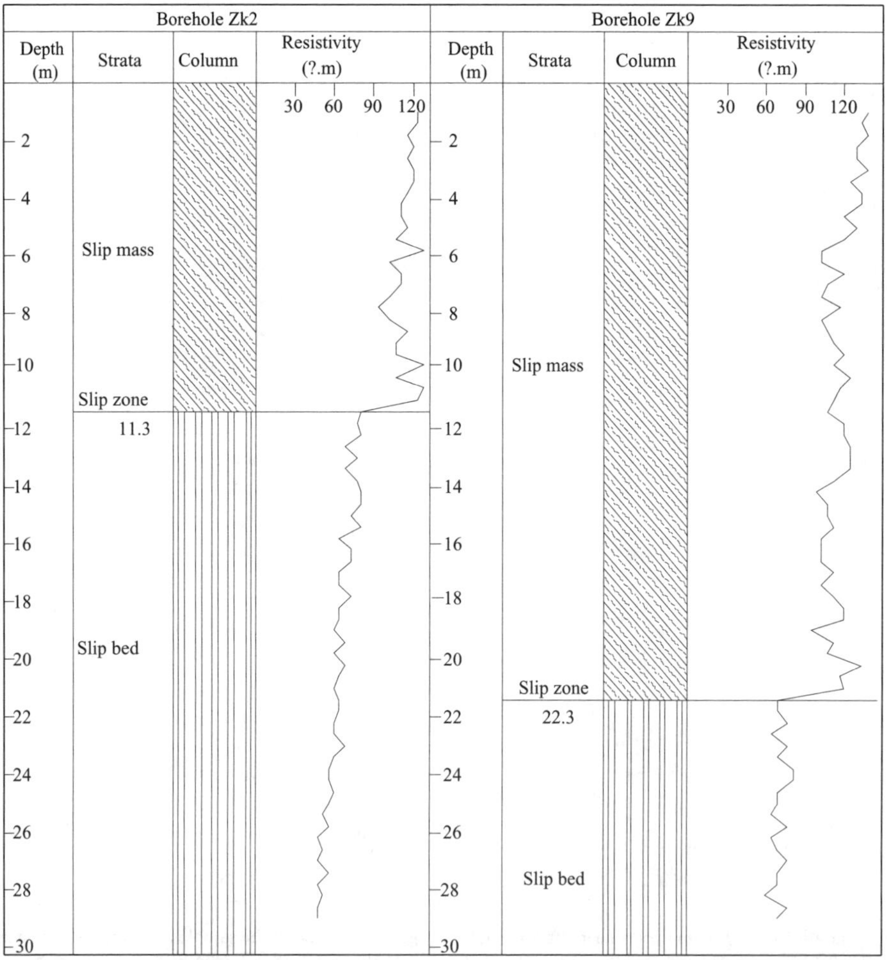

Figure 7. Eclectic resistivity logging in boreholes.

the slip zone of loess landslides. This method could further be used in electric logging to determine the slip zone in boreholes. Figure 7 are the electric logging curves of the two boreholes in the Wohushan landslide near Yan'an City Shanxi Province. The whole holes are loess and the slip zone is not so confident by observing the soil core. From the logging curve, we clearly see the discrepancy between slip mass and slip bed, the slip zone could be confidently confirmed.

5 CONCLUSION

In Chinese loess area, some of old landslides may be reactivated by natural agents or human engineering disturbing, but many engineers are always confusable to determine the slip zone. The main reasons are the invalid drilling skills and unclear of the criteria for slip zone determination. Here suggests some practical and easy operated methods in loess landslide exploration. The above motioned is concluded as the follow aspects:

1. A dual tube auger was designed for the loess landslide drilling, which can collect continuous undisturbed soil core for slip zone determination and laboratory test of soil properties.
2. Determination of slip zone is the most important work in landslide exploration. As observing in megascopic and microscopic scales carefully, the structural differences of the soils among slip mass, slip zone and slip bed are apparent. the physical indexes as water content, liquid limit, plastic indexes and liquid index shows higher values in the slip zone while plastic limit and mean particle diameters is lower in it, therefore, the slip zone of loess landslide can be easily identified.
3. The resistivity has great discrepancy between slip mass and slip bed in loess landslide, the value of slip mass is much higher than that of slip bed. It implies that the steepest point on the curve of resistivity verse depth is corresponding to the position of slip zone. The method applied to borehole electric logging is simple and effective to confirm the depth of the slip zone

ACKNOWLEDGEMENTS

The research work was funded by the projects from National Natural Science Foundation (Project No.40772181) and Communication Bureau of Shaanxi Province. The postgraduate students, Long Jianhui, Fu Yukai and Lei Xiaofeng, attended the field investigations and measurements. Here our thanks are extended to the ones contributing this work.

REFERENCES

Lei Xiangyi. 2001. Geological hazards and human activity on the Loess Plateau. Beijing: Geological Press.
ASTM 2003. *Annual Book of ASTM Standards*, Vol.04.09, West Conshohochen.
Li Xingsheng & Li Tonglu. 1997. The characteristics and analysis of loess landslide in China. *International Symposium on Landslide Hazard Assessment*. 361–366.
Sun Janzhong. 2005. *Loessology*. Hong Kong Archaeological Society Press, Hong Kong.

Testing study on the strength and deformation characteristics of soil in loess landslides

H.J. Liao
Department of Civil Engineering, Xi'an Jiaotong University, Xi'an, China

L.J. Su
School of Civil Engineering, Xi'an University of Architecture and Technology, Xi'an, China

Z.D. Li & Y.B. Pan
Department of Civil Engineering, Xi'an Jiaotong University, Xi'an, China

H. Fukuoka
Disaster Prevention Research Institute, Kyoto University, Uji, Kyoto, Japan

ABSTRACT: Landslides groups in southern highland area of Jingyang are investigated and analyzed. Characteristics of and reasons for the landslides were statistically analyzed. A case history of a large-scale landslide in loess in Dongfeng village of Gaozhuang town in Jingyang is studied. The displacement, maximum width, circumference and volume of the slipping mass were obtained through in situ measurement. Laboratory tests were carried out on soil samples taken from a depth of 2 m below the front edge of the slipping mass. Basic property and shear strength parameters of the loess soil were obtained. These results will benefit further studies on the kinetic mechanism of loess landslide in southern highland area of Jingyang.

1 INTRODUCTION

There are frequently occurred geological disasters in the vast territory of China. Every year, landslides, debris flows, ground collapses and etc. bring enormous economic loss and casualties, especially the landslides (Fig. 1) (Ai, Z.X. 2005). In China, loess is widely distributed, especially in the northwest area, where the loess is the most representative. Various geological disasters, especially the loess landslides, will be encountered inevitably with the Development of the West Regions. The loess landslide disasters are one of the major landslides in special soil (Wu et al. 2002). According to incomplete statistics from 1983~1993, about 200 loess landslides had collapsed only in Guanzhong area of Shaanxi, the direct economic loss caused by which is more than 100,000,000 yuan (Lei, X.Y. 1996). There are over 1000 active landslides at present, which bring serious threats to human's property and life.

The loess is widely distributed in China and landslides in loess are very typical. It is of great importance for preventing landslide to study its strength deformation characteristics and the mechanism of its formation and movement by both theoretical and testing methods. In the southern highland area in Jingyang, Shaanxi, landslide and collapse develop seriously and bring severe threats to human's life and property. Based on a loess landslide in the southern highland area in Jingyang, the causes of landslide were analyzed and physical and mechanical tests were performed and analyzed aiming at providing some information for further study on the mechanism of loess landslide occurred in this area.

■ Ground fissures ■ Ground collapses
■ Debris flows ■ Dilapidation ■ Landslides

Figure 1. Statistics of geological dissasters occurred in 2004.

2 IN SITU INVESTIGATION AND ANALYSIS OF LANDSLIDE GROUP IN SOUTHERN HIGHLAND AREA IN JINGYANG

The southern highland area in Jingyang is situated to the south of Jingyang County and on the south bank of Jinghe River. The area of the southern highland is 180 km² which is 23.1% of the whole county's. The elevation of the highland is 430~500 m. Steep slopes eroded by Jinghe River connect the floodplain at the edge of the southern highland area in Jingyang. The steep slopes on the edge of the highland provide gravity conditions for landslides. Therefore landslides with different scale are often evoked by external factors such as frequent irrigating activities, rainfall and earthquake, etc.

After Weihe River was channelled towards large areas in the southern highland area in Jingyang for irrigation in 1976, landslides occurred more and more frequently. More than 40 landslides had been recorded by 2004, including 7 huge landslides and several collapses. More than 1.60×10^7 m³ of soil and rock cascaded down, 29 were killed and 27 injured, 140 hm² of farmland was damaged, over 10 houses and nearly 200 cave-houses were demolished, and 100 cattle were killed. The total direct economic loss reached about 3.00×10^6 yuan (Wang et al. 2004).

The surface area of the southern highland in Jingyang is quite broad. It is higher in the northwest lower in the southeast. The surface of the highland is 30~90 m higher than the floodplain of Jinghe River. The angles of the slopes lie between 45°~80° (Lei, X.Y. 1995).

Taking Gaozhuang town as the starting point, variations of the angle, height and volume of some previous occurred landslides were statistically analyzed along the southern highland area from east to west. The results are shown in Figure 2. Lines shown in the figure are exponential trend lines of the measured data. It is obvious that the angle, height and volume of the landslides tend to increase from east to west. Especially the volume increases very obviously, which is related to the local geological and topographical conditions.

According to the literature and in situ investigation, the frequency occurrence of landslides is closely related to the structure of loess layer. The loess in southern highland in Jingyang is loose in which vertical joints developed. It is with small cohesion and large permeability, and is easy to be softened by water. The main component of the slopes is paleosol or loess in late and middle Pleistocene. When the soil is dry, it is dense and with a large cohesion and high shearing strength. When it is infiltrated by water, the cohesion and shear strength will decrease sharply and the loess will become soft or flowing plastic form which forms slip planes of landslides. The overall characteristics of the landslides group in southern highland area of Jingyang are as following:

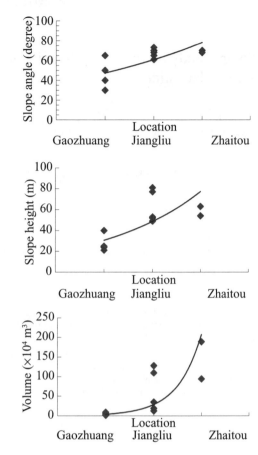

Figure 2. Statistics of landslides from east to west along the southern highland area in Jingyang.

1. Along the boundary of the highland, the landslides group is distributed like a belt due to the effect of topographical characteristics and erosion of Jinghe River.
2. The landslide blocks moved fast and could reach large displacements. According to the author's inspection, the largest displacement of the slope in Dongfeng village of Gaozhuang town may reach 300 m after it slipped from its toe.
3. In southern highland area of Jingyang, the slope stability is low. Old landslides might revive and new landslides might mobilize behind the wall of the old landslides.
4. The sliding mass often cuts the loess layer deeply and slides out from paleosol at the toe of the slope.

5. Rainfall could evoke landslides. Landslides in southern highland area of Jingyang often occurred during or after rainfalls.

According to the in situ inspection, the typical characteristics of the development of loess landslides in southern highland area of Jingyang are as follows: most of them maintain a shape like a round-backed armchair; the wall of the steep slope is almost vertical; the sliding mass moves along the floodplain after it slides out from the slope toe. The sliding mass looks like a tongue and undulant terrain is formed by alternately arranged drumlins and depressions.

A case history of a large-scale loess landslide in Dongfeng village of Gaozhuang town in Jingyang is studied in this paper (Fig. 3). The landslide occurred at 4:40 AM on July 23, 2003. A sliding mass of 53,360 m² moved along the floodplain of Jinghe River. In situ inspection of the landslide was conducted in April 2006. Undisturbed soil sample was taken in order to perform laboratory test.

Based on the measured data, characteristic of the landslide can be obtained. The displacement of the slipping mass along the floodplain of Jinghe River is more than 300 m. The maximum width is 400 m. The surface area covered by the slipping mass is about 106,421 m². The circumference and volume of the sliding mass are 1192 m and 1.28×10^6 m³ respectively.

Figure 3. A photo of the landslide in Dongfeng village.

3 LABORATORY TESTING STUDY

3.1 Basic properties and collapsibility analysis

Basic property tests were conducted on soil samples taken from a depth of 2 m below the front edge of the slipping mass of the landslide in Dongfeng village. The results are shown in Table 1.

Figure 4 shows the results from the particle size distribution test. The percent of particles with diameters larger than 0.075 mm is 27.8%, which is smaller than 50%. The percent of particles with diameters smaller than 0.005 mm is larger than 10%. The plasticity Index I_p is between 3 and 10. Therefore the soil can be classified as loess-like clayey silt. The characteristic of thixotropy and liquefaction of silt is an important factor that causes the landslide group.

Collapsibility is one of the typical characteristics of loess. Double-line collapsibility test was carried out on soil samples taken from the southern highland area in Jingyang and the results are shown in Figure 5. Based on the initial stress level at the depth where the soil samples were taken, the calculated coefficient of collapsibility is 0.03 and the soil can be classified as slightly collapsible loess. It is observed that most of the cracks are due to collapse of the loess. The cracks are located at edge of the depression area caused by collapse of the loess. Depression area was found on farmland a few days before the landslide occurred in Dongfeng village in Jingyang.

3.2 Results from direct shear tests

Both unconsolidated and consolidated direct shear tests were performed under four different vertical pressures of 100 kPa, 200 kPa, 300 kPa and 400 kPa. These two types of tests are suitable for clay whose coefficient of permeability is smaller than 10^{-6} cm/s.

Results of the direct shear tests are shown in Figure 6 and Figure 7. The cohesion and internal friction angle from unconsolidated direct shear tests are 35.74 kPa and 33.8° respectively while those from consolidated direct shear tests are 85.77 kPa and 24.3° respectively. The shear strength parameters obtained from consolidated tests are larger than those obtained from unconsolidated tests, which is in accord with the fact.

Table 1. Basic property test results.

Water content w_n (%)	Density ρ (g/cm³)	Liquid limit w_L	Plastic limit w_p	Specific gravity d_s	Coefficient of collapsibility δ_s	Coefficient of compression a_{1-2} (MPa⁻¹)
6.40	1.80	22	13	2.72	0.03	0.155

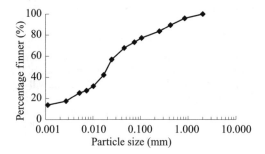

Figure 4. Particle size distribution results.

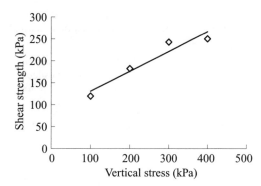

Figure 7. Results of consolidated direct shear test.

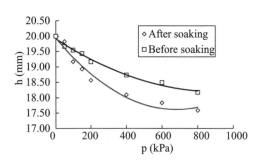

Figure 5. Curves of the collapsibility test.

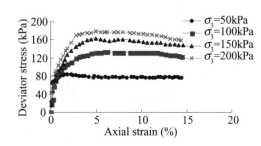

Figure 8. Deviator stress-axial strain curves.

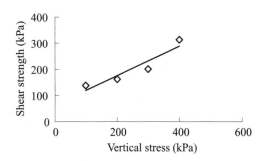

Figure 6. Results of unconsolidated direct shear test.

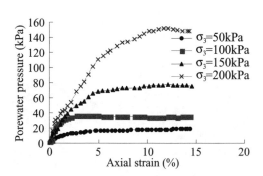

Figure 9. Porewater pressure-axial strain curves.

3.3 Results from triaxial tests

Consolidated undrained (CU) tests were carried out on undisturbed soil samples under confining pressures of 50 kPa, 100 kPa, 150 kPa and 200 kPa. The results are shown in Figures 8 to 10.

During shearing, strain softening was observed after the deviator stresses reached their peaks at the axial strain of 2%~4%. The cohesion and internal friction angle obtained from the failure envelope are 39.54 kPa and 10.4° respectively.

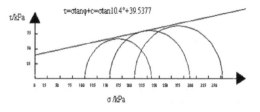

Figure 10. Failure envelope of the CU tests.

The obtained shear strength parameters are smaller than those obtained from consolidated undrained direct shear tests. The reason for this is that the shearing plane is fixed between the lower and upper box in direct shear tests but the weakest plane may not be this fixed plane. The shearing plane in triaxial tests is not fixed and the water drainage is strictly controlled. Therefore the results from triaxial tests are more close to the theoretical value (Yan, L.F. 2004). The shear strength parameters of loess are quite scattered so that different results would be obtained from different type of tests. The shear strength parameters obtained from the above tests are close to the actual value of loess (Wang, J.D. 1999).

4 CONCLUSIONS

The landslide groups in southern highland area of Jingyang were evoked by both internal and external factors. The internal factors include new tectonic movement, topography and geography, lithology and underground hydrology conditions etc. The external factors include irrigation, excavation, and rainfall etc.

Based on the large-scale loess landslide groups in Dongfeng village of Gaozhuang town, in situ observation and a series of laboratory tests have been carried out. Stress-strain curves and shear strength parameters of the loess were obtained. These results will benefit further studies on the kinetic mechanism of loess landslide in southern highland area of Jingyang.

ACKNOWLEDGEMENT

The research is financially supported by Nature Science Foundation of China (50379043). The authors would like to acknowledge the "FY2006 JSPS Invitation Fellowship Program for Research in Japan" for its support.

REFERENCES

Ai, Z.X. 2005. Analysis of landslide evoked by rainfall. *Disaster and Prevention Engineering* (2): 9–11.

Lei, X.Y. 1995. The hazards of loess landslides in the southern tableland of Jingyang county, Shaanxi and their relationship with the channel water into fields. *Journal of Engineering Geology* 3(1): 56–64.

Lei, X.Y. 1996. The study of loess landslide type caused by human activities in Guanzhong of Shaan'xi province. *Hydrology Geology and Engineering Geology* (3): 36–39.

Wang, D.Y., Du, Z.C. & Zhang, M.S. 2004. Geological hazard of cliff collapse, landslide and their occurrence in southern Jingyang county of Shaanxi province. *Bulletin of Soil and Water Conservation* 24(4): 34–37.

Wang, J.D. 1999. *The systematic geological research of typical high-speed loess landslide groups*. Chengdu: Sichuan Science Press.

Wu, W.J. & Wang, N.Q. 2002. Basic types and active features of loess landslide. *The Chinese Journal of Geological Hazard and Control* 13(2): 36–40.

Yan, L.F. 2004. Comparison of contrast between direct box shear test and triaxial test. *Shanxi Architecture* 30(24): 64–65.

Failure mechanism of slipping zone soil of the Qiangjiangping landslide in the Three Gorges reservoir area: A study based on Dead Load test

Xianqi Luo & Ailan Che
School of Naval Architecture, Ocean and Civil Engineering, Shanghai Jiao Tong University, China

Ling Cao & Yuhua Lang
China Three Gorges University Key Laboratory of Geological Hazards on Three Gorges Reservoir Area, Ministry of Education, China

ABSTRACT: The Qianjiangping landslide occurred during the rainy season after the first impoundment of the Three Gorges Reservoir up to 135 m. Rainfall and reservoir water are considered as two main inducements for its failure. The failure mechanism of the Qianjiangping landslide induced by these two factors was explored through Dead Load test (DL test). In the test, consolidation pressures σ_1 and σ_3 were kept unaltered to simulate the deformation behavior and strength characteristics of the landside slipping zone soil while the matrix suction decreasing.

1 INTRODUCTION

The Qianjiangping landslide took place in Shazhenxi Town, Zigui County, Hubei Province, PRC at 0:20 a.m. on July 13, 2003. 129 shelters had been damaged; 1200 men had left homeless; 14 died and 10 were missing till the morning of July 20. This landslide caused a traffic halt, collapse of shelters and workshops, block of the Qinggan River and enormous economic loss. Four enterprises, a silicon company, a brick-fabricating factory, a transportation company, and an architecture and construction company, were destroyed. The rush of the landslide fiercely struck the lower part of Shazhenxi Town with huge waves. Some anchors anchored on the dock were overturned; the vegetation partially demolished; the use value of the land debased.

The Qianjiangping Landslide locates on a slope with southeast trend along the left bank of Qinggan River, which is in Qiangjiangping Village, Shazhenxi Town, Zigui County. The angle of the slope is $35°\tilde{1}5°$, from the top to the bottom. The landslide develops in clastic sedimentary rocks of lower-middle Jurassic Niejiashan formation. In terms of lithology, it consists of medium and thick layer siltstone, silty mudstone and shale. The incline of the terrane is roughly identical with that of the slope, its angle is $35°\tilde{1}9°$, from the top to the bottom, i.e., it is a dip slope. Owing to the influences by regional structure, soft interlayer in the dislocation-mudding zone was more developed. And the attitude of stratum was chiefly stable; its variable scope was $110°\tilde{1}50°\angle15°\tilde{3}0°$. There were many faultages, fissures and soft interlayer in this region, which worked as the intrinsic condition and material basis for forming the landslide.

The impoundment of Three Gorges Reservoir has been planed to carry through in three stages of time: the first stage starts from June 1, 2003, when the reservoir water level is 67 m, to June 15, 2003, the water level

Figure 1. Location of the Qianjiangping landslide.

then rises up to 135 m; the second stage lasts from September 22, 2006 to October 27, the reservoir water level rises from 135 m up to 156 m; the last stage will begin in the year 2009, the water level when finished will go up to 175 m.

The Qiangjiangping landslide is a dip rocky-slope; its geomorphic features, geologic constitution and structure, and physical composition are quite typical in the Three Gorges reservoir area. Features of dip slope and that of stratum structure with soft and hard rock layer spaced in-between are seen on the location of the Qiangjiangping landslide. They provide physical basis for new failures of reservoir landslide. The landslide occurred during the rainy season after the first stage of impoundment of the Three Gorges reservoir, it is natural to associate its potential inducement with the rainfall and reservoir water. It is definite that these two factors are quite common in the Three Gorges Reservoir area, so that it is significant important for later landslide controlling and forecasting in the area.

Figure 2b. View of the Qianjiangping landslide from the upstream side of the Qingganhe River.

2 COMPOSITION OF THE QIANGJIANGPING LANDSLIDE SLIPPING ZONE SOIL

The Qiangjiangping landslide is a dip slope with soft and hard stratum layers spaced in-between. Carbonaceous shale is merely appeared in one of the layers within the extent of the landslide, which thickness is 20–30 cm. It is the primary intercalation of the landslide. Interlamination stick slip during the course of anaphase tectogenesis rebuilds the layer containing carbonaceous shale, and then forms an intercalated crushed dislocation-mudding zone, which worked as the slipping zone for the Qiangjiangping landslide during the rainy season and after the first stage of impoundment of Three Gorges Reservoir.

Such typical slipping zone soil with quite clear distinctions are found in the Qiangjiangping landslide

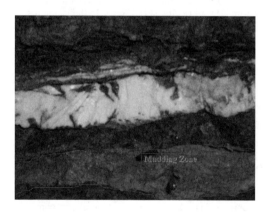

Figure 3. Picture for the intercalated crushed dislocation-mudding zone.

Figure 2a. Overlook of the Qiangjiangping landslide.

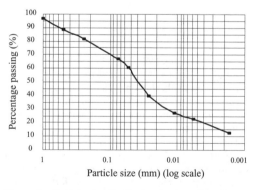

Figure 4. Particle-size distribution for the soil used in DL test.

Table 1. The average index value of its physical property.

Wet density/dry density (g/cm)^3	Natural water content (%)	Specific gravity	Void ratio	Saturation (%)	Liquid limit (%)	Plastic limit (%)
2.03/1.79	13.5	2.70	0.52	70.2	34.7	18.3

in prospecting drift wall mapping, borehole core and field surface outcrop. Since the slipping zone soil mass may be affected, more or less, by water and slide movement, samples for test were collected from No.1 prospecting drift wall mapping in the middle part of hauling area of the landside mass, where most typical slipping zone soil were found (Figure 2).

The intercalated crushed dislocation-mudding zone resulted from partial mylonitization of broken porphyry. Its protolithic composition is described: carbonaceous shale and limestone banding/conglomeration spaced in-between. Cataclasite distributed on the dislocation surface. Its terrane was extruded into flakiness. Agglutination was found between calcareousness and mudding, and weathering was noticeable. Steps and longitudinal scratches were often found. The granulometric composition obtained from particle size analysis on slipping zone soil sample was as follows: particles with its diameter >2 mm count 49 %~53 %; particles with its diameter 2~0.075 mm, 43 %~47 %. Powder cosmids with its diameter <0.075 mm, is only contained about 4%. And Figure 4 is the particle-size distribution for the soil used in DL test. The average index value of its physical property is shown as table 1.

3 CONSOLIDATED-UNDRAINED TEST (CU TEST) OF THE QIANGJIANGPING LANDSLIDE SLIPPING ZONE SOIL

Generally, the Qiangjiangping landslide slipping zone soil is unsaturated, which shear strength usually get from CU test where suction and net confining pressure are under control simultaneously.

3.1 Sample preparation

Before the test, the soils collected were dried and scrunched, and then screened through with a 2 mm screen. Then it was made into wet soil with reference to the liquid limit water content. After 24 hours' moist curing, which served to homogenize water content in the soil, it was chopped into remodeling clay soil samples, with 100 mm height, and 50 mm diameter. The proportion, weight, natural water content, dry density and void ratio were measured respectively (Table 2).

Table 2. Physic parameters of the Qiangjiangping Landslide slipping zone soil.

Unit weight (kN/m^3)	Specific gravity	Water content (%)	Dry density (g/cm^3)	Void ratio
19.6	2.70	19.6	1.67	0.57

3.2 Test instrument

The instrument used for this test was GDS. Unsaturated Triaxial Testing System (GDS UNSAT) produced by GDS in UK. The biggest strongpoint of this instrument stays with its preciseness in measuring the volume of samples; and also their volume changing can be read at any moment during the test by surveying, continuously and automatically, the water-level discrepancy between that in the inner cell and that in the reference tube through a sensitive differential pressure transducer (Zhang and NG 2006).

3.3 Testing program

The test is conducted through 4 individual groups in accordance with different suctions; they are 0 kPa, 50 kPa, 100 kPa and 200 kPa, respectively. In each group, there are 3 samples, which net confining pressure $(\sigma_3 - u_a)$ are, 50 kPa, 100 kPa, and 150 kPa, respectively.

Mohr circles under the same suction are put in the same group; the common tangent of these circles is its failure envelope. And the intercept and slope of the failure envelope are respectively c and φ under different suction (Table 3).

From Table 3, it is noticed that the value of ϕ has no significant changes within the range of suction (0~200 kPa) given in the test. When the given value of suction is 0, it was actually a triaxial test of saturated soil, therefore it is $\phi' = 26.6°$.

Table 2 demonstrates principally a linear relationship between gross cohesion C and suction $u_a - u_w$. Its linear equation is: $c = c' + (u_a - u_w)tg\phi^b$. Where, c' is the intercept of Mohr-Coulomb envelope surface and shear stress axis, when both matrix suction and net normal stress are 0, that is, the intercept of gross cohesion. From the equation, we get: $c' = 15.7$ kPa, $\phi^b = 22°$.

Table 3. c and φ of the Qiangjiangping landslide slipping zone soil under different pressures.

Suction (kPa)	0	50	100	200
c (kPa)	10	40.3	59.1	94
ϕ (°)		26.6	27.2	27.5	28.2

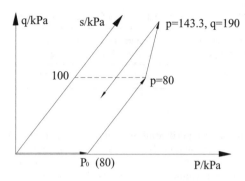

Figure 5. Stress path of the DL test.

4 DEAD LOAD TEST (DL TEST) OF THE QIANGJIANGPING LANDSLIDE SLIPPING ZONE SOIL

During the action of rainfall and reservoir water, the shear breakdown to soil mass owes chiefly to the increase in pore water pressure or, in other words, the decrease in matrix suction. According to Brand (1981) and Chen et al (2004), the decrease in matrix suction will cause debasing effective stress on the potential slip zone of the slope, where the shear strength will be weaken down. Anisotropic triaxial test can simulate the process in which the pore water pressure rises from a negative value till the failure of soil mass. If σ_1 and σ_3 remain unaltered, unsaturated soil test can be a relatively accurate simulation to the failure stress path of the unsaturated-soil slope when couple affected by the rainfall and reservoir water, and therefore the deformation features of the slope soil mass could be examined.

Dead Load test of unsaturated soil is also able to simulate the influences upon the strength and volumetric strain behavior of the soil mass while the matrix suction is decreasing during the process of rainfall and rising level of reservoir water.

4.1 Sample preparation, test instrument and testing program

Sample preparation and test instrument are identical with that in the CU test. And the testing program is as follows:

Figure 5 shows the stress path of the DL test. The test consists of four stages: saturation and consolidation, moisture dehydration, anisotropic consolidation and suction decrease. The next is our test operations.

At first, put samples into the pressure chamber, and give counter pressures to saturate and consolidate them. Then take $p - u_a = p_0$ under control to make sure its value keeps unaltered. When the pressures given from each direction are identical, gradually rises the matrix suction from 0 to a certain value. Anisotropic consolidation begins after the suction becomes stable. The top of each sample are made connected with the probe of the axial pressure sensor while anisotropic consolidation $\sigma_1 = \sigma_3/k_0$ take place. Here coefficient of at-rest earth pressure $k_0 = 1 - \sin\varphi'$ (refer to results in the CU test for the value of φ', namely, $\varphi' = 26.6°$ $k_0 = 0.55$). After the stabilization of anisotropic consolidation, measure the gross volume deformation, axial deformation, lateral deformation and volume change of the specimen, and calculate the water content of the soil mass. The increasing rate of pressure and the criterion to stop the application of suction is referenced (Zhan 2003; Sivakumar 1993; Yang 2005; Fredlund and Rahardjo 1997).

From Geotechnical Test Regulations (SL237–1999) for the criterion to stop the consolidation stage during triaxial test, the tonnage is less than 1 mm^3 within 2 hours.

After the stabilization of consolidation, u_a remains unaltered, u_w increases at each phase, in other words, the matrix suction gradually decreases. During such a process, the pressure of anisotropic consolidation keeps unaltered. It is meaning that gross stress σ_1, σ_3 and deviatoric stress $q = \sigma_1 - \sigma_3$, as Dead Load, constantly acts on the samples. Shear deformation occurs along with the decreasing of matrix suction. While the samples are reaching shear, use the strain controlling mode within "4D UNSATURATED" module of GDS till axial strain ε_a reaches its maximum value, which can be regarded as the sheer failure of the samples.

4.2 Results and discussions

(1) The stage of moisture dehydration

The backpressure and consolidation is given as the initial state of this stage, which is when the tonnage and gross volume deformation is measured after the consolidation; the water content was calculated to 18.50%. The suction load was increased at each individual phase, which target value is 20, 60, 100 kPa, respectively.

Figure 6a shows the variation in the water content during the stage of moisture dehydration. It divided itself into three phases noticeably. With target suction increasing and test time lasting, the water content decreased more and more sharply. At the first phase, when the suction $u_a - u_w = 20$ kPa is stable, the water content is get as $\omega = 17.66\%$. At the second phase, when the suction is stable, the water content is get as $\omega = 16.02\%$, and at the third phase, $u_a - u_w = 100$ kPa, $\omega = 14.95\%$.

Figure 6b shows the volume variations of the sample in the process of moisture dehydration. Similar to the variations in water content, the volume variation curve may also be divided into three phases. The rate of volume change in each phase presents the same trend: gradually getting slower till down to a certain value. The volume decreases in the whole stage. Here its absolute shrinkage η_d is defined as:

$$\eta_d = \frac{|v_d - v_0|}{v_0} \times 100\%$$

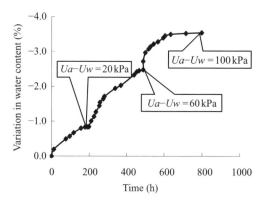

Figure 6a. Variation in water content during the stage of moisture absorption.

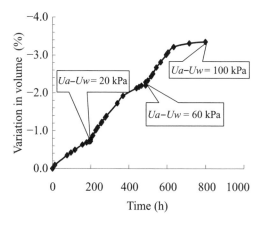

Figure 6b. Variation in volume during the stage of moisture absorption.

Where, v_d is the sample's volume when the suction is stable in the end of the stage of moisture dehydration. The volume shrinkage in this test is 3.35%.

Moreover, the relationship between specific volume $(v = 1 + e)$ and its corresponding suction value is also found, while the suction getting stable in each of the three phases during the stage of moisture dehydration. Figure 7 shows the relationship between the specific volume and $\ln[(s + p_{at})/p_{at}]$ (where s is the value of matrix suction p_{at} is referential pressure, namely, a standard atmospheric pressure). It is shown that the specific volume and suction (within the range from 0 to 100 kPa) is basically of a liner relation, which can be described as the following formula (Alonso 1998):

$$v = v_s - k_s \ln\left(\frac{s + p_{at}}{p_{at}}\right)$$

where v_s and k_s are respectively the intercept and the slope of the relation curve $v - \ln[(s + p_{at})/p_{at}]$ when the value of suction is 0, where $k_s = 0.07$, $v_s = 1.64$.

This result does not testify a suction increase suggested by Gens (1992) & Alonso (1990) for unsaturated soil. The reason may be considered as that the maximum value of suction we discuss here, i.e., 100 kPa is not big enough to reach its yield point to suction. The suction in this test has not gone beyond the elastic stage of soil mass deformation.

(2) The stage of anisotropic consolidation

When the suction becomes stable, $(u_a - u_w) = 100$ kPa, $u_a = 150$ kPa and $u_w = 50$ kPa; then conduct anisotropic consolidation ($p = (\sigma_1 + 2\sigma_3)/3$, $\sigma_3 = k_0\sigma_1$). The stable state after the first stage is taken as the initial state for the stage of anisotropic consolidation. At the end of axial loading, $\sigma_1 - u_a = 270$ kPa, $\sigma_3 - u_a = 80$ kPa, axial deformation of the sample reaches 5.19%. Figure 8 shows the relation curve of $\varepsilon_a - p$ (axial deformation-average net normal stress).

A distinct yield point can be found in the compression curve when the suction is 100 kPa. The average net normal stress at the yield point is 114.6 kPa, and the void ratio e is 0.561. After the yield point, the compression coefficient is $\alpha_v = 0.009$ kPa^{-1}; this data enables to calculate slope deformation in the latter stages.

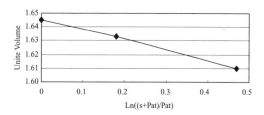

Figure 7. The volume-suction relation curve.

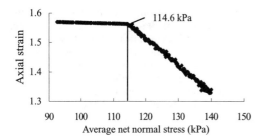

Figure 8. Compression curve during anisotropic consolidation.

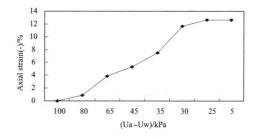

Figure 9a. Axial strain-matrix suction curve.

(3) Suction down to the failure stage

After the stabilization of anisotropic consolidation, DL test is conducted to the stress routine of decreasing matrix suction to the testing slope receiving actions of rainfall and rising reservoir water. That means anisotropic consolidation σ_1, σ_3 remains unaltered, $\sigma_1 = 420$ kPa, $\sigma_3 = 230$ kPa, and deviatoric stress $q = \sigma_1 - \sigma_3$, as Dead Load, continuously acts on the sample. When decrease gradually the matrix suction (100 kPa–80 kPa–60 kPa–40 kPa–30 kPa–20 kPa–5 kPa), the deformation of the sample increase gradually with it. Similarly, the stable state of anisotropic consolidation is taken as the initial state for this stage.

Figures 9a, 9b demonstrate the relationship between axial strain and matrix suction, between volume change and matrix suction. From Figure 10b it shows that unsaturated soil in the Qiangjiangping Landslide demonstrates dilatability in the course of moisture absorption. It can be explained as the influences upon deformation of unsaturated soil by pore water and pore gas. For triaxial test to the route of moisture hydroscopic, during the process of unsaturated soil consolidation, i.e., the process when ambient pressure, back pressure and atmospheric pressure in various phase are reaching its balanced, macroscopic pores crush and die away; and in the shear process, water in microscopic pores redistribute and thus cause the soil mass to dilate.

When matrix suction $u_a - u_w$ finally down to the value of 5 kPa and the axial strain $\varepsilon_a = 12.6\%$, it is shown as Figure 10a that the axial strain-matrix suction curve $\varepsilon_a - (u_a - u_w)$ get to its peak value and becomes even. It is meaning that the soil mass has come into the state of failure. Then the stress condition is as follow:

$u_a - u_w = 5$ kPa $\sigma_1 - u_a = 270$ kPa

$\sigma_3 - u_a = 80$ kPa

Compared these results with ones obtained from the unsaturated soil CU test concerning strength parameters, in the CU test,

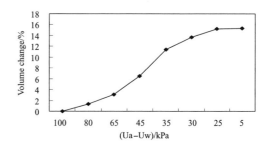

Figure 9b. Volume change-matrix suction curve.

$\tau_f = c' + (\sigma - u_a)_f tg\phi' + (u_a - u_w)_f tg\phi^b$ and

$\phi' = 26.6°$, $\phi^b = 22°$, $c' = 15.7$ kPa,

suppose stress path would not affect shear strength of the soil mass, to substitute the known values of ϕ', ϕ^b and c' in the above formula, then the shearing stress on the shear failure surface of the soil mass is $\tau = 86.96$ kPa, shear strength is $\tau_f = 84.05$ kPa, when $\sigma_1 - u_a = 270$ kPa, $\sigma_3 - u_a = 80$ kPa, $u_a - u_w = 5$ kPa. It is suggested that moisture hydroscopic of the soil mass caused by the rainfall and rising reservoir water (increasing pore water pressure and decreasing matrix suction $u_a - u_w$) would lead to the failure of soil mass.

5 CONCLUSION

The main results are as follows:

1. From the DL test, the following data of the Qiangjiangping landslide slipping zone soil are obtained: in case of the net ambient pressure $p - u_a = 80$ kPa, water content variations and volumetric behavior during the stage of moisture dehydration; through its volumetric behavior, liner relationship between specific volume and suction is:

$$v = v_s - k_s \ln\left(\frac{s+p_{at}}{p_{at}}\right)$$

$$= 1.6448 - 0.0751\left(\frac{s+p_{at}}{p_{at}}\right)$$

2. A distinct yield point has been found in the compression curve when the suction is set as 100 kPa during the stage of anisotropic consolidation. Across the yield point, the compression coefficient is as follow: $\alpha_v = 0.009$ kPa^{-1}.

3. Dilatancy has been demonstrated in the shear process of the Qiangjiangping landslide unsaturated slipping zone soil. The pore water and pore gas have influences upon the deformation behavior of the soil mass. And during the stage of unsaturated soil consolidation, i.e., the process when ambient pressure, back pressure and atmospheric pressure in various phase are reached its balance, macroscopic pores crush and die away; and in the shear process, water in microscopic pores redistribute and thus cause the soil mass to dilate.

4. After anisotropic consolidation, keep its net stress $p - u_a$ and deviatoric stress q unaltered, when matrix suction $u_a - u_w$ finally down to 5 kPa and the axial strain $\varepsilon_a = 12.6\%$, the axial strain-matrix suction curve $\varepsilon_a - (u_a - u_w)$ get to its peak value and becomes even, the soil mass has come into the state of failure. To substitute the known data from the CU test, $\phi' = 26.6°$, $\phi^b = 22°$, $c' = 15.7$ kPa, the following parameters are gotten: $\tau = 86.96$ kPa, $\tau_f = 84.05$ kPa.

REFERENCES

Alonso E.E., 1998. Modeling expansive soil behavior, Proc of the 2nd International Conference of Unsaturated Soils. Beijing.

Alonso E.E., GENS A. & JOSA A., 1990. A constitutive model for partially saturated soils. Geotechnique, 40 (3): 405–430.

Brand, E.W., 1981. Some thoughts on rain-induced slope failures, Proc., 10th Int. Conf. on Soil and Mechanical Foundation Engineering, Stockholm, Sweden, Vol. 3, 373–376.

Chen H., Lee C.F. & Law K.T., 2004. Causative Mechanisms of Rainfall-Induced Fill Slope Failures, Journal of Geotechnical and Geoenvironmrntal Engineering. 130(6): 593–602.

Fredlund D.G. & Rahardjo H., 1997. Unsaturated Soil Mechanics.

Gens A. & Alonso E.E., 1992. A framework for the behavior of unsaturated expansive clays, Canadian Geotechnical Journal, 29: 1013–1032.

Sivakumar V., 1993. A critical state framework for unsaturated soils, Sheffield: University of Sheffield.

Yang Heping & Xiao Duo, 2005. The Influence of Alternate Dry-wet Effect on the Strength Characteristic of Expansive Soils, Journal of Hunan Light Industry College (natural science), 2 (2): 1–5.

Zhan Liang tong, 2003. Field and laboratory study of an unsaturated expansive soil associated with rain-induced slope instability. Hong Kong: The Hong Kong University of Science and Technology.

Zhang Liang tong & Ng, C. W.W., 2006. The Triaxial Test Study on Strength and Deformation of Unsaturated Expansive Soils, Chinese Jounal of Geotechnical Engineering, 28 (2): 196–201.

Post-failure movements of a large slow rock slide in schist near Pos Selim, Malaysia

A.W. Malone
Dept of Earth Sciences, University of Hong Kong, China

A. Hansen
Dept of Spatial Sciences, Curtin University of Technology, Australia

S.R. Hencher
Halcrow China Ltd., Hong Kong; School of Earth & Environment, University of Leeds, UK

C.J.N. Fletcher
Dept of Building & Construction, Hong Kong City University, China

ABSTRACT: This paper describes the results of the monitoring, by total station and photogrammetric surveys, of the movements of a slow compound rock slide from failure in 2003 to December 2006. During this period the head moved downwards more than 21 m. Whilst the rate of displacement is declining slightly year on year, for much of the time the landslide mass is accelerating and then decelerating in surges. Evidence is presented of some correspondence between the timing of the surges and the seasonal rainfall pattern. It is inferred from surface observations that the failure involves sliding at the head and in the upper main body of the landslide on joints roughly orthogonal to the foliation, which dips at a shallow angle into the slope. In the central toe zone the landslide slides up and out on the foliation. The failure, which occupies an area of about 8.5 ha, has reactivated major pre-existing faults which run obliquely through the landslide mass.

1 INTRODUCTION

A landslide occurred in September 2003 during hillside excavation for a new strategic road in mountainous terrain near the Cameron Highlands hill resort in northern Peninsular Malaysia. The site is on the Simpang Pulai—Lojing Highway, close to Longitude 101° 20′ 43″ Latitude 4° 35′ 27″ (Figure 1). Roadworks commenced in 1997 and movements occurred in roadside cut in the vicinity of chainage 23+900. The slope was cut back to a flatter angle but instability persisted. Progressively more extensive slope flattening was undertaken in response to continuing failure until the works reached the ridgeline, 200 m to 260 m above the road. Gross movements occurred in the cut in September 2003 with the formation of a main scarp and associated disruption and the displaced mass has since moved continuously. A study of the landslide was carried out by the authors in 2005 and 2006 (Andrew Malone Ltd, 2007).

2 TOPOGRAPHY, GEOLOGY AND THE LANDSLIDE

The site is on the western hillside of the Gunung Pass ridge which reaches an elevation of 1587 m above sea level. Prior to cutting, the valley sides were densely forested and generally steeper than 30°, with ravines leading down to the deeply-incised River Penoh some 600 m below the ridge.

The cuts were formed largely in nominally 12 m high 1:1 batters and 2 m berms to produce an excavation up to 260 m high and inclined about 33°–35° overall.

The geology of the Gunung Pass area consists of a sequence of sedimentary rocks, probably of Paleozoic age, which have undergone low- to medium-grade dynamic metamorphism. The metasedimentary rocks outcrop in a 4 km-wide shear zone contained within Mesozoic granites. The landslide has taken place in quartz mica schists (Figures 1 & 2) which at the base contain impersistent graphite schist layers

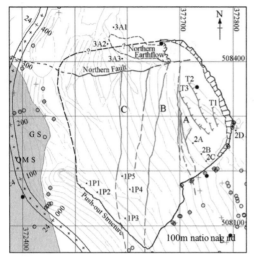

QMS *Quartz Mica Schist Unit* GS *Graphite Schist Unit*
ABC geological faults; T1 T2 T3 transverse cracks
1P1, 2A etc total station survey markers
o points used in the photogrammetric adjustment
+ survey check points • survey targets

Figure 1. Outline of the landslide superimposed on a simplified geological map of the site; and survey points.

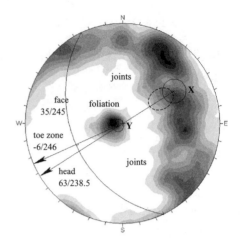

equal angle, lower hemisphere, 541 poles
Shading represents concentration of poles (5% maximum). The circles X & Y are centred on the mean disposition of vectors within 100 m wide blocks of ground (on the centreline) at the highest part of the head (X) and at the toe zone (Y) and enclose >80% of the vector data points.

Figure 3. Stereographic projection of poles to joint planes and foliation and surface displacement vectors.

Figure 2. Several joint sets cutting the low-dipping foliation within the *Quartz Mica Schist Unit*.

less than 30 cm thick. The foliation strikes generally north and the orientation of the excavated face of the hillside is NNW-SSE (Figure 3). The foliation dips at shallow angles towards the east, i.e. into the slope.

The rock sequence at site is cut by sets of pre-existing faults. The most prominent fault set dips steeply towards the E to ESE and three of these faults can be traced across the landslide (Figure 1, faults A, B & C). The fault planes form counterscarps at outcrop on the landslide and have oblique and vertical striations, suggesting distinct phases of slip. The faults also show signs of recent but pre-landslide movement and have been reactivated during landslide movements. The schists at outcrop are highly jointed with typical joint spacing less than 0.5 m (Figure 2). The poles of the joints form a girdle that is roughly orthogonal to the low-dipping foliation (Figure 3). Where unweathered, the schist is generally strong to very strong and most of the rock material currently exposed across the site can be classified as 'slightly' to 'moderately weathered' (British Standards Institution, 1999: Figure 19).

The main surface features of the landslide are the main scarp, the head graben, the north and south flanks, the counterscarps of the oblique faults (A, B & C etc) and a low-angle push-out structure at the toe zone (Figure 1). Neither the northern flank, which is partly concealed beneath an earthflow, nor the toe of the basal slip surface can yet be fully delineated. The term 'toe zone' is therefore used here in preference to the word 'toe'.

The head graben is crossed by multiple high-angle internal shears with counterscarps at outcrop (T1, T2 & T3 etc).

3 DEFORMATION MONITORING BY TOTAL STATION

Monitoring of the landslide has been carried out by the road contractor since October 2003. The work

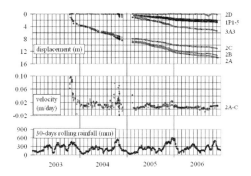

Figure 4. Displacements and velocities from total station monitoring; and rainfall at Stesen Kajicuaca Cameron High-lands.

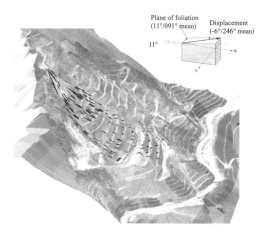

Figure 5. 3D visualisation of the excavated hillside showing the surface displacement vectors (2003–5) to scale. Inset: surface movement of the central toe zone block relative to foliation.

involves nominally weekly measurement, by total station (Sokkia SET5E), of distances and horizontal and vertical angles from base stations west of the road to reflective markers installed on the landslide. The plan co-ordinates and reduced levels of the markers are computed from these data. The magnitude, dip and dip direction of the displacement of each survey marker have been calculated and velocities of movement have been determined. Some of the monitoring data are presented in Figure 4.

Uncertainty in the data may be assessed by examining the reported movements of a marker located above the crown of the landslide (Marker 2D). In contrast to the markers within the landslide, the reported changes in the position of Marker 2D are very small (save for an unexplained excursion in September and October 2006) and no systematic pattern is evident. The variance in horizontal position data to August 2006 (standard deviation of data = 45 mm) is about three times that expected from equipment error alone. The variance in height determination is as expected from equipment error alone.

4 TOPOGRAPHIC SURVEY AND PHOTOGRAMMETRIC MEASUREMENTS

A digital elevation model had been created from topographic survey in November 2003 and another was made photogrammetrically from aerial photographs taken from a helicopter in September 2005. Displacement vectors were constructed from the differences between the two digital elevation models at identifiable features such as the ends of drainage channels and berm edges.

Many of the 150 displacement vectors are shown in Figure 5.

Uncertainty in these measurements is associated with the coordinates of the ground control points used for photogrammetric adjustment and errors in the digital camera system. The photogrammetric survey was compared to the November 2003 survey at 56 survey check points in areas thought not to have moved between 2003 and 2005 (see Figure 1). The error standard deviation is 0.2 m. Uncertainty in the dip and dip direction of the vectors is less than 1° for the longest vector (24.3 m), a possible maximum of 13° for the shortest vector (1.6 m), and an average of 2.7° for the mean vector (8.0 m).

5 POST-FAILURE LANDSLIDE MOVEMENTS

The surface displacement vector data (2003–5) advance our understanding of landslide behaviour. Viewed in plan the vectors are seen to be normal to the slope face contours with lateral extension revealed by radial divergence (north-south spreading), conforming to topography, which takes the form of a subdued ridge. Movements are greater at the head than in the toe zone (compression) and, on any slope face contour, displacements are greater in the north than in the south (rotation). Viewed in cross-section the vectors are seen to plunge at the head of the landslide, to generally lie sub-parallel to the slope in the upper main body and to emerge in the toe zone.

It is instructive to examine the disposition of vectors along the centreline of the displaced mass by means of stereographic projection (Figure 3). The directions of vectors within 100 m wide blocks of ground at the highest part of the head (mean 238.5°) and at the toe zone (mean 246°) closely correspond to the dip direction of the face (mean 245°). The disposition of vectors at the highest part of the head coincides with a concentration of joint planes (Figure 3 – X) and at the central toe zone block corresponds to the attitude of the foliation (Figure 3 – Y and Figure 5 inset).

The vectors reveal significant downslope compression. Compressive strain (defined as the displacement normalized against downslope length) measured on centreline between upper main body (at the elevation of Markers 2A-2C) and the toe zone is about 5% (2003–5). Such compression is evident in small-scale sliding on foliation seen as shear offsetting ('kicking out'), especially in the southern part of the landslide, and by slip on the reactivated faults A, B & C etc. Observed fault slip movements are dextral, increase to the south and are greatest on fault B, where slip at the centreline is 3.5 m.

The total station data give further insights into landslide behaviour. Whilst the overall rate of displacement is declining slightly year on year, for much of the time the displaced mass appears to be either accelerating or decelerating. Five surges are apparent (Figure 4) and comprise an accelerating phase (six to eight weeks) and a decelerating phase (two to three months). The velocity reached during surges at markers 2A-C is generally about 20 mm/day (greater in late 2004).

6 DISCUSSION

The nature of the basal sliding surface(s) is of interest. Evidence is given above of movement at surface stations which is parallel to joint planes at the highest part of the head and to foliation in the central toe zone; slip on foliation is also visible on the ground. It may be inferred, if the effects of non-parallel internal shear and change in landslide thickness are assumed insignificant, that the landslide is sliding on joint planes at the highest part of the head (i.e. at the main scarp 'normal fault') and sliding upwards on foliation in the central toe zone (but oblique to dip, Figure 5 inset).

The vectors plunge steeply at the head and emerge sharply in the toe zone, the profile suggesting a non-circular basal slip surface (Figure 5). The presence of multiple counterscarps in the head graben (T1, T2, T3 etc. Figure 1) may signify curvature of the basal slip surface (Hutchinson, 1988). There are joints disposed to facilitate slip on such a curved surface (Figure 3 – dashed oval). The landslide is probably a compound slide. An educated guess was made about the geometry of the basal slip surface, using the surface station movements and crack patterns, and estimates were made of landslide volume.

It appears that the volume of the landslide is about 2 million m^3.

After failure the landslide decelerated until March 2004 and it has since continued to move, for much of the time accelerating and then decelerating in surges. The timing of the surges generally coincides with peaking in the 30-day rolling rainfall (Figure 3), rainfall being measured at the Stesen Kajicuaca Cameron Highlands raingauge of the Malaysian Meteorological Service, 13 km SSE of the site. The bimodal rainfall pattern shown in Figure 4 is characteristic of an inland climatic regime in peninsular Malaysia. It may be that the landslide is responding to rainfall-induced seasonal rise and fall of groundwater levels. Such fluctuation is manifest by intermittent seepage from the southern toe zone. Other causal factors may have contributed to surges: a surge in late 2004 concurred with the removal of 100,000 m^3 of ground from the northern toe zone of the landslide.

7 CONCLUSIONS

The landslide is a slow rock slide in schist. Failure occurred in September 2003 and by December 2006 the head had moved downwards more than 21 m. The rate of displacement is declining slightly year on year, but for much of the time the landslide mass is accelerating and then decelerating in surges. There is some correspondence between the timing of the surges and the seasonal rainfall pattern. It is likely that the surges are induced by groundwater fluctuations. It may be inferred from surface observations that the failure involves sliding at the head and in the upper main body of the landslide on joints roughly orthogonal to the foliation, which dips at a shallow angle into the slope; in the central toe zone the landslide is sliding up and out on the foliation. The failure, which is probably a compound slide of volume about 2 million m^3, has reactivated major pre-existing faults that run obliquely through the landslide mass.

ACKNOWLEDGEMENTS

The study of the Pos Selim landslide was carried out on behalf of the Slope Engineering Branch of the Public Works Department of Malaysia and facilitated by the road contractor, MTD Construction Sdn Bhd., who supplied the total station data. Prof YQ Chen of the Department of Land Surveying and Geo-informatics, Hong Kong Polytechnic University checked the displacement and velocity calculations and assessed the errors in the total station surveying.

REFERENCES

Andrew Malone Ltd 2007. *Landslide study at Ch 23+800 Simpang Pulai-Lojing Highway, Malaysia*. Report to Minister of Works of Malaysia.

British Standards Institution, 1999. *Code of Practice for Site Investigations BS5930:1999*.

Hansen, A. 2007. *Semi-automated geomorphological mapping applied to landslide hazard analysis.* Ph.D. Thesis, Department of Spatial Sciences, Curtin University of Technology, 281p.

Hutchinson, J.N. 1988. General Report: Morphological and geotechnical parameters of landslides in relation to geology and hydrogeology. In *Proc Fifth International Symposium on Landslides* (C Bonnard ed.) Balkema Rotterdam v1 3–35.

Characteristics of rock failure in metamorphic rock areas, Korea

W. Park & Y. Han
Architectural Engineering Team, Civil Part, Samsung Corporation, Seoul, Korea

S. Jeon
School of Civil, Urban and Geosystem Engineering, Seoul National University, Seoul, Korea

B. Roh
Technical advisor Team, Samsung Corporation, Seoul, Korea

ABSTRACT: The metamorphic rock ranges in Korea very extensively, which has the unique characteristics through crustal movement and complex metamorphic processes for a long time. So the metamorphic rocks have discontinuities such as dislocation and fault as well as weak part such as fractured zone and fault gauge. For these reasons, there exist great potential of collapses, so it is required great caution to construction and design roads or tunnels in this area. This paper describes the past experiences of slope failure during the construction of roads at the area of metamorphic complex, especially Gyong-gy gneiss, and analyzes the characteristics of the failure patterns in order to prevent from loss of the life and property.

1 INTRODUCTION

Length of the Korean Peninsula is about 1,000 km and the area is 223,000 km², especially 70% of the land consists of the mountain. But despite of small area, there exist various rock types and the complex geological structures from Pre-Cambrian to Cenozoic era. As the characteristics of distribution of the rock, metamorphic rock of Cambrian period and plutonic rocks of Mesozoic exist extensively in the middle of the Korean Peninsula. Sedimentary rock and igneous rock which include stratum of Paleozoic and after that period scatter on these bedrock. According to roughly compositions of the rock, the metamorphic rock is 40%, igneous rock is 35% and sedimentary rock is 25%, and in the report of slope failures, mostly rock collapses were happened in metamorphic and sedimentary rock area in the road constructions. Actually, many losses of life and property happened from repetitive slope failure.

Therefore, it is important to understand geological characteristics and collect data of the failure cases of the each rock type in order to avoid natural disaster and failures of artificial structures of rock. For this purpose, this paper described characteristics of failures in metamorphic rock area.

1.1 Geological characteristics of Korea

Figure 1 shows the geological map of South Korea. Most of metamorphic rock of Pre-Cambrian exists as the type of gneiss and schist in the middle region and sporadically puts in the whole country.

As it is well known, metamorphic rocks have many complicated geological structures by the movement of crustal and metamorphism processes for a long time, so they have irregular discontinuities, rock fractured, various scale fault and weak gouge.

In case of the fault in these rocks, the size is not large scale, but considering direction and location of fratures, it is not easy to determine and predict.

Discontinuities such as joints and foliations are very complex and these crevices are infilled by gouge and weak deposit material. So on excavating slope or pit

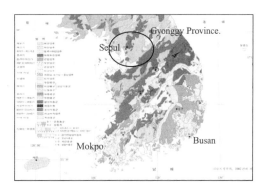

Figure 1. The geological map of Korea.

and tunneling in rock, the failures were frequently happened by the disadvantage geological structures.

The sedimentary rock which is mostly mud stone, shale and sand stone from Cenozoic era ranges in the south-east region, and most of the these rock, except the Pohang's mud stone, are very hard and mass state.

This rock sometimes has some problems when the excavation directions are the same with the dip direction of sedimentary rock. Many engineers know how to design in this rock type, so comparatively the damages are low, but if the failures have happened this area, the loss would be largely.

In case of the igneous rock, even if the rock is intrusive rock or eruptive rock, it can be seen easily in Korea. Representative igneous rock is granite which ranges very largely in Seoul and Kyung sang Province and mostly eruptive rock such as the andesite and tuff exists in south region nearby Mokpo.

The characteristics of this rock failure are that the many failures were happened after cutting the slope by erosion and weathering.

As the rock types, there are unique characteristics of the movements, so for designing the slope or tunnel, it is very important to apply the appropriate geological structures.

1.2 Statistics of the failures

About the slope failures, the important things, except geological problems, are related with the point time of excavation and elapsed time. Figure 2 shows the data of slope failures as the elapsed time of the slope excavation at highway construction site in Korea (You, 1997).

As seen in the graph, though initial collapses were relatively rare, many slope failures occurred shortly after cutting slope and opening to public, totally 85% of the failures were these times.

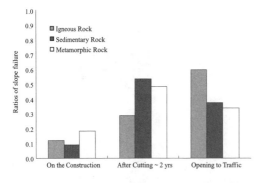

Figure 2. Ratios of slope failure in term of elapsed time from cutting slope for there different rock types in Highway construction sites in Korea (You, 1997).

Many failures in metamorphic rock area happened immediately after cutting, and then as the time goes, failures of sedimentary rock occurred much more. And in case of the igneous rock, the slope became weak by erosion of the rock and fast weathering, therefore swallow face-failure and the scour were easy to happen on the slope. On exposing the slope face, weathering velocity of some ingredients of the rock mass becomes much faster and the rainfall and underground water infiltrate into discontinuities.

These unpredicted collapses brought many losses of the construction period and the additional cost.

The important points of these accidents were that most of the failure could not predict in design. This paper is the case study about investigation of the collapse of soil and rock slope by cutting slope in the metamorphic rock site which is located at Gyong-gy Province and analyses the pattern of the failures.

2 CHARACTERISTIC OF THE SITE

2.1 Geological characteristics

This investigation area was located at the Gyong-gy Province near the Seoul and there was the Route 45 which was a new lying road in the South-North direction, especially most parts of the route passed the mountain area, so major works were tunneling and cutting slope.

As characteristics of the geology, most of rock at this area was metamorphic rock, called Gyong-gy gneiss which is the representative metamorphic rock in South Korea. This rock has unique characteristics of the geological structures through crustal movements and metamorphism processes for a long time. Joints, dislocations, fault and weak part such as fractured zone and fault gauge were existed, and till some of depth, discontinuity surfaces were covered by ferrite oxide and very weak clay, also it is hard to predict the direction of the discontinuity.

Figure 3 is the cut slope with containing the fault which parallels the strike of slope. Discontinuities of these structures were common in this site, because the fault is the same direction with the road. In the other side, upper and side of slope consisted of residual soil and weathered soil, and some of the complex dyke rock. Figure 4 shows the plane failure and wedge failure which was two persistent joint with the line of intersection of the joints daylighting at the rock face.

2.2 Design and construction the slope

In order to design, designers need many data of the site such as geotechnical investigation data and the detailed geological structures. But it is hard to get the all data

Figure 3. The exposed fault in the slope and this fault is the same direction with the Yong-in fault.

Figure 4. Exposed discontinuity which can be collapsed in the slope.

Table 1. Standard Guide for slope's dip.

Condition	Height	Dip	Remarks
Soil	0~5 m	1:1.2	Establish the 1 m ramp
	Above 5 m	1:1.5	per 5 m height
Rock	Weather rock	1:1.0	Establish the 3 m ramp
	Hard rock	1:0.5	per 20 m height

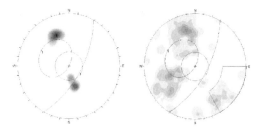

Figure 5. Stereographic Analysis; a) Discontinuities at the first design. b) after cutting.

in case of the construction site of the road, owing to the limit of the investigation, so the standard of the slope design, which need not the precious data, is used very often.

The standard of slope design is practically the same as related organizations of the civil construction, and most of the design criteria are based on the strength of the rock without the rock type. But if the designer used only these guides no concerning the geological information, many problems can be occurred, and a few organizations recommend that the characteristics of the geological structures are considered on the slope, if need.

Table 1 shows the standard guide for design of slope at the Korea Expressway Corporation & Ministry of Construction and Transportation.

This guide can be used comfortably in case of lack of the geotechnical investigation data, but there are many potentialities of the dangerous failure, but in case of being applied to importance cutting slope.

As the briefly examples, the dips of the slope in this investigation site were determined by the standard guide such as Table 1 without considering direction of discontinuity and geological structures.

The Figure 5a shows stereographic analysis of the slope at the initial design. As seen the result of the stereographic analysis, the dangerous block could not be found, but as the face mapping on slope after the slope was finished the excavation, unpredicted fault and joint were existed in the slope. Actually this slope was happened the wedge failure and small scale plan failure by the discontinuities. So in case of design the slope with the standard guide, it is simple to use it, but it is always not safe.

3 CHARACTERISTIC OF FAILURE IN THIS SITE

3.1 *Characteristic of the slope failure*

The rock failures are influenced on many factors such as the discontinuities, strength of rock and the time of construction. In case of the metamorphic rock, there are many collapses after finishing the excavation as result of the investigation. These results were the same of the investigation which was performed by Korea Expressway Corporation.

Table 2 shows results of investigation of the relation rock type which was classified by the rock strength and elapsed time of the excavation. 66% of slope failure happened after finishing excavation, and 85% of the failure occurred in the rock masses which were above strength of weather rocks. These reasons that the more slope was cutting, the much free surface in slope face was enlarged. As it is well-known, the free surface gives the many influences in the slope. By the investigation, the causes of many failures were by the discontinuity of the slope, especially fault and joint.

On the other side, the failures of residual soil or completely weathered soil, which had no geological

Table 2. Number of slope failure.

Rock	On construction	On complete	After project	Total
R.S	3	1	1	5 (15%)
W.R	2	5	–	7 (20%)
S.R/H.R	7	15	1	23 (65%)
Total	12 (34%)	21 (60%)	2 (6%)	35

*R.S is residual soil; W.R is weathered rock; S.R is soft rock; H.R is Hard rock.

structure patterns, were caused by the soil weight, and most of failure shapes were circular type in this part. Figure 6a shows the circular failure, and it is typical circular failure by the weight of soil mass. However geological structures such as dyke rock and foliation are faintly reminded in slope on account of the different weathered speed, so failures in soil occasionally happened like the type of rock mass failures, especially in the metamorphic or sedimentary rock area.

Figure 6b shows the circular failure by dyke rock. Dyke rock which did a role as discontinuity which divided into some soil layers, if this discontinuity was the same direction with the cut slope, circular or other shape failure would be occurred.

As seen the result of the Table 2, soil failures did not often happen, but 2-case happened on the finished excavation and opening the public, the three failures happened the on the construction. In case of weak zone in the slope, the characteristic of the movement was same as soil rather than rock, so it is hard to predict the slope failure in the area where the fractured zone and mixed fault.

As seen the table 2, collapse of 85% is related with rock, and large scaled failures were caused by the fault and gouge which is called Yong-in fault and this discontinuity was put to South-North direction. Even if the fault were so small scale, when the fault was created, nearby the crust was move at once, so short and dangerous discontinuities were extensively existed at this area. And as another reasons, there were fault gouge and infilled materials into the gap of joint.

Figure 7 shows the representative large scale failure before starting to excavate in this area. The height of the mountain is about 70 m and width is 250 m. The main cause of the collapse was fault which existed in slope of lower part, and there was a thin thickness of the fault clay between discontinuities. The strike of fault paralleled with direction of slope, so mostly mode of failures were plan failure.

Due to this failure, 5 m of settlements and over the 50 cm of tension crack were occurred for several weeks. Because this failure was large scale and the

Figure 6. Soil failure; a. The circular failure through the line of least resistance by soil weight. b. The circular failure by dyke.

Figure 7. This picture is large slope failure in the this site. This slope's height is 56 m and the width is 120 m. a. view of the slope. b. Vertical discontinuity exposed by ground settlement at the failure. c. Tension crack on the ridge.

problem of compensation with inhabitants around the site, it was hard to do additional cutting and lots of monetary problems were happened.

The sources of these problems were lack of properly geotechnical investigation, especially investigation of the fault and dangerous discontinuities is not easy and in usually case, small fault does not consider to design.

Table 3 shows classification of the slope failure as shape of the failures and number of the occurrences are referred to table 2.

As above mentioned, because the discontinuity are the main cause, failures are happened the rock which related the discontinuity.

According to the investigation of failures in the site, there were several cases of wedge and plane failure but fortunately the toppling failure did not happen.

On the excavation, some of topping failure blocks were in the slope, but reinforcement methods such as the bolt and rock anchorage were used to prevent from failure through using.

Figure 9 shows the plane failure and wedge failure after cutting and passed after 1year in this site. Height of slope was 25 m and width that collapse occurs was about 50 m.

The main cause was the unexpected fault which of dip of the fault was N40°E/63°SE.

Figure 10 is the stereographic analysis of the Figure 9. As seen the results, the direction of the joint were changed as the construction, and could not reflect the fault at the first design because there were no data about the fault.

As seen the Figure 10, although there were many dangerous blocks in slope at the initial condition, but designer used the standard guide, then the collapsed happened.

Figure 9. View of the failure in the slope: a: Slope shortly after cutting; b: Wedge failure in the slope; c: Fault gouge with 50 cm of width and slickenside in the slope; d: Tensile Crack in the ramp.

Figure 8. Rock face formed by persistent discontinuities: a. plane failure formed by bedding planes parallel to the face with continuous length over the slope (Gneiss on Route 45 near Yong-in); b. wedge failure formed by two intersecting planes dipping out of the face.

Table 3. Types of the slope failure.

Type of failure	Circle failure	Planar failure	Wedge failure	Toppling failure
No. of occurrence	5	19	11	–

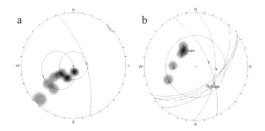

Figure 10. Stereographic analysis of the discontinuities plane at the construction site: a. is state of the discontinuity before digging and b. is result of the face mapping after 1st cutting. Dip of the fault was 63 degrees.

If more accurately investigation data of this site were existed in design, failure could be prevent by using a few rock bolts.

4 CONCLUSION

In this paper, we investigated the characteristics of slope failure in the metamorphic rock area which is called Gyong-gy gneiss. This rock mass has many unpredicted discontinuities from the metamorphism process and crustal movement for along time.

Therefore most of the road and tunnel constructions of this region, many failures were happened by the geological structures. 85% of the slope failures were concerned discontinuities of the rock and most of the failures happened after cutting the slope or opening the public.

As these reasons, the exposed discontinuities were revealed on the slope after excavation. Also in case of the soil parts, some of the failures were induced by the reminded geological structures such as fold and foliation, especially dyke rock.

As the problems of the design, the designer depends on the standard guides of the slope without no considering the characteristic of the geology, even if there are many studies of slope stability.

Therefore, in order to prevent from the rock failure, it is important to collect the more data such as accurate geotechnical investigation which include the characteristic of the geological structures and to evaluate slope stability by the face mapping after cutting, and through the international research on the slope design, the more stable and rationally design method must be established.

REFERENCES

Lee Dae-Sung (edited). 1987. *Geology of Korea*. South Korea.
Hoek, E. & Bray, J. 1977. *Rock Slope Engineering*, 3rd edn, IMM, London.
Hoek, E. & Bray, J. 2004. *Rock Slope Engineering*, 4th edn, IMM, London.
Korea Expressway Corporation. 1996. *Highway slope manual*.
Park, W.S. 2002. *The report of the slope stability in Yongin*, Samsung Corp.
You, B.O. 1997. *A Study on Harzad Rating System and Protective Measures for Rock Slope*, PH.D Dissertation, Dept. of Civil Eng., Hanyang University, Seoul, South Korea.

Shape and size effects of gravel grains on the shear behavior of sandy soils

S.N. Salimi, V. Yazdanjou & A. Hamidi
Tarbiat Moallem University, Tehran, Iran

ABSTRACT: The shear behavior of sandy soils containing gravel particles has been investigated by many researchers. However, the effects of the shape and size of gravel particles have not been particularly evaluated. In the present study the shear strength of sand-gravel mixtures with two different gravel grain sizes and shapes is studied in loose, medium and dense states using large direct shear test. The results of this study indicate that the gravel shape and size has a little effect on the shear strength of sand-gravel mixtures in low gravel contents. Increasing the gravel content to higher values makes this effect more clear. This is more obvious when the gravel particles are no more floating in the sandy soil matrix. The samples containing angular gravel particles generally show higher shear strength and dilation compared to the mixtures containing rounded to sub-rounded gravels. Also the samples containing larger gravel particles usually show more shear strength and dilation compared to the samples with smaller gravel particles in the same gravel content.

1 INTRODUCTION

Fragaszy et al (1990, 1992) introduced a new method to evaluate the shear strength of soils containing oversized particles. This method was based on the assumption that larger particles floating in a matrix of finer grained material do not significantly affect the strength and deformation characteristics of mixture. In other words, while oversized particles are floating in the finer matrix without any contact, the strength and deformation characteristics are controlled by the matrix part alone. However, for higher oversized contents, it is controlled by both the sand matrix and oversized particles. Therefore, in a floating state the behavior of the soil containing oversized particles can be simulated by testing the matrix portion alone, provided that the model specimen is prepared in near field density. The near field density is the density of matrix in vicinity of oversized particles (Fragaszy et al. (1992)). Based on this concept the shear strength of granular soils containing oversized particles reduces by increase in oversized content provided that the relative density of the mixture remains constant.

Yagiz (2001) investigated the effects of the shape and content of gravel particles on the shear strength of fine sandy soils using direct shear tests. It was concluded that the shape and content of gravel particles have important effects on the friction angle of the mixture.

Simoni and Houlsby (2004) performed 87 large direct shear tests on sand-gravel mixtures with different gravel contents. They concluded that increase in gravel content enhances the dilatancy rate and the critical state friction angle.

In the present study a regular set of 84 large direct shear tests were performed using a $300 \times 300 \times 170$ mm direct shear box apparatus to investigate the effects of the shape and size of oversized particles on the shear behavior of sand-gravel mixtures.

2 SAMPLE PREPARATION

A bad graded fine sand with sub-rounded to rounded grains was used as the base soil. Also two gravel types were used as the oversized particles. The first type was river gravel with rounded to sub-rounded grains and the second type was an angular to sub-angular one. Two different gradations with maximum grain size of 12.5 mm and 25 mm were considered for each gravel type. Each one was mixed with the base sand in different contents to prepare sand-gravel mixtures.

The maximum and minimum void ratios of the mixtures were measured according to the ASTM-D4253 and ASTM-D4254. Also the specific gravity of the sand and gravel grains was measured as 2.74 and 2.64 respectively according to ASTM-D854. The dry unit weights of different mixtures in desired relative densities were computed. The weight of the soil required

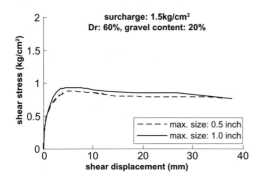

Figure 1. Shows the shear stress-shear displacement curve for two mixtures containing 20% gravel of maximum size of 12.5 and 25 mm.

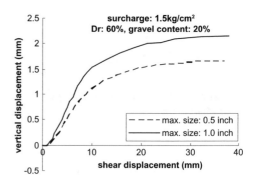

Figure 2. Vertical displacement-shear displacement curves of mixtures containing 20% of different gravel sizes.

to make samples in a specific relative density was selected. The gravel and sand portions were mixed based on the desired weight percents. The whole mixture was divided into three equal fractions and the soil sample was made in three layers by purring the first fraction and to compact it using a metal hammer in order to fill one third of the shear box height. The two other layers were placed using the same procedure.

3 EFFECT OF THE GRAVEL SIZE ON THE SHEAR BEHAVIOR OF SAND-GRAVEL MIXTURES

The tests performed under an overburden pressure of 1.5 kg/cm^2 on samples in a relative density of 60%. As it can be seen the shear stress-shear displacement curves are almost coincided for both mixtures. Although the peak shear strength of mixture containing larger gravel grains is a little higher, the residual strengths of both mixtures are the same. In fact in low gravel contents, the gravel grains are floated in the sand matrix and there is little contact between them. Therefore the shear behavior of the mixture is controlled mainly by the sand portion and the gravel grain size does not play an important role on the shear behavior of the mixture.

The variation of vertical displacement with shear displacement for the above mentioned mixtures is shown in Figure 2. It can be observed that the vertical displacement of the mixture containing gravel grains with maximum size of 25 mm is more than the one containing gravel grains with maximum size of 12.5 mm. It seems that the increase in gravel grain size increases the vertical displacement or the dilation of mixture due to the more vertical movement and slip or topping of the larger gravel particles over each other. The more dilation of the mixtures containing larger gravel

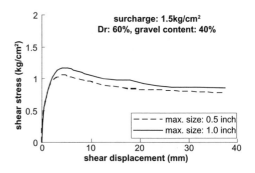

Figure 3. Shear stress-shear displacement curves of mixtures containing 40% of different gravel sizes.

grains is the main reason for higher shear strength in comparison to the mixtures with smaller gravel grains.

Mixtures with 40% and 60% gravel contents were prepared to investigate the effect of gravel size on the shear behavior in higher gravel contents. Figure 3 shows the shear stress-shear displacement curve for samples with 40% gravel content. It can be seen that the difference between the curves has been increased for different gravel grain sizes compared to the curves for mixtures with 20% gravel content. This is mainly due to the increase in the contacts between gravel grains in the mixture in higher gravel content. The contact forces between particles are more for larger gravel grains compared to the smaller ones due to the more contact surface that leads to higher shear strength.

The same comparison was made considering the vertical displacement-shear displacement curves. It was concluded that larger grains cause higher vertical displacement or dilation. The difference between dilation of mixtures with different gravel grain sizes increases with increase in gravel content. Figure 4 indicates this case for the mixtures with 40% gravel content. The differences of shear strengths and dilation of the mixtures with different gravel grain sizes are more obvious when the gravel content is 60% in mixture. This is shown in Figures 5 and 6 which indicate the shear stress-shear displacement and vertical displacement-shear displacement curves for samples with 60% gravel content.

4 EFFECT OF THE GRAVEL SIZE ON THE FRICTION ANGLE OF SAND-GRAVEL MIXTURES

The variation of friction angle for mixtures containing gravel grains of different size prepared in different relative densities is shown in Figure 7. It can be concluded

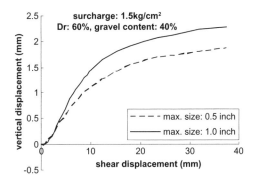

Figure 4. Vertical displacement-shear displacement curves of mixtures containing 40% of different gravel sizes.

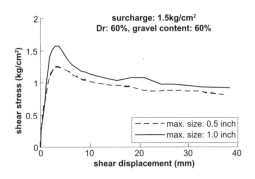

Figure 5. Shear stress-shear displacement curves of mixtures containing 60% of different gravel sizes.

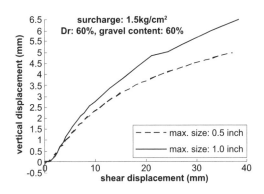

Figure 6. Vertical displacement-shear displacement curves of mixtures containing 60% of different gravel sizes.

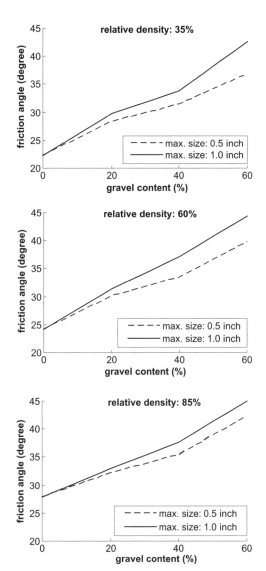

Figure 7. Friction angle of mixtures containing gravel grains with maximum size of 12.5 mm and 25 mm.

that the increase in gravel content leads to the increase of friction angle for all the mixtures. Besides the mixtures containing larger gravel grain particles show higher friction angle. Also the difference between the friction angles of mixtures containing different gravel size increases in higher gravel contents.

5 EFFECT OF THE GRAVEL SHAPE ON THE SHEAR BEHAVIOR OF SAND-GRAVEL MIXTURES

In order to study the influence of gravel shape on the shear behavior of sand-gravel mixtures, a set of direct shear tests have been conducted on mixtures containing angular and rounded gravel grains.

It was observed that in low gravel content of 20%, the shear stress-shear displacement curves of mixtures containing angular and rounded gravel grains are nearly coincided as shown in Figure 8. However, the vertical displacement of the mixture containing angular gravel grains is a little more than the one containing rounded gravel grains as shown in Figure 9. In fact the angularity of gravel grains results in more vertical displacement or dilation. This is due to the increase in overtopping of gravel grains during shearing.

Figures 10 and 11 indicate the shear stress-shear displacement and vertical displacement-shear displacement curves for mixtures with 40% of different gravel shapes. It can be observed that the effect of gravel shape on the peak and residual shear strength and dilation of sand-gravel mixtures are more obvious in higher gravel contents. The more shear strengths of mixtures containing angular gravel grains are mainly due to the higher dilation occurs in these mixtures compared to the mixtures containing rounded gravel grains. The same trend can be seen when the gravel content increases to 60% as shown in Figures 12 and 13.

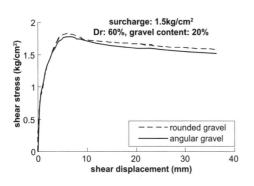

Figure 8. Shear stress-shear displacement curves of mixtures containing 20% of different gravel shapes.

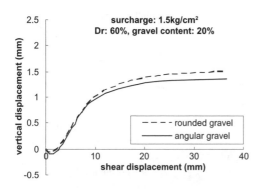

Figure 9. Vertical displacement-shear displacement curves of mixtures containing 20% of different gravel shapes.

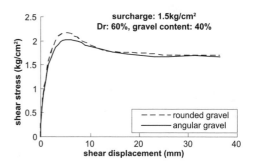

Figure 10. Shear stress-shear displacement curves of mixtures containing 40% of different gravel shapes.

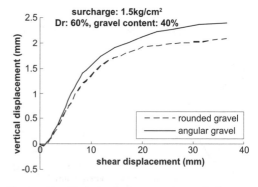

Figure 11. Vertical displacement-shear displacement curves of mixtures containing 40% of different gravel shapes.

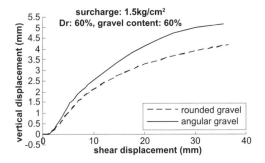

Figure 12. Shear stress-shear displacement curves of mixtures containing 60% of different gravel shapes.

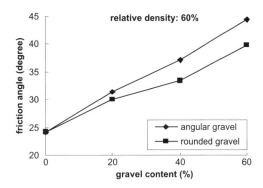

Figure 14. Friction angle of mixtures containing angular and rounded gravel grains.

For mixtures containing angular gravel with maximum grain size of 12.5 mm in a relative density of 60%:

$$\phi = 0.32 Gc + 24.7 \quad R^2 = 0.99 \qquad (2)$$

In the above equations Gc is the gravel content in percent and ϕ is friction angle of the mixture in degrees.

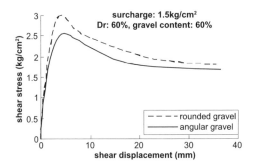

Figure 13. Vertical displacement-shear displacement curves of mixtures containing 60% of different gravel shapes.

6 EFFECT OF THE GRAVEL SHAPE ON THE FRICTION ANGLE OF SAND-GRAVEL MIXTURES

The variation of friction angle for mixtures containing angular and rounded gravel grains in different gravel contents is shown in Figure 14 in a relative density of 60%. The figure shows that increase of the angular gravel content in sandy soil increases the friction angle of the mixture more compared to the rounded gravel.

A relationship can be determined for mixtures containing rounded and angular gravel grains to estimate the friction angle of the mixtures containing certain amount of gravel using a linear regression as shown in equations (1) and (2). These equations may be used to determine the friction angle of sand-gravel mixture with a relative accuracy.

For mixtures containing rounded gravel with maximum grain size of 12.5 mm in a relative density of 60%:

$$\phi = 0.24 Gc + 24.7 \quad R^2 = 0.99 \qquad (1)$$

7 SUMMARY AND CONCLUSION

A regular set of direct shear tests performed on mixtures of sand and different types of gravel grains to investigate the effects of gravel grain shape and size on the shear behavior of sand-gravel mixtures. It was concluded that the addition of each type of gravel grains to the sandy soil increases the shear strength and dilation of the mixture intensively. However, the addition of larger gravel grains increases the shear strength and dilation of mixture more compared to the smaller ones. Also the difference between shear strengths of mixtures containing gravel grains of different size is more obvious in higher gravel contents. The same results obtained for mixtures containing gravel grains with different angularity. Addition of angular gravel grains increases the shear strength and dilation of the mixture more than rounded gravel grains. The effect of gravel grain angularity is more obvious in higher gravel contents. The friction angles of the mixtures containing different amounts of gravel content were also studied and it was understood that the increase in gravel content results in more friction angle. However, the friction angle of the mixtures containing angular gravel grains is higher compared to the one containing rounded gravel grains in equal gravel content. The same result obtained for friction angle of mixtures containing larger gravel grains compared to the mixtures containing smaller gravel grains. For equal gravel content, the friction angle of the mixture containing larger gravel grains is higher than the one containing smaller gravel grains.

REFERENCES

Fragaszy, R.J., Su, W. & Siddiqi, F.H. 1990. Effect of oversized particles on the density of clean granular soils. *Geotechnical Testing Journal* 13(2): 106–114.

Fragaszy, R.J., Su, W., Siddiqi, F.H. & Ho, C.L. 1992. Modeling strength of sandy gravel. *Journal of Geotechnical Engineering Division, ASCE* 118(6): 920–935.

Yagiz, S. 2001. Brief note on the influence of shape and percentage of gravel on the shear strength of sand and gravel mixture. *Bulletin of Engineering Geology and the Environment* 60(4): 321–323.

Simoni, A. & Houlsby, G.T. 2006. The direct shear strength and dilatancy of sand-gravel mixtures. *Geotechnical and Geological Engineering* 24(3): 523–549.

Nonlinear failure envelope of a nonplastic compacted silty-sand

D.D.B. Seely
IGES, Inc., Salt Lake City, Utah, USA

A.C. Trandafir
Department of Geology and Geophysics, University of Utah, Salt Lake City, Utah, USA

ABSTRACT: Accurate evaluation of shear strength of compacted fill materials represents a key issue in performing reliable stability analyses of embankment slopes. In engineering practice, the shear strength of compacted soils is expressed in terms of a linear failure envelope usually derived from consolidated-undrained (CU) triaxial tests with pore water pressure measurements. Experimental evidence indicates, however, that the failure envelope of many soils is not linear, particularly within the range of low effective normal stresses. The present study addresses the nonlinear character of failure envelope of a nonplastic compacted silty-sand. Linear and nonlinear strength functions were fitted to experimental results from CU triaxial tests to derive the strength parameters of linear and nonlinear failure envelopes. For the investigated soil and typical range of mean effective normal stresses of interest in embankment slope design, the nonlinear failure envelope appears to give a more accurate representation of the experimental information compared to the linear strength model.

1 INRODUCTION

The importance of incorporating the nonlinear character of soil failure envelope in slope stability computations has been demonstrated by two dimensional limit equilibrium (LEM) and finite element (FEM) analyses (Trandafir et al. 2000) as well as three dimensional slope stability studies (Jiang et al. 2003). Previous research has shown that failure envelopes of various soils (e.g., silt, sand, compacted clay, and compacted rockfill) exhibit nonlinearity especially within the range of small normal stresses (Penman 1953, Ponce & Bell 1971, Charles & Soares 1984, Day & Axten 1989). In this context, the present study is concerned with the nonlinear character of the failure envelope of a compacted silty-sand representing potential embankment material for earthworks throughout Utah. Experimental data from a series of consolidated-undrained (CU) triaxial tests with pore pressure measurement are utilized to derive both the linear and nonlinear failure envelopes of the analyzed soil, and the results are discussed.

2 PRELIMINARY SOIL TESTING

Preliminary laboratory tests were performed on the soil used in this study for the purpose of geotechnical classification, including liquid and plastic limits (ASTM D 4318-05), and grain size analysis (ASTM D 422-63). A compaction test (ASTM D 698-00) was also conducted in order to obtain the optimum water content associated with the maximum dry unit weight of the material. The results of the grain size analysis are shown in Figure 1, whereas the proportion of clay, silt, and sand characterizing the soil are depicted in Figure 2. Based on the laboratory test results, the analyzed soil classifies as a nonplastic silty-sand (SM) according to the Unified Soil Classification System (USCS) (ASTM D 2487-06). The soil consists of 66.4% sand, 30% silt, and 3.6% clay (Figure 2).

The compaction test yielded an optimum water content of 16.14% and maximum dry unit weight of

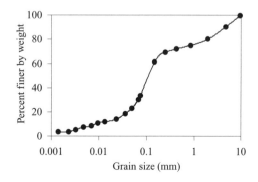

Figure 1. Grain size distribution curve of the silty-sand used in the experimental study.

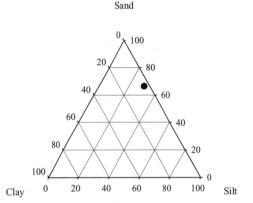

Figure 2. Ternary diagram showing the percentages of sand, silt and clay in the analyzed soil.

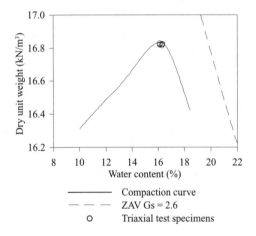

Figure 3. Results of the compaction test performed in accordance with ASTM D698-00. The water content-dry unit weight characteristics of the soil specimens used in the triaxial testing program are plotted on the same curve.

16.82 kN/m³ (Figure 3). These were the target compaction parameters used to prepare the soil samples for triaxial testing.

3 TRIAXIAL TESTING

A total of eight samples were prepared for triaxial testing. The soil was blended with water at a target water content equal to the optimum water content derived from the compaction test, and allowed to attain the steady moisture condition for a minimum of 16 hours prior to compaction. The soil specimens were hand compacted in the laboratory to approximately 100% of the maximum dry unit weight obtained from the compaction test using 254 mm lifts, with the soil being scarified between lifts. The initial dimensions of the specimens were 61.37 mm in diameter and 152.4 mm in length. The samples were then trimmed to a height to diameter ratio between 2:1 and 2.5:1 prior to preparation in the triaxial cell. The moisture-dry unit weight data points of the compacted specimens for triaxial testing plot very close to the point of optimum on the compaction curve, as seen in Figure 3.

The triaxial samples were percolated with CO_2 to aid in saturation prior to the initialization of the triaxial tests as recommended by Rad (1984). The samples were backpressure saturated, ensuring a Skempton's B pore pressure parameter of 0.95 prior to consolidation. A strain rate of 0.06%/min was used for undrained shearing, as derived from the time rate of consolidation data. The triaxial testing program involved CU axial compression (AC) and axial extension (AE) tests on the compacted soil specimens. The experiments were conducted using a state-of-the-art microprocessor controlled fully automated triaxial equipment manufactured by Geocomp Corporation (Dasenbrock 2006). The effective stress paths from triaxial tests are shown in Figure 4.

Seven CU-AC samples were sheared up to 20–25% axial strain with average data sampling every 0.1% strain. The AC specimens demonstrated a dilative response yielding an effective confining stress at failure larger than the initial (consolidation) effective confining stress (Figure 4). Consequently, a CU-AE test (characterized by a contractive tendency of the soil before failure) was included in the experimental program to capture the soil strength behavior within the range of low effective normal stresses (Figure 4). The AE specimen was sheared up to approximately −10% axial strain with data sampling every 0.01% strain.

4 INTERPRETATION OF FAILURE IN TRIAXIAL TESTING OF SILTY SOILS

Failure in dense low-plasticity silty soils subjected to triaxial compression testing is difficult to define due to a continuous dilative behavior of these materials with increasing shear strain. Consequently, a variety of criteria used to define the onset of failure for triaxial tests can be found in the literature (Brandon et al. 2006). Three of these criteria have also been evaluated in the context of the triaxial test results shown in Figure 4, i.e., Skempton's A pore pressure parameter = 0 (Brandon et al. 2006), peak deviator stress $(\sigma_1' - \sigma_3')_{max}$, and peak principal stress ratio $(\sigma_1'/\sigma_3')_{max}$. Figure 5 graphically demonstrates all three of the previously mentioned criteria and their applicability to triaxial test results in this study.

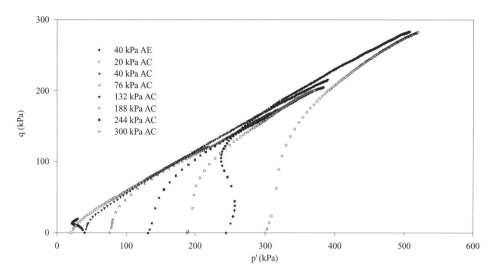

Figure 4. Results of the CU-AC and CU-AE triaxial tests. The values in the legend represent the initial effective consolidation stresses for the triaxial tests.

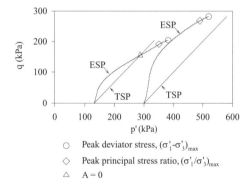

○ Peak deviator stress, $(\sigma'_1-\sigma'_3)_{max}$
◇ Peak principal stress ratio, $(\sigma'_1/\sigma'_3)_{max}$
△ $A = 0$

Figure 5. Graphical representation of various criteria to define the onset of failure in triaxial tests. Both samples in the figure were sheared up to 20% axial strain. The $A = 0$ failure condition occurs where the effective stress path, ESP, and the total stress path, TSP, intersect.

Brandon et al. (2006) recommends the condition $A = 0$ to define the onset of failure in low-plasticity silts. As illustrated in Figure 5, the $A = 0$ condition seems an appropriate choice for the triaxial specimen with an initial effective consolidation stress of 132 kPa. It results in a large quantity of points on the failure envelope described by the effective stress path above the $A = 0$ stress point. However, the $A = 0$ condition is not achieved for some of the triaxial test results in this study (e.g., the triaxial specimen with an initial mean effective consolidation stress of 300 kPa in Figure 5). For this reason, the $A = 0$ criterion was discarded.

The disadvantage of the peak deviator stress $(\sigma'_1-\sigma'_3)_{max}$ criterion applied to the tested soil is that it results in only one point on the failure envelope, i.e., the strength value at the maximum shear strain attained at the end of the triaxial test (Figure 5). Compared to the other analyzed criteria, the $(\sigma'_1-\sigma'_3)_{max}$ criterion will provide the minimum number of experimental data points that can be used to derive the strength parameters characterizing the failure envelope.

The criterion selected to define the onset of failure in this experimental study is the peak principal stress ratio $(\sigma'_1/\sigma'_3)_{max}$. As seen in Figure 5, the stress point associated with $(\sigma'_1/\sigma'_3)_{max}$ is above the stress point corresponding to $A = 0$ failure condition for the sample with an initial effective consolidation of 132 kPa thus providing a smaller number of experimental data points characterizing the failure envelope in this case. However, unlike the $A = 0$ criterion, the $(\sigma'_1/\sigma'_3)_{max}$ failure condition is achieved in all of the performed triaxial tests (thus including the sample with an initial effective confining stress of 300 kPa in Figure 5 that did not achieve the $A = 0$ condition). Furthermore, the $(\sigma'_1/\sigma'_3)_{max}$ criterion appears to provide a significantly larger number of experimental data points describing the failure envelope compared to the $(\sigma'_1-\sigma'_3)_{max}$ criterion. Figure 6 demonstrates the range of experimental data points located on the failure envelope based on $(\sigma'_1/\sigma'_3)_{max}$ failure condition, in relation to axial strain, ε_a, for a triaxial test with an initial effective confining stress of 244 kPa.

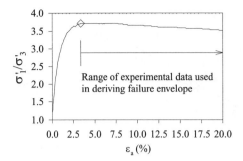

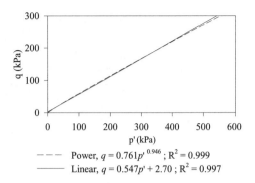

Figure 6. Principal stress ratio plot for sample with an effective confining stress of 244 kPa showing the range of data used in failure envelope regression analysis. The diamond point represents the peak principal stress ratio.

Figure 8. Linear and nonlinear failure envelopes obtained from fitting linear and nonlinear strength functions to the experimental data.

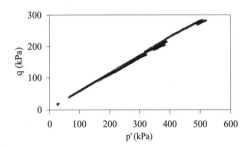

Figure 7. Experimental stress points on the failure envelope according to the principal stress ratio failure definition.

By selecting the $(\sigma'_1/\sigma'_3)_{max}$ criterion to define the onset of failure in triaxial testing, a reasonably large number of experimental stress points describing the failure envelope was obtained. This will allow for a more accurate evaluation of the shear strength parameters characterizing the linear and nonlinear strength functions derived from regression analysis. Figure 7 displays the experimental failure envelope data from triaxial testing according to the $(\sigma'_1/\sigma'_3)_{max}$ failure condition.

5 LINEAR AND NONLINEAR FAILURE ENVELOPES FROM REGRESSION ANALYSIS

A regression analysis was performed using the experimental data set shown in Figure 7. Conventional least squares minimization methods were employed to derive the strength parameters for a typical linear k_f function and a nonlinear empirical power law relationship. The resulting linear and nonlinear failure envelopes derived from the regression analyses are shown in Figure 8. A high coefficient of determination R^2 of 0.997 and 0.999 obtained for the linear and nonlinear regression analyses indicates that both strength functions provide a very good description of the experimental information which in fact covers a range of mean effective normal stresses, p', within 25 to 500 kPa (Figure 7). For the interval of very small mean effective normal stresses (i.e., $p' < 25$ kPa), the nonlinear strength function predicts smaller strength values compared to the linear model (Figure 9).

Pariseau (2007) has shown that both forward and backward extrapolation of a particular linear fit applied to the experimental data can overestimate the available shear strength of the material. This aspect is also demonstrated in this study by the q-axis intercept of the linear failure envelope, unlikely for a cohesionless nonplastic soil such as the analyzed silty-sand (Figure 9), and the deviation of the linear and nonlinear failure envelopes in the range of mean effective normal stresses above 350 kPa (Figure 8).

Reliance on the R^2 value alone to assess the accuracy of a strength function in describing the experimental information should be approached with caution. Pariseau (2007) pointed out that a common but erroneous procedure is to interpret the failure envelope with the greatest R^2 value as the "best fit" failure criterion. Care must be taken in evaluating the "best fit" for the range of normal stresses available from laboratory testing as long as this range does not coincide with the interval of normal stresses for the design problem of interest.

In case of compacted embankment slope applications, the mean effective normal stress typically does not exceed 200 kPa. However, the experimental results in Figure 7 are within a larger p' interval (i.e., 25 to 500 kPa), thus the R^2 parameter characterizing the linear and nonlinear strength functions in Figure 8 corresponds to this large range of mean effective normal

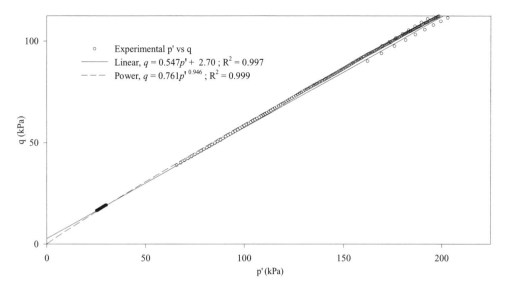

Figure 9. Expanded view of the linear and nonlinear failure envelopes together with the experimental data in the range of small mean effective normal stresses.

stresses. Therefore, in order to investigate the accuracy of strength predictions within the interval of mean effective normal stresses of interest, an approach based on cumulative squares of residuals has been used. The residual for a given p' value was calculated as the difference between the experimentally measured q value and q predicted by the linear or power strength functions in Figure 8. The cumulative squares of residuals was plotted against the mean effective normal stress (Figure 10) to demonstrate the "fit" of the two strength models within the range of stresses applicable to compacted embankment slopes (i.e., 0 to 200 kPa mean effective normal stress). As seen in Figure 10, for p' within 0 to 200 kPa, the nonlinear strength model provides either a comparable or a much lower value of the cumulative squares of residuals compared to the linear strength function. Thus, it may be concluded that for the analyzed compacted silty-sand and for the range of mean effective normal stresses of interest in this study, the failure envelope described by the nonlinear strength function is a better representation of the actual strength of the soil.

6 CONCLUSIONS

Linear and nonlinear strength functions were fitted to laboratory triaxial test data to derive the failure envelope of a compacted silty-sand. For the range of experimental mean effective normal stresses, both linear and nonlinear strength models provided equally

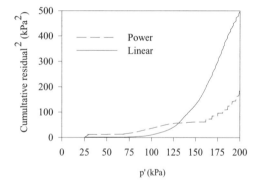

Figure 10. Demonstration of the "best fit" failure criterion for the given range of stresses using the cumulative residual^2. The nonlinear failure criterion provides a more accurate representation of the experimental data, having an overall lower cumulative residual^2 for the given range of stresses.

valid descriptions of the experimental information. However, a closer examination of the accuracy of predictions within the interval of mean effective normal stresses of interest for compacted embankment slopes revealed a better "fit" of the nonlinear strength function to the triaxial data. The derived nonlinear failure envelope also predicted lower strength values than the linear strength model within the range of very small normal stresses uncovered by the experimental information. This departure from linearity is usually associated with smaller computed safety factors in a conventional slope stability analysis based on the non-linear strength model, implying therefore a safer slope design.

ACKNOWLEDGMENTS

Funding for this research was provided by the Undergraduate Research Opportunity Program (UROP) at the University of Utah. The authors would like to thank IGES, Inc. for open access to their extensive laboratory facilities that were used in the present experimental investigation.

REFERENCES

ASTM 2007. Standard D 4318-05: Standard Test Methods for Liquid Limit, Plastic Limit, and Plasticity Index of Soils, *Annual Book of Standards*, Vol. 4, ASTM International, West Conshohocken, 2007.

ASTM 2007. Standard D 422-63: Standard Test Method for Particle-Size Analysis of Soils, *Annual Book of Standards*, Vol. 4, ASTM International, West Conshohocken, 2007.

ASTM 2007. Standard D 698-00: Standard Test Method for Laboratory Compaction Characteristics of Soil Using Standard Effort (12,400 ft-lbf/ft^3 (600 kN-m/m^3)), *Annual Book of Standards*, Vol. 4, ASTM International, West Conshohocken, 2007.

ASTM 2007. Standard D 2487-06: Standard Classification of Soils for Engineering Purposes (Unified Soil Classification System), *Annual Book of Standards*, Vol. 4, ASTM International, West Conshohocken, 2007.

Brandon, T.L., Rose, A.T. & Duncan, J.M. 2006. Drained and undrained strength interpretation for low-plasticity silts. *J. Geotech. Geoenviron. Eng.*, 132(2): 250–257.

Charles, J.A. & Soares, M.M. 1984. Stability of compacted rockfill slopes. *Geotechnique* 34(1): 61–70.

Dasenbrock, D.D. & Hankour, R. 2006. Improved soil property classification through automated triaxial stress path testing. *Proc. GeoCongress,* Atlanta, Feb. 26–Mar. 1 2006.

Day, R.W. & Axten, G.W. 1989. Surficial stability of compacted clay slopes. *J. Geotech. Eng.* 115(4): 577–580.

Jiang, J.C., Baker, R. & Yamagami, T. 2003. The effect of strength envelope nonlinearity on slope stability computations. *Can. Geotech. J.* 40(2): 308–325.

Pariseau, W.G. 2007. Fitting failure criteria to laboratory strength tests. *J. Rock Mech. & Mining Sciences*, 44: 637–646.

Penman, A. 1953. Shear characteristics of saturated silt measured in triaxial compression. *Geotechnique* 15(1): 79–93.

Ponce, V.M. & Bell, J.M. 1971. Sear strength of sand at extremely low pressures. *J. Geotech. Eng.* 9(4): 625–638.

Rad, N.S. & Clough, G.W. 1984. New procedure for saturating sand specimens. *J. Geotech. Eng.* 110(9): 1205–1218.

Trandafir, A., Popescu, M. & Ugai, K. 2001. Two dimensional slope stability analysis by LEM and FEM considering a non-linear failure envelope. *Proc. 40th Annual Conf. of Japan Landslide Society, Maebashi,* August 2001: 219–222.

An investigation of a structurally-controlled rock cut instability at a metro station shaft in Esfahan, Iran

Ali Taheri
Zaminfanavaran Consulting Eng. (ZAFA), Esfahan, Iran

ABSTRACT: On July 29th 2006 at 1.30 A.M, an overall slope failure took place at eastern wall of a 28 m-depth metro station shaft in Esfahan, Iran. The failure caused some serious damage to the city utilities and loss of one man life. The instability occurrence was about 10 month after the shaft completion, where, the shaft walls had been supported by a regular grid of grouted dowels, steel mash and shotcrete.
According to the geotechnical investigation results carried out after the failure, the main cause of the rock mass instability at the shaft wall was a major shear joint existence in the rock mass, trending almost parallel to the slope face and dipping towards the shaft bottom with about 60 degrees angle. This structural feature was unrecognized before the failure incidence.
In this paper, by describing the induced failure mechanism, the dominant role of the existing natural plan of weakness on the rock instability is analyzed and the unsuccessful effect of the used rock supporting system is discussed. It has been concluded in this paper that the possibility of the structurally-controlled rock failures should be taken into account seriously in design and execution of any earth structures, especially in urban areas.

1 INTRODUCTION

The Esfahan Metro Project is a large scale infrastructure project which has attracted international interest form inception as it passed through the city of Esfahan with 2500 years history and architecture and home to a number of highly treasured world heritage sites. On the other hand, the geological and hydrogeological conditions of the project area is various and rather complicated.

The line 1 of the Esfahan subway passes through bedrock at southern part of the city and through soft deposits at the middle and northern parts. The kargar Station shaft instability occurred in the discontinuous bedrock mass due to sliding on a major shear joint, 10 months after completion of the ground excavation and support installation.

In this paper, by describing the geological conditions of the site and explanation of the installed support system of the station cut, the induced failure mechanism has been discussed and conclusions and recommendations are given in order to prevent such instabilities in similar projects.

2 GEOLOGY

2.1 Lithology

The lower Jurassic (Lias) deposits, comprising shale and sandstone alternation, form the bedrock of the project area. These deposits which are denoted as "Shemshak Series" are overlain by a slightly cemented, coarse grained alluvium, which its thickness is around 9 m at the proposed site.

2.2 Structural features

The dominant structural features of the bedrock mass are bedding plane and three systematic joint sets, with the orientations given in Table 1.

There is no visible geological discontinuity in the covering alluvial deposits.

2.3 Groundwater condition

There is a shallow, low transmissible groundwater aquifer at the southern part of the Esfahan city, which has been formed in the alluvial deposits and shallow layers of the bedrock mass. The main recharge source

Table 1. Orientations of discontinuity system in the rock mass.

	Orientation (deg.)	
Discontinuity type	Dip direction	Dip amount
Bedding plane	185–200	35
Joint set 1, J_1	310	85
Joint set 2, J_2	110	75
Joint set 3, J_3	010	60

of this aquifer is water infiltration form green areas, as well as, leakage from water and sewage lines. The groundwater flow direction is generally from south to north, with hydraulic gradient of about 3%. The original groundwater table at the proposed site lies in the range of 6.2 to 8.8 m. bellows the ground surface.

3 GEOTECHNICAL ASPECTS

In order to assess the effect of the geological discontinuities on stability of the proposed rock mass, the orientation of the four distinguished discontinuity sets have been plotted on a streonet, together with the orientation of the shaft cut face (Figure 1).

It is evident from this streoplot that the existing discontinuities were unlikely to be involved in the slope failures. So, the potential slope failure had been considered as non-structurally controlled and the slope stability analyses were performed on the basis of this assumption. This led to design and execution of the following slope supporting system.

In designing the executed supporting system, by assuming the rock mass as an equivalent continuum media, the shear strength of the rock mass had been determined on the basis of the Hoek-Brown failure criterion, as follows (Hoek et al. 2002).

$$\sigma_1' = \sigma_3' + 49(0.0176\sigma_3' + 0.0003)^{0.508} \qquad (1)$$

where σ_1' and σ_3' are the major and minor effective principle stresses at failure, in Mpa.

4 THE INSTALLED SUPPORT SYSTEM

In order to support the proposed station shaft walls, a passive supporting system has been designed and executed during the excavation process. In designing this system, the rock mass had been considered as a equivalent continuous media, with potential failure plane as non-structurally controlled (similar to soil). The wall supporting system was comprised of ϕ32 mm grouted dowels with 12 , 8 and 6 m. lengths, at 2 × 2 m. regular grid, and 30 cm. thickness shotcrete, reinforced by two 10 × 10 × 0.8 cm. weldmeshes (Figure 2).

5 THE FAILURE MECHANISM

When the overall failure of the shaft wall started to take place at 1:30 A.M., there were five persons working inside the shaft, which one of them had not chance to survive. Another worker who was witness of the incident reported that the failure started from bottom of the shaft by mass sliding towards the opposite wall with a loud voice, and followed by toppling of soil, rock and shotcrete blocks from high levels of the wall (Figure 3). It should be added that 10 days before failure incidence, some tension cracks had been formed at upper surface of the shaft, which nobody paid attention to it.

According to the geotechnical investigation results, carried out after the failure incidence, a major shear joint (local fault plane), trending almost parallel to the slope face and dipping towards the shaft bottom with about 60 degrees angle, was the main cause for the failure occurrence. This fault plane which had not been recognized before the failure incidence was filled by a soft and saturated gouge of about 30 cm thickness.

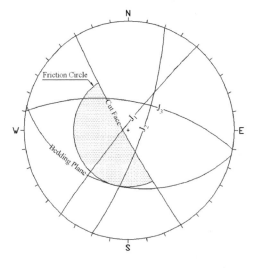

Figure 1. Streoplot of the discontinuity system in the rock mass.

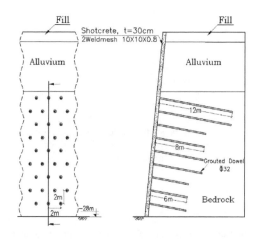

Figure 2. Wall supporting system of the station shaft.

Figure 3. General view of the failed shaft wall.

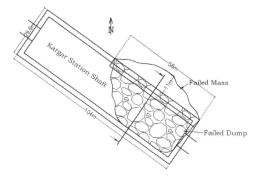

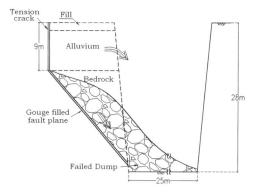

Figure 4. Plan and cross section of the failed shaft wall.

As shown in Figure 4, the induced failure had a dual mechanism; sliding of a discontinuous rock on the existing fault plane and then rotation and toppling of the soil blocks and shotcrete slabs from higher levels of the wall (slide head toppling).

6 THE MAIN REASONS OF THE INSTABILITY

In spite of the supporting system installation for stabilization of the shaft walls, about half of the eastern wall was failed in a few seconds, 10 months after completion of the shaft excavation.

By consideration of the induced failure mechanism and taking into account all the parameters affecting the stability of the slope, the main reasons of the shaft wall instability were reported as follow.

- Design of the wall supporting system by assuming the rock mass as a continuum media with equivalent geomechanical parameters. While, the existing fault plane caused the rock mass to behave as discontinuous, with structurally controlled failure incidence.
- Besides the insufficient supporting system, the poor installation of the grouted dowels intensified the instability problem of the shaft wall. For instance, by inspecting the failure plane after removal of the debris, it was found that just a limited number of the dowels were failed in tension, while, most of them were pulled out from failed or remained rock masses (Figure 5).
- As the failure took place 10 months after completion of the shaft excavation, the sliding of the potentially unstable rock block can be related to increase of groundwater pressure on the fault plane, due to the high water infiltration rate from adjacent green areas in hot summer.

7 FAILURE MECHANISM ANALYSIS

In order to analyze the induced failure mechanism, a windows grogram called "ROCPLANE" has been used. This program is an interactive software tool for assessing the stability of planar sliding blocks in rock slopes. It also allows users to estimate the support capacity required to achieve a specified factor of safety.

To do this analysis, the Mohr-Coulomb shear strength parameters of the fault plane have been determined by performing the Consolidated-Undrained triaxial tests on a number of representative undisturbed samples of the fault gouge material in the laboratory. The mean shear strength parameters determined in this manner are as follows:

- Cohesion, $c' = 0.5$ kg/cm^2,
- Angle of internal friction, $\phi' = 30$ degrees.

By assuming the groundwater pressure distribution as shown in Figure 6, the factor of safety of the so supported slope was found as 1.1, without taking into account the retaining effect of the reinforced shotcrete. In this case, the total retaining force of the installed grouted dowels was found 157.5 ton/m, assuming all the dowels were mobilized simultaneously. While, by factor of safety of 1 the retaining force of the dowels drops to 118.5 ton/m, i.e. 75% of the expected total force, and by taking into account the retaining effect of the reinforced shotcrete, it was found that around 50% of the installed grouted dowels were not mobilized simultaneously with the others, at the verge of the failure.

Figure 5. Pull out of the dowels from the failed rock mass.

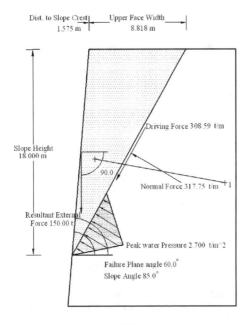

Figure 6. Geometry of the failed slope.

8 CONCLUSIONS AND RECOMMENDATIONS

According to the induced failure mechanism and reasons, the following conclusions and recommendations can be given.

- Regarding the dominant role of the major geological discontinuities on stability of rock slopes, in preliminary designing of the wall supporting systems, the potentiality of structurally controlled failures should be taken into account, especially in urban areas. To do this, a good practice is to carry out some sensitivity analysis by assuming a range of possible orientations and shear strength parameters for the major structures in rock masses.
- To confirm the geological structure of the slope, further geological mapping together with evidence on groundwater and tension cracks, would provide information for review of the situation to decide upon the best means of slope stabilization, in addition to the drainage measures.
- Due to the considerable effect of the supporting systems installation quality on stability of rock masses, the proper execution of the rock supporting designs is highly recommended. In this sense, it is desirable that unlike the studied case, the passive supporting systems be mobilized simultaneously. Otherwise, the resisting capacity of the supporting systems will

be decreased and may not be able to sustain the destructive forces of the unstable rock blocks.
- In order to control the stability of the slopes and efficiency of the ground stabilization systems, it is recommended that to perform some instrumentation and monitors surface movements and subsurface deformations of the ground, during execution and operation of the shafts in the urban areas.

ACKNOWLEDGMENTS

The author would like to record his appreciation of help from Mr. H. Mansoori Broojeni in collection and analysis of the failure data.

REFERENCES

Alamoot Bridge & Building Eng. Co. 2006. Geological report of the failure event at the Kargar Station eastern wall, *Report* No. EURO-ALMT-RO3-RPT-GEO-2003-(Rev. 0), in Persian, Tehran-Iran.

Haraz Rah Consulting Eng. Group. 2006. A study of the Kargar Station eastern wall failure. *Technical report*, in Persian, Tehran-Iran.

Hoek, E. & Bray, J.W. 1981. *Rock Slope Engineering*, 3rd. ed. London: IMM.

Hoek, E., Carranza-Torres, C.T. & Corkum, B. 2002. Hoek-Brown failure criterion-2002 ed., proc. North American Rock Mechanics Society meeting in Toronto in July 2002.

Ortigao, J.A.R. & Sayao, A.S. 2004. *Handbook of Slope Stabilization*. Berlin: Springer.

Simons, N., Menzies, B. & Matthews, M. 2001. *Soil and Rock Slope Engineering*. London: Thomas Telford.

US Army Corps of Engineers. 2003. *Engineering and Design Slope Stability, Manual* No. 1110–2-1902.

Wyllie, D.C. & Mah, C.W. 2004. *Rock Slope Engineering, Civil and Mining*, 4th ed., Based on the 3rd. ed, by Hoek, E. & Bray, J., London: Spon Press.

Zaminfanavaran Consulting Eng. (ZAFA). 2006. *Concluding report of the instability at the Kargar Station eastern wall.* in Persian, Esfahan-Iran.

Yield acceleration of soil slopes with nonlinear strength envelope

A.C. Trandafir
Department of Geology and Geophysics, University of Utah, Salt Lake City, Utah, USA

M.E. Popescu
Department of Civil and Architectural Engineering, Illinois Institute of Technology, Chicago, Illinois, USA

ABSTRACT: This paper presents and discusses the results of a pseudostatic slope stability analysis aiming to address the influence of strength envelope nonlinearity on the computed yield acceleration of a uniform slope in a homogeneous soil. Published linear (Mohr-Coulomb) and nonlinear strength envelopes derived from the same experimental database were employed in the analysis. For the variety of slope geometries analyzed in this study, the nonlinear strength envelope always resulted in a smaller yield acceleration compared to the traditional Mohr-Coulomb strength envelope. The difference between yield accelerations associated with the two failure criteria appears to increase with decreasing slope height and increasing slope inclination. For the same slope geometry, the critical sliding surface corresponding to the yield acceleration for the nonlinear strength envelope is shallower than the critical sliding surface associated with the Mohr-Coulomb strength envelope.

1 INTRODUCTION

Assessment of seismic displacements of slopes based on Newmark sliding block method (Newmark 1965) represents a routine engineering practice. Newmark model consists of one-block translational or rotational mechanism along a rigid-plastic sliding surface which is activated when the ground shaking acceleration exceeds the yield acceleration of the sliding mass. The yield acceleration is defined as the earthquake acceleration required to bring the sliding mass to the limit equilibrium condition corresponding to a safety factor of 1.0. The yield acceleration is the yield coefficient (k_y) multiplied by the gravitational acceleration (g). The yield coefficient (k_y) is obtained from conventional pseudostatic slope stability analyses and depends, among other factors, on the shear strength properties of the material along the sliding surface.

Soil shear strength is generally expressed in terms of Mohr-Coulomb linear failure criterion $\tau = c + \sigma \tan\phi$, with c (cohesion) and ϕ (angle of internal friction) representing the conventional shear strength parameters. There is however considerable experimental evidence showing that the strength envelope of many soils is not linear particularly within the range of small effective normal stresses (e.g., Bishop et al. 1965, Charles and Soares 1984, Atkinson and Farrar 1985, Maksimovic 1989). Furthermore, slope stability analyses conducted for materials exhibiting a nonlinear failure envelope resulted in lower static safety factors compared to those provided by the linear (Mohr-Coulomb) strength envelope (Maksimovic 1979, Charles and Soares 1984, Popescu et al. 2000, Trandafir et al. 2001, Jiang et al. 2003). Given the interdependence between static safety factor and yield acceleration of a potential sliding mass, the importance of considering the strength envelope departure from linearity in seismic slope stability evaluations may be easily inferred; the result may be a smaller value of the computed yield acceleration and therefore a higher vulnerability of the analyzed slope to earthquake induced instability. This paper examines through a parametric study the influence of soil strength envelope nonlinearity on the computed yield acceleration of a uniform slope in a homogeneous soil. Published linear Mohr-Coulomb and nonlinear strength envelopes derived from the same experimental database are employed with pseudostatic slope stability computations to obtain the yield acceleration of the sliding mass.

2 SLOPE GEOMETRY AND SHEAR STRENGTH ENVELOPES

Figure 1 shows the geometric variables (i.e., slope height H and slope inclination parameter b) for the parametric study, and the forces acting on the sliding mass in a pseudostatic slope stability analysis to

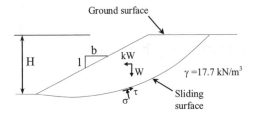

Figure 1. Slope geometry and forces acting on the slide mass.

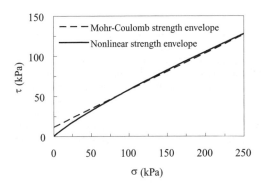

Figure 2. Linear and nonlinear strength envelopes considered in the analysis.

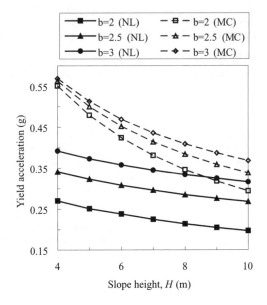

Figure 3. Yield acceleration versus slope height for various slope inclinations; NL—nonlinear strength envelope; MC—Mohr Coulomb strength envelope.

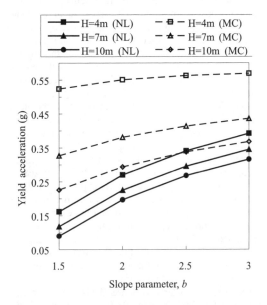

Figure 4. Yield acceleration versus slope inclination for various slope heights; NL—nonlinear strength envelope; MC—Mohr Coulomb strength envelope.

the shear strength estimate in the range of small normal stresses obtained from the projection of the linear strength envelope in the range of larger normal stresses is unsafe.

determine the yield acceleration of a specific slope. The linear (Mohr-Coulomb) and nonlinear strength envelopes used in the analysis are depicted in Figure 2. These envelopes were derived by Jiang et al. (2003) using least-square estimates on experimental results from numerous (i.e., 103) laboratory triaxial tests on heavily compacted Israeli clay. The conventional shear strength parameters characterizing the linear (Mohr-Coulomb) failure envelope (Figure 2) are $c = 11.7\,\text{kPa}$ and $\phi = 24.7°$. The nonlinear strength envelope is described by a power-type relationship $\tau = P_a A(\sigma/P_a)^\delta$ (Baker 2004), with P_a representing the atmospheric pressure, whereas A and δ are the dimensionless strength parameters of the nonlinear strength function. For the nonlinear failure envelope in Figure 2, $A = 0.582$ and $\delta = 0.857$.

By comparing the sums of squares of residuals, Jiang et al. (2003) noticed equally valid descriptions provided by both linear and nonlinear strength models of the available experimental data associated in fact with normal stress levels greater than 35 kPa. It is to be noted that triaxial tests reported by Atkinson and Farrar (1985) indicate that the nonlinear shear strength envelope apparent in the range of small stresses becomes linear in the range of larger normal stresses. Therefore

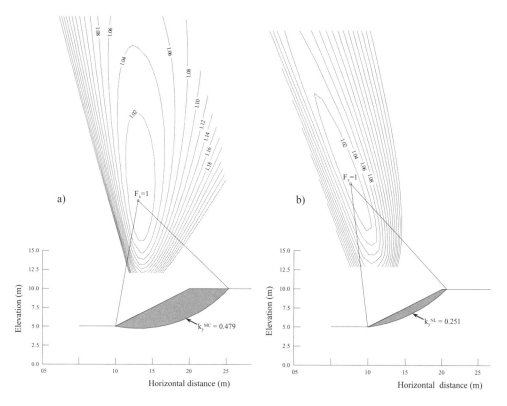

Figure 5. Critical sliding surface corresponding to the yield coefficient of a 5 m high slope with an inclination $b = 2$: (a) linear strength envelope model (MC); (b) nonlinear strength envelope model (NL).

For the particular soil considered in our analyses (Figure 2), the nonlinear strength model predicts smaller strength values than linear (Mohr-Coulomb) strength envelope for normal stresses below 35 kPa. The following section of this paper presents and discusses the results of a parametric study performed to illustrate the effect of the nonlinear failure envelope in the range of very low normal stresses (i.e., <35 kPa) on the yield acceleration.

3 PARAMETRIC STUDY

The effect of strength envelope nonlinearity on the yield acceleration of the sample slope depicted in Figure 1 was studied for various slope heights and inclinations. The yield coefficient (k_y) giving the yield acceleration ($k_y g$) of the slide mass was determined from a pseudostatic slope stability analysis along circular sliding surfaces based on Bishop's simplified method of slices. The analysis was conducted using the Slope/W module of the GEO-SLOPE OFFICE package (GEO-SLOPE International, Ltd., 2002), a software that can handle both linear and nonlinear strength envelopes in a limit-equilibrium slope stability analysis. In Slope/W procedure, the nonlinear strength envelope is introduced as a general data point function consisting of a series of (σ, τ) values characterizing the failure envelope. For each analyzed slice, Slope/W computes the local slope angle (ϕ) and cohesion intercept (c) of the tangent to the nonlinear failure envelope, as a function of the normal stress at the base of the slice (GEO-SLOPE International, Ltd., 2002). Consequently, in slope stability analysis with nonlinear strength envelope using Slope/W module, the linear strength parameters (c, ϕ) are different for each slice.

In limit-equilibrium slope stability analysis, the seismic coefficient (k) associated with a safety factor of 1.0 represents the yield coefficient (k_y) of the sliding mass. Therefore, k was gradually increased in the analysis until the calculated safety factor reached 1.0. For each input k value, the minimum factor of safety was determined by enabling the automatic search option in Slope/W to locate the pole of the critical failure surface across a grid of potential slip circle centers (Figure 5).

Figures 3 and 4 present the yield acceleration versus slope height and, respectively, slope inclination, as computed for the linear (MC) and nonlinear (NL) strength envelope models shown in Figure 2. Overall, the nonlinear strength model resulted in smaller yield acceleration values compared to the linear strength model. The difference between the computed yield accelerations based on the two strength envelope models becomes more pronounced with decreasing slope height (Figure 3). For a slope inclination corresponding to $b = 2$, this difference increased from 33% ($H = 10$ m) to 51% ($H = 4$ m) expressed as percentage of the yield acceleration associated with the linear (MC) strength envelope. Additionally, the yield acceleration given by the NL model is less sensitive to variations in slope height compared to the yield acceleration determined for the MC model. On the other hand, the yield acceleration corresponding to the NL model shows a more abrupt increase with decreasing slope angle compared to the MC based yield acceleration (Figure 4). The difference between the NL and MC yield accelerations increases significantly with increasing slope inclination. For example, this difference increased from 31% ($b = 3$) to 69% ($b = 1.5$) expressed as percentage of the yield acceleration associated with the linear (MC) strength envelope for a slope height $H = 4$ m (Figure 4). The differences in terms of depth and geometry between critical sliding surfaces corresponding to the yield coefficient associated with the NL and MC strength models for a slope with $H = 5$ m and $b = 2$ are illustrated in Figure 5. Apparently, the type of failure envelope has a significant influence on the critical sliding surface. The nonlinear strength model always resulted in a much shallower critical sliding surface compared to the linear (Mohr-Coulomb) strength model.

4 CONCLUSIONS

The parametric study presented in this paper illustrates the significant effect of strength envelope nonlinearity in assessment of seismic slope stability. For slope materials exhibiting nonlinear strength envelope within the range of low normal stresses, an analysis based on the traditional Mohr-Coulomb failure criterion may lead to considerable overestimate of the yield acceleration and therefore to unsafe seismic slope design, if the relevant range of normal stresses along the sliding surface of the investigated slope is overlooked in laboratory evaluations of Mohr-Coulomb strength parameters. The discrepancy between yield accelerations computed using linear and nonlinear strength models becomes larger for smaller height slopes and steeper slopes. This is related to the fact that in smaller height slopes and steeper slopes, the zones of smaller normal stresses which are within the range of pronounced strength envelope nonlinearity have more extent.

REFERENCES

Atkinson J.H. & Farrar D.M. 1985. Stress path tests to measure soil strength parameters for shallow landslips. *Proc. 11th Int. Conf. on Soil Mechanics and Foundation Engineering*, San Francisco: 983–986.

Baker R. 2004. Nonlinear Mohr envelopes based on triaxial data. *J. Geotech. and Geoenv. Eng.*, 130(5): 498–506.

Bishop A.W., Webb D.L. & Lewin P.I. 1965. Undisturbed samples of London Clay from the Ashford Common shaft: Strength effective normal stress relationship. *Geotechnique* 15(1): 1–31.

Charles J.A., & Soares M.M. 1984. The stability of slopes in soils with nonlinear failure envelopes. *Canadian Geotech. J.*, 21: 397–406.

GEO-SLOPE International Ltd. 2002. Computer program SLOPE/W for slope stability analysis. *User's guide*, Version 5, Calgary, Alberta, Canada.

Jiang J.C., Baker R., & Yamagami T. 2003. The effect of strength envelope nonlinearity on slope stability computations. *Can. Geotech. J.* 40(2): 308–325.

Maksimovic M. 1979. Limit equilibrium for nonlinear failure envelope and arbitrary slip surface. *Proc. 3rd Int. Conf. on Numerical Methods in Geomechanics*, Aachen, A.A. Balkema, Rotterdam, Vol. 2: 769–777.

Maksimovic M. 1989. Nonlinear failure envelope for soils. *J. Geotech. Eng.*, 115(4): 581–586.

Newmark N.M. 1965. Effects of earthquakes on dams and embankments, *Géotechnique* 15(2): 139–159.

Popescu M., Ugai K. & Trandafir A. 2000. Linear versus nonlinear failure envelopes in LEM and FEM slope stability analysis. *Proc. 8th Int. Symp. Landslides*, Cardiff, A.A. Balkema, Rotterdam, Vol. 3: 1227–1234.

Trandafir A., Popescu M., & Ugai K. 2001. Two dimensional slope stability analysis by LEM and FEM considering a non-linear failure envelope. *Proc. 40th Annual Conf. of Japan Landslide Society, Maebashi*, August 2001: 219–222.

derlandslides and Engineered Slopes – Chen et al. (eds)
© 2008 Taylor & Francis Group, London, ISBN 978-0-415-41196-7

Evaluation of rockfall hazards along part of Karaj-Chaloos road, Iran

A. Uromeihy
Dept. of Engineering Geology, Tarbiat Modares University, Tehran, Iran

N. Ghazipoor & I. Entezam
Engineering Geology Sect. Geological Survey of Iran, Iran

ABSTRACT: Karaj-Chaloos Road is a vital route that connects Tehran and the southern part of Alborz to the northern part and the resort areas along the Caspian Sea. The road is located in the Central Alborz Mountains where the occurrence of rockfall hazards is always considered to be very high. The event of May 28th 2004 earthquake triggered a large number of rockfalls along the road which caused human casualties and huge damages to the infrastructures along the road. The aim of this paper is to evaluate the potential of rockfalls along part of the route between Pol-e-zangoleh and Marzan-abad. The study area has a very complicated geological a geomorphological conditions. Since most of the slope instabilities along the road are of rockfalls type, rockfalls theory proposed by Evans and Hungr (1993) was used together with CONEFALL program (introduced by Jaboyedoff, 2003) to analyze and to predict the potential of rockfalls along the road. Many factors such rock type, slope morphology, drainage pattern and the seismic activity were considered in this research as main affecting factors. Based on GIS program a rockfall hazard zonation map was prepared. It was found that about 45% of the rout has a high potential of rockfall activity and slope morphology is the most affecting factor for the development of rockfalls. The influences of other factors such as rock type, fault alignment were found to be less.

1 INTRODUCTION

Rockfalls are the most dominant type of slope instabilities in mountainous areas. They can cause a large number of casualties and huge economic losses along road and other infrastructures. Rockfall can be defined as the detachment and falling of rock blocks down the slope of a rock body.

Evaluation of rockfall potential along the road in mountainous areas was the interest of many researchers in recent years. For examples; Aksoy & Ercanoglu, (2006), evaluate the rockfall initiation mechanism along highway in forest preservation and urban settlement areas in Turkey. Batterson et al. (2006), investigate the risk of geological hazards including the rockfalls in Newfoundland, Canada. Wasowski & DelGaudio, (2003), evaluate seismically induced mass movement hazard in Caramanico, Italy. Okura et al. (2000), studied the effect of rockfall volume on run-out distance in Japan.

Karaj-Chaloos Road is a main route that connects the Capital Tehran and the southern part of Alborz Mountain to the northern part along the Caspian Sea. Evaluation of rockfall potential along part of this road between Pol-e-zangoleh and Marzan-abad, with a total length of about 40 kilometers, was considered in this paper. Figure 1, illustrates the location of the study area along the main road. Due to the rough topography features and deep cuts along the road, most of rockfall hazards occurred along this part of the road.

Initiation of rockfall events along the study area can be sourced from different geological, geomorphologic, climatologic and human-related processes. Earthquake also has a clear influence on triggering rockfall occurrence. Recently, the magnitude 6.3 Baladeh Earthquake of May 28th 2004 is reported to cause 28 deaths due to the rockfall along the Chaloos road (Bolourchi et al. 2007). Details of the earthquake mechanism are investigated by Tatar et al. (2007).

Since most of the slope instabilities along the road are of rockfalls type, rockfalls theory proposed by Evans & Hungr (1993) was used together with CONE-FALL program (introduced by Jaboyedoff, 2003) to analyze and predict the potential of rockfalls along the road. Based on GIS data, a rockfall hazard zonation map was also prepared.

2 GEOLOGICAL SETTING

Karaj-Chaloos road with North-South alignment cut through the Alborz Mountain and passes various types

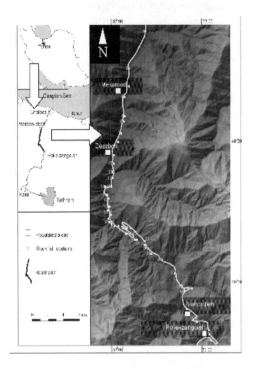

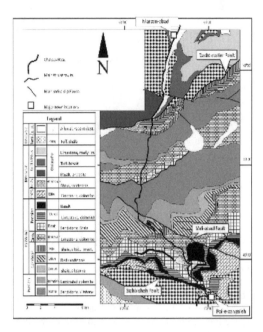

Figure 2. Geological setting of the area between Pol-e-zangoleh and Marzan-abad.

Figure 1. Geographical location of the studied area along the Karaj-Chaloos road.

Table 1. Summary of geological units in the area.

Geology time	Formation's names	Rock type
Pre-Cambrian	Kahar, Soltanieh	Sandstone, siltstone, shale, dolomite
Paleozoic	Zaigon, Lalon, Mila, Mobarak, Dorood, Roteh	Shale, siltstone, sandstone, dolomite, limestone, basalt
Mesozoic	Elika, Shemshak, Chaloos	Limestone, shale, sandstone, coal, tuff, basalt
Cenozoic	Karaj, Hezar-darreh	Tuff, alluvial

of geological features and structures. The geological setting of the site consists of a sequence of sedimentary rocks from Pre-Cambrian to Quaternary. The rocks include laminated layers of sandstone, shale, limestone and dolomite of different ages and some patches of volcanic rocks such as basalt and tuff. A summary of the most dominant geological units are presented in Table 1. The main geological features and rock distribution of the area are also shown on Figure 2.

There are two sets of thrust and reverse faults in the area, (Vahdati, 2001). They include faults with north-west south-east alignment such as Siahbisheh, Vali-abad, and Dona faults. The second sets have north-east south-west alignment named as Mekarood, Dezbon, Mojlar and Marzan-abad faults and they dips generally towards the south. The first and second sets of the faults are generally concentrated to the southern and northern part of the area respectively, (Stoklin, 1974). Examples of rockfall potential along the road are shown on Figures 3a and 3b.

3 CONEFALL THEORY

Conefall theory is a simple way to demonstrate the free movement of a block along a slope. A block can propagate from its source point making an angle ϕ_p with horizontal. The space where a block can propagate from a grid point is located within a cone of slope ϕ_p with a summit placed at the source point. It can be shown that the mean velocity of block is given as a function of the difference in altitude Δh between the cone and the topography. The details of the theory can be found in Evan and Hungr (1993). A computer program named as CONEFALL is designed by Jaboyedoff (2003), to estimate roughly the potential of rockfall prone area. The principle of the program is to define a maximum run-out distance for block propagation regarding the source point. The angle with horizontal of the line joining the stop point and the source ϕ_p may vary according to the slope of topography. Figure 4 shows the relationships between angle ϕ_p and the maximum run-out distance.

(a)

(b)

Figure 3. Potential of rockfall along the Chaloos road.

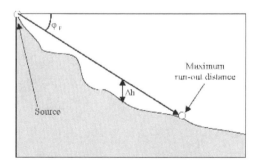

Figure 4. Relationships between angle ϕ_p and the maximum run-out distance (after Jaboyedoff, 2003).

4 ZONATION OF ROCKFALL HAZARD

In order to perform a precise evaluation of rockfall potential, the area was divided into three quadrants named as Mekarood, Dezbon and Pol-e-zangoleh from north to south respectively as shown on Figure 5. The required data for the analysis were collected from 47 stations along the road. The locations of each station were selected regarding their past history of rockfall occurrence and specially their response to the more recent earthquake event of May 28th 2004.

The CONEFALL program was run twice considering two cone angle of 40 and 45 degrees. The cone angles were selected regarding the common slope angle along the road in the area. The results showed that as the cone angle is increased the run-out distribution of block at the base of the slopes decreases. Figure 5 shows a rockfall hazard zonation map of the area. The zonation map was correlated to other sheet data by the aid of GIS program and following results were obtained:

- Potential of rockfall is higher along the alignment of main faults.
- The drainage pattern has no great effect on the development of rockfalls.

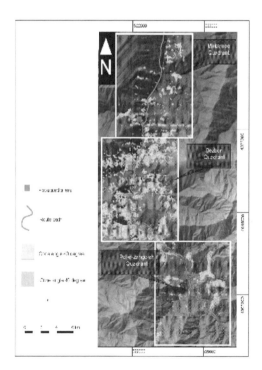

Figure 5. Rockfall hazard zonation map along Chaloos road (between Pol-e-zangoleh and Marzan-abad).

- Rockfall potential is higher where the Pre-Cambrian rocks are outcropped.
- Slope angle of natural slopes and cuts along the road greatly influences the generation of rockfalls.

5 CONCLUSIONS

The CONEFALL theory and program were used to evaluate the potential of rock fall hazards along part of Chaloos road. The rockfall zonation map showed that the rockfall potential is higher in places where the elevation is higher, the cuts are deeper. Also it was found that the development of rockfalls is directly related to the alignment of the main faults and other fractures. The type of rock also showed great influence on the generation of rockfalls. The number of rockfalls in sandstone were distributed among the whole areas while in limestone were restricted to the deep cuts along the road path. The potential of rockfall is highest in Dezbon Quadrant where Pre-Cambrian rocks are dominant. While the distribution of rockfalls in Pol-e-zangoleh Quadrant comes in second place where Paleozoic rocks have greater outcrops. Finally the occurrence of rockfalls in Mekarood Quadrant is lowest in which the topographic reliefs are less and Mesozoic rocks are more frequent.

REFERENCES

Aksoy, H. & Ercanoglu, M. 2006. Determination of rockfall sources in an urban settlement area by using a rule base fuzzy evaluation. *Nat. Hazards Earth Syst. Sci*, 6, 941–954.

Batterson, M.J., McCuaig S. & Taylor, D.M. 2006. Mapping and assessing risk of geological hazards on the Northeast Avalon Peninsula and Humber Valley, Newfoundland. *In Current Research, Newfoundland Department of Natural Resources, Geological Survey, Canada*. Report 06–1, 147–160.

Bolourchi MJ. Entezam I. Mahmoudpour M. & Ansari F. 2007. Investigation of rockfall hazard in Chaloos road. *Report Number 85-5-1 Geological Survey of Iran*. 209 pages, (in Persian).

Evan, S. & Hungr, O. 1993. The assessment of rockfall hazard at the base of talus slopes. *Canadian Geotechnical Journal*, vol. 30, pages: 620–636.

Jaboyedoff, M. 2003, CONEFALL 1.0 user guide, Open report-soft 01, Quanterra, www.quanterra.org, 15 pages.

Jaboyedoff, M. & Labiouse, V. 2003. Preliminary assessment of rockfall hazard based on GIS data. *Technology roadmap for rock mechanism, South African Institute of Mining and Metallurgy*. 575–578.

Okura Y. Kitahara H. Sammori T. & Kawanami A. 2000. The effect of rockfall volume on rumout distance. *Engineering Geology, Elsevier*, vol. 58, Issue 2, 109–210.

Stoklin J. 1974. Northern Iran Alborz Mountain, in Mesozoic-Cenozoic orogenic belts, *Geological Society London*, spec. pub. No. 4.

Tatar M, Jackson J. Hatzfeld D. & Bergman E. 2007. The 2004 May 28 Baladeh earthquake (M_w 6.2) in the Alborz, Iran: overthrusting the South Caspian Basin margin, partitioning of oblique convergence and the seismic hazard of Tehran. *Geophysical Journal International*, vol.170, issue 1, 249–261.

Vahdati F. 2001. Geological description of Marzan-abad map, scale 1:100,000. *Geological Survey of Iran*, Sheet No. 6262.

Wasowski J. & DelGaudio V., 2003. Evaluating Seismically induced mass movement hazard in Caramanico Terme (Italy). *Engineering Geology, Elsevier*, vol. 67, Issues 3–4, 281–296.

Coupled effect of pluviometric regime and soil properties on hydraulic boundary conditions and on slope stability

R. Vassallo & C. Di Maio
Dipartimento di Strutture, Geotecnica, Geologia Applicata, Università degli Studi della Basilicata, Italy

M. Calvello
Dipartimento di Ingegneria Civile, Università degli Studi di Salerno, Italy

ABSTRACT: The case of a landslide in Southern Italy was studied by modeling pore pressure distribution under steady and transient conditions. The goal was to interpret the *in situ* piezometric and inclinometric measurements, and to investigate the effects of the pluviometric regime on the slope stability. Steady state analyses show that the hypothesis of zero pore pressure at the ground surface gives the best fit between theoretical and experimental results, despite the nearly arid climate. The results of the transient state analyses show that the condition of zero pore pressure is appropriate for both rainy and non-rainy periods because of the particular soil properties, i.e. heterogeneity, compressibility and hydraulic conductivity. This condition generates a distribution of pore pressures which can justify the observed horizontal displacements.

1 INTRODUCTION

This paper focuses on the study of an unstable slope located in Tricarico (Fig. 1). The town of Tricarico is situated in Southern Italy, Basilicata Region, in the hill zone of the Lucanian Apennine chain, north of the Basento river. The historical centre rises on a calcarenite plate which overlies a formation of Pliocenic Clay. The newest quarters are mostly located in the areas where a more recent clay formation outcrops and overlies the calcarenites.

This town has always suffered from slope instability problems. In the 1950s a landslide in the Saracena area, north of the town, caused severe damages to buildings and infrastructures. As a consequence, the most part of the community living there had to move to other quarters in the 1960s and 1970s. Nowadays this zone is nearly uninhabited.

In the zone of expansion, west of the town, on the *Carmine* slope, many buildings constructed in the last ten years suffered significant deformations. The most serious damages occurred in some buildings situated north-west, near the line where calcarenites outcrop.

The eastern sector of the town was only affected by local instability problems and fissuring of road pavements.

In October 2002 the University of Basilicata started a study of the above mentioned three zones of Tricarico, funded by the Basilicata Administrative Region. A very extensive geotechnical investigation was carried out, including boreholes, undisturbed sampling, laboratory testing for the determination of index and mechanical properties of soils. *In situ* monitoring of pore pressures and horizontal displacements was carried out over several years.

The analysis presented herein is centred on the interpretation of *in situ* measurements carried out on the Carmine slope, along section AA' reported in Figure 1. Particular attention was devoted to the study of the pore pressure distribution and of its influence on slope stability.

2 IN SITU AND LABORATORY INVESTIGATIONS

As mentioned above, the subsoil of Carmine slope consists of a clay formation overlying a calcarenite formation. In this zone, five continuous coring boreholes were driven, each equipped with two Casagrande piezometers. These boreholes are indicated with the letter "S" in Figure 1. Two further continuous coring boreholes (i.e. S9 and I9 in the same figure) were equipped with full length inclinometer casings. By rapidly extracting small soil specimens from the inner part of remoulded borehole samples and sealing them hermetically (i.e. undrained conditions) it was possible to measure water contents. The obtained *in situ* water content profiles are reported in Figure 2. Five core-destruction boreholes were also driven, each equipped

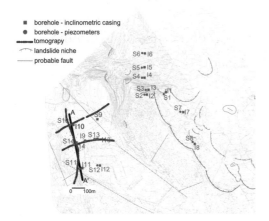

Figure 1. Plan view of the investigated zone.

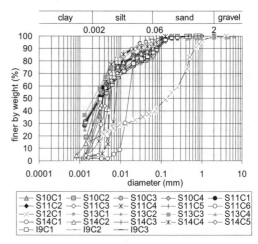

Figure 3. Grain size distribution of soil samples from the Carmine slope.

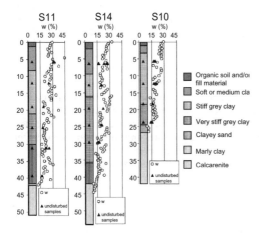

Figure 2. Stratigraphy and water content profiles for boreholes S11, S14 and S10.

with a full length inclinometer, indicated with the letter "I" in Figure 1.

The clayey material is about 42 m thick in boreholes S11 and S14 and 27 m thick in borehole S10, as reported in Figure 2. Verticals S11, S14 and S10 belong to section AA', used to analyze the pore pressure distribution and the stability conditions.

Electrical Resistivity Tomographies (ERT) were also used to recognize the position of the contact between clay and calcarenite where no direct observation was available. A longitudinal tomography crossing verticals S14 and S10 and several transversal tomographies were carried out (Fig. 1) as described in detail by Perrone et al. (2007). A very clear difference of electrical resistivity between an upper and a lower material (associated with clay and calcarenite, respectively) was found and the location of their interface was in good agreement with borehole results. As it will be shown, the integration of geophysical surveys with the geotechnical investigation was quite useful to model the pore pressure distribution of the slope with a good reliability.

Many undisturbed samples were used in the laboratory to measure index properties and to carry out mechanical testing, in particular oedometer, triaxial and direct shear tests. The fine-grained soil ranges from clay with silt to clayey-sandy silt (Fig. 3). In the Casagrande chart this soil is classified as high plasticity inorganic clay. In the activity chart most data points are located between the Na-montmorillonite and the illite lines. Peak friction angle from triaxial tests ranges between 17° and 24° and residual friction angle from direct shear tests ranges between 7° and 14°.

The results of the consolidation stages of oedometric tests were used, under the hypothesis of the Terzaghi one-dimensional model, to estimate values for the coefficient of consolidation c_v and for the hydraulic conductivity k in the axial direction and to investigate the influence of the void ratio, e, on those parameters. The variation of k with e observed experimentally (Fig. 4), along with the significant reduction of porosity with depth indicated by the water content profiles reported in Figure 2, suggested that the clay formation could be divided into two sub-layers. The upper layer, about ten meters thick, can be characterized by an hydraulic conductivity one order of magnitude greater than the lower layer. It will be illustrated in the following how this hypothesis was corroborated by an extensive parametric study of the pore pressure distribution within the slope and by the comparison of calculation results with *in situ* measurements.

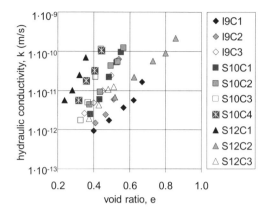

Figure 4. Hydraulic conductivities obtained by interpreting oedometer test data.

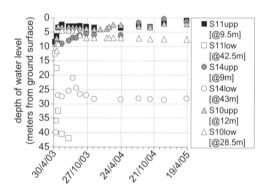

Figure 5. Piezometric measurements.

As it was mentioned above, two Casagrande piezometers were placed at different depths in each of five verticals of the Carmine slope. Piezometric measurements started in May 2003 and were carried out frequently for about two years. Subsequent less frequent measurements were also carried out in 2006 and 2007. Figure 5 shows the depth of the piezometric level from the ground surface as a function of time for the six piezometers considered in this study. One can observe that total heads decrease significantly with depth along the verticals of boreholes S11 and S14. In particular, piezometer S11$_{low}$, after a short initial transient, always resulted empty. Inclinometric readings started in November 2003 and the last measurements were taken in April 2007. The observed displacements of the ground surface since the installation of the inclinometric casings are about 3 cm for borehole I11, 4 cm for borehole I14 and 1 cm for borehole I10. The profiles reported in Figure 6 show that most displacements take place in the upper 10 meters.

3 PORE PRESSURE ANALYSIS

3.1 Steady state analyses

Figure 7 shows the 2-D model used for pore pressure calculations and the position of the verticals S11, S14 and S10 with piezometers. The interface between clay and calcarenites was initially obtained schematically by drawing a straight polyline through the limit found in boreholes S11, S14 and S10 and the points were calcarenites outcrop. However, this choice was validated, at a later stage, by tomography results.

Several analyses were performed using the finite element code SEEP/W (by Geoslope International Ltd), considering different boundary conditions and assuming different values of the anisotropy ratio. For each analysis, the calculated values of water pressures were compared with the values measured at the internal points of the domain.

At first, the numerical simulations were carried out by considering only two homogeneous and isotropic materials: the upper soil (clay) was characterized by a conductivity three orders of magnitude lower than calcarenite. Zero water pressure was assumed at the ground surface. Lateral boundary conditions able to optimize the results for the lower piezometers were found and left unchanged in subsequent analyses. These boundary conditions are also shown in Figure 7.

Then, an heterogeneity within the clay was introduced, based on the observed significant reduction of porosity with depth and on the relationships found experimentally between conductivity and porosity.

An upper layer of altered soil, about 10 m thick, was distinguished from a lower, less permeable, one. The latter was given a conductivity three orders of magnitude lower than calcarenites. A parametric study of the hydraulic properties of the clay formation was carried out, looking for those which could minimize the differences between measured and calculated total heads.

Figure 8a plots the difference Δ between measured and calculated values as a function of the ratio k_{upp}/k_{low} of the conductivities of the two clay layers, considered isotropic. Figure 8b plots the same difference as a function of the anisotropy ratio k_x/k_y of the whole clay layer considered as homogeneous (the direction x of maximum principal conductivity was assumed parallel to the slope). The conditions $k_{upp}/k_{low} = 1$ and $k_x/k_y = 1$ are relative to homogeneous and isotropic clay having a conductivity 1000 times lower than the conductivity k_c of the calcarenite. As the ratio k_{upp}/k_{low} increases, $|\Delta|$ at first significantly decreases and then stays practically constant for a ratio greater than 20. On the other hand, as the ratio k_x/k_y increases, assuming $k_x = 10^{-3} \cdot k_c$, $|\Delta|$ increases. Therefore, as it is also highlighted by the square root of the sum of squares plotted in the same figures, the hypothesis which gives the best agreement with the *in situ* measurements is that of heterogeneous and isotropic clays, with $k_{upp}/k_{low} = 20$. It is worth

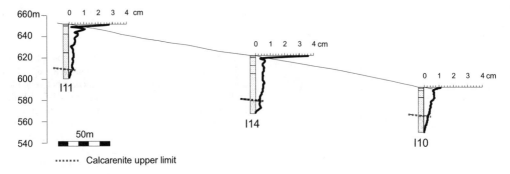

Figure 6. Profiles of horizontal displacements in April 2007 (zero: November 2003).

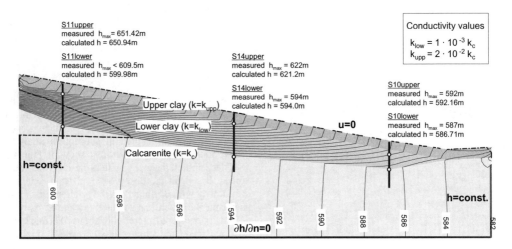

Figure 7. Results of a steady state analysis: optimized solution.

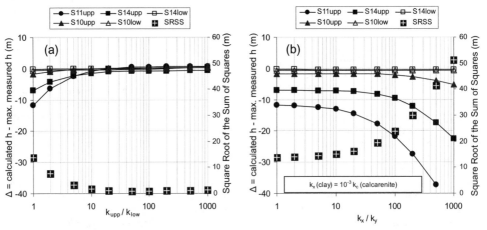

Figure 8. Effect of heterogeneity (a) and anisotropy (b).

noting that this value is in agreement with that coming from oedometer tests and water content profiles. The total head contours and the calculated total head values at the piezometers, versus the measured ones, are also shown in Figure 7 for the final and optimized solution. The very good agreement between calculations and measurements proves the accuracy of the proposed interpretation.

The validity of the chosen lateral boundary conditions was further proved by some analyses. Different boundary conditions were used above the phreatic surface in the calcarenites to account for possible fissuring and for unsaturated flow. The domain was also extended on both sides up to where the calcarenite outcrops and a significant reduction of the hydraulic conductivity above the phreatic surface was taken into account for this soil. These changes did not significantly influence the results at the piezometers and a very good agreement with the measurements was still obtained.

3.2 Transient analyses

As mentioned above, the best agreement between calculations and measurements is obtained with $u = 0$ at the ground surface. Given that the clay is water saturated up to the surface, this condition would certainly derive from a steady rainfall with an intensity greater than the soil hydraulic conductivity. Yet, since the climate is nearly arid in Tricarico, further analyses were carried out to evaluate which other process can make $u = 0$ at the ground surface as the most significant condition. To this aim, the transient flow caused by a cycle of alternating dry periods (normal unit flux $q_n = 0$ at ground surface) and wet periods ($u = 0$ at ground surface) was studied.

Two different steady state cases, i.e. $u = 0$ and $q_n = 0$ at the ground surface, were used as initial condition of the transient analyses. The conductivity of the lower clay was chosen two orders of magnitude higher than the value measured in the laboratory because of in situ large-scale heterogeneity, fissuring, etc. The conductivities of the other two materials were calculated from the same ratios used in steady state analyses. Therefore, the selected values were $k_{upp} = 2 \cdot 10^{-8}$ m/s and $k_{low} = 1 \cdot 10^{-9}$ m/s for intact and altered clay, respectively, and $k_c = 1 \cdot 10^{-6}$ m/s for calcarenite. Upper and lower clays were assumed to have the same compressibility, deduced from oedometer test results. The calcarenite was assumed to have a compressibility significantly lower than the clay one.

A first analysis was carried out considering a 6-month dry period followed by a 6-month wet period, starting from the steady state condition which resulted from $u = 0$ at the ground surface. The evolution of the total head at the ground surface and in the upper piezometer at vertical S11 is reported in Figure 9. The

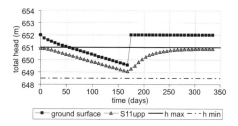

Figure 9. Effect of a 6 months dry period followed by a 6 months wet period. The starting condition is a steady state-distribution of pressures obtained using $u = 0$ at ground surface.

two horizontal lines plotted in this graph represent the minimum and the maximum total head recorded during the period of measurement. Total head at the ground surface starts with a $u = 0$ condition, then decreases as a consequence of the dry period and then immediately returns to a $u = 0$ condition as a consequence of rain. Because of this boundary condition, in the 9.5 m deep-piezometer the total head decreases slowly during the dry period and then returns to it initial value more rapidly during the wet period. In other words, the condition $u = 0$ prevails. Furthermore, total head always stays in the range between the maximum and the minimum measured values. In the other piezometers, the effect of dry periods is even less significant.

Further analyses were carried out using different "rain functions", i.e. different cycles of wet and dry periods, starting both from a wet period- and a dry period- steady state distribution of pore pressures.

These analyses included the case of 4 dry weeks followed by 4 wet weeks and that of 3 dry weeks followed by 1 wet week. The results are reported in Figure 10 for S11$_{upp}$, which was still the piezometer most influenced by the imposed rain functions.

For the analyses which started from a wet period-steady state (Fig. 10a), the total head always fluctuates around values close to that relative to $u = 0$ at the ground surface. On the other hand, the analyses which started from a dry period- steady state (Fig. 10b) indicated that after a few years of alternating wet and dry periods the total head again fluctuates around values close to that relative to $u = 0$ at the ground surface. Therefore, the condition $u = 0$ at the ground surface confirms to be appropriate even during dry periods.

4 STABILITY ANALYSIS

The trend of total head contours shown in Figure 8 is clearly a consequence of the subsoil stratigraphy and

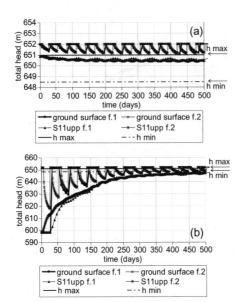

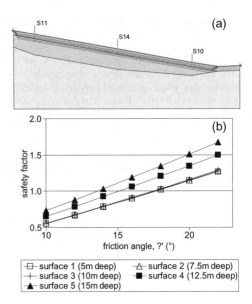

Figure 10. Effect of cyclic successions of wet and dry periods. The starting condition is a steady state- distribution of pressures obtained using: (a) $u = 0$ at ground surface and (b) $q_n = 0$ at ground surface. Rain functions are indicated as f.1 (4 dry weeks followed by 4 wet weeks) and f.2 (3 dry weeks followed by 1 wet week).

Figure 11. Results of stability analyses: (a) tentative slip surfaces; (b) safety factor as a function of friction angle.

of the hydraulic properties which were inferred by a combined use of boreholes, tomography and laboratory measurements. A confirmation of the validity of the chosen model came from the comparison between computed and measured pore pressures.

A detailed study of pore pressure distribution was essential to carry out an accurate slope stability analysis. Figure 11 shows for several hypothetical slip surfaces the safety factor (lower value between those obtained from Janbu and Morgenstern & Price methods) as a function of the friction angle.

The safety factor stays almost constant with the depth of the slip surface in the upper clay layer. On the other hand, it considerably increases with depth in the lower clay layer. This is a clear consequence of the particular pore pressure distribution. In fact, the upper layer has a pore pressure regime quite similar to that of an infinite slope, with lines parallel to the ground surface. Differently, total head contours in the lower clay layer are nearly horizontal, and this results in pressures lower than those which would be predicted for an infinite slope at the same depth. This seems in agreement with the displacements measured during almost four years, which are more relevant in the uppermost 10 meters.

The mobilized friction angle on the slip surface at the interface between upper and lower clay is about 17°. This value is within the range of residual and peak friction angles measured on undisturbed samples extracted from boreholes S11, S14, S10.

5 CONCLUSIONS

The results of pore pressure analyses showed that the interpretation of *in situ* measurements is satisfying if a 10-meters thick altered-clay layer characterized by greater void ratio and conductivity than the lower clay is considered. The condition $u = 0$ was imposed at the ground surface, despite the nearly arid climate of Tricarico. The transient analyses proved that the soil properties of the studied slope, i.e. heterogeneity, along with compressibility and hydraulic conductivity values, make the condition of zero pore pressure appropriate for both rainy and non-rainy periods. The obtained trend of total head contours significantly influences the safety factor, which is lower in the altered clay layer, in agreement with the inclinometric measurements performed over the last 4 years.

REFERENCES

Perrone, A., Piscitelli, S. Lapenna, V., Loperte, A., Di Maio, C. & Vassallo, R. 2007. Electrical resistivity tomography and geotechnical techniques for the stability analysis of the Tricarico landslide. *Thirteenth European Meeting of Environmental and Engineering Geophysics, Instanbul, Turkey, 3–5 September 2007.*

Mechanical characters of relaxing zone of slopes due to excavation

H. Wang
College of Environment and Resources; Fuzhou University, Fuzhou, China

X.P. Liao
Northwest Research Institute of China Railway Engineering Corporation, Lanzhou, China

ABSTRACT: The analysis for relaxing zone is very important for stability analysis of slopes, and it is difficult to confirm the boundary of relaxing zone because of excavation. Based on a typical case study by finite element analysis for excavation, the mechanical character and the distribution regulation for relaxing zone of slopes, especially the variety of differential principal stress and plastic zone, are researched in this paper. The results shows that: the change of differential principal stress tensor which is obtained by subtracting the stress tensors before and after excavated can be used to confirm the range of relaxing zone; and the distribution and development of the negative differential principal stress zone is almost the same as the plastic zone; in the end, the stress attenuation plastic relaxing zone near excavation surface, the original stress zone far from excavation surface and the stress concentration elastic compressing zone between above two are partitioned.

1 GENERAL

The analysis for relaxing zone is very important for stability analysis and design of slopes, and it is attached importance to by researcher for rock mechanics at home and abroad. Recently, with the large-scale development of infrastructure in China, especially hydropower high slopes and the deep cut slopes along expressway, some research results about relaxing zone of slopes can be summarized as follows:

Unloading rockmass mechanics had been founded by Q.L. Ha et al. (2001) by systemically researching anisotropy, strength theories, scale effect, and rheology. Estimate method for excavation disturbed zone of slopes had formed using numerical simulation technique and engineering test method by Q. Sheng (2002), and the relative research result had been used in the Three Gorges Project. The formation principle of unloading zones of high slopes had been analyzed by R.Q. Huang (2001), and the basic regularities of secondary stress distribution due to excavation had been studied. The stress field and the relative displacement field of the deep cut slope had been studied by S.G.Xiao (2003), and the saltation position of displacement or safety factor within influence area due to excavation. The distributing range and the variety character of the relaxing zone in centrifugal model test had been researched, and the displacement value had been used to confirm the width of unloading zones. The excavation disturbed zone in the permanent shiplock slopes of the Three Gorges Project had been studied using investigate, engineering physics exploration, test and monitor by J.H. Deng et al. (2001). The relation between the longitudinal wave speed of rockmass, rockmass deformation modulus and the stress of rockmass had been studied after rock slope excavated using the dynamics theory of wave, and then the change of rockmass deformation had been researched, the thickness of rockmass relaxing band had been predicted.

In general, the research on the relaxation of rock slope till now is based on qualitative analysis and semi-quantitative analysis, and is now in the stage of research independently and express respectively. Some important problems about the relaxation should be deeply studied, such as the production process of relaxing zone, the evaluation method for the reducing effect for strength parameter of rock or soil, and the method to confirm the boundary of relaxing zone because of excavation. To deep into the relative research, the basic concept and physical meaning of relaxation due to excavation are discussed in this paper firstly, and then the mechanics principle of relaxation have been emphatically studied using numerical simulation technique; in the end, the partition standard, confirm method on the relaxing zone have been expounded.

2 THE BASIC CONCEPT OF THE RELAXING ZONE FOR SLOPES DUE TO EXCAVATION

The relative research on the relaxation of rockmass at home and abroad began from the excavation of the tunnel, and the relaxing zone, the compaction zone and the original rock zone usually had been partitioned mainly based on the change of stress for rockmass. Similarly, the relaxation of slope is refer to the behavior or the phenomena about the relaxation of rockmass because of the change of rock stress during excavating. The definition of the relaxing zone of slopes due to excavation can be comprehended as the zone near the excavation surface within which the rockmass stress level reduces, frees, releases or adjusts to a new balance level. And the phenomena of the relaxation are refer to the swell deformation near the excavation surface, the stretch and broaden of rock joints or the produce of new joints, the loosening of rock or soil, the producing of plastic zone, and the failure of slopes locally or holistically.

According to the comparison of theory analysis, numerical simulation and case test, the basal viewpoint on the mechanical character and its engineering effect about the relaxing zone due to excavation have formed, and can be summarized as follows: stress balance with the rockmass has been breached during excavating, then the stress state has been adjusted, so the relaxing zone come to being. During this process, the stress level within the local area beyond the load limitation, the deformation and the failure forms relevantly, and then the physical and the mechanical character have been weakened, and the failure of rock mass has been accelerated.

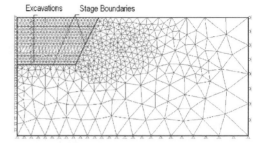

Figure 1. Computing mesh of slope excavation by FEM.

3 THE MECHANICAL CHARTER OF RELAXING ZONE DUE TO EXCAVATION

To study the mechanical charter of relaxing zone of slopes, a finite element method calculation model for slope excavation has been discussed in this paper. This model has been excavated in 8 steps to form a slope of 64 m highness, the slope angle is 63 degree. The elastic-plastic model which use Mohr-Coulomb yield criterion as the yield function has been used to simulate rockmass, and the calculation has been performed with Phase 2 software. This analysis model have some assumed condition, which can been expounded as follows: (1) only considering the gravity field; (2) considering the model as problem of plane strain, (3) joints and ground water are not considered in this model, (4) using strength reduce method to calculate the safety factor of slopes. The computing mesh can be drawn as figure.1, and the relative parameters are listed in table 1.

In this paper, the variety of differential principal stress tensor and the change regulation or distribution of plastic zone are meanly discussed to clarify the mechanical character of the relaxing zone due to excavation. It is necessary to define the differential sigma 1 which is obtained by subtracting the stress tensors and calculating the differential principal stress tensor at each node. And differential sigma 1 is not obtained by simply subtracting the stress magnitudes, since this would not account for changes in principal stress orientation at different stages. The variety of the differential sigma 1 and the plastic zone of slopes during all the process of excavation are showed from Figure 2 to Figure 11.

Firstly, we study the evolvement regulation of the differential sigma 1 and the plastic zone of slopes during excavating. When the 1st and the 2nd excavation step, stress concentration appear near the foot of the slope, but the excavated slope surface is still in the elasticity state, also we can say the excavated slope is in the elastic state. When the 3rd excavation step, stress concentration increase near the foot of the slope, and some area of rockmass is in the plastic state. In the other hand, we can find that the plastic zone is in the upper and the elastic stress concentration zone is in the lower of the foot of the slope, and these two zones are coterminous. We can say the excavated slope is in the plastic state within local area. When from the 4th to the 7th excavation step, the plastic zone expands obviously, and the differential sigma 1 within the plastic zone is negative, near the plastic zone is the elastic stress concentration zone within which the differential sigma 1 is positive, and the area of plastic zone or stress concentration zone all enlarge. We can say the excavated slope is in the extending plastic state. When the 8th excavation step, the plastic zone is continuous, the max shear stain and volume stain develop directionally, and the slip surface come to being (as shown in the figure 12). In this time, the area within which the differential sigma1 is negative or positive are all enlarge, and the change value all increase exquisitely. The safety factor is equal to 1.0. We can say the excavated slope is in the failure state.

502

Table 1. Rockmass parameters of slope excavation model.

Unit weight $\gamma/\text{kN}\cdot\text{m}^{-3}$	Young's modulus E/kPa	Poisson's ratio υ	Tensile strength T/kPa	Cohesion c'/kPa	Cohesion (residual) c'_r/kPa	Friction angle $\phi/°$	Friction angle (residual) $\phi'/°$	Dilation angle $\psi/°$
25	500000	0.3	5	100	85	40	35	20

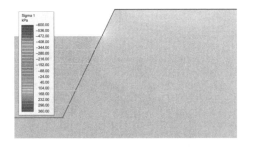

Figure 2. Differential Sigma 1 of the 2nd excavation step.

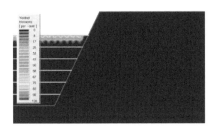

Figure 3. Yield zone of the 2nd excavation step.

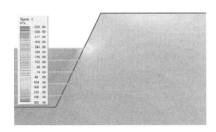

Figure 4. Differential Sigma 1 of the 3rd excavation step.

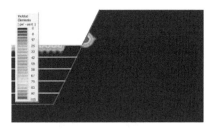

Figure 5. Yield zone of the 3rd excavation step.

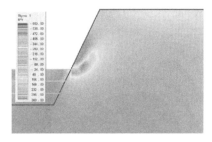

Figure 6. Differential Sigma 1 of the 5th excavation step.

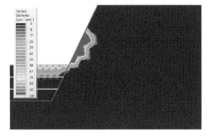

Figure 7. Yield zone of the 5th excavation step.

Secondly, we study the distributing area of the differential sigma 1 and the plastic zone. There are local area near the excavated surface within which the differential sigma 1 is negative, and the final failure surface appears within which. As shown in the figure 12, the minimum differential sigma 1 is near −600 kPa, the minimum value lay near the upper foot of slopes, and the value decrease with the distance to the excavated surface increasing. This zone can be defined as the stress attenuation plastic relaxing zone. Near the plastic relaxing zone, there are local area within which the differential sigma 1 is positive, the maximal differential sigma 1 is near 360 kPa, and the grads of the change value is very great. This zone can be defined as the stress concentration elastic compressing zone,

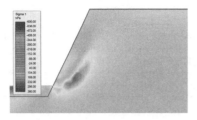

Figure 8. Differential Sigma 1 of the 7th excavation step.

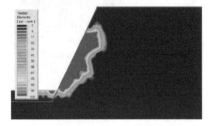

Figure 9. Yield zone of the 7th excavation step.

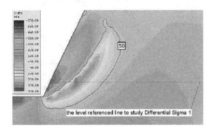

Figure 10. Differential Sigma 1 of the 8th excavation step.

Figure 11. Yield zone of the 8th excavation step.

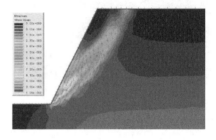

Figure 12. Max shear strain and stress trajectories.

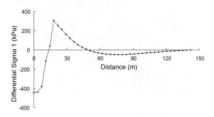

Figure 13. Variety of Differential Sigma 1 in the level referenced line.

and these two zones above transform quickly. With the distance to excavated surface increasing, there are a wide area within which the stress change little, and the deformation are also little. This zone can be defined as the original stress zone. To deeply comprehend the meaning of these three zones, we study the variety of differential sigma 1 in the level referenced line as shown in the figure 13. Within the area that the distance to excavated surface is less than 8 m, the minimum differential sigma 1 is about −450 kPa, showing consuming relaxing effect of rockmass. With the distance increasing to 12 m, the differential sigma 1 transform to zero, this zone is the transition of plastic zone and elastic zone. With the distance increasing to 20 m, the differential sigma 1 increase rapidly to 320 kPa, and then with the distance increasing to 45 m, the differential sigma 1 decrease slowly to zone. This zone shows the stress concentration effect of rockmass. In the area far from excavated surface in which there are a little change of the differential sigma 1, we call it the original stress zone.

According to the analysis expounded above, and combining the case test about tunnel and slope at home and abroad, the stress attenuation plastic relaxing zone near excavation surface, the original stress zone far from excavation surface and the stress concentration elastic compressing zone between above two can be partitioned. The rockmass in the plastic relaxing zone step into plastic state, joints splay and slip within which, and the volume of rock mass increases, the elastic wave speed value decreases, the carrying capacity of relaxing zone weakens, and then the reside strength can be used to simulate their mechanical character. The rockmass in the stress concentration zone is in the elastic always, and the joints in which close yet, the volume of rockmass decrease, the elastic wave speed value increase, and the carrying capacity increase too. For safety considering, the initial carrying capacity evaluation can be used in numerical simulation. The rockmass in the original stress zone is ultimately changeless because it is disturbed a little.

4 THE PARTITION STANDARD AND CONFIRM METHOD OF THE RELAXING ZONE DUE TO EXCAVATION

As summarized above, the differential sigma 1, plastic zone and the weaken of mechanical character are the basic characteristic for the relaxing zone due to excavation, and can be used as the partition standard. It is need to point out that the designer ordinarily lack case test data because of little money or difficulty to test, so we can use the basic regulation of the relaxing zone which are researched from theory analysis, numerical simulation, and case test to establish its partition standard and confirm method. The partition standard of the stress attenuation plastic relaxing zone, the stress concentration elastic compressing zone and the original stress zone is shown in the table 2.

To enhance the veracity and the applicability of this partition principle, it is necessary to build a applied confirm method and operation flow on the relaxing zone. The basic flow during design process to confirm the boundary of relaxing zone of slopes due to excavation is shown in the figure 14.

In the process of confirming the relaxing zone, data processing is the most important flow. In general, the differential sigma 1 is the most representative parameter. Strictly speaking, the field within which the differential sigma 1 is negative can be confirmed as the stress attenuation zone, and the field within which the differential sigma 1 is positive can be confirmed as the stress concentration zone. But considering the computing error of FEM, we recommend 50 kPa as the partition standard. Also we can confirm the field within which the differential sigma 1 is less than −50 kPa can be confirmed as the stress attenuation plastic relaxing zone, and the field within which the

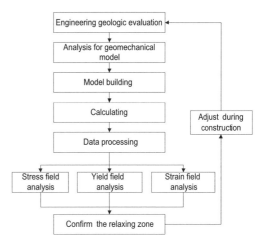

Figure 14. Flow chart to confirm the relaxing zone due to excavation.

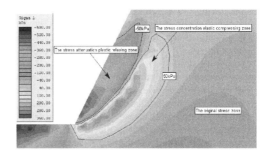

Figure 15. Distribution chart of relaxing zone of slopes due to excavation.

differential sigma 1 is more than 50 kPa can be confirmed as the stress concentration elastic compressing zone; the field between above two is the transitional field, and the field far away these two zone is the original stress zone. The distribution chart of relaxing zone of slopes due to excavation is as shown in figure 15.

From the figure 15, we can see this analysis method can simply confirm the boundary of relaxing zone,

Table 2. Partition principle for relaxing zone of slopes due to excavation.

Partition principle	The stress attenuation plastic relaxing zone	The stress concentration elastic compressing zone	The original stress zone
Differential sigma 1	negative	positive	Approximate equal to zero
Yield state and strain character	Plastic, large strain	Elastic, small strain	Elastic, little strain

and its physical meaning are clear. It is need to point out the plastic zone is almost the same as the distribution and development of the negative differential principal stress zone, and the maximal shear strain and the deformation field in the state of limit equilibrium are very typical, these character can be used as the accessorial criterion.

5 CONCLUSION

Based on the relative research on the relaxing zone for tunnel and slope at home and abroad, the mechanical character of relaxing zone of slopes due to excavation has been studied using numerical simulation, and the partition standard and confirm method of the relaxing zone has been built, the main conclusion can be summed up as follows:

1. The definition, phenomena, mechanical character and its engineering meaning on the relaxing zone for slopes due to excavation has been enucleated.
2. The change of differential principal stress tensor which is obtained by subtracting the stress tensors before and after excavated indicates the essence of relaxing zone by discussing the analysis result of numerical simulation.
3. The distributing and regulation of differential sigma 1, plastic zone and the shear stain can be used to as the partition standard of the relaxing zone.
4. Based on the mechanical character of relaxing zone, the stress attenuation plastic relaxing zone near excavation surface, the original stress zone far from excavation surface and the stress concentration elastic compressing zone between above two are partitioned.

REFERENCES

D.X. Nie. 2004. The study on rock mass deforming paramenters and relaxing thickness of rock high slope. Advance in Earth Sciences. 19 (3); P472–47.

J.H. Deng, Z.F. Li & X.R. Ge. 2001. Disturbed zones and displacement back analysis for rock slopes. Chinese Journal of Rock Mechanics and Engineering. 20 (02); P171–174 [in Chinese].

Q.L. Ha. 2001. Study on the anisotropic unloading rock mass mechanics for the steep-high rock slope of the three Gorges Project permanent shiplock. Chinese Journal of Rock Mechanics and Engineering. 20 (05); P605–610 [in Chinese].

Q. Sheng. 2002. Excavation disturbed zone of deep cutting rock slopes and mechanics behaviour of engineering rock mass. The dissertation for doctor degree of Institute of Rock and Soil Mechanics; The Chinese Academy of Sciences. P48–68 [in Chinese].

R.Q. Huang, F. Lin & D.J. Chen. 2001. Formation mechanism of unloading fracture zone of high slopes and its engineering behaviors. Journal of Engineering Geology. 9 (03); P228–229 [in Chinese].

S.G. Xiao & D.P. Zhou. 2003. Determination and numerical analysis method of relaxation region for cutting slope. Journal of Southwest Jiaotong University. 38 (03); P318–321 [in Chinese].

X.Y. Zhao, H.T. Hu & L.X. Pang. 2005. Study on unloading effect and width of unloading zones in excavating of soil-like material slopes. Chinese Journal of Rock Mechanics and Engineering. 24 (02); P710–711 [in Chinese].

Deformition characteristics and stability evaluation of Ganhaizi landslide in the Dadu River

Yunsheng Wang, Yaoming Sun, Ou Su, Yonghong Luo & Jiuling Zhang
National Key Laboratory, Chengdu University of Technology, Chengdu, China

Chunhong Zhou & Shengfeng Zhang
East China Investigation and Design Institute Under China Hydropower Engineering Consulting Group Corporation, Hangzhou, China

ABSTRACT: Ganhaizi landslide, about 19790×10^4 m^3, is located on the right bank of the Dadu River. There are two cracks and one creep body in the accumulation of Ganhaizi landslide. The potential slippage volume of the accumulation involved by each deformation body is more than 10×10^4 m^3. Cracks 1 and 2 are located in the middle part of Ganhaizi landslide, while the creep body is situated at the south of the landslide body. Several springs are found to the southwest of the staggered platform. The spring flows perennially and keeps limpid in flood period. The accumulation of Ganhaizi landslide is so huge that it may frighten the upstream hydropower project (3 km away). Therefore, its stability evaluation is extremely important. The analysis shows that rainstorm or earthquake will induce Ganhaizi landslide only partly reactive, but the whole landslide is stable. The area of potential instability is small, and it cannot surge or dam Dadu River again. In a word, the accumulation of Ganhaizi landslide has little influence on the hydropower project.

1 INSTRUCTION

The Ganhaizi landslide, situated on the right of the Dadu River, is a giant cut-bedding one, with an accumulation volume about 15950×10^4 m^3. It once dammed Dadu River. The road across the front of the landslide sank apparently. Additionally, there are two deformed cracks and one creep body in Ganhaizi landslide. The characteristics above shows the landslide is partly under the deformation. Therefore, the landslide stability is vital to the upstream hydropower station construction which is just 3 km away.

2 THE CHARACTERISTICS OF THE LANDSLIDE

2.1 The basic characteristics of the landslide

Ganhaizi landslide, on the right bank of Dadu River, lies at downstream of Bawang ditch. The front edge altitude is 1950 m, its back edge altitude is 2540 m, its length is 1580 m from east to west, and its average width is 2070 m from south to north. The area of the landslide is 3190×10^3 m^2, the average thickness is about 50 m, and its volume is about 15950×10^4 m^3.

According to the field investigation (Figure 1), the back edge altitude of Ganhaizi landslide is 2540 m, on top of the back edge is Eluo village platform. The average grade of this platform is about $10°$, the most width of the platform is 700 m, and this platform is primarily covered by Gazha colluvial accumulation. The boundary of landslide at upstream side is new settlement; Its downstream boundary is located in Guan stacked village of Bawang town downstream is a steep to gentle common boundary in land form.

On the basis of the interior terrain of landslide, its relief is descending from north to south. The north (upstream), located in north to Xiaobawang village, is a protuberant platform. This platform extends from the north boundary of landslide to the back edge of landslide, its altitude is between 2130 m to 2200 m, the largest width of this platform is 520 m, and the average gradient is $10°$; The south relief is low, but the area is big, and its area occupies 2/3 to the landslide body surface area. Civilian houses of Xiaobawang village repose on a depression to the south of the platform, and its gradient is comparative gentle.

There is a gentle platform (Xiaobawang platform) in the back edge of Ganhaizi landslide downstream from west to east (Figure 2), whose altitude is about 2200 m and the average gradient is $12°$. Under the altitude of 2200 m, there gradient of the terrain is steep and is about $20°$. From the altitude of 2200 m at Xiaobawang platform to the one of 2050 m at the front edge of landslide is an accumulating platform, its middle part

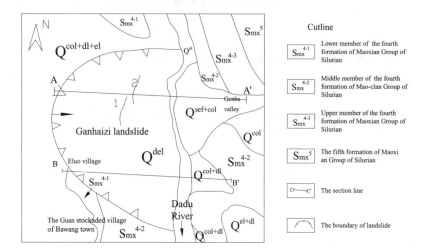

Figure 1. The plan of Ganhaizi landslide.

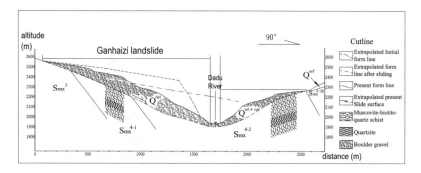

Figure 2. The A-A' section of Ganhaizi landslide.

gradient is about 8°, the front gradient of the platform facing river changes from 30° to 35°.

There is few gullies at the superficial of Ganhaizi landslide, no wide incised gullies.

Associating with clay layer of checked-up accumulation in Bawang (Figure 4), we conclude that Ganhaizi landslide once damed Dadu River, as Figure 2, broken line indicates the conjecture terrain before the landslide slide, dot line indicates the shape of accumulation slide later. The Dadu River undercut and increased the free face in the front edge of slope, owing to channel was narrow, and the right bank was high and steep (the altitude of shoulder of slope is 2400 m), it is easy to check-up Dadu River (Chen De-chuan, 2004). By the time dating, the checked-up accident happened in Pleistocene, and dammed the river for long time. Finally, the river unceasingly corrades the accumulation to form the present terrain (real line).

We find the checked-up accumulation nipped in the middle layer of the sand and pebble of the terrace II downriver of Yan'eryan valley in the left bank of Dadu River, which shows that the landslide took place when the terrace formed. According to regional time-dating, the second-level terrace formed between late Pleistocene to initial Holocene, according to time dating of the clay, Ganhaizi landslide formed 14~17 thousands years age.

2.2 The characteristics of the composition and the slide bed of the landslide

The accumulation of landslide takes on echelon longitudinally, in top of which is a wide-gentle platform, its average gradient is 12° (Figure 3). This platform has thin deluvium in superficial, whose gravels mainly vary from 20 cm to 30 cm. Below the platform, its average gradient is steep, approximately 29°, and the slope

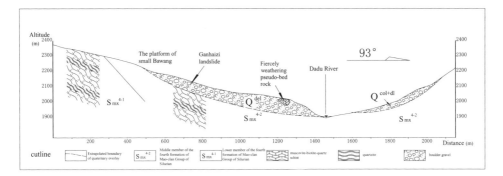

Figure 3. The B-B′ section of Ganhaizi landslide.

Figure 4. Silt-clay layer of checked-up accumulation in the north of Bawang village.

is composed of lots of blocky stones, about 40% to the total, especially the gentle part in the middle of accumulation downstream, its average particle diameter between 50 cm to 70 cm, and the biggest can be reach 3 m. The component of block and debris in the slope is mainly composed by quartzite and muscovite-biotite-quartz schist. The low district downstream side is covered by thicker soil layer, the vegetation and arbor is rich, the soil layer contains few block and broken stone, its particle diameter is between 5 cm to 10 cm, and its lithology is muscovite-biotite-quartz schist and garnet muscovite-biotite schist. The Xiaobawang village is located in this region.

The middle of the Ganhaizi landslide downstream is a protuberant platform, which makes up of broken and blocky stone with silty soil, and partly includes huge block with original rock sequence that weathering and broken fiercely, its lithology is muscovite-biotite-quartz schis. The average particle diameter of block stone in the slope is about 20 to 30 cm, and the largest one reaches 80 cm, its lithology is muscovite-biotite-quartz schist. The surface of the platform is the slope wash, in which the average broken particle diameter is about 10 cm and the slope wash has some stratification which thickness is about 0.5 m. The vegetation on it is poor. Most part of the Ganhaizi landslide is located on the stratum of Smx^{4-1}, Smx^{4-2}, the lithology of which is mainly garnet muscovite-biotite schist, quartzite and has few laminated marble additionally.

The attitude of the bedrock in the district of landslide is N40~60°W, NE∠50°~65° basically, and the whole dip of stratum is steep to upstream, and the intersectant angle to river is 45°. The back cliff of landslide is overlaid with Gazha colluvial and deluvium accumulation in which there appears a few bedrock, the bedrock in the back cliff lithological characteristic is quartzite and garnet muscovite-biotite-quartz schist of Smx^{4-1} and Smx^3. The left (upstream) is covered with Gazha colluvial and deluvium accumulation, the bottom rock is garnet muscovite-biotite-quartz schist and quartzite of Smx^{4-1}. The right (downstream) widely spread bedrock, and its lithology is grey and grayish brown color garnet muscovite-biotite schist, muscovite-biotite-quartz schist and biotite schist of the stratum of Smx^{4-2}.

According to the adit, the belt of slide is composed of clay with gravels, about 30 cm thick. The site test shows the ϕ of the belt is 23°, and C is 0.001 MPa.

3 THE DEFORMED CHARACTERISTICS OF THE LANDSLIDE

The chief deformed characteristic of Ganhaizi landslide is that there are two deformed cracks and one creep body. The crack lies in the middle of Ganhaizi landslide in the right bank of river upstream, the avalanching locates in front of the remained landslide in the left bank of river, and the creep body is on the right bank of Ganhaizi landslide downstream (Figure 5).

No.1 fracturing crack: It is located in the middle part of Ganhaizi landslide towards its upstream, presented echelon. It presents the fracturing crack directed to N30°E near the Daodaoyi, the width is about 20 to 30 cm, the eyeable depth is over 1 m, the length is over 50 m, and this crack vanishes to the Dujize house. The landslide superficial layer sliding partly brings on the crack, and the crack grows in the interior of the landslide, where the average grade is about 20°. There is a scarp in the front of the crack with the distance about 100 to 150 m between them. The grade of the scarp is between 30° and 35°, the height is about 40 to 50 m. This crack is caused by the part slide of the superficiality of landslide. Its extending is long, and its controlled area is big, besides the stability is weak in present, therefore, if this part fails, the estimated volume is 15×10^4 m^3.

No.2 fracturing crack: It is located in the middle part of colluvial and deluvium accumulation of Ganhaizi landslide towards its upstream, extending south to north, and vanishing after prolonging 10 m. This crack mainly developed at the back of the isolated stone, its particle diameter is about 2 to 3 m, and the depth is about 2 m, the width is 5 m. Generally, there is a scarp in front of the crack approximately 5 to 10 m. According to the characteristics of this crack, we know that this crack is caused by the falling isolated rock impinging against the superficial unconsolidated colluvial accumulation or by the landslide superficial deformed, and this fracturing crack controls the deformed area finitely, thus, it has little effect to the stability to the whole landslide (Xu Hua, et al. 2005).

Creep body No.3 (Figure 5) is situated at the south of the landslide body. The back edge situated in the scarp below the platform upstream, extending 50 m ahead northwest, the slide direction is towards northeast, and after sliding it formed a sunk ditch, the width of which is 50 m, and the depth is 40 m, the surface grade is about 30 to 35°. The back cliff of this creep body had been fractured, and formed two staggered level scarp, each scarp highness is about 2 m, and the potential slippage volume is 10×10^4 m^3. We found several springs to the southwest of the staggered scarp, the distance is about 30 m. The volume of each spring outlet is approximately 0.01 l/s. After convergence, the volume is about 0.05 l/s, the water is colorless and tasteless, its temperature is about 15°C. The spring flows perennially, and keeps limpid in flood period.

Accordingly the downstream concave trench formed as early as the trench upstream, thus, the reactivation of this creep body is that Dadu River corroded laterally and earthquake or other geological action induced (Li Xue-ping & Tang Hui-ming 2005).

4 THE STABILITY EVALUATION OF LANDSLIDE

According to the investigation of Ganhaizi landslide and limit equilibrium method, the whole Ganhaizi landslide is stable (K is 1.56 in nature state), but the superficiality may be kept on sliding locally (Chen Dong-liang, et al. 2002), for example, the No.1 fracturing crack upstream, but the potential unstable area is small, and its volume is not big, consequently, the influence to the hydropower station is smaller. The stability of the remained accumulation of Ganhaizi landslide in the left bank of Dadu River is poor. The colluvial accumulation in front edge of Ganhaizi landslide is unconsolidated, and the gradient of free face of main scarp is quite steep, those factors may led to secondary fall or slide below the platform of Genba valley.

5 CONCLUSION

The remained Ganhaizi landslide in the two bank of Dadu River is likely to reactive partly, and the chief deformed characteristic of Ganhaizi landslide is that there are two deformed cracks, one avalanching and one creep body. No.1 fracturing crack extending is long, and the area it controlled is big, thereby, the stability is lower in present. No.2 fracturing crack controls limited area, it has little effect to the stability to the whole landslide. The creep body will react on the function of rainstorm or earthquake, its stability is low.

Analysis above shows that rainstorm and earthquake will induce Ganhaizi landslide part slide or fall secondarily, but the whole stability is well. The area

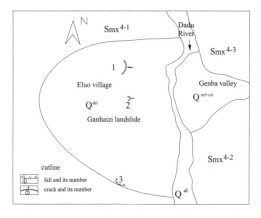

Figure 5. The distribution of deformed phenomena in Ganhaizi landslide.

of potential instability is small, and the volume is little, it cannot surge or dam Dadu River. In a word, the remains accumulation of Ganhaizi landslide has little influence on the hydropower project.

REFERENCES

Chen De-chuan, 2004. Stability study on Bawangshan landslide near ErtanReservior [J]. *Sichuan Water Power*, 9(3), 21–23.

Xu Hua, Li Tian-bin & Xiao Xue-pei, 2005. Formation mechanism and stability prediction of Andu landslide in the Three Gorges Reservoir area [J], *Hydrogeology and Engineering Geology*, 4(4), 28–31.

Li Xue-ping & Tang Hui-ming, 2005. Application on Likelihood Ratio Test Used for Sensitivity Factors of Area Slope Stability [J], *Journal of Yangtze River Scientific Research Institute*, 10(5), 37–48.

Chen Dong-liang, Ying Jing-hao & Yin Guo-sheng, 2002. Analysis on the Forming Cause Mechanism of Loess Landslide in the Area of Crossing Yellow River Project along the Middle Route of South to North Water Transfer Project [J], *Journal of North China Institute of Water Conservancy and Hydroelectric Power*, 4(4), 49–51.

Landslide-prone towns in Daunia (Italy): PS interferometry-based investigation

J. Wasowski & D. Casarano
CNR-IRPI, Italy, Bari, Italy

F. Bovenga & A. Refice
CNR-ISSIA, Italy, Bari, Italy

R. Nutricato & D.O. Nitti
Dipartimento Interateneo di Fisica, Politecnico di Bari, Italy

ABSTRACT: Persistent Scatterers Interferometry (PSI) and satellite radar imagery can be used to detect very slow displacements (mm-cm per year) of targets (PS) exhibiting coherent radar backscattering properties (mainly man-made structures). Here we present results of the PSI application to the Daunia Apennines, which include many hilltop towns affected by landslides. Examples from the towns Casalnuovo Monterotaro and Pietramontecorvino are used to illustrate that the interpretation of PS data on urbanised slopes can be difficult, because their movements may arise from a variety of processes: i) volumetric strains within soils, ii) natural or anthropogenic subsidence or uplift, iii) settlement of engineering structures, iv) deterioration of man-made structures, v) extremely slow slope deformations that may or may not lead to failure. Where true landslide movements are detected, they likely regard long-term post-failure displacements involving clay-rich materials.

1 INTRODUCTION

Since slope failures are a world-problem, an effective approach to landslide hazard reduction seems to be through exploitation of Earth Observation systems, with focus on long term monitoring, early detection and warning (e.g. IGOS GEOHAZARDS 2004; Wasowski et al., 2007a). In particular, space-borne synthetic aperture radar differential interferometry (DInSAR) techniques are useful in landslide investigations, because of their capability to provide wide-area coverage (thousands km^2) and, under suitable conditions, spatially dense information on small ground surface deformations (e.g. Colesanti et al., 2003). Furthermore, advanced multi-pass DInSAR methods such as Persistent Scatterers Interferometry (PSI; Ferretti et al., 2001) and Small Baseline Subset (SBAS; Lanari et al., 2004) overcome the limitations of conventional DInSAR and extend the applicability of radar interferometry from regional to local-scale engineering geology investigations of ground instability (Colesanti & Wasowski, 2006; Ferretti et al., 2006). PSI techniques can be used to detect and monitor very slow displacements of targets (PS) exhibiting coherent radar backscattering properties. Since the presence of slow deformations represents evidence of potential slope failure hazard, PSI results can be used to provide a preliminary distinction between conditions of stability and instability.

However, with the exception of urban landslides, the density of persistent radar targets suitable for interferometric measurements on unstable rural and peri-urban slopes is typically low and this makes difficult PSI analysis, as well as introduces uncertainties in the assessments of true ground motions (e.g. Bovenga et al., 2006; Colesanti & Wasowski 2006). Furthermore, the interpretation of the exact geotechnical significance of millimetric to centimetric yearly displacements currently detectable by PSI can be difficult, because i) such slow ground surface deformations may arise from a wide variety of causes and, therefore, their presence on slopes may not always reflect shear movements or occurrence of landslides, and ii) most radar targets correspond to man-made objects (e.g. houses) and thus their structural behaviour (and ground-foundation interactions) should be taken into account.

In this study we present the results of the application of the SPINUA PSI technique (Bovenga et al., 2004) to landslide-prone towns in the Daunia Apennines (Southern Italy). We expand upon the initial works of Bovenga et al. (2005, 2006) by

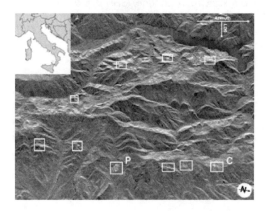

Figure 1. Filtered SAR amplitude image of the Daunia area (28 km × 27 km) showing the predominantly moderate relief hillslope topography. In addition to the towns of Casalnuovo Monterotaro (marked by C) and Pietramontecorvino (P), the white rectangles (few km in width) enclose several other towns selected for the Permanent Scatterers (PS) analysis. Inset shows location of Daunia in Southern Italy.

providing additional analysis of the urban areas of Pietramontecorvino and Casalnuovo Monterotaro (Fig. 1). We also focus on some difficulties in interpreting the exact origin of the PS movements detected by exploiting the European Space Agency (ESA) ERS-1/2 satellites data.

2 BACKGROUND ON PSI

Detailed information on PSI techniques is available in specialised remote sensing publications (e.g. Ferretti et al., 2001), and its practical applicability has already been addressed in engineering geology literature (e.g. Colesanti et al., 2003; Bovenga et al., 2006). Here we mention only some basic aspects of DInSAR and PSI. For a recent comprehensive review of radar-based remote sensing and PSI applications for landslide assessment, the reader is referred to Colesanti & Wasowski (2006).

Space-borne synthetic aperture radars are active microwave systems capable of recording coherently the electromagnetic signal backscattered from the Earth surface. With two or more SAR images acquired over the same area during successive satellite passes it is possible to detect ground surface movements occurred between the SAR acquisitions along the Line Of Sight (LOS) direction. For the ERS satellites (incidence angle 23°), the LOS unit vector has director cosines of about 0.9 (up–down), 0.4 (east–west), and 0.05 (north–south), respectively. Thus, the maximum sensitivity is for vertical displacements.

By using C-band radar data, the detectable velocities range theoretically from about 1 mm/y up to 29 cm/y; however, recent assessments (Crosetto et al., 2007) indicate that displacement rates lower than 1.5–2 mm/y and higher than 15–20 cm/y are hardly measurable in real experimental conditions.

In spite of the above-mentioned limitations, the possibility to measure, relatively quickly, deformations on slope surfaces over wide areas ($10^3 \div 10^6$ km^2), coupled with the regular update capabilities of satellite sensors, opens new perspectives in regional scale detection of slope hazards and monitoring of slow landslides (Colesanti et al. 2003; Colesanti & Wasowski 2004, 2006; Farina et al., 2006; Hilley et al. 2004; Wasowski et al., 2007a).

3 THE STUDY AREA

The area studied is located in the Daunia region (Southern Italy), characterized by gentle hills and low mountains, only locally exceeding 1000 m above sea level. Daunia belongs to the highly deformed area between the frontal thrusts of the Apennine chain and the western-most part of the foredeep (Dazzaro et al., 1988). The chain units are characterised by a series of tectonically deformed flysch formations of pre-Pliocene age.

The clay-rich flysch units are more prone to landsliding, compared to the formations containing higher proportion of lithoid intercalations (sandstones, limestones). The widespread presence of clayey materials with poor geotechnical properties is the underlying cause of landsliding. Furthermore, as a result of the tectonic history of the Apennines, the geological materials are intensely deformed and hence also rock units are susceptible to slope movements. In general, the activity of landslides in the Daunia Apennines is characterized by seasonal remobilisations of slope movements, typically related to rainfall events. Individual meteoric events have been the most frequent triggers of landslides, even though the mean annual rainfall is modest (in the order of 600–700 mm per year).

Although mass movements appear widespread throughout the entire region (Zezza et al., 1994), there are relatively few studies published on landslides in Daunia. The better documented events are concentrated within or in the immediate proximity of the urban areas. In the 1990's there has been an apparent increase in landslide activity in several urban and peri-urban areas. It is probable that the stability of slopes bordering the hilltop towns has gradually worsened because of residential development over recent decades. This has led in some cases to reactivations of pre-existing old landslides. Furthermore, the urban expansion onto marginally stable hillslopes and improper land use has led to increases in damaging first-time failures (Wasowski et al., 2007b).

4 PSI ANALYSIS OF TWO TOWNS

To illustrate the potential of the PSI technique as well as current difficulties in PS data interpretation we present the results concerning two towns: Casalnuovo Monterotaro and Pietramontecorvino (Fig. 1). These two cases are considered representative of the geologic and geomorphologic conditions encountered in other towns of Daunia.

4.1 PSI processing

The initial analysis (Bovenga et al., 2005) involved an area of 28×27 km^2, enclosing 10 towns affected by slope instability problems (Fig. 1). We used a SAR dataset of both descending and ascending ERS-1/2 acquisitions (40 and 35 scenes, respectively) covering the period 1995–1999. The dataset was pre-processed adopting optimized solutions for co-registration, relative calibration, and re-sampling (see Bovenga et al., 2005, 2006 for more details). The analysis was limited to small image windows of size ranging from 5 to 15 km^2 (Fig. 1). The windows enclose towns, excluding vegetated rural areas that typically contain very few PS. This strategy was adopted to ensure an adequate distribution of PS candidates needed for a reliable correction of the atmospheric signal. Such approach was also considered as most practical given that landslides with high socio-economic impact in the study area are those affecting urban centres.

4.2 Interpretation of the PS results: The case of Casalnuovo Monterotaro

Figure 2 shows that the great majority of PS, which falls within the built up area, results to be stable. This is not surprising as the central part of the town develops along a flat topped N-S trending ridge made of the Flysch di Faeto Formation. This Miocene age flysch, consisting of an alternation of marly limestones, calcarenites and marly clays, is relatively less prone to landsliding than the so-called Argille Bentonitiche (clay-rich unit of Miocene age including bentonite clays). The latter crop out at the eastern and western peripheries of the town (Zezza et al., 1994).

There are, however, two small groups of moving PS situated in the eastern periphery of the town, on gently inclined head portion of a local valley (Fig. 2). These PS show low velocity (from -3 to -4 mm/y) movements. The negative sign stands for the displacements away from the radar sensor and, given the ascending satellite acquisition geometry, can be interpreted as

Figure 2. Distribution and average line of sight (LOS) velocity of radar targets (PS marked by black symbols) in Casalnuovo Monterotaro. Negative and positive velocities indicate, respectively, movements away from and towards the sensor; PS within the velocity range -2–2 mm/y are assumed motionless. Background image is a 1997 orthophoto. Locations of relict and recent landslide scarps (dashed and dotted white lines, respectively), are after Zezza et al. (1994). A small watercourse draining the valley at the eastern periphery of the town is shown in light gray. Note also ascending radar satellite acquisition geometry (white arrows).

indicative of either downward (subsidence) or downslope (eastward) movements, or a combination of both. The presence of moving PS in the east-facing head portion of the valley, in the vicinity of old landslide scarps, could suggest landslide origin for the detected movements. However, in cases like this, in situ monitoring data are needed to demonstrate whether the predominant displacements are indeed in downslope or vertical directions.

Nevertheless, inspections of the slope stability conditions in the eastern part of the town (conducted in the recent years for the Department of Civil Protection) revealed that several buildings and retaining walls, including those with moving PS, show signs of distress and have suffered recurrent damage (cracks) since the 1990s. This confirms the reliability of the PS results.

Because the clusters of moving PS are small, they probably point to local site instabilities rather than to true landslide movements. The cut and fill re-shaping of the valley head during the post-second world war development of the town has obliterated the evidence of landslide legacy, but the presence of artificial fill mantling the clay-rich slope substratum could be a cause of the local ground settlements and structure instability. Variations in local drainage conditions and in water input to the slope could also play an important role by inducing volumetric changes in the soil.

4.3 Interpretation of the PS results: The case of Pietramontecorvino

The PS pattern in the Pietramontecorvino area is much different from that of Casalnuovo Monterotaro. Several zones in the town centre and at its northern and southern outskirts include clusters of moving radar targets (Fig. 3). Indeed, Pietra Montecorvino is not a hilltop town, because it develops mostly on a SSE facing slope mantled by large landslides. However, Fig. 3 does not reveal any obvious link between the distribution of moving PS and the landslides. As in the case of the landslides, both moving and stable PS are present also in the slope areas occupied by the Flysch Rosso Formation of pre-Pliocene age (Fig. 3). This clay-rich unit, also referred to as Varicoloured Clays Zezza et al. (1994), is known for its high suceptibility to landsliding.

Pietramontecorvino landslide legacy has been examined on a detailed scale by Zezza et al. (1994), who identified and mapped very old and more recent landslide scarps, quiescent, mappable movements, as well as active and quiescent landslide zones (Fig. 4). However, it is apparent that the pattern of moving and motionless PS bears no specific relation with the distribution of pre-existing landslides. One possible exception can be identified at the southern periphery of the town, where a significant number of slowly moving PS falls within and near the limits of a N-S elongated landslide zone (Fig. 4), interpreted as active by Zezza et al. (1994). Our field inspections showed that the recent landslide activity in this area is conditioned by the erosion activity of a local torrent at the slope base.

However, again, the PS one-dimensional LOS motion data alone are insufficient to resolve the nature of the observed displacements, i.e. whether they represent predominantly vertical or downslope movements or a combination of both. The fissures observed on

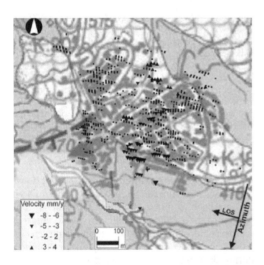

Figure 3. Distribution and average line of sight (LOS) velocity of radar targets (PS marked by black symbols) in the Pietramontecorvino area. Negative and positive velocities indicate, respectively, movements away from and towards the sensor; background geological map from http://www.apat.gov.it: clay-rich Flysch Rosso Formation (in grey colour) of pre-Pliocene age and three old landslides (in white). General downslope direction is to SSE. Note also descending radar satellite acquisition geometry (black arrows).

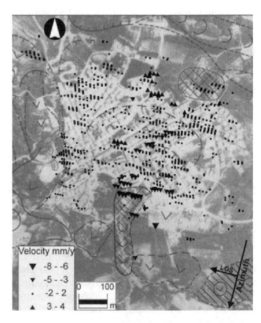

Figure 4. LOS velocity of radar targets (black symbols) in the Pietramontecorvino area. Background image is a 1976 airphoto showing photo-interpreted (after Zezza et al., 1994) landslide features including: old scarps (semicircular barbed lines, dashed if uncertain), quiescent, mappable movements (with Vs and Us standing, respectively, for slides and flows) and active and quiescent landslide zones (respectively crosshatched and hatched areas).

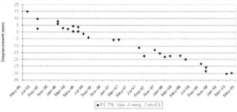

Figure 5. Top: close-up on the southern portion of the town centre with local distribution and average LOS velocity of radar targets (black symbols), superimposed on a recent ortophoto of Pietramontecorvino. White arrow points to a small landslide. Bottom: Displacement time series of the PS located near the slide.

Figure 6. Small rotational slide at the SW corner of Pietramontecorvino football field, whose movements were detected by PS interferometry (cf. Figure. 5). Note semi-circular scarp with minor ground settlement.

some buildings confirm the persistence of the conditions of ground and/or structure instability, but the surface expression of the past landslide events has been substantially altered by recent town development and man activity. Clearly, in situ investigations would be needed to distinguish between the possible causes of the PS movements (e.g. local ground settlements versus true slope movements or structure instability).

In some instances, however, even a simple site inspection can allow for a straightforward interpretation of PS data. This is the case of a single moving PS situated near the SW corner of the town's football field (Fig. 5). Its position coincides with a small rotational slide: we inspected the site only after obtaining the PS results and found a semi-circular scarp with minor settlement of the field ground, as well as a locally rotated gabion (Fig. 6). The field is also guarded by metallic wire-frame sustained by iron poles: these objects could be associated with the PS behaviour and they were locally deformed by the slide. Although the time of the initial failure is not known, the PS displacement time series (Fig. 5) indicate that the movements were taking place in the period covered by radar imagery (1995–1999). Interestingly, this is the fastest (-8 mm/y) moving radar target in Pietramontecorvino. It seems that in this case PSI allowed us to capture the slow post-failure rotation of the slide head, and that the PS motion reflects predominantly sub-vertical displacements.

Nevertheless, interpretations based on a single PS should generally be viewed as a limit case. Clearly, clusters of moving radar targets are needed for a reliable detection of ground instability.

5 DISCUSSION AND CONCLUSIONS

The PSI results indicate that the majority of PS in Casalnuovo Monterotaro are stable; these PS are concentrated within the historical centre of the town. Very slowly moving radar targets are present in two clusters situated at the eastern border of the town. This is in agreement with the observation that displacements related to landslide events or simply to ground instability occur mainly in the peripheries of the Daunia hilltop towns, where the urban expansion is more recent and the man-made structures acting as PS are located close to the potentially unstable slopes (Bovenga et al., 2006).

In the case of Pietramontecorvino the deformations detected by PSI are present not only at the town's peripheries, but also in the centre. Indeed, this town is known for the presence of instability problems affecting many buildings. Although a major part of Pietramontecorvino appears to be sited over a large, very old landslide, we consider unlikely that the slowly moving PS (3–5 mm/y) reflect its activity. Instead, it is suggested that processes such as settlements of engineering structures, volumetric changes and post-failure creep of the clay-rich Pietramontecorvino soils and colluvia, could be responsible for the localised deformations detected by PSI. Nevertheless, a link seems to exist between the moving radar targets and recent landslide activity in the southern periphery of the town (Fig. 5).

Even though in several cases the PS displacement fields show clear evidence of moving objects in urban and peri-urban areas, local knowledge of the investigated area and site inspections are required to interpret the significance of PS motion data and to identify the mechanism of the detected deformations. In general, on slopes, surface displacements over time might be found to be in a downslope direction but such deformations might not necessarily always reflect shear movements or movements leading to shear failure, i.e. to landsliding. With the exception of "natural" PS (e.g. corresponding to rock outcrop targets), without an appropriate in situ investigation, several different interpretations of the very slow PS displacements are possible. Our PS results showing ground surface deformation changes over time on landslide susceptible slopes are very promising. However, the geotechnical parameters and geological boundary uncertainties which control PS displacements need to be investigated and better understood before they can be used directly for landslide hazard/risk zonation or for predicting (warning) of potential instabilities.

ACKNOWLEDGEMENTS

This work was supported in part by the European Community (Contract No. EVGI 2001-00055—Project LEWIS). Images were provided by ESA under the CAT-1 project 2653.

REFERENCES

Bovenga, F., Refice, A., Nutricato, R., Guerriero, L. & Chiaradia, M.T. 2004. SPINUA: a flexible processing chain for ERS / ENVISAT long term interferometry, *Proceedings of ESA-ENVISAT Symposium 2004*, Saltzburg, Austria (CD-ROM).

Bovenga, F., Chiaradia, M.T., Nutricato, R., Refice, A. & Wasowski, J. 2005. On the application of PSI technique to landslide monitoring in the Daunia mountains, Italy, *Proceedings of FRINGE 2005*, ESA-ESRIN, Frascati, Italy (CD-ROM).

Bovenga, F., Nutricato, R., Refice, A. & Wasowski, J. 2006. Application of multi-temporal differential interferometry to slope instability detection in urban/peri-urban areas. *Engineering Geology*, **88** (3–4), 218–239.

Colesanti, C. & Wasowski, J. 2004. Satellite SAR interferometry for wide-area slope hazard detection and site-specific monitoring of slow landslides. *Proc. International Landslide Symposium*—ISL2004 Rio de Janeiro, Brasil, 795–802.

Colesanti, C. & Wasowski, J. 2006. Investigating landslides with satellite Synthetic Aperture Radar (SAR) interferometry. *Engineering Geology*, **88** (3–4), 173–199.

Colesanti, C., Ferretti, A., Prati, C. & Rocca, F. 2003. Monitoring landslides and tectonic motions with the Permanent Scatterers Technique. *Engineering Geology*, **68** (1), 3–14.

Crosetto, M., Agudo, M., Raucoules, D., Bourgine, B., de Michele, M., Le Cozannet, G., Bremmer, C., Veldkamp, J.G., Tragheim, D., Bateson, L. & Engdahl, M. 2007. Validation of Persistent Scatterers Interferometry over a mining test site: results of the PSIC4 project, *Proc. ENVISAT Symposium*, Montreux, Switzerland (CD-ROM).

Dazzaro, L., Di Nocera, S., Pescatore, T., Rapisardi, L., Romeo, M., Russo, B., Senatore, M.R. & Torre, M. 1988. Geologia del margine della catena appenninica trail F. Fortore ed il T. Calaggio (Monti della Daunia—Appennino Meridionale). *Mem. Soc. Geol. It*. 41: 411–422.

Farina, P., Colombo, D., Fumagalli, A., Marks, F. & Moretti, S. 2006. Permanent Scatterers for landslide investigations: outcomes from the ESA-SLAM Project. *Engineering Geology*, **88** (3–4), 200–217.

Ferretti, A., Prati, C. & Rocca, F. 2001. Permanent Scatterers in SAR Interferometry. *IEEE Trans. Geoscience and Remote Sensing*, **39** (1), 8–20.

Ferretti, A., Prati, C, Rocca F. & Wasowski, J. 2006. Satellite interferometry for monitoring ground deformations in the urban environment. *Proc. 10th IAEG Congress*, Nottingham, UK (CD-ROM).

Hilley, G., Burgmann, R., Ferretti, A., Novali, F. & Rocca F. 2004. Dynamics of Slow-Moving Landslides from Permanent Scatterer Analysis. *Science*, 304, 1952–1955.

IGOS GEOHAZARDS 2004. GEOHAZARDS theme report: For the monitoring of our Environment from Space and from Earth. *European Space Agency publicatiom*, 55 p.

Lanari, R., Mora, O., Manunta, M., Mallorqui, J.J., Berardino, P. & Sansosti, E. 2004. A Small Baseline Approach for Investigating Deformations on Full Resolution Differential SAR Interferograms. *IEEE Trans. Geoscience And Remote Sensing*, 42, 1377–1386.

Wasowski, J., Ferretti, A. & Colesanti, C. 2007a. Space-Borne SAR Interferometry for Long Term Monitoring of Slope Instability Hazards. Proceeding of the First North American landslide Conference, Vail, USA. (CD-ROM).

Wasowski, J., Casarano, D. & Lamanna, C. 2007b. Is the current landslide activity in the Daunia region (Italy) controlled by climate or land use change? *Proc. International Conference on "Landslides and Climate Change—Challenges and Solutions"*, Ventnor, UK, 41–49.

Zezza, F., Merenda, L., Bruno, G., Crescenzi, E. & Iovine, G. 1994. Condizioni di instabilità e rischio da frana nei comuni dell'Appennino Dauno Pugliese. *Geologia Applicata e Idrogeologia*, **29**, 77–141.

Basic types and active characteristics of loess landslide in China

Weijiang Wu, Dekai Wang & Xing Su
Geological Hazards Prevention Institute, Gansu Academy of Sciences, Lanzhou, Gansu, China

Nianqin Wang
Department of Geology and Environment Engineering, Xi'an University of Science and Technology, Xi'an, Shanxi, China

ABSTRACT: Loess landslide disasters are widely distributed in loess area in china. Four kinds of basic types can be divided which are homogeneous loess landslide, loess interface landslide, loess-mudstone layer plane landslide and loess-mudstone cutting layer landslide. The characteristics of loess interface landslide and loess-mudstone layer plane landslide are low sliding velocity, short sliding distance and the sliding body has low stability after sliding. The characteristics of homogeneous loess landslide and loess-mudstone cutting layer landslide are high sliding velocity, long sliding distance and the sliding body has high stability after sliding, except its scarp. Basic laws mentioned above have some guiding significance and practical value in economic construction.

1 INTRODUCTION

One of the most basic classification of landslide is based on the main material the sliding body composed of[1]. Loess landslide which is widely distributed in the Loess Plateau in northwest of China is one of the major special soil landslide types. But some large landslides are not composed entirely of loess, and contained a lot of soft rock formed in Neogene, Cretaceous and so on[2][3][4] except for the loess in the different periods. Because they are in loess region some scholars also call them loess landslide. Thus, loess landslides types in the traditional concept are different with the actual situation.

The most of difficult questions are to forecast the sliding distance and hazard areas when deal with the dangerous situation of the landslide. If the types and activity characteristic can not be realized fully, to reduce the disaster lose furthest would be difficult. Generally, loess landslide may collapse, mostly belongs to high-speed or high-speed long-distance landslide. In fact, to different types of loess landslide the difference of sliding and resurrection are quite different. Therefore, it is significance for bring forward the new concept of loess landslide in broad sense and studying the characteristics of different types of loess landslides. This can help to have the correct understanding and forecasting of deformation and damage, activities and hazard areas of landslide.

2 TYPE CHARACTERISTIC

The rock-soil is not only the material base of a landslide developing, but also the problem should be identify firstly in research and prevention of the landslide. Therefore, the classification method based on the material composition is the best method to reflect the basic characteristics of the landslide. Controlling by the regional composition of lithology conditions, in addition to the loess in different periods which develops pure loess landslides, argillaceous rock formation which is widespread distributed in the Neogene, Cretaceous and Jurassic of the Mesozoic and Cenozoic Erathem, is also an easy-sliding stratum because of the poor cementation and low mechanical strength in the Loess Plateau of the northwest in China. So many landslides in the region, especially some large landslides whose sliding body are often composed of loess and the underlying soft rock joint, are mixed type landslide, has gone beyond the traditional concept of the loess landslide which is composed of the pure loess. The concept in broad sense and classification of loess landslide are needed. Its basic characteristics of the various types of loess landslides can be summarized in Table 1.

Comparing to the traditional concept of loess landslide, The new classification is more in line with the actual development of landslide in the loess area and make the concept be more clearly. At the same

Table 1. The basic types and characteristics of loess landslide.

Types of loess landslide		Position of the sliding belt	Constructed profiles
Pure loess landslide	Homogeneous loess landslide	The main sliding surface develops in loess formation.	
	Loess inter-face land-slide	The main sliding surface is in the interface of high water content or saturated loess and mudstone.	
Mixed loess land-slide	Loessmudstone layer plane landslide	The main sliding surface develops in the weak formation and inter-layer.	
	Loessmudstone cutting layer landslide	The main sliding surface cuts the underlying stratum, controlled by joints, fracture and other structural plane.	

time it consider the relationship between development location of the sliding surface and stratum and conducive to understand the law of loess landslide and its prevention.

3 ACTIVITY CHARACTERISTICS

The sliding velocity, sliding distance and other characteristics of loess landslide are very important significance in the landslide disaster prediction, prevention and control. Nowadays the main common forecast methods of landslide are: the slope theory was offered by Austria scholar A.E. Scheidegger[5], the formula was offered by Japanese scholar Moriwaki • width[6], the corresponding formula and the method of the sliding velocity and sliding distance are also given by Jia-zheng Pan Academician, Si-Jing Wang Academician, Yu-shu Fang and other scholars in China[7][8][9]. The statistical models are restricted by many influence factors. So it is difficult to predict the sliding velocity, sliding distance accurately, sometimes results are opposite to the actual situation. For example, in recent years, the volume of Huang-Ci landslide, Jiayouzhan landslide and Bengzhan station landslide occurred at Heifangtai in Gansu Province is above 300×10^4 m^3, but their sliding distance is only about 50 m. The volume of Jiaojia landslide, the Tanhuaguichang factory landslide is generally only 10×10^4–40×10^4 m^3, but the sliding distance is more than 300–500 m. Such phenomena cannot be explained and predicted with Scheidegger's theory or Moriwaki • width formula. There are nearly relationship among sliding velocity, sliding distance and type of loess landslide by the massive loess landslide Statistics survey.

1. The homogeneous loess Landslide generally develops in the steep parts of loess slope, the main sliding surface develops in the homogeneous loess, sliding plane is approximate arc, controlled by vertical joints with the steep scarp (generally around 60°), and the sliding plane is smooth and straight, be conducive to release the sliding energy rapidly, and become a high-speed slide. The sliding distance will be farther while there are open terrain conditions (Figure 1 and 2). Others such as Tianshui Forging Machine landslide, has 80 m high, with sliding distance is nearly 200 m and average sliding velocity is 10 m/s around.
2. Loess interface landslide always develops in loess slopes of Loess Hills Area with the dip angle from

Figure 1. The section of Tanhuaguichang homogeneous loess landslide at Heifangtai, Gansu.

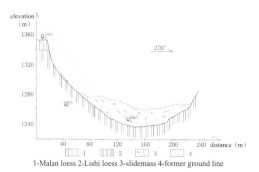

1-Malan loess 2-Lishi loess 3-slidemass 4-former ground line

Figure 2. The section of Qiewa homogeneous loess landslide at Chongxin County, Gansu.

Figure 3. The section of Luoyugou valley loess interface landslide in Tianshui city, Gansu.

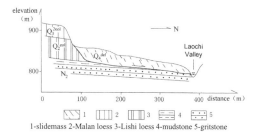

1-slidemass 2-Malan loess 3-Lishi loess 4-mudstone 5-gritstone

Figure 4. The section of Renjiawan loess interface landslide in Lantian county, Shananxi.

10° to 20°, many examples found in Tianshui, Tongchuan, Lantian cites. Malan loess deposits on the ancient topography which is up and down with draping form, and the underlying stratum mainly are formed in Cenozoic. The upper loess has good permeability, the lower mudstone and sandstone have poor permeability. So they formed a double heterojunction structure of the slope. The groundwater got together at interface, so the interface area was wet soft plastic state in a long-term and become the weak-soft structure surface of landslide. Therefore, the loess upper in the slope can slide easily along the weak-soft structure plane in the above area where there are abundant precipitations, forming loess interface landslide (Figure 3 and 4), often occurs during the rainy season.

Because of high water content, low mechanical strength near interface, the sliding velocity of loess interface landslide is low, sliding distance is short. Jiaoping highway toll stations landslide caused by slope cut, the main sliding surface is less than 5°. Since 1993 sliding always accompany with rainy season, massive sliding soil buried and blocked the highways, impact the transportation seriously. It is low-speed sliding landslide[10].

3. Loess-mudstone layer plane landslide occurs in the slope site where the underlying stratum dip angle always from 10° to 30° according with the direction of the slope. The main sliding surface of this kind of landslide develops along the relatively soft formations or interlayer in underlying stratum. Because of the sliding of the bottom sliding mass, the upper loess body forms tension cracks. Although this type landslide have large sliding potential energy, the slope structure, the shape and nature of the main sliding plane determine that its sliding velocity is low and sliding distance is short. Firstly, the dip angle of the main sliding plane is gentle, and its length is large relatively, it is not conducive to sliding. Secondly, the strength of the rock-soil is low, the cohesion has basically released during the long-term shear deformation, sliding body maintains a balance while main stabilizing force is friction of the main sliding plane. Therefore, the cohesion reduced gradually and caused the damage of the rock-soil. On the other hand it can not create a tremendous push force to make the sliding body slide in high-speed and long distance.

Huangci landslide, gas station landslide and pumping stations landslide at Heifangtai region in Gansu, Wenchangge landslide, Qingbaishi landslide, Wushan Dingjiamen landslide at Dajiatai in Lanzhou, all these are this kind of loess landslide with the features of low sliding velocity and short sliding distance. For example, the main sliding process of Huangci landslide spent about 90 minutes, the sliding distance is only 30–60 m although the terrain is open in the front of the landslide (Figure 5 and 6).

4. Loess-mudstone cutting layer landslide develops in the slope areas where the underlying stratum inclination is opposite to the direction of the slope, with steeper slope and greater sliding potential energy. This type landslide Developed through progressive deformation and damage by the effects of gravity and groundwater. Firstly, in the process, because of unloading and rebound, groundwater acting, stress relative concentration at the toe of the landslide, the fracture develops shearing and creeping is relatively strong, shear-creeping phase formed. It results the changing of stress and deformation of upper loess slope, formation of the tensile stress near the top of the landslide, emergence of tension cracks at the back of landslide, formation of tensile paragraph at top. Shear-creeping phase and rupturing phase develop, the stress concentrates in the mudstone of the central part of sliding body, the slope will be stability by the effect of the part of a lock-solid role temporarily. With the time passes, the scale and failure rate of the landslide are increasing and accelerating continuously, the stress of mudstone on the lock-solid phase is concentrated consistently, length is shorter, the entire slope loses its stability and slide completely when reach critical length and failure instantaneously.

According to experimental data, mudstone in lock-solid phase has high cohesion, its peak intensity effect to present brittle failure. Therefore, the sliding body will lose stabilizing force when the failure of the lock-solid phase happen, meanwhile get great corresponding pushing force. Add up the effects of vaporization and a speed cushion when slide in high-speed, form high-speed and long-distance landslide when the terrain is open.

Chana landslide in Longyangxia Reservoir and Shaleshan landslide in Gansu are such kind of landslides. Part of the sliding body of Chana landslide surpasses the Yellow River and arrives at the other side of the bank, sliding distance is more than 2000 m, sliding velocity is 41 m/s based on experience counting[11]. Shaleshan landslide happened on 7 March of 1983. The height between anterior border and posterior border is 321 m (Figure 7 and 8), the furthest sliding

Figure 5. The loess-mudstone layer plane landslide at Taohe River, Gansu.

Figure 7. Saleshan loess-mudstone cutting layer landslide, Gansu Province.

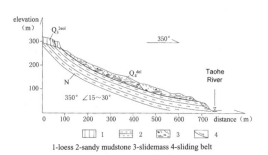

1-loess 2-sandy mudstone 3-slidemass 4-sliding belt

Figure 6. The section of loess-mudstone layer landslide at Taohe River, Gansu.

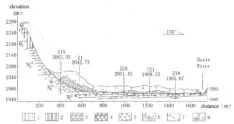

1-Malan loess 2-Lishi loess 3-sandy mudstone 4-conglomerate 5-cobble 6-slidemass 7-former ground line 8-borehole and evelation of ora

Figure 8. The section of Saleshan loess-mudstone cutting layer landslide, Gansu Province.

distance is 1050 m in less than one minute, the average speed is 20 m/s and the quickest can be 32 m/s. It killed 220 people and overwhelmed four villages. Such landslide often cause serious economic losses and people death.

4 THE REVIVAL CHARACTER

According to the investigation data, the stability of sliding body and the scarp of different kinds of the landslide are different and the characters of sliding for the second time or revival are also different.

1. To homogeneous loess landslide, the potential energy has released after sliding in high-speed and long-distance. Its stability is high. And it is always not overall revival or occur local collapse in the scarp sometimes. For example, the stable coefficient of Tianshui Forging Machine landslide is 2.3. Even when the factory was rebuilt the sliding body was cleared 60 m in the toe, the sliding body is still in stable state. But because the scarp of the landslide is 50 m high and 45° in dip angle, its stability is worse and collapse often. Since 1985, the scarp of Panjizhai landslide has slided 8 times and the sliding scale become bigger and bigger. The landslide harms to State Road 310.
2. To loess interface landslide, the sliding potential energy can not release completely because of low sliding velocity and short sliding distance of, sliding body will slide once more when faces with rain, heavy rain, water erosion, or excavation. Some landslide is in perennial creep slowly. And its sliding velocity will be quicker in rainy season or after heavy rain.

    There are 3 times of large scale sliding at Zhangjiabanpo ancient loess interface landslide in Tianshui over the past 50 years. It destroyed 5 houses from 1973 to 1974. And a number of cracks occurred after heavy rain in August of 1990, one of the fissures was 7 cm width and 60 m long, stagger down 40–50 cm. It was sliding again in 2001. The Jiaoshuwan ancient landslide in Tianshui relived during the Autumn in 1990, was seriously harmfal to the city and had to invest heavy fund to control[12]. The revival of landslides at Lantian in Shaanxi belong to such kind.
3. To loess-mudstone layer plane landslide, its sliding distance is short, most of the sliding body remains at main sliding surface, sliding potential energy was not released completely, the sliding body may be in whole or partial revival because of the rainfall, irrigation water or artificial excavation. Sliding distance of gas station landslide at Heifangtai in Gansu in March of 1989 was only 35–45 m, sliding velocity was also low, stability was poor, large-scale revival of landslide occurred in August of 1994 after five years later, damaged the highway. Similarly, Huangci landslide occurred in January of 1995, the revival of large-scale landslide had happened after 11 years later in May of 2006, buried10 yards of residenter.
4. To loess-mudstone cutting layer landslide, Sliding body was accumulated in the flat valley area after the sliding in high-speed and long-distance, sliding potential energy was released completely basically, so the sliding body has high stability, sliding body will be stable even anterior border of the landslide is eroded by water, artificial excavation. That was called ultra-stable by Professor Guangtao Hu[10]. Sliding body of Shaleshan landslide, Chana landslide and Wolong Temple landslide all were in ultra-stable. The stable coefficient of Shaleshan landslide which happened in 1983 was 5.5 using Bishop Law to count. The scarp is high and steep when the landslide slides, develops nearly parallel tensile-tension crack by the effect of sliding strongly at the scarp. So the stability of the scarp is poor and may slide once again. There was big scale sliding at the scarp of Wolong Temple landslide in 1970[13]. There was small scale sliding of the scarp of Shaleshan landslide, and the volume is 300×10^4 m^3 when the large scale sliding happened on 26 March of 1986.

5 CONCLUSION

1. The loess landslide can be divided into four basic types according to rock-soil formation and location of the sliding surface in loess areas: homogeneous loess landslide, loess interface landslide, loess-mudstone layer plane landslide and loess-mudstone cutting layer landslide.
2. Sliding velocity is low and sliding distance is short to the homogeneous loess landslide, sliding body is in a more stable state, may be revival because of the poor stability of the scarp.
3. Sliding velocity is low and sliding distance is short to the loess interface landslide, the sliding body is in unstable state, may revive to slide by the effect of motivating factors.
4. Generally, Sliding velocity is low and sliding distance is short to loess-mudstone layer plane landslide, sliding potential energy has not been released completely, may revive to slide by the effect of motivating factors.
5. The loess-mudstone cutting layer landslide has the characters of high-speed and long-distance, the stability of the sliding body is usually high after the sliding potential energy was released basically, but high-steep scarp is instability, may slide once again.

REFERENCES

[1] Xu, Bang-dong. 2001. Analysis and Prevention of landslide. Beijing: China Railway Press.
[2] Wu Wei-Jiang & Wang Nian-Qin. 2006. Landslide disasters in Gansu. Lanzhou: Lanzhou University Press.
[3] Wu Wei-Jiang & Wang Shou-ying. 1989. Landslide mechanisms of Shaleshan. Landslide Paper anthology in 1987 the National Landslide Colloquium. Chengdu: Sichuan Science and Technology Press.
[4] Wu Wei-Jiang, et al. Characteristics and causes of the Huangci landslide[J]. Journal of Gansu Science, 1996 Supplement 73–78.
[5] Guo Chong-yuan. 1982. Super landslide and the counting of the velocity. Landslide Collection, Part3. Beijing: China Railway Press 87–193.
[6] Moriwaki • Width. 1989. Landscape Forecasting of sliding distance (Nian-Qin Wang translation). Geological and railway roadbed. The third period. 42–47.
[7] Pan Jia-zheng. 1980. Resistance sliding stability of buildings and landslides analysis. Beijing: Water conservancy Press, 120–132.
[8] Wang Si - Jing & Wang Xiao-ning. 1989. Analysis energy of Large-High-speed landslide and disaster prediction. Landslide Paper anthology in 1987 the National Colloquium Landslide. Chengdu: Sichuan Science and Technology Press, 117–124.
[9] Fang Yu-shu. 1993. Research of the large high-speed landslide forecasting. The natural slope stability analysis and Huayingshan slope deformation seminar collection. Beijing: Geological Press. 92–102.
[10] Hu Guang—tao, et al. Landslide dynamics. 1995. Beijing: Geological Press. 36–56, 125–139.
[11] Wang Cheng-hua. 1989. Longyangxia Reservoir Dam hydropower project near large landslide prediction. Landslide Paper anthology In 1987 the National Colloquium Landslide. Chengdu: 190–197 Sichuan Science and Technology Press.
[12] Zhang Shi-wu(c) Wang Nian-Qin, et al. 1996. The basic characteristics and control of Jiaoshuwan landslide in Tianshui. Journal of Gansu Science. Supplement 55–60.
[13] Baoji in Shaanxi Province Irrigation Authority. 1976. The stability and control measures of Wolong Temple landslide. Landslide Collection. Beijing: The Railwa Press, 147–152.

Investigation of a landslide using borehole shear test and ring shear test

Hong Yang
MWH Americas, Inc., Walnut Creek, California, USA

Vernon R. Schaefer & David J. White
Iowa State University, Ames, Iowa, USA

ABSTRACT: This paper presents a case history using the Borehole Shear Test (BST) as the in-situ test method to investigate a landslide involving glacial till in Iowa, USA. Shear strength values of the soil were measured on the walls of boreholes using an in-situ BST. The investigation was supplemented by measuring the residual shear strength values of the soil on remolded soil samples in the laboratory using the Ring Shear Test (RST). Back analyses were also performed on the slope to evaluate the possible shear strength values and slope conditions at failure. The results show that the average shear strength value of the soil obtained by the BST is comparable with the residual strength value obtained by the RST and the shear strength value obtained by back-calculation. The slope failure could have occurred at a water table higher than that measured during the slope investigation, as the failed slope is stable under the currently slope geometry and water table conditions.

1 INTRODUCTION

An important aspect of landslide investigation is to determine and evaluate shear strength of the soils in the field or in the laboratory. One of the many in-situ techniques is the BST which can directly measure soil shear strength parameter values (i.e., the internal friction angle, ϕ', and the cohesion, c'). The BST is essentially a direct shear test that is performed on the wall inside a borehole. Normal stress is applied to the wall of the borehole through a pair of shear plates, and the peak shear stress is measured in-situ separately and concurrently. Thus, the ϕ' and c' values of the soils are determined from the Mohr-Coulomb failure envelope (e.g. Handy & Fox 1967). The BST is normally considered a consolidated-drained test (Demartinecourt and Bauer 1983).

In the laboratory, the RST has been used to determine the residual shear strength of the soils. The RST involves continuously shearing a remolded soil sample to a large displacement using a torsional ring shear apparatus such as the Bromhead ring shear apparatus (e.g. Bromhead 1979; ASTM 2002). Thus, the shearing resistance of the soils at very large displacement represents the soil condition in a landslide with a large movement.

In this study, the BST and RST were used to evaluate the shear strength of the soils in a landslide in Iowa. The shear strength values were also compared with those obtained from back-calculations. Based on the shear strength values and the slope stability analysis results, the possible failure conditions and current stability of the slope are evaluated.

2 SITE CONDITIONS

The landslide is located next to highway E57 near Luther, Iowa, USA. The regional geology indicates that the soils and landscapes near the project site formed in glacial till deposited by the most recent, the Wisconsin glaciations; and the soil deposits are generally up to one hundred meters in thickness. No bedrock outcrops or other soil types were observed during the field investigation. Examination of the aerial photos indicates that the slide occurred sometime between 1994 and 2002. The head scarp and the humps generally appeared old in 2003 when the slide was first investigated; and bushes and vegetations were well grown on the slope surface, as indicated in Figure 1. The surface soil near the scarp appeared relatively loose.

The slope has an overall sloping angle of about 16 degrees (H:V = 3.5:1), a maximum length of 85 m and a maximum height of 23 m. The width of the slope is about 80 m along the highway. The scarp of the slope near the top has a maximum height of 5 m. The slope also has a few cracks near the middle and the toe of the slope (Figure 1). The maximum widths of the cracks are about 0.3 m. There is also a hump near the toe of the slope. A 2-m wide shallow ditch

Figure 1. Slope photograph showing the head scarp and the cracks on the slope.

was located at the toe of slope (beside the highway). The surface features suggested that a relatively large movement had occurred resulting the landslide.

Six boreholes were drilled manually on the slope during a field investigation in 2004. The boreholes were located either near the head scarp, at middle of the slope or near the toe of the slope. The maximum depth of the boreholes was 4.1 m. The boreholes showed that the slope was made of yellowish brown glacial till which was generally soft to medium stiff. Ground water table was observed to be located near the bottom of the boreholes after boring. The ditch at the toe of the slope was wet indicating the water table was shallow near the toe. The water table near the crest of the slope appeared relatively high, which could have been contributed by the agricultural activities in the large area of agricultural land behind the crest of the slope.

3 RESULTS OF FIELD AND LABORATORY INVESTIGATIONS

The BSTs were conducted in the boreholes to obtain the shear strength of the soils. A total of nine BST were performed near or below the ground water table. Shear stresses versus normal stresses for the tests are presented in Figure 2, and the test results are summarized in Table 1. The results show that the ϕ' value

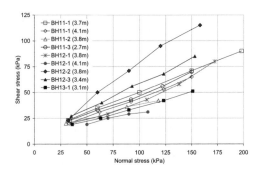

Figure 2. Shear stress versus normal stress for the borehole shear tests.

Table 1. Summary of the borehole shear test results.

Test	Depth (m)	ϕ' (deg.)	c' (kPa)	R^2
BH11-1	3.7	22	11	1.000
BH11-1	4.1	21	7	0.998
BH11-2	3.8	15	12	0.990
BH11-3	2.7	22	9	0.993
BH12-1	3.8	23	1	0.971
BH12-1	4.1	12	10	0.962
BH12-2	3.8	36	4	0.990
BH12-3	3.4	26	10	0.998
BH13-1	3.1	16	8	0.997
Average	-	21.3	7.8	-

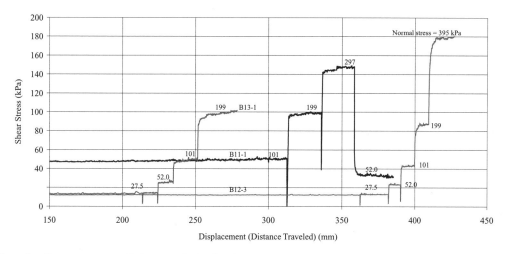

Figure 3. Shear stress versus displacement for the ring shear tests.

Table 2. Summary of the index test results of the soil samples.

Sample No.	Depth (m)	Grain Size Sand (%)	Slit/Clay (%)	Atterberg Limit LL (%)	PI (%)	USCS
BH11-1	3.7	48	52	28	14	CL
BH12-3	3.4	32	68	31	17	CL
BH13-1	3.1	47	53	27	14	CL

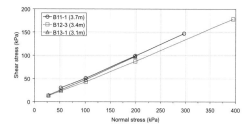

Figure 4. Shear stress versus normal stress for the ring shear tests.

ranged from 12° to 36°, and the c' value varied from 1 to 12 kPa. The average shear strength value was represented by a ϕ' value of 21.3° and a c' value of 7.8 kPa. Basic properties for representative soil samples were also investigated and the results are summarized in Table 2. The results indicate that the glacial till could be classified as sandy, silty clay and had low plasticity.

The RSTs were conducted on three remolded soil samples and the results are presented in Figures 3 and 4. The shear stresses versus displacements at various test stages (various normal stresses) for the samples were plotted to determine the corresponding ultimate shear stresses (Figure 3). Then, the ultimate shear stresses were plotted against the corresponding normal stresses, and the residual shear strength values of the soil samples were obtained (Figure 4). The RST results are also summarized in Table 3, which indicates that the glacial till has residual shear strength values of residual friction angle, ϕ'_r ranging from 24.3° to 26.0° and residual cohesion, c'_r ranging from 0 to 3.8 kPa. The average residual shear strength value could be represented by ϕ'_r value of 25.3° and c'_r value of 1.3 kPa.

4 SLOPE STABILITY ANALYSES AND RESULTS

Slope stability analyses were performed assuming that the soil condition was uniform for the slope. This was because the ground conditions appeared relatively simple based on the site geology and field investigation, and the slope was of relatively small size. The slip surface was assumed circular passing the observed scarp at the top of the slope and a point near the toe of the slope based on the estimated pre-failure slope surface (the area of "loss" of the pre-failure slope near the top equals the area of "gain" of current slope near the toe, as shown in Figure 5). Thus, the centers of the rotations are all located on the line that is perpendicular to and bisects the slope surface. The most probable

Table 3. Summary of the ring shear test results.

Test	Depth (m)	ϕ_r' (deg.)	c_r' (kPa)	R^2
BH11-1	3.7	25.7	3.8	0.9996
BH12-3	3.4	24.3	0.0	0.9995
BH13-1	3.1	26.0	0.1	0.9994
Average	-	25.3	1.3	-

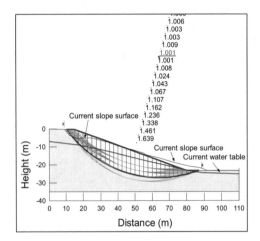

Figure 5. Slope profile, possible slip surfaces and results of slope stability analyses.

Table 4. Summary of shear strength values and factors of safety.

No.	Source	ϕ'	c'	M-P	Bishop	Surface
1	BST (average)	21.3	7.8	1.072	1.069	Pre-failure
2	RST (low)	24.3	0	1.034	1.029	Pre-failure
3	Back-calculated	23.6	0	1.001	0.996	Pre-failure
4	BST (average)	21.3	7.8	1.414	1.411	Current
5	RST (low)	24.3	0	1.393	1.389	Current
6	Back-calculated	18.0	0	1.002	1.000	Current

Note: M-P = Morgenstern-Price method; Bishop = Bishop simplified method; Current water table during investigation was used.

depth of the slip surface was determined by the critical slip surface.

The observed ground water table condition was used in the slope analyses. Both current and pre-failure slope geometry were considered. Back-calculations were also performed to determine the average shear strength values of the soil giving unity factor of safety (FS). Two methods, the Morgenstern-Price method and the Bishop simplified method, were used for the slope stability calculations. The average shear strength values from the BST and the lowest shear strength values from the RST (to be conservative) were used as inputs of soil properties for the analyses.

The results of the slope stability analyses and the corresponding shear strength values are summarized in Table 4. A few observations can be made from the results: 1) FS values obtained from the Morgenstern-Price method and the Bishop simplified method are essentially the same for each of the analyses. 2) FS based on the pre-failure slope surface are generally smaller than the FS based on the current slope surface (i.e., FS of Analyses 1 and 2 are smaller than FS of Analyses 4 and 5, respectively); or shear strength values resulting unity FS based on pre-failure slope surface is higher than those based on current slope surface (i.e., back-calculated shear strength in Analysis 3 is higher than that in Analysis 6). These results are consistent with the fact that the current slope is more stable than the pre-failure slope, as driving force was reduced by the soil "loss" and resistant force was increased by the soil "gain" (Figure 5). 3) The FS values based on the average BST and lowest RST shear strength values are very close indicating that the average BST and RST shear strength values are comparable, and they are also both close to the back-calculated shear strength values based on the pre-failure slope surface. 4) The RST shear strength is slightly higher than that based on back-calculation suggesting that the slide may have occurred under less favorable conditions (such as water table higher than the current one), so that the failure occurred at a higher shear strength (i.e. the RST shear strength), since the RST shear strength represent the ultimate shear resistance of the soil. 5) The FS values are about 1.4 based on current water table and slope surface conditions indicating that the slope is currently stable if the water table is not changed drastically. In fact, no apparent slope movement has been observed since 2004 when the slope was investigated.

5 CONCLUSIONS

The BST and RST were employed for a landslide investigation, and the shear strength values obtained were used in the slope stability analyses. Shear strength values of the soils were also estimated from back-calculations based on the slope giving unity factor of safety. The results show that the RST shear strength value essentially represented the shear strength of the soils at the slope failure, which was also comparable with the average BST shear strength value and the back-calculated shear strength values. The slide most likely has occurred under a higher water table. However, the slope is stable under current water table and ground conditions.

ACKNOWLEDGEMENTS

The Iowa Highway Research Board sponsored this study under contract TR-489. The findings and opinions expressed in this paper are those of the authors and do not necessarily reflect the views of the sponsors and administrations.

REFERENCES

ASTM 2002. D6467. Standard test method for torsional ring shear test to determine drained residual shear strength of cohesive soils. *American Society of Testing and Materials*.

Bromhead, E.N. 1979. A simple ring shear apparatus. *Ground Engineering*, 12 (5), 40–44.

Demartincourt, J.P. Bauer, G.E. 1983. The modified borehole shear device. Geotechnical Testing Journal, ASTM, 6, 24–29.

Handy, R.L. Fox, N.S. 1967. A soil borehole direct shear test device. *Highway Research News, Transportation Research Record*, 27, 42–51.

The importance of geological and geotechnical investigations of landslides occurred at dam reservoirs: Case studies from the Havuzlu and Demirkent Landslides (Artvin Dam, Turkey)

A.B. Yener
General Directorate of State Hydraulic Works (DSI), Ankara, Turkey

S. Durmaz
General Directorate of Mineral Research and Exploration (MTA), Ankara, Turkey

B.M. Demir
General Directorate of Disaster Affairs (AIGM), Ankara, Turkey

ABSTRACT: Landslides, occurring in dam reservoirs, result in a large number of causalities and huge economic losses all over the world. Therefore, regional and moderate scaled landslide investigations around a dam and in its reservoir provides major contribution to the decrease of losses caused by landslide disasters. In this study, moderate-scale susceptibility, hazard and risk assessments of landslides in the reservoir area of the Artvin Dam are introduced with present limitations and review of the precautionary measures. There are two huge landslides in the reservoir area of the Artvin Dam, called Havuzlu and Demirkent landslides. Based on the field investigations, laboratory experiments and evaluations on the previous and most recent data, a susceptibility map of the area was constructed. This map indicated that 88.9% of the existing landslides, including Havuzlu and Demirkent landslides, have high and very high susceptibilities and it is obvious that a risk should be expected in the investigated area. This finding suggests that the crest of the Artvin Dam should be reassessed and raised to compensate possible wave effects. While it is predicted that the reservoir is blocked after a new and huge landslide, and the lake water overruns over the dam. Besides, driving by-pass tunnels at 565 m asl with a length of about 2000 m and a diameter of 10 m could be helpful to derive the rising water in the reservoir. On the other hand, the Yusufeli Dam Power Building and structures located in the upstream area could be damaged by the flood, and therefore the foundation levels of the buildings and structures should be reassessed and the construction project should also be revised. Then, possible choices should be assessed by the control office and the contractor in terms of feasibility and ease of application.

1 INTRODUCTION

1.1 Aim of the study

Landslides, that occur in dam reservoirs, have caused large number of casualties and huge economic losses in all over the world. Therefore, regional and middle scaled landslide investigations around the dam and in its reservoir provides major contribution to the decrease of losses caused by landslide disasters. For instance, in 1963, a major rock slide resulted in a death toll of approximately 2600 in Italy. The slide block moved suddenly into the newly filled reservoir of the Vaiont Dam (Italy), flushing the lake water up and over the dam. The wall of water was over 200 m high as it swept into nearby villages, wiping out everything in its path. The rock slide and the flood could have been readily foreseen if better geological consulting had been carried out before the construction of the dam and reservoir.

This study, discussion of the medium scaled landslides susceptibility, hazard and risk assessments which take place in the Artvin Dam Reservoir, introducing present limitations and review of the precautionary measures were aimed.

1.2 Description of the study site

The Artvin Dam and its hydroelectrical power plant (HEPP) are planned to be constructed on Coruh River, NE Anatolia, Turkey. It is designed as concrete arch dam with a height of 180 m and 332 MW power plant capacity.

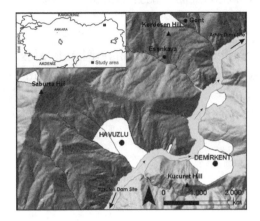

Figure 1. Location map of the investigated area.

Figure 2. Stratigraphic columnar section of the investigated area (Not-to-scale, simplified from Ertunc, 1980).

Different scaled landslides have been identified at the Artvin Dam reservoir site (Ertunc., 1980, 1991, Gunay., 1991). In this study, two large-scaled landslides, called Havuzlu and Demirkent landslides, were investigated and evaluated in the reservoir area (Fig. 1).

1.3 The methods employed

Detailed geological surveys were carried out by using 1/25 000 scaled topographic maps and 1/5 000 scaled geological maps. The landslide characteristics were also evaluated by measurements and observations. Laboratory experiments were conducted on the samples taken from the study area, and a landslide database was obtained. Finally, the lithological, morphological, gradient and aspect maps of the study site were produced using Arcmap, SPSS and Arcview computer codes and landslide susceptibility maps were constructed.

2 GEOLOGY

Liassic Yusufeli formation and the Ýkizdere Magmatites of Upper Eocene—Oligocene crop out in the study area (Baydar et al. 1969, Ertunc. 1980, 1991, Gunay. 1991). In addition, Quaternary terrace, alluvium and slope wash deposits with landslide material widely cover the investigated area (Figure 2).

Liassic Yusufeli formation covering large areas in the middle of the reservoir consists of basic igneous rocks (gabbro, amphibolite), spillite, metabasalt, agglomerate, slate and lithic tuff. The formation is black, dark green and dark gray in color. Due to fracturing, a blocky rock mass is observed in the study area. The strength of the rocks increases with depth.

The Ikizdere magmatites, observed at the middle part of the reservoir and towards downstream, form

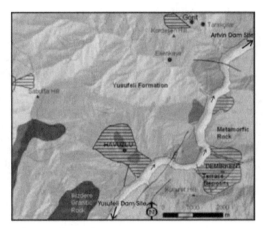

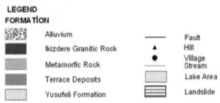

Figure 3. Geological map of the investigated area (adapted from Ertunc 1980, Gunay, 1991).

the main body of a regional batolite which partly penetrated into the Yusufeli formation. The unit is composed of granodiorite-tonalite, adamellite, porphyritic microgranite and granite-gneiss. The Ikizdere magmatites are identified with the presence of granite outcrops in the study area. Light pink colored and coarse grained granitic rocks generally have high strength (Figure 3).

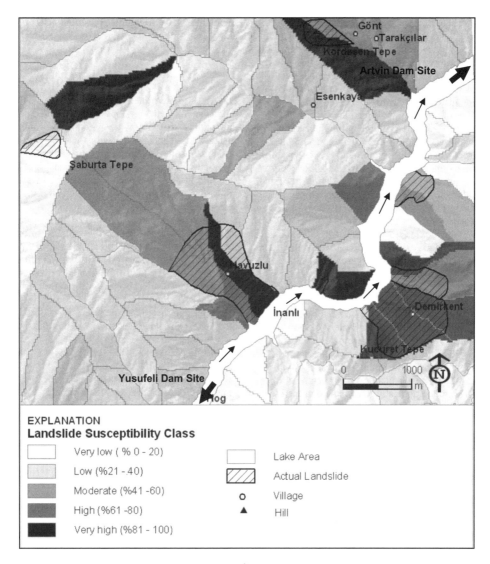

Figure 4. Landslide susceptibility map of the investigated area.

3 LANDSLIDE INVESTIGATIONS IN THE ARTVIN DAM RESERVOIR

To assess landslide susceptibility, hazard and risk in the studied area, firstly, landslide locations were determined through field studies and aerial photo interpretations. Seven different scaled landslides were determined in the reservoir site of the Artvin Dam, with an area of 36.8 km². Areas of landslides vary between 0.3 km² and 1.2 km². Due to the scales of the landslides, the Havuzlu and Demirkent landslides were investigated in detail.

3.1 The Havuzlu landslide

The Havuzlu landslide is located on the NW of the Coruh River, approximately 8 km to the upstream. It takes place below the Saburta Mountain between the two parallel hills trending NW-SE. Based on the investigations, the bedrocks in the region have been identified as phyllite, tuff and gabbro of the Yusufeli formation (Baydar et al. 1969, Ertunc.1980, 1991, Gunay. 1991). It is observed that the landslide developed in the Yusufeli Formation within the loose hill material. Its average gradinet is 22° Some flat areas observed suggest a secondary movement. Head of the landslide could not be observed and a definite boundary of the sliding surface could not also be identified. The thickness of the landslide material varies between 100–150 meters. The laboratory experiments on the samples taken from the study area indicated that the landslide materials are generally clayey gravel (GC).

Volume of the landslide is predicted as 86×10^6 m^3, however, the calculated volume of the reservoir below the Havuzlu landslide is 4×10^6 m^3. The data supplied from field and laboratory studies, susceptibility map and partial movements on the landslide occurred in 1988 indicated that the Havuzlu landslide is still active. For this reason, the reservoir slopes would easily be blocked in case of any partial movement on the landslide towards the reservoir. Besides, even a partial sliding could create a wave at 25–30 m heights where the material slip into the reservoir and 12 m heights around the Artvin Dam is calculated.

3.2 The Demirkent landslide

The Demirkent landslide, which is located on the SE of the Coruh River approximately 6.5 km to the upstream, developed between two parallel hills trending NE-SW. The volume of the landslide is predicted to be 57×10^6 m^3. Maximum thickness is about 120 meters. Although the landslide has an average gradient of 23°, it is topographically flat when compared to its surroundings. Diabase and gabbro of the Yusufeli formation are observed in the area. The landslide material is generally composed of fine grained materials such as green to gray clay, silt and gravel. No evidence of instability was encountered during the field studies, and no mass movement has recently been reported in the area. Due to these reasons, once the reservoir is filled, secondary mass movements are not expected in the Demirkent landslide.

3.3 Landslide susceptibility map of the investigated area

Landslide database including landslide features was established by observations and measurements during the field studies, also the evaluations and data of preceding works were used for this purpose.

Then, the slope unit map of the area was produced by using computer programs such as Arcmap, SPSS, etc. Slope, slope aspect, topographical elevation, shape of slope, lithology, water conditions and vegetation cover were considered as independent input parameters in this study.

Finally, the susceptibility map of the reservoir site of the Artvin Dam and its surroundings was produced using the slope unit map with logistic regression method. Based on this map, 88.9% of the existing landslides, included Havuzlu and Demirkent Landslides, have high and very high susceptibilities (Fig 4).

4 CONCLUSIONS

Due to the determined geological properties of the units that formed the slopes of the landslide areas and sliding materials, the models and geometries of landslides, and susceptibility maps, reasons and possible results of landslides are explained. Also, comments and evaluations on the instability of the area and the possible movements on the investigated landslides are presented. In conclusion, with regard to the field investigations, laboratory studies and evaluations of existing and new data, susceptibility map of the area indicated that 88.9% of existing landslides, including Havuzlu and Demirkent landslides, have high and very high susceptibilities and that a risk of new landslides should be expected in the investigated area. Meanwhile monitoring of the landslides should be carried out, continuously.

It is also predicted that the reservoir will be blocked after a new and huge landslide, causing the overflow of lake water over the dam. Therefore, the crest of the Artvin Dam should be reassessed and raised to compensate possible wave effects. Besides, driving by-pass tunnels at 565 m asl with about 2000 m in length and 10 m in diameter, could be helpful to derive the rising water in the reservoir. On the other hand, as the Yusufeli Dam Power Building and structures located in the upstream could be damaged because of the flood, foundation levels of the buildings and structures should be reviewed and the project should be revised. Available choices should be determined by the control and contractor authorities according to the allowance of the feasibility and the ease of application.

REFERENCES

Baydar, O., Erdogan, A., Topcan, A., Kengil, R., Korkmazer, B., Kaynar, A. & Selim, M. 1969. *Geology of the region between Yusufeli, Ogdem, Madenkoy, Tortum Lake and Ersis.* Ankara: MTA (in Turkish).

Ertunc, A. 1980. *Engineering geological investigation of possible dam sites, reservoirs and tunnel routes at the Coruh Basin..* Ankara: EIE (in Turkish).

Ertunc, A. 1991. Effects of landslides on dam projects along the Coruh River, *Turkey 1. landslide symposium proceedings, Trabzon, 27–29 November 1991.* Trabzon: Karadeniz Technical University Press (in Turkish).

Gunay, S. 1991. Engineering geology report of the Artvin Dam and HEPP project. Artvin: DSI (in Turkish, unpublished).

Yilmaz, S.B., Gulibrahimoglu, I., Yazici, E.N., Yaprak, S., Saraloglu, A., Konak, O., Kose, Z., Cuvalci, F. & Tosun, C.Y. 1998. *Environmental geology and natural resources of Artvin.* Ankara: MTA (in Turkish).

Landslides and Engineered Slopes – Chen et al. (eds)
© 2008 Taylor & Francis Group, London, ISBN 978-0-415-41196-7

An innovative approach combining geological mapping and drilling process monitoring for quantitative assessment of natural terrain hazards

Z.Q. Yue, J. Chen & W. Gao
Department of Civil Engineering, The University of Hong Kong, Hong Kong, China

ABSTRACT: This paper presents an innovative approach to characterize geotechnical conditions of soil and rocks in the ground of hilly natural terrains. The approach combines the conventional geological mapping method and the recent invented Drilling Process Monitoring (DPM) method. The approach can take advantages of the two methods for accurate, economical, quick and flexible applications in hilly natural terrains. Consequently, the approach can offer us quality factual data on the ground conditions for better quantitative assessment of natural terrain hazards.

1 THE ISSUE

Figure 1 is a satellite image showing the hilly natural terrain above the Po Shan Road Landslide in 1972 in Hong Kong Island. Figure 2 is a site photograph showing a front view of the hillside in 2004 when slope maintenance works were being carried out. The 1972 landslide destroyed a 4-story building, collapsed a 12-storey building and killed 67 persons (HKG, 1972).

The Po Shan Road is about 180 mPD (meter above the mean sea level). The failure scarp was up to 300 mPD. The hillside peak is the Victoria Peak of 552 mPD. The hillside has variable slope angles with an average of 33°. Since 1972, the hillside surrounding the hillside has not been developed due to great concerns on the natural terrain hazards.

Prior to any development at a hillside, a quantitative assessment of the hilly natural terrain hazards has to be conducted. A site investigation has to be carried out accordingly (Chen, 2003; Chen et al., 2005; GEO, 1987, 1988, 1994). However, there is an issue.

The issue is how to safely, accurately, economically and environmental-friendly characterize the mechanical conditions of soil and rocks in hilly natural terrains.

The conventional methods for site investigation include trial pits and trenches, engineering geological mapping and aerial photograph interpretation. Modern digital technologies including geographical information system (GIS), digital photogrammetry, mobile mapping systems combining GIS and global positioning system (GPS) can also be used. A natural terrain landslide inventory for the site in GIS can also be established using historical records and aerial photographs.

These methods, however, can reveal only the ground conditions at or near the hillside surface. They

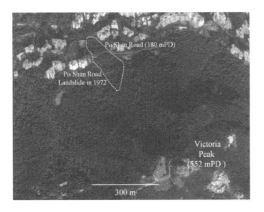

Figure 1. Steep natural hillside above Po Shan Road in Hong Kong Island (modified after http://www.google.com).

Figure 2. A front view of the steep hillside surrounding the landslide.

cannot reveal effectively the ground properties, in particular, the defined mechanical properties at depth. Landslide investigations in Hong Kong and other regions have evidently shown that the ground properties at depth play a dominant role in the occurrence of slope failures and the extrapolation of ground properties from the surface observations can be inaccurate and unreliable (Chen et al. 2007; Lan et al, 2003; Shang et al. 2003; Sheng et al. 2002; Yue et al. 2001; Yue & Lee, 2002; Zheng et al. 2006).

The hillside terrain at Po Shan was formed via a long and natural process. It experienced numerous geological and geotectonic movements, physical and chemical weathering and deterioration. It has colluvium, residual soils, weathered soils and rocks, weak zones, faults and joints. Figure 3 is a site photograph taken in 2006 showing the surface condition of the hillside indicated in Figures 1 and 2. Abundant boulders of weathered volcanic rocks and heavy vegetation and trees were present.

In order to obtain the factual data about ground conditions below the hillside surface (from 0 to 50 m below the ground), conventional sub-surface exploration methods may be adopted. The common method is to drill vertical or inclined boreholes in the ground using hydraulic rotary core drilling machines to obtain disturbed or undisturbed soil samples and rock cores.

However, such drilling or coring can be used in a very limited scope. The conventional ground investigation approach in civil and geotechnical engineering cannot be used effectively and economically for the natural terrain hazard studies. The reasons can be summarized as follows.

The hydraulic rotary core drilling machines are usually heavy and need a drilling lubricant such as water. The water can destabilize the hillside slope that may be marginally stable, which is why air foam is nearly always specified in ground investigation for natural terrain hazard assessment in Hong Kong. The access is difficult. Heavy timber scaffolding is needed to form safe and stable access and drilling platforms. Sometimes, a helicopter may be mobilized to transport the drilling equipment. The drilling time is long. The natural terrain can cover a considerably large area with dense vegetation. Other difficult features are possible to be encountered in the field.

Therefore, innovative approaches for quantitative characterization of the ground conditions and properties must be developed and used for environmentally responsible investigations in natural terrains.

The purpose of this paper is therefore to address the issue with an innovative approach. The approach is to combine and integrate the conventional engineering geological mapping and the recent invented drilling process monitoring (DPM). Details of the DPM can be found in Yue et al. (2001, 2002, 2003, 2004a, 2004b, 2004c, 2006, 2007) and Yue (2004, 2005), Sugawawa (2003) and Lam & Siu (2006). The combined approach can take advantages of the two methods. As a result, it can be applied to quantifying the ground conditions and particularly the spatial distribution of the geomaterial mechanical properties in the 0 to 50 m deep ground of hilly natural terrains. The application can be accurate, economical, quick, flexible and environmental-friendly.

In ensuing, a brief summary of the proposed approach is given. A case study is further used to illustrate the proposed approach. Next, further discussions are made on how to use the approach as a solution of choice to address the issue.

2 THE INNOVATIVE APPROACH

2.1 *Basic principle*

The basic principle of the proposed approach is to use the conventional engineering geological mapping to accurately characterize and record the conditions and properties on the ground surface of hillsides. Such information may include exposed soils and rocks, their physical and mechanical features, discontinuities, seepage, as well as vegetations. The modern digital technologies can further be used to measure and record the information exposed on the ground surface more accurately and extensively.

Based on the mapped information, the DPM is applied to reveal the mechanical properties and strength beneath the ground surface at some selected spots in the hilly natural terrains. The method uses an air-driven rotary-percussive drilling machine with down-the-hole hammer to drill a hole of 100 mm in diameter in the ground and in the meantime uses a drilling process monitor to record the entire drilling

Figure 3. A corner of the steep natural hillside above Po Shan Road in Hong Kong Island with trees and boulders in 2006.

process in real time. From the monitored data, the mechanical strength properties and their spatial distribution along the hole can be obtained. The drilling can be carried out quickly on a simple and light steel scaffolding. The hole can have a depth up to 60 m and a declination or inclination angle variable with respect to the hillside slope.

2.2 Geological mapping

The conventional geological mapping method has been developed and used for many years. Figure 4 shows an example using a geological hammer to map the mechanical properties of soils and rocks exposed on the hillside and a digital photograph to record the measured spot. Other modern digital technologies can also be used. Details can be found in the relevant literatures (e.g., GEO, 1987, 1988, 1994; Chen et al. 2005; Zheng 2006).

2.3 Air-driven rotary-percussive drilling with down-the-hole hammer

As shown in Figure 5, an air-driven (pneumatic) rotary-percussive drilling machine with a down-the-hole (DTH) hammer was being used to drill a declined hole of 100 mm in diameter into the hillside shown in Figure 4. The platform was a steel scaffolding on the hillside. The flush medium is compressed air flow, which has no adverse effect to the groundwater, however, is very noisy and dusty.

The machine was operated by two skilled men. One man was sitting near the controller, operating and powering the rig by steering the supplies of compressed air flow from the main pipe into five dividing pipes. The other was standing near rig and adding or disconnecting the 1 m long drill rods.

2.4 Drilling process monitoring

It is well understood that forming a hole in the ground with an air-driven rotary percussive drilling machine equipped with a DTH hammer requires detaching geomaterial from the cut face with drill bit and removing the detached geomaterial from the hole with flushing air, which in fact is a mechanical failure process of the ground geomaterial.

The DPM is associated with the drilling machine in Figure 5. It automatically and digitally monitors the drilling process and quickly characterizes the mechanical and strength properties of the ground soil and rocks along the hole. It has a hardware for in-situ automatic monitoring and recording of drilling parameters in real time and a software package for analyzing and presenting the monitored digital data in time series and along the hole depth.

The drilling machine in Figure 5 was equipped with a DPM hardware. The hardware device was non-destructively mounted onto the drilling machine. The whole drilling process was monitored in a digital manner and in real time series.

Figure 6 presents an interface from the DPM software showing a typical original DPM data in time series associated with the drilling of a single hole. The graphs under the columns 1 to 7 represent the complete time-histories of the downward and upward thrust pressures, the forward and reverse rotation pressures, the percussion pressure, the bit rotation speed, and the chuck position, respectively. The 1 to 5 pressures represents the five compressed air flows in the five pipes from the controller for pushing the rod downward, lifting the rod upward, rotating the rod clock-wisely, rotating the rod anti-clock-wisely, and punching the DTH hammer and flushing debris at the bit out of the hole, respectively, in real time.

Figure 4. Mapping the properties of soils and rocks on ground surface with geological hammer and digital photographs.

Figure 5. Drilling a declined hole in a hillside with an air-driven rotary-percussive drilling with down-the-hole hammer, which was being monitored with a DPM device.

Furthermore, the depth of hammer bit in the hole in real time can be derived from the original DPM data in Figure 6. The upper graph in Figure 7 shows the complete time-history of bit depth in the hole.

The drilling of the 35 m deep hole started at 11:30:21 and completed at 13:31:20 and used a total of about 2 hours. The advancement of bit into the new ground ended at about 13:08:00. The retrieving of bit from the hole started at about 13:14:00.

Figure 8 plots the depth of bit with respect to the net drilling time, where the associated percussion pressure, downward thrust pressure and the forward rotation pressure are not shown.

The curve of the bit depth versus the net drilling time can be divided into 10 linear zones along the hole of 35.094 m in total length. Each zone has an average drilling rate that is listed in Figure 8. The lower depths of the zones 1 to 10 are 7.624, 15.861, 19.943, 22.421, 23.673, 24.184, 31.644, 32.776, 33.086, and 35.094 m, respectively. Their thicknesses are 7.624, 8.237, 4.082, 2.478, 1.252, 0.511, 7.460, 1.132, 0.310, and 2.008 m, respectively.

The plot in Figure 8 clearly shows the spatial variation of the mechanical strength of the weathered volcanic rock along the drill hole.

2.5 The combined approach

The geological mapping at the hillside found that hillside surface was made of completely decomposed volcanic (CDV) in dry condition, as shown on Figure 9. From Figure 8, it is evident that this CDV soil layer extended about 7.624 m into the ground. Then much stronger weathered or better graded volcanics were encountered in the zones 2 and 3 for 12.32 m thick. Subsequently, weaker zones were present in the zones 4 to 8. Stronger geomaterials were shown again in the bottom two zones 9 and 10.

Since the drilling and monitoring can be carried out quickly and flexibly, many holes can be drilled and monitored in the hillside. The spatial distribution of the geomaterial strength along each hole can be revealed and inter-related.

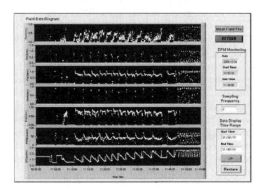

Figure 6. Original DPM data showing the time history of whole drilling process for forming a hole in the hillside.

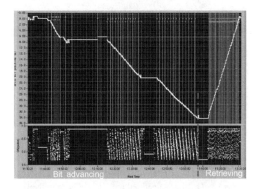

Figure 7. The complete time-history of the hammer bit depth in the hole and the chuck position outside the hole.

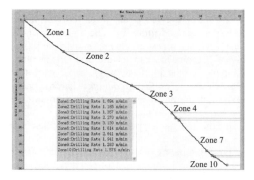

Figure 8. DPM identified mechanical strength zones of constant drilling rates along the hole in the hillside in Figures 6 & 7.

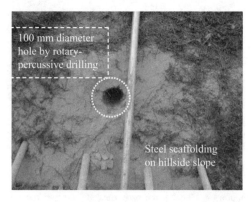

Figure 9. Completely decomposed volcanic soil exposed at the entrance of the declined hole in Figures 6 to 8.

As a result, the two dimensional geological map made on the ground surface can be extended and extrapolated into the hillside for a depth upto the drill hole lengths. Three-dimensional geological map of the hillside can be drawn with high accuracy and reliability.

3 CASE STUDIES

A trial use of the proposed approach was conducted in Hong Kong in 2005. Figure 10 shows the hillside, where both the conventional hydraulic drilling machine and the air-driven rotary-percussive drilling machine were used. Details of the two drilling methods can be observed in Figures 5, 11 and 12.

The hydraulic drilling machine was used to obtain undisturbed soil samples and rock cores from the hillside. The drillhole was declined at an angle of 15 to the horizontal. The time for setting up the timber access and platform was about 10 working days. The time for the drilling of a 40 m depth inclination hole was about five working days.

The steel platform for the rotary-compressive drilling was established about one day and the drilling equipment was mobilized onto the platform by hand. No special access scaffolding was used. The drilling with automatic monitoring for a similar 40 m depth declined hole used about 3 hours.

Figures 4 to 9 and the associated text in Section 2 above are part of the case study.

Furthermore, the rotary-percussive drilling can be carried out for rock cliff up to 50 m high. Figure 13 shows an example of such drilling work in Hong Kong. The drilled holes were used for installation of rock bolts and relief drainage pipes.

4 THE SOLUTION TO THE ISSUE

From the above case studies and discussions, it is evident that conventional ground investigation approaches have various limitations in obtaining clear

Figure 11. Hydraulic drilling on the hillside in Figure 10.

Figure 12. DTH hammer and bit with two flushing holes on above two sliding channels of the rig steel frame.

Figure 10. Hydraulic rotary core and pneumatic rotary-percussive drilling on a steep hillside in Hong Kong.

Figure 13. Air-driven rotary percussive drilling in a cage lifted by a crane to drill a hole in rock cliff in HK.

and reliable pictures and understanding of ground conditions in large hilly natural terrains.

Hydraulic rotary core drilling techniques are usually time-consuming and costly although they can provide undisturbed soil and rock samples. Only a very limited number of rotary boreholes can be carried out, which results in limited factual data and could lead to inaccurate and unreliable ground geotechnical models.

The recently invented DPM method can acquire digital data of the drilling parameters for the real-time process of drilling a hole in the ground. The DPM data are factual data and can identify the variation and distribution of the geomaterial strengths associated with the soils and rocks along the hole depth.

The air-driven rotary percussive drilling equipped with DPM are portable, flexible, reliable and economic in-situ techniques. They can be easily and non-destructively established on natural hillsides with a simple steel platform. The drilling does not cause side effects to the hillside stability and can be considered non-destructive. The drilled hole can further used for soil nailing, rock bolting and drainage pipe installing so that the hillside can also be strengthened after the ground investigation.

Therefore, the proposed approach takes advantages of both the conventional engineering geological mapping method and the drilling process monitoring method in association with the air-driven rotary percussive drilling equipment. As a result, the proposed approach gives an idealized way for accurate, economical, quick and flexible applications in hilly natural terrains. The approach can offer us rich and reliable factual data on the ground conditions for better quantitative assessment of natural terrain hazards.

Returning back to the issue proposed at the beginning about how to conduct the site investigation on the vegetated steep hillside above Po Shan Road, the proposed approach can be a solution of choice. The mapping and DPM can accurately and economically and environmental-friendly characterize mechanical conditions of ground soil and rocks in the hilly natural terrain.

Consequently, the quantitative hazard assessment of the natural terrain can be made using the rich and reliable factual DPM data in addition to the conventional factual mapping data on the hillside surfaces. Mitigation measures such as check dam, flexible barrier and preventive measures such as soil nailing, retaining structures, drainage improvement, and boulder stabilization can be proposed and designed more effectively and environmental-friendly.

ACKNOWLEDGEMENTS

The work described in this paper was substantially supported by a grant from the Research Grant Council of the Hong Kong Special Administrative Region, China (Project No. HKU 7137/03E) and a grant from China Natural Science Foundation (No. 50729904). The authors also thank some in-kind supports from the Geotechnical Engineering Office of the Government of the Hong Kong Special Administration Region, Halcrow China Ltd, Maunsell Geotechnical Services Ltd., Fugro (Hong Kong) Ltd., and Soils and Materials Eng. Co., Ltd. The two coauthors J. Chen and W. Gao also thank The University of Hong Kong for providing financial supports to them during their Ph.D. studies.

REFERENCES

Chen, Z.Y. 2003. *Soil Slope Stability Analysis—Theory, Methods and Programs*, China Water Sources and Hydropower Publisher, Beijing, 560p.

Chen, Z.Y., Wang, X.G., Yang, J., Jia, Z.X. & Wang, Y.J. 2005. *Rock Slope Stability Analysis—Theory, Methods and Programs*, China Water Sources and Hydropower Publisher, Beijing, 890p.

Chen, N.Sh., Yue, Z.Q., Cui, P. & Li, Z.L. 2007. Landslide induced debris flow and its maximum discharge estimation—illustrated with case studies, *Geomorphology*. 84 (1): 44–58.

GEO. 1987. *Guide to Site Investigation*, Geotechnical Engineering Office (GEO), Civil Engineering Department, HK.

GEO. 1988. *Guide to Rock and Soil Descriptions*. Geotechnical Engineering Office (GEO), Civil Engineering Department, HK.

GEO. 1994. *Geotechnical Manual for Slopes* (Second Edition). Geotechnical Engineering Office (GEO), Civil Engineering Department, HK.

GHK. 1972. *Commission of Inquiry into the Rainstorm Disasters, 1972*. Final Report. Hong Kong. Govt. Printer.

Lam, J.S. & Siu, C.K. 2006. *Evaluation of application of drilling process monitoring (DPM) techniques for soil nailing works, GEO Report, No. 189*, Geotechnical Engineering Office, HKSAR Government.

Lan, H.X., Hu, R.L., Yue, Z.Q., Lee, C.F. & Wang, S.J. 2003. Engineering and geological characteristics of granite weathering profiles in South China, *Journal of Asian Earth Sciences*, 21 (4): 353–364.

Shang, Y.J., Yue, Z.Q., Yang, Z.T., Wang, Y.C. & Liu, D.A. 2003. Addressing severe slope failure hazards along Sichuan-Tibet highway in southwestern China, *Episodes, Journal of International Geoscience*, 26 (2): 94–104.

Sheng, Q., Yue, Z.Q., Lee, C.F., Tham, L.G. & Zhou, H. 2002. Estimating the excavation disturbed zone in the permanent shiplock slopes of the Three Gorges Project, China, *International Journal of Rock Mechanics and Mining Sciences*. 39: 165–184.

Sugawawa, J., Yue, Z.Q., Tham, L.G., Lee, C.F. & Law, K.T. 2003. Weathered rock characterization using drilling parameters. *Canadian Geotechnical Journal*, 40 (3): 661–668.

Yue, Z.Q., Lee, C.F. & Pan, X.D. 2001. Landslides and stabilization measures during construction of a national expressway in Chongqing, Southwestern China, *Proceedings of International Conference on Landslides: Causes, Impacts and Countermeasures*, June 17–21, 2001, Davos, Switzerland, pp. 625–633.

Yue, Z.Q., Lee, C.F., Law, K.T., Tham, L.G. & Sugawara, J. 2001. HKU drilling process monitor and its applications to slope stabilization, *Proceedings of Annual Conference—Works Bureau and Ministry of Construction on the Development and Co-operation of the Construction Industry of the Mainland and Hong Kong*, 20–21 September, 2001, Vol. 2, pp. II35–II46.

Yue, Z.Q. & Lee, C.F. 2002. A plane slide that occurred during construction of a national expressway in Chongqing, SW China, *Quarterly Journal of Engineering Geology and Hydrogeology*, 35: 309–316.

Yue, Z.Q., Lee, C.F., Law, K.T., Tham, L.G. & Sugawara, J. 2002. Use of HKU drilling process monitor in soil nailing in slope stabilization, *Chinese Journal of Rock Mechanics and Engineering*, 21 (11): 1685–1690. (in Chinese)

Yue, Z.Q., Guo, J.Y., Tham, L.G. & Lee, C.F. 2003. Application of HKU DPM in automation of geotechnical design and construction, *Proceedings of the First National Congress on Geo-Eng China*. Vol. 1, pp. 147–155. China Communications Press, ISBN 7-114-04724-X.Oct. 22–25, 2003, Beijing, China.

Yue, Z.Q., Lee, C.F., Law, K.T. & Tham, L.G. 2004a. Automatic monitoring of rotary-percussivedrilling for ground characterization—illustrated by a case example in Hong Kong, *International Journal of Rock Mechanics & Mining Science*, 41: 573–612.

Yue, Z.Q., Guo, J.Y., Tham, L.G. & Lee, C.F. 2004b. Drilling process monitoring for ground characterizations during soil nailing in weathered soil slopes, *Proceedings of the 2nd International Conference on Site Characterization* (ISC-2), Porto, Portugal, September 19–22, 2004. Vol. 2, 1219–1224.

Yue, Z.Q., Lee, C.F. & Tham, L.G. 2004c. Automatic drilling process monitoring for rationalizing soil nail design and construction, *Proceedings of the 2004 Annual Seminar of HKIE Geotechnical Division*, Hong Kong, China, May 14, 2004. pp. 217–234.

Yue, Z.Q. 2004. Automatic monitoring of drilling process for optimizing ground anchorage, *Proceedings of the 8th Conference of Chinese Association of Rock Mechanics and Engineering*, Chengdu, Oct. 2004, Science Press, pp. 879–886 (in Chinese).

Yue, Z.Q. 2005. Automatic drilling process monitoring for soil and rock strengths and their spatial distribution in ground. *Proceedings of the 2nd World Forum of Chinese Scholars in Geotechnical Engineering*, Nanjing, China. August 22–22, 2005. pp. 85–90 (in Chinese).

Yue, Z.Q., Gao, W., Chen, J. & Lee, C.F. 2006. Drilling process monitoring for a wealth of extra factual data from drillhole site investigation, *Proceedings of the 10th international congress of the international association of engineering geology*, Nottingham, United Kingdom, 6–10 September 2006. Theme 5—Urban Site Investigation, paper number: 5–746, page 1–10 (CD RAM softcopy).

Yue, Z.Q., Chen, J. & Gao, W. 2007. Automatic drilling process monitoring (DPM) for in-situ characterization of weak rock mass strength with depth, *Proceedings of the 1st Canada-US Rock Mechanics Symposium*, editors: Erik Eberhardt, Doug Stead, & Tom Morrison, Vancouver, Canada, 27–31 May 2007. Taylor & Franics, London. Vol. 1, pp. 199–206.

Zheng, Y.R., Chen, Z.Y., Wang, G.X. & Ling, T.Q. 2006. *Engineering Treatment of Slope & Landslide*, China Communication Press, Beijing.

Types of cutslope failures along Shiyan-Manchuanguan expressway through the Liangyun fracture, Hubei Province

Haiying Zhao & Ren Wang
State Key Laboratory of Geomechanics and Geotechnical Engineering, Institute of Rock and Soil Mechanics, Chinese Academy of Sciences, Wuhan, China
Institute of Rock and Soil Mechanics the chinese Acadeny of Sciences, Wuhan, China

Jianhai Fan & Wei Lin
Headquarters of Shi-Man Expressway Construction, Shiyan, China

ABSTRACT: The Shiyan-Manchuanguan expressway is an important section of the Yinchuan-Wuhan interprovincial Expressway. The road must go through different geological structure units and Liangyun Fracture. Owing to the unique geological conditions and high-steep cut slopes, slope deformation and failure have so frequently happened in the course of construction as to build up serious pressure and threaten to the normal constructing and traffic safety after completing. Through lots of field investigation and study, the slope deformation and failure mainly are two types: landslide and dilapidation. The landslides occur usually in interbed between soft rock and hard rock. The rainstorm between May and October is the peak of slope deformation and failure. In order to relieve or prevent the economic losses caused by the geologic hazards, the paper studied the mechanisms of different landslides.

1 GENERAL SITUATION OF THE PROJECT

The expressway starting from Shiyan in Hubei province to Manchuanguan in Shaanxi Province is an important section of Yinchuan-Wuhan Expressway, one of the eight interprovincial expressways planned. The expressway is designed to begin with Xujiapeng, the end of Xiangfan-Shiyan Expressway, passing through Shiyan, Yun county and Yunxi county, and end up Manchuanguan town in Shaanxi Province. The road extends in SE-NW direction and is 107 km long. There are 107 bridges and 28 tunnels, the percentage accounting for 45.71%. Constructed in November 2004 and completed and opened to traffic in 2008, the road will play a vital role to ensure smooth transportation on the highway north and south. The land along the expressway is higher in north-west than in south-east. With high altitude, its direction is consistent with that of the regional mountains. Most sections of the Shiyan-Manchuanguan expressway lie in the Daba mountainous area south of the Qinling Mountains where the master stratum is metamorphic rocks. This makes it inevitable to go through different geological structure units and Liangyun Fracture. Under multi-stage tectonic movement and weathering, this sequence of metamorphic strata has developed penetrative tectonic foliations, ductile-brittle fracture, brittle fracture, and diplogenetic fold, which does a great damage to the initial rock solidity (Zhang Guo-wei, etc., Che Zi-cheng etc.). Therefore the rock mass gets greatly fragmented and develops joint, cleavage, and weak intercalation. The destructive forms of strata lithology are controlled by different scheduled structural time, deformable system, deformable environment and structural pattern of Liangyun fault. Due to such complex geognostic and geological conditions, the construction of this section of reservoir is harsher. There are 409 slopes in all along the whole line. The height of the excavated slopes in the district is generally from 24 m to 50 m, and the tallest one reaches 70 m or so. And the slopes buckling collapse are extraordinarily serious with 42 landslides and collapsing disasters having occurred. In addition, lots of latent infection excavated slopes may result in buckling deformation and collapse triggered by certain external factors, rainstorm in particular. All in all, landslide and excavated slope treatment has become one of the most important and most serious bottlenecks which will seriously influence the construction schedule of the whole project.

2 REGIONAL GEOLOGIC SETTING

2.1 Topography and physiognomy

The Shiyan-Manchuanguan Expressway is situated in the northwestern mountainous area of Hubei Province, belonging to the southern piedmont of Qinling Mountains. The terrain slopes eastward with distinct rising and falling. Its altitude ranges from 140 m to 1000 m. The Mountains trend northwest, so does the Shiyan-Manchuanguan Expressway. From west to east, the physiognomy is tectonic denudation middling-mountains, low-mountains and hills. According to the genesis and surface configuration, the morphologic feature can be divided into three geomorphic units. They are tectonic denudation midding and low mountains which distribute over K30~K58 and K81~K107, tectonic denudation low-mountains and hills in K0~K29 and K59~K80, erosion and deposit river valleys mostly on either side of the bank of the Han River and her tributary.

2.2 Stratum and Lithology

In terms of chronologic age and lithology of the stratum, there crop out Proterozoic Wudang Group, Sinian, Cambrian, Ordovician, Devonian, Cretaceous, Tertiary and Quaternary (Communicational Planning and Design of Hubei Province, 2003, Second Expressway Survey & Design Institute of China, 2003). The lithology is complex and can be divided into three classes.

Schist: It is the dominating strata, primarily comprising plakite and mica quartz schist of Wudang Group and chlorite schist of Sinian Yaolinghe Group, which accompany partly diabase intrusive body. The surficial rock, that is laminar hypo-hard rock, has been seriously eroded. In contruction area schist is distributing along K0+000~ K58+700 and K75+700~K89+100 where there are local sections of glutenite and quaternary deposit.

Limestone: Distributing along K89+100~K107+300. Middle bedded and heavy bedded dolomie, limestone with shale band and phyllite band that belong to laminar hard rock and hypo-hard rock crop out along K89+100~K100+100, whereas phyllite clipping laminal limestone belonging to hypo-hard rock, which is very soft and seriously eroded, crop out along K100+100~K107+300.

Granulitite: The granulitite of Proterozoic Wudang Group with discontinuously distributed pycnophyllite quartz schist intercalation crops out along K58+700~K75+700.

2.3 Geologic composition

The region of research interest is located at the southern margin of Indo-China orogenic belt south of

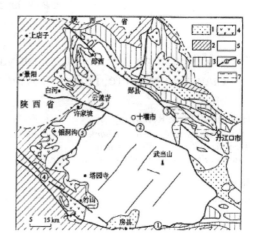

Figure 1. A real structure outline map. 1. Mesozoic-Cainozoic, 2. Palaeozoic, 3. Upper Sinian, 4. Lower Sinian (Yaolinghe Group), 5. Middle-Upper Proterozoic (Wudang Group), 6. Main fault and coding: ① Qingfeng Fault, ② Gonglu Fault, ③ Liangyun Fault, ④ Zhunshan Fault, 7. Danjiangdou Reservoir.

Qinling Mountains and at the northwestern margin of Wudang rise. In this area geological structure is complex, which appears as a series of territorial fracture and fold trending northwest. The road extends on or beside the two sections of Liangyun Fracture (Figure 1). It is the very Liangyun Fracture that determines route selection and architectural competition of the Shiyan-Manchuanguan expressway. It is one active rift from epipleistocene and has no potentiality to give rise to biggish and moderately strong earthquake. Influenced by the tectogenetic movement, the rock in the neighborhood of the fracture is in the poor state of completeness and fragmented, while the attitude of other rock out of the fracture affected zone is stable, extending northward and eastward, and the angle of dip is $40°\sim80°$.

2.4 Weather and hydrology

The research area has abundant atmospheric water. The average rainfall rate per year is 696.1~900.0 mm. Rainfall is concentrated between July and September annually, which accounts for 39~48% of precipitation all year round. In this region, the rainfall is characterized by continuity, concentration and intensity. And surface runoff is developed. There is a wide variety of groundwater, which can be classified into loose stuff intergranular water, clastic rock stuff intergranular water and crack water, basement rock crack water, carbonate rock karstic water according to its occurrence conditions.

2.5 Earthquake

This region is situated in the Jianghan seismic belt of central China's seismic area where the continental seismic activity is weak. Referring to "Zoning map of the ground shock parameters in China" (GB18306–2001), with 10% of exceeding probability during a period of 50 years, the bedrock peak ground accelerations is 0.05 g and the characteristic period is 0.35 s. So the earthquake intensity is VI degree.

3 DISTRIBUTION CHARACTERISTICS AND DEFORMATION MECHANISM OF LIANGYU FRACTURE

3.1 Structural feature of the fracture

From Nanxiang Basin Liangyun Fracture extends in a NW direction with a length of 250 m. By way of Yun county, Yunxi county, and Manchuanguan county, the fault joins with Shanyang fracture extending in an EW direction. The primary fractured surface trend is NE and its angle of dip is 50°–70°. Only in late mesozoic and cainozoic does it change into a SW direction beside marginal basin. Liangyun fracture is composed of a series of parallel or approximately parallel, and synthetic branch reverse thrust fault with different tectonic deformation. On the ground the fault takes a form of linear clough, fault cliff and fault facet developed tens of kilometers long. Moreover, in the stretch zone beside the fault there are a series of concomitant block mountains of the transtorm squeezing zone and fault basins of the transtorm stretch zone. Finally, in the dimension the contemporary basic geomorphic feature behaves as fault morphology (Zhang Guo-wei et al. 2001, Chen Shu-jun et al. 2004).

3.2 Analysis of the deformation mechanism

3.2.1 Tectonic styles of fracture
Liangyun fracture comprises Xinping-Manchuanguan-Yunxi fault and Yunxian-Guanghua fault which are arrayed in right en-echelon formation. Beside the fracture and on the top and the lap joint of the fracture, there are associated with a series of fastigiate fault, shear pull-apart red beds basin, and shear-compressive fold (Zhang Guo-wei, et al. 2001, Chen Shu-jun et al. 2003).

3.2.2 Deformation mechanism of fracture
On the basis of field structure analysis and existing information analysis, it is considered that there are at least three stage metamorphoses in Liangyun fracture belt. They are ductile thrust fault at the early stage, friability strike-slip deformation action at metaphase, ductile abnormal fault at the late stage. Influenced by different deformation mechanism fault at different forming stages, the deformation of faulted structure reflects obvious zoning, which is schistose zone, mylonite zone, cataclastic rock zone, and tectonic breccia zone.

4 ANALYSIS OF TYPES OF SLOPE DEFORMATION AND FAILURE

Because of complex geologic structure and artificial influence, the main types of slope deformation and failure include landslide, collapse or rock fall, which have a detrimental impact on the highway.

4.1 Slope engineering geology

In the geologic hazard distributing graph of Hubei Province, the roadworks lies in the high occurrence zone of landslides and collapse hazard in the northwestern Hubei Province. All of the cut slopes are manual excavation. According to engineering geology investigation from September to November in 2006, there are 409 cut slopes, of which 15 slopes remain uncut and one has been excavated out by a unit in charge of construction. The statistics (393 cut slopes involved) show that at present the percentage of 227 slopes being in the stable condition is 57.76%, the percentage of 83 slopes being in the hypo-stable condition is 21.12%, the percentage of 42 slopes being in the critical state is 10.69%, and the percentage of 4 slopes being in the unstable condition is 10.43%. Table 1 shows the statistics of damage to all the cut slopes along the expressway.

4.2 Landslide

4.2.1 Bedding slip
Within the whole line there are two types of bedding landslide (Table 2), one of which is schist landslide and the other is red sandstone landslide. The stability of different landslide types has huge distinction because of dissimilar characteristics of the structural surface of rock mass.

The red sandstone belongs to soft rock. The bedding surface interpenetration rate generally reaches 100%. And the bedding surface spacing is less than 30 cm, with weak intercalation. This type of slope failure owes to slipping along bedding weak intercalation (Chen Zu-yu et al. 2005, Zhang Xian-gong et al. 2000). Because the inclination of bedding weak intercalation is low and the layer of it is thick, the scale of sliding mass is larger. For instance, No.0728 slope lies in the left of Yunxi county service area and is about 24 m high. The attitude of rock stratification composing the slope is $33°\angle 19°$ and its thickness is

Table 1. Statistics of all cut slopes failure mode.

	Failure mode	Type of slope	Terrain and lithology	Unit/ percentage	Unstable slope Unit
Landslides	Bedding slip	Clockwise parallel slope	Red sandstone	14/3.6	2
			Schist	40/10.2	17
	Insequent slip	Anticlockwise parallel slope	Schist	57/14.5	17
		Steep dip parallel slope	Schist		
	Wedge slip	Clockwise crossing slope	Schist	81/20.6	15
		Anticlockwise crossing slope	Schist	100/25.5	15
	Fossil landslide		clayey soil		3
			Schist		2
	Soil (akin soil) landslide	Soil slope	Clay	35/8.9	7
		Rock & soil slope	Elluvium and schist	4/1.0	1
	Collapse	Limestone slope	Limestone	18/4.6	2
		Crossing slope	Schist	10/2.5	0
		Diabase slope	Diabase	20/5.1	11

Table 2. Distribution range of bedding slip.

Type of slope	Chronologic age	Group	Section
Red sandstone	Cretaceous	Paomagang Group	VII
sthist	Proterozoic	Wudang Group	I~VIII
	Sinian	Yaolinhe Group	VIII, IX

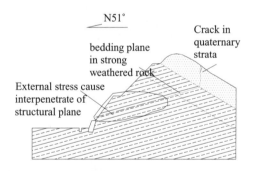

Figure 2. Landslide sketch map of No.0728.

10~40 cm. There are three joints whose attitudes are 5°∠87°, 65°∠77°, 179°∠45° respectively. Excavating the cut slope made its foot unshored and at the same time rainfall infiltration along vertical joint sheared off strata, so large-scale landslide is induced (Figure 2). After the landslide took place, the landslide trailing edge generated many vertical cracks and perforated transversal cracks, as wide as 3 m. Many Golgotha on the sliding mass are pulled apart. The leading edge of the landslide advances to the roadbed median line. This is a large-scale landslide, with the landslide mass 200 m long and 30 m thick.

The slopes of soft and hard interbedding are in the majority of schistous slopes. The bedded plane trending the slope is well developed and badly eroded into quite fragmented rock mass. Owing to filling mud between interlamination, the shear strength of rock mass suffers extreme deviation. In the case of the structural surface complete interpenetration, bigger interbedded spacing and filling mud between interlamination, the slope off which the lower footslope is cut to form free rock mass is easy to come into deformation in the direction of free face under the action of rainstorm. When the slope angle is bigger than schistosity angle, this slope slips along the plane of schistosity and the glide plane is smoothing and flattening (Xu Bang-dong et al. 2001, Chen Zu-yu et al. 2005). If the terrane is cut off a little, large-scale bedding creeping will happen. When the slope angle is smaller than schistosity angle, glide and curve will take place and at last this slope will be going to its yield limit to failure. Take No.0917 slope which is located in the left side of K83+620+860. It has five steps and is 40 m high. The schistosity attitude of rock composing the slope is 38°∠44°. Since the basal slope is destroyed as the slope is excavated, large-scale bedding slip occurs from peak to foot. The drop in level of the landslide is over 60 m and the primary glide direction is 55°. The vertical drop height at the trailing edge of landslide extends to more than 12 m. The leading edge gets over roadbed to pile up at the foot of the opposite slope (Figure 3 and Figure 4). This destroyed type is the chief destructive form, such as No.0837, No.0125, No.0410, and No.0516.

4.2.2 Insequent slip

The countertendency slope and steep dip slope are apt to insequent slip (Table 3). Under the action of self-weight and the crack perpendicular to bedding plane,

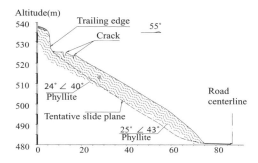

Figure 3. Sectional drawing of No.0917 slope.

Figure 4. Sketch map of the destroyed No.0917.

Table 3. Distribution range of insequent slip.

Type of slope	Chronologic age	Group	lithology	Section
Countertendency parallel slope	proterozoic	Wudang group	plakite Albite schist	III~VIII
Steep dip parallel slope	proterozoic	Wudang group		II~IV

the rock mass collapses in the direction of free face like domino (Chen Zu-yu et al. 2005). As to single formation, flexural deflection similar to cantilever beam occurs. On the ground, there are a series of imbricate parallel cracks taking the shape of "V". When the dump deformation reaches to a certain degree, the maximum buckling distortion of all the strata will be broken to form one perforated failure plane to induce whole slope slipping. No.0615 is a typical slope situated at the left of K52+520~+650, which has five steps and is 40 m high. The natural slope angle is over 50° and the rock mass composing the slope is laminal schist rich in pinal and falls to pieces. As a consequence of continuous rainfall after excavation, the slope slipped from the first step to the third step

Figure 5. Imbricate fissure on the mountaintop of No.0615.

Table 4. Distribution range of wedge landslide.

Type of slope	Chronologic age	Group	lithology	Section
proterozoic		Wudang group	plakite Albite schist Quartz schist	I, V~IX
contertendencyl cross slope	proterozoic	Wudang group		II, IV, V VI, IX

and covered the roadbed. After landslide is over, we can see a score of imbricate parallel cracks range along the natural slope to the mountaintop. The crack extends as wide as 30 cm and the fracture spacing is 1~3 m (Figure 5).

4.2.3 Wedge landslide

Clockwise oblique crossing schist slope and countertendency oblique crossing schist slope are apt to this kind of collapse mode (Table 4).

The failure mode slide badly along the plane composed by bedding plane and two joint planes. No.0704 slope is one typical wedge landslide, which is situated on the left of K59+590~+690 and is 32 m high. This slope is composed of thick layer clipping laminar layer albite-quartz phyllite. The schistosity plane ($36°\angle 62°$) and "X" shear joint plane ($229°\angle 44°$, $65°\angle 83°$) contribute the sliding wedge mass (Figure 6).

4.2.4 Resurgent fossil landslide

Because partial skid resistance of the fossil landslide is excavated under construction, the fossil landslide relives or relives partly in the condition of abundant rainfall. This typical landslide is the left slope at K50+745~K51+103 (Figure 7). As the first step of the slope has been excavated in April 2005, large-scale

Figure 6. Wedge landslide of No.0704.

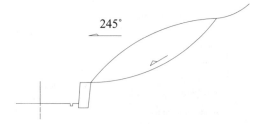

Figure 8. Imbricate fissure of soil landslide.

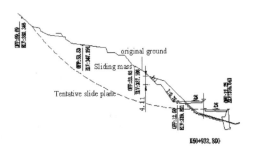

Figure 7. Resurgent fossil landslide at K50+745~K51+103.

Figure 9. Soil landslide.

deeper layer landslide begins to appear. The landslide is about 360 m wide, 100 m long, and 15–50 m thick. The primary slipping direction is NE35°. At present, the leading edge of the fossil landslide has revived and pushed its way in the roadbed and formed obvious deformation, which is threatening the ground to produce the girder. Another example is a revived clay fossil landslide at K71+880~+950.

4.2.5 *Soil (akin soil) landslide*

The formulation of soil (akin soil) landslide is homogeneous. After being excavated, the slope is apt to creeping in the case of rainfall. The glide plane appears circular arc. Failure mode is to pull behind (Figure 8). For instance, the right slope at K43+975~+875 is 8–16 m high and its angle of natural slope is smaller than 10° (Figure 9). After the whole slope slumped, the trailing edge is pulled apart and drop height reaches 8 m. In addition, the slope surfacing has been upheaved and there are many fissures and fracture zone, and at present the last crack has extended to the top of the second slope protection located 15 m far away. The bulge of the leading edge is 4~5 m length and is 15 m

Figure 10. Akin soil landslide.

width. In the middle of the akin soil slope there are slightly weathered rocks and on both sides are fully eroded elluvium and sliderock. In the course of excavation the intrinsic stress equilibrium is broken. At this point rainfall can straightly seep into the slope along manual excavation surface, which leads the slope to buckling collapse. The primary slide face is along the contact zone of hard schist and soft schist. The primary slide direction is not only along the aspect of slope but also along the direction of the slope side (Figure 10).

4.3 Collapse

Within the rock mass of slopes which are diabase slopes, limestone slopes and crosscut schist slopes there do not exist large interpenetrate constructional surface arising slope slumping. Instead there exist many groups of joint and the joint plant is seriously eroded and filled up with clay, so the whole slope is stable. In this case, the collapse mode of the slope is rock failure, and the slope failure is on a small scale.

5 CONCLUSION AND SUGGESTIONS

1. In the region, stratum and geologic structure are simple. Wudang group crops out along the expressway, which is controlled by Liangyun fault zone. The schist of Wudang group is rich in anthrophyllite and the rock beside fault falls to pieces. In this condition there exist landslide and collapse. Landslide is one of the geologic hazards with the highest frequency, largest scale, most serious damage and most complex mechanism.
2. Listing all the slopes in descending order of instability, they are the consequent parallel schist slopes, soil slopes, countertendency parallel schist slopes, and countertendency crossing schist slopes.
3. The mechanism of all the landslides existing along the expressway is that because the foot of the excavated slopes is free, the original stress equilibrium is destroyed and the slope slumps along the direction of the free face. In addition, rainstorm is the dominant factor inducing slope deformation and failure. Most of the slopes are destroyed in the rainstorm from May to October. Moreover, the compact district of landslide is dispersed over the zone of soft rock and hard rock interbedding.
4. Too much explosive is forbidden to use in the course of construction in order to prevent rock mass quality from declining and to protect the slopes against slump. Presplitting blasting is recommended. Besides, it should be excavated step by step while getting done with reinforcement and waterproof in time.

ACKNOWLEDGEMENT

Supported by Project of National Natural Science Funds: Catastrophe mechanism of cascade collection and selective treatment for local debris flows (4067 2193); Self-organized criticality and balance-deviating mechanism of debris flows collapsed deposits (50709035)

REFERENCES

Zhang Guo-wei, Zhang Ben-ren, Yuan Xue-cheng, et al. 2001. *Qinling orogenic belt and continental dynamics.* Beijing: Science press, 221–261.

Xu Bang-dong. 2001. *Landslide analyzing and controlling.* Beijing. China railway publishing house, 140–163.

Chen Zu-yu, Wang Xiao-gang, Yang Jian, et al. 2005. *Rock slope stability analysis.* Beijing: China WaterPower Press, 51–151.

Zhang Xian-gong, Wang Si-jing, Zhang Zhuo-yuan, et al. 2000. China engineering geology. *Beijing.* Science press, 186–216.

Chen Shu-jun, Liu Suo-wang, Yao Yun-sheng, et al. 2004. Research on structural analysis and quaternary slip rate of Liangyun fault. *Journal of geodesy and geodynamics,* 24 (3):60–66.

Chen Shu-jun, Liu Suo-wang, Huang Guang-si, et al. 2003. Report of seismic safety evaluation of Huojuling tunnel, Erdaoya tunnel and Yunling tunnel under the influence of Liangyun fault. Wuhan. Institute of seismology, CEA.

Communicational Planning and Design of Hubei povince. 2003. The specification of engineering geology advanced exploration of the section from Shiyan to Hongyanzi in Hubei province.

Second Expressway Survey & Design Institute of China. 2003.The specification of engineering geology exploration of the section from Hongyanzi to Manchuanguan in Hubei province

*Advances in analytical methods, modeling
and prediction of slope behavior*

Probability limit equilibrium and distinct element modeling of jointed rock slope at northern abutment of Gotvand dam, Iran

M. Aminpoor, A. Noorzad & A.R. Mahboubi
Power and Water University of Technology, Tehran, Iran

ABSTRACT: Rock slope stability analysis has improved significantly in the last few years utilizing new advanced numerical techniques. Application of these new methods have made it possible to realize complexities relating to geometry, material anisotropy, non-linear behavior, in situ stresses and the presence of several coupled processes (e.g. pore pressure, seismic loading, etc.). It is also provided for researchers to understand how initial instability mechanisms become followed by or preceded by creep, progressive deformation and extensive internal disruption of the slope mass. In this paper, it is demonstrated that the numeric distinct element method is useful to perceive some local instable parts of a rock slope due to earthquake loading that it is not possible to anticipate by limit equilibrium analysis.

1 INTRODUCTION

Rock slope stability analyses are routinely performed and directed towards assessing the safe and functional design of excavated slopes (e.g. open pit mining, road cuts, etc.) and/or the equilibrium conditions of natural slopes.

Nowadays, a vast range of slope stability analysis tools exist for both rock and mixed rock-soil slopes; these range from simple infinite slope and planar failure limit equilibrium techniques to sophisticated coupled finite-/distinct-element codes.

Selection of the method of rock slope stability analysis depends on both site conditions and the potential mode of failure, with careful consideration being given to the varying strengths, weaknesses and limitations pertinent to in each methodology.

Given the wide scope of numerical applications available today, it has become essential for the practitioner to fully understand the varying strengths and limitations inherent in each of the different methodologies. For example, limit equilibrium methods still remain the most commonly adopted solution method in rock slope engineering, even though most failures involve complex internal deformation and fracturing which bears little resemblance to the 2-D rigid block assumptions required by most limit equilibrium back-analyses. The factors initiating eventual failure may be complex and not easily allowed for in simple static analysis (Eberhardt, 2003).

2 PROJECT DESCRIPTION: ENGINEERING AND STRUCTURAL GEOLOGY OF GOTVAND DAM SITE

2.1 Dam site geology

The Neogene sediments of the Fars Group (Gachsaran, Mishan and Aghajari Formations) outcrop in the vicinity of the dam site. They include thick evaporitic units (Gachsaran), marl, limestone and alternating sandstone, siltstone and claystone. Plio-pleistocene Bakhtyari conglomerates unconformable overlain these fine-grained sediments. The rock foundation of the dam consists of Aghajari and Bakhtyari Formations.

Figure 1 shows a geological cross-section of the dam site and also cross section of northern abutment of dam. The rock at the site comprises deposits of the Aghajari formation with interbedded claystone, siltstone and sandstone (AJn), overlain by conglomerate of the Bakhtiari formation (BKn). The sandstone and siltstone are indurated and fissured while the claystone appear as heavily over consolidated hard clay. The conglomerate is for the most part indurated and

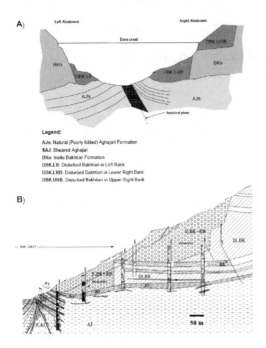

Figure 1. A) Schematic geological cross-section of the dam site B) Cross-section of northern abutment showing different strata and formations.

moderately strong but contains lenses and beds with poor cementation. The rock is massive in the left valley side and rises as a steep cliff some 300 m above the valley floor.

2.1.1 Bakhtyari formation
BK formation is basically formed by conglomerate. Massive conglomerate units, which can be as thick as 100 meters, are only separated by a few sandstone and claystone interlayers. Bedding planes, usually tight, are mostly marked by sandy/silty interbeds, lenses or, in some cases, by the orientation of the elongated pebbles. Pebbles of variable size, from 2 cm up to 30 cm, generally rounded to sub-rounded, are mainly limestone fragments. Chert and sandstone or siltstone pebbles are subordinate. The cement varies from calcareous to siliceous, commonly finely crystallized but sometimes with a coarser, sandy texture. Variable degree of consolidation or washout of sandy bound materials throughout the thick sequence resulted locally in loosely cemented, friable conglomerates.

2.1.2 Aghajari formation
This formation is generally formed by a sequence of brown to gray calcareous sandstone with some interbeded mudstone, marlstone and some siltstones. Thin veins of gypsum are spread out in some horizons of this formation. The sedimentation of this formation is related to the river-flood plane deposits. This formation belongs to late Miocene-Pliocene and composes limey (approx-70%) and silicic (fundamentally Chert-30%) grains. The cementation is generally limy. The thickest and thinnest layers are sandstone and Claystone respectively.

Lateral variations in AJ formation are apparently observed when assessing the boreholes; as in lateral parts, sandstone is changed to siltstone and mudstone. In general, the dominant lithology of the station is formed by siltstone. Sandstone layers are 0.20–0.30 up to 5 m thick. Fine, tight lamination is characteristic for the siltstones and siltstone/sandstone or siltstone/claystone transitions.

2.2 Structural geology: Jointing and disposition of the strata in DBK-URB

One set is the bedding, dipping towards south with a dip angle of 30 to 60° with 45° as a mean (set "B"). The bedding planes, with few exceptions, can be considered as tight.

The second set is perpendicular to the bedding planes (which is commonly offset), dipping around 45° to the north. Spacing is between 1 to 3 m. Their surfaces are generally undulating, while the small scale roughness profiles are variable, from slickenside to rough. Frequently filled with clayey and/or silty material, their width varies between a few millimeters to 30–50 centimeters. Shearing along these discontinuities has been mostly undertaken by the filling material. The conglomerate walls show different conditions, from saw-cut, slightly disturbed to broken matrix and, where greater displacement occurred, mylonite.

Some erratic joints dipping towards north or northeast, are filled with broken conglomerate in a silty clayey matrix.

3 PROBABILITY LIMIT EQUILIBRIUM ANALYSIS

3.1 Application of limit equilibrium analysis in rock slope stability problems

Limit equilibrium techniques are routinely used in the analysis of landslides where translational or rotational movements occur on distinct failure surfaces. Analyses are undertaken to provide either a factor of safety or, through back-analysis, a range of shear strength parameters at failure. In general, these methods are the most commonly adopted solution method in rock slope engineering, even though many failures involve complex internal deformation and fracturing which bears little resemblance to the 2-D rigid block assumptions required by limit equilibrium analyses. However, limit equilibrium analyses may be highly relevant to simple block failure along discontinuities or rock slopes that are heavily fractured or weathered (i.e. behaving like a soil continuum).

For very weak rock, where the intact material strength is of the same magnitude as the induced stresses, the structural geology may not control stability and failure modes such as those observed in soils may occur. These are generally referred to as circular failures, rotational failures or curvilinear slips (Eberhardt, 2003).

Considerable advances in commercially available limit equilibrium computer codes have taken place in recent years. These include:

- Integration of 2-D limit equilibrium codes with finite-element groundwater flow and stress analyses (e.g. Geo-Slope's SIGMA/W, SEEP/W and SLOPE/W—Geo-Slope 2004).
- Development of 3-D limit equilibrium methods (e.g. Hungr et al. 1989; Lam & Fredlund 1993 and SLOPE/W—Geo-Slope 2004).
- Development of probabilistic limit equilibrium techniques (e.g. SWEDGE—Rocscience 2001b; ROCPLANE—Rocscience 2001c).
- Ability to allow for varied support and reinforcement.
- Incorporation of unsaturated soil shear strength criteria.
- Greatly improved visualization, and pre- and post-processing graphics (Stead et al. 2005).

3.2 The SLOPE/W software

SLOPE/W is the leading software product for computing the factor of safety of earth and rock slopes. SLOPE/W, can be used to analyze both simple and complex problems for a variety of slip surface shapes, pore-water pressure conditions, soil properties, analysis methods and loading conditions.

Using limit equilibrium, SLOPE/W can model heterogeneous soil types, complex stratigraphic and slip surface geometry, and variable pore-water pressure conditions using a large selection of soil models. Analyses can be performed using deterministic or probabilistic input parameters. Stresses computed by a finite element stress analysis may be used in addition to the limit equilibrium computations for the most complete slope stability analysis available.

3.3 Probability limit equilibrium analysis of the rock slope at northern abutment of Gotvand dam

There is an ever-increasing interest in looking at stability from a probabilistic point of view. In SLOPE/W simulation, almost all input parameters can be assigned a probability distribution, and Monte Carlo scheme is then used to compute a probability distribution of the resulting safety factors.

As is well known, many natural data sets follow a bell-shaped distribution and measurement of many random variables appear to come from population frequency distributions that are closely approximated by normal probability density function. This is also true for many geotechnical engineering material properties. In this paper, the input data have been considered to have a normal probability distribution function.

For description of uncertainty in a data set, we can use the covariance coefficient that is equal to

$$Cov = \frac{SD}{\overline{X}} \quad (1)$$

where SD is standard deviation and $\overline{X}$ is the mean of data.

Cov coefficient for input parameters of this investigation has been chosen to be 0.1. Therefore, the standard deviation of all input parameters has been considered to be 10% of mean quantity. In order to perform the probability analysis on limit equilibrium results, shear strength parameters (C and φ) and unit weight of all materials have been entered in the program whit a uncertainty of 10%.

A factor of safety is really an index indicating the relative stability of a slope. It dose not imply the actual risk level of the slope, due to the variability of input parameters. With a probabilistic analysis, two useful indices are available to quantify the stability or the risk level of a slope. These tow indices are known as probability of failure and the reliability index.

The reliability index (β) is defined in terms of the mean ($\overline{X}$) and the standard deviation (SD) of the factors of safety as shown in the following equation:

$$\beta = \frac{(\overline{X} - 1.0)}{SD} \quad (2)$$

Table 1. Geotechnical properties of materials of dam site and dam body.

Region	Unit weight (kN/m³)		Cohesion (MPa)		Friction angle (degree)	
	Saturated	Dry	Saturated	Dry	Saturated	Dry
AJn	25	23	0.540	0.500	26	29
S.AJ	24	17.5	0.440	0.500	23	25
BKn	51	35	0.980	0.690	23	25
DBK	40	29	0.390	0.460	23	25
INF	37.7	29	0.832	0.640	24	25
BK^s	42.9	33	1.300	1.000	24	25
Core of dam	20	28	0.100	0.000	19.1	20.1

Table 2. Results of probability limit equilibrium analysis on northern abutment for the time before dam construction.

Analysis type	Slope condition	FS_{mean}	Reliability index	Probability of FS decrease less than 1.5 (%)	Probability of FS decrease less than 1.0 (%)
Static	Natural	2.364	14.173	0	0.00
	Saturated	1.131	1.864	100	3.50
Pseudostatic	Natural	1.666	13.004	0	0.00

This index can also be considered as a way of normalizing the factor of safety with respect to its uncertainty.

The input data for different rock masses in the slope region are listed in Table 1. Also the results of probability limit equilibrium analysis of the rock slope before dam construction considering both static and earthquake event conditions are summarized in Table 2.

It can be seen from the results that earthquake occurrence will not have a considerable effect on the slope stability. This means that total rock slope stability can be reliable during and after earthquake event. But some local instable regions may be occurred due to rock mass jointing that limit equilibrium analysis is not able to demonstrate.

Note that completely saturated condition is a pessimistic state of slope which simulates the intense rainfall and it should be considered as a very improbable state because of depth of slope and also high dip of slope that makes a great runoff.

4 DISCONTINUUM MODELING

4.1 Application of discontinuum modeling in rock slope stability problems

Where a rock slope comprises multiple joint sets, which control the mechanism of failure, then a discontinuum modeling approach may be considered more appropriate. Discontinuum methods treat the problem domain as an assemblage of distinct, interacting bodies or blocks that are subjected to external loads and are expected to undergo significant motion with time. This methodology is collectively referred to as the discrete-element method (DEM).

The development of discrete-element procedures represents an important step in the modeling and understanding of the mechanical behavior of jointed rock masses. Although continuum codes can be modified to accommodate discontinuities, this procedure is often difficult and time consuming. In addition, any inelastic displacements are further limited to elastic orders of magnitude by the analytical principles exploited in developing the solution procedures. In contrast, discontinuum analysis permits sliding along and opening/closure between blocks or particles. The underlying basis of the discrete-element method is that the dynamic equation of equilibrium for each block in the system is formulated and repeatedly solved until the boundary conditions and laws of contact and motion are satisfied. The method thus accounts for complex non-linear interaction phenomena between blocks.

Discontinuum modelling constitutes the most commonly applied numerical approach to rock slope analysis. Several variations of the discrete-element methodology exist:

– distinct-element method;
– discontinuous deformation analysis;
– particle flow codes.

4.2 The UDEC software (Universal Distinct Element Code)

The distinct-element method, developed by Cundall (1971) and described in detail by Hart (1993), was the first to treat a discontinuous rock mass as an assembly of quasi-rigid, and later deformable, blocks interacting through deformable joints of definable stiffness. As such, the numerical model must represent two types of mechanical behavior: that of the discontinuities and that of the solid material.

The dual nature of distinct-element codes, for example UDEC (Itasca 2000), make them particularly well suited to problems that involve jointed rock slopes. On the one hand, they are highly applicable to the modeling of discontinuity-controlled instabilities, allowing two-dimensional analysis of translational mechanisms of slope failure and are capable of simulating large displacements due to slip, or opening, along discontinuities. On the other hand, they are also capable of modeling the deformation and material yielding of the joint-bounded intact rock blocks. This becomes highly relevant for high slopes in weak rock, flexural-topples and other complex modes of rock slope failure (Eberhardt, 2003).

UDEC is a distinct element program for the 2D modeling of jointed rock subjected to quasi-static or dynamic loading conditions. It simulates large displacements (slip and opening) along distinct surfaces in discontinuous medium treated as an assemblage of discrete (convex or concave) polygonal blocks with rounded corners. Discontinuities are treated as boundaries between the blocks. Relative motion along the discontinuities governed by linear and non-linear force-displacement relations for movement in both the normal and shear directions.

4.3 Selection of deformable versus rigid blocks in UDEC simulation of jointed rock masses

An important aspect of a discontinuum analysis is the decision to use rigid blocks or deformableblocks to represent the behavior of intact material. The considerations for rigid versus deformableblocks are discussed in this section.

Early distinct element codes assumed thatblocks were rigid. However, the importance of including block deformability is now recognized, particularly for stability analyses of underground openings and studies of seismic response of buried structures. One of the most obvious reasons to include block deformability in a distinct element analysis is the requirement to represent the "Poisson's ratio effect" of a confined rock mass.

Rock mechanics problems are usually very sensitive to the Poisson's ratio chosen for a rock mass. This is because joints and intact rock are pressure-sensitive: their failure criteria are functions of the confining stress (e.g., the Mohr-Coulomb criterion). Capturing the true Poisson behavior of a jointed rock mass is critical for meaningful numerical modeling.

The effective Poisson's ratio of a rock mass is comprised of two parts: (1) a component due to the jointing; and (2) a component due to the elastic properties of the intact rock. Except at shallow depths or low confining stress levels, the compressibility of the intact rock makes a large contribution to the compressibility of a rock mass as a whole. Thus, the Poisson's ratio of the intact rock has a significant effect on the Poisson's ratio of a jointed rock mass.

A single Poison's ratio, v, is, strictly speaking, defined only for isotropic elastic materials. However, there are only a few jointing patterns which lead to isotropic elastic properties for a rock mass.

Therefore, it is convenient to define a "Poisson effect" that can be used for discussion of anisotropic materials.

The Poisson effect will be defined as the ratio of horizontal-to-vertical stress when a load is applied in the vertical direction and no strain is allowed in the horizontal direction. Plane strain conditions are assumed. As an example, the Poisson effect for an isotropic elastic material is

$$\frac{\sigma_{xx}}{\sigma_{yy}} = \frac{v}{1-v} \quad (3)$$

Consider the Poisson effect produced by joints dipping at various angles. The Poisson effect is a function of the orientation and elastic properties of the joints. Consider the special case shown in Figure 2. A rock mass contains two sets of equally spaced joints dipping at an angle, θ, from the horizontal. The elastic properties of the joints consist of a normal stiffness, k_n, and a shear stiffness, k_s. The contribution of the elastic properties of the intact rock will be examined for the case of $\theta = 45°$. the intact rock will be treated as an isotropic elastic material. The elastic properties of the rock mass as a whole will be derived by adding the compliances of the *jointing* and the intact rock.

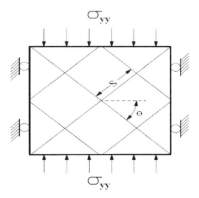

Figure 2. Model for Poisson's effect in rock with joints dipping at angle θ from the horizontal, and with spacing S.

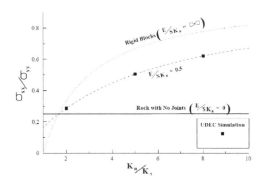

Figure 3. Poisson's effect for rock with two equally spaced joint sets, with $\theta = 45°$ (blocks are deformable with $v = 0.2$).

Table 3. Rock material properties, utilized for distinct element UDEC modeling.

Region	Density (ton/m³)	Bulk modulus (GPa)	Shear modulus (GPa)	Cohesion (kPa)	Friction angle (deg)	Tensile strength (kPa)
D.BK (intact rock)	2.5	10.8	0.500	10,000	45	2,000
BK$_n$ (rock mass)	2.5	3.130	0.500	490	35	0

The compliance matrix due to the two equally spaced sets of joints dipping at 45° is

$$C^{(\text{jointing})} = \frac{1}{2Sk_nk_s}\begin{bmatrix} k_s+k_n & k_s-k_n \\ k_s-k_n & k_s+k_n \end{bmatrix} \quad (4)$$

Thus, the Poisson effect for the rock mass as a whole is

$$\frac{\sigma_{xx}}{\sigma_{yy}} = \frac{[\upsilon(1+\upsilon)]/E + (k_n-k_s)/(2Sk_nk_s)}{[(1+\upsilon)(1-\upsilon)]/E + (k_n+k_s)/(2Sk_nk_s)} \quad (5)$$

Eq. (5) is graphed for several values of the ratio $E/(S.k_n)$ in Figure 3 for the case of $\upsilon = 0.2$. Also plotted are the results of *UDEC* simulations. For low values of $E/(S.k_n)$, the Poisson effect of a rock mass is dominated by the elastic properties of the intact rock. For high values of $E/(S.k_n)$, the Poisson effect is dominated by the jointing.

4.4 Distinct element modeling of the rock slope at northern abutment of Gotvand dam

For discontinuum modeling of the northern abutment rock slope, UDEC code has been used to simulate both intact rock and discontinuity behavior under different conditions.

Table 3 has listed the rock material properties utilized in UDEC model. Also Table 4 has presented discontinuity properties existing in regions used in UDEC model.

In this paper, the slope has simulated into two main regions. First is the disturbed Bakhtyari region and second is the in situ Bakhtyari formation. Disturbed Bakhtyari is a huge rock mass that include the main body of the slope and because of its three main discontinuity sets, it has the principle role in the probable instability occurrence that should be modeled. Therefore, the model has been made by D.BK region that includes of multiple discontinuity sets and BK$_S$ that is made of continuous rock masses.

Three models have been made of rigid, elastic and elastic-plastic blocks and it can be seenv that rigidity

Table 4. Discontinuity properties, utilized for distinct element UDEC modeling.

Discontinuity	Jkn (GPa/m)	Jks (GPa/m)	Cohesion (kPa)	Friction angle (deg)	Dip (deg)	Spacing (m)
Joint set 1	1.7	0.17	20	30	315	1–3
Joint set 2	1.7	0.17	20	30	43	2–4
Bedding	1.3	0.13	20	30	34	1–3

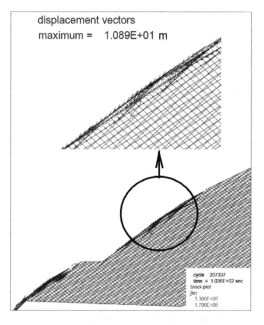

Figure 4. Moving blocks along bedding strata due to earthquake loading, showing displacement vectors.

of blocks has no effect on model results, just as which can be seen from method described in section 4.3.

Built models in distinct element code indicate that slope will have a reliable stability under natural conditions. But earthquake loading can cause some instable planes in surface places that will move down through middle and toe of slope. These surface planes consist of blocks above surface bedding strata. Figure 4 illustrates moving blocks under seismic loading.

5 CONCLUSION

In this study, it is demonstrated that in complex rock slopes under different conditions, some local instable blocks may be formed that simple continuous methods are not able to analyze. As it is observed, limit equilibrium analysis of the current rock slope showed that it will resist under earthquake loading. But discontinuous modeling revealed that surface bedding strata will start to move due to occurrence of earthquake.

Therefore, consideration should be given that limit equilibrium techniques should be used in conjunction with numerical modeling to maximize the advantages of both.

REFERENCES

Eberhardt, E., Kaiser, P.K. & Stead, D. 2002. Numerical analysis of progressive failure in natural rock slopes. In: da Gama, Dinis, e Sousa, Ribeiro (Eds.), EUROCK 2002—Proc. ISRM Int. Symp. on Rock Eng. for Mountainous Regions, Funchal, Madeira. Sociedade Portuguesa de Geotecnia, Lisboa, pp. 145–153.

Eberhardt, E. 2003. Rock Slope Stability Analysis—Utilization of Advanced Numerical Techniques Geological Engineering/Earth and Ocean Sciences, UBC, 6339 Stores Rd:4–5, 8–9 & 23–29.

Geo-Slope, 2000. Geo-Slope Office (Slope/W). Geo-Slope International Ltd., Calgary, Canada.

Itasca, 2004. Itasca Software Products—UDEC. Itasca Consulting Group Inc., Minneapolis.

Mahab Ghodss Consulting Engineers, 2007. Overall Review of Geological and Geotechnical Characteristics of Gotvand Dam Site.

Stead, D., Eberhardt, E. & Coggan, J.S. 2006. Developments in the characterization of complex rock slope deformation and failure using numerical modeling techniques. Engineering Geology 83: 217–235.

Rock block sliding analysis of a highway slope in Portugal

P.G.C. Santarém Andrade & A.L. Almeida Saraiva
Earth's Science Department of Science and Technology Faculty, University of Coimbra, Portugal

ABSTRACT: This work analyses the stability problems that exist on a road slope located in the central part of Portugal, at km 55 of the highway IP3. A joint-study recorded the main characteristics of the joint-sets at two separate sections of the slope. To assess the relative stability and potential unstability in the rock slope the authors used several methods of analysis. A kinematics analysis allowed an identification of the discontinuities which are likely to cause instability problems. The studied slope showed frequent rock block sliding and sometimes wedge failures. The SMR (Romana) system was used to define the relative susceptibility of the slope's rock unstability. The study of the slope's unstability behaviour is concluded with the evaluation of the factor of safety using the Mohr-Coulomb and Barton & Bandis models.

1 INTRODUCTION

This paper analyses a road slope located in the central part of Portugal, showing unstability problems. The most frequent or the worst unstability situations that have occurred were identified. It was important to characterize the different parameters that influenced the rock slope stability. The rock mass classification Slope Mass Rating (SMR) proposed by Romana (1985) system was used to identify the susceptibility of the rock slope to failure. When using the SMR system, the favourability of joint orientations regarding slope orientations must be defined.

For the unstable sectors, the factor of safety (FS) was determined in accordance with the Mohr-Coulomb and Barton & Bandis models.

2 GEOLOGY

The slope is located at km 55 of the IP3 (Itinerário Principal n°3), it is 150 meters long and about 21 meters high (Figure 1). The studied slope is a 75° road cutting slope and has a strike of N80°W. It is situated in the Buçaco Ordovician—Silurian synclinal and is mainly constituted by impure quartzites and also by schists and carbonaceous schists. The later show a black tonality and sometimes present a plastic behaviour. They are very to completely weathered (W4 to W5) and disintegrate easily. The quartzites are the predominant rocks, while schists and carbonaceous schists are mainly located at the slope's toe.

The quartzites are less weathered than the carbonaceous schists. They show whitish, purplish and greyish colours and are constituted by more than 75% of quartz. The quartzites show a granoblastic texture. Other minerals of relative importance are present: white micas (sericite and muscovite), albite, orthoclase, pyrite and iron oxides. The quartzites have pyrite minerals deposits mostly very weathered.

Near to the slope surface the quartzites are more fractured and more weathered (W4) and have a F3 spacing.

The physical and mechanical characteristics of these schists and quartzites can be found on a previous work by Santarém Andrade & Almeida Saraiva (2000).

3 METHODOLOGY

In a stability analysis the discontinuities orientation must be determined initially in order to foresee and prevent situations of unstability.

A graphic kinematics analysis was carried out for the different stability problem situations. Rock blocks slides are frequent and they are related with three major joints families (Figure 1). The occasional wedge failures occur at the main joint intersections.

The rock blocks are originated by the intersections of the main joint sets: stratification joints (N85°E; 75°N), (N85°W; 31°S and N9°E; subvertical).

The rock slope's stability behaviour was studied using a geomechanical classification. A classification adopted from the Romana system (1985, 1996)—Slope Mass Rating (SMR) was used to define the slope's unstable sectors. The discontinuities characteristics, slope's orientation, ground water conditions, rock material strength and rock blocks dimensions are the most important parameters of the SMR system.

Figure 1. Rock blocks slides of the studied slope.

This classification was elaborated from the Bieniawski's Rock Mass Rating (RMR) (1979, 1989). The SMR determination is defined on Equation 1:

$$SMR = RMR - (F_1 \times F_2 \times F_3) + F_4 \quad (1)$$

where F_1, F_2 and F_3 = adjustment factors related with the joint-slope orientation; and F_4 = factor related with the slope's method of excavation. The different factors allowed the RMR correction.

The SMR values vary between 0 and 100 and can be classified in five stability groups (Romana, 1985). The values below 21 can be assumed as completely unstable and the values above 80 are classified as completely stable. The SMR system can be executed on a previous study phase to identify possible unstability situations and to indicate support measures. The SMR classification considers planar and toppling failures. The wedge failures are classified as a special type of plane failures.

The present study only analysed only failures with dimensions superior to 0,005 m³. The determination of the rock blocks volumes included the discontinuity spacing and the angles between the joint sets. The spacing for each discontinuity set was defined as the modal value of the considered intervals for each family.

The analysis of the stability of the slope can be done through deterministic and/or probabilistic methods.

In a first stage the factor of safety (FS) of the unstable sectors was determined in accordance with the Mohr-Coulomb model. The block slides were analysed considering the gravity forces and the factor of safety was obtained by the Equation 2:

$$FS = \frac{(cA + (W \cos \psi) \tan \phi)}{W sen \phi} \quad (2)$$

where FS = factor of safety; c = apparent cohesion; A = area of the block base; W = weight of block; ψ = slope angle and ϕ = friction angle.

The rock block's weight is related with its dimensions and its specific weight.

Unstability problems are generally associated with the presence of water and, in this case, the model of rupture, based on the Mohr-Coulomb equation (Equation 3), includes a perpendicular fracture to the landslide surface, which was verified "in situ". The presence of water causes an accentuated reduction of the rock blocks stability and causes sliding.

$$FS = \frac{(cA + (W \cos \psi - U_w) \tan \phi)}{(W sen \psi + V_w)} \quad (3)$$

where U_w = force related with the water flowing at the surface between the block and its base and V_w = force related with the water filling the subvertical discontinuity plane.

The studied slope corresponds to a structure located on the superficial part of the rock masses. The rock's deformability is not an important aspect.

The friction angle values were obtained through rock joint shear tests. Some of these tests were performed under low normal stresses (smaller than 0.2 MPa) to reproduce as close as possible the situation that occurs "in situ".

To execute the joint shear test several blocks were carefully removed from the slope using picks and hammers not to damage the discontinuity surfaces.

The FS obtained through the Barton & Bandis model (1990) was also determined. This model implies the definition of the JRC (Joint Roughness Coefficient) values.

4 RESULTS

The authors proceeded to evaluate the different stability problems that occur in the sections of the slope. The SMR (Romana) values (Table 2) were defined and compared with the unstability behavior observed "in situ" in the last 6 years (Tables 1–2). From the structural geologic survey and kinematics analysis it was verified that slope unstability is mainly related with rock blocks slides.

The slope was divided in 2 sections 50 and 100 meters long in order to use the SMR classification and compare its results with the "in situ" values. The geologic-structural survey and the laboratorial and "in situ" tests of the geotechnical characterization allowed the definition of the SMR classification parameters.

The SMR ("in situ") value corresponds to the situation of unstability or rupture that was registered during six years.

Table 1. Unstability reported situations and SMR ("in situ") values.

Slope section	Unstability situations	SMR ("in situ")
Slope 1 - a	Plane failure, rock falls and small wedge failures	55
Slope 2 - b	Plane failure, rock falls and small wedge failures, the unstability situations are more frequent than in Slope 1-a	45

Table 2. SMR (Romana) and SMR ("in situ") values.

Slope section	SMR (Romana)	SMR ("in situ")
Slope 1 - a	47	55
Slope 1 - b	34	45

The Slope 1 - a was classified as partially stable and the Slope 1 - b was defined as unstable according to the SMR (Romana) system. The SMR obtained "in situ" for both sections were classified as partially stable.

The differences between SMR ("in situ") and SMR (Romana) are 7 and 8 positive points, confirming that the SMR (Romana) system values can be superior to the observed "in situ" and are conservatives for the time period considered. These differences can be explained by the SMR values, as they were calculated for a longer period of time of the slope's useful life, whereas the situations of unstability analysed in this work were only observed during six years.

The most problematic unstability sites were defined through a kinematics analysis: the Markland test and its refinement by Hocking (1976); and using a geomechanical classification: the SMR (Romana) system. The FS values for these sites were then determined by Mohr-Coulomb and Barton & Bandis models. The FS was established in order to evaluate the adequacy of design.

In table 3 we can observe a great variety of FS results obtained from Equation 1, the ϕ maximum and the ϕ minimum obtained from the rock joint shear tests were respectively 45.5° and 34.8°.

The FS results showed higher values when the apparent cohesion was considered. When it was assumed that the joints do not display any apparent cohesion, an accentuated decrease of the FS (FS < 1) was verified, with values that assumed immediate failure. The FS results for the same range of Joint Roughness Coefficient (JRC) values tend to diminish with the increase of the ϕ.

Table 3. FS results from Equation 1.

Joints safety c	Factor of safety considering c		Factor of safety without considering c	
	$\phi = 45.5°$	$\phi = 34.8°$	$\phi = 45.5°$	$\phi = 34.8°$
Quartzites	3.62–1.54	2.45–1.27	0.27–0.20	0.24–0.19

Table 4. FS results from Equation 3.

Joints	Factor of safety considering c	
	$\phi = 45.5°$	$\phi = 34.8°$
Quartzites	2.15–0.69	1.40–0.54

Table 5. JRC results of the quartzites discontinuities.

Joints	JRC	
	JRC maximum	JRC minimum
Quartzites W2	14	9
Quartzites W3	11	7

The situations of unstability are associated with the presence of water. The occurrence of important rock slope unstability is generally connected with higher rainfall periods. Considering these situations, the FS values were also calculated by Equation 3.

The failure model of Equation 3 approximately includes a perpendicular fracture to the landslide surface, which is similarly verified "in situ".

With the introduction of the influence of water in the calculation of the FS, a clear reduction of these values was verified, as can be observed in table 4, where some FS results are inferior to 1. The results revealed that water action causes an increase of the rock blocks unstability.

The JRC values for the quartzites discontinuities of slope 1 are presented in table 5.

The JRC related FS values were obtained according to the Barton & Bandis model (1990) (Equation 4):

$$FS = \frac{W \cos\psi \left[\tan(JRC \log_{10}(JCS/\sigma_n') + \phi_r)\right]}{W \operatorname{sen}\psi} \quad (4)$$

where JCS = joint compressive strength; σ_n' = normal stress acting on a plane on which the shear strength is mobilized and ϕ_r = residual friction angle.

Table 6. FS results by Equations 4 and 5 for quartzites W2 discontinuities.

Joints	Factor of safety	
	JRC = 14	JRC = 9
Quartzites W2	0.45	0.33

Table 7. FS results by Equations 4 and 5 for quartzites W3 discontinuities.

Joints	Factor of safety	
	JRC = 11	JRC = 7
Quartzites W3	0.37	0.29

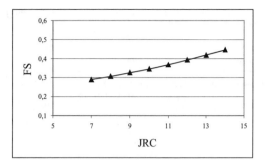

Figure 2. Relation between the FS defined by the Equations 4 and 5 and the JRC values.

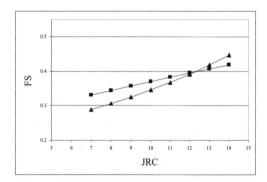

Figure 3. Relation between the FS defined by the Equations 4 and 5 (■ - With scale effect; ▲ - Without scale effect for a joint profile of 0.4 m) and the JRC values.

The blocks slide occurs in the upper part of the rock mass where normal stresses are very reduced, with situations where $(JCS/\sigma'_n) > 100$. For these normal stresses, the Equation 5 allowed a more conservative FS determination based on the Barton & Bandis model (1990). Authors such as Vallejo et al. (2002) consider the Equation 5 very useful to avoid extremely high friction angles obtained from Equation 4 when low normal stresses are used.

$$\phi = \phi_r + 1.7 JRC \qquad (5)$$

The results of the FS based on Equations 4 and 5 were similar to those obtained using the Mohr-Coulomb model when the values of the apparent cohesion were considered null. The results show a direct proportional relationship between the FS and the JRC values, which translates the influence of the increase of roughness in the reduction of the verified displacements. The FS values correspond to situations of instability, with values lower than 1.

The JRC and JCS values tend to decrease with the increase of the discontinuities extension.

To analyse the scale effect on the FS determination, the Equations 6 and 7 defined by Bandis et al. (1981) were used:

$$JRC_n = JRC_0 [L_n/L_0]^{-0,02 JRC_0} \qquad (6)$$

$$JCS_n = JCS_0 [L_n/L_0]^{-0,03 JCS_0} \qquad (7)$$

where JRC_o = joint roughness coefficient for samples with approximately 10 cm of length profile; JCS_o = joint compressive strength for samples with approximately 10 cm of length profile; L_n = joint length profile considered "in situ" and L_0 = sample length profile (10 cm).

In the present study, it was assumed that the sliding rock blocks have approximately 40 cm of length profile and the relation $[L_n/L_0]$ is equal to 4.

An adaptation of the Barton & Bandis criterion, considering the scale effect, is represented in the Equation 8:

$$\tau = \sigma \tan[JRC_n \log_{10}(JCS_n/\sigma'_n) + \phi_r + i_r] \qquad (8)$$

where i_r = angle obtained through the large-scale roughness profiles defined by Hack et al. (2003).

The ratio between L_n and L_0 for the studied joints was not very high. Consequently the FS values with scale correction were not very different from those obtained without scale correction. The Figure 3 shows a distinction, not very accentuated, between the different results. When the scale effect was considered a small increase of the FS for the lower JRC values was verified, while for the higher JRC values a reduction of the FS results was observed.

In order to verify the scale effect on a discontinuity surface of greater extension (approximately 2 meters),

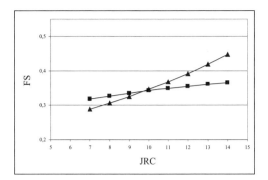

Figure 4. Relation between the FS defined by the Equations 4 and 5 (■ - With scale effect; ▲- Without scale effect for a joint profile of 2.0 m) and the JRC values.

the FS with and without scale effect was determined wile the other parameters values remaining constant (Figure 4).

The results in Figure 4 showed a reduction of the FS calculated with scale effect for the majority of the JRC values. This reduction is more evident for the higher JRC values.

In most of the studied situations of unstability, it was verified a reduction of the tangential resistance and FS values when the relation $[L_n/L_o]$ was increased and the scale effect influence was considered.

5 CONCLUSIONS

The slope sections were defined from partially stable to unstable according to the SMR (Romana) system. The SMR obtained "in situ" was slightly different as both sections were considered as partially stable.

The SMR system must be applied with some limitations and it can be used in rock slopes to define the main types of unstability that can occur as well as their degree of intensity. One of its advantages is to be a relative fast method which prevents the execution of an expensive and time consuming characterization. The SMR and its possible adaptations do not constitute the exclusive methods of rock slopes stability study. Other geomechanical classifications and more accurate methods of analysis should be used.

Despite the similarity of the results, between the FS values obtained through Barton & Bandis (1990) model (Equations 4 and 5) and Equation 5 and those obtained through the Mohr-Coulomb model when the apparent cohesion was considered null, to exclusively establish the FS derived from the JRC values is not recommended. The FS values must be confirmed whenever it is possible, through the results of joint shear tests.

The FS values defined through the Equations 4 and 5 and the Mohr-Coulomb model when the apparent cohesion was considered null seemed very conservative. A more suitable approach was provided by the Mohr-Coulomb model considering the presence of water: the FS results varied between 1.40 and 0.54 for the ϕ minimum.

REFERENCES

Bandis, S.C., Lumsden, A.C. & Barton, N.R. 1981. Experimental studies of scale effects on the shear behaviour of rock joints. *International Journal of Rock Mechanics and Mining Science and Geomech.*, 1 (18): 1–21.

Barton, N.R. & Bandis, S.C. 1990. Review of predictive capabilities of JRC-JCS model in engineering pratice. *Proc. International Conference on Rock Joints, Loen, Norway*: 603–610. Rotterdam: Balkema.

Bieniawski, Z.T. 1979. The geomechanics classification in rock engineering applications. *Proc. 4th Congress International on Rock Mechanics, Montreux, Swiss*: 51–58. Rotterdam: Balkema.

Bieniawski, Z.T. 1989. *Engineering rock mass classification*. Chichester: John Wiley & Sons.

Hack, R., Price, D. & Rengers, N. 2003. A new approach of rock slope stability—a probability classification (SSPC). *Bulletin of the International Association of Engineering Geology and the Environment* 62 (2):167–184.

Hocking, G. 1976. A method for distinguishing between single and double plane sliding of tetrahedral wedges. *International Journal of Rock Mechanics and Mining Science and Geomech.*, 9 (13): 225–226.

Romana, M. 1985. New adjustment ratings for application of Bieniawski classification to slopes. *Int. Symposium on the Role of Rock Mechanics, Zacatecas, México*: 49–53. International Society of Rock Mechanics.

Romana, M. 1996. The SMR geomechanical classification for slopes: A critical ten-years review. *Proc. 8th International Conference and Field Trip on Landslides, Granada, Spain*: 255–267. Rotterdam: Balkema.

Santarém Andrade, P.G.C. & Almeida Saraiva, A.L. 2000. Physical and mechanical characterization of rock material in excavation slopes. *Proc. 7th National Geotechnics Congress, Porto, Portugal*: 311–318. Lisbon: SPG. Porto: FEUP. (in Portuguese).

Vallejo, G., Ferrer L., Ortuño, M., Oteo, C. 2002. *Ingenieria Gelogica*. Madrid: Prentice Hall.

Contribution to the safety evaluation of slopes using long term observation results

J. Barradas
Laboratório Nacional de Engenharia Civil, LNEC, Lisboa, Portugal

ABSTRACT: The existence of patterns of behaviour in some slopes, relating the accumulated precipitation with movements along slip surfaces, is illustrated by the results of the long-term observation of two slopes. The increments of sub-horizontal displacements, along deep slip surfaces, in excavated natural soil slopes, obtained using inclinometers, along more than ten and twenty years, are directly correlated with the accumulated amounts of rain by means of polynomial functions. These relations provide the basis for simple methods of evaluation of the need for improvement of the stability conditions and of the performance of new stabilization works. If they are associated with methods of climatic forecast, an evaluation of the medium and long term safety conditions and forecasts of future behaviour can be done as well. These methods require the availability of medium or long term good quality observation results, but they do not require permanent monitoring or even very frequent measurements.

1 INTRODUCTION

The evaluation of the behaviour and the stability conditions of instrumented slopes after construction are frequently difficult when long term movements are detected. In some cases, the same slip surfaces that existed during construction go on being active during the lifetime of the infrastructure for which the slopes were created or modified. Such movements are usually irregular and irreversible and they are highly dependent on the intensity and distribution of rainfall in the vicinity of the slope, particularly if other causes such as new excavations, other works, leakage of pipes, earthquakes, etc. do not intervene.

We have in these cases slow moving active landslides and many times the most economic solution is to live with the movements associated with the slip surfaces. However this implies a strong capacity of prediction of future movements and trends which usually is not available.

In these cases, models of the same kind as those used in the design of the geotechnical works, although useful for the control of safety, are usually not enough for the establishment of criteria to evaluate the behaviour of the works, in order to manage properly the infrastructure affected by the slope.

Studies presenting correlations of rainfall with the behaviour of slopes deal more frequently with the occurrence of landslides or present comparisons with displacements without any explicit function of correlation (Picarelli & Russo 2004).

This paper deals with cases of excavated natural soil slopes in which there is no apparent progressive acceleration of movements, that are instead irregular (of very different amount from one year to the other), discontinuous and irreversible). It has been found that there are correlations between the increments of sub-horizontal displacements, along deep slip surfaces, obtained using inclinometers, and the accumulated amounts of rain in the vicinity of the slopes. These correlations express a kind of pattern that represents the behaviour of the slope, so long as there are no changes in the stability conditions or other relevant environmental actions (apart from the rainfall).

The cases of two slopes are presented that show the widespread pattern of behaviour (obtained in a set of inclinometers), the change of pattern when a significant change of the stability conditions (improvement in that instance) was achieved and the polynomial function that best represents the correlation. These slopes have been monitored along more than ten and more than twenty years.

Some applications of this method as well as its requirements and main limitations are briefly presented. These uses include a contribution to the safety evaluation of the slopes and to the decision making involved in the management of the slope and the infrastructure affected by it. From this method, some gains in the efficiency of the execution of the slope monitoring can also be obtained.

2 SLOPE IN THE REGION OF LISBON

2.1 Presentation

The enlargement of a motorway at the lower part of a natural slope, 35 m in height, required the execution of its excavation, in 1994–95. The same slope had already been excavated when the motorway was inaugurated (at the beginning of the 1960 decade).

In figure 1, a typical section of the slope is presented, with the geometry after the excavation, the stabilization works executed until 1995 and the inclinometric tubes installed in that section (I3, I6, I7).

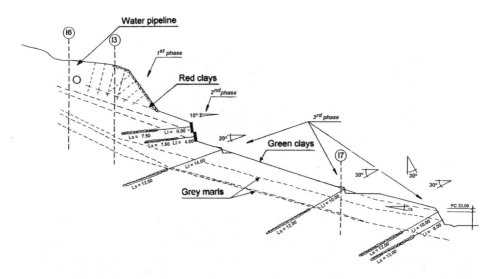

Figure 1. Slope in the region of Lisbon. Cross section with stabilization works.

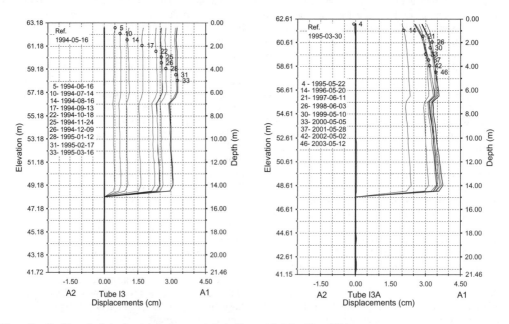

Figure 2. Profiles of sub-horizontal displacements in inclinometric tubes I3 and I3A.

2.2 Observation results and additional works

In figure 2, profiles of sub-horizontal displacements, obtained in inclinometers I3 and I3A, in the direction closer to the steepest direction, are presented. The second of these tubes replaced the first when this became obstructed by excessive distortions.

The displacements in tube I3 were obtained during the execution of a large part of the excavation and stabilization works and those of tube I3A were measured after these works.

Figure 2 shows that the same slip surface was active during the excavation and afterwards.

Since the end of construction, in almost all the inclinometers, from the beginning of the rainy season of the rainiest years, there were increments of the sub-horizontal displacements. In most cases these corresponded to distortions along a slip surface located inside a formation of grey clayey marls (Barradas 2003) which had been already active during the execution of the excavation (see fig. 1).

Figure 3 shows evolutions in time of the resultants of sub-horizontal displacements measured in three pairs of inclinometers (they are in pairs for the same reason as inclinometers I3 and I3A) and of the load in a typical anchorage of those that are affected by the deep ground movements. The similarity of the evolutions is striking, except, in the case of the anchorage, where some decrease in tension after each new increment can be observed.

In 1996 and in 1997, a reinforcement of some anchored works (beams and abutments) was executed, with new anchors (in the first year) and (mostly in 1997) there was an improvement of the deep drainage system (construction of sub-horizontal linear drains, drainage wells and vertical linear drains).

In the calculation of sub-horizontal displacements, a bias error correction (Mikkelsen 2003) was executed.

2.3 Correlation between the observed behaviour, the rainfall and the slope conditions

Figures 4 and 5 show comparisons of the accumulated quantities of rainfall in the rainiest three consecutive months of those years that were more rainy (of the periods 1995–2003 and 1998–2003, respectively) with the increments of sub-horizontal displacements in the rainy season measured in the inclinometers. The results of precipitation published by the Meteorological Institute, for a station in Lisbon (Gago Coutinho) were used. In these figures, each rainy season is identified by the year of its end.

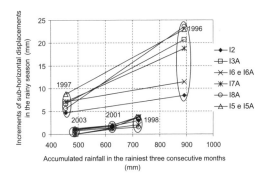

Figure 4. Accumulated rainfall in the rainiest three consecutive months vs. increments of sub-horizontal displacements in the rainy season.

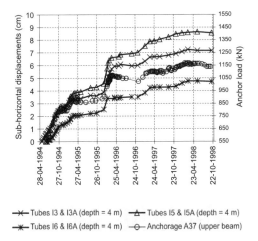

Figure 3. Inclinometric tubes I3, I3A, I5, I5A, I6, I6A and anchorage A37. Evolution of sub-horizontal displacements and anchor loads.

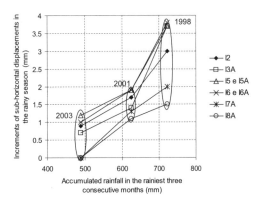

Figure 5. Accumulated rainfall in the rainiest three consecutive months vs. increments of sub-horizontal displacements in the rainy season.

It can be seen that there is a very clear positive correlation between the above mentioned accumulated quantities of rainfall and the referred increments of sub-horizontal displacements in all inclinometric tubes presented, which are all that were installed in the main part of the slope. On the other hand, the consequences of the works implemented in 1997 seam to have been remarkable, in reducing the detrimental effects of the rainiest periods, according to the correlations shown in figure 4.

The results of figures 4 and 5 also suggest:

- a difference in the behaviour pattern between different parts of the slope, since inclinometers I2, I6 and I6A (which had a different behaviour in the first two years) and inclinometers 7A and 8A (which had a different behaviour later) are in parts of the slope (lateral and rear part, the two first and lower part the two last) that are different from the other inclinometers and they had a different response;
- the apparent non-linearity of the relation accumulated rainfall/displacements specially in the case of the upper part of the slope.

2.4 Evaluation of the safety conditions

Figures 4 and 5 show the improvement of the behaviour pattern of the slope after the works executed in 1997. However, due to:

- the irreversible and detrimental character of certain aspects of the behaviour observed in the first service years of the excavated slope (in which permanent increments of displacement occurred, including the upper part of the slope, where an important water pipeline was buried, and affecting a large number of the anchorages);
- the continuation of such behaviour, although attenuated, after the additional works of 1996 and 1997;
- and the state of overload that affected already many anchorages of the slope (probably all of them in certain parts of the slope),

the safety conditions of the slope were deemed insufficient, considering the risks arising from the proximity of two important infra-structures (the motorway at the base of the slope and the water pipeline at the top) and it was decided to study a new solution for the slope (Barradas 2005). This has already been implemented. It included a substantial amount of earthwork that changed the profile of the slope (reducing its height) and the construction of a by-pass for the pipeline.

3 SLOPE IN THE REGION OF COIMBRA

3.1 Presentation

In 1980, during the construction of a section of A1 Motorway, in Coimbra region (200 km to the North of Lisbon), some landslides of considerable size occurred when the toe of a natural slope, about 45 m in height, was excavated.

The upper part of this slope includes colluvium deposits reaching a considerable depth (20–25 m). They consist mainly of clayey silt and silty sand, with some gravel. The weaker materials (clayey silt) had an estimated $\varphi' = 13°$. Underneath and behind the colluvium there are several "in situ" layers of quaternary and cretacic age.

In this case the stabilization works consisted in the placement of rockfill masses and drainage works, including wells and linear drains (Barradas 1996).

The monitoring system included about 20 inclinometric tubes.

3.2 Observation results

Figure 6 shows the evolution in time of sub-horizontal displacements, from 1987 to 2006, in one of the inclinometers in which deep seated movements have been observed: tube 20A.

The deepest level of which results are presented in figure 6 (11.5 m) is immediately above the slip surface intersection with tube 20A.

In the calculation of sub-horizontal displacements, a bias error correction (Mikkelsen 2003) was executed.

3.3 Correlation between the observed behaviour and the rainfall

Figure 7 shows the existing positive correlation between the accumulated rainfall and the increments of sub-horizontal displacements in the rainy season measured in the inclinometer 20A.

The accumulated rainfall (obtained from data of Coimbra meteorological station) is computed in two ways:

- quantities of precipitation occurred in the rainiest three consecutive months of those years that were more rainy;

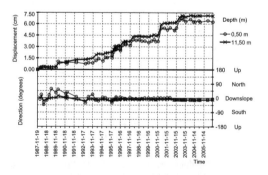

Figure 6. Evolution of sub-horizontal displacements since November 1987 until October 2006, relative to the observation of 1987-11-19.

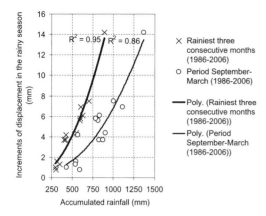

Figure 7. Comparison of accumulated rainfall in the rainiest three consecutive months and in the period September–March with the increments of sub-horizontal displacements in the rainy season, between 1986 and 2006, in tube 20A.

– values accumulated from September of each year to March of the next year (seven months).

In figure 7, the quadratic regression functions for both set of data are presented, as well as the correspondent value of R^2.

When accumulated rainfall of the three rainiest consecutive months is considered a minimum threshold of 300 mm has been found, for movements in the slope to be detected.

No correlations similar or comparable to those that are presented in figure 7 can be obtained, in the case of this slope, comparing piezometric observation (instead of accumulated rainfall) with displacements, given the fact that no frequent measurements of water levels were generally available. On the other hand, even if they existed, they would not allow the same kind of assessment of the behaviour of the slope, since the evolution of the piezometric level is already part of the response of the slope.

In figure 7, it can be seen that:

– The quadratic polynomial function used for the regression expresses well the correlation when accumulated rainfall of the rainiest three consecutive months is considered ($R^2 = 0.95$, for the period 1986–2006).
– The same type of function expresses a worse correlation when the values of accumulated rainfall from September to March of each year are considered ($R^2 = 0.86$, for the period 1986–2006).

The results of R^2 for several types of polynomial functions that were used in the regression of the results of tube 20A, obtained between 1986 and 2006, are presented in table 1.

It is also interesting to remark that, in those years in which there were two periods of three consecutive months that were rather rainy and in which the accumulated rainfall values (for three months) did not differ much from each other (the second value being still above 300 mm), the respective data do not agree so well with the general regression function. There were several of these years before 1986; that is the reason why this year was chosen for the beginning of the period considered so far. Apparently there has been a trend in the region, during the last decades, towards a certain concentration of the rainiest period (when several months are considered). This tendency improves the applicability of the proposed method.

Although the same trend and pattern of behaviour can be observed, in general, in the inclinometric tubes of this slope in which a slip surface was intersected and there are long term movements, the correlation between rainfall and displacements is not always as good as in inclinometer 20A. This can be seen in figure 8, in which a comparison is presented, with the same criteria as above, but in this case for tubes 20A and 17A, and for the period 1994–2006, and only for the rainiest three consecutive months of those years

Table 1. Comparison of accumulated rainfall with the increments of sub-horizontal displacements in the rainy season, between 1986 and 2006, in tube 20A. Values of R^2 for several types of polynomial functions.

Type of polynomial function	Linear	Quadratic	Third degree
Rainiest three consecutive months	0.92	0.95	0.98
Period September–March	0.82	0.86	0.88

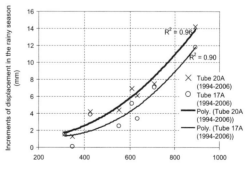

Figure 8. Comparison of accumulated rainfall in the rainiest three consecutive months with the increments of sub-horizontal displacements in the rainy season, between 1994 and 2006, in tube 20A and 17A.

that were rainiest. Tube 17A, for which the displacements measured immediately above the slip surface are used (depth = 12.5 m), is located in the southern part of the slope, in a cross section that is different from the one of tube 20A (both tubes are about 100 m apart). The obtained results of R^2 (see fig. 8) were 0.96 (for tube 20A) and 0.90 (for tube 17A).

The observation results of tube 17A, although important for the part of the slope in which it is located, seam to be of a lesser quality than inclinometer 20A (as is also apparent comparing the displacement profiles of tube 17A, obtained in successive observations, which show some irregularities in the part that is subjected to the largest distortions).

3.4 Evaluation of the safety conditions

The near permanence of the relations between accumulated rainfall and displacements that are illustrated in figures 7 and 8 do not show a trend towards a worsening of the global safety conditions in the period 1986–2006. However they show that the pattern had no trend to an attenuation of movements with time either. So the detrimental effects, such as continuing damages (although irregular and generally at a slow rate) in infra-structures and houses were expected to continue.

The safety conditions were improved in 2006–2007 when a new deep drainage system including wells and automatic pumping was installed. Giving the absence of anchorages, there is no problem of their overload, as was the case in the slope previously presented (in section 2).

4 CONCLUSIONS

From what has been presented, the following conclusions can be drawn:

1. In cases of excavated natural slopes properly instrumented and monitored for long periods, such as those that were presented, it is possible to establish behaviour patterns for the slopes, relating this directly with the accumulated amounts of rainfall in the vicinity of the slopes.
2. For this purpose, the better way that was found to consider the accumulated rainfall was the rainiest three consecutive months and the functions used for regression that better express the correlation are polynomial quadratic.
3. The establishment of such patterns, accompanied by adequate statistical criteria, allows an evaluation, in each new observation year, if there was an improvement or a worsening of the behaviour of the slope and so, indirectly, also an assessment of its stability conditions. The patterns give an overall information related with the real state of the slope, incorporating the influence of aspects that exist usually only in a fragmentary and disperse way (such as those related with the performance of the drainage system and the hydro-geological conditions) or may be even inexistent (such as the loss of strength of the soil subject to progressive distortions along the slip surfaces). The evaluations and assessments that were mentioned will be of great interest in the decision making involved in the management of the slope and the infrastructure affected by it.
4. The quality of the observation results affects, as it would be expected, the quality of the correlations that can be established. For each slope, a choice of the more reliable (and representative) monitoring devices should be done, in order to make use of the proposed method.
5. Associating this method with statistical methods of climatic forecast (distribution of rain) it will be possible to evaluate the medium and long term safety conditions of the slope and to forecast its future behaviour.
6. From the use of this method, some gains in the efficiency of the execution of the slope monitoring can also be obtained.
7. This method requires the availability of medium or long term good quality observation results (including several years in which the accumulated quantities of rainfall were larger then average), but they do not require permanent monitoring or even very frequent measurements (it can be applied with only three or four surveys per year).

REFERENCES

Barradas, J. 1996. Long term behaviour of a natural soil slope excavated at the toe. In *Proc. Seventh International Symposium on Landslides, vol.3*. Trondheim, Norway. Balkema.

Barradas, J. 2003. Behaviour and stabilization of a natural slope excavated at the toe. In *Field Measurements in Geomechanics, Proceedings of the 6th International Symposium FMGM 2003*. Oslo, Norway. Balkema.

Barradas, J. 2005. Comportamento e estabilização da escavação de um talude natural. In *2.as Jornadas Luso-Espanho-las de Geotecnia. Modelação e segurança em Geotecnia—LNEC*. Lisbon, Portugal (in Portuguese).

Mikkelsen, P.E. 2003. Advances in inclinometer data analysis. In *Field Measurements in Geomechanics, Proceedings of the 6th International Symposium FMGM 2003*. Oslo, Norway. Balkema.

Picarelli, L. & Russo, C. 2004. Remarks on the mechanics of slow active landslides and the interaction with man-made works. In *Landslides: Evaluation and stabilization. Proceedings of the 9th International Symposium on Landslides*. Rio de Janeiro, Brazil. Balkema.

Landslides and Engineered Slopes – Chen et al. (eds)
© 2008 Taylor & Francis Group, London, ISBN 978-0-415-41196-7

Delimitation of safety zones by finite element analysis

J. Bojorque, G. De Roeck & J. Maertens
Dept. of Civil Engineering, K.U. Leuven, Leuven, Belgium

ABSTRACT: This paper deals with the determination of safety zones and local minima by using finite element analysis. Used is made of finite element method in order not to constraint the analysis by neither the assumptions in the location of the sliding surface nor in the interslice force function. It is known, that only the most critical failure mechanism and global minimum are evaluated by the strength reduction method, in such approach local minima most of the time are unnoticed. Here, it is proposed that the safety zones and the local minima can be detected by keeping the information generated in the strength reduction process. In addition, the importance of soil properties in the location of the failure mechanism is highlighted. The methodology is presented in an artificial case study and in a real natural slope. The safety zones should be considered in landslide stabilization and remediation.

1 INTRODUCTION

Numerical methods have been recognized as a powerful tool for practical geotechnical applications. These techniques have become important for slope stability analysis, especially when complex stratigraphy and complex soil behaviour are treated. Moreover, the use of the strength reduction method (SRM) to define the stability of a slope has shown some advantages over traditional methods (Griffiths & Lane 1999, Dawson & Drescher 1999), among others. On the other hand, one limitation of the strength reduction approach is that only the most critical failure mechanism and global minimum are evaluated (Cala et al. 2004; Cheng et al. 2007). Therefore, by SRM, local minima most of the time are unnoticed. In engineering practice apart from the global factor of safety and the critical sliding surface associated with it, it is important to detect the local minima.

In classical limit equilibrium analysis, local minima are defined by evaluating different sliding surfaces associated with different safety factors. A single line to characterize the sliding mechanism can mislead the implementation of remedial measures. Safety maps can be generated by using limit equilibrium concepts to detect safety/unsafety areas. Those maps are represented by a series of contour lines along which minimal safety factors are constant (Baker & Leshchinsky 2001, Renaud et al. 2003). However, limit equilibrium methods need to use key assumptions in other to solve the problem. These assumptions and the disregard of the strain-stress relationship limit their use in some extension.

The modified shear strength reduction technique in the framework of Finite Difference Method (Cala et al. 2004), has been proposed in order to detect several sliding surfaces. The technique is based on gradually reducing the strength properties after identification of the first sliding surface. The modified technique needs extra computations and is not applicable for finite element analysis, causing some numerical problems. In this paper, it is proposed that the strength reduction technique used in the framework of finite element method can still given information regarding different failure mechanisms by keeping the different stages generated during the process. Furthermore, the necessity of detect different failure mechanisms that can arise from small changes in soil parameters is highlighted. Used is made of finite element methods in order not to constraint the analysis by neither the assumptions in the location of the sliding surface nor in the interslice force function. Moreover, the advantages of finite element slope stability analysis can be exploited.

Safety zones and local minima are detected in finite element analysis by keeping all the shear zones defined by using different strength reduction coefficients (Bojorque et al. 2007). The safety zones should be considered for the selection and location of remedial measures for landslide stabilization. The verification of the procedure is presented in an artificial case study acquired form literature in which the development of local minima is indicated. The methodology is implemented, as well, in a case study of a natural slope located in Ecuador where the presence of different failure mechanisms should be evaluated.

2 FINITE ELEMENT METHOD—STRENGTH REDUCTION TECHNIQUE

In the framework of finite element slope stability analysis, the Factor of Safety (FoS) is defined as classical methods. Hence, the factor of safety of a slope is defined as the factor by which the shear strength parameters must be divided in order to bring the slope to failure (Griffiths and Lane 1999). When the same strength reduction factor (SRF) is used for both, the cohesion (c) and tangent of the friction angle ($\tan \phi$), Equation 1 holds.

$$SRF = \frac{c}{c_f} = \frac{\tan \phi}{(\tan \phi)_f} \quad (1)$$

where c_f and $(\tan \phi)_f$ represent the factored parameters.

To compute the factor of safety and its associated failure mechanism, finite element method uses the strength reduction method (SRM). In this process, the cohesion and $\tan \phi$ are gradually reduced or increased until non-convergence or convergence, respectively, in the plastic solution is found. This reduction or increase will depend on both, the initial estimation of the strength reduction factor and the stability of the slope. When small changes in the strength reduction factor produce jumps from convergence to non-convergence in the solution, the factor of safety is determined. At this limit, Equation 1 can be re-written as,

$$SRF \text{ at failure} = \text{Factor of Safety} = \frac{\text{available strength}}{\text{critical strength}} \quad (2)$$

It is noticed, based on Equation 2, that if the initial estimation of the safety factor, represented by SRF, is higher than the factor of safety, the SRM will reduce the soil strength parameters by increasing the estimated SRF. Otherwise, if the initial estimation is lower than the factor of safety, the approach needs to increase the strength parameters.

Before performing the SRM, the initial state of stress should be determined. For cases where the slope is stable, FoS $\geq$ 1, the initial state of stress is normally computed by the gravity loading procedure using the slope own-weight. On the contrary, when the slope is unstable, FoS < 1, the K0-procedure is adopted, with a typical value of $K0 = 1 - \sin \phi$. In this last approach, the vertical stress (σ_v) is determine by the weight of the slope and the horizontal stress (σ_h) is obtained from the relationship $K0 = (\sigma_h/\sigma_v)$. Once the initial stresses are computed, the determination of the factor of safety is obtained by the SRM. The SRM will process different calculations for different strength parameters (strength reduction factor), some of them will be higher than the factor of safety and other lower than this value. By retaining and properly visualization of the different steps, it is possible to detect other failure surfaces that can emerge from the computations. The failure mechanism can be identified by the shear strain contour computed from the results of the SRM. Different failure mechanisms associated with different reduced parameters can be detected and incorporated in safety zones by locating different shear zones (Bojorque et al. 2007).

The finite element program PLAXIS using 15-node elements is used for the slope stability analysis based on the strength reduction method (PLAXIS-BV 2004). The soils are modelled as elastic-plastic material with Mohr-Coulomb failure criterion and zerotensile strength. Vertical and horizontal displacements restricted on the base and horizontal displacements restricted on the sides are used as boundary conditions for the analyses.

3 NUMERICAL EXAMPLE

A two slope angles example is presented to compute the safety zones. The geometrical configuration and soil properties are given in Figure 1. This example is taken from the benched slope case presented in (Cala et al. 2004). The lower slope is inclined 45° and the upper one is inclined 40°.

3.1 Limit equilibrium safety map

A comparative analysis is presented with respect to limit equilibrium calculations in order to validate the results. Bishop's limit equilibrium equations using circular slip surface is adopted. Figure 2 shows the critical sliding surface determined by Bishop's method. Besides, the location of the 100 most unsafe slip surfaces are given.

It is worth noticing that two well defined failure mechanisms are presented. One located in the lower slope, in which the critical slip surface is developed

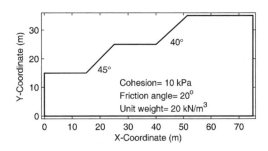

Figure 1. Geometrical configuration and soil properties of the two-angle slope example.

and has a safety factor, associated with it, equal to 0.94. The other mechanism is located in the upper slope with a minimum safety factor corresponding to 0.99. The first 100 sliding surfaces corresponded to factors of safety from 0.94 to 1.06. A better representation of the limit equilibrium results can be given by a safety map (Baker and Leshchinsky 2001; Renaud et al. 2003).

The safety map is constructed by dividing the slope model by a mesh and assigning to each point in a mesh a factor of safety obtained by minimizing the factor of safety between all the slip surfaces going through this point (Renaud et al. 2003). For this example the slope is divided into a rectangular mesh instead of a triangular mesh. For each point in the grid the minimum factor of safety from the nearest slip surface is input. Any limit equilibrium method can be used for the computation of the sliding surfaces, for this example Bishop's method with circular slip surfaces is used. A rectangular mesh spacing of 0.15 m is adopted for the discretization. A filtering value of 1.4 in the safety factor is used to enhance the visualization of the safety map. Figure 3 shows the safety map for the two slope angles example and the critical slip surface is represented by the white hashed line which correspond a factor of safety equal to 0.94.

The darker zones are critical areas where the failure can occur. This map is generated by using 10,000 circular slip surfaces. From this map the determination of safety zones are enhanced by retaining and visualizing all slip surface information.

3.2 Finite element results

For the two slope angles example the FoS determine by FE-SRM is equal to 0.87. This value was computed by performing 60 steps in the SRM. It is noticed that after the fourth step, the critical factor of safety has been reached, the next computations from 5 to 60, are performed to check the stability of the SRM. From step 4 to the last step 60, the plastic points are constant and are located at the lower slope (Fig. 4b). In stage 3, when the strength reduction factor is equal to 0.93, two failure mechanisms are detected (Fig. 4a). If only the last result is considered, as it is typically the case, the second mechanism at the upper slope is unnoticed. This can mislead the implementation of remedial measures. Therefore, the slope stability performance for others strength reduction factors is needed.

In PLAXIS, the initial SRF is computed automatically, this can be a drawback if others values are needed

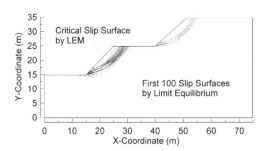

Figure 2. Critical and first 100 most unsafe slip surfaces generated by Bishop's limit equilibrium method. Factor of safety for the critical slip surface equal to 0.94.

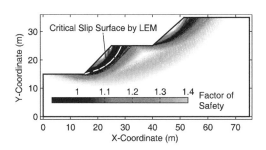

Figure 3. Safety map generated from Bishop's limit equilibrium method. Critical slip surface (white hashed line) correspond to a factor of safety equal to 0.94.

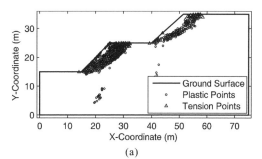

(a)

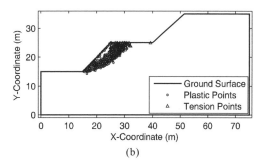

(b)

Figure 4. Plastic and tension points generated during the strength reduction method. (a) Stage 3, strength reduction factor equal to 0.93. (b) Stage 4, strength reduction factor equal to 0.88.

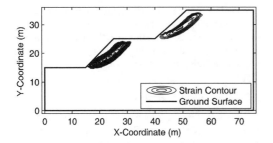

Figure 5. Shear strain contour generated during the strength reduction method at Stage 4, strength reduction factor equal to 0.88.

to be check. For this example, the initial SRF value is 1.03, for step 2 the SRF is reduced to 0.98, and in stage 3 to 0.93. For the example, these SRF values are adequate to detect different potential failures.

When performing a FE-SRM, it is necessary to retain all the different strength reduction computations. This information enables to detect potential unstable mechanisms. By exporting the strains generated at the Gauss points and drawing the shear strain contour for different limits, others failure mechanisms can be detected. Figure 5 shows the shear strain contour computed for stage 4, from this stage and on, plastic points are not more generated at the upper slope (Fig. 4b). By the properly vizualization the location of the two failure zones and the potential unstable zones are located. By this process it is plausible to detect different safety zones in which adequate stabilization measures can be implemented.

Good correspondence between limit equilibrium results (Figs. 2–3) and SRM (Fig. 5) are produced. The safety map generated by Bishop's limit equilibrium method and the safety zones determined by the shear strain contour shows the same failure mechanisms.

For simple slope problems, the determination of safety zones, by shear zones or safety map, is less critical since the critical sliding mechanism and local minima are near to each other. Any slope stabilization action done for the most critical failure mechanism will affect the other mechanisms. The local minima will fall inside the shear strain contour and plastic points developed by FE-SRM. Therefore, it facilitates the slope stability analysis.

4 INFLUENCE OF SOIL PROPERTIES

In this second example, the importance of considering different soil parameters in the failure location is shown. This example is a slope located at km 27+000 at the left hand side of the road Cuenca-Machala in Ecuador. It has an inclination of about 40 degrees. Three main soil units are identified; colluvium, weathered tuff material (Tuff) and the volcanic rock basement. The stratigraphy and cross-section configuration are shown in Figure 6. Soil mechanical parameters are given in Table 1, in which, γ is the unit weight, c is the cohesion, ϕ is the friction angle, E represents the Young's modulus, and ν is the Poisson's ratio. For all the analyses zero-tensile strength is used. Soil parameters were obtained from performing in situ and laboratory tests.

For the colluvium stratum, in which the potential landslide is situated, the cohesion has a value between 9 to 15 kPa and the friction angle varies from 23 to 29°. Even tough, the relative small variation in soil parameters, this changes the location of the failure surface. Figure 7a shows the shear strain contour for the slope having a $c = 9$ kPa and $\phi = 29°$, for those parameters the FoS is equal to 1.27 and the failure surface indicated by the strain contour is located at the upper part of the colluvium. For the same geometrical configuration and by changing the c to 10 kPa and ϕ to 26°, a second failure mechanism arises, this second failure is deeper and it is located at the lower part of the stratum (Fig. 7b). For this case, the computed FoS is equal to 1.19. If now, the $c = 15$ kPa and $\phi = 23°$, the upper failure disappears and only the lower failure is detected (Fig. 7c), having a FoS equal to 1.14.

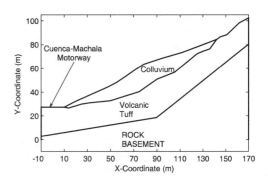

Figure 6. Stratigraphy and geometrical configuration for the potential instability at Cuenca-Machala motorway (km 27+000), Ecuador.

Table 1. Geomechanical properties for the different materials.

Parameter symbol	Units	Colluvium	Material tuff	Rock
γ	kN/m³	16.7	17.6	24.5
c	kPa	9–15	60.0	200
ϕ	degrees	23–29	33.5	40.0
E	MPa	60	100	300
ν	[-]	0.25	0.25	0.40

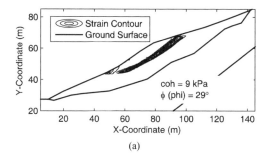

(a)

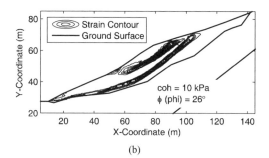

(b)

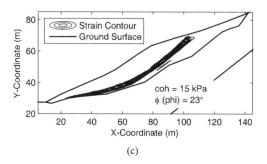

(c)

Figure 7. Shear strain contours generated during the strength reduction method. (a) upper failure, $c = 9\,\text{kPa}$ and $\phi = 29°$, factor of safety equal to 1.27. (b) mixed failure, $c = 10\,\text{kPa}$ and $\phi = 26°$, factor of safety equal to 1.19. (c) lower failure, $c = 15\,\text{kPa}$ and $\phi = 23°$, factor of safety equal to 1.14.

This example shows that for small different soil strength parameters very different failure mechanisms can be developed. To construct safety zones the location of the different failures should be delimitated.

It is worth noticing that by the principal of the strength reduction method, all slopes which have the same strength parameters (c and $\tan \phi$) but are factored by any scalar, will have the same failure mode. In other words, if a slope has parameters c_1, $\tan \phi_1$ and FoS_1, then, if another slope has parameters $c_2 = c_1^* F$ and $\tan \phi_2 = \tan \phi_1^* F$, then the factor of safety FoS_2 is equal to $FoS_1^* F$, and the failure mechanisms is the same. In which, F is any scalar number. This definition is valid for cases where no external load or water forces are presented. Taking this relationship into account, a plot can be draw indicating some of the combination of the soil parameters (c and $\tan \phi$) that will have the same failure mechanism. Figure 8 shows the combination of cohesion, $\tan \phi$ and ϕ, which gives the same failure mechanism and where the factor of safety will be ratio to each other.

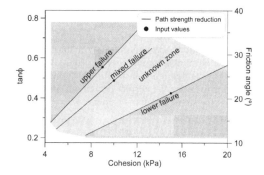

Figure 8. Delimitation of different failure modes depending in soil properties combinations.

The upper, mixed and lower lines represent the factored parameters (c and $\tan \phi$). All the strength parameter combinations that fall above the upper failure line will produce an upper failure. And all the combinations that fall below the lower failure will develop a lower failure. At the "unknown zone" (lack of computations), if the parameters are above the mixed failure line, the failure can be mixed or upper, if the parameters fall below the later line the failure will be mixed or lower. The exact limits between the different mechanisms are not yet defined, further research is still going on to define these limits and reduce the unknown zone.

5 CONCLUSIONS

This study presents a proposal to identify different failure surfaces by retaining the information given in the strength reduction method by different strength reduction factors. Finite element method is used in order not to constraint the analysis by the assumptions employed by others methods regarding the failure. By properly vizualization of the shear strain contour the safety zones are detected. The strength reduction factor used to compute the factor of safety gives local minima for different computations. Good agreement is encountered with the safety map generated by limit equilibrium method.

Small changes in soil properties can cause remarkable changes in the failure mechanism, thus safety zones should be evaluated by performing different soil combinations. By appropriately manipulation of the finite element slope stability information, the full extent of potential slope instability is depicted. This will contribute for the correct design and implementation of the stabilization measures.

ACKNOWLEDGEMENTS

The first author would like to thank for the financial support provided by the K.U. Leuven in the context of the Selective Bilateral Agreement between K.U. Leuven and Latin America.

REFERENCES

Baker, R. & D. Leshchinsky. 2001. Spatial distribution of safety factors. *Journal of geotechnical and geoenvironmental engineering* 127(2), 135–145.

Bojorque, J., G. De Roeck & J. Maertens. 2007. Comments on two-dimensional slope stability analysis by limit equilibrium and strength reduction methods by Cheng et al. doi:10.1016/j.compgeo.2007.04.005 (in press).

Cala, M., J. Flisiak & A. Tajdus. 2004. Slope stability analysis with modified shear strength reduction technique. In W. Lacerda, M. Erlich, S. Fontoura, and A. Sayao (Eds.), *Proc. 9th intern. symp. on Landslides*; *Landslides: Evaluation and Stabilization, Rio de Janeiro, Brazil.*

Cheng, Y.-M., T. Lansivaara & W.-B. Wei. 2007. Two-dimensional slope stability analysis by limit equilibrium and strength reduction methods. *Computers and Geotechnics* 34(3), 137–150.

Dawson, E. & W. Drescher. 1999. Slope stability analysis by strength reduction. *Géotechnique* 49(6), 835–840.

Griffiths, D. & P. Lane. 1999. Slope stability analysis by finite elements. *Géotechnique* 49(3), 387–403.

PLAXIS-BV. 2004. 2d-version 8, finite element code for soil and rock analysis. *Slope Stability Analysis, Delft University of Technology and Plaxis, The Nederlands.*

Renaud, J.-P., M. Anderson, P. Wilkinson, D. Lloyd & D. Muir-Wood. 2003. The importance of visualization of results from slope stability analysis. *Proceedings ICE, Geotechnical Engineering* 156, 27–33.

Laboratory and numerical modelling of the lateral spreading process involving the Orvieto hill (Italy)

F. Bozzano, S. Martino & A. Prestininzi
"Sapienza" University of Rome, Research Centre for Geological Risks (CERI—Valmontone), Rome, Italy

A. Bretschneider
"Sapienza" University of Rome, Department of Earth Sciences, Rome, Italy

ABSTRACT: The results of a study carried out in the cliff rock of the medieval town of Orvieto (Umbria region, Italy) are here presented. The Orvieto hill is involved in a lateral spreading process as a consequence of its geological setting, due to the overlaying of a rigid 60 m thick tuff slab on plastic overconsolidated blue clays. In order to reconstruct the stress path experienced by the Orvieto blue clays during their geological evolution, a not standard laboratory test was performed, by a triaxial device. The obtained function was used to perform a sequential FDM numerical modeling, whose findings are consistent with the actual site conditions. These findings point out the significant contribution of both laboratory tests and numerical modeling in order to reproduce the present stress-strain conditions of the hill involved in the lateral spreading process, and then to better define the engineering-geology model of the Orvieto hill.

1 INTRODUCTION

Lateral spreading processes are widespread in many sites of central Italy. The Orvieto town, located in the southern Umbria region (Figure 1), is one of the most famous examples of medieval towns involved in lateral spreading process.

Lot of paintings and pictures testify that since Middle Ages the Orvieto hill was affected by falls on its cliffs and complex landslides on its slopes; those types of instabilities, induced by lateral spreading processes, affect the Orvieto hill till present time.

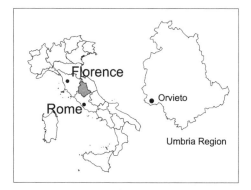

Figure 1. Location of the Orvieto town.

The Orvieto hill has been chosen for human settlements beginning from the Bronze Age until the present, and then there are so many historical buildings, churches and cultural heritage to be preserved.

2 GEOLOGICAL AND GEOMORPHOLOGICAL EVOLUTION

The Orvieto hill is set up by an about 700 m large, 1500 m long and ~40 m thick tuff plate that overlies epiclastic de-posits and blue marine clays 15 m and some hundred meters thick respectively (Figure 2). The blue clays (Argille di base, Conversini et al. 1977, Felicioni, et al. 1995) were deposited during the Middle Pliocene in the marine basin of the Media Valle del Tevere (i.e. the middle Tiber valley) at a depth of 100–200 m below sea level (Lembo Fazio et al. 1982). A regional uplift in Upper Pliocene—Lower Pleistocene caused the end of the clay sedimentation and, in subaerial condition, the partial erosion of the clay succession for a total thickness of 150 m and the gently dipping of the clay top to north-east. The blue clays are overlain by a fluviolacustrine epiclastic deposits about 15 m thick known as the Serie dell'Albornoz. At the hill top there is the Tufo di Orvieto tuff (Nappi et al. 1982), an ignimbritic flow coming from the Vulsini volcanic complex about 336 ky ago. The Tufo di Orvieto tuff is characterised

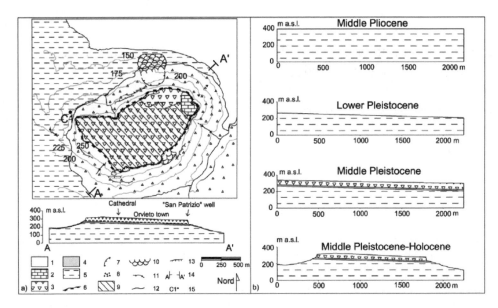

Figure 2. Left, Geological sketch of the Orvieto hill: 1) recent and present alluvia; 2) lacustrine travertines; 3) *Tufo di Orvieto* tuff; 4) *Serie dell'Albornoz* epiclastic deposits; 5) *Argille di base* clays of Orvieto; 6) cliff edges evolving by falls; 7) landslide scarp; 8) debris; 9) structural surface; 10) complex landslide debris; 11) gully; 12) downcutting stream; 13) landslide terrace edge; 14) section trace; 15) sampling point. Right, the A-A′ geological section, from the top to the bottom, in the four main evolutionary steps of the Orvieto hill.

by two lithofacies: the first is lithoid with black scoriae in a reddish matrix, the second one is a poorly cemented grey pozzolana. In the eastern part of the hill outcrops a lens of spongy, well-stratified travertines.

During the Quaternary age, the linear erosion of the rivers caused the isolation of the Orvieto tuff plate from the Vulsini plateau, and now the Orvieto hill is shaped as a mesa-type relief. Considering the morphological profile, the top of the hill is a quasi-planar surface bounded by sub-vertical cliffs slopes with an average height of 40 m; at the base of the tuff plate the slope angle rapidly changes from sub-vertical to about 12°. The gravitational instabilities, both on the cliff walls and on the hill slopes, produced in the past a thick debris belt all around the Orvieto hill.

The main factor that causes the cliff instabilities is the lateral spreading process, due to the overlapping of the tuff plate on the blue clays (Lembo Fazio et al. 1984, Manfredini et al. 1980); falls on the cliff edges are a consequence of this process and of the absence of the horizontal confining pressure due to the river valley erosion. The instabilities of the clayey slope is, instead, mainly due to the rainfall regime (Tommasi et al. 2006).

3 STRESS-STRAIN EVOLUTION AND LABORATORY MODELLING

The geological evolution of the Orvieto hill can be summarized in four main steps: 1) submarine deposition of the blue clays during the Middle Pliocene; 2) emersion and erosion of the highest part of the blue clays during the Lower Pleistocene; 3) deposition of both fluviolacustrine sediments and the tuff during the Middle Pleistocene; 4) lateral release due to the river valley erosion from the Middle Pleistocene to the Holocene (Manfredini et al. 1980).

The discontinuous geological processes responsible for the geological setting and the evolutionary history of the Orvieto hill, induced, on the basal blue clay succession, a complex stress path characterized by loading and unloading conditions. These conditions, in addition to weathering, caused the geomechanical properties of the clay deposits to significantly decay, as proved by the presence of a 15 m thick superficial layer of weathered and softened clays. In the previous studies many authors distinguished these three main layers along the vertical profile of the clay deposit starting from ground level: weathered clay, softened clay, stiff clay (Conversini et al. 1977, Lembo Fazio et al. 1982, Tommasi et al. 1996), about 10 m and 5 m thick respectively along the slope of the Orvieto hill.

In order to reconstruct the stress path experienced by the Orvieto blue clays during their geological evolution, a not standard laboratory test was performed, by a triaxial de-vice. The stress-path experienced by the Orvieto clays has been planned considering the four main steps of the geo-logical evolution of the hill (Tables 1, 2).

The test simulated the stress-strain evolution of a point in the clay slope near the base of the tuff plate, where some cubic clay samples were taken. The target of the not standard triaxial test (NSTT), was to characterize the deform-ability of the blue clays in similar stress-strain conditions experienced during their geological evolution. The stress path of the NSTT has been planned starting from the values of vertical and horizontal stresses (σ_v and σ_h in Table 1), in the sample point at the four evolutionary steps of the geological evolution.

Then a blue clay specimen was prepared in accordance with standard triaxial test procedures. The test was performed using a standard triaxial isotropic-confinement test apparatus in CID configuration, using a loading rate of 0.002 mm/min. During the test, as experimental conditions did not permit to simultaneously change the σ_h and σ_v in the different sections of the simulated stress path, the same path was approximated by a segmented straight line; on this line, the σ'_h were first changed, then the material was let to consolidate and finally the σ'_v were changed in a controlled way. Initially the specimen was isotropically consolidated at a pressure of 8.6×10^5 Pa and brought to a vertical load of 16×10^5 Pa simulating the clay deposition (step 1 of Figure 3).

Subsequently, a vertical unloading process simulated the erosion with a vertical unloading until the minimum vertical stress of 1×10^5 Pa (step 2 of Figure 3). Then the emplacement of the epiclastic deposits and of the ignimbritic flow was simulated with a vertical reloading up to 1×10^6 Pa (step 3 of Figure 3). Simulation of the last section of the stress path (lateral valley erosion, step 4 of Figure 3) was not feasible. This last step of the test, which was expected to simulate zero horizontal load and constant vertical load, would have required an apparatus capable of directly measuring the horizontal deformations of the specimen.

The NSTT took six weeks for completing. The NSTT allowed, reproducing a laboratory modelling of the stress conditions that the clay experienced during its history, to derive the E-σ'_3 correlations, both for the tangent modulus E_{ti} and for the secant modulus E_s, under the tested stress conditions: loading, unloading and reloading. In detail, on the related stress-strain curves, a first roughly rectilinear segment was used to measure the initial tangent modulus (E_{ti}) and a second segment (comprised between the point where the stress-strain curve becomes non-linear and the maximum or minimum vertical stress applied for each σ_3 level) was used to determine the secant modulus (E_s). A good linear correlation between the deformability and the confining pressure has been identified in the first loading step and in the unloading step; otherwise in the reloading step no statistically valid correlation

Table 1. Effective vertical stress calculated in the sample point for each evolutionary steps.

	Lithologies	Thickness m	γ_n kN/m^3	σ'_v kN/m^2
step 1	clays	165	20	1686
step 2	clays	10	20	105
step 3	tuffs	10	20	
	epiclastic deposits	15	17	1005
	clays	50	13	
step 4	tuffs	10	20	
	epiclastic deposits	15	17	1005
	clays	50	13	

Table 2. Evolutionary steps of the Orvieto hill. σ'_v = effective vertical stress; σ'_h = effective horizontal stress; K_0 = lateral stress ratio at rest; $p' = (\sigma'_v + \sigma'_h)/2$; $q = (\sigma'_v - \sigma'_h)/2$.

Steps	σ'_v kN/m^2	σ'_h kN/m^2	K_0	p' kN/m^2	q kN/m^2
I	1686	843	0.5	1264.5	421.5
II	105	63	0.6	84	21
III	1005	603	0.6	804	201
IV	1005	0		502.5	502.5

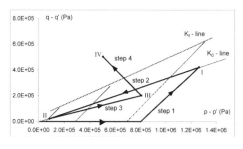

Figure 3. Assessed stress path for not standard triaxial test (NSTT).

can be established, therefore, a single average value was selected for the E_{ti} and E_s moduli. The obtained functions E_{ti} E_s vs. (σ'_3) and the values are summarised in Table 2.

4 NUMERICAL MODELLING

In the second part of this study a numerical modeling of the Orvieto hill by the FDM code FLAC 5.0 (Itasca 2005) was performed. The target was to realize a sequential numerical analysis changing the engineering-geological section and the input data in each simulated evolutionary step.

This was realized inputting in the code, for every step of the sequential numerical model, the geological section used to plan the NSTT, and the linked E-σ'_3 function obtained from the NSTT (Tables 3a, 3b). In detail the engineering-geological model of the cliff was defined by assigning the physical and mechanical parameters obtained from laboratory tests and integrated with literature data to the investigated lithologies.

Moreover a vertical variability of the investigated parameters was assigned in the 2D numerical model; as a con-sequence the parameters E_{ti} and E_s was supposed to change with depth within the same geotechnical unit either discretely (for predetermined stress levels) or continuously (under the law of variation of the parameter) (Tables 3a, 3b).

The stress-strain history of the Orvieto cliff was numerically modeled through engineering-geological sections representing the above-described four evolutionary steps as sketched in Figure 2 and in Figure 4; for the step 4 was adopted the same unloading function E-σ'_3 obtained from the step 2 of the NSTT (Tables 3a, 3b).

The results of the numerical modeling, consistent with the actual site conditions, show in the step 4 (Figure 4-step 4) the presence of an about 30 m thick belt characterized by a G modulus decrease of about two thirds vs. the initial one. An about 15 m-thick portion of this belt corresponds to the Serie dell'Albornoz deposits, while the remaining 15 m represent the shallowest part of the clays, softened by stress loading and reloading according to the data already reported by Conversini et al. 1977, Lembo Fazio et al. 1982 and Tommasi et al. 1996. Sequential numerical modeling of the stress evolution of the investigated lithotypes revealed the existence of a softened portion of the clays which would have not been demonstrated by a conventional simulation approach.

Furthermore was also simulated, in a parametrical way, the decay of the properties of the tuffs joints. The results (Figure 4) show that with the decreasing of the joints properties, new tension cracks can arise in an area extending as far as 140 m from the cliff edge and that at the cliff edges, stresses cluster along well-defined surfaces. These surfaces create a potential wedge failure mode with concentration of tensile stresses.

Table 3a. Values of parameters used in the numerical simulations.

Deposits	γ_n kN/m^3	ten Pa	c Pa	t Pa	ϕ °	ν
Orvieto Tuffs*	13	4.90E+05	9.80E+05		35	0.2
Albornoz Series deposits**	17		5.00E+04	7.15E+04	35	0.3
Clays	20		4.00E+04	8.20E+04	26	0.3

γ_n = natural weight density; ten = tension cut-off; c = cohesion; t = resistance to tensile stress; ϕ = shear resistance angle; n = Poisson ratio.

Table 3b. Values of parameters used in the numerical simulations.

Deposits	σ_3 kPa	$E_{t(\sigma 3')}$ Pa	$E_{s(\sigma 3')}$ Pa
Orvieto Tuffs*		$2 * 10^9$	
Albornoz Series deposits**		$5 * 10^8$	
Clays	$0.55 * 10^5 >$	$45.62(\sigma'_3)$	$8\,(\sigma'_3)$
	$7.9 * 10^5$	$+ 2 * 10^7$	$+ 3 * 10^6$
	$1 * 10^5 < 7.9 * 10^5$	$132.22\,(\sigma'_3) + 4 * 10^7$	
	$1 * 10^5 > 6 * 10^5$	$6.48 * 10^7$	$1.89 * 10^7$

σ_3 = confining pressure range in tests; E_t = tangent Young's modulus; E_s = secant Young's modulus.

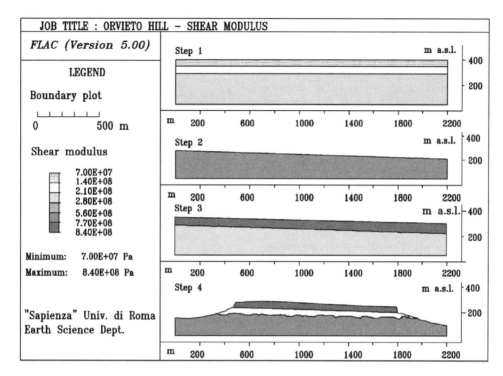

Figure 4. Results of numerical stress-strain analyses of the Orvieto hill. Detail of north-western margin of the tuff plate on the assumption of minimum joint strength, contour of shear stress on vertical plane and plasticity state.

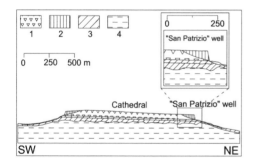

Figure 5. Engineering-geology model of the Orvieto hill: 1) undisturbed tuff plate, 2) released tuffs zone, resulting from the numerical model, 3) softened soils resulted from the numerical model, 4) stiff and undisturbed clay.

5 ENGINEERING-GEOLOGY MODEL

An engineering-geology model for the Orvieto hill was obtained by merging the previously described geological model, laboratory model and numerical model. This engineering-geology model takes into account the stress variations due to the evolution of the hill during its geological history since it summarizes the results of both the laboratory and numerical modeling.

On the basis of the obtained engineering-geology model (Figure 5) four different zones can be distinguished in the Orvieto hill.

The zone 1 corresponds to the tuffs at the hill-top; the zone 2 corresponds to the tuff deposits involved in the horizontal stress reduction as a consequence of the lateral spreading process and according to the numerical modeling. The zone 3 corresponds to both the epiclastic deposits and the softened layer of the basal clays as a consequence of the loading and unloading cycles occurred during the geological evolution and according to the numerical modeling; at last, the zone 4 represents the stiff and undisturbed clay portion of the hill.

6 CONCLUSIONS

An engineering-geology model of the Orvieto hill has been experimentally obtained by reproducing its stress-strain evolution via laboratory and numerical modeling. At this aim a not standard triaxial test was performed and the obtained values for the stiffness parameters were considered as the input for the sequential numerical modeling, per-formed by a FDM approach. The results of the numerical modeling allow to set up an engineering-geology model, that is not only referred to the actual conditions of the site since it takes into account all the geological evolution of the hill. As a consequence, a geomechanical zoning, based on the stress-strain evolution of the Orvieto hill, was obtained, showing the existence of a 15 m thick low-stiffness clayey level, just below the Orvieto tuff plate, which well corresponds to the softened clays recognized by various Authors all along the Orvieto hill slopes.

This study infers that a coupled laboratory and numerical modeling approach, which takes into account the geological evolution of a specific site, has to be regarded as a fundamental tool for better refining an engineering-geology model.

REFERENCES

Barbero, M., Barla, G. & Demarie, G.V. 2004. Applicazione del metodo degli elementi distinti alla dinamica di mezzi discontinui. Rivista Italiana di Geotecnica 3 (2004): 9–24.

Barla, G., Borli Brunetto, M. & Vai, L. 1990. Un esempio di modellazione matematica in rocce tenere: la Rupe di Orvieto. Mir '90, III Ciclo di conferenze di meccanica ed ingegneria delle rocce: 14.1–14.12. Torino.

Conversini, P., Lupi, S., Martini, E., Pialli, G. & Sabatini, P. 1977. Rupe d'Orvieto Indagini geologico-tecniche. Quaderni della Regione Umbria suppl. al n. 15.

Felicioni, G., Martini, E. & Ribaldi, C. 1995. Studio dei centri abitati instabili in Umbria. Articolo estratto dalla Pubblicazione n. 979 del GNDCI-CNR, U.O. 2.17 Rubettino editore.

ITASCA 2005. FLAC 5.0: User manual. Licence number 213-039-0127–16143 (Sapienza—University of Rome, Earth Science Department).

Lembo Fazio, A., Manfredini, G., Sciotti, M. & Totani, G. 1982. Caratteristiche geotecniche dell' argilla di Orvieto. Quaderni dell'Istituto di Arte Mineraria Giugno 1982.

Lembo Fazio, A., Manfredini, M., Ribacchi, R. & Sciotti, M. 1984. Slope failure and cliff instabilities in the Orvieto hill. In Proceedings of the 4th International Sym-posium on Landslides, Toronto, Ont., 16–21 September 1984, Canadian Geological Society, Rexel, Ont. Vol. 2: 115–120.

Nappi, G., Chiodi, M., Rossi, S. & Volponi, E. 1982. L'ignimbrite di Orvieto nel quadro dell'evoluzione vulcanotettonica dei vulsini orientali. Caratteristiche geologiche e tecniche. Boll. Soc. Geol. It., 101: 327–342.

Manfredini, G., Martinetti, S., Ribacchi, R. & Sciotti, M. 1980. Problemi di stabilità della Rupe di Orvieto. Associazione Geotecnica Italiana— XIV Convegno Nazionale di Geotecnica. Firenze 28–31 ottobre 1980; Volume I: 231–246.

Tommasi, P., Pellegrini P., Boldini, D. & Ribacchi, R. 2006. Influence of rainfall regime on hydraulic conditions and movements rates in the overconsolidated clayey slope of the Orvieto hill (central Italy). Can. Geotech. J. 43(2006): 70–86.

Tommasi, P., Ribacchi, R. & Sciotti, M. 1996. Geotechnical aspects in the preservation of the historical town of Orvieto. In Carlo Viggiani (ed.), Geotechnical Engineering for the Preservation of Monuments and Historic Sites; Proc. Int. Symp., Napoli, Italy, 3–4 October 1996. Rotterdam: Balkema, 849–858.

Albano Lake coastal rock slide (Roma, Italy): Geological constraints and numerical modelling

F. Bozzano, C. Esposito, S. Martino & P. Mazzanti
"Sapienza" University of Rome, Department of Earth Sciences and "CERI" Research Center on Geological Risks, Rome, Italy

G. Diano
Geologist

ABSTRACT: The aim of this paper is to back analyze the stress-strain conditions that led to an ancient massive rock slope failure occurred in the inner slope of the Albano lake (Rome, Italy), whose scar and accumulation areas are respectively subaerial and submerged. Detailed geological and geomorphological field surveys, DEM-derived topographical reconstructions of the pre-landslide topography and geomechanical data derived from both in situ and laboratory characterizations, allowed to obtain the engineering-geology model of the slope prior to the landslide occurrence. A numerical modelling by the FDM code FLAC 5.0 was then performed in order to analyze the possible landslide trigger; static, hydrostatic and pseudostatic conditions were taken into account. Static and hydrostatic analyses indicate significant displacements, with the lake level close to the actual one; if an horizontal acceleration higher than 0.02 g is applied, a failure occurs according to a rock-slide kinematic mechanism, strongly influenced by the local geologic-structural setting.

1 INTRODUCTION

The hereby presented case study is framed within a wider research addressed to define the landslide hazard conditions of a peculiar morphologic context such as a lacustrine environment. As a matter of fact, the assessment of landslide-related hazard and risk conditions of coastal areas has to take into account also the "secondary" effects linked to the slope instabilities such as the possible generation of tsunami waves due to the impact of a landsliding mass from a subaerial slope onto the water (Bozzano et al. 2006, Tinti et al. 2005). In this context, the recognition of even past large-sized landslides assumes a relevant role, since they can be considered as geomorphic evidence of the potential occurrence of this kind of processes. Particularly significant is then the comprehension and reconstruction of the stress-strain conditions that led to a massive rock slope failure in order to assess the potential of similar landslides to occur in the present boundary conditions.

For these purposes an effort has to be done in order to collect the most significant data for the reconstruction of an affordable engineering-geology model of the slope before the landslide occurrence. In the presented case the traditional studies (such as geological and geomorphological surveys and geomechanical characterizations) were coupled with the analysis of detailed topographic information on the area obtained through a Digital Elevation Model derived from LiDAR data for the subaerial slope and through a Lacustrine Digital Elevation Model (LDEM) derived from a multi-beam sonar survey (Anzidei et al. 2007, Baiocchi et al. 2007) for the submerged one. This kind of data allowed us to reconstruct a geological cross section and a related engineering-geology section of the slope involved in the recognized massive rock slope failure. The stress-strain analysis on this section was then performed via numerical modelling by taking into account different conditions (static, hydrostatic and pseudostatic analyses) in order to identify and assess both the predisposing and the triggering factors.

2 GEOLOGIC AND GEOMORPHOLOGIC SETTING

The study area is located along the slope which bounds in the southern part the Albano lake. The latter corresponds to the inner slope of a large multiple maar partially filled with water (Fig. 1). This maar is in its turn framed within the Colli Albani volcanic edifice, whose origin can be related to back-arc extensional processes which affected the Thyrrenian coast of central Italy (Funiciello et al. 2003, Giordano et al. 2006).

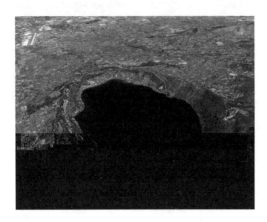

Figure 1. Virtual aerial oblique view of the Albano lake (from the website http://earth.google.it).

The Albano maar activity is linked to the final hydromagmatic phase of the Colli Albani volcanic complex; the corresponding crater is actually the result of the coalescence of at least three recognizable craters which mark different stages of activity. As a result, the geological setting of the inner slopes of the Albano lake, corresponding with the internal walls of the multiple maar, is mainly featured by outcrops of hydromagmatic deposits, which locally overlay lava and scoria deposits related to the previous volcanic phases of the Colli Albani complex. The former are featured by two typical lithofacies (Giordano et al. 2002). The first one represents the most part of the Albano maar deposits and is a plane parallel to low-angle cross-stratified alternation of scoria lapilli beds and ash-rich layers, generally cemented for the zeolitisation. The second lithofacies is represented by massive and chaotic, ash-matrix supported, up to 30 m thick density current deposits, containing up to block-sized xenoliths. In addition, an intracrater facies shaped as a series of tongues that locally drape the internal walls of the Albano maar from the crater rim down to the lake level outcrops with a marked dipslope attitude (De Rita et al. 1986). The subaerial geological frame is completed by talus slope and shore deposits that cover the lowest parts of the subaerial inner slopes.

Many gravity-induced landforms are present along the slopes that bound the lake; based on the geomorphologic surveys performed on purpose, active rock falls (and/or topple) and debris flows with volumes ranging between 10^{-1} and 10^3 m^3 are the most diffuse processes. In addition, the availability of the LDEM allowed us to observe a large number of subaerial landslide detachment areas whose deposit is located below the lake level. Even if the most diffuse and frequent landslides consist of mainly small and medium sized phenomena, geomorphic evidence of at least one past, large landslide is present in the southern slope of the maar.

3 THE ALBANO ROCKSLIDE

As above mentioned one of the most significant gravity-induced landform was identified along the southern sector of the slope which bounds the lake, where a wide, markedly concave slope sector, whose shape is ascribable to the scar area of a large landslide is present (Fig. 2). In addition, the observation of the LDEM allowed us to recognize a huge, convex positive landform just downslope the previously mentioned scar area (Fig. 2). This landform

Figure 2. In this figure a photograph and a 3d rendering are combined to give a view of the whole subaerial-submerged southern slope of the lake, affected by the slope failure.

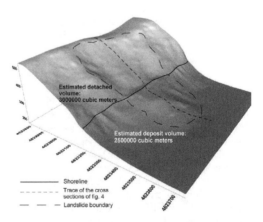

Figure 3. DEM and topographic profile of the investigated slope adopted for the landslide volume assessment.

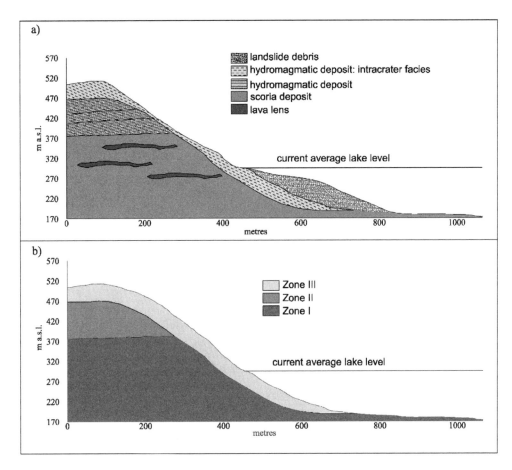

Figure 4. a) Geological cross section of the investigated slope; b) Engineering-geology model adopted for the numerical modeling, based on the hypothesized topography of the slope prior to the failure.

Table 1. Main parameters of the intact rock.

	Zone I	Zone II	Zone III (lava)	Zone III (scoria)
ν	0.25	0.25	0.25	–
$dens_i$ (kg/m³)	1956	1956	2873	–
ϕ_i	28°	28°	35°	–
σt_i (Pa)	$1.38 * 10^6$	$1.38 * 10^6$	$9.58 * 10^6$	–
c_i (Pa)	$2.76 * 10^6$	$2.76 * 10^6$	$1.92 * 10^6$	–
E_i (Pa)	$1.07 * 10^{10}$	$1.07 * 10^{10}$	$8.00 * 10^{10}$	$1.10 * 10^9$
stdv (E_i)	–	–	–	$4.29 * 10^8$
CV	–	–	–	0.39

* ν is the Poisson ratio, ϕ is the friction angle, σt is the tension cut-off, c is the cohesion, E is the Young's modulus, CV is the coefficient of variation; the subscript i indicates a parameter evaluated for the intact rock.

Table 2. Main parameters of the rock mass (discontinuities plus intact rock) derived from the equivalent continuum approach.

	Zone I	Zone II	Zone III
$dens_j$ (kg/m³)	2358	2358	2703
E_j (Pa)	$9.27 * 10^9$	$9.27 * 10^9$	$7.08 * 10^{10}$
ϕ_j	4°	4°	10°
σt_j (Pa)	$1.09 * 10^5$	$1.09 * 10^5$	$8.68 * 10^4$
c_j (Pa)	$7.48 * 10^6$	$7.48 * 10^6$	$4.03 * 10^7$

* See the caption of Table 1 for the symbols; the subscript j indicates a parameter evaluated for the jointed rock mass.

Table 3. Main parameters of the considered joint systems.

	Zone I	Zone II	Zone III
ϕ_j	33°	33°	–
I_j	0°; 17°; 34°	–	–
c_j	0	0	–
RMR	42–44	42–44	52–57
Q	658	658	317

* f and c are respectively the friction angle and the cohesion along the joints, I is the inclination of the joints, RMR and Q are respectively the values of the Bieniawski (1988) and Barton (1988) classifications.

can be interpreted as the corresponding debris accumulation deriving from the slope failure occurred in the present subaerial slope. An indirect confirmation of the relationship between the landforms previously described can be found also in the calculation of their volumes. The estimation of the detached rock volume was performed by means of DEM-derived topographical reconstructions of the hypothesized pre-event morphology and gave as result a volume of about $3 * 10^6$ m^3 (Fig. 3). This volume is quite well comparable with the volume of about $2.5 * 10^6$, estimated for the hypothesized debris accumulation performed on the DEM of the submerged part, especially if some evidence of successive landslides that involved the debris material are taken into account.

The geological setting of the subaerial part of the hereby discussed slope sector is featured by the superimposition of hydromagmatic deposits upon a thick bank of lava lenses and scoria deposits. In particular, detailed geological surveys pointed out that the material actually involved in the rock slope failure is constituted by massive and chaotic, ignimbrite deposits belonging to one of the previously mentioned intracrater facies of the hydromagmatic deposits, characterized by a marked dip slope attitude of the constitutive layers, quite parallel to the slope angle of the topography (Fig. 4a). Based on the morphologic evidence (such as the straight and sharp shape of both the crown and the flanks of the landslide scar, the flat morphology of the topmost part of the debris accumulation and the abrupt slope angle reduction in the lower part of the scar area) and on the geological-structural setting, this landslide can be preliminarily classified as a compound translational rockslide—structurally controlled of type E (block slide with toe breakout), according to Hungr and Evans 2004.

3.1 Numerical modelling

On the basis of the collected geological-geomorphological data, another phase of activity was addressed to better define the above mentioned Albano rockslide, the most relevant evidence of a large-sized event occurred in the study area and then the only constraint to depict a possible scenario in case of a large landslide occurrence. The mechanism of the rock-slide was simulated by a stress-strain numerical modelling. A back analysis was performed in order to analyse the role of the geological setting, of the rock mass properties as well as of the external forces on the rock-slide trigger. A 2D numerical model by the FDM code FLAC 5.0 was obtained starting from the topographical section obtained via LiDAR and multibeam surveys (Baiocchi et al. 2007), respectively for the subaerial and for the submerged slope. A pre-failure slope shape and geological setting were hypothesized on the basis of the morphological features adjacent to the detachment area as well as to the depositional mechanism assumed for the volcanic flow-type deposits. In order to build the engineering-geology section of the slope, three different "geomechanical" zones were distinguished in the above mentioned section (Fig.4b) by considering the geology features and the rock mass (intact rock plus discontinuities) properties derived from in situ surveys, laboratory tests and bibliography data (Tabs. 1–3).

A first zone (Zone I) corresponds to the scoria deposits including lava levels; a second zone (Zone II) corresponds to the hydromagmatic deposits with a sub-horizontal attitude; a third zone (Zone III) corresponds to the hydromagmatic deposits, showing a slope-dipping attitude. Taking into account both geological properties and geomechanical behaviour of the different deposits the following constitutive model were applied to the three zones:

Zone I: elastic constitutive model;
Zone II: Mohr Columb elastic-perfectly plastic constitutive model;
Zone III: Mohr Coulomb elastic-perfectly plastic constitutive model with ubiquitous joints.

The values of the corresponding parameters were evaluated on the basis of a conceptual model of the slope based on field evidences, by considering heterogeneities, anisotropies and/or weak planes within the rock masses. Moreover, the kinematic role of the rock mass discontinuities were also taken into account. In particular, the following rock mass features were attributed to the simulated geological units:

Zone I: heterogeneous (scoria and lava levels) and isotropic material, for the scoria deposits a Gauss-normal distribution of the mechanical parameter values was applied;
Zone II: homogeneous and isotropic material (no possible effects of the actual discontinuities on the rock slide);

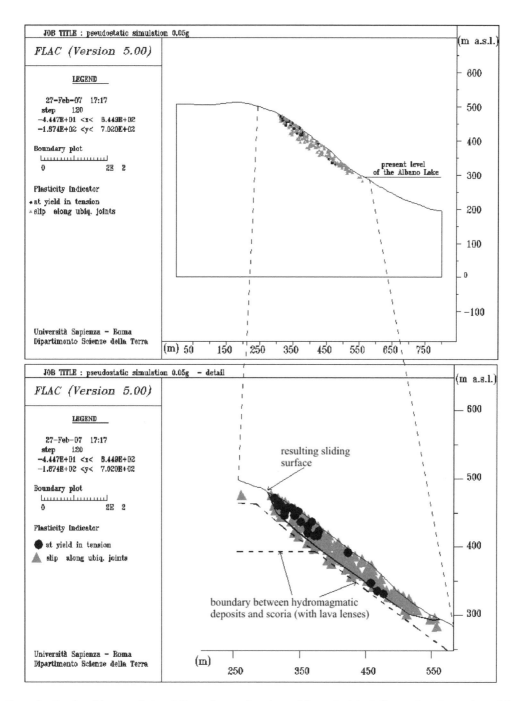

Figure 5. Results of the numerical modeling under pseudostatic conditions: plasticity indicators show the envelope of the failing rock mass that fits well with the actual observed features of the scar area.

Zone III: homogeneous and anisotropic material (possible effects of the actual discontinuities on the rock slide).

For both Zone II and III an equivalent continuum approach (Sitharam et al., 2001) was adopted in order to quantify the geomechanical parameter of the rock mass; the obtained values were derived from both intact rock and joints properties (Esposito et al., 2007). The iso-oriented discontinuities within the rock mass were considered as weak planes for the anisotropic constitutive model.

The back analysis of the rock-slide was performed under three different conditions: i) static analysis: only the weight force of the slope rock mass was applied; ii) hydrostatic analysis: the weight force of both the slope rock mass and the lake water column were applied; iii) pseudostatic analysis: equivalent horizontal accelerations due to the expected earthquakes were applied. In particular, the hydrostatic analysis was performed by supposing different levels of the lake, close to the actual one, in the range 323–268 m a.s.l.; the pseudostatic analysis was performed by applying different horizontal incremental accelerations up to 0.175 g, in agreement with the maximum values for the statistical recurrence time of 475 years (INGV, 2005). The obtained results point out that equilibrium is reached in all the static and hydrostatic conditions; nevertheless, in the hydrostatic conditions, the analysis of both the displacement field and the plasticity state within the slope points out significant displacements, mainly in the detachment area of the rock-slide, if the lake level is assumed to be very close to the actual one. If an horizontal acceleration higher than 0.02 g is applied, a sliding failure occur within the slope-dipping phreatomagmatic deposits at an average depth of about 30 m b.g.l.

The crown area is marked by tension cracks while the break out zone closely corresponds to the morphological plain which can be actually observed at the bottom of the subaerial slope. As a consequence, the resulting landslide mechanism is very consistent with the observed rock-slide scar; moreover, the obtained plasticity state shows sliding along joints within the rock mass, without failures of the intact rock. The latter evidence justifies the kinematic role of the slope-dipping anisotropies within the hydromagmatic deposits (Fig. 5).

4 CONCLUDING REMARKS

The main findings from the numerical back analysis of the Alban rock-slide can be summarised in the following points: i) significant constraints due to the geological setting and, in particular, to the anisotropic conditions of the rock mass, since the plasticity state shows sliding along joints within the rock mass, without failures of the intact rock, thus highlighting the relevant kinematic role of the slope-dipping anisotropies and thus the relevant constraint due to the local geological setting of the slope; ii) critical role of the lake level in predisposing the landslide phenomenon; iii) possible role of the seismic input in the landslide trigger.

Moreover, a back-analysis of the post-failure dynamic was performed for this landslide via the DAN-W software (Hungr 1995), by using a simple basal frictional rheology and considering the sliding mass as a rigid block. In spite of the simple model applied the values of basal friction (about 25°) and reached maximum velocity (about 15 m/s) are comparable with the available literature data for this kind of slope failure.

The above mentioned results can be considered the basis on which successive studies, such as the numerical modelling of the landslide-generated wave propagation, should be developed in order to better define the possible scenarios deriving from the impact of the landslide mass on the lake and the related hazard and risk conditions for the nearshore villages.

REFERENCES

Anzidei, M., Esposito, A. & De Giosa, F. 2007. The dark side of the Albano crater lake. *Annals of Geophysics,* 49: 1275–1287.

Baiocchi, V., Anzidei, M., Esposito, A., Fabiani, U., Pietrantonio, G. & Riguzzi, F. 2007. Integrer bathymetrie et lidar. *Geomatique Expert*, 55: 32–35.

Barton, N. 1988. Rock Mass Classsification and Tunnel Reinforcement Selection Using the Q-System. *American Society of Testing and Materials*, Philadelphia, 59–88.

Bieniawski, Z.T. 1988. The Rock Mass Rating (RMR) System (geomechanics classification) in engineering pratice. In: Louis Kirkaldie, (Ed.), *Rock Classification Systems for Engineering Purposes*, pp 17–34.

Bozzano, F., Chiocci, F.L., Mazzanti, P., Bosman, C., Casalbore, D., Giuliani, R., Martino, S., Prestininzi, A. & Scarascia Mugnozza, G. 2006. Subaerial and submarine characterisation of the landslide responsible for the 1783 Scilla tsunami. EGU 2006, *Geophysical Research Abstracts* 8, 10422.

De Rita, D., Funiciello, R. & Pantosti, D. 1986. Dynamics and evolution of the Albano crater, south of Rome. *Proc. IAVCEI Int. Conf.,* Kagoshima: 502–505.

Esposito, C., Martino, S. & Scarascia Mugnozza, G. 2007. Mountain slope deformations along thrust fronts in jointed limestone: an equivalent continuum modelling approach. *Geomorphology*, 90: 55–72.

Freda, C., Gaeta, M., Karner, D.B., Marra, F., Renne, P.R., Taddeucci, J., Scarlato, P., Christensen, J.N. & Dallai, L. 2006. Eruptive history and petrologic evolution of the Albano multiple maar (Alban Hills, Central Italy). *Bull. Volcanol.* 68: 567–591.

Funiciello, R., Giordano, G. & De Rita, D. 2003. The Albano maar lake (Colli Albani Volcano, Italy): recent volcanic activity and evidence of pre-Roman Age catastrophic lahar events. *Journal of Volcanology and Geothermal Research*, 123: 43–61.

Giordano, G., De Rita, D., Cas, R. & Rodani, S. 2002. Valley pond and ignimbrite veneer deposits in the small-volume phreatomagmatic 'Peperino Albano' basic ignimbrite, Lago Albano maar, Colli Albani volcano, Italy: infuence of topography. *Journal of Volcanology and Geothermal Research*, 118: 131–144.

Giordano, G., De Benedetti, A.A., Diana, A., Diano, G., Gaudioso, F., Marasco, F., Miceli, M., Mollo, S., Cas, R.A.F. & Funiciello, R. 2006. The Colli Albani mafic caldera (Roma, Italy): Stratigraphy, structure and petrology. *Journal of Volcanology and Geothermal Research*, 155: 49–80.

Hungr, O. 1995. A model for the runout analysis of rapid flow slides, Debris flows, and avalanches. *Can. Geotech. J.*, 32: 610–623

Hungr, O. & Evans, S.G. 2004. The occurrence and classification of massive rock slope failure. *Felsbau* 22: 16–23.

INGV 2005. Carta dei valori di pericolosità sismica del territorio nazionale (OPCM 28/04/06 n. 3519), http://zonesis miche.mi.ingv.it/mappa_ps_apr04/griglia002/lazio.html.

Itasca 2005. FLAC, Fast Lagrangian Analysis of Continua, Version 5.0. Itasca Consulting Group, license: DST—"La Sapienza", Roma—serial number: 213–039–0127–16143.

Sitharam, T.G., Sridevi, J., Shimizu, N. 2001. Practical equivalent continuum characterization of jointed rock masses. *Int. J. of Rock Mech. and Min. Sciences* 38: 437–448.

Tinti, S., Manucci, A., Pagnoni, G., Armigliato, A. & Zaniboni, F., 2005. The 30 December 2002 landslide-induced tsunamis in Stromboli: sequence of the events reconstructed from the eyewitness accounts. *Natural Hazards and Earth System Sciences*, 5: 763–775.

Superposition principle for stability analysis of reinforced slopes and its FE validation

F. Cai & K. Ugai
Department of Civil and Environmental Engineering, Gunma University, Kiryu, Japan

ABSTRACT: This paper proposes to use principle of superposition for stability analysis of slopes reinforced with anchors, geosynthetics, or nails. The principle of superposition is implicitly used in Fellenius' method. This paper proposes to use the principle of superposition for Bishop's simplified method. The safety factor calculated using the principle of superposition of anchor-reinforced slopes was compared with the results calculated using finite element method with shear reduction technique (SSR-FEM). The comparison shows that the safety factor calculated using the principle of superposition for stability analysis of anchor-reinforced slopes was consistent well with that calculated with SSR-FEM, a different approach. The principle of superposition can be easily extended to other slope stability analysis methods.

1 INTRODUCTION

The end type of anchorage, where the tendon is grouted below the potential slip surface, has been used to stabilize dangerous slopes to a specified safety factor because of its significant technical advantages resulting in substantial cost savings and reduced construction period. The basic design concept is transferring the resisting tensile forces generated in anchors into the ground through the friction mobilized at their interfaces. The loads are usually developed by the anchorage of the tendon within the soil mass and tensioning at the surface against a bearing plate, concrete pad, or tensioned geosynthetics (Koerner 1984, Greenwood 1985). Some successful applications of this technique to stabilizing slopes have been reported (Tan et al. 1985, Hashimoto 1986, Corona 1996).

Limit equilibrium method, such as Bishop's simplified method, has been commonly used to assess the safety factor of slopes reinforced with anchors, assuming that an anchor only supplies an axial tension to stabilize a slope. The axial tension has been assumed to act at different positions: at the base of the slice where the slip surface intersects the anchor (e.g. Geo-Slope International Ltd. 2000), or at the ground surface where the anchor head is located (e.g. Oasys Ltd. 2001, Yamakawa 1995). Additionally, as described in detail in the following section, the effect of the axial tension on the safety factor is incorrectly considered in some design tools widely used in practice (e.g. Geo-Slope International Ltd. 2000, Oasys Ltd. 2001).

This paper proposes to use principle of superposition for stability analysis of anchor-reinforced slopes. The principle of superposition is implicitly used in Fellenius' method. This paper proposes to use the principle of superposition for Bishop's simplified method. The safety factor calculated using the principle of superposition of anchor-reinforced slopes was compared with the results calculated using finite element method with shear reduction technique (SSR-FEM). The comparison shows that the safety factor calculated using the principle of superposition for stability analysis of anchor-reinforced slopes was consistent well with that of SSR-FEM, a different approach.

2 PRINCIPLE OF SUPERPOSITION

2.1 *Principle of superposition*

For an anchor-reinforced slope shown in Figure 1, if the concept of Fellenius' method was used for the safety factor, the normal force at the base of the slice where the slip surface intersects the anchor is given

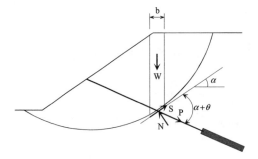

Figure 1. Anchor-reinforced slope.

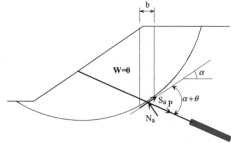

(a) Anchor-reinforced slope of a weightless soil

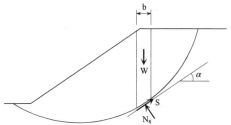

(b) Slope considering the weight of the soil. The shearing resistance at the base of the slice, S, is S_s+S_a.

Figure 2. Anchor-reinforced slope superposed an anchor-reinforced slope of weightless soil and a slope considering the weight of the soil.

by Equation 1 based on the force equilibrium in the direction perpendicular to the base of the slice.

$$N = W \cos\alpha + P \sin(\alpha + \theta) \quad (1)$$

Calculating the shear strength st the base of the slice with the Mohr-Coulomb equation, and using the equilibrium of the resisting moment and the driving moment, we can obtain the safety factor of anchor-reinforced slopes for Fellenius' method as follows.

$$F_S = \frac{\sum_n (cl + W \cos\alpha) \tan\phi + \sum_m P \sin(\alpha + \theta) \tan\phi}{\sum_n W \sin\alpha - \sum_m P \cos(\alpha + \theta)} \quad (2)$$

where n is total number of slices, and m is total number of rows of anchors.

The safety factor of Fellenius' method can also be obtained by principle of superposition, as shown in Figure 2. The normal force at the base of the slice for the anchor-reinforced slope of a weightless soil is given by

$$N_a = P \sin(\alpha + \theta) \quad (3)$$

This normal force allows an additional frictional shearing resistance to be mobilized.

$$S_a = N_a \frac{\tan\phi}{F_s} = P \sin(\alpha + \theta) \frac{\tan\phi}{F_s} \quad (4)$$

The tangential component, i.e., shear force in the anchor, $P\cos(\alpha + \theta)$, only supplied a resisting moment or resisting force to sliding of the slope; however, it had no effect to the normal force and shearing resistance at the base of the slice.

For the slope with the weight of the soil, the normal force on the base of the slice is given by

$$N_s = W \cos\alpha \quad (5)$$

Thus the total mobilized shear strength at the base of the slice can be expressed using the principle of superposition as follows.

$$S = S_s + S_a = \frac{1}{F_s}(cl + N_s \tan\phi) + S_a \quad (6)$$

Equilibrium of the resisting moment and the driving moment is expressed by

$$\sum_n S_s R + \sum_m S_a R = \sum_n WR \sin\alpha - \sum_m P \cos(\alpha + \theta) R \quad (7)$$

Combining Equations 6 and 7, and solving it, we can obtain the same equation as Equation 2 for the safety factor of anchor-reinforced slopes for Fellnius' method.

2.2 Bishop's simplified method

For Bishop's simplified method, the normal force at the base of the slice for the anchor-reinforced slope of a weightless soil is also given by Equation 3, and the total mobilized shear strength along a slice base plane is given by Equation 6 but N_s is obtained as follows.

For the slope considering the weight of the soil, shown in Figure 2(b), the equilibrium of forces in the vertical direction is given by

$$N_S \cos\alpha + S \sin\alpha = W \qquad (8)$$

Equations 8 and 6 give the normal force and shear strength at the base of the slice for the slope considering the weight of the soil.

$$N_S = \frac{F_s W - cl \sin\alpha - P \sin(\alpha+\theta) \sin\alpha \tan\phi}{F_s m_\alpha} \qquad (9)$$

$$S = \frac{cb + W \tan\phi + P \sin(\alpha+\theta) \cos\alpha \tan\phi}{F_s m_\alpha} \qquad (10)$$

where

$$m_\alpha = \cos\alpha + \sin\alpha \frac{\tan\phi}{F_s} \qquad (11)$$

The safety factor of Bishop's simplified method for the anchor-reinforced slopes can be obtained using equilibrium of the resisting moment and the driving moment, and it can be expressed by

$$F_S = \frac{\sum_n (cb + W \tan\phi)/m_\alpha + \sum_m P \sin(\alpha+\theta)\cos\alpha \tan\phi/m_\alpha}{\sum_n W \sin\alpha - \sum_m P \cos(\alpha+\theta)} \qquad (12)$$

Equation 12 implies that an additional resisting moment of the anchor force can be expressed as

$$M_{RF} = \sum_m \frac{PR \sin(\alpha+\theta) \tan\phi}{1 + \tan\alpha \tan\phi/F_s} \qquad (13)$$

The additional resisting moment of the anchor force in BS 8081 (1989) is given by

$$M'_{RF} = \sum_m PR \sin(\alpha+\theta) \tan\phi \qquad (14)$$

Because the angle α of the base of the slice located by anchors is usually larger than zero, the additional resisting moment of the anchor force in BS 8081 is

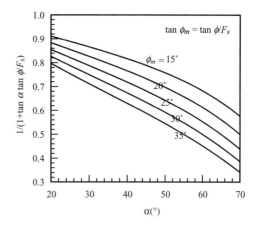

Figure 3. Ratio of additional resisting moment of Equations 13 to that of Equation 14 for a row of anchor.

larger than that given by Equation 13. Equation 14 did not consider the influence of the shearing resistance, induced by the normal force of the anchor force, on the normal force on the base of the slice for the slope with the weight of the soil, as shown in Equation 9. For a row of anchor, the ratio of the additional resisting moment of Equation 13 to that of Equation 14 for a slope reinforced with one row of anchor is shown in Figure 3.

Figure 3 shows that $\alpha+\varphi_m$ was larger, the additional resisting moment calculated by Equation 14 was larger than that by Equation 13. If $\alpha+\varphi_m > 90°$, the ratio of the additional resisting moment of Equation 13 to Equation 14, $1/(1+\tan\alpha \tan\varphi/F_s) < 0.5$.

For Fellenius' method, comparing the equations of the safety factor for anchor-reinforced slopes and slopes with no anchor, we know that Equation 14 gave the same additional resisting moment as that in Equation 2 because the shear strength had not influence on the normal force on the base of the slice.

If the normal force at the base of the slice was calculated directly using the equilibrium of force in the vertical direction, we can obtain

$$N \cos\alpha + S \sin\alpha = W + P \sin\theta \qquad (15)$$

Combining Equation 15 and the shear strength expressed with the Mohr-Coulomb equation, we can obtain the normal force and shear strength. Thus we can obtain the safety factor of Bishop's simplified method. The details can be found in Cai & Ugai (2003a).

$$F_S = \frac{\sum_n (cb + W \tan\phi)/m_\alpha + \sum_m P \sin\theta \tan\phi/m_\alpha}{\sum_n W \sin\alpha - \sum_m P \cos(\alpha+\theta)} \qquad (16)$$

For Equation 16, the additional resisting moment can be given by

$$M''_{RF} = \sum_m \frac{PR \sin\theta \tan\phi}{\cos\alpha + \sin\alpha \tan\phi/F_s} \quad (17)$$

The additional resisting moment in Equation 13 was identical to that given by Cai & Ugai (2003a). The calculated safety factor of anchor-reinforced slopes is consistent with that of SSR-FEM, a different approach. Cai & Ugai (2003a) have compared the safety factor of Equations 12 and 16. In the following section, we compare the safety factor for the Equation 14 and Equation 13 or 12 using the data reported by Cai & Ugai (2003a, 2003b).

3 SLOPE REINFORCED BY A ROW ANCHOR

Figure 4 shows a model slope reinforced by a row of anchors with a height of 8 m; a slope of 1V:1H; and a ground thickness of 10 m and the three-dimensional finite element mesh used for SSR-FEM (Cai & Ugai 2003). Two symmetric boundaries, one vertically passing through the anchor centerline and the other through the soil midway between the anchors, were used, so that the slope analyzed was really stabilized by one row of anchors with the same spacing. A small thickness mesh was used around the anchor to reduce the computational ill-conditioning of the interface elements. The slope was reinforced with only one row anchor for simplicity.

An anchor, consisting by a tendon 6 m long and 32 mm in diameter and a grouted body 6 m long and 90 mm in diameter, was connected to a rigid circular plate 30 cm in diameter at the slope surface. The anchors were installed with a horizontal distance between the slope toe and the anchor head $Lx = 4$ m; an orientation indicated by an angle between the anchor and a horizontal line $\theta = 15°$; and a center-to-center spacing $D_1 = 1.5$ m; unless otherwise stated. The anchor was modeled by finite elements in the axial direction and six wedge elements in the circumferential direction that form a semicircular shape. The tendon was assumed to contact smoothly with the soil. The shear strength of the soil-grouted body interface was the same as that of the soil unless otherwise stated. Table 1 lists the material parameters of the soil, soil-tendon interface, and soil-grouted body interface. Tendon and grouted body was modeled as elastic body with Young's modulus of $2.1 \times 10^5, 2.4 \times 10^4$ MPa, and Poisson's ratio of 0.3 and 0.2 respectively.

The anchor force calculated by SSR-FEM was used for limit equilibrium analyses. Consequently, the safety factor of SSR-FEM and limit equilibrium analyses can be compared with each other.

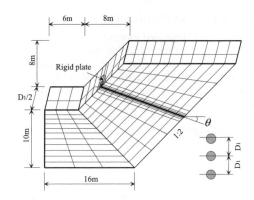

Figure 4. Anchor-reinforced model slope.

Table 1. Material parameters.

Parameter	Soil	Soil-tendon interface	Soil-grout interface
Young's modulus, E(MPa)	200	200	200
Poisson's ratio, ν(-)	0.25	0.25	0.25
Unit weight, γ(kN/m³)	20.0		
Cohesion, c(kPa)	12.0	0.01	12.0
Friction angle, φ(°)	20.0	0.0	20.0
Dilatancy angle, ψ(°)	0.0	0.0	0.0

For the idealized slope shown in Figure 4 with no anchor, SSRFEM give a safety factor of 1.09, which compares well with a value of 1.084 of Bishop's simplified method. The calculated results indicate that the failure mechanism of SSRFEM agrees with the critical slip surface of Bishop's simplified method (Cai & Ugai 2003).

4 RESULTS AND DISCUSSIONS

4.1 Effect of anchor orientation

Figure 5 shows the influence of the anchor orientation on the safety factor in the case of $Lx = 4.0$ m and $D_1 = 1.5$ m. There was an optimal range of the angle $\theta = 7.5°$ to $22.5°$ for the safety factor of the slope considered herein. The safety factor is rather insensitive to the angle θ near the optimum; this is similar to the results for the effect of anchors on the safety factor of infinite slopes reported by Hryciw (1991). Although the difference between the safety factor of SSR-FEM and Equation 16 and BS 8081was smaller than 3% for the slope reinforced with only one row of anchors, the safety factor of Equation 12 was more consistent with the safety factor of SSR-FEM with regards to the optimal range of the angle θ. The critical slip surfaces of the vertical and normal approaches were almost identical to one another.

4.2 Effect of anchor position

Figure 6 shows the influence of the anchor position Lx on the safety factor in the case of $\theta = 15°$ and $D_1 = 1.5$ m. The safety factor of SSR-FEM agreed with those of limit equilibrium methods, and generally the agreement was better between the safety factor of SSR-FEM and Equation 12. The safety factor of Equation 12 was more consistent with the safety factor of SSR-FEM than those of Equation 16 and BS 8081.

4.3 Effect of anchor spacing

Figure 7 shows the influence of the spacing between anchors D_1 on the safety factor in the case of $Lx = 4.0$ m and $\theta = 15°$. The results once again indicate that the safety factor of Equation 12 was more consistent with the safety factor of SSR-FEM than those of Equation 16 and BS 8081: The maximum axial tension and the failure mechanism were almost the same for the range of the spacing considered herein (Cai & Ugai 2003a), so the safety factor increased with the decrease in the spacing. The spacing of anchors used in practice is usually larger than 1.5 m (JGS 4101-2000, 2000).

5 SLOPE REINFORCED BY THREE ROWS OF ANCHORS

When the slope Figure 4 is reinforced with three rows of anchors, we compared the safety factors calculated by two-dimensional SSR-FEM analysis, Equation 12

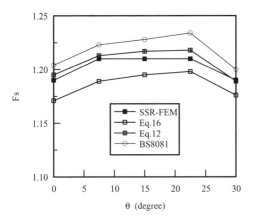

Figure 5. Safety factor versus direction angle ($Lx = 4.0$ m, $D_1 = 1.5$).

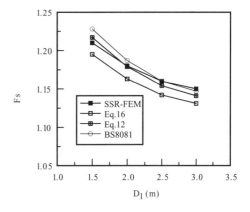

Figure 7. Safety factor versus direction angle ($Lx = 4.0$ m, $D_1 = 1.5$).

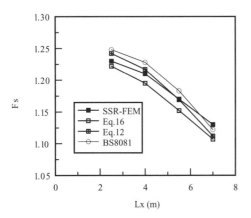

Figure 6. Safety factor versus anchor position ($D_1 = 1.5$ m, $\theta = 15°$).

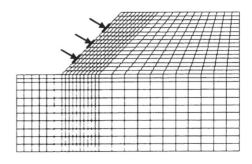

Figure 8. Two-dimensional finite element mesh (Rigid bearing plates of anchors are simulated by assuming that the displacement of such nodes within the extent of the solid black lines on the slope surface is the same).

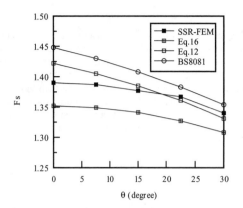

Figure 9. Safety factor of a slope reinforced with three rows of anchors versus angle θ.

(Cai & Ugai 2003b), and Equation 16 and BS 8081. The position of the three rows of anchors is $Lx = 2$, 4, and 6 m, and anchor force is 30 kN/m. The two-dimensional finite element mesh, as shown Figure 8, is used for SSR-FEM analysis.

Figure 9 shows the influence of the anchor orientation on the safety factor of the anchor-reinforced slope. For the slope reinforced with three rows of anchors, the safety factor of Equation 12 was more consistent with the safety factor of SSR-FEM than those of Equation 16 and BS 8081.

6 CONCLUSIONS

This paper proposes to use principle of superposition for stability analysis of anchor-reinforced slopes. The principle of superposition is implicitly used in Fellenius' method. This paper proposes to use the principle of superposition for Bishop's simplified method. As a result, the additional shearing resistance induced by the normal component of the anchor force only was considered in the equilibrium condition for the normal force at the base of the slice. This was different from that in conventional equation (Equation 16) in which the anchor force is directly used in the equilibrium condition in the vertical direction for Bishop's simplified method, and also this is different from BS 8081 in which the influence of the additional shear resistance induced by the normal component of the anchor force was not considered in the equilibrium condition for the normal force at the base of the slice.

The safety factor calculated using the principle of superposition of anchor-reinforced slopes was compared with the results of SSR-FEM. The comparison shows that the safety factor calculated using the principle of superposition for stability analysis of anchor-reinforced slopes was more consistent with that calculated with SSR-FEM, a different approach than the conventional equation and BS 8081.

The principle of superposition can be easily extended to other slope stability analysis methods and slopes reinforced with nails and geosynthetics.

REFERENCES

BS 8081.1989. Ground anchorages. British Standards Institution.

Cai, F. & Ugai, K. 2003a. Reinforcing mechanism of anchors in slopes: a numerical comparison of results of LEM and FEM. Int. J. Numer. Anal. Meth. Geomech., 27: 549–564.

Cai, F. & Ugai, K. 2003b. Numerical comparisons of stabilizing effects of anchors in slopes between limit equilibrium methods and elasto-plastic finite element method. Journal of the Japan Landslide Society, 40(4): 8–14.

Corona, E. P. 1996. Stabilization of excavated slopes in basalt. Proc. 7th Intern. Symp. Landslides, vol. 3, Senneset K (ed.). A.A. Balkema: Rotterdam, 1771–1776.

Geo-Slope International Ltd. 2000. User's guide for SLOPE/W, Version 4. Geo-Slope International Ltd.: Calgary.

Greenwood, J.R. 1985. Geogrids and anchors for slope stabilization-simple method of analysis. Proc. 11th Inter. Conf. Soil Mech. Found. Eng., Vol. 5, San Francisco: 2770–2771.

Hashimoto, I., Kawasaki, K. & Kodera, H. 1986. Cases of landslide control by dead-anchors. Proc. 21st Japan National Conf. Soil Mech. Found. Eng., Sapporo: 1483–1486 (in Japanese).

Hryciw, R.D. 1991. Anchor design for slope stabilization by surface loading. Journal of Geotechnical Engineering ASCE; 117 (8):1260–1274.

JGS 4101–2000. 2000. Standard for Design and Construction of Grouted Anchors. The Japanese Geotechnical Society: Tokyo (in Japanese).

Jewell, R.A. & Wroth, C.P. 1987. Direct shear tests on reinforced sand, Geotechnique, 37(1): 53–68.

Koerner, R.M. 1984. In-situ soil slope stabilization using anchored nets. Proc. Conf. Low Cost and Energy Saving Construction Materials. Envo Publishing Company: Lehigh vally, Pennsylvania: 465–478.

Oasys Ltd. 2001. SLOPE 17, GEO suite for Windows, Version 17.8.0. Oasys Ltd., Part of the ARUP Group: London.

Tan, S.B., Tan, S.L., Yang, K.S. & Chin, Y.K. 1985. Soil improvement methods in Singapore. Third Inter. Geotech. Seminar, Nanyang Technological Institute, Singapore: 249–272.

Yamakawa O. 1995. Simplified design method of slope reinforced with piles and anchors. Ph.D. Thesis, The University of Tokushima (in Japanese).

Soil suction modelling in weathered gneiss affected by landsliding

M. Calvello, L. Cascini & G. Sorbino
Department of Civil Engineering, University of Salerno, Italy

G. Gullà
CNR-IRPI, Rende-CS, Italy

ABSTRACT: The paper presents a numerical procedure to estimate, at a test site, the hydraulic properties and the net rainfall flux across the ground surface, by modelling the suction regime observed by tensiometers. The test site is located in an area affected by landsliding that involves highly heterogenous gneissic soils. Within this area, suction modelling is a fundamental issue as the movements of the landslides are strictly related to transient perched water tables directly linked to rainfall. The numerical procedure adopts an inverse analysis to model the observed suction regime and highlights the role played by the heterogeneity of the soil and by the net flux across the ground surface on the unsaturated flow characteristics at the slope scale.

1 INTRODUCTION

In natural and man-made slopes, soil suction is certainly one of the most important physical variables governing the transfer of energy and water (i.e. infiltration, evaporation and transpiration) between the atmosphere and the soil through the slope surface. Indeed, understanding the mechanisms that control soil suction variations, induced by changes in boundary conditions, and appropriately modelling the soil suction regime are crucial issues for the prediction of the slope behaviour.

In many natural slopes, modelling the soil suction regime may be a quite difficult task, especially when the hydraulic properties of the soils forming the slopes are strongly heterogeneous. In such cases, using the appropriate values of both the hydraulic properties and the boundary condition at the ground surface (i.e. rainfall intensity, evapotranspiration rate) is certainly a crucial point that often calls for experimental and numerical efforts. Among these, non-conventional approaches, analysing the soil suction regime at the slope scale, can be profitably used.

Referring to a particularly complex geolithological context located in Southern Italy, where the landslides movements are strictly related to rainfall, the paper presents an innovative procedure to cope with this issue. Particularly, an inverse numerical analysis is carried out aimed at modelling the suction regime observed by tensiometers at a test site. In the paper, after the description of the geolithological context, the landsliding of the area and the test site, the adopted numerical analysis is illustrated and the obtained results are discussed.

2 GEOLITHOLOGICAL CONTEXT AND LANDSLIDING

The test site is inside a study area of about 7.5 km^2 in the Western Sila Massif (Southern Italy), where several geological and geotechnical investigations have been developed during the time in order to characterise the landsliding affecting the area and the involved soils (Cascini et al. 1992a, 1992b, 1994). As for the soils, weathered gneiss is largely diffused in the area. The grade of weathering of the gneiss was defined following the procedures developed for Hong Kong (GCO 1988) on similar rocks. A simplified version of the adopted classification system (Cascini et al. 1992a, Gullà & Matano 1997) is shown in Table 1.

Table 1. Weathering grade of gneiss (from Cascini et al. 1992a).

Class	Description
VI	Residual and Colluvial soils
V	Completely weathered rock (Saprolitic soil)
IV	Highly weathered rock
III	Moderately weathered rock
II	Slightly weathered rock
I	Fresh rock

This classification, together with detailed geomorphological analyses, allowed the creation of a landslides inventory map where landslide distribution can be fully interpreted according to morphology, tectonics and weathering grade of the outcropping gneiss. The landslides inventory map reveals that the most widespread types of instability phenomena involve heterogenous residual, colluvial and saprolitic soils (classes VI and V). These are characterised by an impulsive kinematism, with lengthy periods of total inactivity, followed by brief phases of sudden reactivations triggered by remarkable increments in pore water pressures induced by rainfall.

Similarly to landsliding that affects the territories of Brasil and Hong Kong (Lacerda 2004, Brand 1984), *in-situ* measurements revealed that the pore water pressures in the landslide bodies are strictly related to transient perched water tables. These last are directly linked to the intensity and duration of rainfall events (Cascini et al. 2006).

Therefore, landslides characterisation calls for the definition of relationships among rainfall, pore water pressure increments in the perched water tables and the triggering phases of landsliding movements. To this end, an adequate knowledge of the suction regime and of the unsaturated flow characteristics is absolutely necessary. An innovative approach to achieve this goal is discussed in the following sections.

3 THE TEST SITE

The test site concerns a small landslide, with a maximum depth of 6 m, in a slope which is part of a large ancient landslide (Fig. 1). The small landslide involves gneiss of classes VI and V, resting on a basement formed by less weathered gneiss of class IV and class III-II (Cascini et al. 1992b, Sorbino 1995, Gullà & Sorbino 1996). As for the landslide debris, the particle size distribution with depth (Gullà & Sorbino 1996) reveals an uneven sequence of soils belonging to classes VI and V (Fig. 2).

The conspicuous number of piezometers installed in the test site—in some cases up to five piezometers per borehole—show, in the landslide debris, the presence of a perched water table with a marked transient behaviour related to the seasonal meteoric events. On the other hand, piezometers located in the basement give remarkably lower piezometric heads with annual or multiannual regime (Fig. 1). According to the data acquired in other sites of the study area (Cascini et al. 2006), this circumstance clearly reveals the presence of two different groundwater regimes in the subsoil: the first in the landslide debris, the other in the basement.

As for the relationship between the perched water table and rainfall, soil suction data were collected

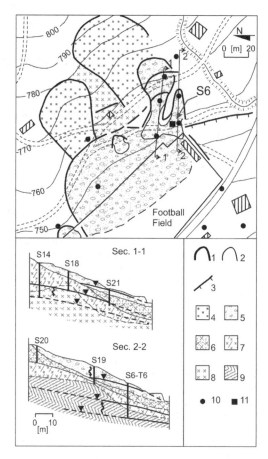

Figure 1. Test site: 1) landslide scarp; 2) limit of filled hollow; 3) terrace; 4) colluvial deposits; 5) gneiss of classes VI and V; 6) gneiss of class V; 7) gneiss of class IV; 8) gneiss of class IV-III; 9) gneiss of class III-II; 10) borehole with piezometer; 11) tensiometers' location; (from Gullà & Sorbino 1996).

within the landslide debris by means of five "Jetfill" tensiometers (Gullà & Sorbino 1996). They were installed, along the same vertical (S6-T6 in Fig. 1), at depths of 0.81 m, 1.45 m, 2.03 m, 2.88 m and 3.58 m below ground surface (Fig. 2). Soil suction measurements from the installation date (May 1993) to May 1995 are shown in Figure 3 together with the daily rainfall (Gullà & Sorbino 1996). Readings at the tensiometers were taken with an average time interval of about one week.

As for the hydraulic properties of the soils forming the landslide debris, they were provided by *in-situ* and laboratory tests. Saturated hydraulic properties were estimated by means of *in-situ* piezometer tests

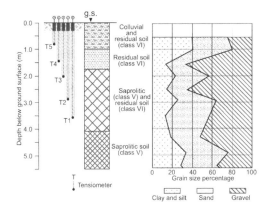

Figure 2. Location of tensiometers, soil profile and grain-size distribution in the borehole S6 (modified after Gullà & Sorbino 1996).

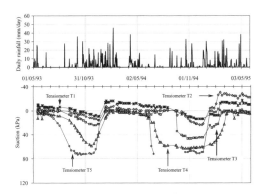

Figure 3. Daily rainfall and soil suction measured by tensiometers (modified after Gullà & Sorbino 1996).

and by permeameter and oedometric tests in laboratory (Sorbino 1995). As for the unsaturated hydraulic properties, they were determined in laboratory on homogenous undisturbed soil samples collected in the gneiss layers belonging to classes VI and V. Figure 4 shows the obtained experimental values of volumetric water content (θ) and hydraulic conductivity (K) against suction, normalized, respectively, by the values of the saturated volumetric water content (θ_s) and of the saturated coefficient of permeability (K_s).

4 NUMERICAL ANALYSIS

4.1 Model definition

In order to address the role played, at the slope scale, by the heterogeneity of the soil and by the net flux across the ground surface on the observed suction

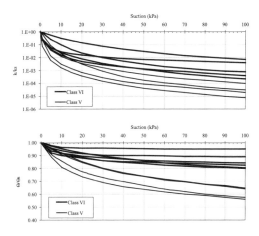

Figure 4. Experimental soil water characteristic curves, normalized by saturated volumetric water content, and hydraulic functions, normalized by saturated coefficient of permeability, for class VI and V soils (modified after Sorbino 1995).

regime, a transient seepage analysis was performed. The domain of the analysis concerns the vertical column of soil where the 5 tensiometers were installed (Fig. 5).

To this aim, two different schemes were adopted. With Scheme 1 the hydraulic properties of the three different soil layers were determined by best fitting the numerical results with the suction measurements at tensiometers T2 to T4. Scheme 2 was used to estimate a reliable range of evapotraspiration rates (i.e. net rainfall at the ground surface) by best fitting the results of the calibrated model with the measurements at tensiometer T5.

The analyses were carried out using the Finite Element code SEEP/W (GeoSlope 2004), which is able to integrate the well known Richards' differential equation governing the transient saturated—unsaturated water flow in soils. The one-dimensional domain was discretised using quadrilateral elements with secondary nodes (Fig. 5). Three different soil layers were considered, which refer to the saprolitic (A), residual (B) and upper colluvial (C) layers.

The time-dependent boundary conditions of the two schemes are different. In the first one, pressure boundary conditions were set at the locations of the lowermost and uppermost tensiometers (T1 and T5) and were assumed equal to their record of measurements. In the second scheme, the lower pressure boundary condition was set at the location of tensio-meter T4; while a transient flux condition, corresponding to estimates of the net daily rainfall intensities based on monthly evapotraspiration rates, was applied at the ground surface (i.e. upper boundary).

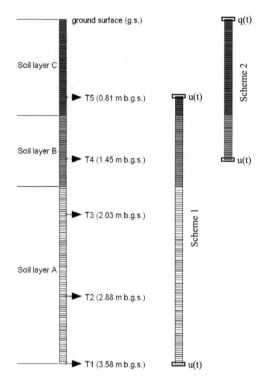

Figure 5. Transient seepage Finite Element model of the vertical column where tensiometers T1 to T5 are installed: mesh and boundary conditions adopted for Scheme 1 and Scheme 2.

For both schemes the transient numerical analysis was carried out with daily time steps covering the two-year long monitoring period of Figure 3. The assumed initial suction distribution was given by a steady-state analysis in which the boundary conditions refer to the first available monitoring data.

For the Scheme 1, the hydraulic properties of the three layers were assumed unknowns, but defined by two analytical relationships. Particularly, the relationship proposed by Van Genuchten (1980) was used for the variation of the volumetric water content against suction, while for the hydraulic conductivity function the relationship proposed by Mualem (1976) was assumed. Both functions are represented by the following:

$$\theta(s) = \frac{\theta_s}{\left[1 + \left(\frac{s}{a}\right)^n\right]^m} \quad (1)$$

$$K(s) = K_s \frac{\left[1 - (As)^{N-1}(1 + (As)^N)^{-M}\right]^2}{\left[1 + (As)^N\right]^{M/2}} \quad (2)$$

where: s = suction; a, n, m, A, N and M = curve fitting parameters. If these last three parameters are assumed dependent from the previous ones through the relations $A = 1/a$; $N = n$; $M = 1 - 1/n$, equations (1) and (2) are completely defined by only five parameters (namely a, n, m, θ_s and K_s).

4.2 Scheme 1

Scheme 1 in Figure 5 was used to calibrate the five unknown parameters defining equations (1) and (2) for each of the three soil layers.

The five independent parameters characterizing each soil layer were determined using an inverse analysis algorithm to minimize the error between the measured suctions at tensiometers T2, T3 and T4 and the corresponding computed results. Figure 6 shows a flowchart of the inverse analysis procedure. A regression analysis was performed to minimize an objective function, which quantifies the fit between computed results and observations. The minimization was attained by the optimization of the input parameters needed to perform the numerical model. If the model fit was not "optimal", the procedure was repeated until the model is optimized. For details on the procedure used and on how to choose the relevant parameters to optimize by inverse analysis see Calvello & Finno (2002, 2004).

Figure 7 shows the model boundary conditions and the comparison between the results of the calibrated model and the measured suction values at tensiometers T2, T3 and T4. The results clearly indicate that the model well reproduces the recorded behaviour both at low and high suction values.

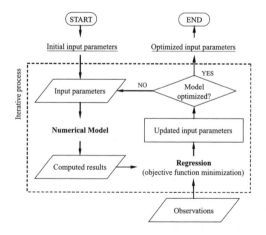

Figure 6. Flow chart of inverse analysis (from Calvello & Finno 2004).

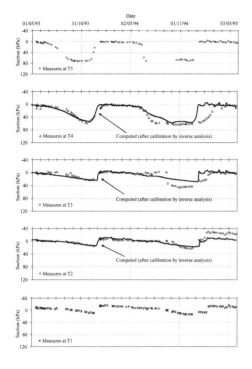

Figure 7. Boundary conditions (tensiometers T1 and T5) and comparison between measured suction values at inner tensiometers (T2, T3 and T4) and computed results after calibration by inverse analysis.

Table 2. Parameters of volumetric water content curve and hydraulic conductivity function before and after calibration.

Layer Parameter	Initial estimates			After calibration		
	A	B	C	A	B	C
θ_s	0.32	0.32	0.35	0.32	0.32	0.35
K_s (m/d)	0.11	0.04	0.04	0.11	**0.70**	**0.24**
a (kPa)	20 (range 1–29)			**6**	**10**	20
n	1.5 (range 1.01–2.13)			1.5	**1.35**	**1.1**
m	0.33 (range 0.18–0.57)			0.33	0.33	0.33

*in **bold** parameters calibrated by the inverse analysis.

The calibrated values of the 15 independent model parameters are reported in Table 2 together with their initial estimates. The initial saturated volumetric water content and coefficient of permeability were set equal to the mean values reported by Sorbino (1995). Initial values of the parameters a, n and m were estimated by best fitting each experimental curve (see Figure 4) with the functions defined by equations (1) and (2).

Only 6 parameters were calibrated by the inverse analysis. The results show some significant differences among the initial and final estimates of these parameters. In particular, the values of the calibrated saturated conductivities for soil layers 2 and 3 are not equal as initially assumed and are higher than their initial estimates. The calibration also lead to a differentiation of the shape of the curves for the three soil layers, as indicated by the different values of the first two curve fitting parameters, a and n.

The comparison between the experimental curves of Figure 4 and the calibrated curves for each soil layer is shown in Figure 8. The results indicate that the calibrated hydraulic conductivity functions well compare with the experimental results, while the calibrated volumetric water content curves show a more rapid desaturation of the soils with suction than the measured ones. This could depend by the presence of largest *in-situ* pore networks that are not adequately represented by the small dimensions of the specimens used in the laboratory tests.

4.3 *Scheme 2*

Scheme 2 defines a domain bounded by the ground surface and the location of tensiometer T4 (Fig. 5). The hydraulic properties of the two soil layers of the model (i.e. layers B and C) were assumed equal to the corresponding calibrated pair of curves in Figure 8. At the lower boundary, a time-dependent pressure condition based on the values measured at T4 was used. At the upper boundary a flux condition was applied,

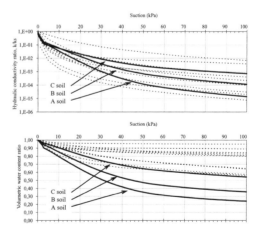

Figure 8. Dimensionless comparison among the experimental soil water characteristic curves and hydraulic functions (dashed lines) and the calibrated curves for soil layers A (lower solid line), B (middle solid line) and C (upper solid line).

based on daily rainfall measures and on the estimates of the evapotraspiration rate.

To this end, the procedure suggested by Tarantino et al. (2002) was adopted. This is based on a comparison between the potential evapotraspiration (E_p) (i.e. the maximum evaporation rate in the case of water availability) and the water limiting evapotraspiration (E_{lim}), which is related to the capacity of the soil to transmit water to the atmosphere. The value of daily evaporation rates for the definition of the flux boundary is then fixed, in the numerical analyses, equal to the lowest value between E_p and E_{lim}. A detailed description of the procedures used for the evaluation of E_p and E_{lim} is reported in Sorbino (2005).

The estimation of E_p was carried out adopting the Thornthwaite's empirical equation (Thornthwaite 1954) and using air temperatures recorded at the test site. The limiting evapotraspiration, E_{lim}, was estimated using the steady-state flow solution proposed by Gardner (1958) for a homogenous soil, which is a function of the saturated and unsaturated hydraulic properties of the soil and of the depth of the water table.

Figure 9 shows the cumulative total rainfall and the cumulative net flux at the ground surface for two different evapotranspiration (EVT) scenarios. The two scenarios refer to two different estimates of the hydraulic properties of the soil between T4 and the ground surface, assumed as homogeneous in the Gardner solution, leading to two different estimates of the limiting evapotranspiration.

In Figure 10 the measured suction values at tensiometer T5 and the computed results are compared for the two different evapotranspiration scenarios. The Figure clearly shows that, when the soil is highly heterogeneous, the estimation of the evapotranspiration rate can lead to inadequate definition of the boundary conditions at the ground surface. On the contrary, the reliability of its estimation can be highly improved when adequate experimental data on suction are available.

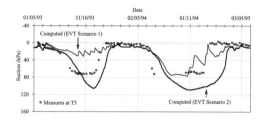

Figure 10. Comparison between measured suction values and computed results at tensiometer T5 for two different estimates of the net rainfall infiltration rate.

5 CONCLUSIONS

In the paper, the *in-situ* soil suction regime is modelled at a test site, to be considered representative of the subsoil conditions for landsliding that affects a large area where weathered gneiss prevails.

The comparison between the model results and a two-year long record of *in-situ* measurements, relative to 5 tensiometers installed along the same vertical, shows that the soil heterogeneity and the net flux across the ground surface play a relevant role in the soil suction regime.

As for the heterogeneity, the laboratory characterization of the hydraulic properties of these soils, albeit accurate, may not be sufficient to define reliable values for the unsaturated flow analysis at the slope scale. To this aim, accurate *in-situ* tensiometric measures may instead be profitably used. As for the net flux across the ground surface, the results underline how the estimation of the evapotranspiration rate is equally relevant. In this case, the tensiometric measurements may be used to define a range of possible evapotranspiration rates.

Tensiometric measurements and the presented procedure may also be useful to address further goals. Among these, forecasting the recharge of perched water tables at site-scale and/or predicting soil suction regime for the landslides involving class VI and V soils that have been mapped over the whole study area of the Western Sila Massif.

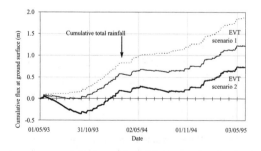

Figure 9. Cumulative total rainfall and net flux at the ground surface for two different estimates of the net rainfall infiltration rate (EVT scenarios 1 and 2).

REFERENCES

Brand, E.W. 1984. Landslides in Southeast Asia: A State-of-the-Art Report. *Proc. 4th ISL*. Toronto, Canada, pp. 17–59.

Calvello, M. & Finno, R.J. 2002. Calibration of soil models by inverse analysis. *Proc. NUMOG VIII, Rome*, pp. 107–113.

Calvello, M. & Finno, R.J. 2004. Selecting parameters to optimize in model calibration by inverse analysis. *Computers and Geotechnics* 31 (5): 411–425.

Cascini, L. Gullà, G. & Sorbino, G. 2006. Groundwater modelling of a weathered gneissic cover. *Canadian Geotechnical Journal* 43 (11): 1153–1166.

Cascini, L. Critelli, S. Di Nocera, S. & Gullà, G. 1992a. A methodological approach to landslide hazard assessment: A case history. *Proc. 6th ISL, Christchurch,* 2: 899–904.

Cascini, L. Critelli, S. Di Nocera, S. & Gullà, G. 1994. Weathering and landsliding in Sila Grande Massif gneiss (Northern Calabria, Italy). *Proc. 7th IAEG, Lisbon,* pp. 1613–1622.

Cascini, L. Critelli, S. Di Nocera, S. Gullà, G. & Matano, F. 1992b. Grado di alterazione e franosità negli gneiss del Massiccio Silano: L'Area di S. Pietro in Guarano. *Geologia Applicata ed Idrogeologia* 27: 49–76.

Gardner, W.R. 1958. Some steady-state solutions of the unsaturated moisture flow equation with application to evaporation from a water table. *Soil Science* (85): 228–232.

GCO 1988. Geoguide 3: guide to rock and soil descriptions. *Geotechnical Control Office (GCO), Civil Engineering Services Department, Hong Kong.*

GEO-SLOPE International Ltd. 2004. Seepage modelling with SEEP/W, user's guide version 6.16. *GEO-SLOPE International Ltd., Calgary, Alberta.*

Gullà, G. & Sorbino, G. 1996. Soil suction measurements in a landslide involving weathered gneiss. *Proc. of the 7th ISL, Trondheim, Norway* 2: 749–754.

Gullà, G. & Matano, F. 1997. Surveys of weathering profile on gneiss cutslopes in Northern Calabria, Italy. *Proc. 8th IAEG, Athens,* pp. 133–138.

Lacerda, W.A., 2004. The behavior of colluvial slopes in a tropical environment. *Proc. 9th ISL. Rio de Janeiro,* pp. 1315–1342.

Mualem, Y. 1976. A new model for predicting the hydraulic conductivity of unsaturated porous media. *Water Resources Res.* 12: 513–522.

Sorbino, G. 1995. Unsaturated hydraulic characteristics of weathered gneissic soils. *Proc. 10th Panam. Conf. Guadalajara,* pp. 25–35.

Sorbino, G. 2005. Numerical modelling of soil suction measurements in pyroclastic soils. Proc. Conf. on Advanced Experimental Unsaturated Soil Mechanics, Trento, pp. 541–547.

Tarantino, A. Mongiovì, L. & Mc Dougall, L.R. 2002. Analysis of hydrological effects of vegetation on slope stability. *Proc. 3rd Int. Conf. on Unsat. Soils, Recife* (2): 749–754.

Thornthwaite, C.W. 1954. A re-examination of the concept and measurement of potential transpiration. *In The measurement of potential evapo-transpiration (ed. J.R. Mather),* 200–209.

Van Genuchten, M.Th. 1980. A closed-form equation for pre-dicting the hydraulic conductivity of unsaturated soil. *Soil Science Society American Journal* 44: 892–898.

Modelling the transient groundwater regime for the displacements analysis of slow-moving active landslides

L. Cascini, M. Calvello & G.M. Grimaldi
Department of Civil Engineering, University of Salerno, Italy

ABSTRACT: Active slow-moving landslides in clayey soils exhibit continuous movements generally related to changing ground water levels. This paper highlights the importance of groundwater modelling for the prediction of these types of movements through the analysis of a well-documented case history: the Porta Cassia landslide on the Orvieto hill in central Italy (e.g. Tommasi et al. 2006). The analysis uses recorded rainfall and pore pressure data to define a reliable model of the transient groundwater regime in the slope, which is then used to derive the time-dependent shear stresses along the main slip surfaces. The displacements at selected points along the slip surface are then computed using a relationship between the local factor of safety and the displacement rate at those points.

1 INTRODUCTION

According to the general framework proposed by Leroueil et al. (1996) for the geotechnical characterisation of slope movements, active landslides move along one or several pre-existing slip surfaces, where the mobilised shear strength corresponds to residual conditions. When most of the landslide shear deformations occur in a narrow shear zone above the main slip surfaces, many authors consider the nature of the deformations as viscous (e.g. Vulliet & Hutter 1988). In this case, the displacements rate is related to changing groundwater levels that affect the shear stress level along the slip surface.

The international literature offers different models to compute landslide displacements from measured pore pressure data (Angeli et al. 1996; Corominas et al. 2005; Mandolini & Urciuoli 1999; Maugeri et al. 2006; Van Asch 2005; Van Asch et al. 2007; Vulliet & Hutter 1988). The main limitation of such models is the need for reliable pore pressure measurements along the slip surfaces. To overcome this limitation, a procedure is proposed which relates the displacement rates along the slip surfaces to the pore pressures computed by a physically-based analysis of the groundwater regime.

The analysis refers to an active translational landslide in stiff clays on the Orvieto hill, in central Italy (Tommasi et al. 2006). The analysis uses recorded rainfall and pore pressure data to define a reliable model of the transient groundwater regime in the slope, which is then used to derive the time-dependent shear stresses along the main slip surfaces. The rate of displacement in a given point along the slip surface at a given time is then computed using one of the phenomenological relationships proposed by Vulliet & Hutter (1988), which relates the displacement rate to the time-dependent local factor of safety (i.e. ratio of the average shear strength to the average shear stress).

2 THE PORTA CASSIA EARTH SLIDE

2.1 Site description

Orvieto is an ancient town in central Italy resting on a tufaceous slab overlying an overconsolidated clayey hill. This town has long been affected by many instability phenomena (Tommasi et al. 1986). Among these, a sudden reactivation occurred in November 1900, when an earth slide movement, successively named Porta Cassia landslide (Figure 1), affected the northern slope of the hill causing lots of damages and interrupting a major road and the Rome-Florence railway (Vinassa de Regny, 1904). The landslide is still active.

With the aim of collecting data for a quantitative confirmation of the mechanism governing the slope evolution (Tommasi et al. 1997), two monitoring programmes started in 1982 and 1996–1998 and thus a comprehensive data set of pore pressure and displacements measurements is now available. Figure 1 shows the map of the Orvieto hill, the boundaries of the Porta Cassia landslide on the northern flank of the slope and the location of the main monitoring stations.

2.2 Analysis of monitoring data

Figure 2 shows the stratigraphy of a representative cross-section of the landslide (cross-section 1–1' from

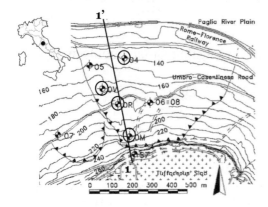

Figure 1. Boundaries of the Porta Cassia landslide on the northern flank of the Orvieto hill and location of the monitoring stations along cross-section 1–1' (from Tommasi et al. 2006).

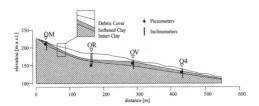

Figure 2. Representative cross-section and location of the monitored piezometers and inclinometers.

Figure 1). The three layers geotechnical section has been drawn using reported stratigraphic data and geotechnical properties and includes: a clayey debris cover, varying in depth from a few meters to 20 meters; a softened clay stratum, whose thickness varies between two and ten meters; an intact overconsolidated clayey bedrock.

The Figure also shows the location of the instrumented boreholes for which piezometric and inclinometric measurements are available (Tommasi et al. 2006). As for the piezometers, OR was located in the lower stratum, at a depth of 31 m b.g.s., and has been regularly monitored since 1988; while OV, OM and O4 were installed within the upper two layers, respectively at 6, 4.5 and 6 m b.g.s., and have been monitored since 1997, 1996 and 1998. As for the inclinometers, OR has been installed in the central part of the slope in 1982, and was since regularly monitored, while OM, OV and O4 were installed in 1996–1998, respectively in the upper, middle and lower part of the slope.

Figure 3 reports the groundwater levels recorded, between 1996 and 2000, by the 4 piezometers shown in Figure 2. The measurements highlight the existence of two different ground water regimes in the upper and

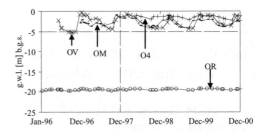

Figure 3. Measured groundwater levels of available piezometers (modified after Tommasi et al. 2006).

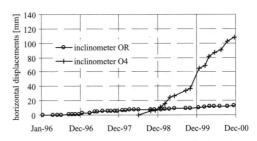

Figure 4. Cumulative horizontal displacements measured by inclinometers OR and O4 along the slip surface (modified after Tommasi et al. 2006).

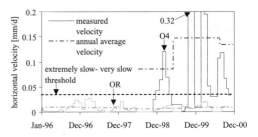

Figure 5. Measured and annual average velocities for the verticals OR and O4 showing two different displacement rates.

lower strata. Indeed, the deeper piezometer, OR, only records modest variations of the groundwater level (less than 1 m); the other three, all installed within the upper strata, show seasonal changes of the groundwater level, with differences of a few meters between the dry and the wet season. The available data suggest that changes of pore water pressures in the debris cover are not significantly influenced by the groundwater regime in the deeper clayey bedrock.

Figure 4 reports the horizontal displacements recorded, between 1996 and 2000, by two of the 4 inclinometrs shown in Figure 2. It is clear that two movements with two different displacement rates exist: a slower one in the upper part of the slope, as recorded by OR at about 20 m b.g.s.; and an order of magnitude faster movement in the lower part of the slope, as recorded by O4 at about 7 m b.g.s.. As shown in Figure 5, which reports both the measured and the annual average velocities of the two movements, O4 records a "very slow" phenomenon while OR an "extremely slow" movement, according to the classification by Cruden & Varnes (1996).

2.3 Landslide classification

The in-situ investigation and the monitoring data indicate that the Porta Cassia landslide can be classified as an active (Leroueil et al. 1996) roto-translational earth slide in clayey material (Varnes, 1978) with very slow maximum measured velocity (Cruden & Varnes, 1996). Within this definition, the term active is here used to mean a phenomenon sliding along one or several pre-existing shear surfaces, where the mobilized shear strength corresponds to residual conditions, without any references to the time and/or the duration of the movements.

3 GROUNDWATER MODEL

3.1 Model description and hypotheses

Figure 6 shows the cross-section and the mesh used in the finite elements model of the transient groundwater flow in the slope. The analysis was performed using the commercial finite element code SEEP/W (GEO-Slope 2004). The mesh only considers the debris cover and the softened clay layers, not explicitly considering the bedrock layer. According to Tommasi et al. (2006) and consistently with the pore pressure measures shown in Figure 3, the flow regime characterizing the stiff clay at depth was assumed not influencing the seasonal changes of pore water pressures inside the slope formed by the softened clay formation and the clayey debris cover.

On the basis of this hypothesis, the lower boundary of the mesh was assumed to be impermeable. As for the left boundary, the mesh ends at the location of the monitored vertical OM. Along this vertical, hydrostatic conditions were assumed with the stationary groundwater level set at the ground surface. These are justified by the presence of a spring at that location (Lembo-Fazio et al. 1984). As for the right boundary, the model section ends at the foot of the slope. The hydrostatic conditions with the ground water level at the ground surface are justified by the presence of a River (Paglia River) at that location. As for the top boundary, rainfall and temperature measures have been used to compute the net rainfall flux to apply at the ground surface. The net rainfall was computed subtracting the evapotraspiration term to the total monthly recorded rainfall, using the Thornthwaite method (Thornthwaite 1948). The frame on the upper right corner of Figure 6 shows the total and net monthly rainfall during the five years considered in the analysis (1996–2000).

As for the hydraulic properties of the two layers, the lower one was assumed to be saturated at all times, while the upper clayey debris cover was assumed to be in partially saturated conditions, in accordance with experimental evidence by Cafaro et al. (2005). Figure 7 shows the hydraulic conductivity function and the soil water characteristic curve for this upper layer. In the graphs, both the hydraulic conductivity and the volumetric water content are divided by their saturated values. The shape of both curves is very close to the shape of the laboratory tests curves reported by Cafaro et al. (2005).

3.2 Results of the analysis

The results of the groundwater model are reported with reference to the two phases of the analysis: the initial stationary conditions and the successive transient

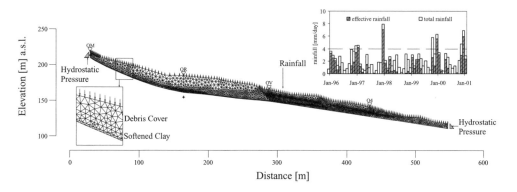

Figure 6. Section of the finite element transient groundwater model of the Porta Cassia landslide and chart of the total and effective rainfall between 1996 and 2000.

analysis. For the former, the calibration of the hydraulic conductivities of the two layers was conducted by minimizing the error between the groundwater level measured and computed at piezometer OV in November 1997. For the latter, a calibration stage and a validation stage have been defined when comparing the numerical results with the piezometric measures of the pore water pressure.

The stationary boundary conditions of the model are the ones described in the previous paragraph for the left, bottom and right boundaries of the mesh (Figure 6), while a null value of flux is assumed at the ground surface. Figure 8 shows the ratio between the saturated hydraulic conductivity of the clayey debris cover (k1) and the softened clay (k2) plotted against the difference between measured and computed groundwater levels at OV (ΔH). The best results, i.e. ΔH null, are obtained for a ratio of about 5.2, indicating that the permeability of the debris cover is about five times larger than the one of the softened clay layer. Tommasi et al. (2006) report differences of about one order of magnitude between the two hydraulic conductivities.

As for the transient analysis, a one-year long calibration stage was defined using the values of pore pressures measured at piezometer OV between November 1996 and October 1997. The ground water levels measured at O4 and all the other measurements at OV were instead used to validate the results of the calibrated model. Figures 9 and 10 show the main results of the analysis with reference to points along the monitored verticals OV, O4 and OR. The extremely satisfactory results of the analysis are confirmed by the values of the coefficients of correlation between computed and measured ground water levels at piezometer OV, respectively equal to 0.981 and 0.874 for the calibration and validation stages. As for the point on the slip surface along vertical OR (Figure 10), a comparison with measured data is not shown because piezometer OR is installed at 31 m b.g.s. and thus well below the slip surface at that location. The numerical results of the calibrated and validated groundwater analysis are, therefore, even

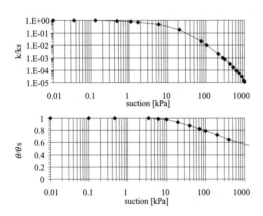

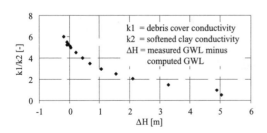

Figure 7. Hydraulic properties of the clayey debris cover: shape of the hydraulic conductivity function and of the soil water characteristic curve.

Figure 8. Stationary analysis: ratio between the saturated hydraulic conductivities of the debris cover and of the softened clay vs. difference between computed and measured groundwater levels at OV.

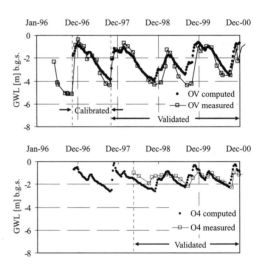

Figure 9. Results of the transient analysis: comparison between computed and measured groundwater levels at piezometers OV and O4.

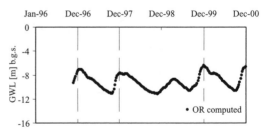

Figure 10. Results of the transient analysis: computed groundwater level on the slip surface along the vertical OR.

Table 1. Values of the calibrated saturated hydraulic conductivities for the two layers.

Layer [-]	Hydraulic conductivity [m/s]
Clayey debris cover	$7 \cdot 10^{-6}$
Softened clay	$1.16 \cdot 10^{-6}$

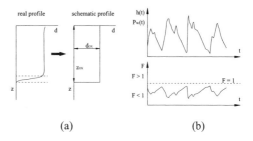

(a) (b)

Figure 11. Schematic representation of the movement along an infinitesimal slip surface and correlation between the computed pore water pressure and "local" factor of safety.

more significant at this location, for which measures of the displacements rate along the slip surface are available.

Table 1 reports the values of the calibrated saturated hydraulic conductivities for the two layers. For both the debris cover and the softened clay layers, the values of the calibrated saturated conductivities are significantly higher than the values reported by Tommasi et al. (2006). These are respectively equal to 10^{-10} and 10^{-9} m/s and were derived from falling head tests in Casagrande piezometers. A possible reason for that is the presence, in the softened clay layer, of oxidized joints (Lembo-Fazio et al. 1984; Tommasi et al. 1986) acting as preferential flow paths. Another factor to consider could be that at times, as reported by Senneset and Sandven (1987), the disturbance induced in the soil by the installation of the piezometers may cause a reduction of the measured coefficients of permeability.

4 DISPLACEMENT MODEL

4.1 Model description and results

The kinematic model, used to compute the time-dependent displacements of the Porta Cassia landslide, employs a phenomenological relationship between the velocity of the slide and the shear stress level acting along the slip surface. The latter, in turn, is a function of the time-dependent pore water pressures computed by the transient groundwater analysis. A detailed description of the model is presented in Calvello et al. (2007). Herein, only the main assumptions of the model and the most significant results are discussed.

Figure 11 shows that the landslide body is assumed to move as a rigid block along a well defined infinitesimal slip surface, at a depth equal to the average depth of the deep creep zone, z_{OX} (Figure 11a). The model is defined as "local" in that the analysis is carried out with reference to the stress state at a single point on the slip surface. The state of stress at the selected point along the slip surface is computed using the infinite-slope scheme, assuming the slip surface parallel to the ground surface and the computed pore water pressure, $P_w(t)$, as the only component of the local factor of safety varying with time (Figure 11b).

The displacements along the slip surface, d_{OX}, are computed introducing a threshold level of shear stress (i.e. the residual shear strength in the Mohr-Coulomb plane) below which there are no movements. Above the threshold, a phenomenological relationship is defined relating the velocity of the movement to the local factor of safety. The following relationship, first proposed by Vulliet and Hutter (1988), was used:

$$v_{OX} = B \cdot \left(\frac{1}{F(t)}\right)^n \Rightarrow d_{OX} = \sum v_{OX} \cdot \Delta t \quad (1)$$

where B and n are phenomenological constants representing, respectively, the typical velocity of the phenomenon and the variation of viscosity with the excess shear stress (Van Asch et al. 2007).

The displacement monitoring data from inclinometer O4 (Figure 4) were used both for calibrating the input parameters of the model and for validating the results of the calibrated model. Particularly, the measures relative to the period July 1998—November 1999 were used for calibrating the model, while the rest of measurements were used to validate it. This was done in order to use some of the observations for evaluating the "predictive capability" of the model.

Figure 12 shows the results of the model with reference to both the calibration and the validation stages. The fit between the numerical results and the observations is almost perfect for the calibration stage and remains extremely positive for the validation stage. Indeed, the coefficient of correlation between computed and measured displacements is 0.995 for the first stage and a 0.953 for the second one. The values of the parameters of the calibrated model are summarized in Table 2, where the creep threshold was set equal to the measured residual friction angle, as determined by direct shear tests by Lembo Fazio et al. (1984).

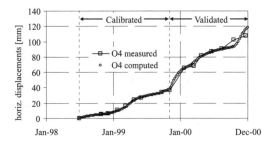

Figure 12. Measured and computer horizontal displacements for the selected point on the slip surface along the vertical O4.

Table 2. Values of the creep threshold and of the coefficients B and n calibrated with the kinematic model for the vertical O4.

Vertical [-]	Creep threshold [°]	Coefficient B [m/d]	Coefficient n [-]
O4	11.3	1.08 e-6	10

5 CONCLUSIONS

This paper underlines the importance of groundwater modelling for the analysis of active slow-moving slides in clayey materials. Key factors of this analysis are rainfall and pore pressure data which can be used to define a reliable model of the transient groundwater regime in the slope. The obtained results can then be used to derive the time-dependent shear stresses along the slip surface. The displacements at selected points along the slip surface can finally be computed using a relationship between the inverse of the time dependent local factor of safety and the displacement rate at those points.

This model has been applied to the Porta Cassia slide, a well-monitored slow-moving active slide in Central Italy. The comparison between the monitoring data and the corresponding computed results is extremely satisfactory for both the groundwater analysis and the displacement model.

The proposed model could be used as a predictive tool to obtain possible displacements scenarios interpreting the evolution of a landslide when rainfall scenarios, based on statistical analysis, are defined. This aspect is particularly relevant when the mechanism governing the landslide evolution is not straightforward. For instance, for the Porta Cassia landslide, one may evaluate whether the sudden reactivation occurred in 1900 was related to exceptional rainfall (i.e. rainfall with a high return period) or if other factors need to be taken into account.

REFERENCES

Angeli, M.G., Gasparetto, P., Menotti, R.M., Pasuto, A. & Silvano, S. 1996. A visco-plastic model for slope analysis applied to a mudslide in Cortina d'Ampezzo, Italy. *Quarterly Journal of Engineering Geology* 29: 233–240.

Cafaro, F., Boldini, D. & Tommasi, P. 2005. Drying behaviour of the Orvieto overconsolidated clay. *International Conference on Problematic Soils, Bilsel and Nalbantoglu eds, Famagusta* 1: 95–102.

Calvello, M., Cascini, L. & Grimaldi, G.M. 2007. Displacements scenarios at site scale of rainfall-controlled slow moving active slides in stiff clays. *Proc. ISSGSR 2007, Int. Symp. on Geotecnical Safety and Risk, Shanghai, China* In Press.

Corominas, J., Moya, J., Ledesma, A., Lloret, A. & Gili, J.A. 2005. Prediction of ground displacements and velocities from groundwater level changes at the Vallcebre landslide (Eastern Pyrenees, Spain). *Landslides* 2: 83–96.

Cruden, D.M. & Varnes, D.J. 1996. Landslide types and processes. *Landslides: investigation and mitigation. Transportation Research Board, Special Report No. 247, National Research Council*, National Academy Press, Washington DC, USA 36–75.

GEO-SLOPE 2004. Seepage modelling with SEEP/W, User's Guide (version 6.16). *GEO-SLOPE International Ltd.* Calgary, Alberta, Canada.

Lembo-Fazio, A., Manfredini, G., Ribacchi, R. & Sciotti, M. 1984. Slope Failure and Cliff Instability in the Orvieto Hill. *Proceedings of the 4th International Symposium on Landslides, Toronto, Canada* 115–120.

Leroueil, S., Locat, J., Vaunat, J., Picarelli, L., Lee, H. & Faure, R. 1996. Geotechnical characterization of slope movements. *Proceedings 7th International Symposium Landslides, Trondheim, Norway* 1: 53–74.

Mandolini, A. & Urciuoli, G. 1999. Previsione dell'evoluzione cinematica dei pendii mediante un procedimento di simulazione statistica. *Rivista Italiana di Geotecnica* 35–42.

Maugeri, M., Motta, E. & Raciti, E. 2006. Mathematical modelling of the landslide occurred at Gagliano Castelferrato (Italy). *Natural Hazards and Earth System Sciences* 6: 133–143.

Senneset, K. & Sandven, R. 1987. Field and laboratory testing—General report. *Proceedings IX European Conference on Soil Mechanics and Foundation Engineering, Dublin, Ireland.*

Thornthwaite, C.W. 1948. An approach toward a rational classification of climate. *Geographical Review* 38: 55–94.

Tommasi, P., Ribacchi, R. & Sciotti, M. 1986. Analisi storica dei dissesti e degli interventi sulla rupe di Orvieto: un ausilio allo studio del'evoluzione della stabilità del centro abitato. *Geologia Applicata ed Idrogeologia* XXI: 99–153.

Tommasi, P., Ribacchi, R. & Sciotti, M. 1997. Slow Movements along the slip surface of the 1900 Porta Cassia landslide in the clayey slope of the Orvieto Hill. *Rivista Italiana di geotecnica* 2: 49–58.

Tommasi, P., Pellegrini, P., Boldini, D. & Ribacchi, R. 2006. Influence of rainfall regime on hydraulic conditions and movement rates in the overconsolidated clayey slope of the Orvieto hill (central Italy). *Canadian geotechnical journal* 43: 70–86.

Van Asch, T. 2005. Modelling the hysteresis in the velocity pattern of slow-moving earth flows: the role of excess pore pressure. *Earth Surface Processes and Landforms* 30: 403–411.

Van Asch, Th.W.J., Van Beek, L.P.H. & Bogaard, T.A. 2007. Problems in predicting the mobility of slow-moving landslides. *Engineering Geology* 91: 46–55.

Varnes, D.J. 1978. Slope movement types and processes. *Landslide Analysis and Control. Transportation Research Board Special Report No. 176,* National Academy of Sciences, Washington DC, USA 11–33.

Vinassa de Regny, P. 1904. Le frane di Orvieto. *Giornale di Geologia Pratica* 1: 110–130.

Vulliet, L. & Hutter, K. 1988. Viscous-type sliding laws for landslides. *Canadian Geotechnical Journal* 25: 467–477.

Numerical modelling of the thermo-mechanical behaviour of soils in catastrophic landslides

F. Cecinato & A. Zervos
School of Civil Engineering & Environment, University of Southampton, UK

E. Veveakis & I. Vardoulakis
Faculty of Applied Science, National Technical University of Athens, Greece

ABSTRACT: A new landslide model is proposed by improving on an existing one, which is able to interpret using a simple 1-D mechanism the post-failure sliding regime of catastrophic landslides and rockslides consisting of a coherent mass sliding on a thin clayey layer. The model takes into account frictional heating and subsequent pore pressure build-up, leading to the vanishing of shear resistance and unconstrained acceleration. First, an existing thermo-elasto-plastic constitutive model for clays is discussed, and modified by re-formulating it in a general stress space and taking into account thermal softening. The soil constitutive model is then employed into an existing landslide model. The resulting model equations are shown to be well-posed, and then are discretised and integrated numerically to back-analyse the final stage of the well-documented case history of Vajont that occurred in Italy in 1963. Finally, the results are used to highlight the possible importance of thermal softening in the development of catastrophic failure.

1 INTRODUCTION

The Vajont landslide of October 9, 1963, has been the subject of numerous geological and geomechanical investigations, due both to its potential contribution to slope stability analysis and to the social and legal implications of the disaster. The landslide moved approximately 2.7×10^8 m³ of rock into an artificial reservoir of about 1.5×10^8 m³, impounding the Vajont deep gorge. The slide moved an 120 m thick (on average) compact rock mass over a front of 1850 m for a maximum slip of 450–500 m (Hendron and Patton, 1985) and at a final slip rate of about 25–30 m/s. The abrupt filling of the reservoir with debris produced a giant wave (4.8×10^7 m³) that propagated up and down the valley, overflowing the dam and wiping out the village of Longarone, located 2 km west.

Habib (1975) proposed that the high slip velocity achieved by the Vajont landslide was due to the conversion of mechanical energy into heat during frictional sliding, which should lead to the "vaporization" of pore water and hence to a cushion of zero friction. Temperature increase in the slipping zone may also have led to pressurization of pore water with the same effect on the shear strength of the slope (Anderson, 1980; Voight and Faust, 1982; Vardoulakis, 2000, 2002). Total loss of strength by thermal pressurization has also been claimed for the Jiufengershan rock and soil avalanche triggered by the Chi-Chi (Taiwan) 1999 earthquake (Chang et al., 2005a, 2005b).

Vardoulakis (2000, 2002) analyzed the pressurization phase of the Vajont slide, when thermal pressurization sets in, during which the slide accelerates rapidly. He proposed a one-degree-of-freedom, frictional pendulum model, employing a Mohr-Coulomb constitutive model for the soil and assuming that frictional heating triggered pore water pressurization inside a shear band of the order of 1 mm. This analysis showed that the catastrophic pressurization phase of the Vajont slide should not have taken more than a few seconds to develop in full.

In this paper we extend the above study by using a more general thermo-elasto-plastic constitutive model, based on the one recently proposed by Laloui et al. (2005). Furthermore we investigate the impact of thermal softening, which some clays exhibit, in the development of the catastrophic mechanism.

In the following, we present in section 2 the landslide model developed by Vardoulakis (2000, 2002), and in section 3 the new thermo-elasto-plastic constitutive model. Section 4 deals with the modification of the landslide model to include this new constitutive law. Finally, in section 5 some computational results are presented and discussed and conclusions are drawn in section 6.

2 LANDSLIDE MODEL

A thermo-poro-mechanical landslide model has been derived by Vardoulakis (2000, 2002) from first principles, and successfully employed to back-analyse the catastrophic sliding phase of the Vajont slide. In landslides of that type, all deformation is assumed to be concentrated on a thin clayey layer with thickness of the order of 1 mm (Tika & Hutchinson 1999, Vardoulakis 2002), overlain by a coherent mass sliding as a rigid block.

The model consists of a set of coupled partial differential equations describing the time evolution of temperature, pore pressure and velocity within the shear-band. As the rigid mass starts sliding, strain and strain-rate softening occur within the shear-band, alongside frictional heating. As soon as the critical temperature for thermoplastic collapse of the soil skeleton is reached, pore pressure build-up takes place leading to catastrophic acceleration.

The equation describing the movement of the rigid block is obtained through moment balance, using a modified slip-circle method. The resulting velocity is the upper boundary condition for the shear-band velocity profile, thus providing the coupling between shearband and sliding mass.

The model consists of the following equations:

$$\begin{cases} \dfrac{\partial \theta}{\partial t} = k_m \dfrac{\partial^2 \theta}{\partial z^2} + \dfrac{D}{j(\rho C)_m} \\ \dfrac{\partial u}{\partial t} = \dfrac{\partial}{\partial z}\left(c_v \dfrac{\partial u}{\partial z}\right) + \lambda_m \dfrac{\partial \theta}{\partial t} \\ \dfrac{dv_d}{dt} = R\omega_0^2 \left(A(\mu_m) + \dfrac{u_d}{p_c(\mu_{mf})}\right). \end{cases} \quad (1.1)$$

The first equation describes the evolution of temperature. The first term on the right-hand side is a diffusion term, where k_m is a constant diffusivity coefficient, and the second one is a heat generation term, where D is the dissipation and is taken equal to the plastic work. The second equation describes the diffusion of excess pore pressure u, where c_v a temperature-dependent consolidation coefficient and λ_m a temperature- and pressure-dependent coefficient determining the generation of excess pore pressure. It is zero until the temperature reaches the critical value for thermal collapse of the soil skeleton. Finally, the third equation describes the time evolution of the slide velocity and is composed of two terms that depend on mobilised friction and pore pressure.

3 THERMOPLASTIC CONSTITUTIVE MODEL

3.1 Problem formulation

A simple constitutive model for clays taking into account thermo-plastic behaviour has been recently developed by Laloui and co-workers (Laloui 2001, Laloui & Cekerevac 2003, Laloui et al., 2005). It was formulated in 'triaxial' stress space and it was used to simulate isothermal isotropic compression for different temperatures and thermal loading paths (Laloui & Cekerevac 2003).

Laloui et al. (2005) added to the isotropic thermal law a deviatoric mechanism based on original Cam-Clay within the framework of multi-mechanism plasticity, which implied the use of two separate plastic multipliers. A simple linear law of thermal softening was also proposed, as experimental evidence summarised by Laloui et al. (2001) shows that the friction coefficient of some soils decreases with increasing temperature (Hicher 1974, Despax 1976, Hueckel et al. 1989).

The above thermoplastic model can be further developed by adopting the isotropic hardening law describing the evolution of apparent preconsolidation stress (Laloui & Cekerevac 2003) and the linear thermal softening law (Laloui et al., 2005). With these ingredients we can re-derive the constitutive equations from a standard plasticity framework, with a general stress space formulation in order to facilitate future generalisation of the model in 2- and 3-D. Furthermore, Modified Cam-Clay will be adopted for the deviatoric behaviour, as this is a widely used model with clear advantages when it comes to numerical implementation. The resulting shape of the yield surface in 3 dimensions is shown in Figure 1.

3.1.1 Thermo-elasticity
The thermoelastic law is derived from the classic Duhamel-Neumann relations (e.g. see Mase, 1970):

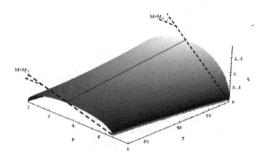

Figure 1. Thermo-plastic yield locus, demonstrating temperature dependence and thermal softening.

$$\dot{\sigma} = \mathbf{D}^{me}\dot{\varepsilon}^{me} + \mathbf{D}^{te}\dot{\theta} \qquad (1.2)$$

where $\mathbf{D}^{me}$ is the stress-dependent elasticity tensor, $\mathbf{D}^{te} = -\beta_s K\delta$ the thermal tensor, K is the bulk modulus, β_s the volumetric thermal expansion coefficient, δ the Kronecker delta, ε the strain and θ the temperature.

3.1.2 Thermo-plasticity

At plastic yielding, employing standard techniques, the above can be rewritten in terms of total strain and temperature rates as:

$$\begin{cases} \dot{\sigma} = \mathbf{D}^{mep}\dot{\varepsilon} + \mathbf{D}^{tep}\dot{\theta} \\ \dot{\sigma}_c = \dfrac{f_{\varepsilon_v^p}f_{p'}}{f_{\sigma_c}}\cdot\dot{\lambda} \end{cases}$$

where σ_c is the preconsolidation stress, p' the mean effective pressure, and $f = f(\sigma_{ij},\varepsilon_v^p,\theta)$ the yield surface derived by substituting the thermal hardening law of Laloui et al. (2003) into the standard expression of the Modified Cam Clay yield locus. Associativity is assumed, and partial derivatives of the yield function with respect to other model variables are denoted by subscripts. In contrast to Laloui et al. (2005), where two different plastic multipliers were defined for isotropic and deviatoric deformation respectively, here we use a single plastic multiplier:

$$\dot{\lambda} = \frac{f_\sigma \mathbf{D}^{me}\dot{\varepsilon} + (f_\theta + 2\mathbf{D}^{te}f_\sigma)\dot{\theta}}{f_\sigma \mathbf{D}^{me}g_\sigma - f_{\varepsilon_v^p}g_p}$$

derived from the consistency condition $df = 0$.

3.2 Numerical results

The above constitutive relations were discretised using a refined explicit scheme with automatic error control (Sloan et al., 2001), controlling independently the applied total strains and the temperature. The response of a soil element to different strain and temperature paths is demonstrated in Figure 2. It is shown how the apparent preconsolidation stress decreases as temperature increases, and how the slope of the plastic loading path changes if temperature and mechanical strains are applied at the same time.

In Figure 3 numerical results are compared to some experimental data on three isothermal isotropic compression paths carried out at different temperatures (Laloui & Cekerevac 2003). The dashed line simulates an isothermal loading path at ambient temperature, while the solid line represents a combined thermo-mechanical loading path to simulate the transition

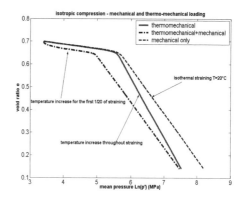

Figure 2. Numerical simulation of some thermo-mechanical isotropic paths. The dashed line represents isothermal loading, the solid line thermo-mechanical loading, the dash-dotted path was obtained by applying a higher temperature rate during the first 5% of total loading, followed by mechanical loading only.

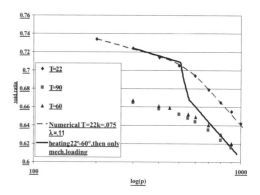

Figure 3. Simulation of isothermal loading and combined thermo-mechanical loading compared to experimental data of isothermal isotropic compression at different temperatures. The black line simulates thermo-mechanical loading until temperature reaches 60°C, followed by mechanical loading only. It can be seen how the state path moves between the two sets of data corresponding to isothermal compression at the respective temperatures.

between two isothermal loading data sets. Good agreement with experimental data points is achieved in both cases.

4 MODIFIED LANDSLIDE MODEL

4.1 Constitutive modifications

The system of equations (1.1) provides a 1-D analysis of the phenomenon, employing a Mohr-Coulomb yield locus as soil constitutive model. To provide a more realistic constitutive assumption, which can be more easily generalised for applications to a wider range of soils, we modify the equations to include the thermoplastic model described above.

The soil is assumed to be at critical state since our analysis starts at incipient failure. Plane strain is also assumed due to the problem geometry. Thus, it is possible to express the dissipation term as

$$D = 2\dot{\lambda}q^2 = 2\dot{\lambda}(Mp')^2$$

where $M = M_0 - g(\theta - \theta_0)$ is linearly dependent on temperature g is the thermal sensitivity and M_0 the critical state parameter at reference temperature. The mean effective pressure p' can be expressed as the difference between the initial effective stress acting on the landslide base p'_0 and the excess pore pressure u:

$$p' = p'_0 - u(z,t)$$

Thus,

$$D = 2\dot{\lambda}(\dot{\theta}, \dot{\varepsilon}_q)\{M(\theta)[p'_0 - u(z,t)]\}^2$$

and

$$\dot{\lambda} = F_1\dot{\theta} + F_2\dot{\varepsilon}_q$$

where:

$$F_1 = \frac{(f_\theta - 2K\beta_s f_p)}{f_p^2 K + 3Gf_q^2 - f_p f_{\varepsilon_v^p}},$$

$$F_2 = \frac{3f_q G}{f_p^2 K + 3Gf_q^2 - f_p f_{\varepsilon_v^p}}$$

and G and K denote the standard stress-dependent elastic moduli.

By substituting the dissipation expression in the heat equations, we finally obtain

$$\frac{\partial \theta}{\partial t} = \frac{k_m}{1 - \frac{2[M(p'_0-u)]^2 F_1}{j(\rho C)_m}} \frac{\partial^2 \theta}{\partial z^2} + \frac{2[M(p'_0-u)]^2 F_2}{j(\rho C)_m - 2[M(p'_0-u)]^2 F_1}\dot{\varepsilon}_q. \quad (1.3)$$

4.2 Well-posedness of heat equation

Expression (1.3) is a diffusion-generation equation for temperature, where, unlike Vardoulakis (2002), the diffusivity varies non-linearly with temperature and pore pressure as

$$D_i = \frac{k_m}{1 - \frac{2[M(\theta)(p'_0-u)]^2 F_1(\theta,u)}{j(\rho C)_m}}.$$

Such an expression can, theoretically at least, assume negative values, which would imply mathematical ill-posedness of the equation and inability to solve it. To ensure that this does not happen and that the problem remains well-posed, the sign of the diffusivity coefficient is calculated for a wide range of values of its parameters and the temperature range $0 < \theta < 1000°C$. Diffusivity proved to be always positive for all values examined of the parameters involved, showing very little variation around the value of 10^{-7} m^2/sec. The parameter ranges used are: $0.25 \leq M_0 \leq 0.85$ and $0.1 \leq \sigma_{c0} \leq 10$ MPa for the critical state parameter and the preconsolidation stress respectively at ambient temperature, $0.003 \leq \varepsilon_v^p \leq 0.5$ for the accumulated plastic volumetric strain, $5 \leq \beta \leq 20$ for the plastic compressibility, $10^{-3} \leq k \leq 5 \cdot 10^{-2}$ for the slope of the elastic compression line, $0.2 \leq \nu \leq 0.45$ for the Poisson's ratio, $0.2 \leq voidr \leq 1.5$ for the void ratio and $0.05 \leq \gamma \leq 0$ for the parameter-γ of the thermoplastic model (Laloui et al., 2003).

4.3 Numerical implementation

Assuming a linear velocity profile within the shearband we obtain the following system of governing equations:

$$\begin{cases} \dfrac{\partial u}{\partial t} = \dfrac{\partial}{\partial z}\left(c_v(\theta)\dfrac{\partial u}{\partial z}\right) + \lambda_m(\theta,u)\dfrac{\partial \theta}{\partial t} \\[6pt] \dfrac{\partial \theta}{\partial t} = D_i(\theta,u)\dfrac{\partial^2 \theta}{\partial z^2} + F'_i(\theta,u)\dfrac{v_d}{d} \\[6pt] \dfrac{dv_d}{dt} = R\omega_0^2\left\{A_0\left[1 - \dfrac{M(\theta_d)}{\tan\varphi_F\sqrt{3 - M(\theta_d)^2}}\right.\right. \\[6pt] \left.\left. + \dfrac{u_d}{p_c(\theta_d)}\right]\right\} \end{cases}$$

The second equation describing the evolution of temperature within the shearband is, unlike its corresponding one of Section 2, nonlinear, since both the diffusivity term and the heat generation term are temperature- and pore pressure-dependent. The third equation describing the movement of the rigid block is modified by the implementation of thermal softening, and coupled to the previous two through the values of temperature, excess pore pressure and velocity v_d at the interface of the shearband with the rigid block.

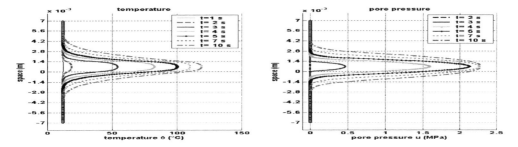

Figure 4. Temperature and pore pressure profile within the shearband and its surroundings, for thermal softening.

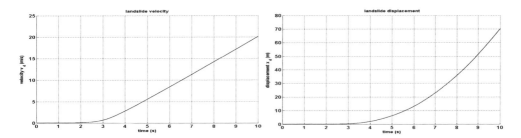

Figure 5. Velocity and displacement profile if thermal softening is included.

The first two nonlinear partial differential equations were discretised using a Finite-Time Centered-Space explicit scheme with a very small time-step $dt = 10^{-4}$ sec, which was found through numerical experimentation to yield stable numerical results. The last equation, which is an ordinary differential equation, was discretised with a standard 4th order Runge-Kutta scheme using the same time-step.

Calculations were performed for a time window of 10 seconds from the initiation of sliding, using the physical parameters of 'section 5' of Vajont landslide (Hendron & Patton 1985, Vardoulakis 2002). Initial conditions for temperature and excess pore pressure are the steady state values, equal to ambient temperature (12°C) and zero respectively.

The 1-D spatial domain is here set to be 10 times the thickness of the shearband (taken 1.4 mm, after Vardoulakis 2002) and assumed to be uniform in hydraulic and geotechnical properties. On the other hand, the shearband is the only area where shear straining, i.e. the mechanism driving energy dissipation and therefore heat production, occurs. Temperature and pore-pressure were computed at each grid-point of the spatial domain and isochrones through the domain were produced at significant time values. The value of the slope of the thermal softening law g representing the 'thermal sensitivity' of the soil proved to be crucial to the order of magnitude of catastrophic evolution of the phenomenon. Other model parameters values were chosen equal to those established by Vardoulakis (2002).

5 RESULTS AND DISCUSSION

In Figure 4 the isochrones of temperature and pore pressure are shown for the case of $g = 10^{-2}$, which corresponds to the average value for the rate of thermal softening found in the literature. It can be seen that temperature inside the shearband reaches 120°C after 10 seconds. This is well below the water vaporisation threshold at the given pressure. The overburden corresponds to an initial effective stress of 2.38 MPa, which is the maximum value that excess pore pressure can reach. It can also be seen that full pressurisation is reached after $t = 7$ s, corresponding to the vanishing of the shear resistance.

In Figure 5 the corresponding computed slide velocity and displacement are plotted. The velocity profile shows negligible increase up to $t = 2$ s, followed by a sudden kick at $t = 3$ s which corresponds with the start of pressurisation due to overtopping of the critical temperature (c.f. Figure 4). These values predict a similar behaviour and magnitude of results to those observed in Vajont, and are in accordance with those calculated by the strain and strain-rate softening model of Vardoulakis (2002).

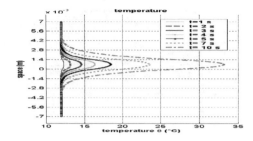 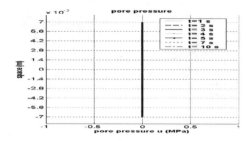

Figure 6. Temperature and pore pressure profile within the shearband and its surroundings, for zero softening.

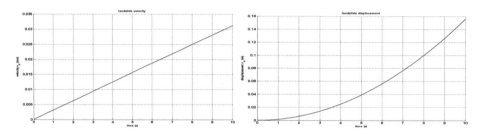

Figure 7. Velocity and displacement profile for zero thermal softening.

In Figure 6 and Figure 7 results are plotted for the case of no thermal softening ($g = 0$). The drastic difference in the computed values shows how sensitive is the timescale of the phenomenon to thermal softening. After 10 seconds the velocity is still of the order of mm/sec, temperature has increased by a few degrees Celsius and pressurisation has not taken place. A much longer time would be needed for catastrophic sliding to be triggered in this case.

The results show that thermal softening acts as a destabilising factor in slides and that its effect can be comparable in magnitude to that of strain and strain-rate softening.

6 CONCLUSIONS

A new catastrophic landslide model has been proposed, able to interpret the frictional heating and pressurisation phase in catastrophic landslides. The model accounts for thermal softening instead of the formerly proposed strain- and strain-rate softening mechanism. The model uses a thermoplastic constitutive model based on Modified Cam-Clay, allowing future generalisation to a wider range of soils and extension to 2-D and 3-D calculations.

The model was discretised and integrated numerically to back-analyse the behaviour of the well-documented case of Vajont landslide. The good accordance of the predicted values with available observations, as well as past calculations, shows that the contribution of thermal softening can be significant in the destabilisation of a slide, and should be taken into account in addition to strain and strain rate softening.

REFERENCES

Anderson, D.L. 1980. *An earthquake induced heat mechanism to explain the loss of strength of large rock and earth slides*. Proceedings of the International Conference on Engineering for Protection From Natural Disasters, pp. 569–580, John Wiley, New York.

Chang, K.J., Taboada, A., Lin, M.-L. & Chen, R.-F. 2005a. *Analysis of landsliding by earthquake shaking using a block-on-slope thermomechanical model: Example of Jiufengershan landslide, central Taiwan*. Eng. Geol., 80, 151–163.

Chang, K.J., Taboada, A. & Chan, Y.C. 2005b. *Geological and morphological study of the Jiufengershan landslide triggered by the Chi-Chi Taiwan earthquake*, Geomorphology. 71, 293–309.

Despax, D. 1976. *Influence de la température sur les propriétés mécaniques des argyles saturées*. Doctoral thesis, Ecole Centrale Paris.

Habib, P. 1975. Production of gaseous pore pressure during rock slides, Rock Mech., 7, 193–197.

Hendron, A.J. & Patton, F.D. 1985. The Vajont slide, a geotechnical analysis based on new geologic observations of the failure surface, Tech. Rep. GL-85–5, U.S. Army Corps of Eng., Washington, D.C.

Hicher, P.Y. 1974. *Etude des proprietes mecaniques des argiles a l' aide d'essais triaxiaux, in uence de la vitesse et de la temperature*. Report, Soil Mech. Lab., Ecole Cent. de Paris, Paris.

Hueckel, T. & Pellegrini, R. 1989. *Modeling of thermal failure of saturated clays*. Numerical models in geomechanics, 81–90, Elsevier.

Laloui, L. 2001. *Thermo-mechanical behaviour of soils*. Environmental geomechanics, 5: 809–843.

Laloui, L. & Cekerevac, C. 2003. *Thermo-plasticity of clays: An isotropic yield mechanism*. Computers and Geotechnics, 30 (8): 649–660.

Laloui, L., Cekerevac, C. & Francois, B. 2005. *Constitutive modelling of the thermo-plastic behaviour of soils*. Revue Europeenne de genie civil, 9 (5–6): 635–650.

Mase, G.E. 1970. *Theory and problems of Continuum Mechanics*. Schaum's outline series, McGraw-Hill.

Sloan, S.W., Abbo, A.J. & Sheng, D. 2001. *Refined explicit integration of elastoplastic models with automatic error control*. Engineering Computations, 18 (1–2): 121–154.

Tika, T.E. & Hutchinson, J.N. 1999. *Ring shear tests on soil from the Vajont landslide slip surface*. Geotechnique, 49, 59–74.

Vardoulakis, I. 2000. *Catastrophic landslides due to frictional heating of the failure plane*. Mechanics of Cohesive-Frictional Materials, 5 (6): 443–467.

Vardoulakis, I. 2002. *Dynamic thermo-poro-mechanical analysis of catastrophic landslides*. Geotechnique, 52 (3): 157–171.

Voight, B. & Faust, C. 1982. *Frictional heat and strength loss in some rapid landslides*. Geotechnique, 32, 43–54.

Some notes on the upper-bound and Sarma's methods with inclined slices for stability analysis

Z.Y. Chen

China Institute of Water Resources and Hydropower Research, Benjing, China

ABSTRACT: In 1997, Donald and Chen proposed a slope stability analysis method that divides the failure mass into a series of slices with inclined interfaces and solves the factor of safety based on the upper-bound theory of plasticity. This approach is capable of producing accurate results provided by the classical slip-line field method. It has also been demonstrated that the method is identical to Sarma's method that has been widely used for rock slope analysis. This paper gives some further examples demonstrating its accuracies and discusses two issues involved in the use of the upper-bound or Sarma's method: (1) the requirement for identifying alternative directions of relative movement between two contiguous slices; (2) treatments when tension occurs in the calculations.

1 INTRODUCTION

Among a variety of limit equilibrium approaches, Sarma's method that divides the failure mass into a number of slices with inclined interfaces has been commended by Hoek (1983, 1987) for rock slope stability analysis as rock mass normally contains at least one set of sub-vertical discontinuities that can be best modeled by the inclined interfaces. Sarma assumes that limit equilibrium conditions prevail on both the base and interfaces and solve the factor of safety by the force equilibrium conditions. Donald and Chen (1997) presented a new approach that established the governing equation based on the virtual work principle. In addition to its elegant analytical formulations, the method enjoys a sound theoretical background supported by the upper-bound theory of Plasticity. A series of testing problems has demonstrated its ability of producing closed-form solutions provided by the slip-line field method using its numerical procedures. In this Paper, we will present an analytical example demonstrating its accuracy.

Practical applications of Sarma's method have encountered the following two problems that will be discussed in this Paper: (1) the requirement for identifying alternative directions of relative movement between two contiguous slices; and (2) treatment when tension occurs in the calculations.

2 THE UPPER-BOUND METHODS

2.1 Background

The method suggested by Sarma (1979) assumes that both on the interfaces and base of a slice the limit equilibrium state applies. The factor of safety F is defined to be a coefficient that reduces the available effective shear strength parameters c and ϕ to new values of c_e and ϕ_e in order to bring the structure to failure,

$$c_e = c/F \quad (1)$$

$$\tan \phi_e = \tan \phi_e / F \quad (2)$$

In the following presentations, the subscription 'e' appeared for all variables would invariably mean that the related c and ϕ values are reduced by Eqs. (1) and (2).

In order to explain the equivalence between Sarma and the upper bound methods, Chen (1999) presented a slope that is composed of two blocks as shown in Figure 1. The factor of safety can be readily solved by establishing force equilibrium equations for the left and right blocks, according to Sarma's concept. However, the problem can be solved in a more efficient way by the Virtual Work Principle.

In this problem, the normal force N and its contribution of shear force $N \tan \phi$ on each of the faces forms

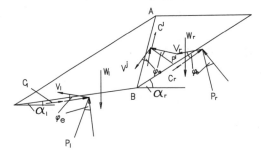

Figure 1. A two-block slope explaining the equivalency between Sarma and the upper-bound methods.

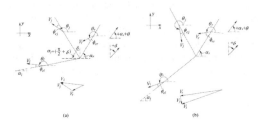

Figure 2. Determining V_r and V_j based on V_l. (a) The left slice moves upward relative to the right one; (b) The left slice moves downward relative to the right one.

a resultant P that inclines at an angle ϕ_e to the normal of the failure plane. If we assign a set of virtual displacements V_l, V_r, V_j, each inclined at an angle of ϕ_e to the shear surface and allow all forces applied on the slope to do work on them. The work done by P_l, P_r, P_j on V_l, V_r, V_j respectively is thus zero. P_l, P_r, P_j, as unknowns, disappear in the work and energy balance equation, which is

$$A_l c_l \cos\phi_{el} V_l + A_r c_r \cos\phi_{er} V_r + A_j c_j \cos\phi_{ej} V_j$$
$$= W_l V_l \cos\rho_l + W_r V_r \cos\rho_r \quad (3)$$

where ρ is the angle between the weight vector and V. Eq. (3) remains only one unknown F which is implied in the subscripts 'e' and is readily obtainable.

The values V_r, V_j can be expressed as a linear function of V_l (Refer to Figure 2),

$$V_r = V_l \frac{\sin(\theta_l - \theta_j)}{\sin(\theta_r - \theta_j)} \quad (4)$$

$$V_j = V_l \frac{\sin(\theta_r - \theta_l)}{\sin(\theta_r - \theta_j)} \quad (5)$$

where θ is the angle of the velocity vector measured from the positive x axis. The superscripts j refers to the variable on the interfaces, and l and r refer to the left and right sides of the interfaces.

Donald and Chen (1997) argued that there are two possible directions of relative movement: the left slice moves upward relative to the right one as shown in Figure 2(a) or downward, in Figure 2(b). Detailed discussions will be given in Section 3.

2.2 Formulations of the upper-bound methods

The governing equation for calculating the factor of safety has been given by Donald and Chen (1997), which is

$$\int_{x_0}^{x_n} \left[(c_e \cos\phi_e - u \sin\phi_e) \sec\alpha - \left(\frac{dW}{dx} + \frac{dT_y}{dx} \right) \right.$$
$$\left. \times \sin(\alpha - \phi_e) - \left(\eta' \frac{dW}{dx} + \frac{dT_x}{dx} \right) \cos(\alpha - \phi_e) \right]$$
$$\times E(x) dx - \int_{x_0}^{x_n} (c_e^j \cos\phi_e^j - u^j \sin\phi_e^j)$$
$$\times L \csc(\alpha - \phi_e - \theta_j) \frac{d\alpha}{dx} E(x) dx + K_i = 0 \quad (6)$$

where

$$E(x) = \kappa \exp\left[-\int_{x_0}^{x} \cot(\alpha - \phi_e - \theta_j) \frac{d\alpha}{d\zeta} d\zeta \right] \quad (7)$$

$$K_i = \prod_{j=1}^{i} \frac{\sin(\alpha_i^l - \phi_{ei}^l - \theta_i^j)}{\sin(\alpha_i^r - \phi_{ei}^r - \theta_i^j)} \quad (8)$$

K_i is s a coefficient accounting for possible discontinuities in α, ϕ_e and c_e, defined as

$$K_i = -\sum_{i=1}^{n} \left(C_e^j \cos\phi_e^j - u^j \sin\phi_e^j \right)_i L_i$$
$$\times \csc\left(\alpha^r - \phi_e^r - \theta_j\right) \sin(\Delta\alpha - \Delta\phi_e)_i E^l(x_i) \quad (9)$$

T_x and T_y are horizontal and vertical surface load respectively, and η refers the horizontal seismic coefficient.

The first term of the left-hand side of Eq. (6) refers to the work done by the external loads and the energy dissipation on the slip surface, while the second term is the energy dissipation on the interfaces between two contiguous slices.

2.3 Theoretical verifications

A number of test examples that have closed-form solutions by the slip-line field method (Sokolovski, 1960) have been performed, confirming that the optimization process of the upper-bound method is capable of creating the critical failure mode of the plastic zone and the minimum factor of safety identical to the theoretical solutions. This paper gives a further example that demonstrates the equivalence between Eq. (6) and the solution given by Sokolovski in an analytical form. On a separate Paper (Chen and Ugai, 2008), we gave a demonstration of the equivalence by the numerical approaches of the upper-bound method.

Figure 4 shows a uniform slope subjected to a vertical surface load q. The weight of the soil mass is neglected. The closed-form solution for the ultimate vertical surface load is provided by Sokolovski (1960) as:

$$q = c \cot\phi \left\{ \frac{1+\sin\phi}{1-\sin\phi} \exp[(\pi - 2\gamma')\tan\phi] - 1 \right\} \quad (10)$$

where γ' is the inclination of the slope surface.

The slip-line filed includes a series of failure surface, each of them consists of two straight lines (e.g., AB and CD) connected by a log-spiral (e.g. BEC).

AB and CD incline at μ to the slop surface and crest respectively. BO and CD incline at μ to OA and OD respectively.

$$\mu = \frac{\pi}{4} - \frac{\phi}{2} \quad (11)$$

The distance between a point on the log-spiral BC and the origin O, as represented by OE with a length L, can be obtained by the equation,

$$L = L_b \exp[-(\delta_b - \delta)\tan\phi] \quad (12)$$

where δ is the angle between OE and the y axis. L_b and L_c are the lengths of OB and OC respectively.

We have

$$L_c = L_b/s \quad (13)$$

where

$$s = \exp\left[\left(\frac{\pi}{2} - \gamma'\right)\tan\phi\right] \quad (14)$$

At any point on BC, we have

$$\alpha - \phi = -\delta \quad (15)$$

and

$$dx = -L \sec\phi \cos\alpha d\delta \quad (16)$$

For a slope at a state of limiting equilibrium, $F = 1$. Calculating the terms in Eq. (6) with the particular conditions defined by equations (11) to (16), as shown in Table 1, we have

$$q(1 - \sin\phi) = (s^2 + 1)c \cos\phi + (s^2 - 1)c \cot\phi \quad (17)$$

It is not difficult to demonstrate that Eq. (10) is identical to Eq. (17). This example validates Eq. (6), which is original in our discipline.

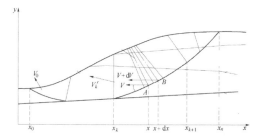

Figure 3. Sketch for the analyses by the energy approach of Sarma's method.

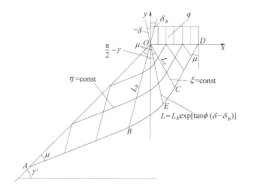

Figure 4. Verifications of Eq. (6) compared to the theoretical solution provided by the slip-line field method.

3 ON THE ALTERNATIVE DIRECTIONS OF THE INTERSLICE FORCES

3.1 *The argument*

When establishing the velocity field, one must examine whether the left slice moves in an upward (Case 1) or downward direction (Case 2), with respect to the right one, as shown respectively in Figure 2(a) and (b).

Failure to do so will cause negative values of V_r or V_j. when Eqs. (4) and (5) are employed. Substituting these negative values into the work and energy balance equation means a violation of Drucker's Postulate. As a consequence, the calculation will lead to absurd results

Table 1. Integrals of various terms on Eq. (9) for example.

Section	$E(x)$	$\int_l (c\cos\phi - u\sin\phi) \sec\alpha E(x)dx$	$\int_l \frac{dT_y}{dx} \sin(\alpha - \phi)E(x)dx$	$\int_l (c^j \cos\phi^j - u^j \sin\phi^j) \times L\cosec(\alpha - \phi - \theta_j) \times \frac{d\alpha}{dx} E(x)dx$
AB	1	$L_c sc \cos\phi$	0	0
BC	$\exp[-(\delta_b - \delta)\tan\phi]$ $= \exp[-(\alpha - \alpha_b)\tan\phi]$	$\frac{1}{2} cL_c \cot\phi(s - s^{-1})$	0	$\frac{1}{2} cL_c \cot\phi(s - s^{-1})$
CD	s^{-1}	$L_c s^{-1} c \cos\phi$	$qL_c(1 - \sin\phi)s^{-1}$	0

indicating that the bigger the cohesion value at the shear surface where the negative velocity develops, the smaller the factor safety. An example of explaining this absurd event was given by Chen (1999).

3.2 The criteria

The criteria for identifying Case 1 and Case 2 and their associated parameters are given by Donald and Chen (1996). They can be summarized as follows.

Case 1. The left slice moves in an upward direction with respect to the right one, i.e. $\theta_l > \theta_r$
In this case, we have

$$\theta_l - \theta_j > 0 \tag{18}$$

and θ_j is defined as

$$\theta_j = \frac{\pi}{2} - \delta + \phi_{ej} \tag{19}$$

Case 2. The left slice moves in a downward direction with respect to the right one. i.e. $\theta_l < \theta_r$
In this case, we have

$$\left. \begin{array}{l} \theta_l - \theta_j < 0 \\ \theta_r - \theta_j > -\pi \end{array} \right\} \tag{20}$$

and θ_j is defined as

$$\theta_j = \frac{3\pi}{2} - \delta - \phi_{ej} \tag{21}$$

Chen et al. (2005) demonstrated that with these conditions, V_r and V_j, calculated by Eqs. (4) and (5), will always be positive.

4 THE TREATMENTS WHEN TENSION OCCURS

4.1 The argument

Once the factor of safety is obtained, all internal forces applied on a slice can be calculated based on the force equilibrium equations. Hoek (1987) noticed that on some occasions, these values may be negative. This means that tensile forces are actually exerted on these faces. This is contradictory to the fundamental principle we have employed in Sarma or the upper-bound methods, which assumes that the frictional shear force is proportional to the normal force on the failure surface, a phenomenon only valid under compressive conditions. In his program SARMA, Hoek (1987) warned that the results may not be applicable. However no further instruction was offered to handle this situation.

4.2 A brief review of Lajtai's failure criteria for rock bridges

It has been understood that the rock mass can more or less provide tensile resistance due to the existence of rock bridges. Lajtai (1969) carried out a theoretical study and a series experiments for rock bridges under enforced shear and confirmed that a rock bridge would exhibit the following three failure modes at different values of normal stress σ_n.

1. Failure in tension
   At low value of σ_n, tension will develop and result in failure, at which shear strength τ_a is determined by

$$\tau_a = \sqrt{\sigma_t(\sigma_t - \sigma_n)} \tag{22}$$

where σ_t is the tensile strength of the rock bridge.

2. Failure in shear
   When σ_n reaches a certain higher level, the rock bridge will fail in shear, at which τ is determined by

$$\tau_a = \frac{1}{2}\sqrt{\frac{(2c_r + \sigma_n \tan\phi_r)^2}{1 + \tan^2\phi_r} - \sigma_n^2} \tag{23}$$

where ϕ_r and c_r are friction angle and cohesion of the intact rock respectively.

3. Failure at ultimate strength

If σ_n exceeds the compressive strength of the rock bridge, the shear strength will be determined based on that developed by crushed rocks with the following criterion.

$$\tau_a = \sigma_n \tan \phi_u \qquad (24)$$

where ϕ_u is the friction angle of crushed rocks.

In case of slope stability analysis, σ_n is on most cases small, compared to σ_t. Therefore Case 1, i.e., failure in tension, will always apply. Since on most slope problems, σ_t is much bigger than σ_n, eq. [22] can be approximated as

$$\tau_a \cong \sigma_t$$

This means that the shear strength of a rock bridge can be regarded as a purely cohesive material, whose cohesion can be approximated to be its tensile strength.

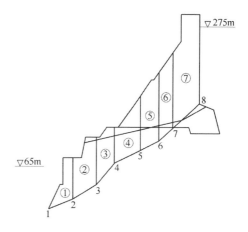

Figure 6. An example that explains the treatments for alternative directions and tensions.

4.3 The suggested method to deal with tensions

Based on the understanding that the rock bridge could behave like a purely cohesive material with a c value approximately equal to its tensile strength, it is suggested that once tensile force appears on the base or interfaces of a slice when Sarma or the upper-bound method is performed, their strength parameters will be substituted with $c = \sigma_t$ and $\phi = 0$, and the problem will be recalculated with these new parameters. The solution is then 'applicable'.

5 AN ILLUSTRATIVE EXAMPLE

Figure 6 shows a sketch of stability analysis for the Three Gorges dam along a potential slip surface consisting of a number of long persistent joints designated 1234567. The failure mode assumes that a part of failure surface cut the concrete dam at 78.

Since the foundation consists of two sets of vertical joints the inter-slice faces were assumed to be vertical when Sarma's method was performed. The shear strength parameters were both assumed to be zero for the part of interface that passes through the concrete dam body. This means that the concrete parts of the slices 4, 5 and 6 are assumed to be a stack of concrete blocks without any international friction and cohesions. However the interfaces between block 6 and 7 is a longitudinal construction joint. The shear strength parameters representing this joint were assigned for this joint.

The shear strength parameters used in the stability analysis are indicated in Table 2 from which one may find that a rapid change in strength parameters at point 7 that transit from the slice base 67 (representing the rock joint) to 78 (representing the concrete). Both the issues of the alternative shear force directions and tensions were raised in this case. In table 2 Alternative 1 employs the correct direction of inter-slice shear forces according to the criteria described in Section 3.2, while Alternative 2 invariably employs Case 1 for all interfaces.

1. The first round of computation

   The calculation employs the original strength parameters as shown in Table 2 for this round.

   Firstly, one may find that the Alternative 1 and 2 gave substantially different factors of safety, which are 2.966 and 2.048 respectively. The latter underestimated the value of F because negative velocities, obtained from Eq. (4), were involved in the computation. Secondly, tensions associated with negative values of the normal inter-slice forces E on interfaces 6 and 7 for Alternative 1 and 2 respectively. This means that the parameters on the interfaces require readjustment.

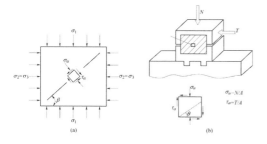

Figure 5. A rock bridge under enforced shear (Lajtai, 1969).

Table 2. The results for the three rounds of computations.

Round	Alternative	Parameters and forces	2	3	4	5	6	7	8	F
1	1	c (kPa)	200	200	200	200	200	3000	0	2.966
		ϕ (°)	35.0	35.0	35.0	35.0	35.0	47.7	0	
		E (9.8 kN)	1331.8	981.5	584.0	70.8	−342.4	3310.4	2112.8	
		X (9.8 kN)	528.0	379.2	−277.5	326.2	492.1	−7618.0	0	
	2	c (kPa)	200	200	200	200	200	3000	0	2.048
		ϕ (°)	35.0	35.0	35.0	35.0	35.0	47.7	0	
		E (9.8 kN)	2103.9	1935.0	1384.3	1212.1	1029.9	−2827.2	2112.14	
		X (9.8 kN)	1027.9	875.1	675.4	808.1	888.2	10782.3	0	
			2	3	4	5	6	7	8	
2	1	c (kPa)	200	200	200	200	200	3000	0	2.912
		ϕ (°)	0	0	0	0	0	47.73	0	
		E (9.8 kN)	1320.5	917.0	395.0	−95.0	−516.1	3231.6	2112.6	
		X (9.8 kN)	241.4	269.0	−203.8	313.1	398.0	−7730.0	0	
	2	c (kPa)	200	200	200	200	200	3000	0	1.966
		ϕ (°)	35.0	35.0	35.0	35.0	35.0	0	0	
		E (9.8 kN)	2219.5	2073.4	1544.1	1402.6	1250.8	−3267.4	2112.1	
		X (9.8 kN)	1112.0	961.0	760.6	909.6	1004.1	9645.9	0	
			2	3	4	5	6	7	8	
3	1	c (kPa)	200	200	200	200	200	0	0	2.419
		ϕ (°)	0	0	0	0	0	0	0	
		E (9.8 kN)	1601.3	1292.0	895.9	525.7	219.7	−362.4	2113.6	
		X (9.8 kN)	290.5	323.8	−245.3	376.8	479.0	0	0	
	2	c (kPa)	200	200	0	200	200	0	0	2.455
		ϕ (°)	35.0	35.0	35.0	35.0	35.0	0	0	
		E (9.8 kN)	1675.9	1412.5	900.9	569.0	264.8	−302.1	2114.9	
		X (9.8 kN)	735.5	581.1	183.9	490.7	522.8	0	0	

2. The second round of computation

In this round, we assigned $\phi = 0$ for interface 2 to 6 for Alternative 1. For Alternative 2 the ϕ value for interface 7 was set to zero. The results contained in Table 2 still exhibit tension on interfaces 6 and 7 for Alternative 1 and 2 respectively. However they are considered to be acceptable since these interfaces do not offer friction. The factors of safety are 2.912 and 1.966 for Alternative 1 and Alternative 2 respectively, which still deviate from each other appreciably.

3. The third round of computation

In order to confirm that the results given by Alternative 2 was wrong; this round of calculation set the value of c to be zero for interface 4 and 7. It can be found in Table 2 that Alternative 2 gave a factor of safety 2.455, greater than 1.966 of the second round of calculation in which c was not zero. This is obviously absurd. On the other hand, Alternative 1 decreased factor of safety from 2.912 to 2.419 by reducing the cohesion from 3000 kPa to zero at interface 7, which can be regarded as a normal situation.

From these calculation results, we may conclude that the correct answer is associated with Alternative 1 and factor of safety is between 2.912 and 2.419 depending on how much the cohesion at interface 7 can be developed. On the other hand, the Morgenstern-Price method gave $F = 2.589$, that is in general agreement to the upper-bound solutions.

6 CONCLUSIONS

This paper provides an example demonstrating that the controlling equation (6) of the upper-bound method proposed by Donald and Chen (1997) is reducible to that provided by the classical slip-line field method. This useful numerical approach for slope stability analysis thus has been offered a more sound theoretical background.

This paper also discussed the two issues commonly encountered when the upper-bound or Sarma's method is performed.

1. The requirement for identifying alternative directions of relative movement between two contiguous slices. Criteria for identifying correct directions have been given.
2. Treatments when tension occurs in the calculations based on the Lajtai's criteria for rock bridge strength criteria.

An illustrative example has been given indicating that a proper handling of these two issues will ensure a correction solution for rock slope stability analysis, especially when the slip surface exhibits rather irregularity in shape and abrupt changes in the strength parameters.

REFERENCES

Chen, W.F. 1975. Limit analysis and soil plasticity. Elsevier.
Chen, Z. and Morgenstern, N.R. 1983. Extensions to the generalized method of slices for stability analysis, Canadian Geotechnical Journal. 20(1): 104–119.
Chen, Z.Y., Wang, X.G., Yang, J., Jia, Z. and Wang, Y.J. 2005. Rock slope stability analysis: theory, method and computer programs, China Water Resources Press., 2005 (in Chinese).
Chen, Z.Y. and Ugai, K. 2008. Limit equilibrium and finite element analysis—a perspective of recent advances. Proceedings of the 10th International Symposium on landslide and engineered slopes. Beijing.
Donald, I. and Chen, Zuyu. 1997. Slope stability analysis by an upper bound plasticity method. Canadian Geotechnical Journal, 34(11): 853–862.
Hoek, E. 1987. General two-dimensional slope stability analysis-Analytical and Computational Methods in Engineering Rock Mechanics. 95, Allen Unwin, London.
Hoek, E. 1983. Strength of jointed rock mass. Geotechnique 33 No.3: 187–223.
Lajtai, E.Z. 1969. Strength of discontinuous rocks in direct shear. Geotechnique, (2): 218–233.
Morgenstern, N.R. and Price, V. 1965. The analysis of the stability of general slip surface, Geotechnique, (15) 1: 79–93.
Sokolovski, V.V. 1960. Statics of soil media. (Translated by Jones DH and Scholfield AN) Butterworth, London.
Sarma, S.K. 1979. Stability analysis of embankments and slopes. ASCE Journal of the Geotechnical Engineering Division. 105(GT12): 1511–1524.
Chen, Z.Y. 1999. The limit analysis for slopes: Theory, methods and applications. Shikoko'99. Vol. 1. 15–30. Balkema.

Slope stability analysis using graphic acquisitions and spreadsheets

L.H. Chen
Beijing Jiaotong University, Beijing, China

Z.Y. Chen & Ping Sun
China Institute of Water Resources and Hydropower Research, Beijing, China

ABSTRACT: This paper describes an approach that performs the conventional slope stability analysis by catching the geometry information through a graphic information acquisition program STAB_E.LSP in Auto-CAD and computing the factor of safety in an MS Excel spread sheet named LOSSAP.XLS. The coordinates of various controlling points of the slope profile can be easily captured by STAB_E.LSP, which are imported to STAB_E.LSP whose VBA facility has allowed a series of subroutines for calculating the weight, pore pressure, strength parameters, etc., by which the factor of safety can be easily obtained.

1 INTRODUCTION

Almost simultaneously with the advent of computers, our profession started exploring its potential for facilitating the conventional limit equilibrium analysis of slope stability problems by computer programs (Whitman and Bailey, 1967, Morgenstern and Price, 1965). However, nowadays the conventional slope stability computation using software is still a luxury undertaking for most geotechnical engineers. Some commercial software is available but with the following limitations:

- It is not accessible to most geotechnical engineers;
- It disclaims any legal liability for the calculation results;
- It requires training to the users, which is not always practically possible. Lack of fully understanding to the program may lead to incorrect calculations.

Using a spread sheet to calculate the factor of safety can be an attractive although its functions may not be as powerful as commercial software. The calculation details in a spread sheet are transparent, self-checkable and ready for further extensions by users. Only a limited amount of coding efforts are required. Therefore theses spread sheets can be shared by geotechnical engineers through the web on a non-commercial basis.

Low and Lee (1997) reported their developments of spreadsheets for calculating the factor of safety by using the method proposed by Chen and Morgenstern (1983). However, their work has not been widely extended perhaps due to the following reasons:

- The calculation still involves a large amount of manual work for determining the material and geometry input data of each slice.
- The software has not been completely open through the web and is therefore not widely applicable to potential users.

A handful approach, which is easy to understand and self-explanatory, is proposed in this paper. The geometrical information of the slope profile and slip surface shown in an AutoCAD image can be automatically captured by a program coded in Lisp language. These data are imported into the spreadsheet, from which the in-built VBA subroutines will calculate the basic data for slope stability analysis, such as weight, pore pressure, shear strength parameters of each slice. The various limit equilibrium methods for slope stability analysis have been arranged in separated sheets, which calculate the factors of safety in a completely transparent format.

This paper describes the technical details of the graphic acquisitions and spreadsheets facilities. The concept of establishing a library of slope stability analysis programs shared through web is also discussed.

2 DATA ACQUISITION FROM A GRAPHIC IMAGE

In the vector based graphic software AutoCAD, the property of each entity, such as point, line, surface, etc., has been stored in its own database. The in-built language AutoLisp can access and operate the entity property in the database directly, which provides a feasible way to catch the geometry information of the slope drawn as an AutoCAD image.

A graphic information acquisition program named STAB_E.LSP was developed by AutoLisp language to determine the input data for slope stability analysis automatically using Polyline in the AutoCAD image (Figure 1). Before applying the program, the line defining the border between two different soil layers, as well as that defining the phreatic surface and the slip surface, should be redrawn by Polylines. After loading the program STAB_E.LSP, the user-computer interrogation will start at the command column (Figure 2). The program requires the user to select Polyline objects interactively by clicking them and inputting some fundamental data such as the soil layer number underlying this line.

The main flowchart of the acquisition program is shown in Figure 1, which includes identifying the Polyline objects and acquiring the coordinates of all the vertex of the Polyline. Function "entsel" of AutoLisp was employed to ask the user to select a Polyline and return the name of the selected object, and Function "entget", for returning a list of the definition data of the entity. An AutoLisp list is a group of related values separated by spaces and contained in parentheses. The first value in the parentheses is called group code. The group code for the entity type is 0. If the value following 0 equals to LWPOLYLINE, a Polyline object is selected. The group code for the vertex coordinates of the LWPOLYLINE entity is 10. By searching through the list, all vertex coordinates of the selected Polyline could be obtained. Figure 2 shows a picture during the process of data acquisition. The captured information is saved in a data file with a default name 'qqq.dat'.

3 FORMULATIONS FOR SLOPE STABILITY ANALYSIS

3.1 Advantages of using the methods with integral and explicit formats

Most slope stability analysis methods involve iterations for factor of safety by employing certain algorithms, which can fall into the following categories.

1. Trial-and-error

This technique employs some randomly selected variables and expects the difference in factors of safety between two consecutive trials is small enough to meet the convergence criterion. For example, the Janbu's generalized method (1973) starts with a series of $t(x)$ (the gradient of shear force on the vertical interfaces) and an assumed value of factor of safety F. The algorithm expects the differences in $t(x)$ for each slice and F between two consecutive trials are all small enough. There is no particular theoretical or practical ground to ensure convergence. This technique is not suitable for a spreadsheet.

2. Consecutive calculations

This technique is normally used for the method that satisfies the force equilibrium method only, such as the method proposed by Corps of Engineers, Lowe and Karafiath. An initially guess of F is assumed and the calculation starts at the top first slice, numbered 1, which determines the inter-slice force G_1 applied on the left side of the slice. Consecutive calculations will enable the determination of $G_2, G_3, \ldots, G_n$ for the subsequent slices. It is expected that G_n to be zero. Otherwise F will be re-assumed until this criterion is met.

It is obvious that G_n is dependent on the slice number n and is therefore unfriendly to the non-linear programming solver provided by Excel, which

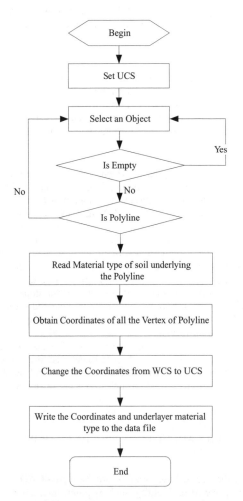

Figure 1. Flow chart of program STAB_E.LSP WCS: World coordinate system UCS: User coordinate system.

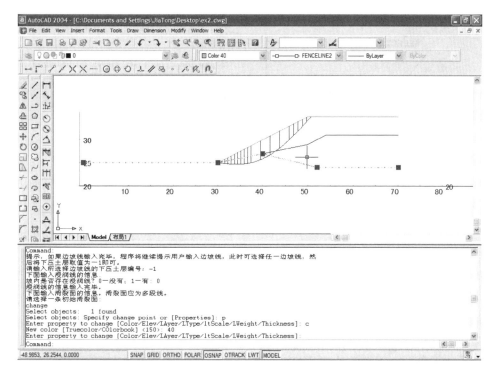

Figure 2. Data acquisition in an AutoCAD interface.

requires the optimized object contained in a fixed cell rather than in a dynamic position. Coding by Excel can be extremely difficult, if not possible.

3. Non-linear regressions

Normally, a generalized slope stability analysis method involves both force and a moment equilibrium equations designated respectively as

$$f(F, \lambda) = 0 \qquad (1)$$

$$M(F, \lambda) = 0 \qquad (2)$$

where F is the factor of safety and λ is a coefficient defining the inter-slice force inclination. A number of researchers (Fredlund, 1967, Spencer, 1967) suggested creating the linear regression curves F_f and F_m by solving Eqs. (1) and (2) respectively associated with different values of λ. F is determined by finding the point of intersection of the two curves (Fig. 3). While this technique is feasible to normal computer programming, although not efficient, it is not easy to be realized in a spreadsheet.

4. Non-linear programming

A variety of non-linear programming techniques are available for solving Eqs. (1) and (2). Among them, Newton-Raphson method is the mostly commonly known and adopted by Morgnestern and Price (1965) as:

$$\Delta F_i = F_{i+1} - F_i = \frac{G_n \frac{\partial M_n}{\partial \lambda} - M_n \frac{\partial G_n}{\partial \lambda}}{\frac{\partial G_n}{\partial \lambda} \frac{\partial M_n}{\partial F} - \frac{\partial G_n}{\partial F} \frac{\partial M_n}{\partial \lambda}} \qquad (3)$$

$$\Delta \lambda_i = \lambda_{i+1} - \lambda_i = \frac{-G_n \frac{\partial M_n}{\partial F} + M_n \frac{\partial G_n}{\partial F}}{\frac{\partial G_n}{\partial \lambda} \frac{\partial M_n}{\partial F} - \frac{\partial G_n}{\partial F} \frac{\partial M_n}{\partial \lambda}} \qquad (4)$$

The iteration terminates when ΔF_i and $\Delta \lambda_i$ meet the convergence criterion.

As a matter of fact, Excel offers a powerful non-linear programming solver, provided that the optimized objects are all contained in fixed cells, free of the slice number n as discussed previously. Our approach to solving this problem is to set n a very large value covering all possible slice number the use might take, say 100. Doing some calculations with nil columns or rows would not be harmful for a spreadsheet, provided that the formulation does not cause overflow problems due to a variable whose value is zero. This means that both $f(F, \lambda)$ and $M(F, \lambda)$ should be preferable in integral and explicit format. By 'integral' we mean that both $f(F, \lambda)$ and $M(F, \lambda)$ should

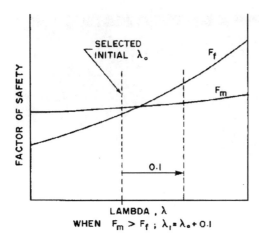

Figure 3. Non-linear regressions for the factor of safety.

eventually lead to manipulations of summing up the valuables contained in a certain column in an Excel sheet, and by 'explicit' we mean that the variables involved in the formulations for f and M should only contain F and λ, free of any intermediate objects, such as the inter-slice force G_i determined through consecutive calculations. It is probably due to this reason that Low and Lee (1997) selected Chen and Morgenstern's formulations for their attempt in creating the first officially reported spreadsheet in the area of slope stability analysis.

3.2 Formulations based on the integral and explicit formats

It is obvious that the formulation for Bishop's simplified method is explicit and involves the summations for the driving and resistant moments of all known variables except the factor of safety. It is suitable for spreadsheets.

In their keynote paper of this Proceedings, Chen and Ugai (2008) summarized the analytical presentations as follows (the explanations for the symbols are abbreviated).

1. For Spencer method, the formulations are:

$$\int_a^b p(x) \sec(\phi_e - \alpha + \beta) dx = 0 \qquad (5)$$

$$\int_a^b p(x) \sec(\phi_e - \alpha + \beta)(x \sin \beta - y \cos \beta) dx = M_e \qquad (6)$$

The two integrations involve all known valuables except F and λ. They are suitable for spreadsheets programming.

2. For the generalized method of slices (Chen & Morgenstern, 1983), the formulations are:

$$\int_a^b p(x) s(x) dx = 0 \qquad (7)$$

$$\int_a^b p(x) s(x) t(x) dx - M_e = 0 \qquad (8)$$

3. For Sarma or the upper-bound method, the formulation is

$$\int_{x_0}^{x_n} \left[(c_e \cos \phi_e - u \sin \phi_e) \sec \alpha - \frac{dW}{dx} \sin(\alpha - \phi_e) \right] \qquad (9)$$

$$E(x) dx - \int_{x_0}^{x_n} (c_e^j \cos \phi_e^j - u^j \sin \phi_e^j) L \csc$$

$$\times (\alpha - \phi_e - \theta_j) \frac{d\alpha}{dx} E(x) dx + K_i = 0 \qquad (10)$$

4 CALCULATING THE FACTOR OF SAFETY THROUGH A SPREAD SHEET

4.1 Structure of the spread sheet

A spread sheet named LOSSAP.XLS is developed for computing the factor of safety by inputting the data contained in 'qqq.dat' and formulating Eqs. (5) to (9). It includes three parts, six worksheets. Worksheet "Finfo" is calculates the basic information of each slice, such as its average unit weight, pore pressure, strength parameters, etc. The second part "Load" works for inputting external loads. The third part includes several sheets that calculate the safety factors using the methods proposed by Bishop, Spencer, Sarma (or Donald & Chen), and Chen and Morgenstern (1983) respectively.

4.2 Calculations for the basic information

Worksheet Finfo has four regions as shown in Figure 5. The left upper region is concerned with the fundamental information including the geotechnical properties of soils, and the water level out of the slope. At the upper right there is a column that reads the input of data 'qqq.dat', which has been saved after performing STAB_E.LSP as described in Section 2. Immediately after reading the geometry information contained in 'qqq.dat', the in built subroutines calculate geometrical and geotechnical information for each slice.

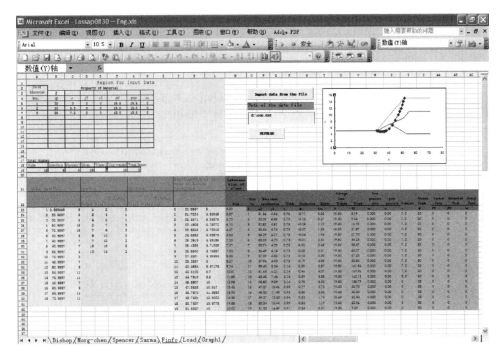

Figure 4. Spreadsheet Finfo.

Figure 5. Spreadsheet for solve FS by Bishop's method.

The width of the slice, inclination at the base of each slice, and the coordinates of the centre of the slice base can be calculated directly by the simple formulas in a spreadsheet. However, determination of the height and average unit weight of the slice, as well as the shear strength parameters and the pore water pressure ratio, require several special sub-routines. These values will be calculated by a user-defined Function named HWM coded by VBA, which is a macro language to improve the flexibility and automatic ability of Office software, such as Word, Excel.

Function HWM first computes the intersection and vertical distance between the center at the base of the slice and each interface line segment. The intersection points are sorted by their ordinates (the value of the y axis), using a technique called 'bubbling algorithm'. This procedure allows the determination of weight of the slice and the shear strength parameters of the slice base, which are stored in Column T to X. Similarly, the pore pressure, normally expressed in terms of pore water pressure coefficient r_u at the base of the slice, can be calculated and stored in Column Y.

4.3 Computing the factor of safety

With the basic information for each slice calculated by 'Finfo' and 'Load', it is possible to compute the factor of safety at the Worksheet 'Morg-Chen', 'Bishop',

'Spencer', and 'Sarma' with the help the non-linear programming facilities provided by Excel.

For Bishop's method, the task is relatively easy. Calculations for sliding and resistant moments of each slice are straight forward, which are contained in column AE and AF respectively. As mentioned previously, our approach to summing up the sliding and resistant moments is to set a very large slice number, $n = 100$, which covers all possible value a user may employ. The formula of cell AF19 is: "= sum (AI24:AI123)−sum (AJ24:AJ123)" (Figure 5). In Solver tool's dialog box, we set cell AE19 as variable, in which an initial value of factor of safety, F, is contained, and AF19, as target. By requiring AF19 to be zero, the final solution for F will be caught immediately by the Solver and updated in cell AE19, as shown in Figure 6.

Low and Lee (1998) described the technique of implementing Chen-Morgenstern method in a spread sheet. The Three-node Gauss-Legendre Quadrature method was employed to solve the integration of the force and moment equilibrium equations (7) and (8).

Figure 6. Pop-up dialog box of solver.

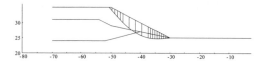

Figure 7. Example of ACADS.

Table 1. The material property of the soil.

No. of soil	$\phi(°)$	c (kPa)	γ_d (kN/m^3)	γ_{sat} (kN/m^3)	r_u
1	38	0	19.5	19.5	0
2	23	5.3	19.5	19.5	0
3	20	7.2	19.5	19.5	0

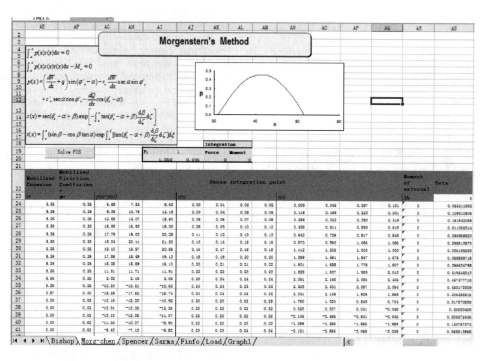

Figure 8. Spreadsheet for solve F by Chen-Morgenstern method.

Similar technique can be used for Spencer method that employs equations (5) and (6).

5 AN ILLUSTRATIVE EXAMPLE

Figure 7 is an example that has been documented by Chen and Ugai (2008) of this Proceedings, designated Example 5. The geotechnical property parameters are shown in Table 1. The minimum factor of safety is 1.366. Using the same critical slip surface, the spreadsheet solution is 1.368 for Chen-Morgenstern method and 1.366 for Spencer method.

Figure 8 shows the calculation procedures and details related to Chen-Morgenstern (1983) method.

6 CONCLUSIONS

The approach described in this Paper allows practitioners to perform routine slope stability analysis work without the need for commercial software and to check the computation details on their own effort. The data acquisition technique in an AutoCAD interface permits a rapid capturing of geometric information, which is processed by the VBA programs in Excel for determining the geometry and geotechnical information of each slice. The integral and explicit forms for slope stability analysis make the procedures of calculating factors of safety in a Spread Sheet very simple and straight forward.

This work requires a limited amount of coding. Therefore, it should be established on a non-commercial basis. The data acquisition program and Spread sheet described in this Paper has been announced at the website: *www.geoeng.iwhr.com/geoeng/download.htm*.

Further work on updating the algorithms and extending its applicability is certainly an encouraging deed, which should also be non-commercial, and carried out by all geotechnical engineers interested in this area. A concept of establishing a website entitled 'Library Of Slope Stability Analysis Programs (LOSSAP)' has been developed. The authors wish it will come to reality in the near future.

REFERENCES

Chen, Z.Y. & Morgenstern, N.R. 1983. Extensions to the generalized method of slices for stability analysis. Canadian Geotechnical J. 20 (1), 104–119.

Chen, Z.Y. & Shao, C.M. 1998. Evaluation of minimum factor of safety in slope stability analysis. Canadian Geotechnical J. 25 (4), 735–748.

Fredlund, D.G. 1984. Analytical methods for slope stability analysis. 4th International symposium on landslides. 209–228.

Janbu, N. 1973. Slope stability computations. Embankment Dam Engineering, 47–86.

Low, B.K. & Lee, S.R. 1998. Slope stability analysis using generalized method of slices. J. Geotechnical and Geoenvironmental Engineering. 124: (4) 350–362.

Morgenstern, N.R. & Price, V. 1965. The analysis of the stability of general slip surface. Geotechnique, 15 (l): 79–93.

Spencer, E. 1967. A method of analysis of embankments assuming parallel inter-slice forces. Geotechnique, 17: 11–26.

Whitman, R.V. & Bailey, W.A. 1967. Use of computers for slope stability analysis. J. Soil Mechs. Fnd. Div. ASCE., Vol. 93: 475–498.

Efficient evaluation of slope stability reliability subject to soil parameter uncertainties using importance sampling

Jianye Ching
National Taiwan University of Science and Technology, Taipei, China

Kok-Kwang Phoon
National University of Singapore, Singapore, China

Yu-Gang Hu
National Taiwan University of Science and Technology, Taipei, China

ABSTRACT: Evaluating the reliability of a slope is a challenging task because the possible slip surface is not known beforehand. Approximate methods via the First-Order Reliability Method (FORM) provide efficient ways of evaluating failure probability of the "most probable" failure surface. The tradeoff is that the failure probability estimates may be biased towards the unconservative side. The Monte Carlo Simulation (MCS) is a viable unbiased way of estimating the failure probability of a slope, but MCS is inefficient for problems with small failure probabilities. This study proposes a novel way based on the importance sampling technique of estimating slope reliability that is unbiased and yet is much more efficient than MCS. In particular, the issue of the specification of the importance sampling Probability Density Function (PDF) will be addressed in detail. An example of slope reliability will be used to demonstrate the implementation of the new method.

1 INTRODUCTION

Slope stability analysis has a long history in geotechnical engineering. In its nature, the analysis of slope stability is obscured by uncertainties, e.g.: material uncertainties and spatial variabilities. The consequence is that even for slopes with nominal safety factor more than 1, they are not necessarily safe. Reliability, namely one minus failure probability, rigorously quantifies such uncertainties. In recent years, analysis methods (e.g.: Chowdhury and Xu (1993), Christian et al. (1994), Low and Tang (1997), Low et al. (1998) and Griffiths and Fenton (2004)) have been proposed to evaluate reliability of slopes. However, evaluating reliability of a slope is not an easy task, primarily due to the lack of knowledge of the failure surface.

A standard procedure of evaluating the reliability of a slope is through Monte Carlo simulation (MCS) (Rubinstein 1981; Ang and Tang 1984). This approach was taken by Griffiths and Fenton (2004) where the MCS method is implemented with a random field model for spatial distribution of shear strengths. The procedure for MCS is straightforward. Let Z denote all uncertain variables in the slope of interest. Without loss of generality, let us assume Z is independent standard Gaussian. In the case that Z is not standard Gaussian, proper transformations can be taken to convert the problem into the standard Gaussian input space. Monte Carlo simulation contains the following steps:

a. Draw Z samples $\{Z_i: i = 1, \ldots, N\}$ from the standard Gaussian PDF.
b. For each sample Z_i, conduct a deterministic slope stability analysis to find the most critical slip surface among all trial surfaces. If the safety factor of the most critical surface is less than 1, the entire slope is considered to fail for that Z_i sample.
c. Repeat Step 2 for $i = 1, \ldots, N$. The average of the failure indicators is simply an estimate of the failure probability of the slope.

Mathematically, the MCS procedure can be summarized as the following equation:

$$P(F) \approx \frac{1}{N} \sum_{i=1}^{N} I\left[\min_{\omega \in \Omega} FS_\omega(Z_i) < 1\right] \equiv P_F^{MCS} \quad (1)$$

where F is the failure event of the entire slope; $I[\cdot]$ is the indicator function; Ω is the set of all trial slip surfaces; ω is one of the trial surface, and FS_ω is the safety factor for that trial surface, which is clearly a function of Z.

Another way of interpreting the overall failure probability in (1) is as follows: let F_ω denote the event $FS_\omega(Z) < 1$, i.e.: the failure event of the trial surface ω. The overall failure event F is simply the union of all individual failure events $\cup_{\omega\in\Omega}F_\omega$. This interpretation is graphically depicted in Figure 1. The overall failure probability is therefore the volume under $f(z)$ (the standard Gaussian PDF) within the F region.

The MCS method provides an unbiased estimate of the actual failure probability. However, it can be very time consuming, especially for slopes with small failure probabilities. This is because the coefficient of variation (c.o.v.; standard deviation divided by mean value) of the MCS estimator P_F^{MCS} is equal to

$$\delta(P_F^{MCS}) = \sqrt{[1-P(F)]/[N \cdot P(F)]}$$
$$\approx \sqrt{1/[N \cdot P_F^{MCS}]} \quad (2)$$

where $\delta(.)$ denotes the c.o.v. Therefore, in order to make the c.o.v. of the MCS estimator to be as small as 30%, it roughly requires $10/P(F)$ MCS samples, i.e.: $10/P(F)$ deterministic slope stability analyses.

A more efficient method based on the first-order reliability method (FORM) was proposed to estimate the failure probability. The idea is to solve for the following optimization problem:

$$\min_{z,\omega\in\Omega} ||z|| \quad \text{such that} \quad FS_\omega(z) = 1 \quad (3)$$

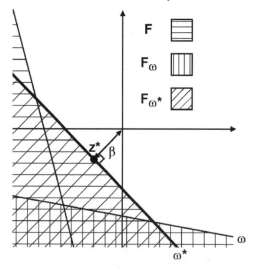

Figure 1. Illustration of various failure regions in the standard Gaussian space.

Note that the slip surface variable ω is augmented into the original FORM optimization problem. Let the solution of the optimization problem be z^* and ω^*, then $\Phi(-||z^*||)$ is the estimated failure probability and ω^* is the "most probable" slip surface. This approach was taken by Low and Tang (1997) and Low et al. (1998) for the evaluation of slope failure probability.

The FORM technique is efficient and convenient because the repetitive deterministic slope stability analyses required by MCS are not needed. However, the tradeoff is that the FORM methods may provide biased and unconservative estimates of the actual overall failure probability (Oka and Wu 1990). This can be seen in Figure 1, where the thick curve indicates the limit state function of the most probable slip surface ω^*. It is clear that the volume under $f(z)$ within the union region $\cup_{\omega\in\Omega}F_\omega$ (the actual overall probability) is always greater than or equal to that within the region F_{ω^*} (the FORM-estimated failure probability) because the latter is a subset of the former.

What is missing in the literature is a technique that can provide unbiased estimate of the actual overall failure probability and yet only requires a small number of repetitive deterministic slope stability analyses. The purpose of this study is to demonstrate that it is possible to implement the importance sampling (IS) technique (Rubinstein 1981; Shinozuka 1983; Melchers 1989) to achieve so. Moreover, it is shown by examples that the c.o.v. of the failure probability estimator made by the IS technique can be as small as 0.2 with only 100 deterministic slope stability analyses for practical range of failure probability. In the paper, the discussion of the IS technique will be made in the context of circular trial surfaces and method of slices although its use is obviously not limited to this scenario. The limitation of the IS technique will also be addressed.

2 METHODS OF SLICES

Some popular methods of determining the safety factor for a given trial slip surface and fixed soil parameters are briefly reviewed. This presentation does not aim to give a complete review of the slope stability methods but just to give enough background for the forthcoming presentation of the importance sampling technique. Moreover, the presentation is limited to methods of slices with circular trial surfaces although the importance sampling technique may apply to general methods and non-circular trial surfaces.

2.1 Ordinary method of slices

As shown in Figure 2, a circular trial surface is under consideration, and the soil parameters Z are fixed

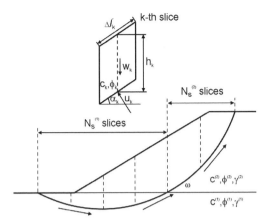

Figure 2. A demonstrative slope.

at some values (e.g.: for MCS, they are fixed their sampled values for each deterministic slope stability analysis). The goal is to determining the safety factor FS_ω of a slip surface ω. A very simplified method that assumes no interacting forces between slices can be used to compute the safety factor:

$$FS_\omega = \sum_{k=1}^{N_s} [c_k \Delta l_k + (W_k \cos(\alpha_k) - u_k \Delta l_k)$$

$$\times \tan(\phi_k)] \bigg/ \sum_{k=1}^{N_s} W_k \sin(\alpha_k) \quad (4)$$

where c_k and ϕ_k are the cohesion and friction angle of the k-th slice; W_k is the total weight of the k-th slice; u_k is the pore water pressure at the middle point of the slice bottom; Δl_k is the length of the slice bottom; α_k is the inclination angle of the slice bottom. Equation (4) is obtained based on the equilibrium equation of the overall moment for all slices about a chosen point. This method of determining the safety factor is called the ordinary method of slices (OMS) (Fellenius 1936).

conservative estimates of safety factors. It is also the only method of slices that does not require iterative calculations to obtain the safety factor. In applications, many trial surfaces are randomly generated, and the trail surface with the smallest safety factor is the critical slip surface. If the safety factor of the critical slip surface is less than 1, the entire slope is then considered as unstable, or failure.

2.2 Simplified Bishop method of slices

The simplified Bishop method of slices (SBMS) (Bishop 1955) takes a different assumption that all inter-slice forces are horizontal. Equilibrium of vertical forces in all slices and the overall moment gives the following expression for the safety factor:

$$FS_\omega = \frac{\sum_{k=1}^{N_s} \left[\frac{c_k \Delta l_k \cos(\alpha_k) + (W_k - u_k \Delta l_k \cos(\alpha_k)) \tan(\phi_k)}{\cos(\alpha_k) + (\sin(\alpha_k) \tan(\phi_k))/FS_\omega} \right]}{\sum_{k=1}^{N_s} W_k \sin(\alpha_k)} \quad (5)$$

The safety factor FS_ω can be found by iteratively solving (5). Again, many trial surfaces are randomly generated, and the trail surface with the smallest safety factor is the critical slip surface. If the safety factor of the critical slip surface is less than 1, the entire slope is then considered failed.

2.3 Transformation to standard Gaussian space

Previously, it is assumed that all uncertain variables are transformed to the standard Gaussian input space for the ease of presentation. To demonstrate how the expressions of safety factors can be transformed into the standard Gaussian space, consider the example in Figure 2, where the trial surface is ω, both soil layers are homogeneous, and the c, ϕ and γ parameters for both layers are uncertain and independent. Note that the first few slices share the same shear strengths and unit weights, and similar for the last few slices. In the case that the OMS is taken, the safety factor can be expressed as:

$$FS_\omega = \frac{\left[\begin{array}{l} c^{(1)} \sum_{k=1}^{N_s^{(1)}} \Delta l_k^{(1)} + \tan(\phi^{(1)}) \cdot \left[\gamma^{(1)} \sum_{k=1}^{N_s^{(1)}} (h_k^{(1)} \cos(\alpha_k^{(1)})) - \sum_{k=1}^{N_s^{(1)}} u_k^{(1)} \Delta l_k^{(1)} \right] \\ + c^{(2)} \sum_{k=1}^{N_s^{(2)}} \Delta l_k^{(2)} + \tan(\phi^{(2)}) \cdot \left[\gamma^{(2)} \sum_{k=1}^{N_s^{(2)}} (h_k^{(2)} \cos(\alpha_k^{(2)})) - \sum_{k=1}^{N_s^{(2)}} u_k^{(2)} \Delta l_k^{(2)} \right] \end{array} \right]}{\gamma^{(1)} \sum_{k=1}^{N_s^{(1)}} h_k^{(1)} \sin(\alpha_k^{(1)}) + \gamma^{(2)} \sum_{k=1}^{N_s^{(2)}} h_k^{(2)} \sin(\alpha_k^{(2)})} \quad (6)$$

The OMS is the simplest method among all methods of slices. It does not satisfy force and moment equilibrium of individual slices, and it usually provides

where $c^{(n)}, \phi^{(n)}$ and $\gamma^{(n)}$ are the cohesion, friction angle and (average) unit weight of the n-th soil layer; $N_s^{(n)}$ is the total number of slices whose bottom sides are within the n-th layer; $h_k^{(n)}$ is the mid-height of the k-th slice in the n-th soil layer.

Now let P be the mapping from a standard Gaussian random variable to the variable of interest, e.g.: if $c^{(1)}$ is Gaussian with mean $= \mu_c^{(1)}$ and standard deviation $= \sigma_c^{(1)}$, one can verify

$$c^{(1)} = p_{c^{(1)}}(Z_{c^{(1)}}) = \mu_{c^{(1)}} + \sigma_{c^{(1)}} \cdot Z_{c^{(1)}} \quad (7)$$

where $Z_c^{(1)}$ is standard Gaussian; if $c^{(1)}$ is lognormal with mean $= \mu_c^{(1)}$ and c.o.v. $= \delta_{c^{(1)}}$,

$$p_{c^{(1)}}(Z_{c^{(1)}}) = \exp\left(\ln\left[\mu_{c^{(1)}}/\sqrt{1+\delta_{c^{(1)}}^2}\right]\right.$$
$$\left. + \sqrt{\ln(1+\delta_{c^{(1)}}^2)} \cdot Z_{c^{(1)}}\right) \quad (8)$$

if $c^{(1)}$ is uniform with lower bound $= l_c^{(1)}$ and upper bound $= u_c^{(1)}$,

$$p_{c^{(1)}}(Z_{c^{(1)}}) = l_{c^{(1)}} + (u_{c^{(1)}} - l_{c^{(1)}}) \cdot \Phi(Z_{c^{(1)}}) \quad (9)$$

where Φ is the cumulative density function (CDF) of standard Gaussian.

Therefore, (6) can be transformed into the following standard Gaussian-space expression:

$$FS_\omega(Z) =$$

$$\left[\underbrace{p_{c^{(1)}}(Z_{c^{(1)}})\sum_{k=1}^{N_s^{(1)}}\Delta l_k^{(1)}}_{a^{(1)}} + \tan(p_{\phi^{(1)}}(Z_{\phi^{(1)}})) \cdot \underbrace{\left[p_{\gamma^{(1)}}(Z_{\gamma^{(1)}})\sum_{k=1}^{N_s^{(1)}}(h_k^{(1)}\cos(\alpha_k^{(1)})) - \sum_{k=1}^{N_s^{(1)}}u_k^{(1)}\Delta l_k^{(1)}\right]}_{b^{(1)}}\right.$$
$$\left. + \underbrace{p_{c^{(2)}}(Z_{c^{(2)}})\sum_{k=1}^{N_s^{(2)}}\Delta l_k^{(2)}}_{a^{(2)}} + \tan(p_{\phi^{(2)}}(Z_{\phi^{(2)}})) \cdot \underbrace{\left[p_{\gamma^{(2)}}(Z_{\gamma^{(2)}})\sum_{k=1}^{N_s^{(2)}}(h_k^{(2)}\cos(\alpha_k^{(2)})) - \sum_{k=1}^{N_s^{(2)}}u_k^{(2)}\Delta l_k^{(2)}\right]}_{b^{(2)}}\right]$$

$$\underbrace{p_{\gamma^{(1)}}(Z_{\gamma^{(1)}})\sum_{k=1}^{N_s^{(1)}}h_k^{(1)}\sin(\alpha_k^{(1)})}_{d^{(1)}} + \underbrace{p_{\gamma^{(2)}}(Z_{\gamma^{(2)}})\sum_{k=1}^{N_s^{(2)}}h_k^{(2)}\sin(\alpha_k^{(2)})}_{d^{(2)}}$$

$$= \frac{[p_{c^{(1)}}(Z_{c^{(1)}})a^{(1)} + p_{c^{(2)}}(Z_{c^{(2)}})a^{(2)} + \tan(p_{\phi^{(1)}}(Z_{\phi^{(1)}})) \cdot b^{(1)} + \tan(p_{\phi^{(2)}}(Z_{\phi^{(2)}})) \cdot b^{(2)}]}{p_{\gamma^{(1)}}(Z_{\gamma^{(1)}})d^{(1)} + p_{\gamma^{(2)}}(Z_{\gamma^{(2)}})d^{(2)}} \quad (10)$$

Note that $a^{(1)}, a^{(2)}, d^{(1)}$ and $d^{(2)}$ are deterministic, while $b^{(1)}$ and $b^{(2)}$ are random. Correlation between input standard Gaussian variables, e.g.: correlation between $Z_c^{(1)}$ and $Z_\phi^{(1)}$, can also be easily handled.

3 IMPORTANCE SAMPLING TECHNIQUE

The inefficiency of MCS is primarily due to the fact that the main support region of $f(z)$, i.e.: standard Gaussian PDF, is quite far from the failure region $\cup_{\omega\in\Omega}F_\omega$, especially when the actual failure probability is small. The consequence is that MCS requires many samples before a failure sample is obtained. The basic idea of the importance sampling method is to adopt a shifted version of $f(z)$ as the so-called importance sampling PDF (IS PDF) $q(z)$. That is,

$$q(z) = \frac{1}{\sqrt{2\pi}^n} \exp\left[-\frac{1}{2}(z-z')^T(z-z')\right] \quad (11)$$

where n is the dimension of Z; z' is the center of the IS PDF, which is chosen by the analyst. The IS technique is based on the following observation:

$$P(F) = E\left(I\left[\min_{\omega\in\Omega} FS_\omega(Z) < 1\right]\right)$$
$$= \int \left(I\left[\min_{\omega\in\Omega} FS_\omega(z) < 1\right] f(z)/q(z)\right) q(z) dz$$
$$\approx \frac{1}{N}\sum_{i=1}^{N} I\left[\min_{\omega\in\Omega} FS_\omega(Z_i) < 1\right] f(Z_i)/q(Z_i) \quad (12)$$

where $\{Z_i : i = 1, \ldots, N\}$ are independent samples from $q(z)$. Therefore, the IS simulation contains the following steps:

a. Draw Z samples $\{Z_i : i = 1, \ldots, N\}$ from $q(z)$.
b. For each sample Z_i, conduct a deterministic slope stability analysis to find the critical slip surface among all trial surfaces. If the safety factor of the most critical surface is less than 1, the slope is considered to fail for that Z_i sample.

c. Repeat Step 2 for $i = 1, \ldots, N$. The IS estimate of the overall failure probability is simply

$$P(F) \approx \frac{1}{N} \sum_{i=1}^{N} I \left[\min_{\omega \in \Omega} FS_\omega(Z_i) < 1 \right]$$

$$\times \exp \left[\frac{1}{2}(Z_i - z')^T (Z_i - z') - \frac{1}{2} Z_i^T Z_i \right]$$

$$\equiv P_F^{IS} \qquad (13)$$

Note that P_F^{IS} is an unbiased estimator of the actual overall failure probability regardless the choice of z'. More importantly, if z' is carefully chosen so that the main support region of $q(z)$ is close to the failure region $\cup_{\omega \in \Omega} F_\omega$, the c.o.v. of P_F^{IS} can be quite small. The c.o.v. of the estimator P_F^{IS} is simply

$$\delta(P_F^{IS}) = \sqrt{Var(P_F^{IS})/E(P_F^{IS})}$$

$$\approx \sqrt{\frac{1}{N} \sum_{i=1}^{N} \left(I \left[\min_{\omega \in \Omega} FS_\omega(Z_i) < 1 \right] \cdot \exp\left[\frac{1}{2}(Z_i - z')^T(Z_i - z') - \frac{1}{2}Z_i^T Z_i \right] - P_F^{IS} \right)^2 } \Big/ P_F^{IS} \qquad (14)$$

3.1 Choice of z'

Although P_F^{IS} is unbiased regardless the choice of z', the choice of z' may seriously affect the efficiency of the IS estimator. An ideal choice of z' is simply the point on the failure limit state surface that is closest to the origin, i.e.: the solution of (3), denoted by z^* (see Figure 1). However, finding z^* is itself a challenging task. Nonetheless, if z' is taken to be a point that is easy to obtain and yet is close to z^*, the resulting IS estimator can still be satisfactory.

In the following, an analytical way of finding such a point that is based on the OMS is presented. Among the slope stability analysis methods, the OMS seems most linear. Moreover, it does not require iterative calculations. Therefore, it is proposed in this research to implement the OMS to find the approximate design point z'. Again, taking the slope in Figure 2 as an example, let the slip surface ω' be a "representative" slip surface, e.g.: the critical slip surface when all uncertain soil parameters are fixed at their expected values.

One can re-write (4) as the following OMS limit state equation:

$$g_{\omega'}^{OMS} = c^{(1)} a^{(1)} + \tan(\phi^{(1)}) \cdot b^{(1)} + c^{(2)} a^{(2)}$$
$$+ \tan(\phi^{(2)}) \cdot b^{(2)} - \gamma^{(1)} d^{(1)} - \gamma^{(2)} d^{(2)} \qquad (15)$$

where $g_{\omega'}^{OMS}$ is the OMS limit state function for the failure event of the representative slip surface ω': if $g_{\omega'}^{OMS} < 0$, the surface fails under the OMS criterion, and vice versa. Let us consider the soil parameters $c^{(1)}, c^{(2)}, \phi^{(1)}, \phi^{(2)}, \gamma^{(1)}, \gamma^{(2)}$ to be uncertain. Note that if $c^{(1)}, c^{(2)}, \tan(\phi^{(1)}), \tan(\phi^{(2)})$ and $\gamma^{(1)}, \gamma^{(2)}$ are independent Gaussian, the limit state function in the standard Gaussian space can be approximated as

$$g_{\omega'}^{OMS}(Z) = (\mu_{c^{(1)}} + \sigma_{c^{(1)}} Z_{c^{(1)}}) a^{(1)}$$
$$+ (\mu_{c^{(2)}} + \sigma_{c^{(2)}} Z_{c^{(2)}}) a^{(2)}$$
$$+ (\mu_{\tan(\phi^{(1)})} + \sigma_{\tan(\phi^{(1)})} Z_{\tan(\phi^{(1)})}) \bar{b}^{(1)}$$
$$+ (\mu_{\tan(\phi^{(2)})} + \sigma_{\tan(\phi^{(2)})} Z_{\tan(\phi^{(2)})}) \bar{b}^{(2)}$$
$$- (\mu_{\gamma^{(1)}} + \sigma_{\gamma^{(1)}} Z_{\gamma^{(1)}}) d^{(1)}$$
$$- (\mu_{\gamma^{(2)}} + \sigma_{\gamma^{(2)}} Z_{\gamma^{(2)}}) d^{(2)} \qquad (16)$$

where μ and σ are the mean value and standard deviation of the subscripted variable;

$$\bar{b}^{(n)} = \mu_{\gamma^{(n)}} \sum_{k=1}^{N_s^{(n)}} (h_k^{(n)} \cos(\alpha_k^{(n)})) - \sum_{k=1}^{N_s^{(n)}} u_k^{(n)} \Delta l_k^{(n)} \qquad (17)$$

Note that (16) is an approximation of (15) because $b^{(1)}$ and $b^{(2)}$ are in fact random, but we have deliberately replaced them by $\bar{b}^{(1)}$ and $\bar{b}^{(2)}$. By doing so, (16) is now a linear function of the standard Gaussian input Z. It is soon clear that the OMS design point in the standard Gaussian space is roughly

$$\left[z_{c^{(1)}}^{OMS*} \; z_{c^{(2)}}^{OMS*} \; z_{\tan(\phi^{(1)})}^{OMS*} \; z_{\tan(\phi^{(2)})}^{OMS*} \; z_{\gamma^{(1)}}^{OMS*} \; z_{\gamma^{(2)}}^{OMS*} \right]$$
$$= \frac{\beta^{OMS}}{\sigma_{g_\omega^{OMS}}} \cdot [-a^{(1)} \sigma_{c^{(1)}} - a^{(2)} \sigma_{c^{(2)}} - \bar{b}^{(1)} \sigma_{\tan(\phi^{(1)})}$$
$$- \bar{b}^{(2)} \sigma_{\tan(\phi^{(2)})} \; d^{(1)} \sigma_{\gamma^{(1)}} \; d^{(2)} \sigma_{\gamma^{(2)}}] \qquad (18)$$

where

$$\beta^{OMS} = \frac{a^{(1)} \mu_{c^{(1)}} + a^{(2)} \mu_{c^{(2)}} + \bar{b}^{(1)} \mu_{\tan(\phi^{(1)})} + \bar{b}^{(2)} \mu_{\tan(\phi^{(2)})} - d^{(1)} \mu_{\gamma^{(1)}} - d^{(2)} \mu_{\gamma^{(2)}}}{\sigma_{g_\omega^{OMS}}}$$

$$\sigma_{g_\omega^{OMS}} = \sqrt{\begin{aligned}&a^{(1)^2}\sigma_{c^{(1)}}^2 + a^{(2)^2}\sigma_{c^{(2)}}^2 + b^{(1)^2}\sigma_{\tan(\phi^{(1)})}^2\\&+ b^{(2)^2}\sigma_{\tan(\phi^{(2)})}^2 + d^{(1)^2}\sigma_{\gamma^{(1)}}^2 + d^{(2)^2}\sigma_{\gamma^{(2)}}^2\end{aligned}}$$

(19)

The above design point is in the standard Gaussian space of $c^{(1)}, c^{(2)}, \tan(\phi^{(1)}), \tan(\phi^{(2)})$ and $\gamma^{(1)}, \gamma^{(2)}$. A simple algebraic operation shows that the design point in the standard Gaussian space of $c^{(1)}, c^{(2)}, \phi^{(1)}, \phi^{(2)}$ and $\gamma^{(1)}, \gamma^{(2)}$ is

$$\begin{bmatrix} z_{c^{(1)}}^{OMS*} \\ z_{c^{(2)}}^{OMS*} \\ z_{\phi^{(1)}}^{OMS*} \\ z_{\phi^{(2)}}^{OMS*} \\ z_{\gamma^{(1)}}^{OMS*} \\ z_{\gamma^{(2)}}^{OMS*} \end{bmatrix} = \begin{bmatrix} -a^{(1)}\sigma_{c^{(1)}}\beta^{OMS}/\sigma_{g_\omega^{OMS}} \\ -a^{(2)}\sigma_{c^{(2)}}\beta^{OMS}/\sigma_{g_\omega^{OMS}} \\ p_{\phi^{(1)}}^{-1}(\tan^{-1}(\mu_{\tan(\phi^{(1)})} - b^{(1)}\sigma_{\tan(\phi^{(1)})}^2\beta^{OMS}/\sigma_{g_\omega^{OMS}})) \\ p_{\phi^{(2)}}^{-1}(\tan^{-1}(\mu_{\tan(\phi^{(2)})} - b^{(2)}\sigma_{\tan(\phi^{(2)})}^2\beta^{OMS}/\sigma_{g_\omega^{OMS}})) \\ d^{(1)}\sigma_{\gamma^{(1)}}\beta^{OMS}/\sigma_{g_\omega^{OMS}} \\ d^{(2)}\sigma_{\gamma^{(2)}}\beta^{OMS}/\sigma_{g_\omega^{OMS}} \end{bmatrix}$$

(20)

where $p_{\phi^{(n)}}$ is the mapping from a standard Gaussian random variable to $\phi^{(n)}$. In this study, the design point defined in (20) will be taken as the center z' of the IS PDF.

Note that the derivations of the design point in (20) are based on the following assumptions: (a) $c^{(1)}, c^{(2)}, \tan(\phi^{(1)}), \tan(\phi^{(2)})$ and $\gamma^{(1)}, \gamma^{(2)}$ are independent Gaussian and (b) the OMS is the adopted method of slices. Those assumptions are taken because the resulting design point z' has an analytical form. Clearly, such a design point z' in general is not the actual design point z' because for the actual application those assumptions may not be true, i.e.: the actual adopted slope stability method may not be the OMS and the soil parameters may not be independent Gaussian. Nonetheless, because the proposed IS approach works regardless of the choice of z', the resulting estimator P_F^{IS} is always unbiased. The only concern is whether the approximate design point z' is close to the actual design point z^*. Empirically, it is found that the approximate design point z' is usually close to the actual design point z^* even the two assumptions are violated in the actual application.

4 PROCEDURE OF PROPOSED APPROACH

The procedure of the proposed IS method is described as follows:

a. Execute an OMS analysis where all uncertain soil parameters are fixed at their mean values to find the representation slip surface ω'.

b. Use (20) to obtain the design point z' for the representative surface ω'.
c. Draw Z samples $\{Z_i: i = 1, \ldots, N\}$ from $q(z)$, a Gaussian PDF centered at z' with unit covariance matrix.
d. For each sample Z_i, conduct a deterministic slope stability analysis (not necessarily the OMS) to find the critical slip surface among all trial surfaces. If the safety factor of the critical surface is less than 1, the slope is considered to fail for that Z_i sample.

e. Repeat Step 2 for $i = 1, \ldots, N$. The IS estimate of the overall failure probability is simply

$$P(F) \approx \frac{1}{N}\sum_{i=1}^{N} I\left[\min_{\omega \in \Omega} FS_\omega(Z_i) < 1\right]$$

$$\times \exp\left[\frac{1}{2}(Z_i - z')^T(Z_i - z') - \frac{1}{2}Z_i^T Z_i\right]$$

$$\equiv P_F^{IS}$$

(21)

5 NUMERICAL EXAMPLE

Consider the following example extracted from the STABL user manual (Siegel 1975): the slope in Figure 3 underlain by a rock layer. The shear strength parameters c and ϕ of the soil are uncertain. It is assumed that tension cracks are present near the top surface of the slope up to a depth of 11 m and that the failure surface cannot propagate into the rock layer. The uncertain cohesion c is lognormally distributed with mean $\mu_c = 76.3$ kN/m² and c.o.v. $\delta_c = 20\%$, while the friction angle is Gaussian with mean $\mu_\phi = 18°$ and standard deviation $\sigma_\phi = 1.8°$. There is a negative correlation of -0.3 between Z_c and Z_ϕ. The simplified Bishop method of slices is taken as the slope stability method for this example.

The MCS method with sample size N = 100,000 is taken to estimate the failure probability of the slope. For each Z sample, one hundred trial surfaces are

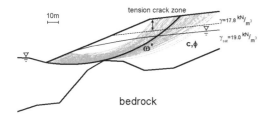

Figure 3. The slope considered in the numerical example.

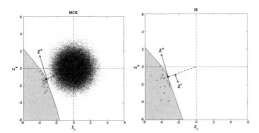

Figure 4. The MCS (left) and IS (right) samples in the standard Gaussian space. The shaded region is failure region F.

randomly generated to form the set of all trial slip surfaces Ω (shown in Figure 3), and the factor of safety is determined by solving (5), then (1) can be employed to find P_F^{MCS}. Figure 4 shows the locations of all MCS samples in the standard Gaussian space, where a dot indicates a non-failure sample, while a 'x' indicates a failure sample. One can see most of the samples are non-failure ones. The analysis shows that P_F^{MCS} is $2.4e^{-4}$ and the c.o.v. of the estimator is 20.4%.

As a verification, the actual failure region F is found by an exhaustive analysis described as follows. First divide the standard Gaussian space into dense grid points. At each grid point, a simplified Bishop stability analysis is taken to see if the entire slope fails. Do so for all grid points to obtain the actual failure region F. The actual design point z^* can be readily identified in Figure 4 as the point on the failure boundary that is closet to the origin.

For the IS technique, the representative surface ω'. (shown in Figure 3) is first found by a single OMS stability analysis where c and ϕ are fixed at their mean values. The approximate design point z' for the representative surface ω'. is determined from (20), and its location is shown in Figure 4. It is clear that z' is fairly close to the actual design point z^* even though the actual analysis method is the simplified Bishop method, not the OMS, and even though in reality c and $\tan(\phi)$ are not independent Gaussian. Following the IS procedure, only 100 samples of Z are drawn from $q(z)$ to obtain P_F^{IS} via (13). Those samples are plotted in Figure 4. It is clear that a large portion of the samples are failure samples. The resulting P_F^{IS} is $2.0e^{-4}$, and its c.o.v. is estimated to be 26.1% via (14). It is clear that the MCS method becomes inefficient when the failure probability gets smaller, while the IS method seems robust on that aspect. In particular, with a total number of 100 samples, the IS method performs satisfactorily even for small failure probability.

6 CONCLUSIONS

A new method based on the importance sampling (IS) technique is proposed to efficiently estimate failure probability of slope stability. The center of the importance sampling probability density function (PDF) can be found by a first-order reliability analysis based on the ordinary method of slices. From the results of the numerical example, it is found that the proposed IS method is more efficient than the Monte Carlo simulation (MCS). The proposed IS method provides unbiased estimates for slope failure probabilities, and yet it requires much less computation than MCS especially for the cases with small failure probabilities. A limitation of the proposed IS method is that the dimension of uncertain soil parameters cannot be too large due to the dimension limitation of the basic IS method. Therefore, the proposed method is not suitable for determining failure probability of slopes with spatial variability. It is left as a future research to overcome this limitation.

REFERENCES

Ang, A.H.S. and Tang, W.H. 1984. Probability Concepts in Engineering Planning and Design, Volume II Decision, Risk, and Reliability, John Wiley and Sons.

Bishop, A.W. 1955. The use of the slip circle in the stability analysis of slopes. Geotechnique, 5, 7–17.

Chowdhury, R.N. and Xu, D.W. 1993. Rational polynomial technique in slope-reliability analysis. ASCE Journal of Geotechnical Engineering, 119(12), 1910–1928.

Christian, J.T., Ladd, C.C. and Baecher, G.B. 1994. Reliability applied to slope stability analysis. ASCE Journal of Geotechnical Engineering, 120(12), 2180–2207.

Fellenius, W. 1936. Calculation of the stability of earth dams. Transactions of 2nd Congress on Large Dams, Washington, DC, 4, 445–462.

Griffiths, D.V. and Fenton, G.A. 2004. Probabilistic slope stability analysis by finite elements. ASCE Journal of Geotechnical and Geoenvironmental Engineering, 130(5), 507–518.

Low, B.K. and Tang, W.H. 1997. Probabilistic slope analysis using Janbu's generalized method of slices. Computers and Geotechnics, 21(2), 121–142.

Low, B.K., Gilbert, R.B. and Wright, S.G. 1998. Slope reliability analysis using generalized method of slices. *ASCE Journal of Geotechnical and Geoenvironmental Engineering*, 124(4), 350–362.

Melchers, R.E. 1989. Importance sampling in structural systems, *Structural Safety*, 6, 3–10.

Oka, Y. and Wu, T.H. 1990. System reliability of slope stability. *ASCE Journal of Geotechnical Engineering*, 116(8), 1185–1189.

Rubinstein, R.Y. 1981. *Simulation and the Monte-Carlo Method*, John Wiley & Sons Inc., New York.

Shinozuka, M. 1983. Basic analysis of structural safety, *ASCE Journal of Structural Engineering*, 109, 721–740.

Siegel, R.A. 1975. *STABL User Manual*, Joint Highway Research Project 75-9, School of Engineering, Purdue University, West Lafayette, Indiana.

Prediction of the flow-like movements of Tessina landslide by SPH model

S. Cola & N. Calabrò
Dept. IMAGE, University of Padova, Italy

M. Pastor
Centro de Estudios y Esperimentación de Obra Públicas (CEDEX) and ETS de Ingenieros de Caminos, Madrid, Spain

ABSTRACT: The SPH (Smoothed particle hydro-dynamics) method is a powerful tool for modelling debris/mud flows, which can be described in terms of local interactions of their constituent parts. The characteristics of the 1992 Tessina landslide make it appropriate for modelling by means of SPH method. In this paper we will investigate the evolution and dynamics of the 1992 Tessina landslide and present future hazard scenarios based on potential volumes of masses mobilised in the future.

1 INTRODUCTION

Over the past decades, flow-like landslides (debris-flow and mud-flow) have been intensively studied, because in their catastrophic form they have caused many victims and serious economic damage around the world. Their distinctive features and evolution are strictly related to the mechanical and rheological properties of the materials involved which are responsible for the long distances travelled and high velocities in some cases reached.

Lately a number of new numerical models have been developed: the prediction of both run-out distances and velocity can notably reduce losses inferred by these phenomena, providing a means for defining and estimating the hazardous areas, and the working out of appropriate design measures.

Most of the current numerical methods are based on either structured (finite elements) or unstructured (finite elements and volume) grids. An alternative approach is that of Cellular Automata (CA) which is appropriate in modelling debris/mud flow (Avolio et al., 2000). But recently another interesting and powerful tool for modelling natural systems has been provided by a new group of "meshless" numerical methods, among which the method of "Smoothed particle hydro-dynamics" (SPH) independently introduced by Lucy (1977) and Gingold and Monaghan (1977) and firstly applied to astrophysical modelling (Benz W., 1990). This method relies on nodes (points) to approximate functions or derivates, discluding any element based information.

In this paper, an in-depth integrated SPH model (Pastor et al, 2001) is proposed to simulate the propagation stage of a slow mud-flow landslide (Takeda, 1994): that of Tessina one, being present for many decades in North-East Italy. Model calibration is obtained from the simulation of the 1992 event and, subsequently, used in forecasting flow expansion during a possible future reactivation.

2 MATHEMATICAL MODEL

The mathematical model is presented starting with the Biot—Zienkiewicz equations for non-linear materials and large deformation problems, from where the depth integrated model can be derived.

It is based on:

i. the balance of mass, which is:

$$\frac{D^s \rho}{Dt} + \rho \, div v^s = 0 \qquad (1)$$

where D^s refers to a material derivative following the soil particles, ρ is the density of the soil and v_s is the velocity of soil skeleton;

ii. the balance of linear momentum, which reads:

$$\rho \frac{D^s v_s}{Dt} = \rho b + div \sigma \qquad (2)$$

where b are the body forces and σ the Cauchy stress tensor.

The model is completed by suitable rheological and kinematical relations relating the stress tensor to the rate of deformation tensor d, and the rate of deformation tensor to the velocity field v_s.

Many landslides have average depths which are small in comparison with their length or width, so it is possible to simplify the 3D propagation model by integrating its equations along the vertical axis. The resulting 2D depth integrated model presents an excellent combination of accuracy and simplicity, providing important information such as velocity of propagation, time in reaching a particular place, depth of the flow at a certain location, etc.

Depth integrated models have been frequently used in the past to model flow-like landslides. It is worth mentioning the pioneering work of Hutter and his co-workers (Savage and Hutter, 1991; Hutter and Koch 1991) or Laigle and Coussot (1997).

The equations for the depth averaged model are obtained by integrating the balance of mass and momentum equations along the vertical axis. The balance of mass results:

$$\frac{\partial}{\partial x_j}(h\bar{v}_j) + \frac{\partial h}{\partial t} = 0 \quad j = 1,2 \tag{3}$$

where h is landslide thickness and $\bar{v}_j$ depth averaged velocity.

Indeed, the balance of momentum, depth integrated equation is:

$$\rho \left[\frac{\partial}{\partial t}(h\bar{v}_j) + \frac{\partial}{\partial x_j}(h\bar{v}_i\bar{v}_j) \right]$$

$$= \frac{\partial}{\partial x_j}(h\bar{\sigma}_{ij}) + t_j^A + t_j^B + \rho b_j h \quad j = 1,2 \tag{4}$$

where the terms t_j^A and t_j^B are the normal stress acting on the surface and bottom respectively.

It is important to note that the above results depend on the rheological model chosen, from which basal friction and depth integrated stress $\bar{\sigma}_{i,j}$ are obtained. In fact it is assumed that the flow at a given point and time, with depth and in-depth averaged velocities known, has the same vertical structure as a uniform steady state flow. In the case of flow—like landslides this model is often referred to as the infinite landslide as it is assumed to have constant thickness and move at constant velocity along a constant slope.

The main advantage of this numerical method is a big reduction of the calculation time compared to a standard finite element code formulated on the Eulerian approach, because the computational grid is separated from the structured terrain mesh used to describe terrain topography.

In the first application of the model to the Tessina phenomenon, the Bingham rheological model was considered sufficiently appropriate to describe the main features of landslide mass behaviour. This is because the soil involved in the slide contains appreciable fraction of fine and plastic material whose behaviour may be controlled prevalently by viscosity.

The shear stress inside a Bingham fluid is depicted by the equation:

$$\tau = \tau_c + \mu_b \frac{du}{dy} \tag{5}$$

where τ and τ_c are the shear stress and the yield strength respectively, μ_b the plastic viscosity and du/dy the shear rate.

3 TESSINA LANDSLIDE

The Tessina landslide, which was first triggered in October 1960, is a complex movement with a source area affected in the upper sector by rotational slides; downhill the slide turns into a mud flow through a steep channel. The landslide develops in the Tessina valley between the altitudes of 1220 m and 625 m a.m.s.l, with total longitudinal extension of nearly 3 km and maximum width of about 500 m. The mud flow passes very close to the village of Funes and stretches downhill as far as the village of Lamosano (Figure 1). The rock types involved in the landslide belong to the Flysch Formation, consisting of a rhythmic alternation of marly—argillaceous and calcarenite layers about 1000–1200 m thick.

During the 1960s, several reactivations, involving about 5 million m³ of material, occurred causing the filling of the Tessina valley with displaced material

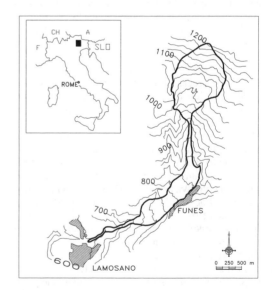

Figure 1. Distribution area of the Tessina landslide.

30–50 m thick. These movements seriously endangered the village of Funes, which is situated on a steep ridge—originally quite high above the river bed, but now at nearly the same level as the mud-flow surface (Dall'Olio et al, 1987).

During April 1992 a rotational slide with 20–30 m deep failure surface, affecting also the Flysch bedrock, caused the collapse of an area 40.000 m² wide, on the left hand-side of the Tessina stream, with an approximate volume of 1 million m³. The movements initially caused the formation of 15 m high scarp, a 100 m displacement of all the unstable mass and the destruction of drainage systems set up some years earlier, but they continued with a certain intensity up to June 1992, causing the mobilization of another 30.000 m² more. The highly fractured and dismembered material from this area was channelled along the river bed where, due to continuous remoulding and increasing of water content, it became more and more fluid, thus generating some small earth flows converging into the main flow body. After these events, the inhabitants of Funes and Lamosano were evacuated.

After this important event, the landslide was not stabilized but its evolution was kept under control by an accurate monitoring activity and an alarm system set up. Some other collapses and mudflows, similar in evolution to the 1992 event but involving smaller portion of soils, were observed with an occurrence of about 3 years. Anyway, the hazard level reached in 1992 was never overcome and the inhabitants were evacuated no more.

4 SIMULATION OF THE 1992 TESSINA LANDSLIDE

4.1 Properties of soils from experimental tests

Laboratory analyses were performed on some remoulded samples collected inside the moved mass along the closing channel near Funes village.

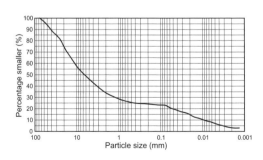

Figure 2. Typical grain size distribution of material involved in mud-flow.

The soil may be classified as a tout—venant with a percentage of clay and silt ranges from 5 to 20% (Figure 2). The fine fraction is composed prevalently of medium-low plasticity inorganic clay (Figure 3).

In order to evaluate the rheological behaviour of the soil composing the mud-flow, some tests with the

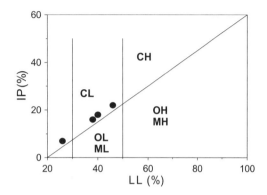

Figure 3. Classification of mud-flow material.

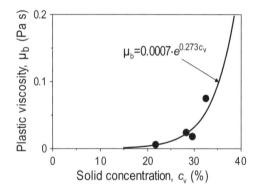

Figure 4. Viscosity curve.

$$\mu_b = 0.0007 \cdot e^{0.273 c_v}$$

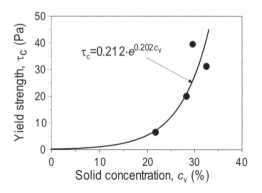

Figure 5. Yield strength curve.

$$\tau_c = 0.212 \cdot e^{0.202 c_v}$$

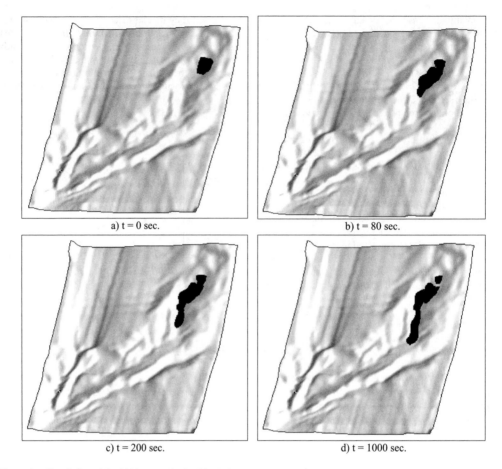

Figure 6. Simulation of the 1992 event obtained by SPH.

FANN V-G rheometer were performed on the fine fraction (passing to n.40 sieve). In accordance with the facilities of this equipment, the strength was recorded at two speeds—i.e. at 300 and 600 r/min. From these two readings the Bingham viscosity and the yield strength were determined.

The measurements were repeated with a solid concentration varying from 20 to 45%: Figures 4 and 5 sketch the experimental data obtained for sample n.3. The rheological properties depend on solid concentration c_v according to the following exponential laws (O' Brien and Julien, 1988; Major and Pierson, 1992; Coussot and Piau, 1994, Sosio et al, 2006):

$$\tau_c = \alpha e^{\beta c_v}$$
$$\mu_b = \gamma e^{\delta c_v}$$
(6)

In Figures 4 and 5 the best fitting equations are also reported. For the concentration analyzed the yield strength and plastic viscosity range from 5 to 40 Pa and from 0.005 to 0.08 Pa·s respectively.

4.2 Numerical modelling

In order to simulate the 1992 event and calibrate model parameters, the flow-like movements of slide were reproduced assuming a topographic mesh of 17464 nodes provided by a 15 × 15 m digital terrain model (DTM). Figure 6a shows the topographical base and initial position of the failed mass, the latter being modelled through 3304 particles with average spaces of 3.5 m.

It is important to note that the analysis with SPH code doesn't account for the triggering of the landslide, when large portions of slope, including rock

Table 1. Assumed rheological properties.

Density (Kg/m^3)	1700
Solid concentration (%)	50
Bingham viscosity (Pa s)	600
Yield strength (Pa)	5200

blocks coupled with debris and mud, move. This initial phase exhausts itself quickly in comparison to the duration of the phenomenon, because thereafter such masses split continuously until reaching complete fluidity. The numerical simulation considered that, the whole mass involved in the landslide detaches instantaneously. In reality the cohesion inside the unstable mass is lost little by little: in future advances of this model, this aspect would be taken into account by assuming initial high viscosity of unstable mass and decreasing it gradually up to a complete fluidity state.

The 1992 landslide was simulated assuming the Bingham constitutive model with viscosity coefficients corresponding to a liquid state of the tested soils, as summarized in Table 1.

The results of the performed simulation are shown in Figure 6. In particular, Figure 6b shows the landslide at the first simulation step (the elapsed time equal to 80 s), that means the masses have just moved away and the mudflow begins to channel correctly into the river valley. Simulations at the elapsed time of 200 s and 1000 s (Figures 6c and 6d) show how the mudflow has canalized in the river valley and here runs correctly.

Comparison between the real event and this simulation shows substantial agreement in the landslide development, even if the numerical analysis gives runout distances shorter than those observed. This result depends on the fact that the SPH code is unable to reproduce recurrent detachments of mass and, without the contribution of other material moving later from the upper zone, the flow stops earlier than in reality. But, nevertheless, the mud path is properly identified and the velocities (4 ÷ 5 mm/s) and thicknesses of the landslide (6 ÷ 7 m) are in agreement with field observations.

5 FUTURE HAZARD SCENARIOUS

The SPH code permitted to develop reliable simulations of future events. To this aim, an accurate and update morphology of the landslide area has been set up using the most updated DTM. As regards the possible instable volume, on the basis of previously observed events, Avolio et al. (2000) indicated an interval from 0.5 to 1.5 million on cubic meters: these limit values represent the minimum mud quantity needed to obtain a landslide event and the worst case of detachment. The detachment location may be anywhere in the superior basin where the monitoring has marked the highest movements: in this case, the most critical area seems to be the zone known as "*Pian de Cice*" (Figure 7), where maximum displacements of about 3 cm/ year were recorded in the last decade.

On the basis of these remarks, three different cases were simulated, concerning the collapse of 0.5, 1 and 1.5 million m^3 of material. To note that the value of 1 million m^3 is more than that experienced in all the previous movements.

Also in this case the material is described by the Bingham rheological model, using the constitutive parameters obtained from the previous calibration.

The simulation results, shown in Figure 8, underline how the mud path is similar to that of the 1992 event. More, even if the involved volume changes, the maximum distance reached for the three cases are more and less equivalent because it is effected mainly by the topographic gradient, which is the same in all three analyses.

Velocities and thickness are different for the three cases (Table 2), the highest value resulting in the simulation of 1.5 million m^3, that is, therefore, the most hazardous case for Lamosano and Funes villages. Although the model can't analyse recurrent collapses, and the real path length can't be estimated exactly, it is possible to suppose that a detachment from this area could involve downhill villages with a very high risk, resulting in a necessary partial or even totally evacuation.

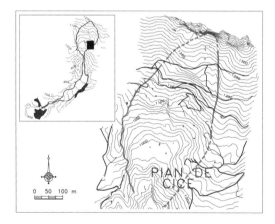

Figure 7. Localization of Pian de Cice hazardous area.

Table 2. Results for future hazardous scenarios simulation.

Volume (m^3)	0.5	1	1.5
Average velocity (mm/s)	2–4	4–6	9–10
Average thickness (m)	11–10	8–11	7–10

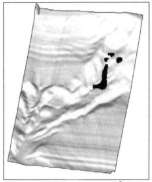 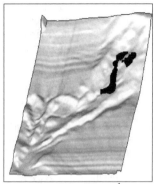 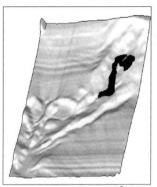

case 1) 0.5 million m³   case 2) 1 million m³   case 3) 1.5 million m³

Figure 8. Simulation of the future scenarios due to mobilization of Pian de Cice area.

6 CONCLUSIONS

Results of first simulations using the depth averaged model developed on the SPH approach are promising. Despite the limitations existing in representing the trigger of the rotational superior slides and the recurrent collapse of different area with this model, the reproduction of landslide path for the 1992 event was appreciable, supporting the assumption that the development of the initial part doesn't influence the simulation of the subsequently run-out phenomenon.

On the basis of this positive result, possible future events, of a magnitude comparable or higher than the previous slide, are simulated, hypothesizing that physical characteristics of the landslide don't change significantly over time. The results obtained show a situation of real risk affecting the hamlets involved in the landslide and underline the need to take adequate countermeasures.

The experience acquired through the Tessina landslide is useful for better identification and definition of potential applications of this procedure in order to forecast areas that may be involved in low—viscous landslides, such mud flows. The SPH model may be successfully used in order to follow the progress of an event, predict its evolution and, finally verify the possible effects of flows on man's intervention.

ACKNOWLEDGEMENTS

This paper is a part of the research Project "MoVeMit: studio e modellazione di movimenti di versante finalizzati alla formulazione di interventi di mitigazione", financed by the funding of the "Fondazione Cassa di Risparmio di Verona, Vicenza, Belluno e Mantova. At least the authors wish to thank the "Fondazione Angelini" of Belluno and the CNR-IRPI of Padova which covenant this activity research.

REFERENCES

Avolio, M.V., Di Gregorio, S., Mantovani, F., Pasuto, A., Rongo, R., Silvano, S. & Spataro, W., 2000. Simulation of the 1992 Tessina landslide by cellular automata model and future hazard scenarios, *Geomorphology*, 2(1): 41–50.

Benz, W., 1990. Smooth particle hydrodynamics: a review, in *The Numerical Modelling of Nonlinear Stellar Pulsatations*, pp. 269–288, Buchler (Ed) Kluwer Academic Publishers.

Coussot, P., 1997. Mudflow Rheology and Dynamics. *IAHR monograph*, Balkema, Rotterdam, 260 pp.

Coussot, P. & Piau, J.M., 1994. On the behaviour of fine mud suspensions. *Rheological Acta* 33, 175–184.

Dall'Olio, L., Ghirotti, M., Iliceto, V. & Semenza, E., 1987. La frana del Tessina, Alpago (Bl). *Atti VI Congresso Ordine. Nazionale Geologi*, Venezia 25–27 settembre 1987, pp. 275–293.

Gingold, R.A. & Monaghan J.J., 1977. Smoothed particle hydrodynamics: theory and application to non-spherical stars. *Monthly Notices of the Royal Astronomical Society.* 81, pp. 375–389.

Gingold, R.A. & Monaghan J.J., 1982. Kernel estimates as a basis for general particle methods in hydrodynamics. *J. Comput. Phys.* 46.

Hutter, K. & Koch, T., 1991. Motion of a granular avalanche in an exponentially curved chute: experiments and theorical predictions. *Phil. Trans. R. Soc. London*, 334, pp. 93–138.

Laigle, D. & Coussot, P., 1977. Numerical modeling of mudflows. *Journal of Hydraulic Engineering, ASCE*, 123(7), pp. 617–623.

Lucy, L.B. 1977. A numerical approach to testing of fusion process. *Astronomical Journal*, Vol. 82, pp. 1013–1024.

Major, J.J. & Pierson, T.C., 1992. Debris flow rheology: experimental analysis of fine-grained slurries. *Water Resources Research* 28, 841–857.

Monaghan J.J. & Gingold, R.A., 1981. Shock simulation by the particle method SPH. *J. Comp. Phys.* 52, pp. 374–389.

Pastor, M., Quecedo, M, Merodo, J.A., Herreros, M.I., González, E. & Mira, P., 2002. *Modelling of debris flows and flow slides*. Revue française de genie civil: Numerical modelling in Geomechanics, vol 6, 1213–1232.

Pastor, M., Sopeña, L., Fernández Merodo, J.A., Quecedo, M., Mira, P., Herreros, M.I. & González, E., 2001. *Modelización de deslizamientos de laderas y olasprovocadas en embalses: aplicaciones prácticas*. V Simposio Nacional sobre Taludes y Laderas Inestables, Madrid, v.2, pp. 431–460.

O'Brien, J.S., Julien, P.Y. & Fullerton, W.T., 1993. Two dimensional water flood and mudflow simulation. *Journal of Hydraulic engineering*, 119, 244–259.

O'Brien, J.S. & Julien, P.Y., 1988. Laboratory analysis of mudflow properties. *Journal of Hydraulic Engineering*, 110, 877–887.

Savage S.B. & Hutter, K., 1991. The dynamics of avalanches og granular materials from initiation to runout. *Part I: Analysis. Acta Mechanica* 86, pp 201–223, Madrid 2001.

Sosio, R., Crosta, G.B., Frattini, P. & Valbuzzi, E., 2006. *Caratterizzazione reologica e modellazione numerica di un Debris flow in ambiente alpino*, Giornale di geologia applicata, v. 3.

Takeda, H.T., Miyama, S.M. & Sekiya, M., 1994. *Numerical simulation of viscous flow by smooth particle hydrodynamics*, Progress of Theoretical Physics, V.92, 939–960.

Applications of the strength reduction finite element method to a gravity dam stability analysis

Q.W. Duan, Z.Y. Chen, Y.J. Wang, J. Yang, & Y. Shao
China Institute of Water Resources and Hydropower Research, Beijing, China

ABSTRACT: This paper examines an example that includes gravity dam stability analysis by using the strength reduction finite element method (SRF), compared with the analytical results of the limit equilibrium approach proposed by Sarma. Extensions of plastic zones along the specified seam, as well as the development of new yield zone in the intact rock, can be clearly visualized as the reduction factor F increases, until a failure criterion, either defined as divergence of the numerical iteration, or rapid increase of displacement at some characteristic points, is reached. The associated values of F are in good agreement with those of Sarma's method for two cases investigated. It shows that strength reduction method is feasible in gravity dam stability analysis along specified or potential slip surfaces.

1 INTRODUCTION

During the past ten years, finite element analysis using the strength reduction approaches (SRF) has been widely used in engineering (Griffths & Lane 1999, Ugai & Leshchinsky, 1995, Matsui & San, 1992). It considers nonlinear constitutive relations of materials and complicated geological conditions of the rock mass. Various complicated loads, such as waterstoring, seepage, and earthquake can be included. The analytical results include the stress region, spatial distribution of deformation, the development process of plastic zone etc., which are much more informative, compared to the conventional limit equilibrium methods.

This paper studies the feasibility of strength reduction method in the deep anti-sliding stability of gravity dams. The results obtained by this study are compared with those obtained by the limit equilibrium method.

2 THE STRENGTH REDUCTION METHOD

Strength reduction finite element method (SRF) is firstly presented by Zienkiewicz in 1975 in the application of elastic-plastic by the Finite Element Analysis. The reduction coefficient is defined as a ratio of the available shear strength of the soil within slope to the actual shear stress developed when the limiting equilibrium state is believed happened, either based on the criterion that the numerical iteration fails or the displacement rate approaches infinity. This definition for safety factor is the same as that proposed in the limit equilibrium method. In the analysis, F, the strength reduction factor, is gradually increased until the structure reaches instability.

$$c' = c/F \tag{1}$$

$$\tan \phi' = \tan \phi / F \tag{2}$$

where c and ϕ are the strength parameters of the material.

Based on the shear strength reduction concept by Zienkiewicz et al. (1975) and the elastic-plastic finite element method, many domestic and foreign scholars developed a variety of research outcomes that have confirmed that SRF can always give calculated factor of safety almost identical to that given by the conventional limit equilibrium methods. In this paper, we present a study that concerns the stability of gravity dam along potential slip surfaces. The analytical results again show good agreements with the conventional approach that uses Sarma's method.

3 THE TEST EXAMPLES

3.1 The gravity dam example

A gravity dam with height of 129 m is shown in Figure 1(a), which contains two soft layers AD and BC intersecting at point B. The inclination of AD is

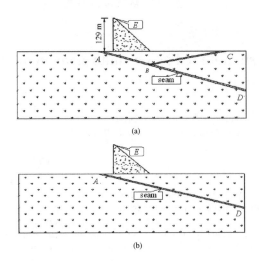

Figure 1. An example of gravity dam stability analysis. (a) Double soft layers, (b) Single soft layer.

Table 1. Mechanical parameters of the dam and abutment.

Material	Deformation Modulus (GPa)	Poisson's ratio	Friction angle	Cohsion (MPa)	Tensile strength (MPa)	Density kN/m^3
Dam body	20	0.167	35	2.0	1.85	24
Bedrock	10	0.26	35	1.3	0	27
Layers	2.5	0.35	19.8	0.115	0	25.6

15° downstream, inclination of BC is 10° upstream. Figure 1(b) shows the dam containing only a weak layer AD as described in Figure 1(a). Water thrust of 128 m height and the water pressure at the ground surface of upstream base are considered. The parameters of dam body and dam base are shown in Table 1. Mohr-Coulomb yield criterion is adopted for the elastic-plastic analysis.

3.2 Calculation results by Sarma method

Figure 2 shows the calculations for double-layer model associated with the safety factor of 2.19, with vertical interfaces.

For the calculation of single layer model illustrated in Figure 3, the method of slices with inclined interfaces is adopted and the initial and critical sliding surfaces of calculation results are shown respectively. The factors of safety with regard to initial and critical sliding surfaces are 4.732 and 3.190.

Figure 2. Analysis by Sarma's method, the double-layer model.

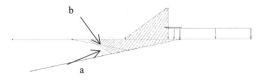

Figure 3. Analysis by Sarma's method, the single-layer model.
NOTE: The arrows 'a' and 'b' refer to the initial and critical slip surfaces respectively.

4 THE DOUBLE - LAYER MODEL BY SDF

4.1 The iteration process

The safety factor for calculation model shown in Figure 4 is 2.12 by the commercial software FLAC. Its corresponding surface is shown in Figure 4. The result by the authors has good agreement with that obtained by Sarma method.

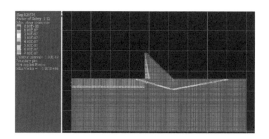

Figure 4. Sliding Plane in ultimate State, $F = 2.12$.

Calculation started with $F = 1.0$, with a results shown in Fig. 5(a), followed by $F = 1.2$, $F = 1.4, \ldots$, until $F = 2.2$ (Figure 6), when the iteration broke.

The calculation results have shown that the critical sliding surface is along the soft layer ABC. In order to study the progressive development of destruction of dam base with the reduction of strength, the plastic zones are listed in Figure 5 with regard to different reduction coefficients (from 1.0 to 2.0 with equal intervals of 0.2, 2.12 and 2.20).

As shown in Figure 5, greater portions of the soft layer become plastic as the reduction coefficient increases. When the reduction coefficient reaches 2.0, the whole weak layer is plastic; however, the dam base has not reached the ultimate Status. When the reduction coefficient increases to 2.12, the whole weak layer is plastic, moreover, a large plastic area appears between the dam site and the turning point, the dam base has been at the ultimate status.

Figure 6 shows the plastic zone at the moment the numerical iteration broke, associated with a factor of safety $F = 2.20$.

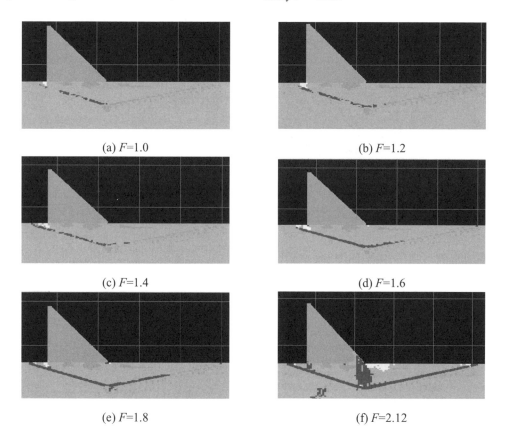

Figure 5. The plastic zones for different reduction coefficients.

4.2 The displacement and iteration features

From the reduction factor of distribution of plastic zone, the relationship can be seen between the increase in the reduction factor and development process of the plastic area. So we can come into conclusion that the appearance of the plastic area is an essential stage from this process from all the plastic zone transfixion to the ultimate Status of dam base. Therefore, it is the ultimate Status when secondary sliding surface appears, state as illustrated in Figure 5.

From Figure 5, one may see the gradual development of plastic zone on the slip surface as F increases. At $F = 2.12$, which corresponds to the solution that Sarma's method offered, plastic yielding developed along the whole seam, and in the meanwhile, a vertical plastic zone developed near the toe. The failure modes of SRF and Sarma coincide with each other.

Table 2 shows information of the iterative convergence times and horizontal displacement of some characteristic points located at the entrance, exit points of the seam and dam crest (designated A, C, and E respectively in Figure 1a) associated with the reduction factor. Figure 7 is for the curves of the reduction factor and horizontal displacement of the point of entry, Figure 8 is for the curves of iterative steps of convergence. From Table 2 we can see, there are different

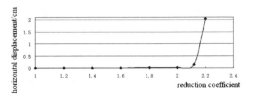

Figure 7. Reduction factor and the curve of the horizontal displacement of the entrancing.

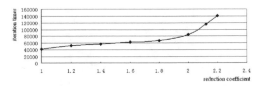

Figure 8. Reduction factor and the curve of iterative step.

feature points, but the extent that horizontal displacement increases as reduction factor increases remains the same level, and there is a reduction factor of 2.12, there exists a mutation and the largest displacement.

As shown in Figure 7, when rapid development of displacement takes place, factor amounts to about 2.1, As shown in Figure 8, with the increase of reduction factor, the iterative steps in tandem increase, corresponding mutations range is from 1.8 to 2.12.

5 THE SINGLE - LAYER MODEL BY SDF

5.1 The iteration process

It can be seen from Figure 9 that as the reduction factor increases, yield length of soft layer increases. When the reduction factor reaches 1.8, the weak layer is in plastic state. There appears a small range of plastic area in the dam, as shows in Figure 9(e). As the reduction factor continues to increase, the plastic area expands. When the reduction factor arrives at the range from 2.8 to 3.0, as shown in Figure 9(j) and (k), a plastic zone between the soft layer and the dam developed in a downstream direction.

When the reduction factor increases to 3.15, there will appear a long plastic zone which tends to the upstream of the dam between the earth surface and plastic area, which implies the dam base has arrived at the ultimate status, as shown in Figure 9(l). The gradual development of the plastic zone and related sliding surface is noticeable as can be seen in Figure 11.

Figure 10 shows the plastic zone at the moment the numerical iteration broke, associated with a factor of safety $F = 3.15$.

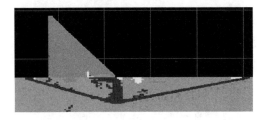

Figure 6. The final stage at which the numerical algorithm broke ($F = 2.20$).

Table 2. Horizontal displacement of feature points in reduction factors.

Reduction factor, F	Displacement, cm			Iteration number
	A	C	E	
1.0	0.6644	0.1571	0.2521	40920
1.2	0.6495	0.1699	0.308	52415
1.4	0.6445	0.1882	0.4007	56524
1.6	0.6840	0.2148	0.5592	62013
1.8	0.9165	0.2544	0.8283	67215
2.0	1.643	0.528	1.542	84727
2.12	15.04	13.41	26.64	114876
2.2	203.8	200.1	379.7	140920

NOTE: The locations of A, C, E are indicated in Figure 1(a).

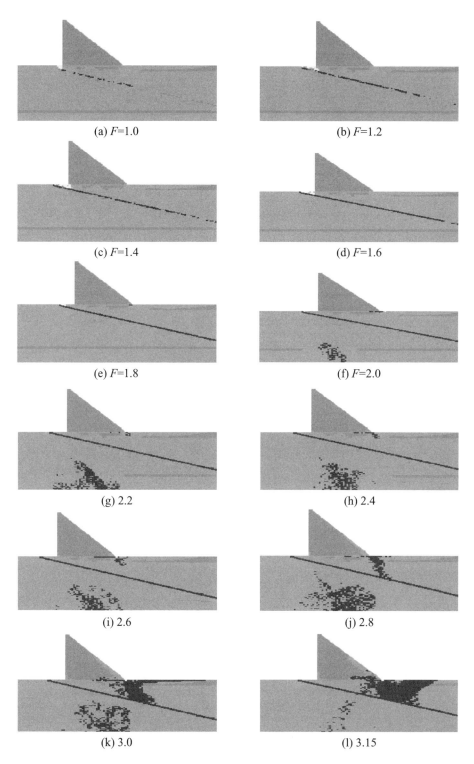

Figure 9. The plastic zones for different reduction coefficients.

Figure 10. The final stage at which the numerical algorithm broke.

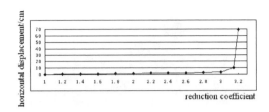

Figure 13. The reduction factor and horizontal displacement curves of dam top.

Figure 11. The single sliding plane's horizontal displacement in ultimate status.

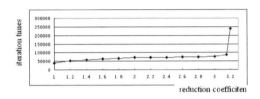

Figure 14. The reduction factor and iterative frequency required to achieve convergence curves.

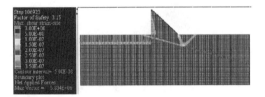

Figure 12. The single sliding plane, at $Fr = 3.15$.

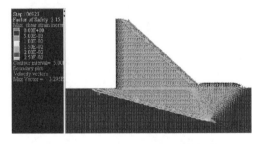

Figure 15. The shear strain rate vector under ultimate status.

5.2 The displacement and iteration features

Figure 11 shows the contour of horizontal displacement under ultimate State. Figure 12 gives the sliding surface under ultimate status, which has good agreement with that shown in Figure 3 with Sarma method.

Figure 13 shows the curves between reduction factor and the horizontal displacement at the dam top, Figure 14 shows the curve between reduction factor and iterative steps required by the convergence of the calculation. As can be seen from Figure 13 and Figure 14, when the reduction coefficient is 3.15, divergence occurs in both the two curves. Figure 15 is shear strain rate vector under the ultimate status, which shows the overall deformation trend of the dam base.

6 CONCLUSIONS

The dam stability analyses of two typical failure modes have been performed by finite difference strength reduction method that confirms the results in good agreement with those by Sarma method.

The strength reduction analysis for the double soft layers calculation model provides the development of plastic zone along the seam with the increase of reduction coefficient. Both SRF and Sarma method gave a factor of safety $F = 2.12$. Different failure criterion might give slightly different results. For this particular problem, judging based on the distribution of plastic zone of dam base may result in a little conservative estimation for the factor of safety, while the use of criterion of displacement divergence can give closer results to that of the limit equilibrium method.

The deep sliding pattern of single soft layer models is obtained by strength reduction method. The main sliding surface can be found in the dam base, together with sub-sliding surfaces passing through dam toe. Not only the factor of safety but also the failure zone are in good agreement with those obtained by Sarma's method.

REFERENCES

Griffths D.V. & Lane P.A. 1999. Slope stability analysis by finite elements. Geotechnique 49(3):387–403.

Matsui, T. & San, K.C. 1992. Finite element slope stability analysis by shear strength reduction technique. Soils and Foundations 32(1):59–70.

Sarma, S.K. 1979. Stability analysis of embankments and slopes. Journal of the Geotechnical Engineering Division, ASCE 105(GT12):1511–1524.

Zienkiewicz, O.C, Humpheson, C. & Lewis RW. 1975. Associated and nonassociated visco-plasticity and plasticity in soil mechanics. Geotechnique. 25(4):671–89.

Ugai, K. & Leshchinsky, D. 1995.Three-dimensional limit equilibrium and finite element analysis: a comparison of results. Soils and Foundations 35(4):1–7.

Study on deformation parameter reduction technique for the strength reduction finite element method

Q.W. Duan & Y.J. Wang
China Institute of Water Resources and Hydropower Research, Beijing, China

P.W. Zhang
Central Research Institute of Building & Construction, Beijing, China

ABSTRACT: Some problems related to stability analyses of slopes and dam foundations against sliding by employing the strength reduction method are discussed in this paper. In view of the fact that the plastic zone appeared during the reduction of strength parameters does not conform to the real case when using the strength reduction method, a relationship between the friction angle, and Poisson's ratio, is proposed, that is, among which is a parameter related to the depth, unit weight and cohesion of geomaterials. According to this inequality, the minimum friction angle can be determined for different depth, cohesion, unit weight and Poisson's ratio. This inequality can also take into account the more general cases, such as zero cohesion, infinite depth of geomaterial.

1 INTRODUCTION

Over the past decades, Finite-element (FE) method (including Discrete Element Method and Finite Difference Method), has been widely used to analyze problems of slope stability and stability of dam foundation against sliding. Compared to limit equilibrium methods, FE methods have advantages of being able to simulate the non-linear stress-strain relationship of geomaterial and complex geometric and boundary conditions, and to take into account construction process, seepage, earthquake and reinforcement. In addition, FE methoda can also predict distribution of stress or displacement, progress of plastic zone. However, FE Methods has a difficulty in estimating the factor of safety similar to the limit equilibrium method, which leads to no essential progress for solving stability problems by using FE methods.

In recent years, FE methods combined with a technique called strength reduction (or shear strength reduction) have been developed to evaluate the structure safety. The factor of safety is just one by which the shear strength parameters are reduced, and the reduced strength parameters make the caollapsed of strucure occur. The FE method combined the shear strength reduction technique called strength reduction FE method in this paper. Griffith (1999) firstly applied the strength reduction FE method to slope stability problems and gave a compartive factor of safety to the the Bishop's Method. However, the strength reduction technique implies that only the shear strength parameters are reduced, simultaneous reduction of deformation parameters, such as Young's modulus, E, and Possion's ratio, μ, are not considered.

Zheng (2002) discussed the necessity of reducing the shear strength and deformation parameters simultaneously when using the strength reduction FE method, and found that when these two kinds of parameters are reduced, the Poisson's ratio, μ, and internal friction angle, φ, should satisfy the following inequality, that is,

$$\sin \varphi \geq 1-2\mu \quad (1)$$

However, Equation 1 was derived based on: (1) a null value of cohesion of geomaterial; and (2) an infinite depth of analysis area below the ground surface. Actually, these two assumptions are not often satisfied, which leads to Equation 1 not be stood. Neglecting the deficiency mentioned above, utlization of Equation 1 may produce some abnormal plastic zones when the strength and deformation parameters are reduced simultaneously. In view of the deficiency embedded in Equation 1, an update of Equation 1 is made in this paper. The validity of the update is also discussed. In addition, the necessity of reducing Young's modulus and its criterion are also discussed for strength reduction FE technique.

2 STRENGTH REDUCTION FE METHOD

The strength reduction technique was first forward by Zien-kiewicz et al. (1975), when performing elasto-plastic analysis for soil stability by using FE method. The concept of strength reduction is defined as a ratio of the shear stress produced by the external loads acting on the structure to the maximum mobilized shear strength of soil. This strength reduction factor determined by this definition is conceptually the same as the facor of safety used by the Bishop's method. The general procedure for determining the factor of safety by using the strength reduction FE method is as following: (1) the material shear strengths are progressively reduced by a reduction factor, F; (2) preforming FE analysis based on the reduced strength parameters until collapse occurs. then he factor, F, is just the factor of safety for the strength reduction FE method.

Obviously, the value F determined has the same meaning as material strength reservation factor (Pan 1980). That is, If material strength parameters are divided by the value of F, the structure system would collapse.

$$\left.\begin{array}{l} f_i = \dfrac{f}{K_f} \\ C_i = \dfrac{C}{K_f} \end{array}\right\} \quad (i = 1, 2, \ldots, n) \qquad (2)$$

where f_i, c_i are the reduction factor at time i.

By combining the concept of strength reduction technique with the elasto-plastic FE method, a number of strength reduction FE methods have been developed, and be applied to preform slope stability analysis and stability analysis of dam foundations against sliding (Li 2000, Ma et al. 1984)

3 RELATIONSHIP BETWEEN FRICTION ANGLE AND POISSON'S RATIO

Generally speaking, the larger values of material strength parameters c and φ are, the larger its elastic modulus is, and the smaller its Poisson's ratio is. For example, the value of Young's modulus of hard rock is on the level of GPa, and the value of Poisson's ratio is not greater than 0.25. By contrast, the Young's modulus of soft rock is on the level of MPa, and the value of Poisson's ratio is above 0.3. So far, the geneal way is to reduce the strength parameters, c and φ, only, and simultaneous reduction of deformation parameters, E and μ, is not reported when using the strength redution FE technique.

For a semi-infinite geomaterial space as shown in Figure 1, is only subjected to gravity, stresses acting on an element at depth, h, below the ground surface, can be defined as follows:

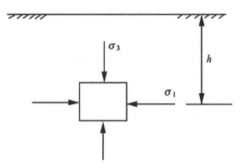

Figure 1. The distribution of stress in semi-infinite rockmass with depth of h.

$$\left.\begin{array}{l} \sigma_1 = \sigma_2 = -K\gamma h \\ \sigma_3 = -\gamma h \end{array}\right\} \qquad (3)$$

where γ is the unit weight of geomaterial, K the lateral pressure coefficient, which is defined as $K = \mu/(1-\mu)$.

If it is assumed that the stress field shown in Figure 1 is satisfied the Mohr-Coulomb criterion, thus

$$(1 + \sin\varphi)\sigma_1 - (1 - \sin\varphi)\sigma_3 \leq 2c\cos\varphi \qquad (4)$$

Substituting Equation 4 into Equation 3 gives

$$-K\gamma h(1 + \sin\varphi) + \gamma h(1 - \sin\varphi) \leq 2c\cos\varphi \qquad (5)$$

Let h in Equation 5 tend to infinite or the cohesion is equal to zero, the inequality $\sin\varphi \geq 1-2\mu$, proposed by Zheng (2002), can be obtained.

Rearrangement of Equation 5 leads to

$$(1 + K)\sin\varphi + \frac{2c}{\gamma h}\cos\varphi \geq 1 - K \qquad (6)$$

and then,

$$\sin(\varphi + \alpha) \geq \frac{1 - K}{\sqrt{(1 + K)^2 + \left(\dfrac{2c}{\gamma h}\right)^2}} \qquad (7)$$

Where,

$$\cos\alpha = \frac{1 + K}{\sqrt{(1 + K)^2 + \left(\dfrac{2c}{\gamma h}\right)^2}} \qquad (8)$$

$$\sin\alpha = \frac{2c/\gamma h}{\sqrt{(1 + K)^2 + \left(\dfrac{2c}{\gamma h}\right)^2}} \qquad (9)$$

It can be from Equations from 6 to 9 obtained that, when the depth h tends to infinite or the cohesion c is zero, then $\alpha = 0$.

By substituting $K = \frac{\mu}{1-\mu}$ into Equation 7, one obtains

$$\sin(\varphi + \alpha)/\cos\alpha \geq (1-2\mu) \quad (10)$$

when $\alpha = 0$, Equation 10 can be returned to Equation 1.

Equation 10 has demonstrated that there is a relationship between friction anger, φ, and Poisson's ratio, μ. When preforming the strength reduction FE analysis, the abnormal plastic that is inconsistent with the real case will occur if $\sin(\varphi+\alpha)/\cos\alpha < (1-2\mu)$, so adjusting Poisson's ratio, μ, according to Equation 10 is necessary to gurantee the unreaonable plastic does not occur.

4 ELASTIC MODULUS REDUCTION THEORY AND METHOD

For the soil whose mechanical characteristics follows E-ν model, The elastic constants accord with hyperbola relation, as shown in Figure 2, where a and b are constants, σ_1 and σ_3 are major and minor principal stresses respectively, ε_1 is the axial strain, E_i is the initial elastic modulus. $(\sigma_1 - \sigma_3)_{ult}$ is the asymptotic line of $(\sigma_1 - \sigma_3)$ when $\varepsilon_1 \to \infty$.

$$\sigma_1 - \sigma_3 = \frac{\varepsilon_1}{a + b\varepsilon_1} \quad (11)$$

According to differential relation, the tangent modulus can be obtained as

$$E_t = \frac{\partial(\sigma_1 - \sigma_3)}{\partial \varepsilon_1} = \frac{a}{(a+b\varepsilon_1)^2} \quad (12)$$

Elastic modulus is derived from two conditions, When $\varepsilon_1 \to \infty$

$$(\sigma_1 - \sigma_3) \to 1/b = (\sigma_1 - \sigma_3)_{ult} \quad (13)$$

When $\varepsilon_1 \to 0$

$$E_t = 1/a = E_i \quad (14)$$

According to Mohr-Coulomb criterion, the shear stress at failure can be obtained by using the following equation,

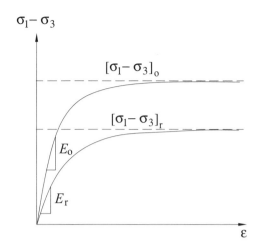

Figure 2. The concept of reducing Young's modulus E based on the hyperbolic stress strain relationship, the subscript 'o' and 'e' stand for the original and reduced variables and parameters.

$$\tau = c + \sigma \tan\varphi \quad (15)$$

$$(\sigma_1 - \sigma_3)_f = \frac{2c\cos\varphi + 2\sigma_3 \sin\varphi}{1 - \sin\varphi} \quad (16)$$

It can be obtained from Qian (1994) that

$$Et = (1 - R_f s)^2 E_i \quad (17)$$

Where, $s = \frac{(\sigma_1-\sigma_3)}{(\sigma_1-\sigma_3)_f}$, $Rf = \frac{(\sigma_1-\sigma_3)_f}{(\sigma_1-\sigma_3)_{ult}}$, which are known as failure ratio,

$$E_i = Kp_a \left(\frac{\sigma_3}{p_a}\right)^n \quad (18)$$

Where, p_a is atmospheric pressure, K and n are constants.

It can be known from Equation 17 that, the initial elastic modulus is invariable when R_f of soil is given. And the elastic modulus of soil is decreased gradually when s is increased, which means that when using the strength reduction FE method to determine the factor of safety of a slope, the value $(\sigma_1 - \sigma_3)$ at Mohr circle shown in Figure 3 is not changed because the stress within slope due to external loading is not changed. When shear strength parameters c and φ are reduced, the corresponding strength determined by the Mohr-Coulomb are also changed as shown in Figure 3, and the stress state of a point within the will approach to the yield state progressively.

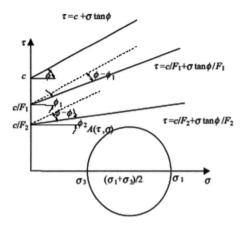

Figure 3. Strength reduction FEM sketch map.

It is also known from Equation 17 that, the elastic modulus of soil is related to the stress state at failure, which means that for a set of material parameter c_1 and φ_1 reduced already by a reduction coefficient F_1, the maximum stress at failure, $(\sigma_1 - \sigma_3)_{ult}$, can be determined by using the Mohr-Coulomb criterion. Therefore, the elastic modulus of materials can be determind by using Equation 17, the detailed procedures are as follows:

1. Determine the elastic modulus corresponding to parameter c and φ by Equation 17

$$E_1 = \left[1 - \frac{(\sigma_1 - \sigma_3)}{(\sigma_1 - \sigma_3)_{ult}}\right]^2 E_i \quad (19)$$

2. Compute c_2 and φ_2 by a given strength reduction coefficient F_2, and determine E_2 as

$$E_2 = \left[1 - F + F\sqrt{\frac{E_1}{E_i}}\right]^2 E_i \quad (20)$$

3. Input E_2, c_2 and φ_2 into FE model to perform FE analysis;
4. If the limit state is reached, then F_2 is the factor of safety obtained by strength reduction FE method, and the calculation terminates, otherwise, increasing the strength reduction coefficient and repeat Step 1 to Step 4.

5 EXAMPLES

Example 1 For a semi-infinite soil, the physical and mechanical parameters are summarized in Table 1.

Table 1. Mechanic parameters of soil in Example 1.

Deformation modulus (GPa)	Poisson's ratio	Friction angle°	Cohesion (MPa)	Tension strength (MPa)	Unit weight kN/m³
10	0.26	35	1.3	0	27

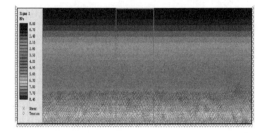

Figure 4. Analysis result without adjusting the Poisson's ratio (F = 2.0).

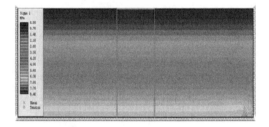

Figure 5. Analysis result with adjusting the Poisson's ratio (F = 2.0).

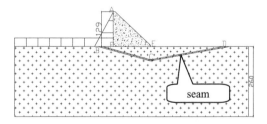

Figure 6. Gravity dam and foundation model.

This special example is designed to illustrate how the adjustment of the Poisson's ratio affect the appearance of plastic zone when strength reduction FE method is applied. For this simple semi-infinite soil mass, all the surrounding boundaries are only moved in the direction of normal to their side and the geomaterial is only subjected to gravity. When using the strength reduction FE method to analyze this problem, the Poisson's ratio is firstly not changed, and the

Table 2. Adapting parameter of dam and foundation.

Material name	Deformation modulus (GPa)	Poisson's ratio	Friction angle°	Cohesion (MPa)	Tension strength (MPa)	Unit gravity kN/m³
Dam	20	0.167	35	2.0	1.85	24
Foundation	10	0.26	35	1.3	0	27
Interlayer	2.5	0.35	19.8	0.115	0	25.6

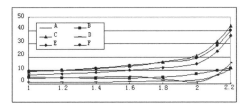

Figure 7. Horizontal displacement-reduction coefficient curve without adjusting Poisson's ratio.

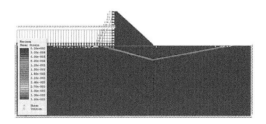

Figure 9. Distribution of plastic zone without adjusting Poisson's ratio ($F = 2.12$).

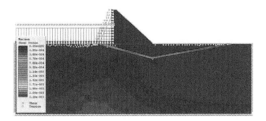

Figure 8. Distribution of plastic zone without adjusting Poisson's ratio ($F = 2.12$).

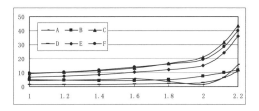

Figure 10. Horizontal displacement-reduction coefficient curve with adjusting Poisson's ratio.

soil below a depth of 200 m is in a plastic state when the reduction factor $F = 2.0$, as shown in Figure 4. However, when both strength parameters and Poisson's ratio are adjusted, espe-cially the Poisson's ratio is changed from 0.26 to 0.29, the plastic zone appeared in Figure 4 disappeared, as shown in Figure 5, which is more identical with the real case.

Example 2 stability analysis of dam foundation against sliding.

This example is focused on a gravity dam with height of 129 m, under which two weak seams BC and CD are present as shown in Figure 6, among them, weak seam BC dips at 15° downstream, and weak seam CD dips upstream at 10°.

For this case, except the gravity is considered in analysis, only water pressure acting on the upper stream side of dam and upstream foundation is considered. The physical and mechanical parameters of geomaterials are summarized in Table 2 the Mohr-Coulomb failure criterion is applied also in analysis, following conditions and parameters mentioned

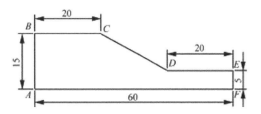

Figure 11. Geometry model of slope (unit: m).

Table 3. Material parameter of slope.

Deformation modulus (GPa)	Poission's ratio	Friction angle°	Cohesion MPa	Unit gramity kN/m³
0.1	0.4	20	0.01	20

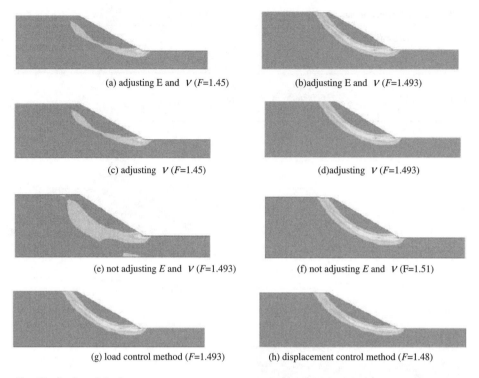

(a) adjusting E and ν (F=1.45)
(b) adjusting E and ν (F=1.493)
(c) adjusting ν (F=1.45)
(d) adjusting ν (F=1.493)
(e) not adjusting E and ν (F=1.493)
(f) not adjusting E and ν (F=1.51)
(g) load control method (F=1.493)
(h) displacement control method (F=1.48)

Figure 12. Distribution of plastic zone.

above, the strength reduction FE method gives a factor of safety of 2.12, which is close to Fos = 2.19 determined by traditional Sarma's method.

For the case of no adjustment of Poisson's ratio, curves of horizontal displacement at characteristic points, A, B, C, D, E, and F, with reduction coefficient are shown in Figure 7. If utilization of the phenomenon that the displacement curve for the characteristic point changes sharply at a vertain value of the reduction coefficient as judging criterion for the occurrence of limit failure state, the factor of safety of Example 2 is 2.12 for the case of no adjusting the Poisson's ratio. The corresponding plastic zone is shown in Figure 8. It can be seen from Figure 8 that, the shear plastic zone occurs not only in weak seams, but also in the deep foundation. actually, the plastic zone appeared in the deep foundation are due to the disaggrement of Poisson's ratio with the reduction strength parameters and Eqigure 9 is violated.

To avoid occurrence of the unreasonable plastic zone, and make Equation 9 stand, the value of Poisson's ratio of foundation material is adjusted to 0.33. the plastic zone in deep foundation as shown in Figure 10 disappear except the weak seams. The curves of displacement at characteristic points with the strength reduction coefficient as shown in Figure 10 shows the factor of safety for the case of reducing Poisson's ratio is 2.1 or so, just a little different rom that for the case of no adjusting the Poisson's ratio. However, characteristics of displacement for the two cases of adjusting and no adjusting the Poisson's ratio are quite different, as shown in Figure from 7 to10, which indicates that Poisson's ratio adjustment during the strength parameters reduction, provides not only a correct estimation of the factor of safety, but also a pertinent evolution of plastic zone.

Example 3

Example 3 is a homogeneous slope with a slope of 26.57°, its geometry is shown in Figure 11, and mechanical parameters for the slope material are listed in Table 3. For the FE model, the bottom boundary is fixed, left and right boundaries can only move in vertical direction.

For this example, three different cases are considered when performing the strength reduction FE analysis, namely, (1) Case 1: adjusting both deformation modulus and Pois-son's ratio; (2) Case 2: adjusting Poisson's ratio only; (3) Case 3: not adjusting deformation modulus and Poisson's ratio. The plastic zones for these three cases are illustrated in Figure 12. It can be found from Figure 12 that, (1) an adjustment of Poisson's ratio plays an imortant role on the distribution of plastic zone; (2) an adjustment of elastic modulus has no minimal influence on distribution of plastic zone and the factor of safety; however, taking much small value of elastic modulus would cause unreasonable factor of safety; (3) no adjustment of elastic moduls and Poisson's ratio will produce abnormal and impractical plastic zone.

6 CONCLUSIONS

Based on the analysis results, the following conclusions can be drawn:

- The previous strength reduction FE method only consider reduction of shear strength parameters, adjustment of Poisson's ratio and Young's modulus are not available. In this paper, an adjustment criterion for Poisson's ratio and Young's modulus, $\sin(\varphi + \alpha)/\cos\alpha \geq (1-2\mu)$, are proposed for propore utlization of strength reduction FE technique.
- Although the strength reduction FE method not considering the Poisson's ratio be adjusted may give the factor of safety in a good agreement with that determined by limit equilibrium method, however, some abnormal and impractical plastic zones appears. The strength reduction FE method with adjusting the Poisson's ratio according to the criterion presented in this paper is proved to provide only an appropriate factor of safety, but also a sound plastic zone.
- The reduction of Young's modulus of soils obeying E-υ hyperbolic model has been studied and its effect on the plastic zones and factors of safety has been discussed.

REFERENCES

PAN J.Z. 1985. *Engineering geology calculation and foundation treatment*. Beijing: Water Resources and Electric Power Press. (in Chinese)

Zheng H. & Li C.G. 2002. *Finite element method for solving the factor of safety*, Chinese Journal of Geotechnical Engineering, V24(5), 626~628.

Griffith D.V. & Lane P.A. 1999. *Slope stability analysis by finite element*. Geotechnique, 49(3), 387~403.

Zienkiewicz O.C., Humpheson C & Lewis R.W. 1975. *Associated and Non-Associated Visco-Plasticity and Plasticity in Soil Mechanics*. Geotechnique, 25(4), 671~689.

LI P.J. 2002. *Analysis on deep anti-sliding of gravity dam, Design of water resources and hydroelectric projects*, 2000, V19(1), 49–52.

MA L., ZHANG Y.M. & ZHANG L.Q. 1984. *Several problems concerning the analysis of stability against deep sliding in gravity dam foundation*. Journal of Hydraulic Engineering, (1), 27–35. (in Chinese)

LUAN M.T., WU Y.J. & NIAN T.K. 2003. *A criterion for evaluating slope stability based on development of plastic zone by shear strength reduction FEM*. Journal of Seismology, 23(3), 1–8. (in Chinese)

ZHAO S.Y., ZHENG Y.R. & DENG W.D. 2003. *Stability analysis of jointed rock slope by strength reduction FEM[J]*. Chinese Journal of Rock Mechanics and Engineering, 22(2), 254–260. (in Chinese)

Stability and movement analyses of slopes using Generalized Limit Equilibrium Method

M. Enoki
Tottori University, Tottori, Japan

B.X. Luong
University of Transport and Communication, Hanoi, Vietnam

ABSTRACT: The authors proposed Generalized Limit Equilibrium Method, which can analyze every type of plasticity problems. This method was developed from the ordinary slice method, but the assumptions on the forces on inter-block planes are more reasonable. By introducing inertia forces caused by the earthquake accelerations, GLEM can treat the dynamic problems. In this paper GLEM is introduced and applied to some plasticity problems including static stability and dynamic behavior. Finally, the applicability of classical theories of plasticity including GLEM is discussed in regard to the assumption of rigid-perfect plastic material.

1 INTRODUCTION

Geotechnical engineers often encounter the dynamic and plastic problems, such as deformations of foundations or slopes subjected by an earthquake or dynamic loading. Newmark's method and the derivative methods are famous to analyze these problems (Newmark 1965). However, they have the common weak point that they cannot obtain the actual solution for the static problems, for example, the Prandtle's solution for the bearing capacity problem.

The authors already proposed the Generalized Limit Equilibrium Method (GLEM), which can analyze every type of plasticity problems such as bearing capacity problems, earth pressure problems and slope stability problems (Enoki et al. 1991a, b).

GLEM can give the approximate solution obtained by Slip Line Method (like Kötter's equation) and the upper bound solution in the strict sense of Limit Analysis Method.

By introducing inertia forces caused by the earthquake accelerations, GLEM can treat the dynamic problems, such as the movement of slopes (Enoki et al. 2005).

In this paper GLEM is introduced and applied to some plasticity problems including static stability and dynamic behavior.

2 STATIC GLEM

2.1 Formulation of static GLEM

The Generalized Limit Equilibrium Method was developed from the ordinary slice method (Enoki et al. 1990), but the failing soil mass is divided into n pieces of triangular or quadrangular blocks, as shown in Fig. 1.

The following equilibrium conditions of every block give $2n$ equations under a given geometry of block system:

Equilibrium toward the direction normal to the block base plane

$$-H_j \cos(\beta_j - \alpha_i) + V_j \sin(\beta_j - \alpha_i) \\ + H_{j-1} \cos(\beta_{j-1} - \alpha_i) - V_{j-1} \sin(\beta_{j-1} - \alpha_i) \\ + M_i g \cos \alpha_i = P_i \quad (1)$$

Equilibrium toward the direction tangential to the block base plane

$$H_j \sin(\beta_j - \alpha_i) + V_j \cos(\beta_j - \alpha_i) \\ - H_{j-1} \sin(\beta_{j-1} - \alpha_i) - V_{j-1} \cos(\beta_{j-1} - \alpha_i) \\ + W_i \sin \alpha_i = T_i \quad (2)$$

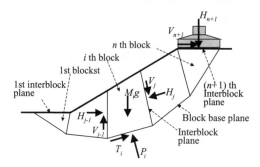

Figure 1. Block and force system used in Generalized Limit Equilibrium Method.

Table 1. Number of variables and equations in static GLEM.

Equations	Equilibrium condition	
	Normal to base plane	n
	Tangential to base plane	n
	Failure condition	Total $4n$
	On base plane	n
	On interblock plane	n
Variables	Forces on block base plane	
	Normal force $P_1,$---, P_n	n
	Tangential force $T_1,$---, T_n	n
	Forces on interblock plane	Total $4n$
	Normal force $H_2,$---, H_n	n
	Tangential force $V_2,$---, V_n	n

The following failure conditions are applied not only to the block base planes but also to the interblock planes, and $2n$ equations are obtained:

Failure condition on the block base plane

$$T_i = P_i \frac{\tan \phi}{F_s} + \frac{c}{F_s} l_i \quad (3)$$

Failure condition on the interblock plane

$$V_j = H_j \frac{\tan \phi}{F_s} + \frac{c}{F_s} l_j \quad (4)$$

Then, total $4n$ equations are obtained. In a bearing capacity problem the variables are normal and tangential forces, $P_1,$ ---, P_n and $T_1,$ ---, T_n on the block base plane, and $H_2,$ ---, H_n and $V_2,$ ---, V_n on the interblock plane. H_{n+1} is the objective force (ultimate bearing capacity Q_u), $F_s = 1$, and H_1 and V_1 are known surcharge.

The number of variables is also $4n$, as shown in Table 1, and the problem is determinate under a given geometry of block system. In the case of slope problem the safety factor F_s becomes a variable instead of H_{n+1}.

Thus, the static plasticity problem is transformed into an optimization problem in which bearing capacity or safety factor is minimized for the geometry of block system with restrained conditions expressed by equations.

This kind of optimization problem can be easily solved by "Solver", additional function of Excel. The readers can download the Excel sheets and FORTRAN source programs free for this analysis from the following web:
http://www.denkishoin.co.jp/index.html
(Only Japanese language for the present)

2.2 The meaning of optimization

In the limit equilibrium method including GLEM, the process of optimization is required. For example, in a bearing capacity problem the ultimate bearing capacity is minimized for the geometry of block system. The physical meaning of this process in GLEM is demonstrated in Figs. 2 and 3.

As shown in Fig. 2, in the initial geometry the stress state of every block (expressed by a Mohr's stress circle) has intersections with two failure lines, failure conditions on block base plane and interblock plane. However, the stress circles exceed the failure lines. The centers of stress circles are not always on σ-axis,

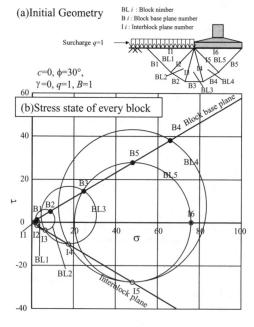

Figure 2. Stress state of every block before optimization in GLEM.

and this means equilibrium of moment of rotation is not satisfied.

As shown in Fig. 3, after the minimization of bearing capacity for the geometry of block system, the stress state of every block hardly exceed the failure lines, and the centers of stress circles are almost always on σ-axis. In GLEM the equilibrium of moment is not used, but it is satisfied as a result.

Thus, the authors believe that GLEM can give the approximate solution obtained by Slip Line Method (like Kötter's equation) and the upper bound solution in the strict sense of Limit Analysis Method. The detailed reason should be referred to Enoki 2007. In the limit equilibrium methods other than GLEM, such as all the slice methods, the failure condition is not adopted on inter-slice planes, then the physical meaning of optimization is not clear and not same with GLEM.

2.3 Illustrative examples of static GLEM

So-called N_q problem, ultimate bearing capacity of unit width foundation on weightless ground ($c=0$, $\phi=30°$) under unit surcharge, is analyzed by static GLEM. The bearing capacity and slip surface obtained are added to Hansen's figure and shown in Figure 4. In this analysis initial slip surface before minimization of bearing capacity is a circle of which center is the left end of foundation and radius is B.

Stability of a slope is analyzed by static GLEM under the assumption of toe failure and the result is shown in Figure 5.

3 DYNAMIC GLEM

3.1 Introduction of inertia forces

To formulate the dynamic GLEM, the horizontal and vertical accelerations of the i-th block, α_{hi} and α_{vi}, are introduced into the static GLEM as inertia forces, $-M_i\alpha_{hi}$ and $-M_i\alpha_{vi}$ using d'Alembert's principle (Enoki et al. 2005). The kinematic condition of the foundation is required to solve the problem. Now, we give the mass of the foundation, M_f, and assume the rough base that means the foundation and the n-th block move as one, as shown in Figure 6.

Then, the following motion equations are obtained for every block instead of Equations 1 and 2 for a dynamic problem under a given geometry of block system:

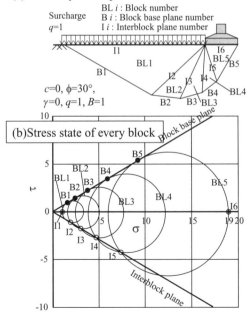

Figure 3. Stress state of every block after optimization in GLEM.

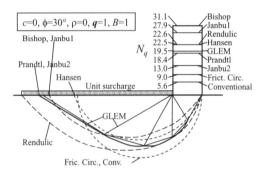

Figure 4. Bearing capacity and slip surface for N_q problem analyzed by static GLEM and other methods (after Hansen 1970).

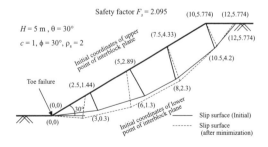

Figure 5. Safety factor and slip surface of slope problem analyzed by static GLEM (Enoki 2007).

673

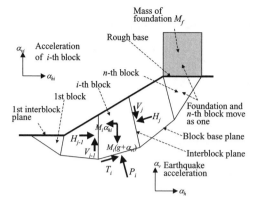

Figure 6. Block and force system used in dynamic GLEM.

Table 2. Number of variables and equations in dynamic GLEM at the instant of failure.

Equations	Equilibrium condition		
	Normal to base plane	n	
	Tangential to base plane	n	Total $4n–1$
	Failure condition		
	On base plane	n	
	On interblock plane	$n–1$	
Variables	Forces on block base plane		
	Normal force $P_1,---,P_n$	n	
	Tangential force $T_1,---,T_n$		Total $4n–1$
	Forces on interblock plane		
	Normal force $H_2,---,H_{n-1}$	$n-1$	
	Tangential force $V_2,---,V_{n-1}$	$n-1$	
	Critical earthquake acceleration		
	Horizontal $\alpha_{h,cr}$	1	

*α_v is given as the earthquake acceleration.

Equilibrium toward the direction normal to the base-block plane,

$$-H_j \cdot \cos(\beta_j - \alpha_i) + V_j \cdot \sin(\beta_j - \alpha_i)$$
$$+ H_{j-1} \cdot \cos(\beta_{j-1} - \alpha_i) - V_{j-1} \cdot \sin(\beta_{j-1} - \alpha_i)$$
$$+ M_i(g + \alpha_{vi}) \cdot \cos \alpha_i - M_i \cdot \alpha_{hi} \cdot \sin \alpha_i = P_i \quad (5)$$

Equilibrium toward the direction tangential to the base-block plane,

$$H_j \cdot \sin(\beta_j - \alpha_i) + V_j \cdot \cos(\beta_j - \alpha_i)$$
$$- H_{j-1} \cdot \sin(\beta_{j-1} - \alpha_i) - V_{j-1} \cdot \cos(\beta_{j-1} - \alpha_i)$$
$$+ M_i(g + \alpha_{vi}) \cdot \sin \alpha_i + M_i \cdot \alpha_{hi} \cdot \cos \alpha_i = T_i \quad (6)$$

The failure conditions, Equation 3 and 4 are still applicable, and $F_s = 1$.

Equations 1 and 2 are a special case of Equations 5 and 6, where every acceleration of block, $\alpha_{h1}, \alpha_{h2}, \ldots,$ $\alpha_{hn}, \alpha_{v1}, \alpha_{v2}, \ldots, \alpha_{vn}$, are equal to zero. As for the horizontal and vertical accelerations of every block, there are two processes in dynamic GLEM.

3.2 The instant of failure

Before and at the instant of failure, every block moves similarly to the base of the ground, or $\alpha_{h1} = \alpha_{h2} = \cdots = \alpha_h$ and $\alpha_{v1} = \alpha_{v2} = \cdots = \alpha_v$.

In this case the objective is to determine the critical acceleration $\alpha_{h,cr}$ that causes the ground to fail under the given α_v, and the problem is still determinate as shown in Table 2.

The critical acceleration $\alpha_{h,cr}$ must be minimized for the geometry of the slip surface.

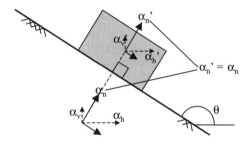

Figure 7. The continuity condition of accelerations.

3.3 After the failure

When the actual earthquake acceleration α_h exceeds $\alpha_{h,cr}$ at $t = t_1$, the ground fails, and this process will begin. In this process every block moves relatively to each other, and all the accelerations $\alpha_{h1}, \alpha_{h2}, \ldots,$ α_{hn} and $\alpha_{v1}, \alpha_{v2}, \ldots, \alpha_{vn}$ are variables. In order to determine these accelerations, the continuity condition of accelerations and the conservation of slip surface are useful.

The continuity condition of accelerations was found by one of the authors, when the motion of the surface layer during an earthquake was analyzed. It says that the components of the accelerations normal to a slip plane, α_n and α'_n, are equal to each other across the slip plane, and Equation 7 is written, where α_h, α_v and $\alpha'_h,$ α'_v are accelerations of one side and another side across the slip plane, respectively, and θ is the inclination of the slip plane, as shown in Figure 7.

$$\alpha'_n = \alpha'_v \cos \theta - \alpha'_h \sin \theta = \alpha_v \cos \theta - \alpha_h \sin \theta = \alpha_n \quad (7)$$

This means the conservation of mass during the motion, and also written as $(\alpha'_v - \alpha_v)/(\alpha'_h - \alpha_h) = \tan \theta$.

Table 3. Number of variables and equations in dynamic GLEM after the failure.

Equations	Equilibrium condition		
	Normal to base plane	n	
	Tangential to base plane	n	
	Failure condition		Total
	On base plane	n	$6n-2$
	On interblock plane	$n-1$	
	Continuity condition of acceleration		
	On base plane	n	
	On interblock plane	$n-1$	
Variables	Forces on block base plane		Total
	Normal force $P_1,\text{---},P_n$	n	$6n-2$
	Tangential force $T_1,\text{---},T_n$	n	
	Forces on interblock plane		
	Normal force $H_2,\text{---},H_{n-1}$	$n-1$	
	Tangential force $V_2,\text{---},V_{n-1}$	$n-1$	
	Acceleration of every block		
	Horizontal $\alpha_{h1},\text{---},\alpha_{hn}$	n	
	Vertical $\alpha_{v1},\text{---},\alpha_{vn}$	n	

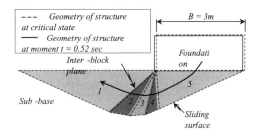

Figure 8. Displacement of block system in foundation problem at moment $t = 0.52$ sec.

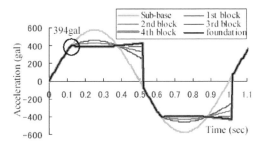

Figure 9. Horizontal accelerations of sub-base, foundation and soil blocks.

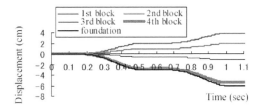

Figure 10. Vertical displacements of foundation and soil blocks.

The conservation of slip surface says that the slip surface remains unchanged from the initial one during a single motion. This is not always necessary to formulate the dynamic GLEM, but convenient in the calculation and realistic for the actual problems.

After applying these conditions, the problem becomes determinate as shown in Table 3, and the accelerations of every block can be determined. The optimization of the acceleration for the geometry of the slip surface is not required in this process.

When the relative velocity of arbitrary block becomes zero at $t = t_2$, the slip stops, as written in Equation 8.

$$\int_{t_1}^{t_2} (\alpha_{hi} - \alpha_h) dt = 0 \tag{8}$$

3.4 Illustrative examples of dynamic GLEM

The behavior of a shallow foundation with a rough base, $M_f = 70$ t and $B = 3$ m, during an earthquake is analyzed by dynamic GLEM. The soil has density of 1.8 t/m³ and strength parameters of $c = 10$ kPa, $\phi = 32°$. Only a horizontal wave $\alpha_h = 5.75 \sin 2\pi t$ (m/s²) is given as an earthquake acceleration. A system of five triangular blocks is used as shown in Figure 8.

After the given acceleration exceeds the critical acceleration $\alpha_{h,cr} = 3.94$ m/s², all the blocks move differently from each other as shown in Figure 9. Within one wave the failure occurred twice, and the settlement of foundation is about 6 mm as shown in Figure 10.

The model test of finite slope was conducted on a shaking table, as shown in Figure 11. Toyoura sand with density of 1.8 t/m³, void ratio $e = 0.8$, moisture content 1%, and strength parameters of $c_{peak} = 1.25$ kPa, $c_{residual} = 0.85$ kPa, $\phi_{peak} = \phi_{residual} = 36.4°$ was used. The slope had height $h = 30$ cm and inclination 50°. Only a horizontal wave $\alpha_h = 10 \sin 12\pi t$ (m/s²) was given as an earthquake acceleration.

The slip surfaces observed in the experiment is compared with the analyzed one in Figure 12. The horizontal accelerations and displacements observed in the experiment are shown in Figure 13. It can be compared with the analytical ones shown in Figure 14. According to the experimental results, the sliding begins to occur from the second wave of input acceleration at a

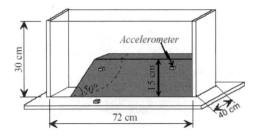

Figure 11. Model test of finite slope on shaking table.

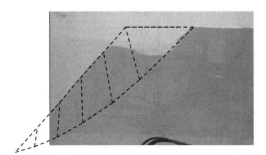

Figure 12. Slip surfaces observed and analyzed.

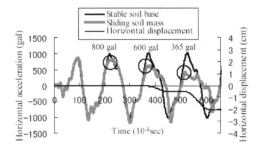

Figure 13. Experimental results of slope motion.

critical acceleration of about 800 gal. Then, the critical acceleration decreases and has the value of about 600 gal corresponding to the third wave, and 365 gal for the following waves.

This fact indicates that for the early stage of the failure, the peak shear strength of the soil are mobilized to resist the sliding, thereafter the strength of the soil will decrease and be at residual value. In the theoretical analysis, with the use of the residual strength of the soil, the sliding begins to occur just from the first wave with the critical acceleration of about 438 gal.

Consequently the residual displacement in horizontal direction of the sliding soil mass obtained by

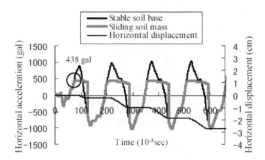

Figure 14. Analytical results of slope motion by dynamic GLEM.

theoretical analysis is a little larger than the experimental one (2.76 cm compared to 2.02 cm).

Although there are still some differences, it can be said that the experimental results have rather good agreement with the analytical ones.

The comparison of experimental result of foundation model on a shaking table with the analytical one showed not so good agreement as in the case of slope.

4 DISCUSSION ON THE APPLICABILITY OF GLEM

The stability analysis based on the classical theories of plasticity such as Slip Line Method and Limit Analysis Method, requires the assumption of the perfect plasticity (with no hardening nor softening) of the material. This assumption can be easily satisfied in the unsaturated medium-dense sand. However, the motion analysis based on the classical theories of plasticity requires the assumption of the rigid-perfect plasticity. This assumption cannot be easily satisfied in the foundation problems where the change in stress in the soil is large, and volumetric compression and elastic deformation cannot be neglected. On the other hand, the change in stress from the initial state till the failure state is small in the slope problems; therefore the "rigid" assumption can be approximately satisfied.

As the static or dynamic GLEM is a kind of the classical theories of plasticity, the applicability may have the same limitation. The authors believe that this is the reason why the analytical results agrees well with the observed one in dynamic slope problems but not agrees in dynamic foundation problems.

5 CONCLUSIONS

The outline of the static and dynamic Generalized Limit Equilibrium Method is introduced in this paper. The static GLEM can give the approximate solution obtained by Slip Line Method and the upper bound solution obtained by Limit Analysis Method.

The procedure of optimizing the external force (earth pressure or bearing capacity) or safety factor in GLEM plays a role to prevent stress state from exceeding the failure criterion. It also plays a role to satisfy the equilibrium of moment of rotation in a block.

Since the static GLEM is formulated as a special case of the dynamic GLEM, where acceleration of every block is equal to zero, there is no essential difference between them. Thus, the authors suppose that the dynamic GLEM gives the approximate solution obtained by the dynamic Slip Line Method and the upper bound solution obtained by the dynamic Limit Analysis Method, although they have not been established yet.

GLEM seems to have the limitation on application, that is, unloading and active problems can be properly analyzed, but loading and passive problems cannot properly be analyzed, because of the assumption of rigid-perfect plasticity of the material.

REFERENCES

Enoki, M., Yagi, N., Yatabe, R. & Ichimoto, E. 1990. Generalized Slice Method for Slope Stability Analysis. *Soils and Foundations* 30(2): 1–13.

Enoki, M., Yagi, N., Yatabe, R. & Ichimoto, E. 1991a. Generalized Limit Equilibrium Method and Its Relation to Slip Line Method. *Soils and Foundations* 31(2): 1–13.

Enoki, M., Yagi, N., Yatabe, R. & Ichimoto, E. 1991b. Relation of Limit Equilibrium Method to Limit Analysis Method. *Soils and Foundations* 31(4): 37–47.

Enoki, M., Luong, X.B., Okabe, N. & Itou, K. 2005. Dynamic Theory of Rigid-Plasticity. In *Soil Dynamics and Earthquake Engineering*: 635–647, London, Elsevier.

Enoki, M. 2007. *Analyses of Stability and Deformation in Soil-like Frictional Material*: Tokyo. Denki-Shoin (in Japanese).

Hansen, J.B. 1970. A Revised and Extended Formula for Bearing Capacity, *Geotekniks Institute, Bull.* 28: 5–11.

Newmark, N.M. 1965. Effect of earthquakes on dams and embankments. Fifth Rankine Lecture Géotechnique 2: 139–160.

Long-term deformation prediction of Tianhuangpin "3.29" landslide based on neural network with annealing simulation method

Faming Zhang, Chenxin Xian, Jian Song & Binyue Guo
Earth Science and Engineering Department of Hohai University, China

Zhiyao Kuai
Geological Department of Chang'an University, China

ABSTRACT: Landslide has already become one of the most dangerous geo-hazards in China, there are a lot of economy loss and personnel casualty. Recently, the safety research on stability analysis or predict to the hazard taken place for landslides is more and more important. But, it is difficult to predict when the sliding will happen accurately since the factors affected on landslide are complex. Thus, in these years, monitor scheme is adopted for many landslides, how to use the monitoring data to predict the landslide will occur is an important problem in geological field. In this paper, based on adopting neural network with simulated annealing method, the model of landslide deformation prediction in long-term was set up, which using simulated annealing method to overcome the disadvantage of BP neural network, furthermore, by using dynamic forecasting technique to reduce the influence of the prophase displacement, it can get better forecast precision. Finally, the Tianhuangpin "3.29 landslide" deformation in long-term was predicted by using the model discussed in this paper, the result is coincident to the real condition of the slope.

1 GENERAL INSTRUCTIONS

Tianhuangpin pump-storage power station is located at Tianhuangpin town in Zhejiang Province, it is 57 km apart from Hangzhou and 180 km from Nanjing. The 329 Landslide is distributed in the left bank of the lower reservoir of the project, it is 430 m from the dam in upstream. The first sliding of the landslide was occurred in the morning of 29, March of 1996, which the 2×10^6 m^3 slide mass was rushed into the reservoir. The further exploration work found the instability slope body was more than 19×10^6 m^3, which slip surface of the landslide is in middle weathered break rock mass. From 1997, the monitoring network was set up. In the past 10 years, there are many monitored data was collected. But, how to evaluate the landslide stability and predict the long-term deformation performance by using the monitored data is an important subject in safety running of the power station.

It is well known that the research work of prediction for landslide are only recent 30 years, the method used in forecasting normally is statistics and experience [Liu 1996, 1998]. In recent years, grey system theory, fuzzy mathematic etc. are used in landslide deformation prediction [Deng 1985, 1987]. When talk about the methods, the errors between monitored and predicted value are companied with the factors affected on landslide, so, it is difficult to forecast the deformation of landslide in long term. The main factor is the monitored data are changed with seepage and weather, In this paper, based on adopting neural network with simulated annealing method, the model of landslide deformation prediction in long-term was set up, which using simulated annealing method to overcome the disadvantage of BP neural network, the prediction result of Tianhuangpin landslide is more accurate with monitored data.

2 GEOLOGICAL CONDITIONS

2.1 Topography and physiognomy

"3.29" landslide of Tianhuangpin power station is located in Tianmu Mountain area, the famous mountain in China, the highest altitude of the landslide is 970 m, the vertical difference of the landslide is more than 400 m. The strike direction of the landslide is S~N, and the slope degree is 40~50°. The width of the landslide is from 15 m (top area) to 600 m (lower part) (Figure 1). All the accumulation of the landslide body is consisted of weathered volcanic rock mass (Figure 2).

Figure 1. "3.29" landslide scope.

Figure 2. Accumulation of "3.29" landslide.

2.2 Stratum and rock mass property

In the landslide distributed area, the rock mass is Jurassic system up series Hangjian group (J_3^{h1}) and Lacun group ($J_3^{L1(d)}$) volcanic rock with intrusive granite-porphyry ($\gamma \pi_5^{2-3}$). The Quaternary loose deposits were distributed on volcanic rock. Huanjian group (J_3^{h1}) is grey tufflava and ryholitical agglomerate. Lacun group ($J_3^{L1(d)}$) is gravel agglomerate.

2.3 Structure

The structure in landslide area is well developed, the main faults are F_{105} and F_{106}. These two faults are 0.4~2.0 m width, which filled with fault mud and mylonite. The joints are closely spaced.

2.4 Hydro-geological property

Atmosphere precipitation is the supply source for underground water, the buried depth of the table is more than 50 m since the joint closed spaced.

2.5 Weathering phenomenon

The depth of complete weathered rock mass subface is 2.0~5.0 m (EL464.90~461.90 m), the maximum depth is 59.50~60.10 m (in EL506.97~506.37 m). The strong weathered is 60.10~70.90 m (EL506.37~495.57 m), and the weak weathered is 40~85 m which under the stress-release interface 10~30 m.

3 MONITORING SCHEME

3.1 Monitoring scheme

After the "3.29" landslide happened, the accumulation of the landslide was clean up by the Tianjian Corporation, the slope was reinforced by taken downwearing and revetment, the stability of the slope is strengthened. From 1999, the monitoring scheme was taken in the surface deformation, slope indicator, surface settlement, rainfall and piezometer etc. The scheme map of landslide monitor is illustrated in Figure 3, which HP is indicated as slope indicator, TP is indicated as horizontal surface displacement and LD is vertical surface displacement.

For surface deformation monitoring, there are 18 monitor sites lain on the surface of the slope, which take 15 days as one cycle. For the interior deformation

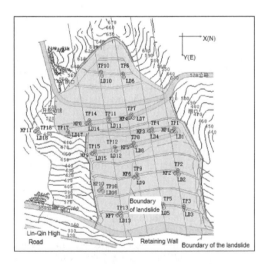

Figure 3. Monitoring scheme of "3.29" landslide.

monitoring of the slide body, there are 11 clinometers installed from 1999.

3.2 Deformation information statistics

It is obvious that the interior deformation especially on the sliding surface may representative the stability of the landslide or slope, so, in this paper, the deformation prediction is only discussed for interior deformation performance. On considering the probability sliding surface of the landslide, the deformation with depth of 10 monitoring holes was considered from 2000. Take an example, monitoring hole 2 (called HP2), installed at 450 m, the depth is 38.5 m. There are 2 abnormal changes at 18 m and 34.5 m in direction A. In direction B, there are 2 obvious variations along the depth at 17.5 m and 37 m. The 8 m depth' deformation changes with time in direction A were illustrated in Figure 4. The curves of deformation along the depth were figured in Figure 5 and Figure 6.

The other monitoring sites of interior deformation were lain out in Table 1.

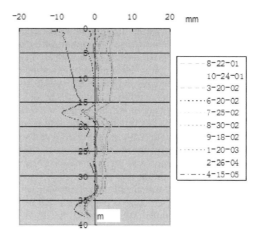

Figure 6. Deformations with depth in direction B of HP2.

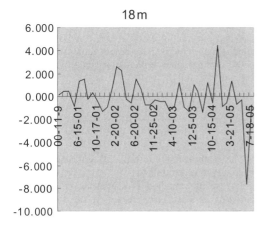

Figure 4. Deformation with time curve of 18 m for HP2.

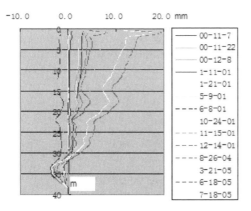

Figure 5. Deformations with depth in direction A of HP2.

Table 1. The maximum exceptional deformation changed in monitoring hole with depth.

Monitoring location	Direction	Time	The depth of obvious deformation variation (m)	The maximum deformation variation (mm)
hp1	A	2004-11-15	8	8.80
		2005-6-18	18	7.63
		2005-6-18	33	3.21
	B	2001-9-18	17.5	6.94
		2003-7-22	25.5	4.22
hp3	A	2003-8-21	21.5	7.74
	B	2002-3-15	21.5	8.08
		2002-3-15	24	3.52
hp4	B	2005-7-18	21	13.13
		2003-9-16	35	15.36
		2005-7-18	42.5	9.75
hp5	A	2002-7-25	19.5	6.19
		2004-1-10	26	4.61
	B	2002-7-25	27	6.96
hp6	A	2003-12-5	18.5	5.86
		2005-4-15	31.5	2.43
	B	2001-6-16	27	4.72
hp7	A	2003-7-22	7	8.76
		2002-2-28	33	1.16
	B	2002-7-25	27.5	4.51
hp9	A	2001-7-27	12.5	5.02
	B	2002-6-20	12.5	14.10
hp10	A	2003-1-20	12.5	5.78
	B	2001-4-13	7	7.06

4 MODEL OF LONG-TERM DEFORMATION PREDICTION

4.1 Prediction model of simulated annealing with neural work

Since the BP neural network technology can't guarantee the convergence at the minimum value, it is difficult to get the best result in optimization process [Sandro 1997], but not the same with BP neural network technology, simulated annealing method can research the optimum result [Kang etc., 1994]. As we known, the study of neural work process is to look for a suitable mapping or optimalizing weight and critical point threshold, so, the simulated annealing method can be used in dealing with the study of neural work process. In this reason, the model of simulated annealing with neural work was set up to forecasting the deformation of landslide.

In the model of simulated annealing with neural work, the network structure is the same as BP neural network, it adopt three layers, but use the simulated annealing method to search the weight and critical point threshold instead of the opposite propagation errors in BP neural work method. So, the weight value and the critical point threshold value is obtained from interval of $[-1, +1]$ by using random search method of simulated annealing, the objective function is:

$$E = \sum_{k=1}^{m} E_k = \sum_{k=1}^{m}\sum_{t=1}^{q} (y_t^k - c_t^k)^2/2 \qquad (1)$$

where y_t^k is output victors, $Y^k = (y_1^k, y_2^k, \ldots, y_q^k)$, $k = 1, 2, \ldots, m$, q is the number of output units; $c_t^k = f(Q_t^k) t = 1, 2, \ldots, q Q_t^k = \sum_{j=1}^{n} v_{jt} b_j^k - r_t t = 1, 2, \ldots, q$, v_{jt} is the weight from middle layer to output layer, r_t is critical point threshold of output layer, b_j^k is activation value of j unit of middle layer, m is the number of specimen (n) is imported unit. More details about neural network method can reference to Zhang's paper [Zhang 1983].

4.2 Process of neural network with simulated annealing method for deformation prediction

4.2.1 Analysis of monitored data
Since affected of monitoring equipments and artificial factors, the monitored data will be unreal sometimes, for these reasons, it is necessary to analyze the data reality before prediction by using neural network method.

4.2.2 Select the specimen of data
The specimen data is used in training process of neural network method for building the rules of deformation changing, so, in obtaining the reasonable rule, the plentiful specimen data is required. It is the best way to use interpolation method for unequal interval data.

4.2.3 Data alternate process
Characteristic function of BP neural network normally used is S (Sigmoid) function the interval of the function value is $(0, 1)$, so, it is necessary to change the monitored data value into $(0, 1)$ interval. The method can be used as following methods.

The first method is normalizing, it changes the import data to normalize to $(0, 1)$ interval. The normalizing formula is:

$$X = X_{min} + \frac{Y - Y_{min}}{Y_{max} - Y_{min}}(X_{max} - X_{min}) \qquad (2)$$

where Y_{max} and Y_{min} are the maximum and the minimum values of monitored data, X_{max} and X_{min} are the normalized values corresponding to Y_{max} and Y_{min}.

When the monitored data were forecasted by BP neural network method, the output values must restore to the original state as the actual data by using the following formula:

$$Y = Y_{min} + \frac{X - X_{min}}{X_{max} - X_{min}}(Y_{max} - Y_{min}) \qquad (3)$$

The symbols in the formula (3) represent the same meanings as in formula (2).

The second method is standardizing, the standardizing process is expressed in the formula (4) to (6).

$$\bar{x} = \frac{\sum_{i=1}^{N} x_i}{N} \qquad (4)$$

$$\sigma = \sqrt{\frac{\sum_{i=1}^{N}(x_i - \bar{x})^2}{N - 1}} \qquad (5)$$

$$S_i = \frac{x_i - \bar{x}}{Std} \qquad (6)$$

where x_i is the import data (the monitored data), $i = 1, 2, \ldots, N$, N is the number of specimen, $\bar{x}$ is the mean value of import data, σ is the standard deviation, S_i is the standardized value of x_i.

The modeling flow chart of BP neural network with simulated annealing method for predicting landslide deformation was showed in Figure 7.

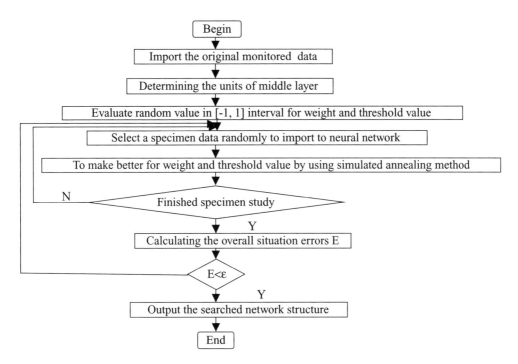

Figure 7. The modeling flow chart of BP neural network with simulated annealing method for predicting landslide deformation.

4.3 Model of BP neural network with simulated annealing method

According to the process discussed in the paper above, the model of BP neural network with simulated annealing method for predicting landslide deformation was set up in this paper. In the model, the import parameters are $(x(t-1), x(t-2), x(t-3), t, L)$, where $x(t-1), x(t-2)$ and $x(t-3)$ are real monitored deformations, t is the period of deformation from initial monitoring to forecasting time, L is the water level of underground water in t moment, suppose the output value is $x(t)$ then, it can be described as function:

$$x_{(t)} = f(x_{(t-1)}, x_{(t-2)}, x_{(t-3)}, t, L) \qquad (7)$$

For the model described above, there are 3 layers, such as import layer, middle layer and output layer, there are 5 nodes in import layer and 1 node in output layer, but, for the middle layer, the middle nodes and the weight and critical point threshold value can be obtained by using simulated annealing method.

Additionally, for reducing the affects of random interfere, the import data are not used the original monitored data directly but adopting totting-up method for original monitored data and set in equal intervals. If the original monitored data is $x^{(0)}$, then, the generated data under one time totting-up is:

$$x^{(1)}(t) = \sum_{m=1}^{t} x^{(0)}(m) = x^{(1)}(t-1) + x^{(0)}(t) \qquad (8)$$

where $x^{(1)}(t)$ is the accumulated deformation value in t moment. When cutting the accumulated data in certain time, training is carried out by using the BP neural network with simulated annealing method, and, after the training is finished, the prediction for deformation can be executed.

5 DEFORMATION PREDICTING FOR "3.29" LANDSLIDE

Take the monitored data before Sept. 15 of 2003 as the training specimen, let the data from Sept. 15 of 2003 to Sept. 15 of 2004 as the checking specimen, then import the training specimen in BP network with simulated annealing structure, the prediction of deformation for interior deformation was showed in Table 2 and Figure 8 to Figure 9. On comparing the prediction results with the monitored data in site, it is easy

Table 2. Comparing with the prediction and monitored data in site of HP1 and HP2.

	hp1				hp2			
	Direction A		Direction B		Direction A		Direction B	
Prediction time	Monitoring value (mm)	Prediction value (mm)	Monitoring value (mm)	Prediction value (mm)	Monitoring value (mm)	Prediction value (mm)	Monitoring value (mm)	Prediction value (mm)
9-15-04	3.19	3.228	4.415	3.804	4.36	5.104	−2.425	−2.211
10-15-04	1.26	1.15	5.255	6.183	2.89	3.251	−2.975	−3.217
11-15-04	6.92	8.043	6.875	6.722	1.12	1.088	−2.705	−3.017
12-15-04	1.07	0.967	4.165	3.586	1.89	1.573	−2.615	−2.961
2-15-05	1.54	1.449	5.925	5.334	1.89	1.55	−2.265	−2.346
3-15-05	3.63	4.051	4.105	4.59	6.13	7.126	−3.325	−3.251
4-15-05	1.36	1.597	2.895	2.382	3.13	2.669	−3.035	−3.581
5-15-05	1.94	1.966	2.915	2.496	1.3	1.114	−2.545	−2.354
6-15-05	−6.13	−5.129	3.115	3.456	10.32	9.406	−3.175	−3.706
7-15-05	1.84	1.689	4.325	4.744	8.3	7.163	−2.995	−2.959
8-15-05		0.926		5.313		3.769		−6.057
9-15-05		1.594		11.12		8.726		−4.715
Average error		9.66%		12.21%		13.50%		9.09%

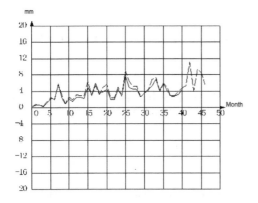

Figure 8. The diagram of deformation predicted with monitored of HP1 in direction B.

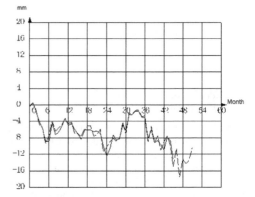

Figure 9. The diagram of deformation predicted with monitored of HP4 in direction B.

to find the maximum error is 16.26%, the maximum absolute error is 0.8 mm, most of the errors are less than 10%, that means the model discussed in this paper is suitable to forecast the future deformation for landslide.

The results for predicting the interior deformation in 3 years, 5yeares and 10 years are indicated in table 3. It is not difficult to get the long term deformation value from the tables. It can be known that the maximum deformation is 26 mm in July, 15 in 2015, the location is in the middle of the landslide, so, the 3.29 landslide is stable in future 10 years.

6 CONCLUSIONS

The neural network methodology possess the function of dealing with high nonlinearity and parallel calculation problems, the simulated annealing method can be used in searching the minimum in overall situation for time-sequence problems, so, it is possible to set up the model for forecasting the value based on deformation monitored data in combining with BP neural network and simulated annealing method. As an example, the Tianhuangpin "3.29 landslide" deformation in long-term was predicted by using the model discussed in this paper, the result is coincident to the real condition of the slope.

Table 3. Deformation predicted in long term of monitoring site.

Prediction time	Hp1 Prediction value (mm)		Hp2 Prediction value (mm)		Hp3 Prediction value (mm)	
	A	B	A	B	A	B
2008-7-15	1.783	8.683	9.046	−3.957	2.275	−12.744
2010-7-15	3.339	10.921	11.506	−5.45	3.514	−14.888
2015-7-15	6.197	16.872	13.476	−8.953	6.728	−18.903

Prediction time	Hp4 Prediction value (mm)		Hp5 Prediction value (mm)		Hp6 Prediction value (mm)	
	A	B	A	B	A	B
2008-7-15	7.181	−17.75	13.875	−29.43	1.093	1.285
2010-7-15	10.532	−21.551	17.654	−40.203	1.445	2.909
2015-7-15	17.584	−27.607	25.981	−57.45	2.824	4.14

Prediction time	Hp7 Prediction value (mm)		Hp9 Prediction value (mm)		Hp10 Prediction value (mm)	
	A	B	A	B	A	B
2008-7-15	7.253	2.452	−2.765	6.011	−3.764	1.525
2010-7-15	9.338	0.35	−3.236	8.864	−4.205	0.764
2015-7-15	13.358	−1.544	−5.816	13.753	−7.272	−3.396

ACKNOWLEDGEMENTS

The research work was financed by Jiangsu Natural Science foundation (No.Bk2006171).

REFERENCES

Brown, E.T. 1987. Analytical and computational methods in engineering rock mechanics. John Wiley (eds), New York.

Deng Julong, 1985. Grey control system, Wuhan: Huazhong Technology University Publishing House.

Deng Julong, 1987. The basic method of grey system. Wuhan: Huazhong Technology University Publishing House.

Kang lishan, Xie yun, You siyong etc., 1994. None numerical parallel arithmetic—simulated annealing method. Beijing: Science Publishing House.

Liu Hangdong, 1996. Prediction Theory and Method of Landslide failure. Zhenzhou: Yellow Water Publishing House.

Liu Hangdong, 1998. Develop stage of slope failure forecasting, The 5th Rock Mechanics and Engineering Congress in China, Beijing: Chinese Sciences and Technology Publishing House.

Sandro Ridolla, Stefano Rovetta & Rodolfo Zunino, 1997. Circular back-propagation Networks for classification, IEEE Transaction on Neural Networks, 8 (1): 84–97.

Zhang liming. 1983. Artificial neural network model and it's application. Shanghai: Fudan University Publishing House.

New models linking piezometric levels and displacements in a landslide

R.M. Faure
Tunnel Study Centre, Bron, France

S. Burlon
Public Works Research Laboratory of Lille, France

J.C. Gress
Company Hydrogéo, Fontaines, France

F. Rojat
Public Works Research Laboratory of Toulouse, France

ABSTRACT: The aim of this article is to present two models linking the piezometric level to the displacements of a landslide. These two relations are validated with the study of the Petit Caporal landslide (Boulogne-sur-Mer, France). The obtained results are very correct since both models manage to follow the evolution of the displacements. These relations stem from two distinct approaches: the first one based on experimental concepts and the second one based on the study of creep. These models are relevant alert systems, able to detect changes in the move of the slide. Their easy use encourages to validate them on other sites and to generalise this kind of studies to other unstable slopes.

1 INTRODUCTION

The constant lack of free building areas may lead to set up constructions and infrastructures on unstable zones. Slow landslides can be dangerous when their speed increases, as this phenomenon often announces an imminent failure. Therefore, models allowing to understand better the behaviour of such slides constitute a major issue for both experts and administrators. The site of "Petit Caporal" (Boulogne-sur-Mer, France) that has been studied for nearly thirty years by the technical services of French State is precisely one of these slow landslides. It is crossed by a national road (RN1) and by gas and water pipes, which implies security problems and important financial interests for the neighbouring population. This landslide is fully tooled up with monitoring systems such as pluviometer, inverse pendulum, piezometers and inclinometers. The data gathered during several years allow building and validating various models linking piezometric levels to displacement speed. Two kinds of models are presented in this paper. The first one bases on empirical and statistical considerations as it aims at determining at best the value of two behaviour parameters. The second one rests on more physical considerations as it insists on creep modelling. The validity of these models and their possible improvements are also discussed.

2 THE "PETIT CAPORAL" LANDSLIDE

This landslide does not concern a very important volume of material. It is about 220 m long and 200 m wide with a regular slope around 5.5°. It is delimited uphill by the RN1 road and downhill by the Liane stream. The global geology appears quite complex. Figure 1 shows the typical longitudinal section that was used to carry out the study.

Two kinds of geological formations can be identified: first the Kimmeridgian stratigraphic formations (upper Jurassic, secondary era) and on the other hand the superficial formations of the Liane Valley (quaternary era). To simplify matters, the reader should keep in mind three main guidelines about the geology of the site:

– the material is globally clayey despite the presence of two strata with more sandy characteristics in which most water flows happen;
– the "Moulin Wibert" clays constitute an impermeable substratum;

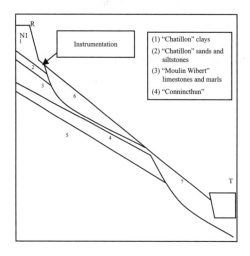

Figure 1. General geology of the "Petit Caporal" site (drawing not to scale).

	c' (kPa)	φ' (°)	γ (kN/m³)
(1) "Chatillon" clays	20	25	20.4
(2) "Chatillon" sands and siltstones	10	30	20.2
(3) "Moulin Wibert" limestones and marls	12	19	20.5
(4) "Connincthun" sands	5	26	20.5
(5) "Moulin Wibert" clays	0	8 (residual)	21
(6) Uphill soils	0	20	20.2
(7) Downhill soils	0	25	20.2

Figure 2. Geotechnical characteristics of the landslide.

– the main surface of failure is located at 6 m depth at the interface between this clay layer and the Connincthun sands.

Movement prediction from rainfall measures is an old challenge; a synthesis was done in (Favre et al., 1992) and very good data were obtained on the full-scale experimental site at Sallèdes in France (Pouget et al, 1994).

(Vuillet et al., 1996) tested new approach using neural networks, with a conclusion saying that a great amount of data is necessary for building an accurate model.

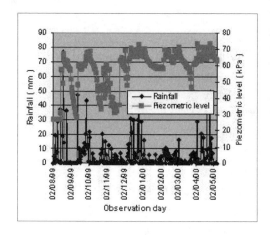

Figure 3. Rainfall and piezometric levels between August, 2nd 1999 and May, 7th 2000.

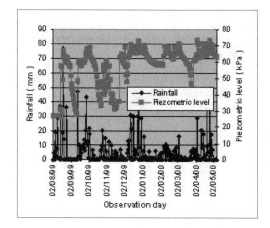

Figure 4. Rainfall and piezometric levels between August, 2nd 2000 and May, 8th 2001.

Coming back to Le Petit Caporal, the models presented in this paper are tested on two observation periods of 280 days, the first one beginning on August, 2nd 1999 and the second one beginning on August, 2nd 2000.

The rainfall and piezometric readings on both periods are presented hereafter on figures 3 and 4.

3 A FIRST MODEL FOR DISPLACEMENT ASSESSMENT

The first model that was used is the one proposed by P. Alphonsi (Alphonsi, 1997) who had carried out similar research programs about the "Clapière" landslide

(Alpes-Maritimes, France). The relationship he had established was as follows:

$$V_{j+1} = V_j \exp(-\beta)\frac{H_{j+1}}{H_j} \quad (1)$$

with:

V_j the displacement speed at day j in mm/day
H_j the piezometric level at day j in kPa
β a coefficient expressing ground sewage.

In fact, it appeared that this model was giving no relevant results in the case of the "Petit Caporal" landslide. Therefore, a second formulation has been elaborated:

$$V_{j+1} = \alpha_1 \exp\left(\frac{H_j}{\alpha_2}\right) \quad (2)$$

with:

V_{j+1} the displacement speed at day j + 1 in mm/day
H_j the piezometric level at day j in kPa
α_1 a scale parameter in mm/day
α_2 a piezometric level of reference, for which the displacement speed is minimal.

This relationship simply expresses that the sliding speed increases when the piezometric levels are higher. It can be used to evaluate the ground displacements over a few days but not the precise daily speed. Indeed, it appeared too ambitious to build a two-parameters model only based on piezometric data to express explicitly the daily speed changes, so equation (2) expresses an average behaviour. The results obtained are compared with the readings from the inverse pendulum that provides cumulated displacements.

This formula was used in two steps. First, the observation period beginning on August, 2nd 1999 allowed determining both α_1 and α_2 parameters. The indicator Δ that was chosen to assess the quality of the correlation is the average of the absolute values of the difference between measured and calculated displacements. Thus, the following values of parameters α_1 and α_2 have been determined: $\alpha_1 = 0.01$ mm/day and $\alpha_2 = 20.5$ kPa, with $\Delta = 2.86$ mm. As shown in figure 5 below, considering that the displacement range is around 50 mm, the obtained correlation is satisfactory.

In a second step, the above-determined values of parameters α_1 and α_2 have been applied to the other observation period. In this case Δ reaches the value of 12.82 mm: the correlation is not very good, as it appears on figure 6.

In fact, this second observation period has to be analysed more carefully, as it seems to indicate a change in the kinematics of the landslide. Indeed, the

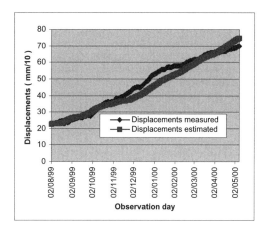

Figure 5. Comparison between observed and calculated displacements over the first observation period.

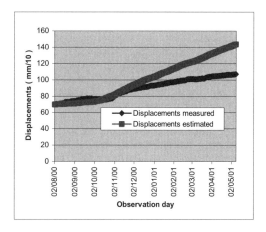

Figure 6. Comparison between observed and calculated displacements over the second observation period.

correlation shows to be very precise during the first 100 days, and then a constant drift is observed as if the slide was reacting differently to piezometric changes. The sudden inflexion in the curve of calculated displacements can be easily explained by a higher—and constant—piezometric level, due to important rainfalls. So the question is to know why the landslide did not accelerate as well with this increase of the water level. The first hypothesis is that the calculation model does not use enough parameters to describe correctly the kinematics of the landslide as a whole. The second hypothesis takes into account that reinforcement works were realized during the observation period. In fact, two test trenches filled with 7500 m³ of granular materials were installed at the bottom of the slide.

Due to their position with respect to the slope, these trenches have no noticeable influence on piezometric levels. However, substituting clays and silts with a sandy material must have increased the global resistance to shear at the level of the failure surface, slowing down the observed displacements. This last hypothesis has to be confirmed through the study of a second calculation method.

Thus, this first simple model, despite its uncertainties, seems to translate very correctly the interactions between piezometric levels and displacements in a slow landslide. Its main advantages are its very simple use and its ability to detect changes in the kinematics of the sliding area. Validating this model on other landslides of the same kind would be of great interest.

4 A SECOND WAY TO ASSESS SLOPE DISPLACEMENTS, BASED ON CREEP MODELLING

The study of creep phenomena usually allows to bring out three kinds of creep, each of them being associated with a characteristic strain speed: no speed for primary creep, constant speed for secondary creep and regularly increasing speed—until failure—for tertiary creep (see Figure 7). However some authors (Ter-Stepanian, 1996 e.g.) contest this characterisation and define four kinds of creep corresponding to four levels of material state and stress conditions.

The creep function that was used in this research had been presented in a previous article (Faure et al., 2002). It describes the three kinds of creep mentioned above and can be written as follows:

$$\delta(s,t) = \lambda(s) + \frac{v_0(s)}{s_1 - s} \cdot \frac{\sinh((s_1 - s)(t + T))}{(\cosh(t + T))^{s_1 - s}} \quad (3)$$

with:

s the tangential stress ratio τ/τ_{max} at the level of the failure surface, with $\tau = W \cdot \sin(\alpha)$ and $\tau_{max} = c' + (W \cdot \cos(\alpha) - u) \cdot \tan(\varphi')$ (u being the pore pressure).
s_1 a reference stress ratio expressing the stress state in the ground;
λ the instant displacement when the material is submitted to the shear stress τ. In this study, λ was supposed equal to 0;
v_0 the initial slope of the creep function;
t a time increment;
T a time variable expressing ground damage.

The parameters T and t above have the dimension of a time. However, the exact time scale they refer to has not been determined yet as it would require complete laboratory creep tests, which was not possible in this study. For the moment, they can be simply considered as time increments necessary to calculation progress. Further research may allow to identify better their real physical meaning and to determine how the scale changes between laboratory tests and a real landslide should be taken into account. Applying the creep function above also requires defining an average cohesion and friction angle at the level of the failure surface in order to assess the evolution of the shear stresses. From the laboratory tests realised on the various kinds of soils encountered in the "Petit Caporal" slope, the following values were inferred: $c' = 5$ kPa and $\varphi' = 25°$. The soil density is: $\gamma = 21$ kN/m^3.

The displacements are estimated with the creep formula (3) by giving t a fixed value and by determining T and s_1 in order to get the best possible correlation. Fixing t comes down to the hypothesis that the observation (and modelling) period is very short in comparison with the complete lifetime of the landslide. It allowed getting round the above-mentioned difficulties on parameters T and t, by using only the transitions from one kind of creep to another (Y-axis) and not the time-dependence (X-axis) of the creep function.

To determine the various parameters, the fitting procedure was similar to the one described in the previous sections. For the first observation period, $\Delta = 2.41$ mm is obtained with $s1 = 0.27$, $T = 6$ and $v_0 = 0.1$. For the second observation period, the same parametering yields $\Delta = 12.90$ mm, which is quite deceptive. Figures 8 and 9 give an overview of the results: it clearly appears that the observed behaviour is very similar to what had been inferred from the

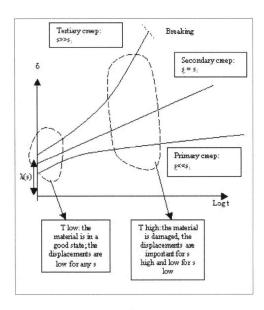

Figure 7. The three kinds of creep.

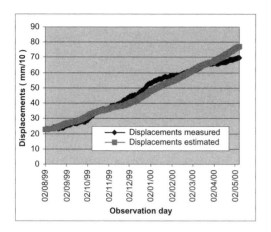

Figure 8. Comparison between observed and calculated displacements over the first observation period.

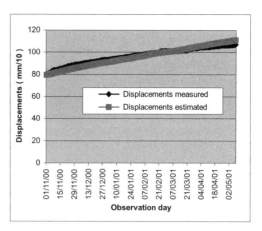

Figure 10. Comparison between observed and calculated displacements after soil strengthening.

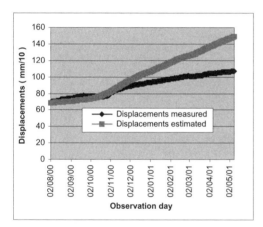

Figure 9. Comparison between observed and calculated displacements over the second observation period.

first calculation method (equation (2)). The change in the kinematics of the landslide during the second observation period seems to be confirmed.

In order to know if a change in the parameters of the model—in particular s_1 and T—could allow a better fitting, the second part of the second observation period has been studied as a third observation period (i.e. from November 1st, 2000 to May 8th, 2001). With the same calculation method as previously, a very good correlation ($\Delta = 1.91$ mm) can be obtained by simply increasing s_1 from 0.27 to 0.33. The results are shown in figure 10. This increase in parameter s_1 clearly indicates an improvement of ground strength. Quite similar results can be obtained by keeping s_1 constant and by increasing τ_{max} (for instance with $\varphi' = 29°$ instead of $25°$), which is logical as equation (3) uses $s_1 - s$. However, it must be emphasized that increasing s_1 or increasing τ_{max} refer to two different physical approaches.

In the first case the reference stress ratio is changed, which means different creep thresholds (primary/secondary/tertiary) and a change in the behaviour of the soil that deforms less for the same stress level.

In the second case the creep curves remain the same, but the maximal shear strength is increased, which reduces the stress level τ/τ_{max}. It is likely that both approaches play a role in the real behaviour of the slope. This third observation period shows that interesting data can be obtained from such a model when trying to quantify ground improvement. For instance, testing the model on a more recent period could have allowed knowing whether the effects of the reinforcement works were still perceptible.

Finally, it seems interesting to give additional comments about the definition of the parameters and variables of the creep function. In the case of the "Petit Caporal" landslide, the variable s directly expresses the influence of the piezometric changes on displacements. It allows determining how the material might be subjected to creep. In the studied cases, the calculated variations of s show that the material does not behave the same according to the season. When the water table is high, $s > s_1$ and tertiary creep is observed. On the contrary, when the water table is low, primary creep happens. Besides, the parameter T expresses the material fatigue. During the 39th Rankine Lecture, S. Leroueil clearly explained this fatigue phenomenon with loading and unloading cycles when the soil is submitted to ground water level variations. The various calculations performed for this study showed that the parameter T does not vary much on the observation

691

period, as small changes in T may lead to important displacements. This observation is in good agreement with the hypothesis that the observation period is rather small in comparison with the total lifetime of the slide. In any case, finding how T varies in function of the stress-strain history of the material, even if of great interest for the improvement of the model, remains rather complex.

5 CONCLUSION

The validity of two formulas expressing the displacements of a landslide from piezometric changes has been tested on the "Petit Caporal" case history. The first relationship shows that a simple formula based on intuitive concepts can allow over a few months a good assessment of the evolution of the slope. The second relationship resting on a creep model yields quite similar results that can be analysed in function of various parameters, in particular s and T.

Both relations constitute interesting tools for control or even forecast (when no fundamental change occurs in the landslide behaviour). They might be used as well to assess the efficiency of reinforcement works on slope motions. Curve fitting before and after reinforcement may yield very useful quantitative data concerning the improvement of the global shear strength. However additional validations on other unstable sites and through laboratory experiments are still necessary. It must be emphasised that at the present time, when security and risk management become more and more sensitive topics, such research works should have good prospects.

REFERENCES

Alfonsi P. 1997. Relation entre les paramètres hydrologiques et la vitesse dans les glissements de terrain. Exemples de la Clapière et de Séchilienne (France). *Revue française de géotechnique, n° 79*, p.3–12.

Faure R.M., Gress J.C. & Rojat F., 2002. An easy to use model for taking in account rainfall in soil displacement. European Slope Stability Symposium, Rybar, Prague.

Favre J.L., Gevreau E. & Durville J.L. 1992. Prévoir l'évolution des mouvements de terrain. *Revue française de géotechnique, n° 59*, p.65–73.

Leroueil S. 2001. Natural slopes and cuts: movement and failure mechanisms. *Geotechnique 51, No. 3*, p.195–243.

Pouget P. & Livet M. 1994. Relations entre la pluviométrie et la piézométrie et les déplacements d'un versant instable (site expérimental de sallèdes, Puy-de-Dôme), *Etudes et Recherches des Laboratoires des Ponts et Chaussées, série géotechnique, GT 57.*

Ter-Stepanian G. 1996. Concentration du fluage avec le temps. *Revue française de géotechnique, n° 74*, p.31–43.

Vuillet L., Cornu T. & Mayoraz F. 1996. Using neural networks to predict slope movements. *Balkema-Rotterdam: Senneset*, p.295–300.

3D slope stability analysis of Rockfill dam in U-shape valley

X.Y. Feng & M.T. Luan
Dalian University of Technology, School of Civil and Hydraulic Engineering, Dalian, China

Z.P. Xu
China Institute of Water Resources and Hydropower Research, Beijing, China

ABSTRACT: By using the method of non-linear elastic-plastic finite element analysis with the software of ABAQUS, the three dimensional slope stability of an earth core rockfill dam constructed in U-shape valley was analyzed. The constitutive model of the materials in the analysis is Mohr-Coulomb model and the factor of safety of the slope is determined by strength reduction method. In the analysis, the development of equivalent plastic strain of dam slope under different strength reduction factors was presented. The final critical state of slope failure is defined as the plastic strain zone run through the slope and the sudden changes of equivalent plastic shear strain and displacement occurred. Compare with the conventional limit equilibrium method, the proposed strength reduction method can not only provide the similar factor of safety, but also the strain and deformation changes of the slope. In addition, in 3D analysis, it can also present a clear spatial sliding surface of the slope.

1 GENERAL INSTRUCTIONS

The slope stability is one of the most important issues in the safety of high rockfill dam. At present time, the conventional methods for conducting slope stability analysis include: limit equilibrium method, limit analysis method, slip line field method, etc. The most popular method in engineering application is the limit equilibrium method. In general, those methods are all based on the theory of limit equilibrium of rigid body. It can neither consider the internal stress-strain relationship of the soil, nor the development of the slope failure and the interaction of different material zones. Besides, it is also necessary to assume the shape of the sliding surface, such as circle, broken line or logarithmic spiral, etc.

The application of FEM in slope stability analysis has no need to assume soil as a rigid body. It can not only meet the requirement of equilibrium, but also take into account of the stress-strain relationship of soil material. It can simulate the procedure of slope failure and the real shape of sliding surface and it is applicable in any complex boundary conditions. The first application of FEM in slope stability analysis was started in 1970s. But at that time, due to the limitation of computation conditions, the error of the computation is relatively large. It was not widely accepted (Zienkiewicz, 1975). By gradually reduce the strength of material in FEM analysis to make the structure in an unstable status, which may finally lead to the unconvergence of the iteration of the computation, the factor of strength reduction at this time will be the factor of safety of the slope (Song, 1997. Griffiths, 1999. Dawson, 1999. Manzari, 2000. Zhao, 2002). In recent years, with the development of computation technology and the progress in numerical analysis of geotechnical engineering, many large scale FEM software which are applicable for soil and rock materials are developed. Most of those software are powerful in pre and post processing, which are helpful in applying FEM analysis to study the stability of various slopes.

2 COMPUTATION METHOD OF STRENGTH REDUCTION

2.1 Yield criteria

The yield criteria employed in the paper is Mohr-Coulomb model. It is widely accepted yield criteria in elastic-plastic analysis. The main defect of the model is the incontinuous edge points in the three dimensional stress space. The Mohr-Coulomb yield criteria is expressed as:

$$\frac{1}{3}I_1 \sin\phi - \left(\cos\theta_\sigma + \frac{1}{\sqrt{3}}\sin\theta_\sigma \sin\phi\right) \\ \times \sqrt{J_2} + c\cos\phi = 0 \qquad (1)$$

where I_1 is the first stress invariant of stress tensor, J_2 is the invariant of deviatoric stress tensor, θ_σ is Lode's angle, c is the cohesion, ϕ is the internal fraction angle (Li, 2006).

In the analysis, the modified Mohr-Coulomb criteria was developed (Abaqus, 2003), where the deviatoric stress space has no edge point, the flow potential is totally smooth and only one flow direction can be developed. For general states of stress the model is more conveniently written in terms of three stress invariants as

$$F = R_{mc}q - p\tan\phi - c = 0 \quad (2)$$

where

$$R_{mc}(\Theta, \phi) = \frac{1}{\sqrt{3}\cos\phi}\sin\left(\Theta + \frac{\pi}{3}\right)$$
$$+ \frac{1}{3}\cos\left(\Theta + \frac{\pi}{3}\right)\tan\phi \quad (3a)$$

Θ is deviatoric polar angle (Chen and Han, 1988).

The flow potential G is chosen as a hyperbolic function in the meridional stress plane and the smooth elliptic function proposed by Menétrey and Willam (1995) in the deviatoric stress plane:

$$G = \sqrt{(\varepsilon c|_0 \tan\psi)^2 + (R_{mw}q)^2} - p\tan\psi \quad (4)$$

where

$$R_{mw}(\Theta, e) = R_{mc}\left(\frac{\pi}{3}, \phi\right)$$
$$\times \frac{4(1-e^2)\cos^2\Theta + (2e-1)^2}{2(1-e^2)\cos\Theta + (2e-1)\sqrt{(1-e^2)\cos^2\Theta + 5e^2 - 4e}} \quad (5)$$

and

$$R_{mc}\left(\frac{\pi}{3}, \phi\right) = \frac{3 - \sin\phi}{6\cos\phi} \quad (6)$$

In the analysis presented in the paper, a certain dilatancy of soil material is considered. If the property of dilatancy of soil material is not considered ($\psi = 0$), the computation results will be too conservative. If the dilatancy of soil material is overestimated ($\psi = \phi$), the deformation will be too large. Therefore, the dilation angle is accepted as half of the friction angle ($\psi = \phi/2$), which may lead to a nonassociated plastic flow and unsymmetric stiffness matrix.

2.2 The principle of strength reduction method

The idea of strength reduction was first introduced by Zienkiewicz et al. in 1975. It has the same concept as Bishop (Bishop, 1955) presented in limit equilibrium method. The definition of the factor of shear strength

Table 1. Material parameters.

Material	E/Mpa	υ	Density kg/m³	Strength index c/KPa	$\phi/°$	$\psi/°$
Rockfill	100	0.3	2200	100	40	20
Core	100	0.3	2160	30	25	12.5

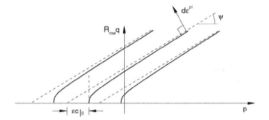

Figure 1. Mohr-Coulomb yield criteria ($p - R_{mc}q$ plane).

reduction can be expressed as (Luan, 2003): under the condition of unchanged external load, the factor of shear strength reduction is the ratio of maximum shear strength of the soil to the real shear stress of the soil. The method of strength reduction is conducted by dividing the original strength indexes of c, ϕ with a factor F_s to get a new strength indexes c', ϕ'. With the new indexes, FEM analysis will be carried out. By gradually increasing the factor F_s to let the slope finally reach to the critical state, then the last F_s will be the minimum factor of safety of the slope (Zhang, 2003). The strength indexes c', ϕ' can be calculated by (7) and (8). Elastic modulus E and Possion's ratio υ are kept unchanged during the computation.

$$c' = \frac{c}{F_s} \quad (7)$$

$$\phi' = \arctan\left(\frac{\tan\phi}{F_s}\right) \quad (8)$$

2.3 Criteria for determining slope failure

In the slope stability analysis by using strength reduction FEM method, the criteria for determining the critical state of slope failure is one of the key points. In practice, some researchers use the un-convergence of calculation as the criteria (Lian, 2001), which means the un-convergence of the calculation represents the stress distribution cannot guarantee the general equilibrium of the slope (Liu, 2005). Ugai (Ugai, 1989) specifies the iteration number of 500 as the limit of un-convergence. Dawson assumes the ratio of imbalance nodal force to the external load greater than 10^{-3} is the criteria of slope failure. Lian Zhenyin (Lian, 2001) considered the character of slope failure should be thorough development of the generalized shear strain from the top of the slope to the bottom of the slope. The appearance and the development of plastic strain in slope indicate that the occurrence of unrecoverable residual deformation. Therefore, the development of plastic strain could essentially represent the procedure of the yield and failure of the slope.

In the analysis presented in this paper, the stability of slope is judged by the conditions of the distribution of equivalent plastic strain. If the plastic strain zone is not yet run through the slope, it means the slope is still stable. Further reduction of the strength indexes of soil material will be conducted. If the plastic zone is run through from the top to the bottom of the slope, it means the slope will be in critical status. The strength reduction factor will be defined as the factor of safety of the slope.

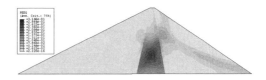

Figure 4. Distribution of plastic strain ($Fs = 1.5$).

Figure 5. Distribution of plastic strain ($Fs = 1.9$).

3 THE APPLICATION OF STRENGTH REDUCTION FEM METHOD

According to the computational method presented above, the 3D slope stability of an earth core rockfill dam constructed in U-shape valley was analyzed by ABAQUS software. The finite element mesh of the computation model is shown in Figure 2. The upstream and downstream slopes of the dam are 1:2 and 1:1.5 respectively. The upstream and downstream slope of central core of is 1:0.2. The constitutive model of rockfill material and earth material is Mohr-Coulomb

Figure 6. Distribution of plastic strain (homogeneous dam).

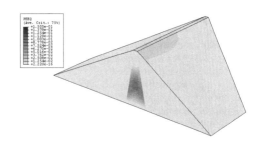

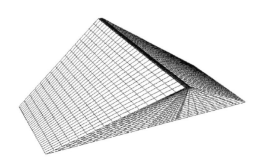

Figure 7. Distribution of plastic strain (3D, original condition).

Figure 2. 3D FEM mesh.

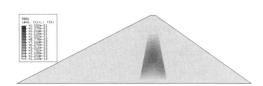

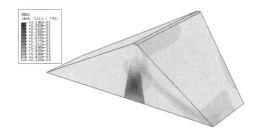

Figure 3. Distribution of plastic strain (original condition).

Figure 8. Distribution of plastic strain (3D, $Fs = 1.5$).

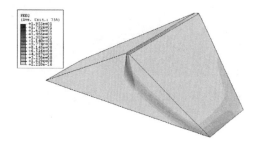

Figure 9. Distribution of plastic strain (3D, $Fs = 1.9$).

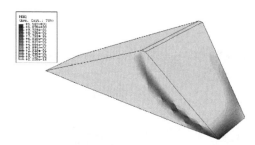

Figure 10. Distribution of plastic strain (homogeneous dam).

model. The parameters of materials are shown in Table 1. The bottom and abutment boundary is treated as fixed boundary. In the analysis, several different strength reduction factors (1.5, 1.7, 1.8, 1.9) were selected for conducting the computation. Besides, the results of the cases in homogeneous dam and central core rockfill dam are presented.

From the computation results, it can be noticed that: under the original condition, the plastic strain zone mainly distributed in the region of central core. No continuous plastic strain zone occurred on downstream slope. When the strength reduction factors reach to 1.5, 1.7 and 1.8, plastic strain zone occurred on downstream slope. But the plastic strain zone not run through the slope. When $Fs = 1.9$, an obvious circular plastic strain zone developed through the top to bottom of the slope. Large deformation occurred at dam crest and downstream toe (as shown in Figure 6 and 10). It can be concluded that the final factor of safety of the slope is 1.9.

4 CONCLUSIONS

From the computation results of the elastic-plastic finite element analysis of the earth core rockfill dam, it can be found that the strength reduction method is applicable in three dimensional slope stability analysis. Compare with the conventional method, the development procedures of plastic strain zone in the slope can be clearly identified. Especially, with the application of three dimensional analysis, the spatial sliding surface of the slope can be presented in the analysis.

In the application of strength reduction method, the developing of plastic strain zone is suggested to be the criteria for determining the slope failure. Besides, the observation of deformation changes of the slope could also be provided as a secondary judgment.

REFERENCES

ABAQUS. 2003. ABAQUS analysis User's manual[M]. inc.
Bishop A.W. 1955. The use of the slip circle in the stability analysis of slopes[J]. Geotechnique, (5):7~17.
Dawson E.M, Roth W.H, Drescher A. 1999. Slope stability analysis by strength reduction[J]. Geotechnique, 49(6):835~840.
Griffiths D.V, Lane P.A. 1999. Slope stability analysis by finite elements[J]. Geotechnique, 49(3):387~403.
Li C.Z, Chen G.X, Fan Y.W. 2006. The analysis of slope stability of strength reduction FEM based on ABAQUS software[J]. Journal of Disaster Prevention and Mitigation Engineering, 26(2):207~212.
Lian Z.Y, Han G.C, Kong X.J. 2001. Stability analysis of excavation by strength reduction FEM[J]. Chinese Journal of Geotechnical Engineering, 23(4):407~411.
Liu Z.Q, Zhou C.Y, Dong L.G, etal. 2005. Slope stability and strengthening analysis by strength reduction FEM[J]. Rock and Soil Mechanics, 26(4):558~561.
Luan M.T, Wu Y.J, Nian T.K. 2003. A criterion for evaluating slope stability based on development of plastic zone by shear strength reduction FEM[J]. Journal of Disaster Prevention and Mitigation Engineering, 23(3):1~8.
Manzari M.T, Nour M.A. 2000. Significance of soil dilatancy in slope stability analysis[J]. Journal of Geotechnique and Geoenvironmental Engineering, America Society of Civil Engineers, 126(1):75~80.
Song E.X. 1997. Finite element analysis of safety factor for soil structure[J]. Chinese Journal of Geotechnical Engineering, 19(2):1~7.
Ugai K.A. 1989. Method of calculation of total factor of safety of slopes by elaso-plastic FEM[J]. Soil and Foundation, 29(2):190~195.
Zhang L.Y, Zheng Y.R, Zhao S.Y, etal. 2003. The feasibility study of strength reduction method with FEM for calculating safety factors of soil slope stability[J]. Journal of Hydraulic Engineering, (1):21~27.
Zhao S.Y, Zheng Y.R, Shi W.M, et al. 2002. Analysis on safety factor by strength reduction FEM[J]. Chinese Journal of Geotechnical Engineering, 24(3):343~346.
Zienkiewicz O.C, Humpeson C, Lewis R.W. 1975. Associated and nonassociated visco-plasicity in soil mechanics[J]. Geotechnique, 25(4):671~689.

3-D finite element analysis of landslide prevention piles

K. Fujisawa, M. Tohei & Y. Ishii
Public Works Research Institute, Tsukuba, Japan

Y. Nakashima & S. Kuraoka
Nippon Koei co., Ltd., Tsukuba, Japan

ABSTRACT: The landslides prevention piles is one of effective countermeasures against landslides. The design of landslides prevention piles is usually conducted by 2-dimentional analysis. Recently, 3-dimentional limit equilibrium analysis such as Hovland method is applied in order to achieve rational design of piles. In this method, stabilizing force (The force to achieve planed factor of safety) is calculated. Then, this force is divided by the number of piles in order compute the load that need to be carried by each pile. It is assumed that stability force is uniformly distributed to the piles. However, landslides have 3-dimentional (3-D) geometries. For instance, thickness of moving body, slip surface angle and material constants for soils are not uniform. Therefore, the load that acts on each pile is different and load may exceed the capacity of piles. For this case, it is possible to apply 3-D finite element analysis that models 3-D geometries of landslide mass and each prevention pile. This paper presents study of 3-D finite element analysis for simulating restraining mechanism of each piles constructed in landslide that is triggered by groundwater. Especially, this paper presents modeling techniques for slip surface using joint element and group of piles with 3-D beam elements. Simulation of actual pile behavior has been performed to validate the model. Possible application for rational design is also discussed.

1 INTRODUCTION

Many landslides that require prevention piles in Japan are those with pre-existing sliding surface. In such cases, three dimensional limit equilibrium methods such as 3-D Janbu's Method and Modified Hovland's Method are applied to reduce amount of piles by taking into account 3-D geometry of landslide body.

It is important that how the total prevention forces obtained from 3-D stability analysis is distributed to each pile. One idea is to assume that the load of piles is uniformly distributed. However, the thickness of landslide body is thick at center, and thin in the edge. In addition, the inclination of sliding surface and material properties of landslide mass are not uniform. As a result, load of piles is different depending on the position of piles, in some piles, load may exceed the capacity.

For this case, it is possible to apply three dimensional finite element method (3-D FEM), but it is necessary to validate whether the 3-D FEM can simulate the real landslide.

The objective of the research described in this paper is to develop a practical three dimensional finite element model for active landslide. A simple elasto-viscoplastic interface model for sliding surface was developed, and each prevention piles were modeled by a beam element. Validation of proposed model was performed by comparison between analysis results and measurement results.

2 MODEL FOR LANDSLIDES PREVENTION PILES

2.1 Outline of analysis model

The model for landslides prevention piles consist of landslide mass, sliding surface, base of landslide, and piles. The landslide moving mass and base are modeled by the solid elements as elastic material. The sliding surface is modeled by the interface (joint) elements as elasto-viscoplastic material is described 2.2. The piles are modeled by the beam elements as elastic material.

2.2 Model for sliding plane

The landslide modeled in this research is the type where soil and rock mass move as a single mass over the pre-existing sliding surface. It is therefore assumed that the landslide mass itself does not fail and modeled by elastic solid elements.

It is thought that the finite element of the sliding surface is solid elements and interface (joint) element.

In this paper, interface (joint) element was chosen for the reasons described as follows.

There are two types of stress concepts, one is the surface traction on the surface of a body, and another is the internal stress in a body.

The former is the normal direction stress (σ_n) on the surface and shear stress (τ_s), and they are called surface traction. On the other hand, the latter is defined as tensor with six components expressed ($\sigma_x, \sigma_y, \sigma_z, \tau_{xy}, \tau_{yz}, \tau_{zx}$), and they are called internal stress.

Stresses that control failure of sliding plane are the surface traction and therefore it is necessary to use joint elements. If solid elements are used, the direction of failure will be different from that of the prescribed sliding surface as described below. The direction of failure base on internal stresses is determined by the Mohr's circle and the failure envelope (Figure 1).

Let us suppose that the direction of principal stress is equal to the direction of gravity. Then we can estimate the direction of failure. The direction of this surface is not necessary in the direction of sliding surface. Especially, if the inclination of sliding surface is gentle, the direction of failure of a solid element may be significantly different from the direction of sliding surface. On the other hand, in case the joint element,

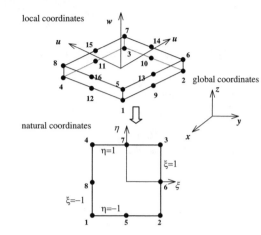

Figure 2. 16-node hexahedron joint element.

the direction of failure is equal to the direction of sliding surface because the direction of plane of the joint element corresponds to sliding surface.

It is thought that a solid element is suitable if the purpose is to simulate the process formation of sliding surface. However, when the sliding surface has already been formed, assuming that the large deformation take place, it is more suitable that the sliding surface is modeled by joint element.

We have been studying the joint element with liner shape functions. If the sliding surface geometry is simple, it is not especially a problem. However large number of joint elements may be required to gain better solution for soil-pile interaction.

So in this paper, modified version of the joint element is introduced. The element has midside nodes and has quadratic shape functions.

As an example, natural coordinates ($\xi \eta$) is introduced for numerical integration of 16-node hexahedron joint elements. Shape functions of eight-node plane serendipity elements are expressed as:

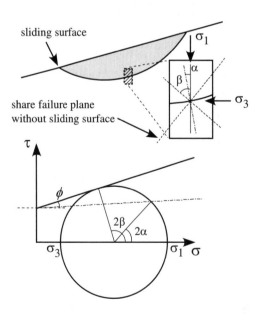

Figure 1. The schematic diagram relationship between the shear failure plane based on internal stress and the angle of sliding surface.

$N_1 = 1/4 (1 - \xi)(1 - \eta)(-1 - \xi - \eta)$

$N_2 = 1/4 (1 + \xi)(1 - \eta)(-1 + \xi - \eta)$

$N_3 = 1/4 (1 + \xi)(1 + \eta)(-1 + \xi + \eta)$

$N_4 = 1/4 (1 - \xi)(1 + \eta)(-1 - \xi + \eta)$

$N_5 = 1/2 (1 - \xi^2)(1 - \eta)$

$N_6 = 1/2 (1 + \xi)(1 - \eta^2)$

$N_7 = 1/2 (1 - \xi^2)(1 + \eta)$

$N_8 = 1/2 (1 - \xi)(1 - \eta^2)$ (1)

This shape function is introduced to translate physical coordinates to natural coordinates.

The relative displacement between the upper surface and lower surface of the joint element is given by

$$\{W\} = [B]\{U\} \quad (2)$$

where $\{W\} = \{\delta u\ \delta v\ \delta w\}^T$ is the relative displacement of 8-node, and $[B]$ is the matrix of 3 by 48 that translates displacements vector $\{U\}$ of 8-node to relative displacements.

The stiffness matrix of the joint element is given by

$$[K] = \int_A [B]^T [D][B]\, dA \quad (3)$$

where $[D]$ is constitutive matrix for joint element.

Also, we introduced 12-node penta type element to model more general shape of landsides.

The elasto-viscoplastic model is based on the fundamental theory developed by Perzyna in 1966, and it is extension of the two dimensional model introduce by Sekiguchi (1990) with modification of yield and plastic potential functions. The model permits both shear and opening failure modes, while compression failure is not allowed. When the interface opens due to tensile failure, displacements in shear are permitted since there is no shear resistance.

When there is no failure, elastic analysis is performed. If failure takes place, viscoplastic flow rule applies. The viscoplastic flow rule is defined for relative displacement, u_{vp} in local coordinates, since stresses are functions of relative displacements,

$$\{\dot{u}_{vp}\} = \gamma \left\langle \Phi\left(\frac{F}{F_0}\right) \right\rangle \frac{\partial Q}{\partial \sigma} \quad (4)$$

Where γ is the fluidity parameter, Q is the plastic potential, F is a yield function, F_0 is a reference value, and Φ is a scalar flow function for $x > 0$ and the notation $<>$ is defined as,

$$\langle \Phi(x) \rangle = \Phi(x) \quad \text{(for } x > 0\text{)}$$
$$\langle \Phi(x) \rangle = 0 \quad \text{(for } x \leq 0\text{)} \quad (5)$$

The rate of relative displacement mainly depends on the choice of functions Q and Φ. When the interface is in shear failure mode under compression, the yield function, F is taken as,

$$F = \sqrt{\tau_{rs}^2 + \tau_{sr}^2} + \sigma_n \tan\phi - c \quad (6)$$

$$F_0 = c \cdot \tan\phi \quad (7)$$

where τ_{rs} and τ_{sr} is shear stress for joint element, σn is normal stress act on joint element plane, ϕ is friction angle.

The plastic potential Q is expressed as,

$$Q = \sqrt{\tau_{rs}^2 + \tau_{sr}^2} \quad (8)$$

Compression is negative such that $F > 0$ in case of shear failure. For tensile failure mode, the yield and plastic potentials, and reference value for normal direction are formulated in terms of tensile strengths. The yield and plastic potentials for the shear direction are the same as Eq. (6) and (7) except that angle of friction and cohesion are set to zero.

2.3 Model for prevention piles

The beam element is introduced for the model of prevention pile. We can choose other type pile model composed of solid elements, if it necessary to assess detailed stress distribution around the piles.

2.4 Model for sliding mechanism

The judgment of shear failure of joint element is performed by subtract the pore water pressure from the normal stress.

The landslide sliding mechanism is expressed by increasing the groundwater level at slopes.

3 APPLICATION

3.1 Outline of the area and measurement

Validation of proposed model for landslide and prevention pile is performed by comparison between analysis and measurement results.

The Arahira Landslide is located in Miyazaki Prefecture in western Japan. It is 120 m long (south-north direction) and 120 m wide (east-west direction). The maximum depth of moving soil mass is about 19 m (Figure 3 and 4).

The geology is composed of phyllite from the Makimine Formation in the Morozuka Group.

Figure 4 is the section of the landslide due to field investigation and exploratory borings.

The sliding surface shape is plane towards the toe of landslide.

The counter measure such as drainage well and prevention piles were designed as the target factor of safety 1.15 (the increase with piles is 0.106).

These were constructed from March through August of 2000.

Strain gauges were attached to the pile by 1 m pitch in the direction of depth to measure bending moment. The measurement is conducted at the frequency about once a Month. Also, the extensometer (S-1, S-2) and the groundwater level meter (W-1, B11-2) were set up. Figure 5 shows the results.

The bending moment of pile increased rapidly from May 19, 2000 (measurement was started) through October 13, 2000. In the same period, the extensometer (S-2) recorded the displacement of 4.5 mm, and the groundwater level meter (W-1, B11-2) recorded increase of water level. The rainfall of total 671 mm was recorded from September 8, 2000 to September 16. Thus, it is thought that the landslide was caused under the influence of increased ground water level, and load acted on the prevention piles.

The pressuremeter test was performed near the pile by 1 m pitch in the direction of depth to investigate deformation property of geomaterial (Table 1).

3.2 Analysis model and procedure

Finite element mesh of Arahira landslide is shown Figure 6, in which the sliding plane is modeled by the joint elements described 2.2, whereas rest of the domain is modeled by 10-node tetrahedron solid elements as elastic material, the prevention piles was modeled by beam element. All of the mesh was made based on borehole investigation and topographical measurement.

The validation analysis was performed in two steps. In the first step, initial stress condition was generated under gravity loads where the ground water level was set to low water level (L.W.L), and piles were not generated at this step. Subsequently, in the second step, the piles were generated and the ground water level was set to high water level (H.W.L) to simulate landslide by inducing plastic failure in joint elements.

The validation analysis was performed by simulating behavior of piles from May 19, 2000 through October 13, 2000 where ground level increased by 3.4 m (from May 27, 2000 to June 20, 2000 at W-1).

3.3 Analysis case and parameter

The parameter for analysis, such as cohesion and the angle of internal friction, modulus of deformation, unit weight, Poisson's ratio is shown Table 2.

Unit weight and Poisson's ratio was set to typical value, the angle of internal friction of joint elements was back-calculated and by assuming that F.S. of sliding plane is 1.0 with 212.5 m of ground water level on September 5, 2000 (W-1).

In the analysis, F.S. was calculated as the ratio of sliding force which is summation of each joint element and resistant force.

The fluidity parameter γ in Eq. (4) was set to 0.025.

The modulus of deformation was set to the mean value of the pressuremeter test (Table 1) from sliding plane to distance of $1/\beta$ in the direction of depth.

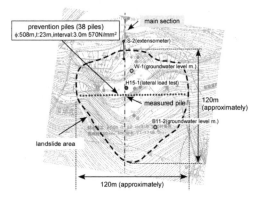

Figure 3. Plan view of Arahira Landslide and location of instruments, piles.

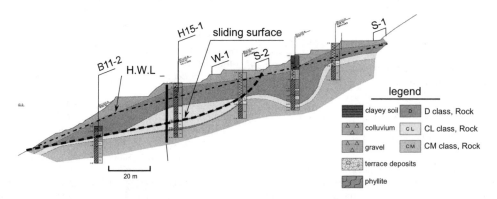

Figure 4. Sectional view of Arahira Landslide.

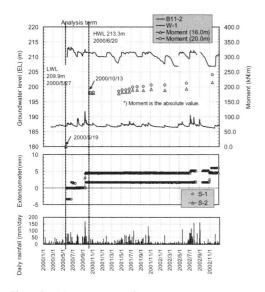

Figure 5. Measurement results.

Table 1 Results of the pressuremeter test.

Depth (m)	Geology	Classification	Modulus of deformation E_0 (kN/m²)
1	Colluvium-		2682
2			5090
3			4305
4			3503
5			5022
6		D	9085
7			8470
8			9297
9			8429
10			7717
11			106890
12			27700
13	Phyllite		18640
14		CL	14750
15			73850
16			20070
17			555090
18	Sliding surface	D	69160
19			22580
20		CL	17230
21		D	24710
22			155390
23		CM	79840
24			129420

The value of β which decides the property of piles is given by

$$\beta = \sqrt[4]{\frac{k_h \cdot d}{4 \cdot E \cdot I}} \tag{9}$$

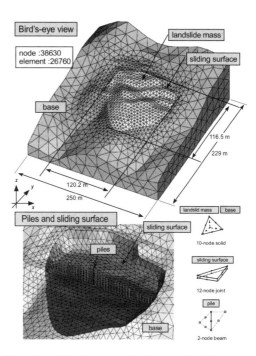

Figure 6. Finite element mesh of Arahira landslide.

Table 2. Parameter of the validation analysis.

	Unit weight	Cohesion	Angle of internal friction	Poisson's ratio
Landslide mass	18 kN/m³	-	-	0.3
Sliding surface		19.0 kN/m²	12.17°	-
Base	18 kN/m³	-	-	0.3

Where k_h is coefficient of horizontal subgrade reaction, d is diameter of piles, E is Young's modulus of steel, I is geometrical moment of inertia for piles section.

3.4 Results of analysis

The displacement and moment distribution in depth of measured pile is shown in figure 7. The bending moment from FEM analysis shows a good agreement with the measurement results. From this figure, the bending moment distribution shows similar peak moments at upper and lower depth from sliding plane. This tendency agrees with measurement results.

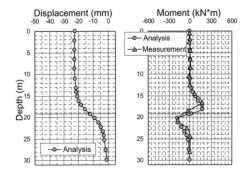

Figure 7. Displacement and moment of measured pile.

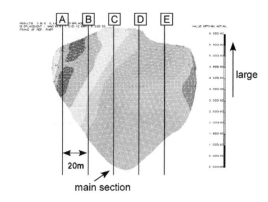

Figure 9. Displacement contour of sliding plane from FEM analysis when there is no pile.

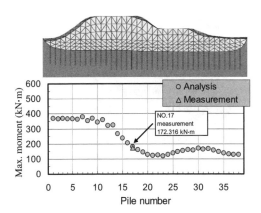

Figure 8. The maximum moment distribution of each pile.

The maximum moment distribution of each pile from FEM analysis and measurement results (pile No.17) are shown in figure 8. The maximum moment is different in each pile. In the left side area of landslide block, the maximum moments of piles are large. On the other hand, in right side area of landslide block, the maximum moments of piles are relatively small.

It is thought that this trend is due to the differences in the thickness of the landslide. However, this reason can not explain the fact that the bending moments of the piles No.1 to 5 are relatively large, where the thickness of left side area is relatively thin.

The reasons for the high moment of piles on the left region may be related to the inclination of sliding plane. Figure 9 shows displacement contour of sliding plane from FEM analysis when there is no pile, and figure 10 shows displacement contour of each section.

From these figures, the inclination of sliding plane of the left side is steeper than that of the right side, causing the left side of the displacement to be larger than that of the right side.

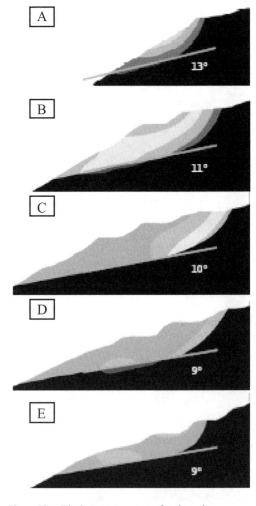

Figure 10. Displacement contour of each section.

4 CONCLUSIONS

This paper presents modeling techniques for slip surface using joint element and group of piles with 3-D beam elements. Simulation of actual pile behavior has been performed to validate the model.

It is thought that the 3-D finite element method enables more reasonable design of piles because it can consider 3-D effects.

REFERENCES

Ghaboussi, J., Wilson, E.L. & Isenberg, J. 1973. Finite elements for rock joints and interfaces, *Jl. Soil Mechs. Dn.*, ASCE, 94, SM3.

Goodman, R.E., Taylor, R.L. & Brekke, T. L. 1968. A model for the Mechanics of Jointed Rock. *Proc. ASCE*, 94 SM3, pp. 637–659.

Kuraoka, S., Ota, K. & Nakashima, Y. 2003. *International Conference On Slope Engineering*, Vol. II, pp. 834–839.

Owen, J & Hinton, E. 1980. Finite Elements Plasticity; Theory & Practice, Pineridge Press, pp. 272–318.

Sekiguchi, K., Rowe, R.K. & Lo, K.Y. 1990. Time step selection for 6 noded non-linear joint element in elasto-viscoplasticity analyses. *Computers and Geotechnics*, Vol. 10, pp. 33–58.

Desai, C., Samtani, N.C. & Vulliet, L. 1995. Constitutive modeling and analysis of creeping slopes. *Journal of Geotechnical Engineering*, Vol. 121, No. 1, pp. 43–56.

Miura, F., Okashige, Y. & Okinaka, H. 1985. Rupture propagation analysis by the three dimensional finite element method with joint element. *Research report University of Yamaguchi, Dept. of Engineering*, Vol. 36, No. 1, pp. 81–87.

Integrated intelligent method for displacement predication of landslide

W. Gao
Institute of Rock and Soil Mechanics, The Chinese Academy of Sciences, Wuhan, China

ABSTRACT: Displacement predication of landslide is very important in the control of landslide disaster. Considering the monotonously increasing character of time series of the landslide displacement, a new intelligent prediction method ton combine the Grey System and the Evolutionary Neural Network (ENN) is proposed here. On the basis of the principles of displacement decomposition, the trend of time series is extracted by the Grey System, while the deviation of the Grey System is approximated by the new ENN proposed. The architecture and algorithm parameters in the new ENN can evolve simultaneously through the modified BP algorithm and Immunized Evolutionary Programming proposed by the author. This new method is applied in the study of Xintan landslide, and the results show that the new method is good in generalization and can predict the displacement of landslide very well.

1 INTRODUCTION

Landslide is a very serious geological disaster. For there are lot of mountains in the west of china, as the progress of West Development Project in China, more and more landslide disasters will be encountered. So, how to control the landslide has become a very important work. To control landslide disaster, the forecasting of landslide is a very powerful method. But the development of landslide is a very complicated dynamic procedure. To describe this system very accurately is very hard. But the measured displacement series can describe the general laws of landslide development. So, some methods (Feng 2000, Gao & Zheng 2000, Huang 1999, Liu & Fan 1992, Shi & Xu 1995) for displacement predication of landslide are proposed. From analysis of those methods (Feng 2000, Gao 2002), we can find that, the neural network method and evolutionary neural network method can describe displacement series more accurately and more easily, and are two better methods. For predication problem, the generalization of neural network is very important. From the theory studies on generalization of neural network, the precision of inner interpolation can be ensured, but the error of outer interpolation is very large. For predication of one monotonously increasing time series, the outer interpolation is need. So, for predication of one monotonously increasing time series, the neural network is not very good. But the landslide displacement is one monotonously increasing time series, so to predict this time series, the new method must be proposed. Generally, the time series of the landslide displacement can be divided into some sections, such as, even section, periodic section and fluctuant section, et al. For different sections, different methods should be taken. But in previous studies using neural network, this problem have not been mentioned. To solve this problem very well, in this paper, an intelligent method combining Grey System and Evolutionary Neural Network is proposed.

2 NEW INTELLIGENT METHOD FOR LANDSLIDE PREDICATION

2.1 Division of time series of the landslide displacement

Supposing u_i is measured time series of landslide, it can be divided as follows,

$$u_i = u_{si} - v_i \quad (1)$$

where u_{si} is the trend section of displacement time series; and v_i is the deviation section.

The previous studies show that (Liu & Fan 1992), the trend of landslide displacement time series can be described by grey system very well. But the deviation section is still a very complicated time series. For describing this time series, the evolutionary neural network is a very suitable method.

2.2 Grey system model for trend section

Generally, the trend section of landslide displacement can be concluded into two types. One type is a kind of monotonously increasing curve that is a concave line. To describe this curve, the GM (1, 1) model in grey system is a very suitable one. Another type is an S shape curve. To describe this curve, the Verhulst model or DGM (2, 1) model in gery system is an very suitable one.

Here, the grey system is only to extract the trend, so its precision is not exigent. And the only objective is that the deviation is not a monotonously increasing curve. The details of the grey system can be found in reference (Liu et al. 1999).

2.3 Evolutionary neural network model for deviation section

As in above description, the precision of grey system is not high, so the deviation section is still a very complicated series. Even if the precision of grey system is high, it is very hard to guarantee that the deviation section is a simple random series. So, to describe the deviation section very well, the neural network is very suitable.

So, to construct a neural network model for modeling displacement time series, the construction of neural network is the main problem to be solved. Because in this problem, the hidden layer construction and input layer construction all must to be confirmed. This problem can be solved by evolutionary algorithm very well. Here, as a primary study, the evolutionary neural network which construction is confirmed by evolutionary algorithm and which weight is confirmed by MBP algorithm is proposed. To make problem simpler and generalization bigger, the three layers neural network is studied. So, here, only the number of input neuron and number of hidden layer neuron are to be confirmed. In MBP algorithm, there are two parameters, iterating step η and inertia parameter α, to be confirmed. These two parameters affected MPB algorithm very seriously. So, these two parameters are all confirmed by evolutionary algorithm. And then, in evolutionary neural network, there are four parameters to be evolved. In order to get the better effect, the new evolutionary algorithm- immunized evolutionary programming (Gao & Zheng 2003) proposed by author is used in evolutionary neural network.

The details of this new evolutionary neural network are given as follows.

1. The search range of input neuron and hidden layer neuron are given firstly. And also the search ranges of two parameters in MBP algorithm are given. And some evolutionary parameters, such as evolutionary generation stop criteria, individual number in one population, the error criteria of evolutionary algorithm, number of output neuron in neural network, iterating stop criteria and iterating error criteria in MBP algorithm are all given.

   It must be pointed out that, to construct the suitable samples, the number of input neuron must be smaller than total number of time series.

2. One network construction is generated by two random numbers in search range of input neuron and hidden layer neuron. And also, one kind of MBP algorithm is created by two random numbers in search range of parameters η and α. And then, one individual can be generated by the four parameters.

   It must be pointed out that, for two numbers of neuron are integer numbers and two MBP parameters are real numbers, so the expressions of one individual must be structural data.

3. To one individual, its fitness value can be gotten by follow steps.

   a. The whole time series of landslide displacement is divided to construct the training samples based on number of input neuron and number of hidden layer neuron. And also, the total number of samples is noted.
   b. The whole learning samples are to be divided into two parts. One part is the training samples, which is to get the non-linear mapping network. The other part is the testing samples, which is to test the generalization of network.
   c. The initial linking weights of network individual are generated randomly.
   d. The iterating step of MBP algorithm is taken as $j = 1$.
   e. This network individual is trained by testing samples, and the square error $E(j)$ is computed, and this error is taken as minimum error of the whole training, $\min E = E(j)$. If $\min E$ is smaller than the error criteria of evolutionary algorithm, then the fitness value is $\min E$. And the computing process is transferred to step (3).
   f. This network individual is trained by training samples. If its training error is smaller than iterating error criteria of MBP algorithm, then the fitness value is also the $\min E$. And the computing process is transferred to step (3).
   g. The whole linking weights are adjusted by MBP algorithm.
   h. $j = j + 1$, and the computing process is transferred to step e.
   i. If j is lager than iterating stop criteria of MBP algorithm, then the fitness value is also the $\min E$. And the computing process is transferred to step (3).

4. If the evolutionary generation reaches its stop criteria or computing error reaches error criteria of evolutionary algorithm, then the algorithm stop. At this time, the best individual in last generation is the searching result.

5. Every individuals in population are mutated. For there are different data types in one individual, the different mutation types are used for each parameter. For numbers of input neuron and hidden layer neuron are integer number, the uniform mutation

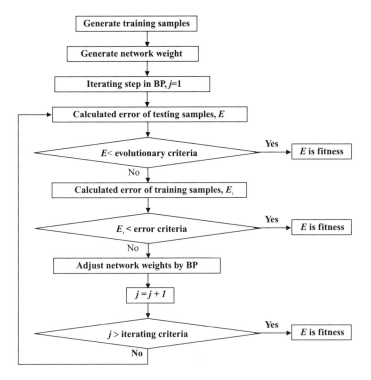

Figure 1. Flow chart of evolutionary neural network.

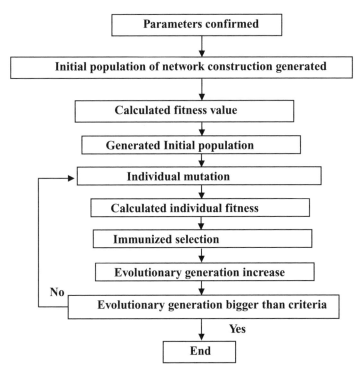

Figure 2. Flow chart of calculated error E.

707

Table 1. Measured displacement and calculated displacement of Xintan landslide.

Time step	Measured ones	Trend of GM (1, 1)	Deviation section	Training samples of ENN	Generalized results	Calculated results
1	0.077					
2	0.092	0.414	0.322	0.677		
3	0.615	0.468	−0.146	1.146		
4	0.65	0.529	−0.120	1.120		
5	0.69	0.599	−0.091	1.091		
6	0.738	0.677	−0.091	1.060		
7	0.846	0.766	−0.080	1.080		
8	0.962	0.866	−0.096	1.096		
9	1.0	0.979	−0.020	1.020		
10	1.03	1.108	0.078	0.922		
11	1.061	1.253	0.192	0.808		
12	1.077	1.417	0.34	0.660		
13	1.1	1.602	0.502	0.498		
14	1.23	1.811	0.581	0.4184	0.460	1.267
15	2.46	2.048	−0.411	1.411	1.382	2.431
16	2.754	2.316	−0.437	1.437	1.359	2.675
17	2.83	2.619	−0.210	1.210	1.268	2.888
18	2.92	2.962	0.042	0.957	1.087	3.049
19	3.46	3.350	−0.109	1.110	1.116	3.467
20	4.00	3.788	−0.211	1.211	1.135	3.923
21	4.25	4.284	0.034	0.966	0.826	4.110
22	4.38	4.844	0.464	0.535	0.559	4.403
23	4.615	5.478			0.585	5.064
24	5.77	6.195			0.804	5.999

is used. For parameters η and α are real numbers, the adaptive Cauchy mutation is used. And then the offspring population is generated.
6. The fitness value of each individual in offspring population is calculated by the method in step (3).
7. The set of offspring population and parent population is selected by selection operation based on thickness, then the new offspring population is generated.
8. The number of evolutionary generation increases 1, then the computing process is transferred to step (4).

From the above algorithm, the four parameters, number of input neuron, number of hidden layer neuron, two parameters η and α in MBP algorithm can be confirmed. So, the optimization neural network for deviation section of landslide displacement series can be gotten.

The flow charts of evolutionary neural network are as Figures 1 and 2.

3 APPLICATION OF NEW INTELLIGENT METHOD IN REAL ENGINEERING EXAMPLE

To verify the above algorithm, the displacement time series of Xintan landslide (Huang 1999) is used.

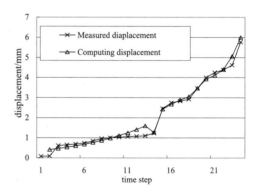

Figure 3. Measured displacement and computing displacement of Xintan landslide.

The displacement time series on key measured point of A_3 is showed as in follow Table 1.

The measured displacements can also be showed in Figure 3. From the Figure, we can see that, the trend of displacement is a kind of concave exponential curve. To model it, the GM (1, 1) model is a suitable one. Also, to test the forecasting capability of our algorithm, the data at step 23 and 24 cannot be used.

According to the model process of GM (1, 1), the GM (1, 1) model of trend section can be gotten as follows.

$$\hat{x}^{(1)}(k+1) = \left(x^{(0)}(1) - \frac{0.379826}{-0.122959}\right) e^{0.122959k}$$
$$+ \frac{0.379826}{-0.122959} \qquad (2)$$

From above model, the trend series of landslide displacement can be gotten. And also the deviation section series can be gotten. The two series are all showed in Table 1. By some pre-disposal to deviation section series, the samples of ENN can be gotten, which is also showed in Table 1.

After computing, we can get the follow results. The number of input node is 12. The number of hiding node is 14. The parameters in MBP algorithm are $\eta = 0.226$ and $\alpha = 0.948$.

By the above results, the computing displacements can be gotten as in Table 1 and Figure 3.

From the above results, the follow conclusions can be drown. The new intelligent algorithm proposed in this paper can reveal the essential rule of landslide displacements. It has not only the good approximated capability, but also the good forecasting capability, and is a very good method for modeling the landslide displacement.

4 CONCLUSIONS

The new intelligent algorithm proposed in this paper can solve the problem of landslide displacement forecasting very well. In this method, based on the principles of displacement decomposition, the trend of displacement time series is extracted by Grey System and the deviation of Grey System is approximated by the new ENN. So, the different method is used to model the different section of displacement. And the whole forecasting capability is very well. At last, one real engineering example is used to verify this new algorithm, and the results are very well.

REFERENCES

Feng, X.T. 2000. *Guide to intelligent rock mechanics*. Beijing: Science Press.

Gao, W. & Zheng, Y.R. 2000. Study on Some Forecasting Methods in Geo-technical Engineering. In *Proc., 6th Conf. of Chinese Rock Mechanics and Rock Engineering Society*: 90–93. Beijing: Science Press.

Gao, W. 2002. Study on neural network method for displacement forecasting in geo-technical engineering. *Geotechnical Engineer* 14(1): 8–12.

Gao, W. & Zheng, Y.R. 2003. An New Evolutionary Back Analysis Method in Geo-technical Engineering. *J. Chinese Rock Mechanics and Rock Engineering* 22(2): 192–196.

Huang, Z.Q. 1999. *Study on non-linear mechanism of slope evolution and forecasting of landslide*. Beijing.

Liu, H.W. & Fan, J.W. 1992. Forecasting and analysis of landslide displacement based on Grey System theory. *J. Chengdu Science and Technology Univ.* (2): 57–64.

Liu, S.F., Guo, T.B., Dang, Y.G., et al 1999. *Grey system theory and its applications*. Beijing: Science Press.

Shi, Y.S. & Xu, D.J. 1995. Application of time series analysis method in slope displacement forecasting. *Rock and Soil Mechanics* 16(4): 1–7.

A new approach to *in Situ* characterization of rock slope discontinuities: The "High-Pulse Poroelasticity Protocol" (HPPP)

Y. Guglielmi, F. Cappa, S. Gaffet & T. Monfret
Geosciences Azur, Sophia Antipolis, France

J. Virieux
LGIT, Grenoble Cedex 9, France

J. Rutqvist & C.F. Tsang
LBNL-ESD, Berkeley, CA, USA

ABSTRACT: The High-Pulse Poroelasticity Protocol is an alternative approach for *in situ* rock-slope-properties estimation. It relies on an innovative probe that allows simultaneous pressure and deformation measurements in boreholes. The method consists of short-duration hydromechanical pulse tests to estimate local hydraulic and mechanical properties of fractures (normal, shear stiffnesses and permeability). Then, a long-term injection induces a large slope deformation, measured at the injection point with the HPPP probe and in the near field with tiltmeters. Fully coupled, hydromechanical, numerical elastic models are then used to match all pressure, deformation, and tilt measurements by adjusting discontinuity properties. Applied to a rock slope of fractured limestone, the Coaraze slope in southern France, the method evidenced a hyperbolic relationship between hydraulic apertures and stiffness of fractures, and enabled an estimation of fracture compressive and shear strengths. Compared to other approaches based on joint roughness analyses, this approach seems less subjective and more accurate.

1 INTRODUCTION

Complex interactions among numerous pre-existing discontinuities primarily determine the potential areas of instability in fractured rock slopes. However, the stability analysis and monitoring of any rock structure faces the major problem of determining both the properties and constitutive stress-displacement models characterizing the discontinuities. Several criteria have been proposed to identify the strength of a natural rock joint (Patton, 1966, Ladanyi & Archambault 1970, Barton & Choubey 1977, Amadei et al., 1998, Saeb & Amadei 1992, Plesha, 1987). In practice, Barton's model is assumed (ISRM, 1978), and in general used based on parameters describing the joint morphology called the JRC (Joint Roughness Coefficient) and the JCS (Joint Compressive Strength). These aparmeters are measurable on joint laboratory samples and *in situ* through profiling methods, the Schmidt hammer method (Aydin & Basu 2005) and visual estimation (Laubscher, 1990). They are then related to joint strength, stiffness, and aperture through empirical relations (Bandis et al. 1983), and introduced into stability models using the Barton-Bandis or the derived Mohr-Coulomb parameters of the jointed rock slope.

This approach is an efficient method for a first estimation of jointed distinct elements and continuum numerical analyses, to assist in the design of tunnel and cavern reinforcement and support. However, applying this method to rock-slope stability analyses is less reliable, because the state of stresses close to the surface is highly heterogeneous and with low values, making the estimation of Mohr-Coulomb parameters (c, cohesion, and ϕ, friction angle) from the derivation of the non-linear Barton's equation less than precise. Moreover, conversion of JRC numbers into the ISRM roughness descriptions is subjective and unambiguous only for some roughness profiles (Barton & Bandis 1990). Measuring joint roughness also requires separating the discontinuity surfaces, which causes cohesion break, damage, and mismatching of the two surfaces, all of which can alter any estimation of joint properties.

Moreover, the mechanical and hydraulic properties of joints are estimated independently. Coupled hydromechanical effects are seldom integrated into slope stability analyses, even though water is often mentioned as a triggering mechanism for failure in rock with well-developed preexisting fractures networks (Erisman & Abele, 2000). The mechanisms operating in the discontinuity behavior are complex, and involve fully coupled hydromechanical effects (Rutqvist & Stephansson, 2003). Compressive effective normal stresses ($\sigma'_n = \sigma_n - P_f$) press opposing discontinuity walls together, resisting sliding motion along the discontinuity surface possibly induced by shear stresses (τ) acting parallel to the discontinuity plane. A reduction in the effective normal stress state leads to the normal opening of discontinuities, inducing a reduction of the inherent shear strength of these discontinuities. When the slope discontinuities are deformed, their hydraulic properties are modified, mainly through dilatancy and the crushing of discontinuities asperities (Gentier et al. 2000). As a consequence, there is a close relationship between the way mechanical and hydrauli properties evolve within discontinuities. This relationship has been studied in the laboratory (Li et al. 2007, Gentier et al. 1997, Olsson & Brown 1993 for examples) but hardly ever in the field.

We present the first results from an innovative geophysical method for *in situ* quantification of hydraulic and mechanical properties within fractures, through coupled pressure/deformation measurements in boreholes combined with surface tiltmeter measurements. This method, called the High-Pulse Poroelasticity Protocol (HPPP), relies on high-accuracy hydromechanical pulse injections analyses, which enable in situ estimation of joint properties. This method is then applied to characterize the stability of a fractured slope. Specifically, we show (1) the principles and the sensitivity of the HPPP test, (2) an example of an *in situ* estimation of joint properties in a natural slope, and (3) the results in terms of joint strength estimation.

To evaluate the usefulness of the HPPP method, it was applied to the Coaraze site on the lower 15 m section of a 40°–60° dipping slope comprised of a thick sequence of fractured limestone (Figure 1a). The slope is cut by 12 parallel bedding - planes, with a 40° trend dipping 45° SE, and two sets of approximately orthogonal, near-vertical faults, with 50°/70° trends dipping 70° to 90° NW and 120°/140° trends dipping 75° to 90° NE (Figure 1b). The mechanical properties of the rock matrix, previously determined from laboratory testing, indicated Young's modulus values ranging from 44.4 to 70 GPa, Poisson ratios of 0.29 to 0.34, and intact-rock permeability of 10^{-17} m². Nine corings were performed at different locations in the slope to investigate local fracture properties. There is a broad range of joint surface JRC values (from 2 to 20) (Figure 1c). In general, fault surfaces display higher average JRC values than bedding planes (respectively 16 and 8). Nevertheless, a large JRC variation occurs on the same fault plane parallel to the direction of the tectonic

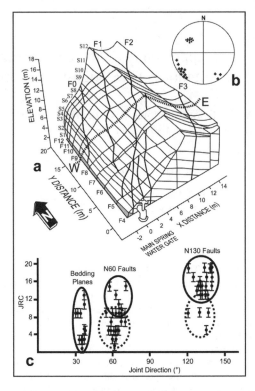

Figure 1. (a) Three-dimensional view of Coaraze fractured slope; (b) Pole plots showing brittle faults and bedding-plane orientations (lower hemisphere); (c) JRC values of slope's fractures.

ridges (low values in dashed circle, Figure 1c) and perpendicular to that direction (high values in continuous circle in Figure 1c). Joint' surfaces are not weathered.

2 THE HIGH PULSE POROELASTICITY PROTOCOL

The "High-Pulse Poroelasticity Protocol" (HPPP) is a new in situ approach developed for a very large broadband geophysical monitoring of rock HM deformations within boreholes. In this protocol, the rock is subjected to a controlled source corresponding to a fast (i.e. few seconds) hydraulic pressure pulse localized within a short injection chamber (0.4 to 3 m), isolated between two inflatable packers in a borehole. In the chamber, measurements conducted with fiber-optic sensors allow simultaneous monitoring of changes (high frequency [120 Hz] and high accuracy) in fluid pressure ($\pm$ 1 kPa) and displacement normal to the walls of the tested joint ($\pm 10^{-7}$ m) (Cappa et al.,

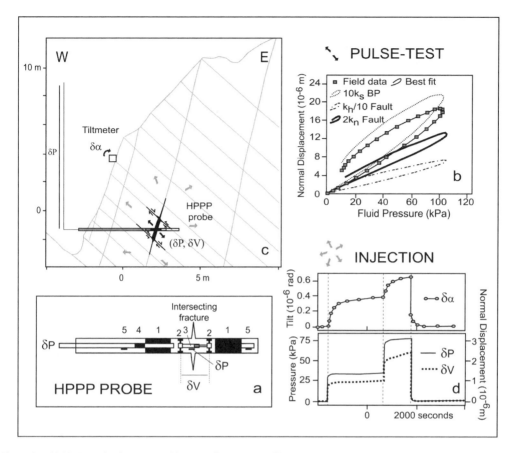

Figure 2. (a) HPPP probe (1: upper and lower packers, 2: extending anchors, 3: displacement sensor, 4: valve, 5: pressure sensors); (b) typical pressure-deformation loop-shaped HPPP curve; (c) in situ installation of HPPP protocol; (d) pressure-deformation-tilt responses to a two-pressure-steps injection experiment.

2006). The displacement sensor is fixed to the borehole walls by two extending anchors located on both sides of the tested discontinuity (Figure 2a).

To conduct the HPPP tests, we increased the fluid pressure in the injection chamber to the required pressure-pulse magnitude using a volumetric water pump, allowing the pressure to increase from 10 to 120 kPa (chosen to be lower than the ambient state of stress on the joint to prevent hydraulic fracturing). The injection chamber is connected to a valve leading to an upstream volume used to perform a pressure pulse. The fluid pressure is first increased upstream of the valve. Thereafter, a pulse is initiated when the valve is opened to allow water to enter the injection chamber. The pressure first increases in the packed-off section, and then it begins to decrease as a result of fluid flow into the fractures and rock.

The transient poroelastic response of the fractures when the pressure pulse is applied in the injection chamber was extensively analyzed through numerical modeling (Cappa et al., 2006). In figure 2b-c, we show an example of a test conducted at the Coaraze slope on a 80° dipping fault. The resulting normal-displacement-versus-pressure curve shows a characteristic loop behavior, in which the paths for pressure increase and decrease are different. The initial rising portion of the curve is highly dependent on the tested fracture's permeability, normal stiffness, and intact rock stiffness close to the fracture walls. For example, a lowering by a factor of 10 in permeability or an increase by a factor of 2 in normal stiffness respectively induce a 50% and 25% lowering of the curve slope (Figure 2b). The falling portion of the curve is influenced by mechanical processes within a larger portion of the surrounding fractures rock. In this example, shear along bedding planes and, as a result, bedding-plane shear stiffness strongly affect the results, an increase by a factor of 10 inducing a

20 to 45% variation in this part of the curve (Figure 2b). Then, if the fracture network geometry and the state of stress are known close to the test, it is possible to estimate the hydraulic permeability (k_h) and the normal stiffness (k_n) of the tested fracture, as well as those of adjacent bedding planes.

The application of HPPP to rock slope stability characterization includes two successive steps:

1. The HPPP tests are conducted to estimate the local poroelastic properties of joints. This is a short-duration test (about 15 seconds long) meant to induce a local deformation of the tested joint and to estimate its local properties. In the example, nine tests were conducted in small diameters (Ø = 0.07 m) boreholes, as well as in five faults and four bedding-planes.
2. An injection test is conducted in one borehole using the HPPP probe while simultaneously monitoring the slope surface tilt (Figure 2c). This is a long-duration test (about 2000 seconds long) meant to induce an extended deformation in the slope and to test slope behavior. In the example, the injection was performed at 5 m depth in the slope, in two increasing pressure steps to produce a clear signal both at the injection and at the near-field tiltmeter points (Figure 2c).

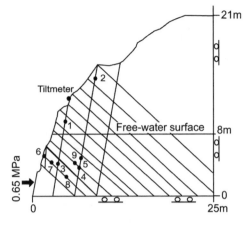

Figure 3. Numerical model of the Coaraze slope (points 1 to 9 are HPPP tests points).

3 RESULTS

Field data were analyzed with different coupled hydromechanical (HM) codes (ROCMAS, FLAC3D, and UDEC). Considering the discontinuities and the rock matrix to be linearly elastic, we developed a 2D model that represented the topography and joint network geometry of the EW cross section (Figure 3). A 25 m × 21 m model was chosen to minimize boundary effects. Stress concentration at the valley base, calculated from large-scale models, was applied to the bottom 2 meters of the topographical left boundary. The remainder of the topographical surface was free to move. Fixed X and fixed Y displacement conditions were imposed on the right and basal boundaries, respectively. All model hydraulic boundaries were impervious. HPPP tests were analyzed one by one. The model was loaded by imposing the time-dependent pressure pulse measured in the injection chamber. Hydraulic and mechanical properties were estimated after matching calculated and measured pressure and normal displacement curves at the point. Second, the injection test was analyzed. The model was loaded by imposing the time-dependent pressure curve measured at Injection point 4 (Figure 3). Properties from all the local tests were introduced as input values within the model and adjusted until both local responses at points 1 to 9 and tilt responses at the slope surface matched the experimental responses.

Close agreement (5 to 25% discrepancy depending on the points) for both slope surface and internal measurements was obtained when a large variability in slope-element properties was introduced into the models. Specifically, we had to define a strong contrast between faults and bedding-plane normal stiffnesses ($k_{nfaults} = 10^{-3} \times k_{nbedding-planes}$) and hydraulic apertures ($b_{hfaults} = 10 \times b_{hbedding-planes}$), There was a highly nonlinear correlation between normal stiffness and the hydraulic aperture b_h of fractures, which was fitted to a hyperbolic function (Figure 4a). This correlation was the result of both the significant property contrast between the faults and bedding planes, and the high variability of properties within a given fracture type. Since there are several parameters that affect the hydraulic aperture of a given fracture (mechanical aperture, roughness, and tortuosity), it is hard to formulate a simple explanation for the high b_h of such a fracture. In the very special case of shallow discontinuities in slopes, it was commonly observed that discontinuities almost parallel to the slope direction and dip (roughly the case of vertical faults at Coaraze) were in general widely opened. We can assume that their tortuosity was low and their roughness small compared to their mechanical aperture. Elsewhere, a few centimetric contacts were observed between the two walls of those discontinuities, as a result of the progressive failure of those contacts due to slope decompression. The small number of contacts, and hence relatively small contact area could explain the relatively low stiffness of discontinuities, which would thus have a large hydraulic aperture.

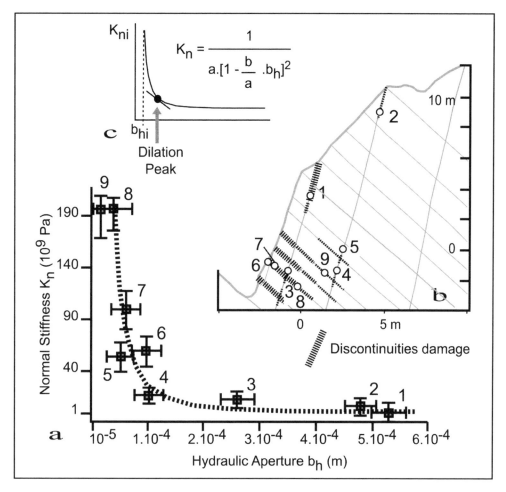

Figure 4. (a) correlation between normal stiffness and hydraulic aperture of discontinuities; (b) localization of damage in the slope; (c) proposed joint model.

Such a mechanism could explain the property variation from Point 3 to 1 on Fault F1, a factor-of-10 lowering of the normal stiffness and a factor-of-2 increase of b_h, respectively. This mechanism could also explain the property variation between Points 4, 5 to 2, a factor-of −2.4 lowering of normal stiffness and a factor-of-10 increase in b_h (Figure 4a). Points 1 and 2 are located on shallow segments of the faults and are more disturbed than points 3, 4, and 5, which are located deeper (Figure 4b). On the other hand, bedding planes slightly affected by decompression effects and submitted to higher compressive stresses (mainly because of their inward dipping orientation) are more closed than faults. In this case, with more contacts between the two walls of the discontinuity, bedding plane would display higher normal stiffnesses and lower b_h. Nevertheless, there is a high variability in bedding-plane properties corresponding to bedding-plane damage characterized by a tenfold increase of the b_h value and a threefold decrease of normal stiffness, from points 8 and 9, located at 2 m and 5 m depths, respectively, to points 6 and 7, located at 1 and 0.5 m depths, respectively (Figure 4a and b).

The shear stiffness of bedding planes and faults was estimated to be 0.1 to 0.02 times lower than normal stiffnesses.

4 DISCUSSION AND CONCLUSION

The HPPP method, based on combining surface tilt with localized internal coupled pressure/deformation measurements, can help to characterize the influence of discontinuities properties on slope destabilization. For a given slope geometry, the HPPP test enables an *in situ* estimation of the hydraulic and mechanical properties of discontinuities at the single fracture scale, as well as the coupling relationship between those key properties. First, an empirical hyperbolic relationship is proposed between normal stiffness and the hydraulic aperture of joints. Second, a rough estimation of fracture normal/shear stiffness ratio of 1 to 5 is given. This agrees well with laboratory studies in which such a relation is obtained from fracture shear tests performed under a state of stresses sufficiently low to favor dilation when sliding along discontinuity planes occurs.

The established permeability-stiffness relationship reduces the number of unknown properties to be introduced in stability numerical models. Thus, those two properties are no longer unknown, but can be used as input data to calculate shear processes happening in natural slopes. Further, the compressive and shear strengths of discontinuities can be carried out as follows. First, the initial normal stiffness K_{ni} and initial hydraulic aperture b_{hi} can be estimated using the hyperbolic normal stiffness-hydraulic aperture relashionship. If we refer to laboratory experiments described in the literature (Bandis et al., 1983), a good fitting of K_{ni} can be obtained, mainly with the uniaxial compressive strength σ_c and bh, and secondly with joint surface roughness. This can be illustrated from equation (1):

$$K_{ni} = \sigma_c / b_h \qquad (1)$$

Second, an estimation of shear strength can be made by considering that the fracture dilation peak is reached at the maximum inflection point of the normal stiffness-hydraulic aperture curve (Figure 4c). The shear stiffness estimated from the measured normal/shear stiffness ratio is then related to the shear strength in Equation (2):

$$K_s = \sigma_n \tan \phi \qquad (2)$$

where σ_n = normal stress and ϕ = friction angle.

In our example, the estimated values for JCS and friction angle are 10^8 Pa and $90°$ (for an average $\sigma_n = 4 \ 10^4$ Pa), respectively. This is in good accordance with values extrapolated with the Barton-Bandis method: JCS = 10^8 Pa, JRC = 19, and residual friction angle = $40°$. The HPPP approach is then particularly effective for it relies on a reduced number of parameters (K_n, K_s and b_h) that can be measured in-situ with the HPPP probe, making the estimation of in situ fracture-strength parameters less subjective. Moreover, further development of the probe will allow direct estimation of K_s through in situ measurement of both normal and tangential displacements of fracture walls.

REFERENCES

Amadei, B., Wobowo, J., Sture, S. & Price, R.H. 1998. Applicability of existing models to predict the behavior of replicas of natural fractures of welded tuff under different boundary conditions. Geotech. Geo. Eng. 16, pp. 79–128.

Aydin, A. & Basu, A. 2005. The Schmidt hammer in rock material characterization. Engineering Geology 81, pp. 1–14.

Bandis, S.C., Lumsden, A.C. & Barton, N.R. 1983. Fundamentals of Rock Joints Deformation. Int. J; Rock Mech. Min. Sci. & Geomech. Abstr. 20, No. 6, pp. 249–268.

Barton, N. & Choubey, V. 1977. Shear strength of rock joints in theory and practice. Int. J. Rock Mech. Sci. & Geomech. Abstr. 10, pp. 1–54.

Barton, N. & Bandis, S. 1990. Review of predictive capabilities of JRC-JCS model in engineering practice. Rock Joints. Barton & Stephansson (eds). Balkema, Rotterdam. ISBN 90 6191 109 5. pp. 603–610.

Cappa F., Guglielmi Y., Rutqvist J., Tsang C.F. & Thoraval A. 2006. Hydromechanical modeling of pulse tests that measure both fluid pressure and fracture-normal displacement at Coaraze Laboratory Site, France. Int. J. Rock. Mech. Min. Sci. 43, pp. 1062–1082.

Erisman T. & Abele G. 2000. Dynamics of rockslides and rockfalls. Springer Verlag, Berlin, 316 p.

Gentier, S.S., Lamontagne, E., Archambault, G. & Riss, J. 1997. Anisotropy of flow in fracture undergoing shear and its relashionship to the direction of shearing and injection pressure. Int. J. Rock Mech. Min. Sci. Geomech. Abstr. 34.

Gentier, S.S., Riss, J., Archambault, G., Flamand, R. & Hopkins, D.L. 2000. Influence of fracture geometry on sheared behaviour. Int. J. Rock Mech. Min. Sci. 37, pp. 161–174.

ISRM 1978. Suggested methods for the quantitative description of discontinuities in rock masses. Int. J. Rock Mech. Min. Sci. & Geomech. Abstr. 15, pp. 319–368.

Landanyi, B. & Archambault, G. 1970. Simulation of shear behaviour of a jointed rock mass. Pro. Of the 11th Symp. On Rock Mech. (AIME), pp. 105–125.

Laubscher, D.H. 1990. A geomechanics classification system for rating of rock mass in mine design. J. South African Inst. Of Mining and Metallurgy 90, No. 10, pp. 257–273.

Li B., Jiang, Y., Koyama, T., Jing, L. & Tanabashi, Y. 2007. Experimental study of the hydro-mechanical behavior of rock joints using a parallel-plate model containing contact areas and artificial fractures. Int. J. of Rock Mech. & Min. Sci., in press.

Olsson, W.A. & Brown, S.R. 1993. Hydromechanical response of a fracture undergoing compression and shear. Int. J. Rock Mech. Min. Sci. Geomech. Abstr. 30, pp. 845–51.

Patton, F.D. 1966b. Multiple Modes of Shear of Failure in Rock. Proc. 1st Cong. Int. Soc. Rock Mech., Lisbon, pp. 509–513.

Plesha, M.E. 1987. Constitutive models for rock discontinuities with dilatancy and surface degradation. Int. J. Numer. Anal. Meth. Geomech. 11, pp. 345–362.

Rutqvist J. & Stephansson O. 2003. The role of hydromechanical coupling in fractured rock engineering. Hydrogeology Journal 11, pp. 7–40.

Saeb, S. & Amadei, B. 1992. Modelling rock joints under shear and normal loading. Int. J. Rock Mech. Min. Sci. Geomech. Abstr. 29, pp. 267–278.

Fuzzy prediction and analysis of landslides

Yong He, Bo Liu, Wen-juan Liu, Fu-qiang Liu & Yong-jian Luan
Qingdao Agricultural University, Shandong Qingdao, China

ABSTRACT: There are internal and external reasons for the occurrence of landslip. Through the analysis on a large number of investigation materials of landslip, the growth inducements of landslip are found out. By using theories in blur maths, we conduct order arrangement of these inducements in accordance their significance, put forward blur judgment rule of landslip estimation and present applicable examples.

1 INTRODUCTION

Landslip, a serious geological disaster, with serious harmfulness, greatly threatens human beings' existence. The inducements of landslip which are complex, include artificial elements and natural elements. A great deal of facts on landslip shows that landslip is often the result of mutual effect of many elements. There is a high degree of obscurity of the relationship between different elements and the action mechanism, so it is hard to present a quantitative mathematic model with common methods. It is one of blur maths' (Kamal et al. 2001, Duzgoren et al. 2002, Furuya et al. 1999, Lee et al. 2002, Mauritsch et al. 2000, Raetzo et al. 2002) strong points to research and deal with the nonclearness of differences of objective things in their transitions. If blur theories can be applied to settle the forecast problem of landslip possibility, it will be helpful for the forecast of landslip accidents and reducing the harm of landslip disasters to the lowest. The writer, together with the leader of the scientific research team, has studied various geological disasters (including landslip, ground sedimentation and so on) for many years, and in the research process, we have accumulated a mass of materials on landslip and the in situ motoring data of landslip. On the base of comprehensive analysis and research, the writer and the scientific research team has summed up many associated elements (Duzgoren et al. 2002, Furuya et al. 1999), using theories in blur maths and making basic principles of fuzzy evaluation method as the guidance, we scientifically set down the subjection degree of each variable in the fuzzy set of landslip and confirm the relative weight of appraisal index elements by use of analysis of hierarchy process. Through the analysis of every associated index of landslip, we find out their subjection degree in the fuzzy set. By use of estimate model, the probability of the landslip's occurrence is determined. In the following part, the writer will make a brief introduction of fuzzy forecast problems of landslip raised by the writer and the scientific team.

2 BASIC PRINCIPLE OF FUZZY FORECAST OF LANDSLIP

2.1 Establishment of landslip forecast index system

At present, the chain of causation model is commonly used to analyze the reasons of various kinds of accidents. The writer and the scientific research team, making abundant materials on landslip and the actual in situ motoring materials as the basis, has formed the landslip forecast index element system which can be seen in Table 1.

2.2 The determination of index weight

The determination of index weight is in accordance with the AHP method (Analytic Hierarchy Process Method) put forward by American operational research master A.L. Saaty in the 1970s.

2.2.1 The determination of index, element and sub-element weight

The lowest layer of the writer's analysis of hierarchy process on landslip forecast is the element collection of subsidiary appraisal index, the middle layer is the element collection of leading appraisal index and the highest layer, also called the target layer, is the stabilization condition of the landslip body. Therefore, the hierarchical structure model is established, which can be seen in Table 1. Then, establish the pairwise comparison matrix (that is judgment matrix)

Table 1. Appraisal index system of landslip forecast.

Target Layer	The middle layer		The bottom layer	Appraisal matrix (poll result of the appraisal team)						
				Weight	Stable	Basically stable	Sub-stable	Critical stage	Unstable	Extremely unstable
	Main appraisal elements and code	Weight	Subsidary appraisal elements and code		V1= 0.000	V2= 0.20	V3= 0.40	V4= 0.60	V5= 0.80	V6= 1.00
Stabilization condition of landslip body	Internal elements A	0.41	Composing of landslip body materials A1	0.44	0	1	6	3	1	0
			Thickness of landslip body A2	0.19	1	2	6	2	0	0
			Form of the leading slipping surface A3	0.37	0	1	4	4	1	1
	External environmental elements B	0.59	Coverage condition of the surface of landslip body B1	0.06	0	2	4	4	1	0
			Free surface condition of landslip body B2	0.20	0	1	5	4	1	0
			Change condition of water table B3	0.08	0	1	4	4	1	1
			Erosion condition of surface water B4	0.13	0	1	3	4	2	1
			Drainage condition of the surface of landslip body B5	0.07	1	2	3	4	1	0
			Fragmentation degree of the surface of landslip body B6	0.05	1	1	5	2	2	0
			Loading condition of the surface of landslip body B7	0.06	1	1	5	3	1	0
			Precipitation condition B8	0.20	0	1	3	4	2	1
			Vibration-bearing condition of the surface of landslip body B9	0.15	1	7	3	0	0	0

according to Saaty's nine-rank rating scale method. See Table 1 to acquire the weight of index, element and sub-element.

2.2.2 Calculation of each level's index weight

First of all, the level simple sequence is conducted. According to judgment matrix, the weight of the simple sequence is calculated. That is to say, to the elements of the last level, calculate this level's weight of all the elements which has association with them. According to the principle of analysis of hierarchy process, through mathematic calculation, the eigenvector which is the result of judgment matrix's corresponding with the maximum eigenvalue λ_{max}, is gained and each element of eigenvector (that is weight value) is the simple sequence result. By use of MATLAB software, the writer calculates the eigenvector and gains the calculation result of level simple sequence of corresponding elements which can be seen in Table 1.

Then, the coincidence test of single taxis is carried on. The main advantage of AHP method is that it can quantify decision-maker's fixed thoughts. Due to the complexity of the appraised objects and experts' diversity and unilateralism on cognition, the nine-rank rating scale can't guarantee the complete coincidence of each judgment matrix. Therefore, the coincidence test must be used to check if there are contradictions between each index's weight and the check step is as follows:

1. Calculate the maximum latent root of judgment matrix λ_{max}

$$\lambda_{max} = \left\{ \sum_{i=1}^{n} [(AW)_i / W_i] \right\} / n \quad (1)$$

In formula (1), λ_{max} is the maximum latent root; n is the linage of the judgment matrix which is also the number of index in the arrangement subsystem; A is the judgment matrix; W is the eigenvector of the judgment matrix; and $(AW)_i$ is the number i element of vector AW gained by the multiplication of the judgment matrix A and eigenvector W.

2. Calculate the coincidence index C_I

$$C_I = (\lambda_{max} - n)/(n - 1) \quad (2)$$

3. Calculate the random coincidence ratio C_R

$$C_R = C_I / R_I \quad (3)$$

R_I is the average random coincidence index. See Table 2.

Table 2. Average random coincidence index.

n	2	3	4	5	6	7
R_I	0.00	0.58	0.90	1.12	1.24	1.32
n	8	9	10	11	12	
R_I	1.41	1.45	1.49	1.52	1.54	

When $C_R \leq 0.10$, the judgment matrix can have satisfying coincidence, otherwise, it should be adjusted.

According to the above calculation methods, carry on the calculation of the coincidence index of judgment matrix in different levels of target—index, index—element and element—sub-element. See the calculation result in Table 1. None of the C_R is more than 0.10 and all the judgment matrix meet the coincidence requirement.

Next, the level total sequence is conducted. Taking advantage of all the results of the level simple sequence in the same level and all the weight of elements in the last level, calculate all the elements' weight value in this level aiming at the total target.

Finally, the coincidence test of the level total sequence is conducted. Like the level simple sequence, the level total sequence also needs the coincidence test. The test result all accord with $C_R \leq 0.10$ and also meets the coincidence requirement.

2.3 Determination of the appraisal element base

The writer chooses six grades to form the appraisal collection V of landslip body's stabilization condition.
V = {V1 (stable), V2 (basically stable), V3 (sub-stable), V4 (critical state), V5 (unstable), V6 (extremely unstable)}.

2.4 Establishment of appraisal matrix

Many experts on landslip and local residents who have experienced the landslip disasters form the appraisal team. After the comprehensive investigation (including drilling materials, experiment materials and motoring materials) towards the landslip bodies that are to be appraised and analysis, poll on conditions of every second-grade index elements which are divided into six appraisal collections are hierarchically conducted. See Table 1 to find out the appraisal result. Form the judgment matrix according to appraisal results. Every element in the judgment matrix is the result gained by the ballot possessed by this element to divide the total ballot.

2.5 Establishment of landslip forecast model

The landslip forecast model established by the writer and the scientific team is as follows:

$$P = \sum_{j=1}^{n} [a_j b_j] \quad (4)$$

In formula (4)

$$a_j = x_j / \sum_{j=1}^{n} x_j \quad (5)$$

In formula (4), when element j is not contained in the objective elements, $b_j = 0$; when element j is contained in the objective elements, $b_j = \lambda_j$. In formula (4) and (1), a_j is the probability coefficient of the number j element; n is the number of the main determined influential elements; x_j is the number of accidents that has relationship with element j; λ_j is the degree of subjection of element j determined by experts in accordance with fuzzy appraisal method. From formula (4), we can find out that the more the accidents caused by certain accident element j, the more the probability coefficient a_j and the element j accounts for a more weight comparing with other elements. At this time, if the element j is contained in the objective conditions of forecast objects, the probability of the landslip occurrence P is higher, which can be seen from the model (formula (1)). The more the objective elements that the landslip causes accidents, the more the items which is not zero and the higher the summation value. Therefore, the possibility of its occurrence is higher.

3 APPLICATION EXAMPLES OF LANDSLIP MODEL

Table 1 is the fuzzy appraisal that the writer and scientific team has made towards the landslip body in Cuobu Mountain. The appraisal team consists of 11 experts. After thorough investigations and analysis of landslip scene of Cuobu Mountain and whole experiments and data of supervision and measurement, these 11 experts make a classification vote for each subsidiary appraisal index in accordance with 6 evaluation collections. After the analysis of fuzzy forecast model, they make a prewarning of landslip disaster of Cuobu Mountain and the landslip occurs sooner. According to Table 1, the calculating process of fuzzy appraisal of landslip body in Cuobu Mountain is.

3.1 Relative calculation of main appraisal element A

The area of A which is composed by three subsidiary appraisal elements is: A = {A1 (Composing of landslip body materials), A2 (Thickness of landslip body), A3 (Form of the leading slipping surface)}. In appraisal systems, the area which is composed by 6 appraisal classifications is V: V = {V1 (stable), V2 (basically stable), V3 (sub-stable), V4 (critical state), V5 (unstable), V6 (extremely unstable)}. According to the subjection degree which is gained by 11 experts' votes, (for example, in Table 1, the appraisal of subsidiary appraisal element A1 (Composing of landslip body materials) is that the vote for stable is 0, the proportion is 0/11 and the subjection degree is 0.00; the vote for basically stable is 1, the proportion is 1/11 and the subjection degree is 0.09; the votes for sub-stable are 6, the proportion is 6/11 and the subjection degree is 0.55; the votes for critical state are 3, the proportion is 3/11 and the subjection degree is 0.27; votes for unstable are 1, the proportion is 1/11 and the subjection degree is 0.09; votes for extremely unstable are 0, the proportion is 0/11 and the subjection degree is 0.00) we establish the appraisal matrix R_A which is composed by three subsidiary appraisal elements as:

$$R_A = \begin{bmatrix} 0.00 & 0.09 & 0.55 & 0.27 & 0.09 & 0.00 \\ 0.09 & 0.18 & 0.55 & 0.18 & 0.00 & 0.00 \\ 0.00 & 0.09 & 0.36 & 0.36 & 0.09 & 0.10 \end{bmatrix}$$

Three subsidiary appraisal elements weight fuzzy vectors A_A which are determined by integrating with existing landslip data and utilizing the analysis of hierarchy process is: $A_A = (0.44, 0.19, 0.37)$. Integrated appraisal B_A of main appraisal element A is: $B_A = A_A R_A = (0.0171, 0.1071, 0.4797, 0.2862, 0.0729, 0.0370)$. The subjection degree of main appraisal element A is calculated by λ_A. Endowing value K = (0.00, 0.20, 0.40, 0.60, 0.80, 1.00) respectively into every appraisal value of main appraisal area A, the subjection degree λ_A of main appraisal element A, $\lambda_A = KB_A^T = 0.4803$.

3.2 Relative calculation of main appraisal element B

The area B which is composed by nine subsidiary appraisal elements is B = {B1 (Coverage condition of the surface of landslip body), B2 (Free surface condition of landslip body), B3 (Change condition of water table), B4 (Erosion condition of surface water), B5 (Drainage condition of the surface of landslip body), B6 (Fragmentation degree of the surface of landslip body), B7 (Loading condition of the surface of landslip body), B8 (Precipitation condition), B9 (Vibration-bearing condition of the surface of landslip body)}. The appraisal area V which is composed by six appraisal classifications in appraisal collection is: V = {V1 (stable), V2 (basically stable), V3 (sub-stable), V4 (critical state), V5 (unstable), V6 (extremely unstable)}. According to the subjection degree which is gained by 11 experts' votes, we establish the appraisal matrix R_B which is composed by nine subsidiary appraisal elements as:

$$R_B = \begin{bmatrix} 0.00 & 0.18 & 0.36 & 0.36 & 0.10 & 0.00 \\ 0.00 & 0.09 & 0.45 & 0.36 & 0.10 & 0.00 \\ 0.00 & 0.09 & 0.36 & 0.36 & 0.09 & 0.10 \\ 0.00 & 0.09 & 0.27 & 0.36 & 0.18 & 0.10 \\ 0.09 & 0.18 & 0.27 & 0.36 & 0.10 & 0.00 \\ 0.09 & 0.09 & 0.46 & 0.18 & 0.18 & 0.00 \\ 0.09 & 0.09 & 0.46 & 0.27 & 0.09 & 0.00 \\ 0.00 & 0.09 & 0.27 & 0.36 & 0.18 & 0.10 \\ 0.09 & 0.64 & 0.27 & 0.00 & 0.00 & 0.00 \end{bmatrix}$$

Nine subsidiary appraisal elements weight fuzzy vectors A_B which are determined by integrating with existing landslip data and utilizing the analysis of hierarchy process is: A_B = (0.06, 0.20, 0.08, 0.13, 0.07, 0.05, 0.06, 0.20, 0.15). Integrated appraisal B_B of main appraisal element B is: $B_B = A_B R_B$ = (0.0297, 0.1842, 0.3395, 0.2916, 0.1140, 0.0410). The subjection degree of main appraisal element B is calculated by λ_B. Endowing value K = (0.00, 0.20, 0.40, 0.60, 0.80, 1.00) respectively into every appraisal value of main appraisal area B, the subjection degree λ_B of main appraisal element B, $\lambda_B = KB_B^T = 0.4798$.

3.3 Forecast of occurrence probability of landslip

According to a great number of landslip data statistical results, we chose probability coefficient value a_1 (that is a_A) = 0.41; a_2 (that is a_B) = 0.59. According to Landslip Forecast Model (formula (4)) we can get the probability P of landslip occurrence of Cuobu Mountain as

$$P = \sum_{j=1}^{n} [a_j b_j] = a_A \lambda_A + a_B \lambda_B = 0.481276$$

Generally when P is more than 0.25, small omen of landslip will appear, if more than 0.50, the probability of landslip occurrence will be much higher. Due to the probability P of landslip occurrence of landslip body in Cuobu Mountain has reached 0.481276 which is near to dangerous value of 0.50, so scientific research team makes a prewarning that landslip body in Cuobu Mountain will slide. It is proved to be right by later facts.

4 CONCLUSIONS

The key of landslip forecast fuzzy appraisal is the determination of subjection degree which depends on experts' experiences in most degree, so the experts who are chosen must have very abundant spot experiences and know the occurring rules of landslip disaster very well. To settle the problems of landslip forecast by fuzzy mathematical methods is to number landslips which are a fuzzy system by mathematical methods. Landslip forecast model which is established by the author and scientific research team with fuzzy mathematical theory is a good method to solve the problem of landslip forecast, with the characteristics of simple model structure, strong pragmatic, fast calculation and good operation. Due to the writer's level, landslip forecast model which is established by the author and scientific research team with fuzzy mathematical theory and the design of its appraisal index must have many blemishes, the choosing of its elements collection is not very proper and reasonable and some important landslip elements maybe be left out, so we hope that you can complement and perfect it in applications.

ACKNOWLEDGMENTS

The research project discussed in this paper has been funded by the National Natural Science Fund of China (No.83200655) and Construction Science Fund of Shandong Province (No.LJK-20020312). Sincere gratitude is presented to them.

REFERENCES

Kamal M., Al-Subhi & Al-Harbi. 2001. Application of the AHP in project management. *International journal of project management*, 19:19–27.

Duzgoren Aydin N.S., Aydin A. & Malpas J. 2002. Reassessment of chemical weathering indices: case study on pyroclastic rocks of Hong Kong. *Engineering Geology*, 63(1/2):99–119.

Furuya G., Sassa K., Hiura H., et al. 1999. Mechanism of creep movement caused by landslide activity and underground erosion in crystalline schist, Shikoku Island, southwestern Japan. *Engineering Geology*, 53(3/4):311–325.

Lee H.S. & Cho T.F. 2002. Hydraulic Characteristics of Rough Fractures in Linear Flow under Normal and Shear Load. *Rock Mechanics and Rock Engineering*, 35(4):299–318.

Mauritsch H.J., Seiberl W., Arndt R., et al. 2000. Geophysical investigations of large landslides in the Carnic Region of southern Austria. *Engineering Geology*, 56 (3–4):373–388.

Raetzo H., Lateltin O., Bollinger D. et al. 2002. Hazard assessment in Switzerland-codes of practice for mass movements. *Bulletin of Engineering Geology and the Environment*, 61:263–268.

LPC methodology as a tool to create real time cartography of the gravitational hazard: Application in the municipality of Menton (Maritimes Alps, France)

M. Hernandez, T. Lebourg & E. Tric
University of Nice—Sophia Antipolis, Géosciences Azur, UMR 6526, Valbonne, France

M. Hernandez & V. Risser
Cabinet Risser, Saint Laurent du Var, France

ABSTRACT: This paper presents the LPC (Landslide Predictive Cartography) methodology for evaluating slope failure in wide study area. This deterministic methodology is based on the limit equilibrium theory and combined with a dynamic hydrogeological model. It assesses, for each time laps, the slope stability in relation with real or modelled climatic events. This model was described and applied on the Menton municipality (Maritimes Alps, France). This studied area is densely populated in the thalweg zone and have very steep slope. The Menton landscape morphology comes from the erosion of the sedimentary formation. The results obtained indicate that, for a 10 m DEM scale resolution and a 20-year return rainfall modelling, 21.4% of the studied area should be unstable and 70.7% of the landslide occurred have a FS inferior to 1.

1 INTRODUCTION

Emergency management planning requires prediction of the damage associated to the landslide occurrence. In this paper a new deterministic model, using dynamic hydrogeological models, is presented: the LPC (Landslide Predictive Cartography) methodology. In order to assess the efficiency of this methodology, it is applied on the Menton area (Maritimes Alps, France) where a landslide mapping is well defined. The origin of this methodology comes from the study of the Rucu Pichincha volcano near the town of Quito in Ecuador (Risser, 2000). Throughout an operational model, this methodology assesses, for each time laps, the soil stability in relation with climatic events.

2 LPC: A SHALLOW LANDSLIDE SLOPE STABILITY MODEL

2.1 Infinite slope stability model

Data is organized by grids depending on the resolution of the Digital Elevation Model (DEM). The LPC methodology uses the limit equilibrium theory to analyse the stability state (Spencer, 1967) for each pixel (figure 1). The slope stability is based on infinite slope model including several simplifying assumptions (Selby et al. 1993).

The equation (1) of the Factor of Safety (FS) is commonly used for the study of the equilibrium conditions of the soil mass. This equation is based on Mohr—

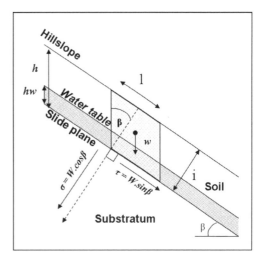

Figure 1. Infinite slope model, where the resisting force holding the block in place is given by W (which the mass * the gravita-tional force) multiplied by the cosines of the slope angle. The driving force is $W\sin\beta$ (inspired by Selby et al. 1993).

Coulomb failure criterion: $\tau_{max} = C' + \sigma' \cdot \tan\varphi'$, where τ_{max} is the maximum shearing stress and σ the normal stress (Costet and Sanglerat, 1981).

$$FS = \frac{C'}{\rho s.g. \sin\beta.\cos\beta} + \frac{(\rho s.h - \rho w.hw(t)) \times \tan\varphi'}{\rho s.h.\tan\beta} \quad (1)$$

where φ' is the effective angle of internal friction (deg), C' is the effective cohesion (KPa), h the vertical soil depth (m) which is calculated with electrical resistivity measurements, $h_w(t)$ the vertical dynamic saturated soil depth (m) depending on the hydrogeological model, ρ_s is the dry soil density (kg.m^{-3}), ρ_w is the density of water (kg.m^{-3}), β is the slope angle (deg), and g is the gravitational acceleration (m.s^{-2}).

A sensitivity study of this equation was realised. The results are similar to those obtained by Borga et al. (2002). The following parameters are quoted in the decreasing order of influence on the safety factor equation: slope angle, effective angle of internal friction, soil depth and groundwater-soil ratio.

In this study the pixel resolution of the DEM is 10 m × 10 m (extrapolation of the DEM 25 × 25 m). The other maps are also organized by grid scale of 10 m. The whole maps are composed by 672 × 530 pixels i.e. 6,720 m × 5,300 m area. Only the Menton municipality zone is well mapped (geological map), hence the surface of the effective stability study zone is 35.6 km^2.

2.2 Hydrogeological model

In order to evaluate the parameter $h_w(t)$, a hydrogeological model plays an important role in the FS calculation. The groundwater response following a rainfall event can be modelled by many hydrologic models like TOPOG (O'Loughlin, 1986), TOPMODEL (Beven, 1997), and DYNWET (Wilson and Gallant, 2000).

The LPC methodology uses its own hydrogeological dynamic model. This model assumes that the shallow subsurface flow has the same behaviour than a perched water table. In case there is a homogenous soil thickness the shallow subsurface flow downslope follows the topographic gradient. But, contrary to the other methods, in case there is a variation in the soil thickness, the subsurface flow follows the substratum gradient. The flow direction is calculated with the D∞ algorithm (Taborton, 1997). The rainfall event could be whether modelled with the Desbordes (1987) methodology whether came from a rain gauge measured data. These two methods provide intensity-duration (I-D) rainfall. The infiltration process is founded on the empirical Horton equations (Horton, 1933). After the infiltration step, the one-dimensional vertical flow in the unsaturated zone is based on the simplified equations used in Fuentes et al. (1992).

Once the accumulation of the water begins at the base of the soil column, the accumulated zone is considered like saturated. The saturated flow is based on the Darcy's equation (Darcy, 1856) and more precisely on the Bernoulli's theorem.

The hydrogeological model depends on the quality of the landuse mapping for the infiltration rate and on the permeability mapping for the water flow in the soil.

In order to have the most critical situation of the slope stability we kept the hydrogeological map with the maximum soil saturation coming from a 20 year-return rainfall modelling with one hour duration (Desbordes, 1987).

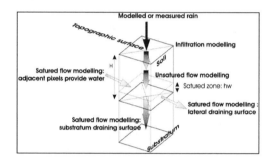

Figure 2. Hydrogeological model.

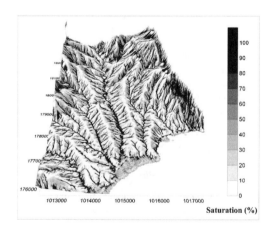

Figure 3. Example application, Menton, (Maritim Alps, France) for a 20-year return rain, 60 min after the beginning of the rainfall. The white zone at the South-East is the Mediterranean Sea. The white zone is Italy.

3 STUDY AREA AND LANDSLIDES DATA

3.1 Localisation of the study area

The Menton municipality is localised in the "Alps Maritimes" (SE, France) touching the Italian border and the Mediterranean Sea. The geomorphology of the studied area is compound by very steep slope. This zone is densely populated in the thalweg zone; the landslide factor is one of the most important risks for this municipality. The landslide data and the geological map are provided by the "C.E.T.E Méditerranée" (In French: "Centre des études techniques de l'équipement", in English: Facility Technical Study Center).

3.2 Geological setting

The field study is localised in the Alps external sedimentary cover units. The figure 4 displays the localisation of the different lithology. The great part of the map shows the importance of the colluviums and eluviums deposit. Theses colluviums and the eluviums are classified according to their geological origin and their thickness. Three levels of thickness corresponding to the vertical soil depth (h) are used: low (0.5 to 1 m), medium (1 to 3 m), and high (more than 3 m). The South-East part of the study area is constituted by Jurassic limestone and dolomite cliffs. In the north part, large outcrops of Cretaceous marl are present. In the central part, Oligocene sandstones and flyschs

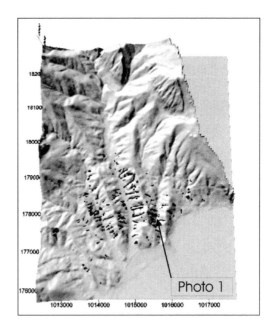

Figure 5. Landslide localisation on the Menton landscape. The black polygons represent the landslides occurred during the period 1999–2000.

show on the surface. The Quaternary formation, constituted of alluvium and colluvium, covers the thalweg and the coast zone.

3.3 Landslides localisation

On the field 191, landslides were counted and mapped. In order to compare the FS map obtained with the LPC methodology and the landslides map, the landslide zones are redrawn keeping only the failure zone and deleting the accumulation zones. The landslides surface corresponds to 1872 pixels i.e. 0.18 km² or 1.34% of the total studied surface. Some of these events caused several deaths. The origin of these landslides can be a natural slope destabilisation or an anthropogenic perturbation (bad management of the embankments). They are mainly localised in the central part of the map, i.e. in the Oligocene formations constituted by flyschs, sandstones, and their colluvium and eluvium.

The photo 1 is an example of the landslide occurrence in the Menton municipality. This destabilisation occurred after a heavy rain estimated as being a 20-year return rainfall. In this example, the steep slope and the anthropogenic action (deforestation and habitations) had support the landslide occurrence.

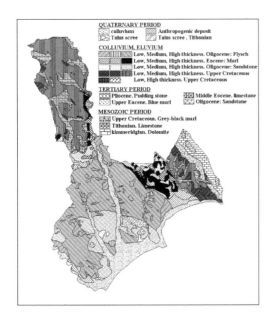

Figure 4. Geological map of the Menton municipality.

Photo 1. Landslide occurred in 2000.

4 RESULTS

Database and field tests were used to obtain the mechanical and hydrogeological parameters. Depending on theses, a "Factor of Safety" map was realised (Figure 6).

According to the FS map, 21.4% of the studied area should be unstable for a 20-year return rainfall (Figure 7).

The realisation of the DEM (1995) is posterior to the landslides database (1999–2000), consequently the DEM shows the altimetric data repartition before the landslides occurrence. In the landslide areas:

- 70.7% of these zones are FS < 1,
- 26.6% are contained between 1 and 1.8,
- 2.7% are Fs > 1.8 (Figure 8).

The FS values superior to 1 can have several different origins: a bad calibration of the model, and/or a landslide mapping problem, and/or unexpected conditions variation (geomorphological, mechanical, hydrogeological, anthropological ... conditions).

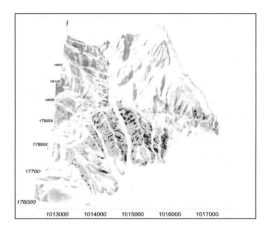

Figure 6. FS map of the Menton municipality.

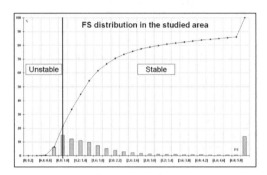

Figure 7. Factor of Safety distribution in the studied area. The line graph represents the cumulated percentage and the bar chart represents the percentage for each class.

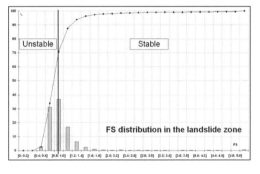

Figure 8. Factor safety distribution in the landslides zones. The line graph represents the cumulated percentage and the bar chart represents the percentage for each class.

In a densely populated zone, in order to have the most efficiency map on a long period of time, a re-evaluation of the several parameters is necessary: (topography, anthropogenic perturbations ...).

5 CONCLUSION

This paper presents the LPC approach for evaluating slope failure in wide study area. This deterministic model was described and applied on the Menton municipality (Maritimes Alps, France). This methodology requires numerous calibrations: mechanical and hydrogeological parameters are necessary. The dynamic hydrogeological modelling is an important aspect of this approach to model the FS map. Nevertheless the database accessibility remains possible and permits to apply this methodology on several types of regions.

The results obtained in this study indicate that, for a 10 m DEM resolution, the LPC methodology is a successful application on the Menton area. Indeed, for a 20-year return rainfall, 70.7% of the landslide occurred have a FS inferior to 1.

A classification of the FS map can gives a hazard map depending on the model. Nevertheless, the expert's evaluation stays an important step to refine on the field and to conclude the hazard maps.

REFERENCES

Beven, K. 1997. Topmodel: A critique. *Hydrological Processes*. 11:1069–1085.

Borga, M., Dalla Fontana, G., Gregoretti, C. & Marchi, L., 2002. Assement of shallow landsliding by using a physically based model of hillslope stability. *Hydrological Processes*, Vol. 16, pp. 2833–2851.

Darcy, H. 1856. *Les fontaines publiques de la ville de Dijon*. V. Dalmont, Paris.

Desbordes, M. 1987. *Contribution à l'analyse et à la modélisation des mécanismes hydrologiques en milieu urbain*. Thèse de Doctorat d'Etat, Université Montpellier 2, Montpellier, France, pp. 242.

Fuentes, C., Haverkamp, R. & Parlange, J.-Y., 1992. Parameter constrains on closed-form soilwater relationships. J. Hydrol., Vol. 134, pp. 117–142.

Horton, R.E., 1933. The role of infiltration in the hydrologic cycle: EOS, Transactions, American Geophysical Union, Vol. 14, pp. 446–460.

Montgomery, D.R. & W.E. Dietrich, 1994. A Physically Based Model for the Topographic Control on Shallow Landsliding, *Water Resources Research*, 30 (4): pp. 1153–117.

Musy, A. & Soutter, M., 1991. *Physique du sol. Laussane*. Presses Polytechniques et Universitaires Romandes, 335p.

O'Loughlin, E.M., 1986. Prediction of surface saturation zones in natural catchments by topographic analysis. *Water Resources Research*, Vol. 22: pp. 794–804.

Risser, V., 2000. *Mouvements de terrains sur le versant oriental du volcan Rucu Pichincha* (Quito, Equateur). Projet Sishilad—EMAAP-Q/INAMHI/IRD. 67p.

Selby, M.J., 1993. *Hillslope material and processes*. Edition Oxford. Second edition. 445p.

Spencer, E., 1967. A method of analysis of the stability of embankments assuming parallel inter-slice forces. *Geotechnique*; Vol. 17. pp. 11–26.

Tarboton, D.G., 1997. A new method for the determination of flow directions and upslope areas in grid digital elevation models. *Water Resources Research*, 33 (2): p. 309–319.

Wilson, J.P. & Gallant, J.C., 2000. *Terrain Analysis—Principles and Applications*. Wiley, New York, pp. 87–132.

Landslides and Engineered Slopes – Chen et al. (eds)
© 2008 Taylor & Francis Group, London, ISBN 978-0-415-41196-7

Back-analyses of a large-scale slope model failure caused by a sudden drawdown of water level

G.W. Jia, Tony L.T. Zhan & Y.M. Chen
MOE Key Laboratory of Soft Soils and Geoenvironmental Engineering, Zhejiang University, Hangzhou, China

ABSTRACT: Slope failure caused by a sudden drawdown of water level was simulated in a large-scale model box (15 m by 5 m by 6.5 m). The model test revealed a retrogressive multiple rotational sliding in a loose silty soil slope. This paper presented the back-analysis work of the large-scale model test. Transient seepage analyses were conducted to simulate the seepage field induced by rapid drawdown. Slope stability analyses were conducted to simulate the retrogressive sliding observed in the model test. The results from the back analyses on the seepage field and slope stability were basically consistent with the observation.

1 INTRODUCTION

Sudden drawdown of water level usually occurs to an earth dam or an embankment along a river or a coastal line due to the climate change, sudden breaching or periodical tide. The sudden drawdown usually affects the stability of the slope on the verge of water and even causes slope failures. Slope failures induced by water level drawdown are often reported (Morgenstern, 1963; Nakamura, 1990; Liao et al, 2005). Jones (1981) investigated the landslides occurred in the vicinity of Roosevelt lake from 1941 to 1953 and found about 30% of the landslide happened due to the drawdown of the reservoir. Nakamura (1990) reported that about 60% of the landslides around reservoirs occurred under a drawdown condition in Japan.

Numerous desk studies have been conducted to investigate the performance of slope subjected to a sudden drawdown (Lane et al. 2000; Zhang et al. 2005; Liao et al. 2006), but most of the previous numerical studies usually focused on a specific assumed slope and no experiments were carried out to verify the analysis results. In these analyses, the slip surface was assumed and the slope failure mode can't be simulated. Because it is difficult to obtain the accurate input data, such as slope geometry, phreatic lines and the slip surface, so few back-analyses of slope failures caused by rapid drawdown have been carried out.

This paper presents the back-analyses work on a large-scale model slope experiment conducted in Zhejiang University. In this paper, transient seepage and limit equilibrium analyses have been conducted for the slope subjected to rapid drawdown. The stability of the slope during the drawdown process is analyzed incorporating the pore water pressures calculated from the transient seepage analyses. The main objective of the analyses is to investigate the seepage field and slope stability of the slope during the rapid drawdown.

2 LARGE-SCALE MODEL TEST SIMULATING THE SUDDEN DRAWDOWN

The large-scale model slope experiment was conducted in Zhejiang University. The model slope was constructed with sandy silt. The geometry of the model slope is illustrated in Figure 1. The initially constructed slope model is 15 m in length, 5 m in width and 6 m in height. The model slope consists of homogenous sloping ground that was 6 m thick upslope and 2 m thick downslope with the slope angle of 45°, giving a net model slope height equal to 4 m. The crest and toe of the slope are 5 m and 6 m long, respectively.

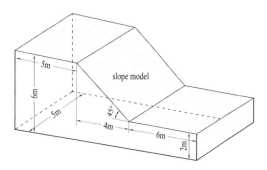

Figure 1. Three dimensional view of the initial slope model.

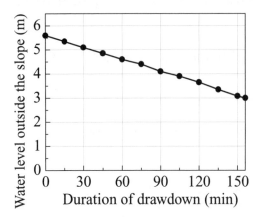

Figure 2. Change of water level outside the slope during the drawdown process.

Figure 3. The retrogressive multiple rotational landslide.

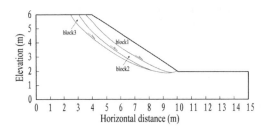

Figure 4. The observed slip surface of the slide.

A water-level control system was developed to simulate the rise and drawdown of water level in the slope model (Zhan et al. 2007). Firstly, the water level inside and outside the slope was elevated by the water-level control system. The elevating program comprised of 6 elevating steps and each elevation increment was 1 m and sustained for a period of typically 24 h. The initial constructed slope surface collapsed during the water level rising process and the slope angle become approximately 33° (Zhan et al, 2007). Secondly, the sudden drawdown of water level outside the slope surface was started after the final step of water rise had lasted for 72 hours. The water level outside the slope was shown in Figure 2. The drawdown rate is about 1 m/s. A typical retrogressive multiple rotational slide occurred to the slope model during the sudden drawdown process (Fig. 3). There were totally three displaced masses taken place in succession and three slip surfaces were observed (Fig. 4).

3 BACK-ANALYSES OF THE SLOPE FAILURE

3.1 Analyses procedures

For investigating the pore water pressure distributions in the model slopes and then on the slope stability, firstly, transient seepage analyses for saturated and unsaturated soils are carried out using a computer program SEEP/W (Geo-slope, 2003). After obtaining the pore water pressure distributions from the transient seepage analyses, limit equilibrium analyses are then carried out to determine the factor of safety of the slopes using SLOPE/W (Geo-slope, 2003). The shear strength of unsaturated soils can be represented by an extended Mohr-Coulomb failure criterion (Fredlund & Rahardjo, 1993).

The finite element mesh of the slope is shown in Figure 5. The downstream boundary subjected to the rapid drawdown is defined as known total head function which is shown in Figure 2. The drawdown process (156 minute) is divided into 156 time step with one minute for one time step in the numerical analyses. The downstream boundaries is specified as review boundary which means that if the total head is smaller than the corresponding elevation the flux will be zero at the node. All the other boundaries are specified as zero flux boundaries. The initial groundwater conditions for transient seepage analyses are established by conducting a steady state analysis, in which the initial water level is specified at the elevation of 5.6 m.

3.2 Hydraulic properties for seepage analyses

To simulate transient seepage in unsaturated soils, it is essential to specify hydraulic parameters including soil-water characteristic curves (SWCC), saturated water permeability and permeability functions for the sandy silt. The soil-water characteristic curve of the silt was obtained from volumetric pressure plate extractor and the results are shown in Figure 6. The saturated permeability of the soil material was calibrated by

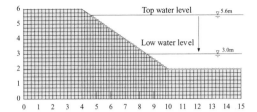

Figure 5. The finite element mesh of model slope for seepage analyses.

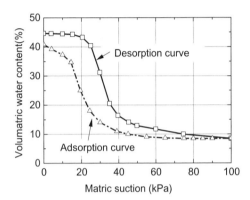

Figure 6. Soil-water characteristic curve for the sandy silt.

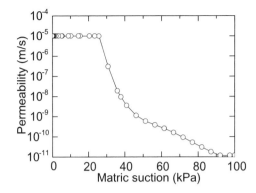

Figure 7. Permeability functions for the sandy silt.

varying the hydraulic permeability until the computed phreatic surface matched the observation during the water level rising process. The calibrated permeability for the soil was 1.0×10^{-5} m/s. The permeability function was predicted from the SWCC by using the method proposed by Fredlund et al. (Fredlund et al, 1993). The calculations were assisted by the program SEEP/W. The estimated permeability function for the soil is shown in Figure 7.

3.3 Approach for analyzing retrogressive slope failure

Limit equilibrium method of slices is commonly used for calculating the factor of safety of a slope. In this study, the Morgenstern-Price method of slices has been adopted for the calculation.

Though there have been many methods for the analysis of slope stability, few numerical investigations on landslides with retrogressive mode is available in the literature. Only few simplified approaches have been proposed to analyze such failures (Saucer 1983; Li et al, 2005). Li et al (2005) assumed that the soil mass above the critical slip surface is slipped down completely and the soil mass below forms a new slope, the slope surface is the critical slip surface. In the analysis, the shear strength resistance of the previous slide debris was neglected. Saucer et al. (1983) divided the retrogressive landslide into several blocks and each successive block combination is considered to move as a coherent unit. In this study, Saucer's method was adopted to analyze the retrogressive landslide.

4 BACK-ANALYSES OF THE SLOPE FAILURE

4.1 Results from transient seepage analyses

Phreatic lines and pore water pressure field for every minute can be obtained from the seepage analyses. Figure 8 shows phreatic lines changing with time during the drawdown process. It can be seen that the phreatic lines inside the slope show a significant delay to the water level outside the slope. It is unfavorable to the stability of the slope. The delayed responses of phreatic lines inside the slope are consistent with the phenomenon that outward seepage was observed on the slope surface during the drawdown process.

In order to verify the back-analysis results, the numerical results of pore-water pressure at the position of Piezometer-4 were compared with the observation. Figure 9 shows the comparison. Due to the matches between the computed and the field-measured pore-water pressure to a certain extent, the results from seepage analyses can be in accordance with the fact.

4.2 Calculation of factor of safety

In this section, the pore-water pressure distributions obtained from the seepage analyses were incorporated into a slope stability analyses to find the factor of safety. It can be seen from Figure 3 that sliding distance

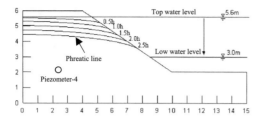

Figure 8. Phreatic lines changing with time during the drawdown process.

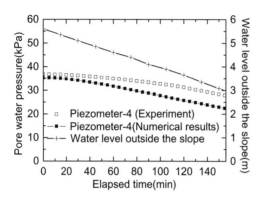

Figure 9. The comparison between the pore-water pressures from numerical analyses and experiment.

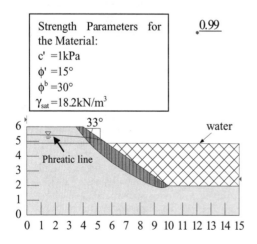

Figure 10. Back analyses of block 1.

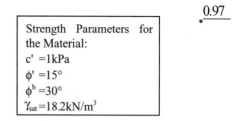

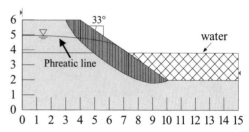

Figure 11. Back analyses of the combination of blcok 1 and block 2.

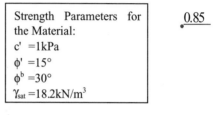

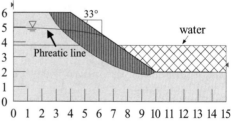

Figure 12. Back analyses of the combination of blcok 1, block 2 and block 3.

for block 2 and block 3 is relatively small to the slope dimensions, so the Saucer's method can be applied in the slope stability analyses. The slope failure surfaces were specified to the three observed slip surfaces observed from the observation windows (Fig. 4).

When the water level outside the slope drawdown for 0.7 m at the 40th minute, the first slide block (named block 1) formed and slide downward with a rapid speed. The factor of safety value corresponding to slip surface 1 calculated by SLOPE/W was shown in Figure 10. In the slope stability analyses, both the matric suctions above the phreatic line and the thrust of water outside the slope surface were taken into account. It can be seen that the factor of safety obtained from the back analysis was slightly lower than 1. The slope safety factor is consistent with the observation.

When the factor of safety was lower than 1, the slope failure initiated.

When the water level was drawdown for 1.7 m at the 100th minute, the second block and the third block initiated nearly at the same time. Figure 11 shows the back analyses of the combination of blcok 1 and block 2. The factor of safety of the combination of blcok 1 and block 2 was slightly lower than 1 when the slip surface 2 is the critical slip surface. Figure 12 shows the back analyses of the combination of blcok 1, block 2 and block 3. The factor of safety of the combination of blcok 1, block 2 and block 3 was 0.85, which is much lower than 1. The difference between the calculated factor of safety and practical slope failure may be caused by the sliding mass stacking at the foot of the slope.

5 CONCLUSIONS

In this paper, the back-analyses of a large-scale slope model failure caused by a sudden drawdown of water level were carried out. The phreatic lines inside the slope show a significant delay relative to the water level outside the slope. The phreatic lines and pore-water pressure of the back-analyses were basically consistent with the observation.

The procedures that the successive sliding blocks combination were assumed to move as a coherent was employed to analyze the retrogressive landslide. The calculated values of factor of safety were consistent with the observation. It is demonstrated that procedure was feasible to analyze retrogressive landslides.

ACKNOWLEDGMENTS

The authors would like to acknowledge the financial support from the Key Technology R& D Program for the eleventh five years provided by the Ministry of Science and Technology of the People's Republic of China.

REFERENCES

Fredlund, D.G. & Rahardjo, H. 1993. *Soil Mechanics for Unsaturated Soils*, New York: John Wiley & Sons, Inc.

GEO-SLOPE International Ltd. 2003. *SEEP/W and SLOPE/W for finite element seepage analysis, vol.5. Users' Manual*. Calgary, Alberta, Canada, 2003.

Lane, P.A. & Griffiths, D.V. 2000. Assessment of stability of slopes under drawdown conditions. *J Geotech Geoenviron Eng, ASCE* 126 (5): 443–450.

Li, S., Yue, Z.Q., Tham, C.F. et al. 2005. Slope failure in under consolidated soft soils during the development of a port in Tianjin, China. Part 2: Analytical study. *Canadian Geotechnical Journal* 42: 166–183.

Liao, H.J., Sheng, Q., Gao, S.H. et al. 2006. Influence of drawdown of reservoir water level on landslide stability. *Chinese Journal of Rock Mechanics and Engineering*, 24 (19): 3454–3458. In Chinese.

Liao, Q.L., Li, X. & Dong, Y.H. 2005. Occurrence, geology and geomorphy characteristics and origin of Qianjiangping landslide in Three Gorges Reservoir area and study on ancient landslide criterion. *Chinese Journal of Rock Mechanics and Engineering* 24 (17): 3146–3153. In Chinese.

Morgenstern, N. 1963. Stability charts for earth slopes during rapid drawdown. *Geotechnique* 13 (2): 121–131.

Nakamura, K. 1990. On reservoir landslide. *Bulletin of Soil and Water Conservation*, 10 (1): 53–64. In Chinese.

Saucer, E.K. 1983. The Denholm landslide, Saskatchewan. Part II: Analysis. *Canadian Geotechnical Journal*, 20: 208–220.

Zhan, L.T., Zhang, W.J. & Chen, Y.M. 2006. Influence of reservoir level change on stability of a silty soil bank. *The fourth international conference on unsaturated soils, ASCE*: 463–472.

Zhan, L.T., Jia, G.W. & Chen, Y.M. 2007. A large scale model test simulating a slope failure caused by a sudden drawdown of water level. *Proceeding of the 3rd Asian conference on unsaturated soils*: 531–536.

Zhang, W.J., Chen, Y.M. & Ling, D.S. 2005. Seepage and stability analysis of bank slopes. *Chinese Journal of Hydraulic Engineering* 36 (12): 1510–1516. In Chinese.

Effect of Guangxi Longtan reservoir on the stability of landslide at Badu station of Nankun railway

Riguang Jiang, Rongguo Meng, Aizhong Bai & Yuliang He
Guangxi Survey Institute of Hydrogeology and Engineering Geology, Liuzhou 545006, China

ABSTRACT: The landslide at the Badu station of Nankun railway is an ancient landslide. The volume of the main landslide is about 3.772 million m^3, and that of the secondary is about 0.8344 million m^3. The landslide was activated by construction of Nankun railway. However, the ten-year monitoring indicates that the landslide did not have any deformation after maintained. If the Longtan reservoir is running, the stored water would have a negative effect on the stability of the landslide. It is necessary to reinforce the existing stabilization engineering, which is calculated and analyzed in terms of the water level change.

Figure 1. Overlook of Badu landslide.

1 CHARACTERISTICS OF THE LANDSLIDE ON BADU STATION OF NANKUN RAILWAY

1. The landslide lies on the bank of Nanpan river with the altitude of about 367 m in the front and 556–568 m in the rear. The main slide body extends at the direction of 145°–175° with the length of about 300 m, the width of 460–570 m, the thickness of about 23 m, the area of 16.40 × 10^4 m^2 and the volume of about 377.20 × 10^4 m^3. Whereas the secondary landslide is about 280–310 m in length, 320–440 m in width, and about 7 m in thickness, with the area of 11.92 × 10^4 m^2 and the volume of about 83.44 × 10^4 m^3 (Figure 1).

    The landslide in the front was badly collapsed, the front and the middle of which were revived. The new body of landslide has the length of about 280 m, the width of 460–570 m, the area of 13.24 × 10^4 m^2, the average thickness of about 25 m, and the volume of about 331.00 × 10^4 m^3. Because of Nanning-Kunming railway built between 1996 and 1997, as well as the change of the geologic condition such as hydrogeological condition, the secondary landslide was revived with the volume of about 83.44 × 10^4 m^3.

2. The body of the landslide is mainly made up of detritus, angular boulder and clayey soil. The old sliding zone is mainly composed of gray green, brown, yellow brown soft-plastic to plastic silty clay containing gravel which was visibly burnished, while the new deformation sliding zone is weak and have no obvious characteristics of slipping. The bedding is made up of light gray, gray green calcareous, siliceous quartz sandstone and gray black mud, shale of the Triassic EC Bianyang Group (T$_2$b). In front of the formation, there are alluvial sand and gravel layers, which indicate that the landslide has moved forward and covered fluvial sediments. Three reverse faults and one translation fault has developed in the bedrock of the slide bed (the width of the broken belt is 10–60 m), and intersected each other.

3. It is an ancient landslide. The emergency management was taken on the main landslide after partially revived, including two rows of anchor piles, ground and underground drainage engineering measures. The controlling measures were adopted on the secondary landslide, such as prestressed chain, cantilever piles, ground and underground drainage. According to monitoring data, there was no deformation observed in the secondary landslide during last 10 years.

2 MAIN OBJECTIVES

It is necessary to evaluate the effect of Longtan reservoir on the stability of the landslide, based on investigation of the Badu station landslide's characteristics. As long as the reservoir runs, the water table will arise and accordingly affect the stability of the landslide. Therefore, the characteristics of the stored water on Longtan reservoir would determine the stability of the landslide, which can be grouped into 5 cases shown as follows (Ministry of Construction of the People's Republic of China, 1999).

Case 1: The present state with the water level of 368 m, including before the building of railway, between after the building of railway and the maintenance of the slide, and after the maintenance of the slide;

Case 2: The present state in case of the flood once every hundred year with the water level of 387.01 m, including before the building of railway, between after the building of railway and the maintenance of the slide, and after the maintenance of the slide;

Case 2–1: The present state in case of the flood once every hundred year, after the slide was maintained, with the water level between 368 and 387.01 m;

Case 3: The reservoir operation after the slide is maintained for the railway building, with the water level of 375 m in the near future and 400 m in the future;

Case 4: The reservoir operation in case of the flood once every hundred year after the slide is maintained for the railway building, with the water level decreasing 375 m from to 392.50 m in the near future and from 407.33 m to 400 m in the future.

Therefore, it can be seen that Cases 1 and 3 are the normal conditions, while Cases 2 and 4 are exceptive.

3 METHODS

3.1 Data collection

The first step is to collect related data obtained from the predecessors. The data include results of various studies and investigation, information on design and construction completion, and ten-year monitoring data. Then do further research to the information.

3.2 Complementary reconnaissance

Comprehensive approaches was used to further identify the basic features of the landslide and to investigate the effectiveness of the accomplished project, such as engineering geological mapping, four adits, trenching, geophysical exploration, large shear test in-situ, drilling, laboratory tests, the rigidity test, water-injecting test and dating of sliding zone (Ministry of Construction of the People's Republic of China, 2002).

3.3 Analytical methods

Both limit equilibrium analysis (Ministry of Construction of the People's Republic of China, 2002) and finite element analysis were used to evaluate the effect of the reservoir (Shibiao Chang et al.1994) on the stability of the landslide.

Figure 2. The sketch map of Badu landslide's profile for evaluating.

3.3.1 Limit equilibrium analysis

Three vertical axial profiles (Profile 1, Profile 2 and Profile 3) were selected as a model. Because the ancient landslide was revived to a new body deformation after the excavation for the railway, the stability of each profile is separately calculated by the following sections: original main landslide (C'-A' section); the original secondary landslide (A-B Section); the whole landslide composed of the main landslide and the secondary landslide (A-A' section); the new deformable body as a result of the railway excavation, in which Profile 1 and Profile 2 are divided into two new deformable bodies (IV_1, IV_2), IV_1 is the B'-A' section, IV_2 is the C'-A' section, and Profile 3 is composed of one new deformable body (IV) (Figure 2).

3.3.2 Finite element analysis

Geological analysis and mode of the Badu Landslide geological characteristics are used to simulate the groundwater flow, slope stress field and deformation field of the Badu landslide in two- and three-dimension in three conditions, including the natural condition, the railway construction condition and the stored water condition, in order to estimate the stability of Badu landslides and to forecast deformation of the landslide.

4 RESULTS AND DISCUSSIONS

It is reasonable to synthetically confirm the parameters of slip soil depending on the lab test with the samples from the adits, the large in-situ shearing test and the result of the anti-count on the untreated condition after railway excavation.

4.1 Limit equilibrium analysis

1. Results of the limit equilibrium analysis show that the ancient landslide was at stable state prior to railway construction.
2. Between the railway construction and the treatment, the landslide was stable or less stable, while it was less stable or instable when the flood took place once every one hundred-year.
3. Treatment increased the stability of the landslide, which was stable or less stable in the flood once every one hundred-year.
4. Because of the reservoir running, the landslide became less stable at diverse extends: the stability of the secondary landslide was less affected, while the new deformable body on the main landslide

was significantly affected. The original main landslide and the original secondary landslide were at steady state, while the whole landslide was stable or less stable. The new deformation body of the main landslide was less stable in Case 3.
5. When the water level drops from 392.5 to 375 m or from 407.33 to 400 m due to the flood once every 100-year, the landslide in every profile was less stable.

Therefore, local parts of the landslide will possibly slide after the reservoir functions.

4.2 Finite element analysis

1. Three-dimension Modflow software was used to simulate the groundwater flow and dynamic characteristics of the Badu landslide in natural condition and on the conditions of water level of 375 m and 400 m in the reservoir. The pressure of pore water and the change of the hydraulic gradient were calculated in terms of the water level change in the reservoir.
2. The result of the Badu landslide stress analysis indicates that the Badu landslide is stable in natural conditions, where there is tensile stress only at local slope surface. The landslide is easily revived by induced factors, such as rainstorm, excavation, and reservoir. There was tensile stress on the excavated slope, and partial parts were damaged. With the excavation of railway, the stress and the tensile stress of the landslide increased in case of heavy rain, and the value of destroyed degree η increased dramatically, which made the landslide plastic damage. After the comprehensive repair, the landslide was generally stable. However, its stability would decline because of Longtan reservoir, especially at the water level up to 400 m, which would make local part be damaged. Therefore, the reinforcement must be carried out in order to ensure the safety of the landslide.
3. The results of the Badu landslide deformation and the stability status analysis indicate that the Badu landslide was stable in natural conditions. It was deformed only in the initial stages, and gradually tended to slow creep state and eventually became stable while the time went by. The displacement of the landslide was less, primarily occurred in the angle of excavated slope after the railway had been excavated. In case of the heavy rain, the displacement increased rapidly (up to 120 mm). Accordingly, the landslide was revived and instable. With the comprehensive repair, the deformation tended to slow down and eventually to stabilize. However, the deformation will increase and the landslide stability will decline because of Longtan reservoir, especially in case of the water level up to 400 m. The displacement will reach 47 mm, mainly in the front edge of the landslide below the water level.

In summary, the reservoir has impacted effects on the landslide as follows.

Firstly, the reservoir increased groundwater water table and caused the change of the natural groundwater flow in the landslide, which would negatively affect the stability of the landslide.

Secondly, after the reservoir run, the project protecting the landside from sliding would play a floating role, which would decrease the stability of the landslide.

Thirdly, the descent of the water level in the reservoir for flooding protection led to higher penetration pressure. This increased the downward stress imposed on the landslide.

Fourthly, the dammed water and the change of the water level acting on the bank reduced the skid resistance, while increased the downward stress.

4.3 Overall merit

1. There is a drainage system on the land surface, which need be repaired in time.
2. The underground drainage works have put into effect in the secondary Landslide, while there are less engineering measures in the main landslide. It is proposed to build underground drainage channel instead of the adits.
3. The project to counteract sliding has been built in the secondary landslide. Some of anti-sliding pegs are slightly shallow in the rock in the main landslide (below the rail), and few prestressed anchors and chain are damaged. If the reservoir is filled to a high level or the high water level drops, the shore will serious collapse. That has a serious effect on the stability of the main landslide. Therefore, it is essential to accomplish the bank protection and reinforcement works before the reservoir runs, in order to protect the landslide and to avoid endangering the Badu part of the Nanning-Kunming railway.

ACKNOWLEDGEMENTS

The authors would like to express their thanks to experts and professors in Longtan Hydropower Development Co., Ltd., Southeast Electric Power Design Institute, Liuzhou Railway Bureau, Railway Survey and Design Institute, the Ministry of Land and Resources, the Ministry of Railways, the former Ministry of Water and Electricity, Guangxi University, Chengdu University of Technology and Chongqing University for their guidance and helps during the reconnaissance, analysis and the modification of the disquisition.

REFERENCES

Ministry of Construction of the People's Republic of China.1999.*Code water resoure and hydropower engineering geological investigation*, GB50287–99. Beijing: China Archicture & Building Press.

Ministry of Construction of the People's Republic of China.2002.*Technical code for building slope engineering*, GB50330–2002. Beijing: China Archicture & Building Press.

Ministry of Construction of the People's Republic of China.2002.*Code for investination of geotichincal engineering*, GB50021–2001.Beijing: China Archicture & Building Press.

Shibiao Chang et al.1994. *Handbook of geologic engineering*, the third edition. Beijing: China Archicture & Building Press.

Application of SSRM in stability analysis of subgrade embankments over sloped weak ground with FLAC3D

Xin Jiang
Key Laboratory of Road & Traffic Engineering of the Ministry of Education, Tongji University, Shanghai, P.R. China

Yanjun Qiu
School of Civil Engineering, Southwest Jiaotong University, Chengdu, P.R. China

Yongxing Wei
No. 2 Civil and Architecture Design Institute, China Railway Eryuan Engineering Group Co. Ltd., Chengdu, P.R. China

Jianming Ling
Key Laboratory of Road & Traffic Engineering of the Ministry of Education, Tongji University, Shanghai, P.R. China

ABSTRACT: Subgrade embankments over sloped weak ground exhibit different deformation behavior due to downhill sliding potential compared to filling over flat ground. The stability analysis of subgrade embankments over sloped weak ground plays an important role in embankment design process. Some numerical simulation software packages including Shear Strength Reduction Method (SSRM) to analysis slope stability were summarized. Based on Finite Difference Method (FDM) software package FLAC3D, 3D stability analysis of embankments over sloped weak ground was performed. The influences of some model geometry parameters, such as the weak subsoil layer longitudinal length and vertical thickness, on potential slip surface behavior and Factor of Safety (FS) were discussed. The research results present the considerable differences between the slip surface behavior and Factors of Safety (FS) of an embankment over sloped weak ground estimated from 2D and 3D numerical calculations. The slip surface behavior depend on the weak subsoil layer longitudinal length and vertical thickness seriously. It seems that FS obtained from 2D calculations may be underestimated when the weak subsoil layer longitudinal length is limited. FS obtained from 2D calculations were lower than from 3D. With the rapid development of computer hardware and available commercial numerical simulation software, application of Shear Strength Reduction Method (SSRM) in 3D seems to be a reasonable alternative to 2D analysis to take the complexity of geology under consideration—especially the presence of thin and weak strata.

1 INTRODUCTION

The surrounding mountains around Sichuan basin produce many geotechnical challenges for highway and railway engineers such as deep cut, high fill, cut-fill transition, sloped weak ground, and other unconventional subgrade design in southwest China. For example, there are about 80 sections and total length 11.8 km sloped weak ground in Yuhuai (Chongqing-Huaihua) railway project, a newly built main railway route in southwest China. Due to its downhill sliding potential, subgrade embankments over sloped ground exhibit different deformation behavior as compared fills over flat ground and the trend will be strengthened when there is a weak layer over sloped ground. So sloped ground could be regarded as the special case of sloped weak ground (the surface weak layer thickness is zero) as illustrated in Fig. 1.

Two major concerns arise in highway and railway construction over sloped weak ground: embankment stability and lateral deformation. Embankment stability may cause severe problems without sufficient design consideration and full understanding of structural mechanism. Literature reported that embankment and pavement failure in railway and highway construction with fills over sloped ground as shown in Fig. 2 and Fig. 3, including sliding failure, pavement surface longitudinal crack due to intolerable uniform deformation. Anti-slide piles or pile net foundation in sloped weak ground were designed in original plan to increase slope stability in Fig. 4.

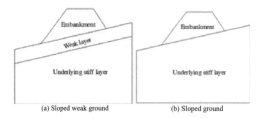

Figure 1. Engineered embankments over natural sloped ground.

Figure 2. Embankment failure (photo by Yongxing WEI).

Figure 3. Pavement longitudinal crack (photo by Xin JIANG).

Figure 4. Anti-slide piles (photo by Yongxing WEI).

Most research efforts on mechanical behavior of embankments over sloped weak ground. Wei (2001) established the concept of engineered fill over sloped ground. Jiang et al. (2002, 2003) simulated deformation and stability behavior of railway embankment over sloped ground using commercial software GEOS-LOPE, based on plane strain FEM and 2D limit equilibrium method (LEM), while Luo et al. (2002) conducted geo-centrifuge testing to investigate deformation behavior, stability characteristics and effective of engineering measures of railway subgrades over sloped ground. You et al. (2002) studied the construction process of railway fill over sloped weak ground in plateau area. Later Jiang et al. (2006, 2007a) explored the stability of railway fills over sloped weak ground using 2D strength reduction method (SRM) with commercial software Plaxis. Qiu et al. (2007) studied the deformation and stability behavior of highway fill over sloped weak ground and influence on pavement responses.

It can be concluded based on literature review that embankments over sloped ground need more research efforts to have a better understanding of the deformation and stability behavior in order to design a sound upper structures, such as pavement or rail track. The stability analysis of embankments over sloped weak ground plays an important role in embankment design process without doubt. Traditional 2D limit equilibrium method and FEM were used to perform the stability analysis in the existing research and the limit longitudinal length of the weak subsoil will be ignored, however. 3D shear strength reduction method (SSRM) was conducted in this paper to investigate potential sliding mode and more realistic value of factor of safety (FS) of embankment over sloped weak ground.

2 BRIEF INTRODUCTION TO SSRM AND SOFTWARE

2.1 Shear strength reduction method

The shear strength reduction method (SSRM) is also known as strength reduction method (SRM). By automatically performing a series of simulations while changing the soil strength properties, the factor of safety (FS) can be found to correspond to the point of stability, and the critical failure (slip) surface can be determined (Griffiths et al., 1999). Actual shear strength properties, cohesion (c) and friction (φ), are reduced for each trial according to the following equations:

$$c^{trial} = c/F_{trial} \quad (1)$$

$$\varphi^{trial} = \arctan(\tan\varphi/F_{trial}) \quad (2)$$

The SSRM provides advantages over traditional limit equilibrium solution as follows (Itasca Consulting Group, Inc, 2005b):

1. Any failure mode develop naturally; there is no need to specify a range of trial surface in advance.
2. No artificial parameters (e.g., functions for inter-slice force angles) need to be given as input.
3. Multiple failure surface (or complex internal yielding) evolve naturally, if the conditions give rise to them.
4. Structural interaction (e.g., rock bolt, soil nail or geogrid) is modeled realistically as fully coupled deforming elements, not simply as equivalent forces.
5. The solution consists of mechanisms that are kinematically feasible. (Note that the limit equilibrium method only considers forces, not kinematics).

2.2 Software including SSRM

Due to the rapid development of computing efficiency, some geotechnical numerical simulation codes were developed to perform the slope stability analysis using SSRM directly. Now FLAC2D, FLAC3D, FLAC/SLOPE developed by Itasca Consulting Group, Inc. in USA, Phase 2 V6.0 developed by Rocscience Inc. in Canada, Plaxis, Plaxis 3D Tunnel and Plaxis 3D Foundation developed by Plaxis B.V. in Netherlands, Z-SOIL.PC developed by Zace Service Ltd. in Swzerland, Geo FEM developed by FINE Ltd. in Czech, MIDAS/GTS developed by POSCO Group in Korea, RFPA (Realistic Failure Process Analysis) developed by Dalian Mechsoft Co., Ltd. and VFEAP developed by Zhejiang University in China are some mainstream numerical simulation codes that can be performed the slope stability analysis using SSRM directly. FLAC2D, FLAC3D and FLAC/SLOPE are developed using finite difference method (FDM) and the others are developed based on finite element method (FEM) among the above codes. More information about these software features and comparisons can be found in Jiang et al. (2007b).

3 FLAC3D NUMERICAL ANALYSIS MODEL

3.1 Geometry dimensions

FLAC3D developed by Itasca Consulting Group, Inc. was chosen to perform analysis in this paper

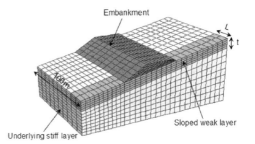

Figure 5. FLAC3D model.

Table 1. Mechanical properties of soil layers.

Soil type	E/MPa	μ	γ/(kN/m³)	c/kPa	φ/°
Embankment	30	0.35	19	25	25
Sloped weak layer	5	0.40	18	10	10
Stiff layer	300	0.25	22	200	50

(Itasca Consulting Group, Inc., 2005a). FALC3D is widely used for slope stability analysis recently (Varela Suarez et al., 2003; Cala et al., 2006). The FLAC3D analysis model can be seen in Fig. 5. In this model, the embankment top width is assumed to be 28 m, embankment height is 5 m, side slope ratio is 1:1.75 (V:H) and the ground inclination ratio is 1:5. The thickness t and the longitudinal length L of the weak subsoil layer were changed to investigate the model geometry design parameter sensitivity. It is noted that the full longitudinal length of embankment is assumed to be 100 m.

3.2 Material properties

Table 1 shows the mechanical properties assumed in the paper for the soil units.

In Table 1, E is elastic modulus, μ is Poisson's ration, γ is soil unit weight, c is cohesive force and φ is inner friction angle. E and μ should be transformed to bulk modulus K and shear modulus G in FLAC3D input process. More information can be found in FLAC3D manual (Itasca Consulting Group, Inc., 2005).

4 RESULTS AND DISCUSSION

The comparison pictures of potential failure mode for the longitudinal length of the weak subsoil $L = 20$ m, 100 m (weak subsoil thickness t is 1 m) are presented in Fig. 6. The FEM computer code, Phase 2 V6.0 was used for 2D slope stability analysis using

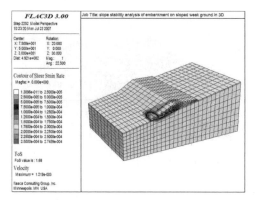

(a) 3D analysis using FLAC3D (L=20m, t=1m)

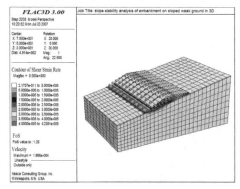

(b) 3D analysis using FLAC3D (L=100m, t=1m)

(c) 2D analysis using Phase2 V6.0 (t=1m)

Figure 6. Potential slip surface behavior (weak layer thickness $t = 1$ m).

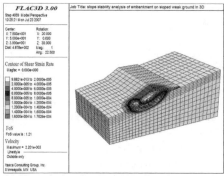

(a) 3D analysis using FLAC3D (L=20m, t=9m)

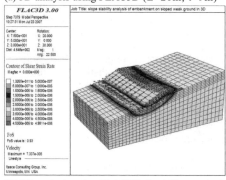

(b) 3D analysis using FLAC3D (L=100m, t=9m)

(c) 2D analysis using Phase2 V6.0 (t=9m)

Figure 7. Potential slip surface behavior ($t = 9$ m).

SSRM. The contours of shear strain rate and direction of velocity vectors are clearly identifying failure surface in FLAC3D. However, the potential failure mode is presented using the maximum shear strain or displacement vectors in Phase 2 V6.0. The failure surface is only local when the longitudinal length of the weak subsoil L is 20 m but the failure surface is turned to be global when the longitudinal length of the weak subsoil L is 100 m. The failure mode of 2D analysis is only close to 3D analysis when the longitudinal length of the weak subsoil L is 100 m. It seems that the failure surface looks like to be a circle line in embankment body and a beeline in weak subsoil layer, in other words the failure slip surface is a composite slip surface.

The comparison pictures of potential failure mode for the longitudinal length of the weak subsoil $L = 20$ m, 100 m (weak subsoil thickness t is 9 m) are presented in Fig. 6. The FEM computer code, Phase 2 V6.0 was used for 2D slope stability analysis using SSRM. The contours of shear strain rate and direction of velocity vectors are clearly identifying failure surface. However, the potential failure mode is presented using the maximum shear strain or displacement vectors in Phase 2 V6.0. The failure surface is only local when the longitudinal length of the weak subsoil L is 20 m but the failure surface is turned to be global when the longitudinal length of the weak subsoil L is 100 m. The failure mode of 2D analysis is only close to 3D analysis when the longitudinal length of the weak subsoil L is 100 m. The failure slip surface is a circle slip surface but not a composite slip surface. The reason is the weak subsoil layer thickness is thicker than before. So reasonable engineering measures should be adopted to improve guidance of embankment design.

Fig. 8 shows the potential slip surface behavior using SLIDE V5.0 developed by Rocscience Inc. based on limit equilibrium method (LEM). It can be seen that slip surface behaviors are similar to those obtained from numerical simulation.

Twenty five series of analysis were performed using FLAC3D and Phase 2 at the same time. Fig. 9 shows 2D and 3D FS values for several longitudinal length of the weak subsoil stratum. The value of FS is constant up to the length of the weak subsoil stratum in 2D analysis. Increasing the weak subsoil layer result in decreasing of FS value. The FS values of 3D analysis are larger than those of 2D analysis obviously especially when the weak subsoil thickness is thin relatively. These means the 2D analysis will magnify the effect of weak subsoil layer because the longitudinal length in 2D analysis is considered as unlimited. FS obtained from 3D calculations slowly tends to the factor of safety value obtained from 2D calculations.

Fig. 10 shows the relationship between FS and sloped weak layer thickness t. It can be seen that increasing the sloped weak subsoil layer thickness will lead to the decreasing of FS value.

Table 2 shows the 2D and 3D FS value comparisons when the weak subsoil layer thickness vary (the longitudinal length L = 20 m). It can be seen the largest FS value difference reach to 0.57. The FS value difference will be relatively flat when the weak layer thickness t increase. In case of the longitudinal length of weak subsoil layer, FS obtained from 2D calculations may be seriously underestimated. Application of SSRM in 3D may produce a reasonable value of FS for most cases.

(a) 2D LEM analysis using Slide V5.0 (t=1m)

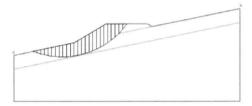

(b) 2D LEM analysis using Slide V5.0 (t=9m)

Figure 8. Slip surface obtained from 2D LEM.

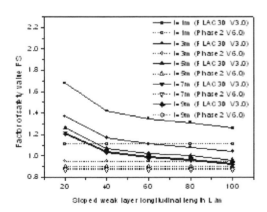

Figure 9. Factor of safety-sloped weak layer longitudinal length L.

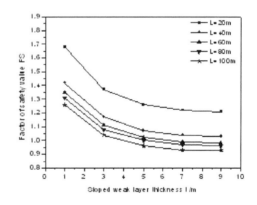

Figure 10. Factor of safety-sloped weak layer thickness t.

Table 2. FS comparisons between 3D model and 2D model.

Weak layer thickness t/m	Factor of safety (FS)		FS difference
	3D model (FLAC3D)	2D model (Phase 2)	
1	1.68	1.11	0.57
3	1.37	0.95	0.42
5	1.26	0.90	0.36
7	1.22	0.88	0.34
9	1.21	0.87	0.34

5 SUMMARY AND CONCLUSIONS

Embankment over sloped weak ground is still a great challenge for geotechnical and highway engineers due to its complexity. Sufficient safety reserve is one major concern in design and construction. Using FLAC3D as the major simulation tool, this paper conducted preliminary research on stability behavior of highway embankments over sloped ground in 3D SSRM. Conclusions can be summarized as follows:

1. It seems that SSRM can be conducted to perform slope stability analysis of embankment over sloped weak ground.
2. Application of SSRM in 3D may produce a reasonable value of FS and potential slip surface behavior.
3. SSRM in 3D often cost more computer time and need more computer hard disk space to store the result files than 2D analysis and the effect of 3D is often considered as an additional safety reserve so one must find a reasonable equilibrium between safety and economy.

ACKNOWLEDGEMENTS

This research is supported by Chinese Postdoctoral Science Foundation (20060390654). The authors are grateful to Mr. Yongsheng ZHU of ITASCA Consulting China Ltd for his valuable assistances.

REFERENCES

Cala, M., Flisiak, J. & Tajdus, A. 2006. Slope stability analysis with FLAC in 2D and 3D. In Pedro Varona & Roger Hart (eds), *FLAC and numerical modeling in geomechanics. Proceedings of the fourth international FLAC symposium. Madrid, Spain, 29–31 May, 2006.* (CD-ROM). Minneapolis: Itasca Consulting Group, Inc.

Griffiths, D.V. & Lane, P.A. 1999. Slope stability analysis by finite elements. *Géotechnique.* 49(3): 387–403.

Itasca Consulting Group, Inc. 2005a. *FLAC3D-Fast Lagrangian Analysis of Continua in 3 Dimensions, Ver.3.0 User's Guide.* Minneapolis: Itasca Consulting Group, Inc.

Itasca Consulting Group, Inc. 2005b. *FLAC/Slope User's Guide.* Minneapolis: Itasca Consulting Group, Inc.

Jiang, X., Wei, Y. & Qiu, Y. 2002. Numerical simulation of subgrade embankment on sloped weak ground, *Journal of Traffic and Transportation Engineering,* 2(3): 41–46.

Jiang, X., Wei, Y. & Qiu, Y. 2003. Stability of subgrade embankment on slope weak ground, *Journal of Traffic and Transportation Engineering,* 3(1): 30–44.

Jiang, X., Qiu, Y. & Wei, Y. 2006. Research on the subgrade embankments engineering on sloped weak ground, *Journal of Railway Engineering Society.* (1): 32–35, 39.

Jiang, X., Qiu, Y. & Wei, Y. 2007a. Engineering behavior analysis of subgrade embankments on sloped weak ground based on strength reduction FEM, *Chinese Journal of Geotechnical Engineering,* 29(4): 622–627.

Jiang, X. & Qiu Y. & Ling, J. 2007b. Comparisons of strength reduction method software for slope stability analysis. *Chinese Journal of Underground Space and Engineering.* (in press).

Luo, Q. & Zhang, L. 2002. Centrifuge testing of railway embankments over sloped weak ground, Research Report, School of Civil Engineering, Southwest Jiaotong University, Chengdu, China.

Qiu, Y., Wei, Y. & Luo, Q. 2007. Highway embankments over sloped ground and influence on pavement responses. *International Conference of Transportation Engineering 2007, Chengdu, China, 22–24 July, China 2007.* ASCE Press. pp1615–1620.

Varela Suarez, A. & Alonso Gonzalez, L.I. 2003. 3D slope stability analysis at Boinas East gold mine. In Richard Brummer, Patrick Andrieux, Christine Detournay & Roger Hart (eds), *FLAC and numerical modeling in geomechanics. Proceedings of the third international FLAC symposium, Sudbury, Ontario, 21–24 October 2003.* A.A. Balkema Publishers.

Wei, Y. 2001. Stability evaluation methods of embankments over sloped weak ground, *Journal of Geological Disaster and Environment Protection,* 12(2): 73–79.

You, C., Zhao, C., Zhang, H. & Liu, H. 2002. Study on construction test of embankment on soft clay of plateau slope, *Chinese Journal of Geotechnical Engineering,* 24(4): 503–508.

Strength parameters from back analysis of slips in two-layer slopes

J.-C. Jiang & T. Yamagami
Department of Civil and Environmental Engineering, The University of Tokushima, Japan

ABSTRACT: For a two-layer slope with a known geometry, unit weight and pore water pressure distribution, the position of critical slip surface from a limit equilibrium stability analysis will remain unique for a given set of $(\lambda_1 = c'_1/\tan\phi'_1, \lambda_2 = c'_2/\tan\phi'_2, \lambda_3 = c'_1/c'_2)$ values regardless of the magnitude of individual strength parameters $(c'_1, \phi'_1, c'_2, \text{ and } \phi'_2)$ of two soil layers. This theoretical relationship between c' and ϕ' and the critical slip surface is used to develop a back analysis method by which the shear strengths of soils can be estimated from a known position of the failure surface observed in the field two-layer slope. In this method, an objective function is first defined to describe the difference between the actual failure surface and theoretical critical slip surface, and then a nonlinear programming technique is used to minimize the difference (i.e. the objective function) so as to determine an optimal solution of $(\lambda_1, \lambda_2, \lambda_3)$ that corresponds to the failure surface. When the magnitude of $(\lambda_1, \lambda_2, \lambda_3)$ are obtained, a unique set of $(c'_1, \phi'_1, c'_2, \phi'_2)$ values can easily be computed by considering the fact that the factor of safety is equal to unity. Results back-calculated from a failure surface involved in a two-layer slope are presented to demonstrate the effectiveness of the proposed back analysis method.

1 INTRODUCTION

Back analysis of slope failures has been a useful tool to determine soil strengths along the slip surface especially in connection with remedial stabilization works. Existing back analyses have been performed only for failed slopes in homogeneous soils. In practice, empirical methods are often used to estimate c and ϕ by assuming one of these parameters and back-calculating the other for a factor of safety of unity (e.g., Duncan and Stark 1992, Japan Road Association, 1999). On the other hand, it has been suggested that the magnitude of both c' and ϕ' can be determined by considering the position of the actual slip surface together with the fact that the factor of safety should be equal to unity (Saito 1980, Li and Zhao 1984, Yamagami & Ueta 1989, Greco 1996, Wesley & Leelaratnam 2001, Jiang & Yamagami, 2006).

When shear failure occurs in slopes consisting of multi-layered soils, stability analysis requires the determination of the strengths of each layer involved. This paper presents a method to back-calculate the strength parameters from the slip surface in a two-layer slope. As estimation of the shear strengths is based on the information provided by a failure surface, the theoretical relationship between the (c', ϕ') and the location of critical slip surface in a two-layer slope is studied. It is shown that when the slope geometry, unit weight and pore water pressure distribution in a two-layer slope are given, the position of the critical failure surface from a limit equilibrium stability analysis will remain unique for a particular set of $(\lambda_1 = c'_1/\tan\phi'_1, \lambda_2 = c'_2/\tan\phi'_2, \lambda_3 = c'_1/c'_2)$ values regardless of the magnitude of individual parameters $(c'_1, \phi'_1, c'_2, \phi'_2)$ of two soils. Based on this finding, a back analysis method is established to determine the four strength parameters of two soil layers involved. In this method, an objective function with three independent variables $(\lambda_1, \lambda_2, \lambda_3)$ is first defined to describe the difference between the actual failure surface and theoretical critical slip surface. Then, this difference (i.e. the objective function) is minimized using a nonlinear programming technique to obtain an optimal solution of $(\lambda_1, \lambda_2, \lambda_3)$ that corresponds to the failure surface. When the magnitude of $(\lambda_1, \lambda_2, \lambda_3)$ are obtained, a unique set of $(c'_1, \phi'_1, c'_2, \phi'_2)$ values can easily be computed by considering that the factor of safety is equal to unity.

The proposed method satisfies the two essential requirements for back analysis: i) the actual failure surface must be consistent with a theoretical critical slip surface, and ii) the factor of safety should be equal to unity. Back analysis satisfying these conditions takes full advantage of the information provided by an existing landslide which represents the field large-scale shear test conducted by nature. Therefore, the (c', ϕ') values obtained in such a way can be expected more reliable.

2 THEORETICAL RELATIONSHIP BETWEEN STRENGTH PARAMETERS AND LOCATION OF CRITICAL SLIP SURFACE

In limit equilibrium methods of slope stability analysis, the factor of safety, F, is commonly defined as

$$F = \frac{\text{shear strength of soil}}{\text{shear stress required for equilibrium}} \qquad (1)$$

The shear strength of the soil is usually described by the Mohr-Coulomb failure criterion as a function of two parameters: cohesion c' and internal friction angle ϕ'. When the soil mass above a slip surface is divided into vertical slices, the definition of F can be expressed as:

$$\sum T = \frac{\sum R_f}{F} = \frac{\sum (c'l + N' \tan\phi')}{F}$$

$$= \sum \left(\frac{c'}{F} l + \frac{\tan\phi'}{F} N'\right) \qquad (2)$$

where R_f -the available shear resistance on the base of a slice, T-the shear force mobilized on the slice base, N'-the effective normal force on the slice base, l-length of the slice base. T and N' can be obtained by solving some or all of the equations of equilibrium. c'/F and $\tan\phi'/F$ in Equation (2) are sometimes called the strength parameters that are necessary only to maintain the slope in limit equilibrium. In other words, F can be defined as the factor by which the shear strength of the soil would be reduced to bring the slope into a state of barely stable equilibrium (Michalowski, 2002).

For a two-layer slope such as the one shown in Figure 1, Equation (2) can be rewritten as

$$\sum T = \sum \left(\frac{c'}{F} l + \frac{\tan\phi'}{F} N'\right)$$

$$= \sum_{AB} \left(\frac{c'_1}{F} l + \frac{\tan\phi'_1}{F} N'\right)$$

$$+ \sum_{BC} \left(\frac{c'_2}{F} l + \frac{\tan\phi'_2}{F} N'\right) \qquad (3)$$

From Equation (3) it follows that

$$\sum T = \sum_{AB} \left(\frac{\bar{c}'_1}{\bar{F}} l + \frac{\tan\bar{\phi}'_1}{\bar{F}} N'\right)$$

$$+ \sum_{BC} \left(\frac{\bar{c}'_2}{\bar{F}} l + \frac{\tan\bar{\phi}'_2}{\bar{F}} N'\right) \qquad (4)$$

where $\bar{c}'_1 = \mu c'_1$, $\tan\bar{\phi}'_1 = \mu\tan\phi'_1$, $\bar{c}'_2 = \mu c'_2$, $\tan\bar{\phi}'_2 = \mu \tan\phi'_2$, and $\bar{F} = \mu F$ (μ is a positive

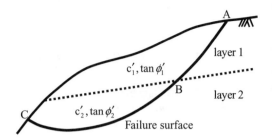

Figure 1. Failure surface in a two-layer slope.

constant). This means that if the strength parameters (c'_1, $\tan\phi'_1$, c'_2, $\tan\phi'_2$) are multiplied by μ, respectively, then the factor of safety for the same slip surface is also multiplied by μ in order to maintain a limit equilibrium state for the same slip surface.

Now let us consider an arbitrary two-layer slope with the (c'_1, $\tan\phi'_1$, c'_2, $\tan\phi'_2$) soils where the critical slip surface location with a minimum factor of safety, denoted by F_0, is known. Equations (3) and (4) indicate that if the soil layers in the slope are substituted by the two artificial materials having the strength parameters ($\bar{c}'_1 = \mu c'_1$, $\tan\bar{\phi}'_1 = \mu\tan\phi'_1$, $\bar{c}'_2 = \mu c'_2$, $\tan\bar{\phi}'_2 = \mu\tan\phi'_2$) and the problem is reanalyzed using the same method of slices, the factor of safety defined for each of the trial slip surfaces will be multiplied by μ, but the critical slip surface with the lowest factor of safety, $\bar{F}_0 (= \mu F_0)$, will remain at the same position.

It is of interest to note that in the cases mentioned above the ratios of the strength parameters keep unchanged although individual strength parameter values of the two soils are different. That is, the following relationships between the (c'_1, $\tan\phi'_1$, c'_2, $\tan\phi'_2$) and $\bar{c}'_1$, $\tan(\bar{\phi}'_1, \bar{c}'_2, \tan\bar{\phi}'_2)$ are held.

$$c'_1 / \tan\phi'_1 = \bar{c}'_1 / \tan\bar{\phi}'_1 \qquad (5.1)$$

$$c'_2 / \tan\phi'_2 = \bar{c}'_2 / \tan\bar{\phi}'_2 \qquad (5.2)$$

$$c'_1 / c'_2 = \bar{c}'_1 / \bar{c}'_2 \qquad (5.3)$$

$$\tan\phi'_1 / \tan\phi'_2 = \tan\bar{\phi}'_1 / \tan\bar{\phi}'_2 \qquad (5.4)$$

When any three of these four ratios are fixed, the rest one will automatically be determined. In other words, only three of the four ratios are independent of each other. For the sake of convenience, the following three ratios are chosen and used in this study.

$$\lambda_1 = c'_1 / \tan\phi'_1 \qquad (6.1)$$

$$\lambda_2 = c'_2 / \tan\phi'_2 \qquad (6.2)$$

$$\lambda_3 = c'_1 / c'_2 \qquad (6.3)$$

Summarizing the above discussions, we can conclude that different combinations of $(c'_1, \phi'_1, c'_2, \phi'_2)$ with a same set of $(\lambda_1, \lambda_2, \lambda_3)$ values will result in the identical critical slip surface. That is, when all other conditions except for (c', ϕ') are the same (the same slope geometry, the unit weight of soils, the pore water pressure distribution), the position of the critical slip surface in a given two-layer slope depends only on a set of $(\lambda_1, \lambda_2, \lambda_3)$ values regardless of the magnitude of individual strength parameters. Moreover, while the value of the minimum factor of safety varies as the strength parameters change from $(c'_1, \tan\phi'_1, c'_2, \tan\phi'_2)$ into $(\bar{c}'_1, \tan\bar{\phi}'_1, \bar{c}'_2, \tan\bar{\phi}'_2)$, the following relationships exist.

$$c'_1/F_0 = \bar{c}'_1/\bar{F}_0, \quad \tan\phi'_1/F_0 = \tan\bar{\phi}'_1/\bar{F}_0 \qquad (7.1)$$

$$c'_2/F_0 = \bar{c}'_2/\bar{F}_0, \quad \tan\phi'_2/F_0 = \tan\bar{\phi}'_2/\bar{F}_0 \qquad (7.2)$$

The above-mentioned relationship between the strength parameters and the location of critical slip surfaces is found directly from the definition of the factor of safety, and therefore, it is commonly valid for any existing method of slices. As an illustrative example, the stability analysis of a two-layer slope such as the one shown in Figure 2 was carried out using the same unit weight and pore water pressure distribution but different strength parameter values. Three different combinations of $(c'_1, \phi'_1, c'_2, \phi'_2)$ associated with a same set of $(\lambda_1, \lambda_2, \lambda_3)$ values were considered. For each of these cases, the critical slip surface was located using the search approach by Baker (1980) which combined the Spencer method (1967) with dynamic programming technique. The critical slip surfaces obtained are illustrated in Figure 2 together with their values of minimum factor of safety (F_{min}). It is seen from Figure 2 that the three critical slip surfaces completely coincide with each other. Also, the relationships shown in Equation (7) are hold for all the cases considered.

When the problem shown in Figure 2 was solved using combinations of $(c'_1, \phi'_1, c'_2, \phi'_2)$ corresponding to different sets of $(\lambda_1, \lambda_2, \lambda_3)$ values, the positions of critical slip surfaces located were also different, as shown in Figure 3. Consequently, it may be concluded that for a given two-layer slope with the known unit weight and pore water pressure distribution, the critical slip surface associated with a particular method of slices corresponds uniquely to a set of $(\lambda_1, \lambda_2, \lambda_3)$ values. This signifies that it is possible to estimate the magnitude of $(\lambda_1, \lambda_2, \lambda_3)$ from a known position of the actual slip surface observed in a field two-layer slope.

3 BACK ANALYSIS PROCEDURE

Figure 4 shows a two-layer slope in which a failure plane, AOB, has occurred. Its factor of safety

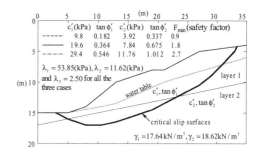

Figure 2. Critical slip surfaces for different (c', ϕ') combinations associated with a same set of $(\lambda_1, \lambda_2, \lambda_3)$ values.

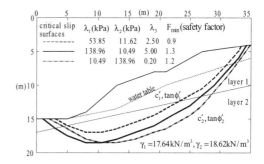

Figure 3. Critical slip surfaces for different sets of $(\lambda_1, \lambda_2, \lambda_3)$ values.

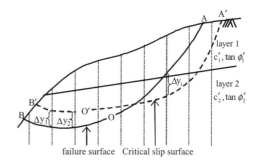

Figure 4. Difference between failure surface and critical slip surface ($DIS = \Sigma|\Delta y_i|$).

is denoted by F_0 which is usually taken to be unity. As mentioned previously, the position of the failure surface can be used to determine the magnitude of $(\lambda_1, \lambda_2, \lambda_3)$ that correspond to the true field values of $(\lambda_{10}, \lambda_{20}, \lambda_{30})$. In other words, by carrying out a conventional stability analysis with $(c'_1, \phi'_1, c'_2, \phi'_2)$ values that satisfies $(\lambda_{10} = c'_1/\tan\phi'_1, \lambda_{20} = c'_2/\tan\phi'_2, \lambda_{30} = c'_1/c'_2)$, it is possible to obtain a critical slip surface that is consistent with the actual slip surface, AOB.

Because the position of the failure surface, AOB, corresponds uniquely to the ($\lambda_{10}, \lambda_{20}, \lambda_{30}$) values, the stability analysis using another set of ($\lambda_1, \lambda_2, \lambda_3$) will yield a critical slip surface, $A'O'B'$ (Figure 4), which is different from AOB. Now let us take note of the difference in the positions of these two slip surfaces, defined by $DIS = \Sigma|\Delta y_i|$, in which Δy_i stands for vertical distance between the failure surface and the critical slip surface at each slice dividing line, as shown in Figure 4. DIS can be regarded to be a function of three variables ($\lambda_1, \lambda_2, \lambda_3$) because the position of critical slip surface varies with the change in values of ($\lambda_1, \lambda_2, \lambda_3$). When this function reaches a minimum (i.e. $DIS = 0.0$), a required solution of ($\lambda_1, \lambda_2, \lambda_3$) is obtained that should be equal to ($\lambda_{10}, \lambda_{20}, \lambda_{30}$), respectively. Therefore, if DIS is defined as an objective function, the present back analysis can be mathematically described as a minimization problem, as shown below.

Minimize $\quad DIS(\lambda_1, \lambda_2, \lambda_3) = |\Delta y_i|$ (8)

Subject to inequality constraints:

$0 \leq \lambda_1 \leq \lambda_{1max}, \quad 0.0 \leq \lambda_2 \leq \lambda_{2max},$

$0.0 \leq \lambda_3 \leq \lambda_{3max}$

where $DIS(\lambda_1, \lambda_2, \lambda_3)$—the objective function, and $\lambda_{1max}, \lambda_{2max}, \lambda_{3max}$ are estimated maximum values of ($\lambda_1, \lambda_2, \lambda_3$), respectively.

The minimization problem shown in Equation (8) can generally be solved by a nonlinear mathematical programming method. The SUMT technique called the interior point method (Jacobi et al., 1972) is used in this paper to transform $DIS(\lambda_1, \lambda_2, \lambda_3)$ into a modified objective function without the constraints. Then, the Nelder and Mead simplex method (1964) is employed to obtain an optimal solution of ($\lambda_1, \lambda_2, \lambda_3$) that gives the minimum value of the objective function. The details of the solution procedure are omitted due to the limitation of space.

A critical slip surface search is needed for each of the sets of ($\lambda_1, \lambda_2, \lambda_3$) in performing the simplex iterations. In this study, we used the search method by Baker (1980) which integrated the Spencer slope stability analysis (1967) with dynamic programming to locate the critical slip surface for each set of ($\lambda_1, \lambda_2, \lambda_3$).

When the optimum solution denoting by ($\lambda_{10}, \lambda_{20}, \lambda_{30}$) is obtained, all the four parameters ($c'_1, \phi'_1, c'_2, \phi'_2$) can be uniquely determined by considering the fact of $F = F_0$. This has been done using the following procedure.

1. Specify a value for one of ($c'_1, \phi'_1, c'_2, \phi'_2$) appropriately and calculate the values of the other three parameters by substituting ($\lambda_{10}, \lambda_{20}, \lambda_{30}$) into the relationships in Equation (6), so that a set of ($c'_1, \phi'_1, c'_2, \phi'_2$) values are obtained.
2. Compute the factor of safety (the Spencer method is used in this study), $\bar{F}_0$, of the failure surface using the ($c'_1, \phi'_1, c'_2, \phi'_2$) values obtained in 1).
3. Determine the magnitude of ($c'_{10}, \phi'_{10}, c'_{20}, \phi'_{20}$) for the factor of safety of $F_0 (=1.0)$ by substituting the ($c'_1, \phi'_1, c'_2, \phi'_2, \bar{F}_0, F_0$) values obtained in 1) and 2) into the relationships in Equation (7).

4 VERIFICATION

The accuracy of the proposed back analysis method was verified using a theoretical (hypothetical) slope failure in a two-layer slope such as the one shown in Figure 5. When the soil unit weight, pore water pressures (zero) and ($c'_1, \phi'_1, c'_2, \phi'_2$) values shown in the figure are given, the critical slip surface with a minimum factor of safety of $F_0 = 1.06$, as shown in the figure, was located using the Baker's (1980) dynamic programming search. It is seen that the slip surface passes through the two soil layers that have totally different strength parameter values.

In order to demonstrate the effectiveness of the proposed method, the slope profile in Figure 5 is now treated as a post-failure problem in which the (critical) failure surface and its factor of safety ($F_0 = 1.06$) are known but the strength parameters of the two soil layers are assumed to be unknown. Thus, the proposed method can be used to back-calculate ($c'_1, \phi'_1, c'_2, \phi'_2,$) values based on the information provided by the "failure" surface in Figure 5, and the results can be compared with their known (correct) values to verify the accuracy of the method.

Numerous tests performed herein have illustrated that the simplex technique provides a systematic and efficient tool to solve the minimization problem as shown in Equation (7). The convergence is reached rapidly in all the cases considered. However, the simplex iterations may converge to a local minimum when

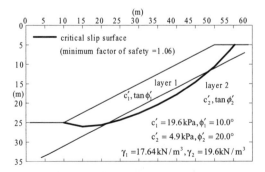

Figure 5. Hypothetical failure in a two-layered slope.

Table 1. Results back analyzed from hypothetical failure (Correct solution: $\lambda_1 = 111.16$ kPa, $\lambda_2 = 13.46$ kPa, $\lambda_3 = 4.0$).

No	Initial values			Back calculated values			Objective function
	λ_1(kPa)	λ_2(kPa)	λ_3	λ_1(kPa)	λ_2(kPa)	λ_3	$\sum |\Delta y_i|$ (m)
1	98.0	9.8	10.0	109.16	14.71		0.0003
2	9.8	9.8	10.0	18.81	27.13		28.167
3	45.0	180.0	5.0	102.49	311.45	13.23	30.667
4	180.0	9.8	4.0	164.71	19.98	4.55	21.367

the objective function is multimodal (Nguyen, 1985). That is, the optimum solution is dependent upon the selection of initial values of $(\lambda_1, \lambda_2, \lambda_3)$. This is also true for the present problem. A number of solution sequences are therefore needed to obtain an overall optimum, each starting from different initial values of $(\lambda_1, \lambda_2, \lambda_3)$. By running the proposed method with different initial values for $(\lambda_1, \lambda_2, \lambda_3)$, the sequences will yield DIS values associated with different $(\lambda_1, \lambda_2, \lambda_3)$ combinations as potential solutions. The smallest one of these DIS values gives a correct (optimal) solution and the corresponding set of $(\lambda_1, \lambda_2, \lambda_3)$ is therefore taken as the required values of $(\lambda_{10}, \lambda_{20}, \lambda_{30})$.

Several different initial values for $(\lambda_1, \lambda_2, \lambda_3)$ were randomly selected for the problem shown in Figure 5, and for each case the back analysis procedure described in Section 3 was executed. Table 1 shows the obtained values of $(\lambda_1, \lambda_2, \lambda_3)$ and DIS as well as the associated initial values for typical four cases. Results in Table 1 indicate that the best solution is obviously given by the $(\lambda_1, \lambda_2, \lambda_3)$ values for the No.1 case which resulted in the smallest difference (lowest DIS value) between the theoretical critical slip surface and the actual failure plane. Using $\lambda_{10} = 109.16$ kPa, $\lambda_{20} = 14.71$ kPa, and $\lambda_{30} = 3.65$, four strength parameters are calculated for $F_0 = 1.06$ and the results were found to be $c'_{10} = 19.40$ kPa, $\phi'_{10} = 10.1°$, $c'_{20} = 5.29$ kPa, and $\phi'_{20} = 19.9°$. These solutions agree well with their correct values.

5 BACK ANALYSIS OF A FAILED SLOPE IN TWO-LAYER SOILS

The proposed method is herein used to back-analyze a rainfall-induced landslide in a two-layer soil slope. Heavy rain fell on the Shikoku area during Typhoon Namtheun (the 10th tropical storm in the western Pacific in 2004). The total precipitation from July 30 to August 2 was more than 2,000 mm in Kisawa village and Kaminaka town, Tokushima Prefecture, Japan. This is several times the normal precipitation for the months of July and August in this area. Hourly precipitation reached more than 120 mm and the highest daily precipitation of 1,317 mm was recorded on 1 August.

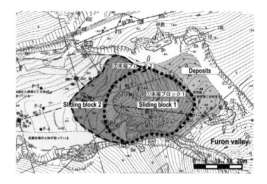

Figure 6. Plan view of moving sliding blocks.

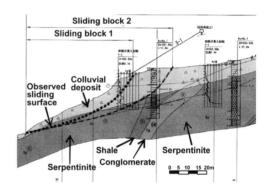

Figure 7. Geological cross section of slope and sliding blocks.

The strong precipitation for the storm was centered on a very narrow area of 5 to 6 km in the east-west direction, and 10 to 20 km in the south-north direction. Within the area, many landslides and debris flows were triggered during the first two days of August (Wang et al. 2005). One of the rainfall-induced landslides was a landslide in Shiraishi town. This slide occurred near the top of the Furon valley in a small residential area in the town. The sliding mass from the covering deposits overlying on weathered bedrock moved downslope and transformed into a large debris flow,

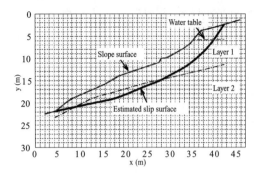

Figure 8. Cross section and failure surface of sliding block 1.

traveling about 800 m. The debris flow destroyed a number of houses which were built on the right side and near the exit of the valley.

Around the top of the valley, two unstable blocks with many cracks were found to move slightly after the landslide. The ground deformation observed by the installed extensometers indicates that the deforming landslides show nearly translational movement. Figures 6 and 7 show a plan view and a representative geological cross section of the landslides. At this site, the slope is underlain by the weathered bedrock that is mainly composed of a serpentinite. The covering materials overlying on the bedrock are colluvial deposits. Field investigation showed that the colluvial deposits are unconsolidated mixtures of clay and weathered fragments of sandstone and claystone. The serpentinite is commonly fissured, crumbling easily.

Bore-hole data and inclinometer observations indicate that the failure surfaces of the both unstable blocks cut through the covering deposits and the serpentinite. As a very large portion of the sliding block 2 is located in the underlying bedrock, the sliding block 1 is chosen and used to back calculate the strength parameters of the two different materials. Figure 8 shows a cross section of the failure surface of sliding block 1. The water table in the figure was determined after the debris flow event using the highest groundwater level observed in bore-holes.

As the sliding block 1 moved slowly, its factor of safety was estimated to be 1.0. A number of combinations of $(\lambda_1, \lambda_2, \lambda_3)$ are appropriately assumed as initial values, and then the proposed method was run for each of the combinations. The back calculated values of $(\lambda_1, \lambda_2, \lambda_3)$ and the magnitude of the corresponding objective functions are summarized in Table 2 for typical four cases.

It is seen from Table 2 that when $\lambda_1 = 14.3$ kPa, $\lambda_2 = 10.1$ kPa, and $\lambda_3 = 1.2$, the difference between the theoretical critical slip surface and the failure plane reaches a minimum of $DIS = 5.6$ m. Thus, these values of $(\lambda_1, \lambda_2, \lambda_3)$ are taken as the correct solution. The four strength parameters are calculated to meet the condition of $F_0 = 1.06$ and the relationships of $(c'_1/\tan\phi'_1 = 14.3$ kPa, $c'_2/\tan\phi'_2 = 10.1$ kPa, $c'_1/c'_2 = 1.2)$. As a result, $c'_1 = 9.7$ kPa, $\phi'_1 = 34.2°$, $c'_2 = 8.3$ kPa, and $\phi'_2 = 37.8°$ are obtained.

6 CONCLUSIONS

It has been shown that when the soil unit weight and pore water pressure distribution in a two-layer slope are given, the location of the critical slip surface from a limit equilibrium stability analysis will remain unique for a particular set of $(\lambda_1 = c'_1/\tan\phi'_1, \lambda_2 = c'_2/\tan\phi'_2, \lambda_3 = c'_1/c'_2)$ values regardless of the magnitude of individual strength parameters $(c'_1, \phi'_1, c'_2, \phi'_2)$. This theoretical relationship between the (c', ϕ') and the critical slip surface is found directly from the definition of the factor of safety and thus is available for any existing limit equilibrium slope stability method.

Based on the findings of the relationship between (c', ϕ') and the critical slip surface, a straightforward back analysis method has been presented to estimate the magnitude of $(\lambda_1, \lambda_2, \lambda_3)$ from a known position of the failure surface observed in the field slope. In the proposed method, back analysis of $(\lambda_1, \lambda_2, \lambda_3)$ is described as a problem to minimize the difference between the actual failure plane and theoretical critical slip surface. Although the optimal solution is affected by initial values, it is possible to obtain a correct combination of $(\lambda_1, \lambda_2, \lambda_3)$ using the solution sequences suggested in this paper. By running the method with

Table 2. Results back-calculated from the slope failure shown in Figure 8.

No	Initial values			Back calculated values			Objective function
	λ_1(kPa)	λ_2(kPa)	λ_3	λ_1(kPa)	λ_2(kPa)	λ_3	$\sum |\Delta y_1|$(m)
1	10.0	10.0	1.0	14.3	10.1	1.2	5.6
2	30.0	120.0	1.0	29.0	47.0	2.7	56.0
3	150.0	40.0	10.0	109.3	74.3	11.4	61.5
4	500.0	250.0	1.5	391.4	253.2	2.4	64.5

different initial values for $(\lambda_1, \lambda_2, \lambda_3)$, several *DIS* values can be obtained as potential solutions. The smallest one among the *DIS* values gives the correct solution of $(\lambda_1, \lambda_2, \lambda_3)$ that corresponds to the actual failure surface. When the magnitude of $(\lambda_1, \lambda_2, \lambda_3)$ is obtained, a unique set of $(c'_1, \phi'_1, c'_2, \phi'_2)$ can easily be determined by considering the fact that the factor of safety of the actual failure surface is equal to unity.

Results for two examples have been presented to demonstrate the proposed method. For the shear failure in a hypothetical slope, back calculated $(c'_1, \phi'_1, c'_2, \phi'_2)$ values of two soil layers agreed well with their correct solution. The back analysis of a rainfall-induced landslide resulted in a set of $(c'_1, \phi'_1, c'_2, \phi'_2)$ that probably reflected the actual situation of the failure and could be used for the remedial work design of the slope. These results show the potential of the proposed method for practical use. Future research planned is to carry out laboratory failure tests for model slopes to further verify the accuracy of the proposed back analysis method.

REFERENCES

Baker, R. 1980. Determination of the critical slip surface in slope stability computations. *Int. J. Numer. and Anal. Meth. in Geomech.*, 4, 333–359.

Duncan, J.M. & Stark, T.D. 1992. Soil strengths from back analysis of slope failures. *Proc. Specialty Conf. Stability and Performance of Slopes and Embankments-II*, ASCE, Berkeley, CA, 1, 890–904.

Greco, V.R. 1996. Back-analysis procedure for failed slopes. *Proc. 7th Int. Symp. on Landslides*. Balkema, 1, 435–440.

Jacoby, S.L.S., Kowalik, J.S. & Pizzo J.T. 1972. Iterative methods for nonlinear optimization problems. Prentice-Hall.

Japan Road Association, 1999. Guidelines for slope stabilization works and slope stability. (in Japanese).

Jiang, J.-C. & Yamagami, T. 2006. Charts for estimating strength parameters from slips in homogeneous slopes. *Computer and Geotechnics*, 33, 294–304.

Li, T.D. & Zhao, Z.S. 1984. A method of back analysis of the shear strength parameters for the first time slide of the slope of fissured clay, *Proc., 4th Inter. Symp. on Landslides*, Toronto, 2, 127–129.

Michalowski, R.L. 2002. Stability charts for uniform slopes. *ASCE J. Geotech. Geoenvir. Engrg.*, 128 (4), 351–355.

Nelder, J.A. and Mead, R. 1964. A simplex method for function minimization. *Computer Journal*, 7, 308–313.

Nguyen, V.U. 1985. Determination of critical slope failure surfaces. *ASCE J. Geotech. Engrg.*, 111, 238–249.

Saito, M. 1980. Reverse calculation method to obtain c and ϕ on a slip surface. *Proc. 3rd Inter. Symp. on Landslides*, New Delhi, 1, 281–284.

Spencer, E. 1967. A method of analysis of the stability of embankments assuming parallel inter-slice forces. *Géotechnique*, 17 (1), 11–26.

Wang, G.H., Suemine, A., Furuya, G., Kaibori, M. & Kyoji Sassa. 2005. Rainstorm-induced landslides at Kisawa village, Tokushima Prefecture, Japan, August 2004, *Landslides*, 2, 235–242.

Wesley, L.D. & Leelaratnam, V. 2001. Shear strength parameters from back-analysis of single slips. *Géotechnique*, 51 (4), 373–374.

Yamagami, T. & Ueta, Y. 1989. Back analysis of average strength parameters for critical slip surfaces. *Proc. Computer and Physical Modelling in Geotech. Engrg.* (eds. A. S. Balasubramaniam, et al.), Balkema: Rotterdam, 53–67.

Development characteristics and mechanism of the Lianhua Temple Landslide, Huaxian County, China

Jia-yun Wang, Mao-sheng Zhang, Chuan-yao Sun & Zhang Rui
Xi'an Institute of Geology and Mineral Resources, Xi'an, China

ABSTRACT: The Lianhua Temple landslide is located in Huaxian county of Weinan City, Shaanxi province, China. On the basis of the analysis of its development characteristics and mechanism of the landslide, it is concluded that the destruction mode is a progressive process of sliding-tension along the joints and gneissosities, and that the contribution factors to the landslide include triggering and controlling factors.

1 INTRODUCTION

The Lianhua Temple landslide is located in Huaxian county of Weinan City, Shaanxi Province, and becomes one of the famous rock landslides with high speed and long slip distance because of its complicated geological background, the unique development characteristics, the perplexing mechanism and the long slip distance etc. Previous historical documents and surveys on the landslide show that the Lianhua Temple landslide was not the so-called avalanche triggered by the intensive earthquake in Huaxian county in 1556, instead, it is a super huge rock landslide with long slip distance and high speed which happened before this earthquake. Based on the engineering geological mapping, pitting and trenching, the authors expounded the terrain and physiognomy, the formation lithology, the characteristics of the tectonic, the rock mass discontinuity, the shape of the landslide, and the landslide bed, and discussed the characteristics of the landslide in the physical structure and the movement. In the end, the authors analyzed the mechanism of the landslide, including the deformation and destruction mode, as well as the triggering and controlling factors.

2 THE GEOLOGICAL CONDITIONS OF THE LANDSLIDE

2.1 *The terrain and physiognomy*

The Lianhua Temple landslide, Huaxian county lies in the north of the Qinling mountains, which are located in the steep transition belt that crosses the Qinling mountains and the Guanzhong basin. The elevation ranges from 300 m to 2500 m, so the hill has large relative elevation difference and steep terrain. The profile of the valleys is in "V" shape. The landslide is adjacent to the Mihu valley on the east, and connected with the Baila valley on the west. At the foot of the mountains are the alluvial and prolvial fans of the Baila valley and the Mihu valley, of which the terrain is flat and wide, while in the landslide deposit zone the terrain is uneven and disorderly.

2.2 *Formation lithology*

The parent rock of the landslide is gneiss in the archean era. Well-developed gneissosities down the slope are favorable to the occurrence of landslide. In the fault zones, rocks are fragments consisted of crushed rock and tectonite. Due to poor shear resistance, the fault zone is the favorable site for shear failure of the slope. The hill at the foot of the mountains is covered by five to six meters loess. Paleosol develops along the hill, and the vertical joints and the joints along the hill are well developed. During the accumulation, loess suffered so much dislocation that about 1 meter loess at the bottom of the loess slope has been immingled with the crushed rock, and that the arc interface, which is formed by the shearing of loess to the crushed rock, can be seen obviously.

2.3 *Tectonic characteristics*

The Lianhua Temple landslide is located in the Baoji-Tongguan fault zones, consisted of East-West or NWW complex folds and faults. Faults experienced several tectonic movements and their types transformed many times. In the early stage, their types were compression or compression-shearing faults. Later, strong tectonic movements transformed the faults into tensional or shearing faults (Xu Ren-chao et al. 1987) Due to multiple tectonic movements, rocks are crushed strongly and tectonic joints down the slope develop, which provide potential slip plane for the landslide.

3 DEVELOPMENT CHARACTERISTICS OF THE LANDSLIDE

3.1 The shape characteristics

The shape of landslide is like a fan, which spreads out from the rear to the middle and the front part (Figure 1). The slip mass is 5 km long from the south to the north, which is adjacent to the Mihu valley on the east and adjacent to the Baila valley on the west. The rear of the landslide is 1 to 1.5 km wide, the middle is 1.5 to 1.8 km wide, and the front part is 1.8 to 3.25 km wide. It covers an area of 7.5 km^2, and its volume is 300 million cubic meter. The head of the landslide is 1300 m above the sea level, and the shear-outlet is 550 m above the sea level. So the relative height difference between them is 750 m, and the one from the shear-outlet to the foot of the hill is 120 m. Therefore, there is high potential energy existing caused by the obvious relative height difference. After landing, the slip mass slides along the direction of 350°.

3.2 The characteristics of rock mass discontinuity

The parent rock of the landslide is gneiss with well-developed gneissosities. The occurrence of the gneissosities is 355°∠34°, which is almost parallel with the hill. And the rock mass discontinuity causes anisotropy and reduces the strength of the rocks. The faults experienced several tectonic movements and their types were transformed many times from the compression or compression-shearing faults in the early stage to tension or shear faults in the late stage. Due to multiple tectonic movements and transformation of faults types, tectonic joints of the rocks are well developed down the hill. The occurrence of the joints in the right sidewall of the landslide is 335°∠40°, and the one in the left sidewall is 0°∠40°. The well-developed gneissosities and tectonic joints along the hill reduce the stability of the slope, in the mean time, provide potential slip plane.

3.3 The characteristics of the landslide bed

When the rock mass slides out of the landslide bed with high speed, only small part of slip mass is left on the landslide bed. So most of the landslide bed is bare, which is 2500 m long and 500 m wide and can be seen obviously. The landslide bed with curving shape is narrow in the rear and wide in the middle. It constringes at the shear-outlet. The shape of the landslide bed is like fold line, and the average gradient of slope is 20° (Figure 2). The sidewall of the landslide bed is gentle in the west and steep in the east, of which the slope angel is greater than 70° in the east and is less than 60° in the west. And in the west, huge rocks scatter on the surface. There are large amount of mixtures of rocks and loess on the landslide bed. In the front, the diameter of the rock blocks is 5–50 cm with certain psephicity. And in the rear part, the diameter is over 1 m with poor psephicity. The slip direction of the slip mass changes several times from 320° to 355° and then to 310°, and the scrape on the right sidewall of the landslide can prove this.

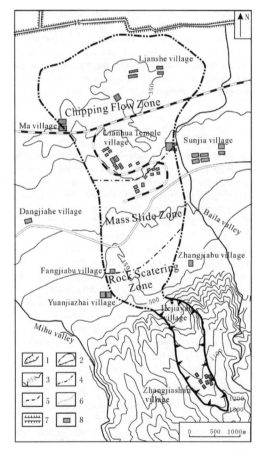

Figure 1. The shape of the Lianhua Temple landslide. 1—the perisporium of the landslide; 2—the boundary of the slip mass; 3—the contour line and altitude; 4—the zoning boundary of the slip mass accumulation; 5—railway; 6—highway; 7—channel; 8—residential area.

3.4 The physical structure characteristics of the landslide

The region of the slip mass accumulation can be divided into three zones: chipping flow zone, mass

3.5 The movement characteristics of the landslide

The large difference between the head and the shear-outlet of the landslide is 750 m. The great height difference causes high potential energy, which can be transformed into kinetic energy for landslide in high speed. During sliding nearly 2500 m along the landslide bed, the rocks impact the landslide bed continuously; at the same time, the potential energy is being converted into kinetic energy. So the accelerating effects like the elastic impulse a near the landslide bed and drop of peak-residual strength (Hu, 1988) appear. The sliding speed of rocks is accelerated and the rocks get high speed when they slide away from the shear-outlet. Because of the alluvial and proluvial deposits at the foot of the mountains, the tremendous impact makes the pore-water pressure in the deposits increase sharply and makes the deposits fluidified when the slip mass gets to the ground. So the resistance decreases and the slip mass can keep sliding in high speed. The friction caused by the slide in high speed makes the pore-water vapored and the air cushion effect forms on the interface between the slip mass and the alluvial and proluvial deposits, which keeps the slip mass slide in high speed. Because of the above reasons, the mass slide body can slide 2.5 km away from the shear-outlet. Due to wide and gentle terrain of the alluvial and proluvial fans and due to the effects of inertia and the pore-water, the mixtures of loess, crushed rocks and chippings flow down along the gentle slope, and the slip distance gets to more than 2 km. The longest slip distance of the slip mass is 5 km away. From the above analysis it can be seen that the landslide has the movement characteristics of high speed, long slip distance, the air cushion effect and the chipping flow effect.

4 THE MECHANISM OF THE LANDSLIDE

4.1 The mode of landslide destruction

Well-developed gneissosities in parent rocks and over-developed tectonic joints for multiple tectonic movements provide the potential slip plane. The occurrence of landslide bed is close to those gneissosities and tectonic joints, which indicates that the landslide is caused by deformation and destruction of sliding-tension of the gneissosities and tectonic joints for gravity. The historical records of Song dynasty indicated that the inhabitants in the mountains said that there were always clouds on the peak in a few years and the rocks sent out sounds when it rained. This is the evidence that the deformation and destruction of the landslide is the progressive process. From the above evidence, it is proved that the deformation and destruction is the progressive process of sliding-tension of the gneissosities and tectonic joints.

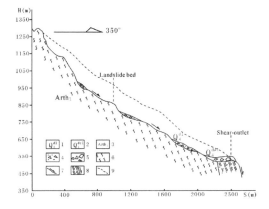

Figure 2. The profile of the landslide bed. 1—the slope deposit in Holocene; 2—the landslide deposit in Holocene; 3—Taihua group in archean; 4—mixture of rock blocks and loess; 5—rock block; 6—gneiss; 7—the landslide bed and slip direction; 8—fault and tectonite; 9—the original terrain line inferred.

slide zone and rock scattering zone. The rock scattering zone is below the shear-outlet. The surface is concave which was the Baiya lake, and it is covered by rock blocks with average diameter of more than 1 m and the largest diameter of 10 m. Below it is the 0.5 to 1 m mixture of loess and rock blocks and the loess has the evidence of extrusion. The mass slide zone is covered by 3–5 m loess. The waved paleosol are visible, though the loess is disturbed. The middle part of the mass slide body is the mixture of rock blocks and loess with several meters to more than tens meters thick. The rock blocks in the mixture have clear angularities and no psephicity. And the diameters of the rock blocks range from several centimeters to several meters. Loess in the mixture is brick red and compact, and it has signs sintered. In partial zones it can be seen that the rock blocks and chippings are turned on the loess. Below them is the gneiss. The gneiss is still continuous in general, and even the gneissosities can be seen, although it experiences strong disturbance. White belt of materials in the gneiss is sinter formed in the high temperature environment caused by slide friction. In the mass slide zone, the bottom boundary can not be seen clearly despite of the largest exposure thickness of over 30 m. The chipping flow zone is located in the gentle terrain to the north of the Yishan temple. Rock blocks in this zone is smaller and the diameters of the rock blocks are from several centimeters to more than tens centimeters, which are immingled with the loess and chippings. The mixtures of loess, chippings and rock blocks move forward several kilometers due to the Inertia and ground water (Figure 3), and the chipping flow zone comes into being.

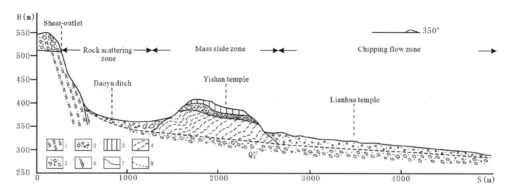

Figure 3. The profile of the slip mass in major slip direction. 1—tectonite; 2—mixture of rock block, chipping and loess; 3—loess; 4—gneiss; 5—proluvial deposit in late pleistocene; 6—the fault; 7—present terrain; 8—the original terrain inferred.

4.2 The triggering factors

4.2.1 Rainfall

The historical records indicate that it always rained and the rocks in the hill sent out sounds a few years before occurrence of the landslide. So the rainfall is the active factor that triggers the happening of the landslide. It includes two aspects: one is the decrease of rocks strength. The rainfall makes the water content of the rocks increase, even saturated. For the materials exchange of water and rocks, the mechanical properties of the rocks are changed, especially the shear resistance is decreased. The decrease of rocks strength causes the decrease of the stability of the slope. The other aspect is the effect of the pore-water pressure. The rainfall causes the joint network saturated and the pore-water pressure increase, so that the horizontal thrust and the buoyancy force to rocks increase. The horizontal thrust causes the increase of the pushing force of rocks, and at the same time, the buoyancy force reduces the sliding friction force, so the increase of pore-water pressure reduces the stability of the slope.

4.2.2 The effect of early earthquakes

The Lianhua Temple landslide is located in the down-faulted belt and earthquake belt of the Fen-Wei basin, and the borders of the basin are controlled by the normal faults with intense movements. In this area the tectonic movements of faults cause earthquakes, so the earthquakes epicenter always distributes along the faults belt. The historical records shows that the Lianhua Temple landslide happened in 1072 before the Huaxian county earthquake (1556), so the landslide has no relationship with this earthquake. While multiple earthquakes appeared in the eastern area of the Guanzhong Basin before the Lianhua Temple landslide (He, 1990). The intense shake changed the structures of the slope and the stress was adjusted several times. The stability of the slope decreased and the possibility of the destruction increased along with it.

4.3 The controlling factors

4.3.1 Great relative height difference and steep terrain

The altitude of the landslide zone ascends sharply from 300 m in the northern Fen-Wei basin to over 2 km in the north of Qinling mountains and the terrain is steep. The great relative height difference (750 m) from the head to the shear-outlet provides enough potential energy for the slip in high speed.

4.3.2 Shear-outlet in crushed fault belt with high free surface

The shear-outlet of the landslide lies in the fault belt and the lithology is made of crushed rocks with inferior cementation and low strength. So the shear-outlet zone becomes prone to destruction. The relative height difference of the free surface is 120 m, which provides favorable condition for slip of the slope.

4.3.3 Well-developed gneissosities and tectonic joints along the slope

The gneissosities is well developed in parent rocks, and the discontinuity of the gneissosities causes anisotropy and reduces the strength. Multiple tectonic movements cause the tectonic joints overdeveloped along the slope, which aggravates the anisotropy of rocks and reduces the stability of the slope. Well-developed gneissosities and tectonic joints along the slope not only reduce the stability of the slope, but also provide the landslide with potential slip plane.

5 CONCLUSIONS

According to the surveys, analyses and studies of the Lianhua Temple landslide, Huaxian county, the following conclusions can be obtained:

1. The Lianhua Temple landslide has the physical structure characteristics of chipping flow, mass slide and rock scattering.
2. The Lianhua Temple landslide has the movement characteristics such as high speed, long slip distance, the air cushion effect and the chipping flow effect.
3. The deformation and destruction of the Lianhua Temple landslide is the progressive process of sliding—tension along the gneissosities and tectonic joints.
4. The triggering factors of the landslide include the effect of early earthquakes and the rainfall, which reduces the strength of rocks and increases the pore-water pressure to cause horizontal thrust and buoyancy force; the controlling factors of the landslide include great relative height difference and steep terrain, shear-outlet in crushed fault belt with high free surface, and well-developed gneissosities and tectonic joints along the slope.
5. Further surveys and studies of the Lianhua Temple landslide are needed to find out the dynamic mechanisms.

REFERENCES

Xu Ren-chao, Zhong Li-xun, Pan Bie-tong & HU Guang-tao. 1987. The Stability Research of Slopes in the Southern Mountainous Area of Shaanxi Province.

Hu Guang-tao. 1988. *Dynamical Landslideology*. Xi'an: Shaanxi Science and Technology Press.

He Ming-jing. 1990. *The Hazards Research of the Earthquakes in Huaxian County*. Xi'an: Shaanxi People Education Press.

Cheng Qian-gong & Peng Jian-bing et al. 1999. *Dynamics of Rock Landslide with High Speed*. Chengdu: the Southwest Communication University Press.

Modeling landslide triggering in layered soils

R. Keersmaekers
Department of Architecture, Provincial University College Limburg, Hasselt, Belgium
Department of Civil Engineering, Building Materials Division, Katholieke Universiteit Leuven, Leuven, Belgium

J. Maertens & D. Van Gemert
Department of Civil Engineering, Building Materials Division, Katholieke Universiteit Leuven, Leuven, Belgium

K. Haelterman
The Authorities of Flanders, Geotechnics Division, Ghent, Belgium

ABSTRACT: The Flemish Government (represented by AMINAL; Administration of Environment, Nature, Land and Water management) commissioned a research project to study the triggering of landslides in the Flemish Ardennes. A few representative sites subject to landslides were studied from a geotechnical point of view. Several sites were selected for geotechnical calculations in order to predict the conditions necessary to trigger a landslide and to verify the predictions with the observations on site. Two of those sites are discussed in this paper. A hypothetical collapse mechanism, possibly responsible for many landslides in the Flemish Ardennes, was numerically verified. The presence of a sand layer (high water permeability) between two clay layers (low water permeability) causes the building up of pore water overpressures, decreasing the effective stresses, eventually resulting in the uplift and/or collapse of the slope. The understanding of the mechanisms responsible for the studied landslides resulted in specific recommendations to prevent future landslides on these sites, but also on similar sites in the region.

1 INTRODUCTION

1.1 Landslides in layered soils: causes and parameters

Insufficient safety of the global stability of a slope can cause a landslide that occurs along a slip surface (circular or not), or a slow displacement (creep) of the entire slope. Landslides can also be triggered by the seepage of ground water out of the slope, causing local erosion and caving.

An insufficient safety of the global stability of the slope may be induced by several parameters. The geometry of the slope is obviously a very important factor. A steeper slope will collapse more rapidly.

The shear strength characteristics of the involved soil layers play an even more important role. The friction angle $\varphi[°]$ and cohesion c [MPa] (cfr. the Mohr-Coulomb soil model) must be determined for every layer when evaluating the global stability of the slope. When checking the safety of a slope which already experienced a landslide, it is best to evaluate the stability also with the residual shear resistance parameters (De Beer, (1979)).

Another important parameter is the piezometric height of the water in the different soil layers, or the ground water level. The shear resistance is determined (in drained circumstances) by the effective stresses in the soil, which are directly correlated with the piezometric heights in the soil layers. When dealing with permeable layers with significant thickness, the groundwater level or the piezometric height is easily measured using water level tubes. The determination of correct piezometric heights becomes much more difficult when thin permeable layers occur within little permeable or impermeable layers. The variation of piezometrc height within an enclosed layer can be very high (i.e. during periods with heavy rainfall). In Figure 1 a hypothetical collapse mechanism caused by high piezometric levels is illustrated.

Seepage of rainwater through the top layer into the sand layer will cause high water overpressures in this permeable layer when the entrance water amount (e.g. true cracks) is bigger than the exit water amount (which is very limited because of the inclusion between two impermeable layers). This high water pressures result in low effective stresses (= total stress minus water

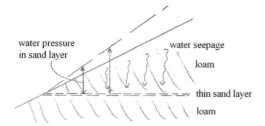

Figure 1. Typical failure mechanism. A permeable (sand) layer is enclosed by little permeable layers (e.g. loam or clay) resulting in high water overpressures and thus low effective stresses in the sand layer, resulting in collapse during heavy rainfall.

pressure) and thus lowers the shear resistance, causing the landslide to occur.

The global stability can also be affected when the load on the slope is altered. This can have a stabilizing or a destabilizing effect. When adding load (buildings, soil deposits, ...) at the top of the slope, this will have a destabilizing effect. Adding load at the toe of the slope will have a stabilizing effect.

1.2 Research aim

The Flemish Government (represented by AMINAL; Administration of Environment, Nature, Land and Water management) commissioned a research project to study the triggering of landslides in the Flemish Ardennes. This particular rolling region in Belgium has a history of landslides and general stability problems of slopes. The project marked out a part of the Flemish Ardennes, then produced an inventory, a classification, a statistical and spatial analysis and a methodology for the production of hazard maps (Van Den Eeckhaut et al. (2005)). Complementary, a few representative sites subject to landslides were studied from a geotechnical point of view. This geotechnical study is the topic of this paper.

Three sites were selected for geotechnical calculations in order to predict the conditions necessary to trigger a landslide and to verify the predictions with the observations on site. Two of those sites are discussed in this paper. The study of each site started with the execution of soil investigation tests (Cone Penetration Tests, borings and triaxial shear tests on representative non disturbed soil samples) to draw a geotechnical profile of the slope, determining the stratification of layers and their geotechnical parameters. Combined with topographical data of the collapsed slope and the assumed profile before the landslide occurred, numerical models of these sites were implemented using two software programmes based on numerical (finite elements method using PLAXIS, www.plaxis.nl) and analytical (SLOPE, www.geo-slope.com) mathematical algorithms.

With these models it is possible to calculate an overall factor of safety to evaluate the stability of the slopes. An hypothetical collapse mechanism as defined in paragraph 1.1 (See Figure 1), possibly responsible for many landslides in the Flemish Ardennes, will be numerically verified. The presence of a sand layer (high water permeability) between two clay layers (low water permeability) causes the building up of pore water overpressures, decreasing the effective stresses, eventually resulting in the collapse of the slope.

Numerical simulations also proved to be very useful to determine the negative influence on the overall slope stability of adding additional loads to the upper slope surface (buildings), the slightly positive influence of vegetation and the negative influence of excavations (swimming pools, ponds) at the bottom of the slope.

The understanding of the mechanisms responsible for the studied landslides results in specific recommendations to prevent future landslides on these sites, but also on similar sites in the region. For a more detailed publication in Dutch of this work see (Keersmaekers et al. (2005)).

2 MODELLING

2.1 Analytical method: SLOPE (Bishop)

The development of calculation methods to check the global stability of slopes with arbitrary shapes and materials with cohesion and friction goes back for decades. Firstly only circular slide surfaces were considered (Figure 2). Later on calculation methods were developed for irregular slide surfaces. Recently, the use of finite element methods is becoming more and more common.

The most used analytical method to evaluate circular slip surfaces is the Bishop method. The soil above

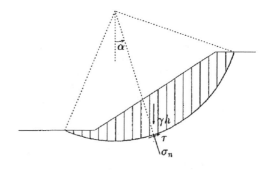

Figure 2. Circular slide surface for analytical calculations conform Bishop. This analytical method divides the slide volume in vertical strips.

762

the slip surface is divided in vertical strips, bounded by vertical surfaces. A shear stress t is mobilized along the slip surface, which is supposed to be a factor SF (safety factor) smaller than the maximum possible shear resistance. A 2D-geometry is assumed.

Verifying the stability of the slope consist of the expression of the momentum equilibrium to a center point of the considered circular slip surface. The calculation of SF is done iteratively (starting with $SF = 1$) and must be done for many center points (i.e. many possible circular slip surfaces) to find the lowest value of the SF.

The software package used in this research project to calculate the Bishop method is Slope.

2.2 Finite element method: PLAXIS

Most calculations in this research project were made with Plaxis, a finite element based software package, especially developed for geotechnical applications. The volume is divided into small elements which are numerically coupled to each other. The stress equilibrium and the soil deformation are described by a system of regular and partial differential equations which are solved numerically. In this way the soil stresses and deformations can be calculated.

The advantage of using finite elements is that the real behavior of the soil is better simulated and that the real stresses occurring in the soil are taken into account. There is for example a clear division between vertical and horizontal stresses which is not the case in analytical methods.

To model the behavior of the soil a so called Mohr-Coulomb model has been used. This model assumes a complete elastic behavior until the shear stress in the element equals the shear resistance. After this point the soil behaves completely plastic.

For the determination of the safety factor (SF as defined above), the so called phi, c (φ, c)-reduction method is used. The shear resistance parameters φ' and c' are reduced in the same way until collapse of the slope. This collapse is verified by the displacement of one or more well chosen physical points.

2.3 Safety factor

When slopes are designed with the analytical Bishop method, a safety factor SF of minimum 1.3 is normally required. The calculation is then based on the momentum equilibrium of a circular slip surface.

The calculation of the safety factor using finite element methods is based on the phi, c (φ, c)-reduction method. Many calculations in the past learned that the overall safety factor, obtained from analytical methods like Bishop do not differ a lot from the safety factor obtained using the phi, c (φ, c)-reduction method (finite elements).

When evaluating the Slope and Plaxis calculations in this project, an overall SF of 1.3 is required to have a save slope. SF-values lower than one indicate slopes that will collapse under the given parameters. SF-values between 1 and 1.3 correspond to slopes with insufficient safety, but will not necessarily collapse.

3 SITE1: SOCCER FIELD "KORTE KEER" AT MAARKEDAL

3.1 Situation

The first site is the football field Korte Keer at Maarkedal (Nukerke, Belgium). The site is located above a very large, deep and old landslide. The football field itself was built on a slope by rectifying the terrain by filling the site with a sandy embankment. From a geotechnical point of view, this added material has similar properties as the top layer of the original slope (see hereafter).

Figure 3 shows the situation of the site Korte Keer. The line defines the topographic profile used to model the geometry of the collapsed and original slope (see also Figure 4). The original profile of the football field had a steep inclination of 1/1.

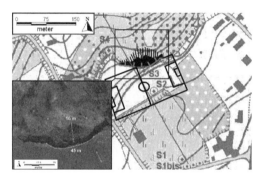

Figure 3. Situation of the site Korte Keer. The line defines the topographic profile used to model the geometry of the collapsed and original slope.

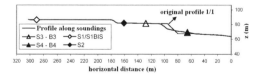

Figure 4. Profile before and after the event of the landslide. Notice the location of the soundings.

3.2 Characteristics of the soil layers

There were five CPT-tests made on site with a 200 kN apparatus, which probed to depths until 25 meters. Also two borings and triaxial tests on undisturbed soil samples were performed. The information obtained from these tests are than used to determine the stratification and characteristics of the different soil layers. The locations of these CPT-tests are given in Figures 3 and 4. Also a piezometric pipe was installed to monitor the variation of the water table over one year.

From the on-site investigations it was concluded that the site consists of two major layers. The top layer is a well permeable sand layer which is partly constructed from sand deposits to rectify the slope and the underlying original sandy material (quaternary origin). Both layers have very similar characteristics and are therefore modeled as one top layer. Under this top layer, a clay layer (from tertiar origin, i.e. "Ieperiaan") is found with an almost horizontal orientation. Table 1 summarizes the main characteristics of both layers.

A Mohr-Coulomb soil model is used to define the two layers in Plaxis. Two models are constructed. The first one with the original slope geometry and one with the present profile. This second profile, representative for the situation after the landslide, aims to determine the present safety of the site (see Figure 5).

3.3 Results of the calculations

3.3.1 Original profile with inclination 1/1
Figure 6 gives the slip surface of the original slope. The safety factor SF is calculated to be 0.561, meaning that the collapse of the site was inevitable. The inclination 1/1 is too steep. The calculation is made with a deep water table, which proves the landslide was not due to water overpressures.

3.3.2 Profile after landslide
The measurements of the water level pipe showed a maximum piezometric height of +73.37 TAW (TAW is the reference level for Belgium) between September and October 2004.

The SF-value obtained was 1.218, meaning the slope will not collapse, but has an insufficient SF according to literature (minimum SF = 1.3). When the piezometric height of the top layer increases to +76 TAW (i.e. 2.63 m higher than in the above situation), the SF-value drops to 0.997, meaning the slope has just reached equilibrium, the landslide can occur at any moment.

The influence of the growth of plants on the site is incorporated by giving the first meter of the top soil layer a cohesion of 5 kN/m². The SF increases

Table 1. Ground characteristics of the different soil layers.

Parameter	Top sand layer		Bottom clay layer	
	Value	Unit	Value	Unit
γ_{drv}	18	kN/m³	19	kN/m³
γ_{wet}	19	kN/m³	19	kN/m³
E-mod	2E+4	kN/m²	1E+4	kN/m²
Poisson v	0.3	[-]	0.35	[-]
Cohesion c	0.0	kN/m²	25	kN/m²
φ	27.5	°	23	°

Figure 6. Slip surface of the original slope. The safety factor SF is 0.561, meaning that the collapse of the site was inevitable.

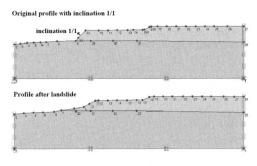

Figure 5. Plaxis geometric models for the original profile and after the landslide occurred.

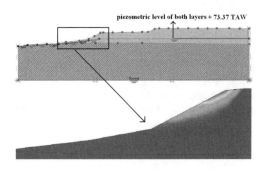

Figure 7. Top: Piezometric heights at +73.37 TAW of both layers. Notice the drop in water level towards the toe of the slope. Bottom: Preferential slip surface, the safety factor SF is 1.218.

from 0.997 to 1.074, which proves the positive effect of vegetation.

Comparison with a Slope calculation (analytical model) resulted in a SF-value of 1.08 (compared to 0.997), which proves that the overall safety factor, obtained from analytical methods like the Bishop method does not differ a lot from de safety factor obtained using the phi, c (φ, c)-reduction method (based on finite elements modeling). Also the slip surface showed a very similar sliding surface, showing a rather shallow collapse of the slope.

Finally a calculation was made to rebuild the football field without the danger of triggering a landslide. The piezometric height of the top layer was chosen one meter below the surface (this can be done by placing a drainage system under the new football terrain and a drainage at the toe of the slope). A minimum inclination of the slope of 12/4 resulted in a SF of 1.333, reaching the minimum required SF-value of 1.3.

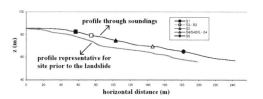

Figure 9. Profile before and after the advent of the landslide. Notice the location of the soundings.

Table 2. Ground characteristics of the different soil layers.

Parameter	Top clay Value	Middle sand Value	Bottom clay Value	Unit
γ_{dry}	18	17	19	kN/m^3
γ_{wet}	18	20	19	kN/m^3
E-mod	1E+4	1.3E+4	1E+4	kN/m^2
Poisson v	0.35	0.30	0.35	[-]
Cohesion c	10	1	20	kN/m^2
φ	25	30	25	°

4 SITE2: SCHERPENBERG RONSE

4.1 Situation

The Scherpenberg is a complex landslide on a slope without buildings. Therefore it is easily accessible for cone penetration test and chosen for this project.

Figure 8 shows the situation of the site Scherpenberg. The dotted line defines the topographic profile used to model the geometry of the original slope. The continuous line defines the topographic profile used to model the geometry of the collapsed slope. Notice the location of the CPT-tests (see Figure 9).

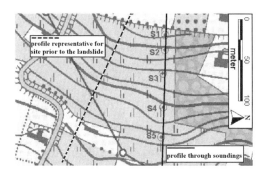

Figure 8. Situation of the site Scherpenberg. The dotted line is representative for the original slope (no landslide has occurred there). The continuous line defines the topographic profile used to model the geometry of the collapsed slope. Notice the location of the soundings.

4.2 Characteristics of the soil layers

There were five CPT-tests made on site with a 200 kN apparatus, which probed to depths until 25 meters. Also two borings were made and triaxial shear tests on undisturbed soil samples were performed. The information obtained from these tests are then used to determine the stratification and characteristics of the different soil layers. The locations of these CPT-tests are given in Figures 8 and 9.

From the on-site investigations it was concluded that the site consists of three major layers. The middle layer is a well permeable sand layer which is enclosed between two low permeable clay layers. Table 2 summarizes the main characteristics of the three layers.

A Mohr-Coulomb soil model is used to define the three layers in Plaxis. Two models are constructed. The first one with the original slope geometry and the second one with the present profile. This second profile, representative for the situation after the landslide, aims to determine the present safety of the site (Figure 10).

4.3 Results of the calculations

4.3.1 Presumed original profile

Figure 11 gives the slip surface of the presumed original slope. The piezometric height of +65 TAW (for the middle and bottom layers, for the top clay layer the piezometric height is assumed one meter under the ground surface) is the maximum value measured on

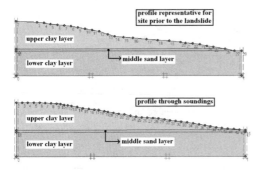

Figure 10. Plaxis geometric models for the presumed original profile and after the landslide occurred.

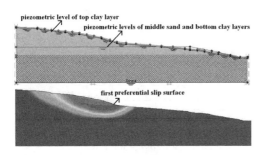

Figure 11. Top: Piezometric heights of the middle and bottom layers at +65 TAW. Bottom: Preferential slip surface at the top of the slope, SF = 1.526.

Figure 12. Piezometric heights of the middle and bottom layers at +68 TAW. Preferential slip surface, SF = 1.026.

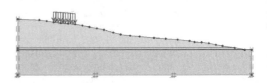

Figure 13. Plaxis geometric model including the load representative for a building at the top of the slope.

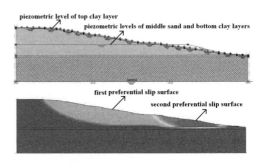

Figure 14. Top: Piezometric heights of the middle and bottom layers at +68 TAW. Bottom: The preferential slip surface shifts from the top to the bottom of the slope, SF = 1.865.

site between November 2004 and January 2005. The safety factor SF is calculated to be 1.526, meaning that the site is safe under these conditions. Notice the preferential slip surface at the top of the slope.

Raising the piezometric level of the middle and bottom layer to +68 TAW results in a SF of 1.026 (meaning the slope has just reached equilibrium, the landslide can occur at any moment), and a different preferential slip surface at the bottom of the slope is obtained (Figure 12). This means that when a landslide will occur, the slope will collapse according to this second slip surface.

When adding a load of 40 kN/m^2 at the top of the slope (representative for a building on a fill layer, Figure 13), for the same conditions of Figure 11 (+65 TAW, SF = 1.526), the SF drops to 1.264. In case the load is 60 kN/m^2, the SF drops to 1.155, proving the negative effect of load at the top of the slope.

4.3.2 Profile after landslide

Figure 14 gives the slip surface of the slope after occurrence of the landslide. The piezometric height for the middle and bottom layers is set at +68 TAW, for the top clay layer the piezometric height is set one meter under the ground surface.

The safety factor SF is calculated to be 1.865, meaning that the site is safe under these conditions (which are unlikely to occur in reality). For lower water levels the slope gave even higher SF's and preferential slip surfaces at the top of the slope. Notice the shift from a high to a low preferential slip surface in Figure 14.

When the piezometric height for the middle and bottom layers is set at +71.5 TAW, SF drops to 1, 212, resulting in a slip surface at the bottom of the slope (cfr. Figure 15). For a TAW level +72.5, SF = 0.952, meaning the slope has just reached equilibrium, the landslide can occur at any moment.

Figure 15 shows the negative influence of a swimming pool at the bottom of the slope. The empty swimming pool is lifted, SF drops from 1.212 to 1.025, respectively without and with swimming pool.

Finally the influence of the decrease of the strength characteristics due to an already occurred landslide is taken in to account. Figure 16 shows the Plaxis

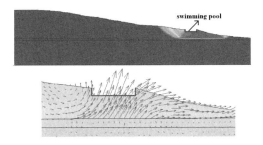

Figure 15. Top: Influence of a swimming pool at the bottom of the slope. Piezometric heights of the middle and bottom layers at +71,5 TAW. Bottom: Incremental displacement vectors: The empty swimming pool is lifted, SF drops from 1.212 to 1.025, respectively without and with swimming pool.

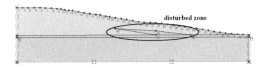

Figure 16. Plaxis geometric model including the presumed disturbed zone (based on the sounding profiles) of the former landslide.

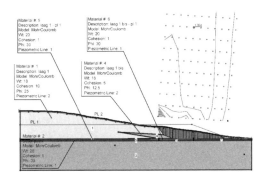

Figure 17. Slope calculation (Bishop method) of a geometric model including the presumed disturbed zone of the former landslide, $SF_{Slope} = 1.094$.

model including the presumed disturbed zone (based on the sounding profiles) due to the occurred landslide. In (De Beer, (1979)), values for the residual shear strength characteristics for quaternary clay are given: $\varphi'_r = 12.5°$ and $c'_r = 5\ kN/m^2$. The piezometric heights of the middle and bottom layers are set at +71,5 TAW. The SF is 1.037 (compared to 1.212 above) proving the negative influence on the present stability of the already disturbed zones in the top layer. This result was verified with Slope, using the Bishop method (Figure 17), giving a value $SF_{Slope} = 1.094$ (almost equal to 1.037 above).

5 CONCLUSIONS

This study demonstrates that the presence of a sand layer between two clay layers may cause the building up of pore water overpressures, decreasing the effective stresses, eventually resulting in the collapse of a slope (cfr. Figure 1).

Numerical simulations proved very useful to determine the negative influence on the overall slope stability of adding additional loads to the upper slope surface (buildings), the positive influence of vegetation and the negative influence of excavations (swimming pools, ponds) at the bottom of the slope.

REFERENCES

De Beer, E. 1979. Historiek van het kanaal Lei-Ieper; Eigenschappen en gedragingen van Ieperiaanse klei. Uittreksel uit het tijdschrift der openbare werken van België nrs 4, 5 en 6.

Keersmaekers, R., Maertens, J. & Van Gemert, D. 2005. Verkennende studie met betrekking tot massabewegingen in de Vlaamse Ardennen. Deel II: Geotechnisch onderzoek van enkele representatieve sites onderhevig aan massabewegingen. Rapport Laboratorium Reyntjens R/30232/04. Rapport in opdracht van de Vlaamse Gemeenschap, AMINAL, afdeling Land.

Van Den Eeckhaut, M., Poesen, J., Verstraeten, G., Vanacker, V., Moeyersons, J., Nyssen, J. & van Beek, L.P.H., 2005. The effectiveness of hillshade maps and expert knowledge in mapping old deep-seated landslides. International journal on Geomorphology 67, p351–363.

Numerical modeling of debris flow kinematics using discrete element method combined with GIS

Hengxing Lan & C. Derek Martin
Dept. Civil & Environmental Engineering, University of Alberta, Edmonton, Alberta, Canada

C.H. Zhou
LREIS, Institute of Geographic Sciences and Natural Resources Research, CAS, Beijing, P.R. China

ABSTRACT: Numerical modeling of debris flow kinematics was performed using the integration of Discrete Element Method and Geographic Information Systems (GIS). A discrete element code was used to investigate the dynamic behavior of debris flows in mountainous terrain. The macroscopic dynamic behavior of the debris flows was evaluated using a spatial averaging method implemented in GIS. The coupling of the discrete element method with GIS technology was evaluated at a debris flow site along a railway corridor in the Rocky Mountains of Western Canada.

1 INSTRUCTION

A debris flow represents a mixture of sediment particles of various sizes and water flowing down a confined or unconfined channel (Hutter et al. 1995). Its physical and mechanical behavior exhibits a distinct discrete nature which is characterized by the internal interaction between solid particles and between solid and fluid in the mixture of the debris flow body. The solid components are characterized by a broad distribution of grain sizes, from boulders to clay particles (Pirulli M. & Giuseppe S. 2007). The characteristics of flow kinematics also vary widely depending on the fluid condition and particle size which result in large challenges in numerical modeling.

Continuum mechanics using depth averaging methods is the current principle approach for the dynamic modeling of debris flows (Chen & Lee 2000; Denlinger & Iversion 2001; McDougall & Hungr 2005). Physically, this approach has difficulties accounting for the microscopic behaviors of debris flow materials in terms of transportation kinematics and collision or friction interaction. Such an approach can not lead to the construction of a clear picture about the debris flow dynamics in an irregular channel (Zhu & Yu 2005).

The Discrete element method is recognized as an effective approach to study the fundamental behaviour of granular materials (Cundall & Strack 1979). Compared with the continuum techniques, it has advantages of allowing us to elucidate the internal properties of debris flow such as internal stress distribution (Zhu & Yu 2005). The microscopic quantities such as velocities of particles and interaction force between particles, ground surface can also be examined. Difficulties exist using the discrete element method in interpreting the output at a macroscopic scale due to the large local divergence during the simulation (Richard et al 2004). A range of methods can be used to extract the useful quantitative macroscopic information for the debris flow kinetics such as spatial averaging or Fourier transform.

In this paper, an integration of discrete element code (PFC3D) and Geographical System Information (GIS) is used for the numerical modeling of debris flow kinematics. PFC3D is used to investigate the microscopic dynamic behavior of debris flow in an irregular ground-surface channel. The macroscopic dynamic behavior of the debris flow is inspected using a spatial averaging method implemented in GIS. The integration methodology is applied to a debris flow site along a major railway near Klapperhorn Mountain in the Canadian Rockies of Western Canada.

2 STUDY AREA

The study area is located on Klapperhorn Mountain near the border between Alberta (AB) and British Columbia (BC), Canada (Fig.1). The site is within the Rocky Mountains Proterozoic Middle Miette Group, which includes feldspathic sandstone, granule and pebble conglomerate, siltstone and argillite (Mountjoy, 1980). Two rail-lines operated by CN (Canadian National Railway), the Albreda and Robson subdivisions run parallel to each other at the base of Klapperhorn Mountain. The 3 km long section of

Figure 1. Study area of Klapperhorn Mountain. Five debris flow drainage basins are identified. Photo shows the overview of study area. The Albreda (upper) and Robson (lower) railway Subdivisions along the base of Klapperhorn are also shown. The main creek in the foreground is at Mile 54.3 of the Albreda Subdivision. The rockshed at Mile 54.7 is visible (From Davies, 2005). Photo in the top left corner show the debris flow source area.

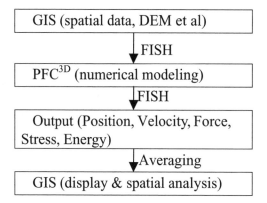

Figure 2. Integration of PFC3D and GIS (Geographic Information System).

both railways has been impacted by debris flow events in the past. Most of these debris flows have reached the track elevation with varying consequences. Figure 2 clearly shows the feature of distinct debris flows in the study area, particularly at mileage 54.3. At this location a major channel has formed and this channel has been engineered to reduce the impact to the track.

Davies et al (2005) classified the study area into six drainage basins based on detailed air photograph interpretation and field survey works. The characteristics of the debris flow drainage basins are described in Davies' work including the feature of debris flow creeks, debris source areas, debris deposition areas and their transportation. In general, the bedrock in upper drainage basin dominates the terrain units. Rockfalls and snow avalanches occurring frequently in this region deposit material at the top of a colluvial cone where most of the debris flows initiate. Debris transport occurs in the unconfined or confined drainage channel whose morphology is usually bedrock controlled. Debris flows deposit debris on the colluvial fans at the end of primary or secondary channels which can be above or below the track level.

3 NUMERICAL SIMULATION OF DEBRIS FLOW KINEMATICS

3.1 Method

The integration of discrete element code (PFC3D) and GIS functions was implemented to facilitate the numerical modeling of debris flow kinematics and to facilitate the interpretation of the results (Fig. 2).

The geospatial data related to the debris flow characteristics are managed using the GIS system, such as debris flow channel topography, debris flow mode (confined or unconfined), source and deposition region delineation. For example, the Digital Elevation Model (DEM) for the debris flow channel topography can be prepared in GIS using TIN format (Triangulated Irregular Network). It is then imported into PFC3D as irregular walls to indicate the mountain surface by writing a FISH function. FISH is an embedded programming language in PFC3D. Figure 3 shows the three dimensional topography of debris flow channel in Klapperhorn Mountain area imported into PFC3D from a digital elevation model generated in GIS.

Numerical modeling of debris flow was performed in PFC3D with the fluid option. This approach ensures that the 3D kinematics of the flowing mass, the force

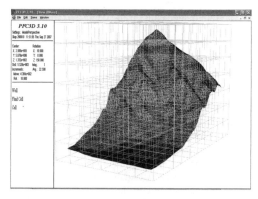

Figure 3. GIS was used to create the digital elevation model (DEM) for the site and export the topography as a triangulated irregular network. These were then imported into PFC3D. The screen shot represents the final DEM in PFC3D covered with fluid domain.

transmission within the granular mass and in particular the lateral expansions are properly accounted for. A "fixed coarse-grid fluid scheme is implemented in PFC3D for particle-fluid coupling simulation. This scheme solves the continuity and Navier-stokes equations for incompressible fluid flow numerically in an Eulerian Cartesian coordinate system, and then derives the pressure and fluid velocity for each fixed grid (or cell) by including the influence of particles, and the corresponding porosity, within each cell. Driving forces from the fluid flow are applied to the particles as body forces. These forces are also added to the fluid equations and cause change in momentum, as reflected by the change in the pressure gradient in the flow direction (Manual of PFC3D). The present formulation in PFC3D only applies to fully saturated, fixed, fluid domains. Interaction is only between the fluid and particles; fluid interaction with walls is not included. Also, fluid boundaries can only be specified for a rectangular fluid domain; arbitrary fluid boundaries cannot be given.

During the simulation, the status of each particle was recorded at a certain time interval, for example every 5 seconds. It includes the position of each particle, velocity, displacement, force, stress, stain rate and energy. Using FISH, this information can be exported to ASCII files and then import into GIS system for interpretation.

Spatial averaging on the numerical result is performed by means of a spatial mean interpolation approach in the spatial analyst extension of GIS. The discrete time series data generated by PFC3D are then converted into various continuous raster surfaces. The raster surfaces could represent the spatial distribution of kinetic property of the flowing mass in terms of velocity, stress or thickness at a certain time step. Figure 4 shows an example of converting particle velocity calculated from PFC3D into a velocity raster surface in GIS. It can be seen that using raster surface, the modeling results from discrete element code can be easily interpreted and provide useful information for the further analysis and prediction.

Five energy and work terms are also be calculated and traced in every PFC3D run including body work, boundary work, strain energy, kinetic energy and friction work. In this paper, we focus on the kinetic energy and friction work. The kinetic energy represents the total kinetic energy of all particles, accounting for both translational and rotational motion. The friction work indicates the total energy dissipated by frictional sliding at all contacts.

3.2 Procedure

Specific aspects must be considered in the debris flow model creation and solution including: the assembly of particle generation and compaction; choice of contact model and material properties; boundary and initial conditions specification; loading, solution and sequential modeling; and finally interpretation of results.

According to the field survey, the boundaries for debris flow material source were identified. Four new boundary walls were created in PFC^{3D} to specify the generation region for the assembly of particles. The particles are statistically uniformly distributed with radius from 0.5 meter to 2.5 meter. Under gravity, the assembly was compacted and reached equilibrium state on the irregular debris channel wall. In this study area, the assembly has an approximate porosity of 0.4. Three different source boundaries were specified to investigate the effect of magnitude of debris flow source indicated by the number of particles. In this study, we generated three different assemblies of particles which have 600 particles, 1604 particles and 7309 particles respectively.

The particles were released by deleting the boundary walls. Different material properties were set before the particles release including contact property (elastic stiffness and Coulomb's friction), damping ratio and fluid conditions. In order to account for the modulus of debris flow mass, a modulus-stiffness scaling relation was employed (Potyondy & Cundall 2004). The normal and shear stiffness of particles are scaled with the particle radii to achieve a constant grain modulus.

A rectangular fluid domain was setup covering the debris flow channel. The fluid pressure boundary condition was assigned based on the assumption that the fluid flows parallel to the slope. So the fluid pressure boundary was applied to ensure the fluid pressure gradient is same as slope gradient.

3.3 Parameter calibration

The input parameters for debris flow numerical simulation cannot be measured easily, since the understanding of rheology or mechanics of the flowing material

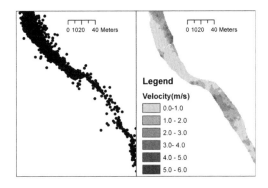

Figure 4. Converting particle velocity calculated from PFC^{3D} into velocity raster in GIS.

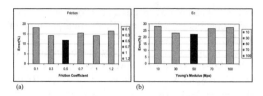

Figure 5. Parameters calibration. (a) Friction coefficient calibration; (b) Contact properties calibration.

is still limited. In discrete element method, the contact force and displacements of a stressed assembly of particles are determined mainly by contact properties, frictional and damping coefficient. The parameter calibration was conducted using a trail-and-error analysis method. It takes into account both the debris flow run out and the shape of debris flow transportation.

From the field work and aerial photo interpretation, the detailed debris flow channel path can be delineated indicating the debris flow initiation, transportation and deposition. With proper parameters, the simulated particles should travel within the defined debris flow channel. The particles dropping out the debris flow channel boundary are considered as wrong particles. The percentage of wrong particle to total particles is defined as a modeling error. Figure 5 show the calibration result for the frictional coefficient and contact properties determined by Young's modulus. The process was conducted using a smaller assembly of particles (600 particles). The calibration results show that using frictional coefficient of 0.5 (friction angle is about 26 degree) and Young's modulus of 50 Mpa, a model could obtain a better result.

The damping coefficient affects the kinematics of the flowing mass in an evident way. However this parameter is not clearly related to any physical mechanism. The realistic value must be obtained by back-analysis of the experimental data. Crosta et al (1991) found that using damping coefficient = 3%, the numerical model can reproduce the experimental flow by Hutter et al (1995) with high satisfactory. So in this study, the damping ratio of 3% was utilized for all the simulations.

4 RESULT AND DISCUSSION

4.1 Results

A real scale three dimensional numerical model was established and performed for debris flow M54.3 in Klapperhorn Mountain study area. A series of pictures in Figure 6 show the typical configuration of flow mass at different phases of debris flow dynamics. The transportation of debris flow mass is highly controlled by the irregular channel topography. The interaction of

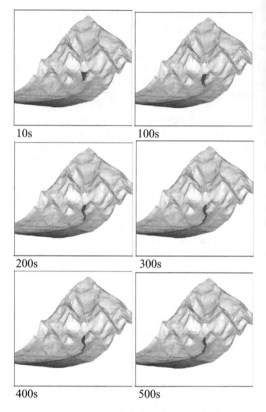

Figure 6. A Series of screenshots at different time indicating the debris flow kinematics.

particles resulted in the various flow mass configurations during the flowing process. A quite compact granular mass configuration dominates the rear part of the flow while the debris flow mass front is more dispersed with time.

The most reprehensive component of the flowing kinematics is the velocity. The macroscopic velocity distribution during the flowing process is shown in Figure 7. A very non-uniform velocity distribution was observed. At the initiation stage, the magnitude of velocity increases gradually away from the source region indicating the kinetic energy propagates downward the slope. Most of the particles obtain a high velocity with a range from 3 m/s to 5 m/s when they reach the central part of the debris flow channel and then decrease gradually when travel into the lower part of the channel.

4.2 Sensitivity analysis for debris flow kinematics

A preliminary analysis of the influence of the numerical parameters on the debris flow kinematics was

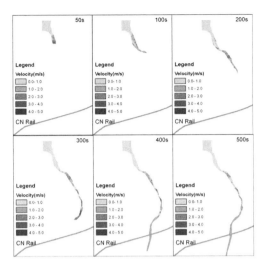

Figure 7. Velocity distribution during the simulation.

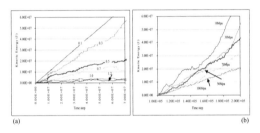

Figure 8. (a) Correlation between kinetic energy and friction coefficient, (b) Correlation between kinetic energy and contact properties determined by Young's Modulus.

performed. The effect of friction coefficient and contact properties on the kinetic energy of flowing mass is shown in Figure 8a. As expected, the results show the total kinetic energy of flowing mass has distinct dependency on the friction coefficient between particles and between particles and walls. The kinetic energy decreases dramatically with the increase of contact friction coefficient. The lower kinetic energy accounts for a lower runout distance for the flowing mass. The high dependency of kinetic energy on the contact stiffness can also be observed from Figure 8b. The contact stiffness is determined by the Young's modulus using a stiffness scaling method. The loose flowing mass with lower Young's modulus tends to have a higher mobility than the dense flowing mass with high Young's modulus. Therefore the effect of contact stiffness can not be neglected for the simulation of flowing mass kinematics. The effect of size of particle and assembly on the kinetic energy and its dissipation was also analyzed. The general trend is that an increase of the size of assembly of particles results in a larger energy dissipation and the larger particle tends to obtain higher kinetic energy during the simulation.

4.3 Velocity profile

Velocity profiles were generated along the longitudinal section along central and left line of debris flow channel (Figure 9). The velocity along both central line and left line was vibrating frequently due to the algorithm PFC uses and the influence of irregular surface. It can be seen for the central line velocity profile, the velocity vibrates strongly at the upper flow path indicating a high particle interaction at the steeper channel portion. In the lower part of flowing path, the velocity of debris mass decreases gradually until it vibrates again when reaching the deposition area. The velocity profile adjacent to the left bank of the channel shows an irregular vibrating character which might indicate the interaction between debris flow mass and the left boundary of channel.

In order to study the effect of topography on the debris flow kinetics, statistical correlation between velocity and slope of flowing path was conducted. It can be found that at a certain time period during the flowing process, the particles traveling along the steeper portion do not necessarily obtain a higher velocity compared to those along the gentle portion. However they show a higher potential to gain the higher kinetic energy since they definitely obtain a higher velocity changing rate (or strain rate) when they travel into a steeper terrain (Figure 10a).

Figure 10b show the velocity profiles along a cross section located at 300 m far away from the debris source at different time. The pattern of the curves at time = 200 s and time = 400 s show a general trend that the velocity at the center of the channel has a higher magnitude but the curves at time = 300 s and time = 500 s show an opposite phenomenon. It, on the other hand reveals that the discrete element method can

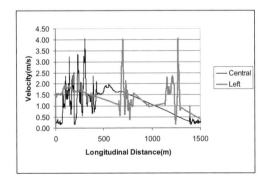

Figure 9. Velocity profile along longitudinal section in central and left line of the debris flow channel.

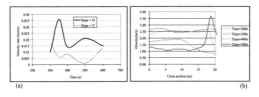

Figure 10. (a) Velocity rate vs. Time for flowing mass traveling on different terrains. (b) Velocity profile at different time along cross section which is 300 meter away from source.

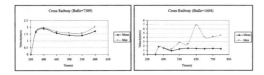

Figure 11. Velocity profile of flowing mass crossing the railway track. (a) large assembly of particles; (b) small assembly of particles.

capture the transient feature of the kinetic properties of the flowing mass.

4.4 Velocity profile crossing railway

A big concern for the debris flow hazard analysis along railway is to predict the velocity or kinematical energy of flowing mass when reaching and crossing the railway track. The maximum and mean velocity profile were prepared using two different simulations (small and large assembly of particles) to indicate the kinetic property of flowing mass crossing the railway track at Albreda subdivision (Figure 11). It can be seen that the mean velocity profiles from the two different simulations show a similar tend. They are primarily changing in the range from 1 m/s to 2 m/s. But the maximum velocity profiles exhibit a large difference. The particles in the smaller assembly tend to get higher maximum velocity due the smaller energy dissipation by interaction of particles. The kinetic energy could be easily calculated when taking into account the mass of flowing mass.

5 CONCLUSIONS

A debris flow composed of complex mixture of materials exhibits a distinct discrete nature in its physical and mechanical behavior. The continuum theory which is widely used for flowing dynamics has difficulties accounting for the internal interaction in the solid-fluid mixture of the debris flow mass.

The modeling the debris flow kinematics using the discrete element method accounts for the microscopic behavior of the flowing mass. PFC3D was used to investigate the microscopic dynamic behavior of debris flows over an irregular ground surface.

The numerical output of discrete element code was easier to interpret when integrated with a GIS system. A time series of continuous raster surfaces was generated using a spatial averaging approach in GIS to represent the macroscopic behavior of the debris flow kinematics, which provided useful information for further analysis and prediction for the debris flow hazard.

The debris flow kinematics is highly dominated by the characteristics of the irregular ground surface which has a direct effect on the internal interaction of the flowing mass. The most reprehensive component of the flowing kinematics is the velocity. The spatial distribution of velocity has been studied in detail for a debris flow in the Klapperhorn Mountain. The potential kinetic energy of the flowing mass reaching or crossing the railway track at the base of the mountain could be evaluated using the predicted velocity magnitude.

ACKNOWLEDGMENTS

This research was supported by the Canadian Railway Ground hazard Research Program http://www.tc.gc.ca/tdc granted by CN, Pacific Railway, Transport Canada and Natural Sciences and Engineering Research Council of Canada, and partly by the National Science Foundation of China (40501055). The authors also wish to acknowledge the contribution of Mr. Michael Davies from BGC Engineering Inc. for providing data and sharing his experience of field survey of debris flow in the study area.

REFERENCES

Chen, H. & Lee, C.F. 2000. Numerical simulation of debris flows. *Can. Geotech. J.,* 37(1): 146–160.

Crosta, G.B., Calvetti, F., Imposimato, S., Roddeman, D., Frattini, P & Agliardi, F. 2001. Granular flows and numerical modeling of landslides. Thematic report. *Debrisfall assessment in mountain catchments for local end-uses* contract No EVG1-Ct-1999-00007.

Cundall, P.A. & Strack, O.D.L. 1979. A discrete numerical model for granular assemblies. *G'eotechnique* 29: 47–65.

Davies, M.R., Froese, D.G. & Cruden, D.M. 2005. Klapperhorn mountain debris flows, Yellowhead Pass, British Columbia. *Proceedings of International Conference on Landslide Risk Management, Vancouver.* D010.

Davies, M.R. 2007. Klapperhorn mountain debris flows, Yellowhead Pass, British Columbia. *MSc Thesis,* University of Alberta.

Denlinger, R.P. & Iverson, R.M. 2001. Flow of variably fluidized granular masses across three-dimensional terrain: 2. Numerical predictions and experimental tests. *J. Geophys. Research*, 106: 553–566.

Hutter, K., Koch, T., Pliiss, C. & Savage, S.B. 1995. The dynamics of avalanches of granular materials from initiation to runout. Part II. Experiments. *Acta Mechanica* 109: 127–165.

Hutter, K., Svendsen, B. & Rickenmann, D. 1996. Debris flow modeling: A review. *Continuum Mech. Thermodyn.* 8: 1–35.

McDougall, S. & Hungr, O. 2005. Dynamic modelling of entrainment in rapid landslides. *Can. Geotech. J.*, 42(5): 1437–1448.

Mountjoy, E.W. 1980. Geology, Mount Robson, West of Sixth Meridian, Alberta-British Columbia. *Geological Survey of Canada. "A" Series Map* 1499A, 1: 250 000.

Pirulli, M. & Giuseppe, S. 2007. PROPAGATION OF DEBRIS FLOWS: COMPARISON OF TWO NUMERICAL MODELS. In V.R. Schaefer, R.L. Schuster & A.K. Turner (eds), *Proceeding of 1st North American Landslide Conference June 3–8, 2007 Vail, Colorado*: 1542–1551 (CD-Rom).

Potyondya, D.O. & Cundallb, P.A. 2004. A bonded-particle model for rock. *International Journal of Rock Mechanics & Mining Sciences* 41: 1329–1364.

Richards, K., Bithell, M., Dove, M. & Hodge, R. 2004. Discrete-element modelling: methods and applications in the environmental sciences, *Phil. Trans. R. Soc. Lond. A*, 362: 1797–1816.

Zhu, H.P. & Yu, A.B. 2005. Steady-state granular flow in a 3D cylindrical hopper with flat bottom: macroscopic analysis. *Granular Matter* 7: 97–107.

Three dimensional simulation of landslide motion and the determination of geotechnical parameters

Yuhua Lang & Xianqi Luo
China Three Gorges University, Key Laboratory of Geological Hazards on Three Gorges Reservoir Area, Ministry of Education, China

Hiroyuki Nakamura
Tokyo University of Agriculture and Technology, Japan

ABSTRACT: A geotechnical model was proposed to do three dimensional simulation of landslide motion. Coulomb's law was used to express the cohesion, friction angle and other geotechnical parameters along the slip surface. The total variation diminishing method was used in data processing to disperse three independent variables: thickness and velocities of x and y directions. A formula on the relationship between static friction angles and dynamic friction angle was proposed on the basis of three dimensional simulations of worldwide landslides.

1 GEOTECHNICAL MODEL

1.1 Introduction

The simulation of landslide motion can be helpful for the prediction of disaster area. Landslide mass can be regarded as a composition of soil particles with internal friction. This composition can be considered as an incompressible viscous fluid then the laws of motion can be expressed by Navier-Stokes equations. And the friction in sliding surface can be expressed by Coulomb's law of friction.

1.2 Laws of mass conservation

Landslide body is divided to grid of right angle as basic calculation cell. The average numerical values of every element are put in the gravity center of each grid. By the laws of mass conservation, the mass volume of flowing into a finite cell is equal to the volume flowing out of the cell in finite time.

$$\frac{\partial u}{\partial x}dx + \frac{\partial v}{\partial y}dy + \frac{\partial w}{\partial z}dz = 0 \quad (1)$$

where u, v, w is the speed along x, y, z directions.
Calculus along vertical direction from the bottom z of mass to the ground surface h in equation (1)

$$\frac{dh}{dt} = \frac{\partial M}{\partial x}dx + \frac{\partial N}{\partial y}dy \quad (2)$$

where $M = \int udz$; $N = \int vdz$; $\int$: calculus from z to h.

1.3 Equation of movement

There are 3 kinds of forces acting on individual grid along x and y directions (Figure 1). Arithmetic elements include stress F_1, internal friction F_2 and slip friction F_3. According to the equation of movement,

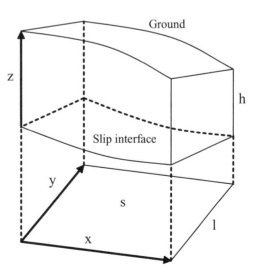

Figure 1. Three dimensional grid of sliding mass.

the internal forces along x direction can be expressed by equations below:

$$F_1 = \rho g h \frac{\partial h}{\partial x} dx \quad (3)$$

$$F_2 = cdx + \rho g h \tan \phi_s \quad (4)$$

$$F_3 = \rho g h \tan \phi_m dx \quad (5)$$

$$\frac{dM}{dt} = \rho g h \frac{\partial h}{\partial x} + cdx + \rho g h \tan \phi_s + \rho g h \tan \phi_m \quad (6)$$

where ρ = density; g = gravity; h = thickness of sliding column; Φ_s = dynamic or apparent friction angle along slip surface; Φ_m = dynamic friction angle in sliding mass.

In formula (6),

$$\frac{dM}{dt} = \frac{\partial M}{\partial t} + u\frac{\partial M}{\partial x} + v\frac{\partial M}{\partial y} \quad (7)$$

With the same pattern, the equations along y direction are:

$$\frac{dN}{dt} = \rho g h \frac{\partial h}{\partial y} + cdy + \rho g h \tan \phi_s + \rho g h \tan \phi_m \quad (8)$$

$$\frac{dN}{dt} = \frac{\partial N}{\partial t} + u\frac{\partial N}{\partial x} + v\frac{\partial N}{\partial y} \quad (9)$$

There are no direct results from equations (2), (6), and (8), they have to be dispersed in data processing.

1.4 Total variation diminishing method

Total Variation Diminishing method (Harlow et al. 1970) is used in the data processing operation to disperse three independent variables: thickness, velocity of x and y directions. Law of mass conservation can be dispersed as formula below:

$$\frac{(h\rho)_{i,j}^{n+1} - (h\rho)_{i,j}^{n}}{\Delta t} + conv(A) = 0 \quad (10)$$

$$Conv(A) =$$

$$\frac{l_{i+1/2,j}(h\rho)_{i+1/2,j}^{n}(u_x)_{i+1/2,j}^{n+1} - l_{i-1/2,j}(h\rho)_{i-1/2,j}^{n}(u_x)_{i-1/2,j}^{n+1}}{s_{i,j}}$$

$$+ \frac{l_{i,j+1/2}(h\rho)_{i+1/2,j}^{n}(u_x)_{i,j+1/2}^{n+1} - l_{i,j-1/2}(h\rho)_{i,j-1/2}^{n}(u_y)_{i,j-1/2}^{n+1}}{s_{i,j}}$$

$$\quad (11)$$

where l = length of a calculation cell, s = bottom area of the cell, i,j are subscript for two calculating steps one after another.

Equation of movement along x direction is:

$$(h\rho)_{i+1/2,j}^{n} \frac{(u_x)_{i+1/2,j}^{n+1} - (u_x)_{i+1/2,j}^{n}}{\Delta t} + conv(x)$$

$$= -g_z(h\rho)_{i+1/2,j}^{n} \frac{H_{i+1,j}^{n} - H_{i,j}^{n}}{\Delta x_{i+1/2,j}}$$

$$- g_z(\rho h)_{i+1/2,j}^{n}(\tan \phi_m)_{xi+1/2,j} - (f_s)_{xi+1/2,j}^{n+1}$$

$$+ (hf_x)_{i+1/2,j}^{n+1} \quad (12)$$

Because the directions of velocity are parallel or vertical to the grid, One time disperse modulus can be calculated as below:

$$conv(x) = (h\rho)_{i+1/2,j}^{n} \frac{l_{i+1,j}(u_x^2)_{i+1,j}^{n} - l_{i,j}(u_x^2)_{i,j}^{n}}{2s_{i+1/2,j}}$$

$$+ (h\rho u_y)_{i+1/2,j}^{n}$$

$$\times \frac{l_{i+1/2,j+1/2}(u_x)_{i+1/2,j+1/2}^{n} - l_{i+1/2,j-1/2}(u_x)_{i+1/2,j-1/2}^{n}}{s_{i+1/2,j}}$$

$$\quad (13)$$

where the resistance along x direction on slip surface is:

$$(f_s)_x = (c + \rho g h^n \phi_s) \frac{u_x^{n+1}}{\sqrt{u_x^2 + u_y^2 + u_z^2}^n} \quad (14)$$

where h^n is the thickness in the n step of calculation, u_x, u_y, u_z is the speed of movement along x, y, z directions.

Equation of movement along y direction is:

$$(h\rho)_{i,j+1/2}^{n} \frac{(u_y)_{i,j+1/2}^{n+1} - (u_y)_{i,j+1/2}^{n}}{\Delta t} + conv(y)$$

$$= -g_z(h\rho)_{i,j+1/2}^{n} \frac{H_{i,j+1}^{n} - H_{i,j}^{n}}{\Delta x_{i,j+1/2}}$$

$$- g_z(\rho h)_{i,j+1/2}^{n}(\tan \phi_m)_{yi,j+1/2}$$

$$- (fs)_{yi,j+1/2}^{n+1} + (hf_y)_{i,j+1/2}^{n+1} \quad (15)$$

$$conv(y) = (h\rho)^n_{i,j+1/2} \frac{l_{i,j+1}(u_y^2)^n_{i,j+1} - l_{i,j}(u_y^2)^n_{i,j}}{2s_{i,j+1/2}}$$

$$+ (h\rho u_x)^n_{i,j+1/2}$$

$$\times \frac{l_{i+1/2,j+1/2}(u_y)^n_{i+1/2,j+1/2} - l_{i-1/2,j+1/2}(u_y)^n_{i-1/2,j+1/2}}{s_{i,j+1/2}} \quad (16)$$

where the resistance along y direction on slip surface is:

$$(f_s)_y = (c + \rho g h^n \phi_s) \frac{u_y^{n+1}}{\sqrt{u_x^2 + u_y^2 + u_z^2}^n} \quad (16)$$

1.5 Structure of the simulation program

Figure 2 is the flow chart of the simulation program. It can be divided to mainly 4 steps: (1) data input and time setting, (2) landform and friction calculation; (3) movement and thickness calculation; (4) data output to other graphical tools.

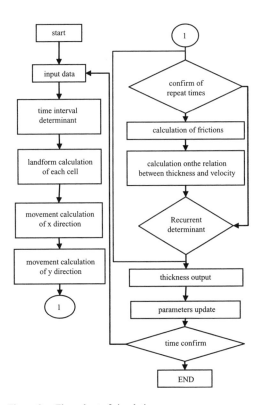

Figure 2. Flow chart of simulation program.

2 DETERMINATION OF GEOTECHNICAL PARAMETERS

2.1 Input & output data

Data required for the program are the 3D data of ground surface and sliding surface; physical properties of the sliding mass such as unit weight, cohesion, coefficient of internal friction, and seismic data etc. The results of calculation are velocity, flow thickness of each grid from which 3D map, longitudinal and traverse sections of each time interval are outputted (Lang & Nakamura, 1998).

The geography data of ground and sliding surface can be obtained from GIS data or other investigation. The static friction angle can be obtained by two dimensional slice methods or soil experiments.

2.2 Reliability of the simulation

Reliability of the model is proved by case studies on 20 worldwide landslides (Table 1). Property parameters, which make the simulation most suitable to the real one, are selected for each landslide.

Take the No.18 Nigawa Landslide as an example of case study. It occurred on Jan. 17th, 1995, by M7.2 earthquake. Its length was 175 m, width 100 m, volume 110,000 m³. Figure 3 shows the calculated

Table 1. Landslides used for calculating coefficient of friction.

No.	Landslide	$\tan \phi_s$	$\tan \phi_m$	$\tan \phi_c$	B	A
1	Kaerikumo	0.132	0.222	0.613	0.353	0.577
2	tukaro	0.087	0.176	0.521	0.264	0.507
3	nata	0.009	0.268	0.344	0.277	0.804
4	Frank	0.012	0.466	0.839	0.479	0.570
5	tuedani	0.044	0.268	0.543	0.312	0.574
6	kinnkoji	0.044	0.268	0.488	0.312	0.639
7	Madison	0.123	0.176	0.445	0.299	0.672
8	nakagi	0.141	0.176	0.613	0.317	0.517
9	Itinomiya	0.087	0.176	0.374	0.264	0.706
10	Kobuki	0.070	0.158	0.404	0.228	0.565
11	Nigorisawa	0.105	0.141	0.268	0.246	0.917
12	Sala	0.141	0.176	0.613	0.317	0.517
13	Matukosi	0.044	0.176	0.543	0.220	0.405
14	Jizukiyama	0.105	0.194	0.466	0.299	0.642
15	Jiaojia	0.141	0.249	0.625	0.390	0.624
16	Shangtan	0.141	0.213	0.625	0.353	0.565
17	Tuedani	0.158	0.213	0.675	0.371	0.550
18	Nigawa	0.044	0.176	0.325	0.220	0.677
19	Takaratuka	0.070	0.087	0.213	0.157	0.741
20	Dangjiacha	0.141	0.194	0.601	0.335	0.557

ϕ_s = dynamic friction angle along slip surface;
ϕ_m = dynamic friction angle in sliding mass;
ϕ_c = static friction angle;
B = $\tan \phi_s + \tan \phi_m$; A = B/$\tan \phi_c$

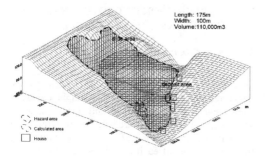

Figure 3. 3D image of hazard area and calculated area of the Nigawa Landslide.

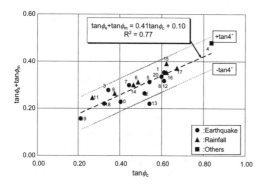

Figure 4. Relation between static coefficient of friction and dynamic coefficient of friction. (Numbers in the figure show the number of landslides in Table 1).

3D image after slide. The calculating mesh is 30 × 30 with a width of 10 m. The calculated area mach the hazard area when $\phi_s = 2.5$ degree, $\phi_m = 10.0$ degree. All other case studies do the same kind of calculation in order to get suitable parameters.

It is generally considered that the dynamic coefficient of friction is around 0.7 times of static one. From the Table 1 we can see that friction on the interface plus the friction inside sliding mass ($\tan \phi_s + \tan \phi_m$) is about 0.4–0.7 times of the static one ($\tan \phi_c$).

2.3 Determination of dynamic friction angle

Figure 4 shows the relation between static coefficient of friction and dynamic coefficient of friction. Numbers in the figure are the same number of landslides in Table 1. The relation line can be summarized to a formula below:

$$\tan \phi_s + \tan \phi_m = 0.41 \tan \phi_c + 0.10 \qquad (17)$$

The correlation modulus of all data is 0.77. It shows a good relationship between the two coefficient. It can be used to calculate the dynamic friction angle instead of getting by soil experiments. Two lines up and down the formula line are dropped to cover all of the data; The range of friction angle can be obtained to simulate maximum and minimum extension of landslide mass.

The different shapes of marks in Figure 4 show the different inducement of landslide. Circle, triangle, and square represent the inducement of earthquake, rainfall, and others respectively. Triangle marks of rainfall induced landslide situated all on the upper side of the formula line. It mains that this kind of landslide has bigger coefficient of friction than that of earthquake induced landslide.

3 PREDICT HAZARD AREA BY SIMULATION

3.1 Process of simulation

The method and process of hazard area prediction are proposed as following: (1) Calculate the critical sliding surface and static friction angle in main section of a landslide; (2) Determine the 3D shape of sliding mass according to the critical sliding surface, geology, and geomorphology of slope; (3) Calculate the maximum and minimum dynamic friction angle by using of the proposed formula; (4) Simulate the movement of landslide by the developed model.

3.2 Recurrence of Tonbi Landslide in Tateyama

Tonbi Landslide was triggered by a M7.1 earthquake on Feb. 26th, 1858 in Tateyama area, Japan. Its 1.23×10^8 m^3 debris formed a nature dam across the valley. The dam broken after several weeks then brought huge damages to the towns of down stream. The simulated area, volume and speed is most similar with occurred one when the coefficient below were used (Figure 5): r = 18 kN/m^3, c = 0.10 kN/m^2, $\phi_s = 3.0°$; $\phi_m = 18.0°$; $\phi_c = 33.0°$.

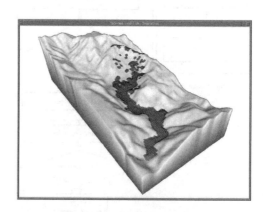

Figure 5. Tonbi Landslide simulation can be used to get parameters of this area for further prediction.

All the parameters in this simulation can be used to do prediction of landslide in the same area.

3.3 Prediction of landslide in Tateyama area

Stability of slopes in this area was investigated and the possibility of sliding was ranked. Siqirou Slope was pointed out as the most unstable slope. Critical sliding surface and other geomorphology data were determined by investigation and 2D calculation.

The simulation result shows that the landslide will dam valley and bring huge damages to down streams.

4 CONCLUSIONS

A three-dimensional movement simulation model is developed to predict the extension of landslide. A formula on the relationship between static friction angle and dynamic friction angle is proposed, based on three-dimensional simulation of worldwide landslides. The process of hazard area prediction of landslide is proposed with a case study in Tateyama area.

REFERENCES

Yuhua Lang & Hiroyuki Nakamura. 1998. Characteristics of slip surface of loess landslides and their hazard area prediction. *J. Japan Landslide Society*, 35–1, 9–18.

Harlow, F.H. & Amsden, A.A. 1970. A numerical fluid dynamics method for all flow speeds. *J. Comp. Phys.*, Vol. 8, 197–213.

Stability analysis and stabilized works of dip bedded rock slopes

Jin-yan Leng & Zhi-dong Jing
The Third Railway Design and Survey Institute Limited Corporation, Tianjin, China

Xiao-ping Liao
Fuzhou University, Fuzhou, China

ABSTRACT: In this paper, a typical equivalent unit was brought forward because of inhomogeneity and discontinuity of bedded rock. Through one engineering case, we obtained orthotropic mechanical parameters of the equivalent unit by a series of numerical tests, and so the bedded rock had the homogeneity and continuity characteristics. Then we simulated and analyzed the instability and stabilized works of some dip bedded rock slopes by finite element method. Finally, a convenient and applied technique to simulate instability and stability works for some kind dip bedded rock slopes was offered.

1 GENERAL INSTRUCTION

With the construction of large-scale infrastructure, especially in the west development and the mountainous construction, there occur many rock slopes. Among them, the dip bedded rock slope is one of the familiar types of slopes.

With the characters of large quantity, wide distribution, diversification, bad condition of engineering geology and instability of works, the dip bedded rock slope is analyzed commonly by the theory of rigid body limit balance. But coupling action between engineering and structure is not considered in it. The analogy technique of engineering geology applied in reinforce works usually, the works depend on experience of engineering mostly, and so either sometimes the works are stronger or reliability inferiority.

Due to the characteristics of inhomogeneity and discontinuity on bedded rock, the typical equivalent unit was brought forward in this paper. By one engineering case, we obtain orthotropic mechanical parameters of equivalent unit by a series of numerical tests, and endow the bedded rocks with homogeneity and continuity characteristics. Then we simulated and analyzed instability and stabilized works of some a dip bedded rock slope by finite element method. And so a convenient and applied technique to simulate instability and stability works for some kind dip bedded rock slopes was offered in this paper.

2 TYPICAL EQUIVALENT UNITS

Bedded rock mass is make up of rocks and structure planes. Typical equivalent unit is the large unit that can compose characteristics of rock and structure planes. It is smaller than engineering scale, whereas larger than structure plane spacing. And so its mechanical characteristics can represents the wholly bedded rock mass mechanical characteristics.

The study rationale of bedded rock mass with "typical equivalent unit" is numerical simulation test on it to study the property of deform and intensive, and obtains

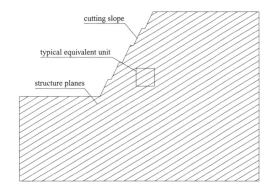

Figure 1. Dip bedded rock slope sketch.

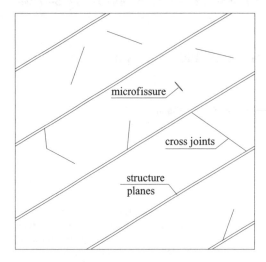

Figure 2. Typical equivalent unit sketch.

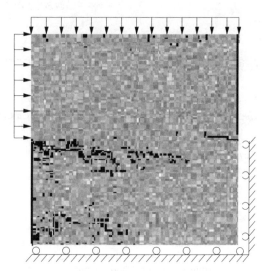

Figure 4. Shear numerical test.

way it can substitute the physical model test that needs long time and large cost.

Based on the characteristics of bedded rock, we simulate and analyze bedded rock slopes with RFPA software. Through uniaxial compressive tests and direct shear tests on parallel bedding plane and vertical bedding plane, the mechanical property of bedding plane are equaled to whole rock mass, and then we can obtain homologize anisotropic mechanics parameter of rock mass. Thereby bedded rock with aspectual discontinuity grew into orthotropic continuous medium, and it supply a convenience way to analysis the stability.

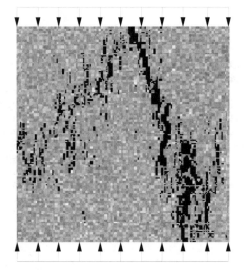

Figure 3. Compressive numerical test.

typical equivalent unit strength of materials parameter; Then the macro layers rock mass is treated as transversely isotropy media, and analysis the stability of bedded rock slope.

4 STABILITY ANALYSES

We obtained orthotropic mechanical parameters of bedded rock by a series of numerical tests. The dip bedded rock was endowed with orthotropic plastic material. Then we analyzed and simulated cutting process of dip bedded rock slope by finite element method, and summarized instability models of dip bedded rock slope. And so we gave the stabilized works corresponding to the instability models. Consequently, a convenient and applied technique to simulate instability and stability works was offered in paper. It was testified that it was a feasible and applicable technique by simulated instability and stability works of an engineering case.

3 NUMERICAL TESTS

As a sort of numerical simulation, numerical tests is a means of simplicity virtual supplementary, and in a

5 ENGINEERING CASE

5.1 Engineering general situation

Certain speedway was in construction, after cutting two-stage aback, one segment in the same direction as layer slope all at once occurred slippage following stratification plane, moved along slope towards about fifteen meters, the slope body structure dismembered to fractures. It's a typical dip bedded rock slope.

This segment lithology contains sandy mud rock and argillaceous sand conglomerated rock, layer thickness is 0.5~2.0 meter, the rock layer inclines to speedway line, dip angle is 15°~20°, the max altitude of slope is 30 meter, slope designed four stages, 1st step ratio of slope is 1:0.75, stage height is 8 meter, ratios of 2nd and 3rd stages are 1:1.0, stage height are 6 meter, per stage have 2 meters platform. It was TBS planting grass operation in 1st and 2nd stage predesigned, 3rd stage was arch framework planting grass operation, 4th stage was triaxiality net planting grass operation.

5.2 Numerical simulation analyses

Procured rock swatch from the 2nd stage top made into 3 samples, 50 mm * 50 mm * 50 mm, to do compression test; from the 3rd stage toe taken disturbed soil sample, did soil quick shear strength, Es, without lateral confinement resist compression test, obtained mechanics parameter (mean) in table 1.

Based on these physical mechanics parameter obtained in lab, utilized RFPA finite element means to do the same direction as layer rock slope proceed resist compression, shear resistant numerical experimentation, obtained this slope typical equivalent unit transversely isotropy.

The mechanical property parameters saw in table1 of the compression test of horizontal and vertical compression test. As per try, drawn the load-displacement curve, stress-strained curve, obtained cell equivalent body compression resistance parameter saw in Table 2.

Based on the structure characteristic of rock slope and familiar destroy form, this model had be done two types shear test: one was shear surface along rock bedding plane, this method could erect sub-macro model of bedding structure independently. The other was erect interlamination structural plane model, erected

Table 1. Parameters of slope rock and structural plane.

	E (Mpa)	υ	φ	σ
Mud-sandstone (strongly weathering)	240	0.31	45	8.3
Weak structure plane	3.2	0.35	10.2	0.055

Table 2. Parameters of slope rock and structural plane.

	Mud-sandstone (strongly weathering)		
	E (Mpa)	υ	φ
Vertical	190	0.33	2.1
Parallel	180	0.34	1.65

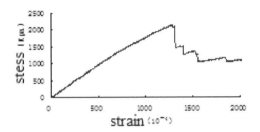

Figure 5. Stress-strain curve about vertical compression on bedding plane.

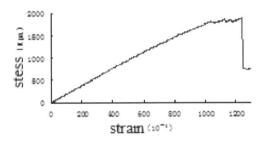

Figure 6. Stress-strain curve about parallel compression on bedding plane.

Table 3. Parameters of shear numerical of slope rock and structural plane.

Mud-sandstone (strongly weathering)		Weak structure plane	
G (Mpa)	τ (Mpa)	G (Mpa)	τ (Mpa)
5.1	1.2	2.7	0.023

sub-macro model for bedded rock body, shorn the rock bedding.

As per numerical test result, obtained typical equivalent unit shear resistant parameter (saw watch table 3), then drawn homologous shear stress-strain curve (saw chart Figure 7 and Figure 8).

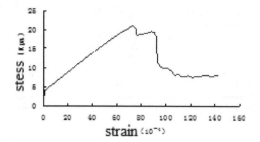

Figure 7. Stress-strain curve about shear compression on bedding plane.

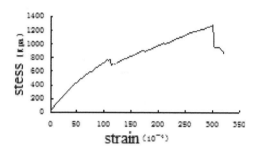

Figure 8. Stress-strain curve about shear vertical compression on bedding plane.

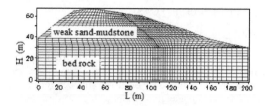

Figure 9. Slope model of finite element method.

Table 4. Jinyin mountain slide numerical calculate achievements.

Maximum shear stress (Mpa)	Maximum tensile stress (Mpa)	Total plastic strain
0.420	0.041	0.0002
0.979	0.129	0.0016

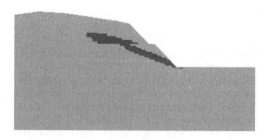

Figure 10. Plastic zone of the 2nd stage excavation.

Through the numerical experiment of this bedded rock body sub-macro model above, obtained this equivalent cell body transversely isotropy physical mechanic property parameter, then used ADINA finite element software, proceeding numerical simulation to the same direction as layer rock slope. Analyses its instability mode, in turn raises reinforce countermeasure.

The primary geological complexion of this slope is 0~32m dip weathering Fractional sand-mudstone, then the under layer is weak sand-mudstone, the interface is weak spot, the obliquity is 20°. Based on the above-mentioned data, and combined pre-designed instance, we establish numerical stimulant model, the figure 9 is discrete cells gridding, mechanical parameters are showed in table 4. We will analysis this bedded rock slope through calculating, undo the changes of slope stress scene, plastic region and other parameters in digging.

Numerical calculation parameters saw table 4.

Based on the calculation result, from plastic zone figure upper can be saw, after the fourth stage excavation, this stage toe appeared plastic zone, extended to slope body middle following inclination 20° bedding angle, formed clearly plane plastic zone strip; after the third stage excavation, this stage toe appeared plastic zone, plastic zone width compared with the fourth stage more augmentation, and extended approach stage top following bedding angle, showed the fact that there was damage trend of following soft stratum bedding (Figure 10).

From tension stress distributive zone figure, after the 4th stage's excavation, stage top appeared small scope tension stress region, magnitude was small too; toe region appeared tension stress region in the small too, furthermore top stress region and toe stress region had run-trough trend; after the 3rd stage excavation, top tension stress region grew bigger, toe tension turned into stress concentration point, the maximum tension stress value was 0.129 Mpa, furthermore developed from top to toe as plastic zone (Figure 11).

Before slope excavation, maximum shear followed linear distribution as depth, shear isolines followed depth adequate distribution from shear stress regularities of distribution. During the 4th stage excavation, shear stress field occurred defluxion, the characteristic approximately was run parallel with line of slope, toe shear stress became overt concentration of about 0.420 Mpa, behaved as toe fail in shear. After the 3rd

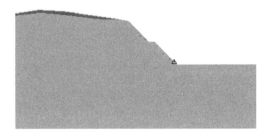

Figure 11. Tensile stress zone of the 2nd stage's excavation.

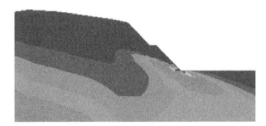

Figure 12. Shear stress zone of the 2nd stage's excavation.

stage excavation, shear stress field kept on defluxion, toe maximum shear concentration turned more apparent, added to about 0.979 Mpa, the characteristic of damage became more apparent too (Figure 12). The bedding rock anisotropy induced large shear, the more larger of aeolotropism was, slope body stress redistributed after excavation, the more bigger of high intensity bedding than low intensity bedding.

Summed up all of upper discuss, after 2nd stage excavation of the same direction as layer rock slope, the distribution and development of plastic zone, tension stress and maximum shear showed that, in conditions without reinforcement work, only two stages excavation could be done. Keeping on excavation simulation, figure had not convergence, showed that the slope had damaged, so approved that this was a limiting state of equilibrium. the simulation conclusion was very anastomotic with realism. It was, aforesaid analog computation parameter equal to project site, by in-house experiments achieved fundamental mechanical parameter, not by amending, adjusting, deleting or filling, fully accounted that its engineering applicability, result veracity and conclusion reliability.

5.3 Numerical simulation

Both simulation conclusion and engineering practice showed that, this slope stability was very difference, needed for fence reinforcement in time.

After the 4th stage excavation, toe site appeared distinct distortion, immediately adopted systematic rock bolt as fence reinforcement, the simulation result clearly showed that the application ameliorated state of stress, far and away reduced plastic zone development (figure 13), thereby leaded the slope could be keeping on excavation.

Kept on excavating the 3rd and 2nd stages, the excavation feet brought distinct distortion and structural plane slack characteristic, plastic zone had followed inclination 20° upswing towards this stage middle, immediately adopted systematic rock bolt as fence reinforcement (if to be necessary, executed excavation and anchor measure simultaneously, body stress state got ameliorate, led this stage toe's plastic zone wane to pimping scope (figure 14).

Proceeding the 1st stage excavation, this stage toe came forth plastic zone, distribution ranges approached its level top; furthermore the interface of the 4th stage bottom and the 3rd stage was on the plastic zone which was augmenting, considering the result of stress adjustment and interface influencing syntheses; for the sake of stabilization of wholly slope, the 1st stage exerted retaining wall as shallow layer fence reinforce, more reduced clearly plastic zone, brought wholly slope body into stable state (Figure 15).

According to correlative slope design specifications, the safety factor responded for 1.2, could we through the means of reduced intensity ascertain for the stability coefficient of side slope. Through the means of repeat calculation, increased and adjusted the 2nd and 3rd stages prestressing force magnitude,

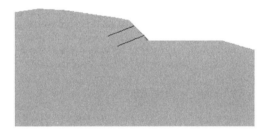

Figure 13. Plastic zone of the 4th stage suspension roof support.

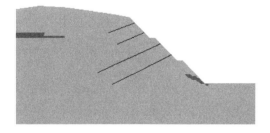

Figure 14. Plastic zone of the 3rd stage excavation.

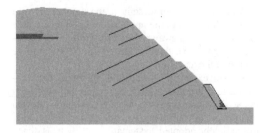

Figure 15. Plastic zone of the 4 stage suspension roof support.

made slope's stability factor satisfy standard requirement, then, resumed practicable strength indexing, by calculation body stress mode gradually ameliorated, plastic zone reduced obviously, brought slope get off to stable state chronically.

After the sliding mass was purged in this project, according to aforesaid fence reinforce measure was proceeded construction, hitherto the slope is stable, so the conclusion was brought of that it was feasible, secure, applied and economical.

6 CONCLUSIONS

Through the analysis of above we drew the several conclusions:

Representative cell equivalent was from macroscopic bedding slope be chosen macroscopic rock mass. Its size was adequately small than this slope project, but was big enough compared joint plane and stratification plane, furthermore it could be used to accurately simulate each stratification plane and per strip joint, and could drew precise conclusion. In this way it was reflected of strength of rock mass and deformation behavior enough.

Through RFPA numerical simulation test of representative cell equivalent, made structural plane lie in wholly rock mass, made discontinuous bedding rock slope into macroscopical isotropic continuous body, in this way simplified this kind questions, at the same time again could reflect rock mass discontinuity and anisotropy characteristic. This way can be used in engineering practice widely.

Used finite element means carried through numerical simulation analyses to flat-lying bedding rock slope stability, put forward homologous work countermeasure, used reduced intensity method or increased bulk density method, ascertained fence reinforced safety factor, founded fence work numerical simulation technology of flat-lying bedding rock slope.

It was important for process distortion of the same direction bedding rock slope construction of that initial stress pressured anchor measure, advanced slope short term and long term stability. As engineering executive physical circumstances, responded in appropriate reinforced scheme.

It was very important of that adopting excavation and reinforcement step by step to prevent stress condition exacerbation of slope body, control slope cutting distortion, long-term stability.

With "Representative cell equivalent" model investigating bedded rock mechanical property, the rationale was: First of all, diverged representative cell equivalent as finite element' require, brought rock cell and stratification plane cell; And then "Representative cell equivalent" proceeded numerical simulation, looked into distortion and intension property of "representative cell equivalent", obtained "equivalent" materials parameters; Be the last treat macro layers rock mass transverse isotropy media, then analyzed engineering stability.

REFERENCES

Chen Zu-yu, Wang Xiao-gang, Yang Jian, et al. 2005. *Rock slope stability analysis—theory, method and programs.* Beijing: China Water Power Press (in Chinese).

Liang Z.Z., Tang C.A., LI H.X. & Zhang Y.B. 2004. Numerical simulation of the 3D failure process in heterogeneous rocks. *Int J Rock Mech Min Sci*, 41(3):419.

Long J.C.S. & Witherspoon P.A. 1985. The relationship of the degree of interconnection to permeability in fractured. *Journal of Geophysical Research.* 90(B4): 3087–3097.

He Man-chao, et al. 2001. A new way of determining mechanical parameters of engineering rock masses. *Chinese J of Rock Mechanics and Engineering.* 20(2):225–229 (in Chinese).

Ling Zheng-zhao. 2005. *Three-dimensional numerical modeling of rock failure process.* Shenyang: Northeastern University (in Chinese).

Xie Wei-hong, Xie He-ping & Zhao-Peng. 1996. Photoelastic simulation of shear strength of rock fractal joints. *Mechanics and Practice.* 18(5):39–43 (in Chinese).

Xu Qiang & Huang Run-qiu. 1996. Discussion on self-organized critical characters in the course of rock failure. *Journal of Geological Hazards and Environment Preservation.* 7(1):25–30 (in Chinese).

Yang Qiang, Chen Xin & Zhou Wei-yuan. 2005. Anisotropic yield criterion for jointed rock masses based on a tow-order damage tensor. *Chinese Journal of Rock Mechanics and Engineering.* 24(8):1275–1282.

Zhang Gui-ke. 2006. *Study on equivalent orthotropic mechanical parameters and yield criterion of jointed rock mass and its engineering application.* Nanjing; Hohai University (in Chinese).

A GIS-supported logistic regression model applied in regional slope stability evaluation

Xueping Li & Huiming Tang
Engineering Faculty, China University of Geosciences, Wuhan, China

Shi Chen
ChangCheng Special Steel Stock Company, Jiangyou, China

ABSTRACT: The logistic regression model is a statistic analytical method, in which there usually are dichotomous variables to be used (two values are given, for example, to the variable "y"). It can provide the results in the form of event probability. Supported by GIS Platform, a logistic regression model for the regional slope stability evaluation was used in evaluation of regional slope stability in new Wushan city site. When the criterion of probability is 0.163, the logistic regression model gives the prediction accuracies as follows: 72.08% for landslide occurrence and 81.44% for no landslide occurrence. As a comparison, when the criterion of probability is 0.274, the Bayesian integrating model gives the prediction accuracies as follows: 79.24% for landslide occurrence and 80.82% for no landslide occurrence and when the criterion of probability is 0.2171, the grouped data-based logistic regression model gives the prediction accuracies as follows: 71.85% for landslide occurrence and 71.70% for no landslide occurrence predicts actual landslide occurrence at accuracy and landslide non-occurrence at 80.82%. Therefore, the logistic regression model is a better election in regional slope stability evaluation.

1 INTRODUCTION

1.1 Study progress

Landslide has characteristics of wide distribution, frequent happening and quick causing, which is one of the main disasters in the world. It has been more than 20 years since the study of regional landslide disaster evaluation and mapping in the world and it has made remarkable progress during the recent years.

The methods of landslide disaster spatial evaluation can be summarized statistics, sensitivity mapping, information and all kinds of experience etc (Tang & Jorg 1998).

Although landslide forecast theories have different evaluation methods, they are based on mechanism analyses, favorableness, and their similarities are to be taken fully consideration of overlapping for various geological environments to evaluate the possibility of causing landslide in certain geological environment and to circle the relatively "dangerous section" in the hazardous area center. Different ways of overlapping are adapted in evaluation process, and the prediction result is expressed by section graphs.

The characteristics of landslide prediction study are intersected and penetrated by various kinds of theories so that different kinds of modern theories such as information theory, systematic theory, fuzzy mathematics, expert system, neural network. Computer technology has been widely applied to landslide spatial analysis (Ohlmacher & Davis 2003).

1.2 Character of Logistic regression model

Logistic regression model is a type of statistic analytical method, which is usually used by dichotomous variables (the variable "y" is given two values). Different from linear regression, Logistic regression is a non-linear model. Parameter estimation is Maximum likelihood estimate (MLE) method. It is proven that the MLE of Logistic model possess consistency, asymptotically efficiency and asymptotically normal in the random sample situation. Logistic regression method can set up models for categorical independent variable, categorical dependent variable (or continuous dependent variable or mix variable), and there is a whole set of criterion to test regression models and the parameters. It provides the result as the form of event probability, and it is undoubtedly a better solution to apply to the regional slope stability evaluation (Wang & Guo 2001). According to the comparison of regression coefficients, the slope stability effect factor can be quantitatively analyzed. With the regression models contrast, the best model on slope stability evaluation can be obtained. For the event probability result is offered, it will be convenient and assured to analyze or predict the result.

2 BASIC RESEARCH METHOD

2.1 Probability formula of Logistic regression

Suppose variable Y is used to represent a sample in a group of independent variable's effect, Y's evaluate rule is

$$\begin{cases} Y = 1 & (right) \\ Y = 0 & (false) \end{cases} \quad (1)$$

The probability of right is P and false is Q, $x_1, x_2, \ldots, x_m$ is to represent m influencing factors to result Y. Logistic regression formula is used to represent the probability of right, which is

$$P = \frac{e^{\beta_0 + \beta_1 x_1 + \cdots + \beta_m x_m}}{1 + e^{\beta_0 + \beta_1 x_1 + \cdots + \beta_m x_m}} \quad (2)$$

For the relational expression $P + Q = 1$. The probability of false will be got according to (2),

$$Q = \frac{1}{1 + e^{\beta_0 + \beta_1 x_1 + \cdots + \beta_m x_m}} \quad (3)$$

Deduced from above formulas, a sample's result and it's relational factors are nonlinear. Define the two probabilities' rate is odds,

$$odds = P/Q = e^{\beta_0 + \beta_1 x_1 + \cdots + \beta_m x_m}$$
$$= e^{\beta_0} \times e^{\beta_1 x_1} \times \cdots e^{\beta_m x_m} \quad (4)$$

The natural logarithm is

$$\ln(P/Q) = \beta_0 + \beta_1 x_1 + \cdots + \beta_m x_m \quad (5)$$

where $\beta_0, \beta_1, \ldots, \beta_m$ are called logistics' regression coefficients. Formula (5) is a nonlinear equation, so the Newton-Raphson Method is applied to solve Logistics' regression coefficient.

2.2 Bayesian statistics inference technology

2.2.1 Events of bayesian formula
Suppose $A_1, \ldots, A_k$ are incompatible events and their sum $\bigcup_{i=1}^{k} A_i$ includes event B, that is $B \subset \bigcup_{i=1}^{k} A_i$, then (Mao 1999, Zhang 1991)

$$p(A_i | B) = \frac{P(A_i)P(B|A_i)}{\sum_{j=1}^{k} P(A_j)P(B|A_j)}, \quad i = 1, 2, \ldots, k \quad (6)$$

2.2.2 Bayesian random variable form
Suppose united distribution density of random variable ξ, η is $p(x, y) = p_\xi(x) f_{\eta|\xi}(y|x)$, $p_\xi(x)$ is brink density of ξ, $f_{\eta|\xi}(y|x)$ is conditional density of η to ξ when $\xi = x$, conditional density of η to ξ, $g_{\xi|\eta}(x|y)$ can be expressed as (when $\eta = y$)

$$g_{\xi|\eta}(x|y) = \frac{p_\xi(x) f_{\eta|\xi}(y|x)}{\int_{-\infty}^{\infty} p_\xi(x) f_{\eta|\xi}(y|x) dx} \quad (7)$$

2.3 Grouped data logistic regression

Suppose independent variable $X = (x_1, x_2, \ldots, x_k)$, which is effect factor of P, is observed L groups results. The jth group is observed n_j and A had occurred m_j. It is possible to calculate the frequency when samples are large. The equation is $f_j = m_j/n_j$. We could use the frequency as group's probability estimate. That is

$$\ln \frac{\hat{p}_j}{1 - \hat{p}_j} = \ln \frac{\hat{f}_j}{1 - \hat{f}_j} = \beta_0 + \beta_1 x_{j1}$$
$$+ \beta_2 x_{j2} + \cdots + \beta_k x_{jk} + \varepsilon_j \quad (8)$$

where x_{jk} is the value of the kth variable in the jth group. ε_j is stochastic error.

The regression model may be calculated with Ordinarily Least Square (OLS). At first, individual data is grouped according to independent variable. Secondly, every group event probability is estimated. Then change every group event probability into logarithm odds which see as dependent variable of linear regression model. All of the dependent variables are categorical variable. Equation (8) has heteroskedastic. There are some methods to eliminate heteroskedastic such as weighted least squares (WLS), Box-Cox transformed method and squared error stability transformed method. All original data including constant are treated through weighted transformed method in WLS model (Dang 1995). After data treated, the model become (9) equation.

$$\left(\frac{1}{S_j}\right) \ln \left(\frac{f_j}{1 - f_j}\right) = \left(\frac{1}{S_j}\right) \beta_0^* + \beta_1^* x_{j1} \left(\frac{1}{S_j}\right)$$
$$+ \cdots + \beta_k^* x_{jk} \left(\frac{1}{S_j}\right) + u_j \quad (9)$$

$$S_j = \hat{V}ar\left(\frac{\varepsilon_j}{p_j(1-p_j)}\right) = \frac{1}{n_j f_j(1-f_j)} \quad (10)$$

where S_j is weight. u_j is transformed residual which has homoscedasticity. The model is also named minimum chi-square estimation.

3 GENERAL INTRODUCTION OF THE APPLIED SITUATION

3.1 Geologic Characteristics of the experimental Area

New city site, Wushan county, Three Gorges is selected as the experimental area. Wushan County lies in the east of Sichuan basin, the north of Daba Mountains, the hinterland of Wushan mountain. The county lies in the center of Three Gorges Reservoir. The water level is 175.4 meter after reservoir impounding. Except some of the inhabiting district of Beimenpo, the majority of the city will be submerged after the reservoir has impounded. New area lies in the joint part of three structures, that is Huaiyang mountainous-type structure west wring reflect arc—Daba mountain arc structure, east of Sichuan belt of folded strata and Sichuan, Hubei, Hunan, Guizhou belt of rise folded structure. The geographical structure is complicated fold and fault. The bedrock is mainly coastal carbonate of Trias, and then the continental sandy argillaceous rock. They include light grey medium thick layer limestone, dolomite and argillaceous dolomite of Jialing River's fourth segment under the Trias system (T_1j^4); grey argillaceous limestone, dark grey argillaceous limestone, dolomite limestone of Badong's first segment, under Trias system (T_2b^1); purple argillaceous limestone contaminating silty argillaceous limestone, silty limestone of the second segment (T_2b^2); grey dolomite limestone, argillaceous dolomite, dark grey medium thick limestone of the third segment (T_2b^3). The featured landslide of new west area is Xiufeng Temple landslide and Sidao Bridge to Dengjiawuchang.

3.2 Slope stability influencing factor

On one hand, slope stability is influenced by the intrinsic factors, such as lithological character, gradient and slope elevation etc. On the other hand, it is influenced by the extrinsic factors, such as rain, human activities etc. Rainfall has not been included in the influencing factor because lack the related data between rainfall and landslide occurrence in the experimental area. The rock type includes T_1j^4, T_2b^1, T_2b^2, T_2b^3. T_1j^4 has the highest intensity so it belongs to firm lithological group. T_2b^1 and T_2b^3 are limestone, marlite which has the medium intensity, so they belong to half-firm lithological group. T_2b^2 has low intensity so it belongs to soft lithological group. In the experimental area, T_2b^1 occupies 17.17% and T_2b^3 occupies 44.85%. If they are put into the same group, it will occupy 2/3 and maybe influence regression analysis precise, so they are treated as different group. The closest distance to effected tectonic line is regarded as geological structure factor. The calculation method is as follows. Each tectonic line above level 4 has been treated single factor logistic regression, and chosen the tectonic line by relevant coefficient R. In the paper, R's limitation is 0.1. The calculation method of the closest distance to effected tectonic line is to search closest distance to the effected tectonic line of each unit. The result of classified group is ≤61 m, 61–122 m, 122–183 m, and >183 m by the singular factor logistic regression coefficient. The detailed selection and classification is as follows.

Table 1. Influencing factors and classification.

Landslide factor	Expression	Classification
Gradient $(^0)$	X_1	≤10, 10–15, 15–20, 20–25, 25–30, 30–35, 35–40, >40
Elevation/(m)	X_2	≤ 150, 150–200, 200–250, 250–300, 300–350, 350–400, 400–450, 450–500, > 500
Slope direction/$(^0)$	X_3	315–45, 45–90, 90–135, 135–180, 180–225, 225–270, 270–315
Lithologic character	X_4	T_1j^4, T_2b^1, T_2b^2, T_2b^3
Slope shape	X_5	Consequent slope 1($\beta < \alpha$), consequent slope 2($\beta > \alpha$), reverse slope, tangential slope
The distance to the closest tectonic line/(m)	X_6	< 61, 61–122, 122–183, >183

Figure 1. Landslide occurrence section in experimental area.
Note: The darkest color is the landslide occurrence section.

3.3 Bayesian statistics inference research method

The study is supported by influence factor database and comprehensively applied two multivariate regression models, trend surface model and Logistic factor model. Combined with GIS technology and statistics technology, regional slope stability has been evaluated. The probability of the first model is regarded as the Bayesian statistic inference's priori information. Factor Logistic model is used to modify trend surface model and to form Bayesian integrating model, and evaluate regional stability.

3.4 Result of the model applied in the experimental area

Supported by MAPGIS, topography data, lithological type data and tectonic line data, etc are input to establish the relevant files. Supported by MAPGIS Secondary Development Database and Visual C++ 6.0, slope stability evaluation system based on Logistic regression model has been established. The experimental area is divided into 14,450 cells with 10 m × 10 m size.

4 RESULT OF THE MODEL APPLIED IN THE EXPERIMENTAL AREA

4.1 The result of six factor Logistic regression model

4.1.1 Regression coefficient
The regression equation is

$$p = \frac{e^{-4.0767+0.2278[x_1=1]+0.5328[x_1=2]+\cdots+0.2550[x_6=3]}}{1+e^{-4.0767+0.2278[x_1=1]+0.5328[x_1=2]+\cdots+0.2550[x_6=3]}} \quad (11)$$

Table 2. Regression coefficient β.

Expression								
$X_1=1$	$X_1=2$	$X_1=3$	$X_1=4$	$X_1=5$	$X_1=6$	$X_1=7$		
β 0.2278	0.5328	0.4490	0.5884	0.3516	0.4342	0.4096		
$X_2=1$	$X_2=2$	$X_2=3$	$X_2=4$	$X_2=5$	$X_2=6$	$X_2=7$	$X_2=8$	
β 0.1207	−0.0270	0.0087	−0.1734	−0.9061	−0.9127	−0.6961	−0.0067	
$X_3=1$	$X_3=2$	$X_3=3$	$X_3=4$	$X_3=5$	$X_3=6$			
β −0.0654	0.2507	0.2698	0.1175	0.0625	0.1394			
$X_4=1$	$X_4=2$	$X_4=3$						
β 1.5995	0.5668	0.8146						
$X_5=1$	$X_5=2$	$X_5=3$						
β 0.6710	0.2294	0.4236						
$X_6=1$	$X_6=2$	$X_6=3$						
β 0.7725	0.5752	0.2550						

β_0 is −4.0767.

4.1.2 Prediction ability
The landslide occurrence accuracy is 72.08%, and non-occurrence accuracy is 81.44%. The criterion of landslide is $Y_0 = 0.163$. If grid probability $P > Y_0$, the grid is judged as the slide will happen, otherwise it will not happen.

4.2 The result of bayesian integrating model

4.2.1 Trend surface model
Trend surface model is a multivariate Logistic regression model which applies coordinate u, v of grid to explain to variable's n times multinomial. The effect of the second multinomial trend surface equation is better than triple multinomial. The result is,

$$p_T = \frac{e^{-2.5682+2.3054\times10^{-2}\mu-1.8320\times10^{-2}v-1.4877\times10^{-4}\mu^2+4.175\times10^{-4}\mu v-3.0117\times10^{-4}v^2}}{1+e^{-2.5682+2.3054\times10^{-2}\mu-1.8320\times10^{-2}v-1.4877\times10^{-4}\mu^2+4.175\times10^{-4}\mu v-3.0117\times10^{-4}v^2}} \quad (12)$$

4.2.2 Integrating model

$$P_B = \frac{1}{1+e^{\ln\frac{1-P_L}{P_L}-\ln\frac{P_T}{1-P_T}}} \quad (13)$$

Where P_L is equation (11).

4.2.3 Prediction ability
The landslide occurrence accuracy is 79.24%, and non-occurrence accuracy is 80.82%. The criterion of landslide is $Y_{B0} = 0.274$.

4.3 The result of grouped data Logistic model

4.3.1 Choice and coding of categorical variable
Three factors have been chosen. They are lithological character factor, elevation factor and distance to the closest tectonic line factor. The categorical variable coding is as follows.

4.3.2 Grouped data Logistic regression equation

$$\hat{P} = \frac{\exp(-4.3976 + 0.8261x_1 + 0.4595x_2 + 0.1286x_3)}{1 + \exp(-4.3976 + 0.8261x_1 + 0.4595x_2 + 0.1286x_3)}$$

(14)

4.3.3 Prediction ability

The landslide occurrence accuracy is 71.85%, and non-occurrence accuracy is 71.70%. The criterion of landslide is $Y_0 = 0.2171$.

5 CONCLUSIONS

1. Bayesian Integrating Model yields the highest landslide occurrence accuracy in experimental area.
2. With support of the system, select 100 m × 100 m, 40 m × 40 m, 4 m × 4 m comparative calculation to test the influence of cell's scale to the calculation result. The result shows that three types scale have the same calculation result.
3. The lithological character is the most sensitivity factor, the elevation is the second sensitivity factor and the distance to effected tectonic line is the third sensitivity factor in six factor Logistic regression model. But the coefficients of these factors are 0.8261, 0.1286 and 0.4595 respectively in grouped data Logistic regression model. The possible cause is qualitative variable coding. The code of 0, 1, 2 and so on imposes amount relative on code in grouped data Logistic regression model. This type coding method maybe effect regression

Figure 2. Landslide occurrence section by six factor Logistic regression model.
Note: The darkest color is the landslide occurrence section.

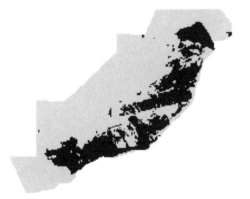

Figure 3. Landslide occurrence section by Bayesian integrating model.
Note: The darkest color is the landslide occurrence section.

Figure 4. Landslide occurrence section by grouped data Logistic regression model.
Note: The darkest color is the landslide occurrence section.

Table 3. Categorical variable coding.

Lithological character X_1	Classification	T_1j^4 0	T_2b^2 1	T_2b^3 2	T_2b^1 3				
The distance to the closest tectonic line/ (m) X_2	Classification	>183 0	122–183 1	61–122 2	≤61 3				
Elevation/(m) X_3	Classification	400–450 0	350–400 1	450–500 2	300–350 3	200–250 4	>500 5	<150 6	250–300 7	150–200 8

precision, but it is not inevitable in grouped data Logistic regression model.
4. GIS plays an important role in the decision-making. On the base of GIS, the paper sets up the slope stability evaluation Logistic model, Bayesian integrating model by employing the Bayesian statistic technique and grouped Logistic model. The results are satisfactory.

REFERENCES

Dang, J.B. 1995. Logistic linear regression model is applied by dealing with qualitative index. *Weaving college foundation course transaction* 8(4): 322–325.

He, X.Q. & Liu, W.Q. 2001. *Application of regression analysis*. Beijing: Renmin University of China press.

Mao, Sh.S. 1999. *Bayesian Statistics*. Beijing: China Statistic Press.

Ohlmacher, G.C. & Davis, J.C. 2003. Using multiple logistic regression and GIS technology to predict landslide hazard in Northeast Kansas, USA. *Engineering Geology* (69) 3–4: 331–343.

Tang, Ch. & Jorg G. 1998. The Principles and Methodology of Landslide Hazard Assessment. *ACTA GEOGRAPHICA SINICA* (53) S:157–159.

Wang, J.Ch. & Guo, Zh.G. 2001. *Logistic regression model—method and application*. Beijing: higher education press.

Wu, X.C. etc. 2002. *The Principles and Methodology of Geographical Information System*. Beijing: electronic industry press.

Zhang, H.L. & Li Zh.X. & Wang R.Ch. Etc. 2000. Application of Bayesian Statistics Inference Techniques Based on GIS to the Evaluation of Habitat Probabilities of Bos Gaurus Readei. *Journal of remote sensing* (4) 1:66–70.

Zhang, Q.R. 1990. Geological Trend Surface Analysis. Beijing: science press.

Zhang, Y.T. & Chen, H.F. 1991. *Bayesian Statistics Inference*. Beijing: science press.

The stability analysis for FaNai landslide in Lubuge hydropower station

Kaide Li, Jin Zhang, Sihe Zhang & Shuming He
Kunming Hydroelectric Investigation Design & Research Institute, State Power Corporation of China

ABSTRACT: The stability of large scale landslide. Fanai in Lubuge hydropower station, with complexed properties, would be the key issue to affect downstream area and villages. So, it is necessary to research, calculate and estimate the stability by any investigation, including drilling holes and test etc. Fanai landslide, located in the reservoir area of Lubuge hydropower station, had been slipped partly which affected by construction and reservoir water. We had lots of detail investigation to estimate the effects on downstream area. Keeping resident safely on the landslide and getting any significant treatments are our purposes.

1 BASIC GEOLOGICAL CONDITION

1.1 Topography

Fanai landslide located on the reservoir left bank with the distance about 1.4 km to 2.2 km from damsite. Upstream valley is open with 15° to 25° of both slopes but downstream is "V" type with slope more than 45°.

The landslide on left bank where valley changed from wide to narrow and flowing direction is S34 W. Riverbed elevation is between 1090 m to 1096 m and the slope height is about 500 m to 600 m. Reservoir water surface altitude is 1130 m.

This landslide locates at 1065 m to 1375 m area. According to the topography, its appearance is just a convex slope or asperity with bluff along trailing edge. Generally, its geomorphic land surface is showing as "arc-chair" shape landslide.

1.2 Stratum

The rock types in this region are:

1. Sandstone, shale intercalated carbon layer or coal shed deposited in Permian age (P_{2l}), outcroped on both river banks and as the formation of part landslide.
2. Sandstone, shale and mudstone deposited in Triassic period (T_{1f}), mainly distributed middle-southern of landslide. Part of this formation outcropped in back slope.
3. Another deposited layer of Triassic period (T_{1y}) composed by limestone with sandstone and shale, distributed on south area and downstream of landslide. Part pf this layers outcropped in back slope and downstream of mountain ridge.

Quaternary covers normally distributed in the gentle slopes and river beds.

1.3 Geological structure

Stratum strike is N°35~65° W and the dip is SW∠30°~40°. Three faults developed in the landslide area showing in Figure 2.

Fault No.1 (F_1) developed in trailing edge of landslide with attitude of N30° E, NW∠85°~90°.

Fault No.2 (F_2) extended from downstream to trailing edge of landslide. Fault No.3 (F_3), with attitude of N20° E, SE∠85°~90°, distributed in the landslide trailing edge too but it approach downstream. Rock Joints in the slide region developed normally.

1.4 Hydrogeology

Groundwaters in this region are pore-water, crack-water & Karst-water etc. Pore-water stored in Quaternary cover layers, crack-water in rock mass and Karst-water in limestone which distributed on back-slope. Because Quaternary colluvium is a lossen layer with strong perviousness, rainfall often discharges quickly.

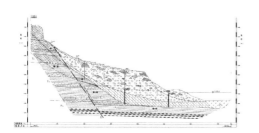

Figure 1. Geological profile of landslide.

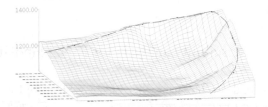

Figure 3. Three-dimensional shape of slip bed.

The leading zone on landslide surface is open and changed narrow backward (Figure 3). Mid surface is showing as gentle and rolling with of depth ranging from 30 m to 100 m. Trailing edge is steep and deep with the depth of 140 m to 170 m. Total sliding direction is N61.5° W. Total volume of Fanai slip mass is about 4.3×10^7 m^3.

2.2 Ancient slide formation mechanism

It is an ancient rock landslide. Sliding plane is controlled by rock bedding surface and structure. Based on physical classification, it is dragging and driving one.

Because of sedimentary layers, such as limestone, fine sandstone and marl dips along the slope; sliding region is shown as uniclinal structure.

According to geological history, this ancient slide formation has been caused by stream corroding and two faults distributed in trailing edge. Whit the subsequent river cutting process, the slope had been deformed backward and created patulous fractures along trailing edge. Moreover, unload and loosed rockmass happen to collapse and slide.

The present landform and surface feature are related to slided movement.

With the reservoir level increased, water softens sedimentary layers and other formation, broken the leading end and lead to slide finally.

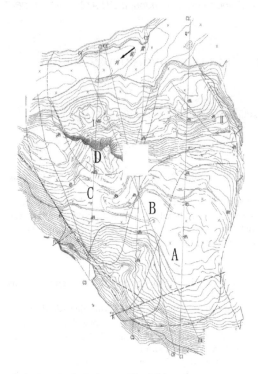

Figure 2. Geological map of landslide.

A few of springs spilled over in this region. Just one spring located on the slopeslide with water discharge about 1 L/min to 2 L/min. The groundwater level is very deep in the back of landslide and the normal depth is 80 m to 110 m. But it gradually become shallower closed to the bank of reservoir. The gradient of water table is too gentle.

2 LANDSLIDE CHARACTERS

2.1 Landslide configuration

This landslide covered about 0.58 km^2 area, 750 m width in leading edge, 690 m to 820 m in middle band and 380 m to 520 m in trailing edge, with longitudinal length of 850 m. The landslide center is max thick with 100 m to 160 m.

Generally, sliding thickness is controlled by the broken types. The mid zone is thicker than the surrounding belt but average thickness is 14.2 m. Because of the soft layer broken by sliding in the leading belt, slipmass is forced forward and accumulative layer about 15 m to 40 m. Moreover, rock layers on both sides and back belt are thicker than the mid area. But average thickness is less than 10 m.

2.3 Developing process of slip mass

This ancient slide was still stabile at the time of preconstruction of Lubuge hydropower project.

In 1979, at the beginning of construction, manual activities, such as living waste water drainage and working load, destroy the primal balance. In 1883, tension cracks were found in No.1 zone (showed as Figure 2), but the slope deformation was not lead to slide.

In November and December 1988, after Lubuge reservoir put into use, it induced to 2.9 and 3.1 earthquake intensity when the storage level arised to 35 m, leading to multi-slides in this region. In the rainy season, from June to September, 1989, the slope had slided again.

Table 1. Accumulated landslide deformation between 2002 to 2003.

Observation point	Horizontal displacement (mm)	Perpendicular displacement (mm)	Observation point	Horizontal displacement (mm)	Perpendicular displacement (mm)	Observation point	Horizontal displacement (mm)	Perpendicular displacement (mm)
C1	43.1	−4.5	N2	20.7	−9.8	P5	8.8	1.9
D1	24.5	0.2	P1	70.7	−0.8	P6	16.5	−1.0
F	22.9	0.6	P3	28.8	−2.3	P7	14.1	−0.1
G	25.4	−5.1	P4	22.0	0.3	P8	12.1	−0.6
H	20.4	−2.9	P4′	16.2	−3.2			

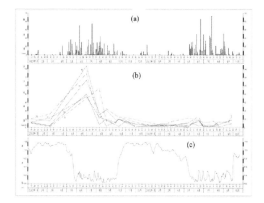

Figure 5. Curve, histogram reflecting to horizontal displacement, rainfall and water level change (2002~2003). (a) Precipitation (b) Displacement vs. time (c) Pool water level vs. time.

Researches for this ancient landslide begin in April, 1989, when there were 13 observation points built on sliding surface. But all of these points had been broken by the sliding.

According to observation datum, deformation rate in leading edge is related to the first storage time and to the strengthen rainfall. During the period from May to June, 1989 and July to August, 1990, this deformation was obvious (Figure 4). In December, 1990, the accumulative horizontal displacement was 32000.4 mm and Perpendicular displacement was 4300.3 mm. These deformations were decreased and tend to stable after September, 1990.

No.1 zone happened to slide on July 13th, 1997, when reservoir emptied. Most fell under the normal pool level (1130 m) and flowed into riverbed. No.2 zone occurred small-scale surface sliding and collapse.

Additionally, land sliding leaded to orbicular cracks in the trailing edge and transverse fissures in mid belt. The land sliding had broken farmer's houses.

At end of 2001, observation network had been rebuilt in active zone for this landslide. We have got a lot of typical information at that time of 2001 to 2003 (Figure 5 & Table 1). These data indicate that landslide deformation is related to rainfall and mutation of reservoir water level.

2.4 *Material of slip mass*

The slip mass is mainly composed of rock block, granule, silty soil and clay, part rock block are still keeping the original rock structure. Rock block are limestone, sandstone and mudstone. Generally, the frame of landslide is a farraginous accumulation with rocks and soil.

2.5 *Formation of sliding belt*

Generally, the sliding belt is mainly composed of soft layers or materials, such as carbon clay and silty soil. Strength of the formation would be decrease while meeting water.

3 FACTORS AFFECTED ON SLIDING

Geological structure, earthquake, groundwater and process of surface water are main factors acted on the slide. Moreover, reservoir water mutation and human activity would other issues induced sliding.

Fanai landslide is an ancient slide. Compositions of slip mass and slid belt, shear strength are key issues for the slide stability. Slide mass is revealing which maxed with silt and stone. It would be happened to slipe when affected by external processes.

According to Chinese seismic zoning and seismological parameter standards, Landslide area intensity is VI and the matching horizontal seismic coefficient is 0.05 g. It would occur to sliding and collapse when earthquake happened. Based on the observation datum of reservoir storage, induced earthquake is the key issue for the slide stability.

Groundwater processing on the slipbody includes hydrostatic pressure, hydrodynamic pressure, load, sofen in slidebelt and contact brushing. While, surface water process includes brush and cut formed air face or imperfect topography leading to collapse.

Additionally, external factors induced sliding includes mutation of storage level and manual excavation. If storage level changes frequently, it would not lead to groundwater drain off in short time. Steep hydraulic gradient would increase hydrodynamic pressure in slidbody. On the other hand, large scale change of storage would induce earthquake, slope cutting and brushing. It would be disadvantaged for the stability of slope. Figure 4 reflected relationships between storage level change, rainfall and horizontal displacement based on observation datum. Road construction had changed pre-stress state of slope, broken the stability of landslide.

4 STABILITY ESTIMATE

4.1 Calculation of stability

4.1.1 Calculation methods

For analysis the slope stability, EMU (Energy Method Upper bound limit analysis) has been used, comparing with Sarma and other methods.

4.1.2 Physical and mechanical parameter

According to test datum and geological analysis, we ascertained physical and mechanical parameter for calculating slide stability (Table 2). Considering basic intensity as VI and the matching horizontal seismic coefficient should be 0.05 g.

4.1.3 Calculation result

Considering any kinds of condition, normal state, earthquake and water sudden drawdown, we have three kinds calculation based on five geological profiles. Calculated results are shown as Table 3, Table 4, Table 5 and Figure 6, Figure 7.

Datum in Table 3 is the calculated results which suppose the unitary body sliding along the bottom. Datum in table 4 and Table 5 are other calculation results suppose the slip mass in the state of stabile limitation.

4.2 Estimate of stability

Based on the calculation of five geological profiles, analysis and comparing results each other, generally, we can get a conclusion that Fanai landslide is steady in natural state, but part of mass, especially the downstream area, is unstable.

Profile C1 shows as stable in all kinds of condition and its safety coefficient is more than 1.2. Safety coefficients of Profile C2, C4 and C5 are more than 1.0 except earthquake happened.

The safety coefficients of Profile C3 close to 1.0. However, considering the following three conditions (natural state, earthquake and water sudden drawdown), stability coefficient by searching and the slide risky area are less than 1.0 except Zone ① in normal state. For the searching process in Profile C3, with deep sliding bed, strong lateral friction drag in Zone ② and Zone ③, the calculation result would be lead to higher than in actual condition. But Zone ② and Zone ③ belong to risky area. Whereas, sliding bed in Zone ① is shallower than in Zone ② and Zone ③, with lower lateral friction drag, so the calculation result would be closed actual condition which belong to unstable area.

Different mediums have various anti-seisms in slip body. Slip bed consist rock with lower anti-seism but materials of sliding body are loose whit high earthquake resistance. Calculation result proved that earthquake would cause holistic land sliding, but not affect on part area. Generally, this dead landslide would be re-active in part area when happen seismic events up to a maximum recorded intensity of VIII.

Sudden drawdown of reservoir level would be the main issues for sliding. Especially, this mutation of water level would affect to the lower belt in downstream area. But middle and higher area would keep stable. Lower belt of C3, C4 and C5 with deep sliding bed, closed to or under the reservoir level, would happen to slide while meeting sudden drawdown of water. According to the observation on 16th to 29th, May 2002, when storage level fell to 16.46 m, sliding horizontal displacement was obvious. But the horizontal displacement would not develop more at that time on 5th to 19th, May 2003, when storage level fell to 12.08 m. According to calculation and observation result, the influenced extent by water sudden drawdown is so limited.

Because it has various stability in different area, calculation results and geological condition proved that Fanai slide can be classify into four zones (A, B, C and D in Figure 2). Zone A is stable but would happen to a little slide. Zone B is a stable area would not happen to

Table 2. Physical and mechanical parameter.

	Friction angle $\Phi(°)$	Cohesion (kPa)	Natural density γ (kN/m^3)	Saturated unit weight γ_{sat} (kN/m^3)
Slip mass	25	35	20	22
Sliding belt	18	28	19	21
Sliding plane	35	500	25	25

Table 3. Stability calculation of unitary landslide.

Computing method	Calculated profile	C1	C2	C3	C4	C5
EMU	Normal state	1.829	1.630	1.177	1.355	1.221
	VI seismic intensity	1.471	1.408	1.025	1.121	1.032
	20 m Sudden drawdown	1.794	1.585	1.124	1.336	1.203
Sarma	Normal state	1.324	1.250	1.014	1.282	1.154
	VI seismic intensity	1.104	1.115	0.892	1.084	0.996
	20 m Sudden drawdown	1.302	1.232	0.984	1.252	1.128
Non-equilibrium traction	Normal state	1.492	1.425	1.110	1.358	1.226
	VI seismic intensity	1.228	1.255	0.970	1.138	1.050
	20 m Sudden drawdown	1.462	1.393	1.066	1.322	1.192

Note: 1. we supposed slide moved along the sliding plane; 2. Considering the storage level as 1130 m and coefficient of seismic force is 0.05.

Table 4. Unitary landslide searching calculation of limiting stability.

Computing method	Calculated profile	C1	C2	C3 (③)	C4	C5
EMU	Normal state	1.593	1.154	0.970	1.355	1.221
	VI seismic intensity	1.317	0.983	0.871	1.121	1.032
	20 m Sudden drawdown	1.573	1.121	0.911	1.336	1.203
Sarma	Normal state	1.486	1.000	0.898	1.282	1.154
	VI seismic intensity	1.248	0.874	0.846	1.084	0.996
	20 m Sudden drawdown	1.464	0.980	0.896	1.252	1.128
Non-equilibrium traction	Normal state	1.540	1.066	0.954	1.358	1.226
	VI seismic intensity	1.280	0.926	0.836	1.138	1.050
	20 m Sudden drawdown	1.508	1.038	0.914	1.322	1.192

Note: 1. We supposed both ends of slip are fixed when automatic searching; 2. Considering the storage level as 1130 m and coefficient of seismic force is 0.05.

Table 5. Upper landslide searching calculation of limiting stability.

Computing method	Calculated profile	C1	C2	C3 (②)	C4	C5	C3 (①)
EMU	Normal state	1.278	1.230	0.966	1.283	1.185	1.012
	VI seismic intensity	1.113	1.093	0.885	1.093	1.027	0.972
	20 m Sudden drawdown	1.268	1.189	0.891	1.240	1.153	
Sarma	Normal state	1.186	1.050	0.874	1.054	0.988	0.972
	VI seismic intensity	1.032	0.932	0.794	0.924	0.870	0.906
	20 m Sudden drawdown	1.176	1.042	0.832	1.036	0.968	
Non-equilibrium traction	Normal state	1.218	1.116	0.916	1.110	1.054	1.014
	VI seismic intensity	1.058	1.022	0.816	0.968	0.920	0.912
	20 m Sudden drawdown	1.208	1.116	0.874	1.084	1.020	

Note: 1. Rearching sliding plane form inner to exterior. 2. Considing the storage level as 1130 m and coefficient of seismic force is 0.05.

large scale slide but would be moved when caused by external effects. Zone C is an unstable area too where it would be happen to large scale sliding if it was caused by external effects. Zone D is another unstable body and would be sliding under other proper conditions. All of these zones are tight each other. One of them slides would be lead to next zone unstably.

5 AFFECTION OF SLIDING

5.1 Affection for reservoir

According to analysis for the stability of landslide, it would be affect on the running of the reservoir. Its movement is small and would not cause to large scale collapse or sudden falling. Sliding activity would

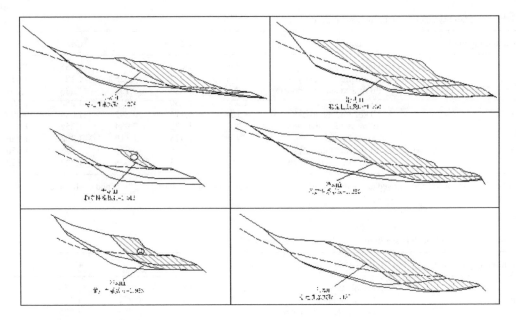

Figure 6. Profiles for unitary landslide searching calculation of limiting stability.

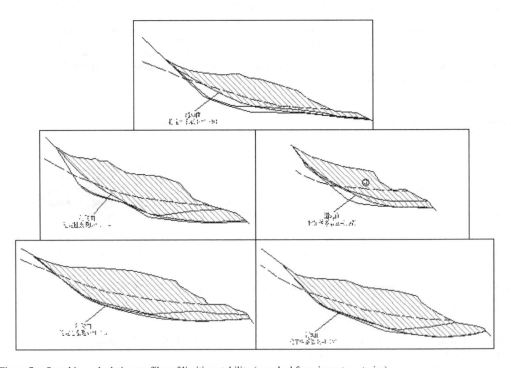

Figure 7. Searching calculation profiles of limiting stability (searched from inner to exterior).

not destroy the reservoir and power station. However, sliding caused by earthquake would lead to reservoir sedimentation.

5.2 *Affection for residents and houses*

More than 100 resident families living on Fanai slide surface. Sliding had been broken resident houses. Especially Zone D, in 1997, surface sliding leaded to cracks in the farmer's soil wall. Sliding in some surface areas is the issue for house and building stability.

6 CONCLUSION AND ADVICE

1. Generally, the landslide would not happen to large scale sliding. According to limitation of stability, Fanai landslide can be classified into 4 zones. Zone C and Zone D may be induced some small slidings but would not influence the running of hydropower station.
2. The groundwater movement would influence the landslide stability to some extent. It would be dangerous if the storage level fell down quickly. Emptying the reservoir quickly for flushing sedimentation is risky behavior. In order to keep slope stably, controlling water falling speed and drawdown are more necessary.
3. Surface sliding is dangerous for the resident building foundation. Human activities are one of another reasons leading further sliding. So it is necessary to emigrate the inhabitant who living on this unstable area.
4. Keeping on observation and inspection for the sliding, building new drainage establishment would be in favor of slope stability and protecting the permanent inhabitant.

Numerical analysis of slope stability influenced by varying water conditions in the reservoir area of the Three Gorges, China

Shaojun Li & Xiating Feng
Institute of Rock and Soil Mechanics, Chinese Academy of Sciences, Wuhan, China

J.A. Knappett
Division of Civil Engineering, University of Dundee, Dundee, Scotland

ABSTRACT: Landslide disasters have become one of the main problems after the Three Gorges impounded. Scholars from all over the world have paid much more attention to the stability of slopes in the reservoir area. Water has been regarded as the main factor which may directly trigger landslides, while the change of water conditions can't be avoided in the slopes in the Three Gorges. The water conditions mentioned in this paper refer to reservoir water level and water table in slope body. They will inevitably deteriorate the slope stability in some degree. The paper aims to investigate their influence on slope stability. A promising numerical analysis technique by shear strength reduction based on finite differential method is introduced here to deal with such a problem. According to the results from a real landslide in the Three Gorges, the curve between the factor of safety and the change of water level is parabolic: the slope stability will be reduced at the first stage of raising reservoir water level and then the factor of safety will be increased after that. However, the landslide will always greatly suffer from the instability when the reservoir water level drops down and the water table rises. It will provide a direct proof for the landslide prevention and remedial designs in this case.

1 INTRODUCTION

The Three Gorges, located on the Yangzi river in Yichang city of Hubei province, China, is the current giantest hydraulic and hydroelectric project in the world. After the impoundment in 2003, about 2490 potential landslides in the reservoir area will be induced according to some materials. All the landslides will be inevitably influenced by varying water conditions, the maximum variable magnitude is up to 30 m. In Japan, about 60 percent of landslides occurred during the sudden drawdown of reservoir water level, while 40 percent are induced by water level rise (Liao et al. 2005). The water condition mentioned in this paper can be divided into two cases, the first one is originated from the impoundment which will change the water level from the toe to the top of the landslides, the other is from the season's rainfall which will alter the water table in the landslide body.

In recent years, many Chinese scholars have done a lot of works related to the slopes stability in the reservoir area of the Three Gorges. Liao et al. (2005) got the relationship between the landslide stability, the permeability coefficient and drawdown speed based on limited equilibrium method using Geo-Slope software. Ding et al. (2004) studied process of the seepage-stress and obtained the deformation trend and failure process in view of water fluctuation for a real landslide based on Flac$^{4.0}$. Zhu et al. (2002) & Liu et al. (2005) discussed the influence by the permeability coefficient of soil under water drawdown based on limited equilibrium method. Tang et al. (2005) & Shi et al. (2004) also focused on the process of water level drawdown and evaluated the stability toward a real landslide respectively. In addition, Liu et al. (2005) analyzed the deformation mechanism when water level raised and calculated the corresponding factor of safety.

In this paper, in view of varying water conditions in the Three Gorges mentioned above, a different and promising numerical method for slope stability analysis called shear strength reduction is to be adopted. Based on a real landslide in the reservoir area, the relationship between factors of safety and varying water condition is obtained with shear strength reduction technique, results will contribute a lot to the stability evaluation and remedial works of slopes in the Three Gorges.

2 SHEAR STRENGTH REDUCTION BASED ON FINITE DIFFERENTIAL METHOD

Factor of safety F is widely adopted for slopes stability analysis. Shear strength reduction technique, integrated with finite element and finite difference, was put forward as early as 1975 by Zienkiewicz et al. (1975), and has been applied by Matsui (1982), Dawson (1999), Zheng et al. (2004) and others. Compared to traditional limit analysis solution, the shear strength reduction technique has a lot of advantages. For example, the critical failure surface can be found automatically, the plastic zone and displacement can also be obtained at the same time. Therefore, it has become a promising method applied in geotechnical slopes.

In order to perform slope stability analysis with shear strength reduction technique, the soil/rock strength should be reduced continually until collapse occurs. In this case a series of trial factors of safety F^{trial} with cohesion C and friction ϕ are to be adjusted according to the following equation:

$$C^{trial} = \frac{1}{F^{trial}} C \qquad (1)$$

$$\phi^{trial} = \frac{1}{F^{trial}} \phi \qquad (2)$$

Factor of safety is to be computed using the explicit-finite difference method in this paper, FLAC. In the computing principal of FLAC, for given element shape functions, the set of algebraic equations solved by FLAC is identical to that solved with the finite element method. The convergence criterion for FLAC is the nodal unbalanced force, the sum of forces acting on a node from its neighbouring elements. Obviously, it is easier for convergence and will be better for the problems with big deformation.

3 CASE STUDY AND NUMERICAL MODEL

3.1 Muzishu landslide in the Three Gorges

Muzishu landslide, locating on the south bank of Yangzi river, is in Zigui county, Chongqing city, China. The landslide has total potential slipping volume of 4.0×10^5 m^3 with the length of 130 m~190 m and the with of 80 m~120 m. According to the material of geological condition, the bedrock, originated from Jurassic system, is mainly made up of sandstone. The soil in landslide body, with the maximum thickness of 25.0 m, consists artificial macadam soil (thickness: 0.5 m~8.0 m), gravelly soil (thickness: 2.0 m~15.0 m), block stone and macadam soil (thickness: 0.5 m~13.0 m). The parameters of soil and bedrock are listed in Table 1 and the typical geological section is shown in Figure 1.

3.2 Numerical model

According to the impounding scheme of the Three Gorges, the simulation of varying water level is divided into these following steps: 125 m, 130 m, 135 m, 140 m, 145 m, 150 m, 155 m, 160 m, 165 m, 170 m, 175 m, 180 m, 185 m.

In order to simulate the change of water table, a coordinate is created as Figure 2, the assumed water tables are almost the same shape and parallel with the initial water table, there is 2 m space between each of the water tables. Final scheme for change of water table is listed below: -6 m, -4 m, -2 m, 0 m, 2 m, 4 m, 6 m.

In the numerical model, the Mohr-Coulomb elasto-plastic model is adopted, properties of soil for computing analysis are selected from the Table 1. Finally a two dimensional numerical computing model is created as shown in Figure 3. However, during the computing procedure, the dynamic seepage and the strength reduction of soil due to water immersion are not taken into account in this paper.

3.3 Computing process

The computing starts from a value of $F^{trial} = 0.8$. According to the Equation 1 and Equation 2, the decreased C^{trial} and ϕ^{trial} will be obtained respectively. Then F^{inc} is set to 0.2, in the next steps, the value of F^{trial} should be increased according to the following formula:

$$F_n^{trial} = F_{n-1}^{trial} + F^{inc} \qquad (3)$$

During the trial computing, if the slope fails, the F^{inc} should be reduced to $1/5 F^{inc}$, then repeat the computing process until the F^{inc} is less than a preset value $\varepsilon = 0.001$. At this time, the slope is in the limit equilibrium state and the corresponding F^{trial} is the final factor of safety, and the failure surface can be also obtained finally.

4 RESULTS AND ANALYSIS

With a view to the landslide field condition and computing schemes mentioned above, the technique of shear strength reduction based on finite differential is performed by FLAC software, a series of factors of safety are calculated as shown in Figure 4 and Figure 5.

Obviously, the relationship curve between reservoir water level and factor of safety F is similar to parabola. Firstly F is equal to 1.04 for the moisture slope with no water table and the reservoir water level is 0 m. However, as the water level rises from 125 m to 135 m~145 m, F decreases from 0.95 to 0.90. Then F will increase gradually until to 0.98 with the reservoir water level 175 m. After that, F will increase

Table 1. Properties of gravelly soil in Muzishu landslide.

Physical properties					
Moisture content ω(%)	Bulk density (kN/m³)		Liquid limit Wl(%)	Plastic limit Wp(%)	Plasticity index Ip(%)
	Wet	Dry			
18.8~21.9 *20.1	19.6~20.9 *20.3	16.3~17.6 *17.0	25.6~29.5 *17.8	14.8~18.0 *16.9	10.2~11.5 *10.8

Mechanical properties			
Consolidated fast shear		Residual shear	
Cohesive strength C(kPa)	Internal friction friction angle φ(degree)	Cohesive strength C'(kPa)	Internal friction friction φ'(degree)
6.0~18.0 *13.0	12.0~25.0 *19.3	0~6.0 #4.0	10.0~25.0 #18.0

*Mean value of five samples, #mean value of three samples.

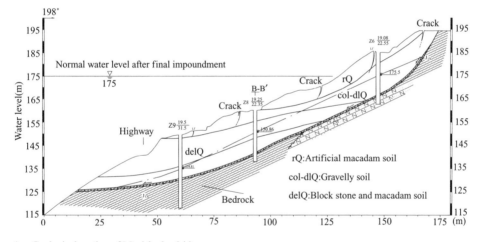

Figure 1. Geological section of Muzishu landside.

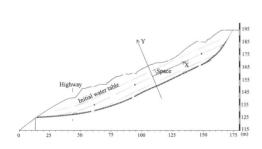

Figure 2. Scheme for change of water table.

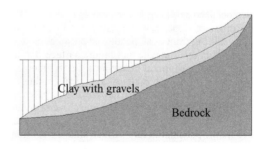

Figure 3. Computing model of Muzishu landslide.

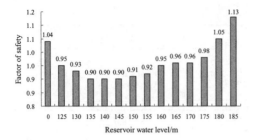

Figure 4. Relationship between the reservoir water level and factor of safety, Muzishu landslide.

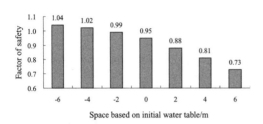

Figure 5. Relationship between the water table in landslide body and factor of safety, Muzishu landslide.

rapidly. The result indicates the natural slope is almost in the sate of limit equilibrium, at the first stage of raising reservoir water level, the slope is in unstable condition, this is why many landslides have occurred during this time. Then after the water level rises over 145 m, the slope stability will be improved. On the contrary, if the reservoir water drops down, the slope stability will be greatly reduced, and it will possibly induce the landslide.

On the other hand, as the water table in landslide body increases as shown in Figure 5, the factor of safety will be greatly decreased from 1.04 to 0.73, the landslide will definitely be triggered. This condition is more dangerous and probably takes place due to heavy rains.

At the same time, the slope deformation (plastic zone and displacement vector) of each condition is also obtained as shown in Figure 6 and Figure 7. For the moisture slope, it is presumed that there are no water level in front of slope toe and no water table in landslide body. The potential failure surface doesn't run through as shown in Figure 6(a). On this moment FOS is 1.04, the slope is almost in critical condition.

The results of plastic zone indicate that the potential failure surface varies with the change of water conditions. When the reservoir water level is lower than 145 m, the potential failure surface almost runs from the slope top to toe, Figure 6(b). However, as the reservoir water level rises, the potential failure surface will

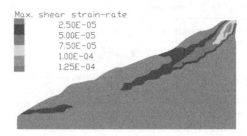

(a) Moisture slope (No reservoir water level and water table)

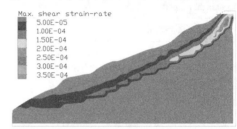

(b) Reservoir water level 145m

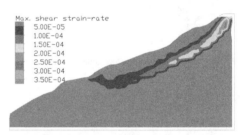

(c) Reservoir water level 155m

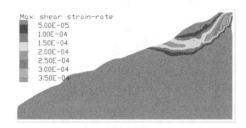

(d) Reservoir water level 175m

Figure 6. Distribution of plastic zone (potential failure surface) under different reservoir water levels.

move up toward the slope top gradually, see Figure 6(c) and 6(d).

Great attention should be paid to the condition of water level 175 m. In this case, although it seems the whole slope is stable, the failure surface still locates on the top of the slope. This part will definitely slip due to some deteriorating external factors, such as heavy rain

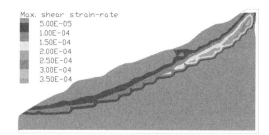

Figure 7. Distribution of plastic zone (potential failure surface) with water table +6 m as shown in Figure 2.

and sudden drawdown of reservoir water level. Therefore, some reinforcing measures such as anti-sliding pile and anchor cables should be taken to control the potential failure.

5 CONCLUSION

In this study, the analysis of slope stability influenced by varying water condition is presented based on shear strength reduction technique, from the case study of Muzishu landslide, some conclusions can be drawn as follows:

1. Shear strength reduction technique based on finite differential is an effective method for slope numerical analysis, through which both the factor of safety and failure surface can be obtained.
2. During the rise of reservoir water level, the factor of safety will decrease at the first stage, the slope maybe fail in this condition. While F will increase when the reservoir water reaches a certain level which is 145 m in Muzishu landslide. On the other hand, the slope will always suffer from the serious instability when decreasing the reservoir water level and raising the water table in the slope body.
3. After the impoundment in the Three Gorges, plastic zone and potential failure surfaces will move up toward the top of slope. Much more attention should be paid to this phenomenon. Particularly after the final impounding level of 175 m, some effective reinforcing measures, such as anchor cables and anti-sliding piles, are essential in those potential dangerous parts.

ACKNOWLEDGEMENT

Financial support from the Pilot Project of Knowledge Innovation Program of the Chinese Academy of Sciences under Grant no. KJCX2-YW-L01 and the Special Funds for Major State Basic Research Project under Grant no. 2002CB412708 are gratefully acknowledged.

REFERENCES

Liu Caihua, Chen Congxin & Feng Xia-ting. 2005. Study on mechanism of slope instability due to reservoir water level rise, *Rock and Soil Mechanics*, Vol.26(5), 769–773.

Dawson, E.M., Roth, W.H. & Drescher, A. 1999. Slope stability analysis by strength reduction, *Geotechnique*, 49, No.6, 835–840.

Zhu Donglin, Ren Guangming & Nie Dexin. 2002. Effecting and forecasting of landslide stability with the change of reservoir water level, *Hydrogeology and Engineering Geology*, No.3, 6–9.

Liao Hongjian, Sheng Qian & Gao Shihang. 2005. Influence of drawdown of reservoir water level on landslide stability, *Chinese Journal of Rock Mechanics and Engineering*, Vol.24(19), 3454–3458.

Tang Huiming & Zhang Guangcheng. 2005. Study on slope stability during reservoir water level falling. *Rock and Soil Mechanics*, Vol.26 (SUP.2), 11–15.

Matsui, T. & San, K.C. 1992. Finite element slope stability analysis by shear strength reduction technique. *Soils and Foundations*, 32, No.1, 59–70.

Shi Weiming & Zheng Yinren. 2004. Stability evaluation of landslide under reservoir water level drawdown condition. *Geotechnical Investigation & Surveying*, No.1, 27–35.

Liu Xinxi, Xia Yuanyou & Lian Cao. 2005. Research on method of landslide stability valuation during sudden drawdown of reservoir level, *Rock and Soil Mechanics*, Vol.26(9), 1427–1431.

Ding Xiuli, Fu Jing & Zhang Qihua. 2004. Stability analysis of landslide in the south end of Fengjie highway bridge with fluctuation of water level of Three Gorges reservoir, *Chinese Journal of Rock Mechanics and Engineering*, Vol.23(17), 2913–2919.

Zheng Yinren & Zhao Shangyi. 2004. Application of strength reduction FEM in soil and rock slope, *Chinese Journal of Rock Mechanics and Engineering*, Vol.23(19), 3381–3388.

Zienkiewicz, O.C., Humpheson, C. & Lewis, R.W. 1975. Associated and non-associated visco-plasticity and plasticity in oil mechanics. Geotechnique 25, No. 4, 671–689.

Landslides and Engineered Slopes – Chen et al. (eds)
© 2008 Taylor & Francis Group, London, ISBN 978-0-415-41196-7

A numerical study of interaction between rock bolt and rock mass

X.P. Li & S.M. He
Key Laboratory of Mountain Hazards and Surface Process, Chinese Academy of Science, Institute of Mountain Hazards and Environment, CAS, Chengdu, China

ABSTRACT: In order to improve the design of rock bolts, it is necessary to have a good understanding of the behavior of rock bolts in rock masses. Several analytical models of describing the interaction among of the rock bolts, the grout medium and the rock mass have been proposed in literatures. In this paper, a numerical simulation procedure was developed for a rock bolt subjected to a concentrated load in pullout tests. The study was concentrated on the failure at the interface between the bolt and the grout. The commercial code FLAC3D was used to analyze the coupling and decoupling phenomenon at the bolt-grout interface. The interface provided by FLAC3D which is characterized by Coulomb sliding and/or tensile separation is proved the premise for simulating the problem. On the basis of a numerical parametric study, the effects of dilation angle and confining pressure on the behavior of the interface were clarified and the failure mechanism on the interface was identified.

1 INSTRUCTIONS

Rock bolts have been widely used to reinforce rock slopes, hydro dams and underground works such as tunnels and mine workings for a long time. Several analytical methods are proposed for bolting design (Farmer, 1975; Freeman, 1978; Li & Stillborg 1999; Cai et al. 2004). However, the interaction mechanism of the rock bolt and the rock mass is not well understood, and the bolting design is still empirical so far.

Pullout tests are currently used to examine the anchoring capacity of rock bolts. Unlike bolts in situ, bolts in a pullout test only have an anchor length. Therefore the anchorage mechanism of bolts and rock masses can be simply investigated through the study on the pullout test. When a fully grouted bolt is subjected to a tension load, failure may occur at the bolt-grout interface, in the grout medium or at the grout-rock interface, depending on which one is the weakest (Li & Stillborg 1999). In this study we concentrate on the failure at the interface between the bolt and the grout.

To investigate the coupling and decoupling behavior of the rock bolt in pullout tests, numerical analysis has been carried out using the explicit finite difference code FLAC3D for better understanding the interface bonding failure phenomena (FLAC3D, 2003). From a series of numerical experiments, it is found that the dilation angle of the interface and confining pressure on the bolt have significant effects on the coupling and decoupling of the interface.

2 OVERVIEW OF PREVIOUS WORK

When a fully grouted bolt is subjected to a tension load, the shear stress along the bolt before decoupling was expressed by Hawkers & Evansas (1951) as:

$$\tau_x = \tau_0 e^{-\frac{Ax}{d}} \quad (1)$$

where τ_0 is the shear stress at the loading end; d is the diameter of the bolt; A is a material parameter which describes the interaction properties of the rock bolt and grout medium.

The Equation (1) can describe the exponentially attenuation of the shear stress at the bolt interface, but it is too simple to represent the effect of the grout and surrounding rock mass on the stress distribution. In Farmer's work (1975), the attenuation of the shear stress was expressed as:

$$\tau_b = \frac{\alpha}{2}\sigma_{b0} e^{-2\alpha \frac{x}{d_b}} \quad (2)$$

where

$$\alpha^2 = \frac{2G_r G_g}{E_b \left[G_r \ln\left(\frac{d_g}{d_b}\right) + G_g \ln\left(\frac{d_0}{d_g}\right) \right]}$$

$$G_r = \frac{E_r}{2(1+\nu_r)}, \quad G_g = \frac{E_g}{2(1+\nu_g)}$$

where σ_{b0} is the axial stress of the bolt at the loading point, E_b is Young's modulus of the bolt steel, E_r is Young's modulus of the rock mass, E_g is Young's modulus of the grout, ν_r is Poisson's ratio of the rock mass, ν_g is Poisson's ratio of the grout, d_g is the diameter of the borehole, and d_0 is the diameter of a circle in the rock outside which the influence of the bolt disappears.

Cai et al. (2004) proposed a model based on the improved shear-lag theory, in which the shear stress distribution along the bolt before decoupling was expressed as:

$$\tau_b(x) = \tau_0 \cosh[\alpha(L-x)]/\cosh(\alpha L) \quad (3)$$

where

$$\alpha = \sqrt{H(1/(A_b E_b) + 1/(E_b A_b))}$$

where τ_0 is the shear stress at the loading end; L is the length of the bolt. H is a material parameter which describes the interaction properties of the rock bolt, grout and the surrounding rock mass. E_b and A_b are the Young's modulus and cross-sectional area of the bolt. Fig. 1 demonstrates the constitutive law of the interface media used by Cai et al.

In general, the shear strength of an interface in coupling stage comprises three components: adhesion/cohesion, mechanical interlock and friction. At this stage, the deformation of the bolt and the surrounding rock mass is compatible along the interface. If the interface medium is ruptured, slippage may take place, which is termed the decoupling behavior. The shear strength of the interface decreases during this process. The shear strength after interface decoupling is called the residual shear strength in this paper. Based on experimental results, Li & Stillborg (1999) proposed a model for the shear stress along a fully grouted bolt illustrated in Figure 2. This model concludes the decoupling along the interface. The stressed in different sections of the bolt is described as follows:

$$\tau_b(x) = 0 \quad \text{when } 0 \leq x < x_0 \quad (4)$$

$$\tau_b(x) = s_r \quad \text{when } x_0 \leq x < x_1$$

$$\tau_b(x) = \omega s_p + \frac{x - x_1}{\Delta}(1 - \omega)s_p \quad \text{when } x_1 \leq x < x_2$$

$$\tau_b(x) = s_p e^{-2\alpha \frac{(x-x_2)}{d_b}} \quad \text{when } x > x_2$$

where $\Delta = x_2/x_1$, and $\omega = s_r/s_p$, the ratio of the residual shear strength to the peak shear strength; x_1 and x_2 are shown in Figure 2.

Numerical experimental results revealed that a confining pressure influences the strength of the interface dramatically (Moosavi et al. 2005). So Cai et al. (2004) proposed a Mohr-Coulomb law to describe the decoupling behavior of the rock bolt and the rock mass. The shear strength of the interface is expressed as

$$\tau_m = c + \sigma_{nb} \tan \varphi \quad (5)$$

where φ and c are the friction angle and cohesion of the interface, σ_{nb} is the normal stress perpendicular to the rock bolt. The constitutive law of the interface media is demonstrated in Fig. 2, where u_{max} is the ultimate coupling shear displacement of the interface. The parameters u_{max} can be obtained by experiment or by analytical calculation from τ_m and k_{ini}.

3 NUMERICAL MODELING PROCEDURE

3.1 Case study

This paper deals with the numerical study of interaction mechanism of rock bolts and the grout medium at the interface. A pullout test conducted by Stillborg

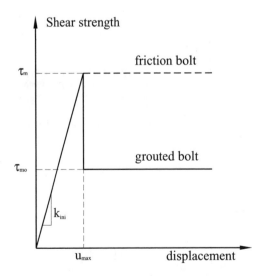

Figure 1. Distribution of shear stress along a fully grouted rock bolt subjected to an axial load (Cai et al. 2004).

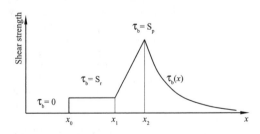

Figure 2. Relationship of the shear stress and displacement at rock bolt interface (Li & Stillborg, 1999).

(1994) is considered in this study. In his test, a 3 m long rebar with a diameter of 20 mm was grouted within two identical concrete blocks. The length of the bolt in each block was 1.5 m. One block was fixed to the ground and the other was pulled. The bolt was pulled out without rupture, indicating that decoupling failure of the interface occurred along the entire length of the bolt. The maximum pullout load was 180 kN. Other necessary parameters are indicated in Table 1.

This problem has many parameters: bolt and borehole geometry, material properties of bolt, grout and rock mass, confining pressure and interface properties etc. The objective of this work is not to consider the influence of all these parameters but to check if a numerical analysis using the finite difference approach can describe correctly the coupling and decoupling mechanisms observed in pullout tests. Therefore the problem is focused on the shear stress and axial load distribution along the bolt in coupling and decoupling stages. In addition, the effect of confining pressure and interface dilation on the shear strength of the interface is studied.

The analysis is carried out using the computer code FLAC3D (Fast Lagrangian Analysis of Continua) which is a commercially available finite difference explicit program.

The bolt, grout and surrounding rock mass (which is replaced by concrete block in pullout test) are all modeled by the elastic model encoded in FLAC3D code. The elastic bulk modulus K and shear modulus G which are input material properties in numerical analysis can be calculated from Young's modulus E and Poisson's ratio v by the following equations:

$$K = \frac{E}{3(1-2v)} \quad (6)$$

$$G = \frac{E}{2(1+v)} \quad (7)$$

3.2 Interface modeling

In the case of a rough bolt, modeling the interface between the grout and the bolt is invariably an integral part of the analysis. In the case of bolt—grout interaction, the interface is considered stiff compared to the surrounding grout medium, but it can slip and may be open in response to the loading. Joints with zero thickness are more suitable for simulating the frictional behavior at the interface between the bolt and the grout medium.

The interface model shown in Figure 3 has been used to simulate the grout/bolt contact described by Coulomb law. The logic contact for either side of the interface is similar in nature to the interface used in the distinct element method.

The spring in the tangential direction and the slider (Figure 3) represent the Coulomb shear-strength criterion. The spring in the normal direction and the limit strength represent the normal contact. The interface has a cohesion c, a friction angle φ, a dilation angle δ, a normal stiffness K_n, a shear stiffness K_s and a tensile strength T. The value of interface properties is given in Table 1.

3.3 Mesh modeling and boundary conditions

In general, fine meshes give more precise results in finite element or finite difference analysis than coarse meshes. But too fine meshes will cause dramatically prolonged cycling time without evident improvement on the results. Figure 4 shows the mesh and boundary conditions retained for this analysis. The model is constructed based on the pullout test referred to in

Table 1. Material properties and geometries adopted in numerical analysis.

Material	Properties	
Rock	Elastic modulus (GPa)	45
	Poisson's ratio	0.25
Grout medium	Elastic modulus (GPa)	30
	Poisson's ratio	0.25
	Thickness (mm)	7.5
Steel bolt	Elastic modulus (GPa)	210
	Poisson's ratio	0.3
	Length (m)	0.5
	Diameter (mm)	20
Interface	Shear stiffness (GPa)	100
	Normal stiffness (GPa)	100
	Poisson's ratio	0.25
	Cohesion (MPa)	13
	Friction angle (°)	30
	Dilation angle (°)	0, 6, 10
	Decoupled cohesion (MPa)	0
	Confining pressure (MPa)	3, 6

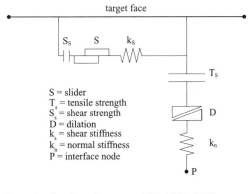

S = slider
T_s = tensile strength
S_s = shear strength
D = dilation
k_s = shear stiffness
k_n = normal stiffness
P = interface node

Figure 3. Interface element used (FLAC3D, 2003).

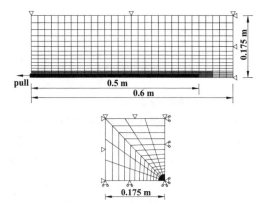

Figure 4. Mesh used and boundary conditions.

3.1, but the length of the bolt is shortened to 0.5 m. This can save cycling time without effects on the analysis results. The mesh size is fine near the bolt where deformations and stress gradients are concentrated. In order to minimize boundary effects, the widths of rock mass mesh are set to five bolt diameter in radius and at the far end of the bolt. As a general rule for the boundary conditions, the top, back and right surfaces are assumed to be fixed, the front and bottom surfaces are assumed to be fixed in horizontal and vertical directions respectively and the left surface is free.

3.4 Simulation procedure with FISH routine

The internal routine of FLAC3D code (FISH) was developed to control the calculation in cycling process and to record the results for plotting and charting. The shear stress, normal stress and shear displacement on the interface can all be read out from the memory by some FISH routines. In addition, FISH functions may be called from several places in the FLAC3D program while it is executing. By a subroutine developed with FISH language, the properties of the interface can be automatically changed when the shear displacement reach a special value. Therefore, the constitutive law of the interface media shown in Fig. 1 can be achieved in numerical analysis. To approximate the actual load conditions on the interface, the following two simulation procedure steps are adopted:

1. The boundary conditions of the two longitudinal sections of the bolt are set to be free. Both of the section surfaces are subjected to a homogeneous pressure to simulate the effect of confining pressure on the bolt. Then, the system is cycled to equilibrium.
2. The two longitudinal sections of the bolt are fixed in radial direction and a pull pressure σ_0 is applied on the bolt end. Then, the system is cycled to equilibrium again.

To simulate the decoupling between the bolt and the grout medium, the cohesion of the interface is set to zero when the shear displacement exceed u_{max} which can be calculated by relation shown in Fig. 1.

4 RESULTS AND DISCUSSION

Numerical studies are performed for different dilation angles and confining pressures and the results of fully coupling and partially decoupling for dilation angle $= 6.0°$, confining pressure $= 6.0$ MPa are presented. Figure 5(a) shows the axial displacement contour of the system in fully coupling state, and it indicates clearly that the displacement of the bolt and the grout medium is consistent at the interface. Figure 5(b) shows a clear decoupling section in blue. Figure 6 present the shear strain increment contour of the system and the decoupling mechanism can also be shown from the discontinuity. Figure 7 shows the shear

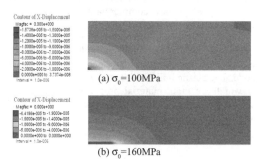

Figure 5. The displacement contour for (a) fully coupling state and (b) partially decoupling state.

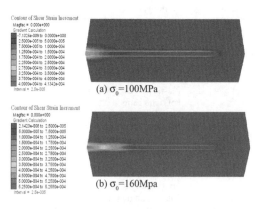

Figure 6. Shear strain increment contours of (a) fully coupling state and (b) partially decoupling state.

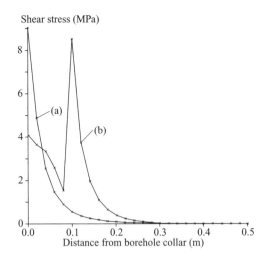
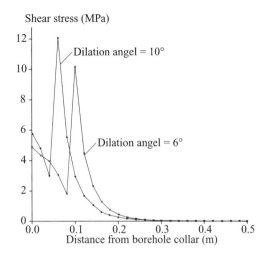

Figure 7. Shear stress distribution along the bolt for (a) $\sigma_0 = 100$ MPa, (b) $\sigma_0 = 160$ MPa.

Figure 8. Distribution of shear stress along the bolt interface for dilation angle = $66°$ and $10°$.

Table 2. Pull pressures σ_0(MPa) for the initial decoupling at the interface.

Dilation angle (°)	Confining pressure (MPa)	
	3	6
0	53	102
6	53	102
10	53	102

stress distribution along the bolt. It can be seen that the curves for two different pull pressures have different forms because decoupling has occurred on the interface of bolt-rock in curve (b). It is clearly indicated that the shear stress decrease exponentially from the point of loading or decoupling front to the far end of the bolt in coupling section.

Table 2 shows the pull pressures of the initial decoupling occur at the interface for different dilation angles and confining pressures. It can be shown from Table 2 that (1) the initial decoupling load of the bolt is independent of the dilation angle; and (2) increasing of confining pressure can add to the coupling capability of the interface. The explanation for this phenomenon may lies in that the dilation only takes effect in the decoupling section in which the shear dilation occurs with the shear displacement. Fig. 8 shows the shear stress distributions of the bolt-grout interface for dilation angle = $6°$ and $10°$ at the same pull pressure $\sigma_0 = 160$ MPa. It indicate that greater dilation angle can mobilize greater shear strength of the interface and lead to a shorter decoupling length at same pull load.

5 CONCLUSIONS

A numerical simulation procedure is proposed to describe the interaction mechanism between bolt and grout medium for fully grouted rock bolt. In spite of the simplicity of the procedure, the coupling and partially decoupling states at the interface of the bolt have been obtained. Specially, different dilation angles and confining pressures have been checked in the paper. The results of these simulations have shown the following:

1. The shear stress decrease exponentially from the point of loading or decoupling front to the far end of the bolt in coupling section.
2. Dilation angle has no effect on the shear stress distribution before decoupling and will not affect the bonding capability, but it has significant effect on the anchorage in decoupling section. Moreover, greater dilation angle can mobilize greater shear strength of the interface and lead to a shorter decoupling length at same pull load.
3. An increased confining pressure can effectively improve the bonding capability and the load bearing capability of the bolt.

Finally, the obtained results validate the simulation procedure using FLAC3D code to simulate the interaction between bolt and grout of fully grouted rock bolts. Such approaches can thus be used to analyse the more complex failure mechanism of grouted bolt, such as failure in grout medium and rock mass.

ACKNOWLEDGEMENTS

The study presented is supported by the open fund of Key laboratory of Mountain Hazards and Surface Process, Chinese Academy of Science and the China National Natural Science No. 40572158.

REFERENCES

Cai, Y. et al. 2004. A rock bolt and rock mass interaction model. *Int J Rock Mech Min Sci Geomech Abstr* 41: 1055–1067.
Farmer, I.W. 1975. Stress distribution along a resin grouted rock anchor. *Int J Rock Mech Min Sci Geomech Abstr* 12: 347–351.
Freeman, T.J. 1978. The behaviour of fully-bonded rock bolts in the Kielder experimental tunnel. *Tunnels and Tunnelling* June: 37–40.
Li, C. & Stillborg, B. 1999. Analytical models for rock bolts. *Int J Rock Mech Min Sci Geomech Abstr* 36: 1013–1029.
Stillborg, B. 1994. *Professional users handbook for rock bolting, 2nded. Trans.* Germany: Trans Tech Publications.
FLAC3D, 2003. *Fast Lagrangian analysis of continua, version 2.1.* Itasca Consulting Group.
Moosavi, M. et al. 2005. Bond of cement grouted reinforcing bars under constant radial pressure. *Cement & concrete composites* 27: 103–109.
Hawkes, J.M. & Evans, R.H. 1951. Bond stresses in reinforced concrete columns and beams. *Journal of the Institute of Structural Engineers* 24(10): 323–327.

Macroscopic effects of rock slopes before and after grouting of joint planes

Hang Lin, Ping Cao, Jiang-Teng Li & Xue-liang Jiang
School of Resources & Safety Engineering, Central South University, Changsha, China

ABSTRACT: Studies on slope reinforcement methods are important for geotechnical engineering. All of the reinforcement methods can be divided into two groups, the passive reinforcement methods and the active reinforcement method. Grouting is an active reinforcement method. It is to inject a cement grout mixture into the rock mass to reduce the permeability of the rock mass and improve the mechanical properties of the rock mass itself. In order to find differences of parameters between the rock mass before and after grouting, the comparative experiments of direct shear were done that indicates that the shear strength and shear stiffness of rock samples are improved greatly after grouting. However, no the whole slope must be grouted. The effective grouting areas (EGA) can be defined by relative calculations. The three dimensional explicit finite difference code, FLAC^{3D}, was adopted to study the differences of mechanical properties of the rock mass before and after grouting. Comparative analysis shows that stress field and displacement field of the rock mass become more symmetrical and continuous after grouting. The tensile stress area reduces and the displacement bifurcation near joint planes vanishes, both are good to the stability of slope.

1 INTRODUCTION

Cement grouting constitutes one of the major techniques used for stabilization in civil engineering. It is widely applied in tunnel engineering (Hoek, 2000, Eberhardt, 2002, Varol et al. 2006, Yesilnacar, 2003, Turkmen et al. 2003). It consists in injecting into the rock mass a cement grout mixture under controlled pressures and volumes. The main results expected from this process are a reduction in permeability and an improvement in terms of mechanical properties. But it is seldom applied in slope engineering, with the reason that the mechanism of grouting is still not well known. Xu (2006) has studied the effects of grouting to rock mass experimentally. In his studies, grouting is done in all the joint planes in samples, but slope is large in size, it is not realistic or practical to grout in all the areas. So it is necessary to decide which joint planes should be grouted and which should not. In present paper, firstly we try to find out the effective grouting area (EGA), then study the macroscopic effect such as stress and deformation of rock jointed slope before and after grouting by the numerical calculation software FLAC^{3D}. Numerical calculation results are integrals of reflections in small units, whose behavior is exactly based on the classical mechanical theory, so we think the numerical analysis is a good way, at least better than the theory deduction based on the simple models, to study the effect of huge body, just as long as the simulation models are close to the reality situation. And these methods are recently widely applied in analysis for the geotechnical engineering (Grgic et al. 2003, Exadaktylos et al. 2002, Guo et al. 2002, Boidy et al. 2002, Waltham et al. 2004).

2 TEST

Joint plane, if loaded in two directions with normal stress and shear stress, will arise one or several different directions of deformation, which exhibits as the closure and slippage deformation. In order to describe the discipline of deformation, we can use stress-displacement relationship curves, whose slope can reflect the shear stiffness K_{ss} and shear stiffness of rock mass. The stress-displacement relationship is described with their increment,

$$d\sigma_{ij} = K \cdot d\delta_{ij} \quad (1)$$

where K is the stiffness matrix.

Substituting σ_{ij} with τ_s and σ_n, δ_{ij} with δ_s and δ_n, we can obtain,

$$\begin{Bmatrix} d\tau_s \\ d\sigma_n \end{Bmatrix} = \begin{Bmatrix} K_{ss} & K_{sn} \\ K_{ns} & K_{nn} \end{Bmatrix} \begin{Bmatrix} d\delta_s \\ d\delta_n \end{Bmatrix} \quad (2)$$

where τ_s is shear stress; σ_n is normal stress; δ_s is shear displacement; δ_n is normal displacement; K_{ss} is shear stiffness coefficient, which indicates the influence of δ_s to τ_s; K_{sn} indicates the influence of σ_n to τ_s; K_{ns} is dilation stiffness coefficient, which indicates the influence of δ_s to σ_n; K_{nn} is the normal stiffness coefficient, which indicates the influence of δ_n to σ_n.

For macroscopic body, effect of normal displacement to shear stress can be ignored (Xia, 2002), which means $K_{sn} = 0$, then eq. (1) can be evolved to,

$$d\tau_s = K_{ss}d\delta_s \qquad (3)$$

and average shear stiffness is,

$$\overline{K_{ss}} = \int_\delta d\tau_s \Big/ \int_\delta d\delta_s \qquad (4)$$

2.1 Materials

Rock samples are uniform in the size of 25 cm × 20 cm × 10 cm. In the tests, M20 cement mortar is chosen to simulate rock, while M7.5 for joint plane, whose strength is lower than rock block. Thickness of grouting is 20 mm, weight proportioning of cement and sand is 1:4.74.

2.2 Results and discussion

Normal stress changes from 0.2 MPa to 0.8 MPa in the test. The curves of relationship between shear stress and shear displacement for rock sample with and without grouting in joint plane are showed in Figure 1 and Figure 2.

Figure 1 shows that, each curve slope changes in the phase prior to peak stress, which indicates that the stiffness K_{ss} changes at the same time. When rock sample begins to slip, value of K_{ss} in is small. During the procedure of slippage, the protuberant fractions meet closer, and the phenomenon of shear dilation occurs, which in turn causes shear stress reaching

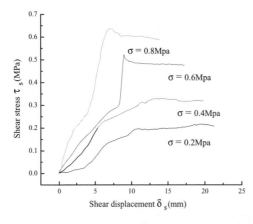

Figure 1. Shear stress and shear displacement before grouting.

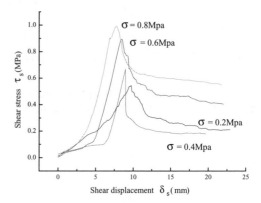

Figure 2. Shear stress and shear displacement after grouting.

peak value. Under combined loads of compressive stress and shear stress, joint plane becomes smooth during slipping, the fluctuation value of peak strength reduces, and the strength of joint plane comes to the residual strength phase. With the increasing of normal stress, the peak shear resistance stress increases. Shear resisting strengths of the four samples are 0.22 MPa, 0.33 MPa, 0.52 MPa and 0.64 MPa for the normal stress of 0.2 MPa, 0.4 MPa, 0.6 MPa, and 0.8 MPa, respectively.

Relationship of shear stress and shear displacement for samples after grouting is shown in Figure 2. The linear elastic states are more obvious and curves become steeper when compared with the curves of samples without grouting in Figure 1. After curves reach peak of stress, they decrease quickly which is very different from that of Figure 1, and exhibits the stiffness characteristic. Shear resistance stress for different normal stresses are 0.54 MPa, 0.66 MPa, 0.89 MPa and 0.99 MPa, respectively, which are larger in magnitude than that before grouting. After grouting, average stiffness increases and joint plane changes from frictional joint plane to stiffness joint plane.

3 EFFECTIVE GROUTING AREA

It is not practical and realistic to grout in all the joint planes of slope, so it is necessary to find the effective grouting areas (EGA). We make the study by building the calculation model, showed in Figure 3, with assumption that the slope face strikes parallel to the underlying exfoliation surface and hence the slope can be analyzed by means of a two dimensional model. Natural angle is β_2, thickness of slope is d, dip angle of excavated slope is α, joint plane inclination is β_1, distance between B_i and C is L_i. The stability of slope is controlled by joint plane. Studies are done for the situation of $\alpha > \beta_1$, and makes the assumption that

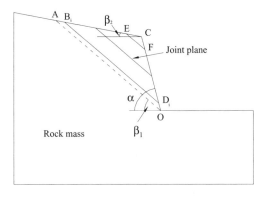

Figure 3. Theory deduction model.

failure mode of slope is shear strength failure with unstable block slipping along joint plane.

For plane B_iD_i, the safety factor of slope is calculated as follow,

$$F_s = \frac{S^{BD}}{\tau^{BD}} = \frac{F_k^i}{F_x^i} \qquad (5)$$

where F_k^i is resisting force of block B_iCD_i, F_x^i is driving force of block B_iCD_i.

Relationships of parameters are,

$$|D_iC| = \frac{L_i \sin(\beta_1 - \beta_2)}{\sin(\alpha - \beta_1)} \qquad (6)$$

$$G_{B_iCD_i} = \frac{1}{2}\gamma \cdot d|B_iC| \cdot |D_iC| \sin(\alpha - \beta_2) \qquad (7)$$

where $G_{B_iCD_i}$ is the gravity of B_iCD_i, γ is the average unit weight of B_iCD_i.

Then we can obtain,

$$F_k^i = G_{B_iCD_i} \cos\beta_1 \tan\phi_j + \frac{c_j L_i \sin(\alpha - \beta_2) \cdot d}{\sin(\alpha - \beta_1)} \qquad (8)$$

$$F_x^i = G_{B_iCD_i} \sin\beta_1 \qquad (9)$$

where ϕ_j is friction angle of joint plane, c_j is cohesion of joint plane.

Substituting F_k^i and F_x^i in eqs. (5) with eqs. (8) and (9), yields,

$$F_s = \frac{G_{B_iCD_i} \cos\beta_1 \tan\phi_j + \frac{c_j L_i \sin(\alpha - \beta_2)}{\sin(\alpha - \beta_1)}}{G_{B_iCD_i} \sin\beta_1} \qquad (10)$$

If $F_s = 1$, block B_iCD_i is in the critical state of stability, then the critical slippage length can be calculated as follow, according to eq. (10),

$$L_{cr} = \frac{2c_j}{\gamma \sin(\beta_1 - \beta_2)(\sin\beta_1 - \cos\beta_1 \tan\phi_j)} \qquad (11)$$

Supposing that $|AC| = L_0$, the following statements can be obtained,

a. if $L_{cr} > L_0$, slope is in stable state;
b. if $L_{cr} = L_0$, slope is in critical stability state;
c. if $L_{cr} < L_0$, slope is in unstable state.

Then the EGA can be obtained from L_{cr} to L_0. After grouted, c and ϕ of rock mass increase, and L_{cr} also increase according to eq. (11). If L_{cr} reaches the condition of (a), slippage along joint plane is controlled, which means that the slope is in stable state.

4 NUMERICAL SIMULATION

4.1 Simulation technique

Macroscopic effect of rock slope is simulated by three dimensions explicit finite difference code, FLAC^{3D}. For given element shape functions, the set of algebraic equations solved by FLAC^{3D} is identical to that solved with the finite element method. However, in FLAC^{3D}, equations are solved using dynamic relaxation (Dawson et al.1999), an explicit, time-marching procedure in which the full dynamic equations of motion are integrated step by step. Static solutions are obtained by including damping terms that gradually remove kinetic energy from the system. The convergence criterion of calculation is the unbalance force ratio which is defined by the ratio of nodes' average force to the maximum unbalance force reaches 10^{-7}.

4.2 Model

Models for jointed slope are divided into two groups: (1) before grouting; (2) after grouting. Three joint planes are preset in the slope. The elasto-plastic block element with lower strength is chosen to simulate joint plane, and the inclination of joint plane is 40° with 0.1 meters thickness. Plane strain model is built with 9 795 triangular elements and 3 293 grids, shown in Figure 4. Size of slope are shown in Figure 5, and rock parameters are shown in Table 1. Natural angle is 10°, slope angle after excavation is 75°. Gravity load is performed as an initial free mechanical state. Critical slippage length can be obtained by eq. (11). Then we can obtain, $L_{cr} = 12.71$ m, $L_0 = 54.26$ m, $L_1 = 9.51$ m and $L_2 = 29.09$ m. It is obvious that $L_{cr} < L_2 < L_0$, so the second and third joint planes are in EGA.

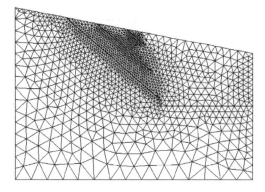

Figure 4. Numerical model.

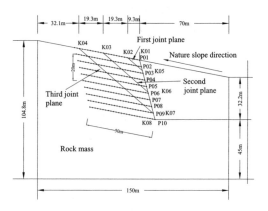

Figure 5. Monitoring points.

4.3 Monitoring points

Failure of rock mass is close related to its inner structure plane and stress field. Perturbation induced by excavation can cause deformation of rock mass, which is nonlinear mechanical procedure. In order to reflect the nonlinear procedure of displacement in different time steps, monitoring points from K01 to K08 are located in slope. These displacements change with time steps, and we call them dynamic displacement. After excavation, the displacements of particles will converge to certain values, then we call them static displacements. Monitoring points from P01 to P10 are located in the slope body to record the static displacements of rock particles. All the monitoring points are shown in Figure 5.

4.4 Results and discussion

Static total displacements of slope before and after grouting are shown in Figure 6. Before grouting, displacements of monitoring points are divided into three groups (three bifurcation groups along monitoring lines), but the magnitudes of bifurcation are small. The bifurcation occurs at the places of joint planes. Values of displacement reduce gradually from surface of slope to inner place. After grouting, monitoring displacements in each monitoring line are different from that before grouting. In the same monitoring line, displacements of each monitoring line (except line P01 and line P02) change fluently without mutation phenomenon. The displacements reduce gradually from surface of slope to inner place, which is the same as that before grouting. But displacements of some monitoring lines (line P01 and line P02) still remain bifurcation, which is because the first joint plane has not been grouted.

Figure 7 shows the relationships of dynamic total displacements and calculation steps. For monitoring points located on the top of slope, the magnitudes of displacements at the same step reduce gradually from surface to the inner place in sequence of K01 → K02 → K03 → K04. At their convergence phases, the peak values of displacement as well as curve's slope reduce in the same sequence. For points on the surface of slope, values of displacement as well as curve's slope reduce in the sequence of K05 → K06 → K07 → K08. Among all the monitoring points, K01 has the largest magnitude of displacement, so it is suitable to choose K01 as the characteristic point for deformation monitoring. Besides, displacement of K08 stays at the same value in most part of calculation procedure, with the reason that it is under the joint plane and little affected by slippage of block along joint plane. After grouting, magnitudes of displacements reduce greatly, which is because grouting strengthens the joint plane. For the monitoring point on the top of slope, displacements reduce in the sequence K01 → K02 → K03 → K04. For monitoring points on the surface of slope, values of displacement increase in sequence of K06 → K07 → K08, which is different from the slope before grouting. This is because rock mass become uniform and continuous after grouting. But the displacement of K08 remains the same after grouting when compared to that before grouting, which results from that K08 is at the toe of slope, and the toe of slope remain stable before grouting, so grouting done in joint plane affect little the displacement of K08. After grouting joint planes change from weakness plane to stiffness plane with the increasing of vertical stress. And we can also see that, the slope of each curve becomes steeper than that before grouting, which indicates the stiffness improvement.

Figure 8 shows the maximum principal stress of slope before and after grouting. Maximum principal stress is an important factor to determine the state of element. According to Mohr-Coulomb criterion, when

Table 1. Calculation parameters.

Material	Unit weight (kN/m³)	Elastic modulus (GPa)	Poissons' ratio	Cohesion (kPa)	Friction angle (°)	Tensile strength (MPa)
Rock mass	25	3.0	0.20	500	35	1.00
Joint plane	17	0.01	0.30	120	20	0.05
Grouting	20	0.1	0.20	240	23	1.00

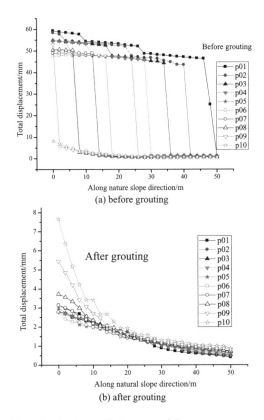

Figure 6. Static total displacement of slope.

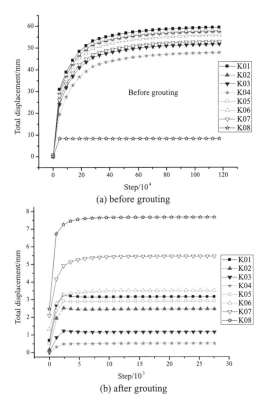

Figure 7. Dynamic total displacement of slope.

certain principal stress states of a particle have reached the failure envelope, the unbalance force of this particle can then be transferred into adjacent particles, which contribute to the increase of its unbalance force, in turn, the sliding mechanism may be initiated. If the failure points connect to form a sliding plane, the unstable rock mass will slip. After excavation, most parts of slope are in stable state, the maximum principal stress in these places is compressive. But at the places near joint plane, the stress is tensile and concentrated, in turn, it causes hidden factor for the instability of slope because rock mass can not resist great tensile stress. After grouting, physical parameters of joint plane are improved, and the shear resisting strength and shear resisting stiffness increase, which result in stress fields in slope becoming more continuous. The magnitudes of stress near joint plane become smaller, with the phenomenon of bifurcation in joint plane reduced, and stability, uniformity of slope improved.

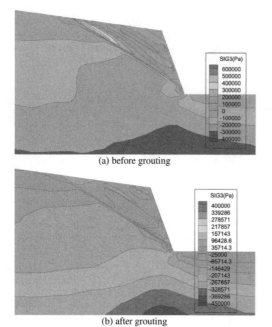

Figure 8. The maximum principal stress.

5 CONCLUSION

1. Direct shear tests for rock samples before and after grouting are done, results of tests show that, before grouting, joint plane belongs to frictional joint plane, its shear mechanical characteristic is influenced by the waviness degree and occlusive degree, and the residual stress descends little compared to peak stress; After grouting, strength and stiffness of rock sample increase.
2. Theory models are founded for the slopes with its angle α greater than joint plane angle β, critical slippage length L_{cr} for joint slope is deducted, then the effective grouting areas are conformed.
3. Numerical simulation models are founded for excavated slope before and after grouting by FLAC^{3D}, results of calculations show that, for slope before grouting, displacements fields of rock mass redistribute during the excavation procedure, and there are many tensile areas near joint planes which lead to the tensile crannies in slope. The magnitude of displacement and deformation are large near joint plane. The bifurcations of displacements happen when the monitoring lines passing joint planes, stress distributions are asymmetry in rock mass, which influence the stability of slope; For slope after grouting, displacement field and stress field become smooth, large magnitudes of deformations near joint plane are controlled. At the same time, the tensile crannies on the top of slope reduce, which is good to the stability of slope.

REFERENCES

A.C. Waltham & G.M. Swift. 2004. Bearing capacity of rock over mined cavities in Nottingham. *Engineering Geology* (75):15–31.

Ahmet Varol & Sueyman Dalg. 2006. Grouting applications in the Istanbul metro, Turkey. *Tunnelling and Underground Space Technology* (21):602–612.

Dragan Grgic, Francoise Homand & Dashnor Hoxha. 2003. A short- and long-term rheological model to understand the collapses of iron mines in Lorraine. *France.Computer and Geotechnics* (30):557–570.

E. Boidy, A. Bouvard & F. Pellet. 2002. Back analysis of time-dependent behaviour of a test gallery in claystone. *Tunnelling and Underground Space Technology* (17):415–424.

E.M. Dawson, W.H. Roth & A. Drescher. 1999. Slope Stability analysis by strength reduction. *Geotechnique* 49 (6):835–840.

Erik Eberhardt. 2002. *Rock Slope Stability Analysis Utilization of Advanced Numerical Techniques*. ETH zurich, Switzerland: Swiss Federal Institute of Technology.

Evert Hoek. 2000. *Rock Engineering*. North Vancouver: Evert Hoek Consulting Engineering Inc.

G.E. Exadaktylos & M.C. Stavropoulou. 2002. A closed-form elastic solution for stresses and displacements around tunnels. *International Journal of Rock Mechanics & Mining Sciences* (39):905–916.

M.I. Yesilnacar. 2003. Grouting applications in the Sanliurfa tunnels of GAP, Turkey. *Tunnelling and Underground Space Technology* (18):321–330.

R. Guo & P. Thompson. 2002. Influences of changes in mechanical properties of an overcored sample on the far-field stress calculation. *International Journal of Rock Mechanics & Mining Sciences* (39):1153–1166.

Sedat Turkmen & Nuri Ozguzel. 2003. Grouting a tunnel cave-in from the surface a case study on Kurtkula irrigation tunnel, Turkey. *Tunnelling and Underground Space Technology* (18):365–375.

Xia Cai-chu & Sun Zong-qi. 2002. *The Mechanics of Engineering Joint Rock Mass*. Shanghai: Tongji university Press (in Chinese).

Xu Wan-zhong, Cao Ping, Peng Zhen-bin, et al. 2006. Analysis of Mechanical Mechanism of Shear Characteristic of Bolted Layered Slope Considering Grouting Factor of Weak Plane. *Chinese Journal of Rock Mechanics and Engineering* 25 (7):1475–1480 (in Chinese).

Xu Wan-zhong, Peng Zhen-bing, Hu Yi-fu, et al. 2006. Simulation Experiment Study on The Treatment of Bolting and Grouting in Rock Slope Reinforcement. *China Railway Science* 27 (4):11–16 (in Chinese).

Two- and three-dimensional analysis of a fossil landslide with FLAC

X.L. Liu
College of Environmental Science and Engineering, Ocean University of China, Qingdao, China

J.H. Deng
College of Water Resources and Hydropower, Sichuan University, Chengdu, China

ABSTRACT: Xietan landslide, a fossil landslide located in Zigui County, Hubei Province, China, is one of the important geo-hazards at the northern bank of Yangtze River in the Three Gorges Reservoir areas. The landslide is mainly composed of sliding mass, slip soil (the weak layer) and bedrock. Slip soil of the landslide, containing coarse particles up to 40% (by weight), is very typical in the Three Gorges Reservoir areas. In this paper, first, the shear strength parameters of the slip soils of Xietan landslide were analyzed by numerical computing. Then with the FLAC program, the stress and deformation characteristics of the fossil landslide at the reservoir water level of 139 m caused by impoundment were simulated using two- and three-dimensional model respectively. According to comparisons of deformation between the monitoring data and the numerical computing data of the two- and three-dimensional simulations, the followings can be concluded. (1) Shear strength properties of slip soils have important influence on landslide stability analysis. The shear strength parameters of the slip soils containing coarse particles of Xietan landslide used for numerical simulation are acceptable. (2) Results of the numerical analysis of the three-dimensional model agree better with the actual condition of the landslide than that of the two-dimensional model. (3) It is feasible to analyze landslides by numerical method, for example by the FLAC program, provided that the reasonable parameters of the landslide for computing can be gained.

1 INTRODUCTION

Xietan landslide, a fossil landslide located in Zigui County, Hubei Province, China, may reactivate to become one of the geo-hazards at the northern bank of Yangtze River due to impoundment of the Three Gorges Reservoir. The main longitudinal section is about 800 meters long, and the averaged width of the transverse section is about 260 meters. The fossil landslide is mainly composed of sliding mass, slip soil (the weak layer) and bedrock. The sliding mass is about 10 to 40 meters thick. Thickness of the slip soil ranges from about 0.5 to 3.0 meters. The whole landslide bulks about 624 ten thousand cubic meters.

Permeability parameter of the sliding mass is 4×10^{-2} cm/s and that is 3×10^{-7} cm/s for the slip soil. The slip soil can be regarded as the aquifuge in the landslide. There is no consistent groundwater table in the landslide. Perched water exists on the localized aquifuge of the slip soils.

Shear strength properties of slip soils have great influence on stability of landslides. Slip soil of Xietan landslide is very typical in the Three Gorges Reservoir areas, containing gravels up to 40% (by weight). In view of little knowledge about the shear strength characteristics of slip soils containing coarse particles, investigation of the shear strength properties of this kind of slip soils is important for numerical analysis of the fossil landslide.

In this paper, shear strength properties of slip soils with coarse particles of Xietan landslide were investigated first. Then the stress and deformation characteristics of the landslide were analyzed according to the numerical computing results and the in situ monitoring data.

2 SHEAR STRENGTH PROPERTIES OF SLIP SOILS OF XIETAN LANDSLIDE

2.1 *Index properties of the slip soils*

The slip soil of Xietan landslide is mainly composed of sandy/silty clay containing gravels of sub-angular and sub-rounded fragments of sandstone, mudstone and shale, etc. Gravel content of the slip soil ranges from 20% to 40% by weight, with 2–60 mm size. A typical particle size distribution (PSD) curve is plotted in Fig. 1 as curve 3.

The average specific gravity of the slip soil is 2.78. Average density of the undisturbed slip soil is about 2237 kg/m³. Natural water content of the slip soil

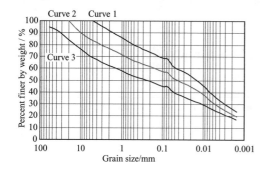

Figure 1. Particle size distribution curves of different series of tests.

Table 1. Shear strength tests on slip soils of Xietan landslide.

Test series (gravel size)	Type of shear test	Maximum particle size	Coarse content (by weight)
1 (2–5 mm)	DST-slow (60 mm box)	5 mm	8%
	Triaxial tests CU (38 mm diameter)	5 mm	
2 (2–20 mm)	Triaxial tests CU/CD (101 mm diameter)	20 mm	23%
3 (2–60 mm)	Multi-reversal LDST (500 mm box)	60 mm	38%

ranges from about 12% to 16%, resulting in saturation of the in situ slip soil ranges from 85% to 95%.

2.2 Shear strength properties of the slip soils

In order to investigate the shear strength properties of the slip soils containing gravels of Xietan landslide, three series of tests were carried out, as indicated in Table 1. Direct shear tests (DST), consolidated undrained/drained (CU/CD) triaxial tests, and the multi-reversal large direct shear tests (LDST) were performed with specimens of different gravel content and size. Remolded specimens were used for series tests 1 and the undisturbed samples were used for series tests 2 to 3. The PSD curves of the first to third series of tests have been plotted in Fig. 1, corresponding to curve 1 to 3 respectively.

The third series of tests were non-standard multi-reversal direct shear tests on one block sample using a large direct shear box (500 mm × 500 mm × 400 mm). Prior to tests, the block sample had been sheared once under the unconsolidated undrained condition. The block sample was consolidated first under normal stress, and then was sheared at the speed of 0.7 mm/min to remain undrained. The block sample was sheared only one time corresponding to one different normal stress. In the whole tests, the same block sample was sheared for 4 times corresponding to 4 different consolidation normal stresses. From the test program, we knew that before consolidation each time, there had been a pre-existing sheared surface in the block sample.

Results of the total shear strength (the consolidated undrained shear strength) tests of series one to three are presented in Fig. 2, from which it was found that the total shear strength parameters of d < 60 mm (d represents the gravel diameter) series were very similar to that of the d < 20 mm series. Taking the existing sheared surface into account for the d < 60 mm series of tests, it could be inferred that shear strength of the large undisturbed sample was greater than that of the small ones in the second series of tests, and the shear strength of the second series of tests could be regarded as the lower bound of the shear strength of the in situ slip soils. Comparing the PSD curves of these three series, it can be expected that the increasing

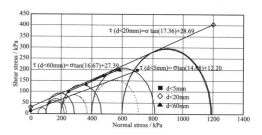

Figure 2. Total shear strength results of series tests of one to three.

Table 2. Parameters of the landslide used for numerical simulations.

Components of the landslide	Bedrock	Slip soils	Sliding mass	
Young's modulus/ GPa	25	0.03	0.08	
Poisson's ratio	0.25	0.38	0.35	
Drained cohesion/ kPa	—	29.29	Natural	10
			Saturated	2
Drained frictional angle/degree	—	21.34	40	
Density/kg/m³				
Natural	2670	2237	2070	
Saturated	2670	2275	2250	

proportion of coarse particles and their sizes result in a significant increase in shear strength of slip soils (Liu et al. 2006).

According to the analysis above, it was acceptable to consider the second set of test results as the shear strength parameters of the slip soils used for numerical simulations of Xietan landslide. The Young's modulus and the Poisson's ratio were also estimated from the CD triaxial tests in the second series, as presented in Table 2.

3 NUMERICAL ANALYSIS OF XIETAN LANDSLIDE

3.1 Numerical models

The original Yangtze River water level is 76 m which has little effect on analysis of the Xietan landslide. Considering availability of the monitoring data, the landslide at long term reservoir water level of 139 m due to impoundment of the Three Gorges Reservoir was simulated by numerical method.

The relative position of the water level line of 139 m and the location of the bore ZK1 for inclination monitor can be referred to Fig. 3.

The landslide was simulated with FLAC (Fast Lagrange Analysis of Continua), using small-strain mode. The numerical model can be referred to Fig. 3 for 2-dimensional (2D) and Fig. 4 for 3-dimensional (3D) analysis. The main longitudinal profile of the 3D model was used to represent the 2D landslide, as indicated in Fig. 3.

Horizontal displacements were fixed for nodes along the left and right boundaries (in 3D model, horizontal displacements of nodes along the front and back boundaries were also fixed) while both horizontal and vertical displacements were fixed along the bottom boundary of the model. The sliding mass and the slip soils were modeled as a linear elastic-perfectly plastic

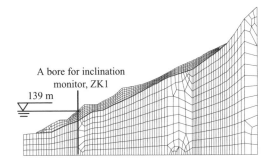

Figure 3. 2D numerical model of Xietan landslide (Main longitudinal section).

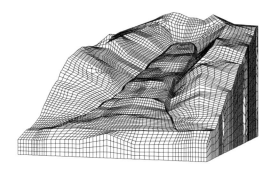

Figure 4. 3D numerical model of Xietan landslide.

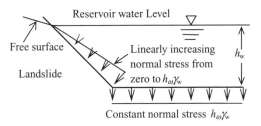

Figure 5. Sketch for considering influence of the reservoir water level.

material, together with the Mohr-Coulomb yield criterion and the non-associated flow rule with the dilation angle of zero. Rock-base of the landslide was modeled as a linear elastic material. All the parameters used in the numerical model have been presented in Table 2.

Influence of the reservoir water was considered by the way of Griffiths and Lane (1999). As shown in Fig. 5, with water pressures acting on the landslide surface in normal directions, the pore pressure at a point was estimated as the product of the unit weight of water γ_w and the vertical distance of the point beneath the free surface.

3.2 Numerical result analysis

Stress state of Xietan landslide in the situation of 139 m reservoir water level for 2D and 3D model has been shown in Fig. 6. Results of the main profile of the 3D model were given to represent the 3D analysis ones. Only elements near the toe of the landslide were given for comparison. Most of the slip soil elements yielded due to the shear stress. The stress state distribution showed that the main anti-sliding zone was the toe of the landslide.

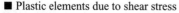
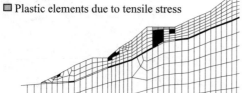

(a) Stress state distribution of the 2D model

(b) Stress state distribution on the main longitudinal profile of the 3D model

Figure 6. Stress state distribution of Xietan landslide at the reservoir water level of 139 m.

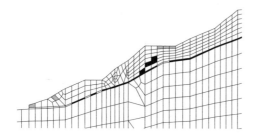

Figure 7. Comparison of the computing and monitoring data.

A bore for inclination monitor, ZK1 in the landslide, has been indicated in Fig. 3. According to the in situ monitoring data, the actual displacements of the landslide at the location of ZK1 can be gained. Monitoring and the 2D and 3D numerical computing horizontal displacements of the landslide at location of ZK1 for the reservoir water level of 139 m are indicated in Fig. 7. Here, the monitoring data 1 was collected on Nov., 5, 2005 and the monitoring data 2 on Oct., 7, 2006. Comparison of the two series of monitoring data showed that deformation of ZK1 in the landslide had been unchangeable after Nov., 5, 2005 at the reservoir water level of 139 m. So the monitoring data indicated in Fig. 7 can be regarded as the long term deformation of ZK1 at the reservoir water level of 139 m. In view of the long term deformation of the landslide computed with the FLAC program, the monitoring and computing data can be compared.

According to horizontal displacements shown in Fig. 7, we found that all the horizontal displacements of the sliding mass were similar to that of the slip soils, which implied that deformation of the slip soils dominated the whole sliding trend of the landslide. Therefore, determination of the shear strength properties of the slip soils is important for simulating the landslide properly.

Figure 7 has obviously shown that computing results of the 3D model agreed better with the monitoring data than that of the 2D model, which implied that the 3D analysis could reflect the actual characteristics of the landslide. At the same time, good agreement of the 3D computing data with the monitoring data also indicated that the used parameters of the slip soils with gravels were reasonable for Xietan landslide numerical simulation.

4 CONCLUSIONS

Shear strength parameters of the slip soils containing coarse particles of Xietan landslide were investigated by a series of tests. Two- and three-dimensional simulations of the landslide were performed by the FLAC program. Some conclusions can be inferred as follows.

1. Shear strength properties of the slip soils have important influence on stability analysis of landslides. The acceptable shear strength parameters of slip soils containing coarse particles of Xietan landslide have been concluded by tests, which can provide some experience for the similar landslides analysis.
2. For numerical simulations of landslide, results of 3D model agree better with the actual characteristics of the landslide than that of the 2D model.
3. It is feasible to analyze landslides by numerical method, for example by the FLAC program, provided that the reasonable parameters of the landslide can be gained.

ACKNOWLEDGEMENTS

Financial support from National Science Foundation of China under grant 40702044 and the permission of using the FLAC program from Institute of Rock and Soil Mechanics, the Chinese Academy of Science are greatly appreciated.

REFERENCES

Griffiths, D.V. & Lane, P.A. 1999. Slope stability analysis by finite elements. *Geotechnique* 49(3): 387–403.

Liu X.L., Loo, H. & Min, H. et al. 2006. Shear Strength of Slip Soils Containing Coarse Particles of Xietan Landslide. *Geotechnical Special Publication* 151: 142–149.

Application of the coupled thin-layer element in forecasting the behaviors of landslide with weak intercalated layers

Y.L. Luo & H. Peng
School of Civil and Architectural Engineering, Wuhan University, Wuhan, China

ABSTRACT: The distribution and seepage-stress coupled interaction of Weak Intercalated Layers (WIL) in landslide are key factors resulting in the landslide disaster. Forecasting the behaviors of landslide with the WIL and mitigating the disaster maximumly are very important. But the studies on the coupled interaction of WIL are seldom reported. So this paper proposed a Coupled Thin-Layer Element (CTE) based on the original Goodman's joint to model both the stress-deformation behaviour and water pressure response of the WIL. An elastic viscoplastic constitutive law was adopted to model the creeping behavior and then a FE program of CTE was developed to analyze the Dayeping landslide. The calculated groundwater free surface is consistent well with the in-situ testing data. The distributions of displacements and stresses suggest that the landslide may slide along the WIL at the elevation of 230 m to 370 m. In the end, the strength reduction method was used to calculate the stabilities for CTE, Solid Element (SE) and Desai Element (DE) used to model the WIL. The results indicate that SE and DE can't simulate the tension fracture damage or pore pressure, although the minimum safety factors (F_{min}) obtained, 1.22 and 1.27, are acceptable. CTE makes up the above faults. It can comprehensively model the true status of stresses in the landslide and the F_{min} obtained is only 1.19, which is lower than those for SE and De, but larger than the allowable safety factor 1.15. Thus CTE is superior to the other elements and can be used in forecasting the behaviors of landslide with WIL correctly.

1 INTRODUCTION

Since the 20th century, with the increasing of population, the extensions of human activities and the influences of engineering activities on geological environment disturbances, the landslide disaster has become the most frequent and serious geological hazards. It is well known that the distribution and seepage-stress coupled interaction of the weak intercalated layers (WIL) in landslide are the key factors resulting in the landslide disaster (Oda, Masanobu 1986). Thus modelling the coupled interaction of WIL correctly is the key forecasting the behaviours of landslide, but the coupled interaction of WIL is seldom considered. So this research is very important and meaningful.

At present, there are two methods modelling the WIL: (1) solid element (SE) (Ding Xiuli, Fu Jing & Zhang Qihua, 2004, Chai Junrui & Li Shouyi 2004). Because the mechanical behaviours of materials on the top and bottom of WIL are very different, and the relative sliding, opening, even the tension fracture damage, may occur between the top and bottom of WIL. Thus the solid element can't model the WIL well; in addition, the solid elements nearby the WIL must be refined to ensure the computational precisions, the corresponding quantity of nodes will increase rapidly, finally the coupled analysis efficiency will slow down seriously. (2) Goodman element or Desai element (DE) (Goodman, R.E., Taylor, R.L. & Brekke, T.L., 1968, Desai, C.S., Zaman, M.M. & Lightner, J.G., et al. 1984). The conception of Goodman element is clear, and it can model the sliding and opening displacement of contact interface well. But its main problems are the randomicity adopting normal stiffness and the difficulty deciding reasonable stiffness; though the DE introduces the interpenetration control method, which overcomes the faults of the original Goodman joint element, it is short of theoretical bases on determining calculation parameters (Luan Maotian & Wu Yajun 2004). In addition, the Goodman and Desai element can't consider the pore pressure, so they can't directly apply to model the seepage-stress coupled interaction of WIL, they can only take the seepage force or pore pressure as the external load (Desai, C.S., Samtani, N.C., et al. 1995, Samtani, N.C., Desai, C.S., et al. 1996), then a single stress analysis is carried out for the WIL. So it can be seen that the two methods will occur large errors and affect the validity forecasting the behaviours of landslide, and some improvements need to do for the above elements.

This paper proposed a new type of 2D four nodes seepage-stress coupled thin-layer element (CTE) based on the original Goodman's joint and Biot's consolidation differential equations. An elastic viscoplastic constitutive model was adopted, and a FE program was developed to verify the new model. In the end, it was applied to analyze the coupled effect of Dayeping landslide.

2 FE MODEL

2.1 Biot consolidation theories

The proposed coupled thin-layer element is shown in figure 1. Every node has three freedoms: u, v and p. Two key assumptions are made here: (1) there are only two components of stress σ_n, τ_s in the coupled element; (2) the velocity of seepage flow can only occur in the tangential direction of the element, that is, $p_1 = p_3$, $p_2 = p_4$. The well known Biot's consolidation equations for the coupled thin-layer element are as follows (Small, J.C., Booker, J.R., Davis, E.H., 1976):

$$\sigma'_n = \sigma_n - p \qquad (1)$$

$$\tau'_s = \tau_s \qquad (2)$$

$$P_k = -H^*_{kl}\delta_l \qquad (3)$$

$$\frac{\partial v_x}{\partial x} = -\frac{\partial \theta}{\partial t} = -\frac{1}{h}\frac{\partial u_n}{\partial t} \qquad (4)$$

$$v_x = -\frac{k_x}{r_w}\frac{\partial p}{\partial x} \qquad (5)$$

where $P = \{\tau_s \; \sigma'_n\}^T$; $\delta_l = \{u \; v\}^T$, u, v represent the relative shear and normal displacement components along the coupled element, respectively.

2.2 FE approximation

Introduce the stress and seepage boundary conditions, into the stress equilibrium equations and water continuity equation, the finite element equations of CTE can be derived by using the Virtual Work principle as follows:

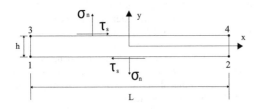

Figure 1. Coupled thin-layer element model (local coordinate system)

$$\int_V \sigma'_i \delta u_i dV + \int_V p\delta u_n dV + \int_S T_i \delta u'_i dS = 0 \qquad (6)$$

$$-\int_V \frac{k_x h}{r_w}\frac{\partial p}{\partial x}\frac{\partial \delta p}{\partial x}dV - \int_V \frac{\partial u_n}{\partial t}\delta p dV = 0 \qquad (7)$$

where δu_i and $\delta u'_i$ represent any arbitrarily change in relative and absolute displacements, respectively; δu_n is the arbitrarily change in the relative normal displacement along the coupled element; T_i are the components of traction along the boundary S.

To obtain an approximate solution of Eqns (6) and (7), it is necessary to first relate the displacements and water pressures to the nodal values. According to the Goodman's joint formulation, the displacements and pore pressures adopt the linear interpolation function. The variables $u, p, u_n, \sigma, \partial p/\partial x$ in the global coordinate system can be expressed

$$u = BR\delta_u \qquad (8)$$

$$p = a^T\delta_p \qquad (9)$$

$$u_n = v^{top} - v^{bottom} = V^T R \delta_u \qquad (10)$$

$$\frac{\partial p}{\partial x} = \frac{\partial}{\partial x}(a^T \delta_p) = E_x \delta_p \qquad (11)$$

$$\sigma' = -Du \qquad (12)$$

$$B = \begin{bmatrix} -N_1 & 0 & -N_2 & 0 & N_1 & 0 & N_2 & 0 \\ 0 & -N_1 & 0 & -N_2 & 0 & N_1 & 0 & N_2 \end{bmatrix} \qquad (13)$$

$$N_1 = N_3 = \frac{1}{2}\left(1 - \frac{2x}{L}\right) \qquad (14)$$

$$N_2 = N_4 = \frac{1}{2}\left(1 + \frac{2x}{L}\right) \qquad (15)$$

$$\delta_u = [u_1 \; v_1 \; u_2 \; v_2 \; u_3 \; v_3 \; u_4 \; v_4]^T \qquad (16)$$

$$a^T = \frac{1}{2}[N_1 \; N_2 \; N_1 \; N_2] \qquad (17)$$

$$\delta_p = [p_1 \; p_2 \; p_3 \; p_4]^T \qquad (18)$$

$$V^T = [0 \; -N_1 \; 0 \; -N_2 \; 0 \; N_1 \; 0 \; N_2] \qquad (19)$$

$$E_x = \frac{1}{2}\left[\frac{\partial N_1}{\partial x} \; \frac{\partial N_2}{\partial x} \; \frac{\partial N_1}{\partial x} \; \frac{\partial N_2}{\partial x}\right] \qquad (20)$$

D is the constitutive matrix; D_e is the elastic matrix, the elastic viscoplastic matrix D_{vp} will be discussed later.

$$D_e = \begin{bmatrix} \lambda_s & 0 \\ 0 & \lambda_n \end{bmatrix} \quad (21)$$

where λ_s, λ_n are the unit shear and normal stiffnesses (Desai, C.S., Zaman, M.M., Lightner, J.G., et al. 1984).

R is the transform matrix; r is the direction cosine matrix, θ is the angle between the global X axis and the local x axis (Zhu Bofang 1998)

$$R = \text{diag}\,[r\ \ r\ \ r\ \ r] \quad (22)$$

$$r = \begin{bmatrix} \cos\theta & \sin\theta \\ -\sin\theta & \cos\theta \end{bmatrix} \quad (23)$$

If Eqns (8~10) and (12) are substituted into Eqn (6) then

$$K\delta_u - L^T \delta_p - m = 0 \quad (24)$$

$$K = \int_L R^T B^T DBR dx \quad (25)$$

$$L^T = \int_L R^T V a^T dx \quad (26)$$

$$m = \int_S R^T N^T T dS \quad (27)$$

$$N = \begin{bmatrix} N_1 & 0 & N_2 & 0 & N_1 & 0 & N_2 & 0 \\ 0 & N_1 & 0 & N_2 & 0 & N_1 & 0 & N_2 \end{bmatrix} \quad (28)$$

If Eqns (9~11) are substituted into Eqn (7) then

$$-L\frac{\partial \delta_u}{\partial t} - \Phi \delta_p = 0 \quad (29)$$

$$\Phi = \frac{k_x h}{r_w} \int_A E_x^T E_x dx \quad (30)$$

Eqns (24) and (29) can be solved by the increment of displacements $\Delta\delta_u$ and pore pressures $\Delta\delta_p$. Therefore we are solving a set of coupled equations,

$$\begin{bmatrix} K & -L^T \\ -L & -\alpha \Delta t \Phi \end{bmatrix} \begin{Bmatrix} \Delta\delta_u \\ \Delta\delta_p \end{Bmatrix} = \begin{Bmatrix} \Delta m \\ \Delta t \Phi p_t \end{Bmatrix} \quad (31)$$

where p_t are the pore pressures at time t, Δ denotes the increments from time t to $t + \Delta t$; α is the integrating factor, its value is related to the integration scheme and ranges from 0 to 1, a value of 0.5 was used here to ensure unconditional stability of the analysis with respect to time (Booker, J.R. & Small, J.C. 1975).

2.3 Constitutive law

Rock slope always behaves slow and continuous sliding under the load of gravity and seepage force, this characteristic is called creep (Desai, C.S., Samtani, N.C., et al., 1995). So this paper adopted the elastic viscoplastic constitutive law based on the Perzyna theory (Perzyna, P. 1966, Liu Baoguo & Qiao Chunsheng 2004). Suppose the relative displacements consist of two parts: elastic and viscoplastic displacements, the elastic displacements u_e are independent of time and they can be decided by the elastic constitutive equations; and yet the viscoplastic displacements u_{vp} are dependent of time, they can be decided by the Perzyna viscoplastic constitutive equations. The increments of the total relative displacements in Δt^n

$$\Delta u^n = \Delta u_e^n + \Delta u_{vp}^n \quad (32)$$

The increments of effective stresses in Δt^n

$$\Delta \sigma_n' = D_e \Delta u_e^n = D_e(\Delta u^n - \Delta u_{vp}^n) \quad (33)$$

The increments of viscoplastic displacements based on the *Euler* time integration method

$$\Delta u_{vp}^n = \Delta t^n \left[(1-\Theta)\dot{u}_{vp}^n + \Theta \dot{u}_{vp}^{n+1}\right] \quad (34)$$

Expand the $\dot{u}_{vp}^{n+1}$ using the *Taylor* series

$$\dot{u}_{vp}^{n+1} = \dot{u}_{vp}^n + \left[\partial \dot{u}_{vp}^n/\partial \sigma'\right]^n \Delta \sigma_n' = \dot{u}_{vp}^n + H^n \Delta \sigma_n' \quad (35)$$

If Eqns (34~35) are substituted into Eqn (31) then

$$\Delta \sigma_n' = D_{vp}^n (B^n \Delta \delta_u^n - \Delta t^n \dot{u}_{vp}^n) \quad (36)$$

$$D_{vp}^n = \frac{D_e}{1 + D_e \Delta t^n \Theta H^n} = \left[(D_e)^{-1} + \Theta \Delta t^n H^n\right]^{-1} \quad (37)$$

where $\Theta = 0.0, 0.5, 1.0$. If the H^n is decided, the elastic viscoplastic matrix D_{vp} will be decided, and the H^n can be decided by the Perzyna viscoplastic theory as follows. The relative viscoplastic displacements can be expressed by the displacements rate

$$\dot{u}_{vp} = \gamma \langle \Phi\left(\frac{F}{F_0}\right) \rangle \frac{\partial Q}{\partial \sigma'} \quad (38)$$

$$\langle \Phi\left(\frac{F}{F_0}\right) \rangle = \begin{cases} \Phi(F/F_0) & F/F_0 > 0 \\ 0 & F/F_0 \leq 0 \end{cases} \quad (39)$$

γ represents the flow parameter, it can be decided by rheological test of interface; Φ represents the function of viscoplastic displacements rate.

$$\Phi(F/F_0) = (F/F_0)^n \tag{40}$$

where n represents the material parameter; F represents the material yield criterion; F_0 is the reference value of F (Perzyna, P. 1966); this paper adopted the Mohr-Coulomb yield criterion and associated flow rule

$$F = Q = \tau_s + \sigma'_n \tan\varphi - c \tag{41}$$

$$F_0 = c\cos\varphi \tag{42}$$

where c represents the cohesion of interface; φ represents the frictional angle (Swoboda, G., Mertz and W., Beer, G., 1987). If Eqns (38~42) are substituted into Eqn (35), the D_{vp} in local coordinate can be obtained.

3 EXAMPLES

A relevant FE program based on the above derivation was developed on the secondary development platform of Abaqus (Hibbitte, Karlsson, Sorenson INC., 2002), then it was applied to the coupled analysis of the Dayeping landslide.

The Dayeping landslide lies on the Loushui river, its main material compositions are nubby dolostone and few clays. The sliding zone is the layered compression crush zone, whose original rockmass is the dolostone fragment, because of the integrated effect of groundwater and special tectonic zone, the dolostone fragments weather seriously, dissolute and scour to form the WIL, and then the WIL results in the sliding. The main sliding direction is $S70°W$, the plane shape of the landslide is round-backed armchair, its elevation is from 230 m to 480 m, and the dip angle of slide surface is between 18° and 26°.

The upriver groundwater level in the landslide is constant, the elevation is 500 m, the downriver normal water level is 294 m, and the dead water level is 235 m. The material zone and finite element mesh of the landslide section 1 are shown in figure 1 and 2, respectively, and the thicknesses of WIL 1 and 2 are both 20 cm. The indexes n of the WIL and general rock are 3 and 2, respectively, other material parameters are shown in table 1. The WIL adopted the coupled element proposed in this paper to model, and the general rock adopted the 2D four nodes coupled element Cpe4p to model (Hibbitte, Karlsson, Sorenson, INC. 2002). The quantity of elements and nodes is 3615 and 3765, respectively. The constitutive law of all materials is elastic viscoplastic model based on the Perzyna theory, and the yield criterion is the Mohr-Coulomb criterion,

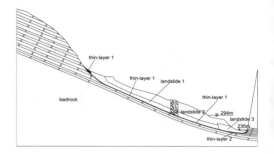

Figure 2. Material zone of landslide section 1.

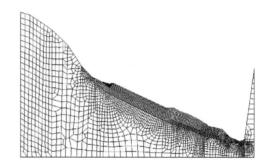

Figure 3. Finite element mesh of landslide section 1.

the corresponding flow rule is associated. Boundary conditions: the x and y displacements at the bottom are restricted, and it is impermeable; the x displacement on the left vertical plane is restricted; the upriver and downriver water level are 500 m and 235 m, respectively. Because it is short of seepage-stress coupled test data, this paper doesn't consider the relationship between permeability coefficient, void ratio and stresses, strains in analysis, suppose the permeability coefficient is constant. The groundwater free surface comparisons between the coupled analysis and in-situ testing data of section 1 and 2 are shown in figure 4, the displacement vector and contour of section 1 are shown in figure 5, and the stresses contours of section 1 are shown in figure 6.

It is seen in Figure 4 that the groundwater free surface often passes through the WIL, whether or not simulating the coupled interaction of WIL correctly matters the location of critical slip surface. Through comparisons between the coupled analysis and in-situ testing data, we found that the results are consistent well on the drill holes, and some discrepancies are also observed at the top bedrock in figure (a) and stepwise WIL in figure (b), the reasons may be few drill holes and the errors in the course of fitting the in-situ testing data, we think that these discrepancies are reasonable,

Table 1. Material parameters.

Material	ρ/(kN/m³)	E/(MPa)	μ	k/(m/s)	e	c/(kPa)	$\varphi/(°)$	$\gamma/(\min^{-1})$
Bedrock	26.5	8000.0	0.26	4.98×10^{-6}	0.012	300.0	50.0	0.00015
Thin-layer 1	19.5	λ_s(MPa/m) 7.28	λ_n(MPa/m) 133.0	0.6(shear) 0.0(normal)	0.220	30.0	19.8	0.053
Landslide 1	22.0	10.0	0.35	2.3×10^{-7}	0.201	55.0	25.0	0.00089
Landslide 2	21.0	9.8	0.33	2.0×10^{-3}	0.195	58.0	27.0	0.00095
Landslide 3	20.5	9.5	0.31	3.2×10^{-4}	0.180	60.0	31.0	0.0015
Thin-layer 2	19.8	λ_s(MPa/m) 5.34	λ_n(MPa/m) 105.0	2.0(shear) 0.0(normal)	0.257	28.0	20.3	0.063

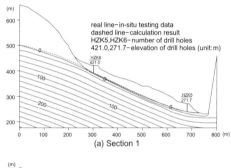

(a) Section 1

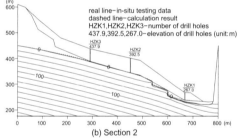

(b) Section 2

Figure 4. Groundwater free surface comparisons.

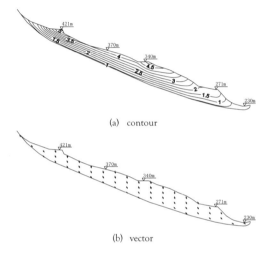

Figure 5. Displacement of landslide section 1 (unit:m).

and they are in the permission range of engineering. Therefore adopting the proposed CTE to model the seepage-stress coupled interaction of WIL is feasible and effective.

The displacement contour shown in figure 5(a) indicates that the displacement on the surface of landslide is very large, and it decreases gradually with the increase of depth. In particular, the maximum occurs on the first and second platform, the elevation is 421 m and 340 m, respectively. Figure 5(b) shows that the displacement vector has the trend of sliding along with the WIL, especially at the elevation from 230 m to 370 m, Thus it can be primarily concluded that the landslide will occur partial broken-line sliding along with the WIL at the elevation from 230 m to 370 m.

It is seen in figure 6(a) that the σ_x is the tensile stress nearby the rear end of landslide, the maximum of the tensile stress is 0.27 Mpa, this may result in the decrease of the rear end's stability. Figure 6(b) shows that the σ_y is the compressive stress in the whole landslide, the maximum is 1.82 Mpa, it lies on the first platform, whose elevation is 421 m. Figure 6(c) shows that the maximum of the τ_{xy} nearby the WIL is 0.3 Mpa, it lies in the front of the landslide, its elevation is 230 m, and its direction is upward along the WIL, the reason is that the landslide has the trend of sliding along the WIL, and yet the bedrock resists sliding, especially in the front of the landslide. The τ_{xy} far from the WIL makes the landslide tend to slide downwards, and the maximum is -0.25 Mpa ("$-$" represents the direction is opposite to the stated positive direction), it lies on the surface of landslide, its elevation is about 380 m.

Finally, this paper adopted the strength reduction method to analyze the stability under three cases of CTE, SE and DE modeling the WIL. The cases of

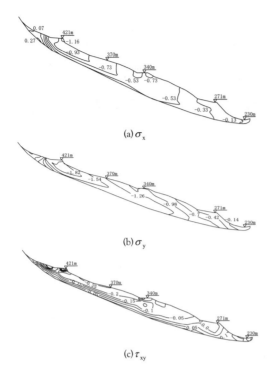

Figure 6. Stress contours of landslide section 1 (unit: MPa).

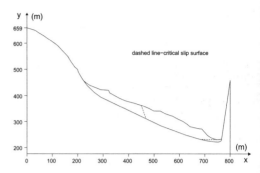

Figure 7. Critical slip surface.

Table 2. Minimum safety factors.

Coupled thin-layer element	Solid element	Desai element
1.19	1.22	1.27

CTE and SE considered the coupled effect, and yet the case of DE only carried out stress analysis, the pore pressure was considered to be the external load. The critical slide surface under the case of CTE is shown in figure 7, and the minimum safety factors are shown in table 2.

The results indicate that the critical slide surfaces of three cases are consistent basically, the elevation ranges from 230 m to 370 m, but the minimum safety factors change evidently, the minimum safety factor of CTE is the smallest, and yet that of DE is the largest. The case of CTE considers the coupled interaction of landslide and WIL comprehensively, so it is close to the true status of stresses in the landslide; The SE only considers the coupled interaction of landslide partly, but the most important affecting factor, that is, the coupled effect of WIL is ignored, so it results in the increase of safety factor; And yet the DE doesn't consider the above coupled interactions at all, it only adopts the discoupling style to solve the coupled problem, thus the corresponding error is large, and the safety factor also increases.

4 CONCLUSIONS

1. A new type of coupled thin-layer element based on Goodman's joint has been presented, which is based on Biot's well known consolidation theory, it can be modeled both the stress-deformation behaviours and water pressure response of WIL, and it also considered the time-dependent behaviors of WIL. So the proposed element is fully coupled.
2. A FE program of the CTE was developed to analyze the Dayeping landslide, the calculated groundwater free surface is consistent well with the in-situ pore pressure testing data. It indicates that the proposed coupled element is right.
3. The strength reduction method was adopted to appraise the stability of landslide. The results indicate that the SE and DE can't model the tension fracture damage or pore pressure of WIL, the corresponding results are inclined to safe. And yet the coupled element makes up the above faults, it can model the relative slide, tension fracture damage and pore pressure, therefore its advantages are very evident. The corresponding calculation is close to the true status of stresses in the landslide, the F_{min} is 1.19, which is evidently smaller than those of SE and DE (1.22, 1.27), but it is larger than the allowable safety factor 1.15, the landslide is still stable. So the proposed coupled element is superior to the other elements; it can forecast the stability of landslide with WIL correctly, and the results also

provide a theoretical basis for the landslide disaster prevention.
4. The proposed CTE model is a completely coupled model, it can consider the relationship between permeability coefficient, void ratio and stresses, strains, but due to be short of seepage-stress coupled testing data, the coupled analysis in the Dayeping landslide only supposed the permeability coefficient is constant, this affects the results to some extent. So developing seepage-stress coupled test equipments and researching the relationship between the permeability coefficient, void ratio and the stresses, strains are the basically but urgent problems in rock seepage-stress coupled research.

REFERENCES

Booker, J.R. & Small, J.C. 1975. An investigation of the stability of numerical solutions of Biot's equations of consolidation [J]. *International Journal of Solids and Structures*, 11(7–8): 907–917.

Chai Junrui, Li Shouyi. 2004. Coupling analysis of seepage and stress fields in Xietan Landslide in Three Gorges region [J]. *Chinese Journal of Rock Mechanics and Engineering*, 23(8): 1280–1284.

Desai, C.S., Samtani, N.C., et al. 1995. Constitutive modeling and analysis of creeping slopes [J]. *Journal of Geotechnical Engineering*, 121(1): 43–56.

Desai, C.S., Zaman, M.M., Lightner, J.G., et al. 1984. Thin-layer element for interfaces and joints [J]. *International Journal for Numerical and Analytical Methods in Geomechanics*, 8: 19–43.

Ding Xiuli, Fu Jing & Zhang Qihua. 2004. Stability analysis of landslide in the south end of Fengjie highway bridge with fluctuation of water level of three Gorges reservoir [J]. *Chinese Journal of Rock Mechanics and Engineering*, 23(17): 2913–2919.

Goodman, R.E., Taylor, R.L. & Brekke, T.L. 1968. A model for the mechanics of jointed rock [J]. *Journal of the Soil Mechanics and Foundations Division Proceedings of the American Society of Civil Engineers*, 94(SM3): 637–659.

Hibbitte, Karlsson, Sorenson, INC. 2002. *ABAQUS/Standard user's manual*[S].

Liu Baoguo & Qiao Chunsheng. 2004. Back analysis of visco-plastic model parameters of rock mass [J]. *Engineering Mechanics*, 21(4): 118–122.

Luan Maotian & Wu Yajun. 2004. A nonlinear elasto-perfectly plastic model of interface element for soil-structure interaction and its applications [J]. *Rock and Soil Mechanics*, 25(4): 507–513.

Oda Masanobu. 1986. Equivalent continuum model for coupled stress and fluid flow analysis in jointed rock masses [J]. *Water Resources Research*, 22(13): 1845–1856.

Perzyna, P. 1966. Fundamental problems in viscoplasticity [J]. *Advances in Applied Mechanics*, 9: 243–377.

Samtani, N.C., Desai, C.S., et al. 1996. An interface model to describe viscoplastic behavior [J]. *International Journal for Numerical and Analytical Methods in Geomechanics*, 20: 231–252.

Small, J.C., Booker, J.R. & Davis, E.H. 1976. Elasto-plastic consolidation of soil [J]. *International Journal of Solids and Structures*, 12: 431–448.

Swoboda, G., Mertz, W. & Beer, G. 1987. Rheological analysis of tunnel excavations by means of coupled finite element-boundary element analysis [J]. *International Journal for Numerical and Analytical Methods in Geomechanics*, 11(2): 115–129.

Zhu Bofang. 1998. *The finite element method theory and applications* (2nd edition) [M]. Beijing: China Waterpower Press, 597–599.

Numerical modelling of a rock avalanche laboratory experiment in the framework of the "Rockslidetec" alpine project

I. Manzella
Rock Mechanics Laboratory, Swiss Federal Institute of Technology, Lausanne, Switzerland

M. Pirulli
Politecnico di Torino, Turin, Italy

M. Naaim
Cemagref, Grenoble, France

J. F. Serratrice
CETE Méditerranée, Aix en Provence, France

V. Labiouse
Rock Mechanics Laboratory, Swiss Federal Institute of Technology, Lausanne, Switzerland

ABSTRACT: This paper illustrates the results of a collaboration born among the authors, in the framework of the INTERREG IIIA "Rockslidetec" project, with the aim of studying rock avalanche propagation by means of physical and numerical modelling. The EPFL rock mechanics laboratory has carried out an experimental campaign. Tests consist in releasing unconstrained gravel flows on an inclined panel. One of the experiments has been simulated by the codes developed by the partners. The Cemagref code and RASH3D (Politecnico di Torino) are based on a continuum mechanics approach; EPAN3D (CETE) on a discrete approach. Cemagref results, although rather far from observations made on deposit morphology, reproduce quite well the travel distance, without any parameter calibration. RASH3D reproduces well the velocity of propagation of the rear and front of the mass, but less the final deposit shape. EPAN3D reproduces very well the final deposit but the calculations are based on the calibration of 11 parameters.

1 INTRODUCTION

Large rock avalanches or sturzstorms, as called by Heim (1932), involve volumes of more than one million of cubic metres of material and are rare but catastrophic events. They engage a great amount of energy and exhibit a much greater mobility than could be predicted using frictional models (Hungr et al. 2001). For these reasons the most effective way to prevent them to cause damages and victims would be to forecast their runout and consequently to define areas that could be affected by their occurrence. On the other hand, there are still a large number of unknown factors and numerical codes are still too sensible to some geomechanical parameters like friction angle. Moreover, since rock avalanches are luckily not so frequent, there is a lack of well known real cases on which to base proper back analysis studies or code validations.

The INTERREG IIIA "Rockslidetec" project (2003–2006), among Italian, French and Swiss partners, has born to partly fill up these lacunas: its main goal is to develop methodological devices to define areas that could be affected by rock avalanche phenomena. The project is organised in three sections:

– Action A: realization of an itemized list of historical rockslide events;
– Action B: rockslide geometrical characterization and determination of instable volumes;
– Action C: study and modelling of rockslide runout with comparison of different codes.

In the action C framework the EPFL rock mechanics laboratory has carried out a research by means of a physical model to determine the main parameters that influence propagation. Even if the quantitative interpretation of the results is not straightforward, owing to the difficulty in matching the scaling laws, laboratory experiments are very helpful for the phenomenological study and the assessment of relevant physical parameters as well as for numerical model validation (Manzella & Labiouse, 2007a).

For this reason test results constitute a base of experimental comparison for the validation of the numerical codes developed by the partners of the project. In particular in the action C framework one specific experiment has been used as a benchmark by Cemagref (Grenoble, France), Politecnico di Torino (Turin, Italy) and CETE Méditerranée (Aix en Provence, France). The results of the simulations have been useful to compare the three codes and underline their specific characteristics.

In this paper the experimental set-up and the three codes are described in details, then results of the exercise on the benchmark are shown and the comparison among the codes is illustrated.

2 LABORATORY TESTS

Tests mainly consist in simulating a rock avalanche releasing an unconstrained flow of granular material on an inclined panel. The experimental set-up (Fig. 1) mainly consists of two rectangular panels (3 m × 2 m) joined by a hinge. The first panel is fixed horizontally on a concrete floor slab and the second one can change its inclination. A wooden cuboidal container measuring 0.20 m height × 0.40 m width × 0.65 m length is filled with different amount of material and placed on the tilting panel (Manzella and Labiouse 2007a, b). The box is opened in an almost instantaneous way using a spring-loaded bottom gate and the material is released directly onto the slip surface.

Each test is filmed by a digital high speed camera placed at a height of about 5 m over the horizontal panel.

The parameters varied during the whole experimental campaign are:

- material volume (10, 20, 30, 40 litres);
- releasing height (1, 1.5, 2 m);
- slope angle (30°, 37, 5°, 45°);
- nature of released material: Aquarium gravel of two different grain size distribution, Hostun sand and small bricks;
- the number of releases (40 litres in one or in two consecutive releases of 20 litres)
- base friction coefficient (high roughness using wood and low roughness using forex, a light PVC sheet, to cover the model boards).

In this paper only one set of parameters is considered.

2.1 Measurements

The displacement and the velocity of the mass front during sliding have been evaluated through image analysis using the software WINanalyze. Run-out, length, height and width of the final deposit have been manually measured after each test.

Volume, deposit geometry and morphology can be evaluated by means of the fringes projection method. This is an innovative method which consists in projecting alternate lines of dark and light (fringes) on the deposit surface. When fringes are projected on a planar surface, they are straight and equally spaced, whereas on a rough surface they are distorted and this distortion is related to the shape of the object (Desmangles, 2003). Consequently, it is possible to retrieve the information on the deposit thickness deriving it from the departure from straightness of the fringes. The precision obtained using this method with the mentioned test set-up is approximately 4 mm, which is in the order of the maximum grain diameter of the gravel used (D = 0.5 – 3 mm).

2.2 Benchmark test

The test on which the partners worked has the following characteristics: a release of 20 litres of aquarium gravel from 1 m height on a 45° inclined panel.

Aquarium gravel is a material used for aquarium decoration; this material is suitable for laboratory experiments since it is treated during production in

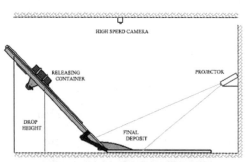

Figure 1. The experimental setup.

order to reduce to minimum the quantity of dust. It has a quite homogeneous grain size distribution with a diameter D = 0.5 − 3 mm. Its unit weight is of 14.3 kN/m³. Its internal static friction angle is $\phi_i = 34°$ and the one at the base between the gravel and the panel surface is $\phi_b = 32°$.

Thanks to the fringes projection method and the film registered with the high speed camera it has been possible to furnish to the partners the digital elevation model of the final deposit and images of the flow each 1/60 second.

3 NUMERICAL MODELS

Three numerical codes are considered in the present paper. Two of them are based on a continuum mechanics approach while the third, EPAN3D, on a discrete element method.

RASH3D (Politecnico di Torino) and the code developed by Cemagref treat the moving mass as a homogeneous continuum, assuming that both depth and length of the flowing mass are usually large if compared with the characteristic dimension of the particles involved in the movement. Under this assumption, it becomes possible to replace the real moving mixture of solid and fluid phases by an "equivalent" fluid, whose rheological properties have to approximate the behaviour of the real mixture (Hungr 1995). By this way, the dynamic behaviour of the flowing mass can be described by the mass and momentum conservation laws. Further, assuming that the vertical structure of the flow is much smaller than its characteristic length, the codes integrate the balance equations in depth, obtaining the so-called depth-averaged continuum flow models (Savage and Hutter 1989). The general system of equation to be solved is as follows:

$$\begin{cases} \dfrac{\partial h}{\partial t} + \dfrac{\partial(\overline{v_x}h)}{\partial x} + \dfrac{\partial(\overline{v_y}h)}{\partial y} = 0 \\[6pt] \rho\left(\dfrac{\partial(\overline{v_x}h)}{\partial t} + \dfrac{\partial(\overline{v_x^2}h)}{\partial x} + \dfrac{\partial(\overline{v_xv_y}h)}{\partial y}\right) \\[6pt] = -\dfrac{\partial(\overline{\sigma_{xx}}h)}{\partial x} + \tau_{zx(z=b)} + \rho g_x h \\[6pt] \rho\left(\dfrac{\partial(\overline{v_y}h)}{\partial t} + \dfrac{\partial(\overline{v_yv_x}h)}{\partial x} + \dfrac{\partial(\overline{v_y^2}h)}{\partial y}\right) \\[6pt] = -\dfrac{\partial(\overline{\sigma_{yy}}h)}{\partial y} + \tau_{zy(z=b)} + \rho g_y h \end{cases} \quad (1)$$

where $\overline{v} = (\overline{v}\hat{x}, \overline{v}\hat{y})$ denotes the depth-averaged flow velocity in a reference frame (x,y,z) linked to the topography, h the fluid depth, ρ the mass density, g_x, g_y the projection of the gravity vector along the x and y directions, $\sigma_{xx}, \sigma_{yy}, \tau_{zx}, \tau_{zy}$ the normal and shear stresses components and t the time.

Adopting the above described approach, changes of the mechanical behaviour within the flow are ignored and the complex rheology of the flowing material is incorporated in a single term, describing the frictional stress that develops at the interface between the flowing material and the rough surface.

On the other hand a model such as EPAN3D (CETE Méditerranée) using a discrete element method (DEM) assumes that the mass consists of separate, discrete particles and that the moving mass is the result of the interaction of the particles. The forces acting on each particle are computed taking into account the initial configuration and the relevant physical laws. According to Cundall (1988) any particle may interact with any other particle and there are no limits on particle displacements and rotations.

In the case of rock avalanche modelling, particles interact between them through friction and the equations of the conservation of mass, of energy and of quantity of momentum are used. The velocity distribution field and the thickness of the mass in movement are obtained by the computation of the dynamics of a discretized deformable mass on a grid.

3.1 Cemagref

In Cemagref model the domain is restricted to rock avalanches made of dry and cohesion-less grains. The Bagnold's profile is adopted to represent the variation of velocity within the depth of the flow. A friction model issued from the recent progress on granular material flows (Pouliquen, 1999) is chosen to represent the momentum loss at the substratum level. The pressure distribution inside the flowing material is considered as isotropic, which means that the earth pressure coefficient is taken equal to 1. Afterwards, the shallow water equations were extended to large rock avalanche flows over a complex topography. The variables of the equations are the depth and the velocities. The system of equations is written in curvilinear coordinates. The mathematical conservative equations system is solved by a Godunov numerical scheme using a finite element unstructured mesh. This kind of mesh allows taking into account the complexity of the mountain terrain especially near singularities. To increase the accuracy of the solver a linear variation of the height and mean velocities inside each cell is considered. In order to avoid numerical instabilities, the gradients of these variables were limited according to the criterion defined by Van Leer (1979). Each component of this model was severely tested using both analytical solutions and laboratory experiments.

3.2 RASH3D

The RASH3D code (Pirulli, 2005) originates from a pre-existing model (SHWCIN) based on the classical finite volume approach for solving hyperbolic systems using the concept of cell centred conservative quantities, developed by Audusse et al. (2000) and Bristeau et al. (2001) to compute Saint-Venant equation in hydraulic problems.

An extension of SHWCIN for simulating dry granular flows using a kinetic scheme was initially introduced by Mangeney-Castelnau et al. (2003). Pirulli (2005) proposed further modifications to SHWCIN to reduce observed mesh-dependency problems, permit simulation of motion across irregular 3D terrain, incorporate the influence of internal strength and allow the selection of more than one possible basal resistance relationship.

As input data the code requires the digital elevation model of the studied area, the identification of the boundary of the source area and the geometry of the initial volume.

As for the rheological characteristics of the flowing mass, three different rheologies are implemented in RASH3D at the present time:

1. the simple frictional rheology, based on a constant friction angle φ, which implies a constant ratio of the shear stress to the normal stress. Shear forces, τ, are independent of velocity.

$$\tau = \gamma h \tan \varphi \quad (2)$$

where γ = unit weight, φ = basal friction angle and h = flow depth.

2. the Voellmy flow relation, which consists of a turbulent term, v^2/g, accounting for velocity-dependent friction losses, and a Coulomb or basal friction term for describing the stopping mechanism. The resulting basal shear stress is given by the following equation:

$$\tau = \gamma h \tan \varphi + \frac{\gamma \bar{v}^2}{\xi} \quad (3)$$

where $\bar{v}$ = the mean flow velocity, ξ = turbulence coefficient; the others symbols are similar as in equation [2].

3. the quadratic rheology, where the total friction is provided by the following expression:

$$\tau = \gamma h \tau_y + \frac{k \eta \bar{v}}{8h} + \frac{n_{td} \bar{v}^2}{h^\beta} \quad (4)$$

where τ is the shear stress, τ_y is the Bingham yield stress, η is the Bingham viscosity, n_{td} is the equivalent Manning's coefficient for turbulent and dispersive shear stress components, k is the flow resistance parameter.

The first and the second terms on the right hand side of equation [4] are, respectively, the yield term and the viscous term as defined in the Bingham equation. The last term represents the turbulence contribution (O'Brien et al. 1993).

The RASH3D code has been widely validated simulating laboratory tests (e.g. Mangeney et al., 2003; Pirulli, 2005) and back analysing real cases (e.g. Pirulli et al., 2007; Pirulli and Mangeney, 2007).

Time for calculation of each numerical analysis cannot be univocally defined but it is mainly a function of the digital elevation model discretization.

Further developments of RASH3D include the numerical implementation of entrainment of material along the path of propagation and the arrangement of a more user friendly interface.

3.3 EPAN3D

The objective of EPAN3D (CETE Méditerranée) is to simulate a rock mass of large volume, its propagation along a mountain slope then the accumulation and the spreading out in the valley. The unstable volume of rock is discretized in small elements of volume. After failure, these elements slide on the digital terrain model (DTM), until they find equilibrium at the bottom of the valley where they accumulate and spread out. The moving rock cluster is permanently associated with morphology of the slope by updating the DTM at each calculation step. The energy is dissipated primarily through friction between the sliding elements and at the base.

Two local evolution laws are used for modelling. A first law fixes the ratio of the friction $\mu = \text{tg}\phi$ against the instantaneous velocity v of a sliding element. It takes the form:

$$v \leq v_0 \quad \mu = \mu_0$$
$$v > v_0 \quad \mu = \mu_0 + (\mu_1 + \mu_0)$$
$$\times \left[1 - \exp\left(-\frac{(v - v_0)}{v_1}\right)\right] \quad (5)$$

where $\mu_0 = \text{tg}\phi_0$, $\mu_1 = \text{tg}\phi_1$, v_0 et v_1 are parameters ($\mu_0 \leq \mu_1$ and $v_0 < v_1$).

Friction increases asymptotically with velocity towards a limit value μ_1. On the other hand it remains equal to a minimal value μ_0 when v is lower than v_0. This formulation is inspired by the laws suggested by Pouliquen (1999).

Another law describes the evolution of the shape and the volume of each sliding element along its path. Each elementary volume is supposed to take the shape of a paraboloid with elliptic section whose shape can be parameterized by means of an elongation modulus and a spreading coefficient. The volume of the sliding elements also evolves according to a bulking law that depends on the distance covered from failure. Finally, the initial unit weight of the rock is given. On the whole, calculations are based on ten mechanical parameters.

After having discretized the unstable rock volume in small elements, a calculation "by spreading" (par épandage) is carried out by successive time steps: at a given time, the trajectory of a sliding element within the rock cluster is determined by linear and angular momentum balance conservation equations applied to the local tangent plan covered by the element. Thus, calculations by spreading allow following the position of all the sliding elements at every time and making the form of the topographic surface evolve during the movement by update of the DTM. Therefore this process provides a method of modelling the propagation and the spreading of the rock cluster. Furthermore it gives at every moment the position of all its elements, the extent and shape of the cluster, from the time of release until its complete stabilization.

EPAN3D is designed to study various scenarios of rupture or ruptures by successive stages (step-by-step scenarios).

4 SIMULATION RESULTS

To simulate the benchmark test the partners have used the following data depending on their codes:

– For his code, based on a "frictional" model, Cemagref used a friction coefficient equal to 0.62 corresponding to the tangent of the friction angle at the base ($\phi_b = 32°$). No calibration has been made.

– The RASH3D simulation was also based on a model of the type "frictional". The basal friction angle was fixed at 42°, after calibration on the results of the laboratory test.
– For the EPAN3D code the released mass was divided in 1911 elementary volumes arranged in seven layers according to a cubic network. A back-analysis of the laboratory test allowed a calibration of the parameters of the constitutive model, such as $\gamma = 16.5$ kN/m^3, $\phi_0 = 11°$, $\phi_1 = 44.5°$, $v_0 = 0.1$ m/s and $v_1 = 0.5$ m/s.

Digitalization of the geometry of the model (an inclined panel at 45° degrees and a horizontal panel) as well as the definition of the zone of departure (0.25 m × 0.20 m × 0.4 m box filled with 20 litres of granular material) was in charge of the teams, but there should not be any differences.

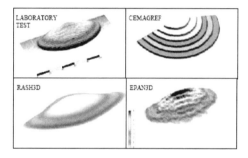

Figure 3. 3D view of the observed deposit and the codes simulations. Data are presented at the same scale.

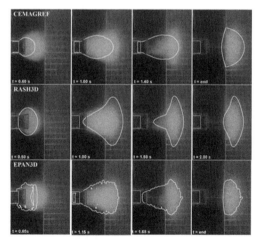

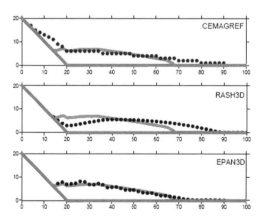

Figure 2. Longitudinal section along the symmetry axis of the observed deposit (light grey) and codes simulations (black points).

Figure 4. Comparison between the mass propagation and spreading observed in laboratory (images in background) and those modelled by the three codes. The outline represented in white concerns the 1 mm thickness contour line.

Table 1. Qualitative comparison among the codes.

Model Analysis		Cemagref frictional without calibration	RASH3D frictional back-analysis	EPAN3D pouliquen back-analysis
Number of parameters		1	1	11
Propagation	Runout	+/−	+/−	+
	Velocity	−	+	−
Deposit Dimensions	Longitudinal ext.	−	+/−	+
	Lateral extension	+/−	−	+
	Height	+/−	+/−	+
Deposit Morphology	General shape	−	−	+
	Front	−	+/−	+/−
	Rear	−	−	+

In Figures 2, 3 and 4 the simulation results are shown.

Although relatively far from observations made on the extension and the morphology of the gravel deposit, the results obtained by Cemagref are relevant in terms of runout distance, if it is taken into account that no parameter calibration has been made (class A prediction). As can be seen from Figure 4, the calculation made by RASH3D reproduces well the velocity of propagation of the rear and front of the mass, but less the final deposit shape, especially in the transversal direction. Besides, the chosen angle of friction is far from the experimental value. Observing Figures 2 and 3 it is found that the deposit morphology and dimensions are better simulated by the model EPAN3D, but it has to be taken into account that they result from a back-analysis and from the calibration of 11 parameters, not all with physical meaning.

In Table 1 the predictions of the three codes are qualitatively compared with the observations made during the laboratory test. The symbols +, +/− and − mean respectively good, average, imperfect adequacy, for the characteristic indicated on the corresponding line.

5 CONCLUSIONS

In the INTERREG IIIA "Rockslidetec" framework a test carried out at the LMR-EPFL has been used as a benchmark for three different codes developed by the project partners. Cemagref method, commonly used for snow avalanches, is based on continuum mechanics approach and a "frictional" model; RASH3D, is also based on a continuum mechanics approach and, in the case considered, a model of the type "frictional" is used as well; EPAN3D is based on the discrete elements method. A back-analysis of the laboratory test allowed a calibration of respectively the friction angle for RASH3D and the eleven parameters of the developed constitutive model for EPAN3D.

The simulated test consisted in an unconstrained flow of 20 litres of aquarium gravel, released from 1 m height on a slope inclined at 45°.

Cemagref reproduces quite well the runout distance, without any parameter calibration (class A prediction). RASH3D reproduces well the velocity of propagation of the rear and front of the mass, but less the final deposit, in particular in the transversal direction. Besides, the chosen angle of friction differs from the experimental value. EPAN3D is better in reproducing deposit morphology and dimensions, but calculations are based on a back-analysis and on the calibration of 11 parameters.

On the other hand, since only one test has been used for this work of comparison, it is important to put the above-mentioned statements into perspective. Only a comparison of the codes among several test configurations would allow a proper analysis.

This study underlines the importance of the validation of numerical models on well-defined laboratory tests since this contributes to understand advantages and limits of the codes, improving their development and their use and consequently enhancing forecast of rock avalanche propagation.

ACKNOWLEDGEMENTS

The authors thank the Canton of Valais, the OFEG, the SECO, the Regione Valle d'Aosta, the Région Rhône-Alpes and the departments of Isère and Savoie for funding. Further acknowledgements go to the Institut de Physique du Globe de Paris for having put the SHWCIN code at disposition and to Professor Claudio Scavia for fruitful discussions.

REFERENCES

Audusse, E., Bristeau, M.O. & Perthame, B.T. 2000. Kinetic schemes for Saint-Venant equations with source terms on unstructured grids. *INRIA Report 3989*, National Institute for Research and Computational Sciences and Control, LeChesnay, France.

Bristeau, M.O., Coussin, B. & Perthame, B. 2001. Boundary conditions for the shallow water equations solved by kinetic schemes. *INRIA Rep. 4282*, National Institute for Research and Computational Sciences and Control, LeChesnay, France.

Cundall, P.A. 1988. Formulation of a three dimensional distinct element model-Part I. A scheme to detect and represent contacts in a system composed of many polyhedral blocks. *Int. J. Rock Mech. Min. Sci.*, 25: 107–116.

Desmangles, A.I. 2003. Extension of the fringe projection method to large object for shape and deformation measurement. *Ph.D. thesis no 2734*, Ecole Polytechnique Fédérale de Lausanne, CH.

Heim, A. 1932. Bergsturz und Menschenleben. *Frets und Wasmuth*

Hungr, O., Evans, S.G., Bovis, M. & Hutchinson, J.N. 2001. Review of the classification of landslides of the flow type. *Environmental and Engineering Geoscience*, VII: 221–238.

Hungr, O. 1995. A model for the runout analysis of rapid flow slides, debris flows, and avalanches. *Canadian Geotechnical Journal*, 32(4): 610–623.

Mangeney-Castelnau, A., Vilotte, J.P., Bristeau, O., Perthame, B., Bouchut, F., Simeoni, C. & Yerneni, S. 2003. Numerical modeling of avalanches based on Saint Venant equations using a kinetic scheme. *Journal of Geophysical Research*, 108(B11): 2527.

Manzella, I. & Labiouse, V. 2007a. Qualitative analysis of rock avalanches propagation by means of physical modelling of non-constrained gravel flows. *Rock Mech. and Roch Engineering Journal* (DO10.1007/s00603-007-0134-y).

Manzella, I. & Labiouse, V. 2007b. Rock avalanches: experimental study of the main parameters influencing propagation. *Proc. 11th ISRM*, 9–13 July, 2007, Lisbon, 1: 657–660.

O'Brien, J.S., Julien, P.Y. & Fullerton, W.T. 1993. Two-dimensional water flood and mudflow simulation. *Journal of Hydrological Engineering*, 119(2): 244–261.

Pirulli, M. 2005. Numerical modelling of landslide runout, a continuum mechanics approach. *PhD Thesis*, Department of Structural and Geotechnical Engineering, Politecnico di Torino, Italy.

Pirulli, M. & Mangeney, A. 2007. Results of back-analysis of the propagation of rock avalanches as a function of the assumed rheology. *Rock Mech. and Roch Engineering Journal* (DO10.1007/s00603-007-0143-x).

Pirulli, M., Bristeau, M.O., Mangeney, A. & Scavia, C. 2007. The effect of the earth pressure coefficients on the runout of granular material. *Environmental modelling & software*, 22: 1437–1454.

Pouliquen, O. 1999. Scaling laws in granular flows down rough inclined planes. *Physics of Fluids*, 11(3): 542–548.

Savage, S.B. & Hutter, K. 1989. The motion of a finite mass of granular material down a rough incline. *Journal of Fluid Mechanics*, 199: 177–215.

Serratrice, J.F. 2006. Modélisation des grands éboulements rocheux par épandage. Application aux sites de la Clapière (Alpes Maritimes) et de Séchilienne (Isère). Bulletin des laboratoires des ponts et chaussées, no 263–264, juillet-août 2006: 53–69.

Van Leer, B. 1979. Towards the ultimate conservative difference scheme, V, *J. Comp. Phys.*, 32: 101–136.

Three-dimensional slope stability analysis by means of limit equilibrium method

Shingo Morimasa & Kinya Miura
Toyohashi University of Technology, Aichi, Japan

ABSTRACT: We have conducted a series of stability analyses of cohesive soil slopes with limit equilibrium method to clarify the effect of three-dimensional shapes of sliding mass and slope on the slope stability. In the analyses, straight and curved slopes consisting of frictionless soil were examined; the curved slopes were defined with concentric circles instead of straight lines, and sliding surfaces were defined with elliptic revolution. Through the analytical investigations some important features of three-dimensional slope stability were found. Even straight slope with finite width has a critical sliding depth with notably higher stability factor than that of with infinite width which shows the lowest stability at infinite depth as in ordinary two-dimensional analysis. Curved slopes have their own critical sliding width depending on its curvature. We updated the stability diagram presented by Terzaghi with three-dimensional effects taken into consideration.

1 INTRODUCTION

Usually soil slope is modeled within two-dimensional condition, and the slope stability against sliding is analyzed in plane strain condition. In the analysis the slope is simplified to be straight and the width of sliding mass is assumed to be infinite. The modeling of slope in two-dimensional condition with the simplifications leads to lower slope stability; that is, lower safety factor is obtained, compared with that in three-dimensional condition. Thus, stability analysis conducted in two-dimensional condition is apparently in conservative design side, and has been accepted in design practices. In reality, however, configuration of soil slope is not straight but curved, and sliding mass has a finite width not as in two-dimensional analysis. The effects of three-dimensional shape of soil slope and sliding mass are notably large and not negligible in the evaluation of the slope stability.

In this study we have conducted a series of three-dimensional slope stability analyses on three types of soil slopes: Straight Slope, Concave Curved Slope and Convex Curved Slope. All the slopes considered are assumed to consist of homogeneous frictionless cohesive soil. In the analysis, sliding surfaces are modeled with spheroid, elliptic revolutions. The sliding mass was divided into thin slices, and driving and resisting moments were numerically calculated within the limit equilibrium method. Through the calculation of the driving and resisting moments of the possible sliding surfaces, the critical sliding surfaces are determined; therefore, the shape, size and location of the sliding surface at critical condition are obtained, and associated stability factor of the slope is calculated. The calculation results are summarized into several charts; some are for the description of critical sliding surface, and the others are for stability factors presented by Terzaghi. Finally the three-dimensional effects of slope configuration and sliding surface are discussed, and the important role of three-dimensional analysis in slope stability evaluation is presented.

In the Terzaghi's text he made a critical comment on the role of stability chart as follows: "Owing to the complexity of field conditions and to the important difference between the assumed and the real mechanical properties of soils, no theory of stability can be more than a means of making a rough estimate of the available resistance against sliding. If a method of computation is simple, we can readily judge the practical consequences of various deviations from the basic assumption and modify our decisions accordingly. ... ". We hope that the update of the Terzaghi's stability chart we make in this paper, contributes to the reasonable evaluation of slope stability and the engineering judgement in three-dimensional condition.

2 ANALYSIS METHOD

2.1 Spheroidal sliding surface

Figure 1 shows a cohesive straight slope with an inclination of β and an assumed three-dimensional sliding surface and sliding mass. The y-axis is situated in the direction of slope width, and another horizontal axis, x-axis, and vertical axis, z-axis, both are perpendicular to the y-axis as shown in the figure; Ugai (1985) explained that the sliding surface must have a shape of revolving solid in accordance with the variation method presented by Baker et al. (1997), provided that the equilibrium of the forces on the sliding mass is satisfied, and the factor of safety against sliding is minimized. The revolving solid is defined by the following equation:

$$(x - x_r)^2 + (z - z_r)^2 = r^2(y) \quad (1)$$

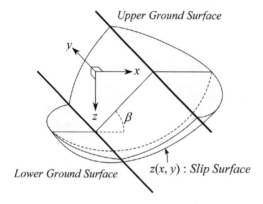

Figure 1. Three-dimensional sliding mass assumed in cohesive straight slope.

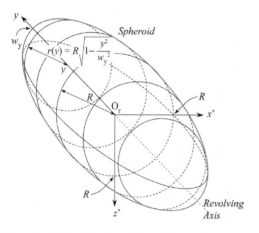

Figure 2. Spheroid for determining sliding surface.

where $O_r(x_r, z_r)$ is the revolving axis, and $r(y)$, a function of y, is the radius of the revolving solid. From Eq. 1 we employed a spheroid as a sliding surface as shown in Fig. 2, because spheroid can be considered an appropriate shape of sliding surface and is easy to analyze mathematically. The sliding surface is determined by the following equation:

$$\frac{(x - x_r)^2}{R^2} + \frac{y^2}{w_y^2} + \frac{(z - z_r)^2}{R^2} = 1 \quad (2)$$

where R is the radius of circle in vertical x-z plane, and w_y is size of the spheroid in the y-direction. The radius of circle in the plane parallel with x-z plane can be calculated as

$$r(y) = R\sqrt{1 - (y/w_y)^2} \quad (3)$$

2.2 Slope shape

Ordinary slopes can be geometrically roughly classified into three types as shown in Fig. 3: (a) Straight Slope, (b) Concave Curved Slope, and (c) Convex Curved Slope. While Straight Slope has two-dimensional configuration, the other two types of Curved Slopes have three-dimensional configuration. Toe and shoulder lines are straight and parallel each other in Straight Slope, and are concentric circles in Curved Slopes. All the slopes are symmetrical with respect to x-z plane and defined by the following parameters: slope angle β and slope height H. Curved Slope has one more parameter, normalized curvature radius R_t/H; the value R_t represents the radius of toe line.

2.3 Analysis method

Figure 4 shows a Straight Slope in plan view and in section. The shaded part in the figure indicates a sliding mass whose width is $2w_{sld}$. The symmetrical slope shape can cause a symmetrical slide, which allows us to evaluate slope stability by computing only half of the sliding mass, $0 \le y \le w_{sld}$. Calculating the ratio of the resisting moment M_r to the driving moment M_d, we obtain the factor of safety F_S of the slope with respect to sliding along the sliding surface

$$F_S = \frac{M_r}{M_d} = \frac{c}{\gamma H} \cdot \frac{H \int_0^{w_{sld}} L(y) r(y) \sqrt{(dr/dy)^2 + 1}\, dy}{\int_0^{w_{sld}} G'_{z'}(y) dy} \quad (4)$$

where c is the cohesion of the soil, γ the unit weight, $G'_z(y)$ geometrical moment of area with respect to z'-axis of a slice at y (see Fig. 4), and $L(y)$ and $r(y)$ is the length and the radius of sliding arc of the slice,

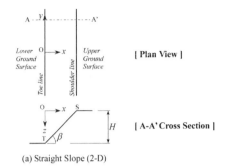

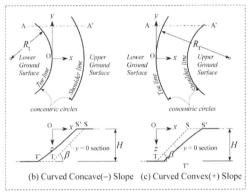

(a) Straight Slope (2-D)

(b) Curved Concave(−) Slope (c) Curved Convex(+) Slope

Figure 3. Three types of slope shape.

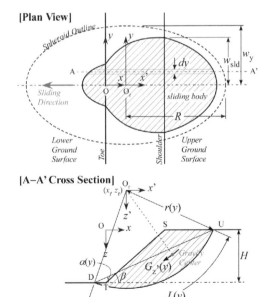

Figure 4. Straight Slope in plan view and in section.

respectively; the values of $G'_z(y), L(y)$ and $r(y)$ can be calculated mathematically. Differentiating Eq. 3 by y, we obtain

$$\frac{dr}{dy} = -\frac{R \cdot y}{w_y\sqrt{w_y^2 - y^2}} \qquad (5)$$

From Eq. 4, the stability factor N_S devised by Terzaghi is

$$N_S = \frac{\gamma H}{c'} = \frac{\gamma H}{c/F_S} = \frac{H \int_0^{w_{sld}} L(y) r(y) \sqrt{(dr/dy)^2 + 1}\,dy}{\int_0^{w_{sld}} G'_{z'}(y)\,dy} \qquad (6)$$

The mathematical integration in Eq. 6 is difficult analytically. The sliding mass is, therefore, divided into several slices and the value N_S is calculated by means of Simpson's rule of numerical integration. This investigation must be repeated for different revolving axis O_r in each analysis condition by trial and error. The stability factor of the slope is equivalent to the smallest value of the factor of safety F_S thus obtained. Two types of Curved Slopes can be analyzed in the similar manner.

In the above calculations, we assume that the sliding mass is rigid body and moves toward the direction perpendicular to the y-axis. For this reason, no share forces is assumed to generated in the y-direction on the sliding surface at sliding condition, that is, the shearing resistance works only in the x- and z-directions as shown in Fig. 5. It should be noted that shear and normal stresses on the interfaces between the slices both have no influence on the calculation of the driving and resisting moments as long as frictionless and cohesive soil slopes are concerned.

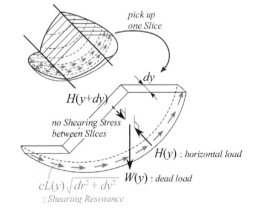

Figure 5. Force acting on slice of sliding mass.

2.4 Analysis conditions

Each slope is analyzed against two types of failure: toe failure and base failure. The depth of firm stratum on which slope rests is expressed by the depth factor

$$n_D = \frac{D}{H} \quad (D = n_D \cdot H) \tag{7}$$

where D is the depth of firm stratum from the upper ground surface. While analysis on Straight Slope has two parameters, slope angle β and assumed sliding width w_{sld}, that on Curved Slopes have one more parameter, normalized curvature radius R_t/H.

3 CALCULATION RESULTS AND DISCUSSION

3.1 Stability of straight slope in two- and three-dimensional condition

3.1.1 End effect

Figure 6(a) to (f) shows the relations between normalized sliding width w_{sld}/H and stability factor N_{S_3D} for straight slopes with different slope angle β. In order to investigate the effect of sliding width on stability of slope, stability factor N_{S_2D} obtained in two-dimensional analysis is plotted on right-hand side of the graphs. Comparing these values on the same slope condition, we can find that N_{S_3D} is always higher than N_{S_2D} regardless of assumed sliding width. It has been indicated that the shearing resistance on the ends of both sides of sliding mass in three-dimensional sliding surface contribute to the increase in stability of slope. The end effect becomes notable with a decrease in sliding width and/or in firm stratum depth.

In Fig. 7(a) to (e), the values of N_{S_3D} are plotted against the slope angle β for different values of w_{sld}/H. It was confirmed that Fig. 7(f), plotted in the same manner for N_{S_2D}, was identical with Terzaghi's stability diagram. Comparing Fig. 7(e) and (f), we can recognize that the end effect is effective even

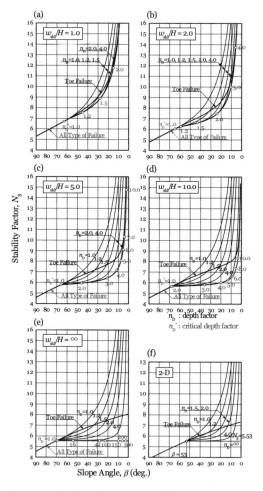

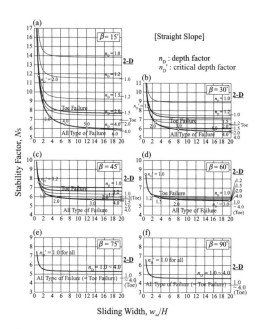

Figure 6. Relations between assumed normalized sliding width w_{sld}/H and stability factor N_S for different slope angle β and depth factor n_D (straight slope).

Figure 7. Stability diagrams for straight slope that are relations between slope angle β and stability factor N_S for different normalized sliding width w_{sld}/H and depth factor n_D; (a) to (e) are for three-dimensional condition; (e) is for almost infinite sliding width; (f) is for two-dimensional condition.

in the case of sliding mass with infinite width in three-dimensional condition.

3.1.2 Critical depth of sliding surface and critical stability factor

Calculating stability factor for sliding mass with arbitrary depth, we can obtain the relations as shown in Fig. 8. While the stability factor in two-dimensional condition decreases with increasing depth, that in three-dimensional condition has the minimum value. The corresponding depth is the critical depth to which the sliding surface can develop. It is natural that the appropriate critical depth and the corresponding critical stability factor N'_S should exist according to the sliding width. The values of N'_S are plotted in Fig. 6 and 7 (all type of failure). We can thus find the minimum stability factor without knowing depth factor, which leads to an error on the safe side.

3.2 Stability of curved slope

3.2.1 Critical width of sliding surface

Figure 9 shows the relations between assumed normalized sliding width w_{sld}/H and stability factor of curved slope for different slope angle β. Each figure contains the values of stability factor against toe failure and those of critical stability factor against base failure for different normalized curvature radius R_t/H; these values for straight slope are also plotted in the figure by broken lines. It can be seen from the figure that each curved slope has the minimum stability factor, which indicates the existence of critical sliding width to which the sliding mass can expand.

3.2.2 Influence of curvature on stability of slope

It is found in Fig. 9 that both types of curved slopes indicate higher critical stability factors than

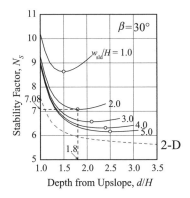

Figure 8. Relation between normalized depth d/H of sliding mass and stability factor N_S for different sliding width (slope angle $\beta = 30°$).

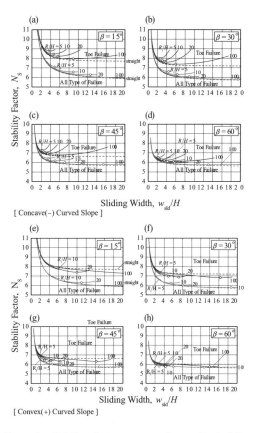

Figure 9. Relations between assumed normalized sliding width w_{sld}/H and stability factor N_S against toe failure and critical stability factor N'_S against base failure for different slope angle β and depth factor n_D; (a) to (d) are for concave curved slope; (e) to (h) are for convex curved slope.

straight slope. On the other hand, for stability factors against toe failure, different behaviors appear depending on the type of curvature; compared with straight slope, concave type has higher value and convex type has smaller value. Incidentally, the difference in stability factor between curved and straight slopes becomes negligible with narrow sliding width and large curvature.

The effect of curvature radius R_t/H on critical sliding width w'_{sld} and the corresponding stability factors are shown in Fig. 10 for different slope angle β. It is found from the figure that the increase in width of critical sliding mass and the decrease in the corresponding stability factors are almost in proportion to increase in the logarithm of curvature radius.

As a result of above discussion, stability diagram for curved slope was finally obtained for different curvature radius and sliding width as shown in Fig. 11.

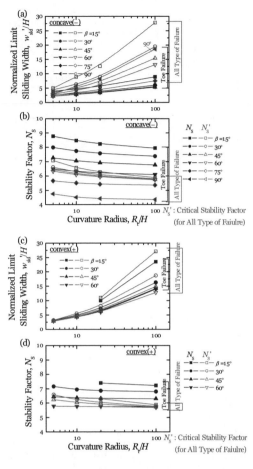

Figure 10. Relations between curvature radius and limit sliding width, and corresponding stability factor; (a) and (b) are for concave curved slope; (c) and (d) are for convex curved slope.

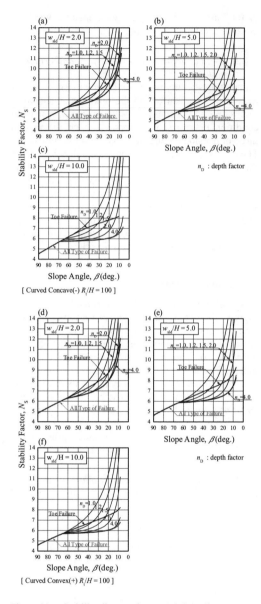

Figure 11. Stability diagram for curved slope (for curvature radius $R_t/H = 100$); (a) to (c) are for concave curved slope; (d) to (f) are for convex curved slope.

4 CONCLUSIONS

To clarify the effect of three-dimensional shapes of sliding mass and slope on stability of the slope, we conducted a series of stability analysis of straight and curved slopes by means of limit equilibrium method. And we presented several slope stability diagrams which make it possible to estimate the slope stability considering three-dimensional effects of slope and sliding surface. The following points were found as a summary of this study.

— In the analysis of slope stability with finite width of sliding mass in three-dimensional condition, end effect always increased the stability consistently both in straight and curved slopes. Sliding surface does not always touch firm base even in the case of base failure, but critical depth in accordance with assumed sliding width exists.

— Though stability factor decreases with an increase of sliding width in the case of straight slope, stability factor becomes minimal at a particular width of sliding mass in the case of curved slopes. At the critical condition depth of sliding surface as well as sliding width are determined as functions of the curvature. As long as toe failure is concerned, stability factor was less in convex slopes than those in straight and concave slopes with common slope angle. In the case of base failure with small slope angles, stability factor becomes smallest in straight slopes with common slope angle.

REFERENCES

K. Terzaghi, *Theoretical Soil Mechanics*, *Chapter IX: Stability of Slopes*: pp. 144–181.

Ugai, K. 1985, Three-Dimensional Stability Analysis of Vertical Cohesive Slopes, *JGS Soils and Foundations*, Vol.25, No.3: pp. 41–48.

R. Baker and M. Garber 1997, Variational approach to slope stability, *9th ICSMFE*, Vol.2: pp. 9–12.

Embankment basal stability analysis using shear strength reduction finite element method

Atsushi Nakamura
FORUM8 Co. Ltd., Meguro-ku, Tokyo, Japan

Fei Cai & Keizo Ugai
Gunma University, Kiryu, Gunma, Japan

ABSTRACT: The stability evaluation of the slope commonly uses limit equilibrium methods, for example, Fellenius' method in Japan. The limit equilibrium methods have two limitations. (1) It is necessary to divide the sliding mass into slices, and to set up some additional equations to make the problem statically determinate. (2) It is necessary to search sliding surface with a minimum safety factor. Shear strength reduction finite element method (SSR-FEM) can solve the limitations of the limit equilibrium methods. This paper inspects superiority of the slope stability analysis by SSR-FEM. And, as a model case, the failure of base ground of the embankment is considered. Comparing the calculated results of SSR-FEM with limit equilibrium method, we point out the problems inherent in limit equilibrium method for the model case when the slip surface is assumed to be circle. We propose an approach to use SSR-FEM for slope stability analysis in practical design.

1 INTRODUCTION

The stability evaluation of the slope commonly uses limit equilibrium methods, for example, Fellenius' method, in most design codes and standards in Japan. The limit equilibrium methods have two limitations. (1) It is necessary to divide the sliding mass into slices, and to set up some additional equations to make the problem statically determinate. (2) It is necessary to search sliding surface with a minimum safety factor. Shear strength reduction finite element method (SSR-FEM) can solve the limitations of the limit equilibrium methods.

SSR-FEM can obtain the safety factor and slip surface without the analyzer to assume any particular shape about the slip surface. SSR-FEM begins to be used to analyze the slope stability though it is not commonly in the design. The limit equilibrium methods are still used in almost cases. It is because the limit equilibrium methods are simple. However, the limit equilibrium methods are not fully verified for the following items. (1) Whether it is correct to assume that the slip surface is one circular arc. (2) How wide is the range of the grid for the center of slip circular arc to search the minimum safety factor.

Then, embankment base failure of the support ground of the fill was taken up as a model case in this report. When the support ground consists of clayey soil layer and bearing stratum, the thickness of the clayey soil layer is changed, and the slip surface are analyzed using SSR-FEM. Using these obtained slip surfaces, we arranges the extent that the slip surface become a circular arc or a non-circular arc. It is the main one of objectives t the stability examination that uses the SSR-FEM from the stability examination by the limit equilibrium method in this report.

2 SSR-FEM FOR STABILITY ANALYSIS

2.1 Basic concept of SSR-FEM

In the finite element method with shear strength reduction technique (SSR-FFM), a non-associated elasto-plastic constitutive law is adopted, where the Mohr-Coulomb yield criterion is used to define the yield function.

$$f = \frac{\sigma_1 - \sigma_2}{2} - c' \cos\phi' - \frac{\sigma_1 - \sigma_2}{2} \sin\phi' \tag{1}$$

and the Drucker-Prager criterion to define the plastic potential.

$$g = -\alpha I_1 + \sqrt{J_2} - \kappa \tag{2}$$

Where

$$\alpha = \frac{\tan\psi}{\sqrt{9 + 12\tan^2\psi}}, \quad \kappa = \frac{3c'}{\sqrt{9 + 12\tan^2\psi}} \tag{3}$$

In Equation 1 and 2, c', ϕ', and ψ are the effective cohesion, friction angle, and dilatancy angle, respectively. I_1 and J_2 are the first invariant of the effective stress, and the second invariants of the deviatoric stress, respectively. σ_1 and σ_3 are the major and minor principal effective stress, respectively.

The global safety factor of slope in SSR FEM identical to the one in limit equilibrium methods. The reduced strength parameters c'_F and ϕ'_F are defined by

$$c'_F = c'/F, \phi'_F = \tan^{-1}(\tan\phi'/F) \quad (4)$$

In SSRFEM, firstly, the initial stresses in slope are computed using the elastic finite element analysis. The vector of externally nodal forces consists of three parts: (1) surface force; (2) body force (total unit weight of soils); and (3) pore water pressure. Secondly, stresses and strains are calculated by the elasto-plastic finite element analysis, where the reduced shear strength criterion. The shear strength reduction factor F is initially selected to be so small, for example 0.01, that the shear strength is large enough to keep the slope in elastic stage. Stresses at some Gaussian points reach the yielding condition with the shear strength reduction factor F in Equation 3 increased gradually. When the stress at anyone Gaussian point reaches the yielding condition, the increment of the shear strength reduction factor will make stresses at more Gaussian points reach the yielding condition because of the residual force induced by the decrease in the shear strength soils.

The shear strength reduction factor F increases incrementally until the global failure of the slope reaches, which means that the finite element calculation diverges under a physically real convergence criterion. The lowest factor of safety of slope lies between the shear strength reduction factor F at which the iteration limit is reached, and the immediately previous one. The procedure described hereby can predict the factor of safety within one loop, and can be easily implemented in a computing code.

One of the main advantages of SSRFEM is that the safety factor emerges naturally from the analysis without the user having to commit to any particular from of the mechanism a priori. When the slope stability is evaluated with the effective stress method, the pore water pressure is usually predicted with the finite element seepage analysis or Biot's consolidation theory. If the same finite element mesh is used for the seepage or consolidation analysis and SSRFEM, the water pressure, predicted in the seepage or consolidation analysis, can be directly used in SSRFEM. This can simplify the slope stability analysis, and consider more accurately the influence of the seepage force.

2.2 Predominance of SSRFEM

This example was firstly reported by Arai and Tagyo and concerns a layered slope (Figure 3), where a layer of low resistance is interposed between two layers of higher shear strength. The material properties of various layers are listed in Table 1. Arai and Tagyo used the conjugate gradient method to search for the critical slip surface and used simplified Janbu's method to calculate the safety factor, and obtained the lowest safety factor of 0.405. This problem has also been examined by Sridevi and Deep using the random-search technique RST-2, by Greco using pattern search and Monte Carlo method, and by Malkawi et al. using

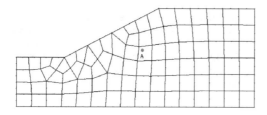

Figure 1. A point.

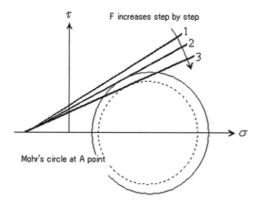

Figure 2. Mohr's circle at A point.

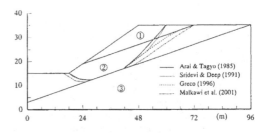

Figure 3. Slope in example.

Monte Carlo method (random walking). The slip surfaces located by various investigators are of significant difference. In this study, the problem is analyzed and Table 2 summarizes the results. Figure 4 shows the slip surface determined by SSRFEM. It is clear that the slip surface by SSRFEM is closer to that located by Greco.

Table 1. Material properties for example.

Layer	ϕ'(Deg)	c'(kN/m^2)	γ(kN/m^2)
1	12	29.4	18.82
2	5	9.8	18.82
3	40	294.0	18.82

Table 2. Minimum safety factors for example.

Method for safety factor	Method for slip surface	Safety factor
Arai and Tagyo (1985)		
Simplified Janbu	Conjugate gradient	0.405, 0.430 ※
Sridevi and Deep (1992)		
Simplified Janbu	RST-2	0.401, 0.423 ※
Greco (1996)		
Spencer	Pattern search	0.388
Spencer	Monte Carlo	0.388
Malkawai et al. (2001)		
Spencer	Monte Carlo	0.401
Rocscience Inc. (2002)		
Spencer	Random search	0.401
Simplified Janbu	Random search	0.410, 0.434 ※
Kim et al. (2002)		
Spencer	Random search	0.44
Lower-bound	Automatic	0.40
Upper-bound	Automatic	0.45
Cai et al. (2003)		
SSRFEM ($\psi = 0$)	Automatic	0.417
SSRFEM ($\psi = \phi$)	Automatic	0.423

※ Corrected the safety factor calculated by simplified Janbu.

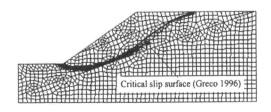

Figure 4. Slip surface by SSRFEM for example.

3 APPLICATION OF SSR-FEM

3.1 Embankment basal failure model

This paper compared limit equilibrium methods (Bishop's simplified method) and SSR-FEM ($\psi = \phi$) about embankment base failure of the support ground of the fill. Figure 5 shows the model embankment. The embankment was assumed 20.0 m in width of levee crown, 10.0 m in height (h1). In two levels of strata, the upper layer was soft, the lower layer was strong. The thickness (h2) of the upper layer was changed from 2.0 m, 4.0 m, 6.0 m, 8.0 m, to 10.0 m, to compare the results of both analytical methods.

Material properties is show in Table 3. The necessary matrix of a plasticity calculation was defined using Young's modulus and Poisson's ratio. The influence of these coefficients before the failure is great, but the influence on the total safety factor is very small (Zienkiewicz et al.1975, Griffiths and Lane 1999).

Therefore, we assumed Young's modulus of 2×10^4 kN/m^2, Poisson's ratio of 0.3 regardless of the soils. As for the embankment, $\varphi = 35.0$ degrees, the adhesive strength of the soft layer assumed 35 kN/m^2, the adhesive strength of the bearing stratum 100 kN/m^2.

3.2 Analysis result

3.2.1 h2 = 2.0 m

Figure 6(a) shows an analysis result of h2 = 2.0 m. In SSR-FEM, the slip surface occurred in soft layer base and was a non-circular arc. In contrast, circular slip was searched by the limit equilibrium method located at the embankment slope. The sliding mechanics was different. As for the safety factor, SSR-FEM gave 1.24, and the limit equilibrium method gave 1.21.

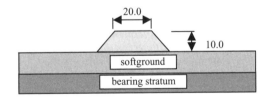

Figure 5. Embankment basal destruction model.

Table 3. Material Properties for Test Model.

Layer	ϕ (Deg)	c(kN/m^2)
Embankment	35.0	0.00
Soft ground	0.0	35.0
Bearing stratum	0.0	100.0

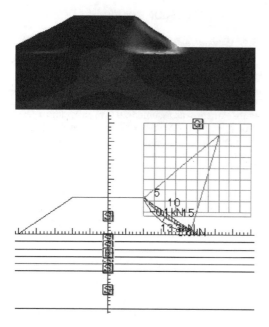

Figure 6(a). Result of SSRFEM and LEM h2 = 2.0 m.

3.2.2 h2 = 4.0 m
Figure 6(b) shows an analysis result of h2 = 4.0 m. In SSR-FEM and limit equilibrium method, the slip surface occurs in soft layer base and become a non-circular arc. As for the safety factor, SSR-FEM gave 1.16, and limit equilibrium method gave 1.22, and some differences occurred between the two methods.

3.2.3 h2 = 6.0 m
Figure 6(c) shows an analysis result of h2 = 6.0 m. In SSR-FEM and limit equilibrium method, the slip surface occurs in soft layer base. As for the safety factor, SSR-FEM gave 1.16, and LEM gave 1.22.

3.2.4 h2 = 8.0 m
Figure 6(d) shows an analysis result of h2 = 6.0 m. In SSR-FEM and limit equilibrium method, the slip surface occurs in soft layer base. As for the safety factor is both 1.10.

3.2.5 h2 = 10.0 m
Figure 6(e) shows an analysis result of h2 = 10.0 m. In SSR-FEM and limit equilibrium method, the slip surface occurs in soft layer base. As for the safety factor is both 1.08.

3.3 The comparison of the analysis result

Table 4 shows the results of both methods. Figure 7 shows the results between the safety factor and the ratio of thickness (h2) of the soft layer and embankment height (h1 = 10.0 m).

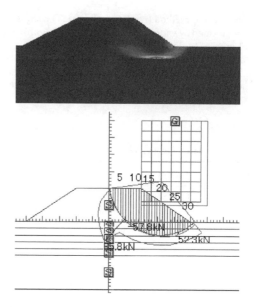

Figure 6(b). Result of SSRFEM and LEM h2 = 4.0 m.

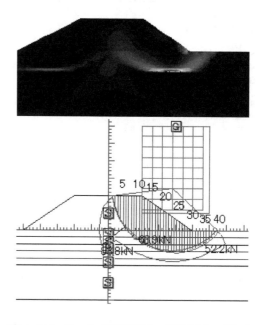

Figure 6(c). Result of SSR-FEM and LEM (h2 = 6.0 m).

Figure shows that both analysis results consistent with each other when the thickness of the soft layer is as above 6.0 m (h2/h1 ≥ 0.6).

However, in the situation that a soft layer is thin, a difference occurs in SSRFEM and limit equilibrium method.

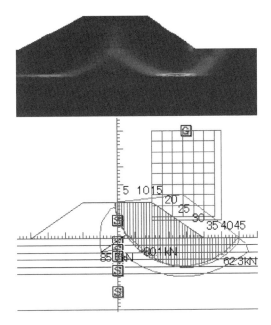

Figure 6(d). Result of SSR-FEM and LEM (h2 = 8.0 m).

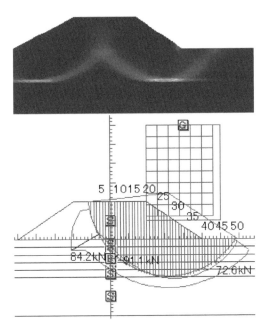

Figure 6(e). Result of SSR-FEM and LEM (h2 = 10.0 m).

When a soft layer is very thin h2 = 2.0 m (h2/h1 = 0.2), in SSR-FEM, bottom failure occurs. On the other hand, embankment slope failure occurs by limit equilibrium method. The calculated failure mechanics was different with each other.

Table 4. An analysis result table every layer thickness.

h2(m)	2.00	4.00	6.00	8.00	10.00
h2/h1	0.20	0.40	0.60	0.80	1.00
SSR	1.211	1.165	1.132	1.097	1.075
LEM	1.240	1.220	1.130	1.100	1.080

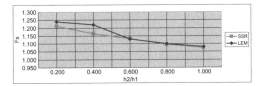

Figure 7. A figure of comparison.

Slip surface in the straight line occurs to go along the soft layer lower part from a result of SSR-FEM. In other words this case becomes the non-circular arc.

The limit equilibrium method assumes that it is circular arc sliding. As a result, it is thought that a safety factor searched the lowest circular arc sliding. However, the limit equilibrium method was not able to search the smallest safety factor and the sliding shape, because a safety factor of SSR-FEM is 1.21, and a safety factor of the limit equilibrium method is 1.24.

In h2 = 4.0 m, embankment base failure occurs both in SSRFEM and limit equilibrium method. The slip surface occurs in soft layer base and become a non-circular arc. As for the safety factor, the latter was 1.16 former 1.22. Among both, some differences occur. In h2 = 6.0 m are regarded as non-circular arc about the sliding shape in SSR-FEM about 8.0, 10.0 m as follows. But, the ratio that the tangent holds shrinks as a soft layer thickens. Therefore, for a safety factor, is approximately equal.

Table 5 shows the result that measured the length of the straight line of the non-circular arc in SSR-FEM.

According to the list, the length of the straight line gets longer so that the thickness of the soft layer increases, but the ratio for the layer thickness understands a thing becoming small. In other words the glide plane suffers from the shape that is near to an arc so that a soft layer is thick.

Figure 8 is a result of non-arc sliding by the limit equilibrium method that considered a straight line calculated by SSR-FEM about layer thickness 4.0 m. It was 1.17, and the safety factor almost equal to that of SSR-FEM.

Table 5. Length of the straight line every layer thickness.

h2(m)	2.00	4.00	6.00	8.00	10.00
L1 (m)	6.3	6.3	7.0	7.5	7.5
L1/h2	3.15	1.57	1.16	0.93	0.75

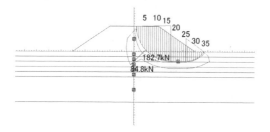

Figure 8. A non-arc by limit equilibrium method slips.

4 CONCLUSION

By this report, we compared an analysis result of SSR-FEM with the limit equilibrium method. With that in mind, we wanted to grasp such soil layer condition that sliding surface became the non-arc from the analysis result of SSR-FEM.

It is necessary to give the center of slip circular act by the judgment of the designer after having supposed sliding shape to be an arc by the limit equilibrium method. Therefore, the limit equilibrium method cannot cope with a non-arc even if the minimum safety factor was able to search the circular act.

SSR-FEM does not have to assume an examination condition, in a given condition, there is the big advantage that it is slippery, and can search shape to become the minimum safety factor.

When the thickness of the layer was thin, SSR-FEM could search a non-arc as sliding line associated with the minimum safety factor, and, in the example, the difference with the limit equilibrium method became great.

By this report, SSR-FEM analyses were performed to find the parameter which was ascertained whether the sliding shape was non-circular. The soft layer thickness was changed and a series of analysis was carried out.

Although the series of analyses was performed, a good parameter was not found to indicate the layer thickness of the soft layer under which non-arc sliding should occur. A future search should be performed.

At present, we advise that the sliding shape was calculated using SSR-FEM, the safety factor of the calculated slip surface was evaluated using limit equilibrium method wanted by design codes and standards.

REFERENCES

Arai, K. & Tagyo, K (1985): Detemination of noncircular slip surface giving the minimum factor of safaty in slop stability analysis. Soils and Foundation., The Japanese Geotechnical Society, Vol 1.25, No.1, pp.43–51.

Sridevi, B. & Deep, K. (1992): Application of globaloptimization technique to slop-stability analysis, Proc. 6th Inter. Symp. on Landslides, pp.573–578.

Greco, V. R. 1996. Efficient Monte Carlo technique for locating critical slip surface, J. Geotech. Eng. Div., ASCE, 122 (7): 517–525.

Malkawi, A.I.H., Hassan, W.F. & Sarma, S.K. (2001): Global search method for locating general slip surface using Monte Carlo technigues, J. Geotech. Geoenviron. Eng., Vol.127, No.8, pp.668–698.

Rockscience Inc. (2002): Verification manual for Slide, 2D limit equilibrium slop stability for soil and rock slopes, Version 4.0, Rocksicence Inc., Canada

Kim, J., Salgado, R. & Lee, J. (2002): Stability analysis of complex soil slopes using limit analysis, J. Geotch. Gepenvion. Eng., Vol. 128, No. 7, pp.547–557.

Cai, F. & Wakai, A. (2003): Use of finite element method in landslide analysis., The Japan Landslid Society, Vol. 40, No. 3. pp.76–80.

Griffiths, D.V. & Lane, P.A. (1999): Slope stability analysis by finite elements, Geotechnique, Vol. 49, No., pp. 387– 403.

Zienkiewicz, O.C., Humpheson, C. & Lewis, R.W. (1975): Associated and non-associated visco-plasticity and plasticity in soul mechanics, Geotechnique, Vol. 25, No. 4, pp.671–689.

Cai, F. & Ugai, K. (2001): The suggestion of the method to evaluate a total safety factor by elasto-plastic FEM of the slope having the measures mechanic. Soils and Foundation., Tokyo, 149 (4): 16–18 (in Japanese).

Back analysis based on SOM-RST system

H. Owladeghaffari
Department of Mining & Metallurgical Engineering, Amirkabir University of Technology, Tehran, Iran

H. Aghababaei
Faculty of Mining Engineering, Sahand University of Technology, Tabriz, Iran

ABSTRACT: This paper describes application of information granulation theory, on the back analysis of "Jeffrey mine- southeast wall-Quebec". In this manner, using a combining of Self Organizing Map (SOM) and Rough Set Theory (RST), crisp and rough granules are obtained. Balancing of crisp granules and sub rough granules is rendered in close-open iteration. Combining of hard and soft computing, namely Finite Difference Method (FDM) and computational intelligence and taking in to account missing information are two main benefits of the proposed method. As a practical example, reverse analysis on the failure of the southeast wall-Jeffrey mine—is accomplished.

1 INTRODUCTION

Back analysis is a reverse procedure, which is to solve the external load or partial material parameters, based on the known deformation and stresses at limited points and the partially known material parameters (Zhu & Zhao, 2004).

In most geotechnical engineering problems, it is often necessary to know the in situ field, material mechanical parameters, and even the mechanical model, by utilizing the monitored physical information such as deformation, strain, stress, and pressure during construction. The back analysis method, based on the required input physical deformation, can be divided in to deformation back analysis method, stress back analysis method and coupled back analysis method (Sakurai et al, 2003). The complex feature of rock mass, associated structures and the difficulty of interpretation of the interaction of excavation and rock mass are the main reasons to deploying of back analysis methods in rock engineering. Generally, two main procedures of back analysis methods have been pointed in the literature: inverse and direct approach (Cividini et al, 1989). In reverse method, the mathematical formalization is inverse of ordinary analysis. In this process, the number of monitored data is more than the unknown parameters, so that by utilizing of optimization techniques, the unknown parameters are obtained. The main advantage of such method is an independent spirit in regard of the iteration operations, which causes to reducing of calculation time, sensibly. Direct approach, is a process associated with the cycling optimization (aim is to minimize the fitness function). Application of direct approach, to evaluation of problem based non-linearity and employing of heuristic optimization methods such Ant Colony (ACO), Particle Swarm (PSO), and Simulated Annealing (SA) are two main results. So, back analysis based on interactions of hard computing and soft computing methods can be settled in direct approach. As an initial over view, 1–1 mapping methods can be supposed as hard computing such analytical, numerical or hybrid methods, whereas soft computing methods (SC), as main part of not 1–1 mapping methods, is a coalition of methodologies which are tolerant of imprecision, uncertainty and partial truth when such tolerance serves to achieve better performance, higher autonomy, greater tractability, lower solution cost. The principles members of the coalition are fuzzy logic (FL), neuro computing (NC), evolutionary computing (EC), probabilistic computing (PC), chaotic computing (CC), and machine learning (Zadeh, 1994).

Utilizing of soft computing methods besides hard computing methods, such support vector machine (Feng et al, 2004) and neural network (Pichler et al, 2003), in the reverse procedures have been experienced, successfully. Advancing of soft computing methods, under the general new approach, namely information granulation theory, has been opened new horizons on the knowledge discovery data bases. Information granules are collections of entities that are arranged due to their similarity, functional adjacency, or indiscernibility relation. The process of forming information granules is referred to as information granulation (Zadeh, 1997). There are many approaches to construction of IG, for example Self

Organizing Map (SOM) network, Fuzzy c-means (FCM), rough set theory (RST). The granulation level, depend on the requirements of the project. The smaller IGs come from more detailed processing. On the other hand, because of complex innate feature of information in real world and to deal with vagueness, adopting of fuzzy and rough analysis or the combination form of them is necessary. In fact, because of being the complex innate of crisp, rough or fuzzy attributes, in the natural world, extraction of fuzzy or rough information inside the crisp granules, can give a detailed granulation. To develop back analysis based new soft computing approaches and taking in to account of a mathematical tool for the analysis of a vague description of objects, as well as rough set theory, a combining of self organizing map-neural network-(SOM), and rough set theory (RST) associated with the hard computing methods is proposed.

The rough set theory introduced by Pawlak (1982, 1991) has often proved to be an excellent mathematical tool for the analysis of a vague description of object. The adjective vague, referring to the quality of information, means inconsistency, or ambiguity which follows from information granulation. The rough set philosophy is based on the assumption that with every object of the universe there is associated a certain amount of information, expressed by means of some attributes used for object description. The indiscernibility relation (similarity), which is a mathematical basis of the rough set theory, induces a partition of the universe in to blocks of indiscernible objects, called elementary sets, which can be used to build knowledge about a real or abstract world. Precise condition rules can be extracted from a discernibility matrix. The rest of paper has been organized as follow: a part from details, section 2 covers a brief on the RST and SOM and in section 3 the details of the proposed algorithm has been presented. Finally, we discuss an example in which we show how the proposed technique behaves.

2 RST & SOM

Here, we present some preliminaries of self organizing feature map—neural network- and rough set theory that are relevant to next section.

2.1 Self organizing map-neural network

Kohonen's (1987) SOM algorithm has been well renowned as an ideal candidate for classifying input data in an unsupervised learning way.

Figure 1 shows Kohonen's SOM topology, where two layers such as input layer and mapping layer are working together to organize input data into an appropriate number of clusters (or groups).

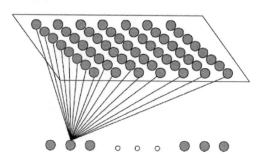

Input layer – each node is a vector representing N information

Figure 1. Kohonen's SOM.

For supervised learning, both input and output are necessary for training the neural network, while the unsupervised learning needs only the inputs. In unsupervised learning, neural network models adjust their weights so that input data can be organized in accordance with statistical properties embedded in input data. Kohonen's SOM includes two layers such as input layer and mapping layer, in the shape of a one or two-dimensional grid. The number of nodes in the input layer is equal to the number of features associated with input data. The mapping (output) layer acts as a distribution layer. Each node of the mapping layer or output node also has the same number of features as there are input nodes. Both layers are fully connected and each connection is given an adjustable weight. Furthermore, each output node of the mapping layer is restricted smaller distance around the cluster center.

For a Kohonen's SOM, suppose that the number of output nodes is m, the number of input nodes is n, and $w_i = (w_{i1}, w_{i2}, \ldots, w_{im})$ ($1 \leq i \leq m$) is the connection weight vector corresponding to output node i. w_i can be viewed as a center of the cluster i. whenever new input data $x = (x_1, x_2, \ldots, x_n)$ is presented to SOM during the training phase, the output value for output node i is computed by square of the Euclidean distance denoted by o_i between x and w_i, as shown in Equation 1:

$$o_i = (d_i)^2 = \| x - w_j \|^2 = \sum_{j=2}^{n}(x_j - w_{ij})^2,$$

$$1 \leq i \leq m \qquad (1)$$

If the node i^* satisfies Equation 2 then it is declared as a winner node:

$$(d_{i^*})^2 = \min o_i, \quad 1 \leq i \leq m \qquad (2)$$

Adjustable output nodes including the winning node i^* and its neighbor nodes are determined by the neighborhood size of the winning node $i^*(|\Omega_{i^*}|)$. Subsequently, connection weights of the adjustable nodes are all updated. The learning rule of SOM is shown in Equation 3:

$$\Delta w_{ij} = \eta(x_j - w_{ij}), i \in |\Omega_{i^*}|, \quad 1 \leq j \leq n \quad (3)$$

Where η is a learning rate. To achieve a better convergence, η and the neighborhood size of the winning node, should be decreased gradually with learning time.

To achieve a better convergence, and the neighborhood size of the winning node, should be decreased gradually with learning time. SOM has been successfully employed in different fields of applied science. Specially, in geomechanics, for example, in clustering of lugeon data (Shahriar & Owladeghaffari, 2007) and joint sets (Sirat & Talbot, 2001).

2.2 Rough set theory

The rough set theory introduced by Pawlak (Pawlak, 1991) has often proved to be an excellent mathematical tool for the analysis of a vague description of object. The adjective vague referring to the quality of information means inconsistency, or ambiguity which follows from information granulation. The rough set philosophy is based on the assumption that with every object of the universe there is associated a certain amount of information, expressed by means of some attributes used for object description. The indiscernibility relation (similarity), which is a mathematical basis of the rough set theory, induces a partition of the universe in to blocks of indiscernible objects, called elementary sets, which can be used to build knowledge about a real or abstract world. Precise condition rules can be extracted from a discernibility matrix.

An information system is a pair $S = \langle U, A \rangle$, where U is a nonempty finite set called the universe and A is a nonempty finite set of attributes. An attribute a can be regarded as a function from the domain U to some value set Va. An information system can be represented as an attribute-value table, in which rows are labeled by objects of the universe and columns by attributes. With every subset of attributes $B \subseteq A$, one can easily associate an equivalence relation I_B on U:

$$I_B = \{(x, y) \in U : \text{for every } a \in B, a(x) = a(y)\}$$

Then, $I_B = \cap_{a \in B} I_a$.

If $X \subseteq U$, the sets $\{x \in U : [x]_B \subseteq X\}$ and $\{x \in U : [x]_B \cap X \neq \varphi\}$, where $[x]_B$ denotes the equivalence class of the object $x \in U$ relative to IB, are called the B-lower and the B-upper approximation of X in S and denoted by $\underline{B}X$ and $\overline{B}X$, respectively. It may be observed that $\underline{B}X$ is the greatest B-definable set contained in X and $\overline{B}X$ is the smallest B-definable set containing X To constitute reducts of the system, it will be necessary to obtain irreducible but essential parts of the knowledge encoded by the given information system. So one is, in effect, looking for the maximal sets of attributes taken from the initial set (A, say) that induce the same partition on the domain as A. In other words, the essence of the information remains intact, and superfluous attributes are removed. Consider $U = \{x1, x2, \ldots, xn\}$ and $A = \{a1, a2, \ldots, an\}$ in the information system $S = \langle U, A \rangle$. By the discernibility matrix M(S) of S is meant an n*n matrix such that $c_{ij} = \{a \in A : a(x_i) \neq a(x_j)\}$

A discernibilty function fs is a function of m Boolean variables a1,..., am corresponding to attributes a1,..., am, respectively, and defined as follows:

$$f_s(a_1, \ldots, a_m)$$
$$= \wedge \{\vee(c_{ij}) : 1 \leq i, j \leq n, j \prec i, c_{ij} \neq \varphi\}$$

where $\vee(c_{ij})$ is the disjunction of all variables with a. $a \in c_{ij}$ (Pal et al, 2004). With such discriminant matrix the appropriate rules are elicited.

The existing induction algorithms use one of the following strategies:

a. Generation of a minimal set of rules covering all objects from a decision table;
b. Generation of an exhaustive set of rules consisting of all possible rules for a decision table;
c. Generation of a set of 'strong' decision rules, even partly discriminant, covering relatively many objects each but not necessarily all objects from the decision table (Greco et al, 2001). In this study we use first approach in combining with initial crisp granules, adaptively.

3 THE PROPOSED METHOD BASED ON SOM-RST SYSTEM AND HARD COMPUTING METHODS

Figure 2 shows a general procedure, in which the information granulation theory accompanies by a predefined project based rock engineering design. After determining of constraints and the associated rock engineering regards, the initial granulation of information as well as numerical (data base) or linguistic format is accomplished. Developing of modeling instruments based on IGs, whether in independent or associated shape with hard computing methods (such fuzzy finite element, fuzzy boundary element, stochastic finite element...) are new challenges in the current discussion.

Thus, one can employ such method as a new mythology in designing of rock engineering flowcharts. The main benefits are considering of the roles of the expert's experiences and educations, the missing or vagueness information and utilizing of advantages of

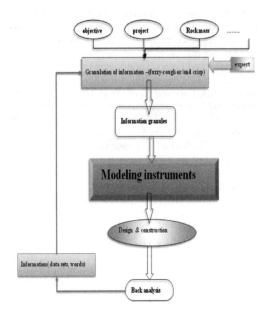

Figure 2. A general methodology for back analysis based on information granulation theory.

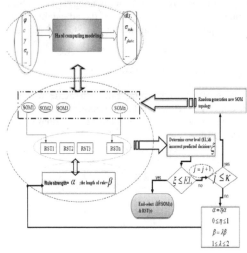

Figure 3. A proposed procedure based on hard computing and SOM-RST system.

soft and hard computing methods. A new advantages of the mentioned method, is to considering of the outward changes in *"words format"* and *"calculation with words and perception" (CWP)*.

Under mentioned methodology and to develop modeling instruments, authors have (are) proposed (proposing) new algorithms with neural networks, fuzzy logic (possibility theory), rough set theory, and meta heuristic optimization methods which are accompanied by close-open world idea. Random selection of initial precise granules can be set as "close world" assumption (CWA). But in many applications, the assumption of complete information is not feasible (CWA) nor realistic, and only cannot be used. In such cases, an open world assumption (OWA), where information not known by an agent is assumed to be unknown, is often accepted. The aim of open-world is to achieve complete knowledge of the universe by set of classified rules or by modifying rules. The close-open iteration accomplishes the balancing of crisp and sub rough granules by some random selection of initial granules and increment (or decreasing) of supporting rules, gradually. Figure 3, shows one of our proposed algorithms, apart from contributing of FL and free-derivative optimization methods. One may employ other shape of open world assumption's implementation. For instance to raise the quality of approximation by much more categories, but the optimum numbers of such scaling could be approached by an algorithm, so that an adaptive disceritization besides balancing of the initial granules with open world is accomplished.

First step is to collect data sets by using 1-1 modeling (here, hard computing methods as one of the main part of the direct modeling methods are selected). Applying of SOM as a preprocessing step and discretization tool is second process. For continuous valued attributes, the feature space needs to be discretized for defining indiscernibilty relations and equivalence classes. We discretize each feature in to three levels by SOM: *"low, medium, and high"*; finer discretization may lead to better accuracy at the cost of a higher computational load. Because of the generated rules by a rough set are coarse and therefore need to be fine-tuned, here, we have used the preprocessing step on data set to crisp granulation by SOM, so that extraction of best initial granules and then rough granules is rendered in close-open iteration. In open world iteration phase, we use a simple idea associated with human's cognition of the surround world: *"simplicity of rules"* (dominant to the problem) whether in numbers, applied operators or/and the length of rules. Here, we take in to account three main parameters in rules generation: length of rules, strength of object and number of rules. By setting the adaptive threshold error level and the number of close-open iteration, stability of the algorithm is guaranteed. After obtain best rules, the monitored data set as real decision parts are compared with the extracted rules and best conditional parts are picked up. These parts, associated with the reduced data sets by SOM, are the approximation of the wanted parameters as well as back analysis results.

4 AN ILLUSTRATIVE EXAMPLE: FAILURE ON THE SOUTHEAST WALL- JEFFREY MINE

This section describes how one can acquired approximated values of the effective parameters under a reverse analysis. To this aim and by using finite difference method, as well as FLAC (), the failure on the southeast wall in Jeffrey mine (Quebec) has been evaluated.

The Jeffrey mine is located at Asbestos; Québec As Asbestos fiber is mined from ultra basic host rocks dominated by periditotes, dunites and serpantites. The rocks mass is intersected by several thick shear zones and smaller scale discontinuous. Strength and deformability for the rock materials vary widely from very soft and weak to moderato stiff and strong rock. In 1970, the slope height was 180 m plus 60 m of overburden (clay, silt, & sand). Failure occurred in relatively fractured, serpantinized peridotite with a major shear zone present. Sliding in the overburden was first observed, followed by local wedge failure and finally a major slide in 1971, involving some 33 million ton of rock (figure 4). Failure was believed to have standard in the weak shear zone which then leads to failure of upper portions of the slope movement rates of the order of 1500 m/month were recorded. After this failure, some reactivation of the 1971 of the 1971 year failure occurred (Sjoberg, 1996).

By using finite difference method under *FLAC4*, as a hard computing method, almost 30 different models on the mentioned slope and by considering of Mohr-Cloumb plastic law have been constructed. The aim was to detect the material properties, while the slope movement rate satisfies the monitored records, approximately. Table 1 shows the part of the obtained results while 8 effective parameters changes.

The performance of $3 * 3$ SOM ($n_x = 3, n_y = 3$; Matrix of neurons $-n_x \cdot n_y$ determines the size of 2D SOM.) on the maximum velocity vectors of the tested data have been depicted in figures 5 and 6 (The discritization procedure using $3 * 3$ SOM on the attribute has been rendered). This step by accounting of different structures of SOM and in interaction with RST was iterated. The effective rough set procedure parameters were selected as follows: *minimum rule strength: 60%; maximum length of rule:2 and maximum number of rules: 5.* It must be noticed that decreasing of the rules is accomplished, gradually (depend on the error level).

Error level, here, is settled in true classified test data (percentage). Since total of analysis is low (30), we set $n = 1$, $EL = 80\%$ and $\kappa = 2$ *(maximum closed-iterations for any incensement of rules).*

After four closed-open iterations, algorithm satisfies conditions of the test data. Figure 7 illustrates the extracted rules which are matched with the recorded data by the proposed algorithm.

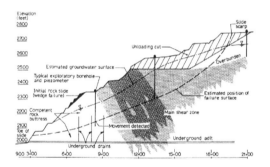

Figure 4. Cross section through the Jeffry mine showing the 1971 failure surface (Sjoberg, 1996).

Table 1. Subset of the obtained results by FLAC4- Cp (kg/cm^2): coherent of Periditotets; Sz: shear zone; T: tensile strength (kg/cm^2); TMD: total max. Displacement (m); MVV: max. Velocity vector (m/s).

No	Cp*10^5	Phi p	Cb	Phi b	C sz	Phi sz	T p	T b	TMD	MVV
1	2.00	25	3.00E + 05	35	500	15	428901.4	42844.4	2.41E-05	4.46E-14
2	3.20	35	2.20E + 05	25	1000	5	428901.4	42844.4	5.82E-06	9.86E-16
3	2.50	30	3.50E + 05	40	1500	10	428901.4	42844.4	2.66E-06	1.84E-14
4	2.75	25	3.75E + 05	35	1500	15	428901.4	42844.4	4.27E-06	1.68E-22
5	3.00	35	4.00E + 05	45	1000	20	428901.4	42844.4	2.52E-08	2.06E-22
6	2.00	25	3.00E + 05	35	1500	15	428901.4	42844.4	2.61E-06	4.24E-14
7	2.00	25	3.00E + 05	35	1500	15	1.00E + 06	6.80E + 05	2.20E-11	4.10E-22
8	2.00	25	3.00E + 05	35	1500	5	1.00E + 06	6.80E + 05	2.58E-06	1.73E-13
9	2.00	25	3.00E + 05	35	500	15	2.71E + 06	1.13E + 06	7.80E-16	1.92E-22
10	2.00	25	3.00E + 05	35	800	7	2.71E + 06	1.13E + 06	3.71E-08	4.95E-13
11	2.00	25	3.00E + 05	35	800	8	2.71E + 06	1.13E + 06	1.76E-07	1.70E-15
12	2.00	25	3.00E + 05	35	500	8	2.71E + 06	1.13E + 06	1.16E-06	8.14E-16

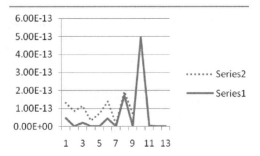

Figure 5. The performance of 3 ∗ 3 SOM on the maximum velocity vectors of the tested data (series 1: real; series 2: deduced data set).

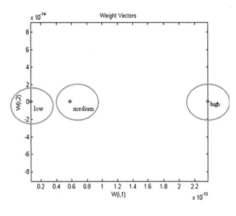

Figure 6. Discritization on the TMV in three clusters: low (3), medium (2), and high (1) by SOM.

```
Rule 1.   (cb<=220000.000000)   =>  (mvv at
most 1);
Rule 2.   (phib<=25.035000)   =>  (mvv at most
1);
Rule 3.   (csz<=999.790000)   =>  (mvv at most
1);
Rule 4.   (phisz<=5.035400)   &
(tb<=42844.000000)   =>  (mvv at most 1);
```

Figure 7. Part of the extracted rules by the proposed algorithm.

So, we could extract some best rules where the decision parts of them satisfy the monitoring rate of the slope. Thus, Application of such system under regular monitoring data set can be rendered. The main point in such data is to granulate the real decisions and compare with the deduced decision parts rules.

5 CONCLUSIONS

The role of uncertainty and vague information in geomechnaic analysis is undeniable feature. Indeed, with developing of new approaches in information theory and computational intelligence, as well as soft computing approaches, it is necessary to consider these approaches within and inside of current and conventional analysis, especially in geomechanic field. Under this idea and to fining best information granules which are picked off inside each other, close-open worlds (cycle) procedure has been proposed. Utilizing this algorithm in interaction with finite difference method and inferring back analysis results are main advantages of our method. So, we could obtain a becoming approximation of the parameters (in failure of southeast wall of Jeffry mine -1971). It must be notice that application of such process ensues a (or some) reduct set of the attributes where the reduct set(s) can be supposed as soft sensitive analysis on the model. Decreasing of time consuming and extractions of effective parameters, behind the core(s)-most effective parameters- of attributes (in condition parts) are other benefits of our model.

REFERENCES

Cividini, A., Jurino, L., Gioda, G. 1989. Some aspects of characterization problems in geomechanics, Int J Rock Mech Min Sci. 18: 487–503.

Feng, X.T., Zhao, H., Li, S. 2004. A new displacement back analysis to identify mechanical geomaterial parameters based on hybrid intelligent methodology. Int J Numer Anal Meth Geomch 28: 1141–1165.

Greco, S., Matarazzo, B., Slowinski, R. 2001. Rough sets theory for multi criteria decision analysis, European Journal of Operational Research 129: 1–47.

Kohonen, T. 1987. Self-organization and associate memory. 2nd ed. Springer —Verlag. Berlin .

Pal, K.S., Polkowski, L.A., Skowron, A. 2004. Rough-neural computing: techniques for computing with words. springer-verlag.

Pawlak, Z. 1982. Rough sets. Int J Comput Inform Sci 11: 341–356.

Pawlak, Z. 1991. Rough sets: theoretical aspects reasoning about data. Kluwer academic. Boston.

Pichler, B., Lackner, R., Mang, H.A. 2003. Back analysis of model parameters in geotechnical engineering by means of soft computing. Int J Numer Anal Meth Geomch 57: 1943–1978.

Sakurai, S., Akutagawa, S., Takeuchi, Shimiziu, N. 2003. Back analysis for tunnel engineering as a modern observational method. Tunneling & underground space technology 18 (2–3): 185–196.

Shahriar, K., Owladeghaffari, H. 2007. Permeality analysis using BPF, SOM & ANFIS. 1st Canada–U.S. Rock Mechanics Symposium (in print).

Sirat, M., Talbot, C.J. 2001. Application of artificial neural networks to fracture analysis at the Aspo HRL, Sweden:fracture sets classification. Int J Rock Mech Min Sci. 38: 621–639.

Sjoberg, J. 1996. Technical report: large scale slope stability in open pit mining. LULEA University. ISSN 0349-3571.

Zadeh, L.A. 1997. Toward a theory of fuzzy information granulation and its centrality in human reasoning and fuzzy logic. Fuzzy sets and systems 90: 111–127.

Zane, L.A. 1994. Fuzzy logic, neural networks, and soft computing. Commun ACM vol. 37: 77–84.

Zhu, W., Zhao, J. 2004. Stability analysis and modeling of underground excavation in fractured rocks. Elsevier.

Temporal prediction in landslides – Understanding the Saito effect

D.N. Petley
Department of Geography, University of Durham, Durham, UK

D.J. Petley
School of Engineering, University of Warwick, Coventry, UK

R.J. Allison
Department of Geography, University of Sussex, Sussex, UK

ABSTRACT: Of all natural hazards, landslides offer the best potential for prediction of the time of an event, and indeed successful predictions of the time of large-scale slope collapse have now been made for a number of landslides. In general these predictions are based upon the use of the approach first proposed by Saito in which the inverse of landslide strain rate is plotted against time, and the time of final failure determined by fitting a straight line to the data, and extrapolating the time at which $v^{-1} = 0$. In this paper, an investigation of Saito linearity is undertaken using laboratory testing. Using undisturbed samples of materials from the basal shear region of landslides, stress path testing has been undertaken. 38 mm diameter samples have been tested using the pore pressure reinflation (sometimes termed the field) stress path, in which the deviatoric and the total mean stresses are kept constant, whilst pore pressure is elevated. This triggers failure, during which the development of axial strain with time is monitored. The results clearly demonstrate that Saito linearity is the result of brittle deformation processes associated with the formation of the shear surface within the landslide mass. In landslides in which no shear surface forms, the linear trend is not observed. The data suggest that the state and rate dependent friction model, which has been implicated as a mechanism in some studies, is not the cause of this behavior. Our tests show that the linear phase of deformation in v^{-1}—time space is associated with stress concentration during the late stage of shear surface formation. These results provide enhanced understanding of the processes of shear surface formation, and the circumstances under which Saito linearity can be observed.

1 INTRODUCTION

1.1 Context

Landslides represent one of the most destructive of geological phenomenon, being responsible for over 100,000 fatalities in the period 1990–2006 (Petley et al. 2005a). The impact in terms of economic loss and damage is undoubtedly substantial, possibly reaching as much as 2–3% of GDP per annum in many mountainous developing countries (Brabb 1991). As a result, a host of techniques have been developed to permit the analysis of the stability of a slope and to allow suitable engineering designs to be devised to mitigate the hazard. These techniques have proven to be effective, such that the incidence of landslide fatalities in many developed countries is now low (Petley et al. 2005a). However, such approaches are generally expensive and are often environmentally insensitive. In mountainous developing countries, the provision of fully engineered solutions to the large of majority slope threats is just not viable given the limited resources, the abundance of dangerous slopes, and the likelihood of environmental degradation. In many of the more developed countries the use of engineering solutions is often opposed due to the incumbent damage to the environment. However, at the same time there is an increasing lack of tolerance of risk. Thus, there is an increasing emphasis being placed on the development of non-engineered approaches to slope problems, and in particular to the development of reliable and robust techniques that permit the prediction of the time of final failure of a slope, allowing the implementation of an effective warning system. In this paper, we seek to investigate the most widely used predictive tool, that first proposed by Saito (1965), in which the inverse of velocity is used to estimate the time of final failure.

1.2 The "Saito" approach to the prediction of failure

Saito (1965) first suggested that the pattern of change in the rate of movement of a landslide can be used to predict the time of final failure. This was based on an observation that the inverse of velocity (v^{-1}) defines a straight line when plotted against time in the period leading up to failure. The time of final failure can thus be predicted by extrapolating the linear trend to the time when $v^{-1} = 0$, i.e. the rate of displacement is infinite. This approach has been used successfully on a number of occasions (Zvelebil 1984; Suwa 1991; Hungr and Kent 1995; Rose and Hungr 2007). However, unfortunately there has also been a number of occasions when this methodology has proven to be unsuccessful (Hungr et al. 2005). The major problem is that to date the actual mechanisms that cause the linear trend in v^{-1}—time (t) space are not well understood. Use of this technique should only be attempted when these processes have been explained.

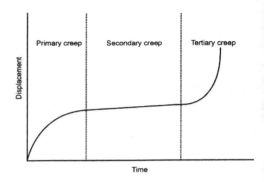

Figure 1. The generalized form of the three phase movement model.

2 EXPLORING THE SAITO METHOD

2.1 Landslide movement

It is well established that during the development of first time failure in landslides three distinct "phases" of movement (Fig. 1) are often. This pattern of movement is often referred to as "three phase creep", although in this context the term "creep" may be a misnomer. In the three phase model, the key types of movement are:-

1. *Primary* movement, which represents the initiation of displacement. Primary creep is characterized by an initially high, but declining, displacement rate.
2. *Secondary* movement, in which the rate of movement is low. The actual movement rate often fluctuates according to the stress state of the landslide, and may stop altogether at times.
3. *Tertiary* movement, in which the displacement rate increases rapidly as the landslide accelerates to final failure. It is during the tertiary movement phase that Saito linearity is observed.

The form of the three phase movement model is the result of the interaction of strain hardening and strain softening processes. During primary creep, strain hardening dominates, resulting in reductions in strain rate with time. In secondary creep, strain hardening and strain softening are in an approximate balance, although it seems likely that the evolution of this phase reflects a slow but progressive change in dominance from hardening to softening mechanisms. Thus, it is likely that the gradient of the displacement—time graph passes through an inflection point during this phase. In tertiary creep, strain softening processes dominate, although exactly how this occurs is contentious.

2.2 The applicability of the Saito approach

Petley et al. (2002) explored the general applicability of the Saito technique by examining the development of the final movement phase of a number of landslides. It was clearly shown that Saito linearity is not seen in all landslides. The linear trend was only observed in failures in which the controlling processes were brittle, i.e. in which final failure occurred through the formation of a new shear surface. Where the controlling processes were not brittle (i.e. they did not rely on the formation of a new shear surface), such as when the movement occurs as a result of ductile deformation, flow processes or sliding along a pre-existing, (comparatively) weak surface, a non-linear trend was noted in v^{-1}—t space. Thus, Petley et al. (2002) and Kilburn and Petley (2003) proposed that the Saito technique is only applicable to landslides in which a shear surface is being formed. This is entirely consistent with the model for "progressive failure" (i.e. the failure of a landslide without mobilizing the peak strength of the basal materials) proposed by Bjerrum (1967). Here, first time failure of landslides in materials that show distinctly different peak and residual strengths (i.e. that show brittleness) was envisaged to occur through the generation of a shear surface that grows due to stress concentration at the tip of the cracks. Shear surface growth could occur because the stress concentration exceeded peak strength. Such a model is consistent with the observations of Petley et al. (2002) in terms of the importance of brittle processes.

Within the geophysical community a different explanation is often advanced to explain three phase creep and Saito linearity. This is a concept of rate- and state-dependent friction, in which the resistance to movement is considered to reduce as strain rate increases (Helmstetter et al. 2003; Helmstetter and Sornette 2004). Such a model is not consistent with the model proposed by Bjerrum (1967). Clearly there is a need to investigate this behavior further both to determine which of the crack growth and the state- and rate-dependent friction models is correct and, more importantly, to provide a better basis for the application of the Saito technique to the provision of warning systems. As full scale monitoring of the failure of landslides is problematic, the best way to approach this is to use laboratory experimentation.

3 METHODOLOGY

To investigate the behavior of materials in the basal regime of a landslide we have undertaken an experimental study on samples of the Foxmould. This is a unit within an Upper Cretaceous greensand formation found on the south coast of England that consists of a yellowish-brown, weakly-bonded, glauconitic sandstone. In terms of landslides the significance of this material is that it is the unit containing the shear zone of the controlling rotational component of the Black Ven landslide complex in southern England (Brunsden 1969; Petley et al. 2005b). In this programme of tests we have undertaken 'pore pressure reinflation' (field stress path) tests, in which the sample is consolidated and then subjected to drained shear to a pre-determined stress/strain state. The key part of the test occurs at this point, when the sample is subjected to increasing pore pressure under constant total stress conditions. In these tests we have used 38 mm × 76 mm cylindrical undisturbed cores of Foxmould, prepared using a soils lathe. Testing has been undertaken in a standard GDS stress path cell (Menzies 1988).

In most of the tests reported here we have consolidated the sample along a K_0 stress path as this produces the most realistic stress and strain state (Table 1), although two PPR tests were undertaken after isotropic consolidation. We have undertaken five monotonic drained compression (MDC) experiments at a displacement rate of 0.05 mm/hour in order to define the peak and residual strength envelopes. Six pore pressure reinflation experiments have been completed, in which pore pressure was increased until the sample failed. In four of these (PPRU100, 200, 400, 1000) pore pressure reinflation was undertaken from the K_0 stress path. An additional two experiments (PPRD100 and PPRI100) were conducted in which the samples were isotropically consolidated to p′ = 100 kPa before being subjected to drained shear at 0.004 mm min^{-1} to a deviatoric stress of 105 kPa, at which point pore pressure was increased at 10 kPa h^{-1} until failure occurred.

Table 1. Experimental parameters for the Foxmould tests.

Experiment	State	Type	p'_0	PPR rate
K_0	Undisturbed	K_0	n/a	n/a
MDC100	Undisturbed	MDC	100	n/a
MDC200	Undisturbed	MDC	200	n/a
MDC400	Undisturbed	MDC	400	n/a
MDC600	Undisturbed	MDC	600	n/a
MDC1000	Undisturbed	MDC	1000	n/a
PPRU100	Undisturbed	PPR	100	10 kPa h^{-1}
PPRU200	Undisturbed	PPR	200	10 kPa h^{-1}
PPRU400	Undisturbed	PPR	400	10 kPa h^{-1}
PPRU1000	Undisturbed	PPR	1000	10 kPa h^{-1}
PPRD100	Remoulded	PPR	100	10 kPa h^{1}
PPRI100	Undisturbed	PPR	100	10 kPa h^{-1}

NB. PPR = pore pressure reinflation.

These latter two tests were briefly described in Petley et al. (2005a), although a detailed analysis was not provided.

4 RESULTS

4.1 Conventional tests

Clear peak and residual strength envelopes were defined by the MDC tests, both having a curved form in q-p′ space (Fig. 2). A Mohr's circle analysis showed that the Foxmould has a comparatively low low cohesion intercept (c′ = 35 kPa), but with a peak strength effective angle of internal friction of 34°, which is typical for a weakly-cemented sand. The residual strength envelope indicates no cohesion and an effective angle of friction of 29°. At failure each sample displayed a set of conjugate shear surfaces.

4.2 Pore pressure reinflation tests

Experiments PPRD100 and PPRU100 were undertaken to compare the behavior of an undisturbed and a remoulded sample of the Foxmould. At the start of both experiments the density of the samples was the same. Isotropic consolidation was used for these two samples as the different K_0 stress paths that the samples would follow would prevent comparison between them. The results of the PPR portion of these tests are shown in Fig. 3.

It is clear that the shape of the axial strain—time plot is quite different for these two samples. First, the undisturbed sample failed much later than did the remoulded sample. This is unsurprising given that it was much weaker than was the undisturbed

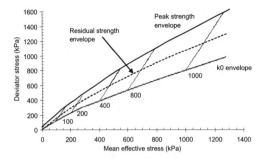

Figure 2. Stress path diagram for the MDC and K_0 stress path experiments, showing the peak, residual and K_0 lines.

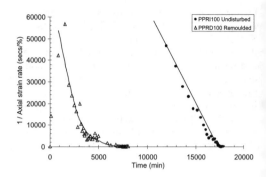

Figure 4. 1/Axial strain rate—time plot for the isotropically consolidated PPR tests.

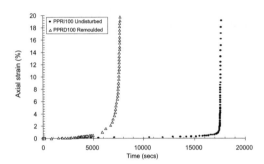

Figure 3. Axial strain—time plot for the isotropically consolidated PPR tests.

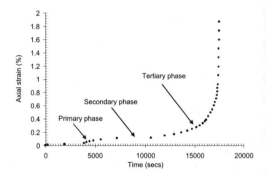

Figure 5. Axial strain—time plot for the first 2% of axial strain for the undisturbed isotropically consolidated PPR test, showing three phase creep.

sample. Note also though that the undisturbed sample fails over a much shorter time period than did the remoulded sample. This is evident in the Saito plot as the undisturbed sample shows the linear trend whilst the undisturbed sample shows the non-linear trend described by Petley et al. (2002) (Fig. 4). This suggests that the linear trend is associated with the presence of bonding in the sample, as this is the key difference between the samples, as suggested by Petley et al. (2005b).

Interestingly, in a plot of the development of axial strain against time, the undisturbed sample clearly shows the three phase movement pattern (Fig. 5). This is entirely consistent with the observed patterns of movement in real landslide systems, suggesting that this type of stress path testing is suitable for the simulation of landslide movements.

The stress paths for the four PPR experiments are shown in Fig. 6. In each case the deviatoric stress was successfully controlled until failure was initiated. In every case final failure occurred at a mean effective stress value slightly lower than that for the conventional tests—i.e. the experiments seem to imply that the PPR failure envelope has a slightly higher angle

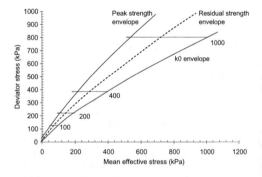

Figure 6. Stress path diagram for the PPR experiments, showing the peak, residual and K_0 lines. Sample numbers (value of p'_0) are indicated.

of internal friction than does the conventional failure envelope.

The pattern of axial strain against time for the 1000 kPa PPR experiment is shown in Figs. 7 and 8. Once again a clear primary movement phase is

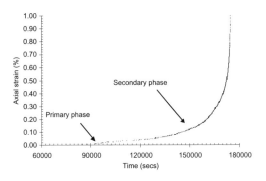

Figure 7. Axial strain against time (0–1% strain) for the 1000 kPa experiment showing the primary and secondary movement phases.

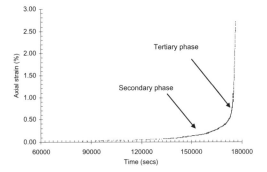

Figure 8. Axial strain against time (0 to 3% strain) for the 1000 kPa experiment showing the secondary and tertiary movement phases.

observed (Fig. 7), which is consistent with that seen in the isotropically consolidated sample described above. The final failure is rapid as per the isotropic sample, although in this case the point of onset of tertiary movement is not easy to determine (Fig. 8). Note that in this experiment, as with the others described here, failure occurs at a surprisingly low value of axial strain (<3%).

The Saito plot for this 1000 kPa test is interesting (Fig. 9). Here, the primary movement phase is clearly evident, and there appears to be a sharp termination into what is apparently the secondary phase. During this secondary movement period the Saito trend is non-linear, resembling the trend seen for the remoulded sample.

However, a detailed examination of the later stages of the experiment shows that the sample did appear to display a clear linear Saito trend (Figure 10). The onset of this linearity occurred at about 14,000 seconds, which is consistent with that indicated by Fig. 8.

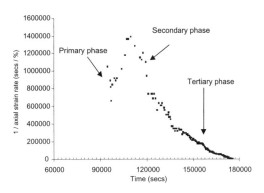

Figure 9. Saito plot for the 1000 kPa experiment showing the primary, secondary and tertiary movement phases. Note the non-linear form of the secondary movement phase.

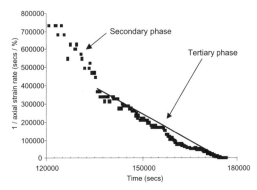

Figure 10. Saito plot for the 1000 kPa experiment showing the onset of the tertiary movement phase.

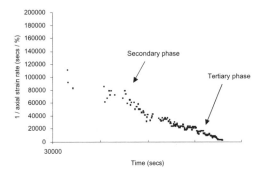

Figure 11. Saito plot for the 200 kPa experiment showing the secondary and tertiary movement phases. Note the non-linear form of the secondary movement phase and the lack of a discernable primary movement phase.

The other experiments showed only a secondary and tertiary movement phase, with no primary movement (Fig. 11). This is probably because this phase of movement could not be resolved by the equipment in use here, given the small levels of strain involved. In each case a very clear non-linear secondary creep phase was evident in the Saito plot, followed by a linear tertiary phase.

5 DISCUSSION

The experiments described here clearly demonstrate that Saito linearity in $v^{-1} - t$ space can be replicated experimentally in the stress path cell. These experiments have been undertaken on a material that is known to fail in a brittle manner through the generation of a conjugate pair of shear surfaces. In the experiments in which the samples had intact bonding the Saito plots clearly demonstrated a linear trend in $v^{-1} - t$ space. In the experiment in which the bonding had been destroyed through remoulding no linear trend was evident. In this case the sample displayed the same non-linear (exponential) trend that Petley et al. (2002 and 2005b) described. This appears to be consistent with the suggestion made by Petley et al. (2002) and Kilburn and Petley (2003) that it is brittle crack growth processes that are responsible for the linearity in $v^{-1} - t$ space first observed by Saito (1965). These results do not appear to be consistent with the model proposed by Helmstetter et al. (2003) and Helmstetter and Sornette (2004) that state- and rate-dependent friction is responsible for the observed behavior. In particular, the state- and rate-dependent friction model does not seem to be able to explain why linearity would be present in undisturbed samples but not present in samples that are identical except for the destruction of their interparticle bonds.

Thus, the Saito technique for the prediction of failure is valid where failure of the slope occurs through brittle processes (i.e. shear surface development). Where movement occurs through a different mechanism the Saito approach cannot be applied. On the other hand, if linearity is seen then it is clear that shear surface formation is occurring.

Bjerrum (1967) suggested that the process of progressive failure in a slope can be associated with the development of the shear surface of the landslide. This is a brittle cracking mechanism. The experimental data presented here support the Bjerrum (1967) model. Interestingly, with slight modification the Bjerrum model might also explain why Saito linearity is observed (Fig. 12). Here, Bjerrum's growing crack is modified by the influence of a tension crack (alternatively, the shear surface moving towards the surface will also achieve the same effect). This creates a zone of stress concentration on the unsheared material. As the area of this zone tends towards zero, the shear stress acting upon it tends to infinity, explaining the observed acceleration behavior.

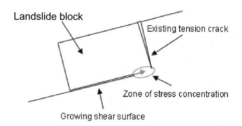

Figure 12. Modified Bjerrum (1967) model to account for Saito linearity.

6 KEY CONCLUSIONS

1. The Saito technique is valid when, and only when, the landslide is brittle;
2. The Bjerrum (1967) model of progressive failure is consistent with the Saito technique, and potentially provides an explanation for it through stress concentration.

REFERENCES

Bjerrum, L. 1967. Progressive failure in slopes of overconsolidated plastic clay and clay shales. *Journal of the Soil Mechanics and Foundations Division, Proceedings of the American Society of Civil Engineers* 93 (SM5): 2–49.

Brabb, E. 1991. The world landslide problem. *Episodes* 14: 52–61.

Brunsden, D. 1969. Moving cliffs in Black Ven. *Geographical Magazine* 41(5): 372–374.

Helmstetter, A., D. Sornette, J.-R. Grasso, J.V., Andersen, S. Gluzman, S. & Pisarenko, S. 2003. Slider-block friction model for landslides: implication for prediction of mountain collapse. *Journal of Geophysical Research*, 108(B2): 409–432.

Helmstetter, A. & Sornette, D. 2004. Slider block friction model for landslides; application to Vaiont and La Clapiere landslides. *Journal of Geophysical Research*, 109(B2): 210–225.

Hungr, O. & Kent, A. 1995. Coal mine waste dump failures in British Columbia, Canada. *Landslide News*, 9: 26–28.

Hungr, O., Corominas, J., and Eberhardt, E. 2005. Estimating landslide motion mechanism, travel distance and velocity. In: Hungr, O., Fell, R., Couture, R., and Eberhardt, E. *Landslide Risk Management*. Taylor and Francis, London, pp. 99–128.

Kilburn, C.J. & Petley, D.N. 2003. Forecasting giant, catastrophic slope collapse: lessons from Vajont, Northern Italy. *Geomorphology*, 54(1–2): 21–32.

Menzies, B.K. 1988. A Computer Controlled Hydraulic Triaxial Testing System. In: Donaghue, R.T., Chaney, R.C. and Silver, M. (eds), *Advanced Triaxial Testing of Soil and Rock, ASTM STP 977*, American Society for Testing and Materials, Philadelphia, USA, pp. 82–94.

Petley, D.N., Bulmer, M.H.K. & Murphy, W. 2002. Patterns of movement in rotational and translational landslides. *Geology*, 30(8), 719–722.

Petley, D.N., Dunning, S.A. & Rosser, N.J. 2005a. The analysis of global landslide risk through the creation of a database of worldwide landslide fatalities. In: Hungr, O. Fell, R., Couture, R., and Eberhardt, E. *Landslide Risk Management*, A.T. Balkema, Amsterdam, 367–374.

Petley, D.N., Higuchi, T., Petley, D.J., Bulmer, M.H. & Carey, J. 2005b. The development of progressive landslide failure in cohesive materials. *Geology* 33(3): 201–204.

Rose, N.D. & Hungr, O. 2007. Forecasting potential rock slope failure in open pit mines using the inverse-velocity method. *International Journal of Rock Mechanics and Mining Sciences* 44: 308–320.

Saito, M. 1965. Forecasting the time of occurrence of a slope failure. *Proceedings of the 6th International Conference on Soil Mechanics and Foundation Engineering* 2: 537–541.

Suwa, H. 1991. Visually observed failure of rock slope in Japan. *Landslide News* 5: 8–10.

Zvelebil, J. 1984. Time prediction of a rockfall from a sandstone rock slope. In: *Proceedings of the IVth International Symposium on Landslides* 3: 93–95.

3D landslide run out modelling using the Particle Flow Code PFC^{3D}

R. Poisel & A. Preh
Institute for Engineering Geology, Vienna University of Technology, Vienna, Austria

ABSTRACT: Rockfalls are modelled as the movements of single rock blocks over a surface or as the movement of a viscous mass over a surface (e.g. DAN). In reality a mass of discrete, interacting rock blocks is moving downslope. Thus the program PFC (Particle Flow Code) based on the Distinct Element Method was modified in order to model rock mass falls realistically in 3 dimensions based on physical relations. PFC models the movement and interaction of circular (2D) or spherical (3D) particles and wall elements using the laws of motion and of force-displacement. In the course of the calculation the contacts between particles and particles or particles and walls are detected automatically. The particles may be bonded together at their contact points, and the bondage can break due to an impact. For realistic modelling of the run out a viscous damping routine in case of a particle—wall contact was introduced. Numerical drop tests and back analyses of several rock mass falls provided appropriate damping factors. Thus, the movement types bouncing, sliding, rolling and free falling of single rock blocks and the interaction between the blocks occurring in a rock mass fall can be modelled realistically by using the adapted code of PFC. The application of this method is demonstrated by the example of Aaknes (Western Norway).

1 INTRODUCTION

Landslide run outs are modelled as the movements of single rock blocks over a surface or as the movement of a viscous mass over a surface (e.g. DAN). In reality a mass of discrete, interacting rock blocks is moving downslope. Thus the program PFC (Particle Flow Code) based on the Distinct Element Method was modified in order to model landslide run outs realistically in 3 dimensions based on physical relations.

PFC models the movement and interaction of circular (2D) or spherical (3D) particles and wall elements using the laws of motion and of force-displacement. In the course of the calculation the contacts between particles and particles or particles and walls are detected automatically. The particles may be bonded together at their contact points, thus modelling a solid, and the bondage can break due to an impact. Thus, PFC can simulate not only failure mechanisms of rock slopes, but also the run out of a detached and fractured rock mass (Poisel & Roth 2004).

Rock mass falls can be modelled as an "All Ball model" and as a "Ball Wall model". An "All Ball model" simulates the slope as an assembly of balls bonded together. The simulation shows the failure mechanism of the slope due to gravity (Poisel & Preh 2004). After detachment of the moving mass, the run out is modelled automatically.

In the "Ball Wall model" the underlying bedrock is simulated by linear (2D) and planar (3D) wall elements (Roth 2003). Therefore, an estimate or a model of the failure mechanism of the slope (Preh 2004) and of the detachment mechanism is needed as an input parameter. However, in the "Ball Wall model" the detached mass can be modelled, using more and smaller balls with the same computational effort in order to approach reality better.

2 RUNOUT RELEVANT PARAMETERS

According to observations in nature, several kinds of movements of the rock fall process (Broilli 1974) have to be distinguished during the computation (Bozzolo 1987):

- free falling,
- bouncing,
- rolling and
- sliding.

In order to achieve an appropriate simulation of these different kinds of movements by PFC, some modifications have been necessary using the implemented programming language Fish.

2.1 Free falling

In order to model the free falling of blocks, neither the acceleration nor the velocity (ignoring the air resistance) is to be reduced during fall as a consequence of mechanical damping.

PFC applies a local, non-viscous damping proportional to acceleration to the movement of every single particle as a default. The local damping used in PFC is similar to that described by Cundall (1987). A damping-force term is added to the equations of motion, so that the damped equations of motion can be written

$$F_{(i)} + F_{(i)}^d = M_{(i)} A_{(i)}; \quad i = 1 \ldots 6 \quad (1)$$

$$M_{(i)} A_{(i)} = \begin{cases} m\ddot{x}_{(i)} & \text{for } i = 1 \ldots 3; \\ I\dot{\omega}_{(i-3)} & \text{for } i = 4 \ldots 6 \end{cases} \quad (2)$$

where $F_{(i)}$, $M_{(i)}$, and $A_{(i)}$ are the generalized force, mass, and acceleration components, I is the principal moment of inertia, $\dot{\omega}$ is the angular acceleration and $\ddot{x}$ is the translational acceleration; $F_{(i)}$ includes the contribution from the gravity force; and $F_{(i)}^d$ is the damping force

$$F_{(i)}^d = -\alpha |F_{(i)}| \text{sign}(v_{(i)}) \quad i = 1 \ldots 6 \quad (3)$$

expressed in terms of the generalized velocity

$$v_{(i)} = \begin{cases} \dot{x}_{(i)} & \text{for } i = 1 \ldots 3; \\ \omega_{(i-3)} & \text{for } i = 4 \ldots 6. \end{cases} \quad (4)$$

The damping force is controlled by the damping constant α, whose default value is 0.7 and which can be separately specified for each particle.

This damping model is the best suited for a quick calculation of equilibrium. There arises, however, the disadvantage of the movements of the particles being damped as well. Therefore, the local damping has been deactivated for all kinds of particle movements.

2.2 Bouncing

Elastic and plastic deformations occur in the contact zone during the impact of a block. Both the kinetic energy of the bouncing block and the rebound height are reduced by the deformation work. The reduction of the velocity caused by the impact is modelled with the help of a viscous damping model integrated in PFC.

The viscous damping model used in PFC introduces normal and shear dashpots at each contact (Fig. 1). A damping force, D_i ($i = n$: normal, s: shear), is added

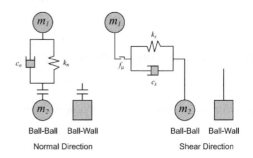

Figure 1. Viscous damping activated at a contact with the linear contact model (Itasca 1999).

to the contact force, of which the normal and shear components are given by

$$D_i = C_i \cdot |V_i| \quad (5)$$

where C_i ($i = n$: normal, s: shear) is the damping constant, V_i ($i = n$: normal, s: shear) is the relative velocity at contact, and the damping force acts to oppose motion. The damping constant is not specified directly; instead, the critical damping ratio β_i ($i = n$: normal, s: shear) is specified, and the damping constant satisfies

$$C_i = \beta_i \cdot C_i^{crit} \quad (6)$$

where C_i^{crit} is the critical damping constant, which is given by

$$C_i^{crit} = 2 m \omega_i = 2\sqrt{mk_i} \quad (7)$$

where ω_i ($i = n$: normal, s: shear) is the natural frequency of the undamped system, k_i ($i = n$: normal, s: shear) is the contact tangent stiffness, and m is the effective system mass.

In rock fall programs, the rebound height of blocks touching the bedrock is calculated using restitution coefficients. The restitution coefficient R_i ($i = n$: normal, s: shear) is defined as the ratio of the contact velocity before and after the impact and can be defined as

$$R_i = \frac{v_i^f}{v_i^i} \quad (8)$$

where v_i^f ($i = n$: normal, s: shear) is the velocity of the block after impact and v_i^i ($i = n$: normal, s: shear) is the velocity of the object before impact. The relation between the restitution coefficient R_i and the critical damping ratio β_i can be estimated by simulating drop tests (Preh & Poisel 2007).

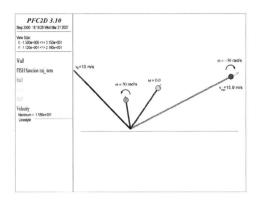

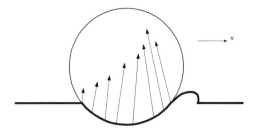

Figure 3. Deformation of the surface and distribution of contact stresses.

Figure 2. Rebound angle influenced by the particle spin (black line—rebound course; green, yellow and red balls—particle position after rebound).

Spin has an impact on both the direction and the velocity of the rebounding block. Therefore, it is essential to consider the spinning when modelling the run outs of rock falls. PFC determines the motion of each single particle by the resultant force and moment vectors acting upon it, and describes it in terms of the translational motion of a point in the particle and the rotational motion of the particle (Equations 1 and 2). Figure 2 depicts the flight trajectories of three particles bouncing at different spins.

Furthermore, with PFC the interaction of friction and spin is considered, since the influence of the spin increases with the increase of frictional resistance.

By modelling rock mass falls, it was shown to be necessary to distinguish between ball-ball contacts and ball-wall contacts. This was done by using the programming language Fish.

2.3 Rolling

The most important run out relevant effect is rolling resistance, because it is known that pure rolling of blocks in the model leads to more extensive run outs than observed in nature.

The rolling resistance is caused by the deformation of the rolling body and/or the deformation of the ground (Fig. 3) and depends strongly on the ground and the block material.

Due to these deformations, the distribution of contact stresses between the ground and the block is asymmetric (Fig. 4). Replacing the contact stresses by equivalent static contact forces results in a normal force N, which is shifted forward by the distance of c_{rr}, and a friction force F_{rr}, opposing the direction of the movement.

The deceleration of the angular velocity caused by the rolling resistance is calculated using conservation

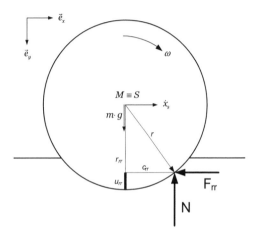

Figure 4. Calculation of the rolling resistance.

of translational momentum (Equation 9) and angular momentum (Equation 10).

$$m \cdot \ddot{x}_s = -F_{rr} \qquad (9)$$

$$-I \cdot \dot{\omega}_{rr} = M_{rr}, \quad I_{sphere} = \frac{2}{5} \cdot m \cdot r^2 \qquad (10)$$

where M_{rr} is the resulting moment caused by the rolling resistance, I is the principal moment of inertia and ω_{rr} is the angular deceleration.

The kinematic link is established by the condition of pure rolling (Equation 11).

$$\ddot{x}_s = \dot{\omega} \cdot r \qquad (11)$$

The angular acceleration is defined by a finite difference relation in order to express the increment of the angular velocity per time increment (Equation 12). Thus, the friction force F_{rri} is defined by the conservation of momentum

$$F_{rr} = -m \cdot \frac{\Delta \omega_{rr}}{\Delta t} \cdot r \qquad (12)$$

Equation 10 and equation 12 yield

$$-\frac{2}{5} \cdot m \cdot r^2 \cdot \frac{\Delta \omega_{rr}}{\Delta t} = F_{rr} \cdot r_{rr} - N \cdot c_{rr}$$

$$-\frac{2}{5} \cdot m \cdot r^2 \cdot \frac{\Delta \omega_{rr}}{\Delta t} = -m \cdot \frac{\Delta \omega_{rr}}{\Delta t} \cdot r \cdot r_{rr}$$
$$- m \cdot g \cdot c_{rr}. \qquad (13)$$

Therefore, the angular deceleration is

$$\Delta \omega_{rr} = \frac{-g \cdot c_{rr}}{r \cdot \left(r_{rr} - \frac{2}{5} \cdot r\right)} \cdot \Delta t$$

$$r_{rr} = \sqrt{r^2 - c_{rr}^2}. \qquad (14)$$

The rolling resistance is implemented by adding the calculated increment of the angular velocity to the angular velocity calculated automatically by PFC at every time step (Equation 14).

$$\omega_i^{(t)} = \omega_i^{(t)} + \Delta \omega_{rr,i} \qquad (15)$$

According to these considerations, the rolling resistance is an eccentricity c_{rr} or sag function u_{rr}. The deeper the block sags, the greater is the rolling resistance $\Delta \omega_{rr}$. In classical mechanics, the rolling resistance is a function of the ratio of the eccentricity c_{rr} to the radius r.

$$\mu_r = \frac{c_{rr}}{r} [-] \qquad (16)$$

This means that spherical blocks of different sizes have the same run out for the same rolling resistance coefficient.

In nature, however, it can be observed that large blocks generally have a longer run out than smaller ones. Therefore, according to the damping model described, the run out is calibrated by the sag u_{rr}.

Calculations carried out by PFC, using the model of rolling resistance described above and modelling a detached rock mass as an irregular assembly of particles of two different sizes ($r_1 = 0.8$ m, $r_2 = 1.6$ m) employing the same sag of $u_{rr} = 25$ cm for both particle sizes have shown that the larger particles have a longer run out than smaller ones and that within the deposit mass the smaller particles (r = 0.8 m) rest at the bottom and the larger particles (r = 1.6 m) at the top (Preh & Poisel 2007). This model behaviour corresponds closely to observations in nature.

2.4 Sliding

Sliding is calculated by the slip model implemented in PFC without any further adaptation.

3 THE AAKNES ROCK SLOPE FAILURE

Large rock slope failures are common events in the inner fjord areas of Western Norway and represent one of the most serious natural hazards in Norway. Rock avalanches and related tsunamis (Harbitz et al. 1993) have caused serious disasters and during the last 100 years more than 170 people have lost their lives in Western Norway. The Tafjord disaster occurred in 1934 when 3 million m³ rock mass dropped into the fjord. The tsunami generated by the avalanche reached a maximum of 62 m above sea level, several inhabited villages along the fjord were destroyed and 41 people were killed (Blikra et al. 2005).

The unstable rock slope at Aaknes (Fig. 5) is situated in a steep mountain slope in Storfjorden and is built up by gneisses with an overall dipping foliation parallel to the slope (Tveten et al. 1998). The slide planes are probably following weak zones along the foliation planes. Geophysical data from 2D resistivity, refraction and reflection seismics and penetrating radar indicate that the slide is covering an area of some 800.000 m² maximum and that the thickness of the unstable area is between 40 and 140 m (Blikra et al. 2007).

The Aaknes/Tafjord project was initiated in 2004 with the aim of investigating, monitoring and providing early warning of the unstable areas at Aaknes and Hegguraksla in Tafjord in More og Romsdal County. The monitoring system includes extensometers, lasers, GPS, a total station, ground based radar and borehole instrumentation (Blikra et al. 2007). The displacement velocities are in the order of 3–10 cm/year at present.

Figure 5. The Aaknes rock slope. Unstable area marked by dashed line. Photo: Th. Sausgruber.

The geometry and structure of the failure is complex and the instable area seems to be composed of several individual blocks.

Based on the relatively frequent slide events documented in the fjord areas, it has been estimated that a flank collapse in this part of the fjord system in the order of 1–8 million m³ may have a probability of less than 1 event /1.000 years (Blikra et al. 2005).

3.1 Numerical investigations of the Aaknes run out using PFC

The mapping of geological structures dividing the moving area into separate blocks (Blikra et al. 2005), displacement vectors showing different displacement directions of these blocks as well as the wavy morphology of sliding planes of already occurred slides in the surrounding area led to the conclusion, that the slope failure will take place rather as a fall of a mass of blocks than as a slide of a coherent mass. Thus PFC was chosen to simulate the Aaknes rock slope failure numerically.

The digital terrain models (DTMs) provided by the Geological Survey of Norway were used to generate the wall elements simulating the detachment and the terrain surface of the smallest scenario (Fig. 6) and the detached rock volume (some 5 million m³) was modelled by 2583 particles (balls) with $r_{min} = 5$ m and $r_{max} = 7.5$ m (Fig. 7).

First it was assumed that the moving mass is fractured completely due to the sliding displacements before the run out starts. Thus the run out process of completely unbonded particles was started by deleting the wall elements above the detached rock volume using the PFC routine described above.

Figure 8 shows the position of the unbonded particles at the moment the first particle hits the water

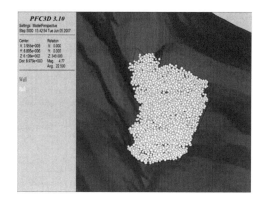

Figure 7. Detached rock volume modelled by 2583 particles.

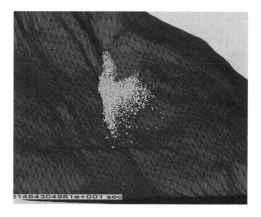

Figure 8. Unbonded material, distribution of particles at the moment the first particle hits the water surface.

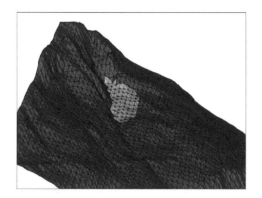

Figure 6. Detachment and terrain surface of the smallest scenario modelled by wall elements.

surface, figure 9 the accumulated mass of particles hitting the water surface over time.

In a second approach it was assumed that the moving mass is composed of several blocks built up by an assemblage of particles. Thus bonds with a shear strength of 1 GN and with a tension strength of 1 GN were introduced at the particle contact points. These bonds partly broke during the following run out.

Figure 10 shows the position of the initially completely bonded particles at the moment the first particle hits the water surface, figure 11 the accumulated mass of particles hitting the water surface over time.

Comparison of Figures 8–11 reveals completely different distributions in space as well as in time. The impacts of the unbonded material are restricted more or less to the channel on the orographic right side of the moving mass while the initially completely bonded material runs down distributed over the whole width

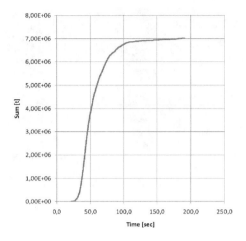

Figure 9. Unbonded material, accumulated mass of balls hitting the water surface over time.

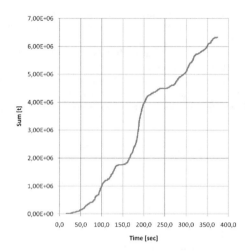

Figure 11. Bonded material, accumulated mass of balls hitting the water surface over time.

Figure 10. Bonded material, distribution of particles at the moment the first particle hits the water surface.

of the sliding mass. The run out of the unbonded material lasts only some 100 seconds, the run out of the initially bonded material some 400 seconds.

These results were transformed into a smoothed particle hydrodynamics (SPH) based depth integrated numerical model with coupling between the solid and the fluid phases by Pastor (2007) in order to simulate the tsunami caused by the rock mass run out.

4 CONCLUSIONS

With the help of the adapted PFC code it is possible to create a mechanically correct model of rock mass falls in 3D. The adapted code was used to model the run out of the Aaknes rock slide in Storfjorden (Western Norway) by unbonded and by bonded particles. The simulations showed different particle distributions in space and time which are expected to cause different tsunamis. Thus investigations of the sliding rock mass disintegration as a function of sliding displacements are essential.

REFERENCES

Blikra, L.H., Longva, O., Harbitz, C.B. & Lovholt, F. 2005. Quantification of rock avalanche and tsunami hazard in Storfjorden, western Norway. In K. Senneset, K. Flaate, J.O. Larsen (eds.), *Landslides and avalanches. Proc. of the 11th Int. Conf. and Field Trip on Landslides (ICFL), Norway, September 1–10, 2005*: 57–63. London: Taylor and Francis.

Blikra, L.H., Jogerud, K., Hole, J. & Bergeng, T. 2007. Åknes/Tafjord prosejktet. Status og framdrift for overvaking og beredskap. Report 01–2007 Åknes Tafjord prosjektet (in Norwegian).

Bozzolo, D. 1987. Ein mathematisches Modell zur Beschreibung der Dynamik von Steinschlag. *Dissertation Nr. 8490 an der ETH Zürich*.

Broilli, L. 1974. Ein Felssturz im Großversuch. *Rock Mechanics*, Suppl. 3: 69–78.

Cundall, P.A. 1987. Distinct Element Models of Rock and Soil Structure. In E.T. Brown (ed.), *Analytical and Computational Methods in Engineering Rock Mechanics*: Ch. 4, 129–163. London: Allen & Unwin.

Harbitz, C.B., Pedersen, G. & Gjevik, B. 1993. Numerical simulations of large water waves due to landslides. *J. of Hydraulic Engineering* 119(12): 1325–1342.

Hoek, E. 1987. Rockfall—A program in basic for the analysis of rockfalls from slopes. Dept. Civil Eng., University of Toronto, Toronto.

Itasca 1999. PFC^{2D} (Particle Flow Code in 2 Dimensions) User's Guide. Itasca Consulting Group, Inc., Minneapolis.

Pastor, M. 2007. Numerical simulation of the tsunamis caused by the Aaknes rock slide due to run outs modelled by PFC. Unpublished report.

Poisel, R. & Preh, A. 2004. Rock slope initial failure mechanisms and their mechanical models. *Felsbau* 22: 40–45.

Poisel, R. & Roth, W. 2004. Run Out Models of Rock Slope Failures. *Felsbau* 22: 46–50.

Preh, A. 2004. Modellierung des Verhaltens von Massenbewegungen bei großen Verschiebungen mit Hilfe des Particle Flow Codes. *PhD Dissertation,* Inst. for Eng. Geology, Vienna University of Technology.

Preh, A. & Poisel, R. 2007. 3D modelling of rock mass falls using the Particle Flow Code PFC^{3D}. *Proceedings of the 11th Congress of the International Society for Rock Mechanics, Lisbon, July 9–13, 2007. Specialized Session S01—Rockfall—Mechanism and Hazard Assessment.*

Roth, W. 2003. Dreidimensionale numerische Simulation von Felsmassenstürzen mittels der Methode der Distinkten Elemente (PFC). *PhD Dissertation,* Inst. for Eng. Geology, Vienna University of Technology.

Spang, R.M. & Rautenstrauch, R.W. 1988. Empirical and mathematical approaches to rockfall protection and their practical applications. *Proceedings of the 5th International Symposium on Landslides, Lausanne, 1988:* Vol. II, 1237–1243.

Tveten, E., Lutro, O. & Thorsnes, T. 1998. Bedrock map Alesund. 1:250,000. Geological Survey of Norway.

Double-row anti-sliding piles: Analysis based on a spatial framework structure

Tonghui Qian & Huiming Tang
Faculty of Engineering, China University of Geosciences, Wuhan, Hubei, China

ABSTRACT: Since the single-row anti-sliding piles could not able to provide an efficient bracing structure under larger sliding-force, the double-row anti-sliding piles, which have the advantages of less displacement in the top of the piles and large anti-force, can be used to reduce the pile deformation. However, the existing calculation models of double-row piles ignored the interaction features of piles, beams and soils, and the calculation results based on the models are not accurate enough. In this paper, a spatial force model of the double-row anti-sliding piles was presented. When taking double-row anti-sliding piles as a single-layer multi-span frame affected by piles, beams and soils, the spatial synergetics interaction between top ring beam and linking beam, as well as the affection of soil-arch on the frame were analyzed. A deformation equation was established according to pile border condition, continuous deformation, static force balance and deformation coordination relationship between top ring beam and pile. An analysis program was developed on the basis of FEM and the theory of Winkler Elastic Foundation Beam. Finally, the model was applied in the practice of Three-Gorge Reservoir Prevention and Treatment project, and the result shows it is suitable for similar engineering.

1 INSTRUCTION

Anti-sliding piles are important retaining structure for landslide prevention. Since single-row anti-sliding piles is not capable of ensuring good anti-sliding effect under larger sliding-force, the double-row anti-sliding piles based on spatial framework structure, which have the advantages of less displacement in the top of the piles and large anti-force, are developed as a new retaining structure, and can be used to reduce the piles deformation. Due to the complicated stress mechanism and imperfect design and calculation, researches about double-row anti-sliding piles in large landslide prevention are just limited. Xu (1998, 2004) calculated the rigid frame anti-sliding piles by Finite Difference Method according to the theory of elastic foundation beam. Xiong (2002) supposed double-row piles as separate cantilever piles and discussed the distribution of anti-sliding push-force on different piles. Zhou (2005) presented an action model for calculation of soils between piles on the front piles. However, the previous calculation models of double-row piles ignored the deformation coordination action between linking beams and ring beams, the interaction between piles and beams. Further more the calculation results based on the models are not accurate enough.

In this paper, a spatial force model of the double-row anti-sliding pile is presented. Taking double-row anti-sliding pile as a single-layer multi-span frame affected by piles, beams and soils, the spatial synergetics interaction between top ring beam and linking beam, as well as the affection of soil-arch on the frame are analyzed. A deformation equation is established according to pile border condition, continuous deformation, static force balance and deformation coordination relationship between top ring beam and pile. An analysis program is developed based on FEM and the theory of Winkler Elastic Foundation Beam. Finally, the model is applied in the practice of Three-Gorge Reservoir Prevention and Treatment project, and the result shows it is suitable for similar engineering.

2 CALCULATION METHOD

2.1 *Presumption*

For the double-row anti-sliding piles under large push-force, the push-force acts on the back-row piles, showed in Figure 1. The height of pile above sliding surface is h and the whole height of pile is H. Simplify the foundation resistance on front and back

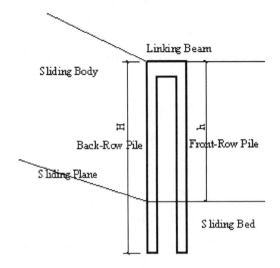

Figure 1. The section of double-row anti-sliding piles.

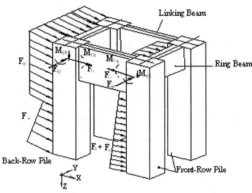

Figure 2. Calculation sketch for the double-row anti-sliding piles.

rows as elastic supporting and the horizontal resistance coefficient is calculated by m method. The basic presumptions are listed below (SS & DI, 1983).

1. Piles, beams and linking beams satisfy the superposition principle of force and displacement.
2. Soils below the sliding plane are thought as Winkler Discrete Linear Spring.
3. Tensile strength above the soils is 0 (Jiang, 2004).
4. The horizontal resistance coefficient is expressed as $k = k_0 + mz$, m is the increasing proportional factor of horizontal resistance coefficient with depths and is of different values for upper and below soils.
5. When the ratio of pile space and pile diameter or pile width (b/B) is smaller than 8, soil arch effects on front row should be considered.

2.2 The mechanics model

The push-force and active soil pressure distribute along the piles length and the soil arch effects of F_a and F_b are shown in Figure 2.

Under the interaction of push-force (F_p), active soil pressure (E_a), axial force (F_{li}) from linking beam to piles, bending moment M_{la} from linking beams to front-row piles, bending moment M_{lh} from linking beams to back-row piles, axial force F_i and bending moment M_{qq} from front-rows ring beam on piles, axial force F_i and bending moment M_{qh} from back-rows ring beam on piles, active soil pressure F_a and soil arch force F_b, the horizontal displacements on piles top are expressed by u_{qi} and u_{hi} angles by θ_{qi}, θ_{hi}.

$$\begin{cases} u_{qi} = u_{qi}(F_p, F_{li}, F_{qi}, M_{lq}, M_{qq}, F_a, F_b) \\ \theta_{qi} = \theta_{qi}(F_p, F_{li}, F_{qi}, M_{lq}, M_{qq}, F_a, F_b) \end{cases} \quad (1)$$

$$\begin{cases} u_{hi} = u_{hi}(F_p, F_{li}, F_{qi}, M_{lh}, M_{qh}) \\ \theta_{hi} = \theta_{hi}(F_p, F_{li}, F_{qi}, M_{lh}, M_{qh}) \end{cases} \quad (2)$$

According to the basic presumptions, the anti-sliding piles are linear elastomer and satisfy the superposition principles. So the above equations (1) and (2) can be modified as following:

$$\begin{cases} u_{qi} = u_{pi} + u_{bi} + u_{qi} - \delta^u_{PFii}(F_{li}) \\ \qquad + \delta^u_{PMii}(M_{lq} - M_{qq}) \\ \theta_{qi} = \theta_{pi} + \theta_{bi} + \theta_{qi} - \delta^\theta_{PFii}(F_{li}) \\ \qquad + \delta^\theta_{PMii}(M_{lq} - M_{qq}) \end{cases} \quad (3)$$

$$\begin{cases} u_{hi} = u_{pi} + \delta^u_{PFii}(F_{lh}) - \delta^u_{PMii}(M_{qh} + M_{lh}) \\ \theta_{hi} = \theta_{pi} + \delta^\theta_{PFii}(F_{lh}) - \delta^\theta_{PMii}(M_{qh} + M_{lh}) \end{cases} \quad (4)$$

Where, u_{pi} and θ_{pi} are the displacement and turning angle of the pile top in the vertical direction under the condition of only push-force existing, δ^u_{PFii} and δ^θ_{PFii} are the displacement and turning angle of the pile top in the vertical direction under the condition of only axial force existing, and δ^u_{PMii} and δ^θ_{PMii} are the displacement and turning angle of the pile top in the vertical direction under the condition of only bending force moment existing.

2.3 Stress analysis of ring beam

Under the interaction of spatial synergetics, the restraint of ring beam provides a certain horizontal force and bending moment to the anti-sliding piles, which

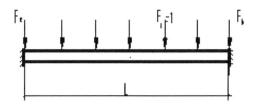

Figure 3. Calculation sketch for ring beam.

increases the stability of anti-sliding piles and reduces the trends of deformation.

Basic presumptions:

1. The crunodes at two ends of ring beams are quite rigid, and seen as the fixed bearing ends.
2. Within the span of ring beams, the piles exert horizontal force on ring beams, and the stress and displacement satisfy the superposition principle.
3. The torque from piles to ring beam is ignored.

The calculation sketch of ring beam is shown in Figure 3.
The displacement can be calculated by

$$u_i = \sum_{j=1}^{n} \delta_{QFij} F_j \quad (5)$$

Where δ_{QFij} is the horizontal displacement of ring beam at point i when only existing horizontal force $F_j = 1$.

2.4 The spatial analysis method

After establishing of the calculation model for the double-row and ring beam, the inner force and deformation can be calculated by the deformation collaborative relationship between boundary condition, deformation continuity of supporting points and static force balance condition (Xu, 1982, Bao, 1998).

1. Based on deformation collaborative relationship, the displacement of double-row piles and ring beams are the same at joint i.

$$u_{pi} + u_{bi} + u_{qi} - \delta^u_{PFii}(F_{li}) + \delta^u_{PMii}(M_{lq} - M_{qq})$$

$$= \sum_{j=1}^{n} \delta_{QFij} F_{qj} \quad (6)$$

$$u_{pi} + \delta^u_{PFii}(F_{lh}) - \delta^u_{PMii}(M_{qh} + M_{lh})$$

$$= \sum_{j=1}^{n} \delta_{qfij} F_{hj} \quad (7)$$

2. Because of the linking beam, the bending moment on pile top is transmitted to linking beam and changed to the axial force, which is only horizontal force on pile top without turning angle. The turning angle of pile top is 0.

$$\theta_{pi} + \theta_{bi} + \theta_{qi} - \delta^\theta_{PFii}(F_{li}) + \delta^\theta_{PMii}(M_{lq} - M_{qq}) = 0 \quad (8)$$

$$\theta_{pi} + \delta^\theta_{PFii}(F_{lh}) - \delta^\theta_{PMii}(M_{qh} + M_{lh}) = 0 \quad (9)$$

$$\sum_{j=1}^{n} \delta_{qfij} F_{hj} = \sum_{j=1}^{n} \delta_{QFij} F_{qj} \quad (10)$$

3. The effect of turning angle of the linking beam is ignored. Under the lateral force the both front and back row anti-sliding piles are connected by linking beam. So the pile top deformation of both front and back row are the same.
   It must be noticed that the flexibility coefficient of back-row piles is different from that of front-row, and the same thing maybe to the ring beams of double rows.
   The flexibility coefficient of ring beams is

$$\delta_{GFij} = \frac{b_i^2 x_i^2}{6EIL^2} \left[3a_j - \left(1 + \frac{2a_j}{L}\right) x_i \right] (i \leq j) \quad (11)$$

where, E is elastic modulus and I is section moment of inertia, b_j, x_i, a_j, L, are presented in Figure 4.

4. The additional stress produced by soil arch between front-row and back-row piles is Fb. When $0 < b < 8B$ (Zhou, 2005),

$$F_b = \frac{2}{\pi} \left(\frac{bB}{b^2 + B^2} + \arctan \frac{B}{b} \right) q_b \quad (12)$$

q_b is the anti-force that soil exerted on the back-row piles.

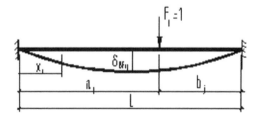

Figure 4. Calculation sketch for the flexibility coefficient of the ring beam.

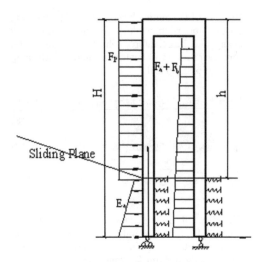

Figure 5. Finite element calculation model of double-row.

The active soil pressure between front-row and back-row piles is F_a. When $0 < b < 8B$ (Zhou, 2005),

$$\beta(b/B) = 2b/b_0 - (b/b_0)^2 \qquad (13)$$

5. Through equation (3) to (13), the $7n$ unknown forces (F_{li}, F_a, F_b, M_{lq}, M_{qq}, M_{lk}) and M_{qh} could be figured out.

3 FEM CALCULATION MODEL FOR DOUBLE-ROW ANTI-SLIDING PILES

Figure 5 show, opposed that the soils below the sliding plane are continuous spring (Winkler Assumption), push-force behind piles is distributed as the pattern of uniform load, the upper part of the pile above sliding plane is a beam, and the part of the pile below sliding plane is an elastic foundation beam. The rigidity coefficient can be calculated by simplifying the horizontal foundation resistance on each joint of double-row piles. The distribution patterns of push-force and active soil pressure on back-row piles are determined by the properties and thickness of sliding body while the active soil pressure on front-row piles (F_a) and the additional soil pressure (F_b) can be gained by equations (12) and (13). Based on deformation collaborative relationship, there are distributed force and moment on pile top. Considering the static balance of retaining structure, the following equation is obtained according to the method of direct rigidity coefficient expressed in FEM (Zhu, 1998)

$$\{F\} = [K]\{u\} \qquad (14)$$

Where $\{F\}$ is load vector, $\{u\}$ is displacement vector and $\{K\}$ is whole rigidity matrix.

4 CASE STUDY

A FEM program was develop for analyzing double-row anti-sliding piles based on the principles and methods above, by which the inner force and deformation could be calculated.

Huangtupo Landslide Prevention Engineering Project is located in the programming area for Badong Country in Three Gorges Reservoir Region. According to geological investigation, the potential sliding plane is of the shape of trumpet with front-edge width of 210 m, middle width of 340 m, back-edge width of 400 m and average portrait length of 280 m. From back edge to front edge, the fragile degree of soils reduces and the soil permeability increases. Three types of anti-sliding piles were adopted: single-row cantilever piles, pre-stress anchored piles and double-row anti-sliding piles. In order to compare the anti-sliding effect, the rigidity is equivalent to square double-row and single-row piles. The bearing condition at pile bottom is gemel. Using **m** calculation method, 280 KN/m surplus push-forces are distributed in a form of rectangle on the back-row piles. The calculation parameters for piles are listed in Table 1.

The bending moment diagrams for three types of anti-sliding piles are shown in Figure 5 to Figure 8. Figures 6, 7, 8, show that the bending moment, stress and displacement of double-row anti-sliding piles are smaller obviously than those of single-row cantilever piles and the pre-stress anchored piles when their rigidity is with the same value, and the distribution of bending moment of double-row anti-sliding piles are more reasonable. The distribution of inner forces between front-row and back-row in double-row anti-sliding piles are different, and the maximal value of bending moment located on the 1/3 of pile body in back-rows is larger distinctly than that in front-row. This means that the back-row piles are subject to larger outer forces for the anti-sliding piles due to the spatial effect. It is also shown that the maximal displacement of back-rows is smaller than that of front-rows. Deformation gradually decreases from pile top to bottom and the inner force of linking beam is larger. That means that the spatial interaction has influence on double-row anti-sliding piles. In addition, the lateral deformation of double-row piles is smaller than that of single-row cantilever piles but close to that of pre-stress anchored piles. Thus, double-row anti-sliding piles can reduce the deformation effectively.

Table 1. Calculation parameters for piles.

	Basic parameters	Single-row cantilever pile	Pre-stress anchored pile	Double-row anti-sliding pile	Remarks
Pile Size	Length (m)	28.000	28.000	28.000	
	Embedded Depth (m)	5.000	5.000	5.000	
	Width (m)	2.800	2.500	2.000	The same equivalent rigidity
	Height (m)	3.200	3.000	2.500	
	Spacing (m)	6.000	6.000	Spacing 6.000 Distance 6.000	
			Horizontal Tension 965 KN	Linking Beam 1.0 × 2.0 (width × height)	
Sliding Plan	C (KPa)	$C_d = 22$ $C_s = 25$ $\phi_d = 19.1$ $\phi_s = 15.8$	Sliding Body	C (kPa)	$C_d = 220$ $C_s = 180$ $\phi_d = 30$ $\phi_s = 25$
	$\phi (°)$			$\phi (°)$	
	ρm (kN/m³)		$\rho_d = 19.7, \rho_s = 20.2$		

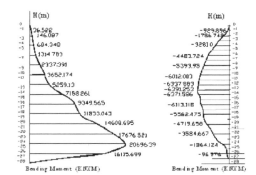

Figure 6. Bending moment of single-row cantilever pile.

Figure 7. Bending moment of anchored—pile.

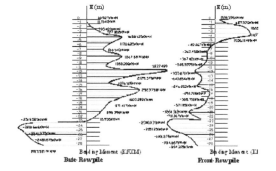

Figure 8. Bending moment of double-row anti-sliding piles.

5 CONCLUSIONS

1. The stress of double-row anti-sliding piles is different from that of single-row cantilever piles and pre-stress anchored piles. Bending moment of pile top is not 0 for both of the front-row and back-row piles. The maximal bending moment of double-row anti-sliding piles is smaller than that of single-row cantilever piles and pre-stress anchored piles.
2. The maximal side displacement of double-row anti-sliding piles is smaller than that of single row cantilever piles but close to pre-stress anchored piles. The displacement of front-row and back-row in double-row piles is close and inner force of linking beam is larger. This means that the deformation of double-row piles is restrained greatly due to the spatial action, which enhances the ability of anti-deformation. In addition, the spatial effects are existing among the anti-sliding piles, ring beams and linking beams.
3. Double-row anti-sliding pile has the advantages of symmetrical stress, small bending moment and less displacement so that it is suitable for similar projects.
4. Pile, beam and soil were thought as an integral part in this paper and the effect of soil arch was also considered. The calculation results showed with the clarity that the calculation model for double-row anti-sliding piles based on the spatial framework structure are reasonable and practicable.

REFERENCES

Xu Fenghe & Wang Jinsheng. 1988. Design and Construction of Anti-sliding Rigid Framework in Luoyixi Project. *Proceedings of Landslides*. Beijing: China Railway Publishing House.

Min Shunnan & Xu Fenghe. 1990. The Research of Chair-shaped Piling Wall in Shirongxi. *Proceedings of Landslides*. Beijing: China Railway Publishing House.

Jiang Chusheng. 2004. Computation of Inner Forces for Chair-shaped Anti-sliding Piles. *Roadbed Engineering*, 112 (01): 57–59.

Xiong Zhiwen, Ma Hui & Zhu Haidong. 2002. Force Distribution of Buried Double-row Anti-sliding Piles, *Roadbed Engineering*, 102(03): 5–10.

Zhou Cuiying, Liu Zuoqiu & Shang Wei. 2005. A New Mode for Calculation of Double-row Anti-sliding Piles, *Rock and Soil Mechanics*, 26(03): 441–444.

Centrifuge modeling of rainfall-induced failure process of soil slope

J.Y. Qian
State Key Laboratory of Hydroscience and Engineering, Tsinghua University, Beijing, China

A.X. Wang
Department of Civil and Environment Engineering, University of Science and Technology Beijing, China

G. Zhang & J.-M. Zhang
State Key Laboratory of Hydroscience and Engineering, Tsinghua University, China

ABSTRACT: Two centrifuge model tests were carried out to investigate the rainfall-induced deformation and failure process of Beijing Olympic Forest Park. A new rainfall generator was developed to realize rainfalls during the centrifugal model tests. The deformation process of slopes subjected to a heavy rain was measured using an image-based measurement system. The tests results indicated that the water content of the soils has a significant effect on the failure of slopes: the slope with high initial water content is more inclined to fail. The displacements of the slope, both in horizontal and vertical directions, increased with increasing rainfall; the area where significant displacements occurred also enlarged. The deformation of the slope was significantly dependent on the wetting front of the slope during a heavy rain.

1 INTRODUCTION

Soil slopes are inclined to fail due to the rainfall application in the rainy season. The failures of theses slopes claim many human lives and cause serious economic loss. Therefore, a study is needed to explore a better understanding of the mechanism of failure of slopes subjected to rainfall; this can help to diminish these disasters so far as possible (Kimura, T., et al. 1991, Tamate, S & Takahashi A. 1998).

A 50-meter-high soil hill was completed in Beijing Olympic Forest Park, China, in 2007. Millions cubic meters of soil was constructed within only 108 days. Significantly, the Olympic Games 2008 will be held is pluvial August .Therefore, to evaluate the stability level of the high soil hill of the Beijing Olympic Forest Park, behavior of deformation and failure process of a soil slope under different rainfall conditions should be concerned.

In this paper, two centrifugal model tests are described to investigate the process and mechanism of failure of soil slopes subjected to a heavy rainfall; the influence of the initial water content and density of the soil are also preliminarily discussed (Liu, D., et al. 2006).

2 DEVICES

The model tests were conducted using the TH-50 g-t geotechnical centrifuge machine. The model container for tests is 500 mm long, 200 mm wide and 500 mm high. Two transparent lucite windows are installed on both sides of the model container.

A new rainfall generator was developed to realize rainfalls during the centrifuge model tests. The device mainly consists of three parts: the water supply module, the water dispersing module, and the calibration module. Water is stored in a water tank that is fixed on the radial beam of the centrifuge machine before the test. It is transported to the dispersing module through the plastic pipes once an instruction is accepted during the test. The dispersing module is used to simulate the rainfall. It is made up of a lucite container with two copper pipes and a layer of semi-permeable cloth beneath the copper pipes (Figure 1). The calibration module is used to determine the rainfall intensity during the test; it is made up of a few measuring cups.

The rainfall is slantwise relative to the fills because the radial velocities are significantly different between the rainfall generator and the slope. Moreover, the rainfall direction is significantly influenced by the wind due to rapid circumrotation of the centrifuge machine. A series of technical measures are used to assure a homogeneous rain on the slope, including: 1) the rainfall generator is made a bit wider than the container; 2) the lucite windows are cut in a given way with an angle of 45° to keep the rain away from the blocking of the lucite windows; 3) two pieces of wind-proof cloth are fixed outside the container to prevent the wind from the rainfall (Figure 2).

An image-record and displacement measurement system was used to record the images of soils during centrifuge model tests. 48 frames of image can be captured per second. An image-correlation analysis algorithm was used to determine the displacement vectors of soil without disturbing the soil itself (Zhang, G. et al. 2006, Mu, T. et al. 2006). The displacement history of an arbitrary point on the soils can be measured with sub-pixel accuracy. Only a colorful region with a random distribution is needed for this system; such a region can be obtained by embedding white particles in the lateral side of the soils. For the test conditions in this paper, the measurement accuracy can reach 0.02 mm based on the model dimension.

3 CENTRIFUGE MODEL TEST

3.1 Soils

The soils used in the centrifuge tests were taken directly from the soil hill of Beijing Olympic Forest Park. The average grain size of the soil is 0.03 mm. The plastic limit and liquid limit are 5% and 18%, respectively; thus the plastic index is 13.

3.2 Models

The soils were compacted to a layer with 6 cm in thickness in the container until the total thickness reached 36 cm. A slope with the inclination of 45° was formed by cutting out the redundant soil. A six-cm-high horizontal soil layer under the slope was set to diminish the influence of the bottom container plate on the deformation of the slope. Moreover, the silicone oil was painted on both sides of the container to decrease the friction between slope and container sides. The images of slope were recorded using the image-based measurement system with which the displacement field and its change can be furthermore obtained.

The model slope was installed on the centrifuge machine and the centrifugal acceleration gradually increased to 50 g, a centrifugal acceleration that was maintained during raining tests. The rainfall started after the deformation of the slope due to increasing centrifugal acceleration became stable. The intensity of rain produced was 15 mm/min based on the model dimension. Two different centrifuge model tests were conducted by varying the dry density and initial water content of the soil (Table 1).

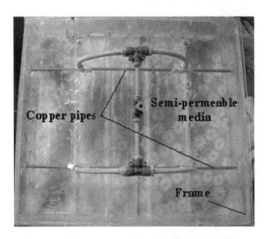

Figure 1. The dispersing module of rainfall generator.

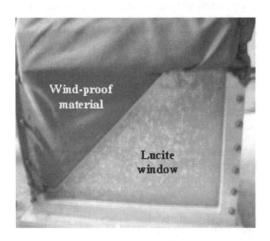

Figure 2. Photograph of model container.

4 RESULT

4.1 The behavior of the failed slopes

The drier slope (No. 1 test) failed when the rainfall was 64 mm, while the wetter one (No. 2 test) failed when

Table 1. The soils and conditions of two tests.

No.	Dry density	Initial water content	Rainfall
1	1.55 g/cm^3	10%	64 mm
2	1.4 g/cm^3	14%	12 mm

the rainfall was only 12 mm. This indicated that the slopes with higher water content were more inclined to fail during a heavy rain. The failure behavior of the two slopes was also significantly different (Figures 3–4). A shallow slip surface was washed out in the wetter slope, while a relative deep landslide occurred in the drier one. Several cracks also appeared on the top of the drier slope, while the top soils of the wetter one were washed out. The toes of the fills were deposited by the silts washed out by the rain; this implied that the surface runoff appeared during the tests.

Figure 5 shows the residue deformation due to rainfall. It can be seen that the horizontal displacement and settlement of the drier slope were significantly larger than those of the wetter one. It should be noted that all the displacements of the slope are based on model dimension. The increasing flexibility of the soil due to increase of water content may be one of the reasons of different displacement. Moreover, the rainfall influence depth, within which the significant deformation occurred, of the drier slope was larger than that of the wetter one. This would explain the different failure behaviors of the two slopes. Thus, it can be concluded that the water content of the soils has a significant effect on the failure of slopes.

4.2 Deformation and failure process

The drier slope (No.1 test) was selected to discuss the deformation and failure process of a slope due to rainfall. The downward movement of the wetting front can be determined using the recorded photos (Figure 7). It can be seen that the shape of wetting front was

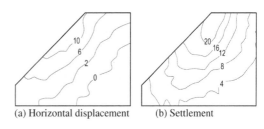

Figure 5. Contour lines of residue deformation of the drier slope (No. 1 test). unit, mm.

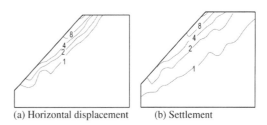

Figure 6. Contour lines of residue deformation of the wetter slope (No. 2 test). unit, mm.

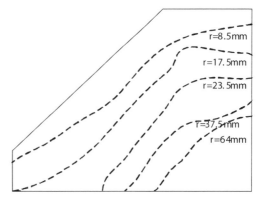

Figure 7. Movement of wetting front obtained from photographs in test No. 1. r, rainfall.

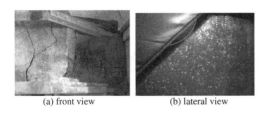

Figure 3. Photographs of the failed drier slope (No. 1 test).

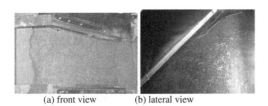

Figure 4. Photographs of the failed wetter slope (No. 2 test).

approximately parallel to the slope. And the displacements were fairly small in the area without rainwater penetrated.

Figure 8 shows the images and displacement contour lines of the slope at several typical times. It can be seen that the cracks occurred when the rainfall reached 17.5 mm. The horizontal and vertical displacements both increased with increasing rainfall. The area where significant displacements occurred also enlarged with increasing rainfall; this exhibited a significant relation to the development of wetting front. It can be noticed

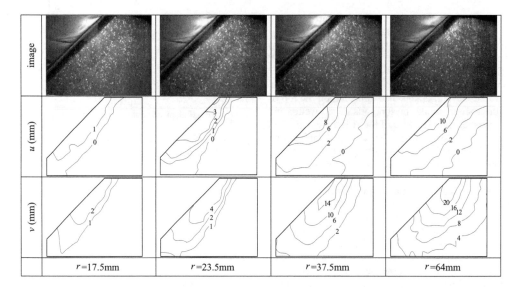

Figure 8. Images and displacement fields under different rainfall of the drier slope (No.1. test). u, horizontal displacement; v, vertical displacement; r, rainfall.

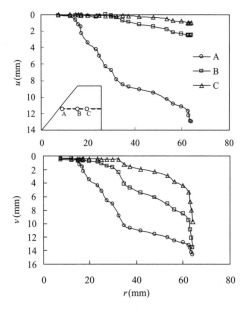

Figure 9. Displacement histories of typical points of the drier slope (No.1 test). u, horizontal displacement; v, vertical displacement; r, rainfall.

that the shape of the zero contour lines of the horizontal displacement was similar to that of the wet front.

In the slopes, three typical points were selected to discuss the deformation process of the slope (Fig. 9).

At the beginning of the rainfall, small settlements occurred at all the three points. Their difference became significant when the rainfall reached 17.5 mm. The significant deformation occurred in the borders of the slope firstly (point A); it developed to the internal with increasing rainfall (point B & C). This is because that the rain water penetrated the slope from border to the internal (Figure 7). When the rainfall reached 60 mm, the horizontal displacement near the slope border (Point A) became straight nearly. In other words, the displacement of the slope increased rapidly at this time; this indicated the failure of the slope occurred. The vertical displacement also increased with increasing rainfall. When the rainfall got to 60 mm, unlike horizontal displacement, the vertical displacements of all the three points exhibited an significant increments, this may be because a significant settlement occurred at the bottom of the slope.

5 CONCLUSIONS

A new rainfall generator was developed to realize rainfalls during the centrifugal model tests. Using the equipment, two centrifugal model tests were carried out to investigate the deformation and failure process of slopes subjected to a heavy rain and the influence of the initial water content of fills.

The tests results indicated that the water content of the soils has a significant effect on the failure of slopes. The slope with high initial water content is more inclined to fail, with a relatively shallow slip surface washed out.

The displacements of the slope, both in horizontal and vertical directions, increased with increasing rainfall; the area where significant displacements occurred also enlarged. The shape of the zero contour lines of horizontal displacement is similar to that of the wet front; this exhibits a significant relation between the deformation and wetting front of the slope.

ACKNOWLEDGEMENTS

The project is supported by National Basic Research Program of China (973 Program)(No. 2007CB714108) and National Natural Science Foundation of China (No. 50679033). The support is gratefully acknowledged.

REFERENCES

Kimura, T., et al. 1991. Failure of fills due to rain fall. *Centrifuge'91*: 509–516.

Tamate, S & Takahashi A. 1998. Slope stability test. *Centrifuge'98*: 1077–1082

Liu, D., et al. 2006. Model study on the irreducible water saturation by centrifuge experiments. *Journal of Petroleum Science and Engineering*, 53, 77–82.

Zhang, G., Liang, D. & Zhang, J-M. 2006. Image analysis measurement of soil particle movement during a soil—structure interface test. *Computers and Geotechnics,* Vol. 33 (4–5): 248–259.

Mu, T., Zhang, G. & Zhang, J-M. 2006. Centrifuge modeling of progressive failure of soil slope. *International Conference on Physical Modeling*, Vol. 1: 373–377.

A GIS-based method for predicting the location, magnitude and occurrence time of landslides using a three-dimensional deterministic model

C. Qiu, T. Esaki & Y. Mitani
Institute of Environmental Systems, Kyushu University, Fukuoka, Japan

M. Xie
Civil and Environmental Engineering School, The University of Science and Technology Beijing, China

ABSTRACT: To predict the location, magnitude and occurrence time of landslides that occur in wide mountainous area, a new GIS-based three-dimensional method is proposed. A GIS raster-based 3D slope stability analysis model is delivered to calculate the 3D safety factor in regional landslide assessment. In order to locate the potential failures, a Monte Carlo technique is used by means of minimizing the 3D safety factor through an iterative procedure, based on a simulation of ellipsoid for the 3D shape of slip surfaces. A new GIS-based model is developed by coupling a dynamic rainfall-infiltration model with the GIS-based 3D model to quantify the varying safety factors during rainfall infiltration. All proposed methods are applied by developing a GIS-based comprehensive system. The effectiveness of the method and the practicality of the developed system are verified by a practical application for landslide-prone area in Japan.

1 INTRODUCTION

Any landslide hazard predicting work that is commonly used for hazard precaution should give answers to three key questions: (1) the magnitude, (2) the location and (3) the time of failure occurrence. The factors that determine a landslide can be summarized as falling into two categories: quasi-static variables and dynamic variables (Dai & Lee 2001). The quasi-static variables, such as geology, terrain, and geotechnical properties, contribute to landslide susceptibility and determine the location and the magnitude of the failure. The dynamic variables, such as rainfall and earthquake, tend to trigger landslides in an area of given susceptibility and thus have a direct impact on the time of occurrence of the landslide. It is therefore necessary to evaluate the effects of both the quasi-static variables and the dynamic variables on slope stability (Wu & Sidle 1995, Crosta 1998).

Once the slope geometry and sub soil conditions have been determined, the stability of a slope may be assessed using the deterministic methodologies. The deterministic models can be one-dimensional (1D), two-dimensional (2D), or three-dimensional (3D). In case of a wide mountainous area where the geometry varies significantly and the material properties are highly inhomogeneous, a 3D analysis may become necessary. Although many 3D models have been proposed in the literature (Hovland 1977, Chen & Chameau 1983, Hungr 1987, Huang et al. 2002), however, applications of the 3D models are generally limited for site-specific slopes, but seldom being used for a wide area due to the difficulties in processing and managing a vast amount of complex information of natural slope, and in identifying unknown slip surface(s).

On the other hand, concerning the effect of the dynamic variables on slope stability, the influence of rainfall on landslide has been a subject of research for many years (Al-Homoud et al. 1999, Gasmo et al. 2000, Kim et al. 2004). When a wide natural area is an object of study, much research has traditionally been focused on the relationship between the probability of landslide occurrence and the rainfall threshold, in which the rainfall threshold, defined as the rainfall required to cause one landslide in an area observed during past rainfall events, was used to predict the occurrence of slope failure. An intrinsic hypothesis underlying such methods is that the landslides analyzed exhibit consistent behavior. Therefore, a model established for a specific site is generally unreliable for application to a different site due to variations in geotechnical conditions. Furthermore, such methods utilize rainfall as the only factor to provide warning and do not take other important variables influencing landslide, such as geology, geometry, and groundwater, into account. Because rainfall-induced landslides occur as the result of the infiltration of water from intense rainfalls, it is reasonable to evaluate this type of slope failure using physically based models to simulate the transient hydrological and geotechnical processes responsible for slope stability (Wu & Sidle 1995).

To efficiently analyze a regional landslide hazard which relates a large number of spatial data, a useful tool for processing spatial data is essential. Recently, the Geographic Information System (GIS), with its excellent spatial data processing capacity, has attracted great attention in landslide hazard assessment. If a GIS-based deterministic model can be established, the computing power provided by GIS will significantly revolutionize the 3D slope stability analyses in two ways: (1) the 3D models can be generated precisely and automatically, based on the GIS data for topography and stratum; and (2) a large number of slip surfaces can be analyzed, making it possible to locate the critical slip surfaces with a high degree of reliability. Integrating a conventional column-based 3D model with GIS raster data, the authors have proposed a GIS-based 3D slope stability analysis model to calculate the 3D safety factor of a predefined failure mass and to further locate the critical slip surface by means of minimizing the 3D safety factor through a Monte Carlo simulation (Xie et al. 2003, Qiu et al. 2007).

This paper describes the development of the proposed model into a new GIS-based 3D model by incorporating an infiltration model and taking account of geomechanical changes of soil strength during rainfall in the calculation. Using this model, the location and the shape of critical slip surfaces can be identified through a random search procedure by varying the geometrical parameters of the ellipsoid that is used to simulate the shape of the slip surface. Furthermore, the time of occurrence of failures can be forecast by mapping the changing distribution of safety factors during rainfall. A GIS system is developed to efficiently implement all of the computational procedures as well as the data preparation and the result visualization. The effectiveness of the proposed method was verified by a practical application for a landslide-prone area in Japan.

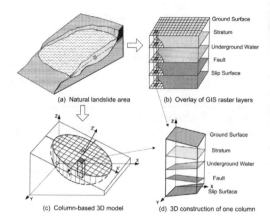

Figure 1. GIS-based 3D slope stability analysis model.

and then converted to GIS raster layers through spatial interpolation, as shown in Figure 1b. Each raster layer represents a certain type of information such as topography, stratum, or groundwater. The superposed construction of the strata can be presented by overlay of the multiple raster layers. The overlay extends the 2D raster layers to a third dimension, which makes the application of 3D engineering models possible. Here, the GIS raster model is manipulated to accommodate the traditional column-based 3D model by discretization of the study mass, as shown in Figure 1c. Slope stability analysis-related information for each soil column, such as topography, strata, groundwater, faults, can thus be obtained from the corresponding cell of raster layers (Fig. 1d) and used to calculate a safety factor.

Integrating Hovland's (1977) 3D model and the GIS raster-based database, the safety factor can be derived from the horizontal force equilibrium in the direction of sliding (Xie et al. 2003):

$$SF_{3D} = \frac{\sum_J \sum_I (cA + W\cos\theta \tan\phi)\cos\theta_{Avr}}{\sum_J \sum_I W \sin\theta_{Avr} \cos\theta_{Avr}} \quad (1)$$

where SF_{3D} = 3D safety factor of the slope; c = cohesion (kN/m^2); A = area of the slip surface (m^2); W = weight of one soil column (kN); ϕ = friction angle (°); θ = inclination of the slip surface (°); θ_{Avr} = angle between the direction of movement and the horizontal plane (°); and J, I = numbers of rows and columns of the cell within the range of failure mass.

To identify critical slip surfaces from the study area, a random search is performed by means of a minimization of the 3D safety factor. The shape of the slip surface is assumed to be the lower part of an ellipsoid

2 A NEW COUPLING RAINFALL-SLOPE STABILITY ANALYSIS MODEL

2.1 The GIS-based 3D model and identification of critical slip surface

The mechanism underlying the proposed GIS-based 3D model is illustrated by Figure 1. A potentially sliding mass, which will be the study object, is shown in Figure 1a. Initially, all the discrete investigation data for the study site can be represented as a number of GIS vector layers (e.g. a line layer of contour), and will vary in depth, dip angle, and direction of inclination. The direction of inclination is calculated as the mode value from the dip direction of all raster cells in the range of the sliding mass. The uncertain parameters, such as dip angle and the a, b, and c axes, are specified by Monte Carlo simulation from the range they may take. For each cell that is taken as the central point of a randomly chosen trial ellipsoid from the range of sliding mass, a minimum safety factor and a critical slip surface can be obtained after enough time of calculations.

2.2 Coupling the slope stability analysis model with a rainfall-infiltration model

Infiltration is a function of soil properties, rainfall and local settings. There are numerous models formulated on the basis of soil characteristics that have been proposed to evaluate infiltration. The widely used Green & Ampt model (1911) is adopted in this study. In this model, the infiltration rate at any time t is calculated by

$$f(t) = K_s + K_s \frac{\psi_f(\theta_s - \theta_i)}{F} \quad (2)$$

The expression of $F(t)$ can be stated as follows:

$$t = t_p + \frac{1}{K_s}\left[F - F_p + \psi_f(\theta_s - \theta_i)\ln(n)\right] \quad (3)$$

$$n = \frac{\psi_f(\theta_s - \theta_i) + F_p}{\psi_f(\theta_s - \theta_i) + F} \quad (4)$$

t_n and F_n can be calculated from the following equations:

$$t_p = \frac{F_p}{P} \quad (5)$$

$$F_p = \frac{\psi_f K_s (\theta_s - \theta_i)}{P - K_s} \quad (6)$$

where $f(t)$ = infiltration rate (m/h) at time t (h); K_s = soil saturated hydraulic conductivity (m/h); ψ_f = matrix suction at the wetting front (m); F = cumulative amount of infiltrated water (m); θ_s = soil saturated volumetric water content; θ_i = initial soil volumetric water content; t_n = time when water begins to pond on the soil surface (h); F_n = amount of water that infiltrates before water begins to pond at the surface (m); and P = rainfall rate (m/h).

It should be noted that a necessary condition for Equation 2 to Equation 6 is that the rainfall rate must be greater than the soil hydraulic conductivity. If the rainfall rate is not greater than the potential infiltration rate ($P \leq K_s$) or no surface ponding occurs ($t \leq t_n$), then all rainfall will infiltrate into the soil without runoff, and the actual infiltration rate is equal to the rainfall rate:

$$f(t) = P \quad (7)$$

$$F = Pt \quad (8)$$

It is widely known that rainfall causes a rise of groundwater level as well as an increase in pore water pressure that results in slope failure. However, in many situations where shallow failures are concerned, it has been noted that the failure was attributed to the advance of a wetting front into the slope instead of a rise of the groundwater level (Cho & Lee 2002). The wetting front causes a reduction in soil suction (or negative pore pressure) and an increase in the weight of soil per unit volume. These result in a process in which soil resistant strength decreases while total stress increases, until failure occurs on the potential failure surface where equilibrium cannot be sustained.

Under these conditions, there are four possible situations regarding the slip surface that can be anticipated, and four models are thus proposed to calculate the corresponding safety factors (Fig. 2).

Model 1: The slip surface forms in the unsaturated zone between the wetting front that is advancing from the ground surface and the groundwater table (Fig. 2a). In this situation, the horizontal resistance force F_1 and the horizontal sliding force F_2 acting on the slip surface can be calculated using Equation 9 and Equation 10, respectively:

$$F_1 = \{c'_i A + [\gamma_i z + (\gamma_{sat} - \gamma_i)H_w]$$
$$\times \cos\theta \tan\phi'\}\cos\theta_{Avr} \quad (9)$$

$$F_2 = [\gamma_i z + (\gamma_{sat} - \gamma_i)H_w]\sin\theta_{Avr}\cos\theta_{Avr} \quad (10)$$

Model 2: The slip surface forms in the saturated zone between the ground surface and the wetting front that is advancing from the ground surface (Fig. 2b). In this situation, the horizontal resistance force F_1 and the horizontal sliding force F_2 can be calculated using Equation 11 and Equation 12, respectively:

$$F_1 = \left[c'_w A + (\gamma_{sat} z \cos\theta - u_w)\tan\phi'\right]$$
$$\times \cos\theta_{Avr} \quad (11)$$

$$F_2 = \gamma_{sat} z \sin\theta_{Avr} \cos\theta_{Avr} \quad (12)$$

Model 3: The slip surface forms in the saturated zone under the groundwater table and the wetting front that has reached the groundwater table (Fig. 2c). The horizontal resistance force F_1 and the horizontal sliding force F_2 can be calculated using Equation 11 and Equation 12, respectively.

Model 4: The slip surface forms in the saturated zone under the groundwater table and the unsaturated

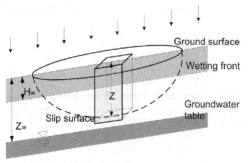

(a) Model1: slip surface forms between the wetting front and the groundwater table

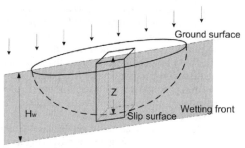

(b) Model2: slip surface forms between the ground surface and the wetting front

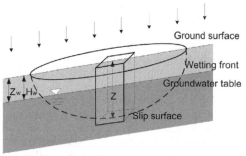

(c) Model3: slip surface forms under the groundwater table

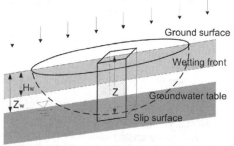

(d) Model4: slip surface forms under the groundwater table and the unsaturated zone exists

Figure 2. Four models for rainfall-induced slope stability analysis.

zone exists between the wetting front and the groundwater table (Fig. 2d). In this situation, the horizontal resistance force F_1 and the horizontal sliding force F_2 can be calculated using Equation 13 and Equation 14, respectively:

$$F_1 = \{c'_w A + [(\gamma_i(z_w - H_w) + \gamma_{sat}(H_w + z - z_w)) \times \cos\theta - u_w] \tan\phi'\} \cos\theta_{Avr} \quad (13)$$

$$F_2 = [\gamma_i (z_w - H_w) + \gamma_{sat} (H_w + z - z_w)] \times \sin\theta_{Avr} \cos\theta_{Avr} \quad (14)$$

Assuming that the vertical sides of each soil column are frictionless, the 3D safety factor can thus be calculated by summing F_1 and F_2 of all soil columns of failure mass:

$$SF_{3D} = \frac{\sum_J \sum_I F_1}{\sum_J \sum_I F_2} \quad (15)$$

where SF_{3D} = 3D safety factor of the slope; F_1 = horizontal resistance force (kN); F_2 = horizontal sliding force (kN); γ_{sat} = saturated unit weight of soil (kN/m³); γ_i = initial unit weight of soil (kN/m³); c'_i = initial effective cohesion of soil (kN/m²); c'_w = saturated effective cohesion of soil (kN/m²); ϕ' = effective friction of soil (°); z_w = depth of the wetting front (m); H_w = depth of the groundwater table (m); z = depth of the slip surface (m); A = area of the slip surface of the soil column (m²); θ = inclination of the slip surface (°); θ_{Avr} = dip angle of the main sliding direction (°); and J, I = numbers of rows and columns of the cell in the range of failure mass.

3 COMPUTATIONAL COMPLEMENTATION

Figure 3 illustrates the computational procedures for estimation of rainfall-induced slope stability. For a given rainfall intensity, the permeation depth at a specified time can be predicted by the infiltration model described above. At the same time, taking a cell of the raster dataset to be the central point of an ellipsoid, a trial slip surface can be formed using the Monte Carlo simulation model. A safety factor is calculated subsequently by the 3D slope stability model, depending on the relationship between the permeation depth and the location of the slip surface after many trials, the critical slip surface and the associated safety factor, which is the minimum of all the trial calculations, can be obtained. The safety factor is then saved to a point dataset as the safety factor of the raster cell. Also, if the safety factor is < 1, the information for the corresponding critical slip surface, such as location, shape, and volume, will be

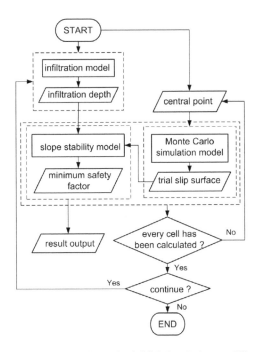

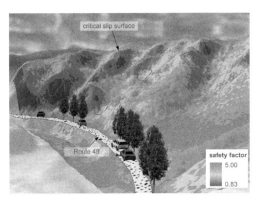

Figure 4. 3D view of the distribution of 3D safety factors with critical slip surfaces.

Figure 3. Flow chart of rainfall-induced slope stability caculation.

saved in the format of a polygon. In this way, the safety factor at a specified time can be calculated for all raster cells inside the slide mass range. As a result, the time of occurrence of a landslide can be predicted by means of the decrease of the safety factor along with the amount of rainfall, and the location and the magnitude of the potential slip surfaces can be known from the polygon dataset of critical slip surfaces.

All of the above calculations have been incorporated within a GIS-based system, called the 3D slope stability analysis system (3DSSAS), which is programmed by Visual Basic language using Microsoft component object model technology (COM). In 3DSSAS, ArcObjects, the framework that forms the foundation of the ArcGIS™ applications (GIS software developed by ESRI) with more than 2000 COM-based components has been used.

4 A PRACTICAL APPLICATION

The Goto section of National Route 49 is located near the city of Iwaki in Japan, about 10 km westward, where slope disasters have happened frequently. To forecast slope failure and to provide support for deciding on suitable countermeasures, the system described here was used for landslide mapping.

The bedrock of the study area is represented by Mesozoic granites, which is hard in a fresh state but often heavily weathered to form deep residual deposits. The soil cover in the area, with an average depth of about 2 m, is composed mainly of colluvium and residual deposits formed by weathering of granites and accumulation of debris as result of landslide activity. It is well known that landslides are common in colluvium and residual soils, particularly during periods of intense rainfall. The well-developed internal drainage of the soils is conducive to water infiltration, subsequent reduction in pore-water tension, and consequent sliding. According to the characteristics of the geological formation, the possible collapse mode was considered to be a shallow slope failure with a sliding surface along the boundary surface of the bedrock or inside the soil layer.

The surface digital elevation model (DEM) data in grid form with 2 m mesh size was produced from airborne laser scanning. The geomechanical and hydraulic parameters of the soil surface consisted of information from a specific geotechnical field investigation, triaxial test results of soil samples collected from three spots, and related literature, as $c'_i = 14.5$ (kN/m^2), $c'_w = 3.2$ (kN/m^2), $\phi' = 40.4$ (°), $\gamma_i = 15.3$ (kN/m^3), $\gamma_w = 16.5$ (kN/m^3), $Ks = 5.2$ (cm/h), $\psi_f = 40.8$ (cm), $\theta_s - \theta_i = 0.31$, respectively. In addition, in a range of 370 × 520 m, detailed data for soil depth, with a precision of one value per 10 m^2, were obtained by a special surveying method using a probe stick. This range is used as the analytic range of this study. The soil depth data is then interpolated using the Gausses method of the Kriging to create a raster data with 2 m mesh size to match mesh size of the DEM data. The bedrock surface was then abstracted as a GIS raster dataset by subtraction of the surface elevation and the soil depth.

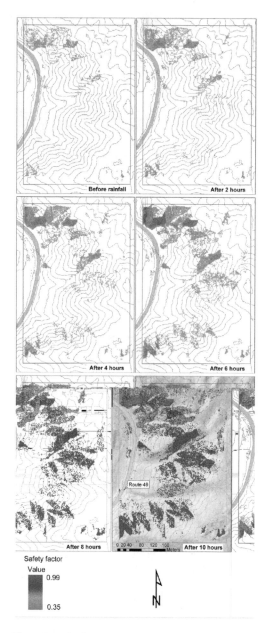

Figure 5. Distribution maps of 3D safety factors changing over time.

of a randomly produced ellipsoid is lower than the boundary surface of the bedrock, the confined surface of the bedrock will be prioritized for selection as one part of the assumed slip surface. After 100 trial calculations for each raster cell taken as the central point of a trial ellipsoid in the study range, finally, the critical slip surfaces were identified, and the variation of safety factors over time was mapped. A 3D view of the distribution of the critical slip surfaces that have a safety factor < 1 is shown in Figure 4. The critical slip surfaces are seen frequently in parts where the slope angle is around 35°–50°. Figure 5 illustrates six distribution maps of the safety factors changing over time. From these maps, a high correlation between rainfall and the decrease of the safety factor can be recognized.

5 CONCLUSIONS

By combining an infiltration model with an improved column-based 3D slope stability analysis model, a new GIS-based 3D model has been developed for evaluating the variation of surficial slope stability during a rainfall event. A GIS-based program has been developed to carry out all the processes, including data handling, calculation, and expression of the results. The introduction of the GIS technology enables both a convenient management of enormous amounts of slope-related data and an easy implementation of complicated calculations.

The suitability of the proposed method for large complex and data-limited natural terrains was demonstrated by a practical case study, in which the distribution of the safety factors changing over time and the location of the critical slip surfaces were mapped. Such a spatio-temporal hazard map of rainfall-induced landslide has profound implications for the identification of the resultant processes that control slope stability and specification of remedial and stabilization measures.

REFERENCES

Al-Homoud, A.S., Prior, G. & Awad, A. 1999. Modeling the effect of rainfall on instabilities of slopes along highways. *Environmental Geology* 37 (4): 317–325.

Chen, R.H. & Chameau, J.L. 1983. Three-dimensional limit equilibrium analysis of slopes. *Geotechnique, Institution of Civil Engineers (ICE)* 33 (1): 31–40.

Cho, S.E. & Lee, S.R. 2002. Evaluation of surficial stability for homogeneous slopes considering rainfall characteristics. *Journal of Geotechnical and Geoenvironmental Engineering* 128 (9): 756–763.

Crosta, G. 1998. Regionalization of rainfall thresholds: an aid to landslide hazard evaluation. *Environmental Geology* 35 (2–3): 131–145.

Dai, F. & Lee, C. 2001. Terrain-based mapping of landslide susceptibility using a geographical information system: a case study. *Canadian Geotechnical Journal* 38: 911–923.

For the study area, a uniform rainfall event with duration of 10 h and an intensity of 8 cm/h was assumed, and the method proposed above was applied. The detection of the critical slip surface was achieved through trial searching and calculation of the 3D safety factor. In the trial searching process, if the lower part

Gasmo, J.M., Rahardjo, H. & Leong, E.C. 2000. Infiltration effects on stability of a residual soil slope. *Computers and Geotechnics* 26: 145–165.

Green, W.H. & Ampt, G.A. 1911. Studies of soil physics: I. The flow of air and water through soils. *J Agric Sci* 4:1–24.

Hovland, H.J. 1977. Three-dimensional slope stability analysis method. *Journal of the Geotechnical Engineering, Division Proceedings of the American Society of Civil Engineers, ASCE* 103 (GT9): 971–986.

Huang, C., Tsai, C. & Chen, Y. 2002. Generalized method for three-dimensional slope stability analysis. *Journal of Geotechnical and Geoenvironmental Engineering, ASCE* 128 (10): 836–848.

Hungr, O. 1987. An extension of Bishop's simplified method of slope stability analysis to three dimensions. *Geotechnique* 37 (1): 113–117.

Kim, J., Jeong, S., Park, S. & Sharma, J. 2004. Influence of rainfall-induced wetting on the stability of slopes in weathered soils. *Engineering Geology* 75: 251–262.

Qiu, C., Xie, M. & Esaki, T. 2007. Application of GIS technique in three-dimensional slope stability analysis, In Z. Yao & M. Yuan (Eds), *International Symposium on Computational Mechanics, China, Springer Press*: 703–712.

Wu, W. & Sidle, R.C. 1995. A distributed slope stability model for steep forested basins. *Water Resources Research* 31 (8): 2097–2110.

Xie, M., Esaki, T., Zhou, G. & Mitani, Y. 2003. GIS-based 3D critical slope stability analysis and landslide hazard assessment. *Journal of Geotechnical and Geoenvironmental Engineering (ASCE)* 129 (12): 1109–1118.

Application of a rockfall hazard rating system in rock slope cuts along a mountain road of South Western Saudi Arabia

B.H. Sadagah
King Abdulaziz University, Jeddah, Saudi Arabia

ABSTRACT: Mountain roads play a vital role in the development of southwestern Saudi Arabia, as they were built in a difficult terrain. This part of the country is characterized by high-rising and steep slope mountains. This forms a natural obstacle to man-made road alignment and the engineered slope-cuts. Abha-Al-Darb mountain road of almost 50 km long lies at one of the harshest terrains in Saudi Arabia. A 12 km portion of such mountain road, in this study, that lies along sharp cliff suffers from frequent rock falls, mainly in rainy seasons, in addition to various types of slope failures. The rock masses are mainly schist of high grade metamorphism. The foliation of the schistosity dipping towards the rock slope cuts, causing a frequent occurrence of mainly rockfalls. The man-made rock slope cuts are $70° - 90°$, and reach up to 40 m height. Many of these slopes are dangerous and potentially unsafe due to rockfalls. The source of rock blocks is from upper slope elevations. Absence of ditches and meshes aggravate the conditions under the slopes. In this research, the rock masses were studied in order to identify the decisive parameters of rockfalls, and applied the Rockfall Hazard Rating System on such steep man-made and natural rock slopes. Parameters such as slope angle, restitution coefficient, rolling friction coefficient, bounce coefficient, trajectories, effect of block size and geometry were studied. Colorado Rockfall Simulation Program (CRSP) modeled the most important rolling factors on rockfalls. According to this study, remedial measures were suggested according to the site conditions and the dominant parameters utilizing the latest technology to arresting fallen blocks.

1 INTRODUCTION

Rockfall occurrences along road cuts create considerable risk for human injury and property damage, posing problems for transportation across the mountain road. Consequences of rockfall include impact damage to pavement from falling rocks, rocks on roads posing hazards to commuters, pedestrians, road closure, and environmental impact due to collisions with vehicles (Moore, 1986, Wyllie & Norrish, 1996). Consequently, as the demand for rockfall protection increases (Flatland, 1993).

Rockfalls occur when rock or debris is shed from a road cut or nearby steep slope by processes such as planar sliding, wedge failure, toppling, differential weathering, and raveling onto the catchment and/or road (Norrish & Wyllie, 1996, Sadagah, 1989). Characterization of rockfall hazard along road cuts is necessary for identifying hazard level and prioritizing remediation measures. The characterization includes attributes such as vehicular traffic patterns, roadway geometry, and rock slope geometry (Wyllie & Norrish, 1996). However, a considerable attention was pointed to characterize the role of geology in the rockfall process. Characterizing geology is important, because this factor and its relation to the roadway controls whether material is available to be shed as a rockfall. Approaches to incorporating geology in hazard assessment have included 1) defining hazard by association with rock type (Hadjin, 2002, Vandewater et al. 2005), 2) ascertaining whether geologic discontinuities are oriented favorably or unfavorably with respect to promoting rockfall (Abbott et al., 1998), and 3) describing the rockfall process for rock and soil slopes (Lowell & Morin, 2000).

The goal of this study is to investigate 1) the role of geological setting in causing rockfalls, 2) applying the Rockfall Hazard Rating System on selected dangerous part of a road cut along the mountainous road to evaluate the safety conditions, 3) apply the latest used technique Colorado Rockfall Simulation Program (CRSP) to model the incidents of rockfall failures, and 4) provide the transportation departments and future investigators with correlations and recommended remedial measures that could be used to consider likely rockfall modes and block sizes for a prospective road cuts in similar geologic characteristics.

2 STUDY AREA

The study area is located in the southwestern part of Saudi Arabia. In order to perform the development plans, the Ministry of Transportation has to design, construct, and execute roads in the mountainous highland areas. The purpose of these urban highways are (i) to connect the road networks located on the plateau on the mountainous areas with the lowland areas across the Red Sea escarpment by creating these descents (Figure 1), to render service to residential areas and commuters, (ii) to improve effectively the developing economic and social life, (iii) to improve an absolute solution to transportation and trade problems between the western, south western and southern regions, and (iv) to supplement the existing road network which by the end of 70's decade became insufficient for the high volume of traffic generated by the modernization of the country.

2.1 Location and physiography

The topography in the western, south western, and southern regions of the kingdom along the Red Sea coast is hilly and rugged. These mountains are bounded by sharp cliffs especially in the western parts. These cliffs form the so-called Red Sea escarpment and they reach an altitude of more than 3,000 meter above sea level at the southern regions of the escarpment, this altitude decreasing northwards. Along the Red Sea escarpment lie many descents, especially in the western and south western parts of the mountainous area. In this region, the contour line elevation indicate that the escarpment edge generally has a drop of about 900 meters along a horizontal distance of 700 meters, which forms a slope angle of >45° at the escarpment edge to about 35° at the bottom of the valley, overall. These slope angles decrease toward the mouth of the valley. Locally much greater slopes are found in the field areas of study.

One of the existing descents is Al-Dilaa descent (Figure 2); it is located along Wadi Dilaa on the south fringe of Abha city, and has constructed in a difficult terrain, this descent connects directly the cities of Abha and Jizan.

The descent is located along the edge of the highland forming the eastern part of what is called Arabian Shield. This area is formed of difficult and inaccessible terrains mostly of hard rocks, generally to moderately weathered, with slightly to moderately jointed rock masses, and moderate to steep rock slopes that are covered in variable amounts of soils and debris. The characteristics are different from one place to another according to the changes in rock type, lithology, and origin. The rocks in this descent are mainly igneous and metamorphic in origin. The slope angle of the rocks is as mentioned above.

2.2 Climate

The climate in the study area (western highlands) ranges from hot and dry to humid and rainy. The daily and seasonal temperature variations are significant by about 15°.

Because the areas of this study lie on escarpment which rise to heights of 3,000 meters and above, they are very similar to the mediterranean region. In winter the temperature falls to below zero at nights. The average annual temperatures are about 13.7°.

The winds in the study area come from the Ethiopian highlands across the Red Sea.

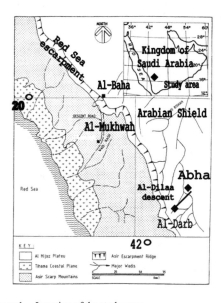

Figure 1. Location of the study area.

Figure 2. The entrance of Al-Dilaa descent at Abha city, where the escarpment edge is shown.

2.3 General setting

The Al-Dilaa descent lies at north of Wadi Dilaa. It connects Abha city at the escarpment with Al-Darb town at downhill. The project begin at Abha city where the upper part of the plateau (chainage km 0.0 to km 14.24); this is where the study area is, and extend for 50 km to Al-Darb town.

The descent starts at the edge of the Red Sea escarpment on a plateau at an elevation of 2,250 meters above sea level and continues down to the beginning of the Dilaa valley at an elevation of about 1,300 meters above sea level. The valley sides at the north end of the descent are very steep because of the sharp decrease of the elevation at the escarpment. The valley runs almost NNW-SSE, meandering between the mountain slopes. The north side of the valley viewed from the valley floor rises sharply up to 500 meters above the valley, and then rises another 250 meters in more gentle slopes. At the south end of the valley the mountain foothills start at an elevation of about 600 meters.

3 BRIEF GEOLOGICAL DESCRIPTION

3.1 Introduction

The study area is located along the escarpment edge of Al-Dilaa descent. It is located in the south western part of the Arabian Shield. Al-Dilaa descent is bounded by Longitudes 42° 29′ 21″ and 42° 31′ 18″ E and Latitudes 18° 9′ 20″ and 18° 11′ 42″ N. the descent is easily accessible; Al-Dilaa descent is forming the south fringe of Abha city.

The Arabian Shield is an extensive occurrence of Precambrian crystalline and metamorphic rocks in the western part of the Arabian Peninsula. It covers an area of 610,000 km^2 and is separated from the African Shield in the west by the Red Sea graben about 25–30 Ma ago (Stoeser and Camp, 1985). Stratigraphic units of the Shield are given by Schmidt et al. (1973) and Greenwood et al. (1980). They are grouped from older to younger in the following lithostratigraphic assemblages: (1) biotite schist and amphibolites; (2) metabasalt-greywacke-chert; (3) metabasalt-andesite; and felsic volcanic rocks.

3.2 Brief geologic description

The rocks at Al-Dilaa descent is partially metamorphosed intrusive plutonic rocks. The tonalite forms a large suite of rocks that intruded layered rocks of the Bahah group (Greenwood et al., 1982; and Stoeser, et al., 1984). The tonalite is light to medium greenish grey and generally medium to coarse grained, fine grained locally. In the field, tonalite is weakly to strongly foliated and has gneissic layering defined by biotite and hornblende concentrations.

A major phase of igneous activity culminated in the formation of monzogranite, granodiorite, diorite and gabbro, renamed collectively by Greenwood (1985) as biotite monzogranite. The field observations show that the contacts are not sharp, and the monzogranite is partially intruded by the granodiorite. Greenwood (1985) grouped the diorite and the gabbro as one unit. Field observations show that the granodiorite is younger than the monzogranite and older than the diorite. A schematic cross section at the Al-Dilaa descent is given in Fig. 3.

The monzogranite at the descent crest is pink, coarse grained, massive to weakly cleaved rock, locally prophyritic. This granite body forms a mixed-rock complex with the diorite, granodiorite and gabbro. The granodiorite is light to dark grey coarse grained and partially gneissed. The diorite is fine grained to medium grained, massive to weakly cleaved, medium to dark greenish grey. The gabbro forms a small plutonic body, massive to moderately cleaved, equigranular, medium to coarse grained.

3.3 Geology and erosion effect on rock jointing

As the Arabian Shield was uplifted 25–30 Ma ago, and because of the intrusive nature of the rocks forming the mountains, the western side of the escarpment is very steep toward the Red Sea and forms a natural obstacle for communications. This escarpment is not sharp all over the continental divide; the different intensities of the erosional cycles and the mechanical properties of the rocks certainly affect their response to erosion processes. Linking the crest with the low lands is a difficult engineering problem.

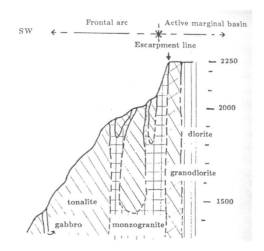

Figure 3. Cross section of the rocks located at Al-Dilaa descent.

The strength and the degree of jointing in the rocks influenced by the processes and cause the altitude of the rock mass to decrease, even to retreat back the escarpment divide line along the high terrains. Al-Dilaa descent is a good example of this process.

The escarpment edge line of Al-Dilaa descent is sharp due to (i) the high strength of the rock materials, (ii) less fractured rocks along the escarpment line (diorite and granite).

4 ROCK MASSES

4.1 Field data acquisition

The main target of the engineering geological study was (1) to record the existing engineering geological, geodynamic and rock conditions of the mapped areas, (2) to demonstrate the suitability of the designed slopes and tendency of rock slope stability, and (3) to map the past and probable future active hazards to the road and to the other existing engineering structures. These data were then to be processed as a case history for creating a safer road design in other areas of similar rock types.

The road was divided into sections and stations as the geological and geotechnical characteristics of the rocks and engineering structures varied along the motorways.

Classical geological mapping pf all the present geological elements was carried out, in addition to mapping of the geotechnical properties of the existing rocks. The engineering geological description of the rocks was based upon the systems of ISRM (1978), Geological Society of London (1977), and Geotechnical Control Office (1984). CSIR, NGI classification systems by Barton et al. (1974) and Bieniawski (1976) were applied in assessing the rock conditions in every section (cut slopes, tunnels, and bridges) along the road alignment at descent.

The collected data comprised (1) for rock slopes; joint survey, joint spacing survey, and joint roughness. A vertical profile was measured for each section along the road cut alignment. In addition, the dimensions and the geometry of each rock slope cut was registered and the spacing between the rock bolts and their number was measured; (2) For every studied section, measurements were made on discontinuities using the Schmidt hammer, shear tests, and point load were covering all rock types along the road; (3) Landslide hazards, failures, and damage were observed and recorded; (4) The water level in wells, seepage and wet areas were marked; and (5) non-destructive test such as density and sonic pulse velocity were done for different rock types. Example of a studied section will be given below.

5 ENGINEERING GEOLOGY OF THE FIELD AREAS

5.1 Introduction

Al-Dilaa descent is generally of steep slopes descending from an escarpment edge, and dissected by the drainage patterns. The descent mountainous is strongly fissured in the upper areas of Wadi Al-Dilaa. The road alignment runs through and area which has a difference in elevation of 1,000 m. most of the problems associated with the construction of the roads arose from (i) the steep escarpment slope, (ii) the drainage patterns, (iii) the unstable natural slopes and (iv) the need to force the route corridors through highly schistosed rock masses.

5.2 Geotechnical parameters

At al-Dilaa descent, the weathering grade of tonalities is generally grade II (slightly weathered), but in some localities the rocks are moderately to highly weathered; in these sections these grades are associated with the existence of groundwater seepage and landslides. The higher value of the weathering grade is associated with the low value for Schmidt hammer rebound number (29). At the higher elevations of the escarpment, the weathering grade is classified as II (slightly weathered).

The engineering solution specified for a rock mass containing weak and weathered rock is shotcrete. The thickness of the shotcrete is 5 cm as setted by Ministry of Transportation, in recognition of the active modes of weathering in progress in the mountainous areas.

The mode of failure is controlled by the orientation of the joints with respect to the loading direction, and by the spacing of the joints in relation with respect to the dimensions of the loaded area. Therefore, joints spacing was measured at the example given.

JRC value for each joint set of the studied rock masses was calculated in addition to the friction angle Φ_b for every rock type, in order to calculate the peak shear strength of the rock types.

The results shows that tonalite rocks shows the highest shear strength value, however these rocks have low JRC values in the field. The granites and the granodiorites show shear strength values lower than the tonalite although the granites and granodiorites have higher JRC values in the field.

Degree of jointing in the study area being related to (1) rock type, (2) geological history, and (3) mechanical properties of the rock. The critical height of steep slopes is dependent not only on the intact strength but also on the orientation and strength of issues and joints.

At Dilaa descent, the granitic rock masses at the edge are widely jointed, the tonalite rocks are closely jointed and schistosed. Although the rock masses at Al-Dilaa descent are classified as low and highly jointed, the rocks vary locally in the intensity of jointing. The general cohesion and hence the stability of the rock masses close to the escarpment edge are likely to be higher than those of the tonalite. Accordingly, the slopes of the granitic rocks are higher in elevation and in the slope angle the rock slopes of the tonalities.

Toppling is another type of rock slope failure. It is controlled by the ratio between the height to the base of the rock blocks, weight of the blocks, the angle of friction and the angle of the joint plane where the block will topple. Toppled blocks and loose boulders eventually accumulate in the form of scree. These screes may be small but can contain big pieces of rock which have become detached from the rock mass and which have fallen as an individual piece.

At Al-Dilaa descent, the toppled rock boulders are not restricted to any particular rock type. Many loose blocks have fallen from above the support of slope face; and generally accumulating in the gullies where they could be forming the basis of a debris flow or a landslide. At Al-Dilaa descent, the free fall of the blocks is triggered mainly by the rain fall and could cause a road closure for clearance. The fallen boulders make holes in the road pavement.

6 MODELING OF THE ROCKFALLS

6.1 Using Rockfall Hazard Rating System

From above description of the geotechnical characteristics at the study area, an example was taken (Fig. 4). The RHRS shows a high values at this section which indicate the necessity of quick solution to take the appropriate remedial measures.

6.2 Rockfall events

Incidents of rockfalls reported frequently along the higher parts from 2260 to 2220 and drop on elevation 2200 m to 2020 m of the descent road, especially after rainfall. The fallen rock blocks vary in number, size and shape according to the technical characteristics of the rock material, rebound coefficient, slope surface roughness. The fracture information, including information on size, roughness and other parameters were measured in the field and exported for analyses to CRSP version 4.0 created by Jones et al. (2000). Results are given in Figures 5 and 6 assuming that the fallen rock blocks are of 1 m diameter.

The results show that the velocity and bounce increase after about 37 meters from the maximum elevation of the slope profile. Further, it reaches the road with a high value of velocity and bounce height. If the diameters of the blocks are smaller, the blocks will have a higher velocity and bounce height. The results shows that as the start point of y-coordinate of rockfall are further high above the hit point (road), the kinetic energy bounce height, and velocity increase. This will cause more damage to the road.

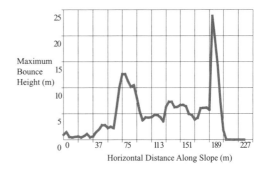

Figure 5. Horizontal distance along slope versus max. bounce height along failure path.

Figure 4. Arial photo showing the steep slopes above the descent highway.

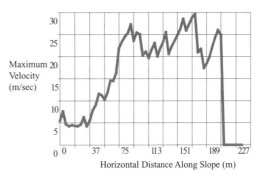

Figure 6. Horizontal distance along slope versus max. velocity along failure path.

7 RECOMMENDATIONS

According to the results of applying HRHS system and CRSP program based on the geotechnical characteristics of the rock masses, it is highly recommended to install a wire mesh at different elevations on the slopes parallel to the road alignment. Elevations where the wire meshes to be installed depend on the results of 1) number of blocks stopped along the profile, which will be provided by CRSP, 2) size of the blocks, and 3) slope profile and roughness.

8 CONCLUSIONS

The Arabian Shield was uplifted 25–30 Ma ago, since then it has been subjected to severe erosional processes leading to the removal of about 1 km of rock, which is greatly decreases the quality of the rock masses.

The nature, the mode of formation and the mechanical properties of the rocks at the topmost elevation along the escarpment are the major factors responsible for shaping the morphology of the descent.

Some of the fallen blocks could fall behind the road and fall in the valley.

CRSP program shows a significant ability to perform simulation of the rockfalls, in addition to back analyses of the unseen rockfall events.

REFERENCES

Abbott, B., Bruce, I., Savigny, W., Keegan, T. & Oboni, F. 1998. Application of a new methodology for the management of rockfall risk along a railway, In Moore, D.P., Hunger, O. (eds.), 8th Congress of the International Association of Engineering Geologists: Vancouver, BC, A.A. Balkema, Netherlands, pp. 1201–1208.

Barton, N., Lien, R. & Lunde, J. 1974. Engineering classification of rock masses for the design of tunnel support. Norwegian Geotech. Inst. Publ. 106, 48p.

Bieniawski, Z.T. 1976. Rock mass classification in rock engineering. Proc. Symp. Exploration for Rock Engineering, Johannesberg, Vol. 1, pp. 97–106.

Flatland, R. 1993. Application of the Rockfall Hazard Rating System to the rock slopes adjacent to US50 and state route 28 in the east side of Lake Tahoe, Nevada: Unpublished M.S. thesis, University of Nevada, Mackay School of Mines, Reno, 316p.

Geological Society of London. 1977. The description of rock masses for engineering purposes. Report by the Geological Society Engineering Group Working Party. Q. J. Engng. Geol., Vol. 10, No. 4, pp. 355–388.

Geotechnical Control Office. 1984. Geotechnical manual for slopes. Engineering Development Department, Hong Kong, 295p.

Greenwood, W.R. 1985. Geologic map of Abha quadrangle, sheet 18F, Kingdom of Saudi Arabia: GM-75C. Saudi Arabian Directorate General of Mineral Resources Geoscience Map GM-77A, p. 27, scale 1:250,000.

Greenwood, W.R., Anderson, R.E., Fleck, R.J. & Roberts, R.J. 1980. Precambrian geologic history and plate tectonic of the Arabian Shield: Saudi Arabian Directorate General of Mineral Resources Bull. 24, 25p.

Greenwood, W.R., Stoeser, D.B., Fleck, R.J. & Stacey, J.S. 1982. Proterozoic island-arc complexes and tectonic belts in the southern part of the Arabian Shield, Kingdom of Saudi Arabia: Saudi Arabian Directorate General of Mineral Resources Open-file Report USGS-OF-02-8, 46p.

Hadjin, D.J. 2002. New York state Department of transportation rock slope rating procedure and rockfall assessment: Transportation Research Record, No. 1786, pp. 60–68.

International Society for Rock Mechanics. 1978. Suggested methods for determining the strength of rock materials in triaxial compression. ISRM Commission on Standardization of Laboratory and Field Tests. Intnl. J. Rock Mech. Min. Sci. & Geomech. Abstr. Vol. 15, No. 2, pp. 47–51.

Jones, C.L., Higgins, J.D. & Andrew, R.D. 2000. Colorado rockfall simulation program version 4.0 for windows manual. 127p.

Moore, H.L. 1986. Wedge rockfalls along Tennessee highways in the Appalachian region: their occurrence and correction. Bulletin Association of Engineering Geologists. Vol. 23, No. 4, pp. 441–460.

Norrish, N.I. & Wyllie, D.C. 1996. Rock slope stability analysis. In Turner, A.K. and Schuster, R.L. (eds.), Landslides: Investigation and Mitigation: Transportation Research Board Special Report 247, National Research Council, Washington, DC, pp. 391–425.

Sadagah, B.H. 1989. Engineering geological maps for road design and construction in Saudi Arabia. Unpublished PhD. Thesis, Imperial College of Science, Technology and Medicine, University of London. 417p.

Schmidt, D.L., Hadely, D.G., Greenwood, W.R., Gonzalez, L., Coleman, R.G. & Brown, G.F. 1973. Stratigraphy and tectonism of the southern part of the Precambrian Shield of Saudi Arabia: Saudi Arabian Directorate General of Mineral Resources Bull. 8, 13p.

Stoeser, D.B., Stacey, J.S., Greenwood, W.R. & Fischer, L.B. 1984. U/Pb zircon geochronology of southern portion of the Nubian mobile belt and Pan-African continental collision in the Saudi Arabian Shield. Saudi Arabian Directorate General of Mineral Resources Technical Record USGS-TR-04-5, 88p.

Stoeser, D.B. & Camp, V.E. 1985. Pan-African microplate accretion of the Arabian Shield. Bulletin of the Geological Society of America, Vol. 96, pp. 817–826.

Vandewater, C.J., Dunne, W.M., Mauldon, M., Drumm, E.C. & Bateman, V. 2005. Classifying and assessing the geologic contribution to rockfall hazard. Journal of Environmental & Engineering Geosciences, Vol. 11, No. 2, pp. 141–154.

Wyllie, D.C. & Norrish, N.I. 1996. Stabilization of rock slope. In Turner, A.K. and Schuster, R.L. (ed), Landslides: Investigation and Mitigation: Transportation Research Board Special Report 247, National Research Council, Washington, DC, pp. 474–504.

Model tests of collapse of unsaturated slopes in rainfall

N. Sakai
National Research Institute for Earth Science and Disaster Prevention (NIED), Tsukuba, Japan

S. Sakajo
Kiso-Jiban Consultants Co., Ltd., Tokyo, Japan

ABSTRACT: The authors conducted two model tests of sandy slope collapse by rainfall in 1G field. The mechanism was investigated very precisely by installing many sensors. The saturation degrees, surface displacement and ground deformation were monitored very well. Through these test results, it was approved that slope collapse could be predicted prior to the failure by monitoring surface displacements and saturation degrees. However, it is not easy in case that the initial water flow was made in the slope before rainfall start. Furthermore, a numerical simulation was conducted coupling seepage analysis and non-linear deformation analysis alternatively, based on the finite element method. The limitation and applicability of the analysis were discussed.

1 INTRODUCTION

Landslide is one of severe natural disasters all over the world. Actually, this can be divided into the 3 research fields in general. 1) The first one is landslide, where is currently sliding in the mountain, with its sliding zone is specified by geologists based on soil investigations. 2) The second one is land collapse by the rainfall or earthquake. 3) The third one is mud water flow mixing soils and gravels. The three fields have been studied individually by the different research groups. The first one has been studied mainly by the geologists. The second one has been studied by a few soil engineers and researchers. The third one has been studied mainly by the national institutes, universities and governmental researchers.

The authors have studied the second one mechanically and experimentally for these years, because of believing this study could contribute to the third field in the future. Slope collapse is very important because it occurs quite often every year and the repairs are very costly, in not the local but also urban areas all over Japan. As the prediction service of slope collapse, several information systems are available by different organizations, which is based on rainfall volume or water infiltration to the ground, although actually it is difficult to predict as shown in the photograph 1. As seen in this photograph, only particular places are damaged in the mountainous area with same rainfall. Therefore, for example, NIED (National Research Institute for Earth Science and Disaster Prevention) provides the real time rainfall information service through the internet of the past record of slope collapses.

The authors have studied this slope collapse from the points of water and deformation to establish a

Photo 1. Typical slope collapse by rainfall.

proper prediction method based on a rational sol investigation. Their challenging numerical analysis has been developed on finite element method, based on seepage analysis and deformation analysis. The numerical procedure could simulate at displacement of slope by rain at a site of slope in Hiroshima Japan (Sasahara, Sakai, Sun & Sakajo, 2006). This is very unique because many researches focus on either water (Tohari, Nishigaki & Komatsu, 2007, Orence, Shimoma, Maeda & Towhata, 2004) or deformation (Sasahara, Ebihara, Tsunaki & Tsunakai, 1996). Furthermore, the numerical procedure could explain the difference of deformations by the different rainfall patterns (Sasahara Kurihara, Sakai, Sun & Sakajo 2007). A further necessary aspect is, of course, to know the more precise mechanism of slope collapse. Then, the authors conducted a series of model tests, which is limited to the sandy slopes in the laboratory. The mechanism of displacement prior to the slope collapse was focused from the points of water and deformations. Then, two different patterns of deformations were observed, which is quite interesting to explain the collapse mechanisms on sandy slope prior to failure.

2 MODEL TESTING

2.1 Test device and soil material

Figure 1 shows model test container with sensors, which were to monitor the deformations, changes of water contents and pressures precisely in order to investigate the mechanism of slope collapse in realizing the different conditions. Size of container is with a height of 60 cm and a length of 200 cm. Slope angle is 45 degrees, which is quite steep.

Moisture content gauges (VW1~VW6) and water pressure gauges (PW1 & PW2) were installed. Water level gauges (1~10) were also installed. Displacements are measured with laser sensor (L1 & L2) dial gauges (D1 & D2) and cupper thin wall (G1~G4 & G5~G8).

About the used soil material, its fineness content is 3.1% and sand content is 96.9%. Particle size of sand particle is 0.211 mm of D50, with the uniformity factor of 1.946. The maximum dry density is 1.571 (g/cm^3) and the minimum dry density is 1.240 (g/cm^3). Then it was compacted to be 1.34 (g/cm^3), which is loose.

The suction pressure and water content curves on wetting and drying processes obtained as shown in Figure 2. Two curves in absorption and drainage of water were tested, which are not much different each

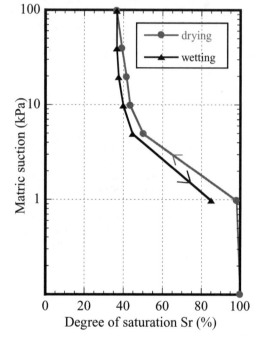

Figure 2. Suction and saturation curve.

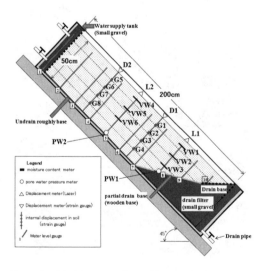

Figure 1. Model test container with sensors.

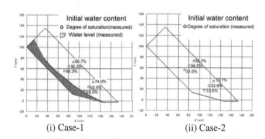

Figure 3. Initial water table for two cases.

other. The air entry suction pressure 1 to 3 kPa. The coefficient of permeability of saturated soil is 2.0×10^{-2} cm/sec.

2.2 Collapse procedure

Two model tests of slope collapse by rainfall were explained. One is under rainfall and another is with initial ground water flow prior to rainfall. The former is Case-1 and the later is. Case-2. Case-2 had a ground water flow and rainfall, where rainfall started after the first 9,000 sec with ground water flow. Rainfall intensity for the both cases is 150 mm/h, provided from water pipes with small holes at the searing. Figure 2 shows the initial water tables differences at rainfall start. The initial water flow in case-2 was made by supplying from water tank at the top of container until to form 10 cm high water table from the bottom of container. And, from many preliminary trials of testing, gravels were installed on the tip with two stepped gentle slopes, to avoid the failure stopping.

2.3 Test results

Figures 4 & 5 show test results of Case-1 and Case-2, with three sets of figures. From the top, the first figure show the saturation degree (VW1 to VW3) and water pressure curves (P1) on elapsed time, for the section located at the lower portion of the slope. The second figure shows the saturation degree (VW4 to VW6)

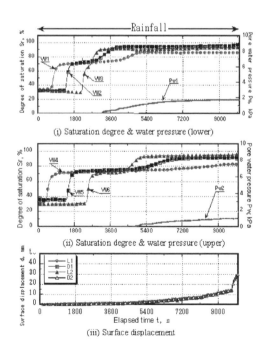

Figure 4. Monitoring data (Case-1).

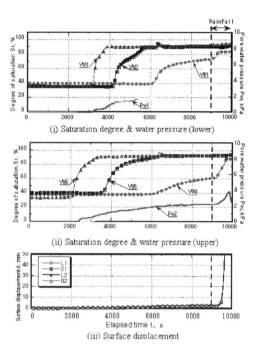

Figure 5. Monitoring data (Case-2).

and water pressure curves (P2) on elapsed time, for the section located at the upper portion of the slope.

The last figure shows surface displacements along the slope measured with laser sensors (L1 & L2) and with strain gauges (D1 & D2).

<Case-1>

From the top figure in Figure 4, the saturation degrees increased on the stepped curve with three stages at the each depth VW1, VW2, VW3 at the lower position of slope. VW1 at the highest position began to increase earlier at the first saturation stage than VW3 at the lowest position. VW2 comes between them. But this order of VW1 and VW3 became inversed at the second and third saturation stages because of the water penetrated from rainfall rose up to near the slope surface. The water pressure PW1 increased from the third stage of water saturation, which shows the delay of water table formation in the ground from rainfall.

From the second figure, the water saturation also increased on the stepped curve with three stages at the each depth VW4, VW5, VW6 at the upper portion of slope. The order of saturation magnitudes at VW4, VW5 and VW6 changes on elapsed time with the same manner of VW1, VW2 and VW3. The water pressure PW2 increased from the third saturation stage as well as PW1 by the same reason. The all sensors at the lower position of slope increased earlier than those at the upper position because penetrated water easily gathers at the down slope.

Meanwhile, from the third figure, the displacement on the slope surface increased corresponding to the water pressure increased gradually. All the sensors showed the almost same values. The first small increase was observed at the time almost 5,800 sec passed. The second large increase was observed at the time 9,800 sec passed. Therefore, two turning points of displacement increase were observed on the curve in Case-1. Then it finally yielded a slope failure after the second turning point, where only PW2 at the upper position of the slope. It suggested that the propagation of water table to upstream caused a failure.

<Case-2>
Case-2 has water flow in the slope for the first 9,000 sec. Therefore, water table was already made before rainfall. From the top figure in Figure 5, the saturation degree increased on the stepped curve with two stages at the each depth VW1, VW2, VW3 at the upper position of slope. On the contrary with Case-1, VW3 at the lowest position began to increase earlier at the first saturation stage than VW1 at the highest position. The position is, the higher the higher magnitude is. VW3 at the lowest position is always higher than VW2 and VW1 at the higher positions, although VW2 and VW1 increased to catch up VW3 gradually. The water pressure PW1 increased from the first saturation stage. This is very different from Case-1. Unfortunately, the measurement of PW1was stopped by accident.

From the second figure, the water saturation degree also increased on the stepped curve with two stages at the each depth VW4, VW5, VW6 at the upper position of sloe. The order of saturation magnitudes of VW4, VW5 and VW6 are same with VW1, VW2 and VW3. The water pressure PW2 increased gradually from the first stage of saturation degree as well as PW1. But PW2 showed a peak then to be decreased at the final stage, where soil might be fully saturated and softened by failure.

Meanwhile, the displacements on the slope surface started to increase corresponding at 9,000 sec passed, when the rainfall started. Then it shortly became large drastically to yield slope failure. This is much larger than Case-1. By comparing these two cases, it was found that the initial seepage is quite influencing to the slope failure mechanism.

2.4 *Deformation and mechanism*

Two different slope deformation processes to collapses were obtained with the same conditions of slope angle and rainfall. In order to investigate slope collapses from the point deformation of slope, the authors measured lateral ground displacements to the depth. Figs. 6 and 7 show the lateral displacements at G1~G4 at the lower position of slope and G5~G8 at the upper positions of slopes for Case-1 and Case-2 respectively.

Figure 6 shows the result of Case-1. It can be seen that the lateral displacements deformed like a bending column. From this, it was accumulated increasingly from the bottom around the depth of 20 cm to the surface on time passing. Soil was getting much weaker at the surface side by the rainfall and this slope collapse seems to be sallow slope failure. At the time 5,400 sec passed, lateral displacements started to increase drastically to the failure. Comparing the lower with the upper position, G1~G4 showed smaller displacements at the depth below 20 cm than G5~G8, because G1~G4 at the lower position is located near at the end of slope, where meets another flat slope.

Figure 7 shows the results of Case-2. There are a big difference between G1~G4 at the lower position and G5~G8 at the upper position. In the position, the lateral displacements deformed like a bending column as well as Case-1. However, the lateral displacement was

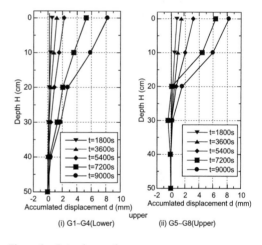

Figure 6. Lateral ground.

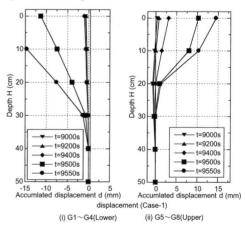

Figure 7. Lateral ground displacement (Case-2).

not like this at the upper position. It shows an inverse trend of displacement increase at the upper position because the water table increase from the bottom and waterfall increase from rainfall yielded complicated weakness of soils in the slope. Focusing the lower position, lateral displacements seems rather small until the time 9,000 sec passed and then it started to increase drastically to the failure. The depth of occurrence of lateral displacement is almost same at the depth upper 30 cm, although Case-2 had the lager slope failure than Case-1.

3 NUMERICAL ANALYSIS

Furthermore, the authors have conducted to simulate these results. As aforementioned, the failure mechanism is complicated, coupling water behavior with deformation. The authors have proposed the new numerical method, which could couple seepage analysis and deformation analysis on the finite element method (Drager-Prager model considering strength reduction with saturation). Most of numerical analysis has not used actual rain records although it is very important to judge stability of slopes. The deformation varies very much by the changes of rain fall volume and sequence. If water content increases by rainfall and seepage, the loss of strength generates the deformation. The used soil parameters are tabulated in Table 1. The suction and saturation curve was employed as shown in Figure 2.

The computation results of Case-1 is shown in Figs. 8 to 10. Figure 8 shows the deformations at the time 1,800 and 10,800 sec passed, respectively. Figure 9 shows the saturation degree at the time 1,800 and 10,800 sec passed, respectively. Figure 10 shows the computed lateral displacement at G1~G4 and G5~G8.

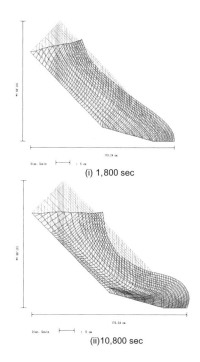

Figure 8. Computed deformation (Case-1).

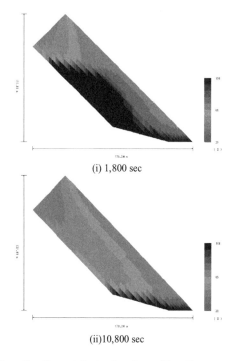

Figure 9. Computed saturation degree (Case-1).

Table 1. Soil parameter used in the analysis.

Parameter	Values
The coefficient of permeability, ks	0.0199 (cm/sec)
Unsaturated properties	See Figure 2
Saturated volumetric water content, θs	0.49
Unit weight, γ	1.85 (gf/cm^3)
Specific gravity, Gs	2.73
Frictional angle, ϕ	35 (degree)
Void ratio, e	1.02
Cu with suction Su (Sasahara et al., 2006)	Cu (Su) = 0.2816 * Su$^{0.433}$
Young's Modulus, E	200 (kgf/cm^2)
Poisson's ratio, ν	0.3

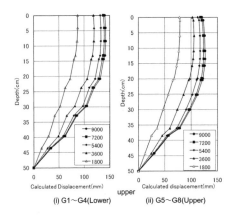

(i) G1~G4(Lower)   (ii) G5~G8(Upper)

Figure 10. Computed lateral displacement (Case-1).

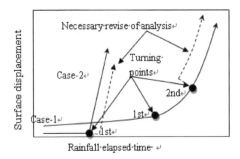

Figure 11. Schematic displacement and time passing.

4 REMARKS

Through the results of two model tests including numerical analysis, the clear difference between Case-1 and Case-2 can be summarized on the relations between displacement and rainfall elapsed time as shown in Figure 11.

As seen in this figure, Case-1 has two turning points to increase the surface displacements but Case-2 has only one. In Case-1, it is easy to predict slope collapse by monitoring the increase of displacement between the first and second turning points. However in Case-2, it seems not easy to predict. If a water flow with a certain water table was made in the slope prior to the slope failure by some reasons, monitoring displacement for prediction of slope collapse could not work for a sudden slope failure. In such cases, the results of Case-2 suggests the importance of monitoring saturation degree and water pressure instead of displacement monitoring. One of author has proposed a water content monitoring widely based on internet techniques in the mountainous area (Okayasu, Hadano, Inoue, Mitsuoka & Sakajo, 2007).

5 CONCLUSIONS

Two experimental model tests were conducted in a sandy steep slope with angle of 40 degrees. Focusing the saturation degree and deformation, the authors investigated the mechanism of the slope collapse precisely. The main conclusions were obtained as follows.

1. If the saturation increases by rainfall, the deformation of ground of slope increases. Generally, the surface displacement should be monitored because it is weakened more than the lower portions.
2. Saturation degrees increase with two or three stepped stages prior to slope collapse. The final stage is where saturation degree is nearly 90%. To find this stage must be the key to the prediction.
3. From Case-2, slope might be collapsed drastically where water flow was made to prior to rainfall stat. In this case, a rather large slope failure will be occurred by softening of soils.
4. The used numerical analysis could explain the deformation and seepage by rain in general, although there are some limitations. It will be useful to simulate the mechanism of slope collapse by rainfall.

REFERENCES

Okayasu, T., Hadano, R., Inoue, E., Mitsuoka, M. & Sakajo, S. 2007. Possibility of application of distributed agricultural information network to geotechnical field. *The 43rd Annual Conference of JGS*, Nagoya, pp.2031–2032 (in Japanese).

Orense, R.P., Shimoma, S., Maeda, K. & Towhata, I. 2004. Instrumented Model Slope Failure due to Water Seepage. *Journal of Natural Disaster Science*, 26 (1), 15–26.

Sakai, N., Yamakoshi, T., Kuriara, J., Sasahara, K. & Morita, K. 2006. Modeling and the deformation characteristics of slope of weather granite. *Annual conference of Japan Society of Erosion Control Engineering*, Wakayama, pp.128–129 (in Japanese).

Sasahara, K., Ebihara, K. & Tsunaki, R. 1996. Experimental study on mechanism of steep slope failure. *Journal of land slide*, 32 (4), pp.1–8 (in Japanese).

Sasahara, K., Kurihara, J., Sakai, N., Sun, Y. & Sakajo, S. 2007. Prediction of surface displacement and collapse critical time on slopes with rain penetration by a FEM analytical technique. *Proc. The 43rd Annual Conference of JGS*. Nagoya, pp.22–23 (in Japanese).

Sasahara, K., Sakai, N., Sun, Y. & Sakajo, S. 2006. Proposal of slope failure analysis on 2D FEM. *The 42nd Annual Conference of JGS*, Kagoshima, pp.2191–2192 (in Japanese).

Tohari, A., Nishigaki, M. & Komatsu, M. 2007. Laboratory rainfall-induced slope failure with moisture content measurement. *Journal of Geotechnical and Geo-Environmental Engineering*, 333 (5), pp.575–587.

Calibration of a rheological model for debris flow hazard mitigation in the Campania region

A. Scotto di Santolo & A. Evangelista
Department of Geotechnical Engineering, University of Naples Federico II, Naples, Italy

ABSTRACT: The paper reports the results of the back-analyses of the propagation of debris flows in pyroclastic (deposits in the Campania region, Italy. 57 well-documented case histories were analyzed using the 2-3D DAN-W) code (Hungr, 2003) with two different rheological models: the frictional model and the Voellmy model. The latter produces the more consistent results in terms of total runout, debris spread and distribution as well as velocity data. The results show that it may be possible to model past events reasonably accurately using the Voellmy model. Although it is difficult to make predictions about future landslides, the calibrated model could be used to predict their propagation if the detachment area and the morphology are known.

1 INTRODUCTION

Rapid long runout landslides represent a difficult challenge in hazard studies because they pose a risk to areas situated a considerable distance from the source. The prediction of runout distance, flow velocity and depth (hereafter referred to as dynamic parameters) are necessary for designing protective measures and are a key requirement for the delineation of the hazard zone. The best existing prediction methods rely on empirical relationships between volume, travel distance, angle of reach or Fahrböschung etc. (e.g. Scheidegger, 1972, Corominas, 1996, Scotto di Santolo, 2000; Fannin & Wise 2001) but they do not make it possible to predict all the dynamic parameters.

This paper is concerned with different kinds of debris flows that have occurred in the pyroclastic deposits of the Campania region (southern Italy). According to other researchers (Di Crescenzo & Santo, 2003; Scotto di Santolo, 2000), three main types of landslides have been detected: un-channelled, channelled and mixed debris flows. 57 well-documented case histories which were suitable for back-analysis were selected. Each of the case histories was analyzed in 3D dimensions using DAN-W (Dynamic Analysis of Landslides, Hungr, 2003) with two different rheological models: the frictional model and the Voellmy model (1955).

The results of each analysis were evaluated by matching the following parameters to the values as determined from maps and on site survey: total horizontal distance (or runout) and flow velocities. Unfortunately, only in a few cases do we have information about the depth of the deposition fan.

The paper reports calibration procedure of the most suitable rheological model for the analyzed debris flows.

2 CASE HISTORIES

2.1 Geotechnical properties of the pyroclastic deposits

57 debris flows that took place in the Campania region between 1973 and 1998, which are well-documented in the literature (Del Prete et al., 1998; Scotto di Santolo, 2000; Di Crescenzo and Santo, 2005) were analyzed (figure 1). The events occurred in the following zones:

1. The Phlegrean Field of which the city of Naples is a part;
2. The carbonatic ridge comprising the Sorrento peninsula (Monti Lattari) and the mountains of Sarno-Quindici.

The soil concerned regards the most recent pyroclastic deposits deriving from the volcanic activity of the Phlegrean Fields in zone 1 and from the volcanic activity of mount Somma/Vesuvius in zone 2. The substratum underlying the above-mentioned cover is of the same volcanic nature in zone 1 and of a carbonatic nature in zone 2. In granulometric terms, the unstable cover consists of sandy silts or slightly clayey or gravelly silty sands. The clayey part is slightly plastic though only in the Vesuvian deposits. The gravelly part mainly consists of pumices and, to a lesser extent, of scoriae and lapilli. The particles are mainly siliceous. Their structure is amorphous and porous;

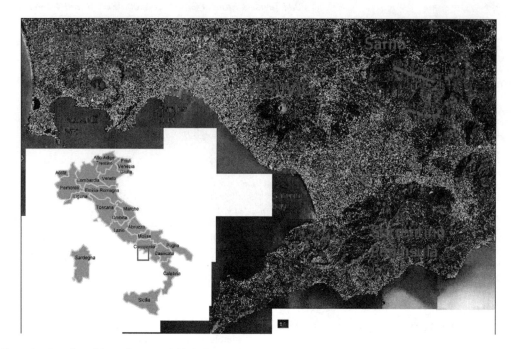

Figure 1. Location of the study area and debris flows.

there is a double porosity system inter and intra particle (not connected to the surface). The specific gravity of solids is 2.54. Porosity exceeds 70% for Vesuvian deposits while it is lower for the Phlegrean deposits. The cover is partially saturated; the level of saturation varies according to weather conditions.

Mechanical behaviour is extremely variable; the peak friction angle depends on stress, density and is a function of the degree of saturation. The ultimate friction angle of the material varies between 35° and 37°; the higher value relates to the Vesuvian deposits (Picarelli et al., 2006). From a phenomenological perspective, it can be observed that the shear strength of non-saturated deposits, in terms of total stress, gives rise to un intercepted cohesion, known as apparent cohesion, which increases with suction according to Fredlund and co-workers criteria (1978) (Scotto di Santolo, 2000b; Picarelli et al., 2006). Mechanical behaviour in conditions of partial saturation still requires more detailed research. Edometric compression tests and direct shear tests revealed that, at a constant total stress, the reduction in suction always leads to structural collapse and a reduction in strength (Scotto di Santolo et al., 2000).

The reduction in shear strength due to the infiltration of rainwater is considered to be the triggering mechanism of the landslide phenomena considered in this paper. When rainwater seeps into the pyroclastic cover, not only does it increase the level of saturation of the soil, thereby causing the existing suction to diminish, but it also, and simultaneously, creates more unfavourable conditions of water circulation in the portion of soil closest to the surface according to in situ suction monitoring (Scotto di Santolo, 2000; Scotto di Santolo & Evangelista, 2004; Evangelista et al., 2007).

The post failure behaviour (or mobility) of the landslide is conditioned by several factors such as the geometry of the slope, the mechanical properties of the deposits cover (porosity, grain size distributions) and the characteristic of rainfall before and after the triggering. This is clearly demonstrated by the comparison between the channelled flows and the flows on the open slope (unchannelled). In the former, the presence of the channel allows a higher concentration of water and therefore greater fluidification of the landslide mass.

According to some authors, the triggering mechanism is attributed to the development of an undrained mechanism (static liquefaction) (e.g. Sassa, 1988). However, in the opinion of the present authors, there are still uncertainties regarding the role played by liquefaction during the triggering phase of unsaturated deposits; it is believed that it contributes to subsequent fluidification (Eckersley, 1980).

2.2 Geometrical and dynamic characteristics of landslides

An inventory of 57 debris flows was collected containing geomorphological, geotechnical and dynamic data. In table 1 some of the available information was reported. The landslides were divided into three distinct types: un-channelled, channelled and mixed debris flows (Figure 2). For each flow, the following information was evaluated according to Figure 3:

- morphometric data: (slope angle of the crown and the sliding and deposition zone and of active zone α_a, the extent of the area, thickness, the difference in height between the crown and the toe of landslide H, trigger volume V)
- dynamic data (runout L)
- unstable deposit type and bedrock
- geotechnical properties.

30% of the landslides were of the channelled type, 28% were un-channelled while 42% were of a mixed typology. The angle of reach (tan (H/L)) was related to the volume of the landslide and to the travel distance as reported in the literature. As volume increases, so does mobility according to other observed flows (Corominas, 1996; Finlay et al., 1999). However this relationship appears to be scattered. The relationship with H was plotted in figure 4a and was selected due to the higher degree of correlation. The angle of reach decreases with H.

Landslides in the Phlegrean Fields are characterized by H less than 200 m and have a high angle of

a)

b)

Figure 2. Main types of landslides detected: a) un-channelled (Nocera, 2005); b) mixed (Pozzano, 1997) debris flows.

Table 1. Debris Flow inventory.

ID	Commune	Date	path	altitude of the crown m (a.s.l.)	L (m)	H (m)	α_a (°)
1	Bracigliano	5.5.1998	M	850	3300	630	38.7
2	Bracigliano	5.5.1998	M	850	3695	610	45.0
3	Bracigliano	5.5.1998	I	975	3590	720	38.3
4	Castellammare di Stabia	22.2.1986	VA	200	177.5	78	29.1
5	Castellammare di Stabia	10.1.1997	M	465	828	460	41.4
6	Castellammare di Stabia	10.1.1997	M	500	127.5	80	41.8
7	Castellammare di Stabia	10.1.1997	M	525	110	100	49.6
8	Corbara	10.1.1997	I	525	310	170	32.8
9	Corbara	10.1.1997	M	840	1145	565	35.2
10	Corbara	10.1.1997	I	725	855	450	36.1
11	Gragnano	2.1.1971	VA	315	325	193	36.7
12	Gragnano	10.1.1997	VA	265	242.5	130	39.3
13	Gragnano	10.1.1997	VA	285	102.5	65	37.3
14	Massa Lubrense	16.1.1973	VA	460	340	240	37.3
15	Moschiano	10.1.1997	M	600	760	280	36.3
16	Moschiano	10.1.1997	VA	520	630	195	31.8
17	Napoli	16.12.2005	VA	145	55	64	39.3
18	Napoli	4.3.2005	VA	250	252.32	195	51.7
19	Napoli	4.3.2005	VA	250	320	205	48.8
20	Napoli	4.3.2005	VA	255	265	190	42.8
21	Napoli	10.1.1997	M	420	250	250	50.9
22	Pagani	10.1.1997	VA	186	226	127	34.8
23	Palma Campania	22.2.1986	M	300	419	221	38.1
24	Piano di Sorrento	10.1.1997	VA	575	255	230	52.1
25	Quindici	10.1.1997	M	560	776	258	40.8
26	Quindici	5.5.1998	M	850	2715	640	38.7
27	Quindici	5.5.1998	M	925	2850	730	34.2
28	Quindici	10.1.1997	M	525	417.5	260	37.1
29	Quindici	10.1.1997	M	640	589	220	37.4
30	Quindici	5.5.1998	M	875	1870	650	41.5
31	Quindici	5.5.1998	I	850	1680	570	26.6
32	Quindici	10.1.1997	I	515	405	225	36.3
33	Quindici	5.5.1998	I	725	2722	540	36.9
34	Quindici	5.5.1998	M	900	2460	655	46.8
35	Quindici	5.5.1998	M	700	1660	525	58.9
36	Quindici	5.5.1998	M	525	783.2	287	33.8
37	Quindici	5.5.1998	I	775	1715	565	32.0
38	Salerno	4.3.2005	VA	395	497.33	290	33.1
39	Sarno	5.5.1998	I	625	1670	580	45.0
40	Sarno	5.5.1998	I	875	3495	835	33.8
41	Sarno	5.5.1998	I	775	3520	735	26.6
42	Sarno	5.5.1998	I	775	2810	735	41.0
43	Sarno	5.5.1998	I	900	3175	860	63.4
44	Sarno	5.5.1998	M	750	2190	735	33.7
45	Sarno	5.5.1998	M	775	2760	705	40.0
46	Sarno	5.5.1998	I	975	3270	895	26.6
47	Sarno	5.5.1998	M	875	2980	800	36.9
48	Sarno	5.5.1998	I	950	3560	880	45.0
49	S. Egidio del M.te Albino	10.1.1997	M	320	610	235	33.4
50	Siano	5.5.1998	I	575	1061.5	260	45.0
51	Siano	5.5.1998	M	575	1440	445	39.8
52	Siano	5.5.1998	M	475	1215	425	33.8
53	Trasaella	11.12.1996	VA	375	127.5	100	39.8
54	Vico Equense	10.1.1997	M	425	210	135	40.4
55	Vico Equense	10.1.1997	VA	480	230	145	35.1
56	Vico Equense	23.1.1966	I	225	237.5	130	43.6
57	Vico Equense	10.1.1997	M	375	235	125	30.8

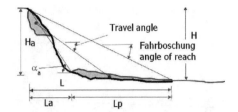

Figure 3. Definition of term.

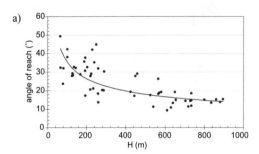

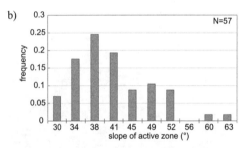

Figure 4. Relation between morphometric data.

reach. Those in the Vesuvian area have a lower angle of reach where H is less than 400 m. For higher levels of H (the mountains of Sarno), the angle of reach is still lower and decreases with H until a stationary value. Therefore the debris flows with H higher than 400 m displaying greater mobility depending on the fluidification of the mass (usually for channelled kind). In figure 4b the slope of the active zone α_a (crown and sliding zone) were also reported. This slope is greater than 30° with a maximum frequency of about 38°.

For each landslide the longitudinal and transversal profiles were plotted (scale 1/5000 or 1/2000) and were then used for the numerical analyses with the calculation program DAN-W (Hungr, 2003).

3 NUMERICAL ANALYSES

3.1 Introduction

DAN-W is a windows-based program used to model the post-failure motion of rapid landslides (Hungr, 1995). It is based on shallow flow assumptions ($H \ll L$, with H flow depth normal to the base). This dynamic model is based on the Lagrangian solution of St. Venant's equation. This equation can be derived by applying the conservation of momentum to thin slices of flowing mass which are perpendicular to the base of the flow. These "boundary blocks" divide the slide mass into n "mass elements" of constant volume.

The following input data were used: the trigger volume, the geometry of the slope (2D), the wide of the channel and the constant erosion depth. The code gave as an output the velocity and thickness of the sliding mass both along the slope and at a pre-specified location along the path as functions of time.

Eight rheologies are available in DAN-W (Hungr, 1995). In this study, however, only two rheological models were analyzed:

- *The Friction model*, where the flow resistant term was controlled by the effective normal stress on the base of the boundary block;
- *The Voellmy fluid model* (1955), where the resistance is a function of a friction term and a turbulent term (Hungr, 1995).

The basal flow resistance term is governed by the rheology of the material and can be expressed thus:

$$\tau = \gamma H_i \left(\cos \alpha_i + \frac{a_c}{g} \right) \cdot (1 - r_u) \cdot \tan \varphi \qquad (1)$$

for friction flow

$$\tau = \gamma H_i \left(\cos \alpha_i + \frac{a_c}{g} \right) \mu + \gamma \frac{v^2}{\xi} \qquad (2)$$

for a Voellmy fluid

where γ is the unit weight of the flowing material, $a_c = v_i^2/R$ is the centrifugal acceleration resulting from the vertical curvature of the flow path R, r_u is the pore-pressure coefficient (ratio of pore pressure to total normal stress at base of boundary block), φ is the friction angle; μ is a friction coefficient and ξ is a turbulence coefficient with dimensions of [m/s^2].

3.2 Calibration procedure

As already mentioned, the friction model and the Voellmy models were selected from the rheological models in the calculation program. The model was calibrated with reference to:

- distance L, calculated horizontally between the edge of the mound and the crown;
- velocity of the front v_f;
- and, only in certain cases, with respect to the height of the deposit.

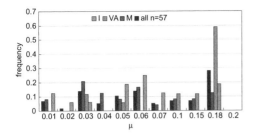

Figure 5. Results of the back analyses with Voellmy model ($\xi = 100$ m/s^2).

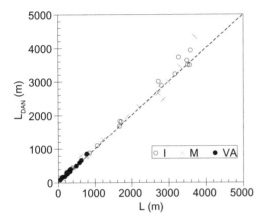

Figure 6. Comparison between numerical runout L_{DAN} and measured ones L with Voellmy model ($\mu = 0.03$ and $\xi = 100$ m/s^2).

With regard to velocity, it should be noted that there is no on site data, only estimates obtained from the back analysis of the damage caused by the flows that took place in the area of Sarno-Quindici (Faella and Nigro, 2003) or from semi-empirical expressions (Zanchetta et al., 2004).

In the analyses undertaken for the study, the model was considered valid on the basis of a comparison of the distance ($L_{DAN} \geq L$). The second parameter for calculation was the maximum speed of the front which was set as no greater than 12 m/s, in the absence of experimental data.

The results obtained were analysed together and the flows were then separated according to the specific route (channelled flow, un channelled flows or a mixture of the two).

3.3 Results of numerical analyses

Of the two models that were analysed in this study, Voellmy's model proved to be better for interpreting the phenomena involved. The use of this rheological model made it possible to consider the dissipative effects of the chaotic motion of water and solid grains together.

It was also observed that with the frictional model, velocities were significantly overestimated (Hungr, 1995). Moreover, Voellmy's model has been widely confirmed in similar contexts (Hungr & Evans, 1996; Fiorillo et al., 2001; Revellino et al., 2004) also using three-dimensional geometric models (Pirulli, 2004; Mc Dougall & Hungr, 2005) both in terms of distances travelled and in terms of the velocity of the volumes mobilized.

The interval of variation of the parameters ξ and μ already used for the landslides in Campania was acquired from bibliographical research. The analyses were carried out by letting μ vary over an interval of values ranging between 0.01 and 0.2 and letting ξ equal to 100 and 200 m/s^2. Figure 5 shows the range of values of the parameter μ, having set ξ at 100 m/s^2 so that L_{DAN} was quite equal to L. It was observed that, for the channelled flows, the most frequent value of μ was 0.06, while the value for the flows on open slopes was 0.18 and was 0.03 for mixed. These values correspond to the observation of greater mobility of channelled landslides compared to landslides on open slopes for which dissipative phenomena occurred, linked to the widening of the transversal section along the path. The two values chosen for parameter ξ lead to practically coincident distances travelled, but at significantly different velocities. The best fit μ value varies in a wide range. The frequency is higher in particular for three values of μ: 0.03, 0.06 and 0.18. Thinking to a possible prevision of this phenomenon through these values for safety results the study was carried out with a value of 0.03 for μ and 100 m/s^2 for ξ.

Figure 6 compares the observed distances and the distances calculated using these parameters for channelled, unchannelled and mixed flows. For flows on open slopes (modest value of L), it was observed that the friction coefficient μ had a reduced influence. For channelled flows, on the other hand, a perfect coincidence was observed between the measured and calculated results for L up to 1500 m. For greater lengths, the model displays a slight tendency to overestimate the theoretical routes (according to the results in Figure 4). In table 2 the synthesis of the results were reported.

Figure 7 shows the velocity profiles of the front and rear, obtained with DAN using the above-mentioned model for the three kind of path analyzed. It can be seen that the range of the velocities reaches a maximum at around the trigger area and then decreases almost with exponential law for the all data. Of course the used model overestimates the distance for unchannelled and channelled ones. For this reason it could be better using the Voellmy model with three

Table 2. Synthesis of the results with Voellmy model ($\mu = 0.03$ and $\xi = 100$ m/s^2).

L (m)	L$_{DAN}$/L	
	Mean value	Dev. ST
0 ÷ 500	1.18	0.14
500 ÷ 1000	1.06	0.09
1000 ÷ 4000	1.04	0.06

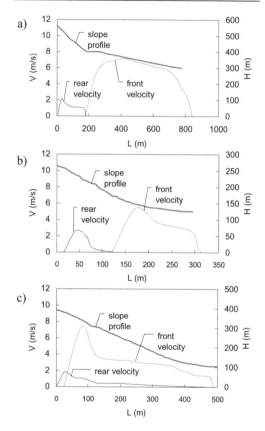

Figure 7. Slope and velocity profile calculated with DAN: a) Channeled; b) Un-channeled; c) Mixed.

different values of friction coefficient according to figure 5. The velocity profile is very useful for predicting the dynamic parameters of future events and therefore for mitigating risk in similar zones.

4 CONCLUSION

Rapid long runout landslides represent a difficult challenge in hazard studies because they endanger areas situated far from the source. Predictions of the runout distance, flow velocity and depth are necessary for planning and designing protective measures and are a key requirement for delineating the hazard zone. 57 well-documented case histories which were suitable for back-analysis were selected. Each of the case histories was analyzed in 3D using DAN-W (Dynamic Analysis of Landslides, Hungr, 2003) with two different rheological models: the frictional model and the Voellmy model. The results of each analysis were assessed by matching the following parameters to the values as determined from maps and on site survey: total horizontal distance (runout) and flow velocities. The Voellmy model produces the most consistent results in terms of total runout, debris spread and distribution, as well as velocity data.

The results show that past events can be modelled with reasonable accuracy using the Voellmy model; however, it is still difficult to make accurate predictions concerning the most likely runout.

REFERENCES

Ayotte, D. & Hungr, O. 2000. Calibration of a runout prediction model for debris-flows and avalanches. Wieczorek G, Naeser (eds) Debris flows hazard mitigation. Balkema, Rotterdam, pp 505–514.
Corominas, J. 1996. The angle of reach as a mobility index for small and large landslides. Can Geotech J 33:260–271.
Di Crescenzo, & Santo, A. 2005. Debris slides-rapid earth flows in the carbonate massifs of the Campania region (southern Italy): morphological and morfometric data for evaluating triggering susceptibility. Geomorphology, 66, pp. 255–276.
Del Prete, M., Guadagno, F.M. & Hawkins, A.B. 1998. Preliminary report on the landslide of 5 May 1998. Bull Eng Geol Environ 57:113–129.
Eckersley, D. 1990. Instrumented laboratory flowslides. Geotecnique J., 40, pp. 873–885.
Evangelista, A. & Scotto di Santolo, A. 2001. Mechanical Behaviour of unsaturated pyroclastic soil. Proc. Landslides: Causes, Impacts and Countermeasures, Davos-Switzerland, June 2001.
Evangelista, A. & Scotto di Santolo, A. 2004. *Analysis and field monitoring of slope stability in unsaturated pyroclastic soil slopes in Napoli, Italy*. Proc. 5th Int. Conf. on Case Histories in Geotechnical Engineering, New York 2004.
Faella, C. & Nigro, E. 2003. Dynamic impact of debris flows on the constructions during the hydrogeological disaster in Campania 1998: failure mechanical models and evaluation of the impact velocity. Proc. Int. Conf. FSM, Naples, 1, pp.
Fannin, R.J. & Wise, M.P. 2001. An empirical-statistical model for debris flow travel distance. Can Geotech J, 38, pp. 982–994.
Fiorillo, F., Guadagno, F.M., Aquino, S. & De Blasio, A. 2001. The December Cervinara landslides: further debris flows in the pyroclastic deposits of Campania (southern Italy). Bull Eng Geol Environ, 60: 171–184.

Fredlund, D.G., Morgenstern, N.R. & Widger R.A. 1978. *The shear strength of unsaturated soils.* Canadian Geotech. J., 15: 313–321.

Hungr, O. 1995. A model for the runout analysis of rapid flow slides, debris flows, and avalanches. Canadian Geotechnical Journal, 32: 610–623.

Hungr, O. & Evans, S.G. 1996. Rock avalanche runout prediction using a dynamic model. Proc. Landslides, Senneset (ed.), Rotterdam: Balkema, 233–238.

McDougall, S.D. & Hungr, O. 2005. Dynamic modelling of entrainment in rapid Landslides. Can. Geotech. J. 42: 1437–1448.

Picarelli, L., Evangelista, A., Rolandi, G., Paone, A., Nicotera, M.V., Olivares, L., Scotto di Santolo, A., Lampitiello, S. & Rolandi, M. 2006. *Mechanical properties of pyroclastic soils in Campania Region. Proc. Int. Conf. on Natural soils, Singapore*, 2006. Rotterdam: Balkema.

Revellino, P., Hungr, O., Guadagno, F.M. & Evans, S.G. 2004. Velocity and runout prediction of destructive debris flows and debris avalanches in pyroclastic deposits, Campania region, *Italy. Environmental Geology*, 45: 295–311.

Sassa, K. 1988. Geotechnical model for the motion of landslides. *Proc 5th Int. Symp. on Landslides* 1: 33–55.

Scheidegger, A.E. 1973. On the prediction of the reach and velocity of catastrophic landslides. *Rock Mech* 5: 231–236.

Scotto di Santolo, A. 2000a. Analisi geotecnica dei fenomeni franosi nelle coltri piroclastiche della provincia di Napoli. *Tesi di dottorato XII ciclo Consorzio Università di Napoli e Roma*, 1/2000.

Scotto di Santolo, A. 2000b. Analysis of a steep slope in unsaturated pyroclastic soils. *Proc. Asian Conference on Unsaturated Soils*, Singapore, 569–574. Rotterdam: Balkema.

Scotto di Santolo, A. 2002. *Le colate rapide.* Hevelius Ed. srl, Benevento 2002, ISBN 88-86977-42-5.

Voellmy, 1955. Uber die Zerstorungskraft von lawinen. Bauzeitung, Jahrgang 73, 212–285.

Zanchetta, G., Sulpizio, R., Pareschi, M.T., Leoni, F.M. & Santacroce, R. 2004. Characteristic of May 5–6 1998 volcaniclastic debris flow in the Sarno Area (Campania Southern Italy): relationship to structural damage and hazard zonation. *J. of Volcanology and Geothermal R.* 133, pp. 377–393.

Optical fiber sensing technology used in landslide monitoring

Yan-xin Shi
Geo-Detection Laboratory, Ministry of Education of China, China University of Geosciences, Beijing, China

Qing Zhang & Xian-wei Meng
Centre for Hydrogeology and Environmental Geology, CGS, Baoding, China

ABSTRACT: Using the distributed optical fiber sensing technology in landslide monitoring, we can obtain the main specialty of the landside and can improve the monitoring efficiency. In the paper, we introduced the theory of FBG and BOTDR, respectively discussed their applied method, then, put forward a notion, BOTDR combined with FBG to monitor landslide. In Canlian landslide, we laid the monitoring optical fiber on the entire landside and used BOTDR to obtain the outline information of it. We installed FBG at the certain essential spots, the strain fissures, to obtain their strain information. Thus we can monitor the landside from dot to line and future to surface. Finally we obtained the completed strain information of the landslide.

1 INTRODUCTION

The distributed optical fiber sensing technology is a new technology in the domain of project survey. The optical fiber sensor uses the light as the carrier and transmission medium of information. It has a lot of advantage, such as anti-electromagnetic interference, anti-corrosive, high sensitivity, quick responds, light weight, small volume, variable shape, wide transport bandwidth, distributed survey and so on. It has widespread application in the aspect of online dynamic monitoring such as the health monitoring of the high-rise construction, the intelligent building, the bridge, the highway and so on. We applied the distributed optical fiber sensor in the landslide monitoring in 2004 and has obtained the good effect (Han, Z.Y. & Xue, X.Q. 2005).

2 THE PRINCIPLE

The optical fiber sensing technology measured the change of certain parameters of the transmission light in fiber (for example, intensity, phase, frequency, polarization condition and so on) to realize the measurement of environment parameter. The distributed optical fiber sensing technology has become the most promising technology by its advantage (multiply, distribution and long distance transmission). It is the development trend of optical fiber sensing technology The optical fiber Bragg grating sensor (FBG) and the Brillouin time domain reflection sensor (BOTDR) are two kind of the most representative distributed optical fiber sensing technology.

2.1 FBG

FBG is one kind of the distributed optical fiber sensor but the information it gained in the distance is not continuous. The refractive index of Bragg grating is periodically changed. If the period of FBG is not same, its reflection light wave length is also different. When this kind of optical fiber with the Bragg grating is under the stretch or compression or its temperature changes, its period will change, thus the reflected light wave length also will change. We measured the change of the reflected light wave length and knew the strain or the temperature (Chen, Y. 2003). The theory of FBG is shown in Figure 1.

Arranging certain FBG sensors on an optical fiber, we used the multiplying technology (TDM, WDM and so on) to construct the distributed sensing network. Thus we can simultaneously monitor multi-spots in the wide range, moreover may reduce the equipment number of monitor system, the length of

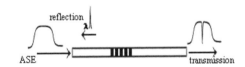

Figure 1. The theory of FBG.

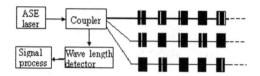

Figure 2. The distributed FBG monitoring system.

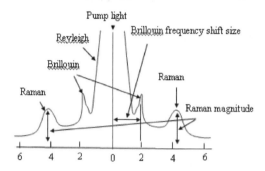

Figure 3. The backward scattered spectrum in optical fiber.

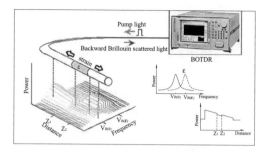

Figure 4. The theory of BOTDR strain measurement.

transmission fiber and reduce the cost of monitoring project. The distributed FBG monitoring system is shown in Figure 2.

The typical measurement accuracy of FBG Sensor is high at about 0.001%, and the typical distance resolution is also high at about 1 cm as determined by the grating length. In addition, measurement systems that can measure strain in real-time have been realized. These advantages have led to the FBG being applied to the precise strain measurement of structures, such as bridges and tunnel health monitoring. When FBG sensors were used in strain monitoring, mainly used at roughly identified positions where large deformations have occurred. Based on the FBG characteristic, when we used it in it landslide monitoring, we mainly used it to monitor the change of the landslide backyard or the known crack in real-time.

2.2 BOTDR

The Brillioun optical time domain reflectometer utilizes the characteristic that the spectrum and power of backward Brilliouin scattered light is correlative to the external environment (temperature, strain and so on). The backward scattered spectrum in optical fiber is shown in Fig 3.

In general, the Brillouin scattered light is shifted in frequency. When fiber materials are affected by temperature or strain, the Brillouin frequency shift size will change. Therefore, we can measure the shift size of backward Brilliouin scattered light to realize the distributed temperature and strain measurement (Shi B et al. 2004).

When the ambient temperature change is smaller than or equal to 5°C, if longitudinal strain ε occurs in the optical fiber, the Brillouin frequency shift VB changes in proportion to that strain. This relation can be expressed as:

$$V_B(\varepsilon) = V_B(0) + C\varepsilon \qquad (1)$$

where $V_B(\varepsilon)$ is the Brillouin frequency shift size when longitudinal strain ε occurs in the fiber, $V_B(0)$ is the Brillouin frequency shift size when no longitudinal strain occurs in the fiber, C is the strain coefficient, approximately equals to 50 MHz/$\mu\varepsilon$. The theory of BOTDR strain measurement is shown in Fig. 4.

The optical fiber which the BOTDR sensing technology used can be embedded in the substrate structure by the random form but does not affect the performance because it may be curving for thin and soft. Only measured the power and frequency of the Brillouin scattered light in the sensing optical fiber, we could obtain the distributed strain and temperature on the fiber. The biggest merit of this sensing technology is that the optical fiber not only is the sensing part but also is the transmission medium. The technology belongs to the distributed monitor technology. We may utilize it to realize long distance and uninterrupted monitor and easy to build the network with the optical fiber transmission system to realize remote-measurement and control of the system. So embedded the optical fiber in the landside mass as the neural network, we could implement the real-time monitor from the line to the surface of the landslide.

3 THE APPLICATION IN LANDSLIDE MONITOR

3.1 The approach to landslide monitoring by optical fiber sensing technology

The distributed optical fiber sensing technology has the widespread application in many domains by its unique merit. If utilize the distributed optical fiber sensing technology in the landslide monitoring, we must discuss the following questions (Wang, A.J. et al. 2006).

The most important is how the optical fiber be embedded in the landside mass. Only when the optical fiber and the monitored object were coupled very well in together, the information we obtained from the fiber is real effective.

Next, the distributed optical fiber sensing technology took the optical fiber as the sensing part, so the choice of fiber type also is a monitoring key.

Furthermore, it is also a question that how cement the fiber and how build the optical fiber monitoring network in order to obtain the comprehensive monitoring information of the landside mass from the spot to the line and further to the surface.

The natural landside mass mostly is natural rock and soil mass which has not passed through the artificial change and mostly is covered the loose quaternary system. If we want to use the optical fiber to monitor this type of landside mass, we must choose an appropriate construction plan and craft. Perhaps we can carry on artificial processing to the landside mass, or we can directly use the construction on the landside mass or other constructions influenced by the landside mass (for example, the concrete drain and stair, the slope control project, the road and the bridge etc.) to fix and cement the optical fiber. For the landside mass which has processed, we can choose the representative section plane on it to fix and cement the optical fiber. Thus, we can really obtain the strain information of the landside mass through the strain of the optical fiber laid on it.

In the landslide monitor, the optical fiber choice is more important. We can enhance the monitor life by choosing the appropriate optical fiber. At present there are two kinds of fiber usually used in monitor: the bare fiber and the tight tube fiber. The monitoring sensitivity of the bare fiber is very high, but its measuring range is small and easy to break off and laying the bare fiber is difficult. So the bare fiber adapts in the small strain monitoring. The tight tube optical fiber is composed by the core, the envelope, the painting layer and the protective tube. It has some merits: the strong inoxidizability, the good waterproof performance, the slightly big measuring range and not easy breaking off. Used the tight tube fiber in monitor is advantageous to the construction and can enhance the monitor life. Therefore we often choose the tight tube fiber to monitor the strain of the landslide.

Generally arranging the optical fiber network has two forms: unidimensional network and two-dimensional network. The former, namely, the optical fiber continuously from bottom to top is made the snake-shaped arrangement along the landslide body. This type of network is suited to monitor the strain changing in a direction. The latter, firstly, the optical fiber continuously from bottom to top is arranged along the landslide body, then, continuously from left to right (or from right to left) is made the snake-shaped arrangement along the horizontal direction. This type of network may monitor the strain changing in two directions. When laying the optical fiber, we should according to the characteristic of the work area to determine the arrangement form of the optical fiber network.

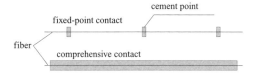

Figure 5. The laying method of optical fiber.

In the actual monitor project, laying the optical fiber has two methods (shown in Figure 5): the comprehensive contact type and the fixed-point contact type.

The monitor motive of the comprehensive contact type is roundly monitoring the distortion and the monitor object is the entire landside mass. The monitor motive of the fixed-point contact type is mainly monitoring the distortion of the crack, the stress concentration area and so on. The monitor object is the latent distortion point on the landslide. When laid the optical fiber, we should according to the special details of the landslide and the monitor motive to select the different method.

3.2 Combined FBG and BOTDR to monitor landslide

FBG and BOTDR are two kind of optical fiber sensing technology and have advantage and disadvantage respectively. The sensitivity of FBG sensor is high that can measure the strain extremely accurately. We can use some FBG sensors to compose the FBG sensing network to realize the distributed monitor, but the sensing array which is used to response the surrounding is set in advance, we must measure these discrete distributed sensing spot respectively. Therefore the monitor flexibility of FBG is low. The BOTDR sensing part is the optical fiber. We may use it to realize the long distance, the uninterrupted and distributed monitor. But for its technical limit, the measurement distance resolution of BOTDR is only achieved 1 m.

If we combined FBG and BOTDR to monitor the landslide, we would offset the insufficiency of them. We laid the optical fiber in the entire landside mass and used BOTDR technology to obtain the outline information of the entire landside mass. We installed the FBG sensor on the essential distort spot (distortion crack) in the landside mass and utilized its high measurement sensitivity characteristic to obtain the strain of the certain essential spots in landslide. Thus, unified FBG and BOTDR to monitor landslide, we overcome the shortcoming of BOTDR (measurement resolution is not high) and offset the insufficiency of FBG (only realize the separate measurement). Through this method, we may realize the landslide monitor from the spot to the line and further to the surface and obtain the completed strain information of landside mass.

3.3 Application example

The Canlian landslide is located in the heartland of Wushan in Chongqing. It is river and valley slope terrain. Although the government has carried on the remedy against it, there are some obvious distortions in the underside of the landslide. Based on this situation, we have been using BOTDR technology to monitor the landslide from August 2004. In October 2006, we have installed the FBG strain sensor in its key distortion spot.

The monitoring area is composed by two parts. On the upside, we made grooves on the concrete trellis which forms the control project and laid the optical fiber in it. We arranged the fiber network with two-dimensional form. On the underside, there is a cement stair passed through the landslide. We made grooves on the right of the stair and laid the optical fiber in it. The fiber was laid with the section plane form. The optical fiber network is shown in Figure 6. We used the optical fiber strain analyzer AQ8603 which made in Japan to monitor the distributed strain of the landslide. On 92–93 m of the stair, we installed the FBG strain sensor. The monitoring equipment we used is FBG demodulator which made by ourselves.

Figure 8 showed the distributed strain along the optical fiber which laid on the underside of Cailian landslide by BOTDR monitored. Obviously, from the figure we found 4 obvious high strain sections along the optical fiber and the strain was symmetrical (because the optical fiber was laid symmetrically). C1 was corresponded to the 92–93 m of the section plane. C2 was corresponded to the 142–143 m of the section

Figure 6. The optical fiber laying network in Canlian landslide.

Figure 7. The FBG installation in field.

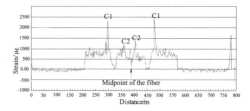

Figure 8. BOTDR monitored distributed strain on the underside of Canlian landslide.

plane. We found there were obvious tension and shear fracture through the macro-survey. FBG monitor also showed that the distort was extended too. From October 2006 to August 2007, the strain of the FBG test point was changed from 37.65 $\mu\varepsilon$ to 51.57 $\mu\varepsilon$. Thus, we have obtained the strain distributed along the section plane of the landside mass and strengthened the monitor to the essential distortion spot.

4 CONCLUSION

It's just starts that FBG and BOTDR distributed optical fiber sensing technology were applied in the landslide monitoring domain. We have obtained certain effect in Canlian landslide monitor. But we still needed to practice, summarize and perfect the approach to monitoring landslide with FBG and BOTDR, such as the fiber embedment craft, the fiber network arrangement principle, the FBG sensor choosing and installing and so on.

FBG and BOTDR unified to monitor landslide may realize comprehensive monitor from the spot to the line and further to the surface and may obtain more complete strain information of the landside mass. The method is one kind of perfect monitoring plan. But at present, we still used two set of instrument systems to realize FBG and BOTDR strain-measurement, the FBG monitoring network and the BOTDR monitoring network are two independent networks. We have not really unified these two kinds of technique. It also is a development direction that how to unify these two kinds of technique in the landslide monitoring domain.

REFERENCES

Han, Z.Y., Xue, X.Q., 2005. Status and development trend of monitoring technology for geological hazards. *The Chinese journal of geological hazard and control*, 16(3): 138–141.

Chen, Y., 2003. The study on the measurement of strain in the buildings by using fiber grating sensors. *Dissertation for the master degree in engineering of Nankai university*.

Shi, B, Xu, H.Z., Zhang, D., et al., 2004. Feasibility study on application of BOTDR to health monitoring for large infrastructure engineering. *Chinese journal of rock mechanics and engineering*, 23(3): 493–499.

Wang, A.J., Zhang, J.Y., et al., 2006. Application of BOTDR in monitoring of Canlian landslide in WuShan. *The national fifth conference of geological hazard and control*, Chongqing, 299–305.

Finite element analysis of flow failure of Tailings dam and embankments

R. Singh
Department of Civil Engineering, Indian Institute of Technology, Kharagpur, India

D. Mitra
Department of Civil Engineering, National Institute of Technology, Warangal, India

D. Roy
Department of Civil Engineering, Indian Institute of Technology, Kharagpur, India

ABSTRACT: Undrained shear strength of cohesionless soils is often estimated from back analyses of dams and embankments that have endured static undrained loading with various degrees of distress. The existing procedures for back analysis assume isotropic material behavior. However, saturated cohesionless soils exhibit strong inherent anisotropy during undrained loading. Consequently, the undrained shear strength of granular soils depends on the angle between the major principal direction and the direction of deposition. Using an anisotropic procedure for back analysis, correlations have been developed in this study between normalized Standard Penetration Test (SPT) blow count or cone penetration resistance and anisotropic undrained shear strength for different modes of loading (e.g. compression, simple shear and extension) from back analysis of pre-failure geometries of 28 static flow failure case histories. To illustrate the efficacy of these correlations, finite element analyses were carried out for three earth embankments using the shear strength parameters obtained from the correlations proposed and the computed deformations were compared with the observations. The comparison indicated a reasonable agreement.

1 INTRODUCTION

Undrained shear strength, s_u, of cohesionless soils is often estimated from back analysis of dams and embankments that underwent various degrees of distress due to undrained loading (Olson and Stark 2003). Although the existing analytical procedures are based on the assumption of inherently isotropic material behavior, untrained mechanical response of saturated cohesionless soils is inherently anisotropic (Vaid et al. 1990 and Yoshiminie 1998). A framework has been proposed recently accounting for inherently anisotropic undrained behavior of granular soils (Singh and Roy 2006). In this paper, 28 case histories involving static undrained response of dams and embankments have been back-analyzed using the proposed procedure and to develop a set of correlations between normalized cone tip resistance, q_{c1}, and Standard Penetration Test (SPT) blow count, $(N_1)_{60}$, and peak undrained shear strength ratio, s_u/σ_v' for various modes of deformation, e.g., plane strain compression and extension, and simple shear. To illustrate the application of the correlations, pre-failure geometries of two embankments that underwent static undrained loading were analyzed using shear strengths obtained from the proposed correlations.

2 ANALYTICAL PROCEDURE

Undrained shear strength of cohesionless soils tends to be higher when a sample is loaded in the direction of deposition than that for loading in any other direction. Such behaviour is referred to as inherent anisotropy. Inherently anisotropic undrained behaviour of cohesionless soils can be approximated testing on 84 undrained undisturbed frozen samples from seven sites of Canada for various types of loading. The behaviour is numerically approximated by Equation 1.

$$(s_u/\sigma_v')/(s_u/\sigma_v')_{\text{TXC}} = 0.402 \times \cos 2\theta + 0.598 \quad (1)$$

where θ is the angle between the direction of deposition and that of the effective major principal stress. The relationship, presented in Fig. 1, was developed

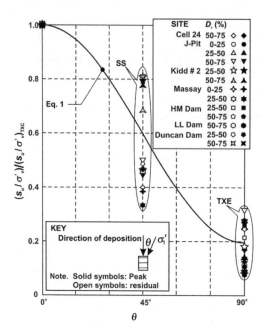

Figure 1. Shear strength ratio for undisturbed sand (modified from Singh and Roy 2006).

using data from undrained monotonic laboratory tests on undisturbed (frozen) samples (Vaid et al. 1996). It is apparent from Fig. 1 that s_u/σ'_v at phase transformation in triaxial compression (denoted with subscript "TXC"), is higher than those in simple shear (SS) and triaxial extension (TXE). It also appears that there is no systematic dependence of s_u/σ'_v on the relative density of the deposit.

The following procedure was used for back analysis to estimate the peak s_u/σ'_v:

- A trial slip surface was first assumed through the zones of lowest penetration resistance in the pre failure configuration.
- The undeformed section of the embankment above the assumed slip surface was divided into a number of vertical slices.
- The mobilized undrained shear strength was assumed to be governed by the vertical effective stress for the undeformed configuration.
- Soils with $q_{c1} \geq 6.5$ MPa or $(N_1)_{60} \geq 12$ and those above water table were assigned drained values of friction angle depending on their q_{c1} or $(N_1)_{60}$.
- The stability analyses were performed using software package XSTABL Version 5.2 (1994) and the GLE method assuming the mobilization of s_u at the base of the slip surface according to

Eq. [1] for until obtaining a value of $(s_u/\sigma'_v)_{TXC}$ by trial and error that gave a factor of safety of 1.
- The above steps were repeated for other trial surfaces until obtaining the minimum value of $(s_u/\sigma'_v)_{TXC}$.

3 CORRELATIONS FOR s_u/σ'_v

Pre-failure geometries of 28 embankments (Table 1) that failed due to static undrained loading were back analyzed using the procedure outlined in the preceding section. The results of these back analyses are summarized in Table 2 for anisotropic. Also included are the results obtained from conventional isotropic back analyses for comparison.

It appears from the results that the undrained shear strengths from isotropic back analyses are in general similar to the anisotropic undrained shear strengths for simple shear loading.

Table 1. Case histories.

Embankment	q_{c1}, $(N_1)_{60}$	Reference
Aberfan Tip No.4	2.2, 5.2	Lucia 1981
Aberfan Tip No.7	2.2, 5.2	Lucia 1981
Asele Embankment	3.8, 7.0	Konard and Watts 1995
Bofokeng Tailings	2.2, 7.4	Lucia 1981
Calaveras Dam	4.4, 8.0	Hazen 1920
Fording South Spoil	1.3, 2.1	Dawson et al. 1998
Copper Tailings Dam	2.1, 6.6	Lucia 1981
Fort Peck Dam	2.6, 6.5	Casagrande 1976
Fraser River Delta	2.9, 5.3	Chillarige et al. 1997
Gypsum Tailings Dam	1.4, 4.7	Lucia 1981
Greenhills Cougar 7	1.2, 2.4	Dawson et al. 1998
Helsinki Harbour	2.8, 6.0	Andresen and Bjerrum 1968
Hoedekenskerke Dyke	4.7, 8.5	Koppejan et al. 1948
Jamuna Bridge Site	3.2, 7.5	Yoshimine et al. 1999
Lake Ackerman Road	3.3, 7.0	Hryciw et al. 1990
Merriespruit Tailings	1.1, 3.2	Fourie and Papageoriou 2001
Mississippi River Bank	3.2, 6.8	Senour and Turnbull 1948
Nerlerk Slide1	2.6, 6.8	Sladen et al. 1985
Nerlerk Slide2	1.5, 3.6	Sladen et al. 1985
Nerlerk Slide3	1.5, 3.6	Sladen et al. 1985
North Dyke	2.3, 4.5	Olson and Stark 2000
Quintette Marmot	1.3, 2.8	Dawson et al. 1998
Sullivan Mine	1.8, 3.7	Davies 1999
Tar Island Dyke	1.2, 3.0	Mittal and Hardy 1977
Trondhiem Harbour	2.5, 6.0	Andresen and Bjerrum 1968
Vlietepolder, Zeeland	2.8, 7.5	Koppejan et al. 1948
Western US Tailings	0.8, 3.0	Davies et al. 2002
Wilheminapolder	2.8, 7.5	Koppejan et al. 1948

Table 2. Pre liquefaction S_u/σ'_v.

Embankment	Anisotropic			Isotropic
	TXC	SS	TXE	
Aberfan Tip No.4	0.55	0.31	0.07	0.33
Aberfan Tip No.7	0.55	0.31	0.07	0.33
Asele Embankment	0.58	0.30	0.06	0.20
Bofokeng Tailings	0.53	0.28	0.06	0.17
Calaveras Dam	0.64	0.31	0.07	0.28
Fording South Spoil	0.34	0.19	0.04	0.19
Copper Tailings Dam	0.49	0.27	0.06	0.33
Fort Peck Dam	0.56	0.31	0.07	0.14
Fraser River Delta	0.53	0.30	0.06	0.16
Gypsum Tailings Dam	0.40	0.22	0.05	0.20
Greenhills Cougar 7	0.34	0.19	0.04	0.19
Helsinki Harbour	0.58	0.30	0.06	0.24
Hoedekenskerke Dyke	0.70	0.39	0.08	0.25
Jamuna Bridge Site	0.61	0.34	0.07	0.20
Lake Ackerman Road	0.55	0.31	0.07	0.25
Merriespruit Tailings	0.32	0.18	0.04	0.12
Mississippi R. Bank	0.52	0.29	0.06	0.27
Nerlerk Slide1	0.54	0.28	0.06	0.16
Nerlerk Slide2	0.33	0.18	0.04	0.16
Nerlerk Slide3	0.32	0.18	0.04	0.14
North Dyke	0.49	0.26	0.05	0.25
Quintette Marmot	0.39	0.21	0.05	0.19
Sullivan Mine	0.34	0.19	0.04	0.19
Tar Island Dyke	0.41	0.23	0.05	0.20
Trondhiem Harbour	0.35	0.20	0.04	0.19
Vlietepolder, Zeeland	0.58	0.30	0.06	0.28
Western US Tailings	0.26	0.15	0.03	0.12
Wilheminapolder	0.39	0.22	0.05	0.12

The pre-failure s_u/σ'_v, listed in Table 2, are plotted in Figures 2 and 3 against q_{c1} and $(N_1)_{60}$, respectively. Following correlations were developed:

$$(s_u/\sigma'_v)_{TXC} = 0.272 \times q_{c1}^{0.223} \quad (2)$$

$$(s_u/\sigma'_v)_{SS} = 0.189 \times q_{c1}^{0.145} \quad (3)$$

$$(s_u/\sigma'_v)_{TXE} = 0.033 \times q_{c1}^{0.209} \quad (4)$$

$$(s_u/\sigma'_v)_{TXC} = 0.123 \times (N_1)_{60}^{0.239} \quad (5)$$

$$(s_u/\sigma'v)_{SS} = 0.116 \times (N_1)_{60}^{0.134} \quad (6)$$

$$(s_u/\sigma'_v)_{TXE} = 0.029 \times (N_1)_{60}^{0.116} \quad (7)$$

The r^2 values for Eqs. (2), (3) and (4) were 0.76, 0.72 and 0.71, respectively, and those for Eqs. (5), (6) and (7) were 0.80, 0.77 and 0.77, respectively.

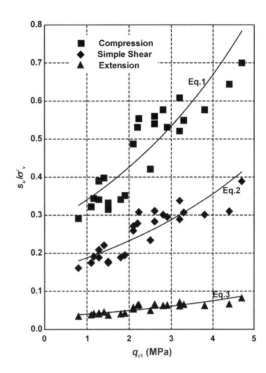

Figure 2. Pre failure $s_u/\sigma'_v - q_{c1}$ relationships.

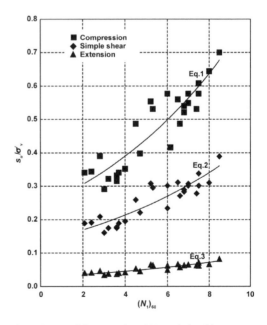

Figure 3. Pre failure $s_u/\sigma'_v - (N_1)_{60}$ relationships.

4 FINITE ELEMENT MODELLING

In order to illustrate the efficacy of the proposed correlations, two embankments that underwent static flow failure were analyzed using finite element package PLAXIS software Version 8.2 (2002). The flow failure case histories are described in the following subsections.

4.1 Tailings Dam No. 2, South America

This is an upstream constructed tailing dam constructed in 1990s. The starter dam for this structure was constructed to a maximum height of about 10 m, of silty sand with fines contents in excess of 30%. A filter facing on the upstream face of the starter dam, together with basal finger drains below the starter dam (drain rock wrapped in geotextile) were included given that the starter dam would not function effectively as a toe drain without such drainage measures. The dam is founded primarily on bedrock of relatively low hydraulic conductivity, with some relatively fine-grained alluvium in the valley bottom. Overall, foundation conditions are such that effective natural under-drainage is likely not present. The undeformed and deformed geometry of Tailings Dam No. 2 is shown in Fig. 4. The more details of the failure are given in Davies (2002).

4.2 Tailings Dam No. 3, South America

This is an upstream constructed tailing dam constructed in 1990s. The dam is constructed across a relatively steep-walled valley with bedrock exposed on the valley slopes and alluvium in its base. The starter dam was constructed of silty sand and gravel (maximum fines content 30%), and included a zone along its upstream face, and along its base, of clean (fines content <5%) sand and gravel. The starter dam was constructed to a maximum height of about 15 m. The undeformed and deformed geometry of Tailings Dam No. 3 is shown in Fig. 5. The more details of the failure are given in Davies (2002).

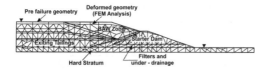

Figure 4. Undeformed and deformed geometry of Dam No. 2.

4.3 Bakacka embankment, turkey

Nearly 100 m long section of the embankment fill, including the northern lane of the highway, slipped and spread into the valley on the north of the highway. The embankment was constructed nearly 40-years-ago using burrowed material from nearby cuts, and had an approximate height of 50 m with an inclination of 2:3. The embankment failure occurred during 12 November 1999 Du'zce earthquake due to the intense near field ground motion. The undeformed and deformed geometry of Tailings Dam No. 2 is shown in Fig. 6. The more details of the failure are given in Bakir and Akis (2005).

4.4 Analysis procedure for flow failure cases

The step by step finite element procedure for the analysis of flow failure cases are given below:

- The flow failure of dams or embankments is analyzed using the PLAXIS Version 8.2 software in which deformations were calculated using the Mohr's Coulomb failure criteria.
- Firstly, drained analysis was carried out using isotropic properties.
- Soil properties were estimated using McGregor and Duncan (1998).
- The angle of the direction of vertical effective stresses with respect to horizontal within the liquefiable layer were estimated from the analysis using isotropic soil properties.

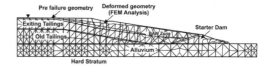

Figure 5. Undeformed and deformed geometry of Dam No. 3.

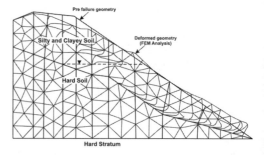

Figure 6. Undeformed and deformed geometry of Bakacka Embankment.

- The liquefiable layer was divided into the various vertical slices with respect to angle of the direction of vertical effective stresses with respect to horizontal.
- Elastic modulus of soil layers were estimated by assuming Poisson's ratio 0.35 and modulus of rigidity in between 10000 to 15000.
- The undrained soil properties within the liqufiable soil layer were estimated below the water table from propose anisotropic correlation and above the water table the value of angle of internal friction were consider in between 30 to 35 degree.
- Using these above properties, the deformation of the following cases were estimated using the finite element analysis.

4.5 *Analysis procedure for seismic failure cases*

The finite element procedure for seismic failure cases is same as flow failure cases except that the earthquake time history records available at dam sites were used in the analysis.

5 RESULTS

A framework proposed for undrained analysis of dams or embankments, which considers inherently anisotropic soil behavior in an approximate manner. In this paper, twenty-eight case histories documenting undrained distress of dams or embankments were back analyzed using the proposed framework and based on the results of these back analyses, correlation was developed between the pre liquefaction shear strength ratio, s_u/σ'_v, and normalized penetration resistances, q_{c1} and $(N_1)_{60}$.

The deformation from finite element analysis using proposed pre failure relationships for anisotropic undrained analyses (Fig. 2) are shown in Table 3. The results from finite element analyses using proposed anisotropic undrained shear strength are showing lesser vertical deformation than the observed vertical deformation.

Table 3. Results of finite element analysis.

Embankment	Vertical displacement Observed	Anisotropic (FEM analysis)
Tailings Dam No. 2	–	0.316
Tailings Dam No. 3	–	1.16
Bakacka Embankment	7.95	7.85

REFERENCES

Andresen, A. and Bjerrum, L. (1968). Slides in subaqueousslopes in loose sand and silt, Norwegian Geotechnical Institute Publication No. 81: 1–9.

Bakir, B.S. and Akis, E. (2005). Analysis of highway embankment failure associated with the 1999 Duzce, Turkey earthquake. J. of Soil Dynamics and Earthquake Engrg., 25: 251–260.

Casagrande, A. (1976). Liquefaction and cyclic deformation of sand: a critical review, Harvard Soil Mechanics Series No. 88, Harvard University Cambridge, MA.

Chillarige, A.V., Morgenstern, N.R., Robertson, P.K. and Christian, H.A. (1997). Seabed instability due to flow liquefaction in the Fraser River delta, Canadian Geotech. J., 34: 520–533.

Dawson, R.F., Morgenstern, N.R. and Stokes, A.W. (1998). Liquefaction flowslides in rocky mountain coal mine waste dumps, Canadian Geotech. J., 35: 328–343.

Davies, M.P. (1999). Peizocone technology for the geoenvironmental characterization of mine tailings, Ph.D. Thesis, University of British Columbia, Canada.

Davies, M.P., McRoberts, E.C. and Martin, T.E. (2002). A tail of four upstream Tailings dam, Proc., Tailings Dams 2002, ASDSO/USCOLD, Los Vegas.

Davies, M.P., McRoberts, E.C. and Martin, T.E. (2002). Static liquefaction of tailings—Fundamentals and case histories, Proc., Tailings Dams 2002, ASDSO/USCOLD, Los Vegas.

Fourie, A.B. and Papageoriou, G. (2001). Defining an appropriate steady state line for Merriespruit gold tailings, Canadian Geotech. J., 38: 695–706.

Hazen, A. (1920). Hydraulic fill dams, Transactions of the American Society of Civil Engineers, Paper No. 1458, 1713–1821.

Hryciw, R.D., Vitton, S. and Thomann, T.G. (1990). Liquefaction and flow failure during seismic exploration, J. Geotech. Engrg., 116: 1881–1899.

Interactive Software Designs, Inc. (1994). XSTABL: An integrated slope stability analysis program for personal computers, Reference Manual.

Konard, J.M. and Watts, B.D. (1995). Undrained shear strength for liquefaction flow failure analysis, Canadian Geotech. J., 32: 783–794.

Koppejan, A.W., van Wamelen, B.M. and Weinberg, L.J.H. (1948). Coastal flow slides in the Dutch province of Zeeland, Proc., 2nd Int. Conf. Of Soil Mechanics and Foundation Engineering, June, 21–30, Netherlands, 89–96. Rotterdam: Balkema.

Lucia, P.C. (1981). Review of experiences with the flow failures of tailings dams and waste impoundments, Ph.D. Thesis, University of California, Berkeley, Calif.

McGregor, J.A., and Duncan, J.M. (1998). Performance and use of the standard penetration test in geotechnical engineering practice, Report, Virginia Tech, Blacksburg, Virginia, USA.

Mittal, H.K. and Hardy, R.M. (1977). Geotechnical aspects of a tar sand tailings dyke, Proc., Conf. On Geotechnical Practice for disposal of solid waste materials, ASCE Specially Conf. Of the Geotechnical Engineering Division, Vol. 1, 327–347.

Olson, S.M., Stark, T.D., Walton, W.H. and Castro, G. (2000). Static liquefaction flow failure of the North Dike of Wachusett Dam. J. of Geotech. and Geoenviro. Engrg., 126: 1184–1193.

Olson, S.M. and Stark, T.D. (2003). Yield strength ratio and liquefaction analysis of slopes and embankments. J. of Geotech. and Geoenviro. Engrg., 129: 727–737.

PLAXIS software 2D- Version 8.2 (2002). Edited by Delft University of Technology and PLAXIS b.v., Netherlands.

Singh, R. and Roy, D. (2006). Anisotropic undrained back analysis of embankments, Proc., Int. Conf. Geoshanghai2006 "Advances in Earth Structures: Research to Practice", June 6–8, Shanghai, China, pp. 225–230.

Senour, C. and Turnbull, W.J. (1948). A study of foundation failures at a river bank revetment. Proc., 2nd Int. Conf. On Soil Mechanics and Foundation Engrg., Vol. 7, 117–121.

Sladen, J.A., D'Hollander, R.D., Krahn, J. and Mitchell, D.E. (1985). Back analysis of the Nerlerk berm liquefaction slides, Canadian Geotech. J., 22: 579–588.

Vaid, Y.P., Chung, E.K.F. and Keurbis, R.H. (1990). Stress path and steady state, Canadian Geotech. J., 27: 1–7.

Vaid, Y.P., Sivathayalan, S., Eliadorani and Uthayakumar, M. (1996). Laboratory testing at University of British Columbia, CANLEX Report, University of Alberta, Edmonton, Canada.

Yoshimine, M., Ishihara, K. and Vargas, W. (1998). Effects of principal stress direction and intermediate principal stress on undrained shear behavior of sand. Soils and Foundations, 38: 179–188.

Yoshimine, M., Robertson, P.K. and Wride, C.E. (1999). Undrained shear strength of clean sand to trigger flow liquefaction. Canadian Geotech. J., 36: 891–906.

Landslide model test to investigate the spreading range of debris according to rainfall intensity

Y.S. Song, B.G. Chae & Y.C. Cho
Korea Institute of Geoscience & Mineral Resources (KIGAM), Daejeon, Korea

Y.S. Seo
Chung-Buk National University, Chungju, Korea

ABSTRACT: Landslide model experiments by considering rainfall intensity were performed to investigate and predict the spreading range of debris. The model flume and the rainfall simulator were designed and manufactured firstly, and a series of model experiments were performed. The model experiments were performed with changing the rainfall intensity from 150 mm/hr to 250 mm/hr. In these experiments, the angle of slope inclination is 25° and the relative density of slope soils is 35%. In order to measure the pore water pressure in a slope, the deformation of slope surface, and the spreading range of debris during experiments, the instrumentation was installed in the slope. As a result of instrumentation, the pore water pressure is increased rapidly at the time of landslides, and the pore water pressure in a slope and the scale of landslide are increased with increasing the rainfall intensity. In addition, the shape of spreading range of debris looks like a bulged pan, and the spreading range is increased rapidly in its early stage and then increased gradually. The increasing velocity of spreading range is influenced by the rainfall intensity, and the final spreading area after heavy rainfall depends on the rainfall intensity and the rainfall duration time.

1 INTRODUCTION

The annual rainfall of Korea ranges from 1,100 mm to 1,400 mm, most of which concentrates in wet season beginning at June through September. The majority of landslides in Korea occurs in wet season and depends highly on heavy rainfall (Park et al., 2006). The human casualties due to natural disasters during 10 year in the 1990s were about 140, and 16% of them were injured by the landslides. The most important factor causing landslides in Korea is the heavy rainfall. However, the relationship between landslides and rainfall such as hourly rainfall, cumulative rainfall, and so on is not yet proved clearly.

In case of Hong Kong, the analysis method considering the antecedent rainfall proposed by Lumb (1975) was used dominantly in the past, while the analysis method considering the rainfall intensity proposed by Brand (1985) has been used in recent years. Kim et al. (1991) studied the occurrence mechanism of landslides using a record of correlating rainfall with landslides in Korea. As the result of this study, landslides were influenced by both the cumulative rainfall and the rainfall intensity. In addition, Chae et al. (2006) presented that the landslides are typical transitional slides at the triggering position, and changed into debris flows as they move down slope. However, the studies of the spreading range of debris by the landslides have not been conducted intensively.

In this study, therefore, landslide model experiments according to the rainfall intensity are performed to investigate the spreading range of debris by the landslides. To do this, a model flume and a rainfall simulator are designed and manufactured firstly, and monitoring instruments are installed to measure a pore water pressure in a slope and a deformation of the slope surface during experiments. The model flume is designed by consideration of the landslides characteristics occurred in Korea as debris flows. The rainfall simulator controls the rainfall intensity. A series of the model experiments according to the rainfall intensity are performed to find out the occurrence characteristics of landslides and predict the spreading range of debris.

2 LANDSLIDE MODEL EXPERIMENT

2.1 Model test equipment

The model test equipment is designed and manufactured to measure a spreading range of the run out distance occurred by debris flows. The test equipment can be divided into three parts: a model flume, a rainfall simulator, and monitoring systems, as shown schematically in Figure 1. The plate for measuring run out distance is attached at the toe part of the model flume to investigate the spreading process and range of run out distance.

The rainfall simulator consists of water sprinkling system, rainfall controller and water tank. It is devised to control the rainfall intensity by a computer. Also, the monitoring systems consist of pore water pressure meters, a data logger, marked pins and digital cameras. The pore water pressure in a slope can be measured by the pore water pressure meter installed on the bottom of the model flume.

In Figure 1, the model flume can be divided into three parts: toe part, slope part and crest part. The length of each part is 0.5 m, 1.5 m and 0.3 m, respectively, and the height is 0.5 m. The model flume is devised to control the slope angle ranged from 0° to 40°. The angle of slope can be controlled by use of the height difference between crest part and toe part. The model flume is made of steel frames and high strength glass plates. The high strength glass plate is used on the front side of the model flume to observe the deformation of slope with the naked eye directly. The bottom surface of the model flume is made with unevenness in order to prevent the sliding at the interface between soils and model flume. The run out distance plate is installed at the edge of toe part to observe the process of spreading the debris triggered by landslides.

The rainfall simulator composed of water sprinkling device and a pedestal. The water sprinkling device is made of PVC pipe, and its dimension is 3.1 m long and 0.9 m wide. The water sprinkling device is consisted of two rows of pipes, and nozzles having inner diameters of 1.5 mm and 2.0 mm are attached to pipes at equal spaces. The water sprinkling device is produced for controlling the rainfall intensity ranged from 100 mm/hr to 1,000 mm/hr. The pedestal having 2.5 m high and 1.8 m long is made of steel bar and can make a move and control the supporting height.

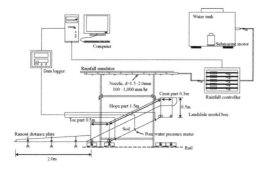

Figure 1. Schematic diagram of the landslide model test equipment.

2.2 Monitoring instruments

In order to measure the pore water pressure in a slope, the deformation of slope surface, and the spreading range of debris due to heavy rainfall, an instrumentation system is designed and installed. The pore water pressure meter is PL1M model made by Senzors Co., USA with 2.5 cm diameter and 9 cm high. To measure the pore water pressure in a slope, three pore water pressure meters are installed on the bottom at the slope part of model flume. The measured pore water pressure is recorded automatically at 10 seconds' intervals through the data logger. The data logger is Geologger 515 made by Data Electronics Co., Australia, and has ten channels. The marked pins are installed on the center of slope surface at 20 cm intervals to measure the deformation of slope. The deformation of slope is measured by the photographing of marked pins using digital and video cameras.

2.3 Soil slope

The soils used in the experiment are Jumunjin standard sands, which are made and widely used for laboratory tests in Korea. To make the homogeneous soil slope, the standard sands are put in the flume and fallen free in the height of 75 cm using a funnel of 1 cm in diameter. The depth of the slope formed in this manner is 30 cm. The relative density of soil slope is 35%, and the internal friction angle obtained from the consolidated drained triaxial test (CD test) is 36.5° (Song, 2004).

2.4 Experiment process

The model experiment of debris flow is performed as follows;

1. Control the angle of slope in model flume
2. Control the rainfall intensity of rainfall simulator
3. Install the pore water pressure meter at the bottom of model flume
4. Make the model soil slope using Jumunjin standard sand, and install the marked pins
5. Develop the debris flow due to rainfall sprinkling
6. Measure the deformation of slope surface, the pore water pressure in a slope, and the spreading range of debris

Table 1. Cases of the model test.

No.	Rainfall intensity (mm/hr)	Slope angle (°)	Relative density (%)
SL-1	250	25	35
SL-2	200	25	35
SL-3	150	25	35

Through this experiment, the spreading range of debris can be able to investigate according to the rainfall intensity. Table 1 shows the planning of model tests according to rainfall intensities; three cases of model test are performed with different rainfall intensities. As shown in Table 1, the relative density of soil slope is 35%, and the angle of slope is 25°. Three rainfall intensities, 250 mm/hr, 200 mm/hr and 150 mm/hr, are applied.

3 RESULTS AND DISCUSSION

3.1 Results of model experiments

To investigate the spreading range of debris caused by landslides according to the rainfall intensity, the model experiments of landslides are performed with changing the rainfall intensities on the same condition of soil slopes. The pore water pressure, the deformation of slope surface and the spreading range of debris are measured during the model experiments.

Figures 2 to 5 show the results of model experiments in the case of SL-2 with the rainfall intensity of 200 mm/hr. Figure 2 shows the variation of pore

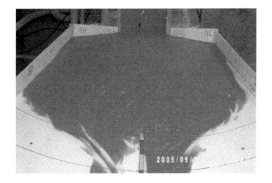

Figure 4. Spreading range of debris on the plate induced by landslide.

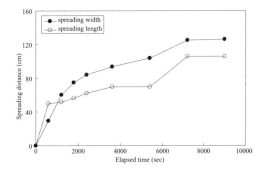

Figure 5. Variation of spreading width and length according to elapsed time.

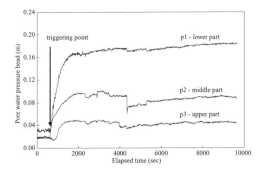

Figure 2. Variation of pore water pressure on the slope according to elapsed time.

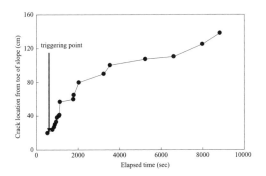

Figure 3. Occurring position of tension crack on the slope surface according to elapsed time.

water pressures in the soil slope with elapsed time. In Figure 2, p1, p2, and p3 indicate the pore water pressures at the toe, middle, and top of slope parts, respectively. The maximum pore water pressure is occurred at the toe of slope part, while the minimum pore water pressure is occurred at the top of slope part. The pore water pressure is kept constant in its early stage and then increased suddenly. Moriwaki et al. (2004) presented that the pore water pressure in slopes is increased suddenly at the landslides occurrence time. Therefore, the rapid increasing the pore water pressure as shown in Figure 2 means the occurrence of landslides.

Figure 3 shows the occurring position of tension cracks on the slope surface triggered by rainfall. The tension cracks are occurred firstly at the toe part of slope and then translated toward the crest part. That is, the landslides are occurred firstly at the toe part, and the scale of landslides is expanded gradually when the cracks are translated toward the crest part. It knows that the tension cracks on the slope surface are

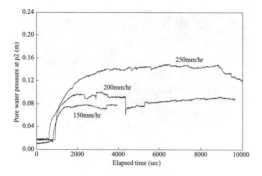

Figure 6. Variation of pore water pressure at the middle part of slope according to rainfall intensity.

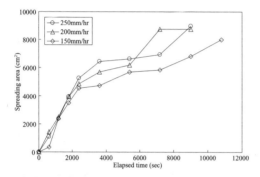

Figure 7. Variation of spreading area of debris according to rainfall intensity.

occurred when the pore water pressure are increased suddenly.

Figure 4 shows the photograph of the spreading range of debris caused by landslides. This Figure indicates that the shape of the spreading range of debris looks likes a bulged pan. Figures 5 show the spreading width and length of debris according to the elapsed time. The increasing tendency of spreading width is similar to that of spreading length, while the spreading width is larger than the spreading length. The spreading length and width is increased rapidly in the early stage and then increased gradually.

3.2 Comparison analysis according to the rainfall intensity

Figure 6 shows the change of pore water pressure with elapsed time according to various rainfall intensities. The data of pore water pressure are obtained from p2 measured at the middle part of slope. The pore water pressure in the slope is increased with increasing the rainfall intensity, and the pore water pressure is increased rapidly at the occurrence time of landslides. The occurrence time of landslides is measured about 700 to 900 seconds. Also, the translational velocity of tension cracks toward the crest of slope is increased with increasing the rainfall intensity. It means that the scale of landslides is increased with increasing the rainfall intensity.

Figure 7 shows the spreading area of debris caused by landslides according to the rainfall intensity. The spreading area of debris is rapidly increased with independent of rainfall intensity at the early period of rainfall. But, the difference of spreading areas according to the rainfall intensity is gradually increased after the passage of a certain time, and the spreading area is increased with increasing the rainfall intensity. Therefore, the spreading area of debris is influenced by not only the rainfall intensity but also the rainfall duration time.

4 CONCLUSIONS AND SUMMARY

In order to investigate the run out distance by the debris flow, the landslide model experiments according to the rainfall intensity were performed. To do this, the model flume and the rainfall simulator were designed and manufactured, and various monitoring instruments were applied in the modeled slope. Based on the results of model experiments, the occurrence characteristics of landslides were analyzed and the run out distance of debris was investigated. The following conclusions could be drawn;

1. The pore water pressure in the slope is increased suddenly at the occurrence time of landslides, and the tension cracks on the slope surface are occurred at that time.
2. The shape of spreading range of debris looks like bulged pan. In addition, the spreading length and width is increased rapidly in its early stage and then increased gradually.
3. The pore water pressure in the slope and the scale of landslides are increased with increasing the rainfall intensity.
4. The increasing velocity of spreading range is influenced by the rainfall intensity, and the final spreading area after heavy rainfall depends on the rainfall intensity and the rainfall duration time.

5 ACKNOWLEDGMENTS

This research was supported by a grant (NEMA-06-NH-04) from the Natural Hazard Mitigation Research Group, National Emergency Management Agency.

REFERENCES

Brand, E.W. 1985 Predicting the performance of residual soil slopes. *Proc. 11th Inter. Conf. on Soil Mech. and Found. Eng. San Francisco. USA.* 2541–2573.

Chae, B.G. Kim, W.Y. Cho, Y.C. Kim, K.S. Lee, C.O. & Song, Y.S. 2006 Probabilistic prediction of debris flow on natural terrain. *Proc. East Asia Landslides Symposium. Daejeon. Korea.* 144–153.

Kim, S.K. Hong, W.P. & Kim, Y.M. 1991 Prediction of rainfall-triggered landslides in Korea. *Proc. 6th Inter. Symp. on Landslides. Christchurch. New Zealand.* 2: 989–994.

Lumb, P. 1975 Slope failure in Hong Kong. *Journal of Engineering Geology.* 8: 31–65.

Moriwaki, H. Inokuchi, T. Hattanji, T. Sassa, K. Ochiai, H. & Wang, G. 2004 Failure processes in a full-scale landslide experiment using a rainfall simulator. *Landslides.* 1: 277–288.

Park, D.K. Oh, J.R. Kim, T.H. & Park, J.H. 2006 Slope-related disasters and management system in Korea. *Proc. East Asia Landslides Symposium. Daejeon. Korea.* 35–46.

Song, Y.S. 2004. *Design methods of the slopes reinforced by earth retention system.* Doctoral Thesis. Chung-Ang University. Korea.

Occurrence mechanism of rockslide at the time of the Chuetsu earthquake in 2004 – A dynamic response analysis by using a simple cyclic loading model

N. Tanaka & S. Abe
Okuyama Boring Co., Ltd., Yokote City, Japan

A. Wakai, H. Kawabata & M. Genda
Gunma University, Kiryu City, Japan

H. Yoshimatsu
SABO Technical Center (STC), Chiyoda-ku, Tokyo, Japan

ABSTRACT: Triggered by an earthquake that occurred in the Chuetsu area, Niigata Prefecture, Japan, in 2004, a large number of landslides occurred in the mountains where Tertiary sedimentary rock is distributed. Three-dimensional dynamic response analysis (FEM) using a simple cyclic loading model (UW model) that takes into account the shear strength and dynamic deformation characteristics of soil was conducted and revealed that high horizontal acceleration occurred on the mountain tops and high shear stress occurred in the valleys. On the basis of those analytical results, this paper discussed the mechanism of a large-scale rockslide that occurred at Hitotsu-minesawa.

1 INTRODUCTION

There are a very small number of earthquake-induced recurrent landslides or rockslides recorded so far in Japan except for some such events observed in parts of the Tohoku District. However, when the Niigata Chuetsu Earthquake (M = 6.8) hit in 2004, many recurrent landslides and rockslides occurred in areas where Tertiary formation is distributed. In particular, the Hitotsu-minesawa area in the mountains located near the epicenter was subject to a large-scale rockslide whose slip surface was 150 m deep. Abe et al. (1997, 2005) pointed out that rockslides of this type that occurred in conjugation with the Niigata Chuetsu Earthquake in 2004 and other earthquakes that have occurred in the past in the area of the Tohoku District where Tertiary formation is distributed have many common factors, one of which is that these landslides are prone to occur in ridge or questa topography. Therefore, clarification of the occurrence mechanism of Tertiary formation landslides triggered by inland earthquakes has the potential to play a pivotal role in hazard evaluation of future landslide disasters that follow earthquakes.

This paper focuses on the Hitotsu-minesawa landslide, which was a large-scale rockslide that occurred on the ridge topography following the Niigata Chuetsu Earthquake in 2004, and reviews the relationship between the behavior of mountain slopes and the occurrence of landslides during earthquakes based on seismic response analysis using three-dimensional dynamic elasto-plastic FEM.

2 OUTLINE OF THE LANDSLIDE

The Niigata Chuetsu Earthquake (hereinafter the Chuetsu Earthquake) that occurred in the Chuetsu area, Niigata Prefecture, in 2004 was a direct-hit earthquake that recorded a magnitude of 6.8 and a maximum acceleration of 1,500 gal (K-NET). The rockslide investigated in our paper occurred in the Hitotsu-minesawa area, located in the mountains about 10 km northeast of the epicenter (Figure 1).

The landslide that occurred at Hitotsu-minesawa transversely severed a ridge protruding west to westsouthwest (Figures 2 and 3). The landslide scarp formed a collapse zone 20 to 30 m in width (b in Figures 2 and 3) that stretched from a ridge area (a in Figures 2 and 3) to a paddy area until it reached an approximately 10 m high landslide scarp formed at the northern edge of the landslide area (e in Figure 2).

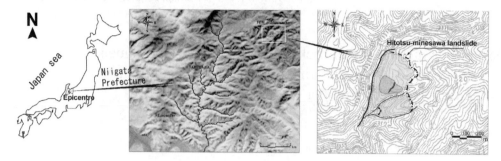

Figure 1. Location map and location of analysis.

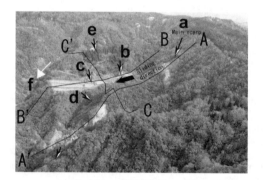

Figure 2. Overall view of Hitotsu-minesawa (from the side) (Photo by Haraguchi, 2004).

Figure 4. Crack occurring in a stream down the ridge (white arrow).

Figure 3. Entire view of Hitotsu-minesawa (from the front) (Photo by Haraguchi, 2004).

A crevice with an approximately 7 m level difference is seen cracking the paddy field in the intermediate part of the landslide area for about 80 m (c in Figures 2 and 3). A few landslide dams were also formed at Hitotsu-minesawa, located at the toe of the landslide area. The movement range of the landslide was about 500 m in width, about 350 m in length, and about 150 m in the difference of elevation from the crown part to the front end (Figure 2).

The local geology is a sandstone-dominant stratigraphic alternation of sandstone and mudstone from the Oligocene-Neogene. The bottom layer seen in the landslide area is presumed to be black mudstone, which is the rock partly exposed at Hitotsu-minesawa, the toe of the landslide area. The layer above it is silty sandstone, the core of the moving mass of the landslide, which compresses thin muddy layers. Coarse sandstone about 5 m in thickness, sandwiched by silty sandstone, is distributed in the paddy field area. At the topmost part of the landslide scarp is seen a more than 10 m-thick layer of black mudstone. Farther above the landslide scarp is seen conglomerate comprising coarse particles. The strike slope of the layers is roughly N15°E and 0 to 5°W.

On both banks of a stream located at the border of the paddy part and the ridge part of the landslide zone (d in Figures 2 and 3) is exposed in series silty sandstone hard enough to defy crushing by hammer impact. These rock masses have cracks in the north-south direction at intervals of 1 to 10 m. These cracks show no entry of tree roots or discoloration and are therefore judged to be fresh. Some of them are open by about 5 cm (Figure 4).

Hitotsu-minesawa, the toe of the landslide, has some characteristic geologic features, including a series of rock masses standing 3 m upright with slickenside, a maximum 10 m high upheaval at the bottom of the stream, and some landslide dams formed in association with the upheaval (Figure 5). These are formed by buckling or folding of the rock mass at the toe following the rockslide. The slip surface has a well-developed bedding and is presumed to linearly continue from down the landslide scarp at the head to Hitotsu-minesawa along the bedding plane due to the presence of thin layers of soft rocks such as black mudstone. If so, the thickness of the moving zone should reach 100 to 150 m (Figure 6).

Figure 5. Rock mass with protruding slickensides at Hitotsu-minesawa, the toe of the landslide.

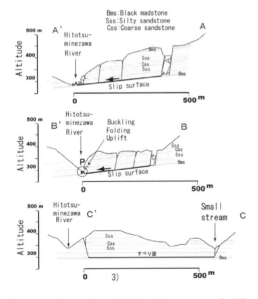

Figure 6. Geologic longitudinal and transverse section (See Photo 1 for A-A', B-B' and C-C').

3 SEISMIC RESPONSE ANALYSIS USING THREE-DIMENSIONAL DYNAMIC ELASTO-PLASTIC FEM

3.1 *Elasto-plastic constitutive model and basic equation*

A number of reports have been issued on three-dimensional analyses of slope failure in the mountainous areas induced by seismic motion. For example, Mizuyama et al. (2002) reviewed the seismic motion distribution by focusing on the salient shapes of the slope, and Asano et al. (2006) studied the topographic effect of the mountain body by seismic motion and its effect on collapse by assuming a three-dimensional elastic and perfectly plastic model.

The elasto-plastic finite element method produces results of a variety of characteristics depending on the constitutive equation used in the method. For the kind of rockslide that occurred at Hitotsu-minesawa, the subject of this paper, it is necessary to use an elasto-plastic constitutive model that can appropriately express the kinetic properties of soil by seismic motion.

With respect to such landslide events, the authors have been engaged in a fundamental review based on 3D dynamic elasto-plastic FEM to realize seismic response analysis for the purpose of predicting wide-area damage in the mountainous areas. To be specific, our achievements include filter output of response acceleration waveforms (Wakai et al., 2005), an attempt to identify wide-area ground physical properties (Tanaka et al., 2006), and development of specifications for a program to conduct input and output data with GIS and realize omission of complicated pre- and post-processing unique to FEM (Wakai et al., 2006). For a constitutive model used in the 3D dynamic elasto-plastic FEM, a simple cyclic loading constitutive model (UW model) was used that can consider both the dynamic deformation characteristics of soil (that is, the relationship involving shear elastic modulus G, shear strain τ, and damping ratio h-τ) and shear strength (adhesive strength c and internal frictional angle ϕ) by Wakai et al. (2004). What characterizes this constitutive model is the use of the hyperbolic stress strain relationship, which is closer to the actual kinetic property of soil, as the skeleton curve in combination with the unique hysteresis loop curve to solve the problem of excessive damping ratio pointed out in a corrected HD model. The hysteresis loop curve can express the dynamic deformation characteristics that match the general h-τ relationship of soil by appropriately providing two parameters (n and $b_{\tau G0}$). In our analysis, the h-τ relationship of Ishihara (1976) was used for reference. As a result of the above formulation, we can realize dynamic deformation characteristics that are closer to the actual soil and reproduce the shear failure of the ground to an excellent degree of precision, based on the Mohr-Coulomb standard.

The basic equation for the elasto-plastic constitutive model described above is expressed as follows. In a dynamic elasto-plastic FEM, the following equation of motion is solved by the time integration algorithm based on Newmark's β method.

$$[M]\{\ddot{u}\} + [C]\{\dot{u}\} + \{P\} = -[M]\{\ddot{U}\} \qquad (1)$$

$\{P\}$ is a nodal point vector equivalent to internal stress, and $\{P\} = [K]\{u\}$ in a linear elastic body. $[M]$, $[C]$ and $[K]$ are respectively mass, attenuation, and initial rigidity matrix. $\{u\}$ and $\{U\}$ are relative displacement (at each position) and absolute displacement vector (of the foundation bed), respectively. The attenuation matrix was derived based on the assumption of Rayleigh attenuation.

$$[C] = \alpha[M] + \beta[K] \qquad (2)$$

This time two values, α and β, that lead to an damping ratio of 3% at a frequency of 0.5 Hz and 5.0 Hz, are assumed, which is 0.171 and 0.00174, respectively.

Refer to Wakai et al. (2004) for the details of formulation based on an elasto-plastic constitutive model.

3.2 Analytical model

The size of one finite element for the seismic response analysis of the Hitotsu-minesawa landslide is set to 25 m × 25 m, and the area measuring 1.5 km × 1.5 km in the neighborhood of the subject area, including Hitotsu-minesawa, was analyzed. When the topography of the subject area was converted to finite element meshes, the raster elevation data obtained from GIS were used.

For the depth direction, element division was conducted assuming the ground to be composed of three kinds of material. They are, in descending order from near the top of the surface, the subsurface, weathered layer and fresh part. The fresh part alone is divided into two elements, and a total of four divisions were made in the depth direction. This analytical model is shown in Figure 7.

For the ground constants, the authors decided to choose the appropriate constants out of the ground survey reports compiled by us for many construction projects conducted in the subject area. The ground constants used for the analysis are tabulated in Table 1.

3.3 Seismic wave input in the basement

Figure 1 shows the observation location of a seismic waveform used as a reference for our determination

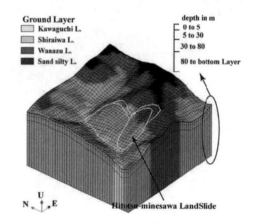

Figure 7. Finite element model for Hitotsu-minesawa.

Table 1. Ground constants used for analysis.

General legend	Ground model (depth in m)	Unit volumetric weight γ[kN/m³]	Young's modulus E[kN/m²]	Poisson ratio ν	Adhesive force C[kN/m²]	Internal frictional angle φ[°]
Sand silty layer	0 to 5	17.658	117720	0.40	9.81	30
	5 to 30	18.639	1285110	0.35	98.10	35
	30 to 80	19.620	5209110	0.30	981.00	40
	80 to bottom layer	19.620	5209110	0.30	981.00	40
Wanazu layer and Shiraiwa layer	0 to 5	17.658	117720	0.40	9.81	30
	5 to 30	19.620	1353780	0.35	98.10	35
	30 to 80	19.620	5209110	0.30	981.00	40
	80 to bottom layer	19.620	5209110	0.30	981.00	40
Kawaguchi layer	0 to 5	19.620	127530	0.40	9.81	30
	5 to 30	19.620	1353780	0.35	98.10	35
	30 to 80	21.582	6925860	0.30	981.00	40
	80 to bottom layer	21.582	6925860	0.30	981.00	40

of the input seismic motion (Takezawa in Yamakoshi Village) and the location of the epicenter of the Chuetsu Earthquake. There are many ways of selecting the input seismic waveforms used for analysis (two components of NS and EW). In our case, before we determined the input seismic waveforms, we conducted FEM analysis of the area around Takezawa using the assumed input seismic motion and adjusted the input seismic waveforms on a trial-and-error basis such that the response acceleration, which is the analysis result, at Takezawa is close to agreeing with the observed strong motion record.

Figure 8 shows the NS and EW components of the input seismic waveforms (which were input to the location 91 m above sea level) thus determined for our analysis.

3.4 Analysis results

The analysis results are put together in Figures 9 through 11. They represent, respectively, the maximum horizontal acceleration on the ground surface during the earthquake, residual horizontal displacement on the ground surface (both are the combined values of the NS and EW components) and the maximum shear stress in the earthquake in the subsurface elements.

Note that the response acceleration resulting from those dynamic elasto-plastic FEM analyses have been put through a low pass filter with a cut-off frequency of 10 Hz to remove the high-frequency components.

As Figure 9 indicates, amplification of acceleration is very much apparent at topographically sharp points or near the ridges. It is shown that acceleration of over 1,000 gal particularly affected the ridge part of the area where the landslide studied in this paper occurred. Horizontal acceleration of over 1,000 gal also occurred at the ridge, shown in A-A' in Figure 2. High values of residual displacement are also

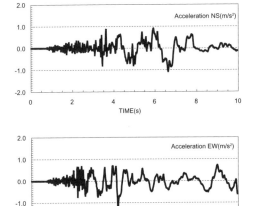

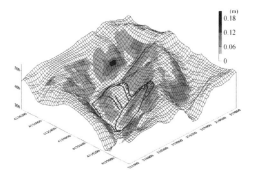

Figure 8. Input seismic waveforms (91 m above sea level).

Figure 10. Result of dynamic response analysis (residual horizontal displacement).

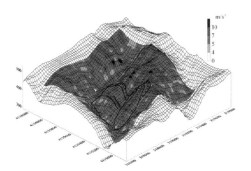

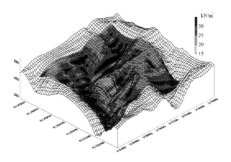

Figure 9. Results of dynamic response analysis (maximum horizontal acceleration).

Figure 11. Results of dynamic response analysis (maximum shear stress).

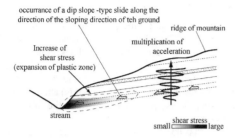

Figure 12. Schematic diagram representing the mechanism of slope failure triggered by the earthquake.

recorded at locations subjected to higher horizontal acceleration (Figure 9). On the other hand, according to Figure 11, a larger shear force of over 30 kN/m² affected Hitotsu-minesawa and the mountain streams at both ends of the A-A′ measuring line that pinches the ridge.

More specifically, a landslide occurs at a topography where high horizontal acceleration occurs at a ridge and high shear force affects the stream part, which corresponds to the dip slope of the ground (Figure 12).

Figure 12 Schematic diagram representing the mechanism of slope failure triggered by the earthquake.

4 DISCUSSION AND SUMMARY

The seismic response analysis using dynamic elasto-plastic FEM revealed that high horizontal acceleration and displacement occurred at salient topography such as at the head of the ridge and that, on the other hand, the large shear force influenced the stream part. These results agree very closely with the phenomenon by which cracks in the north-south direction occurred at intervals of 1 to 10 m over the exposed bedrock of the stream, which is the border between the ridge part and the paddy part (Figure 4). The phenomenon by which high horizontal acceleration occurs at a ridge has previously been reported by Asano et al. (2006) and other researchers, but our analysis results reveal the additional finding that the mountain stream part is subjected to a high shear force. In other words, when the Chuetsu Earthquake occurred, its force severely shook the Hitotsu-minesawa ridge and at the same time caused a high shear force at the stream at the foot of the ridge, and eventually a landslide occurred that took the form of a dip slope-type slide in the direction of slope of the ground. This is how the Hitotsu-minesawa rockslide occurred. A characteristic of the Hitotsu-minesawa rockslide is that the direction of the ridge protrusion matches the direction of the sloping direction of the ground and that the rockslide occurred in the topography where the ridge, which has a stream on either side of its foot, is open on both sides. Thus, it appears that the ridge itself was severed transversely when the land mass slipped. This judgment is corroborated by the report of Abe et al. (2005) that described the presence of a landslide very similar to the Hitotsu-minesawa rockslide, or an earthquake-triggered Tertiary formation landslide that occurred on the ridge part of the mountain.

We are currently making a stability evaluation of mountain slopes by dynamic elasto-plastic FEM analysis for the purpose of corroborating earthquake-induced rockslides at the ridge from a viewpoint of dynamics.

REFERENCES

Abe, S. & Takahashi, A. 1997. Landslide processes during earthquake in the Green Tuff area in the Tohoku district. *Jour. Japan Soc. Eng. Geol* Vol.38 No.5:265–279 (in Japanese).

Abe, S., Takahashi, A., Ogita, S. & Yoshimatsu, H. 2005. Earthquake induced landslides in the Tertiary rock area in Japan. *International Symposium Landslide hazard in Orogenic zone from the Himalaya to island arc in Asia*:397–406.

Asano, S., Ochiai, H., Kurokawa, U., Okada, Y. 2006. Topographic effects on earthquake motion that trigger landslides. *Journal of the Japan Landslide Society* Vol.42 No.6:457–466 (in Japanese).

Ishihara, K. 1976. *The Foundation of Soil Dynamics*: 196–206, Kajima Institute Publishing Co., Ltd.

K-NET: http://www.kyoshin.bosai.go.jp/k-net/

Mizuyama, T., Matsumura, K., Tsuchiya, S., Takahashi, M. & Yang, W. 2002. Evaluation of topographic effects on seismic failures of model slopes through dynamic response analysis. *Congress publication of Interprevent 2002 in the pacific rim* Vol.1:59–66.

Tanaka, N., Wakai, A., Kanto, K. & Ito, H. 2006. Soil constants for seismic evaluation of Mid-Niigata area and their sensibility analysis. *Proceeding of 45rd Conference of Japan Landslide Society*:293–296 (in Japanese).

Wakai, A. & Ugai, K. 2004. A simple constitutive model for the seismic analysis of slopes and its applications. *Soils and Foundations* Vol.44 No.4:83–97.

Wakkai, A., Watanabe, T., Kawabata, H. & Kanto, K. 2005. Fundamental study for development of mountains-area seismic intensity prediction system used on FEM without advanced computer environment. *Proc. Geo-Kanto 2005 JGS*:145–148 (in Japanese).

Wakkai, A., Kawabata, H., Watanabe, T., Ahang, T., Kanto, K. & Tanaka, N. 2006. Mountains-area seismic intensity prediction system based on FEM without excellent computers. Proceeding of 45rd Conference of Japan Landslide Society. 297–300. (in Japanese).

Analysis for progressive failure of the Senise landslide based on Cosserat continuum model

H.X. Tang
Department of Civil and Hydraulic Engineering, Dalian University of Technology, Dalian, China

ABSTRACT: Based on pressure-dependent elastoplastic Cosserat continuum model, progressive failure phenomena of the Senise slope occurred in the excavation processes, characterized by strain localization due to strain softening, are numerically simulated. Numerical results indicate the inability of classical continuum model in simulating the whole failure progress, while the capability and performance of Cosserat continuum model in keeping the well-posedness of the boundary value problems with strain softening behavior incorporated and in completing simulation of the whole failure progress.

Keywords: Progressive failure, Cosserat continuum model, strain localization, strain softening.

1 INTRODUCTION

As strain softening constitutive behavior is incorporated into a computational model in the frame of classical plastic continuum theories, the initial and boundary value problem of the model will become ill-posed, resulting in pathologically mesh-dependent solutions. Furthermore, the energy dissipated at strain softening is incorrectly predicted to be zero, and the finite element solutions converge to incorrect, physically meaningless ones as the element mesh is refined. To correctly simulate strain localization phenomena characterized by occurrence and severe development of the deformation localized into narrow bands of intense irreversible straining caused by strain softening, it is required to introduce some type of regularization mechanism into the classical continuum model to preserve the well-posedness of the localization problem.

One of the radical approaches to introduce the regularization mechanism into the model is to utilize the Cosserat micro-polar continuum theory, in which high-order continuum structures are introduced. As two dimensional problems are concerned, a rotational degree of freedom with the rotation axis orthogonal to the 2D plane, micro-curvatures as spatial derivatives of the rotational degree of freedom, coupled stresses energetically conjugate to the micro-curvatures and the material parameter defined as internal length scale are introduced in the Cosserat continuum.

Among the work, which utilize the Cosserat continuum model as the regularization approach to analyze strain localization problems, are contributions of Muhlhaus (1987, 1989), de Borst et al (1991, 1993), Tejchman et al (1993, 1996), Steinmann (1994, 1999), Manzari (2004), Li and Tang (2005).

de Borst (1991, 1993) formulated the von-Mises elastoplastic model within the framework of the Cosserat continuum, and further extended to the pressure dependent J_2 flow model. Manzari (2004) formulated a micropolar elastoplastic model for soils and employed a series of finite element analyses to demonstrate the use of a micropolar continuum in overcoming the numerical difficulties encountered in application of finite element method in classical continuum when non-associated yield criterion is incorporated in the constitutive model. But the strain softening behavior, which may occur in the geotechnical engineering, was not detected because of the four nodded displacement-based quadrilateral isoparametric element interpolation approximation being used.

Li and Tang (2005) formulated a consistent algorithm of the pressure-dependent elastoplastic model in the framework of Cosserat continuum theory, i.e. the return mapping algorithm for the integration of the rate constitutive equation and the closed form of the consistent elastoplastic tangent modulus matrix.

Based on Li and Tang's pressure-dependent elastoplastic Cosserat continuum model, progressive failure phenomena of Senise landslide occurred in the excavation processes in Italy, characterized by strain localization due to strain softening, are numerically simulated. At the same time, the numerical results based on classical continuum model indicate its inability in simulating the whole failure progress of the slope.

2 PRESSSURE-DEPENDENT ELASTOPLASTIC COSSERAT CONTINUUM MODEL

Each material point in the two dimensional Cosserat continuum has three degrees-of-freedom, i.e. two translational degrees-of-freedom u_x, u_y and one rotational degree-of-freedom ω_z with the rotation axis orthogonal to the two dimensional plane,

$$\boldsymbol{u} = [u_x \; u_y \; \omega_z]^T \quad (1)$$

Correspondingly, the strain and stress vectors are defined as

$$\boldsymbol{\varepsilon} = [\varepsilon_{xx} \; \varepsilon_{yy} \; \varepsilon_{zz} \; \varepsilon_{xy} \; \varepsilon_{yx} \; \kappa_{zx}l_c \; \kappa_{zy}l_c]^T \quad (2)$$

$$\boldsymbol{\sigma} = [\sigma_{xx} \; \sigma_{yy} \; \sigma_{zz} \; \sigma_{xy} \; \sigma_{yx} m_{zx}/l_c m_{zy}/l_c]^T \quad (3)$$

where κ_{zx}, κ_{zy} are introduced as micro-curvatures in Cosserat theory, m_{zx}, m_{zy} are the couple stresses conjugate to the curvatures κ_{zx}, κ_{zy} (Fig. 1), l_c is defined as the internal length scale.

The relation between strain components and displacement components and the equilibrium equations can be written in matrix—vector forms as

$$\boldsymbol{\varepsilon} = \mathbf{L}\boldsymbol{u} \quad (4)$$

$$\mathbf{L}^T\boldsymbol{\sigma} + \mathbf{f} = \mathbf{0} \quad (5)$$

in which the operator matrix

$$\mathbf{L}^T = \begin{bmatrix} \frac{\partial}{\partial x} & 0 & 0 & 0 & \frac{\partial}{\partial y} & 0 & 0 \\ 0 & \frac{\partial}{\partial y} & 0 & \frac{\partial}{\partial x} & 0 & 0 & 0 \\ 0 & 0 & 0 & -1 & 1 & l_c\frac{\partial}{\partial x} & l_c\frac{\partial}{\partial y} \end{bmatrix} \quad (6)$$

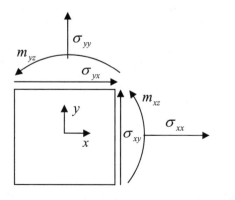

Figure 1. Stress and couple-stress in a two-dimensional Cosserat continuum.

It is assumed that the strain vector $\boldsymbol{\varepsilon}$ is additively decomposed into both the elastic and the plastic parts, i.e. $\boldsymbol{\varepsilon}_e$ and $\boldsymbol{\varepsilon}_p$, and the elastic strain vector $\boldsymbol{\varepsilon}_e$ is linearly related to the stress vector,

$$\boldsymbol{\sigma} = \mathbf{D}_e \boldsymbol{\varepsilon}_e \quad (7)$$

in which the elastic modulus matrix $\mathbf{D}_e$ for isotropic media can be given in the form

$$\mathbf{D}_e = \begin{bmatrix} \lambda+2G & \lambda & \lambda & 0 & 0 & 0 & 0 \\ \lambda & \lambda+2G & \lambda & 0 & 0 & 0 & 0 \\ \lambda & \lambda & \lambda+2G & 0 & 0 & 0 & 0 \\ 0 & 0 & 0 & G+G_c & G-G_c & 0 & 0 \\ 0 & 0 & 0 & G-G_c & G+G_c & 0 & 0 \\ 0 & 0 & 0 & 0 & 0 & 2G & 0 \\ 0 & 0 & 0 & 0 & 0 & 0 & 2G \end{bmatrix} \quad (8)$$

with the Lame constant $\lambda = 2G\upsilon/(1-2\upsilon)$, G and υ are the shear modulus and Poisson's ratio in the classical sense, while G_c is introduced as the Cosserat shear modulus.

In the framework of Cosserat continuum theory, a consistent algorithm of the pressure-dependent elastoplastic model, i.e. the return mapping algorithm for the integration of the rate constitutive equation and the closed form of the consistent elastoplastic tangent modulus matrix can be formulated. The details of the model will not be recited here and the readers can reference Li and Tang (2005)'s work.

3 INTRODUCTION OF THE SENISE LANDSLIDE

According to Troncone (2005), on 26 July 1986 a landslide of great dimensions occurred at Timpone hill in Senise, which is a village located about 70 km from Potenza, in southern Italy. As a result of this event eight people died, and several buildings were destroyed or badly damaged. A reactivation of the landslide occurred on 6 September 1986: it caused further movement of the soil mass.

This case study was analyzed by Troncone (2005), who provided exhaustive documentation of the landslide. After the first catastrophic event and before the reactivation of September 1986, a site investigation consisting of boreholes, penetration tests and conventional laboratory tests was performed. On the basis of the available data, the main features of the landslide mechanism were highlighted in the aforesaid studies: the location of the slip surface was detected, and the laboratory tests showed that the soils involved in the

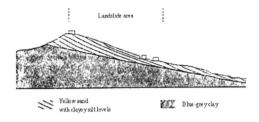

Figure 2. Schematic geological section of Senise hill.

Figure 3. Clayey silt level within sand formation.

movement were characterized by a pronounced strain-softening behavior. A summary of these investigated aspects will be reported below.

A schematics geological section of Senise hill is shown in Fig. 2. The subsoil is essentially constituted of the Plio-Pleistocene Aliano formation, consisting of yellowish sand with varying grain size, interbedded by clayey silt levels. These levels, dipping downslope, have thickness ranging from some centimeters to several decimeters and an inclination of about 18°, as can be seen from Fig. 3. The sand is very dense, and characterized by a significant degree of diagenesis due to calcareous cementation. Fig. 2 also reveals the presence of a basic formation consisting of blue-grey clay. A water table was found at a depth of about 23 m from the ground level, at the slope toe. It therefore was well below the soil mass involved in the landslide occurred after a period of very scarce rain.

The strength properties of the soils under consideration were obtained from consolidated-drained triaxial tests and direct shear tests. As shown in Fig. 4, one peculiar aspect of these soils is their mechanical instability, pointed out by a pronounced strain-softening behavior after the peak is reached. It revealed that both the main and the secondary sliding surfaces were located in thin clayey silt layers present in the sand formation by exploration, ant the kinematics of the Senise landslide was essentially a translational sliding: the main failure surface developed largely along a thin clayey silt layer interbedded within a slightly

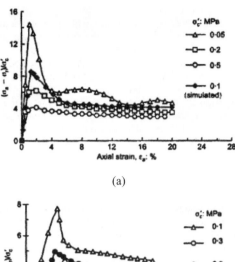

(a)

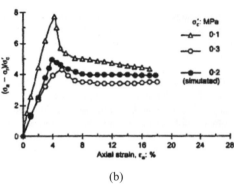

(b)

Figure 4. Experimental and simulated results of triaxial tests: (a) sand; (b) clayey silt.

cemented sand formation. Both these materials exhibited a pronounced strain-softening behavior during the laboratory tests.

4 STABILITY ANALYSES FOR THE SENISE LANDSLIDE

Very different interpretations of the landslide were provided on the basis of stability analyses conducted using the limit equilibrium method. Nevertheless, all the authors agree on the fact that the event was triggered, or at least influenced, by the deep excavations carried out at the toe of the slope before the construction of some of the damaged buildings. The cuts were essentially vertical, and were sustained by reinforced concrete retaining walls.

Using the traditional limit equilibrium approach, Troncone (2005) performed the stability analysis of the Senise slope. By considering a potential slip surface close to that which occurred (Fig. 5), the safety factor SF, computed using Sarma's method in conjunction with the peak strength parameters, is 1.73. By contrast, using the residual strength, the slope is unstable with a safety factor of about 0.6. This should imply that a progressive failure really occurred, with the operational strength along the sliding surface varying between the peak and the residual values.

The application of the finite element method in capturing the development of the yield zone and the effect of progressive failure in strain-softening soils has been illustrated by many authors. One of these studies employed an elasto-plastic constitutive model associated with a law of Mohr-Coulomb type, in which the softening behavior of the material is accounted for by the progressive reduction of the strength parameters with the accumulated deviatoric plastic strains. The solution based on this approach may be affected by a lack of convergence and may depend strongly on the mesh adopted, in terms of both the size and the orientation of the elements. Therefore an analysis of this type requires particular care, especially when the location and propagation of the shear zones cannot be predetermined on the basis of some evidence.

In order to overcome the numerical difficulties of the classical continuum when the strain-softening behavior incorporated, Troncone (2005) used the elasto-viscoplastic model with strain-softening to analyze the stability of the slope. It was found that this model was very effective in capturing the failure process that occurred.

Based on Li and Tang's pressure-dependent elasto-plastic Cosserat continuum model, progressive failure phenomena of Senise landslide occurred in the excavation processes in Italy, characterized by strain localization due to strain softening are numerically simulated. At the same time, the numerical results based on classical continuum model indicate its inability in simulating the whole failure progress of the slope.

According to Troncone's analyses, the soil parameters used in the analyses are reported in Table 1. The soil dilatancy angle has been assumed to be nil, implying a non-associated flow rule with zero volume change during yield. k^p_{shear} is set to zero to account for the brittle behavior experienced by the soils during the laboratory tests. The parameters G_c and l_c, which are related to Cosserat continuum model, are chosen to correspond with the general rule. In order to take into account the strain-softening behavior of the soil, the stability analysis of the slope has been conducted using the FEM. Fig. 6 shows the mesh adopted in the calculation, and it consists of the eight nodded quadratic quadrangle elements.

The initial stress state within the slope, before the excavation, has been reproduced by progressively increasing the gravity acceleration up to the value of 9.81 m/s^2, under the assumption that all the soils exhibit elastic perfectly plastic behavior with the

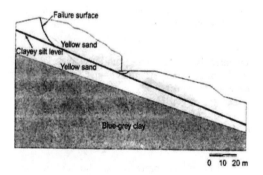

Figure 5. Failure surface considered in stability analysis performed using limit equilibrium method.

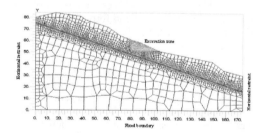

Figure 6. Mesh and boundary conditions used in finite element analysis.

Table 1. Soil parameters used in the analysis.

	γ	E	ν	c'_p	φ'_p	c'_r	φ'_r	ψ	k^{p*}_{shear}	k^r_{shear}	G_c	l_c
	kN/m^3	kPa		kPa	°	kPa	°	°	%	%	kPa	m
Yellow sand	20	70000	0.25	37	43	0	35	0	0	4	14000	0.15
Clayey silt level	20	25000	0.25	15	30	0	12	0	0	4	5000	0.15
Blue-grey clay	20	70000	0.25	150	31	150	31	0	0	0	14000	0.15

* "p" means peak, "r" means residual. For example, k^p_{shear} means the peak shear strain.

Drucker-Prager failure criterion. At the end of the gravity loading, the corresponding displacements and strains have been reset to zero. A linear-elastic element simulating the retaining wall has been inserted at the slope toe, without affecting the stress state within the slope. The excavation process is simulated by removing the elements at the same level in 9 increments from the excavation zone of the initial finite element mesh, and Mana's method (1981) is used to calculate the excavation loading.

At first, analysis has been performed assuming an elasto-plastic strain-softening model with the Drucker-Prager plastic law for all the soils involved. Fig. 7 shows the evolution of the effective plastic strain occurring during the last excavation steps. It can be seen that the plastic strain field develops with a clear localization within the clayey silt level. However, the whole development of the failure surface can not be completed. As the excavation process continues, the classical finite element numerical solution faces significant difficulties with the number of negative eigenvalues in the system stiffness matrix increasing and the numerical calculation can not be carried out any more. Therefore the numerical analysis based on this elasto-plastic model with strain softening is not fully effective in capturing the failure process that occurs in the present study.

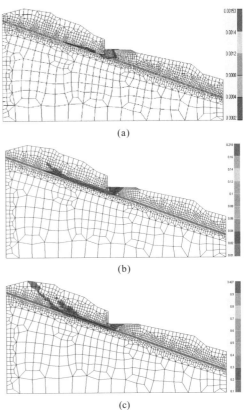

Figure 8. Effective plastic strain distribution in the slope during the excavation process given by Cosserat continuum model: (a) after the removal of the 7th layer; (b) after the removal of the 8th layer; (c) after the removal of the 9th layer.

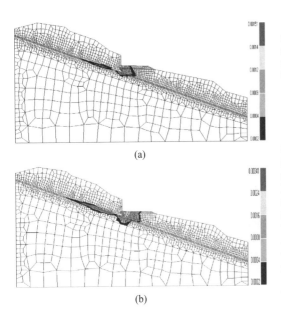

Figure 7. Effective plastic strain distribution in the slope during the excavation process given by classical continuum model: (a) after the removal of the 7th layer; (b) after the removal of the 8th layer.

The calculations have been repeated using the elasto-plastic Cosserat continuum model with strain-softening. Fig. 8 shows the evolution of the more significant effective plastic strain occurring during the last excavation steps. As can be seen that the plastic strain are concentrated in the clayey silt level, and as the excavation process continues they propagate up the slope extending also to the sand formation and, finally cause the collapse of the slope when the excavation is completed. Fig. 9 shows the corresponding deformation configuration at the end of excavation. The position of the failure surface at both the top and the toe of the slope corresponds with that of Troncone's analyses and that actually observed. Therefore the numerical analysis based on Cosserat elasto-plastic model with strain softening is effective in capturing the failure process that occurs in the present study.

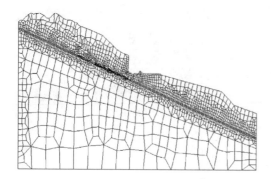

Figure 9. Deformation configuration of the slope after the removal of the 9th layer with Cosserat continuum.

5 CONCLUSIONS

Based on pressure-dependent elastoplastic Cosserat continuum model, progressive failure phenomena of the Senise slope occurred in the excavation processes, characterized by strain localization due to a pronounced strain softening, are numerically simulated. Numerical results indicate the inability of classical continuum model in simulating the whole failure progress, while the capability and performance of Cosserat continuum model in keeping the well-posedness of the boundary value problems with strain softening behavior incorporated and in completing simulation of the whole failure progress.

ACKNOWLEDGEMENTS

The author is pleased to acknowledge the support of this work by the National Natural Science Foundation of China through contract/grant numbers 50278012, 10672033 and the National Key Basic Research and Development Program (973 Program) through contract number 2002CB412709.

REFERENCES

de Borst R., Sluys L.J., 1991. Localization in a Cosserat continuum under static and dynamic loading conditions. Comput. Methods Appl. Mech. Engrg., 90: 805–827.

de Borst R., 1991. Simulation of srtain localization:a reappraisal of the cosserat continuum. Engineering computations, (8): 317–332.

de Borst R., 1993. A generalization of J2-flow theory for polar continue. Computer methods in applied mechanics and engineering, 103: 347–362.

Li Xikui, Tang Hongxiang, 2005. A consistent return mapping algorithm for pressure-dependent elastoplastic Cosserat continua and modeling of strain localization. Computers & Structures, 83 (1): 1–10.

Mana A.I., Clough G.W., 1981. Prediction of movements for braced cuts in clay. Journal of Geotechnical Engineering Division, 107: 759–778.

Manzari M.T., 2004. Application of micropolar plasticity to post failure analysis in geomechanics, Int. J. Numer. Anal. Meth. Geomech., 28: 1011–1032.

Muhlhaus H.B., Vardoulakis I., 1987. The thickness of shear band in granular materials. Geotechnique, 37 (3): 271–283.

Muhlhaus H.B., 1989. Application of Cosserat theory in numerical solution of limit load problems. Ing. Arch., 59: 124–137.

Steinmann Paul, 1994. A micropolar theory of finite deformation and finite rotation multiplicative elastoplasticity. Int. J. Solids Structures, 31: 1063–1084.

Steinmann Paul, 1999. Theory and numerics of ductile micropolar elastoplastic damage. Int. J. Numer. Methods Eng., 38: 583–606.

Tejchman J., Wu W., 1993. Numerical study on patterning of shear bands in a Cosserat continuum. Acta Mechanica, 99: 61–74.

Tejchman J., Bauer E., 1996. Numerical simulation of shear band formation with a polar hypoplastic constitutive model. Computers and geotechnics, 19: 221–244.

Troncone A., 2005. Numerical analysis of a landslide in soils with strain-softening behaviour. Geotechnique, 55 (8): 585–596.

Large-scale deformation of the La Clapière landslide and its numerical modelling (Saint-Etienne de Tinée, France)

E. Tric, T. Lebourg & H. Jomard
University of Nice, Sophia Antipolis, Géosciences Azur, UMR 6526, Valbonne, France

ABSTRACT: The large-scale deformation of high mountain slopes finds its origin in many phenomena from which time-constants are very different. Gravitational effect, tectonics forces, water infiltration are generally the principal causes. However, it is always very difficult to distinguish which cause is dominating and which are their effect respective. The numerical approach offers the possibility of testing some of these causes, in particular the gravitational effect. Then, a two-dimensional numerical experimentation with ADELI code was carried out to determine the effect of gravitational force on mechanical behaviour of the "la Clapière" area. The results show that gravitational instability is possible for values of cohesion and angle of internal friction respectively lower than à 2,00 ± 0,01 Mpa and 27,69° ± 0,14°. These values are compatible with the geomechanical parameters proposed by Gunzburger (2001) from measurements directly done on the site. The numerical results show also a good agreement between the calculated deformation, the actual morphology of the site and geophysical data obtained by resistivity investigation. The deformation leads to destabilisation of the massif by a regressive evolution of the landslide from the bottom up to 1800 m, which is actually the top of Ma Clapière landslide. This progression of the deformation concerns only a depth of around 150 ± 50 m, which could be correlated to the sliding surface like suggested by resistivity investigations (Lebourg et al. 2005; Jomard et al. 2007).

1 INTRODUCTION

For the study of the deformation or the stability analysis of high mountain slopes, knowledge of the large-scale mechanical properties is important. Different approaches exist to estimate those parameters like seismic measurements (Eberhart-Philips et al., 1995; Brueckl and Parotudis, 2001), geotechnical measurements like 'Schmidt hammer tests' (Bieniawski, 1989; Gunzburger, 2001; Gunzburger and Merrien-Soukatchoff, 2002) or mechanical test in laboratory (Lebourg et al. 2004). However, these different observational analysis have not the potential to describe the processes involved on the generation and development of a creeping rock mass out of an originally compact rock. A promising supplement could be the geomechanical modelling of the structures of the creeping or sliding rock masses. Thus, geomechanical parameters may be determined by fitting the geomechanical model to the structures as determined from geophysical exploration (seismic or electrical). This approach rapidly evolving due to the increasing computer power allows also to define stress-strain state of the slope with use of the relevant non-linear mathematical models of the soils and rigorous consideration of all acting forces as well as the variation of mechanical properties with depth (Savage et al. 2000; Brueckl & Parotudis, 2001, Merrien-Soukatchoff et al. 2001, Eberhardt et al. 2004).

In this contribution, we present an application of the numerical approach to the "La Clapière" Landslide (Alpes-Maritimes, France). Indeed, if numerous studies have been realised on this site (hydrological, geologic, tectonic, topographic, ...) (Follacci, 1987, Ivaldi et al. 1991, Compagnon et al. 1997; Guglielmi et al. 2000, 2002, 2003, Gunzburger & Laumonier, 2002), very few numerical studies have been done (Quenot, 2000; Merrien-Soukatchoff et al. 2001, Gunzburger et al. 2002, Cappa et al. 2004) and, to our knowledge, no study has been devoted on the localisation of the deformation and its temporal evolution as a function of the geomechanical parameters and their relation with the gravitational destabilisation of the slope.

2 THE "LA CLAPIÈRE" LANDSLIDE

"La Clapière" landslide is a large unstable slope located in the South-eastern part of France, in the Alps, about 80 km North of Nice city. This landslide which mobilizes a huge volume (55×10^6 m^3) of metamorphic slope in the Mercantour massif (Follacci et al. 1988) is developed on the left bank of Tinee Valley and

Figure 1. The "La Clapière" landslide in 1998 (photo from CETE de Nice).

affect a slope that culminate at 3000 meters, between 1100 and 1800 meters of altitude (figure 1).

It is bordered on its northwestern side by the Tenibres river and to its southeastern sibe by the Rabuons river, flowing into the Tinée river. A large rupture has been identified since the beginning of the century: in 1936 the wrenching at the top of the landslide was already quite visible. In the seventies, the movements became more continuous and a monitoring of the site is working since 1982 (Follaci, 1987, 1988). For a few years, a distance monitoring has been combined with other types of investigations: hydrogeological studies (Compagnon et al. 1997; Guglielmi et al. 2000, Cappa et al. 2004), remote sensing (Casson et al. 2003) and subsurface geophysical investigations (Lebourg et al. 2003, 2005, Jomard et al. 2007a, 2007b). Different types of geometric, mechanic or hydromechanic models have been derived from these data sets (Merrien-Soukatchoff et al. 2001; Cappa et al. 2004) but none of them visualise the morphology of the main slip surface geometry. However, Jomard et al. (2007b) have recently investigated by electrical method the vertical structure of the upper part of the NE lobe of the La Clapière landslide. Their results shows a very strong decreasing of the resistivity to a depth of approximately 90 m which could be associated to the main shear surface of this landslide for this altitude.

In this area, the basement unit is composed of migmatitic paragneisses (Anelle Formation) and orthogneisses (Iglière Formation) bearing a strong Hercynian foliation (Figure 2). This foliation is normally oriented 115° E, 70° NE but, in the la Clapière area, it is progressively rotated to a subhorizontal attitude by a structure called the Clapière fold (Fabri and Cappa, 2001; Gunzburger and Laumonier, 2002). The axis of the fold is 120° E, 15° NW; the axial plane is 140° E, 40° SW. The Clapière landslide occurs in the upper/short limb of the fold (Figure 2).

Two origins of the La Clapière fold are actually proposed. It is either tectonic (Gunzburger and

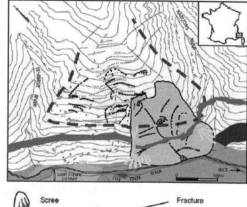

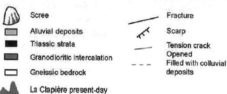

Figure 2. Geomorphological context of La Clapière landslide from Cappa et al. 2004.

Laumonier, 2002, Delteil et al. 2003) or gravitational (Follacci, 1987, 1988). However, the numerical approach done by Merrien-Soukatchoff et al. (2001) shows that gravitational toppling proposed by Follacci (1987) led to be reconsidered. The "La Clapière" slope itself is affected by a lot of tectonic discontinuities. A complete description can be read in Guglielmi et al. (2000), Gunzburger and Laumonier (2002) and Cappa et al. (2004).

Cappa et al. (2004) have tested numerically the influence of the location and the amount of water infiltration on the hydromechanical behaviour of La Clapière slope. Their results showed that most destabilizing area is located in the middle part of the slope. However, due to the internal complexity of the slope hydrogeology, especially the existing perched aquifer in the decompression toppled zone directly connected to the landslide, the most dangerous infiltration inflow is not necessarily that on the landslide area.

3 MECHANICAL MODELLING

The 2D numerical modelling of the "La Clapière" landslide has been done with the computer code ADELI developed by R. Hassani and J. Chery (Hassani, 1994, Hassani et al. 1997). This code is based on a finite element method for space discretization associated with a dynamic relaxation method for time discretization. This method is described in Hassani et al. 1997. It leads to a set of nonlinear equations:

$$M\ddot{u} = F_{int}(u,\dot{u},t) + F_{ext}(u,t) + F_c(u,\dot{u},t)$$

where the vectors F_{int} and F_{ext} are the internal and external nodal forces, F_c is the vector of contact reaction, u, $\dot{u}$ and $\ddot{u}$ are the vectors of nodal displacements, nodal velocities, and nodal accelerations, respectively. M is a fictitious mass matrix. The quasi-static solution is reached when the inertial regularising term $M\ddot{u}$ is negligible compared to the external forces. The numerical divergence of the result means that rupture is appeared and involves a destabilising of the solid mass.

For each numerical experiment presented in this work, several mesh have been tested (6000, 10000, 15000 and 20000 elements) and 5×10^4 to 1×10^6 times steps were used for a total duration of 100 or 10000 years. The lengths of the time steps are then 52 minutes to 17 hours for the duration of 100 years, and 87 hours to 72 days for the duration of 10000 years.

3.1 Geometry and boundary conditions

The geometry used is two-dimensional. The initial state topography is taken from the most ancient available map (dated from 1933) in order to eliminate as far as possible recent slope deformations. The velocity boundary conditions are shown in figure 3. The gravity is the only force.

3.2 Mechanical parameters

The "La Clapière" slope is constituted of migmatitic paragneisses and migmatitic orthogneisses. Those rocks bear a strong Hercyninan foliation, and are therefore highly anisotropic. However, numerical modelling has first been undertaken with isotropic rheological properties. The rocks are also affected by a great number of fractures of various scales and origins, which play a major role in rock mass strength deterioration and deformability increase. To take them into account, a method consist to homogenise the rock mass by using the RMR methodology (Bieniawski, 1989; Gunzburger & Merrien-Soukatchoff, 2002) to bring to an Equivalent Continuous Medium (ECM).

The mechanical parameters of the ECM obtained by this approach for "La Clapière" site are: Young's Modulus, $E = 6,4$ Mpa, cohesion, $c = 210$ kPa, angle of internal friction, $\phi = 29°$, Poisson's ratio, $v = 0,3$ and density, $\rho = 2400$ kg/m³ (Gunzburger, 2001; Gunzburger & Merrien-Soukatchoff, 2002):

These parameters are considered in our study as the reference parameters. In our numerical approach, we model the mechanical behaviour using an elasto-plastic pressure dependent law for which the failure criterion is the Drucker-Prager one (Desai & Siriwardane, 1984). We consider in this study a homogeneous rock mass.

From the five parameters given previously, two have been tested systematically: the cohesion (c) and the friction angle (ϕ). The others were kept constant. Three tests have been performed.

- Test 1: We tested the couple (c, ϕ) from the reference couple (210 kPa, 29°) in order to determine the critical couple for which the destabilisation appears.
- Test 2: The cohesion (c) is constant and we research the critical value of ϕ.
- Test 3: The friction angle is constant and we research the critical value of c.

4 NUMERICAL RESULTS

The results of the test 1 are given in table 1. They are identical whatever the used mesh and the applied duration. They show a critical couple (c, ϕ) equal to 95,0% of the reference couple, i.e. $(1,995.10^5$ Pa, 27,55°) from which the numerical solution does not converge.

All the converged solutions, which correspond consequently to a static solution, are characterised by three plastic deformation zones. The first one is located at the base of the slope up to 1400 meters of altitude and

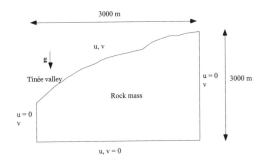

Figure 3. Geometry of the model and boundary conditions. u et v are the horizontal and vertical component of the velocity respectively. Only the gravity (g) is present.

Table 1. Numerical results for the test 1.

c (kPa), ϕ	% of the reference couple	Numerical solution
210, 29.0°	100	Convergence
204, 28.1°	97	Convergence
201, 27.8°	96	Convergence
199, 27.5°	95	Divergence
197, 27.2°	94	Divergence

on a depth about 200 m. The second one is located along the slope between 1400 m and 1800 m of altitude and below the surface (about 100 m). The third zone is at the top of the slop (figure 4a).

When the couple (c, ϕ) is equal or lower to the critical couple, the divergence of the numerical experiment is due to a strong plastic deformation located at the bottom of the slope (figure 4b). This plastic deformation leads to the destabilisation of the massif by a regressive evolution of the landslide from the bottom to the top of the slope. This deformation propagates up to 1800 m, which is actually the top of the "La Clapière" landslide. This progression of the deformation concerns only a depth of around 100–130 m. It is interesting to note that the two altitudes given previously (1400, 1800 m) correspond to a change of the slope in the topography of 1933. Do these results suggest that the distribution of the deformation is controlled by these changes of the slope? The question remains open and deserves to be to study more attentively.

Table 2. Numerical results obtained for different tests.

Test	Constant	Variable	Numerical solution	
			Convergence	Divergence
1		(c, ϕ)	$\geq 96\%$	$\leq 95\%$
2	c	ϕ	$\geq 96\%$	$\leq 95\%$
3	c	ϕ	$\geq 65\%$	$\leq 62\%$

The whole of the results, obtained in the different tests, is resumed in the Table 2. We observe that the instability seems to be mainly controlled by the friction angle. Indeed, when the cohesion is kept constant and equal to 210 kPa, the divergence of the calculation is observed from a value of internal friction angle identical to those obtained when the variable is the couple (c, ϕ). Nevertheless, whatever the used variable, the observed deformation and its temporal and spatial evolution is the same.

5 CONCLUSION

Different mechanical numerical models have been performed to better locate the deformation in the rock mass and determine critical geomechanical parameters from which the landslide occurs in the particular case of "La Clapière". Although the solid mass was regarded as mechanically homogeneous, the first conclusions of this study are encouraging and are:

- A destabilisation of the rock mass is possible with the geomechanical parameters obtained by Y. Gunzburger (2001). Thus, the combination of these mechanical parameters, the topography and the gravity force is completely sufficient to cause a destabilisation of the solid mass.
- Our numerical experiments allow to locate the deformation and his evolution. We observe systematically a good correlation between the calculated deformation and the current morphology of the slope. Indeed, in both cases, the disorders are for an altitude ranging between 1100 m and 1800 m and with a maximum deformation located at the bottom of slope.
- The depth of the deformation is between 80 and 200 m according to the altitude, which is in good agreement with the geophysical results obtained by electrical prospecting in previous studies (Lebourg et al. 2005; Jomard et al. 2007)
- Our results suggest an important role of the changes of the slope in the evolution of the deformation.

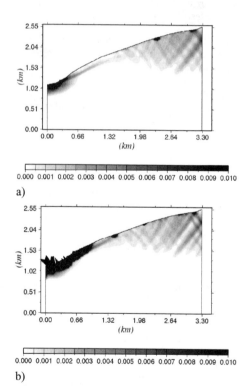

Figure 4. Cumulated plastic deformation obtained for two simulations: (a) the couple (c, ϕ) corresponds to the reference couple (100%). The solution is converged, (b) The couple corresponds to 94% of the reference couple. The solution is not converged.

REFERENCES

Bieniawski Z.T. 1989. Engineering Rock Mass Classifications, J. Wiley.

Brueckl E. & Parotidis M. 2001. Estimation of large-scale mechanical properties of a large landslide on the basis of seismic results. *Int. J. Rock Mech. Sci. & Mining Sci.* 1365–1369.

Cappa F., Guglielmi Y., Soukatchoff V. M., Mudry J., Bertrand, C. & Charmoille, A. 2004. Hydromechanical modelling of a large moving rock slope inferred from slope levelling coupled to spring long-term hydrochemical monitoring: example of the La Clapière landslide (Southern Alps, France). *Journal of Hydrology* 291: 67–90.

Casson B., Delacourt C., Baratoux D. & Allemand P. 2003. Seventeen years of the La Clapière landslide evolution analysed from ortho-rectified aerials photographs. Engineering Geology, 68, 123–139.

Compagnon F., Guglielmi, Y., Mudry, J. & Follacci, J-P. & Ivaldi, J-P. 1997. Chemical and isotopic natural tracing of seepage waters in an important landslide: example from La Clapière landslide (Alpes-Maritimes, France). *C. R. Acad. Sci. Paris*, no. 325: 565–570.

Delteil J., Stephan J-F. & Attal, M. 2003. Control of permian and triassic faults on Alpine basement deformation in the Argentera massif (external southern French Alps). Bull. Soc. Géol. Fr. (174) 55–70.

Desai C.S. & Siriwardane H.J. 1984. Constitutive laws for engineering materials, with emphasis on geologic materials, 457 pp., Prentice-Hall, Englewood Cliffs N.J.

Eberhardt-Philips D. Stanley W.D. Rodriguez B.D. & Lutter W. J. 1995. Surface seismic and electrical methods to detect fluids related to faulting. *JGR* 100: 12, 919–12, 936.

Fabri O. & Cappa F. 2001. Apport de l'analyse structurale à la compréhension de la dégradation du glissement de la La Clapière, Massif du Mercantour, Alpes Maritimes, *S. Spé. Soc., Géol. Fr.*, 13–14.

Follacci J-P. 1987. Les mouvements du versant de la Clapière à Saint-Etienne de Tinée (Alpes—Maritimes). *Bull. Liais. Labo. P. et Cha.* 150–151.

Follacci J.P., Guardia, P. & Ivaldi J.P. 1988. Le glissement de la Clapière (Alpes Maritimes, France) dans son cadre géodynamique. *Compte rendu du 5ème symposium international sur les glissements de terrain*, p1323–1327.

Guglielmi Y. & Bertrand C. Compagnon F. Follacci, J.P. Mudry, J. 2000. Acquisition of water chemistry in a mobile fissured basement massif: its role in the hydrogeological knowledge of the La Clapière Landslide (Mercantour massif, southern alps, France). *Journal of hydrology*, 229, pp 138–148.

Guglielmi Y. et al. Hydrogeochemistry: an investigation tool to evaluate infitration into large moving rock masses (Case study of the La Clapière and Séchilienne alpine landslides). *Bull Eng Geol Env*, vol. 61, pp 311–324, 2002.

Guglielmi., Y. Vengeon, J.M. Bertrand, C. Mudry, J. Follacci J.P. & Giraud, A. 2003. Hydrogeochemistry: an investigation tool to evaluate infiltration into large moving rock masses (case study of La Clapière and Séchilienne alpine landslides). *Bulletin of Engineering Geology and the Environment*, no. 61, pp.311–324.

Gunzburger Y. 2001. Apports de l'analyse de la fracturation et de la modélisation numérique à l'étude du versant instable de La Clapière (Saint-Etienne-de-Tinée, Alpes-Maritimes"), Mémoire de DEA PAE3S, LAEGO, Ecole des Mines de Nancy.

Gunzburger Y., Merrien-Soukatchoff V. & Guglielmi, Y. Mechanical influence of the last deglaciation on the initiation of the "La Clapière" slope instability (southern french alps), 5th European conference on numerical methods in geotechnical engineering. Paris, France, 4–6.

Gunzburger Y. & Laumonier B. 2002. origine tectonique du pli supportant le glissement de terrain de La Clapière (Nord-Ouest du massif de l'Argentera—Mercantour, Alpes du Sud, France) d'après l'analyse de la fracturation. *C. R. Géosciences* 334: 415–422.

Hassani R. 1994. Modélisation numérique de la déformation des systèmes geologiques" Thèse de l'Université de Montpellier II.

Hassani R., Jongmans D. & Chéry J. 1997. Study of plate deformation and stress in subduction processes using two-dimensional numerical models. *J.G.R.* 102: 17, 951–17, 965.

Ivaldi J-P., Guardia P. Follacci J-P. & Terramorsi, S. 1991. Plis de couverture en échelon et failles de second ordre associés à un décrochement dextre de socle sur le bord nord-ouest de l'Argentera (Alpes-Maritimes, France). *C.R. Acad. Sci. Paris, série II*, 313: 361–368.

Jomard H., Lebourg T. & Tric E. 2007a. Identification of the gravitational discontinuity in weathered gneiss by geophysical survey: La Clapière Landslide (France). *Applied Geophysics*, 62: 47–57.

Jomard J., Lebourg T., Binet S., Tric E. & Hernandez M. 2007b. Characterisation of an internal slope movement structure by hydrogeophysical surveying. *Terra Nova*, 19 (1): 48–57.

Lebourg T., Tric E., Guglielmi Y., Cappa F., Charmoille A. & Bouissou S. 2003. Geophysical survey to understand failure mechanisms involved on Deep Seated Landslides. *EGS*, Nice.

Lebourg T., Binet S., Tric E., Jomard H. & El Bedoui, S. 2005. Geophysical survey to estimate the 3D sliding surface and the 4D evolution of the water pressure on part of a Deep Seated Landslide. Terra Nova, 17, 399–406.

Merrien-Soukatchoff V., Quenot Y. & Guglielmi Y. 2001. Modélisation par éléments distincts du phénomène de fauchage gravitaire. Application au glissement de La Clapière (Saint-Etienne de Tinée, Alpes-Maritimes) 95/96, 133–142.

Quenot X. 2000. Etude du glissement de La Clapière. Modélisation du phénomène de rupture, Mémoire de DEA, DEA PAES3S, Ecole Doctorale PROMEMA, Institut national polytechnique de Lorraine, Nancy.

Savage W.Z., Baum R.L., Morrissey M.M. & Arndt B.P. 2000. Finite-element analysis of the Woodway landslide, Washington, U. S. Geological Survey Bulletin 2180, 1–9.

A novel complex valued neuron model for landslide assessment

Kanishka Tyagi, Vaibhav Jindal & Vipunj Kumar
Govind Ballabh Pant University of Agriculture and Technology, Pantnagar, India

ABSTRACT: Landslides are widely spread natural calamity on Earth and is a general term used to describe the down-slope of soil, rock and organic material under the influence of gravity. In this paper we propose a novel method for landslide assessment with the help of input variables that have direct physical significance, using a Novel Neuron Model Approach that works in complex domain and is an extension of the self-look up table approach based Counterpropagation neuron model. Many researchers have investigated landslide assessment problem using conventional neuron models but they suffer from there own limitations of extensive computation and dependency on the pre conditioning of the data being used. The novel part of the paper is that the computation land investigation data is done by converting it into its equivalent complex number form. Architecture of CVCPN, Data validation, Processing of land slide investigation data, and the result of network with various land investigation data have been discussed.

1 INTRODUCTION

There are a number of potentially damaging natural phenomenon, which are termed 'geo-hazards' and slope instability is one of them. Engineers often speak of landslides when the mechanics of movement may be any one of the manifestations of slope instability. In geological term landslide is defined as the displacement of rock masses and the soil along the sloping surface from their normal position under the influence of when structural stability of rock mass is disturbed. This downward movement of the consolidated and unconsolidated rock matter is termed as landslide. Excess quantity of water contents makes the rock weak and more mobile. The intensity and magnitude of these movements varies greatly depending upon the extent and amount of slope, and the type of soil.

These are frequently responsible for considerable losses and they are subsequently considered among the most serious geologic hazards, which plague many parts of the world. Landslides are among the major hydro-geological hazards that affect large parts of India, especially the Himalayas, the North-eastern hill ranges, the Western Ghats, the Nilgiris, the Eastern Ghats and the Vindhyas.

Landslide hazard assessment of existing landslide areas, in which the most probable patterns and mechanisms of future movements are the recurrence of past patterns and mechanisms, is often seen as requiring the recognition of those patterns. Areas of different hazard intensities are mapped according to past activities.

In this study a novel complex valued counterpropagation neural networks which is an extension of already in use real valued counterpropagation network has been used as a means of assessing potential slope failure hazards on the southern slopes of ENE trending ridge near Malpa village situated 43 Km. from Dharchula in Himalayan region of India.

Till now complex valued neural network was actively used in backpropagation algorithm with complex weights and complex valued neuron-activation functions, though this network suffers with many limitations like it is slow with a large set of data, lack of bounded and analytic complex nonlinear activation functions in complex plane (Silverman, 1975). Several approaches have been suggested to process the complex data using backpropagation algorithm (Yadav, 2005). Due to strong power of generalization and simplicity in calculations, Counterpropagation network has an advantage over backpropagation network. Here the complex number theory is successfully applied on forward only Counterpropagation network. Many algorithms have been developed in recent years that work on neural computations techniques with complex values (Hirose, 2003). Many of them have applied complex values on backpropagation algorithm using multilayer or multiplicative neurons (Yadav, 2005). Computation of various Geological parameters by using complex numbers theory and complex valued Counterpropagation network is another plausible approach, which we have explored in this paper.

Section 2 describes the types of failure, Section 3 describes the network architecture of the Complex-Valued Counterpropagation Network (CVCPN). Section 4 shows the evaluation methodology used to process the geological parameters. Section 5 describes the experimental results obtained from the network as well as the possible future work.

2 TYPES OF FAILURE

The type of movement of the slope forming material depends upon the inherent characteristics of the material, geometry of slope and on factors causing movements. These can be broadly described below:

Rock fall: When the movement of slope material is vertically downward due to gravity and not along any plane it is termed a fall. It can be a rock fall if a rock block gets detached from the main rock mass due to presence of at least three sets of adversely oriented joints and here is undercutting of slope by some eroding agency. On steep rocky slope at high altitude having vertical or steeply dipping open joints, whose strike is parallel or subparallel to slope large block of rock may fill due to frost wedging as a result of freezing of water filling the gaping joints. These filled rock masses may ultimately get detached from the main rock mass and come down as rock fall.

Debris fall: When debris is accumulated on a slope it stands for sometime but when more material is added or if the shear resistance between the debris/rock contacts is reduced due to uplift pressure caused by rain the debris starts moving along this place. Some debris may pose perpetual danger to the roads.

Creep: A creep is slow (0.01–1 m/year) movement in over burden material or weathered rock on gentle slope without any defined surface of failure. The movement is parallel to slope and the reduced with depth. The creep is caused when the slope material is cohesive and there is variation in diurnal temperature. The rate of movement depends on rainfall, and the motion is assisted by cyclic swelling and shrinkage.

Mudflow: As against creep the rate of movement in case of low is very fast. In mudflow the movement can be as high as more than 1 m/sec if the rate is slow (less than 10 m/day) it can be termed as mudslide. The mudslide generally takes place on slope steeper than 5°, but a mudflow can take place on a gentler slope. In mudflow the water/soil ration is high and the rate of motion is usually frequent and sudden. It can take place after heavy downpour or after collapse of steep hill slope.

Slide: Earth slide involves loose and consolidated rock matters however there is more significant and rapid movement. Sometimes, it is even be very sudden and violent causing enormous damage.

Slump slides: These are rotational failures with little vertical displacement in thick overburden material or in weathered and soft rocks. Slump failures are rare in hard rocks. However, in certain special setup guided by joints, step like movement planes are developed as a result of stress releases and these cracks are called glide cracks. The head region of a slump is characterized by steep escarpments, where tension cracks are developed which are generally concentric and parallel to the main scrap. The foot region of the slump slide is marked by a zone of tension and uplift, and a bulge is formed in this region. Due to different rate of movement in the central portion and due to resistance created by the firm strata on the periphery, en-echelon cracks are developed in the periphery region on the sides of the slide.

Block slide: These are very common in loose deposits overlying glacial till or where very low dipping rock beds have bedding clay seams which are plastic in nature and the overlying rock is dissected by steeply dipping joints which dissect the rock into blocks. The movement in this case is transnational.

Planar failure and Wedge failure: In jointed rocks the stability of slopes depends upon the nature of joints, their altitude and shear friction along them. If the joints are adversely oriented they offer planes for the movement of overlying rocks. If it is failure along a single plane, it is called a 'planar failure'. If however, two or more set of joints form a wedge and the dip of the line of contact of two planes is towards the valley at angle more than the angle of shear friction along those planes it may result as wedge failure.

Internal causes: This includes such causes which tend to reduce the shearing strength of the rock. The steeping of the slope, water content of the mass, its mineralogical composition and structural features are important factors which will define the stability or otherwise of a given landmass.

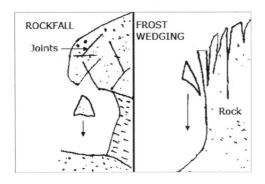

Figure 1. Showing rock slide due to failure at foliation planes.

3 NETWORK ARCHITECTURE

In its simplest version, Counterpropagation network (CPN) is able to perform vector to vector mapping

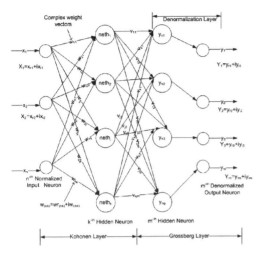

Figure 2. Complex valued counterpropagation network.

similar to heteroassociative memory networks. The complex numbers are treated as vector real and imaginary parts are given equal significance thus following a real-valued approach [Tyagi, 2007]. The weights used in the network are also taken in complex form.

The CVCPN functions in the recall mode as a nearest match look-up table. The difference from the usual table look-up is that the weight vectors are obtained by the training algorithm, rather than in an *adhoc* manner. The input vector $x^r + ix^i$ finds the weight vector $w_m^r + iw_m^i$ that is its closest match among k vectors available in the hidden layer. Weights $v_m^r + iv_m^i$ which are fanning out from the winning mth Kohonen's neuron are updated. The output given by the network is the statistical averages of the similar kinds of input given in the training period.

The input layer contain n cells for distributing training vector x, Kohonen layer with k cells produces output neth. This neth being a real number is compared with other k-1 cells and minimum neth is declare as winner. A fan-in structure, from all input cells is thus created to the winning cell in Kohonen layer. During training stage each complex value is fed into Kohonen network for the self organizing classification. The classification is done separately on real and imaginary part of input values by a predefined algorithm which has been investigated by Tyagi et al.

3.1 Training of input vector in Kohonen layer

We assume that $w^r(t)$ and $w^i(t)$ are the real and imaginary components of Kohonen layer weight vector.

After updating they become $w^r(t+1)$ and $w^i(t+1)$ respectively. Then,

$$w^r(t+1) = w^r(t) + \alpha_t(x^r - w^r(t)) \quad (1)$$

$$w^i(t+1) = w^i(t) + \alpha_t(x^i - w^i(t)) \quad (2)$$

The network is trained until there is no significant change in the updated weight and the old weight i.e.

$$w^r(t+1) = w^r(t) \quad (3)$$

$$w^i(t+1) = w^i(t) \quad (4)$$

In that case the real and imaginary value of the input vector is copied on to its corresponding weight vector. When this situation arises the training is said to be completed and this is the stopping condition for the network. The weight vectors have now settled near the centroid of each cluster. These clusters are different for real and imaginary parts.

Euclidean distance method or the nearest neighbor method is found to be most suitable. The reason for taking the minimum distance (and not the maximum) is that the distance which is minimum is closest to a particular cluster; we then update the weights in such a manner so as to bring it more closely to the cluster thus moving that value to the centroid of the cluster.

We know that if x_1 and x_2 are two complex vectors, then the Euclidean distance between them is given by

$$|x_1 - x_2| = \sqrt{(x_1^r - x_2^r)^2 + (x_1^i - x_2^i)^2} \quad (5)$$

If n be the total input neurons and k be the total number of hidden layer neurons, then let $x_1, x_2 \ldots x_n$ are the complex input given in the Instar layer. w_{11}^r, $w_{12}^r \ldots w_{1k}^r$ are the real and $w_{11}^i, w_{12}^i \ldots w_{1k}^i$ be the imaginary components of the weights connecting the input x_1 and the k'^{th} neurons of Kohonen layer, similarly other weights are connecting input neuron and the hidden layer. The Euclidean distance between x_1 and w_{11} then is:

$$x_1 = x_1^r + ix_1^i \quad (6)$$

$$w_{11} = w_{11}^r + iw_{11}^i \quad (7)$$

$$x_1 - w_{11} = (x_1^r - w_{11}^r) + i(x_1^i - w_{11}^i) \quad (8)$$

$$|x_1 - w_{11}| = \sqrt{(x_1^r - w_1^r)^2 + (x_1^i - w_1^i)^2} \quad (9)$$

Equation 9 is the Euclidean distance between any two points in a complex plane. For a particular set of values of input vectors $x_1, x_2, x_3 \ldots x_u \ldots x_n$ we calculate the following real valued quantity.

$$neth_1 = \sum_{u=1}^{n} \sqrt{[(x_u^r - w_{u1}^r)^2 + (x_u^i - w_{u1}^i)^2]} \quad (10)$$

Similarly

$$neth_2 = \sum_{u=1}^{n} \sqrt{[(x_u^r - w_{u1}^r)^2 + (x_u^i - w_{u1}^i)^2]} \quad (11)$$

Generalizing it we get;

$$neth_k = \sum_{u=1}^{n} \sqrt{[(x_u^r - w_{u1}^r)^2 + (x_u^i - w_{u1}^i)^2]} \quad (12)$$

The minimum of all neth ($neth_1, neth_2 \ldots neth_k$) is calculated and is considered as winner for that set of values of x.

Figure 3 gives a pictorial representation of how the clustering process takes place after the complete training of network. For the sake of simplicity we have shown clustering only for the real part of input vectors, similar clustering takes place for the imaginary part also. Here A, B … E are the input vectors (real numbers) and $W_{rA}, W_{rB} \ldots W_{rE}$ are the real components of weight vectors. It is only after the training that these weight vectors are oriented at the center of the circle. This is what we mean when we say that the weights are adjusted to the centroid of the cluster. The bold line from origin to each cluster is the weight vector and is the mean of all values of the vectors coming in that circle. Observe that cluster 3 is not properly clustered as orientation of vector D (which should be in cluster 2) is into this cluster is same is the case with cluster 4 where vector C wrongly directing towards cluster 4. This is a practical situation and is the cause of error and misclassification. This can be removed by either with more training or proper selection of α and β. The clusters are spread out and are not confined in a single quadrant.

3.2 Training the outstar layer

Once the training in Kohonen layer is completed and all the input vector have been clustered (both real and imaginary) the vector x now makes a *fan out* connection from Kohonen layer to the Grossberg layer and only weights in the Grossberg layer are updated which are connected to the winning neuron. After the training stage, outputs of the CVCPN are separate in real and imaginary parts. They are again combined after the output layer to give the complex output. If each input vector in a cluster maps to a different output vector, then the outstar learning procedure will enable the outstar to reproduce the average of those output vectors when any member of the class is presented to the inputs of the CVCPN. A stuck vector problem is seen if we generate the weights in the range of $[-1\ 1]$. Also to avoid the condition of orthogonality of randomly generated weight vectors we have used weight vectors in the range of $[0\ 1]$.

4 EVALUATION METHODOLOGY

4.1 Study area

For assessing the various soil features, southern slope of ENE trending ridge near Malpa village (Darchula, Uttaranchal, India) in the main range of the central crystalline at the latitude of 30°01′55″ and longitude 80°45′07″ is considered. The rock fall lies between two major tectonic plates, i.e. Vaikrita thrust and trans Himadri fault, both trending NW-SE. The slided mass consisted of massive quartzite interbedded with thin bands of garnet bearing sricicite schist dipping homoclinally at a step angle of 60°–70°. The failure surface of the north east slided portion shows a series of parallel foliation planes. Near vertical joints, perpendicular to the foliation plane, constitute the face of the hazardous area. The precipitation in the preceding days increase the total water content in the open fractures and decrease the cohesion and shearing resistance in the rocks, thus leading to rock slide. The falling rocks reached the base elevation of approx. 2200 m from elevation of 3000 m with a high velocity.

4.2 Data processing and conditioning

In land slide assessment the data is typically a series of analog values that has to be kept within certain

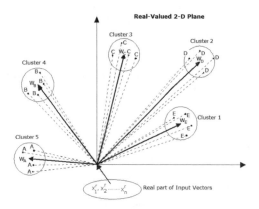

Figure 3. Clustering process in CVCPN.

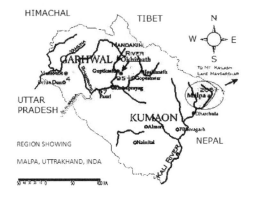

Figure 4. Encircled area is under consideration.

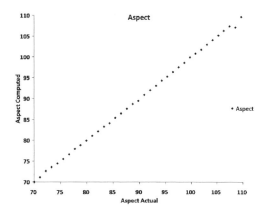

Figure 5. Scattering diagram for aspect.

ranges. The data considered are the various geological parameters (soil aspect, slope, precipitation, elevation) that contribute towards Land sliding. Complex Valued Counterpropagation Network (CVCPN) is trained to recognize optimal values for such data, and make necessary adjustments to control valves, whenever necessary. The data used is for the upper Himalayan region of Malpa. The data conditioning is an important step while using CVCPN. To convert the data in complex number form we first normalize the data (which are real numbers) and then arrange them in such way that the complex part is the next geological parameter from the real part in that complex number. i.e. if a + ib is the complex number then b is the next geological parameter after a. Similarly other parameters is also converted on this form and then feed into the complex CPN.

After training is done in this fashion and during the testing phase the data is again converted into the real number form and de-normalized.

5 RESULTS AND DISCUSSIONS

For each problem, Intel (M) (Celeron Mobile), 1.5 GHz, CPU with 256 MB RAM is used for simulation work with MATLAB® 7.0 as the simulation software. Training and testing data were normalized wherever necessary. It is observed that the proposed model exhibits a more efficient learning in each case. It is due to the reason that in the Kohonen Layer of CVCPN unsupervised and in Grossberg Layer supervised learning is taking place. It is actually the combination of two independent layer of different learning rule that makes up the CPN and this basic structure is preserved in our proposed model also.

Figure shows 4 different graphs obtained between the actual and the computed value of various geological parameters.

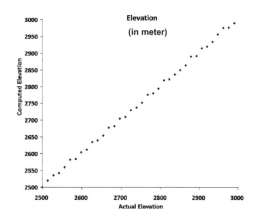

Figure 6. Scattering diagram for elevation.

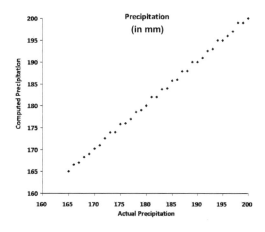

Figure 7. Scattering diagram for precipitation.

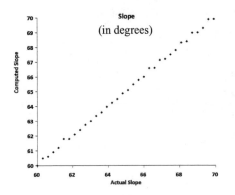

Figure 8. Scattering diagram for slope.

Table 1. Error values for various geological parameters.

S. no.	Geological parameter	Error
1	Aspect	0.0079
2	Elevation	0.0487
3	Precipitation	0.0684
4	Slope	0.0361

From the various graphs obtained it is observed that the minimum scattering is obtained with the aspect, thereby concluding that it is the most important geological parameter in landslide assessment. Total number of hidden neurons are taken as 80. Training and testing time is approximately 40 seconds, number of parameters are 28. Learning rate α varies from 0.9–0.1 and β varies from 0.1 to 0.5. The variations of these learning parameters are exponential in nature contrary to linear change that is more common while using CPN network with real numbers.

CONCLUSIONS

By reducing the number of testing set and still maintaining the required accuracy, reducing the number of hidden layers thus reducing the complexity from hardware point of view and circuit realization are some of the future areas which are still needed to be explored with forecasting using complex valued neural networks.

REFERENCES

Akira Hirose. 2003. *Complex-Valued Neural Network: Theories and Applications*, World Scientific.

Cai F. & Ugai K. 2004. Numerical Analysis of Rainfall Effects on Slope Stability, *International Journal of Geomechanics ASCE*.

Punamia B.C, Jain & Jain., 1998. *Soil mechanics and foundations* Laxmi publication.

Silverman H. 1975. *Complex Variables*, Houghton, Newark, USA.

Subramanya K. 1984. *Engineering Hydrology*, Tata McGraw Hills Publication.

Thandaveswara B.S. & Sajikumar N. 2000, Classification of River Basins using Artificial Neural Network, *Journal of hydrologic engineering*: 290–297.

Tyagi K., Mishra D. & Kalra P.K. 2007. A Novel Complex Valued Counterpropagation Network, *IEEE Proceedings on Computational Intelligence and Data Mining*, pp. 81–87.

Yadav A., Mishra D., Ray S. & Kalra P.K. 2005. Representation of Complex-Valued Neural Networks: A Real-Valued Approach, *IEEE Proceedings on intelligent sensing and information processing*, pp. 331–335.

Prediction of slope behavior for deforming railway embankments

V.V. Vinogradov, Yu.K. Frolovsky, A. Al. Zaitsev & I.V. Ivanchenko
Track and Track Facilities Department, Moscow State University of Railway Engineering, Moscow, Russia

ABSTRACT: The results of the experimental and theoretical researches of slope behavior for deforming railway embankments are presented in the paper. The criteria of similarity for modeling of the serviceability of railway embankments on geotechnical centrifuges under the vibro-dynamical loads from a train are given. In the paper the basic ways of reinforcement and stabilization of potentially-dangerous and deforming embankments of railways are shown, requirements to which are developed on the basis of results of centrifugal modeling of slope stability of a range of embankments, both on real prototypes, and on the basis of development of the generalized parameters of embankments according to the statistical data of the railway network of the Russian Federation. One of the most perspective and economically attractive technical decisions is use for these purposes MSE walls, gabion walls which have received the big distribution abroad and the beginnings 90th years began to be applied widely in Russia in civil and transport construction. For this purpose modern methods of physical modeling and full scale monitoring on centrifuge are used. Among the last the method of centrifuge modeling which allows providing similarity of the processes occurring on small models of soil constructions, to changes of conditions of natural prototypes of a subgrade has the special importance.

1 THEORETICAL INVESTIGATION OF SLOPE BEHAVIOR

The most significant deformations of separate objects of the railway subgrade are failures of stability of their slopes.

Any object of the subgrade, including an intensively operated railway embankment together with external, natural influences and internal loads at studying stability of slopes, is regarded as an open dynamic system. If to treat a parameter of stability of such an embankment in a general way as interdependence between the factors holding a slope and the factors, leading to its failures, it can be written down as:

$$K_s(t) = f[\tau_o(t), \gamma_d(t), U_1(t), U_2(t), U_3(t), t] \quad (1)$$

where K_s = coefficient of stability; t_0 = generalized shear strength of soil; γ_d = unit weight of dry soil; $U_1(t)$ = the function of external loads (external loads on an embankment from constructions, the weight of the track upper structure and the loads from the train); $U_2(t)$ = the function of natural factors (atmospheric precipitations, temperature, wind, flooding); $U_3(t)$ = the function of internal factors (pressure from the weight of the soil, moisture, pore pressure).

The tasks of research of slope stability of the subgrade (as well as other geotechnical constructions) can be divided into two groups: the problems which aim at studying the physical nature of processes of infringement of stability of slopes (they can be carried out to check the hypotheses or various analytical decisions) and tasks of applied character when a definite construction with a definite goal to estimate the slope stability is studied.

At centrifugal modeling all physical nature of the process of infringement of stability of slopes remains unchanged and even in most cases the geometrical similarity of blocks of displacement is provided. Therefore the experimental estimation on the models of slope stability can supplement or replace in very many cases theoretical methods of estimation. Appropriate theoretical preparation is necessary for providing of the experiment. The development of criteria equations of the process and their analysis, as well as the subsequent analysis of the results of experiments with the help of received criteria.

In analysis of slope stability of railway subgrade in Russia the well-known method of calculation of stability by G.M. Shakhunyants is usually used according to which the stability of the slope is estimated by the coefficient of stability.

The necessary criterion of similarity for modeling spontaneous deformation of a slope in a period of time, using the differential equation of deformability of the kind is calculated as:

$$\frac{\partial z}{\partial t} = R\frac{\partial^2 z}{\partial y^2} + \mu y \frac{\partial z}{\partial y} + \mu z \quad (2)$$

where y and z = coordinates of points of a slope; R = the factor of deformability dependent on viscous properties of the ground; m = the coefficient of proportionality between the speed and the parameter of displacement of a particle of the soil.

Having lowered the intermediate calculations, we obtain the general criterion of similarity of the process in question as:

$$\pi = \frac{\pi R t^2}{L^2} = idem \quad (3)$$

where L = general linear size; t = time.

The received criteria can be used for planning experiments and the analysis of results of modeling and their transferring on natural objects.

For practical use the following criteria of similarity are recommended:

$$\pi_1 = \frac{P_u s \cdot P \cdot W}{L^3 \cdot \gamma_d \cdot \tau_0} = idem;$$

$$\pi_2 = P_{us}^2 \cdot P^2 \cdot \left(\frac{W}{L^3 \cdot \gamma_d \cdot \tau_0}\right) I \left(\frac{W}{L^3 \cdot \gamma_d \cdot \tau_0}\right) II$$
$$= idem;$$

$$\pi_3 = \frac{P_{us}^2}{L^4} \cdot \frac{P \cdot W}{(\gamma_d \cdot \varphi \cdot c)_n (\gamma_d \cdot \varphi \cdot c)_c} = idem$$

$$\pi_4 = \frac{R \cdot t}{L^2} = idem \quad (4)$$

where P_{us} and P = external factors from the weight of the upper structure and the influence of a train; W = moisture of soil; γ_d = density of dry soil; t_0 = the generalized resistance to the shift of the soil.

In the second equation (4) index I designate characteristics of a subgrade, index II for foundation. In the third equation of the system (4) index n characterizes properties of draining ground of ballast deepening and loops of an exploited embankment, index c = cohesive soil embankments.

The distinctive feature of modeling of stability of slopes of embankments of railways is the necessity to take dynamic influence of train pressure into account.

The specifics of dynamic influence of train pressure on the subgrade are expressed in the following:

– Influence of external dynamic factors from passing trains on the ground layer is shown in occurrence of dynamic pressure, elastic fluctuations and vibrations; the size and an orientation of these influences is various in space and time;
– Owing to fluctuations and vibrations arise inertial and dispersive forces, the size and direction of which is also various in space and time;
– Influence of dynamic pressure on the ground layer is marked in the appearance of irreversible and convertible processes; these processes are stipulated at vibrations by infringement of structural reactions between particles and their units (for embankments these reactions are restored to a certain extent after deliberate condensation during construction); there is also a transformation of the cohesive water and clearing of immobilized water that reduces the resistance of the soil to a shift, as physically cohesive water possesses ability to resist shifting, free pore water, on the contrary, possesses ability of greasing; at vibrations and significant elastic settlements there is also the general time dilatation of the soil and as a consequence of it:
– Decrease of strength characteristics; all these phenomena and processes are shown non-uniformly in space and time;
– Duration of continuous influence of dynamic pressure is of great importance, as thus irreversible processes start to prevail above convertible in view of decrease in bearing capacity of the ground and resistance to shift that leads to its more intensive deformability; in this connection it is necessary to take into account the character and duration of the actions of dynamic influence on the model of the basic platform of subgrade, their cinematic communication with the model of the prototype.
– Influence of dynamic external factors should be such as to answer natural conditions of influence at which many factors, previously not taken into account or partially used by static calculations reveal themselves, factors: various microstructures of the soil, their natural physical properties—friction, cohesion, shear strength, molecular links of particles of soil and water, chemical cementation of structure, etc.;
– External dynamic influence is transferred to the subgrade through a ballast layer, the length of this transfer is such, that the problem can be considered as a 2D problem.

All the specifics should be whenever possible taken into account at the creation of dynamic model of train pressure and definition of operating conditions with it.

At modeling of dynamic influence of train pressure on the subgrade three variants of its cinematic communication with model of the subgrade can be realized: the model of pressure can transfer dynamic influences through models of rails, at a level of a ballast layer and at a level of the basic platform of the subgrade.

It is by process of elimination received; that it is the most expedient to carry out the application of dynamic model of train pressure at a level of the basic platform of model of the subgrade since modeling of the elements of the upper structure of a track is practically impossible because of their small sizes on the model.

The first condition is the preservation on the model in the same points and in the same moments of time of sizes of natural dynamic pressure, i.e. (i = 1, 2, 3 ...). This condition will be executed, if at a level of the basic platform of a subgrade the stresses for the model and the prototype are equal:

Criteria of similarity which are included in the basic criterion of similarity of N.A.Nasedkin are used:

$$\frac{gl}{\omega^2} = idem \quad (5)$$

where ω = speed of influence on the soil; g = earth gravity; l = the linear size.

The second criterion is similar to the criterion of Frude, at centrifugal modeling:

An effective way of research slope stability of railway embankments is the method of physical modeling on geotechnical centrifuges (Yakovleva, T.G. Ivanov, D.I. 1980).

It has been proved (Shakhunyants, G.M. 1953), that the only and necessary condition, providing equal characteristics of soil strength of nature and models in the same points, in the same moments of time, is $A_m = A_n$ and $v_m = v_n$.

For the creation on the level of the model of the basic platform of subgrade of the dynamic pressure, close to real, the special transmitting device was created.

This dynamic model is executed on MIIT centrifuge as a set of elastic beams of variable section from the material with the corresponding module of elasticity e.g. textolite (Figure 1).

Force *Pm* (see Figure1) equal from 0 up to 16 kgf can be transferred to every beam, meanwhile the dynamic pressures are created in the sections under forces *Pm* and vibrations can be transferred onto the subgrade under frequency from 0 to 200 Hz.

The efficiency of dynamic pressure has been tested in experiments where the settlements of the basic platform of model under certain pressure were calculated—Figure 2.

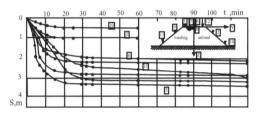

Figure 2. Settlements of the embankment model with and without dynamic loading.

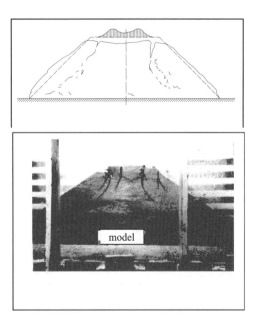

Figure 3. Settlements of the embankment model with and without dynamic loading.

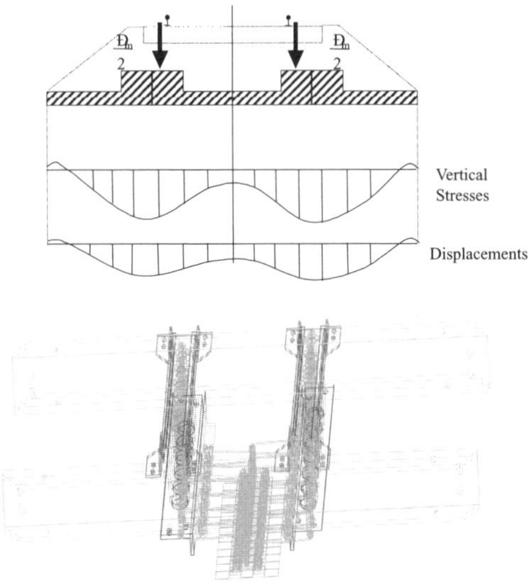

Figure 1. Scheme of model load and construction view.

For the estimation of efficiency of centrifugal modeling stability of objects of railway embankments, experiments on modeling embankments of different height have been executed.

It has been experimentally established, that at steeper slopes of model of an embankment, the weight of the soil in failure mass on slopes is more that corresponds to the natural data.

Also it has been tested, whether the cylindrical form of the surface of failure is reproduced when modeling. The studying of formation of the failure surface on models has shown, that is reproduced—Figure 3.

When modeling the dependence of influence of moisture w and the height of embankments h on the value of destroying dynamic pressure of a train.

2 PRACTICE OF SLOPE BEHAVIOR PREDICTION FOR DEFORMING RAILWAY EMBANKMENTS AND REINFORSING CONSTRUCTIONS

On the basis of the results of centrifugal modeling of stability and behavior of slopes of some embankments both on real prototypes, and on the basis of development of the generalized parameters of embankments on the statistical data of Russian railways the number of effective ways of reinforcement and stabilization of potentially-dangerous and deforming embankments of railways and a number of normative documents on reinforcement of the subgrade of railways has been developed (Zaitsev A.A., Frolovsky Y.K. 2000).

In the Figure 4 the photo of one of the embankments with reinforced by the gabion walls the design of which has been developed on the basis of technical instructions is presented.

The MSE walls and gravitation gabion structures at geotechnical practice of reconstruction of the railway track objects (subgrade) are occupy a big role that time in Russian railways

The full-scale monitoring and physical modeling on the geotechnical centrifuge are main directions of researches conducted in the Moscow State University of Railway Engineering.

Monitoring of the object of the subgrade on 103 km of the Moscow-Aleksandrov line executed: for the evaluation of changing a condition of geotechnical structures in the exploitation time; for the qualification of terms of changing of its condition; for the forecast of the behavior and determinations of life-time.

The several cycles of studies was realized on the object, during which was supporting-geodetic network, executed of the observation and geodetic measurements, winnowed functional diagnostic of the subgrade—a determination vibro-dynamic parameters of soil vibrations from train loading.

For the realization of supervision the displacement and revealing the change tense deformed conditions at different parts of the wall (gabion structure) were provided the bookmark an anchors and winnowed geodetic measurements (Figure 5).

Vertical and horizontal displacement of gabion structures defined comparatively pawned of the anchors and with the supporting geodetic network.

Practically the geodetic removal gabion walls comprise of it:

– Utter measurement of the gabion levels height on top for each row with step 1 meter.
– Measurement of the height for anchors (accessories by diameter 16 mm and length 1 meter) with card face of gabion wall.

Analysis of gabion wall deformations executed on results of measurements settlements of the anchors and measurements, executed as of measurements of the level surface orientations.

Executed analysis of the measurements of the anchors displacement mortgaged with the card face gabion walls is indicate of the happened settlements

Figure 4. Reinforcing of the subgrade on line Aleksandrov-Balakirevo (Nothern Railway).

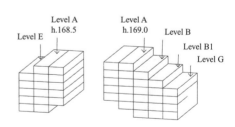

Figure 5. Scheme of the levels for left and right side of gabion walls.

Table 1. Results of the utter measurements of the gabion levels height (Right side).

Points*	1 mm	3 mm	7 mm	17 mm	29 mm
Level A			10	24	13
Level B	−18				12
Level B1	10	19			
Level G			10	11	10

* Points in meters from the construction beginnings.

Table 2. Results of the utter measurements of the gabion levels height (Left side).

Points**	0 mm	1 mm	2 mm	4 mm	10 mm
Level A	31	23	32	20	
Level E	32			36	21

** Points in meters from the construction beginnings.

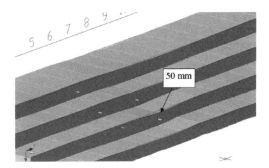

Figure 6. Correlation of the design and the actial positions of the wall. Settlements on right side.

of the construction (walls) with following maximum values:

– Maximum values a settlements for construction of the reinforcement on the right slope of embankment and for anchor, which establishing on the level "A" (in the cross-section 25 meters from the beginnings), has formed 9 mms; average values the deformation do not exceed 3 mms (settlements)
– Maximum values a settlements for the constructions of the reinforcement of the left slope of embankments and for anchor, which establishing on the level "D", has formed 6 mms; average values the deformation do not exceed 4 mms (settlements)

The results of the utter measurements of the gabion surface are shown of the lumpy settlements of the constructions. The maximum values of the settlements are given in the Table 1 and Table 2.

Analysis of the actual position of the gabion walls for design mark has shown certain updating actual height in the construction period.

So for the right side the maximum excess of the actual mark on the design height has formed, for example for the level "A" excess is 96 mms

The results of full scale monitoring have shown small precipitation of separate elements of the wall construction and non-uniform changing a position in different parts of gabion walls. However revealed displacements are not critical for the operation of the construction.

For an estimation of operational working capacity of designs on stabilization and reinforcing of a maintained railway subgrade with use of gabion structures, their physical modeling on geotechnical centrifuge of MIIT's was spent.

According to the preliminary specifications on modelling solved following problems:

– Researches of reasons and conditions of a straining of subgrade (based on the analysis of the information on plant of a subgrade);
– Estimations of working capacity of construction (at a stage of construction and maintenance);
– Predictions of a deformation property of a subgrade;
– Estimations of reliability of use of the accepted computational models;
– The comparative analysis of outcomes of modelling and numerical calculations of the is tense-deformed condition of a subgrade;
– Development of recommendations on perfecting designs of amplification and modes of their construction.

Application of a method of centrifugal modeling has allowed to research and validate a deformation property of various plants of a subgrade and designs of their reinforcing on the small models fulfilled in scale from 1: 25 till 1:50.

For reinforcing of embankment with height up to 9.0 m the two-lane adapted to the use of electric power section Moscow-Aleksandrov Moscow Railways have been developed technical instuctions with use of gabion gravitational walls.

The embankment intersects river Gorelyi Krest with the constant stream flow. There is a water-absorbent pipe here an arch type combined of concrete blocks on a cement slurry, with an orifice a breadth of 3,2 m and 4,5 m. In the location of a water-absorbent building (a southern leg) are observed by altitude a slope of embankment of a bank.

In a upper part the embankment is strengthened by metal columns and the armoring. The breadth of the basic platform is made narrower. The ballast is showered on the slope. Embankment slopes have the overestimated steepness. From a southern leg of a bank the silted section of locality is arranged. The marsh has the mixed feed—atmospheric, superficial and subsoil waters. Average depth varies within the limits of 0,3—0,6 m. Exist 5 ringlets which waters unload in a marsh. The water condition has the mixed type of a feed which develops due to atmospheric precipitation and subsoil waters. The basic quantity of flow of the river (about 60%) happens due to atmospheric precipitation. Because of small depth a river valley the soil feed has smaller value. The hydrological condition of the river is tightly connected with climatic singularities of territory. It represents a strongly pronounced spring high water on which share 60–70% of an annual flow are necessary. A bottom river Gorelyi Krest generally it is silty, covered by vegetation.

Hence, it is possible to state, that outcomes of modeling and their calculated analysis has shown, that origin of not stabilized condition of an embankment if to the foundation combined by soils with low filtration ability, exterior loading will be fast enclosed is possible. As a result of its operation owing to formation of a gauge pressure in void water the angle of an interior friction of the soils, largely determining a load-carrying capacity of usual soils of the foundation, is not realized. Thus, the load-carrying capacity of the foundation will be defined only by values specific cohesion its soils, and it can appear insufficient. Therefore, emersion of strains of an embankment, because of deficiency of its overall stability is possible.

In this connection, it is necessary to note necessity of the account of possible change of state of a reinforced embankment under construction and the subsequent maintenance as the time of consolidation of the waterlogged clayey grounds of the foundation can exceed considerably a time of construction of designs of its reinforcing gravitational gabion walls.

3 CONCLUSIONS

The carried out experimental and theoretical investigations of prediction of slope behavior for deforming railway embankments of have shown that the forms of surfaces of slope displacement of subgrade depend on the dynamic pressure of the train on the basic of the subgrade.

The specially created for such purposes device placed on the centrifuge facility has allowed to etimate the levels of destroying pressures of a train.

For check of working capacity and an estimation of a use reliability of new technology with use of gabion structures for stabilization and reinforcing of subgrade of railways have been worked out experimental researches: full scale monitoring and physical modeling on the geotechnical centrifuge of MIIT.

The results of full scale monitoring have shown small precipitation of separate elements of the wall construction and non-uniform changing a position in different parts of gabion walls. However revealed displacements are not critical for the operation of the construction.

Results of physical modeling and its analysis have allowed to evaluate operational working capacity of the offered the designs with use gabion and MSE walls. It is necessary to note necessity of the account of possible change of state of a reinforced embankment under construction and the subsequent maintenance as the time of consolidation of the waterlogged clayey grounds of the foundation can exceed considerably a time of construction of designs of its reinforcing gravitational gabion walls.

For an estimation of a deformation property of natural railway subgrade objects after their reinforcing by gabion structures was developed the program of geotechnical and geodesic monitoring.

On an instance of embankment of 103 km Moscow-Alexandrov Moscow Railway, reinforced by gravitational gabion wall, are performed works on monitoring of an condition of the gabion structures and definition vibro-dynamic parameters of soils under loadings from a rolling-stock in a cycle 2006–2007 years.

For improvement of quality and effectiveness of realization of researches by a method of centrifugal modeling on geotechnical centrifuge of MIIT it was equipped by the modern control equipment and the automated laboratory complex for definition of physic mechanical properties of soils. There are begun works on working off of conditions of use of control and measuring system at modeling subgrade objects of railways.

In the further researches we project to continue works on monitoring and modeling of subgrade objects of railways both under static and under dynamic loadings, reinforced by gabion structures; to carry out researches with application of the automated laboratory complex for compiling certificates of soils which will be used at numerical calculations of embankment.

REFERENCES

Yakovleva, T.G. & Ivanov, D.I. 1980. *Modeling of toughness and stability of railway subgrade*. Moscow: Transport.

Shakhunyants, G.M. 1953. Subgrade of railways. M.: Transjeldorizdat. 827.

Pokrovsky G. Y. & Fedorov I. S. 1975. *Centrifugal testing in the construction industry*. Vols. 1 & 2. English translation by Building Research Establishment Library Translation Service of monographs originally published in Russian, Watford.

Vinogradov V.V., Yakovleva T.G., Frolovsky Y.K. & Zaitsev A.Al. 2002. Centrifugal modeling of the railway embankments with reinforcement by the various reinforced earth constructions. *Proceedings of the International Conference on Physical Modeling in Geotechnics*, St.John's, Newfoundland, Canada, 10–12 July: 987–991.

Vinogradov V.V., Yakovleva T.G., Frolovsky Y.K. & Zaitsev A.Al. 2005. Evaluation of slope stability of railway embankments. *Proc. of the 16th International Conference on Soil Mechanics and Geotechnical Engineering*, Osaka, Japan.

Vinogradov V.V., Yakovleva T.G., Frolovsky Y.K. & Zaitsev A.Al. 2006 Physical modeling of railway embankments on peat foundations. *Proc. of the 6th International Conference on Physical Modelling in Geotechnics*—6th ICPMGE'06, Hong Kong.

Zaitsev A.A., Mnushkin M.G. & Vlasov A.N. 2001. Object-oriented programming in application to engineering methods for slope stability analysis. *Proceedings of the 8th International Conference on Enhancement and Promotion of Computational Methods in Engineering and Science (EPMESC'VIII)*, Shanghai, China, 9p

Ministry of Railway Transport, 1991. *Technical instructions on reinforcement and stabilizations an embankments on strong foundation of the soil reinforced constructions*. 101p

Ministry of Railway Transport. 1998. *Technical instructions on reinforcement of embankments with using an gabions*. 140p

Yakovleva, T.G. 1992. Principle bases of using reinforced soil for reinforcement of subgrade. *Transactions of Moscow Railway Engineering Institute.* 844: 45–58.

Yakovleva, T.G. & Vinogradov, V.V. & Frolovsky, Y.K. 1997. Methods of stabilization of the embankments by the constructions from the reinforced soils. *Put i putevoe hoziaystvo*. 1: 7–11.

Frolovsky, Y.K & Zaitsev A.A. 2000. Database "Information card" for analogues of railway embankments. Proc. of the 3rd International Conference on Advances of Computer Methods in Geotechnical and Geoenviromental Engineering, Moscow: Balkema.

Yoo, N.J. & Ko, H.Y. 1991. Centrifuge modeling of reinforced earth retaining walls. *Proc. of the International Conference Centrifuge*, Boulder: Balkema.

Finite element simulation for the collapse of a dip slope during 2004 Mid Niigata Prefecture earthquake in Japan

A. Wakai & K. Ugai
Gunma University, Japan

A. Onoue
Nagaoka National College of Technology, Japan

K. Higuchi
Kuroiwa Survey and Design Office Co., Ltd., Japan

S. Kuroda
Institute for Rural Engineering, JAPAN

ABSTRACT: Many landslides in mountain area occurred during 2004 Mid Niigata Prefecture Earthquake in Japan. In this paper, numerical simulations for the collapse of a dip slope by the 2-D dynamic elasto-plastic finite element method is reported. In the case, an upper mudstone has slid along the bedding plane. To simulate such a catastrophic failure, i.e., a long distance traveling failure, it is very important to consider the strain-softening characteristics of the slip surface precisely. In the analysis, the material parameters for the thin sand seam put along the bedding plane were determined by the cyclic direct shear tests of undisturbed block samples. As a result, the observed phenomena could be simulated by the analysis appropriately.

1 INTRODUCTION

In engineering point of view, it should be emphasized that the following two kinds of slope failure can be distinguished definitely. The fist one is a large deformation, where the sliding mass stops after the earthquake. In such a case, the degree of the deformation often attracts attention for the seismic design of adjacent structures. The second one is a catastrophic failure that is often called as a collapse. In such a case, the sliding mass continues moving even after the earthquake as far as there are no obstacles on the way. It means that the sliding mass will not be supported after the earthquake, although it was supported statically before the earthquake. The reduction of the shear resistance along the slip surface during the earthquake may be one of the main causes for the induction of such a long distance traveling failure. In most of previous studies based on the dynamic elasto-plastic FEM, the slope failure of the latter type has not been treated, because of the difficulty of the modeling.

In this paper, numerical simulations for the collapse of a dip slope by the 2-D dynamic elasto-plastic finite element method is reported. In the case, an upper mudstone has slid along the bedding plane. To simulate such a catastrophic failure, i.e., a long distance traveling failure, it is very important to consider the strain-softening characteristics of the slip surface precisely. In the analysis, a newly proposed constitutive model that is extended from the previous model (Wakai & Ugai, 2004) is adopted to consider such a behavior. The material parameters for the thin sand seam put along the bedding plane are determined by the cyclic direct shear tests of undis-turbed block samples. As a result, the observed phe-nomena are simulated by the analysis.

2 ANALYTICAL MODEL

2.1 *Yokowatashi landslide*

Many landslides in mountain area occurred during 2004 Mid Niigata Prefecture Earthquake in Japan. In this paper, one dip slope failure along the Shinano River, Yokowatashi landslide, is simulated numerically as an example of natural slope failure. In the analysis, the dynamic elasto-plastic FEM is applied for reproducing the long distance slide taking strain softening and decrease in strength due to cyclic shear loading into consideration.

As seen in the geological map shown in Figure 1, the Shinano River joins the Uono River at Kawaguchi town near the epicenter of the earthquake and flows toward almost north from Shiroiwa, Nagaoka city after scraping the right coast at Yokowatashi, Ojiya city. The river encroaches on the attacking rock slopes ranging from Shiroiwa to Myoken, Na-gaoka city along the right coast. Neogene deposits, such as Shiroiwa layer, and Uonuma layer sedi-mented by early Diluvium distribute as hills and ter-races from Kawaguchi town to Nagaoka city along the Shinano in this area. These layers fold several times forming the Higashiyama hill areas and the axis directions of these anticlines and synclines are almost South-North in direction. The Shinano locates to the west of the west end anticline axis. The slopes inclining toward west and facing to the river are dip slopes from geological structure view point.

Figure 2 (a) is a photograph taken from a north-west direction. As seen in the figure, a part of the upper Shiroiwa layer and the surface earth with high trees remain as they were on the bedding plane at far end

(a) Whole of collapse

(b) Undisturbed block sample for laboratory tests

Figure 2. Yokowatashi landslide.

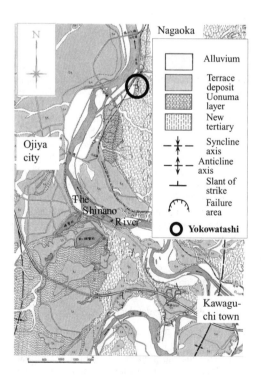

Figure 1. Geological map around Ojiya city.

of this picture. The remaining upper Shiroiwa layer of soft silt rock exposes its side face. The other part of the upper Shiroiwa layer which made up the opposite side of the slid area is visible on site. The portion of the upper Shiroiwa layer between them had covered the planer tectonic dip surface which is clearly seen in the picture, and it has slid more than 72 m to the west toward the Shinano River. The inclination of the bedding plane facing to almost west is approximately 22°. The thickness of the slid Shiroiwa block at the south end is about 4 m and those of earth on the block ranges from 20 cm to 1 m. The height of upper Shiroiwa layer remaining at the north side is about 2.5 m with earth cover of 60 cm thick near the ridge of the slope. Figure 2(b) is a close picture showing soil sampling for the laboratory tests. A thin seam layer of 5–10 mm thick was sandwiched between the upper and lower Shiroiwa layers. The material of the sand seam is tuff sand. Both Shiroiwa layers were gray, weathered, and changed their color to brown up to about 8 cm inside from the boundary of the sand seam. The details of Yokowatashi landslide has already been reported by Onoue et al. (2006).

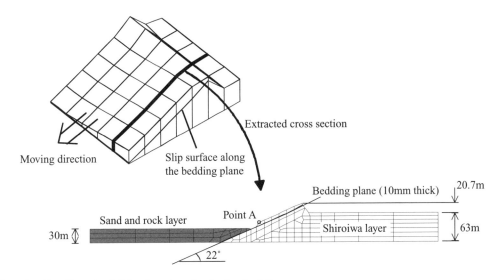

Figure 3. Two dimensional finite element meshes for the simulation.

Table 1. Input parameters.

	Layer	Shiriowa	Sand seam	Sand and gravel
Basic parameters	Young's modulus $E(kN/m^2)$	100000	30000	30000
	Poisson's ratio v	0.3	0.3	0.3
	Cohension $c(kN/m^2)$	–	24	0
	Internal friction angle ϕ(deg)	–	30.9	35
	Dilatancy angle ψ(deg)	–	0	0
	$b \cdot \gamma_{G0}$	–	8.0	18
	n	–	1.40	1.35
	Unit weight $\gamma(kN/m^3)$	20.0	18.0	18.0
Strain Softening parameters	Residual strength τ_π/τ_{f0}	–	0.30	–
	A	–	4.0	–

2.2 Analytical model

As aforementioned, the thickness of the slid soft rock plus earth with trees was thin at the north end and it was thick at the south end. A two dimensional numerical

Figure 4. Test specimen consisting of upper and lower Shiroiwa layers with sand seam in between.

analysis was focused on the cross section of the slid slope with its medium thickness. The finite element mesh consisting of eight nodes per each element were shown in Figure 3. The time history of the response at **Point A** in the figure will be mentioned later. The upper and lower soft rock layers were assumed to be elastic material and the sandwiched tuff sand layer was assumed to be elasto-plastic material having a thickness of 10 mm taking strain softening into consideration. The surface soil at the foot of the slope was assumed to be sand and gravel spreading down to the Shinano.

The basic concept of the elasto-plastic model used here is the same as the cyclic loading model originally proposed by Wakai and Ugai (2004). The undrained strength parameters, c and ϕ, which specify the upper

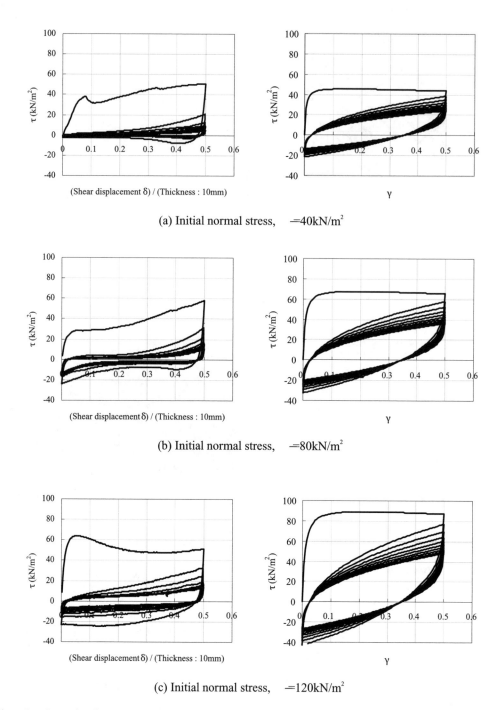

Figure 5. Comparison between tested (left column) and simulated (right column) hysteresis loops.

asymptotic line of the hyperbolic skeleton curve of their model was modified as the decreasing functions of accumulated plastic strain γ^p to incorporate the strain softening characteristics (Wakai et al. 2005) and used here. The shear strength during earthquake is given as,

$$\tau_f = \tau_{f0} + \frac{\tau_{fr} - \tau_{f0}}{A + \gamma^p}\gamma^p \qquad (1)$$

where the initial strength is denoted as,

$$\tau_{f0} = c \cdot \cos\phi + \left(\frac{\sigma_1 + \sigma_3}{2}\right)_{initial} \times \sin\phi \qquad (2)$$

The shear stiffness ratio, G_0, is also assumed to decrease in proportion to the decrease of shear strength. The cyclic loading model disregarding strain softening was used for the sand and gravel layer. The constants of Rayleigh damping were assumed to be basically $\alpha = 0.171$ and $\beta = 0.00174$ which are equivalent to a damping ratio of about 3% for a vibration period of 0.2 through 2.0 s. The material properties used in the analysis were summarized in Table 1.

Figure 4 shows an intact sample consisting of the upper and lower Shiroiwa soft rocks and the tuff sand seam in between. The sample was subjected to the cyclic direct shear test under the constant volume condition. Figure 5 compares the simulated hysteresis loop and the tested loop of each specimen during cyclic loading. The axis of abscissas is written in strain. In the figures for the tests, the strain is defined as the horizontal displacement divided by 10 mm, which corresponds to the thickness of the sandwiched layer. Although they don't perfectly coincide, they are roughly similar to each other for the respective consolidation pressure.

3 ANALYTICAL RESULTS

3.1 *Time histories of response*

The acceleration record in EW direction observed at Takezawa (Figure 6) was used in analyses. Two cases of analysis were conducted to examine the influence of seismic intensity on the inducement of sliding. One was the analysis for which the observed acceleration record was input as it was at the base of the analysis area, and the other was the one for which the acceleration amplitude was compressed to one half that of the observed wave was input.

Figure 7 shows the time histories of horizontal displacement at the foot of the slope, namely **Point A**, in Figure 3. As seen in this figure, the slope does not fail in the case where the acceleration amplitude is compressed to one half that of the actual wave record. Contrarily the large-scale slope failure occurs in the case of actual acceleration amplitude. The sliding amount in horizontal direction is 20 m at t = 40 sec and almost 65 m at t = 50 sec. Since the shear strength became smaller than the shear stress induced only by the self weight of upper Shiroiwa layer, continuous

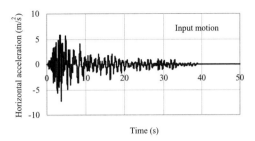

Figure 6. Acceleration record at Takezawa (EW).

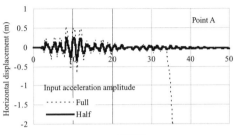

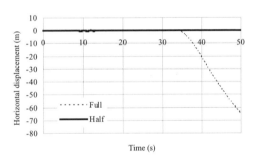

Figure 7. Time history of horizontal displacement of upper Shiroiwa layer.

sliding on the bedding plane started at an elapsed time of about 35 s.

Figure 8 shows the relationship between the mean shear stress and the mean accumulated shear strain of the sand seam, both of which were averaged in all over the sandwiched layer. The accumulated plastic strain converged at $\gamma = 1.5$ and $\tau = 474$ kN/m² in the case of half acceleration amplitude. On the contrary, the shear stress decreases continuously with increasing γ for the actual wave case.

3.2 *Deformation increasing endlessly*

The residual displacement at 0, 10, 20, 30, 40 and 50 seconds after the beginning of the seismic motion,

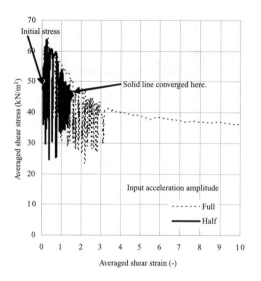

Figure 8. Relationship between shear stress and shear strain of the sand layer during earthquake.

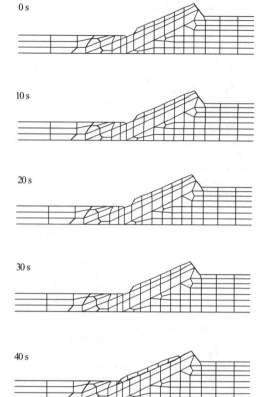

Figure 9. Movement of the sliding mass on the bedding plane during the earthquake.

in the case where the observed acceleration record is input, was shown in Figure 9. The long distance sliding of the upper Shiroiwa layer along the bedding plane can be seen discontinuously at the sand seam in this cross section. The large-scale slide occurred on site was thus reproduced quantitatively through the present analysis.

3.3 Newly proposed index for seismic stability

At the time of the collapse of the slope, the total value of maximum shear resistance mobilized on the sliding plane ΣT_s begins to be less than the total value of the static sliding force corresponding to the gravity of the sliding block ΣR_f. An appropriate stability index for descriptions of this phenomenon can be denoted as the following equation.

$$F_d = \frac{\sum R_f}{\sum T_s} \qquad (3)$$

The time histories of the proposed index obtained in the analyses are shown in Figure 10. The moment when the value of the index becomes 1.0 accords with a moment of the collapse. This index will be extremely convenient for explanation.

4 CONCLUSIONS

The cause of the slope failure at Yokowatashi was examined based on the cyclic shear properties of the sandwiched material on the bedding plane and its long distance sliding was appropriately simulated by the dynamic elasto-plastic FEM. The following con-clusions were obtained from the present work;

1. There exists a tuff sand seam at the bedding plane of Yokowatashi dip slope structure.
2. The shear strength and stiffness of the sand decrease markedly with increasing number of cycles and the sand seam finally lose its shear strength enough to support just the self weight of soils above it.
3. The long sliding distance of the slope failure was reproduced through the elasto-plastic dynamic finite element analysis taking the cyclic shear properties of the sand into consideration.

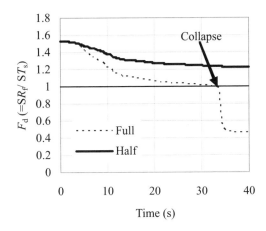

Figure 10. Time histories of the proposed index.

4. A simple stability index for descriptions of the collapse of the slope was proposed. It will be extremely convenient for explanation.

ACKNOWLEDGEMENTS

This study was conducted in "Research project for utilizing advanced technologies in agriculture, for-estry and fisheries" supported by the Ministry of Agriculture, Forestry and Fisheries of Japan. The au-thors wish to thank for the great support.

REFERENCES

Griffiths, D.V. & Prevost, J.N. 1988. Two- and three-dimensional dynamic finite element analyses of the Long Valley Dam, *Geotechnique*, Vol. 38, No. 3, pp.367–388.

Iai, S., Ichii, K., Sato, Y. & Kuwazima, R. 1999. Earthquake response analysis of a high embankment on an existing hill slope. *2nd International Conference on Earthquake Geotechnical Engineering*, pp. 697–702, Lisboa, Portugal.

Onoue, A., Wakai, A., Ugai, K., Higuchi, K., Fukutake, K., Hotta, H. & Kuroda, S. 2006. Slope failures at Yokowatashi and Nagaoka College of Technology due to the 2004 Niigata-ken Chuetsu Earthquake and their analytical considerations. *Soils and Foundations*, Vol. 46, No. 6, pp. 751–764.

Toki, K., Miura, F. & Oguni, Y. 1985. Dynamic slope stability analyses with a non-linear finite element method, *Earthquake Engineering and Structural Dynamics*, Vol. 13, pp. 151–171.

Ugai, K., Wakai, A. & Ida, H. 1996. Static and dynamic analyses of slopes by the 3-D elasto-plastic FEM, *Proc. of 7th International Symposium on Landslides*, pp.1413–1416, Trondheim, Norway.

Uzuoka, R. 2000. Analytical study on the mechanical behavior and prediction of soil liquefaction and flow. A doctoral dissertation of Gifu University, Japan (in Japanese), pp.161–170.

Woodward, P.K. & Griffiths, D.V. 1994. Non-linear dynamic analysis of the Long Valley Dam. *Computer Methods and Advances in Geomechanics*, Balkema, pp.1005–1010.

Wakai, A. & Ugai, K. 2004. A simple constitutive model for the seismic analysis of slopes and its applications. *Soils and Foundations*, Vol. 44, No. 4, pp. 83–97.

Wakai, A., Kamai, T. & Ugai, K. 2005. Finite element simulation of a landfill collapse in Takamachi Housing complex (in Japanese). *Proc. simpojium on Safeness and performance evaluation of ground for hausing, JGS*, pp. 25–30.

Sensitivity of stability parameters for soil slopes: An analysis based on the shear strength reduction method

Ren Wang, Xin-zhi Wang, Qing-shan Meng & Bo Hu
State Key Laboratory of Geomechanics and Geotechnical Engineering, Institute of Rock and Soil Mechanics, Chinese Academy of Sciences, Wuhan, China

ABSTRACT: The stability of soil slope is analyzed through explicit Lagrangian finite-difference method based on the strength reduction method. The predominance of this method is discussed. It is a fast method. Dilatation angle, cohesion and internal friction angle are taken into consideration in analyzing of soil slope. It is concluded that the safety factor will become bigger as the dilatation angle, the cohesion and the internal friction angle, while the plastic area and the slip plane will have fluctuation. The results can be used in practical slope design.

1 INTRODUCTION

There are many reasons causing slope failure such as strength reduction of rock and soil mass, loading, slope excavation and so on. But for a natural slope, the key factor is the strength reduction of rock and soil mass. Therefore, the Strength Reduction Method can be used in slope analyzing work. Explicit Lagrangian Finite-Difference Method is popularly used in geomechanical analyzing work. Because it has nonlinear constitutive functions and can be applicable for any complex boundary condition (Hnang Run-qiu & XU Qiang 1995.). Failure surface must be assumed in analyzing slope stability using Limiting Equilibrium Method. However, the FLAC based on Strength Reduction Method does not need assumption of failure surface, and can educe factor of safety, and display the contour and location of failure surface. So this method is widespread in geoengineering.

Slope stability comes down to many parameters such as gravity of soil, dilatation angle, cohesion and internal friction angle. The single parameter is analyzed to reveal the most important influence to the slope stability. The results can be used in slope design.

2 CALCULATION MODEL

The analysis model is an examine originating from Australia Computer Society designed by B. Donald and P. Giam in 1987. The slope consists of homogeneous soil. The shape and parameters of the soil slope are displayed in figure 1 and table 1. The safety factor is recommended for 1.0. The most possible slip plane is showed in figure 1. This example has been widely accepted to checkout the validity of analysis program. The FLAC program based on Strength Reduction Method reduces the shear strength of soil through dividing the cohesion and internal friction

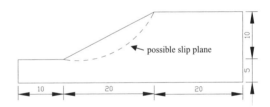

Figure 1. Shape of slope and the possible slip plane (m).

Table 1. Mechanical parameter of slope.

c (KPa)	$\phi(°)$	γ (kN/m³)	E (kPa)	μ
3.0	19.6	20.0	1.0e4	0.25

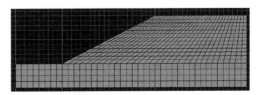

Figure 2. Slope model and lattice division.

angle by the same discount coefficient (Chi Shichun & Guan Li-jun 2004) (formula 1~3).

$$c' = \frac{c}{F_{sr}} \quad (1)$$

$$\phi' = \tan^{-1}\frac{\tan\phi}{F_{sr}} \quad (2)$$

$$\psi' = \tan^{-1}\frac{\tan\psi}{F_{sr}} \quad (3)$$

F_{sr}—discount coefficient

The slope is analyzed using the FLAC program to validate the adequacy of the method of strength reduction.

3 BOUNDARY CONDITION AND RESULTS

In the FLAC2D analysis model, the bottom is fixed, side surface has normal restraint and sloping surface is free. Material properties are given in table 1. The elastic-plastic constitutive equation and Mohr-Coulomb model are employed. Large deformation is considered to adjust the grid coordination. The safety factor of the slope is 0.99 comparing to some traditional limiting equilibrium method listed on table 2. The results of judging program are showed on table 3.

The result of using strength reduction method is quite close to true value. Plastic and tensioned area distribution, shear strain rate and displacement vectograph are almost the same to the judging program. It can be considered reliable in analyzing slope stability.

Table 2. Results of calculation cases.

Analytic method	Safety factor F			
	Mean	Standard deviation	Min.F	Max.F
ALL	0.991	0.031	0.94	1.08
Bishop	0.993	0.015	0.96	1.03
Janbu	0.978	0.041	0.94	1.04

Table 3. Results of judging program.

Donald	SSA (Baker)	STAB (Chen)	GWEDGEM	EMU	Fredlud
1.00	1.00	0.991	1.00	1.00	0.99

4 INFLUENCE TO STABILITY OF SLOPE

4.1 Dilatation angle

For dense sand and over consolidated clay soil, dilatation is observed in shear process. There is explicit peak strength on the stress-strain curve. Firstly, stress and stain increases simultaneously to the peak value, and then strain continues to increase but stress diminishes to residual strength. Dilatation occurs before the peak strength and terminates to material failure. Soil becomes loose and shear strength reduces because of dilatation. So the dilatation angle should not be ignored in the slope stability calculation. Adopting non-associated flow rule which Ignores dilatation will underestimate the shear strength of soil in the stability analysis of slope based on elastic-plastic model (Zienkiewicz O.C., Humpheson C. & Lewis R.W. 1975). But the bearing capacity of soil will be over-evaluated adopting associated flow rule (Zhang Peiwen & Chen Zu-yu 2004). Therefore, the influence of dilatation angle in FLAC analyses based on Strength Reduction Method should be studied particularly.

In order to study the influence of dilatation to slope stability, series of numerical analysis have been done which dilatation angle ψ varies from 0° to 17° and other parameters remain the same. The results display on table 4 and figure 3 that the safety factor has a linear

Table 4. Dilatation angle and safety factor.

Dilatation angle (degree) ψ	Factor of safety F
0	0.96
3	0.97
5	0.98
8	0.99
10	0.99
12	1.00
15	1.01
17	1.01

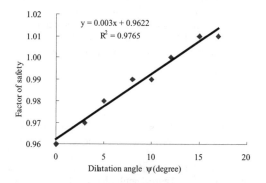

Figure 3. Safety factor vs dilatation angle.

relationship with dilatation angle. The formula can be described as follows:

$$F = 0.003\psi + 0.9622 \quad (4)$$

Safety factor and dilatation angle have good linear correlation (figure 3) which $R^2 = 0.9765$. Safety factor slightly increases as the dilatation angle but not significantly. When dilatation angle increases, the plastic area spreads from slope crest to bottom, displacement increases at the bottom of the slope and the location of slip surface almost remain stable.

4.2 Cohesion

Cohesion which is one of the strength parameters represents the cementation of soil. The cohesion of sand is considered zero. We consider the slope consists of sand soil which its cohesion is zero. The slope is unstable. Safety factor is only 0.72. Plastic area and failure concentrates on slope surface. If diminish the angle of slope, the stability of slope will be improved. We let the cohesion value varies from 0 to 35 KPa, analyses the stability of the slope and find that the safety factor has a linear relationship with cohesion value. The formula can be described as follows:

$$F = 0.0467c + 0.8474 \quad (5)$$

The safety factor increases linearly with cohesion, gliding mass enlarges, failure surface shifts down and becomes concave-down from a plane, its radian gradually enlarges and drops back on the slope crest, slip band extends to deep soil layer. Plastic area distributes around the slip band. Most of the slope crest has tensile stress.

Table 5. Calculation parameters and results.

	c (KPa)	$\phi(°)$	$\psi(°)$	γ (kN/m^3)	F
case1	0	19.6	10	2000	0.72
case2	1	19.6	10	2000	0.84
case3	2	19.6	10	2000	0.92
case4	3	19.6	10	2000	0.99
case5	6	19.6	10	2000	1.18
case6	9	19.6	10	2000	1.33
case7	12	19.6	10	2000	1.47
case8	15	19.6	10	2000	1.61
case9	18	19.6	10	2000	1.74
case10	21	19.6	10	2000	1.86
case11	25	19.6	10	2000	2.02
case12	30	19.6	10	2000	2.21
case13	35	19.6	10	2000	2.4

Figure 4. Shear strain rate and displacement vectograph (c = 0 KPa).

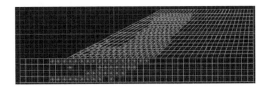

Figure 5. Plastic and tensioned area distribution (c = 0 KPa).

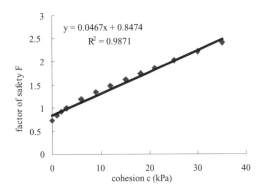

Figure 6. Safety factor vs dilatation angle.

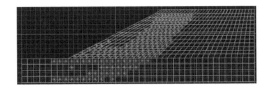

Figure 7. Plastic and tensioned area distribution (c = 12 KPa).

4.3 Internal friction angle

The shear strength of soil is related to internal friction angle which influences the stability of soil slope. The relation between internal friction angle and safety factor of soil slope has been studied in order to make out the regularity. The analyses model has been presented on figure 1 & figure 2. Internal friction angle varies from 0° to 25° and the other parameters are fixed. The parameters and results are displayed on table 6.

The results show that failure occurs when $\phi = 0°$ (figures 10~11). Slip band is concave-down deep at the bottom of the slope. When internal friction angle increases slip band rises close to the slope surface and plastic zone concentrates to gliding surface (figures 12~15). Safety factor increases linearly with the internal friction angle. The formula 6 and figure 9 given below suggest that internal friction angle has a great importance to stability of soil slope. The impact index is 0.454.

$$F = 0.454\phi + 0.1114 \tag{6}$$

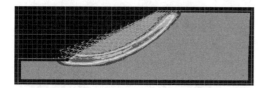

Figure 8. Shear strain rate and displacement vectograph (c = 12 KPa).

Table 6. Calculation parameters and results.

	C (KPa)	Φ(°)	Ψ(°)	γ(kN/m³)	F
case1	3	0	10	2000	0.09
case2	3	5	10	2000	0.36
case3	3	10	10	2000	0.58
case4	3	15	10	2000	0.79
case5	3	20	10	2000	1.01
case6	3	25	10	2000	1.24

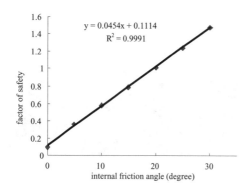

Figure 9. Safety factor vs internal friction angle.

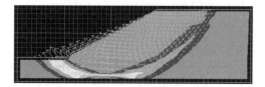

Figure 10. Shear strain rate and displacement vectograph ($\phi = 0°$).

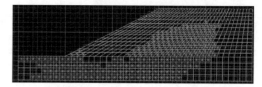

Figure 11. Plastic and tensioned area distribution of slope ($\phi = 0°$).

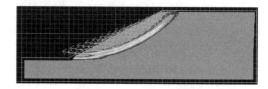

Figure 12. Shear strain rate and displacement vectograph ($\phi = 10°$).

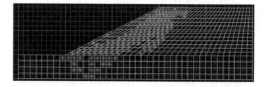

Figure 13. Plastic and tensioned area distribution of slope ($\phi = 10°$).

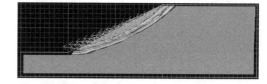

Figure 14. Shear strain rate and displacement vectograph ($\phi = 30°$).

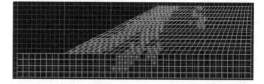

Figure 15. Plastic and tensioned area distribution of slope ($\phi = 30°$).

5 CONCLUSIONS

The stability of soil slope is one of the hot problems because of the harmfulness and its complexity. We analyses the stability of the given soil slope through Lagrangian Finite-Diference Method based on Strength Reduction Method which is proved to be a good method. Through analysis some suggestions can be made that might be helpful to slope design.

Dilatation slightly influences the stability of soil slope that the safety factor slightly increases with it. When dilatation angle increases, the plastic area spreads from slope crest to bottom, displacement increases at the bottom of the slope and the location of slip surface almost remain stable. Variation of dilatation does not impact the location of slip plane. The safety factor increases linearly with cohesion, gliding mass enlarges, failure surface shifts down and becomes concave-down from a plane, its radian gradually enlarges and drops back on the slope crest, slip band extends to deep soil layer. Plastic area distributes around the slip band. Most of the slope crest has tensile stress.

When the internal friction angle is close to zero, soil slope must be unstable. Slip band is concave-down deep at the bottom of the slope. When internal friction angle increases slip band rises close to the slope surface and plastic zone concentrates to gliding surface. Safety factor increases linearly with the internal friction angle. The formula 6 and figure 9 suggest that internal friction angle has a great importance to stability of soil slope. The impact index is 0.454.

In slope prevention project designers can estimate the location of slip plane, the thickness and dimensions of slip band according to cohesion and internal friction of soil so that can select proper length of anti-slide piles or inject grout to enhance the stability of slope.

ACKNOWLEDGEMENTS

This work was funded by the National Science Foundation of China (No. 50639010), the Chinese Ministry of Science & Technology projects (No. 2006BAB19B03), the Knowledge Innovation Program of the Chinese Academy of Sciences (O712041Q01) to Wang. Authors would like to thanks State Key Laboratory of Geomechanics and Geotechnical Engineering. We thank the anonymous reviewer and Chairman Organizing Committee of the 10th ISL Prof. Zuyu Chen for constructive comments and suggestions.

REFERENCES

Chi Shi-chun & Guan Li-jun. 2004. Slope stability analysis by Lagrangian diference method based on shear strength reduction[J]. *Chinese Journal of Geotechnical Engineering*. 26(1): 42–46.

Huang Run-qiu & Xu Qiang. 1995. Application of explicit lagrangian finite—difference methodin rock slope engineering. *Chinese Journal of Rock Mechanics and Engineering*, 14(4): 346–354.

Zhang Pei-wen & Chen Zu-yu. 2004. Finite element method for solving safety factor of slope stabilily[J]. *Rock and Soil Mechanics*, 25(11): 1757–1760.

Zienkiewicz O.C., Humpheson C. & Lewis R.W. 1975. Associated and non-associated visco-plasticity and plasticity in soil mechanics[J]. *Geotechnique*, 25(4): 671–689.

Back analysis of unsaturated parameters and numerical seepage simulation of the Shuping landslide in Three Gorges reservoir area

Shimei Wang, Huawei Zhang, Yeming Zhang & Jun Zheng
Key Laboratory on Geological Hazards of Three Gorges Reservoir Area, Ministry of Education of China, Three Gorges University, Yichang, Hubei, China

ABSTRACT: Obvious ground deformation and displacement occurred in Shuping Landslide in Three Gorges reservoir in June 2003, when the reservoir level reached to EL.135 m. Rainfall and the change of groundwater seepage caused by raise of reservoir water level are the main reasons of landslide deformation. Groundwater movement in landslide mass is a kind of unsaturated and unstable seepage flow, which should be analyzed by the theory of unsaturated unstable seepage. Before the analysis, the determination of parameters is the primary problem. Based on the observed rainfall and water level of drill holes in the landslide area in July 2005, orthogonal optimizing method was used to conduct back analysis for getting the seepage parameters. After that, numerical simulation of groundwater seepage in Shuping landslide mass was carried out by considering real situation of reservoir water level changes. The variation of groundwater seepage in landslide area with the reservoir water level changes was analyzed. The research results will provide useful information for predicting landslide stability under the action of reservoir water level.

1 INTRODUCTION

The impoundment of Three Gorges reservoir has brought a great impact on deformation and stability of reservoir bank. In June 2003, after one month of the reservoir water level reached to EL.135 m, a large scale high-speed landslide with the volume of 24 million cubic meters occurred at Qianjiangping, Shazhenxi town, Zigui County in the reservoir area[1]. At the same time, obvious deformation and displacement was found in Shuping landslide area, where it is only 3 km away from Qianjiangping landslide area. It can be predicted that a great numbers of landslides areas which already have certain deformation will be unstable, and many new potential unstable slopes will appear. This situation will bring a great threat on the lives and properties of residents in the reservoir area.

The variation of groundwater seepage due to the change of reservoir water level and rainfall are the essential reasons of slope deformation. For simulating the variation of groundwater seepage under the condition of rainfall and reservoir water level changes, numerical analysis by using the software of SEEP/W of Geo-slope from Canada, which is based on the non-saturated seepage theory were conducted. In the analysis, the orthogonal optimizing method was employed to design the testing schemes. Back analysis was carried out to get the seepage parameters based on the observed rainfall and groundwater table from drill holes in July 2005. After that, numerical simulation of groundwater seepage in Shuping landslide mass was conducted according to the real situations of reservoir water level changes. The variation of seepage field in landslide mass with the changes of reservoir water level was got from the analysis.

2 BACKGROUND OF SHUPING LANDSLIDE

Shuping landslide area is located on the south bank of the main stream of Yangze River, reservoir area of Three Gorges Project, Shazhenxi town, Zigui County, Hubei province. It is 3 km away from the upstream Qianjiangping Landslide area and 47 km away from the downstream Three Gorges Dam (Figure 1).

Figure 1. The location of Shuping landslide.

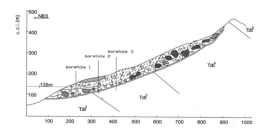

Figure 2. The typical geological profile of Shuping landslide.

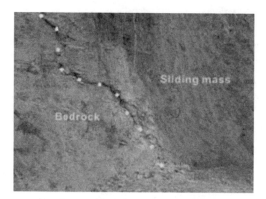

Figure 3. Shear zone in the lateral edge of the slope.

Figure 4. Muddy river water at the exit of the landslide.

The Shuping landslide mass is 700~800 m long and 700~900 m wide. The elevation of landslide exit is 65~68 m and the elevation of trailing edge is 450~500 m. Its thickness is 40~70 m, the area is 525,000 m^2 and the volume is 26 million m^3. It is a multi-period massive landslide with complicate material composition. According to geological investigation and borehole data, the material composition and the stratum structure can be roughly classified into following layers: 1) plough horizon, 2) slope wash, 3) landslide accumulation; 4) sliding zone; 5) bedrock. The typical geological profile is shown in Figure 2.

The water level of Three Gorges Reservoir was raised to 135 m in June, 2003. From October to November in 2003, deformation was occurred at many places of Shuping landslide area. The observable deformations were mainly represented by ground cracks and house cracking. New shear movement was found at the side edge of the landslide mass (Figure 3). From February to March in 2004, large area of muddy water appeared in the river near the shearing exit of the landslide mass (Figure 4).

3 LAYOUT OF TEXT

Consider the obvious deformation occurred in Shuping landslide area after the Three Gorges reservoir impoundment, for observing the development of landslide deformation and understanding the relationship between landslide deformation with rainfall, reservoir water level and groundwater table as well as the relationship between groundwater table with rainfall, reservoir water level, Yichang Institute of Geology and Minerals Resources under China Geological Investigation Bureau has conducted a series of monitoring works, including surface deformation, deep displacement, ground temperature, groundwater table, rainfall and reservoir level. The following is a brief introduction of groundwater and rainfall monitoring and the relationship between them[2]. Then the monitoring data of groundwater table and rainfall were used to conduct back analysis of the unsaturated seepage parameters of Shuping landslide mass.

3.1 Monitoring instruments and allocation

Groundwater level was measured by WS-1040, an automatic dynamic groundwater monitoring device made by the Institute of Hydrogeology and Engineering Geology Technology under China Geological Investigation Bureau. In order to understand the impacts of rainfall on ground water table, a hydrological observation hole was drilled at EL. 180 m in the leading edge of the landslide. The observation is automatic and the frequency of data acquisition is 2 hours.

The rainfall observation is conducted by using N-68 automatic pluviometer, which is made in Japan. It was installed on the roof of residents outside the landslide area. From the observed data, the peak of rainfall all occurred between May to August. At the same period, the reservoir water level was kept at EL.135 m. So the observed data in July 2005 were selected for

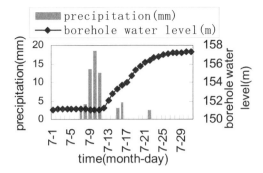

Figure 5. Groundwater table and rainfall in July, 2005.

analyzing the relationship between groundwater table and rainfall.

3.2 The relationship between rainfall and groundwater table

According to the data of rainfall and the groundwater table from July 1 to July 31, 2005, the curve of groundwater table and rainfall can be drawn as Figure 5. From the figure, a continuous rainfall was occurred from July 7~12. Two days after the rainfall, groundwater level in the drill hole was raised from El.151 m to El.157 m. Then, the water level became stable. The fact has shown that the rainfall has a significant impact on groundwater table in Shuping landslide area. Great changes of groundwater table will be happened under the condition of concentrated rainfall in flood season. But it will lag behind the rainfall. Normally, the ground water table will be raised two or three days after concentrated rainfall.

4 BACK ANALYSIS OF THE UNSATURATED SEEPAGE PARAMETER OF SHUPING LANDSLIDE

The coefficient of saturated and unsaturated permeability of landslide mass is the necessary parameters for conducting saturated and unsaturated unstable seepage analysis. The permeability of saturate soil can be gotten from the conventional laboratory permeability test (the results are shown in table 1). But due to the complexity of experimental conditions, the unsaturated permeability of soil is rather difficult to obtain. Normally, the parameters are estimated through soil-water characteristic curve. Based on the data of rainfall and groundwater level of the observation drill hole, by using the soil water characteristic curve V-G model and the corresponding unsaturated permeability function, the parameters of unsaturated seepage can be obtained by back analysis.[3–8]

4.1 Brief introduction of V-G model

Based on measured soil water characteristic curve, Van Genuchten presented the V-G model in 1980. The formula of the model is:

$$\theta_w = \theta_r + \frac{\theta_s - \theta_r}{\left[1 + \left(\frac{\psi}{\alpha}\right)^n\right]^m} \quad (1)$$

Where, θ is the volumetric water content, θ_s is saturated water content, θ_r is the residual water content, and $\psi = \mu_a - \mu_w$ is matric suction, α, n and m are respectively nonlinear regression coefficients, here $m = 1 - 1/n$.

From formula (1), based on the relationship between unsaturated permeability function and the soil water characteristic curve, Van Genuchten gave expression of unsaturated permeability as:

$$k = \frac{k_s \{1 - [\alpha(\mu_a - \mu_w)]^{n-1}[1 + [\alpha(\mu_a - \mu_w)^n]^{-m}]\}^2}{[1 + [\alpha(\mu_a - \mu_w)]^n]^{m/2}} \quad (2)$$

Where, K is the permeability function of unsaturated soil, κ_s is the permeability coefficient of saturated soil. The unsaturated parameters α and n will be gotten through back analysis.

Table 1. The permeability of saturated soil and rock mass in Shuping landslide.

Material	Sliding mass 1	Sliding mass 2	Sliding zone	Bedrock
Saturated permeability (10^{-5} cm/s)	2.75	6.0	0.05	impermeable

Table 2. Factor and level programs.

Level program		Level program 1	Level program 2	Level program 3
Gliding mass 1	n_1	2	2.5	3
	α_1	0.01	0.015	0.03
Gliding mass 2	n_2	7	10	12
	α_2	0.02	0.025	0.03
Sliding zone	n_3	1	1.5	2
	α_3	0.005	0.01	0.015

Table 3. Orthogonal experiment parameters.

No.	n_1 (A)	α_1 (B)	n_2 (C)	α_2 (D)	n_3 (E)	α_3 (F)	/
1	1 (2.0)	1 (0.01)	1 (7)	1 (0.02)	1 (1)	1 (0.005)	1
2	1	2 (0.015)	2 (10)	2 (0.025)	2 (1.5)	2 (0.01)	2
3	1	3 (0.03)	3 (12)	3 (0.03)	3 (2)	3 (0.015)	3
4	2 (2.5)	1	1	2	2	3	3
5	2	2	2	3	3	1	1
6	2	3	3	1	1	2	2
7	3 (3.0)	1	2	1	3	2	3
8	3	2	3	2	1	3	1
9	3	3	1	3	2	1	2
10	1	1	3	3	2	2	1
11	1	2	1	1	3	3	2
12	1	3	2	2	1	1	3
13	2	1	2	3	1	3	2
14	2	2	3	1	2	1	3
15	2	3	1	2	3	2	1
16	3	1	3	2	3	1	2
17	3	2	1	3	1	2	3
18	3	3	2	1	2	3	1

4.2 Orthogonal optimizing test scheme

According to Ven-Genuchten model, the unsaturated properties of each material can be represented by the two parameters α and n. Considering the there kinds of materials in Shuping Landslide, e.g. sliding mass, sliding zone and bedrock, there are 6 parameters. By using the rainfall and groundwater table data in July 2005, orthogonal optimizing method was employed to design the test schemes, and the unsaturated seepage computation was conducted for getting the parameters[9–10]. By taking the reference of the empirical value of the similar landslides, the level program for each group of parameters is shown in table 2. According to the requirement of permutation and combination of orthogonal table, only 18 tests are needed. The test arrangement is shown in Table 3.

4.3 Back analysis of seepage

4.3.1 Model and boundary condition

The geological profile of the computation section is shown in Figure 2. There were three kinds of material in seepage analysis model, e.g. gliding mass, sliding zone and bedrock. The whole section was divided into 3,855 elements and 4,010 nodes (Figure 6).

For the numerical model, the boundary condition of the left and right edge is water head boundary. The left side water head takes the top elevation of the slope and the right side water head is 135 m, which is the water level of Three Gorges reservoir. The upper boundary is a flux boundary, which takes the real values of the rainfall infiltration during July 5 to July 25. The bottom of the model is an impermeable boundary.

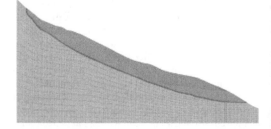

Figure 6. Mesh of seepage analysis model.

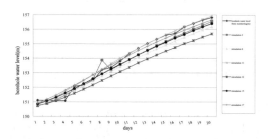

Figure 7. Observed groundwater table and calculated groundwater table.

4.3.2 Computation results and the analysis

According to the 18 groups of parameter listed in Tab. 3, the variation of groundwater level of the observation point in the landslide mass along with the infiltration of rainfall was analyzed respectively. The computation results and the observed changes of

Table 4. Parameters from back analysis.

Material	Gliding mass 1		Gliding mass 2		Sliding zone	
Parameter	n_1	α_1	n_2	α_2	n_3	α_3
Value	2.0	0.015	7	0.02	2	0.015

groundwater level are shown in Figure 7. By comparing the computation results and the observed values, the parameter of the 11th group was the best one. It can be considered as the parameters from the back analysis, which is shown in Tab. 4.

5 NUMERICAL SIMULATION OF SEEPAGE FIELD UNDER DIFFERENT RESERVOIR

5.1 Model, boundary condition and initial condition

The profile of the model was the same as Figure 6. The left boundary is the water head with the elevation of top of the slope. The right boundary is the actual reservoir water level. The bottom is impermeable boundary. The reservoir level and observed groundwater table of landslide mass before reservoir impoundment were taken as the initial condition of the analysis.

5.2 The variation of the Three Gorges Reservoir water level

The actual variations of the reservoir water level were as follows:

1. Initial impoundment period: From April, 2003, the reservoir level was raised from 68 m to 135 m. From EL.68 m to EL.77 m, it was natural impoundment. From 77 m to 135 m, it was forced impoundment, when reservoir water level was rapidly raised in 27 days. After that, the reservoir water level was kept at 135 m, and then slowly raised to EL.145 m.
2. Normal operation period (water level fluctuated between EL.145 m to EL.175 m): Every year, reservoir water level will be raised from El.145 m in June. During middle July to the end of August, the reservoir water level fluctuate between EL.160 m to EL.145 m by the situations of flood. In September, the water level is kept stable at EL.145 m. From October to early November, the water level was raised to EL.175 m, and then kept this level until beginning of next year. During January to the end of May of the next year, the water level falls slowly to EL.145 m again.

To simplify the calculation, the changes of the Three Gorges reservoir water level were generalized into

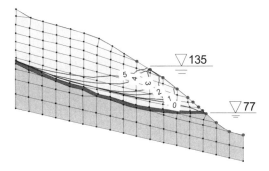

Figure 8. Changes of groundwater table when reservoir water level rose form EL.77 m to EL.135 m.
(Note: 0, 1, 2, 3, 4, 5 represent the groundwater table at the initial time and the time after 12, 27, 40, 100, 200 days).

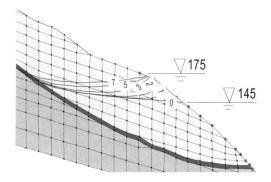

Figure 9. Changes of groundwater table when reservoir water level rose form EL.145 m to EL.175 m.
(Note: 0, 1, 2, 3, 5, 7 represent the groundwater table at the initial time and the time after 20, 40, 60, 100, 200 days).

three typical conditions: First, reservoir water level was raised from El.77 m to EL.135 m within 27 days, and then kept at EL.135 m until the stabilization of groundwater level. Second, the reservoir water level was raised from EL.145 m to EL.175 m within 40 days, and then kept at EL.175 m until the stabilization of groundwater level. Third, reservoir water level was fallen from EL.175 m to EL.145 m within 110 days, and then kept at 175 m until the stabilization of groundwater level.

5.3 Results of numerical simulation

According to the above three typical conditions, numerical simulations of groundwater seepage in landslide area were carried out. The results were shown in Figure8~Figure10.

From the computation results, the groundwater table changes with the fluctuation of the water level.

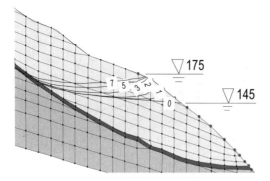

Figure 10. Changes of groundwater table when reservoir water level rose form EL.175 m to EL.145 m.
(Note: 0, 1, 2, 3, 5 represent the groundwater table at the initial time and the time after 50, 110, 150, 100, 200 days).

But the changes of groundwater table are much lag behind for the changes of reservoir water level. The faster of the reservoir water level changes, the longer of the time for ground water table lag behind. Figure 8 and Figure 9 show that when reservoir water level was raised to EL.77 m and EL.135 m, groundwater table are still far from the stable level. At this time, groundwater table is lower than reservoir water level. The seepage direction is from reservoir to the landslide mass (as shown in Figure 8). When reservoir water level falls, the groundwater table in the landslide mass will also drop down. As the fall down speed of reservoir water level is very slow, the change of groundwater table in landslide mass is almost synchronized with the reservoir water level. Only a little time lag occurred.

6 CONCLUSIONS

1. Based on the monitoring groundwater table with the rainfall in July, 2005, by applying Van-Genuchten model and using SEEP/W of GEO-SLOPE, back analysis was conducted for getting the unsaturated seepage parameters of Shuping landslide mass. The parameters have provided sound support for the numerical simulation of groundwater seepage in Shuping landslide area.
2. From the numerical simulation of the seepage field of Suping landslide area under the condition of fluctuation of Three Gorges Reservoir water level, it can be found that: the groundwater table varies with the changes of the reservoir water level. But the changes of groundwater table lag behind the changes of reservoir water level. The faster of the reservoir water level changes, the longer of the time lag, vice versa.

Based on the above characteristics, it can be concluded that: during the initial period of rapid rising of reservoir water level, the rise of groundwater table lag behind, water flow from reservoir into landslide mass and the seepage force toward inside of the slope, which will be in favor of the stability of landslide mass. Whereas, during the initial period of reservoir level quickly drop down, as the water inside the slope has no time to drain, high values of hydraulic gradient will be occurred in landslide mass, which will be unfavorable to the stability of the slope. That is main reason for most of landslides failure occurred during the initial period of reservoir water level drop down.

REFERENCES

Fa-Wu Wang, Ye-Ming Zhang & Zhi-Tao Huo. The July 14, 2003 Qianjiangping landslide, Three Gorges Reservoir, China. Landslides (2004) 1:157–162.

Zhang Huawei. Real-time monitoring and stability forecasting of Shuping Landslide in the Three Gorges Reservoir. Three Gorges University master's thesis, 2005.

Wu Mengxi & Gao Lianshi. Saturated-unsaturated soil unstable seepage numerical analysis. Water Conservancy Journal, 1999, (12): 38–42.

Ye Weimin, Qian Lixin, Baiyun & Chen Bao. Forecasting permeability coefficient of Shanghai saturated soft soil by soil-water characteristic curve. Journal of Geotechnical Engineering, 2005, 27 (11): 1262–1265.

Zhang Hua, Chen Shanxiong & Chen Shouyi. Numerical simulation of unsaturated soil infiltration. Rock Mechanics, 2003, 24 (5): 715–718.

Qi Guoqin & Huang Yunqiu, China. The general mathematical model of soil-water characteristic curve. Journal of Engineering Geology, 2004, 12(2):182–186.

Zhang Yingke, Zan Huiping & Huang Yi. Unsaturated soil seepage function equation. Xi'an Science and Technology University Journal, 2001, 17 (2): 174–177.

Song Xuejun, Chen Yu & Shi Jiying. The application of optimizing design on the super-network division. Journal of Harbin Industrial University, 2002, 34 (2): 265–269.

Xue Maogen, Gu Mingtong & Lin Jieren. The fuzzy analysis of multi-objective optimization orthogonal design of the ship main parameters. Boating Engineering, 1994, 4:20–27.

Slope failure criterion: A modification based on strength reduction technique

Y.G. Wang & R. Jing
Gansu Provincial Communications Planning, Survey & Design Institute, Lanzhou, China

W.Z. Ren
Key Laboratory of Rock and Soil Mechanics, Chinese Academy of Sciences, Wuhan, China

Z.C. Wang
Department of Civil Engineering, University of Calgary, Alberta, Canada

ABSTRACT: The FLAC method based on finite difference technique can simulate the behaviour of slope geomaterials which undergo plastic flow when their yield limits are reached. Thus it can forecast the deformation after slope failure, so it is widely used in slope stability analysis. This paper explored the exiting criteria of slope failure and relevant limitations. Considering the fact that plastic shear strain catastrophe occurs when the points in the critical slice change from a limit equilibrium state to plastic flow after yielding, this paper focused on the changing rate of plastic shear strain of crucial points in the critical slice with the strength reduction coefficient and made a modification for the slope failure criterion: when the catastrophe of the changing rate of plastic shear strain occurs, the corresponding state is a critical instability state and the relevant strength reduction coefficient is a critical factor of safety of the slope. The case analysis in this paper validated that the modified slope failure criterion is rational and feasible.

1 INTRODUCTION

The FLAC method based on finite difference technique can simulate the behaviour of slope geomaterials which undergo plastic flow when their yield limits are reached, and thus forecast the deformation after slope failure. Besides, the slope can have several yield surfaces at the same time (Cala M., 2003), so it is widely used in slope stability analysis. At present, there are three kinds of methods to judge slope failure in critical slice: the first one is to judge slope failure according to deformation characteristics, for example, shear strain method in broad sense, crucial point displacement method, etc. (Lian, 2001, Zheng, 2002), the second one is to judge according to slope stress distributing condition, for example, whether existing the run-through of plastic zone (Luan, 2003); the third one is to judge slope failure according to convergence condition in numerical calculation (Griffiths D.V. 1999, Zheng, 2002). Different researchers adopt different methods and until now no unified criterion comes into being.

There exist some problems more or less in the above three methods. In practice, broad sense shear strain includes not only plastic strain, but elastic strain, so the judgment of the expansion of plastic area and shear deformation area according to these physical parameters is not so rational and accurate; the crucial point displacement method takes the changing curve of strength reduction coefficients with crucial points as analyzing object and takes strength reduction coefficient in the turning point of displacement changing rate as the indication of the beginning of slope failure (Ge, 2003). However, there are two problems need to be solved: ① the displacement changing rate of the crucial points with strength reduction coefficient may increase at any time, so how much change of displacement can serve as a criterion is uncertain; ② the changing curve of displacement and strength reduction coefficient can display different changing effect due to the difference of greatest displacement, so how much reduction coefficient is proper in calculation is not clear; the run-through of plastic zone in slope is the necessary but not the sufficient condition for slope failure. Besides, whether large and infinite expanding deformation and displacement in plastic zone occurs should also be taken into consideration (Zhao, 2005). The geomaterials can still have certain load-carrying capacity when it comes into plastic condition, so simply taking the run-through of plastic zone as sole judging criterion is not rational; Taking the non-convergence condition in numerical calculation as judging criterion of critical instability state of the slope is closely related with numerical calculation technique, the quality of the model and the convergence criterion, so the finite element calculation may sometimes get into the control of minimum factor of safety in part of the slope. And the criterion on the calculation of convergence also needs further discussion.

Considering the close relation between rock and soil plastic failure and the occurrence, expansion and distribution of plastic zone, and the ability of plastic shear strain to memorize and describe the developing and failure evolving process of plastic zone, this paper puts forward the modified slope failure criterion to estimate slope stability with finite difference numerical method based on strength reduction and taking changing rate of plastic shear strain with strength reduction coefficient in the run-through of plastic zone as research object.

2 THE DEFINITION OF SAFETY COEFFICIENT BASED ON STRENGTH REDUCTION

Strength reduction technique is a preferable way to solve the factor of safety of the slope with numerical calculation technique and is more and more widely applied at present. In 1975, Zienkiewicz and his fellows first brought forward the definition of shear strength reduction coefficient in geotechnical elaplastic finite element numerical analysis and hereby determined strength reserve factor of safety is consistent with stable safety coefficient in definition which is presented by Bishop in limit equilibrium methods (Zienkiewicz, 1975). Dawson's research show that the factor of safety derived from numerical method based on strength reduction technique is very close with the factor of safety from limit equilibrium methods which more strictly meets the need of equilibrium condition (Dawson, 1999~2000). Moreover, Zheng (2004) further reveals that the critical slices based on these two methods are very similar.

Shear strength reduction coefficient can be defined as follows: under the condition of consistent outer load, the coefficient is the ratio of maximum shear strength inside slope to practical shear stress produced by outer load within slope. The practical shear stress produced by outer load should be equal to the shear strength after reduction according to practical strength index. In finite difference numerical calculation based on strength reduction, for a certain point on the shear plane, supposing normal and shear stress are σ and τ, so according to the definition of factor of safety given by Bishop in limit equilibrium method and considering the shear strength at this point, Mohr-Coulomb strength criterion can be presented as follows:

$$\tau_f = c + \sigma \tan \phi \quad (1)$$

So the factor of safety of this point on this prearranged shear plane can be presented as:

$$F_s = \frac{\tau_f}{\tau} = \frac{c + \sigma \tan \phi}{\tau} \quad (2)$$

Supposing that no shear failure occurs at this time, the shear stress is equal to shear strength brought into play in practice, namely:

$$\tau = \frac{\tau_f}{F_s} = \frac{c + \sigma \tan \phi}{F_s} = c_l + \sigma \tan \phi_l \quad (3)$$

Therefore, shear strength brought into play in practice equals to shear strength index after reduction. Shear strength indexes after reduction are as follows:

$$c_l = \frac{c}{F_s}, \quad \phi_l = \arctan\left(\frac{\tan \phi}{F_s}\right) \quad (4)$$

In this sense, F_s can be considered as strength reduction coefficient, namely slope whole stable factor of safety or strength reserve factor of safety.

3 IMPROVED SLOPE FAILURE CRITERION AND ITS ADVANTAGES

Based on the analysis of existing criterion of slope failure, this paper combines shear strain methods and stress condition methods and puts forward the modified slope failure criterion: firstly, judging whether there exists the run-through of plastic zone, because the run-through of the plastic zone is necessary condition of slope failure. Slope failure necessarily needs a continuous sliding plane, and the run-through of plastic zone is the minimum shear strength area within the slope which provides conditions for the movement of sliding mass. So this is the first concern when judging whether the slope is in critical state; Secondly, taking the changing rate of plastic shear strain of the crucial points on sliding plane with strength reduction coefficient as research object, when the catastrophe of the changing rate of plastic shear strain occurs, the corresponding state is critical instability state of the slope and the relevant strength reduction coefficient is the critical factor of safety of the slope.

Suppose: (x', y') are the points below and near the sliding plane; (x, y) are the monitoring points on the sliding plane; u_x, u_y, u'_x, u'_y are the displacements of monitoring points respectively; t is the distance between the two points and meets the condition:

$|t_x| \leq 1, \quad |t_y| \leq 1;$

Making $|k| = |\Delta u_y|/|\Delta u_x|$, Δu_x, Δu_y are the increment of displacement respectively.
According to the definition of shear strain:

$$\Delta \varepsilon_{xy} = \Delta \left(\frac{\partial u_x}{\partial y} + \frac{\partial u_y}{\partial x} \right) \approx \Delta \left(\frac{u_x - u'_x}{t_y} + \frac{u_y - u'_y}{t_x} \right)$$

$$= \frac{\Delta u_x - \Delta u'_x}{t_y} + \frac{\Delta u_y - \Delta u'_y}{t_x} = \frac{\Delta u_x}{t_y} + \frac{\Delta u_y}{t_x}$$

$$\approx \left(\frac{1}{t_y} + \frac{k}{t_x} \right) \Delta u_x \geq (1+k) \Delta u_x \quad (5)$$

Because (x', y') are the points below and near the sliding plane respectively, Δu_x, Δu_y are relatively small and can be neglected. From the equation (5), we can clearly see that after the run-through of plastic zone, considering the catastrophe of shear strain changing rate of crucial points on sliding plane as the criterion of slope failure is better than considering the catastrophe of displacement changing rate of crucial points as the criterion, for the former catastrophe value increases $1 + k$ times, which is more obvious and can be effectively avoid the gradual influence of displacement changing rate as well. Therefore, the improved slope failure criterion is more practical.

Which needs explanation is that for general whole slip, critical points on sliding plane are usually selected on shear outlet of main section in landslide, and shear strain should be calculated from the beginning of run-through of plastic zone. Compared with above mentioned slope failure criteria, the improved ones possess the following characteristics:

1. The improved slope failure criterion fully consider the fact that there exists catastrophe of shear strain from limit equilibrium to shear flow after plastic yielding which greatly reduces the uncertainty of criterion. And there is no need to consider the proper reduced coefficient. With the increase of reduction coefficient, as long as there is an obvious turning of plastic shear strain changing ratio of crucial points, certain judgment can be made; otherwise, in the process of incessant expansion of plastic zone, the plastic strain and displacement changing ratio usually appear several inflexion points, which results in too much subjectivity in determining the factor of safety.
2. Taking the run-through of plastic zone as precondition of judgment can meet the basic conditions of slope failure; Meanwhile, judging critical state with the changing ratio of plastic shear strain of crucial points as research object can fully consider anti-sliding potential after the run-through of plastic zone.
3. The selection of critical points on sliding plane avoids some blindness. Because taking changing ratio of plastic shear strain after the rull-through of plastic zone as research object, according to practical condition of the run-through of plastic zone, one or more critical points can be selected to monitor and judge.
4. The existing method with the run-through of plastic zone as judging criteria is a specific example of this paper. Namely, the condition of instant catastrophe of plastic shear strain of critical points after the run-through of plastic zone, also the strength reduction coefficient of the run-through of plastic zone happens to be the corresponding reduction coefficient when the changing ratio of plastic shear strain appears turning point, hence the strength reduction coefficient at this time is factor of safety, and the run-through of plastic zone at this time means the appearance of critical state of slope failure.

4 PROJECT ANALYSIS

4.1 The creation of numerical calculation model

This paper takes the example of HanJiaYa landslide in Xiangshi Expressway to analyze slope stability in natural condition and its strain characteristics with finite difference numerical method based on strength reduction combined with above mentioned improved slope failure criterion.

According to survey report, main section (I-I) is selected for analysis. In FLAC^{3D}, plane strain model is created through controlling deformation of two-directions in front and back of the model. The calculation adopts Mohr-Coulomb yield criterion and non-associated flow law. The angle of the dilatancy is regarded as zero. Numerical calculation model is presented by figure 1~2.

The landslide has all together 9322 units and 3188 nodes. The horizontal direction is presented as x, which is 420 m long, and the vertical direction is presented as y, and the strike along slope is presented as z. The bottom of the model is fixed boundary, and other boundary is restricted by displacement of relevant directions. Physical and mechanical parameters are presented in table 1.

4.2 Slope stability analysis under natural state

According to the idea of "improved slope failure criterion", when the run-through of plastic zone appears in landslide, 1#:2# points on sliding plane round shear outlet in rock mass 5 are selected as crucial points on

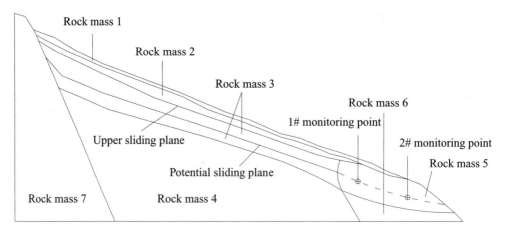

Figure 1. Terrane distribution of main section I-I.

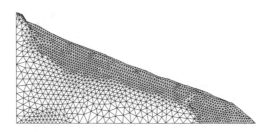

Figure 2. Terrane distribution of main section I-I.

Table 1. Mechanical parameters of rock layers.

Rock mass	Weight KN/m^3	Cohesion c/Kpa	Frictional angle $\varphi/°$	Elasticity modulus E/Gpa	Poisson ratio μ
Rock mass 1	21.2	18	9	0.006	0.2
Rock mass 2	23.2	70	20	2.4	0.245
Rock mass 3	29.3	100	22	6.87	0.261
Rock mass 4	29.7	150	25	17.6	0.283
Rock mass 5	22	50	20	3.32	0.175
Rock mass 6	25	70	21	8.35	0.185
Rock mass 7	27.3	100	22	12.2	0.21

sliding plane and the changing ratio of plastic shear strain with strength reduction coefficient hereafter is analyzed to determine the relevant critical instability state and the factor of safety. Selecting the above two points as monitoring points is based on the following consideration: Looking from the distribution of plastic zone, nephogram of shear strain ratio and displacement vector, the shear outlet is in front of landslide and these two critical points are all on sliding plane, especially 2# monitoring point. The main anti-sliding section is also round the shear outlet of the slope. The plastic shear strain catastrophe will mean the instability of anti-sliding section at last and the occurrence of slope failure of greater scope. So these two points are selected as crucial points in study.

Using shear strength parameters which correspond to a series of strength reduction coefficients (the space between strength reduction coefficient is 0.01) to simulate landslide can obtain slope stress, strain and deformation results. Figure 3 is the vector diagram of displacement distribution which is not strength reduced, namely in natural state.

The figure shows that there is no large scale run-through plastic zone in natural state. The displacement between rock mass 7 in landslide trailing edge and the area around interface of double-layer landslide is relatively large, so slow creep deformation occurs in the double-layer landslide which exerts certain thrusting force to front rock mass 5 and yielding elements come into being in part of the area around the interface of front rock mass 5. This part of slope in front of the landslide become main anti-sliding area of the whole landslide.

We can know from figure 4 that the sliding plane in under layer of the slope is in plastic state. In other words, at this very time the run-through of plastic zone is formed and prerequisite for slope failure is provided; Figure 5, vector digram of displacement distribution shows that relatively large deformations mainly concentrate on interface of landslide trailing edge and rock

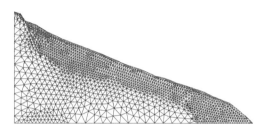

Figure 3. Vector diagram of displacement distribution when strength reduction coefficient is 1.0.

Figure 4. Nephogram of slope shear strain rate when the run-through of plastic zone is formed ($F_s = 1.8$).

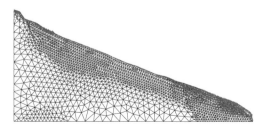

Figure 5. Vector diagram of displacement distribution when the run-through of plastic zone is formed ($F_s = 1.8$).

mass 5 in front of slope toe of steep bedding rock structure. At trailing edge forms tension crack and double-layer landslide creep slides down slowly along with sliding plane. Under residual thrusting force of double-layer landslide, creep deformation of reverse rock-layer continuously occurs in rock mass 5, which results in continuous yielding failure. The slope keeps on sliding down until the whole landslide forms large scale run-through of plastic zone, And the sliding plane in rock mass 5 expands to full run-through.

Analysis of shear strain variation rate of two monitoring points after run-through of plastic zone is presented in figure 6. The variation curve of plastic shear strain with strength reduction coefficient shows that when reduction coefficient reaches 1.20, relatively obvious turning points of plastic shear strain of critical points appear, which can be considered as factor of safety, namely in critical instability state. Thus potential anti-sliding bearing capacity is brought into full consideration.

According to shear strain rate nephogram figure 7 and displacement vector figure 8 when slope is in critical instability state ($F_S = 1.20$), whole rock mass above the underlayer of sliding plane slides down and residual thrusting force acts on rock mass 5 and results in greater deformation. When landslide is in critical instability state, the rock mass 5 in landslide front has been in critical state of yielding failure. Seeing from the variation curve of plastic shear strain of 2# monitoring point around shear outlet, plastic strain acutely increases at this time. It is resulted by the landslide changing from limit equilibrium state to shear flow state after yielding. Great displacement sliding failure occurs in landslide above under layer sliding plane.

According to sliding plane confirmed by nephogram of shear strain variation rate when the slope is in critical instability state, limit equilibrium method is adopted to calculate the stability of the landslide. The factor of safety with Morgenstern-Price method is 1.143, which is 4.75% different from calculation result with failure criterion from this paper. It is obvious that the failure criterion of this paper is rational and feasible and can meet the needs of engineering projects.

5 CONCLUSION

Firstly, this paper deeply discusses the existing slope failure criterion and corresponding problems in numerical calculation and brings forward "improved slope failure criterion": on the basis of the run-through of plastic zone in slope mass, this paper focuses on the variation rate of plastic shear strain of the crucial points on sliding plane with strength reduction coefficient. When the catastrophe of the changing rate of plastic shear strain occurs, the corresponding state

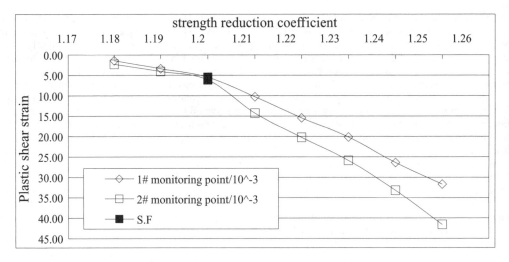

Figure 6. Variation curves of plastic shear strain of two monitoring points along with strength reduction coefficient.

Figure 7. Nephogram of slope shear strain rate when slope is in critical instability state ($F_s = 1.2$).

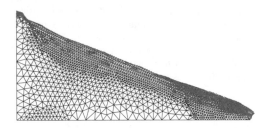

Figure 8. Vector diagram of displacement distribution when slope is in critical instability state ($F_s = 1.2$).

is critical instability state and the relevant strength reduction coefficient is critical factor of safety of the slope.

Secondly, According to "improved slope failure criterion", this paper determines landslide critical instability state and its stable factor of safety and compares with the calculation results with limit equilibrium method. The result of such comparison shows that the improved slope failure criterion suggested by this paper is rational and feasible and can basically meet the needs of practical cases.

REFERENCES

Cala M. & Flisiak J. 2003. Complex geology slope stability analysis by shear strength reduction [A]. *Flac and numerical modeling in geomechanics*, Richard Brummer & Patrick Andrieux, 99–102. A.A. Balkema.

Lian Zhen-ying. 2001. *Studies on some problems of three-dimensional FEM numerical analysis in foundation pit engineering*[D]. Dalian, Dalian University of Technology.

Zheng Hong, Li Chun-guang, LI Zhuo-fen, et al. 2002. Finite element method for solving the factor of safety. *Chinese Journal of Geotechnical Engineering*, 24 (4):323–328.

Luan Mao-tian, Wu Ya-jun & Nian Ting-kai. 2003. A criterion for evaluating slope stability based on development of plastic zone by shear strength reduction FEM. *Chinese Journal of Disaster Prevention and Mitigation Engineering*, 23 (2):1–8.

Griffiths D.V. & Lane P.A. 1999. Slope stability analysis by finite elements. *Geotechnique*, 49 (2):387–403.

Zhang Ying-ren, Zhao Shang-yi & Zhang Lu-yu. 2002. Slope stability analysis by strength reduction FEM. *Chinese Journal of Engineering Science*, 4 (10):57–61.

Ge Xiu-run, Ren Jian-xi, Li Chun-guang, et al. 2003. 3D-FEM analysis of deep sliding stability of 3# dam foundation of left power house of the Three Gorges Project. *Chinese Journal of Geotechnical Engineering*, 25 (3):389–394.

Zhao Shang-yi, Zheng Ying-ren & Zhang Yu-fang. 2005. Study on slope failure criterion in strength reduction finite element method. *Chinese Journal of Rock and Soil Mechanics*, 26 (2): 332–336.

Zienkiewicz O.C., Humpheson C & Lewis R.W. 1975. Associated and Non-Associated Visco-Plasticity and Plasticity in Soil Mechanics. *Geotechnique*, 25 (4):671–689.

Matsui T. & San K.C. 1992. Finite element slope stability analysis by shear strength reduction technique. *Soils and Foundations*, 32 (1):59–70.

Dawson E.M., Roth W.H. & Drescher A. 1999. Slope stability analysis by strength reduction, *Geotechnique*, 49 (6):835–840.

Dawson E., Motamed F., Nesarajah S. et al. 2000. Geotechnical stability analysis by strength reduction[A]. In: Grifiths D.V. ed. *Geotechnical Special Publication: Slope Stability 2000-Proceedings of Sessions of Geo-Denver 2000[C].[s.l.], [s.n.]*, 99–113.

Dawson E., You K. & Park Y. 2000. Strength-reduction stability analysis of rock slopes using the Hoek-Brown failure criterion[A] In: Labuz J.F. ed. *Geotechnical Special Publication: Trends in Rock Mechanics[C]. [s.l.], [s.n.]*, 65–77.

Zheng Hong, Liu De-fu & Luo Xian-qi. 2004. Determination of potential slide line of slopes based on deformation analysis. *Chinese Journal of Rock Mechanics and Engineering*, 23 (6):708–716.

Wang Yong-gang. 2006. *The gradual destruction mechanical model and time-dependent deformation analysis of landslide with double sliding planes and reverse rocklayer*[D]. Wuhan, Institute of rock and soil mechanics, Chinese academy of sciences.

Unsaturated seepage analysis for a reservoir landslide during impounding

J.B. Wei & J.H. Deng
State Key Laboratory of Hydraulics and Mountain River Engineering, College of Water Resources and Hydropower, Sichuan University, Chengdu, China

L.G. Tham & C.F. Lee
Department of Civil Engineering, The University of Hong Kong, Hong Kong, China

ABSTRACT: Xietan landslide is a pre-existing landslide in the Three Gorges Reservoir area and is constituted of the slide body, the slip zone, the sliding-disturbed zone, and the bedrock from up to down. In hydrogeology, the slide body, the slip zone and the sliding-disturbed zone can be regarded as porous continuous medium, while the permeable bedrock is fractured rock. In this paper, an equivalent continuum model was used to model all strata, and a finite element model was established to analyze the unsaturated seepage of Xietan landslide. Then the back-propagation neural network was used to substitute finite element seepage analysis. The saturated permeability coefficient of every stratum was optimized by genetic algorithm based on monitored data during reservoir impounding. The saturated permeability coefficient of the slide body by back analysis is 4.89×10^{-2} cm/s, which is consistent with field testing results (1.78×10^{-2} cm/s to 3.2×10^{-2} cm/s); the saturated permeability coefficient of the slip zone by back analysis is 4.66×10^{-5} cm/s, which is very larger than indoor testing results (2.74×10^{-7} cm/s to 5.73×10^{-7} cm/s). The scale effect of soil samples was reflected by this difference.

1 INTRODUCTION

For reservoir landslide, the temporal and spatial distribution of pore water pressure in the landslide with the change of reservoir water level is of great important to the displacement and stability of landslide. At present, phreatic surface is usually as the base of stability analysis of reservoir landslide, but the methods adopted for determining the ground water tables are mostly empirical and varied from one project to another (Zheng et al. 2004). Along with the development of unsaturated seepage theory, unsaturated seepage analysis method has been widely used in the seepage flow analysis of landslide (Peng et al. 2002; Zhang et al. 2003), but related researches mostly assume that the slide body of landslide is permeable and the slip zone and below of landslide is impermeable. The research of pore water pressure of landslide with complex hydrogeological structure under the condition of reservoir fluctuations is not considered in many jobs, and the comparative analysis on measurement and calculation is few also.

The primary work of unsaturated seepage of landslide is determining the unsaturated hydraulic parameters of every stratum, which include the soil-water characteristic curve (SWCC) and permeability function of unsaturated soil. The soil-water characteristic curve is usually obtained by tests or empirical procedures, and the unsaturated permeability function can be predicted by using the saturated permeability coefficient and the soil-water characteristic curve. However, the result of seepage calculation using test parameters usually differ with the measurement data because the test parameters are affected by sample location, test method and scale effect. Though the back analysis method is an effective way to determinate the parameters of geotechnical materials, it is mostly used in the displacement back analysis of slope or tunnel, the application of back analysis method on seepage analysis of landslide is not so common.

Take the Xietan landslide in the Three Gorges Reservoir area as an example, the finite element model of unsaturated seepage of Xietan landslide is established in the first, and then the saturated permeability coefficient of every stratum is obtained by back analysis method using BP network and genetic algorithm based on monitoring data during impounding, at last the obtained permeability coefficients are used in the finite element model to model the change of groundwater level during impounding. Through comparing the results of numerical modeling and measurement, the feasibility of above method is discussed.

2 ANALYSING THEORY

2.1 Unsaturated seepage modeling

Based on Darcy's law and continuity principle of mass conservation, the governing differential equation of two-dimensional unsaturated seepage can be derived as below (Lei, 1988):

$$\frac{\partial}{\partial x}\left(k_x \frac{\partial H}{\partial x}\right) + \frac{\partial}{\partial y}\left(k_y \frac{\partial H}{\partial y}\right) + Q = \frac{\partial \theta}{\partial t} \quad (1)$$

where H = pressure head; k_x = permeability coefficient in the x-direction; k_y = permeability coefficient in the y-direction; Q = applied boundary flux; θ = volumetric water content; t = time.

The main hydraulic parameters needed in unsaturated seepage analysis include the soil-water characteristic curve and permeability function. In this paper, the soil-water characteristic curve of every stratum refers to Fredlund empirical procedure (Fredlund & Xing, 1994a), as written below:

$$\theta = \left(1 - \frac{\ln(1 + \psi/\psi_r)}{\ln(1 + 1000000/\psi_r)}\right) \\ \times \frac{\theta_s}{\{\ln[e + (\psi/a)^n]\}^m} \quad (2)$$

where θ = volumetric water content; θ_s = saturated volumetric water content; ψ = suction; ψ_r = the suction corresponding to the residual water content; a = air-entry value of the soil; n and m are curve fitting parameters.

The unsaturated permeability function is predicted by using the saturated permeability coefficient and the soil-water characteristic curve (Fredlund et al., 1994b):

$$k(\psi) = k_s \int_{\ln(\psi)}^{b} \frac{\theta(e^y) - \theta(\psi)}{e^y} \theta'(e^y) \, dy \Big/ \\ \int_{\ln(\psi_{aev})}^{b} \frac{\theta(e^y) - \theta_s}{e^y} \theta'(e^y) \, dy \quad (3)$$

where $k(\psi)$ = calculated permeability coefficient for a specified suction ψ; k_s = saturated permeability coefficient; $b = \ln(1000000)$; ψ_{aev} = air-entry value of the soil; θ' = derivative of formula (1).

2.2 BP network

BP network (back-propagation neural network) is one of the most widely used artificial neural networks because of its powerful function-mapping ability (Deng et al. 2001). Through training, BP network can find the potential mapping relation between input and output values.

BP network is constituted by one input layer (X), one output layer (Y) and one or more hidden layer (Z). Each layer contains several nodes, and the nodes are interconnected via weighted links, as shown in Figure 1 (Rogers, 1997). The training process of BP network contains forward propagate of input values, backward propagate of errors and adjust of link values. Input values propagate forward to the hidden layer nodes via weighted links, and the input values in hidden layer transform into output values through transfer function and propagate forward to the output layer via weighted links. The errors between desired outputs and actual outputs are propagated backward through original interconnected links. Then the link values are adjusted according to given learning rate (α) and momentum term (β), and the output values in output layer are adjusted. When the error values between desired outputs and actual outputs of all patterns are less than given tolerance (ε), the training of BP network is finished.

The process of BP network simulating FEM unsaturated seepage is as follows. At first, a BP network should be constructed. The number of nodes (n) in the input layer is the same as the number of parameters to be solved, the number of nodes (m) in the output layer is the same as the number of monitored ground water level, the number of nodes (p) in the hidden layer can be specified either manually or by an optimization method. Usually, only one hidden layer is needed. Secondly, a number of training patterns should be created to train the network. The input valves of training patterns can be created by random method or by orthogonal method, the desired output values of training patterns are the corresponding calculated results

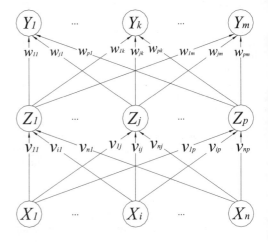

Figure 1. Structure of BP network.

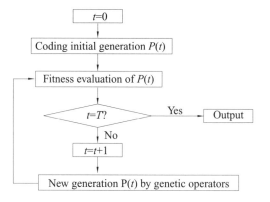

Figure 2. Flow chart of genetic algorithm.

by FEM. Lastly, a number of testing patterns should be created to test the mapping ability of trained network. The construction of testing pattern is the same as training pattern.

2.3 Genetic algorithm

Genetic algorithm is an optimization algorithm that simulated the natural selection and natural genetics, and codes optimization variables as a binary string. The process of genetic algorithm is showed in Figure 2 (Goldberg, 1989). The initial generation is created by random method. Further generations are created by repeated application of genetic operations (reproduction, crossover and mutation), which favor the survival of the fitness of old generations. When the time of reproduction reached a given number (T), the reproduction is stopped, and the individual with highest fitness is the final results.

3 PERMEABILITY COEFFICIENT BACK ANALYSIS OF XIETAN LANDSLIDE

3.1 Unsaturated seepage modeling

Xietan landslide is a preexisting landslide that locates in the reservoir area of the Three Gorges Project, 48.2 km upstream of the Three Gorges dam. The strata of Xietan landslide can be divided into four layers from up to down. The first layer is slide body, which is mainly composed of silt and gravels of siltstone, limestone, and quartz sandstone or shale origin. It is loosely structured and quite permeable. The second layer is slip zone, which is around 2–3 m thick and composed of deep grey or light green clay containing coarse gravels. Its permeability is lower, so it's a relative impermeable stratum. The third layer is sliding-disturbed zone, which is composed of firm clay with coarse gravels of sub-rounded shape. Its permeability is relatively good. The bottom layer is bedrock; the upper part of bedrock is relatively permeable because of fracture.

A 2D finite element software SEEP/W is used for the unsaturated seepage analysis. In the seepage model, sliding-disturbed zone and permeable bedrock are combined into one stratum, namely slip bed, and equivalent continuum model is used to model all strata. Monitoring data shows that the groundwater level in the rear of the landslide is not affected by the reservoir water level during impounding period, so the calculating range is limited in the middle-front part of the landslide, as illustrated in Figure 3.

The monitored groundwater level before filling is applied as initial condition and the monitored reservoir level during filling is applied as boundary condition in seepage analysis. The calculation steps and corresponding reservoir level and monitored borehole water level in ZK9 is listed in Table 1.

The input parameters are the soil-water characteristic curve, saturated permeability coefficient and the permeability function. In this paper, the saturated permeability coefficient is as an undetermined parameter, the soil-water characteristic curve and the permeability function is obtained by empirical formula (2) and

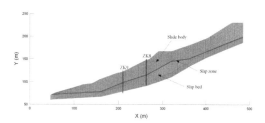

Figure 3. Unsaturated seepage model of Xietan landslide.

Table 1. Steps of unsaturated seepage modeling and monitoring data.

Step	Time	Reservoir level (m)	Water level in ZK9 (m)
0	2003–5-24 19:00	80.90	80.60
1	2003–5-25 7:00	82.43	80.86
2	2003–5-26 7:00	85.04	82.36
3	2003–5-27 7:00	88.10	84.23
4	2003–5-28 7:00	91.26	87.36
5	2003–5-29 7:00	94.48	91.25
6	2003–5-30 7:00	97.82	95.33
7	2003–6-1 7:00	105.92	105.66
8	2003–6-2 7:00	109.75	109.65
9	2003–6-3 7:00	113.06	113.02
10	2003–6-4 7:00	115.60	115.61

(3) respectively. The parameters of the soil-water characteristic curve are obtained by laboratory and field testing results and are listed in Table 2.

3.2 BP network modeling of unsaturated seepage

Based on tested and provisional calculated results, the range of saturated permeability coefficient of every stratum used in BP network is, slide body ($3 \times 10^{-2} - 8 \times 10^{-2}$ cm/s), slip zone ($2 \times 10^{-5} - 7 \times 10^{-5}$ cm/s) and slip bed ($5 \times 10^{-4} - 1 \times 10^{-2}$ cm). A total of 12 training patterns and 4 testing patterns are created. The input values of training patterns and testing patterns are the saturated permeability coefficient of every stratum as illustrated in Table 3. The corresponding output values are the stepped groundwater level in the location ZK9 calculated by finite element method.

A single hidden layer BP network is used; the number of neurons in the input layer, hidden layer and output layer is 3, 7 and 10 respectively. The basic parameters of BP network are, $\alpha = 0.6$, $\beta = 0.5$, $\varepsilon = 0.03$. After the network has been trained, the testing patterns are used to test the mapping ability of the network. The maximum error of output values between FEM and the trained BP network is 0.13 m, so the mapping ability of BP network is demonstrated.

3.3 Genetic algorithm optimizing of permeability coefficient

In genetic algorithm, the probability of an individual reproducing to new generation depends on its fitness. Fitness is transferred from the value of objective function. The smaller the value of objective function is, the higher the fitness. The objective function in this paper is:

$$\min f = \sum_{i=1}^{n} (H_i - h_i)^2 \quad (4)$$

where H_i = the groundwater level by monitor or by FEM; h_i = the groundwater level by BP network.

The basic parameters of genetic algorithm are: crossover probability = 0.6, mutation probability = 0.05, time of reproductions = 2000. At first, the calculated data by FEM of testing patterns are used to inversely predict the saturated permeability coefficient of every stratum through genetic algorithm. The results of inversive prediction are listed in Table 4. The comparison of saturated permeability coefficient between inversive prediction values and actual values listed in Table 3 shows that the error of slide body and slip zone is very little, while the error of slip bed is slightly large. This is because the change range of saturated permeability coefficient of slip bed in training patterns is larger than the other strata. Then the monitoring data of ZK9 during filling period are used to predict the saturated permeability coefficient of every stratum. The results are listed in Table 4 also.

The inversion parameters are input in the FEM model to model the seepage of Xietan landslide during impounding period. The calculated and monitored groundwater level is compared in Figure 4. The two curves coincide very well.

4 COMPARISON OF PERMEABILITY COEFFICIENT BETWEEN BACK ANALYSIS AND TESTS

The inversion permeability coefficient of slide body is 4.89×10^{-2} cm, while the field testing result is $1.78 \times 10^{-2} - 3.2 \times 10^{-2}$ cm (Wei et al. 2007). The inversion and testing values are in the same magnitude. The field test method is double ring infiltration test. Big gravels (above 200 mm) in slide body are avoided in field test because of the size limitation of test instrument. Because the spatial change of permeability coefficient

Table 2. Parameters of SWCC.

Stratum	θ_s	a (kPa)	n	m
Slide body	0.28	100	0.4	0.25
Slip zone	0.31	1000	0.5	0.7
Slip bed	0.25	800	0.3	0.5

Table 3. Input parameters of training patterns and testing patterns.

Pattern no.	Permeability coefficient (cm·s^{-1})		
	Slide body	Slip zone	Slip bed
Training patterns			
1	5×10^{-2}	4×10^{-5}	1×10^{-2}
2	5×10^{-2}	4×10^{-5}	1×10^{-3}
3	8×10^{-2}	4×10^{-5}	1×10^{-3}
4	3×10^{-2}	4×10^{-5}	5×10^{-4}
5	8×10^{-2}	7×10^{-5}	1×10^{-3}
6	3×10^{-2}	7×10^{-5}	5×10^{-4}
7	3×10^{-2}	7×10^{-5}	1×10^{-2}
8	5×10^{-2}	7×10^{-5}	5×10^{-4}
9	3×10^{-2}	2×10^{-5}	1×10^{-3}
10	5×10^{-2}	2×10^{-5}	1×10^{-2}
11	5×10^{-2}	2×10^{-5}	1×10^{-3}
12	8×10^{-2}	2×10^{-5}	5×10^{-4}
Testing patterns			
1	8×10^{-2}	2×10^{-5}	1×10^{-3}
2	5×10^{-2}	4×10^{-5}	5×10^{-4}
3	8×10^{-2}	4×10^{-5}	5×10^{-4}
4	5×10^{-2}	7×10^{-5}	1×10^{-3}

Table 4. Back analysis results of permeability coefficient.

Testing pattern no.	Permeability coefficient (cm·s^{-1})			min f (m^2)
	Slide body	Slip zone	Slip bed	
1	8.09×10^{-2}	2.13×10^{-5}	1.17×10^{-3}	0.0377
2	5.09×10^{-2}	3.95×10^{-5}	5.63×10^{-4}	0.098
3	8.09×10^{-2}	4.0×10^{-5}	3.33×10^{-4}	0.087
4	5.01×10^{-2}	7.01×10^{-5}	8.98×10^{-4}	0.178
monitored water level	4.89×10^{-2}	4.66×10^{-5}	5.46×10^{-4}	2.6

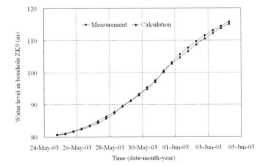

Figure 4. Comparison between calculated and monitored water level in borehole ZK9.

is taken into consideration, the inversion permeability coefficient can reflect the permeability characteristics more exactly.

The inversion permeability coefficient of slip zone is 4.66×10^{-5} cm, while the lab testing result is $2.74 \times 10^{-7} - 5.73 \times 10^{-7}$ cm. There is a big difference between inversion value and testing value. The main reasons of this difference lie in two aspects. At first, remolded samples are used in laboratory permeability test. The sample size is 62.8 mm in diameter and 20 mm in height and the gravels above 2 mm are eliminated from the specimens. Scale effect is exists in the test. While inversion value is the synthetic permeability coefficient of the total slip zone, the non-uniform thickness and discontinuity of slip zone are reflected in the inversion value, so the inversion permeability coefficient of slip zone is larger than test results. And then, due to the limitation of unsaturated seepage modeling method, the fracture seepage in permeable bedrock is modeled by equivalent continuum model, which has a certain impact on the calculated results.

Though the unsaturated seepage modeling method of landslide with complicated hydraulic structure is not perfect now, adopting the inversion parameters and seepage analysis method suggested in this paper, the calculated result is agreement with monitored result, so the feasibility of above method is demonstrated.

5 CONCLUSIONS

1. The infiltrative strata of Xietan landslide are reduced to three strata such as slide body, slip zone and slip bed through combining sliding-disturbed zone and permeable bedrock into one stratum. Then equivalent continuum model is used to model every stratum, and finite element model of unsaturated seepage analysis of Xietan landslide is established.
2. The application of BP network and genetic algorithm in the back analysis of the hydraulic parameters of landslide is discussed, and the saturated permeability coefficient of every stratum is calculated by back analysis based on monitored data.
3. The inversion parameters are used in the FEM model to model the seepage of Xietan landslide during impounding period. The calculated result is agreement with monitored data.
4. The inversion permeability coefficient of slip zone is great larger than laboratory tested result because of scale effect. In this paper, the inversion parameter can reflect the synthetic hydraulic characteristic of the landslide more exactly.

ACKNOWLEDGEMENTS

Financial support from Chinese Postdoctoral Science Foundation under grant 20070410387, the Special Fund for Major State Basic Research Project under grant 2002CB412702 and Research Grant Council of Hong Kong under grant HKU7015/02ERGC are greatly appreciated.

REFERENCES

Deng, J.H. & Lee, C.F. 2001. Displacement back analysis for a steep slop in the Three Gorges Project site. *International Joural of Rock Mechanics and Mining Sciences* 38: 259–268.

Fredlund, D.G. & Xing, A. 1994a. Equations for the soil-water characteristic curve. *Canadian Geotechnical Journal* 31: 521–532.

Fredlund, D.G., Xing, A. & Huang S. 1994b. Predicting the permeability function for unsaturated soils using the soil-water characteristic curve. *Canadian Geotechnical Journal* 31: 533–546.

Goldberg, D.G. 1989. *Genetic algorithms in search, optimization and machine learning*. Reading, Mass: Addison-Wesley.

Lei, Z.D. 1988. *Soil hydrodynamics*. Beijing: T singhua Press.

Peng, H., Chen, S.F. & Chen, S.H. 2002. Analysis on unsaturated seepage and optimization of seepage control for Dayantang landslide in Shuibuya Project. *Chinese Journal of Rock Mechanics and Engineering*, 21: 1027–1033 (in Chinese).

Rogers, J. 1997. *Object-Oriented Neural Networks in C++*. London: Academic Press, Inc.

Wei, J.B., Deng, J.H., Tham, L.G. & Lee, C.F. 2007. Field tests of saturated and unsaturated hydraulic parameters of gravelly soil in Xietan landslide. *Rock and Soil Mechanics* 28: 327–330 (in Chinese).

Zhang, P.W., Liu, D.F., Huang, D.H. & Song, Y.P. 2003. Saturated-unsaturated unsteady seepage flow numerical simulation. *Rock and Soil Mechanics*, 24: 927–930 (in Chinese).

Zheng, Y.L., Shi, W.M. & Kong, W.X. 2004. Calculation of seepage forces and phreatic surface under drawdown conditions. *Chinese Journal of Rock Mechanics and Engineering* 23: 3203–3210 (in Chinese).

A simple compaction control method for slope construction

L.D. Wesley
University of Auckland, Auckland, New Zealand

ABSTRACT: Controlling the quality of compacted fill using dry density and water content can be difficult for the following reasons. Firstly, some soils are highly variable, with rapid and random changes in optimum water content and maximum dry density. Secondly, some soils do not show peak dry density and optimum water content. Thirdly, some soils are highly sensitive and conventional compaction control may be inappropriate. The above observations are particularly true of residual soils, whose formation processes produce quite different properties from sedimentary soils. An alternative compaction control method is presented, making use of shear strength and air voids in place of water content and dry density. By using these parameters, variations in soil properties are taken account of, and the same specification applies regardless of these variations.

1 INTRODUCTION

Compaction of clays and silts has normally been controlled by specifying limits of water content and dry density. While this method has had almost universal acceptance, it is not without deficiencies, and attempts have been made over the years to develop alternative methods. Some of the difficulties in using the conventional method are described in the following sections, and an alternative method is described based on undrained shear strength and air voids.

2 LIMITATIONS OF NORMAL COMPACTION CONTROL METHOD

2.1 Conventional compaction control

Conventional behaviour and compaction tests are well known and only brief comments will be made here. Typical results of compaction tests on a clay are shown in Figure 1. Also plotted on the graph is the zero air voids line; this indicates the maximum value the dry density can have at any particular water content. An important point to recognise is that there is no such thing as a unique "optimum water content" for a particular soil. Almost any value of optimum water content is possible; it simply depends on the compactive effort involved in the compaction process. In Figure 1, the optimum water content from the Modified Compaction Test is about 8% below that for the Standard Compaction Test. This is normal for moderate to high plasticity clays. It is worth noting also that for moderate to high plasticity clays the optimum water content from the Standard Proctor Compaction Test is normally very close to the Plastic Limit.

This procedure is satisfactory with many soils, but certainly not all. Among "problematic" soil characteristics that make the application of this traditional method difficult are the following.

a. Many soils, especially residual soils, are far from homogeneous; their properties change rapidly over

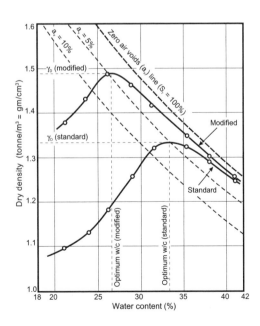

Figure 1. Typical compaction tests on clay.

short distances, so that the optimum water content may vary over a wide range.

b. Standard compaction tests on some residual soils do not show a peak dry density and thus no optimum water content.

c. Many residual soils are highly structured, and have natural water contents close to, or even above their Liquid Limit. Compaction processes destroy the natural structure of these soils, and may turn them into very soft materials, making compaction impossible without substantial drying of the soils.

These aspects are described in more detail in the following sections.

2.2 Highly variable soils

Figure 2 shows the results of compaction tests on soils from two relatively small construction sites in the Auckland region of New Zealand. The soils involved at each site were of the same geological origin. The upper diagram shows tests on soils derived from the weathering of basaltic lava flows and ash deposits. The lower diagram shows tests on soils derived from Pleistocene deposits, some of which may have undergone weathering since deposition. Despite the common geological origin at each site, there is clearly a very large variation in the type of soil, as reflected in the compaction curves shown in Figure 2. Determination of the optimum water content in this situation is rather meaningless, as it will only be valid for the particular location from which the sample was taken.

2.3 Soils without clear optimum water contents

There are some unusual soils that do not conform to the behaviour illustrated in Figure 1. Their compaction curves do not show clear peaks of dry density and thus do not indicate optimum water contents. The best known soil type that displays this behaviour are clays that contain a high proportion of the clay mineral allophone; this is a very unusual clay mineral derived from the weathering of volcanic ash. The non-crystalline nature of volcanic ash results in a very different weathering process to that which produces "conventional" clay minerals, and produces a poorly crystalline, very fine grained mineral. Allophane clays are characterised by generally good engineering properties despite extremely high water contents and Atterberg limits. Water contents between 100% and 200% are common. Another of their distinctive characteristics is that drying causes irreversible changes to their properties. In their natural state they are very fine-grained, and are of moderate plasticity. When air or oven dried, their plasticity is greatly reduced. In many cases, oven drying changes them into non-plastic slightly sandy silts.

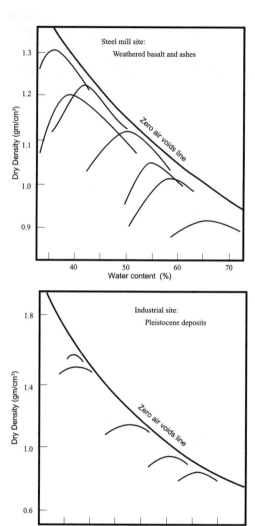

Figure 2. Compaction test results from two small construction sites (after Pickens, 1980).

Figure 3 shows typical results of compaction tests on two clays having high allophane contents. Three different procedures have been followed in obtaining these curves. The curves labelled "natural" have been obtained by progressively drying the soil from its natural state. The other curves have been obtained by re-wetting the soil after it has been both air dried or oven dried. Separate samples were used for each point on the curves, so that the results do not reflect any influence from repeated compaction.

In the case of the second sample, an additional curve was obtained by only partially air-drying the soil, to a

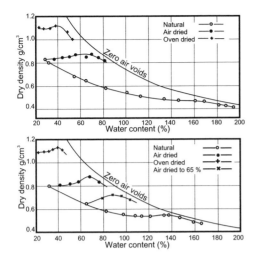

Figure 3. Compaction tests from allophone clays.

water content of only 65%. The number of points on these curves indicates that a large volume of soil was required to carry out the tests. For a full description of the properties of allophone clays see Wesley (2002).

2.4 Soils that soften during compaction

Some soils become softer during the compaction process. Many soils, especially residual soils, are "structured", that is they have some form of weak bonds between their particles. In some cases, they consist of partially weathered parent material, which makes them appear competent, but which is really a fragile, sensitive material. When remoulded the structure is broken down and the soil becomes softer. It is important to recognise, therefore, that compaction of a soil may have two effects, as follows:

a. "Densifying" the soil, that is pressing the particles closer together and squeezing out any air trapped between the particles.
b. Remoulding the soil, causing it to soften. Nearly all natural soils lose some strength when remoulded, and compaction is a form of remoulding. The compaction process destroys the natural structure of the soil, especially any bonds between particles. This is usually accompanied by release of water trapped within the structure, or between the particles, adding to the softening process.

With non-structured soils, which are those of low or negligible sensitivity, the effect of compaction will be purely to force the particles closer together, and thus increase the density and strength of the soil. With highly structured soils, which are normally those of high sensitivity, the effect of compaction will be to destroy the bonds or other structural influences present and thus weaken the soil.

Figure 4 illustrates the softening effect that compaction has on a number of volcanic ash soils in Japan. Tests have been done on a range of volcanic ash soils in which strength measurements have been made after the soils have been compacted using varying compactive efforts. The strength has been measured using a cone penetrometer test, giving the "cone index" as an indication of strength. The compactive effort has been varied by changing the number of blows of the compaction rammer, from a minimum of 5 blows, a maximum of 120 blows. It is seen that most of the samples become softer as the blow count increases. Kanto loam Sample A shows a consistent strength increase until the blow count reaches about 70, beyond which a decrease in strength occurs. Kanto loam Samples B and C show a small initial increase in strength followed by a large loss of strength. This softening effect produced by higher compactive effort is often referred to as "over-compaction".

For sensitive or structured soils at their natural water content, there is thus an "optimum compactive effort" which will produce the strongest fill, as indicated in Figure 4. Kanto loam Samples A, B and C show reasonably clear "optimum compactive efforts" while the remainder show only that the optimum compactive effort is the minimum required to produce a tight fill.

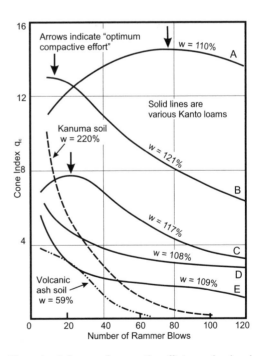

Figure 4. Influence of compactive effort on volcanic ash soils (after Kuno et al., 1978).

3 ALTERNATIVE COMPACTION CONTROL USING SHEAR STRENGTH AND AIR VOIDS

The principal objectives in compacting soil are normally to create a fill of high strength and low compressibility, and in the case of water retaining fills, of low permeability. It is also desirable that the fill will not significantly soften with time as a result of exposure to rainfall, or other weather effects. In adopting the traditional control method it is assumed that by aiming for maximum density the above objectives will be achieved. This is not automatically true, and there is no reason why other parameters will not achieve the intended objectives equally well. Undrained shear strength and air voids are suitable alternative parameters, and are more directly related to the intended properties of the fill. The use of this method was developed in New Zealand in the 1960s and 1970s, and is described by Pickens (1980). The basis for using undrained shear strength is illustrated in Figure 5.

The figure shows the results of a Standard Proctor compaction test on clay, during which measurements of undrained strength have been made in addition to the normal measurements of density and water content. The measurements have been made using both a hand shear vane and unconfined compressive strength measurements on samples of the compacted soil. The two strength measurements give somewhat different results.

It is seen that at the optimum water content the undrained shear strength is about 160 kPa from the unconfined tests and about 230 kPa from the vane tests. Conventional specifications may allow water contents 2 or 3% greater then optimum, in which case the comparable shear strength values would be about 120 and 180 kPa. Thus to obtain a fill with comparable properties to those obtained with conventional control methods, specifying a minimum undrained shear strength in the range of about 150 to 200 kPa would be appropriate. In effect, this would put an upper limit on the water content at which the soil could be compacted.

It is apparent that the required shear strength could be achieved by compacting the soil in a very dry state, which would generally be undesirable, as dry fills may soften and swell excessively when exposed to rainfall. To prevent the soil from being too dry a second parameter is specified, namely the air voids in the soil. Figure 1 indicates that at optimum water content the air voids in the soil is generally about 5%. If the soil is compacted 2 to 3% drier than the optimum water content corresponding to the compaction effort being used, the air voids may be as much as 8 or 10%. Thus to prevent the soil from being compacted too dry an upper limit is placed on the air voids, normally in the range of 8% to 10%.

Figure 6 illustrates how this method of controlling compaction relates to the traditional method. The zero air voids line is always the upper limit of the dry density for any particular water content, and thus applies to both methods. The traditional method involves an upper and lower limit on water content, and a lower limit on dry density, and thus encloses the area shown in the figure.

The alternative method involves an upper limit on water content, corresponding to the minimum permissible shear strength, and a line parallel to the zero air voids line representing the upper limit of air voids. There is no specific lower limit of water content, but the air voids limit prevents the soil from being too dry.

Experience has shown that suitable limits for the two control parameters are as follows:

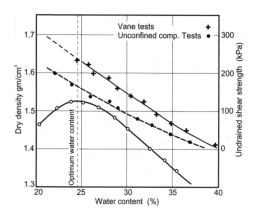

Figure 5. Standard Proctor compaction test showing measurements of undrained shear strength.

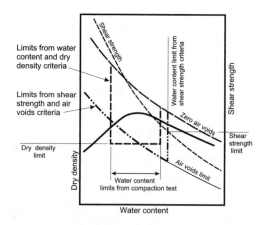

Figure 6. Compaction control using undrained shear strength and air voids.

Undrained shear strength (hand vane values): Not less than 150 kPa (average of 10 tests) Minimum single value: 120 kPa.

Air voids (for "normal" soils): Not greater than 8%. For some unusual soils, such as allophone clays, this value is too low and 12% would often be more appropriate.

These values have been found to be very satisfactory in producing firm, high quality fills. The undrained shear strength can be measured in situ by a hand shear vane, or by taking samples for unconfined compression tests. It can also be measured by empirical penetrometer tests that have been correlated with undrained shear strength. The hand shear vane and examples of hand operated dynamic and static penetrometers are shown in Figure 7. The hand shear vane is both the most useful (since it is a direct measure of undrained shear strength) and the simplest of these methods.

The air voids can only be determined by measuring the density and water content in the usual way. The author's experience has been mainly in temperate or wet tropical climates, where it is often the case that the soil is too wet and the undrained shear strength criteria is difficult to meet while the air voids requirement is easily achieved. This means that the quality control consists essentially of checking only the shear strength. With the hand shear vane, or the hand penetrometers, this checking can be done as the compaction operation proceeds. This is a big advantage over the conventional compaction control method of measuring density and water content, as the results of these tests are often only available the day after the measurements are made. This would not be the case with nuclear methods of control.

While the criteria above are suitable for a wide range of compaction operations, there are some situations where other properties may be more important, and the criteria can be adjusted accordingly. For example, the core of an earth dam built on compressible foundations, or in a seismic zone, may need to be plastic, or ductile, to allow for possible deformations in the dam. This can be achieved by adopting a lower undrained shear strength; a value between about 70 kPa and 90 kPa would produce a reasonably plastic material, assuming the clay is of moderate to high plasticity. For a clay embankment being built for a new highway, it may be desirable that the layers closest to the surface (on which the pavement itself will be constructed) have a higher strength than those deeper down. This could be achieved by increasing the required undrained shear strength to say 200 kPa.

It will be evident from the account given above that this method of compaction control does not actually require compaction tests at all. However, it is still useful to carry out compaction tests to determine the degree of drying, or wetting, that may be needed to bring the soil to a state appropriate for compaction.

4 COMMENTS ON SOILS THAT SOFTEN DURING COMPACTION

As mentioned earlier, many residual soils are "structured", and when remoulded by the compaction process the soil becomes softer, as shown in Figure 4. The natural water content of such soils is generally much higher then the optimum water content of the soil when tested in the normal way. When planning to construct fills with soils of this type it is important to have a clear understanding of two factors. Firstly, it is important to understand the compaction behaviour of the soil by evaluating it with appropriate laboratory testing and field trials, and secondly it is necessary to appreciate clearly what properties are required in the compacted fill. Taking account of both the compaction characteristics of the soil and the required properties of the fill, appropriate compaction criteria and procedures can be adopted. This may necessitate choosing between two options as follows:

a. drying the soil and using normal compactive effort to produce a high quality fill, with a shear strength not less than 150 kPa (ie close to its normal optimum water content),
b. accepting that substantial drying is not feasible because of weather conditions, and adopting a much lower compactive effort, and possibly a lower shear strength, so that the soil can be effectively compacted at, or close to, its natural water content.

Some fills do not require high undrained shear strength. Examples are highway embankments of low to moderate height, and embankments or benched platforms to support light structures. Cores of earth dams are another example. In these situations, undrained

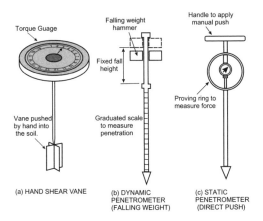

Figure 7. Hand shear vane and hand penetrometers.

shear strengths as low as 80 kPa or less may be acceptable. If this is the case, the need to dry the soil can be avoided by adopting the second option given above.

To determine the feasibility of this second option, it is desirable to conduct trials involving the excavation, transport, and compaction of the soil. In this way the "optimum compactive effort" discussed earlier in relation to Figure 4 can be determined. Excavation, transport, and spreading should be carried out in such a way that disturbance and remoulding of the soil is kept to a minimum. In other words the natural structure and strength of the soil should be retained as much as practical. The compaction operation should similarly be conducted so that remoulding the soil is minimised. Light, tracked equipment is likely to be most appropriate, and the compaction process may consist essentially of "squeezing" intact fragments of soil together to form a uniform fill. For this purpose only a few passes of the compaction equipment is likely to be preferable to a large number of passes; the latter may progressively soften the soil and not make it any more compact or stronger.

5 CONCLUSIONS

To handle the difficulties that "problematic" soils present to earthworks and compaction control, it is necessary to be flexible in the way the soil is evaluated and the way in which the compaction process is carried out and monitored. The use of the standard method involving water content and dry density as control parameters is simply impractical for many residual soils. The alternative method described here using undrained shear strength and air voids provides an appropriate alternative, which, while avoiding the difficulties associated with the standard method, still produces a fill of equally good engineering properties. It is widely used in New Zealand, and is accepted and described in the local code of practice for earthworks. It has also been used on a number of major earthworks operations in Southeast Asia.

To summarise, the advantages of the shear strength and air voids control method are as follows:

- Large variations in soil properties present no difficulty in applying the method. The same specification limits apply regardless of the variations.
- Field control is more direct as the value of the undrained shear strength is known immediately the measurements are made.
- The specification is easily varied to produce fills with particular properties needed in special situations.

The method also allows for the specification to be related more closely to the requirements of the fill itself. The undrained shear strength can be lowered or raised as necessary, depending on whether a relatively soft plastic fill or a high strength fill is required.

ACKNOWLEDGEMENTS

The author wished to acknowledge that the method described here has been developed by a number of people and agencies in New Zealand. The New Zealand Ministry of Works, and the Consulting firm Tonkin and Taylor Ltd have probably played the dominant role.

REFERENCES

Kuno, G., Shinoki, R., Kondo, T. & Tsuchiya, C. 1978. On the construction methods of a motorway embankment by a sensitive volcanic clay. *Proc. Conf. on Clay Fills*, London. 149–156.

Pickens, G.A. 1980. Alternative Compaction Specifications for Non-uniform Fill Materials. *Proc. Third Australia-New Zealand Conf. on Geomechanics*, Wellington 1. 231–235.

Wesley, L.D. 2002. Geotechnical characterisation and properties of allophone clays. *Proc. International Workshop on Characterisation and Engineering Properties of Natural Soils*. Singapore, Dec. 2002.

Numerical analysis of soil-arch effect of anti-slide piles

Yuanyou Xia, Xiaoyan Zheng & Rui Rui
School of Architecture & Civil Engineering, Wuhan University of Technology, Wuhan, Hubei, China

ABSTRACT: This paper aims at solving the problems in the calculation and design of anti-slide piles presently. The basic principle of soil arching is expatiated, FLAC3D is applied to build the 3D model of anti-slide piles, gliding masses and anchorage zone, and the numerical simulation is carried out. Through numerical tests, the soil arching along the whole pile is studied. According to the displacement contours, the effects are mainly appears in the soil under and among the piles. The effect under different depths of thrust and gliding masses is also studied. The results can provide referential information for engineering design and theoretical study of anti-slide piles.

1 INTRODUCTION

As collapse, failure and debris flow, landslide is also a natural disaster commonly encountered in mountain area construction. As landslide happens, traffic interruption, river obstruction, factory and mine destruction, as well as village and town buried are usually accompanied, which will cause serious damage. Therefore, the study on landslide prevention has always been a concern for all countries. Compared with traditional measures of landslide prevention, such as drainage, load reducing and retaining wall, anti-slide pile engineering is the most convenient and efficient way, so it is widely used in engineering practice. Presently, in the calculation and design of anti-slide pile, many assumptions are made for pile-soil interaction. As a result, the load transfer of pile-soil and soil-arch effect along the whole pile cannot be faithfully reflected, which causes a conservative design and a waste of money. So, it has theoretical and realistic meaning to study the soil-arch effect in anti-slide pile.

Blocking effect, such as traffic jams, compaction of granular materials, crowds jams when panic and so on, is unavoidable in the array of all systems. It is the root of soil-arch effect (Garcimartin, 2004). As far back as 1884, an English scientist, Roberts discovered that the pressure acting on barn bottom attained the maximum value and kept constant when the accumulation height of the grain reached a certain value, and he called this phenomenon soil effect. In 1943, using "Trap door" test, Terzaghi (1943) explicated the phenomenon of stress transfer between soil mass and the rigid edge beside, and he called this "soil-arch effect" for the first time. Smid, Novosad, Matuttis, Vanel, Lawrence etc. studied the stress transfer in granular media, and obtained some useful results for soil-arch effect mechanics (Low, 1994, Lawrence, 2002).

Although soil-arch effect plays an important role in various geotechnical engineering, such as tunnels and protecting piles of foundation pits, the definitions of it are not the same in different structure (Zhang, 2004). For anti-slide pile engineering, soil-arch effect transfers the stress that soil subjects to piles. Hence, it can be defined as a phenomenon of shear stress transfer (Chi, 2004). In 1980, Padfield (1980) mentioned arching during his research of riverbank stability. Wu Zishu studied the mechanism of soil-arch effect, and presented the minimum embedded soil depth and the ultimate arch span (Wu, 1995). Based on the research of Liang, Zeng, Chen, and Martin, Zhang Jian-xun etc (Zhang, 2004, Nogami, 1991, Hesham, 2000) found that the major mechanisms for stabilizing soil due to lateral movements by passive piles was soil-arch effect. And then they used a finite element analysis software package, Plaxis 8.1, to study the causes of soil-arch effect in passive piles.

So far, in the field of geotechnical engineering, although the monitoring data, experimental models and theory relating to soil-arch effect become more and more, the interaction between passive piles and moving soil mass is a complex problem that hasn't been sufficiently studied as yet, much less the research on pile-soil interaction in soil slope. So it is necessary to further the study of soil-arch effect problem of passive piles.

2 FORMATION PRINCIPLE OF SOIL-ARCH EFFECT

In slope engineering, after anti-slide pile construction, soil mass between piles is tending to move outside the slope when the deformation of anti-slide piles itself occurs to restrain slope displacement. The trend further develops after the excavations of the soil before piles. Because the lateral displacement of anti-slide pile is less than the slope's, the soil mass after piles in a partial area continuously extrudes piles and uneven earth pressure is generated. The displacement of soil mass between piles is different due to the pile restrain: the displacement of soil mass by piles is small, while the displacement of soil mass away from piles is big. Under this condition, "wedging" function (Wu, 1995, Zhang, 2004, Zhou, 2004) is occurred due to the utilization of shear resistance of soil mass before and after piles, which is also known as soil-arch effect.

3 SIMULATION CALCULATION

3.1 Procedure for anti-slide pile simulations

In this paper, 3D computational model in FLAC3D is applied to simplify the problem, and the example in paper which named Design and Calculation of Pre-Stress Anchor Slide-Resistant Pile is adopted. By symmetry, the part between two piles is considered. Constraint in the z-direction is defined along the model bottom; constraint in the x-direction is defined along the opposite sides of the model; constraint in the y-direction is defined along the front and rear slip surface of the model; stress boundary constraint, whose value is lateral self weight of the gliding mass, is applied above the slip surfaces. As shown in Figure.1, the depth of slip surface is 10 m; rectangle pile with cross section of 1×1.5 m is adopted, and its length is 16 m; pile spacing is 4 m; 4 m in y-direction of the front and rear slip surfaces are considered; calculated depth is 26 m.

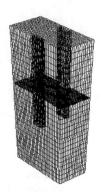

Figure 2. 3D meshing graph.

According to the conditions above, the model is meshed. And "interface element" is used to simulate the slip surface and pile-soil interface. The meshing figure is as shown in Figure 2.

3.2 Numerical test analysis of soil-arch effect

Soil-arch effect, which is the main mode of action in pile-soil interaction, has a bearing on whether anti-slide pile can give its full play to resist slide. There are many factors that can influence soil-arch effect, such as pile-soil interface parameters, thrust, physical mechanic parameters of gliding mass and anchorage zone, as well as the anchorage depth, size and strength of piles. The stress and deformation of piles is different, as a result, the development degree of soil-arch effect is not the same either. This numerical test is carried out under different thrusts to analyze the soil-arch effect of gliding mass.

Design data:

Accumulative formation of soil, block stone, Macadam are above the slip surface and their deformation increases from the top downward, $\gamma_1 = 1900\,\text{kg/m}^3$, $\varphi_1 = 26°$. Mudstone strongly effloresced and shale, which can be considered as compacted soil layer, are below the slip surface, $\gamma_1 = 2100\,\text{kg/m}^3$, $\varphi_1 = 42°10'$. Thrust is as a triangular distribution.

3.2.1 Stress analysis

Arching effect, which is an inherited nature of particulate material, forms a stable structure to prevent particles from flowing when blocking. The soil-arch effect of anti-slide pile is complexity. Because the displacement of soil mass after piles caused by thrust is blocked by piles, blocking effect and extrusion occur. And stress redistributes due to the extrusion. The direction of principal stress changes to the direction

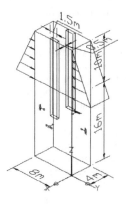

Figure 1. 3D model graph (unit:m).

Table 1. Parameter choosing.

Categories	Thickness	Density/ kg/m³	E/Pa	Poisson's ratio	C/Pa	$\varphi/°$	K/Pa	G/Pa
Detritus	10	1900	1.49E+07	0.3	1.00E+03	26	1.24E+07	5.71E+06
Strong weathered mudstone and shale	16	2100	1.87E+07	0.4	3.10E+04	42.17	3.11E+07	6.67E+06
Pile		2500	2.60E+10	0.167			1.30E+10	1.11E+10

that approximately parallel with the tie of the circular arch between two contiguous piles. Therefore, an arciform compacted complex is formed, and some of the thrust is transferred to anti-slide piles. A certain external force corresponds to a certain structure, that is to say, if the force is not the same, the structure formed will be different either. As to gliding mass with anti-slide piles, the thrust is different in the depth direction. As a result, the soil arch is varied with the depth, and that is why soil-arch effect can not be well reflected by 2D. In this test, the maximum principal stress contour sketches of each plane under different thrust are mapped through simulation calculation.

Through calculation, the variation of displacement and stress on plane $z = 0$, $z = -2$, $z = -4$, $z = -6$, $z = -8$ is simulated. For easy of display, the stress of pile elements is not protracted in some contour sketches, while the omitting will not affect the research. The contours of principle stress under the condition that thrust is 400 kN/m are adopted to analyze.

The contour sketches is as shown in Figure 3. It reveal some laws. The color of the upper part, which corresponds to the soil mass before piles, is homogeneous, and the stress value it indicates is small. The form of the lower part, which corresponds to the soil mass after piles, looks like an anticlinal fold: with the increase of the depth, an "inner core" with high stress appears in the middle; and as the depth increases to −8 m, an arching area appears between and below the two piles, and the contours bending relieves. The contours of maximum principle stress varied with depth fully demonstrate the difference of soil-arch effect in depth. The manifestation of soil-arch effect can be divided into 2 types: gliding mass edges forward with pressure and friction force due to its movement towards the pile is blocked, and soil-arch effect is characterized by tractive force of soil mass; as the thrust & depth increase to a certain degree, the relative displacement between soil and pile increases, the direction of principle stress deflects, and the principle stress in the middle increases continuously, and finally, a "inner core" with high stress generated. With the increase of thrust & displacement, the stress

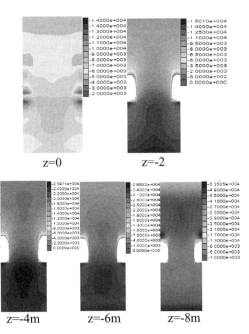

Figure 3. Contours of maximum principle stress of each plane when thrust is 400kN/m.

of the "inner core" increases and gradually expands to surrounding area, some of the soil mass between and after piles is squeezed, and finally, an arching compaction area is formed. The contours of $z = -8$ m is the most obvious, and the principle stress is nearly the same in the compaction area.

Both of the two types of soil-arch effect come into being in all thrust and depth. In this test, numerical simulations are done when thrust is 100 kN/m, 200 kN/m, 400 kN/m and 800 kN/m. It can be seen that, soil-arch effect is not actually formed when the thrust is small, and pile-soil cooperation is characterized by tractive force due to pile blocking and friction. As thrust & depth increases, the relative displacement between soil and pile increases, and then a compacted soil arch is formed in the middle of the soil mass after piles to resist the thrust. By this time, soil-arch effect is the main manifestation of pile-soil cooperation.

3.2.2 Deformation analysis

The calculation results are introduced to SURFER to process. A section per 2 m, the y-direction strain sketches of each section under different thrust are obtained. And the section $z = -8\,\text{m}$ is taken to analyze.

In Figure 4, all of the contours have the trend of decreasing from negative to positive; the contour sketches twist in the positive direction in the middle, and the bent degree is the biggest between the piles with gradual sparse contours. From the y-direction strain sketch of each section under different thrust, the contours of the soil mass after piles become more and more compact, and the bend between the piles become less with the increase of the depth. This phenomenon shows that, the amount of deformation of the soil mass after piles is bigger than the soil mass before, as a result, the soil mass after piles is squeezed; and with the increase of depth, the extrusion become more and more obvious. The displacement on the axle is bigger than the gliding mass near the piles, but along the direction of before-pile, the displacement becomes congruous, and the difference disappears when it reaches a certain position before piles. From the calculation result, the influence sphere caused by deformation difference is smaller than the pile spacing, which indicates the pile spacing chosen is proper at the same time.

From the analysis above, the movement caused by thrust of the soil mass after piles is blocked by piles, the soil mass is squeezed, and soil-arch effect is functional. From the contour sketches, soil-arch effect mainly appears in the soil mass after and between piles. Soil-arch effect of soil mass after piles is characterized by a compact arch zone due to the deflection of principle stress, while soil-arch effect between piles is realized by frictional force on pile-soil interface and tractive force. With the increase of depth, the bend of contours between piles becomes smaller and smaller until reversed bend appears, while the contours of the soil mass after piles become more and more compact, which means soil-arch effect is more obvious: the relative displacement between soil and piles reduces to a microscopic value, the thrust soil mass subjects caused by the gliding mass after piles is very small, and soil arch transfers the thrust to piles. Reversed bend appears due to the deformation of the soil mass before piles. Meanwhile, the friction on pile-soil interface is very small.

The displacement data calculated is plotted: the maximum displacement of the gliding mass varied with depth is as shown in Figure 5; the relative displacement between piles and soil is as shown in Figure 6; the maximum displacement of the gliding mass and relative displacement varied with thrust is as shown in Figure 7; the displacement of piles varied with depth is as shown in Figure 8.

From Figures 5–8, the maximum displacement of gliding mass under different thrust decreases linearly with the increase of depth in general, while the relative displacement in the superficial part is small, and the

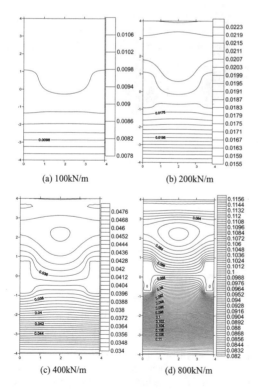

Figure 4. Displacement contours of plane $z = -8\,\text{m}$.

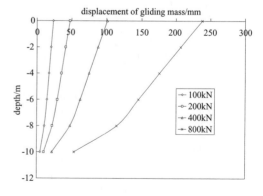

Figure 5. Maximum displacement of gliding mass varies with depth.

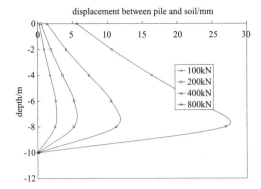

Figure 6. Relative displacement between pile and soil varies with depth.

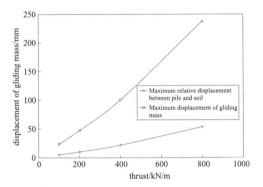

Figure 7. Maximum displacement of gliding mass and relative displacement between pile and soil vary with thrust.

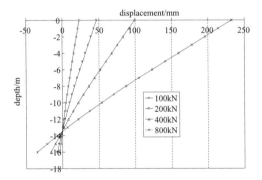

Figure 8. Pile displacement varies with depth.

value increases with the increase of depth, but when the depth exceeds −8 m, the value decreases. The triangular distribution of thrust has much to do with this phenomenon. Meanwhile, from Figure 8, it can be seen that the piles turn about the point −13.15. Form the maximum displacement of gliding mass and the relationship between thrust and relative displacement, the displacement increases linearly with the increase of thrust.

4 CONCLUSIONS

In view of the anti-slide pile design features, the three-dimensional model is created and the parameter numerical test is studied by using the rock & soil engineering three-dimensional numerical calculation software FLAC3D. The pile's total depth of earth-arch effect is studied and the establishment of the earth-arch is analyzed too.

Concluding the above study, we may state the following:

– The movement caused by thrust of the soil mass after piles is blocked by piles, the soil mass is squeezed, and soil-arch effect is functional. From the contour sketches, soil-arch effect mainly appears in the soil mass after and between piles.
– As the thrust & depth increase to a certain degree, the relative displacement between soil and pile increases, the direction of principle stress deflects, and the principle stress in the middle increases continuously, and finally, a "inner core" with high stress generated. With the increase of thrust & displacement, the stress of the "inner core" increases and gradually expands to surrounding area, some of the soil mass between and after piles is squeezed, and finally, an arching compaction area is formed.
– Both of the two types of soil-arch effect come into being in all thrust and depth. Sliding different depth of the soil arching effect of different forms and different thrust of soil arching different forms. It can be seen that, soil-arch effect is not actually formed when the thrust is small, and pile-soil cooperation is characterized by tractive force due to pile blocking and friction. As thrust and depth increases, the relative displacement between soil and pile increases, and then a compacted soil arch is formed in the middle of the soil mass after piles to resist the thrust. By this time, soil-arch effect is the main manifestation of pile-soil cooperation.

REFERENCES

Chi Yue-jun & Song Er xiang et al. 2003. Experimental study on stress distribution of composite foundation with rigid piles. *Rock and Soil Mechanics*, 24 (3):339–343.

Garcimarti, A., Zuriguel, I. & Maza, D. 2004. Jamming in granular matter *Proceeding of 8th Experimental Chaos Conference*: 279–288.

Hesham, M., Naggar, E.I. & Kevin, J. Bentley. 2000. Dynamic analysis laterally loaded piles and dynamic p-y curves. *Canadian Geotechnical Journal*, 37: 1166–1183.

Lawrence. 2002. *The mechanism of load transfer in granular materials utilizing tactile pressure sensor*. University of Massachusetts Lowell.

Low, B.K., Tang, S.K & hhoa, V. 1994. Arching in piled embankments. *Journal of Geotechnical Engineering*, 120(11): 1917–1937.

No. 2 Survey and Design Institute. 1983. *Design and Calculation of Pre-Stress Anchor Slide-Resistant Pile*. Beijing: China Railway Press.

Nogami, T., Jones, H.W. & Mosher, R.L. 1991. Seismic response analysis of pile-supported structure: assessment of commonly used approximations. *Proceedings of 2nd international conference: Recent advancements of geotechnical earthquake engineering and Soil dynamics*, [s.l.]: [s.n.] 931–940.

Richard, L.H. 1985. The arch in soil arching. *Journal of Geotechnical Engineering*, 111(3): 302–318.

Terzaghi, K. 1943. *Theoretical soil mechanics*. New York: John Wiley and Sons.

Wang Nian-xiang. 2000. A summary of interaction between passive pile and soil mass. *Journal of Nanjing Hydraulic Research Institute*, 3: 69–76.

Wu Zi-shu, Zhang Li-min & Hu Ding. 1995. Studies on the Mechanism of Arching Action in Loess. *Journal of ChengDu University of science and technology*, 83 (2): 15–19.

Zhang Fan-rong. 2004. The Studies of design and Construction Technique for Anti-Slide Structures. Ms D Thesis. Southwest Jiaotong University.

Zhang Jian-xun, Chen Fu-quan & Jian Hong-yu. 2004. Numerical analysis of soil arching effects in passive piles. *Rock and Soil Mechanics*, 25 (2): 174–184.

Zhou De-pei, Xiao Shi-yu & Xia, xiong. 2004. Discussion on rational spacing between adjacent anti-slide piles in some cutting slope projects. *Chinese Journal of Geotechnical Engineering*, 26 (1): 132–135.

Determination of the critical slip surface based on stress distributions from FEM

Daping Xiao & Chunqiu Wu
Institute of Foundation Engineering, China Academy of Building Research, Beijing, China

Hong Yang
MWH Americas, Inc., Walnut Creek, California, USA

ABSTRACT: A method is proposed to extend the classic slip line theory and determine the critical slip surface of a slope based on stress distribution from Finite Element Method (FEM). In the proposed method, the slip line field in the slope is determined from the orientation relationship between the slip lines and the principal stress tracing lines. Factors of Safety (FS) are then calculated for each slip line and the critical slip surface is identified to be the one with the minimum FS. The method has taken into account the possible existence of both elastic and plastic zones in the slope and the effect of different flow rules. The validity and applicability of the method are demonstrated using case studies of both a homogeneous slope and a slope with a weak layer.

1 INTRODUCTION

Evaluation of slope stability based on the results of FEM has long been studied in geotechnical engineering. The evaluation involves two interrelated aspects, namely the calculation of a FS for the slope and determination of the critical slip surface. The key issue is to determine the critical slip surface based on the FEM results, which can be solved based on the slip line theory. The slip line method in geotechnical engineering is used to determine the stress distribution of the plastic zone and the ultimate load based on equilibrium equations, failure criteria and boundary conditions. Once the shapes of the slip lines are determined, the critical slip surface is identified as the slip line corresponding to the minimum FS.

The slip line field indicates the stress state either along the slip lines or in the localized plastic zones adjacent to loading. The slip line field does not indicate the stress state of the entire domain of slope stability problem. Therefore, it is necessary to extend the slip line field to the entire domain of slope in order to apply the slip line method to slope stability analysis. The extended stress field should fully comply with both limit equilibrium and failure criteria. To achieve this, an appropriate approach is to determine the slip line field from the information of a stress field generated from elasto-plastic finite element analyses.

Zhen (2000) studied the mathematic model needed to satisfy the critical slip surface and proposed a differential equation which is the equation for slip lines in the classic slip line theory. Zhang et al. (2003) proposed the concept of potential slip line field based on the extension of the classic slip line theory. In addition to these previous studies, the current study developed a numerical modeling method to determine the critical slip surface of a slope based on extended slip line theory and FEM results. The numerical method is capable of solving slope stability problems involving both elastic and plastic zones and different materials using associated or non-associated flow rules for any slope geometry and boundary conditions. The numerical method, therefore, extends the applicability of the slip line method.

2 CLASSIC SLIP LINE THEORY AND ITS EXTENSION

2.1 Classic slip line theory

Under plane strain and plastic state, two intersecting shear failure planes exist in every point of the plane. By connecting these shear planes at all points, two clusters of failure lines can be obtained, namely the α- and β- slip lines, as shown in Figure 1. The tangent of any point on a slip line is the direction of the slip line.

The direction of any point on the slip line is related to the tracing line of principal stress. Two clusters of orthogonal curves (i.e. the so-called tracing lines of principle stresses) can be obtained by connecting the

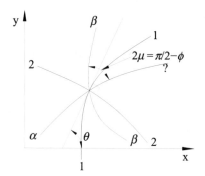

Figure 1. Slip lines and principal stress tracing lines.

fragments representing the direction of principle stress (i.e. curves 1-1 and 2-2 as shown in Figure 1). The analysis of this paper assumes that the tracing line of major principle stress, σ_1, is the baseline; the α-line has a clockwise sharp angle to the baseline, and the β-line has a counter-clockwise sharp angle to the baseline.

For a Mohr-Coulomb material, the angle between the two clusters of slip lines is $2\mu = \pi/\pi 2 - \phi$, and the angle between the slip line and tracing line of principle stress is $\mu = \pi/4 - \phi/2$, in which ϕ is the friction angle of material (Gong and Shi 2000). When using classic slip line theory to determine stresses, the curve coordinates (S_α, S_β) constituted by the two clusters of slip lines can be solved from the following equations:

α-line: $\dfrac{dy}{dx} = \tan(\theta - \mu)$ (1)

β-line: $\dfrac{dy}{dx} = \tan(\theta + \mu)$ (2)

$$-\sin(2\mu)\dfrac{\partial P}{\partial S_\alpha} + 2R\dfrac{\partial \theta}{\partial S_\alpha}$$
$$+ \gamma \left[\sin(2\mu)\dfrac{\partial x}{\partial S_\alpha} + \cos(2\mu)\dfrac{\partial y}{\partial S_\alpha}\right] = 0 \quad (3)$$

$$\sin(2\mu)\dfrac{\partial P}{\partial S_\beta} + 2R\dfrac{\partial \theta}{\partial S_\beta}$$
$$+ \gamma \left[\sin(-2\mu)\dfrac{\partial x}{\partial S_\beta} + \cos(2\mu)\dfrac{\partial y}{\partial S_\beta}\right] = 0 \quad (4)$$

where P is the averaged stress; R is the radii of stress circle, and $R = P\sin\phi + c\cos\phi$ for M-C material; γ is unit weight; θ is the angle between major principle stress σ_1 and x-axis.

The nonlinear differential Equations (1) through (4) for the slip lines with unknown curve coordinates have to be solved using numerical methods, as well as through an iteration method (Gong 1990).

2.2 Potential slip line field in elastic zone

Classic slip line theory was developed based on plastic mechanics and applies only to the plastic zone. The elasto-plastic finite element analysis incorporates both plastic and elastic zones. In order to use FE analysis for determining the critical slip surface, the concept of slip line must be extended to the elastic zone by introducing the concept of potential slip line field. Corresponding to the definition of slip lines, the potential slip lines are defined as the two clusters of the most dangerous shearing lines (i.e., the weakest surfaces for shear resistance) (Zhang et al. 2003). The μ' angle between the potential slip line and the primary principle stress can be determined by:

$$\mu' = \dfrac{\pi}{4} - \dfrac{\theta_{max}}{2} \quad (5)$$

in which $\theta_{max} = \arcsin\dfrac{\frac{\sigma_1 - \sigma_3}{2}}{\frac{\sigma_1 + \sigma_3}{2} + \cot\phi}$; $\theta_{max} = \phi$ when the point reaches it strength limit, and $\theta_{max} < \phi$ when the point is in elastic state.

2.3 Slip line field with consideration of non-associated flow rule

Constitutive relationships are not considered in classic slip line theory. In classic slip line theory, failure criteria constitute only a constraint. However, the classic slip line theory inexplicitly comprises the assumption of associated flow rule for material, namely that the dilation angle, ψ, equals the internal friction angle, ϕ. For geomaterials conforming to the Mohr-Coulomb failure criteria, the effect of dilation will be amplified when the associated flow rule is adopted. Many field observations and laboratory testing have indicated that dilation angles for real soils are much smaller than internal friction angles (Kabilamany and Ishihara 1990). Davis (1968) found that the velocity characteristic lines (i.e. the slip lines) no longer coincide with the characteristic lines of principle stresses when the dilation angle becomes unequal to the internal friction angle. The slip lines no longer satisfy Equations (1) and (2). For this case, the angle between the slip line and major principle stress becomes:

$$\mu' = \dfrac{\pi}{4} - \dfrac{\psi}{2} \quad (6)$$

Davis further found that the normal stress σ^* and shear stress τ^* on slip line has the relationship of:

$$F = \tau^* - c^* - \sigma_n^* \tan \phi^* = 0 \quad (7)$$

where the strength parameters c^* and ϕ^* have similar meanings with those in the Mohr-Coulomb criteria with the following relationship:

$$\frac{c^*}{c} = \frac{\tan \phi^*}{\tan \phi} = \frac{\cos \psi \cos \phi}{1 - \sin \psi \sin \phi} \quad (8)$$

As a result, non-associated flow rule can be considered in the slip line field for geomaterials. For FE analysis using non-associated flow rule, the slip line field obtained from a FE-generated stress field automatically considers the non-associated flow rule.

3 NUMERICAL MODELING OF SLIP LINE FIELD BASED ON FE ANALYSIS

With the incorporation of potential slip lines, the two clusters of slip lines can be uniformly represented by Equations (9) and (10):

$$\alpha\text{-line:} \quad \frac{dy}{dx} = \tan(\theta - \mu') \quad (9)$$

$$\beta\text{-line:} \quad \frac{dy}{dx} = \tan(\theta + \mu') \quad (10)$$

In this paper, rather than solving the nonlinear differential functions (1) through (4), the slip line field is obtained from Equations (9) and (10). Firstly, θ and μ' at each discrete point is obtained from FE results of the stress field, and the direction of slip lines can be determined from Equations (9) and (10). Since FE analysis considers equilibrium, geometric and physical functions, and elastic and plastic zones; FE analysis more realistically and reasonably predicts the potential slip line clusters as compared with classic slip line theory.

As shown in Figure 2(a), the entire range of potential slope failures is divided horizontally and vertically into discrete points for searching slip line paths. The initial horizontal spacing b and vertical spacing d depend on requirements and limitations of computation accuracy and capacity. The spacing can be further divided using the bisection method, if needed. For the case shown in Figure 2(b), the horizontal spacing for the search has to be reduced.

From the calculated direction field of slip lines, the slip line at any point of the exit section can be easily traced (see Figure 3) and the slip line field passing through the exit section can be obtained.

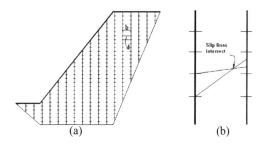

Figure 2. Search for slip line field. (a) Discretization of the slope; (b) Case showing spacing needs to be reduced for searching.

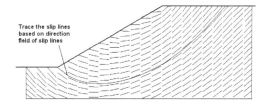

Figure 3. Slip line traced from the slip line direction field.

4 DETERMINATION OF CRITICAL SLIP LINE

On the basis of slip line theory, two clusters of slip lines (i.e. the α and β clusters) can be obtained from FE analysis, as shown in Figure 4. The clusters comprising the critical slip line can be easily identified based on engineering experience. For the case shown in Figure 4, the α-slip line field has the actual slip lines. The critical slip line is that corresponding to the minimum FS, which can be calculated by the following equation:

$$FS = \frac{\sum_n \int_{S_i} (\sigma \tan \phi + c) ds}{\sum_n \int_{S_i} \tau ds} \quad (11)$$

where n is the number of sections searched for the slip lines, and σ and τ are normal and shear stresses on the slip line, respectively.

5 CASE STUDIES FOR VERIFICATION

A program for identifying the critical slip line has been develop, which is used in association with FE results obtained from the commercial software ANSYS. The effectiveness of the proposed method was verified by the following case studies.

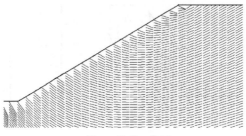

(1) α-slip line field

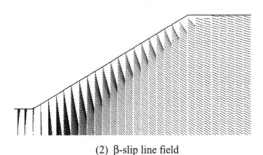

(2) β-slip line field

Figure 4. α- and β-slip line fields.

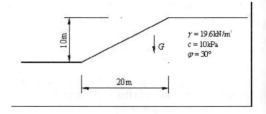

Figure 5. Slope geometry and material properties for case study 1.

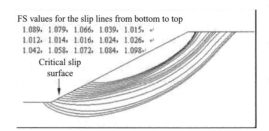

Figure 6. Slip line field for case study 1.

5.1 Case study 1—homogeneous slope

Figure 5 shows the soil properties and geometry of a homogeneous slope. This example has been previously studied by Zhang et al. (2002) using FEM and by Wang and Yin (2002) using limit analysis and limit equilibrium methods. The results of these former studies are compared with the results from this study.

Assuming the dilation angle of the soil is $\psi = 0$ and using the non-associate flow rule and strength reduction method, the stress distributions and slip surfaces in the slope under critical state, as determined using ANSYS, are shown in Figure 6. The FS calculated using FEM (i.e., the coefficient of strength reduction under limit state) was 1.98, which is the same value calculated by Zhang et al. (2002). Meanwhile, the FS calculated by Wang and Yin (2002) was 1.97 using limit analysis and 2.03 using Morgenstern-Price method based on limit equilibrium analysis. This agreement indicates that the calculation using finite element method in this study is correct, since the results are close enough with those obtained using other methods by other researchers.

The FS for each slip surface was calculated using Equation (11) and is shown in Figure 6. The strength parameters are based on strength reduction. The figure shows that the FS initially decreases and then increases as the slip surfaces go up from the bottom to the top. The minimum FS is 1.012 with the slip surface located at the middle, which is the critical slip surface. The minimum FS is close to unity, which indicates the effectiveness of searching for the critical slip surface within the slip line clusters.

Figure 7 shows the relationship between the distribution of plastic zones and the distribution of slip lines. The distributions of slip lines and plastic zones are very similar. The critical slip surface (the 6th from the bottom) just passes through the plastic zone of the slope, which indicates the location of the slip surface is reasonable. Also, the four slip surfaces (5th to 8th) with FS values close to unity are all located in the plastic zone, which indicates that the slope can fail along any of the slip surfaces. This suggests that the slip occurs in a zone rather than along a single surface. This zone is referred to as the shear zone and has been documented with many field observations.

5.2 Slope with a weak layer

Figure 8 shows a slope with a 2 m thick weak layer below the toe of the slope. The geometry of the slope and the soil properties are shown in the Figure 8. The FS value was calculated to be 1.24 using the strength reduction method assuming the materials conform to the non-associated flow rule.

The slip surfaces calculated from limit analysis and FEM are shown in Figure 9. The FS values were calculated from Equation (11). The slip surface with the

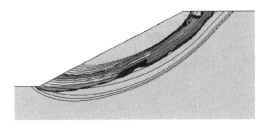

Figure 7. Relationship of the locations for the plastic zone and slip lines for case study 1.

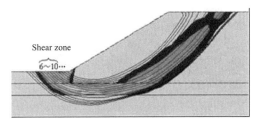

Figure 10. Relationship of the locations for the plastic zone and slip lines for case study 2.

minimum FS value (the 6th from left to right) of 1.029 is the critical slip surface. The FS is close to unity indicating the effectiveness of the search method for the slip surfaces. It can be seen from the figure that, because of the existence of the weak layer, the shape of the slip surface is more like a logarithmic spiral curve rather than a circular curve which is commonly observed or a homogeneous slope.

The relative locations of the plastic zones and the slip surfaces at limit state condition are shown in Figure 10. The shapes of the slip surfaces match the distribution of the plastic zones well. The critical slip surface and those slip surfaces with FS close to unity (6th to 10th form the left to the right) are all located in the plastic zone, which forms the shear zone, as mentioned previously.

Another case example of slope with a weak layer was also studied using both associated flow rule and non-associated flow rule (Wu 2004). The critical slip surfaces calculated are all in the weak layers with the FS close to unity. The results are not presented here due to the length limit of the paper.

6 CONCLUSIONS

This paper extends the classic slip line theory and presents a method to determine the critical slip surface based on stress distribution using FEM. The method determines the slip surfaces using the relationship of the slip line and principal stress tracking line to calculate the FS for each slip surface. The slip surface with the minimum FS is the critical slip surface. Both elastic and plastic zones of a slope are considered, and different flow rules were also adopted. The method was proved to be effective through case examples for both homogeneous slope and a slope with a weak layer.

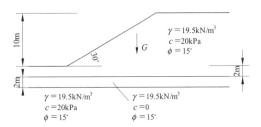

Figure 8. Slope geometry and material properties for case study 2.

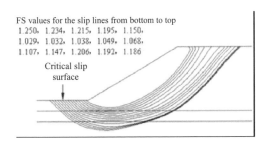

Figure 9. Slip line field for case study 2.

REFERENCES

Davis, E.H. 1968. Theories of plasticity and the failure of soil masses. In Lee I. K. eds., *Soil Mechanics*, Butterworth, London, p. 341–380.

Gong, X.N. 1990. *Soil Plastic Mechanics*. Press of Zhejiang University, Hangzhou, China (in Chinese).

Gong, X.N. Shi, M.D. 2000. Numerical solutions for slip line field, in *Computer Analysis for Geomechanics*. China Construction Press. p. 193–214 (in Chinese).

Kabilamany, K. Ishihara, K. 1990. Stress dilatancy and hardening laws for rigid granular model of sand. *Journal of Soil Dynamics and Earthquake Engineering*, 9(2), 66–77.

Wang, Y.J., Yin, J.H. 2002. Slope stability analysis using a method with associated and non-associated flow rules. *Proceedings of International Conference on Soil Mechanics and Geotechnical Engineering*.

Wu, C.Q. 2004. *Nonlinear Finite Element Method in Soil Stability Problems*. Ph.D. dissertation, Wuhan University (in Chinese).

Zhang, G.X., Liu, X.H., Wei, W. 2003. Two-dimensional slip surface of slope and stability analysis using elasto-plastic FEM. *Chinese Railway Journal*, 25(2), 79–83 (in Chinese).

Zhang, L.Y., Shi, W. M., Zhen, Y. R. 2002. Slope stability analysis using FEM under plane strain conditions. *Chinese Geotechnical Engineering Journal*, 24(4), 487–490 (in Chinese).

Zhen, H. 2000. *Several Types of Non-linear Problems in Geomechanics*. Ph.D. dissertation, China Academy of Science (in Chinese).

Effect of drainage facilities using 3D seepage flow analysis reflecting hydro-geological structure with aspect cracks in a landslide – Example of analysis in OODAIRA Landslide area

Masao Yamada
Japan Conservation Engineers Co., Ltd., Japan

Keizo Ugai
Gunma University, Japan

ABSTRACT: Stratum water and crack water through cracks of bedding planes and of joint surfaces give a great influence on the amount of displacement of moving mass in a landslide area. In this paper, at first, we examined the three-dimensional modeling of seepage flow analysis reflecting aspect cracks and then calculated water pressure distribution specifically by melting snow water in order to grasp a change of water pressure distribution of before-and-after construction of drainage facilities of well and drainage pipe. On that occasion, we made accurate modeling structure of various cracks and drainage facilities of well and drainage pipe. Using the results derived by the seepage flow analysis, we carried out a three-dimensional stability analysis. Based on the result of the stability analysis, we evaluated the stability of the landslide slope and the effect of drainage facilities.

In OODAIRA Landslide area where landslides were reactivated by melting snow, by applying the above mentioned analysis, we were able to obtain the acceptable results that matches field situation between fluctuation of water pressure and displacement of moving mass. We were then able to simulate effectiveness of drainage facilities.

1 INTRODUCTION

Groundwater consists of stratum water and crack water. The hydro-geological structure where groundwater flows in crack zones and stratums affects landslide displacement.

For the purpose of examining how the flow of the ground water vary along the direction and continuity of element in the hydro-geological structure, we carried out a three-dimensional seepage flow analysis by setting the slide surface and joint with crack as aspect crack, and collecting boring pipe and drainage pipe as line crack.

The characteristics of our three-dimensional seepage flow analysis are as follows.

At first, we examined a three-dimensional modeling of seepage flow analysis reflecting not only aspect cracks of the slide surface and joint with crack but also high-permeability zone.

We made modeling of the slide surface and joint with crack as a joint element and high-permeability zone as a series of solid elements with high-permeability. Also, we made modeling of well facility maintaining its form and size as much as possible and of collecting boring pipe and drainage pipe as line crack element.

It is difficult to set up the coefficient of permeability of ground layers and cracks in order for three-dimensional model with cracks to match observed water level. So at first, we set the coefficient of ground permeability by using both two-dimensional steady analysis and two-dimensional unsteady analysis that is a sensitivity analysis for melting snow water. Secondly, we set the coefficient of permeability of the slide surface, joint with crack and high-permeability zone by using three-dimensional unsteady analysis.

In this paper, in OODAIRA Landslide area where landslides were reactivated by melting snow, we carried out a three-dimensional seepage flow analysis reflecting the hydro-geological structure with aspects cracks. We then examined not only the validity of a three-dimensional seepage flow analysis model by comparing investigation between analysis value and observed value and but also examined effectiveness of before-and-after construction of drainage facilities in melting snow period.

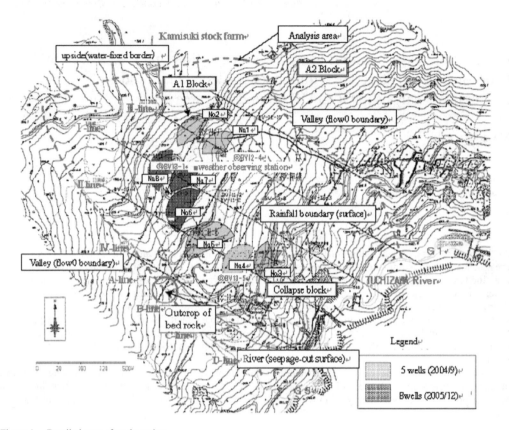

Figure 1. Detailed map of study region.

2 OUTLINE OF OODAIRA LANDSLIDE AREA

2.1 Outline of study area

OODAIRA Landslide area is located in 8 kilometers away from OTARI village in NAGANO prefecture in NNW direction. Its geometry is a large horseshoe geometry that the northern end is flat in KAMISUKI stock farm and TUCHIZAWA River runs along the southern end.

A1 block where the landslide movement is active exists in the designated landslide prevention area (see Figure 1).

From middle to upper-side of A1 block is a gentle slope where slope angle is 20–30° with patches of flat terrain in some places. In the lower-side, the steep escarpment with slope angle of 40–50° is seen toward TUCHIZAWA River.

The steep escarpment forms a collapse area. This collapse caused by a heavy rainfall occured in the lower part of the slope of A1 block in 1995 and the collapse area was further expanded by melting snow in 1999.

2.2 Geological structure

Geological structure in the landslide area consists of alternate strata of sandstone and mudstone of KURUMA group, and pyroclastic materials are found in the raising place such as the crest upside and KAMISUKI stock farm.

Alternate strata of sandstone and mudstone is relatively hard in the stream bed of TUCHIZAWA River and is crushed considerably on the slope of both sides of the river, particularly interbedded coal-seam group is in rock-fragmented state and weathered.

The strike of bedding plane which were observed in the scarp zone and the observation well runs in about EW direction and the angle of bedding plane is 20–40° and agrees with the stratum angle seen on the slope of TUCHIZAWA River.

Looking at the failed slope in the lower side of A1 block, there is a layer interface between sandstone-mudstone that is in fragmented rock state and contains relatively fresh sandstone-mudstone although has multiple crack. The height of layer interface comes gradually down from north to south.

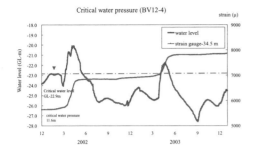

Figure 2. Critical water pressure (BV12–4).

The strike and angle of bedding plane of the layer interface agrees with those of KURUMA group (E-W30°–40° S). That is to say, it is considered that the landslide has dip slope structure regulated by bedding plane of bed rock from left-side wall to end site. This is reinforced by boring survey and some boring observation. Existing of regulatory aspect is showed by reading of aerial photograph and homogenious slope of N-S32° E in right-side wall of the landslide runs straight.

2.3 Subsurface hydrology and activity of the landslide

An example of relationship between landslide and ground water flow is shown in Figure 2 (BV12–4).

From the figure, it is clear that the strain at GL-22.9 m on the strain-gauge change significantly since 6 March 2002.

In melting snow period many strain gauges indicated noticeable accumulation of strain. Observation by GPS is performed since 2003 and the landslide movement was recorded in melting snow period. The variation volume of the strain in melting snow period tends to calm down yearly by the increase in the number of well facilities. The direction of the landslide movement agrees with the direction of the longitudinal field line.

From the past observation, rainfall in non-snow period does not exceed critical water level. Only in melting snow period the landslide movement becomes active.

Namely, melting snow affects the increase of water pressure rather than rainfall.

3 THREE-DIMENSIONAL FEM SEEPAGE FLOW ANALYSIS

3.1 Modeling of three-dimensional hydro-geological structure

From the result of past survey, supply resources of the groundwater include seepage from scarp zone, right-side wall (joint with crack) and its outer high-permeability zone, inflow from upper slope, infiltration from ground surface, and so on. And it is known that the left-side wall and the outer ground have capped rock structure and bedding plane is inclined toward the right-side wall (see Figure 3).

The landslide area has underground hydrology characteristics where melting snow and rainfall produces groundwater with pressure through the slide surface. We made modeling of hydro-geological structure by connecting levels of stratal layers at bore holes sites.

Details of a method for making stratal structure model from boring survey data is reported in YAMADA (2006).

When we carry out FEM seepage flow analysis, it is necessary that the slide surface, joint with crack, drainage facilities (well, drainage pipe) are expressed by appropriate elements.

During our analysis, we kept in mind to make modeling aspect cracks and well facilities to maintain exact form and size as much as possible.

3.2 Setting of the slide surface and joint with crack

The slide surface is a continuous plane of main displacement shear planes and a zone with constant thickness (SHIN 1995).

We will treat the slide surface as aspect crack physically where ground water flows mostly along plane and in FEM analysis, the slide surface is expressed as joint element with constant thickness. Joint element with thickness is shown in Figure 4. Joint element with thickness has only value of thickness as a parameter (although in Figure 4 nodal point seems apart, but in fact nodal point is overlapped).

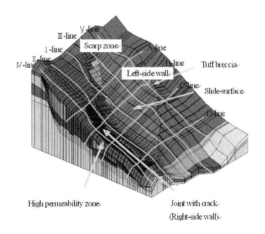

Figure 3. Analytical model.

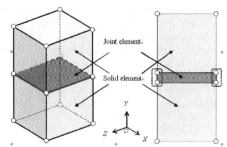

Figure 4. Joint element with thickness.

Table 1. Coefficient of permeability of geologic layer and cracks.

Material name	Coefficient of permeability (m/s)	Coefficient of permeability (m/day)	
Colluvial deposit	5.00E-04	4.32E+01	
Colluvial deposit	1.00E-04	8.64E+00	Setting by 2D back-analysis
Sandstone I	1.00E-05	8.64E-01	
Sandstone II	5.00E-06	4.32E-01	
Sandstone III	1.00E-06	8.64E-02	
Shale I	5.00E-07	4.32E-02	
Shale II	1.00E-07	8.64E-03	
Slip surface	2.00E-03	1.73E+02	Additional setting in 3D analysis
Predominant joint	2.00E-04	1.73E+01	
High permeability	1.00E-04	8.64E+00	

4 SETTING OF THE COEFFICIENT OF GROUND PERMEABILITY AND CRACK ELEMENT

4.1 Setting of ground the coefficient of permeability

Boundary conditions of analysis area were set as follows:

1. Upper side as fixed boundary condition that keeps constant water pressure,
2. River as seep boundary aspect condition,
3. Valley as boundary condition where the flow volume is 0.

Because parameter used for seepage flow analysis is not test value, we gave that unsaturated parameter, specific storage coefficient, saturated volume water content and minimum water holding capacity are constant and specific storage coefficient $= 1 \times 10^{-0.4}$, saturated volume water content $= 0.4$, minimum water holding capacity $= 0$ in order to improve computational performance.

Coefficient of permeability of ground layer is shown as Table 1.

4.2 Setting of crack element

Thickness and coefficient of permeability of the slide surface were respectively given with 15 cm, 2×10^{-3} m/s for joint element. Also thickness joint element with crack is decided as about 5 cm and its coefficient of permeability were given as 2×10^{-4} m/s because it is considered its coefficient of permeability is smaller than the slide surface.

Because the coefficient of permeability in high-permeability zone is small than the coefficient of permeability of joint with crack (2×10^{-4} m/s) and is larger than the coefficient of permeability of the ground ($5 \times 10^{-6} - 5 \times 10^{-4}$ m/s), hence we gave the coefficient of permeability of high- permeability zone as 1×10^{-4} m/s.

5 TREATMENT OF DRAINAGE FACILITIES IN 3D FEM ANALYSIS

5.1 Setting of well

We made modeling of 8 wells (No.1–No.8), drainage pipes in order to reproduce drainage facilities.

Although the well is essentially in circular form, modeling is very simplified if the well is modeled as square section, because solid element of the ground is square form.

As a result of comparing investigation about circular section and square section of equal areas, we were able to obtain same results in case both sections. Therefore we made modeling of the well as square section.

5.2 Setting of collecting boring pipe and drainage pipe

We made modeling of collecting boring pipe and drainage pipe as a line crack model. A line crack model has permeability only along crack direction and there is no outflow from one-dimensional crack element to surrounding ground element.

In the landslide area, volume of melting snow is considerable and the distribution of melting snow volume is not uniform and because collected water volume from collecting boring pipe is different for each setting direction. Therefore we gave the collecting boring pipe coefficient of permeability that is appropriate to observed collecting volume of well.

Also because the gradient of drainage pipe that connects wells and discharges groundwater to surface is low, we modeled drainage pipe as line crack that section areas are different as collecting boring pipe.

6 RESULT OF 3D SEEPAGE FLOW ANALYSIS AND ITS VALIDATION

6.1 3D steady seepage flow analysis of hydro-geological structure model

We carried out a three-dimensional steady seepage flow analysis for the following cases.

1. In case of setting the slide surface and joint with crack,
2. In case of setting high-permeability zone with joint,
3. In case of setting all aspect cracks.

Decreasing values of pore water pressure in II-line are largest where the slide surface and joint plane are crossing. The elevation of this part is lowest among all lines and the gradient of this part is lowest among all lines.

As a result of the survey, we set high-permeability zone in between A-line and B-line as an upper-site and D-line as lower-site. But effective area of high-permeability zone is seen in lower side of the middle point of A-line and B-line.

This is the inflow point of groundwater thorough rock in upper-side of the slope and surface water through crack. In case of field model setting all aspect cracks decrease values of pore water pressure is largest in between II-line and IV-line.

Because slide surface becomes deeper toward right-side and right-side is made of hard bed rock, hydro-geological structure that groundwater collects in between II-line and IV-line reflects field hydro-geological regulated condition.

6.2 3D unsteady seepage flow analysis in before-and-after construction of drainage facilities

We examined a three-dimensional seepage flow analysis before-and-after construction of drainage facilities in melting snow period.

5 wells were completed in September 2004 and additional 3 wells were completed in December 2005 (8 wells in total).

Both analysis values in BV13-1 and BV13-5 mostly match to the chronological water level. Also the decrease of pore water pressure after construction of the well in BV13-1 is larger than that of BV13-5.

It can be said that analysis result can explain variation of chronological water level against melting snow volume, considering that melting snow volume is not uniform in reality and modeling of geological structure in the field has restriction while observation data can contain errors to some extent.

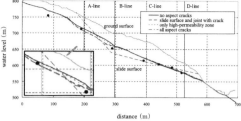
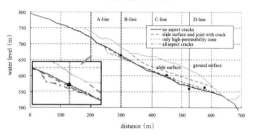

Figure 5. Results of three-dimensional steady analysis.

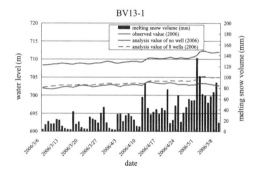
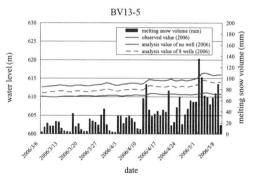

Figure 6. Effect of well facilities on amount of melted snow during the melting season of 2006.

7 EVALUATION OF SLOPE SAFETY USING 3D STABILITY ANALYSIS

7.1 3D analysis method

UGAI (1987) proposed a three-dimensional stability analysis by the simplified Janbu method applicable to slope that has arbitrary terrain and slide surface profile. We evaluated slope safety by calculating three-dimensional safety factor using the three-dimensional Janbu method for chronological water pressure that is derived from a three-dimensional seepage flow analysis.

When performing a three-dimensional stability analysis it is necessary to use the profile of the surface and slide surface, three-dimensional mesh data of water pressure and shear strength of the soil. Water pressure plane can be provided by an isopleth line map of water pressure produced from water pressure data of the element derived by FEM analysis. We made three-dimensional mesh data of ground surface, slide surface and water pressure plane by using GIS software. Here, mesh interspace is 5 m reflecting microtopography of the landslide block.

7.2 Shear strength of slide surface

Angle of shear resistance is derived as $\varphi' = 8.53°$ from the result of slide surface shear test. The landslide area has a peculiar geological structure that the slide surface is regulated by and inclined toward the right-side wall. And so we treat the resistance part of the slide surface as cohesion c'.

The landslide has been reactivated by accumulated strain due to rapid increase of melting snow volume since 6 March 2002. As a result of calculating c' where safety factor of the landslide slope in 6 March 2002 is defined as 1.0, c' is derived as c'=18.11 kPa.

7.3 Evaluation of drainage facilities

In the previous section we calculated safety factor of the slope in the three different periods. When time frame differs, melting snow volume also varies. So we calculated safety factor of the slope by using the same melting snow volume in order to evaluate the levels of construction effects of the drainage facilities.

Figure 7 shows the result of calculating safety factor of the slope for case A, case B and case C by using the same melting snow volume from 1 March 2006 to 9 May 2006.

Safety factor of the slope with 5 wells increases by 0.01–0.05 compared to that with no wells. Safety factor of the slope with 8 wells further increases by 0.004–0.005.

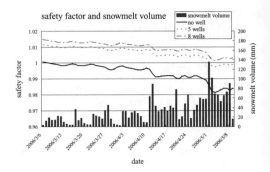

Figure 7. Effect of drainage facilities (calculated using data from the melting season of 2006).

8 CONCLUSION

We carried out a three-dimensional seepage flow analysis considering cracks for OODAIRA Landslide reactivated during melting snow period and verified availability of a three-dimensional seepage flow analysis reflecting hydro-geological structure with aspect cracks.

We presented it was possible to reproduce the rising height of pore water pressure in three-dimensional steady and unsteady analyses. Especially we were able to derive the result that matches to observation value by using a model reflecting real hydro-geological structure model. We carried out a three-dimensional stability analysis based on positional information of water pressure plane derived from a three-dimensional seepage flow analysis and were able to derive the result suitable to variation of strain value.

However, we could not absolutely confirm analysis value to observed value, though we could reproduce water pressure by the analyses. The reasons can be explained as follows.

1. There are in no small part observation bore hole of water level that is unable correspond to theoretical variation of melting snow volume. For it is inherently unreasonable to use melting snow of only one place when chronological data of melting snow is different each place.
2. We can hardly reproduce water pressure distribution by correcting the coefficient of permeability and hydro-geological section in two-dimensional analysis. However, there is a limit to correct the coefficient of permeability because water flow toward cross-section occurs in a three-dimensional analysis.

In the future, if we can calculate safety factor of the slope using water pressure observed by automatic observation system, it will be possible to explain construction effect of drainage facilities in real time by using easy-to-understand evaluation term of safety factor.

REFERENCES

UGAI, K. (1987). Three-dimensional slope stability analysis by simplified Janbu method, landslide, Vol. 24, No. 3, pp. 8–14.

SHIN, J. (1995). Landslide engineering – new topics —, pp. 3–9, sankaido.

YAMADA, M., YAMAZAKI, T., YAMASAKI. (2000). Relationship between groundwater flow and landslides, landslide, Vol. 36, No. 4, pp. 22–31.

YAMADA, M. SHINGO, S. (2006). Construction of landslide GIS, landslide, Vol. 42, No. 4, pp. 51–62.

Back analysis of soil parameters: A case study on monitored displacement of foundation pits

B. Yan, X.T. Peng & X.S. Xu
School of Civil Engineering & Architecture, University of Ji'nan, Ji'nan, Shangdong, China

ABSTRACT: The elastic-reaction method has been widely used in practical engineering for its convenience and validity. The solutions calculated with the elastic-reaction method are effective on condition that the m-values and proportional coefficients of lateral subgrade reaction are consonant with the field soils. The befitting m-values are impossibly obtained in laboratory soil tests. The paper presented an effective back-analysis model to obtain befitting m-values. Since practical m-values are found, the lateral displacements of foundation pits are well predicted.

1 INTRODUCTION

With high buildings rising, retaining and protecting of foundation excavation have been an important phase in constructions in crowded cities. Piles in rows, diaphragms, soil nailing walls, sail anchors and cement-soil walls and so on are often used in deep excavation engineering (JGJ120-99) (GB5007-2002). All academic solutions, including bearing forces or displacements of sides of foundation pits, should be accurate to ensure the security of engineering and surroundings around foundation pits. One of the major problems facing us is that whether soil parameters accord well with the fields or not (Arai et al 1983, 1986).

The paper presents a sort of finite elements method (FEM) adopted in calculating internal forces and lateral displacements for piles in row or diaphragms (Milligan, 1983). The method has been widely used in practical engineering for its convenience and validity. The solutions calculated with the FEM are effective on condition that the m-values, proportional coefficients of lateral subgrade reaction, are consonant with the field soils. The befitting m-values are impossibly obtained in laboratory soil tests, but in fields. Field data, such as lateral displacements or internal forces of bearing piles, are utilized to estimate the m-value to comport with practical fields (Arai et al 1984, 1987). That is back-analysis, which is a valuable method to get befitting parameters, m-values.

2 THE MODEL OF BACK-ANALYSIS

2.1 The objective function of back-analysis

Based on field data, the objective function is expressed as

$$J = \min \sum_{i=1}^{N} \left(\frac{s_i}{s_i^t} - 1 \right)^2 \tag{1}$$

where N = total of monitoring points; s_i = calculated lateral displacement at ith monitoring point; s_i^t = monitored lateral displacement at ith monitoring point.

Figure 1 shows the analysis of search soil parameters, m-values.

2.2 The model for the elastic-reaction method

Based on the Technical Specification for Retaining and Protecting Building Foundation Excavation of China (JGJ120-99), Figure 2 shows the horizontal loads, active earth press, exerted on the bearing piles. The active earth press above bottom of the pit can be estimated with Rankin or Coulomb's Earth Press Theory.

For crushed stone soil and sandy soil above the ground water surface, the standard active earth press is formulated as,

$$e_{ajk} = \sigma_{ajk} k_{ai} - 2c_{ik}\sqrt{k_{ai}} \tag{2}$$

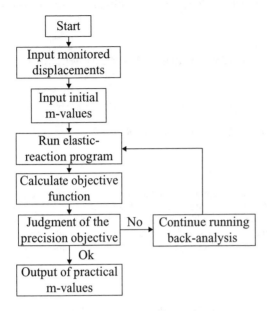

Figure 1. Process of back-analysis.

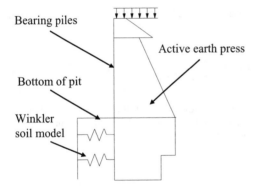

Figure 2. The model for the elastic-reaction method.

under the ground water surface, it is formulated as,

$$e_{ajk} = \sigma_{ajk}k_{ai} - 2c_{ik}\sqrt{k_{ai}}$$
$$+ [(z_j - h_{wa}) - (m_j - h_{wa})\eta_{wa}k_{ai}]\gamma_w \quad (3)$$

for silt and clay, the same formula in (2), where σ_{ajk} = vertical standard stress outside of foundation pits. It can be shown below,

$$\sigma_{ajk} = \sigma_{rk} + \sigma_{0k} + \sigma_{1k} \quad (4)$$

where

$$\sigma_{rk} = r_{mj}z_j$$
$$\sigma_{0k} = q_0 \quad (5)$$
$$\sigma_{1k} = q_1\frac{b_0}{b_0 + 2b_1}$$

All the functions mentioned above may be well known by consulting JGJ120-99.

The soils inside of pits simulate the Winkler soil model for bearing piles. The model is composed of lumped horizontal springs. It's assumed that the modulus of the springs rise at a proportional coefficient, value, along the piles from the top down.

2.3 Back-analysis of m-value

The back-analysis is actually a mathematical optimization problem. The simplex method is used in the paper. Based on Eq. (2), the monitored lateral displacement, s_i^t, should be all-sided and accurate along the piles. Before back analyzing practical m-values, initial values will be estimated by one's experience. The calculated lateral displacements, s_i, may be estimated with the elastic-reaction method program. Obviously, s_i is often unequal to s_i^t. After running the program to cycle time again, that s_i deviates s_i^t is getting smaller and smaller. When the objective value, J, is much smaller than a satisfying value (ε) at that moment, the m-values back-analyzed are well consonant with practical soils.

3 APPLICATION IN PRACTICAL ENGINEERING

3.1 Outline of the project

Figure 3 shows a shallower foundation pit, with 90.0 m in length, 54.0 m in width and 5.0 m in depth, including

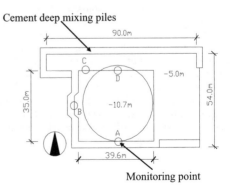

Figure 3. Plane of the foundation pit.

a deeper pit, with 39.6 m in length, 35.0 m in width and 10.7 m in depth. The water plane is 1.5 m below the ground level. The artesian aquifer with 14 m hydraulic head is found at the —18.0 m elevation.

The bracing system is composed of piles in row, top beams on the top of bearing piles and internal struts. The piles in row are formed with bored cast-in-place piles, with its diameter 1000 mm. The piles tip arrive −24.5 m elevation. To prevent ground water from seeping into the deep foundation pit, water tight screens, with a diameter of 600–700 mm, are set near the bearing piles outside. The shallower pit is consolidated with cement deep mixing piles, with 4.0 m in width, 15 m in depth, which can also act as water tight screens.

Figure 4 shows a geological profile composed of five distinct soil layers. Table 1 shows their corresponding index properties.

In the procedure of construction, the ground water and artesian aquifer are availably controlled at 1.0 m under the bottom of the pit.

3.2 Back-analysis of m-value in practical engineering

The m-value used in the elastic reaction method is traditionally estimated one value, that is to say, several soil layers are treated as one same layer. The simple soil parameters are feasible on condition that the behaviors of soil layers are approximately the same. Only approximate soil layers can be united so that the back-analyzed results are minimally influenced. Compare "one united soil layer" with "approximately united soil layers" as follows.

(1) One united soil layer

Figure 3 gives the positions of monitoring points, such as A, B, C, D. Lateral displacements at monitoring points, s_i^t which can be obtained by deflection inclinometers, are practical displacements when the pit is excavated at the level of −10.7 m. Introducing field data into Eq. (1), the practical m-values are searched by back-analysis program.

Table 2 shows the solutions back-analyzed. When the passive earth press areas are supposed to be the same soils, the lateral displacements back-analyzed have relation to the total monitoring points.

(2) Approximate soil layers united

In Figure 4, the soil layer ① and ② belong to softer soil, uniting both of them. The Soil layer ③, ④ and ⑤ belong to stiffer soil, uniting three of them. Now, five layers become two layers. Table 3 shows the solutions back-analyzed in the case.

Conclusions back-analyzed are as follows:
The monitoring points should be set as many as possible. Since behaviors of soil layers in the field have large distinctness, unite five layers into two layers but one. The top layer is softer soil and the bottom one is stiffer layer. Obviously, the top layer has a smaller m-value and the bottom layer has a larger one. By means of the m-value back-analyzed, the lateral displacements calculated with the elastic-reaction method are well consonant with displacements monitored.

Table 2. m-value back-calculated and comparing lateral displacements back-calculated with lateral displacements monitored for one united soil layer.

Monitoring points	m-value calculated (kN/m⁴)	Displacements calculated (mm)	Displacements monitored (mm)
A	3300	26.42	22.92
B	2170	29.20	27.20
C	3200	18.11	21.11
D	2600	24.44	27.94

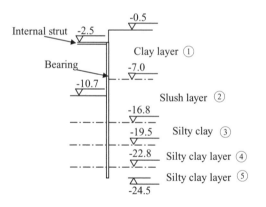

Figure 4. Geological profile.

Table 1. Index properties for soil layers.

Soil layers	Height (m)	ω %	I_p	I_L	γ	φ	c
Clay layer	6.5	35.2	17.9	0.81	18.7	10.6	24.3
Slush layer	9.8	43.4	20.2	1.05	17.9	9.5	21.3
Silty clay	2.7	27.7	12.0	0.89	19.4	15.2	24.8
Silty clay	3.3	27.4	14.5	0.60	19.7	16.3	50.6
Silty clay	11.8	22.5	11.6	0.52	20.8	21.2	56.0

Table 3. The m-values back-calculated and comparing lateral displacements back-calculated with displacements monitored for approximate soil layers united.

Monitoring points	m-value calculated (kN/m^4) Above	Under	Displacements calculated (mm)	Displacements monitored (mm)
A	1020	4500	25.01	22.92
B	1300	3850	27.60	27.20
C	1040	4160	19.61	21.11
D	880	4660	27.94	27.94

Table 4. Comparing lateral displacements predicted with lateral displacements monitored.

Monitoring points	Displacements predicted (mm)	Displacements monitored (mm)
A	18.92	22.92
D	25.94	27.94

4 FORECASTING THE LATERAL DISPLACEMENT AT THE NEXT STAGE

The lateral displacements change with the schedule of construction. Those future lateral displacements known beforehand can effectively direct the next stage of construction. Based on displacements monitored at monitoring points A and D at a certain depth of the pit, befitting m-values has been reached. Then the lateral displacements of the pit at the design elevation will be calculated by means of the elastic-reaction method.

Table 4 shows that the predicted lateral displacements are almost identical to the monitored one.

5 CONCLUSIONS

The elastic-reaction method has been widely used in practical engineering for its convenience and validity. The solutions calculated with the method are effective on condition that the m-values, proportional coefficients of lateral subgrade reaction, are consonant with the field soils. The paper presents an effective back-analysis model. Only approximate soil layers can be united so that the results back-analyzed are influenced in a minor degree. Monitoring points should be set as many as possible to ensure reliable results. Since befitting m-values are found, the lateral displacements for foundation pits can be well predicted.

ACKNOWLEDGMENTS

The supports from Natural Science Foundation of Shandong Province (G0637) and University of Jinan (B0421) are appreciated.

REFERENCES

Arai, K., Ohta, H. & Yasui, T. 1983. Simple optimization techniques for everlasting deformation module from field observations. *Soils and Foundations*. 23 (3):107~113.

Arai, K., Ohta, H., Kojima, K. & Wakasugi, M. 1986. Application of back-analysis to several test embankments on soft clay deposits. *Soils and Foundations*. 26(2):60~72.

Arai, K., Ohta, H. & Kojima, K. 1987. Estimation of nonlinear constitutive parameters based on monitored movement of subsoil under consolidation. *Soils and Foundations*. 27(1):35~49.

Arai, K., Ohta, H. & Kojima, K. 1984. Estimation of soil parameters based on monitored movement of foundations. 24(4):95~108.

China Academy of Building Research. 1999. *Technical Specification for Retaining and Protection of Building Foundation Excavations*. Beijing: China Architecture & building press.

Milligan, G.W.E. 1983. Soil deformation near anchored sheet-pile walls. *Geotechnique*No.1: 44~55.

Ministry of Construction of the PRC. 2002 *Code for design of building foundation*. Beijing: China Architecture & building press.

3D finite element analysis on progressive failure of slope due to rainfall

G.L. Ye
Department of Civil Engineering, Shanghai Jiaotong University, China

F. Zhang
Department of Civil Engineering, Nagoya Institute of Technology, Japan

A. Yashima
Department of Civil Engineering, Gifu University, Japan

ABSTRACT: A soft rock slope in Tokai-Hokuriku Expressway of Japan failed due to a heavy rain. The three-stage failure procedure indicates that it was a typical progressive failure. In this study, by using a modified elastoplastic model with Matsuoka-Nakai failure criterion, which not only can describe the strain-hardening and strain-softening behavior of soft rock but also can take into consideration the influence of the intermediate principle stress, a 3D soil-water coupled finite element analysis is conducted to investigate the mechanics of the progressive failure. The 3D shape of the slope, the geological conditions and the initial ground water level are delicately considered during 3D modeling. The change of the ground water level during rainfall is simulated by increasing the water heads of elements from initial level to the ground surface. In the discussions of the calculation results, the characteristics of slope failure, such as the development of shear strain, the deformation of ground, the propagation of shear band and the failure zone are discussed in detail. It is found that 3D soil-water coupled analyses based on the modified elastoplastic model can simulate the progressive failure of a slope to an engineering acceptable accuracy.

1 INTRODUCTION

Slope failure induced by heavy rain is a very important but difficult problem for geotechnical engineers. Progressive failure, which is a typical time-dependent behavior, is often observed in slope failure. Clarifying the mechanical behaviors of progressive failure is indispensable for predicting when and where the slope failure will occur. From the viewpoint of soil mechanics, the soil skeleton-pore water coupled behavior and/or the inherent viscosity of soil skeleton of the soft rock should be considered when dealing with long-term stability of a soft-rock slope concerning rainfall. Accordingly, a soil-water coupled numerical analysis and/or an elasto-viscoplastic model is necessary for simulation.

In most cases, the failure pattern is strongly related with the 3D profile of the slope. However, numerical simulation on progressive failure by 3D FEM is seldom conducted due to the complexity of the numerical modeling such as the geometry and geological mesh modeling, determination of material parameters, and above all, the time consumption of calculation. Especially in the case considering the effect of the movement of underground water, careful ground survey is necessary to acquire a satisfactory accuracy of the analysis.

In this paper, a slope failure due to heavy rain is simulated by 3D soil-water coupled finite element analysis. The observed failure was a typical progressive failure and the heavy rainfall is the main reason for the failure. Zhang et al. (2001) conducted a 2D soil-water coupled finite element analysis on this slope, which is based on an elastoplastic model proposed by Oka and Adachi (1985).

The model proposed by Oka and Adachi (1985), however, has a Cam-Clay type failure criterion, which as had been pointed out that it could not predict correctly the strength of geomaterials under general loading condition except axisymmetrical condition. Zhang et al. (2003) proposed a modified model by introducing Matsuoka-Nakai failure criterion (SMP failure criterion) to that model. In current paper, based on the modified model, 3D soil-water coupled finite element analysis is conducted to investigate the progressive failure of the slope. The aim of the research is to try to establish an applicable way to assess the progressive failure of slope under 3D topographic condition.

2 CASE HISTORY OF SLOPE FAILURE

2.1 General description of the slope failure

The failed slope locates between the Mino Interchange and Minami Interchange of Tokai-Hokuriku Expressway in Japan. The failure happened on September 22, 1999, which was occasionally recorded by a journalist who was on another failure site near the slope. The video showed that it was a typical progressive failure. Figure 1 shows the sequence of the observed progressive failure, which is a three-stage failure, taking about 3 hours from the first stage to the final stage. The detailed failure process is described as following: About at 9:22 on Sept. 22, 1999, some cracks appeared in the concrete-block frame wall and a great deal of water gushed out of the drainage work. At 9:45, the first stage failure occurred, the failure area was 45 m wide and 50 m long. The second failure occurred at 12:20, the failure area expanded to 120 meters wide and 70 meters long. The final failure occurred at 13:10, continued from the second failure till the crest of the slope, the final failure area was 120 m wide and 125 m long, with fallen soil of almost 110,000 m^3 in volume.

The maximum hourly rainfall depth, the 24-hour rainfall depth and the 9-day rainfall depth before the failure amounted to 63 mm, 182 mm, and 619 mm, respectively. These rainfall depths were recorded at AMeDAS Mino observatory station of Japan Meteorological Agency (JMA), where is very near to the site of the slope failure. The failure happened 6 hours after the maximum hourly rainfall depth was reached. It is reported (Tamura and Matsuka, 1999) that excessive pore-water pressure in a ground of slope usually started to increase at the time 10 hours after a heavy

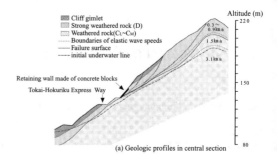

Figure 2. Geologic profile and its modeling of the central section plane of the slope.

rain reached peak value within 20~30 hours, depending on the permeability of the ground. Therefore it is reasonable to think that the increase of the pore-water pressure due to the heavy rain is a main reason for the failure. Since the failure happened six years after its completion, deterioration in the strength of the ground may be possible. In present analysis, however, only the effect of rainfall is considered.

2.2 Geologic conditions of the slope

As plot in Figure 2, the geologic information from field investigations of the slope is summarized as following:

1. Top layer of the ground is strongly weathered rock that is about 5 m in depth on summit and a maximum depth of 17 m in the middle-low part of the failed slope. This layer, classified as D-grade rock, is composed of weathered clay and fresh fragments of rock, with an elastic-wave velocity of 0.3~0.9 km/s. The failure surface formed within this layer. (2) The layer beneath the top layer is a weathered rock with a depth of 20 m~25 m and an elastic-wave velocity of 1.4~1.6 km/s, classified as CL~CM-grade rock.

3 NUMERICAL ANALYSIS CONDITIONS

In the formulation of soil-water coupled analysis, the finite element method is used for the spatial discretization of the equilibrium equation, while the finite difference method is used for the spatial discretization of the pore-water pressure in the continuity equation.

Figure 3 shows the finite element mesh adopted in the analysis of the slope. The size of the ground is 300 m in width, 310 m in length and 214 m in height. The numbers of the node and 8-node isoparametric element are 8190 and 7000 respectively.

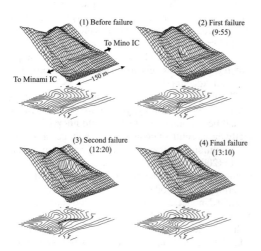

Figure 1. Observed progressive failure sequence of the slope.

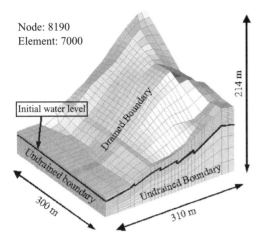

Figure 3. 3D mesh model.

3.1 Boundary conditions

For displacement, it is fixed at the bottom in both horizontal and vertical directions, only the normal direction of the side boundaries is fixed.

For drainage condition, the ground surface is permeable while the others are impermeable.

3.2 Initial conditions

Due to the difficulty to determine the initial stress field for a soft rock ground, the initial stress fields in both calculations are simply assumed to be a gravitational field. Initial total water head, is directly given by the value of underground water level measured from borehole tests in such a way that the underground water level of the elements within one column are the same, and that those elements above the underground water level are also saturated and possess the same value as those of the elements below the underground water level.

In the calculation, the first stage is to calculate the initial stress field; then a prescribed increment of total water head is applied in the second stage and then the calculations are continued to simulate the change of E.P.W.P with a time interval of 2 sec/step.

Figure 4 shows the central cross-section of 3D mesh. Because the initial mean effective stress near the ground surface is very small in the initial stress field calculation, some soil elements near the surface may have already failed at initial stress state, if the stress ratio at residual state is given in the same value disregard the magnitude of its confining stress. This is not coincident with the real situation. According to the laboratory tests (Adachi, T. and Ogawa, T., 1980), the stress ratio at residual state R_f ($R_f : (d\sigma_1/d\sigma_3)_f$ at residual state in a conventional triaxial compression test) is dependent on the confining stress. R_f takes a relatively large value at small confining stress. For this reason, the value of R_f of the strong weathered rock will be given in 3 different values under different initial confining stress in such a way that, the closer the place is to the surface, the larger the R_f will be. By this treatment, it is possible to overcome the discrepancy mentioned above.

3.3 Material parameters

Figure 5 shows the comparison of stress-strain-dilatancy relationships obtained from the theory under drained triaxial and plane-strain compressive/extensive conditions. It shows that the peak stress difference ($q = \sigma_1 - \sigma_3$) of plane-strain condition is larger than those of conventional triaxial compression condition and the volumetric strain of plane-strain condition is less than those of conventional compression condition. In the figure, the material parameters of the

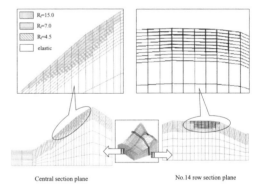

Figure 4. Central section and No.14 row section planes of the 3D mesh.

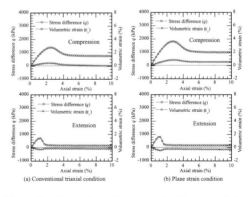

(a) Conventional triaxial condition (b) Plane strain condition

Figure 5. Comparison of stress-strain-dilatancy relationships under different loading conditions ($\sigma_3' = 200$ kPa).

Table 1. Material parameters of slope.

Parameter	Unit	Weathered rock ($C_L \sim C_M$)	Strong weathered Rock (D)
E	(kPa)	$3.0 * 10^6$	$1.0 * 10^6$
ν		0.25	0.33
γ	(kN/m³)	15.7 (γ')	25.6 (γ_{sat})
k	(m/sec)	$1.0 * 10^{-6}$	$1.0 * 10^{-6}$
G'		100.0	100.0
τ		0.050	0.050
b	(kPa)	2000	1000
σ_{mb}	(kPa)	20000	10000
α		1.00	1.00
D		−0.30	−0.30
R_f		+∞ (Elastic)	Upper 15.0 Middle 7.0 Bottom 4.5

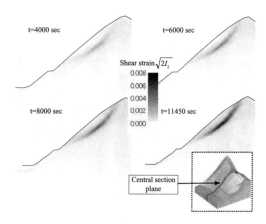

Figure 6. Development of shear strain in the slope with time (central section plane).

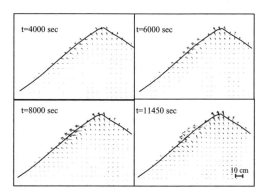

Figure 7. Development of displacement vector in the slope with time (central section plane).

rock are the same as those of strong weathered rock listed in Table 1 except R_f that takes a value of 5.0. The physical meanings and detailed description of the way to determine these parameters can be referred to with the reference (Zhang et al., 2003).

In the numerical calculation, the change of pore-water pressure due to a rainfall is simulated under the assumption that after a heavy rain, the total water head increases from the initial underground water level to ground surface. And the increment of the pressure water head in each element column is applied within six hours.

4 RESULTS AND DISCUSSION

4.1 Investigation on overall behavior of ground

Figure 6 shows the changes in the distribution of the shear strain ($\sqrt{2I_2}$, I_2: second invariant of deviatoric plastic strain tensor). It can be seen that he shear strains developed very slowly in the first 8000 seconds, after then, the shear strains turned to develop rapidly until total failure. A relatively large shear strain occurred at the middle of the slope. Then a shear zone, in which larger shear strain developed, propagated from local area to neighboring areas and finally formed a shear band. The shape of the shear band coincided with the boundary between the strongly weathered rock (D) and rather hard rock ($C_L \sim C_M$). It is found that if the maximum shear strain within the shear band reached a certain value, the failure abruptly happened along the shear band.

Figure 7 shows the change of displacement vectors of the ground with time. The ground moves upward at first 4000 seconds due to the floating force from the rising of the underground water, then it turned to move downhill along the shear band when the shear strain reached a certain value, as shown in Figure 6. Therefore, based on the direction of the displacement of the ground, it is possible to judge whether a slope is in the danger of failure.

Figure 8 shows the comparison of the predicted shear band and the observed failure surfaces. In reality, the slope failed in three stages, taking about 3 hours from the first stage to the final stage. Due to the limitation of finite element method that based on the continuum theory, however, the multi-stage failure cannot be reproduced in simulation. However, the calculated shear band coincides well with the observed final failure surface. Therefore, it can be said that present analysis can simulate the progressive failure of a cut slope to a reasonable accuracy. On the other hand, from the development of shear band, we can judge the degree of the progressive failure.

4.2 Investigation on the behavior of individual element

Figure 9 shows the behavior of element 3, where the formation of the shear band was ignited. Fluctuations were seen in the stress-strain relations and strain rate with time, that is, strain hardening occurs at first, and then softening occurs and followed by succeeding hardening again and fails at last. The reason of the fluctuation can be explained in following way: as strain softening develops in an element, dilatancy occurs while the supply of pore water cannot be achieved simultaneously, resulting in an increase of effective confining pressure in the element. On the other hand, strain softening may happen in some neighboring elements, resulting in some redistribution of stresses due to the lose of strengths, these redistributed stresses have to be resisted by the element. As the result, the element will experience strain hardening again. The repetition of above process results in the fluctuation. In the figures, stress path and stress history path represent the trails of stress state and stress history state in different times. The stress state may overpass the residual strength line in overconsolidated state while stress history state cannot overpass the residual strength line. If it reaches the line, it means that the soil reached residual failure state.

The stress path and the stress-history path of element 3 are also showed in Figure 9. It can be seen that the shear stress and the shear stress-history are decreasing throughout the whole process. The same tendency can be seen for the mean effective stress. When the stress history overpasses the residual strength line, the element fails.

Figure 10 shows the behavior of element 7, which locates outside of the shear band. Different from the element located on shear band, the element did not

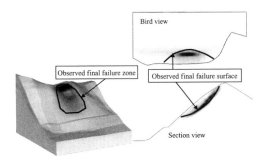

Figure 8. Comparison of the calculated shear zone and the observed failure zone.

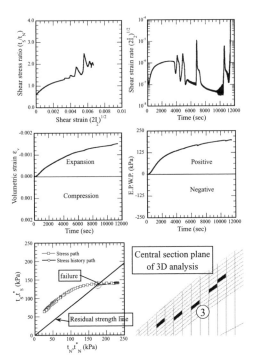

Figure 9. Behavior of element locate on shear band (in the central section plane).

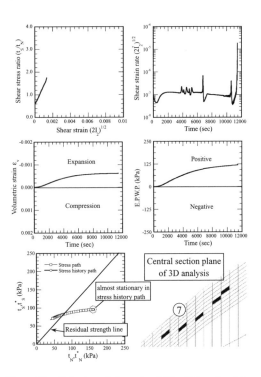

Figure 10. Behavior of element outside of shear band (in the central section plane).

show fluctuation in stress-strain relations. And the magnitudes of the fluctuation in shear strain rates are almost the same, which indicates that the soil-water interaction of the element outside the shear band is not so violent as those on the shear band. Needless to say, the predicted stress history paths show the same result that element did not fail.

Above discussion shows that the behaviors of elements inside and outside of shear band are different. It can be concluded that accurate prediction of stress-dilatancy relation will play a very important rule in describing the progressive failure of a slope related to soil-water coupled problem. This is why a modified elastoplastic model (Zhang et al., 2003) is adopted in the analysis.

Figure 11 shows the development of shear strain in the elements along the shear band of the slope in the middle plane of the failure zone. From the change of both the strains and strain rates, it can be seen that the accelerating stage first occurs at element 3, and then it propagates toward two sides.

Figure 12 shows the development of the shear strain in the elements along the shear band in an orthogonal vertical section. From the change of both the strains and strain rates, it can be seen that the accelerating stage first occurred at element H3 and H4, and then it propagated toward two sides, leftward in the sequence

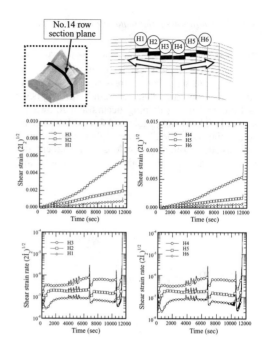

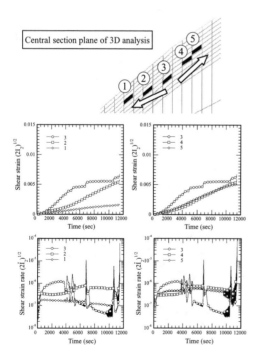

Figure 11. Propagation of failure in the central section plane.

Figure 12. Propagation of failure in horizontal plane.

of element H2 to H1 and rightward in the sequence of element H4 to H5. Therefore, a progressive failure in two orthogonal vertical planes, a typical 3D behavior, can be clearly simulated. From this figure, it can be seen that 3D calculation can give a more comprehensive description of the mechanical behavior of progressive failure than 2D calculation can do.

5 CONCLUSIONS

In this paper, based on an elastoplastic model considering the influence of intermediate stress and strain softening, 3D soil-water coupled finite element-finite difference analysis was conducted to simulate a large-scale slope failure due to heavy rain. From the analyses, the following conclusions can be given:

1. By adopting 3D soil-water coupled finite element methods based on a suitable elastoplastic model for soft rock, and properly 3D modeling, the progressive failure of a real slope due to heavy rainfall can be simulated to an engineering acceptable accuracy.
2. From the stress-strain relations and the time changes of strain rate in soil elements along the shear band, it is found that the formation and the

propagation of the shear band were directly caused by the strain-softening behavior of the ground. 3D analysis is able to describe the propagation in two orthogonal vertical planes, being much more realistic than 2D analysis. In other words, 3D analysis can give a more comprehensive description of the progressive failure of a slope.
3. An overall failure of a slope was ignited by the propagation of shear band within the ground of slope.
4. Accurate prediction of stress-dilatancy relation will play a very important rule in describing the progressive failure of a slope related to soil-water coupled problem.
5. Present analysis cannot simulate the process of multi-stage progressive failure observed in the real failure event, the predicted failure surface coincides only with the observed failure surface at final stage. This is due to the discrepancy that in continuum solid mechanics, medium is considered to be continuum while in reality, after its first-stage failure, the failed soils had already gone away from the slope.

REFERENCES

Adachi, T. & Ogawa, T. (1980): Mechanical properties and failure criterion of soft sedimentary rock, *Proceedings of JSCE*, 295, 51–62 (in Japanese).

Oka, F. & Adachi, T. (1985): A constitutive equation of geologic materials with memory, *Proc. 5th Int. Conf. on Numerical Method in Geomechanics*, Balkema, 1, 293–300.

Tamura, E. & Matsuka, S. (1999): Automatic measurement of pore water pressure in the hard-rock slope and the sliding weathered-rock slope-field survey in mountainous region in Shikoku Island, Japan. *Proc. of Int. Conf. on Slope Stability Engineering (IS-Shikoku), Matsuyama, Japan*, Balkema, 1, 135–140.

Zhang, F., Yashima, A., Sawada, K., Sumi, T., Adachi, T. & Oka, F. (2001): Numerical analysis of large-scale slope failure, *Computational Mechanics—New Frontiers for New Millennium (APCOM'01)*, Australia, Elsevier, 1, 527–532.

Zhang, F., Yashima, A., Ye, G.L., Adachi, T. & Oka, F. (2003): An elastoplastic strain-softening constitutive model for soft rock considering the influence of intermediate stress, *Soils and Foundations*, 43 (5), 107–117.

Block-group method for rock slope stability analysis

Zixin Zhang, Ying Xu & Hao Wu
Department of Geotechnical Engineering, School of Civil Engineering, Tongji University, Shanghai, China
Key Laboratory of Geotechnical & Underground Engineering, Ministry of Education, Tongji University, Shanghai, China

ABSTRACT: This paper provides an extension for the stereo-analytical method, called 'the block-group method'. The stereo-analytical method can consider both convex and concave blocks, but a group of blocks, if considered as a whole, might indeed be unstable and more dangerous compared with individual key blocks. It is revealed that the block group must satisfy the conditions: the block group contains at least one key block; neighboring finite blocks have common face with the key block or key block-group. Block-group types can be determined with the stereo-analytical block method. Then the kinetic energy law is implemented to analyze the movement behavior of each key block or key block-group. Finally, a rock slope project is demonstrated to prove that the block-group method yields more realistic results than the basic key block method.

1 INTRODUCTION

Rock mass is the total in-situ medium containing bedding planes, cracks, faults, joints, folds and other structural features. Rock slope, which is divided into many rock blocks by various discontinuities, is of static equilibrium state in nature stress field. But some rock blocks may slide along structural features or free falling, when rock slope is excavated and disturbed. Even, large-scale rock blocks gradually move. To a great extent, rock slope stability depends on mechanical characteristic, geometrical characteristic and distribute orderliness of discontinuities. Now, many analysis methods of discontinuous mechanics have been put forward and developed, such as limit equilibrium method, key-block method, numerical analysis method, and so on.

The key block method has been widely used in many projects for conducting quick analysis of rock mass media stability. The method is base on two hypothesizes: rock blocks are rigid and run through the whole study rock mass project. Implement of the key block method has been manifested into two main forms: the vector technique by Warburton (1981) and the graphical technique developed by Goodman & Shi (1985). Zhang & Kulatilake (2003) put forward a new stereo-analytical method, which is a combination of the stereo-graphic method and analytical method, having applicability to both convex and concave blocks. There are three main steps in the key block method. The first is to describe the block geometry and define if the blocks are finite or infinite. The second is to identify which finite block intersects the excavation can form a removable block. The third step is to perform to stability analysis to distinguish the stable and unstable blocks.

However, the key block method only considers the key blocks. If no such blocks exist, the method concludes that the rock mass is stable. Yet in reality, group of blocks if considered as a whole, may indeed unstable. Few authors have actually attempted to improve the key block method. Wibowo (1997) sought to take the secondary key blocks into account for a key block analysis based on the Goodman-Shi method. Xu & Wang. (2000) studied the most dangerous sliding-block combination of rock by enlightening search of information with precedence of tree-like structure depth. Yarahmadi & Verdel (2003) proposes "key-group method", that considers not only individual but also groups of collapsible blocks into an iterative and progressive analysis of the stability of discontinuous rock slopes in 2D. To take intra-group forces into consideration, Yarahmadi & Verdel (2005) also implemented the Sarma method within the key-group method to generate a Sarma-based key-group method for rock slope reliability anlyses in 2D. This paper proposes "block-group method", which can analyze key block-groups and non-movable block-groups. Furthermore, the kinetic energy law is implemented to determine the movement behavior of blocks in key block-groups.

2 BLOCK-GROUP METHOD

2.1 Grouping technique of block-group

It is not impossible to consider all possible groups of two or more blocks to produce block-groups, yet a number of combinations would quickly limit that consists of defining the conditions required for assembling block-group. Firstly, the key blocks must be the base blocks for assembling block-group, because the stability of key blocks is the precondition of the stability of total rock slope. Secondly, there are neighboring finite blocks having common joint with key block or key block-group. Thirdly, after the common joint is connected as a whole, the stereo-analytical block method is adopted to analyze stability of block-group.

In order to operate and analyze easily, signed numbering method is adopted to show the topology between block and joints. "+1" denotes that block is in the upper half-space of the joint. "−1" denotes that block is in the lower half-space of the joint. "0" denotes that block does not include the joint.

Figure 1 shows one joint rock slope, which has three joints J_1, J_2 and J_3. The vertical face is free face. Left half-space of the free face is excavation pyramid (EP). Right half-space of the free face is space pyramid (SP). According to the key block method, block 1 is a key block and block 2 is a finite block. Their signed numbering codes are $D_1 = [1\ 0\ -1]$ and $D_2 = [1\ -1\ 1]$ respectively. The finite block 2 is a neighboring block of the key block 1. They have the common joint 3. In other words, block 1 and block 2 are in opposing half-space of joint 3. Correspondingly, the number pair "+1" and "−1" is shown in the third position of their signed numbering codes. When we hide joint 3, joint rock slope only has joint 1 and joint 2. According to the key block method, the big block is a key block. It means that we can assemble the block 1 and the block 2 to be a bigger key block-group 12. Here a sign "∝" is used to operate D_1 and D_2 to express D_{12}. The sign

Table 1. The ∝ operations.

∝	−1	0	1
−1	−1	−1	0
0	−1	0	1
1	0	1	1

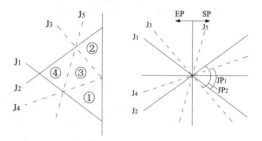

Figure 2. One joint rock slope.

"∝" denotes the operations shown in table 1. So, the code of bigger key block-group can be expressed as:

$$D_{12} = D_1 \propto D_2 = [1\ -1\ 0] \qquad (1)$$

2.2 Progressive grouping

The goal of block-group method is to analyze larger hazardous unstable group of blocks instead of just a few key blocks. The block-group method consists first of analyzing and confirming the key blocks. Then identify if there are neighboring finite blocks for key block or key block-group exist. Subsequently, stability analysis for block-group is performed according to the stereo-analytical block method. For instance, the process of progressive grouping is as follows:

Five joints J_1, J_2, J_3, J_4, J_5 and a free face divided rock slope into four joint blocks. Their codes are $D_1 = [1\ 0\ 0\ -1\ 0]$, $D_2 = [0\ -1\ 1\ 0\ 0]$, $D_3 = [0\ 0\ -1\ 1\ -1]$ and $D_4 = [1\ -1\ 0\ 0\ 1]$ respectively (Figure 2). Based on the key block method, block 1 and block 2 are key blocks. Block 3 is a tapered block. Block 4 is a finite block. So we can get block-groups from block 1 and block 2 respectively.

For key block 1, joint 1 and joint 4 are composing joints. But the number pair of "+1" and "−1" does not appear in the first position in the codes of block 1 and other blocks. In other words, there is no neighboring finite block having common joint 1 with block 1. So we do not need to hide joint 1 to assemble block-group. The number pair of "+1" and "−1" appears in the forth position in the codes of block 1 and block 3. So block 1 and block 3 can be assembled into a block-group. Then the joint 4 can be hidden. According to

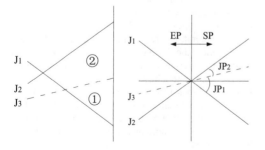

Figure 1. One joint rock slope.

the key block method, the block 13 is a non-movable block-group (Figure 3). The code of block-group 13 can be expressed as:

$$D_{13} = D_1 \propto D_3 = [1\ 0\ -1\ 0\ -1] \quad (2)$$

For block-group 13, joint 1, joint 3 and joint 5 are composing joints. But the number pair of "+1" and "−1" does not appear in the first position of block codes. So we do not need to hide joint 1 to assemble block-group. The number pair of "+1" and "−1" appear in the third position in the codes of block-group13 and block 2. So the block-group 13 and the block 2 can be assembled into block-group123. Then the joint 3 can be hidden. According to the key block method, the block-group 123 is a key block-group (Figure 4). Its code is $D_{123} = D_{12} \propto D_3 = [1\ -1\ 0\ 0\ -1]$.

Also the number pair of "+1" and "−1" appear in the fifth position in the codes of block-group13 and block 4. So the block-group 13 and the block 4 can be assembled into block-group134. According to the key block method, after the joint 5 is hidden, the block-group 134 is a non-movable block-group (Figure 5). Its code is $D_{134} = D_{13} \propto D_4 = [1\ -1\ -1\ 0\ 0]$.

For the key block-group 123, finite block 4 is a neighboring block. The key block 123 and finite block 4 can be assembled into block-group 1234 (Figure 6). For the non-movable block-group 134, key

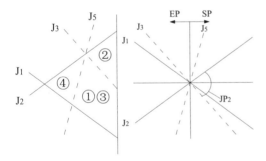

Figure 3. Non-movable block-group 13.

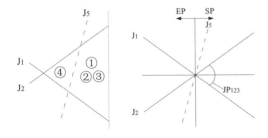

Figure 4. Key block-group 123.

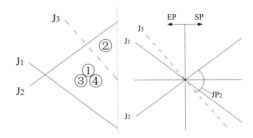

Figure 5. Non-movable block-group 134.

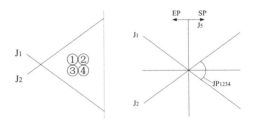

Figure 6. Key block-group 1234.

block 2 is a neighboring block. The non-movable block 134 and finite block 2 can also be assembled into block-group 1234. According to the key block method, the block-group 1234 is a key block-group. Its code is $D_{1234} = D_{123} \propto D_4 = [1\ -1\ 0\ 0\ 0]$.

The grouping steps for key block 2 is almost the same as that for key block 1 and the results are as follows: non-movable block-group 23, key block-group 123, non-movable block-group 234 and key block group 1234.

3 MOVEMENT BEHAVIOR OF BLOCKS FOR KEY BLOCK-GROUP

3.1 The theory of movement behavior analysis

Due to the key block-group is considered as a whole, the theory of key block method is difficult to analyze the movement behavior of blocks in key block-group. The blocks are assumed rigid blocks. So we implement the kinetic energy law to achieve the aim. Movement behavior of any block must be meet kinetic energy law:

$$E_k = W \quad (3)$$

Where E_k is the kinetic energy, W is the force work for the block.

E_k can be denoted as equation:

$$E_k = \frac{1}{2}m\dot{u}_0^2 + \frac{1}{2}m\dot{v}_0^2 + \frac{1}{2}J\dot{w}_0^2 \quad (4)$$

W can be denoted as equation:

$$W = W_p + W_l + W_b \tag{5}$$

where m represents mass of block. J represents rotate inertia. The variable u_0 and v_0 are the rigid body translation of a specific point (x_0, y_0). The variable w_0 is the rotation angle of the block with the rotation center at (x_0, y_0). W_p is the work of point force. W_l is the work of line force. W_b is the work of body force.

3.2 Displacement transformation formula of block

For a single block, the displacement of arbitrary point of block can be ascertained based on one reference point of this block. Generally, the shape center of block is used to be the reference point. The displacements (u, v) of any point (x, y) of a block can be represented by three displacement variables (u_0, v_0, w_0). The displacements of any point can be written as

$$u = u_0 - (y - y_0)w_0 \tag{6}$$

$$v = v_0 + (x - x_0)w_0 \tag{7}$$

3.3 Point force work

When a block moves in a key bock-group, it certainly will contact other blocks. The point contact may occur. The point loading force affects the movement of the blocks. The point force (P_x, P_y) acts on point (x, y) of a block. The work of point force is:

$$\begin{aligned}W_p &= P_x u + P_y v \\ &= P_x[u_0 - (y - y_0)w_0] + P_y[v_0 + (x - x_0)w_0]\end{aligned} \tag{8}$$

3.4 Line force work

The surface contacts between blocks engender line force, such as friction. Assume the force is distributed on a straight line from point (x_1, y_1) to point (x_2, y_2). The equation of the force line is

$$x = (x_2 - x_1)t + x_1 \tag{9}$$

$$y = (y_2 - y_1)t + y_1, \quad 0 \le t \le 1 \tag{10}$$

The length of this line segment is

$$l = \sqrt{(x_2 - x_1)^2 + (y_2 - y_1)^2} \tag{11}$$

The force is

$$F_x = F_x(t) \tag{12}$$

$$F_y = F_y(t), \quad 0 \le t \le 1 \tag{13}$$

where t is a variant along the loading line. The work of the line force is

$$W_l = \int_0^1 [F_x(t)u + F_y(t)v] l \, dt \tag{14}$$

When the line force $(F_x(t), F_y(t)) = (F_x, F_y)$ is constant, the work is derived as follows:

$$W_l = F_x \int_0^1 u l \, dt + F_y \int_0^1 v l \, dt \tag{15}$$

In order to computer the integration, the elements of the integration have to be computed:

$$\int_0^1 (x - x_0) dt = \int_0^1 ((x_2 - x_1)t + (x_1 - x_0)) dt$$
$$= \frac{1}{2}(x_2 + x_1 - 2x_0) \tag{16}$$

$$\int_0^1 (y - y_0) dt = \frac{1}{2}(y_2 + y_1 - 2y_0) \tag{17}$$

Therefore

$$W_l = l \left[u_0 F_x + v_0 F_y - \frac{1}{2} w_0 (y_2 + y_1 - 2y_0) F_x \right.$$
$$\left. + \frac{1}{2} w_0 (x_2 + x_1 - 2x_0) F_y \right] \tag{18}$$

3.5 Body force work

Assuming that (f_x, f_y) is the constant body force acting on the body of the block, (x_G, y_G) is the center of gravity of this block. Then

$$x_G = \frac{S_x}{S} \tag{19}$$

$$y_G = \frac{S_y}{S} \tag{20}$$

where

$$S = \iint dx dy \tag{21}$$

$$S_x = \iint x \, dx dy \tag{22}$$

$$S_y = \iint y \, dx dy \tag{23}$$

The work of the constant body force (f_x, f_y) is

$$W_b = \iint (f_x u + f_y v) dx dy \tag{24}$$

where (u, v) is the displacement of gravity center of body. When the block is homogeneous, the gravity center is the same as shape center. So the work of body force is

$$W_b = S(f_x u_0 + f_y v_0) \quad (25)$$

3.6 Dynamic analysis

Based on the formula (4), (8), (18), (25), the formula (3) can be rewritten as:

$$\frac{1}{2}m\dot{u}_0^2 + \frac{1}{2}m\dot{v}_0^2 + \frac{1}{2}J\dot{w}_0^2$$
$$= (P_x + lF_x + Sf_x)u_0 + (P_y + lF_y + Sf_y)v_0$$
$$+ [P_x(y_0 - y) + P_y(x - x_0) - \frac{1}{2}lF_x$$
$$(y_2 + y_1 - 2y_0) + \frac{1}{2}lF_y(x_2 + x_1 - 2x_0)]w_0 \quad (26)$$

It can be written as

$$\frac{1}{2}m\dot{u}_0^2 - (P_x + lF_x + Sf_x)u_0 = 0$$

$$\frac{1}{2}m\dot{v}_0^2 - (P_y + lF_y + Sf_y)v_0 = 0$$

$$\frac{1}{2}J\dot{w}_0^2 - [P_x(y_0 - y) + P_y(x - x_0)$$
$$- \frac{1}{2}lF_x(y_2 + y_1 - 2y_0)$$
$$+ \frac{1}{2}lF_y(x_2 + x_1 - 2x_0)]w_0 = 0 \quad (27)$$

The result of displacements (u_0, v_0, w_0) of a block can be got through analysis. So the movement behavior of blocks in a key block-group is also determined.

For a block in key block-group, there are several forces caused by counteraction with neighbor blocks. Combine force F and combine moment M are engendered for shape center of block by these force loading. If F and M are not equal to zero, the unequal F and M make block move meeting kinetic energy theory. Because large displacement is involved in stability analysis, the block position and contacts will change following time steps. We have to know all the pairs of blocks which will possibly meet together during the next step. And the counteract force must be got according to a definite method. The method is defined in the conventional ways like Distinct Element Method. For the key block-group, calculation is done for all blocks based on time step iteration. Comparing with distinct element method and other discontinuity analysis methods, the block-group method not only considers a single block but also group of blocks. The method can quickly determine the movement behavior and failure mode of blocks in key block-group. The main advantage of using the block-group method over the other discontinuous analysis methods lies in its processing speed.

4 BLOCK-GROUP METHOD STEP

Based on grouping technique of block-group and movement behavior analysis of blocks in a key block-group, the block-group method steps are illustrated in Figure 7.

5 VALIDITY OF BLOCK-GROUP METHOD

Shown in Figure 8 is an example of a discontinuity rock slope. It consists of four joints J_1, J_2, J_3, J_4 and one free face. The mechanical properties are as follows:

$$\rho = 2500 \text{kg/m}^3, \quad c = 0.03 \text{MPa}, \quad \varphi = 27°$$

where ρ is the rock density, c is the cohesion of the discontinuities, and φ is the friction angle of the discontinuities. According to the key block method, rock slope includes key block 1, tapered block 2, non-movable block 3, finite block 4, finite block 5 and five infinite blocks. However, Based on the block-group method, the biggest key block-group is key block-group 123 (Figure 9). The comparison between the

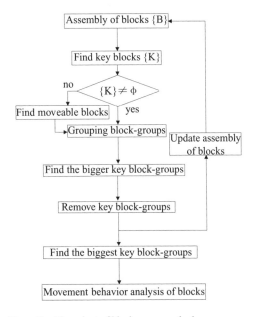

Figure 7. Flow chart of block-group method.

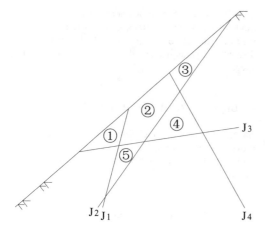

Figure 8. Discontinuity rock slope.

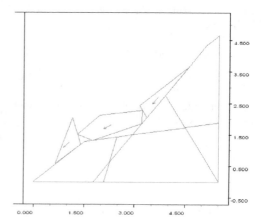

Figure 10. Failure mode of rock slope provided by UDEC.

Table 2. Material properties used in the numerical analysis with UDEC.

Property	Value	Units
Blocks		
Density	2500	Kg/m^3
Bulk modulus	12	GPa
Shear modulus	8	GPa
Discontinuities		
Joint normal stiffness	25.3	GPA/m
Joint shear stiffness	8.5	GPa/m
Cohesion	0.03	MPa
Tensile strength	0.01	MPa
Friction angle	27	Degree

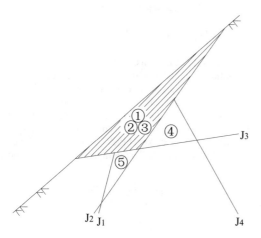

Figure 9. Key block-group 123.

block-group method and the distinct element analysis performed using UDEC is conducted (Figure 10). Table 2 displays the properties used in the numerical analysis with UDEC. We find they make a good agreement. So we can consider block-group 123 as the most hazardous area of rock slope. This proved that the block-group method yields more realistic results than the basic key block method.

6 CONCLUSIONS

Owing to taking a greater number of blocks into account in the stability analysis and more realistic, the block-group method proposed in this paper is superior to the key block method for stability analysis of rock slope. In addition, kinetic energy law is developed to determine the movement behavior of blocks in the key block-group. Compared with other discontinuity numerical analysis methods, the main advantage of using the block-group method lies in its processing speed and can quickly confirm the most hazardous area and failure mode of joint rock slope.

REFERENCES

Cundall, P.A. 1971. A computer model for simulating progressive large scale movements in blocky systems. *Proceeding of the symposium of the international society of rock mechanics. Nancy, France* 1:2–8.

Goodman, R.E. & Shi, G. 1985. *Block theory and its Application to Rock Engineering*. Prentice-Hall: New Jersey.

Shi, G.H. 1988. *Discontinuous deformation analysis new numerical model for the static and dynamics of block system*. Berkeley: Department of civil engineering, University of California.

Warburton, P.M. 1981. Vector stability analysis of an arbitrary polyhedral rock block with any number of free faces. *International Journal of Rock Mechanics and Mining Science & Geomechanics* 18:415–427.

Wibowo, Johannes L. 1997. Consideration of secondary block in key-group analysis. *International Journal of Rock Mechanics and Mining Science & Geomechanics* 34:3–4.

Xu, Mingyi & Wang, Weiming. 2000. Research on the dangerous sliding-block combination of rock slopes. *Rock and soil mechanics* 21(2):148–151.

Yarahmadi Bafghi, A.R. & Verdel, T. 2003. The key-group method. *International Journal for Numerical and Analytical Methods in Geomechanics* 27:495–511.

Yarahmadi Bafghi, A.R. & Verdel, T. 2005. Sarma-based key-group method for rock slope reliability analyses. *International Journal for Numerical and Analytical Methods in Geomechanics* 29:1019–1043.

Zhang, Zixin & Kulatilake, P.H.S.W. 2003. A new stereo-analytical method for determination of removal blocks in discontinuous rock masses. *International Journal for Numerical and Analytical Methods in Geomechanics* 27: 791–811.

Quantitative study on the classification of unloading zones of high slopes

Da Zheng & Run-Qiu Huang
State Key Laboratory of Geohazards Prevention, Chengdu University of Technology, Chengdu, China

ABSTRACT: Unload and unloading zones of high rock slope are familiar phenomenon in southwest area of China, which influence stability of rock slope and other rock engineering. On the basis of analyzing the actual way of classification of unloading zones, the paper puts forward a divisiory way of adopting crack rate, opening crack rate and summation of crack width as quantitative indexes according to formation mechanism and geological exhibition of unloading zones. After large numbers of locale measurement of cracks, analytical result indicates that there is a good corresponding connection between the three indexes and unloading degree, and it is feasible to regard the three indexes as quantitative standard of classification of unloading zone. Furthermore, combining to qualitative geological characteristics of unloading zones, the paper presents suggestions for dividing unloading zones of actual projects.

1 GENERAL INSTRUCTIONS

Unloading of the rock mass slopes is that due to river erosion or manpower excavation free faces' occurrence destroys initial stress balance of the rock mass, as leads to stress releasing at the superficial layer of the slopes and unloading rebound. During this process a new rupture system is formed. These constitute rock mass relaxation and tensile fracture phenomena at the shallow slopes (Huang et al, 1994).

The slope unloading is a common dynamical phenomenon for the high rock slopes in the southwest area of China. Unloading breaks the whole rock mass, and reduce the qualities of the rock mass. At the same time, the channels are formed due to weathering force and exogenic process of underground. Therefore, unloading is meaningful for analyzing the stability of the rock slopes and related rock engineering. In the region of engineering geology and rock mechanics the researches on unloading and unloading zones are always concerned.

Unloading is usually divided into strong, weak and centralized ones. Strong unloading and weak unloading are familiar. In recent years the centralized unloading is a special phenomenon for the high rock slopes in the gorges of southwestern China. The paper mainly discusses strong unloading and weak unloading.

Forming mechanism of the unloading zone, and classification of the unloading zone and engineering unloading characteristics of the rock mass have been researched by many scholars (Huang, 2000; Zhang, 1993; Li, 2001; Wu, 2001). Their findings have been accepted by practical engineering, and the criterion for estimating unloading zones is established by the corresponding exploration survey code (Lu, 1995). However, due to obvious difference of the geological conditions and complex unloading there is uncertain to divide unloading zones in practical operations. Therefore, the geological index and methods to divide unloading zones still wait for more discussions.

In the paper several kinds of unloading divisions are studied. And some basic rules have been found according to many field investigations. Hereby corresponding suggests on geological basis and quantitative index to divide unloading zones are brought forward.

2 IDENTIFYING AND DIVIDING METHODS FOR ACTUAL SLOPE UNLOADING ZONES

Evaluation and division on the slope unloading zones are firstly based on the unloading phenomenon, crack opening, filling condition, weathering of the crack wall and underground in the light of the criterion offered by the codes. Then farther check for the locale evaluation of the unloading zones is made in terms of the wave velocity of the adit walls.

At present, there is no uniform standard for dividing the unloading zones. Hou (2000) researched changing characteristics of unloading crack growth density and opening degree with adit depth. And in strong unloading rock mass cracks are at large open, and full with secondary mud, width of which ranges from 0.5 cm to 1.5 cm. Crack width of the weak unloading rock mass is commonly less than 1 cm, and much full with calcareous and muddy matter (Hou, 2000). Besides the above indexes, due to unloading the rock structure is relaxed, and the rock mass loses the integrality, and the structure plane becomes loose, as results in enhancing water transmitting ability of the rock mass. Therefore, coefficient of permeability obtained by forced water tests reflects the conditions of rock mass's crack growth, joint opening and unloading. It is used to divide the unloading zones of the rock mass. The Code For Water Resources and Hydropower Engineering Geological Investigation in China (GB5087–99) has regulated classification of rock mass permeability in detail, as indicates that the penetrability of rock mass is related with opening degree of the cracks. Quantitative index of the rock mass unloading zones has been applied into practical projects. For example, longitudinal wave velocity and crack opening degree are used for dividing the unloading zone of the dam abutment, Xiaowan Hydropower Station. In addition, opening degree of centralized unloading crack, its growth density, rock mass structure and sound wave velocity of the rock mass are considered as the standard for dividing the unloading zones of Xiluodu Hydropower Station in Jinsha River.

3 SELECTION OF QUANTITATIVE DIVISION OF SLOPE UNLOADING ZONES

Researches show that slope rock mass unloading immediately gives rise to the results that the rock mass in shallow layers is relaxed, original structure planes open, secondary cracks are generated, and even rock mass failures (Ju, 2000). The data statistic indicates that number of rock slope cracks always decreases with increase of the horizontal depth far from the slope surface. And the number and opening degree are stable when the horizontal length is more than a fixed value.

So the paper thinks that identifying and dividing the unloading zones mainly depends on number, opening width of the cracks, and relaxation of the rock mass structure. Therefore, we select change of crack ratio, opening crack ratio, crack opening degree or secondary mud width full with the cracks to describe unloading conditions of the rock mass, and divide unloading zones.

Crack ratio is the crack number per unit length along the measured adit wall. Opening crack ratio is the number of opening cracks per unit length measured. Degree of crack opening is described using "crack width sum". The three indexes are all quantitative ones. In practical engineering, the indexes can be obtained through careful measurements, thus are exercisable.

4 APPLICATION AND DISCUSSION

Using these quantitative indexes, we divide the unloading zones for certain Hydropower Station.

4.1 Geological general situations

The dam of the Hydropower Station is located in the lower braches of Jinshan River. The river valley is narrow and the bank slope is steep, with anisomerous V shape. It is composed of the layer from upper Emeishan basalt of Permian System. The consequent slopes can be seen in the left bank, and its landform shows a ladder shape. In the right bank the slope is steep. There are preferable engineering geological conditions in the dam area. The rock structure in the region of the dam mainly develops in the interlaminar, disturbed belt, small fault and rockbed crack in the basalt. The disturbed belt with good connectivity goes through the whole dam region, which composes the basic structural frame of the rock structure in the area of the dam. The intrastratal disturbed belt with gently dip angles popularly grows in the basalt, which is an important part of the rock structure in the region of the dam.

4.2 Division of the slope unloading zones

Both quantitative and qualitative methods are combined for dividing the unloading zones in the area of the dam. The quantitative indexes used in the paper are crack ratio, opening crack ratio and sum of crack width. These indexes are measured by setting the measured lines along the adit wall.

In order to achieve full statistical data, the paper amply measured more than 40-cm-long cracks over ten adits, and obtained many data about the cracks. At the same time, the data are analyzed to divide unloading zones. For example, Adit 39 is used as a case. Fig. 1 describes changing rule of crack ratio and opening crack ratio with the adit depth. Changing rule of crack

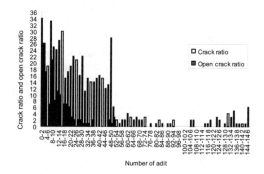

Figure 1. Curve of crack rate and opening crack rate with depth in the adit 39.

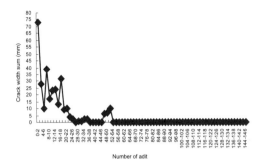

Figure 2. Curve of summation of crack width with depth in the adit 39.

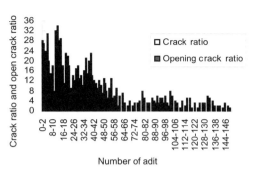

Figure 3. Curve of crack rate and opening crack rate with depth in the adit 57.

width sum with the adit depth can be seen in Fig. 2. Crack ratio and opening crack ratio decrease with increase of the adit depth. At the position from 52 m to 54 m, the two parameters rapidly fall at the same time. The results indicate that the crack is close, and the rock mass no longer is affected by unloading when the adit depth is more than 54 m. The three indexes above are accordant for describing slope unloading. In the light of the results 54 m is considered as unloading depth of the adit.

It should be pointed out that the marked relationship between crack ratio and unloading is related with the properties of the basalt. The basalt under no unloading conditions contains many hidden cracks. As the basalt is affected by unloading the hidden cracks emerge.

According to quantitative parameters we again discuss division of strong unloading and weak one. Fig. 1 and Fig. 2 show that only one of the three indexes well changes at the position from 20 m to 22 m deep in the adit. For 0 m to 20 m, the crack ratio is 25 lines per 2 m, opening crack ratio is 15 lines per 2 m, and opening crack lines account for 60 percent. The crack width sum is up to 26.8 mm per 2 m. The results indicates that from 0 m to 21 m the rock mass is under strong unloading zones. From 21 m to 54 m deep in the adit, crack ratio and opening crack ratio evidently fall, opening crack lines is less than 10 percent of the total number, and crack width sum is averagely 0.84 mm per 2 m. Therefore, the segment from 21 m to 54 m is considered as weak unloading zone. As the adit depth is more than 54 m, crack width sum is zero. This shows that unloading no longer affect the segment more than 54 m.

Other adits' unloading is divided according to the same method. Fig. 3 and Fig. 4 describe that crack ratio and crack width sum for Adit 57 change with adit depth. The conclusion is consistent with Adit 39. The strong unloading ranges from 0 m to 56 m, and weak one changes from 56 m to 102 m.

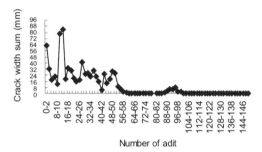

Figure 4. Curve of summation of crack width with depth in the adit 57.

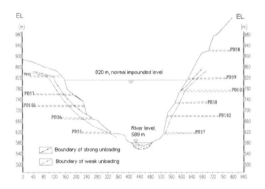

Figure 5. Classification result of unloading zones at I-I section of some hydropower's dam site.

The results of unloading division using field empirical method and the above quantitative method are basically accordant. Therefore, crack ratio, opening crack ratio and crack width sum can be used as geological quantitative indexes for dividing unloading zones of the high slopes' rock mass.

Adopting the above methods we divide unloading zones of all adits in the Profile I-I, distribution of which can be seen in Fig. 5.

Table 1. Index of quantification on classification of unloading zone in dam site of some hydropower.

Unloading zone	Strong unloading zone	Weak unloading zone
Crack ratio (/2 m)	20~25	10~15
Opening crack ratio (/2 m)	10~15	5~8
Crack width sum (mm/2 m)	15~20	1~5
Unloading characteristics	Rock mass is loose, cracks and secondary mud appear, structural cracks are widely open, and full width widely open, and full with debris and secondary mud. Steep dip angles are found	Unloading cracks well develop with small width, and no mud fills the cracks
Rock character	Basalt	

On the basis of the findings, suggest schemes for dividing unloading zones of the damsite are list in Table 1 in terms of geological qualitative analysis and analogue of Xiaowan, Xiluodu and Xiangjiaba Hydropower Stations' experiences.

5 CONCLUSIONS

1. Unloading of the rock mass currently represents opening of the cracks and relaxation of the rock mass structure. The researches in the paper indicate that increase of crack number is an important characteristic of unloading. It, altogether with crack opening, shows like changing rules with increase of adit depth.
2. On the basis of research findings about rock mass unloading, changes of crack ratio, opening crack ratio, and crack opening degree (crack width sum) are selected to describe unloading degree of rock masses. The changes of the three indexes are highly consistent with adit depth.
3. The three parameters are used to divide unloading zones in the practical engineering regions. The results are basically consistent with those divided by empirical geology engineers. Therefore, crack ratio, opening crack ratio and crack width sum can be considered as geological quantitative index and basis for dividing unloading zones of high slope rock mass.

REFERENCES

Huang R.Q., Zhang Z.Y. & Wang Shitian. 1994, Research on rock structure epigenetic reformation. *Hydrogeology and Engineering Geology*, 1994, 21 (4): 17–21. (in Chinese).

Huang Runqiu, Lin Feng, Chen Deji, et al. Formation mechanism of unloading fracture zone of high slopes and its engineering behaviors. *Journal of Engineering Geology*, 2001, 9 (3): 227–232 (in Chinese).

Ren Guangming, et al. A Quantitiative Study on the Classification of Unloading Zones of Rock mass Slope. Journal of ChengDu University of Technology, 2003, 8: 235–238. (in Chinese).

Nie Dexin, Han Aiguo & Ju Guanghong. Study on Integrated Zoning of Weathering Degree of Rock mass. *Journal of Engineering Geology*, 2002, 10 (1): 20–25. (in Chinese).

WU Gang. The present and expectation of the study to mechanism of engineering rock broke unloading. *The Journal of Engineering Geology*, 2001, 9 (2): 174–181. (in Chinese).

The Ministry of Water Resources of People's Republic of China. The Code For Water Resources and Hydropower Engineering Geological Investigation in China (GB5087–99). Beijing: China Plan Press, 1999.

Hou Zhibin. Unloading characteristics of sand-shale rock mass in the pivot area of Chezhuang Hydraulic Engineering. *Shanxi Hydropower*, 2000, 16 (3): 4–9. (in Chinese).

Lu Hong. Engineering geological survey of the high arch dam. *Yunnan hydroelectric technique*, 1995 (4): 4–15. (in Chinese).

Huang Runqiu, Wang Shitian, Hu Xiewen et al. Study on the main Engineering Geological Problems of Xiaowan high arch dam. Chengdu: Southwest Jiao-Tong University Press, 1996. (in Chinese).

National Laboratory of Geohazards Prevention of Chengdu University of Technology. Study on the rock mass qualities and selection of the base surface in the dam site of Xiluodu Hydropower Station. Chengdu: Chengdu University of Technology, 2001. (in Chinese).

Ju Guanghong. Study on engineering geology of weathered unloading zones for deep down-cutting gorge granite in Laxiwa Hydropower Station, Yellow River. Chengdu: Master thesis of Chengdu University of Technology. 2002. (in Chinese).

Investigations on the accuracy of the simplified Bishop method

D.Y. Zhu
School of Civil and Water Resources Engineering, Hefei University of Technology, Hefei, China

ABSTRACT: This paper describes the reason why the simplified Bishop method always gives factors of safety of circular slip surfaces in good agreement with those given by the rigorous methods. The absence of vertical interslice force in the factor of safety equation in this method only means a summation regarding the vertical interslice forces has been neglected. An appropriate set of vertical interslice forces could be found that: (1) allows this term to be zero, (2) renders the satisfaction of the horizontal equilibrium condition, and (3) gives the same factor of safety. In other words, the simplified Bishop method implicitly satisfies the rigorous limit equilibrium conditions.
Key words: simplified Bishop method, limit equilibrium, slope stability, factor of safety

1 INTRODUCTION

The simplified Bishop method (Bishop, 1955) has been widely used in slope stability analysis and is regarded as the best method of limit equilibrium for calculating the factors of safety of circular slip surfaces. In this method, the interslice forces are assumed to be horizontal, or the vertical interslice forces are neglected, the vertical force equilibrium and the moment equilibrium about the centre of the circular slip surfaces are satisfied, but the horizontal force equilibrium is not considered. Thus, the simplified Bishop method is still regarded as one of the non-rigorous limit equilibrium methods of slices. However, the simplified Bishop method always gives factors of safety of circular slip surfaces in good agreement with those given by other rigorous methods of slices such as the Morgenstern-Price method (Morgenstern and Price, 1965) and the Spencer method (Spencer, 1967). In the profession of geotechnical engineering, the simplified Bishop method has been accepted as the accurate method of slices, although it dos not satisfy all the limit equilibrium conditions (Duncan, 1996). The problem why the simplified Bishop method is so accurate has puzzled the profession for over 50 years. This study attempts to give a theoretical study on this issue.

2 FUNDAMENTALS

A sliding body about a circular slip surface of radius R is divided into n vertical slices, as shown in Fig. 1. For the ith slice, the width is b_i, the angle of base is α_i, the weight is W_i, the horizontal interslice forces are E_i and E_{i+1}, the vertical interslice forces are X_i and X_{i+1}, the normal force at the base is E_i, the shear resistance at the base is T_i, the pore water pressure at the base is u_i, the effective internal friction angle and cohesion are ϕ'_i and c'_i, respectively. The factor of safety along the slip surface is F_s.

According to Mohr-Column failure criterion and the principle of effective stress:

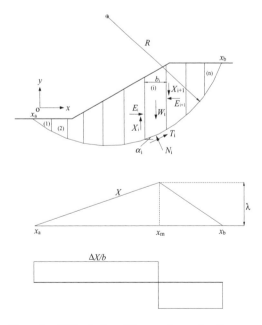

Figure 1. Sliding body and forces acting on the slice.

$$T_i = (N_i - u_i b_i \sec \alpha_i) \tan \phi'_i / F_s + c'_i b_i \sec \alpha_i / F_s \quad (1)$$

Consider the vertical force equilibrium condition

$$N_i \cos \alpha_i + T_i \sin \alpha_i = W_i + X_{i+1} - X_i \quad (2)$$

From equations (1) and (2), one obtains

$$N_i = (W_i + X_{i+1} - X_i + u_i b_i \tan \alpha_i \tan \phi'_i / F_s$$
$$- c'_i b_i \tan \alpha_i / F_s) / m_{\alpha i} \quad (3)$$

$$T_i = [(W_i + X_{i+1} - X_i - u_i b_i) \tan \phi'_i / F_s$$
$$+ c'_i b_i / F_s] / m_{\alpha i} \quad (4)$$

$$m_{\alpha i} = \cos \alpha_i + \sin \alpha_i \tan \phi'_i / F_s \quad (5)$$

Consider the moment equilibrium with respect to the centre of circular slip surface:

$$\sum_{i=1}^{n} T_i R = \sum_{i=1}^{n} W_i R \sin \alpha_i$$

Substituting equations (3) and (4) into the above equation gives

$$F_s = \frac{\sum_{i=1}^{n} [(W_i + X_{i+1} - X_i - u_i b_i) \tan \phi'_i + c'_i b_i] / m_{\alpha i}}{\sum_{i=1}^{n} W_i \sin \alpha_i} \quad (6)$$

The simplified Bishop method assumes that the contribution of vertical interslice forces to the factor of safety is neglected, hence

$$F_s = \frac{\sum_{i=1}^{n} [(W_i - u_i b_i) \tan \phi'_i + c'_i b_i] / m_{\alpha i}}{\sum_{i=1}^{n} W_i \sin \alpha_i} \quad (7)$$

3 SATISFACTION OF HORIZONTAL FORCE EQUILIBRIUM

The vertical interslice forces do not appear in the factor of safety equation of the simplified Bishop method. This does not mean the vertical interslice forces should be zero. In fact, from equation (6), we can see if the following equation holds

$$\sum_{i=1}^{n} (X_{i+1} - X_i) \cdot \tan \phi'_i / m_{\alpha i} = 0 \quad (8)$$

equation (6) is identical to equation (7).

In other words, if a set of vertical interslice force exist that satisfy equation (8), the factor of safety is still identical to that of the simplified Bishop method. Furthermore, if that set of vertical interslice forces are so selected that the horizontal force equilibrium condition is satisfied, then the factor of safety computed by equation (6) or (7) corresponds to that of rigorous method of slices since all the three equilibrium conditions are completely satisfied. The key issue is to find such a set of vertical interslice forces that not only satisfy the horizontal force equilibrium condition but also equation (8).

Consider the horizontal force equilibrium for the whole sliding body:

$$\sum_{i=1}^{n} (T_i \cos \alpha_i - N_i \sin \alpha_i) = 0$$

and substitute equations (3) and (4) into the above equation, one obtains:

$$\sum_{i=1}^{n} \left[(W_i + X_{i+1} - X_i)(\cos \alpha_i \tan \phi'_i / F_s - \sin \alpha_i) \right.$$
$$\left. - (u_i b_i \tan \phi'_i - c'_i b_i)/(F_s \cos \alpha_i) \right] / m_{\alpha i} = 0$$
$$(9)$$

Now what remains is to find the distribution of vertical interslice forces X that simultaneously satisfy equations (8) and (9).

Similar to the Correia method (Correia, 1988), the distribution of vertical interslice forces is of the form as follows

$$X = \lambda f(x) \quad (10)$$

where $f(x)$ is interslice force function and λ is the scaling factor. The difference between the vertical interslice forces on the two sides of the slice is

$$X_{i+1} - X_i = \Delta X_i = \lambda f'_i b_i \quad (11)$$

Substitute of equation (11) into (8) leads to

$$\sum_{i=1}^{n} f'_i b_i \cdot \tan \phi'_i / m_{\alpha i} = 0 \quad (12)$$

The value of λ is determined according to the horizontal force equilibrium. Substitute of equation (11) into (9) gives

$$\lambda = \frac{\sum_{i=1}^{n} Ai / m_{\alpha i}}{\sum_{i=1}^{n} f'_i b_i / m_\alpha} \quad (13)$$

in which

$$A_i = -W_i(\cos\alpha_i \tan\phi'_i/F_s - \sin\alpha_i)$$
$$+ (u_i b_i \tan\phi'_i - c'_i b_i)/(F_s \cos\alpha_i) \qquad (14)$$

For a circular slip surface, the interslice force function $f(x)$ satisfying equation (12) can be found, giving the factor of safety that is identical to that of the simplified Bishop method. The value of λ is then determined from equation (13), resulting in the distribution of the vertical interslice force satisfying the horizontal force equilibrium condition.

For simplicity, the interslice force function $f(x)$ can be chosen as (see Fig. 1)

$$f(x) = \begin{cases} \dfrac{x - x_a}{x_m - x_a} & x_a \leq x \leq x_m \\ \dfrac{x_b - x}{x_b - x_m} & x_m < x \leq x_b \end{cases} \qquad (15)$$

Thus

$$f'(x) = \begin{cases} \dfrac{1}{x_m - x_a} & x_a \leq x \leq x_m \\ -\dfrac{1}{x_b - x_m} & x_m < x \leq x_b \end{cases} \qquad (16)$$

Since the simplified Bishop method implicitly satisfies all the complete equilibrium conditions through proper selection of vertical interslice force distribution, it is not strange that the simplified Bishop method has the accuracy that compares with that of other rigorous methods of slices.

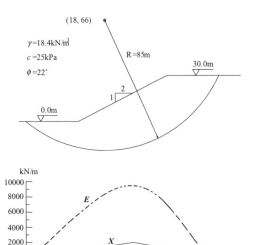

Figure 2. Slope and interslice forces distribution.

4 EXAMPLE

A homogeneous slope has a circular slip surface, as shown in Fig. 2. The unit weight is 18.4 kN/m³, the cohesion and the internal friction angle are 25 kPa and 22° respectively.

The factor of safety is 2.08 computed by the simplified Bishop method. When the interslice force function $f(x)$ is chosen as equation (15) with $x_m = 36.9$, equation (12) is satisfied. From equation (13), the scaling factor λ is calculated as 1988.62. The distribution of the vertical and horizontal interslice forces is shown in Fig. 2.

5 CONCLUSION

The simplified Bishop method, which neglects the vertical interslice forces and takes no account of the horizontal force equilibrium condition, has accuracy as high as the rigorous limit equilibrium methods of slices. This has been confirmed by experiences of many practitioners, but not been explained theoretically. This study shows that the factor of safety equation of the simplified Bishop method does not require that the vertical interslice forces be zero, but a term designated by Eq. (8), be zero. This condition can be realized by assigning an appropriate distribution of vertical interslice forces that satisfies the horizontal force equilibrium condition simultaneously, and lead to a factor of safety identical to that of the simplified Bishop method. Since the approach satisfies the complete equilibrium conditions, this method has accuracy so high that can compares with the Morgenstern-Price method and the Spencer method.

REFERENCES

Bishop, A.W. The use of the slip circle in the stability analysis of earth slopes. Géotechnique 1955, 5(1): 7–17.

Correia, R.M. A limit equilibrium method of slope stability analysis. Proc. 5th Int. Symp. Landslides, Lausanne, 1988, 595–598.

Duncan, J.M. State of the art: limit equilibrium and finite-element analysis of slopes. J. Geotech. Engrg., ASCE, 1996, 122(7): 577–596.

Morgenstern, N.R., and Price, V.E. The analysis of the stability of general slip surfaces. Géotechnique, 1965, 15(1): 79–93.

Spencer, E. A method of analysis of the stability of embankments assuming parallel interslice forces. Géotechnique, 1967, 17(1): 11–26.

Zhu, D.Y., Lee, C.F. and Jiang, H.D. Generalized framework of limit equilibrium methods for slope stability analysis. Géotechnique, 2003, 53(4): 377–395.

Author index

Abdullah, C.H. 1873
Abe, S. 939
Abolmasov, B. 1939
Aceto, L. 373
Adler, R.F. 1991
Aghababaei, H. 857
Akhundjanov, A.M. 1247
Akin, M. 1647
Al. Zaitsev, A. 963
Allison, R.J. 865
Almeida Saraiva, A.L. 561
Alonso, E.E. 67
Aminpoor, M. 553
Anarbaev, E.M. 1475
Araiba, K. 1321
Asano, S. 1375, 1439
Auray, G. 1721

Bach, D. 1991
Badv, K. 1489
Bai, A. 737
Bai, J. 1071, 1077
Bai, S. 1067
Barradas, J. 567
Baum, R.L. 179
Bazarov, Sh.B. 1247
Becker, R. 1083
Bell, R. 1083
Bencardino, M. 365
Beniston, M. 217
Bhandari, R.C. 231
Bianchi, M. 1233
Bica, A.V.D. 293
Bièvre, G. 395
Bigarré, P. 331, 1925
Blikra, L.H. 1089
Bojorque, J. 573
Bomont, S. 1713
Bonnard, Ch. 217
Bottiglieri, O. 1525
Bovenga, F. 513
Bozzano, F. 579, 585, 1381, 1389, 1905
Bressani, L.A. 293
Bretschneider, A. 579
Burghaus, S. 1083

Burlon, S. 687
Burns, S.F. 1979

Cai, D. 1817
Cai, F. 593, 851, 1455
Cai, J.T. 1659
Cai, Y. 1165
Cai, Y.J. 2019
Calabrò, N. 647
Calcaterra, D. 299, 1095, 1397
Calò, F. 299, 1095
Calvello, M. 495, 599, 607
Cao, G.J. 2027
Cao, L. 449
Cao, P. 815
Cappa, F. 331, 711
Casarano, D. 513
Cascini, L. 599, 607, 1103, 1893
Castellanos Abella, E. 1879
Catalano, E. 1433
Ceccucci, M. 1111
Cecinato, F. 615
Cha, K.S. 1763
Chae, B.G. 933
Chai, H.J. 1675, 1799
Chamra, S. 1653
Chang, C.H. 1559
Chang, M. 1493
Che, A. 307, 449
Chen, C.Y. 313
Chen, G.F. 2101
Chen, H. 2005
Chen, J. 535
Chen, K. 1501
Chen, L.H. 631
Chen, S. 789
Chen, S.F. 1559, 2019
Chen, X.D. 1507
Chen, Y.H. 1639
Chen, Y.M. 731
Chen, Z. 25, 2011
Chen, Z.Y. 623, 631, 655
Cheng, Q. 1133, 1985
Cheng, Q.G. 1119
Cheng, Y. 1127

Chi, S.Y. 1639
Ching, J. 639
Cho, Y.C. 933
Choi, Y.S. 1633
Chou, Y.H. 1493
Chu, B.L. 1409
Clément, C. 331, 1143
Clarke, D. 1553
Coe, J.A. 1899
Coelho Netto, A.L. 409
Cola, S. 647
Colombo, A. 1233
Conversini, P. 1609
Corominas, J. 237, 1517
Cotecchia, F. 1525
Cottaz, Y. 319
Couture, R. 1151
Critelli, S. 373
Crovelli, R.A. 1899
Cucchi, A. 1233
Cui, P. 263
Cui, X. 1481

Dahlhaus, P.G. 267
Dai, W.J. 1367
Damiano, E. 1157
Dang, L.C. 1195
Daryono, M.R. 1615
De Roeck, G. 573
Delacourt, C. 395
Demir, B.M. 531
Deng, J.H. 821, 999
Deparis, J. 323
Derevenets, Ph.N. 1225
Dhungana, I. 1463
Di Crescenzo, G. 1951
Di Maio, C. 495
Di Martire, D. 299
Di Nocera, S. 1103
Diano, G. 585
Díez, J.A. 347
Dijkstra, T.A. 1553
Dima, C. 1397
Ding, J. 1533
Ding, X.L. 1367
Dixon, N. 1553
Doanth, B.T. 1475

Dong, S.M. 1675
Du, L. 1775
Du, Y. 1781
Duan, Q.W. 655, 663
Dünner, C. 331, 1143, 1925
Durmaz, S. 531
Dyer, M. 1309, 1965
Dykes, A.P. 339

Eberhardt, E. 39
Effendiantz, L. 323
Eisenberger, C.N. 293
El Bedoui, S. 421, 1273
Ellis, W.L. 179
Emami, K. 1489
Enoki, M. 671
Entezam, I. 491
Eom, Y.H. 1297
Esaki, T. 893
Esposito, C. 355, 585, 1905
Estaire, J. 347
Evangelista, A. 913

Fan, J. 543
Fan, X.-M. 1681
Fasani, G.B. 355, 1905
Faure, R.M. 319, 687, 1721
Fei, L.Y. 1639
Feng, H. 1863
Feng, M. 2061
Feng, S. 307
Feng, X. 803
Feng, X.Y. 693
Ferlisi, S. 1103, 1893
Flentje, P. 267
Fletcher, C.J.N. 457
Floris, M. 1905
Fonseca, A.P. 409
Fornaro, G. 1103
Franz, J. 1165
Fricout, B. 323
Frolovsky, Y.K. 963
Fujisawa, K. 697
Fukuoka, H. 443

Gaeta, M. 1381
Gaffet, S. 711
Galzerano, C.M. 1095
Gao, Q. 1795, 2075
Gao, W. 535, 705
Garcia, C.F. 1463
Garcia, R.A.C. 1831
García de la Oliva, J.L. 1731
Gattinoni, P. 1539
Genda, M. 939

Gens, A. 67
Ghanbari, E. 1405
Gharouni-Nik, M. 1725
Ghazanfari, M. 1911
Ghazipoor, N. 491
Giannico, C. 1233
Gillarduzzi, A. 1547
Glade, T. 1083
Glendinning, S. 1553
Gon, L.S. 1571
Gong, X.N. 1659
González, D.A. 237
González-Gallego, J. 1731
Gori, P.L. 1173
Grasso, E. 1397
Greco, R. 1157
Grelle, G. 365
Gress, J.C. 361, 687
Grimaldi, G.M. 607
Guadagno, F.M. 365
Guglielmi, Y. 331, 711, 1273
Gui, M.W. 1559, 1737
Gullà, G. 373, 599
Gunzburger, Y. 1143
Guo, B. 679
Guo, H.X. 1507

Haelterman, K. 761
Halliday, G.S. 1565
Hamidi, A. 469
Han, J. 1071, 1077
Han, K.K. 1737
Han, Y. 463
Hansen, A. 457
Harp, E.L. 1447
Hassani, H. 1911
Haugen, E.D. 381
He, H. 1839
He, M. 1329
He, S. 795, 1745
He, S.M. 809
He, Y. 719, 737
Hencher, S. 427
Hencher, S.R. 457
Heng, C. 1825
Hernandez, M. 421, 725, 725
Highland, L.M. 1173
Higuchi, K. 971
Ho, K.K.S. 243, 1581, 1595, 1769
Hong, W.P. 1751
Hou, S. 1067
Hsu, H. 1621
Hsu, S.C. 1409

Hu, B. 979
Hu, G.T. 1119
Hu, X.W. 249
Hu, Y. 1481
Hu, Y.G. 639
Huang, G.W. 1367
Huang, R. 389, 403, 1355
Huang, R.Q. 1051, 1177, 2037
Huat, B.B.K. 1919
Huffman, G.J. 1991
Hughes, D.A.B. 1553
Hughes, P.N. 1553
Hui, T.H.H. 1769
Hungr, O. 145
Huo, Z.T. 1321

Il, M.Y. 1571
Ishii, Y. 697
Ivanchenko, I.V. 963

Jafari, M.K. 1427
Jamaludin, S. 1919
Janik, M. 1083
Jeon, S. 463
Jia, G.W. 731
Jiang, J.-C. 747
Jiang, L. 1481
Jiang, L.-W. 1681
Jiang, R. 737
Jiang, X. 741
Jiang, X.L. 815
Jiang, Y.H. 1335
Jibson, R.W. 1447
Jin, W. 1361
Jindal, V. 957
Jing, R. 991
Jing, Z.D. 783
Jomard, H. 421, 951
Jongmans, D. 323, 395
Joo, Y.S. 1763
Ju, L.I. 1571
Ju, N. 403
Jung, S.H. 257
Jung, Y.B. 1633
Jworchan, I. 1757

Kang, X.B. 1689
Kao, T.C. 193
Kawabata, H. 939
Kaynia, A.M. 381
Kazama, M. 1241
Keersmaekers, R. 761
Kim, B.S. 257
Kim, H.W. 1183
Kim, J.H. 1297

Kim, M. 1811
Kim, N.K. 1763
Kim, S.K. 1763
Kim, T.H. 1751, 1763
Klein, E. 1925
Knappett, J.A. 803
Kniess, U. 395
Krishna Murty, V.V.R.S. 231
Krummel, H. 1083
Ku, C.Y. 1639
Kuai, Z. 679
Kuhlmann, H. 1083
Kumar, V. 957
Kuraoka, S. 697
Kuroda, S. 971
Kwong, A.K.L. 1575

Labiouse, V. 835
Lacerda, W.A. 409
LaHusen, R.G. 179
Lan, H. 769
Lang, Y. 449, 777
Lang, Y.H. 1189
Lau, T.M.F. 243, 1581, 1595, 1769
Law, R.P.H. 415
Lebourg, T. 421, 725, 951, 1273
Ledesma, A. 237
Lee, C. 1297
Lee, C.F. 999, 1575
Lee, J.F. 1639
Lee, J.S. 1297
Lee, K.S. 427
Lee, S.G. 257, 427
Lee, W.C. 313
Leng, J.Y. 783
Lenti, L. 1389
Leong, E.C. 67
Li, B.R. 1493
Li, D. 1985
Li, D.L. 1805
Li, H. 1775
Li, H.F. 1315
Li, H.G. 1805
Li, H.P. 1799
Li, J.S. 249
Li, J.T. 815
Li, K. 795
Li, S. 803
Li, T. 435
Li, T.B. 1781
Li, T.F. 1195, 2055
Li, X. 789, 1205, 1745

Li, X.P. 809
Li, Z.C. 1335
Li, Z.D. 443
Lian, J.F. 2055
Liang, S.Y. 1189
Liao, H.J. 443
Liao, X.P. 501, 783
Lin, C.C. 1409
Lin, H. 815
Lin, M.L. 193
Lin, W. 543
Lin, X. 435
Lin, Z.G. 129
Ling, J. 741
Liu, B. 719
Liu, C.Z. 2055
Liu, F.Q. 719
Liu, H. 1421, 1775
Liu, H.C. 1493
Liu, J. 1415, 1415
Liu, J.F. 1933, 2001
Liu, M. 389
Liu, P. 1681
Liu, S.T. 1211
Liu, W.J. 719
Liu, X.L. 821
Liu, Y.H. 2055
Liu, Z. 1421
Lloret, A. 67
Loew, S. 39
Lokin, P. 1939
Longobardi, V. 1095
Lu, G. 1067
Lu, G.P. 1689
Lu, S. 1071, 1077
Lu, X.B. 263
Luan, M.T. 693
Luan, Y.J. 719
Lui, J.F. 1933
Luo, L. 2019
Luo, X. 307, 449, 777
Luo, Y. 507
Luo, Y.L. 827
Luong, B.X. 671
Lupo, G. 365
Lv, M.J. 1805

Ma, H. 2061
Ma, H.M. 1217
Ma, Y. 1689
Maertens, J. 573, 761
Mahboubi, A.R. 553
Mahdavifar, M. 1427
Malone, A.W. 457

Manzella, I. 835
Maranto, G. 1111
Martin, C.D. 3, 769
Martín, R. 1517
Martini, E. 1609
Martino, S. 579, 585, 1381, 1389
Mastroviti, G. 1111
Mathy, A. 323
Matsiy, S.I. 1225
Matsuura, S. 1439
Mazengarb, C. 267
Mazzanti, P. 585, 1381, 1905
McDougall, S. 145
Meisina, C. 1233
Mendes, J. 1553
Meng, Q.S. 979
Meng, R. 737
Meng, X.W. 921
Meric, O. 323
Merrien-Soukatchoff, V. 1143
Meyer, H.J. 1979
Mihalinec, Z. 1587
Minardo, A. 1157
Miner, A.S. 267
Mitani, Y. 893
Mitra, D. 927
Miura, K. 843
Mizuhashi, M. 53
Mohamed, A. 1873
Monfret, T. 711
Montagna, A. 1381
Monterisi, L. 1525
Moreno Robles, J. 1731
Morgenstern, N.R. 3
Mori, T. 1241
Morimasa, S. 843
Mugnozza, G.S. 355
Muhidinov, D.Z. 1475

Naaim, M. 835
Nadim, C. 1925
Nakamura, A. 851
Nakamura, H. 777
Nakashima, Y. 697
Nancey, A. 1721
Ng, C.W.W. 159, 415
Nicoletti, P.G. 1433
Nitti, D.O. 513
Niyazov, R.A. 1247
Noorzad, A. 553
Notti, D. 1233
Nunes Bandeira A.P. 1887
Nutricato, R. 513

O'Brien, A. 1757
Ochiai, H. 1253, 1375
Okada, Y. 1253
Okamoto, T. 1439
Olalla, C. 347
Olivares, L. 203, 1157, 1951
Oliveira, S.C. 1831
Omar, H. 1919
Onoue, A. 971
Orengo, Y. 395
Ortolan, Ž. 1587
Ou, G.Q. 1933, 2001
Owladeghaffari, H. 857

Paciello, A. 1389
Pagano, L. 1259
Pan, H.L. 2001, 2001
Pan, Y.B. 443
Pang, K.K. 1595
Pardo de Santayana, F. 1731
Parise, M. 275, 299, 1095
Park, D.C. 427
Park, H.D. 1633
Park, H.J. 1943
Park, J.S. 1763
Park, W. 463
Pastor, M. 647
Pathier, E. 395
Paulsen, H. 1083
Pecci, M. 355
Peduto, D. 1103
Peng, H. 827
Peng, X.M. 1321
Peng, X.T. 1031
Perri, F. 373
Petley, D.J. 865
Petley, D.N. 865, 1265
Phoon, K.K. 639
Piao, C. 1283
Picarelli, L. 1157, 1301, 1951
Pinheiro, R.J.B. 293
Pirulli, M. 145, 835
Pisciotta, G. 1103
Poisel, R. 873
Popescu, M.E. 487, 1787
Powrie, W. 1553
Preh, A. 873
Prestininzi, A. 579, 1381
Pun, W.K. 85

Qian, J.Y. 887
Qian, T. 881
Qiao, J. 1959, 1995
Qiu, C. 893, 1343

Qiu, Y. 741
Quental Coutinho R. 1887

Rahardjo, H. 67
Ramondini, M. 299, 1095
Raymond, P. 1603
Redaelli, M. 1309, 1965
Refice, A. 513
Reid, M.E. 179
Reis, E. 1831
Ren, W.Z. 991
Renalier, F. 395
Revellino, P. 365
Rezaur, R.B. 67
Rianna, G. 1259
Riopel, S. 1151
Risser, V. 725
Ritzkowski, C.M. 1475
Rizakalla, E. 1757
Roh, B. 463
Rojat, F. 687
Rollins, G.C. 1811
Rosser, N.J. 1265
Rossi, D. 1721
Roy, D. 927
Rui, R. 1011
Rui, Z. 755
Rutqvist, J. 711

Sadagah, B.H. 901
Sakai, H. 1277
Sakai, N. 907
Sakajo, S. 907
Salciarini, D. 1609
Salimi, S.N. 469
Santaloia, F. 1525
Santarém Andrade, P.G.C. 561
Santo, A. 1951
Sarah, D. 1615
Savio, G. 1233
Sawatparnich, A. 1971
Scarascia Mugnozza, G. 1389
Scavia, C. 145
Schaefer, V.R. 525, 1787
Schauerte, W. 1083
Schulz, W.H. 1447
Schwartz, S. 395
Scotto di Santolo, A. 913
Seely, D.D.B. 475
Selkirk-Bell, J.M. 267
Sento, N. 1241
Seo, Y.S. 933
Serratrice, J.F. 835
Shao, Y. 655

Shen, J.M. 1595
Shi, B. 1283
Shi, L.L. 1959
Shi, Y.X. 921
Shimomura, T. 53
Shou, K. 1621
Singh, R. 927
Singh, V.K. 281
Smethurst, J. 1553
Soares, J.M.D. 293
Song, E.X. 1507
Song, J. 679
Song, X. 1177
Song, Y.S. 933, 1751
Sorbino, G. 599
Srinivasa Gopalan, P. 231
Stedile, L. 355
Strom, A.L. 285
Su, L.J. 443
Su, O. 507
Su, T.W. 1639
Su, X. 519
Suemine, A. 1667
Sui, H. 1283
Suk, O.T. 1571
Sun, C.-Y. 755
Sun, H.W. 1581, 1769
Sun, J.C. 1627
Sun, P. 631
Sun, S. 1289, 2075
Sun, Y. 507, 1329
Sunitsakul, J. 1971
Sunwoo, C. 1633

Tacher, L. 217
Taheri, A. 481
Takahashi, C. 1455
Takeuchi, A. 1321
Tamburi, P. 1609
Tan, C.H. 1639
Tanaka, N. 939
Tang, C. 2055
Tang, H. 789, 881
Tang, H.M. 249
Tang, H.X. 945
Tang, W.H. 415
Tang, X. 1795, 1863
Tensay, B.G. 1475
Terranova, C. 1095
Tham, L.G. 999
Theule, J.I. 1979
Thiebes, B. 1083
Tian, H. 1329
Tiwari, B. 1463
Tohari, A. 1615
Tohei, M. 697

Toll, D.G. 1553
Tommasi, P. 203
Topal, T. 1647
Tortoioli, L. 1609
Tosa, S. 1349
Towhata, I. 53
Trandafir, A.C. 475, 487
Tric, E. 725, 951
Truong, Q.H. 1297
Tsang, C.F. 711
Tsui, H.M. 1581
Tyagi, K. 957

Ugai, K. 25, 593, 851, 971, 1023, 1455
Um, J.G. 1943
Urciuoli, G. 85, 1259, 1301
Uromeihy, A. 491
Utili, S. 1309, 1965
Uzuoka, R. 1241

Van Gemert, D. 761
van Westen, C.J. 1879
Vanicek, I. 1653
Vardoulakis, I. 615
Vassallo, R. 495
Vázquez-Suñé, E. 1517
Veveakis, E. 615
Villemin, T. 323, 395
Vinale, F. 1259
Vinogradov, V.V. 963
Virieux, J. 711
Vitolo, E. 1893

Wakai, A. 939, 971
Wang, A.X. 887
Wang, B. 1283
Wang, D. 519
Wang, F.W. 1321, 1469
Wang, G. 1321, 1667, 1855, 1863, 2061
Wang, G.H. 1469
Wang, G.Q. 1627
Wang, H. 501
Wang, H.B. 1469
Wang, H.S. 1493
Wang, J. 1067, 1415
Wang, J.J. 1799
Wang, J.-Y. 755
Wang, K.L. 193
Wang, L. 1367
Wang, M.Y. 1659, 1659
Wang, N. 519
Wang, R. 543, 979
Wang, S. 985, 1355, 2075
Wang, S.T. 1675

Wang, S.Y. 263
Wang, X.B. 1781
Wang, X.Z. 979
Wang, Y. 507, 2061
Wang, Y.G. 991
Wang, Y.J. 655, 663
Wang, Z. 1985
Wang, Z.C. 991
Wang, Z.F. 1343
Wang, Z.H. 2111
Wang, Z.Q. 1315
Wang, Z.W. 1211
Wasowski, J. 513
Watson, A.D. 39
Webb, S. 1475
Wei, G. 1283
Wei, J.B. 999
Wei, Y. 741
Wen, M.S. 2055
Wesley, L.D. 1005
White, D.J. 525
Woo, I. 1943
Wu, B. 1621
Wu, C. 1017
Wu, H. 1043
Wu, M.J. 1335
Wu, S.R. 1469
Wu, A.W. 519, 1475
Wu, X. 1329
Wu, Y. 1745
Wu, Z. 1355

Xia, Y. 1011
Xian, C. 679
Xiang, R. 1825
Xiao, D. 1017
Xiao, W. 1817
Xie, M. 893
Xie, M.W. 1343
Xie, X.R. 2019
Xu, H. 1659, 1863
Xu, M. 1689
Xu, N. 1329
Xu, Q. 1675, 1681
Xu, X.S. 1031
Xu, Y. 1043
Xu, Z.J. 129
Xu, Z.P. 693

Yamada, M. 1023, 1349
Yamagami, T. 747
Yan, B. 1031
Yan, E.C. 1805
Yan, M. 389, 1355
Yan, Z. 1421
Yang, G. 389

Yang, H. 525, 1017, 1811, 1991
Yang, J. 655
Yang, W.T. 1367
Yang, X. 1775
Yang, Z. 1817, 1995
Yao, A. 1825
Yashima, A. 1035
Yazdanjou, V. 469
Ye, G.L. 1035
Yener, A.B. 531
Yin, J.-H. 1361
Yin, Y. 2005
Yin, Y.P. 2089
Yoon, H.K. 1297
Yoshimatsu, H. 939
Yoshiya, O. 307
You, Y. 1933, 2001
Yuan, L. 1481
Yuan, P.J. 1335
Yue, Z.Q. 535, 1697

Zeni, L. 1157
Zervos, A. 615
Zêzere, J.L. 1831
Zhan, T.L.T. 731
Zhang, D. 1283
Zhang, F. 679, 1035, 1067
Zhang, G. 887
Zhang, H. 985
Zhang, J. 507, 795, 1289
Zhang, J.-M. 887
Zhang, K. 2005
Zhang, L. 1839
Zhang, L.M. 1205, 1315, 1703
Zhang, M.S. 129
Zhang, M.-S. 755
Zhang, P.W. 663
Zhang, Q. 921, 1367
Zhang, S. 507, 795
Zhang, X.Y. 1367
Zhang, Y. 985, 1355
Zhang, Y.M. 1321
Zhang, Y.S. 1493
Zhang, Z. 1043, 1825
Zhang, Z.P. 1217
Zhang, Z.Y. 1675
Zhao, H. 543
Zhao, J. 403
Zhao, J.B. 1847
Zhao, M. 2019
Zhao, S. 1177
Zhao, Y. 2011
Zheng, B. 1289
Zheng, D. 1051

Zheng, G.D. 1189
Zheng, J. 985, 1855
Zheng, X. 1011
Zhou, C. 507
Zhou, C.H. 769
Zhou, G.D. 415

Zhou, J.P. 2101
Zhou, P. 1067
Zhu, B. 1289
Zhu, D.Y. 1055
Zhu, H.B. 2027
Zhu, H.-H. 1361

Zhu, J.G. 1799
Zielinski, M. 1309
Zingariello, M.C. 1259
Zolfaghari, M.R. 1427
Zou, L. 1863
Zucca, F. 1233